CLAUDE A. VILLEE
HARVARD UNIVERSITY

WARREN F. WALKER, JR.
OBERLIN COLLEGE

ROBERT D. BARNES
GETTYSBURG COLLEGE

FOURTH EDITION

GENERAL ZOOLOGY

W. B. SAUNDERS COMPANY · PHILADELPHIA · LONDON · TORONTO

W. B. Saunders Company: West Washington Square
Philadelphia, Pa.   19105

1 St. Anne's Road
Eastbourne, East Sussex BN21 3UN, England

833 Oxford Street
Toronto, M8Z 5T9, Canada

General Zoology   ISBN 0-7216-9038-6

Print No: 9   8   7   6

This book is dedicated
with respect and affection
to
Tyler Buchenau,
who served with distinction as College Editor
of the W. B. Saunders Company
and helped and encouraged
each of us to develop as writers

Preface

It would appear to be customary in the Preface of a revised edition of a textbook to state that "this is a thorough revision." With this new edition of General Zoology that statement is true to an unusual extent. We have been joined by a new co-author, Dr. Robert Barnes of Gettysburg College; we have extensively rearranged the order of the subjects considered; and we have a wealth of new illustrative material in addition to revising much of the text. The survey of the animal phyla has been moved to Part Three since we believe it can be better appreciated following a study of general animal structure and function.

The field of zoology, along with all of the biologic sciences, has grown enormously in extent and complexity in the last few decades, and promises to continue to grow in the future. To deal with this vast array of knowledge, some college courses in zoology are based upon a thorough examination of certain representative animals. Other courses center around discussions of broad biologic principles. Each of these approaches has obvious merits, and in writing this text we have tried to blend the two. Neither method can be carried to extreme, for one cannot hope to teach principles without concrete examples, nor can one teach animal types without the intellectual synthesis provided by an understanding of biologic principles.

The special task of anyone writing a textbook is to select with care the topics to be discussed and to present a clear picture of the subject without giving an overwhelming mass of detail. In preparing this text we have endeavored to summarize the factual and theoretical knowledge that constitutes modern zoologic science. However, an appreciation of any science requires not only a grasp of the product of the science—the facts which have been gained in that field—but also an insight into the processes by which such knowledge is acquired. The methods of science are introduced in Chapter 1, and throughout the text examples of experimental work are presented to illustrate modern methods in zoologic science. This text probably includes some material that the instructor will have neither the time nor the inclination to consider in his course. Each instructor, naturally, emphasizes those topics he considers most important; the text provides the interested student with an opportunity to read about subjects which may be omitted or considered only briefly in the lectures and laboratory exercises. In discussing the many subjects which comprise modern zoology, we have tried to distinguish between fact and theory and to cite some of the problems that remain for future zoologists to solve. The conclusions presented and the inferences drawn represent, to the best of our knowledge and ability, the current interpretation of the relevant observations and experiments.

The introductory chapter describes the subsciences of zoology, scientific method and the sources of scientific information. The general concepts basic to a study of the form and function of both invertebrate and vertebrate animals are presented in Part One. Chapters 2 and 3 introduce the chemical and physical concepts needed for an understanding of the dynamic, molecular aspects of cells and tissues. The chemistry and physics relevant to the discussion are not presented separately but are introduced as needed to provide an understanding of the biologic material. The nature of enzymes, their role in cellular physiology, and the principles of bioenergetics are discussed in Chapter 4. Both vertebrate and invertebrate animals have had to solve the same major problems in order to survive, and an examination of their physiologic mechanisms shows that they have much in common. The principles of nutrition, digestion, circulation, respiration, excretion, protection, sensation, locomotion, irritability, and integration are discussed in Chapter 5 to provide a general background for the discussions of the animal types that follow. The general aspects of reproduction—gametogenesis, fertilization, and development and its control—are considered in Chapter 6. The chapters on genetics and evolution have been included in Part One since they are more pertinent to that section dealing with general biologic principles.

Parts of the book have been extensively reorganized. Vertebrate form and function are now considered in Part Two, for they follow naturally from the discussion of general concepts in Part One. Mammalian structure and physiology are emphasized, but sufficient information is given on the lower vertebrates so that mammals can be understood in evolutionary perspective. Much of the material previously in a separate chapter on the frog has been integrated into these chapters. An early consideration of the form and function of vertebrates, which are the organisms most familiar to students, introduces concepts that can be built upon in the discussions of the animal phyla that follow in Part Three. Many modifications have been made throughout this section to improve clarity of presentation and to keep the material current with recent advances in the subject. Notable changes have been made in discussions of vitamins, hormones, breathing mechanisms, hemostasis, immunity, urine formation, and the chemistry of vision. The Annotated References have been updated; few before 1960 are cited unless they are classics.

Part Three, the survey of animal phyla, has two goals. We have attempted to provide the student with an appreciation of the great diversity of animal life and also to provide an understanding of adaptive change in the evolutionary history of animals. To achieve these ends we have in our coverage of each of the major metazoan phyla abandoned the older device of utilizing type animals and have employed instead a general approach. Adaptive morphology and physiology have been strongly emphasized, and processes such as locomotion, nutrition, and gas exchange, which most conspicuously reflect adaptations to various modes of existence and habitats, have received the greatest attention. We have sought to avoid excessive anatomic detail and to introduce only what we have felt to be essential terminology. Throughout, the survey has also been used to illustrate and reemphasize many zoologic principles.

As in previous editions, the chapters on the vertebrate classes (Chapters 30 to 34) emphasize the adaptations of each class to its mode of life, and the evolutionary interrelationships of the various groups. Changes have been made in the discussion of aquatic adaptations, flight, and temperature regulation, among others, to clarify and to keep the presentation current. The chapter on the evolution of man has been completely rewritten and follows the discussion of mammals.

The final section of the book, Part Four, is devoted to a discussion of animals and their environment. It opens with a new chapter on animal behavior and continues with a discussion of ecologic principles and their practical implications. Autecology and synecology are discussed in separate chapters, and the section ends with a new chapter on human ecology.

We have decided to delete the study questions at the end of each chapter. The small number that can be provided are probably too general to be very useful to either student or teacher. The illustrations in this book are either original or have been redrawn especially for this edition to provide greater clarity. We are deeply grateful for the care and artistry with which Susan Heller, Ellen Cole and William Osborn prepared the line drawings. We are indebted to Aldine Publishing Company, Vernon G. Applegate, Robert S. Bailey, Betty M. Barnes, Russell J. Barrnett, Angus Bellairs, Kurt Benirschke, C. M. Bogert, S. H. Camp and Company, Austin H. Clark, Allan D. Cruikshank, H. R. Duncker, Earl R. Edmiston, Alfred Eisenstadt, Frank Essapian, Don Fawcett, D. Fraser, W. H. Freeman Company, Golden Book Encyclopedia of Natural Science, J. N. Hamlet, Harcourt Brace Jovanovich, Harper & Row, Fritz Goro, C. Lynn Haywood, A. A. Knopf, Herbert Lang, Daniel Mazia, Jacques Millot, James W. Moffett, Peter Morrison, W. W. Norton and Company, Jean Luc Perret, J. D. Pye, Col. N. Rankin, E. S. Ross, P. R. Russell, Hugh Spencer, E. P. Walker, L. W. Walker, Time-Life Books, and Weidenfeld and Nicholson, who have kindly permitted us to use certain of their drawings and photographs.

Our special thanks are due to the members of the staff of the W. B. Saunders Company who have given so liberally of their time and care in preparing this revised edition. We want to express our further thanks to Alison Bowen, Kathleen Callinan, Janet Loring and Suzanne Villee for their assistance in preparing the index.

CLAUDE A. VILLEE
WARREN F. WALKER
ROBERT D. BARNES

Contents

37 SYNECOLOGY: COMMUNITIES, BIOMES AND LIFE ZONES 838

38 HUMAN ECOLOGY 873

PART ONE

GENERAL CONCEPTS

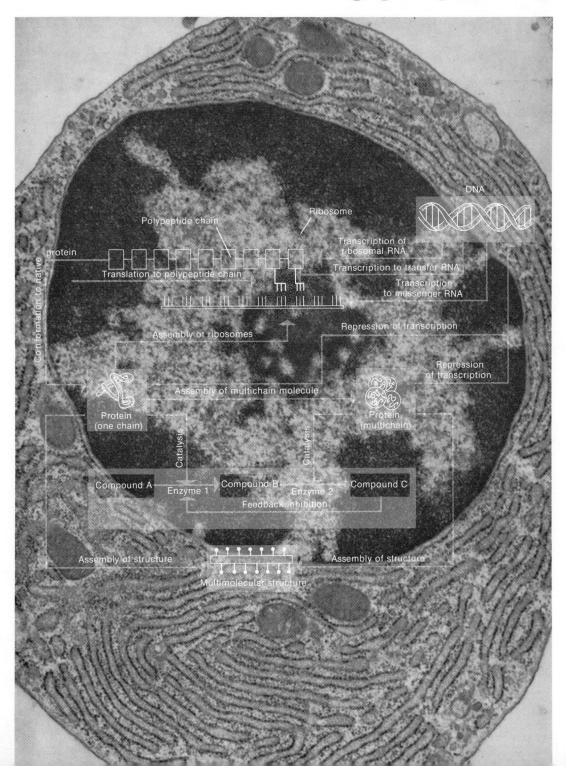

1. Introduction

1.1 ZOOLOGY AND ITS SUBSCIENCES

Zoology is one of the biologic sciences, the one dealing with the many different aspects of animal life. Since a "zoo" is a collection of animals, one could easily guess that "zoology" deals with animals. A visit to a zoo, interesting though it is, can barely begin to suggest the enormous variety of animals that are living today. (There are about one million different kinds of animals!) In addition to the ones living at present, a host of other kinds of animals have lived in past ages but are now extinct.

Modern zoology concerns itself with much more than the simple recognition and classification of the many kinds of animals. It includes the study of the structure, function and embryonic development of each part of an animal's body; of the nutrition, health and behavior of animals; of their heredity and evolution; and of their relations to the physical environment and to the plants and other animals of that region.

Enough facts about animals and their ways are known to fill a whole library of books, and more information appears every year from the intensive researches of zoologists in the field and in the laboratory. No zoologist today can know more than a small fraction of this enormous body of knowledge. Zoology is now much too broad a subject to be treated thoroughly in a single textbook or to be encompassed by a single scientist. Most zoologists are specialists in some limited phase of the subject—in one of the subdivisions of zoology. The sciences of **anatomy, physiology** and **embryology** deal with the structure, function and development, respectively, of an animal. Each of these may be further subdivided according to the kind of animal investigated, e.g., invertebrate physiology, arthropod physiology, insect physiology or comparative physiology. **Parasitology** deals with those forms of life that live in or on and at the expense of other organisms. **Cytology** is concerned with the structure, composition and function of cells and their parts, and **histology** is the science of the structure, function and composition of tissues. The science of **genetics** investigates the mode of transmission of characteristics from one generation to the next and is closely related to the science of **evolution,** which studies the way in which new species of animals arise and how the present kinds of animals are related by descent to previous animals. The study of the classification of organisms, both animals and plants, is called **taxonomy.** The science of **ecology** is concerned with the relations of a group of organisms to their environment, including both the physical factors and the other forms of life which provide food or shelter for them, compete with them in some way, or prey upon them.

Some zoologists specialize in the study of one group of animals. There are **mammalogists, ornithologists, herpetologists** and **ichthyologists** who study mammals, birds, reptiles and amphibians, and fishes, respectively; **entomologists,** who investigate insects; **protozoologists,** who study the single-celled animals, and so on.

In recent years advances in chemistry and physics have made possible quantitative studies of the molecular structures and events underlying biologic processes. The term **molecular biology** has been applied to analyses of gene structure and function and genic control of the synthesis of enzymes and other proteins, studies of subcellular structures and their roles in regulatory processes within the cell, investigations of the mechanisms underlying cellular differentiation, and analyses of the molecular basis of evolution by comparative studies of the molecular structure of specific proteins—enzymes, hormones, cytochromes, hemoglobins—in different species.

The science of zoology thus includes both a tremendous body of facts and theories about animals and the means for learning more. The ultimate source of each fact is in some carefully controlled observation or experiment made by a zoologist. In earlier times, some scientists kept their discoveries to themselves, but there is now a strong tradition that scientific discoveries are public property and should be freely published. In a scientific publication a man must do more than simply say that he has made some particular discovery; he must give all of the relevant details of the means by which the discovery was made so that others can repeat the observation. It is this criterion of *repeatability* that makes us accept a certain observation or experiment as representing a true fact; observations that cannot be repeated by competent investigators are discarded.

When a scientist has made some new observation, or carried out a series of experiments that add to our knowledge in a field, he writes a report, called a "paper," in which he describes his methods in sufficient detail so that another worker can repeat them, gives the results of his observations, discusses the conclusions to be drawn from them, perhaps formulates a theory to explain them or discusses how they are explained by a previous theory, and finally indicates the place of these new facts in their particular field of science. The knowledge that his discovery will be subjected to the keen scrutiny of his colleagues is a strong stimulus for repeating the observa-

tions or experiments carefully before publishing them. He then submits his paper for publication in one of the professional journals in the particular field of his discovery. There are several thousand zoological journals published all over the world. Some of the more important American ones are the *Journal of Experimental Zoology, Journal of Cellular and Comparative Physiology, Biological Bulletin, Physiological Zoology, American Journal of Physiology, Anatomical Record, Ecology* and the journals devoted to research on a particular group of animals, such as the *Journal of Mammalogy.* The paper is read by one or more of the board of editors of the journal, all of whom are experts in the field. If it is approved, it is published and becomes part of "the literature" of the subject.

At one time, when there were fewer journals, it might have been possible for one man to read them all each month as they appeared, but this is obviously impossible now. Journals such as *Biological Abstracts* assist the hard-pressed zoologist by publishing, classified by fields, short summaries or abstracts of each paper published, giving the facts found, the conclusion reached, and an exact reference to the journal in which the full report appears. A considerable number of journals devoted solely to reviewing the newer developments in particular fields of science have sprung up in the past 40 years; some of these are *Physiological Reviews, Quarterly Review of Biology, Nutrition Reviews, Annual Review of Physiology* and *Vitamins and Hormones.* The new fact or theory thus becomes widely known through publication in the appropriate professional journal and by reference in abstract and review journals and eventually may become a sentence or two in a textbook.

The professional societies of zoologists and the various special branches of zoology have annual meetings at which new discoveries may be reported. Two of the largest annual meetings are those of the American Institute of Biological Sciences and the Federation of American Societies for Experimental Biology. There are, in addition, national and international gatherings, called **symposia,** of specialists in a given field to discuss the newer findings and the present status of the knowledge in that field. For example, the discussions of the Cold Spring Harbor Symposia in

Quantitative Biology, held each June at the Long Island Biological Laboratory in Cold Spring Harbor, are published and provide an excellent review of some particular field. A different subject is discussed each year.

1.2 THE SCIENTIFIC METHOD

The ultimate aim of each science is to reduce the apparent complexity of natural phenomena to simple, fundamental ideas and relations, to discover all the facts, and the relationships among them. The essence of the scientific method is the posing of questions and the search for answers, but they must be "scientific" questions, arising from observations and experiments, and "scientific" answers, ones that are testable by further observation and experiment. The Danish physicist Niels Bohr put it this way, "the task of science is both to extend the range of our experience and to reduce it to order."

There is, however, no single "scientific method," no regular, infallible sequence of events which will reveal scientific truths. Different scientists go about their work in different ways. George Sarton, in the *Study of the History of Science,* points out that "Even as all kinds of men are needed to build up a community, even so we need all kinds of scientists to develop science in every possible direction. Some are very sharp and narrow-minded, others broad-minded and superficial. Many scientists, like Hannibal, know how to conquer, but not how to use their victories. Others are colonizers rather than explorers. Others are pedagogues. Others want to measure everything more accurately than it was measured before. This may lead them to the making of fundamental discoveries, or they may fail, and be looked upon as insufferable pedants."

The ultimate source of all the facts of science is careful, close observation and experiment, free of bias and done as quantitatively as possible. The observations or experiments may then be analyzed, or simplified into their constituent parts, so that some sort of order can be brought into the observed phenomena. Then the parts can be reassembled and their interactions made clear. On the basis of these observations, the scientist constructs a **hypothesis,** a trial idea about the nature of the observation, or about the connections between a chain of events, or even about cause and effect relationships between different events. It is in this ability to see through a mass of data and construct a reasonable hypothesis to explain their relationships that scientists differ most and that true genius shows itself.

The role of a hypothesis is to penetrate beyond the immediate data and place it into a new, larger context, so that we can interpret the unknown in terms of the known. There is no sharp distinction between the usage of the words "hypothesis" and "theory," but the latter has, in general, the connotation of greater certainty than a hypothesis. A **theory** is a conceptual scheme which tries to explain the observed phenomena and the relationships between them, so as to bring into one structure the observations and hypotheses of several different fields. The theory of evolution, for example, provides a conceptual scheme into which fit a host of observations and hypotheses from paleontology, anatomy, physiology, biochemistry, genetics and other allied sciences.

A good theory correlates many previously separate facts into a logical, easily understood framework. The theory, by arranging the facts properly, suggests new relationships between the individual facts, and suggests further experiments or observations which might be made to test these relationships. It may predict new phenomena that will be observed under certain circumstances and finally may provide the solution for practical problems. A good theory should be simple and should not require a separate proviso to explain each fact; it should be flexible, able to grow and to undergo modifications in the light of new data. A theory is not discarded because of the existence of some isolated fact which contradicts it, but only because some other theory is better able to explain all of the known data.

Once a hypothesis has been established, the rules of formal logic can be applied to deduce certain consequences. In physics, and to a lesser extent in the biologic sciences, the hypotheses and deductions can be stated in mathematical terms, and far-reaching conclusions may be deduced. From these inferences, one can predict the results of other observations and experiments. Each hypothesis is ultimately kept, amended or discarded on the basis of its ability to make valid predictions. A hypothesis must be subject to some sort of experimental test—i.e., it must make a prediction that can be verified in some way—or it is mere speculation. Conversely, unless a prediction follows as the logical outgrowth of some theory it is no more than a guess.

The finding of results contrary to those predicted by the hypothesis causes the investigator, after he has assured himself of the validity of his observation, either to discard the hypothesis or to change it to account for both the original data and the new data. Hypotheses are constantly being refined and elaborated. There are few scientists who would regard any hypothesis, no matter how many times it may have been tested, as a statement of absolute and universal truth. It is rather regarded as the best available approximation to the truth for some finite range of circumstances. For example, the Law of the Conservation of Matter was widely adhered to until the work of Einstein showed that it had to be modified to allow for the possible interconversion of matter and energy.

Ideally, the scientific method consists of making careful observations and arranging these observations so as to bring order into the phenomena. Then one postulates a hypothesis or conceptual scheme which will explain the facts at hand and make predictions about the results of further experiments or observations. Sciences differ widely in the extent to which prediction is possible, and the biologic sciences have been held by some to be not truly "scientific," for they are not completely predictable. However, even physics, which is generally regarded as the most scientific of the sciences, is far from completely predictable.

The history of science shows that, although many scientists have made their discoveries by following the precepts of the ideal scientific method, there have been occasions on which important and far-reaching theories have resulted from making incorrect conclusions from erroneous postulates, or from the misinterpretation of an improperly controlled experiment! There are instances in which, in retrospect, it seems clear that all the evidence for the formulation of the correct theory was known, yet no scientist put the proper two and two together. And there are other instances in which scientists have been able to establish the correct theory despite an abundance of seemingly contradictory evidence.

In most scientific studies one of the ultimate goals is to explain the cause of some phenomenon, but the hard-and-fast proof that a cause and effect relationship exists between two events is really very difficult to obtain. If the circumstances leading to a certain event always have a certain factor in common in a variety of cases, that factor may be the cause of the event. The difficulty, of course, lies in making sure that the factor under consideration is the *only* one common to all the cases. It would be wrong, for example, to conclude from the observation that drinking Scotch and soda, bourbon and soda, and rye and soda all produce intoxication, that soda is the only factor in common and therefore is the cause of the intoxication. This method of discovering the common factor in a series of cases that may be the cause of the event (known as the **method of agreement**) can seldom be used as a valid proof because of this difficulty of being sure that it is indeed the only common factor. The simple observation that all people suffering from beriberi have diets which are low in thiamine is not proof that a deficiency of this vitamin causes the disease, for there may be many other factors in common.

Experiments based on the **method of difference** provide another way of elucidating cause and effect relations. If two sets of circumstances differ in only one factor, and the one containing the factor leads to an event and the other does not, the factor may be considered the cause of the event. For example, if two groups of rats are fed diets which are identical except that one contains all the vitamins and the second contains all but thiamine, and if the first group grows normally but the second fails to grow and ultimately develops polyneuritis, this would be a strong suggestion (but would not be acceptable as absolute proof) that polyneuritis, or beriberi in rats, is caused by a deficiency of thiamine. By using an inbred strain of rats that are as alike as possible in inherited traits, and by using littermates (brothers and sisters) of this strain, one could make certain that there were no hereditary differences between the controls (the ones getting the complete diet) and the experimentals (the ones getting the thiamine-deficient diet). One might postulate that the thiamine-free diet does not have as attractive a taste as the one with thiamine, and the experimental animals simply eat less food, fail to grow, and develop the deficiency symptoms because they are partially starved. This source of error can be avoided by "pair-feeding," by pairing in some arbitrary way each control and experimental animal, then weighing the food eaten each day by each experimental animal and giving only that much food to the corresponding control member of the pair.

One of the more useful methods of detecting cause and effect relationships is the **method of concomitant variation.** If a variation in the amount of one given factor produces a parallel variation in the effect, the factor may be the cause. Thus, if several groups of rats were given diets with varying amounts of thiamine, and if the amount of protection against beriberi varied directly with the amount of thiamine in the diet, one could be reasonably sure that thiamine deficiency is the cause of beriberi.

It must be emphasized that it is seldom that we can be more than "reasonably sure" that X is the cause of Y. As more experiments and observations lead to the same result, the probability increases that X is the cause of Y. When experiments or observations can be made quantitative, when their results can be counted or measured in some way, the methods of statistical analysis provide a means for calculating the probability that Y follows X simply as a matter of chance. Scientists are usually satisfied that there is some sort of cause and effect relationship between X and Y if they can show that there is less than one chance in a hundred that the observed X-Y relationship could be due to chance alone. A statistical analysis of a set of data can never give a flat yes or no to a question; it can state only that something is very probable or very improbable. It can also tell an investigator approximately how many more times he must repeat the experiment to show with a given probability that Y is caused by X.

The proper design of experiments is a science in itself, and one for which only general rules can be made. In all experiments, the scientist must ever be on his guard against bias in himself, bias in the subject, bias in his instruments and bias in the way the experiment is designed.

Each experiment must include the proper **control group** (indeed some experiments require several kinds of control groups). The control group is one treated exactly like the experimental group in all respects but one, the factor whose effect is being tested. The use of controls in medical experiments raises the difficult question of the moral justification of withholding treatment from a patient who might be benefited by it. If there is sufficient evidence that one treatment is indeed better than another, a physician would hardly be justified in further experimentation. However, the medical literature is full of treatments now known to be useless or even detrimental, which were used for many years, only to be abandoned finally as experience showed that they were ineffective and that the evidence which had originally suggested their use was improperly controlled. There is a time in the development of any new treatment when the medical profession is not only morally justified, but really morally required, to do carefully controlled tests on human beings to be sure that the new treatment is better than the former one.

In medical testing it is not sufficient simply to give a treatment to one group of patients and not to give it to another, for it is widely known that there is a strong psychologic effect in simply giving a treatment of any sort. For example, a group of students in a large western university served as subjects for a test of the hypothesis that daily doses of extra amounts of vitamin C might help to prevent colds. This grew out of the observation that people who drink lots of fruit juices seem to have fewer colds. The group receiving the vitamin C showed a 65 per cent reduction in the number of colds contracted during the winter in which they received treatment as compared to the previous winter when they had no treatment. There were enough students in the group (208) to make this result statistically significant. In the absence of controls, one would have been led to the conclusion that vitamin C does help to prevent colds. A second group of students were given "placebos," pills identical in size, shape, color and taste to the vitamin C pills but without any vitamin C. The students were not told who was getting vitamin C and who was not; they only knew they were getting pills that might help to prevent colds. The group getting the placebos reported that they had a 63 per cent reduction in the number of colds! This controlled experiment thus shows that vitamin C had nothing to do with the decrease in the number of colds and that the reductions reported in both groups were either psychologic effects or simply the result of a lesser amount of cold virus on the campus that year. There have been reports that other substances, called bioflavonoids, present in fruit juices may have some effect in protecting against the common cold. Comparable carefully controlled experiments are needed to substantiate this report.

1.3 HISTORY OF ZOOLOGY

Man's interest in animals is probably somewhat older than the human race, for the ape-men and man-apes that preceded him in evolution undoubtedly learned at an early time which animals were dangerous, which could be hunted for food, clothing or shelter, where these were to be found, and so on. Some of prehistoric man's impressions of the contemporary animals have survived in the cave paintings of France and Spain (Fig. 1.1). Some animals were regarded as good or evil spirits. Later man decorated pottery, tools, cloth and other objects with animal figures.

The early Egyptians had a wealth of knowledge about animals and had domesticated cattle, sheep, pigs, cats, geese and ducks. The Greek philosophers of the fifth and sixth centuries B.C., Anaximander, Xenophanes, Empedocles and others, spec-

Figure 1.1   Paintings by Upper Paleolithic man from the wall of the cavern at Lascaux, Dordogne, France. (Photo by Windels Montignac.) (Villee: Biology, 6th ed.)

ulated on the origin of the animals of the earth. One of the earliest classifications of animals is found in a Greek medical book of this time which classifies animals primarily as to whether or not they are edible. Aristotle (384–322 B.C.) was one of the greatest Greek philosophers and wrote on many topics. His *Historia animalium* contains a great deal of information about the animals of Greece and the nearby regions of Asia Minor. Aristotle's descriptions are quite good and are recognizable as those of particular animals living today. The breadth and depth of his zoologic interests are impressive—he made a careful study of the development of the chick and of the breeding of sharks and bees, and he had notions about the functions of the human organs, some of which, not too surprisingly, were quite wrong. He presented an elaborate theory that animals have gradually evolved, based on a metaphysical belief that nature strives to change from the simple and imperfect to the more complex and perfect. His contributions to logic, such as the development of the system of inductive reasoning from specific observations to a generalization which explains them all, have been of inestimable value to all branches of science.

The Greek physician, Galen (131–201 A.D.), was one of the first to do experiments and dissections of animals to determine structure and functions (Fig. 1.2). The first experimental physiologist, he made some notable discoveries on the functions of the brain and nerves and demonstrated that arteries carry blood and not air. His descriptions of the human body were the unquestioned authority for some 1300 years, even though they contained some remarkable errors, being based on dissections of pigs and monkeys rather than of human bodies. Pliny (23–79 A.D.) and others in succeeding centuries compiled encyclopedias (Pliny's *Natural History* was a 37 volume work) regarding the kinds of animals and where they lived, which are remarkable mixtures of fact and fiction. Some of the ones written in the Middle Ages were called "bestiaries." The zoologic books written in the Middle Ages are, almost without exception, copied from Aristotle, Galen and Pliny; no original observations were made to corroborate or refute the accuracy of these authorities.

Figure 1.2   Galen (131–201 A.D.). The founder of experimental physiology. (Courtesy of J.A.M.A., from the series Medical Greats.)

The Renaissance in science began slowly with scholars such as Roger Bacon (1214–1294) and Albertus Magnus (1206–1280) who were interested in all branches of natural science and philosophy. The genius Leonardo da Vinci (1452–1519) was an anatomist and physiologist as well as a painter, engineer and inventor. He made many original observations in zoology, some of which came to light only much later, when his notebooks were deciphered.

One of the first to question the authority of Galen's descriptions of human anatomy was the Belgian, Andreas Vesalius (1514–1564), who was professor at the University of Padua in Italy (Fig. 1.3). By actual dissections and detailed, clear drawings of what he saw, Vesalius revealed many of the inaccuracies in Galen's descriptions of the human body. He published his observations and illustrations in *De Humani corporis fabrica* (On the Structure of the Human Body) in 1543. Since Vesalius dared to reject the authority of Galen, he was the object of much adverse criticism and was finally forced to leave his professorial post.

Just as Vesalius had emphasized the importance of relying on original observation rather than on authority in anatomy, so did

Figure 1.3 Andreas Vesalius (1514–1564). In his *De Humani corporis fabrica*, he established the basis for modern anatomy. (From Garrison: History of Medicine. 4th ed. Philadelphia, W. B. Saunders Co., 1929.)

William Harvey (1578–1657) in physiology (Fig. 1.4). Harvey, an English physician, received his medical training at the University of Padua, where Vesalius had taught. He returned to England and investigated the circulation of the blood. In 1628 he published *Exercitatio anatomica de motu cordis et sanguinis in animalibus* (Anatomical Studies on the Motion of the Heart and Blood in Animals). At that time blood was believed to be generated in the liver from food and to pass just once to the organs of the body where it was used up. The heart was believed to be nonmuscular and to be expanded passively by the inflowing blood. Harvey described, from direct observations on animals, how first the atria (auricles) and then the ventricles fill and empty by muscular contraction. He showed by experiment that when an artery is cut blood spurts from it in rhythm with the beating of the heart, and that when a vein is clamped it becomes full of blood on the side away from the heart and empty on the side toward the heart. He showed that valves in the veins permit blood to flow toward the heart but not in the reverse direction (Fig. 1.5). From these experiments he concluded that blood is carried away from the heart in arteries and back to the heart in veins. Furthermore,

by measuring how much blood is delivered by each beat of the heart, and by measuring the number of heartbeats per minute, he could calculate the total flow of blood through the heart per minute or hour. This he found to be so great that it could not be generated anew in the liver but must be recirculated, used over and over again. This was the first quantitative physiologic argument. He inferred that there must be small vessels connecting arteries and veins to complete the circular path of the blood but, lacking a microscope, he was unable to see them. In later years he made a careful study of the development of the chick, published in 1651 as *Exercitationes de generatione animalium.* In this he postulated that mammals, like the chick, develop from an egg.

The development of the compound microscope by the Janssens in 1590 and by Galileo in 1610 provided the means for attacking many problems in zoology and botany. Robert Hooke (1635–1703), Marcello Malpighi (1628–1694), Antony van Leeuwenhoek (1632–1723) and Jan Swammerdam (1637–1680) were some of the first microscopists. They studied the fine structure of plant and animal tissues. Hooke was the first to describe the presence of "cells"

Figure 1.4 William Harvey (1578–1657). He proved the circulation of the blood by the first quantitative physiologic demonstration. (From Garrison: History of Medicine. 4th ed. Philadelphia, W. B. Saunders Co., 1929.)

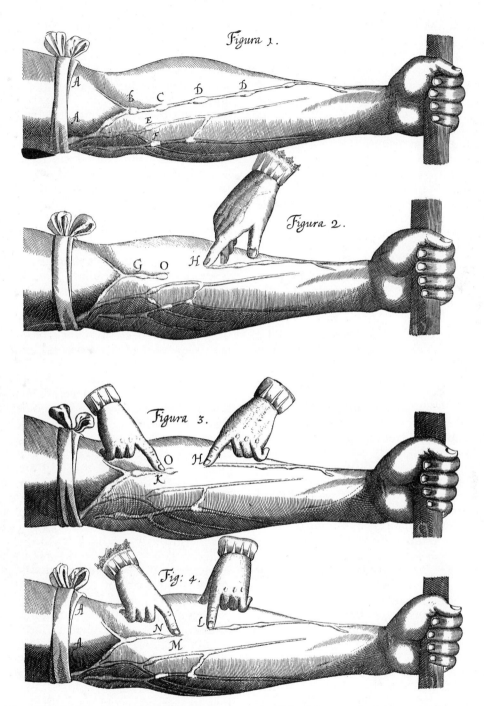

Figure 1.5 Harvey's illustrations to demonstrate the direction of blood flow. *1*, The formation of "knots" at valves. *2*, Stripping a portion of a vessel to show that there is no back flow. *3* and *4*, Demonstration of valvular blocking of blood flow. (From Harvey: Exercitatio Anatomica de Motu Cordis et Sanguinis in Animalibus. Translation by C. D. Leake: Anatomical Studies on the Motion of the Heart and Blood, Springfield, Ill., Charle · C Thomas, 1931.)

in plant tissue, Leeuwenhoek was the first to describe bacteria, protozoa and sperm, and Malpighi was the first to describe the capillaries connecting arteries with veins. The light microscope has been modified and improved greatly in the past century, and man's ability to see the fine structure of cells has been greatly extended by the invention of the phase microscope and of the electron microscope. The latter, with good resolution at magnifications as great as 80,000 to 100,000 diameters, has revealed a whole new level of complexity in the structure of all kinds of cells.

John Ray (1627–1705) and Linnaeus (Karl von Linné) (1707–1778) brought order into the classification of animals and plants and devised the binomial system (two names, genus and species) for the scientific naming of the kinds of animals and plants. Linnaeus first used this binomial system consistently in the tenth edition of his *Systema naturae* (1758).

Contributions to our understanding of the embryonic development of animals were made by Fabricius, the professor of Anatomy at Padua who taught William Harvey, and by Harvey, Malpighi, and Kaspar Wolff (1759). Wolff proposed the theory of epigenesis, the concept that there is a gradual differentiation of structure during development from a relatively structureless egg. Karl Ernst von Baer (1792–1876) established the theory of germ layers and emphasized the need for comparative studies of development in different animals.

Following William Harvey, physiology was advanced by René Descartes (1596–1650), who was a philosopher rather than an experimenter. He believed that "animal spirits" are generated in the heart, stored in the brain, and pass through the nerves to the muscles, causing contraction or relaxation, according to their quantity. Charles Bell (1774–1842) and François Magendie (1783–1855) made notable contributions to our understanding of the function of the brain and spinal nerves. Johannes Müller (1801–1858) studied the properties of nerves and capillaries; his textbook of physiology stimulated a great deal of interest and research in the field. Claude Bernard (1813–1878) was one of the great advocates of experimental physiology, and contributed significantly to our understanding of the role of the liver, heart, brain and placenta.

Henry Bowditch (1840–1911) discovered the "all-or-none" principle of the contraction of heart muscle and established the first laboratory for teaching physiology in the United States. Ernest Starling (1866–1927) made many contributions to the physiology of circulation and the nature of lymph and with William Bayliss (1866–1924) elucidated the hormonal control of the function of the pancreas.

The Scottish anatomist John Hunter (1728–1793) and the French anatomist Georges Cuvier (1769–1832) were pioneers in the field of comparative anatomy, studying the same structure in different animals. Richard Owen (1804–1892) developed the concepts of homology and analogy. Cuvier was one of the first to study the structure of fossils as well as of living animals and is credited with founding the science of paleontology. Cuvier believed strongly in the unchanging nature of species and carried on bitter debates with Lamarck, who in 1809 proposed a theory of evolution based on the idea of the inheritance of acquired characters.

One of the most important and fruitful concepts in biology is the **cell theory,** which has gradually grown since Robert Hooke first saw, with the newly invented microscope, the dead cell walls in a piece of cork. The French biologist René Dutrochet clearly stated in 1824 that "all organic tissues are actually globular cells of exceeding smallness, which appear to be united only by simple adhesive forces; thus all tissues, all animal organs are actually only a cellular tissue variously modified." Dutrochet recognized that growth is the result of the increase in the volume of individual cells and of the addition of new cells. Two Germans, botanist M. J. Schleiden and zoologist Theodor Schwann, studied many different plant and animal tissues and are generally credited with formulating the cell theory, for they showed that cells are the units of structure in plants and animals, and that organisms are aggregates of cells arranged according to definite laws. The presence of a nucleus within the cell, now recognized as an almost universal feature of cells, was first described by Robert Brown in 1831.

Zoology, along with the other biologic sciences, has expanded at a tremendous rate in the past century, with the establishment of the subsciences of cytology, embryology, genetics, evolution, biochemistry, biophysics, endocrinology and ecology. The discoveries and new techniques of chemistry and physics have made possible new approaches to the biologic sciences that have attracted the attention of many biologists. So many men have contributed to the growth of zoology in this past century that only a few in each field can be mentioned: Mendel, deVries, Morgan and Bridges in genetics, Darwin, Dobzhansky, Wright and Goldschmidt in evolution, and Harrison and Spemann in embryology. Many others will be mentioned as these subjects are discussed in detail in the text.

The establishment and growth of the marine biologic laboratories, such as the ones at Naples (Italy), Woods Hole (Mass.), Pacific Grove (Calif.), Friday Harbor (Wash.) and elsewhere, have played an important role in fostering research in zoologic sciences. There are comparable stations for the study of fresh-water biology, such as the one at Douglas Lake, Michigan.

1.4 APPLICATIONS OF ZOOLOGY

Some of the practical uses of a knowledge of zoology will become apparent as the student proceeds through this text. Zoology is basic in many ways to the fields of medicine and public health, agriculture, conservation and to certain of the social sciences. There are esthetic values in the study of zoology, for a knowledge of the structure and functions of the major types of animals will greatly increase the pleasure of a stroll in the woods or an excursion along the seashore. Trips to zoos, aquariums and museums are also rewarding in the glimpses they give of the host of different kinds of animals. Many of these are beautifully colored and shaped, graceful or amusing to watch, but all will mean more to a person equipped with the basic knowledge of zoology which enables him to recognize them and to understand the ways in which they are adapted to survive in their native habitat.

ANNOTATED REFERENCES

Beveridge, W. I. B.: The Art of Scientific Investigation. New York, W. W. Norton & Co., 1957. One of the best over-all surveys of the scientific method.

Cannon, W. B.: The Way of an Investigator. New York, W. W. Norton & Co., 1945. An autobiography with many interesting anecdotes illustrating the application of the scientific method to medical research.

Conant, J. B.: Science and Common Sense. New Haven, Yale University Press, 1951. A general presentation of the methods of science; one of the classics in this field.

Feibleman, J. K.: Testing Hypotheses by Experiment. Perspect. Biol. Med., 4:91, 1960. An excellent, brief discussion of scientific methods.

Gabriel, M. L., and S. Fogel: Great Experiments in Biology. New York, Prentice-Hall, Inc., 1955. By extensive quotations from the original papers, traces the development of some of the basic concepts of zoology.

Guthrie, D.: A History of Medicine. Philadelphia, J. B. Lippincott Co., 1946. Describes the early phases of the development of anatomy and physiology along with other medical sciences.

Sedgwick, W. T., H. V. Tyler, and R. P. Bigelow: A Short History of Science. New York, Macmillan, 1939. A classic survey of the early development of the sciences in general.

Singer, C.: A History of Biology. Revised ed. New York, Abelard-Schumann, 1959. A general history of the biological sciences.

Wilson, E. B.: An Introduction to Scientific Research. New York, McGraw-Hill Book Co., 1952. An excellent discussion in nontechnical terms of the methods of science and some of the factors to be considered in planning and carrying out scientific investigation.

2. The Physical and Chemical Basis of Life

Most biologists are agreed that all the varied phenomena of life are ultimately explainable in terms of the same physical and chemical principles which define nonliving systems. The idea that there are no fundamental differences between living and nonliving things is sometimes called the **mechanistic theory of life.** A corollary of this is that when enough is known of the chemistry and physics of vital phenomena it may be possible to synthesize living matter. An opposite view, widely held by biologists until the present century, stated that some unique force, not explainable in terms of physics and chemistry, is associated with and controls life. The view that living and nonliving systems are basically different and obey different laws is called **vitalism.** Many of the phenomena that appeared to be so mysterious when first discovered have subsequently proved to be understandable without invoking a unique life force, and it is reasonable to suppose that future research will show that other aspects of life can also be explained by physical and chemical principles.

To differentiate the living from the nonliving and then to separate the living into plants and animals are difficult to do sharply and clearly. Organisms such as cats, clams and cicadas are clearly recognizable as animals, but sponges, for example, were considered to be plants until well into the nineteenth century, and there are single-celled organisms which, even today, are called animals by zoologists, plants by botanists, and protista by others. Even the line between living and nonliving is indistinct, for the viruses, too small to be seen with an ordinary light microscope, can be considered either the simplest living things or very complex, but nonliving, organic chemicals.

2.1 CHARACTERISTICS OF LIVING THINGS

All living things have, to a greater or lesser degree, the properties of specific organization, irritability, movement, metabolism, growth, reproduction and adaptation.

Organization. Each kind of living organism is recognized by its characteristic form and appearance; the adult organism usually has a characteristic size. Nonliving things generally have much more variable shapes and sizes. Living things are not homogeneous, but are made of different parts, each with special functions; thus the bodies of animals and plants are characterized by a specific, complex organization.

The structural and functional unit of both animals and plants is the **cell.** It is the simplest bit of living matter that can exist independently and exhibit all the characteristics of life. The processes of the entire organism are the sum of the coordinated functions of its constituent cells. These cellular units vary considerably in size, shape and function. Some of the smallest animals have bodies made of a single cell; the body of a man, in contrast, is made of countless billions of cells fitted together.

The cell itself has a specific organization and each kind of cell has a characteristic size and shape by means of which it can be recognized. A typical cell, such as a liver cell (Fig. 2.1), is polygonal in shape, with a **plasma membrane** separating the living substance from the surroundings. Almost without exception, each cell has a **nucleus,** typically spherical or ovoid in shape, which is separated from the rest of the cell by a **nuclear membrane.** The nucleus, as we shall see later, has a major role in controlling and regulating the cell's activities. It contains the hereditary units or **genes.** A cell experimentally deprived of its nucleus usually dies in a short time; even if it survives for several days it is unable to reproduce.

Irritability. Living things are irritable; they respond to stimuli, to physical or chemical changes in their immediate surroundings. Stimuli which are effective in evoking a response in most animals and plants are changes in light (either in its color, intensity or direction), temperature, pressure, sound, and in the chemical composition of the earth, water, or air surrounding the animal. In man and other complex animals, certain cells of the body are highly specialized to respond to certain types of stimuli: the rods and cones in the retina of the eye respond to light, certain cells in the nose and in the taste buds of the tongue respond to chemical stimuli, and special groups of cells in the skin respond to changes in temperature or pressure. In lower animals such specialized cells may be absent, but the whole organism responds to any one of a variety of stimuli. Single-celled animals such as the ameba will respond by moving toward or away from heat or cold, certain chemical substances, or the touch of a microneedle. Indeed, many of the cells of higher animals have a similar generalized sensitivity.

Movement. A third characteristic of living things is their ability to move. The movement of most animals is quite obvious —they wiggle, swim, run or fly. The movement of plants is much slower and less obvious, but is present nonetheless. A few animals—sponges, corals, hydroids, oysters, certain parasites—do not move from place to place, but most of these have microscopic, hairlike projections from the cells, called **cilia** or **flagella,** to move their surroundings past their bodies and thus bring food and other necessities of life to themselves. The movement of an animal body may be the result of muscular contraction,

Figure 2.1   Schematic drawing of a generalized animal cell.

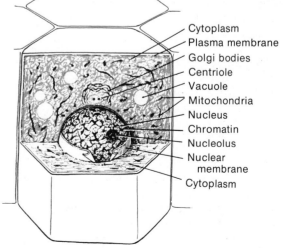

Cytoplasm
Plasma membrane
Golgi bodies
Centriole
Vacuole
Mitochondria
Nucleus
Chromatin
Nucleolus
Nuclear membrane
Cytoplasm

of the beating of cilia or flagella, or of the slow oozing of a mass of cell substance (known as ameboid motion).

Metabolism. All living things carry on a wide variety of chemical reactions, the sum of which we call **metabolism.** There is no way of observing the occurrence of most of these chemical reactions without the aid of special apparatus such as respirometers to measure oxygen utilization and carbon dioxide production and thermometers to measure heat production. Elaborate physical and chemical equipment and substances labeled with radioactive or stable isotopes are used to trace in detail the paths of metabolism and their respective quantitative importance to the animal or plant under investigation. Such studies have shown that all cells are constantly taking in new substances, altering them chemically in a multitude of ways, building new cell components, and transforming the potential energy of some of the molecules taken in into kinetic energy and heat. The large molecules— proteins, fats, carbohydrates and others— are broken down stepwise to yield energy and simpler substances. The never-ending flow of energy within a cell, from one cell to another and from one organism to the next, is the very essence of life, one of the unique and characteristic attributes of living things. The rate of metabolism is affected by temperature, age, sex, general health and nutrition, by hormones, and by many other factors. The study of energy transformations in living organisms is termed **bioenergetics.**

Those metabolic processes in which simpler substances are combined to form more complex substances and which result in the storage of energy and the production of new cellular materials are termed **anabolic.** The opposite processes, in which complex substances are broken down to release energy, are called **catabolic.** Both types of metabolism occur continuously and are intricately interdependent so that they become, in practice, difficult to distinguish. Complex compounds of one sort may be broken down and their parts recombined in new ways to yield new compounds. The interconversions of carbohydrates, fats and proteins that occur continuously in most animal cells are examples of combined anabolic and catabolic processes. Furthermore, the synthesis of most molecules requires energy, so that some catabolic processes must occur to supply the energy to drive the anabolic reactions of these syntheses.

Growth. Both plants and animals grow; nonliving things do not. The increase in mass may be brought about by an increase in the *size* of the individual cells, by an increase in the *number* of cells or both. An increase in cell size may occur by the simple uptake of water, but this is not generally considered to be growth. The term **growth** is restricted to those processes which increase the amount of living substance of the body, measured by the amount of nitrogen, protein or nucleic acid present. Growth may be uniform in the several parts of an organism, or, perhaps more commonly, growth is differential, greater in some parts than in others, so that the body proportions change as growth occurs.

Growth may occur throughout the life span of an organism or may be restricted to a part of it. One of the truly remarkable aspects of the process is that each organ continues to function while undergoing growth.

Reproduction. If there is any one characteristic that can be said to be the *sine qua non* of life, it is the ability to reproduce. Since individual animals grow old and die, the survival of the species depends upon the replacing of these individuals by new ones. Although at one time worms were believed to arise from horse hairs in a trough of water, maggots from decaying meat and frogs from the mud of the Nile, we now know that each can come only from previously existing ones. One of the fundamental tenets of biology is that "all life comes only from living things."

The classic experiment disproving the **spontaneous generation of life** was performed by an Italian, Francesco Redi, about 1680. Redi proved that maggots do not come from decaying meat by this simple experiment: He placed a piece of meat in each of three jars, leaving one uncovered, covering the second with a piece of fine gauze, and covering the third with parchment. All three pieces of meat decayed but maggots appeared on only the meat in the uncovered jar. A few maggots appeared on the gauze of the second jar, but not on the meat, and no maggots were found on the meat covered by parchment. Redi thus demonstrated that the maggots did not come from the decaying meat, but hatched from eggs laid by blow-flies attracted by the smell of the decaying meat. Further observations showed that the maggots develop into flies which in turn lay more eggs. Louis Pasteur, about two hundred years later, showed that bacteria do not arise by spontaneous generation but only from previously existing bacteria. The submicroscopic filtrable viruses do not arise from nonviral material by spontaneous generation; the multiplication of viruses requires the presence of previously existing viruses.

The process of reproduction may be as simple as the splitting of one individual into two. In most animals, however, it involves the production of specialized eggs and sperm which unite to form the zygote or fertilized egg, from which the new organism develops. In some animals, the liver flukes for example, reproduction involves several quite different forms, each of which gives rise to the next in succession until the cycle is completed and the adult reappears.

Adaptation. To survive, an animal or plant must be able to adapt to its surroundings. Each particular species can achieve adaptation either by seeking out a suitable environment or by undergoing modifications to make it more fitted to its present surroundings. This ability to adapt is a further characteristic of all living things. Adaptation may involve immediate changes which depend upon the irritability of cells, or it may be the result of a long-term process of mutation and selection. It is obvious that no single kind of organism can adapt to all the conceivable kinds of environment; hence there will be certain areas where it cannot survive. The list of factors which may limit the distribution of a species is almost endless: water, light, temperature, food, predators, competitors, parasites, and so on.

2.2 THE ORGANIZATION OF MATTER: ATOMS AND MOLECULES

To get a more complete idea of what living matter is and what it can do, we must consider not only the easily visible macroscopic aspects of life and those that can be seen under the light or electron microscopes, but also those molecular patterns that lie far beyond the range of the microscope. To appreciate the latter requires some understanding of certain basic principles of physics and chemistry.

Matter and Energy. The universe consists of two fundamental components—matter and energy—which are related by the famous Einstein equation, $E = mc^2$, where $E =$ energy, $m =$ mass and $c =$ the velocity of light, which is a constant. This equation provides the theoretical basis for the conversion of matter to energy that occurs in an atomic bomb or nuclear reactor. In the familiar, everyday world, however, matter and energy are separate and distinguishable. Matter occupies space and has weight, and energy is the ability to produce a change or motion in matter—the ability to do work. Energy may take the form of heat, light, electricity, motion or chemical energy. We shall return (p. 59) to the kinds of energy transformations of importance in biological phenomena.

Atoms. Regardless of the form—gaseous, liquid or solid—that matter may assume, it is always composed of units called **atoms.** In nature there are 92 different kinds of atoms, ranging from hydrogen, the smallest, to uranium, the largest. In addition to these naturally occurring atoms, there are ten or more larger than uranium that have been man-made in a cyclotron or nuclear reactor. All of them, natural and synthetic, are much smaller than the tiniest particle visible under the microscope. In fact, no one has ever seen an atom; atomic structure and properties have been inferred from experiments made with many types of elaborate apparatus.

Once the atom was believed to be the ultimate, smallest unit of matter. Now physicists know that atoms are divisible into even smaller particles organized around a central core, as our solar system of planets is organized around the sun. The exact number and kind of these particles and their arrangement in the atom are the subject of continuing research. For our purposes we need consider only three types: **electrons,** which have a negative electric charge and an extremely small mass or weight; **protons,** which have a positive electric charge and are about 1800 times as heavy as electrons; and **neutrons,** which have no electrical charge, but have essentially the same mass as protons.

The center of the atom, corresponding in position to the sun in our solar system, is the **nucleus**, containing protons and neutrons; it comprises almost the total mass of the atom. Just as the solar system is mostly empty space, so is the atom, in which electrons move in circular or other paths in the empty space around the nucleus. Each type of atom has a characteristic number of electrons circling the nucleus and characteristic numbers of protons and neutrons in the nucleus (Fig. 2.2). In all atoms, the number of protons in the nucleus equals the number of electrons circling around it, so that the atom as a whole is in a state of electrical neutrality. There are only a few kinds of particles other than protons, neutrons and electrons; the different kinds of matter reflect differences in the number and arrangement of these basic particles. Living systems are composed of exactly the same kinds of atoms, with the same kind of atomic structure, as nonliving systems.

Elements. An **element** is a substance composed of atoms all of which have the same number of protons in the atomic nucleus and therefore the same number of electrons circling in the orbits. A few elements can be found in nature as such—gold, silver, iron and copper, for example—but most elements have a strong tendency to unite with other elements to form compounds.

The unique aliveness of living things does not depend upon the presence of some rare or unique element. Four elements, carbon, oxygen, hydrogen and nitrogen, make up about 96 per cent of the material in the human body. Potassium, sulfur, calcium and phosphorus are four other elements usually present to the extent of 1 per cent or more each. Since bone is largely composed of calcium and phosphorus, the amount of these elements is much greater in a bony animal than in a completely soft-bodied one. Smaller amounts of sodium, chlorine, iron, iodine, magnesium, copper, manganese, cobalt, zinc and a few others complete the list. The unique aliveness of living matter does not depend on the presence of some rare or unique element, for these same elements are abundant in the atmosphere, in the sea and in the earth's crust. The phenomenon of life depends, instead, upon

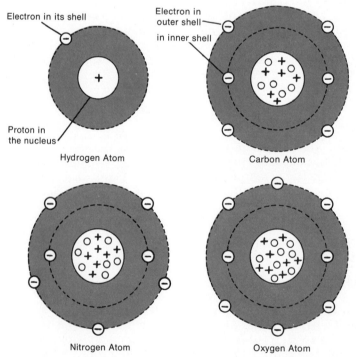

Hydrogen Atom

Carbon Atom

Nitrogen Atom

Oxygen Atom

Figure 2.2   Diagrams of the structure of the atoms of the four chief elements of living matter: hydrogen, carbon, nitrogen and oxygen. The symbols used are $\bigcirc$, neutron; +, proton; $\ominus$, electron.

the complexity of the interrelationships of these common, abundant elements.

For convenience in writing chemical formulas and reactions, chemists have assigned to each of the elements a symbol, usually the first letter of the name of the element: O, oxygen; H, hydrogen; C, carbon; N, nitrogen. A second letter is added to the symbol of those elements with the same initial letter: Ca, calcium; Co, cobalt; Cl, chlorine; Cu, copper; Na, sodium (Latin *natrium*).

Ions. The chemical properties of an element are determined primarily by the number and arrangement of electrons revolving in the outermost orbit around the atomic nucleus and to a lesser extent by the number of electrons in the inner orbits. These, in turn, depend upon the number and kind of particles, protons and neutrons, in the nucleus.

The number of electrons in the outermost shell varies from zero to eight in different kinds of atoms (Fig. 2.2). If there are zero or eight electrons in the outer shell the element is chemically inert and will not readily combine with other elements. When there are fewer than eight electrons, the atom tends to lose or gain electrons in an attempt to achieve an outer orbit of eight electrons. Since the number of protons in the nucleus is not changed, this loss or gain of electrons produces an atom with a net positive or negative charge. Such electrically charged atoms are known as **ions**. Atoms with one, two or three electrons in the outer shell tend to lose them to other atoms and become positively charged ions because of the excess protons in the nucleus (e.g., Na^+, sodium ion; Ca^{++}, calcium ion). These are termed **cations** because they migrate to the cathode of an electrolytic cell. Atoms with five, six or seven electrons in the outer shell tend to gain electrons from other atoms and become negatively charged ions or **anions** (e.g., Cl^-, chloride ion). Anions migrate to the anode or positively charged electrode of an electrolytic cell. Because they bear opposite electric charges, anions and cations are attracted to each other and unite forming an ionic or electrostatic bond. Atoms such as carbon, which have four electrons in the outer orbit, neither lose nor gain electrons, but share them with adjacent atoms.

Isotopes. Physical research has shown that most elements are composed of two or more kinds of atoms, which differ in the number of neutrons in the atomic nucleus. The different kinds of atoms of an element are called **isotopes** (iso = equal, tope = place), because they occupy the same place in the periodic table of the elements. All the isotopes of a given element have the same number of electrons circling the atomic nucleus.

Although the isotopes of a given element have the same chemical properties, they can be differentiated physically. Some of them are radioactive and can be detected and measured by an instrument such as the **Geiger counter** by the kind and amount of radiation they emit. Others can be differentiated by the slight difference in the mass of the atoms caused by an extra neutron in the nucleus.

Substances containing ^{15}N (heavy nitrogen) instead of ^{14}N, the usual isotope, or ^2H (heavy hydrogen, deuterium) instead of ^1H, will have a greater mass detectable with a **mass spectrometer**.

A tremendous insight has been gained into the details of the metabolic activities of cells by preparing some substance — sugar, for example — labeled with radioactive carbon (^{11}C or ^{14}C) or heavy carbon (^{13}C) in place of ordinary carbon (^{12}C). The labeled substances are fed or injected into an animal or plant, or cells are incubated in a solution containing the tracer, and the labeled products resulting from the cell's or organism's normal metabolic processes are then isolated and identified. By such experiments it has been possible to trace step by step the sequence of reactions undergone by a given compound and to determine the form or forms in which the labeled atoms finally leave the cell or organism. The rate of formation of bone, for example, and the effect of vitamin D and parathyroid hormone on this process, can be studied with the aid of radioactive calcium, ^{45}Ca. Many biologic problems that could not be attacked in any other way can be solved by this method.

Analysis of the human body reveals that it contains about 50 per cent carbon, 20 per cent oxygen, 10 per cent hydrogen, 9 per cent nitrogen, 4 per cent calcium, 2.5 per cent phosphorus (P), 1 per cent potassium (K), 0.8 per cent sulfur (S), 0.4 per cent sodium (Na), and 0.4 per cent chlorine (Cl). Analyses of other animals would yield comparable results. Such analyses are not very informative unless the animal has some unusual element. Tunicates, for example, are unusual in that they contain a large amount of the element vanadium (V).

2.3 CHEMICAL COMPOUNDS

Most elements are present in living material as chemical compounds, rather than as free elements. A **chemical compound** is a substance composed of two or more different kinds of atoms or ions joined together; hence they can be decomposed or split apart into two or more simpler substances. The amounts of the elements in a given compound are always present in a definite proportion by weight. This reflects the fact that the atoms are joined together by chemical bonds in a precise way to form the compound. The assemblage of atoms held together by chemical bonds is called a **molecule.** A molecule is the smallest particle of a compound that has the composition and properties of a larger portion of it. A molecule is made of two or more atoms — which may be the same, as in a molecule of oxygen, O_2, or nitrogen, N_2 — or they may be atoms of different elements. The properties of a chemical compound are usually quite different from the properties of its constituent elements. Each water molecule, for example, contains two atoms of hydrogen and one atom of oxygen, and the chemical properties of water are quite different from those of either hydrogen or oxygen. Chemists state this fact by writing the chemical formula of water as H_2O. A **chemical formula** gives, in the chemist's shorthand, the kinds of atoms present in a molecule and their relative proportion.

A molecule of table salt, or sodium chloride, is made of one atom of sodium and one of chlorine. A molecule of the simple sugar, glucose, is composed of three kinds of atoms: six carbon atoms, twelve hydrogen atoms and six oxygen atoms; its formula in chemical shorthand is $C_6H_{12}O_6$. Any larger portion of glucose — a gram or kilogram — will also contain carbon, hydrogen and oxygen in these same proportions. If it were possible to divide a gram of glucose in half, then divide the half in half, then the quarters in half, and so on, eventually the glucose would be subdivided into its con-

stituent molecules. Each of these molecules would have the same composition and properties as the original gram. However, if a glucose molecule were divided, the parts no longer would be sugar, but would be carbon, hydrogen and oxygen atoms, and these would have quite different properties.

The weight of any single atom or molecule is exceedingly small, much too small to be expressed in terms of grams or micrograms. These weights are expressed in terms of the atomic weight unit, the **dalton,** approximately the weight of a proton or neutron. An atom of the lightest element, hydrogen, weighs one dalton, a carbon atom weighs 12 daltons and oxygen has an atomic weight of 16 daltons. The **molecular weight** of a compound is the sum of the atomic weights of its constituent atoms. Thus the molecular weight of water, H_2O, is $(2 \times 1) + 16$, or 18 daltons. The molecular weight of glucose, $C_6H_{12}O_6$, is $(6 \times 12) + (12 \times 1) + (6 \times 16)$, or 180 daltons.

A large part of each cell is simply **water.** In an animal such as man, the water content ranges from about 20 per cent in bone to 85 per cent in brain cells. The water content is greater in embryonic and young cells and decreases as aging occurs. About 70 per cent of our total body weight is water; as much as 95 per cent of a jellyfish is water. Water has a number of important functions in living systems. Most of the other chemicals present are dissolved in it; they must be dissolved in water in order to react. Water also dissolves the waste products of metabolism and assists in their removal. Water has a great capacity for absorbing heat with a minimal change in its own temperature; thus, it protects the living material against sudden thermal changes. Since water absorbs a large amount of heat as it changes from a liquid to a gas, the mammalian body can dissipate excess heat by the evaporation of sweat. Water's high heat conductivity makes possible the even distribution of heat throughout the body. Finally, water has an important function as a lubricant. It is present in body fluids wherever one organ rubs against another and in joints where one bone moves on another.

A **mixture** contains two or more kinds of atoms or molecules which may be present in varying proportions. Air is a mixture of oxygen, nitrogen, carbon dioxide and water vapor, plus certain rare gases such as argon. The proportions of these constituents may vary widely. Thus, in contrast to a pure compound, which has a fixed ratio of its constituents and definite chemical and physical properties, a mixture has properties which vary with the relative abundance of its constituents.

To learn more about the constituents of cells, biochemists have used very sensitive analytical techniques and have taken great pains to preserve the extremely labile substances present in these enormously complicated systems. To prevent the disappearance of certain substances it is necessary to quick-freeze a bit of excised tissue, or even a whole small animal, by dropping it directly into liquid air. Biochemical research has made it abundantly clear that the composition of every cell is constantly changing, that the cell constituents are in a "dynamic state." There is a continuous synthesis of large, energy-rich molecules and continual decomposition of these into smaller, energy-poor ones. Some of the most important compounds are present in cells only in extremely minute amounts at any given time, although the total amount formed and used in a 24 hour period may be quite large. An appreciation of this may be gained from the following consideration: when substances undergo chemical reactions in sequence (and almost all the reactions of importance biologically are sequences or "cycles"), such as $A \rightarrow B \rightarrow C \rightarrow D$, the rate of the whole process is controlled by the rate of the slowest reaction in the chain. For example, if reaction $A \rightarrow B$ is 10 times as fast as $B \rightarrow C$, and if $C \rightarrow D$ is 100 times as fast as $B \rightarrow C$, then the least reactive substance, B, will tend to accumulate and the most reactive one, C, will be present in the smallest amount. For this reason many of the most active and important substances are present in cells in extremely minute amounts. This, coupled with their chemical instability, has made their detection and isolation difficult. There are probably many such intermediate compounds that remain to be discovered.

The compounds present in cells are of two main types: **inorganic** and **organic**. The latter include all the compounds (other than carbonates) that contain the element carbon. The element carbon is able to form a much wider variety of compounds than any other element because the outer shell of the carbon atom contains four electrons, which can be shared in a number of different ways with adjacent atoms. At one time it was believed that organic compounds were uniquely different from other chemical substances and that they could be produced only by living matter. This hypothesis was disproved when the German chemist Wöhler succeeded in 1828 in synthesizing urea (one of the waste products found in human urine) from the inorganic compounds ammonium sulfate and potassium cyanate. Since that time thousands of organic compounds have been synthesized, some of which are quite complex molecules of great biologic importance such as vitamins, hormones, antibiotics and drugs.

2.4 ELECTROLYTES: ACIDS, BASES AND SALTS

The inorganic compounds important in living systems are acids, bases and salts. An **acid** is a compound which releases hydrogen ions (H^+) when dissolved in water.* Acids turn blue litmus paper to red and have a sour taste. Hydrochloric (HCl) and sulfuric (H_2SO_4) are examples of inorganic acids; lactic ($C_3H_6O_3$) from sour milk and acetic ($CH_3 \cdot COOH$) from vinegar are two common organic acids. A **base** is a compound which releases hydroxyl ions (OH^-) when dissolved in water. Bases turn red litmus paper blue. Common inorganic bases include sodium hydroxide ($NaOH$) and ammonium hydroxide (NH_4OH).

For convenience in stating the degree of acidity or alkalinity of a fluid, the hydrogen ion concentration may be expressed in terms of pH, the negative logarithm of the hydrogen ion concentration. On this scale, a neutral solution has a pH of 7 (its hydrogen ion concentration is 0.0000001 or 10^{-7} molar), alkaline solutions have pH's ranging from 7 to 14 (the pH of 1 M NaOH), and acids have pH's from 7 to 0 (the pH of 1 M HCl). Most animal cells are neither strongly acid nor alkaline but contain a mixture of

---

* Other definitions of acid and base, such as those by Brønsted or Lewis, may be useful in understanding certain more complex reactions. Brønsted defines an acid as a proton donor and a base as a proton acceptor.

acidic and basic substances; their pH is about 7.0. Any considerable change in the pH of a cell is inconsistent with life. Since the scale is a logarithmic one, a solution with a pH of 6 has a hydrogen ion concentration 10 times as great as that of one with a pH of 7.

When an acid and a base are mixed, the hydrogen ion of the acid unites with the hydroxyl ion of the base to form a molecule of water (H_2O). The remainder of the acid (anion) combines with the rest of the base (cation) to form a **salt.** For example, hydrochloric acid (HCl) reacts with sodium hydroxide (NaOH) to form water and sodium chloride (NaCl) or common table salt:

$$H^+Cl^- + Na^+OH^- \rightarrow H_2O + Na^+Cl^-$$

A salt may be defined as a compound in which the hydrogen atom of an acid is replaced by some metal.

When a salt, an acid or a base is dissolved in water it separates into its constituent ions. These charged particles can conduct an electric current; hence these substances are known as **electrolytes.** Sugars, alcohols and the many other substances which do not separate into charged particles when dissolved, and therefore do not conduct an electric current, are called **nonelectrolytes.**

Cells and extracellular fluids contain a variety of mineral salts, of which sodium, potassium, calcium and magnesium are the chief cations (positively charged ions) and chloride, bicarbonate, phosphate and sulfate are the important anions (negatively charged ions). The body fluids of land vertebrates resemble sea water in the kinds of salts present and in their relative proportions, but the total concentration of salts is only about one-fifth as great as in sea water. Most biologists now believe that life originated in the sea. The cells of those early organisms became adapted to function optimally in the presence of this pattern of salts. As larger animals evolved and developed body fluids, this pattern of salts was maintained, even as some of the descendants migrated into fresh water or onto the land, for any marked change in the kinds of salts present would inhibit certain enzymes and place that kind of animal at a marked disadvantage in the competition for survival.

Some animals have evolved kidneys and other excretory organs that selectively retain or secrete certain ions, thus leading to body fluids with somewhat different relative concentrations of salts. The concentration of each ion is determined by the relative rates of its uptake and excretion by the organism.

Although the concentration of salts in cells and in the body fluids is small, this amount is of great importance for normal cell functioning. The concentrations of the respective cations and anions are kept remarkably constant under normal conditions; any marked change results in impaired function and finally in death. A great many of the enzymes which mediate the chemical reactions occurring in the body require one or another of these ions — for example, magnesium, manganese, cobalt, potassium — as cofactors. These enzymes are unable to function in the absence of the ion. Normal nerve function requires a certain concentration of calcium in the body fluids; a decrease in this results in convulsions and death. Normal muscle contraction requires certain amounts of calcium, potassium and sodium. If a frog heart, for example, is removed from the body and placed in a solution of sodium chloride, it soon stops beating and remains in the relaxed state. If placed in a solution of potassium chloride, or in a mixture of sodium and calcium chloride, it ceases beating in the contracted condition. But if it is placed in a solution of the three salts in proper proportion it will continue to beat for hours. Under the proper conditions, the strength of the heartbeat is proportional to the concentration of calcium ions in the fluid bathing the heart; this method is sensitive enough to be used to measure the concentration of calcium ions. In addition to these several specific effects of particular cations, mineral salts serve an important function in maintaining the osmotic relationships between each cell and its environment.

2.5 ORGANIC COMPOUNDS OF BIOLOGIC IMPORTANCE

The major types of organic substances present in cells are the carbohydrates, proteins, fats, nucleic acids and steroids. Some of these are required for the structural integrity of the cell, others to supply energy for its functioning, and still others are of prime importance in regulating metabolism within the cell. **Carbohydrates** and **lipids** are the chief sources of chemical energy in almost every form of life; **proteins** are structural elements but are of even greater importance as catalysts (enzymes) and regulators of cellular processes. **Nucleic acids** are of prime importance in the storage and transfer of information used in the synthesis of specific proteins and other molecules.

The basic pattern of the types of substances, and even their relative proportions, is remarkably similar for cells from the various parts of the body and for cells from different animals. A bit of human liver and the substance of an ameba each contain about 80 per cent water, 12 per cent protein, 2 per cent nucleic acid, 5 per cent fat, 1 per cent carbohydrate and a fraction of 1 per cent of steroids and other substances. Certain specialized cells, of course, have unique patterns of chemical constituents; the brain, for example, is rich in certain kinds of fats.

Carbohydrates. The simplest of the organic substances are the carbohydrates—the sugars, starches and celluloses—which contain carbon, hydrogen and oxygen in a ratio of 1 C : 2 H : 1 O. Carbohydrates are found in all living cells, usually in relatively small amounts, and are important as readily available sources of energy. Both **glucose** (also known as dextrose) and **fructose** (also called levulose) are simple sugars with the formula $C_6H_{12}O_6$. However, the arrangement of the atoms within the two molecules is different and the two sugars have somewhat different chemical properties and quite different physiologic roles. Such differences in the molecular configurations of substances with the same chemical formula are frequently found in organic chemistry. Chemists indicate the molecular configuration of a substance by a **structural formula** in which the atoms are represented by their symbols—C, H, O, etc.—and the chemical bonds or forces which hold the atoms together are indicated by lines. Hydrogen has one such bond; oxygen, two; nitrogen, three; and carbon, four. The structural formulas of glucose and fructose are compared in Figure 2.3. Note that the lower four carbon atoms in the two sugars have identical groups; only the upper two show differences.

Carbon atoms can unite with each other as well as with many other kinds of atoms and form an almost infinite variety of compounds. Carbon atoms linked together may form long chains (as in fatty acids), branched chains (certain amino acids), rings (purines and pyrimidines) and complex rings (steroids). Molecules are in fact three-dimensional structures, not simple two-dimensional ones as these formulas suggest. There are more complex ways of representing the third dimension of the molecule. Since the properties of the compound depend in part on the exact nature of its three-dimensional structure, its **conformation,** such three-dimensional formulas are helpful in understanding the intimate relations between molecular structure and function.

Glucose is the only simple sugar which occurs in any quantity in the cells and body fluids of both vertebrates and invertebrates. The other carbohydrates eaten by vertebrates are converted to glucose in the liver. Glucose is an indispensable component of mammalian blood and is normally present in a concentration of about 0.1 per cent. No particular harm results from a simple increase in the concentration of glucose in the body fluids, but when the concentration is reduced to 0.04 per cent or less, the brain cells become hyperirritable. They discharge nerve impulses which result in muscular twitches, convulsions, and finally unconsciousness and death. Brain cells use glucose as their prime metabolic fuel, and a certain minimum concentration of glucose in the blood is required to supply this. A complex physiologic control mechanism, which operates like the "feedback" controls of electronic devices and which involves

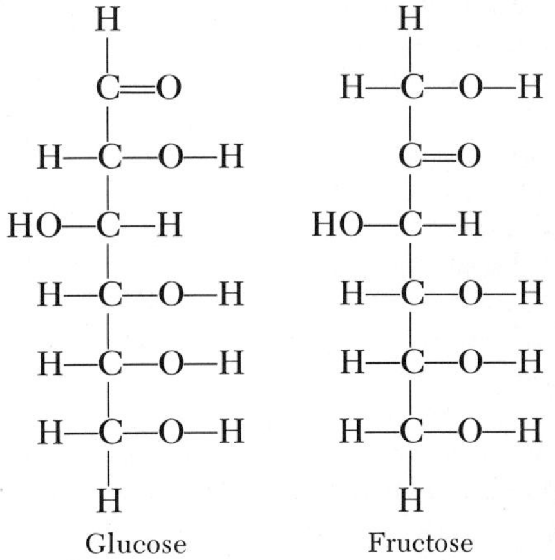

Figure 2.3 Structural formulas of two simple sugars.

the liver, pancreas, pituitary and adrenal glands, maintains the proper concentration of glucose in the blood.

The double sugars, with the formula $C_{12}H_{22}O_{11}$, consist of two molecules of simple sugar joined by the removal of a molecule of water. **Sucrose,** or table sugar, is a combination of glucose and fructose. Other common double sugars are **maltose,** composed of two molecules of glucose, and **lactose,** composed of glucose and galactose. Lactose, found in the milk of all mammals, is an important item in the diet of the young of these forms. Fructose, the sweetest of the common sugars, is more than 10 times sweeter than lactose; sucrose is intermediate.

Most animal cells contain some **glycogen** or animal starch, the molecules of which are made of a very large number—thousands—of molecules of glucose joined together by the removal of an H from one and an OH from the next. Glycogen is the form in which animal cells typically store carbohydrate for use as an energy source in cell metabolism. The glycogen molecules within a living cell are constantly being built up and broken down. Glucose and other simple sugars are not a suitable storage form of carbohydrate for, being soluble, they readily pass out of the cells. The molecules of glycogen, which are much larger and less soluble, cannot pass through the plasma membrane. Glycogen is typically stored intracellularly as microscopic granules, which can be made visible by special stains. Glycogen is readily converted into small molecules such as glucose-phosphate (p. 71) to be metabolized within the cell.

Cellulose, also composed of hundreds of molecules of glucose, is an insoluble carbohydrate which is a major constituent of the tough outer wall of plant cells. The bonds joining the glucose molecules in cellulose are different from those joining the glucoses in glycogen and are not split by the amylases that digest glycogen and starch.

Glucosamine and galactosamine, nitrogen-containing derivatives of the sugars glucose and galactose, are important constituents of supporting substances such as connective tissue fibers, cartilage and chitin, a constituent of the hard outer shell of insects, spiders and crabs.

Carbohydrates serve as a readily available fuel to supply energy for metabolic processes. Glucose is metabolized to carbon dioxide and water with the release of energy. A few carbohydrates combine with proteins (**glycoproteins**) or lipids (**glycolipids**) to serve as structural components of certain cells. **Ribose** and **deoxyribose** are five-carbon sugars that are components of ribonucleic acid (RNA) and deoxyribonucleic acid (DNA).

Fats. The term **fat,** or lipid, refers to a heterogeneous group of compounds which share the property of being soluble in chloroform, ether or benzene, but are only very sparingly soluble in water. True fats are composed of carbon, hydrogen and oxygen, but have much less oxygen than carbon. Fats have a greasy or oily consistency; some, such as beef tallow fat or bacon fat, are solid at ordinary temperatures, others, such as olive oil or cod liver oil, are liquid. Each molecule of fat is composed of one molecule of **glycerol** and three molecules of **fatty acid.** All such neutral fats, termed **triglycerides,** contain glycerol but may differ in the kinds of fatty acids present. Fatty acids are long chains of carbon atoms with a carboxyl group (—COOH) at one end. All fatty acids in nature have an even number of carbon atoms—palmitic has 16 and stearic, 18. Fatty acids with one or more double bonds are called **unsaturated.** Oleic acid has 18 carbons and one double bond (and hence has two less hydrogen atoms than stearic). A fat common in beef tallow, tristearin, $C_{57}H_{110}O_6$, has three molecules of stearic acid and one of glycerol (Fig. 2.4). Fats containing unsaturated fatty acids are usually liquid at room temperature, whereas **saturated** fats, such as tristearin, are solids.

Fats are important as fuels and as structural components of cells, especially cell membranes. Glycogen or starch is readily converted to glucose and metabolized to release energy quickly; the carbohydrates serve as short-term sources of energy. Fats yield more than twice as much energy per gram as do carbohydrates and thus are a more economical form for the storage of food reserves. Carbohydrates can be transformed by the body into fats and stored in this form—a restatement of the generally known fact that starches and sugars are "fattening." The reverse may also occur to some slight extent: Parts of the fat molecule may be converted into glucose and other

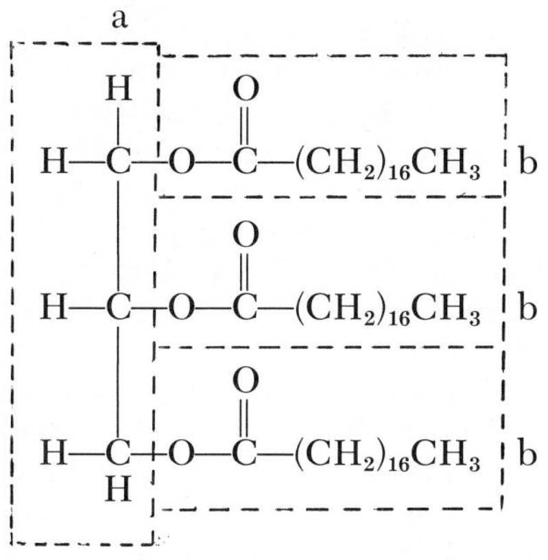

Figure 2.4 Structural formula of tristearin, a fat composed of glycerol (*a*) and three molecules of stearic acid (*b*). In the formula, $(CH_2)_{16}$ represents a chain of sixteen carbon atoms joined in a line, $-\overset{\displaystyle H}{\underset{\displaystyle H}{C}}-\overset{\displaystyle H}{\underset{\displaystyle H}{C}}-\ldots$, to each of which are attached two hydrogen atoms. (Villee: Biology, 6th ed.)

carbohydrates. This has been shown by preparing fatty acids or glycerol labeled with radioactive or heavy carbon, feeding or injecting these into a rat or dog, and then isolating glucose from the blood or glycogen from the liver and demonstrating that these molecules now contain some of the labeled carbon atoms.

Fats are important structural elements of the body. The plasma membrane around each cell and the nuclear membrane contain fatty substances as important constituents, and the myelin sheath around the nerve fibers (p. 56) has a high lipid content. Animals store fat as globules within the cells of adipose tissue. The layer of adipose tissue just under the skin serves as an insulator against the loss of body heat. Women tend to have a thicker layer of adipose tissue than do men and thus should be more tolerant to cold. Whales, which live in cold water and have no insulating hair, have an especially thick layer of fat (blubber) just under the skin for this purpose. The subcutaneous fat in man keeps the skin firm.

The fat deposits are not simply long-term stores of foodstuff used only in starvation, but are constantly being used and re-formed. Studies with labeled fatty acids showed that mice replace half of their stored fats in seven days.

Besides the true fats, composed of glycerol and fatty acids, lipids include several related substances that contain components such as phosphorus, choline and sugars in addition to fatty acids. The **phospholipids** are important constituents of the membranes of plant and animal cells in general and of nerve cells in particular. The fatty acid portion of the phospholipid molecule is **hydrophobic,** not soluble in water. The other portion, composed of glycerol, phosphate and a nitrogenous base such as choline, is ionized and readily water-soluble. For this reason, phospholipid molecules in a film tend to be oriented with the polar, water soluble portion pointing one way and the nonpolar, fatty acid portion pointing the other. This configuration appears to underlie the three-layered unit membrane structure of protein:phospholipid:protein.

Beeswax, lanolin and other **waxes** contain a fatty acid plus an alcohol other than glycerol. **Cerebrosides,** as their name indicates, are fatty substances found especially in nerve tissue. They contain galactose, long chain fatty acids, and a long chain amino alcohol, sphingosine. The metabolic roles of these special fats are not clear at present.

Steroids. Steroids are complex molecules containing carbon atoms arranged in four interlocking rings, three of which contain six carbon atoms each and the fourth of which contains five. Vitamin D, male and female sex hormones, the adrenal cortical hormones, bile salts and cholesterol are examples of steroids. **Cholesterol** (Fig. 2.5) is an important structural component of nervous tissue and other tissues, and the steroid hormones are of great importance in regulating certain aspects of metabolism.

Proteins. Proteins differ from carbohy-

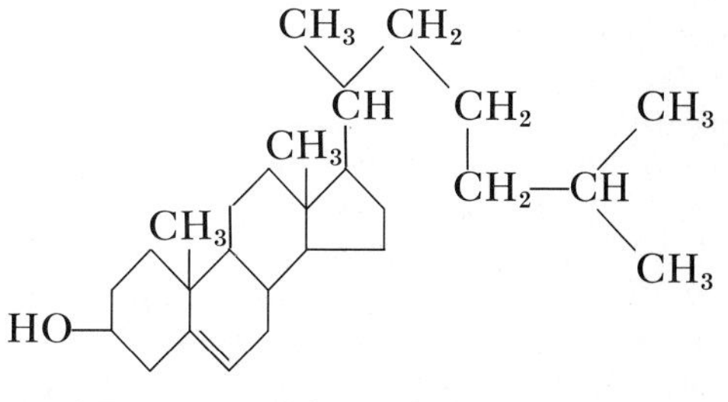

Figure 2.5 Structural formula of a sterol, cholesterol.

drates and true fats in that they contain nitrogen in addition to carbon, hydrogen and oxygen. Proteins typically contain sulfur and phosphorus also. Proteins are among the largest molecules present in cells and share with nucleic acids the distinction of great complexity and variety. Hemoglobin, the red pigment found in the blood of all vertebrates and many invertebrates, has the formula $C_{3032}H_{4816}O_{872}N_{780}S_8Fe_4$ (Fe is the symbol for iron). Although the hemoglobin molecule is enormous compared to a glucose or triolein molecule, it is only a small- to medium-sized protein. Many, indeed most, of the proteins within a cell are **enzymes,** biological catalysts which control the rates of the many chemical processes of the cell.

Protein molecules are made of simpler components, the **amino acids,** some 30 or more of which are known. Since each protein contains hundreds of amino acids, present in a certain proportion and in a particular order, an almost infinite variety of protein molecules is possible. Analytical tools such as the amino acid analyzer permit one to determine the sequence of the amino acids in a given protein molecule. **Insulin,** the hormone secreted by the pancreas and used in the treatment of diabetes, was the first protein whose structure was elucidated. Work culminating in 1957 revealed the sequence of the 124 amino acids that comprise the molecules of the enzyme ribonuclease, secreted by the pancreas.

It is possible to distinguish several different levels of organization in the protein molecule. The first level is the so-called **primary structure** which depends upon the sequence of amino acids in the **polypeptide chain.** This sequence, as we shall see, is determined in turn by the sequence of nucleotides in the RNA and DNA of the nucleus of the cell. A second level of organization of protein molecules involves the coiling of the polypeptide chain into a helix or into some other regular configuration. The polypeptide chains ordinarily do not lie out flat in a protein molecule, but they undergo coiling to yield a three-dimensional structure. One of the common secondary structures in protein molecules is the α-**helix,** which involves a spiral formation of the basic polypeptide chain. The α-helix is a very uniform geometric structure with 3.6 amino acids occupying each turn of the helix. The helical structure is determined and maintained by the formation of hydrogen bonds between amino acid residues in successive turns of the spiral. A third level of structure of protein molecules involves the folding of the peptide chain upon itself to form globular proteins. Again, weak bonds such as hydrogen, ionic and hydrophobic bonds form between one part of the peptide chain and another part, so that the chain is folded in a specific fashion to give a specific overall structure of the protein molecule. Covalent bonds such as disulfide bonds (—S—S—) are important in the tertiary structure of many proteins. The biological activity of a protein depends in large part on the specific tertiary structure which is held together by these bonds. When a protein is heated or treated with any of a variety of chemicals, the tertiary structure is lost. The coiled peptide chains unfold to give a random configuration accompanied by a loss of the biological activity of the protein. This change is termed **denaturation.**

Each cell contains hundreds of different proteins and each kind of cell contains some proteins which are unique to it. There is evidence that each species of animal and plant has certain proteins which are different from those of all other species. The degree of similarity of the proteins of two species is a measure of their evolutionary relationship. The **theory of species specificity** states that each species has a characteristic pattern of its constituent proteins and that this pattern differs at least slightly from that of related species and more markedly from those of more distantly related species. Because of the interactions of unlike proteins, grafts of tissue removed from one animal will usually not grow when implanted on a host of a different species, but degenerate and are sloughed off by the host. Indeed, even grafts made between members of the same species will usually not grow, but only grafts between genetically identical donors and hosts—identical twins or members of a closely inbred strain.

Amino acids differ in the number and arrangement of their constituent atoms, but all contain an **amino group** (NH_2) and a **carboxyl group** (COOH). The amino group enables the amino acid to act as a base and combine with acids; the carboxyl group enables it to combine with bases. For this reason, amino acids and proteins are important biological "buffers" and resist changes in acidity or alkalinity. Protein molecules are built up by linkages, **peptide bonds,** between the amino group of one amino acid and the carboxyl group of the adjacent one (Fig. 2.6). Pure amino acids have a rather sweet taste. The proteins eaten by an animal are not incorporated directly into its cells but are first digested to the constituent amino acids to enter the cell. Subsequently each cell combines the amino acids into the proteins which are characteristic of that cell. Thus, a man eats beef proteins in a steak, but breaks them down to amino acids in the process of digestion, then rebuilds them as human proteins—human liver proteins, human muscle proteins, and so on.

Proteins and amino acids may serve as energy sources in addition to their structural and enzymatic roles. The amino group is removed by an enzymatic reaction, **deamination,** and then the remaining carbon skeleton enters the same metabolic paths as glucose and fatty acids and eventually is converted to carbon dioxide and water by the Krebs tricarboxylic acid cycle (p. 68) and associated paths. The amino group is excreted as ammonia, urea, uric acid or some other nitrogenous compound, depending on the kind of animal. In prolonged fasting, after the supply of carbohydrates and fats has been exhausted, the cellular proteins may be used as a source of energy.

Animal cells can synthesize some, but not all, of the different kinds of amino acids; different species differ in their synthetic abilities. Man, for example, is apparently unable to synthesize eight of these; they must either be supplied in the food eaten or perhaps synthesized by the bacteria present in the intestine. Plant cells apparently can synthesize all the amino acids. The ones which an animal cannot synthesize, but must obtain in its diet, are called **essential amino acids.** It must be kept in mind that these are no more essential for protein synthesis than any other amino acid, but are simply essential constituents of *the diet,* without which the animal fails to grow and eventually dies.

Nucleic Acids. Nucleic acids are complex molecules, larger than most proteins, and contain carbon, oxygen, hydrogen, nitrogen and phosphorus. They were first isolated by Miescher in 1870 from the nuclei of pus cells and gained their name from the fact that they are acidic and were first identified in nuclei. There are two classes of nucleic acid—one containing ribose and called **ribose nucleic acid** or **RNA,** and one containing deoxyribose and called **deoxyribonucleic acid** or **DNA.** There are many different kinds of RNA and DNA which differ in their structural details and in their metabolic functions. DNA occurs in the chromosomes in the nucleus of the cell and, in much smaller amounts, in mitochondria and chloroplasts. It is the primary repository for biological information. RNA is present in the nucleus, especially in the nucleolus, in the ribosomes, and in lesser amounts in other parts of the cell.

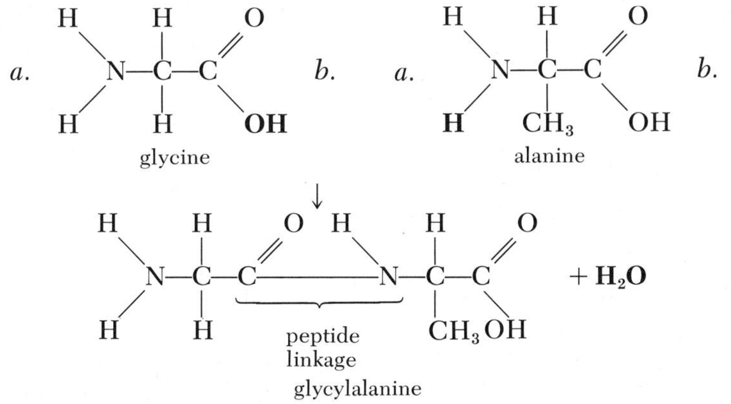

Figure 2.6 Structural formulas of the amino acids glycine and alanine, showing, (*a*) the amino group and, (*b*) the acid (carboxyl) group. These are joined in a peptide linkage to form glycylalanine by the removal of water.

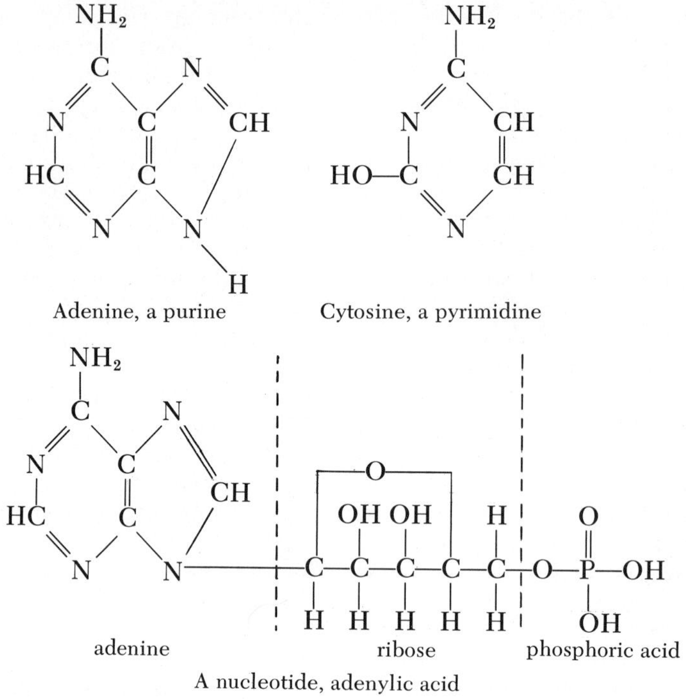

Figure 2.7 Structural formulas of a purine, adenine; a pyrimidine, cytosine; and a nucleotide, adenylic acid. (Villee: Biology, 6th ed.)

Nucleic acids are composed of units, called **nucleotides,** each of which contains a nitrogenous base, a five-carbon sugar and phosphoric acid. Two types of nitrogenous bases, purines and pyrimidines, are present in nucleic acids (Fig. 2.7). RNA contains the purines **adenine** and **guanine** and the pyrimidines **cytosine** and **uracil,** together with the pentose, ribose, and phosphoric acid. DNA contains adenine and guanine, cytosine and the pyrimidine **thymine,** together with deoxyribose and phosphoric acid. The molecules of nucleic acids are made of linear chains of nucleotides, each of which is attached to the next by bonds between the sugar part of one and the phosphoric acid of the next. The specificity of the nucleic acid resides in the specific sequence of the four kinds of nucleotides present in the chain. For example, CCGA-TTA might represent a segment of a DNA molecule, with C = cytosine, G = guanine, A = adenine and T = thymine.

An enormous mass of evidence now indicates that DNA is responsible for the specificity and chemical properties of the **genes,** the units of heredity. There are several kinds of RNA, each of which plays a specific role in the biosynthesis of specific proteins by the cell (p. 189).

2.6 PHYSICAL CHARACTERISTICS OF CELLULAR CONSTITUENTS

The properties of cell constituents depend not only on the kinds and quantities of substances present, but on their physical state as well. A mixture of a substance with water, or other liquid, may result in a true solution, a suspension or a colloidal solution. These are differentiated by the size of the dispersed particles. In a **true solution,** the ions or molecules of the dissolved substance (called the **solute**) are of extremely small size, less than 0.0001 micron in diameter. The solute particles are either ions or small molecules dispersed among the molecules of the dissolving liquid (called the **solvent**). A true solution is transparent and has a higher boiling point and a lower freezing point than pure water. Most acids, bases, salts and some nonelectrolytes, such as sugars, form true solutions in water.

The dispersed particles in a **suspension,** in contrast, are much larger (greater than 0.1 micron) and are composed of aggregations of many molecules. They tend to settle out if the suspension is allowed to stand. Muddy water, for example, contains

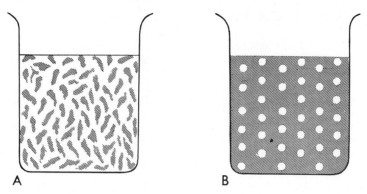

Figure 2.8   Diagram of a colloidal solution as (A) a sol and (B) a gel. The sol contains water as the continuous phase in which the colloidal particles (dark rods) are dispersed. In the gel the colloidal particles have coalesced to form a continuous lacy network in which the water droplets (light circles) are dispersed.

particles of clay in suspension. Suspensions are opaque rather than transparent, and have the same boiling and freezing points as pure water.

A **colloidal solution** contains particles intermediate in size between those of a true solution and a suspension, particles from 0.001 to 0.1 micron in diameter. A colloidal solution, or colloid, is transparent or translucent, has about the same boiling and freezing points as pure water, and is stable; it does not tend to separate into its constituent parts on standing. The particles of a colloidal solution may have a positive or a negative charge, but usually they all have the same charge and tend to repel each other. The presence of the charge is a factor which tends to keep the particles dispersed. A colloidal solution has the unique property of changing from a liquid state, or **sol,** to a solid or semisolid state or **gel** (Fig. 2.8). When a package of gelatin is dissolved in hot water, the particles of gelatin (a protein) are dispersed through the water and a liquid colloidal solution, a sol, results. As the gelatin cools, the gelatin particles aggregate and become the continuous phase, the water particles become dispersed as small droplets in the gelatin and a semisolid gel results. The gel can be converted back to a sol by reheating. The gelatin-water mixture is a liquid sol when it consists of particles of gelatin dispersed in water and a solid gel when the droplets of water are dispersed in gelatin. The sol-gel change in a substance may be effected by changing the temperature, the pH or the salt concentration or by mechanical agitation (whipping cream, for example). The change is reversible, but if the system is subjected to large changes of temperature, acidity, alkalinity or salt concentration, the colloidal solution is destroyed; the particles aggregate to form larger particles and settle out.

Many of the properties of colloids are a result of the enormous amount of surface area between the dispersed particles and the liquid phase. For example, a cube 1 cm. on each edge has a total surface area of 6 sq. cm., but an equal volume of material divided into particles 0.01 micron on an edge has a total surface area of 6,000,000 sq. cm. Many chemical reactions occur only at a surface, and for this reason a colloidal system is a much better medium for chemical reactions than any other type of mixture.

Many of the unique properties of the cell's contents follow from the fact that it is a colloidal system composed of protein molecules in water. The protein molecules are too large to form a true solution in water and too small to settle out. The substance of the cell is constantly and rapidly changing from sol to gel and back; one portion of a cell may be a sol while others are gels. The constant, rapid change from sol to gel is one expression of the "aliveness" of the cell. Any extreme of temperature, acidity or alkalinity, or the presence of certain chemicals will cause an irreversible change to the gel or sol state and the cellular contents are no longer alive. Each cell contains a large amount of water—80 per cent of muscle is water, for example—yet, because the water is part of a colloidal system, bound to the proteins present, muscle itself can become quite solid during contraction. Shortly after death muscle undergoes *rigor mortis*, an irreversible change to the gel state.

ANNOTATED REFERENCES

Baker, J. J. W., and G. E. Allen: Matter, Energy and Life. Reading, Mass., Addison-Wesley, 1965. A presentation of thermodynamic principles and their application to studies of living systems.

Grunwald, E., and R. H. Johnsen: Atoms, Molecules and Chemical Change. Englewood Cliffs, N.J., Prentice-Hall, Inc., 1960. An introductory treatment of basic physics and chemistry.

White, E. H.: Chemical Background for the Biological Sciences. Englewood Cliffs, N.J., Prentice-Hall, Inc., 1964. Interesting presentations of the structure of atoms and molecules and the nature of chemical reactions plus a variety of topics related to the subjects of this chapter.

For more detailed discussions of acids, bases, salts and organic compounds and of the physical and chemical principles relating to life processes, consult one of the standard texts of college chemistry such as:

Jones, M. M., J. T. Netterville, D. O. Johnston and J. L. Wood: Chemistry, Man and Society. Philadelphia, W. B. Saunders Co., 1972.

Lee, G. L., H. O. Van Orden and R. O. Ragsdale: General and Organic Chemistry. Philadelphia, W. B. Saunders Co., 1971.

Watt, G. W., L. F. Hatch and J. J. Lagowski: Chemistry. New York, W. W. Norton & Co., 1964.

The Scientific American publishes excellent discussions of certain topics related to biology. Several hundred of these have been reprinted as "offprints" by Wm. Freeman Co., San Francisco. They are too numerous to cite individually but comprise a rich source of collateral reading.

3. Cells and Tissues

The living substance of all animals is organized into units called cells. Each cell contains a nucleus and is surrounded by a plasma membrane. Mammalian red blood cells lose their nucleus in the process of maturation, and a few types of cells such as those of skeletal muscles have several nuclei per cell, but these are rare exceptions to the general rule of one nucleus per cell. In the simplest animals, the **protozoa,** all of the living material is found within a single plasma membrane. These animals may be considered to be unicellular, i.e., single-celled, or acellular, with bodies not divided into cells. Many protozoa have a high degree of specialization of form and function within this single cell (Fig. 3.1), and the single cell may be quite large, larger than certain multicellular, more complex organisms. Thus, it would be wrong to infer that a single-celled animal is necessarily smaller or less complex than a many-celled animal.

3.1 THE CELL AND ITS CONTENTS

The term "cell" was applied by Robert Hooke, some 300 years ago, to the small, boxlike cavities he saw when he examined cork and other plant material under the newly-invented compound microscope. The important part of the cell, we now realize, is not the cellulose wall seen by Hooke, but the cell contents. In 1839 the Bohemian physiologist Purkinje introduced the term "protoplasm" for the living material of the cell.

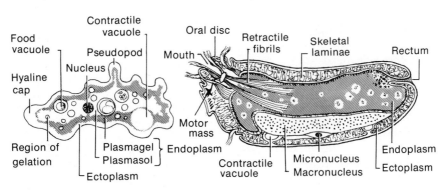

Figure 3.1 Diagrams of an ameba (left) and *Epidinium* (right) to illustrate the range in complexity of the single-celled animals.

As our knowledge of cell structure and function has increased, it has become clear that the living contents of the cell comprise an incredibly complex system of heterogeneous parts. Purkinje's term "protoplasm" has no clear meaning in a chemical or physical sense, but it may be used to refer to all the organized constituents of a cell.

In this same year, 1839, a German botanist, Matthias Schleiden, and Theodor Schwann, his fellow countryman and a zoologist, formulated the generalization which has since developed into the **cell theory:** The bodies of all plants and animals are composed of cells, the fundamental units of life. The cell is both the structural and functional unit in all organisms, the fundamental unit possessing all the characteristics of living things. A further generalization, first clearly stated by Virchow in 1855, is that new cells can come into existence only by the division of previously existing cells. The corollary of this, that all cells living today can trace their ancestry back to the earliest living things, was stated by August Weismann about 1880.

The bodies of higher animals are made of many cells, differing in size, shape and functions. A group of cells which are similar in form and specialized to perform one or more particular functions is called a **tissue.** A tissue may contain nonliving cell products in addition to the cells themselves. A group of tissues may be associated into an **organ,** and organs into **organ systems.** For example, the vertebrate digestive system is composed of a number of organs: esophagus, stomach, intestine, liver, pancreas, and so on. Each organ, such as the stomach, contains several kinds of tissue—epithelium, muscle, connective tissue, nerves—and each tissue is made of many, perhaps millions, of cells.

A single-celled animal placed in the proper environment will survive, grow and eventually divide. For most single-celled animals, a drop of sea water or pond water will provide the environment required. It is more difficult to culture cells removed from a multicellular animal—a man, chick or frog. This was first accomplished in 1907 by Ross Harrison of Yale, who was able to grow cells from a salamander in a drop of nutrient medium containing blood plasma. Since then, many different kinds of cells from animals and plants have been cultured in vitro,* and many important facts about cell physiology have been revealed in this way.

3.2 THE PLASMA MEMBRANE

The outer surface of each cell is bounded by a delicate, elastic covering, an integral functional part of the cell, termed the **plasma membrane.** This is of prime importance in regulating the contents of the cell, for all nutrients entering the cell and all waste products or secretions leaving it must pass through this membrane. It hinders the entrance of certain substances and facilitates the entrance of others. Cells are almost invariably surrounded by a watery medium. This might be the fresh or salt water in which a small organism lives, the tissue sap of a higher plant, or the plasma or extracellular fluid of one of the higher animals.

The plasma membrane behaves as though it has ultramicroscopic pores through which certain substances pass. The size of these pores determines the maximal size of the molecule that can pass through the membrane. Factors other than molecular size, such as the electric charge, if any, carried by the particles, the number of water molecules, if any, bound to the surface of the particle, and the solubility of the particle in lipids, may also be important in determining whether or not the substance will pass through the membrane. The membrane appears to be a three-layer sandwich about 120 Ångstrom units thick. The inner and outer layers, each some 30 Ångstrom units thick, are protein and surround a middle layer of phospholipid molecules 60 Ångstrom units thick (Fig. 3.2). This concept of the cell membrane, derived from chemical and physical studies, agrees with high resolution electron micrographs of the plasma membrane, which depict a three-layered structure. The plasma membranes of animal, plant and bacterial cells and the membranes of a variety of subcellular organelles all appear to have a similar structure. This pattern of protein-lipid-protein in membranes, termed the **unit membrane,** now appears to be a fundamental structural unit of widespread occurrence.

Nearly all plant cells have, in addition to

---

* In vitro, Latin *in glass.* The cells are removed from the animal body and incubated in glass vessels.

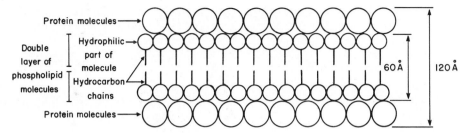

Figure 3.2 Diagram of the molecular architecture of biological membranes such as the plasma membrane or the membrane around the mitochondrion. (Villee: Biology, 6th ed.)

the plasma membrane, a thick cell wall made of cellulose. This nonliving wall, lying outside the plasma membrane, is secreted by the cell substance. It is pierced by fine holes, through which substances may pass, and the cytoplasm of one cell may connect with that of adjacent cells. These tough, firm cell walls provide support to the plant body.

3.3 THE NUCLEUS AND ITS FUNCTIONS

The **nucleus** of the cell is usually spherical or ovoid. It may have a fixed position in the center of the cell or at one side, or it may be moved around as the cell moves and changes shape. The nucleus is separated from the cytoplasm by a nuclear membrane which controls the movement of materials into and out of the nucleus (Fig. 3.3). The electron microscope reveals that the nuclear membrane is double-layered and that there are extremely fine channels through the nuclear membrane through which the nuclear contents and cytoplasm are continuous (Fig. 3.4).

The nucleus is an important center of control and is required for growth and for cell division. An ameba can survive for

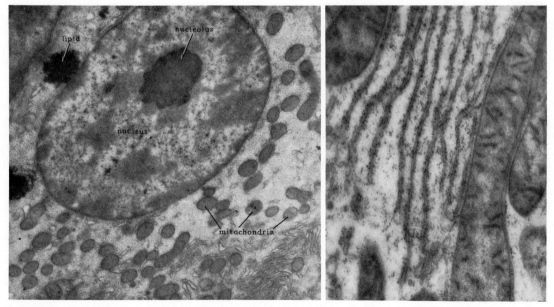

Figure 3.3 *A*, Electron micrograph of the nucleus and surrounding cytoplasm of a frog liver cell. The spaghetti-like strands of the endoplasmic reticulum are visible in the lower right corner. Magnified 16,500 times. *B*, High power electron micrograph of mitochondria and endoplasmic reticulum within a rat liver cell. Granules of ribonucleoprotein (ribosomes) are seen on the strands of endoplasmic reticulum and structures with double membranes are evident within the mitochondria in the upper left corner and on the right. Magnified 65,000 times. (Electron micrographs courtesy of Dr. Don Fawcett.) (Villee: Biology, 6th ed.)

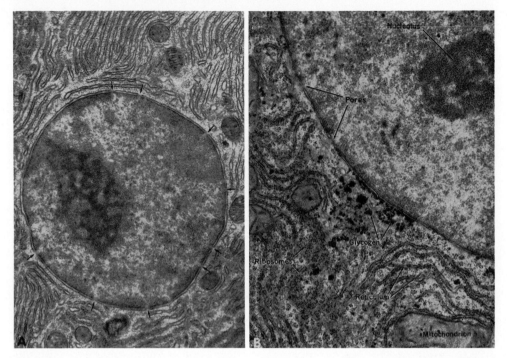

Figure 3.4 Electron micrographs showing pores in the nuclear membrane. The endoplasmic reticulum is evident in both pictures; *B*, shows the ribosomes on the endoplasmic reticulum. *A*, Magnified 20,000 times; *B*, magnified 50,000 times. (Electron micrographs courtesy of Drs. Don W. Fawcett and Keith R. Porter.)

many days after the nucleus has been removed by a microsurgical operation. To demonstrate that it is the absence of the nucleus, not the operation itself, that causes the ensuing death, one can perform a **sham operation.** A microneedle is inserted into an ameba and moved around inside the cell to simulate the operation of removing the nucleus, but the needle is withdrawn without actually removing the nucleus. An ameba subjected to this sham operation will recover, grow and divide. A controlled experiment such as this, in which two amebas are subjected to the same operative trauma and the one with the nucleus lives whereas the one without the nucleus dies, provides strong evidence of the vital role of the nucleus in regulating the metabolic processes that underlie growth and cell division.

A classic demonstration of the role of the nucleus in the control of cell growth is provided by the experiments of Hämmerling with the single-celled plant *Acetabularia mediterranea*. This marine alga, 4 to 5 cm. long, is mushroom-shaped, with "roots" and a stalk surmounted by a flattened, disc-shaped umbrella. The single nucleus is located near the base of the stalk. Hämmerling cut across the stalk (Fig. 3.5) and found that the lower part containing the nucleus could live and regenerate an umbrella. The upper part eventually died without regenerating a stalk and roots. In further experiments, Hämmerling first severed the stalk just above the nucleus, then made a second cut just below the umbrella. The section of stalk thus isolated, when replaced in sea water, was able to grow a partial or complete umbrella. This might seem to show that a nucleus is not necessary for regeneration. However, when Hämmerling cut off this second umbrella the stalk was unable to form a new one. From experiments such as these, Hämmerling concluded that the nucleus supplies some substance necessary for umbrella formation. This substance passes up the stalk and instigates umbrella growth. In the experiments described here, some of this substance remained in the stalk after the initial cuts, enough to produce one new umbrella. After that amount of "umbrella substance" was exhausted by the regeneration of an

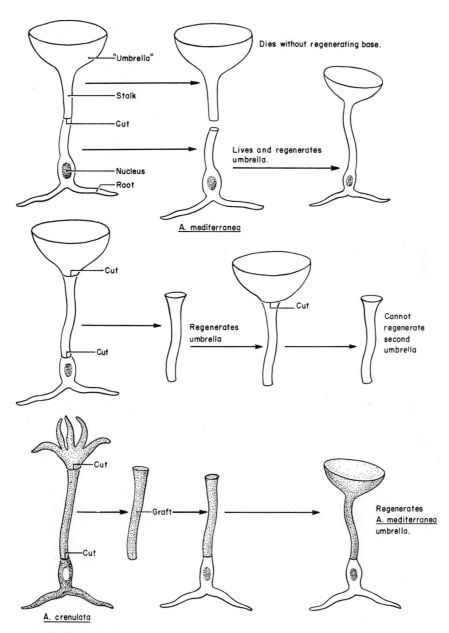

Figure 3.5 Hämmerling's experimental demonstration of the production of an umbrella-regenerating substance by the nucleus of the alga, *Acetabularia*. Lower line, when a stalk from *A. crenulata* is grafted onto the base (white) of an *A. mediterranea* plant, the stalk regenerates an umbrella whose shape is that characteristic of *A. mediterranea* plants. (Villee: Biology, 6th ed.)

umbrella, no second regeneration was possible in the absence of a nucleus.

A second species, *Acetabularia crenulata*, has a branched instead of a disc-shaped umbrella. When a piece of *crenulata* stalk (without a nucleus) is grafted onto the base of a *mediterranea* plant (containing a *mediterranea* nucleus) a new umbrella will develop at the top of the stalk. The shape of the umbrella is determined not by the species of the stalk but by the species of the base (Fig. 3.5, lower part). The nucleus, through the action of its genes, provides the specific information that controls the type of umbrella that is regenerated and can override the tendency of the stalk to form an umbrella characteristic of its own species.

When a cell has been killed by fixation with the proper chemicals and then stained with the appropriate dyes, several structures are visible within the nucleus (Fig. 3.6). Within the semifluid ground substance are suspended a fixed number of extended, linear, threadlike bodies called **chromosomes,** composed of DNA and proteins and containing the units of heredity, the **genes.** In a stained section of a nondividing cell the chromosomes typically appear as an irregular network of dark-staining threads and granules termed **chromatin.** Just prior to nuclear division these strands condense into compact, rod-shaped chromosomes which are subsequently distributed to the two daughter cells in exactly equal numbers. Each type of organism has a characteristic number of chromosomes present in each of its constituent cells. The fruit fly has eight chromosomes, sorghum has 10, the garden pea, 14, corn, 20, the toad, 22, the tomato, 24, the cherry, 32, the rat, 42, man, 46, the potato, 48, the goat, 60 and the duck 80. The somatic cells of higher plants and animals each contain two of each kind of chromosome. A cell with two complete sets of chromosomes is said to be **diploid.** Sperm and egg cells, which have only one of each kind of chromosome, one full set of chromosomes, are said to be **haploid.** They have just half as many chromosomes as the somatic cells of that same species. When the egg is fertilized by the sperm the two haploid sets of chromosomes are joined and the diploid number is restored.

The nucleus also contains one or more small spherical bodies, **nucleoli,** which are difficult to see in a living cell with an ordi-

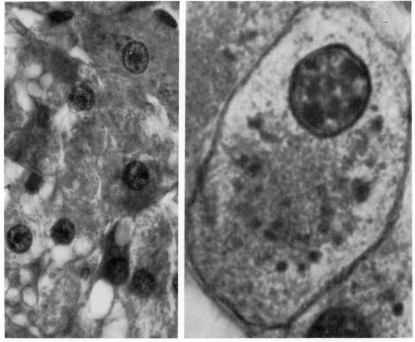

Figure 3.6   Tissue sections of human adrenal gland stained to show cellular details; left, magnified 600 times; right, magnified 1500 times. (Courtesy of Dr. Kurt Benirschke.)

nary light microscope but are evident by phase microscopy. The cells of any particular animal have the same number of nucleoli. The nucleolus disappears when a cell is about to divide and reappears after division is complete. If the nucleolus is destroyed by carefully localized ultraviolet or x-irradiation, cell division is inhibited. This does not occur in control experiments in which regions of the nucleus other than the nucleolus are irradiated. The nucleolus plays a key role in the synthesis of the ribonucleic acid constituents of ribosomes.

3.4 CYTOPLASMIC ORGANELLES

Two small, dark-staining cylindrical bodies, called **centrioles,** are found in the cytoplasm near the nucleus of animal cells. With the electron microscope each centriole is revealed to be a hollow cylinder with a wall in which are embedded nine parallel, longitudinally oriented groups of microtubules, with three tubules in each group. The cylinders of the two centrioles are typically oriented with their long axes perpendicular to each other. The centrioles play a role in cell division in determining the location of the spindle fibers on which the chromosomes move (p. 42). It would appear, however, that centrioles are not essential for cell division, for plant cells are able to divide without them.

Those cells that bear cilia on their exposed surfaces have a structure at the base of each cilium, the **basal body,** that resembles the centriole in the presence of nine parallel tubules. Each cilium contains nine peripherally located longitudinal filaments and two centrally located ones. Like centrioles, basal bodies can duplicate themselves.

The cytoplasm may contain droplets of fat and crystals or granules of protein or glycogen which are simply stored for future use. In addition, it contains the metabolically active cell organelles, **mitochondria, endoplasmic reticulum** and **Golgi bodies.**

Mitochondria range in size from 0.2 to 5 microns and in shape from spheres to rods and threads (Fig. 3.3). Their number may range from just a few to more than 1000 per cell. When living cells are examined, their mitochondria appear to move, change shape and size, fuse with other mitochondria to form longer structures, or cleave to form smaller ones. Each mitochondrion is bounded by a double membrane, an outer smooth membrane and an inner one folded into parallel plates that extend into the central cavity and may fuse with folds from the opposite side. Each of these plates is a unit membrane composed of a middle double layer of phospholipid molecules with a layer of protein molecules on each side. These shelflike inner folds, termed **cristae,** contain the enzymes of the electron transmitter system, of prime importance in converting the potential energy of foodstuffs into biologically useful energy for cellular activities. The mitochondria have been dubbed the "powerhouses" of the cell.

Those cells especially active in the synthesis of proteins are crowded with the membranous labyrinth of the **endoplasmic reticulum;** other cells may have only a scanty supply. In a thin section these appear as a profusion of spaghetti-like tubular strands, but in three dimensions are sheetlike membranes. Some of the endoplasmic reticulum is granular and has bound to it a profusion of **ribosomes,** small nucleoprotein particles on which protein synthesis occurs. The remainder of the endoplasmic reticulum is agranular and consists of smooth sheets believed to play some role in cellular secretion.

The tightly packed sheets of endoplasmic reticulum may form tubules some 50 to 100 nanometers in diameter. In other regions of the cell, the cavities of the endoplasmic reticulum may be expanded, forming flattened sacs called **cisternae.** The membranes of the endoplasmic reticulum divide the cytoplasm into a multitude of compartments in which different groups of enzymatic reactions may occur. The endoplasmic reticulum serves a further function as a system for the transport of substrates and products through the cytoplasm to the exterior of the cell and to the nucleus.

Golgi bodies, found in all cells except mature sperm and red blood cells, consist of an irregular network of canals. In the electron microscope (Fig. 3.7) Golgi bodies appear as parallel arrays of membranes without granules. The canals may be distended in certain regions to form small vesicles or vacuoles filled with material. The Golgi bodies are usually concentrated in the part of the cytoplasm near the centrioles and appear to have a role in the production or

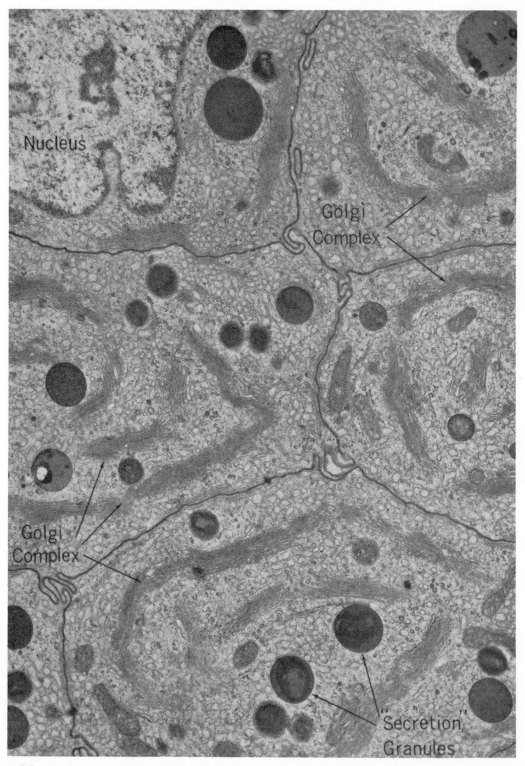

Figure 3.7 Electron micrograph of rabbit epididymis showing the extensive Golgi complex, evident in the parallel arrays of the membranes. Magnification, 9500×. (From Fawcett, D. W.: The Cell. Philadelphia, W. B. Saunders Company, 1966.)

storage of secretions. They may have the appearance of granules, rods, threads or canals.

Lysosomes are intracellular organelles about the size of mitochondria but less dense. They are membrane-bounded structures that contain a variety of enzymes capable of hydrolyzing the macromolecular constituents of the cell. In the intact cell these enzymes are segregated within the lysosome, presumably to prevent their digesting the contents of the cell. Rupture of the lysosome membrane releases the enzymes and accounts, at least in part, for the lysis of dead cells and the resorption of cells such as those in the tail of a tadpole during metamorphosis.

The cytoplasm of certain cells, chiefly those of lower animals, contains **vacuoles,** bubble-like cavities filled with fluid and separated from the rest of the cytoplasm by a vacuolar membrane. Most protozoa, and the endoderm cells of coelenterates and flatworms, have **food vacuoles** in which food is digested. Digestive enzymes are secreted from the cytoplasm into the cavity of the vacuole, the food is digested, and the products of digestion are absorbed through the vacuolar membrane into the cytoplasm. The protozoa living in fresh water have the problem of eliminating the water which enters the cell constantly by osmosis (p. 49). These forms have evolved **contractile vacuoles,** which alternately fill with water from the adjacent cytoplasm and then eject the water to the surrounding environment.

Most animal cells are quite small, too small to be seen with the naked eye. The diameter of the human red blood cell is about 7.5 microns (a micron is 0.001 mm.), but most animal cells have diameters ranging from 10 to 50 microns. There are a few species of giant amebas with cells about 1 mm. in diameter. The largest cells are the yolk-filled eggs of birds and sharks. The egg cell of a large bird such as a turkey or goose may be several centimeters across. Only the yolk of a bird's egg is the true egg cell; the egg white and shell are noncellular material secreted by the bird's oviduct as the egg passes through it.

The limit of the size of a cell is set by the physical fact that, as a sphere gets larger, its surface increases as the square of the radius but its volume increases as the cube of the radius. The metabolic activities of the cell are roughly proportional to cell volume. These activities require nutrients and oxygen, and release carbon dioxide and other wastes which must enter and leave the cell through its surface. The upper limit of cell size is reached when the surface area can no longer provide for the entrance of enough raw materials and the exit of enough waste products for cell metabolism to proceed normally. The limiting size of the cell will depend on its shape and its rate of metabolism. When this limit is reached the cell must either stop growing or divide.

Much has been learned in recent years of the role each of these particles plays in the economy of the cell. Cells are homogenized in special glass grinding tubes to break the cell membrane and release the intracellular structures. Then, by subjecting the homogenate to increasing amounts of centrifugal force in an ultracentrifuge, first the nuclei, then the mitochondria, and finally the microsomes can be sedimented separately. When these sedimented particles are examined in the electron microscope, they are found to have the same structure exhibited by comparable structures in the intact cell. The separated particles can then be suspended in suitable incubation media and their metabolism can be studied. Such separated mitochondria and microsomes will carry out many biochemical reactions, and much is now known about the functions of each of these particles. The liquid left after the homogenate has been subjected to high centrifugal force to sediment the microsomes contains many other enzymes which apparently exist in the cell more or less free in the ground substance.

Where, you may ask, is life localized—in the mitochondria? in the microsomes? in the ground substance? The answer, of course, is that life is not a function of any single one of these parts, but of the whole integrated system of many component parts, organized in the proper spatial relationship and interdependent on one another in a great variety of ways.

3.5 MITOSIS

Because of the limitation on the size of individual cells, growth is accomplished largely by an increase in the *number* of cells. When a single-celled protozoan divides, the resulting two cells are separate individuals, members of a new generation. In multicellular animals, cell division results in an increase in the number of cells per individual. The process of cell division, called **mitosis,** is extremely regular and ensures the qualitatively and quantitatively equal distribution of the hereditary factors between the two resulting daughter cells. Mitotic divisions occur during embryonic development and growth, in the replacement of cells that wear out, such as blood cells, skin, the intestinal lining, and so on, and in the repair of injuries.

When a dividing cell is stained and examined under the microscope, dark-staining bodies, called **chromosomes,** are visible within the nucleus. Each consists of a central thread, the **chromonema,** along which lie the **chromomeres** — small, beadlike, dark-staining swellings. It has been suggested that the chromomeres are, or contain, the genes, for breeding experiments have shown clearly that these hereditary units lie within the chromosome in a linear order. However, the correlation between chromomeres and genes is not regular; some chromomeres contain several genes and some genes have been located between chromomeres. Several theories have been formulated to account for these swellings of the chromosomes, but at present their true significance is not clear.

Each chromosome has, at a fixed point along its length, a small clear circular zone called a **centromere** that controls the movement of the chromosome during cell division. As the chromosome becomes shorter and thicker just before cell division occurs, the centromere region becomes accentuated and appears as a constriction.

Chromosomes occur in pairs; the 46 chromosomes of each human cell consist of two of each of 23 different kinds. The chromosomes differ in length and shape and in the presence of identifying knobs or constrictions along their length (Fig. 3.8). In most animals, the morphologic features of the chromosomes are distinct enough so that one can identify the individual pairs.

Cell division must be an extremely exact process to ensure that each daughter cell receives exactly the right number and kind of chromosomes. If we tamper experimentally with the mechanism of cell division, and the resulting cells receive more or less than the proper number of chromosomes, marked abnormalities of growth, and perhaps the death of these cells, will follow. Mitosis may be defined as the regular process of cell division by which each of the two daughter cells receives exactly the same number and the same kind of chromosomes that the parent cell contained. Although each chromosome appears to split longitudinally into two halves, each original chromosome remains

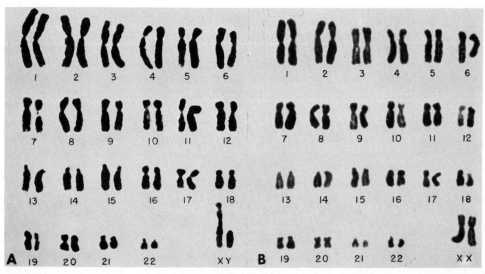

Figure 3.8   Human chromosomes. *A*, Normal male. *B*, Normal female. (Photographs courtesy of Dr. Melvin Grumbach.)

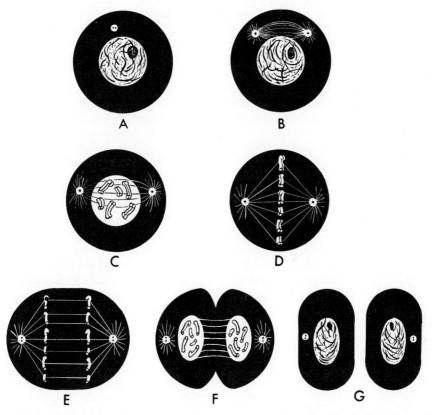

Figure 3.9 Mitosis in a cell of a hypothetical animal with a diploid number of six (haploid number = 3); one pair of chromosomes is short, one pair is long and hooked, and one pair is long and knobbed. *A*, Resting stage. *B*, Early prophase, centriole divided and chromosomes appearing. *C*, Later prophase, centrioles at poles, chromosomes shortened and visibly double. *D*, Metaphase, chromosomes arranged on the equator of the spindle. *E*, Anaphase, chromosomes migrating toward the poles. *F*, Telophase, nuclear membranes formed; chromosomes elongating; cytoplasmic division beginning. *G*, Daughter cells, resting phase.

intact and brings about the synthesis of an exact replica of itself immediately adjacent to it. The new chromosome is made, sometime before the visible mitotic process begins, from raw materials present in the nucleus. When the process is complete, the original and the new chromosomes separate and become incorporated into different daughter cells. The role of the complicated mitotic machinery is to separate the "original" and "replica" chromosomes and deliver them to opposite ends of the dividing cell so they will become incorporated into different daughter cells.

The term mitosis in a strict sense refers to the division of the nucleus into two daughter nuclei and the term **cytokinesis** is applied to the division of the cytoplasm to form two daughter cells, each containing a daughter nucleus. Nuclear division and cytoplasmic division, although almost invariably well synchronized and coordinated, are separate and distinct processes.

Each mitotic division is a continuous process, with each stage merging imperceptibly into the next one. For descriptive purposes biologists have divided it into four stages: **prophase, metaphase, anaphase** and **telophase** (Fig. 3.9). Between mitoses a cell is said to be in the resting stage. It is difficult to visualize from a description or diagram of mitosis, or from examining a fixed and stained slide of cells, just how active a process cell division is. Motion pictures made by phase microscopy reveal that a cell undergoing division bulges and changes shape like a gunny sack filled with a dozen unfriendly cats.

Prophase. The chromatin threads condense and the chromosomes appear as a tangled mass of coiled threads within the nucleus. Early in prophase the threads are stretched maximally so that the individual chromomeres are visible. Later in prophase the chromosomes shorten and thicken and the chromomeres lie so close together that individual ones cannot be distinguished. The reduplication of the chromosomes has occurred previously and in many species of animals the double nature of each chromosome is apparent.

Early in prophase the centrioles migrate to opposite sides of the cell. Between the separating centrioles a spindle forms. The **spindle** is composed of protein threads with properties similar to those of the contractile proteins of muscle fibrils. The protein threads of the spindle are arranged like two cones base to base, broad at the center or equator of the cell and narrowing to a point at either end or pole. With a microneedle attached to a micromanipulator the spindle can be moved as a unit from one part of the cell to another. By appropriate techniques spindles may be isolated from dividing cells (Fig. 3.10). At the end of prophase, the centrioles have gone to the opposite poles of the cell, the spindle has formed between them and the chromosomes have become short and thick.

Metaphase. When the chromosomes are fully contracted and appear as short, dark-staining rods, the nuclear membrane dis-

appears and the chromosomes line up in the equatorial plane of the spindle. The short period during which the chromosomes are in this equatorial plane is known as the metaphase. This is much shorter than the prophase; although times for different cells vary considerably, the prophase lasts from 30 to 60 minutes or more, and the metaphase lasts only two to six minutes.

During the metaphase the centromere of each chromosome divides and the two chromatids become completely separate daughter chromosomes. The division of the centromeres occurs simultaneously in all the chromosomes, under the control of some as yet unknown mechanism. The daughter centromeres begin to move apart, marking the beginning of anaphase.

Anaphase. The chromosomes separate (Fig. 3.11), and one of the daughter chromosomes goes to each pole. The period during which the separating chromosomes move from the equatorial plate to the poles, known as the anaphase, lasts some three to 15 minutes. The spindle fibers apparently act as guide rails along which the chromosomes move toward the poles. Without such guide rails the chromosomes would merely be pushed randomly apart and many would fail to be incorporated into the proper daughter nucleus. The mechanism moving

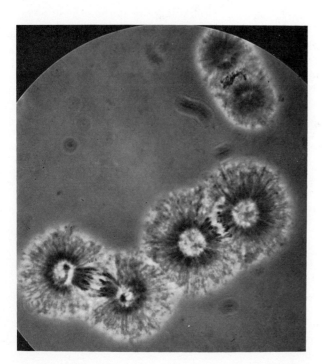

Figure 3.10 Photomicrograph of the mitotic apparatus isolated from dividing cells of a sea urchin embryo. Each mitotic apparatus includes spindle fibers, asters and chromosomes. A metaphase figure appears in the upper right and two anaphase figures below. (Courtesy of Daniel Mazia.) (Villee: Biology, 6th ed.)

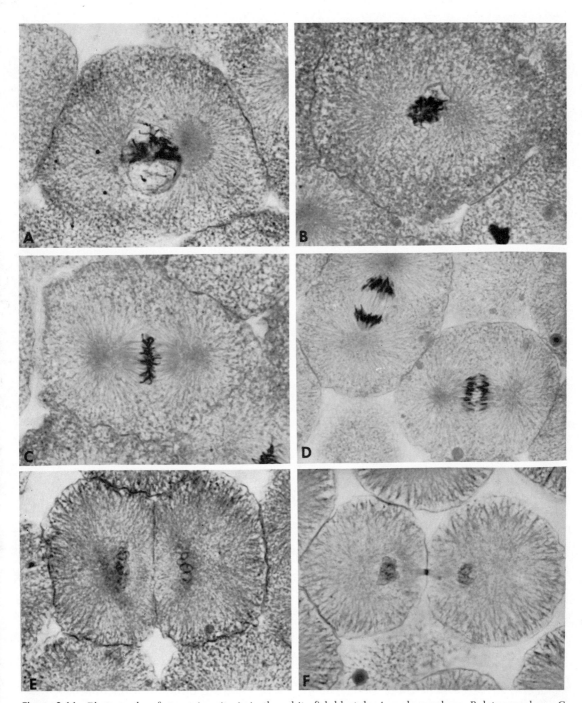

Figure 3.11 Photographs of stages in mitosis in the white fish blastula. A, early prophase; B, later prophase; C, metaphase; D, two cells in early and late anaphase respectively; E, early telophase; and F, late telophase. The dark spot connecting the two cells in E is the remainder of the spindle. (Photographs courtesy of Dr. Susumu Ito.)

the chromosomes apart is not clear. Experiments suggest that some substance between the chromosomes takes up water, swells, and pushes the chromosomes apart. Other experiments indicate that some of the spindle fibers are contractile and can pull the chromosomes toward the poles. The chromosomes moving toward the poles usually assume a **V** shape with the centromere at the apex pointing toward the pole. It appears that whatever force moves the chromosome to the pole is applied at the centromere. A chromosome that lacks a centromere, perhaps as a result of exposure to x-radiation, does not move at all in mitosis.

Telophase. When the chromosomes have reached the poles of the cell, the last phase of mitosis, telophase, begins. Several processes occur simultaneously in this period: a nuclear membrane forms around the group of chromosomes at each pole, the chromosomes elongate, stain less darkly, and return to the resting condition in which only irregular chromatin threads are visible, and the cytoplasm of the cell begins to divide. Division of the cytoplasm is accomplished in animal cells by the formation of a furrow which circles the cell at the equatorial plate and gradually deepens until the two halves of the cell are separated as independent daughter cells. The events of telophase require some 30 to 60 minutes for their completion.

The mitotic process results in the formation of two daughter cells from a single parent cell. Each daughter cell receives exactly the same number and kind of chromosomes, and of the units of heredity (genes) contained in these chromosomes, that were present in the parent cell. Since all the cells of the body are formed by mitosis from a single fertilized egg, each cell has the same number and kind of chromosomes, and the same number and kind of genes, as every other cell.

The speed and frequency of cell division vary greatly from tissue to tissue and from one animal to another. In the early stages of embryonic development, there may be only 30 minutes or so between successive cell divisions. In certain adult tissues, notably the nervous system, mitoses are extremely rare. In other adult tissues, such as the red bone marrow, where red blood cells are produced, mitotic divisions must occur frequently to supply the 10,000,000 red blood cells each human being produces every second of the day and night.

3.6 REGULATION OF MITOSIS

The factors which initiate and control cell division are not known exactly. The possible role of the ratio of cell surface to cell volume was discussed previously (p. 39). The ratio of *nuclear* surface to *nuclear* volume may also be important. Since normal cell function requires the transport of substances back and forth through the nuclear membrane, growth will eventually result in a state in which the area of the nuclear membrane is insufficient to meet the demands of the volume of cytoplasm. Cell division, by splitting the volume of cytoplasm into two parts and increasing the area of nuclear membrane, will restore optimal conditions. There is some evidence to suggest that the chromosomes may release a substance or substances which initiate, first, the nuclear events of prophase and metaphase and, secondly, the reactions in the cytoplasm which form a cleavage furrow and bring about the division of the cytoplasm.

Another theory postulates the initiation of mitosis by a "cell division hormone." The mitoses of the cells of an egg undergoing cleavage occur simultaneously, which suggests that a periodically released hormone may control these divisions.

3.7 THE STUDY OF CELLULAR ACTIVITIES

Despite great differences in size, shape and location in the body, all cells have many metabolic activities in common. The myriad enzymes in each cell enable it to release energy by converting sugars, fats and proteins to carbon dioxide and water. Each cell synthesizes its own structural proteins and enzymes. Superimposed on this basic pattern of metabolism common to all cells may be other activities peculiar to each type of cell. For example, muscle cells have special contractile proteins, **myosin** and **actin**; particular digestive enzymes are produced by the cells lining the stomach and intestine; and the cells of the pituitary, adrenal and thyroid glands manufacture characteristic hormones.

Each of the many ways of studying cellular activity provides useful information about cell morphology and physiology. Living cells suspended in a drop of fluid can be examined under an ordinary microscope or with one equipped with **phase contrast lenses.** In this way one can study the movement of an ameba or a white blood cell, or the beating of the cilia on a paramecium. Cells from a many-celled animal — a frog, chick or man — can be grown by "**tissue culture**" for observation over a long period of time. A complex nutritive medium, made of blood plasma, an extract of embryonic tissues, and a mixture of vitamins, is prepared and sterilized. A drop of this is placed in a cavity on a special microslide, the cells to be cultured are added aseptically, and the cavity is sealed with a glass cover slip. After a few days the cells have exhausted one or more of the nutritive materials and must be transferred again to a fresh drop of medium. Cells transferred regularly in this fashion will grow indefinitely.

Cell morphology may be studied by using a bit of tissue that has been killed quickly with a special "fixative," then sliced with a machine called a microtome, and stained with special dyes. The stained thin slices, mounted on a glass slide and covered with a glass cover slip, are then ready for examination under the microscope. The nucleus, mitochondria and other specialized parts of the cell differ chemically and will combine with different dyes and be stained characteristic colors (Fig. 3.6). For observation in the electron microscope a bit of tissue is fixed with osmic acid, mounted in acrylic plastic for cutting in extremely thin sections, and then placed on a fine grid to be inserted into the path of the electron beam. Both light microscopy and electron microscopy have revealed many details about cell structure (Fig. 3.12).

Some clue as to the location and functioning of enzymes within cells can be ob-

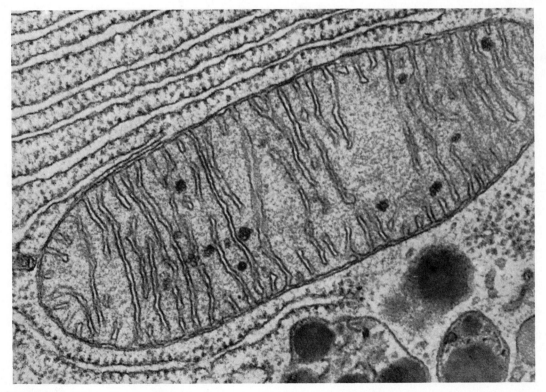

Figure 3.12 Electron micrograph of a typical mitochondrion from the pancreas of a bat, showing the cristae, matrix, and matrix granules. Endoplasmic reticulum is seen at the upper left and some lysosomes at the lower right. ×79,000. (Courtesy of K. R. Porter.)

tained by **histochemical** studies, in which a tissue is fixed and sliced thin by methods which do not destroy enzymic activity. Then the proper chemical substrate for the enzyme is provided and, after a specified period of incubation, some substance is added which will form a colored compound with one of the products of the reaction mediated by the enzyme. The regions of the cell which have the greatest enzymic activity will have the largest amount of the colored substance (Fig. 3.13). Methods have been devised which permit the demonstration and localization of a wide variety of enzymes. Such studies have given interesting insights into the details of cell function.

Another method of investigating cell function is to measure, by special microchemical analyses, the amounts of chemical used up or produced as a bit of tissue is incubated in an appropriate medium. In such experiments much has been learned of the roles in cell metabolism of vitamins, hormones and other chemicals by adding these substances one by one and observing the resulting effects.

Every living cell, whether it is an individual unicellular animal or a single component of a multicellular one, must be supplied constantly with nutrients and oxygen. These materials are constantly being metabolized—used up—as the cell goes about its business of releasing energy from the nutrients to provide for its myriad activities. Some of the substances required by the cell are brought to it and taken in by complex active processes which require the expenditure of energy by the cell. Other substances are brought to the cell by the simpler, more easily understood physical process of **diffusion.** To understand this process, so important in many biologic phenomena, we must first consider some of the basic physical concepts of energy and molecular motion.

3.8 ENERGY

Energy may be defined as the ability to do work, to produce a change in matter. It may take the form of heat, light, electricity, motion or chemical energy. Physicists recognize two kinds of energy: **potential energy,** the capacity to do work owing to the position or state of a body, and **kinetic energy,** the capacity to do work posessed by a body because of its motion. A rock at the top of a hill has potential energy; as it rolls downhill the potential energy is converted to kinetic energy.

Energy derived ultimately from solar radiation, trapped by photosynthetic processes in plants, is stored in the molecules of foodstuffs as the chemical energy of the bonds connecting their constituent atoms. This chemical energy is a kind of potential

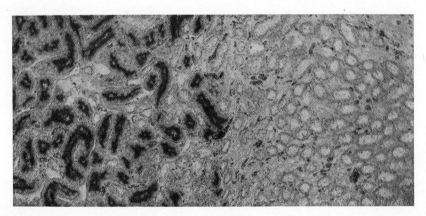

Figure 3.13 Histochemical demonstration of the location of the enzyme alkaline phosphatase within the cells of the rat's kidney. The tissue is carefully fixed and sectioned by methods which do not destroy the enzyme's activity. The tissue section is incubated at the proper pH with a naphthyl phosphate. Some hydrolysis of the naphthyl phosphate occurs wherever the phosphatase enzyme is located. The naphthol released by the action of the enzyme couples with a diazonium salt to form an intensely blue, insoluble azo dye which remains at the site of the enzymatic activity. The photomicrograph thus reveals the sites of phosphatase activity, i.e., the sites at which the azo dye is deposited. The cells of the proximal convoluted tubules (left) have a lot of enzyme; those of the loop of Henle (right) have little or no activity. (Courtesy of R. J. Barnett.) (Villee: Biology, 6th ed.)

TABLE 3.1 ENERGY TRANSFORMATIONS IN CELLS

Transformation	Type of Cell
Chemical energy to electrical energy	Nerve, brain
Sound to electrical energy	Inner ear
Light to chemical energy	Chloroplast
Light to electrical energy	Retina of eye
Chemical energy to osmotic energy	Kidney
Chemical energy to mechanical energy	Muscle cell, ciliated epithelium
Chemical energy to radiant energy	Luminescent organ of firefly
Chemical energy to electrical energy	Sense organs of taste and smell

energy. When these food molecules are taken within a cell, chemical reactions occur which change this potential energy into heat, light, motion or some other kind of kinetic energy. Light is a kind of kinetic energy that may be thought of as the movement of photons or light quanta. All forms of energy are at least partially interconvertible, and living cells constantly transform potential energy into kinetic energy or the reverse (Table 3.1). If the conditions are suitably controlled, the amount of energy entering and leaving any given system can be measured and compared. Such experiments have shown that energy is neither created nor destroyed, but simply transformed from one form to another. This is an expression of one of the fundamental laws of physics, the Law of the Conservation of Energy. Living things as well as nonliving systems obey this law.

3.9 MOLECULAR MOTION

The constituent molecules of all substances are constantly in motion. Despite the fact that wood, stone and steel seem very solid, their component molecules vibrate continuously within a very restricted space. The prime difference between solids, liquids and gases is the freedom of movement of the molecules present. The molecules of a solid are very closely packed and the forces of attraction between the molecules permit them to vibrate but not to move around. In the liquid state the molecules are somewhat farther apart and the intermolecular forces are weaker, so that the molecules can move about with considerable freedom. The molecules in the gaseous state are so far apart that the intermolecular forces are negligible and molecular movement is restricted only by external barriers. Molecular movement in all three states of matter is the result of the inherent heat energy of the molecules, the kinetic energy determined by the temperature of the system. By increasing this **molecular kinetic energy,** one can change matter from one state to another. When ice is heated it becomes water, and when water is heated it is converted to water vapor.

If a drop of water is examined under the microscope, the motion of its molecules is not evident. If a drop of India ink (which contains fine carbon particles) is added, the carbon particles move continually in aimless zigzag paths, for they are constantly being bumped by water molecules and the recoil from this bump imparts the motion to the carbon particle. The motion of such small particles is called **brownian movement,** after Robert Brown, an English botanist, who first observed the motion of pollen grains in a drop of water.

3.10 DIFFUSION

Molecules in a liquid or gaseous state will diffuse, that is, move in all directions until they are spread evenly throughout the space available. **Diffusion** may be defined as the movement of molecules from a region of high concentration to one of lower concentration brought about by their kinetic energy. The rate of diffusion is a function of the size of the molecule and the temperature. If a bit of sugar is placed in a beaker of water, the sugar will dissolve and the individual sugar molecules will diffuse and come to be distributed evenly throughout the liquid (Fig. 3.14). Each molecule tends to move in a straight line until it collides with another molecule or the side of the container; then it rebounds and moves in another direction. By this random movement of molecules, the sugar eventually becomes evenly distributed throughout the water in the beaker. This could be demonstrated by tasting drops of liquid

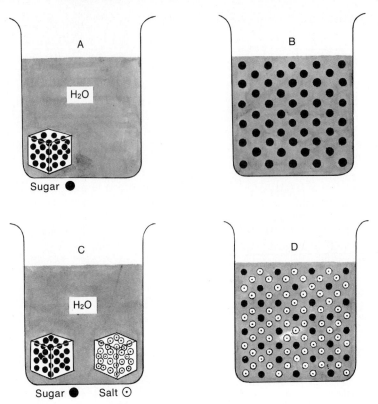

Figure 3.14   Diffusion. When a cube of sugar is placed in water (A) it dissolves and its molecules become uniformly distributed throughout the water as a result of the molecular motion of both sugar and water molecules (B). When lumps of sugar and salt are placed in water (C), each type of molecule diffuses independently of the other and both salt and sugar become uniformly distributed in the water (D).

taken from different parts of the beaker. If a colored dye is used in place of sugar, the process of diffusion can be observed directly. The molecules of sugar or dye continue to move after they have become evenly distributed throughout the liquid in the container; however, as fast as some molecules move from left to right, others move from right to left, so that an equilibrium is maintained.

Any number of substances will diffuse independently of each other. If a lump of salt is placed in one part of a beaker of water and a lump of sugar in another, the molecules of each will diffuse independently of the other and each drop of water in the beaker will eventually have some salt and some sugar molecules.

The rate of movement of a single molecule is several hundred meters per second, but each molecule can go only a fraction of a millimicron before it bumps into another molecule and rebounds. Thus the progress of a molecule in a straight line is quite slow. Diffusion is quite rapid over short distances but it takes a long time—days and even weeks—for a substance to diffuse a distance measured in centimeters. This fact has important biologic implications, for it places a sharp limit on the number of molecules of oxygen and nutrients that can reach an organism by diffusion alone. Only a very small organism that requires relatively few molecules per second can survive if it remains in one place and allows molecules to come to it by diffusion. A larger organism must have some means of moving to a new region or some means of stirring its environment to bring molecules to it. As a third alternative it may live in some spot where the environment is constantly moving past it—in a river, for example, or in the intertidal region at the seashore. The larger land plants have solved this problem by developing an extensively branched system of roots which can tap a large area of the surrounding environment for the needed raw materials.

3.11 EXCHANGES OF MATERIAL BETWEEN CELL AND ENVIRONMENT

All nutrients and waste products must pass through the plasma membrane to enter or leave the cell. Cells are almost invariably surrounded by a watery medium—the fresh or salt water in which an organism lives, the tissue sap of a higher plant, or the plasma or extracellular fluid of a higher animal. In general, only dissolved substances can pass through the plasma membrane, but not all dissolved substances penetrate the plasma membrane with equal facility. The membrane behaves as though it had ultramicroscopic pores through which substances pass, and these pores, like the holes in a sieve, determine the maximum size of molecule that can pass. Factors other than simple molecular size, such as the electric charge, if any, of the diffusing particle, the number of water molecules bound to the diffusing particle and its solubility in fatty substances, may also be important in determining whether or not the substance can pass through the plasma membrane.

A membrane is said to be **permeable** if it will permit any substance to pass through, **impermeable** if it will allow no substance to pass, and **differentially permeable** if it will allow some but not all substances to diffuse through. The nuclear and plasma membranes of all cells and the membranes surrounding food and contractile vacuoles are differentially permeable membranes. Permeability is a property of the *membrane,* not of the diffusing substance.

The diffusion of a dissolved substance through a differentially permeable membrane is known as **dialysis.** If a pouch made of collodion, cellophane or parchment is filled with a sugar solution and placed in a beaker of water, the sugar molecules will dialyze through the membrane (if the pores are large enough) and eventually the concentration of sugar molecules in the water outside the pouch will equal that within the pouch. The molecules then continue to diffuse but there is no net change in concentration for the rates in the two directions are equal.

A different type of diffusion is observed if a membrane is prepared with smaller pores, so that it is permeable to the small water molecules but not to the larger sugar molecules. A pouch may be prepared of a membrane with these properties and filled with a sugar solution. The pouch is then fitted with a cork and glass tube and placed in a beaker of water so that the levels of fluid inside and outside of the pouch are the same. The sugar molecules cannot pass through the membrane and so must remain inside the pouch. The water molecules diffuse through the membrane and mix with the sugar solution, so that the level of fluid within the pouch rises. The liquid within the pouch is 5 per cent sugar, and therefore only 95 per cent water; the liquid outside the membrane is 100 per cent water. The water molecules are moving in both directions through the membrane but there is a greater movement from the region of higher concentration (100 per cent, outside the pouch) to the region of lower concentration (95 per cent, within the pouch). This diffusion of water or solvent molecules through a membrane is called **osmosis,** and is illustrated diagrammatically in Figure 3.15.

If an amount of water equal to that originally present in the pouch enters, the solution in the pouch will be diluted to 2.5 per cent sugar and 97.5 per cent water, but the concentration of water outside the pouch will still exceed that inside and osmosis will continue. An equilibrium is reached when the water in the glass tube rises to

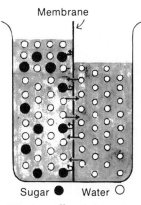

Figure 3.15 Diagram illustrating osmosis. When a solution of sugar in water is separated from pure water by a semipermeable membrane which allows water but not the larger sugar molecules to pass through, there is a net movement of water molecules through the membrane to the sugar solution. The water molecules are diffusing from a region of higher concentration (pure water) to a region of lower concentration (the sugar solution).

a height such that the weight of the water in the tube exerts a pressure just equal to the tendency of the water to enter the pouch. Osmosis then occurs with equal speed in both directions through the differentially permeable membrane and there will be no net change in the amount of water in the pouch. The pressure of the column of water is called the **osmotic pressure** of the sugar solution. The osmotic pressure results from the tendency of the water molecules to pass through the differentially permeable membrane and equalize the concentration of water molecules on its two sides. A more concentrated sugar solution would have a greater osmotic pressure and would "draw" water to a higher level in the tube. A 10 per cent sugar solution would cause water to rise approximately twice as high in the tube as a 5 per cent solution.

It is evident from this discussion that dialysis and osmosis are simply two special forms of diffusion. Diffusion is the general term for the movement of molecules from a region of high concentration to a region of lower concentration, brought about by their kinetic energy. Dialysis is the diffusion of dissolved molecules (solutes) through a differentially permeable membrane, and osmosis is the diffusion of solvent molecules through a differentially permeable membrane. In biologic systems the solvent molecules are almost universally water.

The salts, sugars and other substances dissolved in the fluid within each cell give the intracellular fluid a certain osmotic pressure. When the cell is placed in a fluid with the same osmotic pressure as that of its intracellular fluid, there is no net entrance or exit of water, and the cell neither swells nor shrinks. Such a fluid is said to be **isotonic** or isosmotic with the intracellular fluid of the cell. Normally, the blood plasma and body fluids are isosmotic with the intracellular fluids of the body cells. If the environmental fluid contains more dissolved substances than the fluid within the cell, water will tend to pass out of the cell and the cell shrinks. Such a fluid is said to be **hypertonic** to the cell. If the environmental fluid has a lower concentration of dissolved substances than the fluid in the cell, water tends to pass into the cell and the cell swells. This fluid is said to be **hypotonic** to the cell. A solution of 0.9 per cent sodium chloride, 0.9 gm. per 100 ml. of water, sometimes

loosely called "physiological saline," is isotonic to human cells.

A cell placed in a solution that is not isotonic with it may adjust to the changed environment by undergoing a change in its water content, so that it eventually achieves the same concentration of solutes as in the environment. Many cells have the ability to pump water or certain solute molecules into or out of the cell and in this way can maintain an osmotic pressure that differs from that of the surrounding medium. Amebae, paramecia and other protozoa that live in pond water, which is very hypotonic to their intracellular fluid, have evolved **contractile vacuoles** (see Fig. 3.1) which collect water from the interior of the cell and pump it to the outside. Without such a mechanism the cells would quickly burst from the water entering the cell.

The power of certain cells to accumulate selectively certain kinds of molecules from the environmental fluid is truly phenomenal. Human cells (and those of vertebrates in general) can accumulate amino acids so that the concentration within the cell is two to 50 times that in the extracellular fluid. Cells also have a much higher concentration of potassium and magnesium, and a lower concentration of sodium, than the environmental fluids. Certain primitive chordates, the tunicates (Chap. 29), can accumulate vanadium so that the concentration inside the cell is some 2,000,000 times that in the surrounding sea water. Seaweeds have a comparable ability to accumulate iodine. The transfer of water or of solutes in or out of the cell against a concentration gradient is physical work and requires the expenditure of energy. Some active physiologic process is required to perform these transfers; hence a cell can move molecules against a gradient only as long as it is alive. If a cell is treated with some metabolic poison, such as cyanide, it quickly loses its ability to maintain concentration differences on the two sides of its plasma membrane.

3.12 TISSUES

In the evolution of both plants and animals, one of the major trends has been toward the structural and functional specialization of cells. The cells comprising the body of one of the higher animals are not all alike, but are differentiated and specialized to perform certain functions more efficiently than an unspecialized animal cell could. This specialization has also had the effect of making the several parts of the body interdependent, so that an injury to, or the destruction of, cells in one part of the body may result in the death of the whole organism. The advantages of specialization are so great that they more than outweigh the disadvantages. The cells of the body which are similarly specialized are known as a tissue. A **tissue** may be defined as a group or layer of similarly specialized cells which together perform certain special functions. The study of the structure and arrangement of tissues is known as **histology.** Each tissue is composed of cells which have a characteristic shape, size and arrangement; the different types of tissue of the vertebrate body are readily recognized when examined microscopically. Certain tissues are composed of nonliving cell products in addition to the cells; connective tissue contains many fibers in addition to the fibroblasts or connective tissue cells, and bone and cartilage are made largely of proteins and salts secreted by the bone or cartilage cells.

The cells of a multicellular animal such as man may be classified in six major groups, each of which has several subgroups. These are epithelial, connective, muscular, blood, nervous and reproductive tissues.

Epithelial Tissues. Epithelial tissues are composed of cells which form a continuous layer or sheet covering the surface of the body or lining cavities within the body. There is usually a noncellular **basement membrane** underlying the sheet of epithelial cells. The epithelial cells in the skin of vertebrates are usually connected by small cytoplasmic processes or bridges. The epithelia of the body protect the underlying cells from mechanical injury, from harmful chemicals and bacteria, and from desiccation. The epithelial lining of the digestive tract absorbs water and nutrients for use in the body. The lining of the digestive tract and a variety of other epithelia produce and give off a wide spectrum of substances. Some of these are used elsewhere in the body and others are waste products which must be eliminated. Since the entire body is covered by an epithelium, all of the sensory stimuli must pass through some epithelium to reach the specific receptors for those

stimuli. The functions of epithelia are thus protection, absorption, secretion and sensation. The lining of the digestive tract, windpipe, lungs, kidney tubules and urinary bladder, and the outer layer of the skin are some familiar examples of epithelial tissues.

The cells in epithelial tissues may be flat, cuboidal or columnar in shape, they may be arranged in a single layer or in many layers, and they may have fine hairs or cilia on the free surface. On the basis of these structural characteristics epithelia are subdivided into the following groups.

Squamous epithelium is made of thin flattened cells the shape of flagstones or tiles (Fig. 3.16). It is found on the surface of the skin and the lining of the mouth, esophagus and vagina. The endothelium lining the cavity of blood vessels and the mesothelium lining the coelom are squamous epithelia. In the lower animals the skin is usually covered with a single layer of squamous epithelium, but in man and the higher animals the outer layer of the skin consists of stratified squamous epithelium, made of several layers of these flat cells.

The kidney tubules are lined with **cuboidal epithelium,** made of cells that are cube-shaped and look like dice (Fig. 3.16). Many other parts of the body, such as the stomach and intestines, are lined by cells that are taller than they are wide. An epithelium composed of such elongated, pillar-like cells is known as **columnar epithelium** (Fig. 3.16). Columnar epithelium may be simple, consisting of a single layer of cells, or stratified, composed of several layers of cells.

Either cuboidal or columnar epithelial cells may have cilia on their free surface. Ciliated cuboidal epithelium is found in the sperm ducts of earthworms and other animals, and ciliated columnar epithelium lines the ducts of the respiratory system of man and other air-breathing vertebrates. The rhythmic, concerted beating of the cilia moves solid particles in one direction through the ducts. Epithelial cells, usually columnar ones, may be specialized to receive stimuli. The groups of cells in the taste buds of the tongue or the olfactory epithelium in the nose are examples of sensory epithelium. Columnar or cuboidal epithelia may also be specialized for secreting certain products such as milk, wax, saliva, perspiration or mucus. The outer

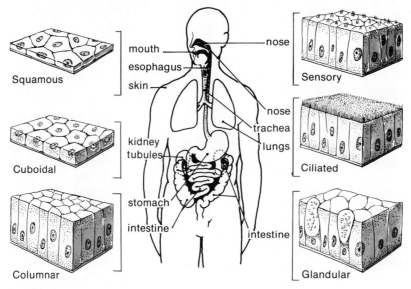

Figure 3.16 Diagram of the types of epithelial tissue and their location in the body.

epithelium of most worms secretes a thin, continuous, noncellular protective layer, called the **cuticle,** which covers the entire body. Insects, spiders, crabs and other arthropods secrete a cuticle which may be quite thick and strengthened with deposits of chitin and salts. The hard protective shells of oysters and snails, composed of calcium carbonate, are secreted by epithelial cells in the mantles of these animals.

Connective Tissues. The connective tissues—bone, cartilage, tendons, ligaments, fibrous connective tissue and adipose tissue—support and bind together the other tissues and organs. Connective tissue cells characteristically secrete a nonliving material called the **matrix,** and the nature and function of each connective tissue is determined primarily by the nature of this intercellular matrix. The actual connective tissue cells may form only a small and inconspicuous part of the tissue. The matrix, rather than the connective tissue cells themselves, does the actual connecting and supporting.

Fibrous connective tissue consists of a thick, interlacing, matted network of fibers in which are distributed the cells that secreted the fibers (Fig. 3.17). There are three types of fibrous connective tissue, widely distributed throughout the body, which bind skin to muscle, muscle to bone, and so on. These include very delicate **reticular fibers;** thick, tough, unbranched,

flexible, but relatively inelastic **collagen fibers;** and long, branched **elastic fibers. Adipose tissue** is rich in fat cells, specialized connective tissue cells which store large quantities of fat in a vacuole in the cytoplasm. Ligaments and tendons are specialized fibrous connective tissues. **Tendons** are composed of thick, closely packed bundles of collagen fibers, which form flexible cables that connect a muscle to a bone or to another muscle. A **ligament** is fundamentally similar in constitution to a tendon and connects one bone to another. An especially thick mat of fibrous connective tissue is located in the lower layer of the skin of most vertebrates; when this is chemically treated —"tanned"—it becomes leather.

Connective tissue fibers contain a protein called **collagen.** Treating the fibers with hot water converts some of the collagen into the soluble protein **gelatin.** The amino acid compositions of gelatin and collagen are nearly identical. Because so much connective tissue is present, nearly one-third of all the protein in the human body is collagen. The collagen units comprising the fibers consist of a helix made of three peptide chains joined by hydrogen bonds.

The supporting skeleton of vertebrates is composed of cartilage or bone. **Cartilage** appears as the supporting skeleton in the embryonic stages of all vertebrates, but is largely replaced in the adult by bone in all but sharks and rays. Cartilage can be felt in man as the supporting framework of the pinna of the ear (the external ear flap) or the tip of the nose. It is made of a firm but elastic matrix secreted by cartilage cells which become embedded in the matrix (Fig. 3.17). These cartilage cells are alive; they may secrete collagenous fibers or elastic fibers to strengthen the cartilage.

Bone consists of a dense matrix composed of proteins and calcium salts identical with the mineral hydroxyapatite, $Ca_3(PO_4)_2 \cdot CaCO_3$. About 65 per cent of the bone is made of this mineral. Bone cells (osteoblasts) remain alive and secrete a bony matrix, both protein and calcium salts, throughout life. The osteoblasts become surrounded and trapped by their own secretion and remain in microscopic cavities (lacunae) in the bone as living osteocytes (Fig. 3.17). The protein is laid down as minute fibers which contribute strength and

resiliency and the mineral salts contribute hardness to bone.

At the surface of each bone is a thin fibrous layer called the **periosteum** (peri, around; osteum, bone) to which the muscles are attached by tendons. The periosteum contains cells, some of which differentiate into osteoblasts and secrete protein and salts to bring about growth and repair. Most bones are not solid but have a marrow cavity in the center. The apparently solid matrix of the bone is pierced by many microscopic channels (haversian canals) in which lie blood vessels and nerves to supply the bone cells. The bony matrix is deposited, usually in concentric rings or lamellae, around these haversian canals. Each bone cell is connected to the adjacent bone cells and to the haversian canals by cellular extensions occupying minute canals (canaliculi) in the matrix. The bone cells obtain oxygen and raw materials and eliminate wastes by way of

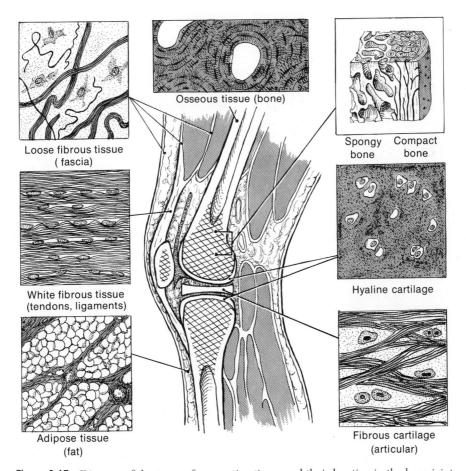

Loose fibrous tissue (fascia)

Osseous tissue (bone)

Spongy bone   Compact bone

White fibrous tissue (tendons, ligaments)

Hyaline cartilage

Adipose tissue (fat)

Fibrous cartilage (articular)

Figure 3.17 Diagram of the types of connective tissue and their location in the knee joint.

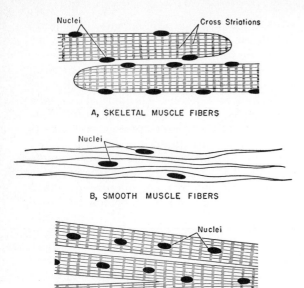

A, SKELETAL MUSCLE FIBERS

B, SMOOTH MUSCLE FIBERS

C, CARDIAC MUSCLE FIBERS

Figure 3.18 Types of muscle tissue. (Villee: Biology, 6th ed.)

these canaliculi. The details of the architecture of a bone can be observed by grinding a slice of bone extremely thin and mounting it on a slide for inspection under a microscope. Bone contains not only bone-secreting cells, but also bone-destroying cells. By the action of these two types of cells, the shape of a bone may be altered to resist changing stresses and strains. Bone formation and destruction is regulated by the availability of calcium and phosphate, by the presence of vitamin D, and by the hormones thyrocalcitonin and parathyrin, secreted by the thyroid and parathyroid glands. The marrow cavity of the bone may contain yellow marrow (largely a fat depot) or red marrow, the tissue in which red and certain white blood cells are formed.

Muscular Tissues. The movements of most animals result from the contraction of elongated, cylindrical or spindle-shaped cells, each of which contains many microscopic, elongate, parallel, contractile fibers called **myofibrils,** composed of the proteins myosin and actin. Muscle cells perform mechanical work by contracting — by getting shorter and thicker; they are unable to do work by pushing. Three types of muscle tissue are found in vertebrates: skeletal, cardiac and smooth (Fig. 3.18). **Cardiac muscle** is found only in the walls of the heart; **smooth muscle** in the walls of the

digestive tract, the urinary and genital tracts, and the walls of arteries and veins; and **skeletal muscle** makes up the muscle masses which are attached to and move the bones of the body. Cardiac and skeletal muscle cells are among the exceptions to the rule that cells have but one nucleus; each of these cells has many nuclei. The nuclei of skeletal muscle cells have an unusual position, at the periphery of the cell, just below the plasma membrane. Skeletal muscle cells are extremely long, two or more centimeters in length; indeed, some investigators believe that the muscle cells extend from one end of the muscle to the other, so that their length is equal to that of the muscle. Muscle fibers range in thickness from 10 to 100 microns; continued, strenuous muscle activity increases the thickness of the fiber. The myofibrils of skeletal and cardiac muscle have alternate dark and light cross bands or **striations.** These striations appear to have some fundamental role in contraction, during which the dark band remains constant but the light band shortens. The contraction of skeletal muscles is generally voluntary, under the control of the will, that of cardiac and smooth muscles is involuntary. Cardiac muscle cells are striated but have centrally located nuclei. Smooth muscle cells are not striated, have pointed ends, and have centrally

TABLE 3.2 COMPARISON OF VERTEBRATE MUSCLE TISSUES

	Skeletal	Smooth	Cardiac
Location	Attached to skeleton	Walls of viscera: stomach, intestines, etc.	Wall of heart
Shape of fiber	Elongate, cylindrical, blunt ends	Elongate, spindle-shaped, pointed ends	Elongate, cylindrical; fibers branch and fuse
Number of nuclei per cell	Many	One	Many
Position of nuclei	Peripheral	Central	Central
Cross striations	Present	Absent	Present
Speed of contraction	Most rapid	Slowest	Intermediate
Ability to remain contracted	Least	Greatest	Intermediate
Type of control	Voluntary	Involuntary	Involuntary

located nuclei. Smooth muscle contracts slowly but can remain contracted for long periods of time. The voluntary muscles of some invertebrates, such as the ones which close the shell of an oyster, are smooth muscles. Striated muscles can contract very rapidly but cannot remain contracted; a striated muscle fiber must relax and rest before it is able to contract again. The muscles of insects, spiders, crabs and other arthropods have cross striations and contract very rapidly. The distinguishing features of the three types of muscle are summarized in Table 3.2.

Vascular Tissues. The **blood,** composed of a liquid part—**plasma**—and of several types of **formed elements**—red cells, white cells and platelets—may be classified as a separate type of tissue or as one kind of connective tissue. The latter classification is based on the fact that blood cells and connective tissue cells originate from similar cells; however, the adult cells are quite dif-ferent in structure and function. The **red cells** of vertebrates contain the red pigment hemoglobin, which has the property of com-bining easily and reversibly with oxygen. Oxygen, combined as oxyhemoglobin, is transported to the cells of the body in the red cells. Mammalian red cells are flattened, biconcave discs without a nucleus; those of other vertebrates are more typical cells with an oval shape and a nucleus.

The five different kinds of **white blood cells**—lymphocytes, monocytes, neutro-phils, eosinophils and basophils (Fig. 3.19)—lack hemoglobin but move around and engulf bacteria. They can slip through the walls of blood vessels and enter the tissues of the body to engulf bacteria there. The fluid plasma transports a great variety of substances from one part of the body to another. Some of the substances transported are in solution, others are bound to one or another of the plasma proteins. The plasma of vertebrates is a light yellow color; in

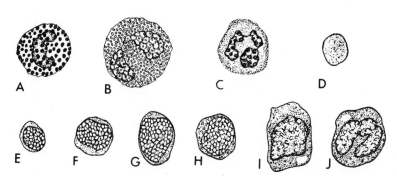

Figure 3.19 Types of white blood cells. *A,* basophil; *B,* eosinophil; *C,* neutrophil; *E–H,* a variety of lymphocytes; *I* and *J,* monocytes; *D,* a red blood cell drawn to the same scale. (Villee: Biology, 6th ed.)

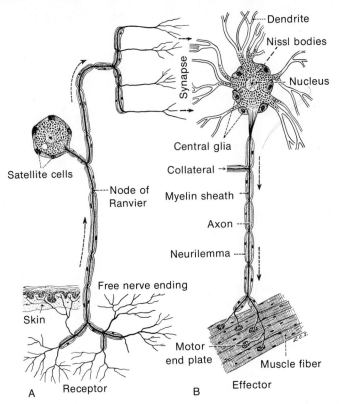

Figure 3.20   Diagrams of (A) an afferent neuron and (B) an efferent neuron. The arrows indicate the direction of the normal nerve impulse. (King and Showers: Human Anatomy and Physiology, 5th ed.)

certain invertebrates the oxygen-carrying pigment is not localized in cells, but is dissolved in the plasma and colors it red or blue. **Platelets** are small fragments broken off from cells in the bone marrow; they play a role in the clotting of blood (Chap. 13).

Nervous Tissues. Cells specialized for the reception of stimuli and the transmission of impulses are called **neurons.** A neuron typically has an enlarged cell body, containing the nucleus, and two or more cytoplasmic processes, the nerve fibers, along which the nerve impulse travels to the next neuron (Fig. 3.20). Nerve fibers vary in width from a few microns to 30 or 40 microns and in length from a millimeter or two to a meter or more. Two types of nerve fibers are distinguished: **axons,** which transmit impulses away from the cell body, and **dendrites,** which transmit them to the cell body. The junction between the axon of one neuron and the dendrite of the next neuron in the chain is called a **synapse.** At the synapse the axon and dendrite do not

actually touch; there is a small gap between the two. Transmission of an impulse across the synapse involves a different mechanism from that moving an impulse along the nerve fiber. An impulse can travel across the synapse only from an axon to a dendrite; thus, the synapse serves as a valve to prevent the backflow of impulses. Neurons show widely diverse patterns in shape of the cell body and in number and length of dendrites and axons.

The cell bodies of neurons commonly occur in groups; there are columns of cell bodies in the spinal cord, sheets of cell bodies over the surface of parts of the brain, nodules of cell bodies ("nuclei") within the brain, and the ganglia of the cranial and spinal nerves. A **ganglion** is a group of nerve cell bodies located outside the central nervous system. A nerve consists of a group of axons and dendrites bound together by connective tissue. Each nerve fiber—axon or dendrite—is surrounded by one or two sheaths, a **neurilemma** and/or a **myelin**

sheath. The neurilemma is a delicate, transparent, tubelike membrane made of cells which envelop the fiber. The myelin sheath is made of noncellular, fatty material which forms a glistening white coat between the fiber and neurilemma. The myelin sheath is interrupted at fairly regular intervals along the nerve by constrictions called the nodes of Ranvier. Nerve fibers are either "medullated" and have a thick myelin sheath, or "nonmedullated" and have an extremely thin myelin sheath. Nerve fibers in the brain and spinal cord have a myelin sheath but no neurilemma; those in the autonomic nerves to the viscera, and the nerves of many invertebrates, are nonmyelinated and have a very thin or no myelin sheath but a neurilemma. The nerves to the skin and skeletal muscles of vertebrates have both a myelin sheath and a neurilemma surrounding them.

Nervous tissue contains, in addition to neurons, several different kinds of supporting cells called **neuroglia.** These have many cytoplasmic processes, and the cells and their processes form an extremely dense supporting framework in which the neurons are suspended. The neuroglia are believed to separate and insulate adjacent neurons, so that nerve impulses can pass from one neuron to the next only over the synapse, where the neuroglial barrier is incomplete.

Reproductive Tissues. The **egg cells** (ova) formed in the ovary of the female and the **sperm cells** produced by the testes of the male constitute the reproductive tissues — cells specially modified for the production of offspring (Fig. 3.21). Egg cells are generally spherical or oval and are nonmotile. A typical egg has a large nucleus, called the germinal vesicle, and a variable amount of yolk in the cytoplasm. Shark and bird eggs have enormous amounts of yolk which provides nourishment for the developing embryo until it hatches from the shell. Sperm cells are small and modified for motility. A typical sperm has a long **tail,** the beating of which propels the sperm to its meeting and union with the egg. The **head** of the sperm contains the nucleus surrounded by a thin film of cytoplasm. The tail is connected to the head by a short **middle piece.** An **axial filament,** formed by the centriole in the middle piece, extends to the tip of the tail. Most of the cytoplasm is sloughed off as the sperm matures; this decreases the weight of the sperm and perhaps renders it more motile.

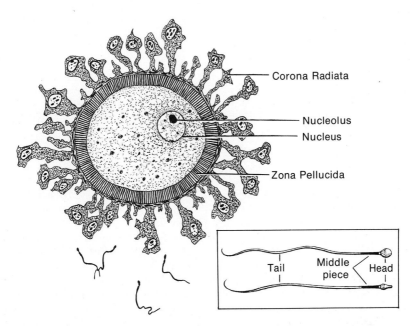

Figure 3.21 Human egg and sperm, magnified 400 times. *Inset,* side and top views of a sperm, magnified about 2000 times. The egg is surrounded by other cells which form the corona radiata.

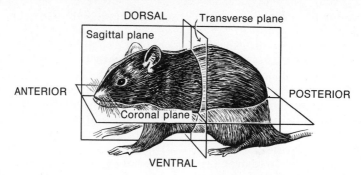

DORSAL   Transverse plane

Sagittal plane

ANTERIOR   POSTERIOR

Coronal plane

VENTRAL

Figure 3.22 Diagram to illustrate transverse, sagittal and frontal or coronal planes in a bilaterally symmetrical animal.

3.13 BODY PLAN AND SYMMETRY

To refer to the regions of an animal body, zoologists use the term **anterior** for the head end and **posterior** for the tail end; the back side is the **dorsal** side and the belly side is the **ventral** side. The midline of the body is **medial** and the sides are **lateral**. The part of a structure nearer the point of reference is **proximal**, the part farther away is **distal**.

A body is symmetrical if it can be cut into two equivalent halves. A few kinds of protozoa can be cut into two equal halves by any plane through the center; they are said to have **spherical symmetry**. Coelenterates and echinoderms have **radial symmetry**; they can be cut into two equal halves by any plane which includes the axis extending from top to bottom through the center. In

such animals a top and bottom side can be distinguished. Most other animals are **bilaterally symmetrical**, and can be cut into two equivalent halves only by a plane passing from anterior to posterior and from the dorsal to ventral sides in the midline. In such a bilaterally symmetrical animal, three types of planes or cuts can be made to get different views: sagittal, frontal and transverse (Fig. 3.22). A **sagittal section** is one made by cutting in the median vertical plane; thus it includes the anterior-posterior axis and the dorsoventral axis but is at right angles to the right-left axis. A **frontal section** is at right angles to a sagittal section and includes the anterior-posterior axis and the right-left axis, but is perpendicular to the dorsoventral axis. **Transverse sections** are cut at right angles to the anterior-posterior axis and include a dorsoventral and a right-left axis.

ANNOTATED REFERENCES

Bloom, W., and D. Fawcett: Textbook of Histology. 9th ed. Philadelphia, W. B. Saunders Co., 1968. One of the standard texts of histology with many superb illustrations of the structure of tissues.

DeRobertis, E. D. P., W. W. Nowinski and F. A. Saez: Cell Biology. 5th ed. Philadelphia, W. B. Saunders Co., 1970. Contains a detailed presentation of the structure and properties of a wide variety of cells.

Fawcett, D.: The Cell: Its Organelles and Inclusions. Philadelphia, W. B. Saunders Co., 1966. An excellent discussion of cell structure illustrated by electron micrographs of outstanding quality.

Hall, T. S.: A Source Book in Animal Biology. New York, McGraw-Hill Book Co., 1951. The development of the cell theory is presented in an interesting fashion by means of long quotations from some of the original scientific papers.

Kennedy, D. (Ed.): The Living Cell: Readings from the Scientific American. San Francisco, W. H. Freeman & Co., 1965. A rich collection of articles about the cell.

4. Cell Metabolism

All cells have complex, efficient systems for transforming one type of energy into another by appropriate chemical reactions. Three major types of energy transformations can be distinguished in the biological world (Fig. 4.1). In the first, the radiant energy of sunlight is captured by the green pigment **chlorophyll,** present in green plants, and is transformed by the process of **photosynthesis** into chemical energy. This is used to synthesize carbohydrates and other complex molecules from carbon dioxide and water. The radiant energy of sunlight, a form of **kinetic energy,** is transformed into a type of **potential energy.** The chemical energy is stored in the molecules of carbohydrates and other foodstuffs as the energy of the bonds which connect its constituent atoms.

In a second type of energy transformation, the chemical energy of carbohydrates and other molecules is transformed by the process termed **cellular respiration** into the biologically useful energy of energy-rich phosphate bonds. This kind of energy transformation occurs in the mitochondrion. A third type of energy transformation occurs when the chemical energy of these energy-rich phosphate bonds is utilized by the cell to do work—the mechanical work of muscu-

Figure 4.1 Energy transformations in the biological world. (1) The radiant energy of sunlight is transformed in photosynthesis into chemical energy in the bonds of organic compounds. (2) The chemical energy of organic compounds is transformed during cellular respiration into biologically useful energy, the energy-rich phosphate bonds of ATP and other compounds. (3) The chemical energy of energy-rich phosphate bonds is utilized in cells to do mechanical, electrical, osmotic or chemical work. Finally energy flows to the environment as heat in the "entropy sink." (Villee: Biology, 6th ed.)

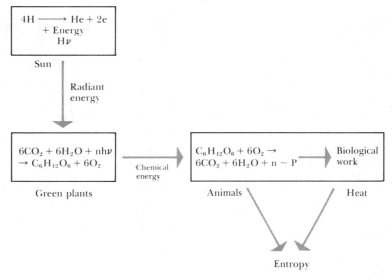

lar contraction, the electrical work of conducting a nerve impulse, the osmotic work of moving molecules against a gradient, or the chemical work of synthesizing molecules for growth. As these transformations occur, energy finally flows to the environment and is dissipated as heat. Plants and animals have evolved some remarkably effective **energy transducers,** such as chloroplasts and mitochondria, to carry out these processes, together with control mechanisms to regulate the transducers and enable the cells to adjust to variations in environmental conditions.

The chemical reactions of cells, which provide for their growth, irritability, movement, maintenance and repair, and their reproduction, are collectively termed **metabolism.** The metabolic activities of animal, plant and bacterial cells are remarkably similar, despite the apparent differences of the organisms themselves. In all cells, sugars and related substances are continually being metabolized, via a large number of intermediate compounds, to water and carbon dioxide. During these conversions the energy of the sugar molecule is made available to drive other processes in the cell.

4.1 ENTROPY AND ENERGY

The branch of physics that deals with energy and its transformations, **thermodynamics,** is based on certain relatively simple principles which are universally applicable to chemical processes whether these occur in living or nonliving systems.

Laws of Thermodynamics. Under experimentally controlled conditions, the amount of energy entering and leaving any system may be measured and compared. It is always found that energy is neither created nor destroyed but is only transformed from one form to another. This is an expression of the first law of thermodynamics, sometimes called the **law of the conservation of energy:** the total energy of any system and its surroundings remains constant. As any given system undergoes a change from its initial state to its final state it may absorb energy from the surroundings or deliver energy to the surroundings. The difference in the energy content of the system in its initial and final state must be just equalled by a corresponding change in the energy

content of the surroundings. Heat is a convenient form in which energy may be measured which is why the study of energy has been called thermodynamics, i.e., heat dynamics. Heat, however, is not a useful way of transferring energy in biological systems, for living organisms are basically **isothermal** (equal temperature). There is no significant temperature difference between the different parts of the cell or between the different cells in a tissue. Stated in another way, cells do not act as heat engines. They have no means of allowing heat to flow from a warmer to a cooler body.

The **second law of thermodynamics** may be stated briefly as "the entropy of the universe increases." **Entropy** may be defined as a randomized state of energy that is unavailable to do work. The second law may be phrased that "physical and chemical processes proceed in such a way that the entropy of the system becomes maximal." Entropy then is a measure of randomness or disorder. In almost all energy transformations there is a loss of some heat to the surroundings, and since heat involves the random motion of molecules, such heat losses increase the entropy of the surroundings. Living organisms and their component cells are highly organized and thus have little entropy. They preserve this low entropy state by increasing the entropy of their surroundings. You increase the entropy of your surroundings when you eat a candy bar and convert its glucose to carbon dioxide and water and return them to the surroundings.

The force that drives all processes is the tendency of the system to reach the condition of maximum entropy. Heat is either given up or absorbed by the system from the surroundings to allow the system plus its surroundings to reach the state of maximum entropy. The changes of heat and entropy are related by a third dimension of energy termed **free energy.** Free energy may be visualized as that component of the total energy of a system which is available to do work under isothermal conditions. Entropy and free energy are related inversely; as entropy increases during an irreversible process, the amount of free energy decreases. All physical and chemical processes proceed with a decline in free energy until they reach an equilibrium in which the free energy of the system is at a minimum and the entropy is at a maximum. Free energy is useful energy; entropy is degraded, useless energy.

4.2 CHEMICAL REACTIONS

A chemical reaction is a change involving the molecular structure of one or more substances; matter is changed from one substance, with its characteristic properties, to another, with new properties, and energy is released or absorbed. Hydrochloric acid (HCl), for example, reacts with the base, sodium hydroxide (NaOH) to yield water (H_2O) and the salt, sodium chloride (NaCl); in the process energy is released as heat. The chemical properties of HCl and NaOH are very different from those of NaCl and H_2O. In chemical shorthand a plus sign connects the symbols of the reacting substances, HCl and NaOH, and the products, NaCl and H_2O. An arrow indicates the direction of the reaction:

$$HCl + NaOH \rightarrow NaCl + H_2O + energy\,(heat)$$

Most chemical reactions are reversible and this reversibility is indicated by a double arrow: $\rightleftharpoons$.

Atoms are neither destroyed nor created in the course of a chemical reaction; thus the sum of each kind of atom on one side of the arrow must equal the sum of that kind of atom on the other side. This is an expression of one of the basic laws of physics, the **Law of the Conservation of Matter.** The direction of a reversible reaction is determined by the energy relations of the several chemicals involved, their relative concentrations, and their solubility.

For each reaction the thermodynamic equilibrium constant, K, expresses the chemical equilibrium reached by the system. Thus for the reaction $A + B \rightleftharpoons C + D$,

$$K = \frac{(C) \times (D)}{(A) \times (B)}$$

The equilibrium constant for any given reaction is determined by the tendency of the reaction components to reach maximum entropy, or minimum free energy, for the system. The equilibrium constant, K, is related mathematically to the change in free energy of the components of the reaction: $\Delta G = -RT \ln K$. R is the gas constant, T the absolute temperature, and ln K is the natural logarithm of the equilibrium constant. The symbol ΔG represents the standard free energy change (i.e., the gain or loss of free energy in calories as one mole of reactant is converted to one mole [*] of product).

Examination of this equation reveals that when the equilibrium constant, K, is high, the standard free energy change, ΔG, is negative. Such a reaction will proceed with a decrease in free energy. However, when the equilibrium constant is very small the reaction does not go far in the direction of completion, and the free energy change is positive. It is necessary to put energy into the system to transform one mole of reactant into one mole of product. When the equilibrium constant is 1, then the change in free energy is zero and the reaction is freely reversible.

The rate at which a chemical reaction occurs is determined by a number of factors, one of which is temperature. Each increase of 10° C. approximately doubles the rate of most reactions. This is true of biological processes as well as reactions in a chemist's test tube. This is further evidence that the chemical reactions of living things are fundamentally similar to those of nonliving ones. Whether or not an enzyme or any other catalyst is present in the system has no effect on either the equilibrium constant, K, or the free energy change, ΔG, undergone. Enzymes and other catalysts simply speed up the rate at which the system approaches equilibrium but do not change the equilibrium point itself.

The unit of energy most widely used in biological systems is the **Calorie,** which is the amount of heat required to raise one kilogram of water one degree centigrade (strictly speaking, from 14.5° to 15.5° C.). Other forms of energy — radiant, chemical, electrical, the energy of motion or position — can be converted to heat and measured by their effect in raising the temperature of water.

Reactions which have a high equilibrium constant, K, and a negative standard free

---

[*] A mole or gram molecular weight of a compound is defined as the amount of that substance equal to its molecular weight in grams. One mole of glucose is 180 grams of glucose. In calculating K the concentrations of all reactants and products are expressed as moles per liter. By convention the parentheses around each symbol imply its molar concentration.

energy change, ΔG, are said to be **exergonic.** A reaction with a very low equilibrium constant and therefore a positive standard free energy change is not spontaneous and will not go to completion under standard conditions, unless energy is supplied to it. Such processes are called **endergonic** processes. In biological systems, the endergonic processes must be coupled with some exergonic process so that the exergonic process delivers the required energy to the endergonic process. In such coupled systems, the endergonic process can occur only if the decrease in free energy of the exergonic process to which it is coupled is larger than the gain in free energy of the endergonic process.

Catalysis.   Many of the substances that are rapidly metabolized by living cells are remarkably inert outside the body. A glucose solution, for example, will keep indefinitely in a bottle if it is kept free of bacteria and molds. It must be subjected to high temperature or to the action of strong acids or bases before it will decompose. Living cells cannot utilize conditions as extreme as these, for the cell itself would be destroyed long before the glucose, yet glucose is rapidly decomposed within cytoplasm at ordinary temperatures and pressures and in a solution which is neither acidic nor basic. The reactions within the cell are brought about by special agents known as **enzymes,** which belong to the class of substances known as catalysts. A **catalyst** is an agent which affects the velocity of a chemical reaction without altering its end point and without being used up in the course of the reaction. The list of substances which may serve as a catalyst in one or more reactions is long indeed. Water is an excellent catalyst for many reactions. Pure, dry hydrogen gas and dry chlorine gas do not react when mixed, but if a slight trace of water is present they react with explosive violence to form hydrogen chloride. Metals such as iron, nickel, platinum and palladium, when ground into a fine powder, are widely used as catalysts in industrial processes such as the hydrogenation of cottonseed and other vegetable oils to make margarine or the cracking of petroleum to make gasoline. A minute amount of catalyst will speed up the reaction of vast quantities of reactants, for the molecules of catalyst are not exhausted in the reaction but are used again and again.

4.3   ENZYMES

Enzymes are protein catalysts, produced by living cells. They regulate the speed and specificity of the thousands of chemical reactions that occur within cells. Although enzymes are synthesized within cells, they do not have to be inside a cell to act as a catalyst. Many enzymes have been extracted from cells with their activity unimpaired. They can then be purified and crystallized and their catalytic abilities can be studied. Enzyme-controlled reactions are basic to all the phenomena of life: respiration, growth, muscle contraction, nerve conduction, photosynthesis, nitrogen fixation, deamination, digestion and so on. There is no need to postulate some mysterious vital force to account for these phenomena.

Properties of Enzymes.   All the enzymes isolated and crystallized to date have been found to be proteins. Enzymes are usually named by adding the suffix -*ase* to the name of the substance acted upon, e.g., sucrose is acted upon by the enzyme **sucrase.** Enzymes are usually colorless, but they may be yellow, green, blue, brown or red. Most enzymes are soluble in water or dilute salt solution, but some, for example the enzymes present in the mitochondria, are bound together by lipoprotein (a phospholipid-protein complex) and are insoluble in water.

The catalytic ability of some enzymes is truly phenomenal. For example, one molecule of the iron-containing enzyme **catalase,** extracted from beef liver, will bring about the decomposition of 5,000,000 molecules of hydrogen peroxide (H_2O_2) per minute at 0° C. The substance acted upon by an enzyme is known as its **substrate;** thus, hydrogen peroxide is the substrate of the enzyme catalase.

The number of molecules of substrate acted upon by a molecule of enzyme per minute is called the **turnover number** of the enzyme. The turnover number of catalase is thus 5,000,000. Most enzymes have high turnover numbers, thus they can be very effective although present in the cell in relatively minute amounts. Hydrogen peroxide is a poisonous substance produced as a byproduct in a number of enzyme reactions. Catalase protects the cell by destroying the peroxide.

Hydrogen peroxide can be split by iron

atoms alone, but it would take 300 *years* for an iron atom to split the same number of molecules of H_2O_2 that a molecule of catalase, which contains one iron atom, splits in one *second*. This example of the evolution of a catalyst emphasizes one of the prime characteristics of enzymes—they are very efficient catalysts.

Although enzymes in general catalyze specific reactions, they do differ in the number of kinds of substrates they will attack. **Urease** is an example of an enzyme which is absolutely specific. Urease decomposes urea to ammonia and carbon dioxide and will attack no substance other than urea. Most enzymes are not quite so specific, and will attack several closely related substances. **Peroxidase,** for example, will decompose several different peroxides in addition to hydrogen peroxide. A few enzymes are specific only in requiring that the substrate have a certain kind of chemical bond. The **lipase** secreted by the pancreas will split the ester bonds connecting the glycerol and fatty acids of a wide variety of fats.

In theory, enzyme-controlled reactions are reversible; the enzyme does not determine the direction of the reaction but simply accelerates the rate at which the reaction reaches equilibrium. The classic example of this is the action of the enzyme lipase on the splitting of fat, or union of glycerol and fatty acids. If one begins with a fat, the enzyme catalyzes the splitting of this to give some glycerol and fatty acids. If one begins with a mixture of fatty acids and glycerol, the enzyme catalyzes the synthesis of some fat. When either system has operated long enough, the same equilibrium mixture of fat, glycerol and fatty acid is reached:

$$\text{fat} \rightleftharpoons \text{glycerol} + 3 \text{ fatty acids}$$

The equilibrium point is determined by complex thermodynamic principles, which will not be discussed. Since reactions give off energy when going in one direction, it is obvious that an equivalent amount of energy in the proper form must be supplied to make the reaction go in the opposite direction.

To drive an energy-requiring reaction, some energy-yielding reaction must occur at about the same time. In most biologic systems, energy-yielding reactions result in the synthesis of "energy-rich" phosphate esters, such as the terminal bonds of **adenosine triphosphate** (abbreviated as ATP). The energy of these energy-rich bonds is then available for the conduction of an impulse, the contraction of a muscle, the synthesis of complex molecules, and so on, much as the energy of a storage battery made by a generator is available for light, heat or running a motor. Biochemists use the term "coupled reactions" for two reactions which must occur together so that one can furnish the energy, or one of the reactants, needed by the other.

Enzymes generally work in teams in the cell, with the product of one enzyme-controlled reaction serving as the substrate for the next. We can picture the inside of a cell as a factory with many different assembly lines (and disassembly lines) operating simultaneously. Each of these assembly lines is composed of a number of enzymes, each of which catalyzes the reaction by which one substance is converted into a second. This second substance is passed along to the next enzyme, which converts it into a third, and so on along the line. From germinating barley seeds one can extract two enzymes that convert starch to glucose. The first, amylase, splits starch to maltose and the second, maltase, splits the double sugar maltose to two molecules of the single sugar glucose. Eleven different enzymes, working consecutively, are required to convert glucose to lactic acid. The same series of 11 enzymes is found in human cells, in green leaves and in bacteria.

Some enzymes, such as pepsin and urease, have been found to consist solely of protein. Many others, however, consist of two parts: one is protein (called the **apoenzyme**) and the other (called a **coenzyme**) is some smaller organic molecule, usually containing phosphate. Coenzymes can usually be separated from their enzymes and, when analyzed, have proved to contain some vitamin—thiamine, niacin, riboflavin, etc.—as part of the molecule. This finding has led to the generalization that *all vitamins function as parts of coenzymes in the cell.* Neither the apoenzyme nor the coenzyme alone has catalytic properties; only when the two are combined is activity evident. Certain enzymes require for activity, in addition to a coenzyme, the presence of one or more ions. Magnesium (Mg^{++}) is required for the activity of several of the enzymes in the chain which converts glucose to lactic acid. **Salivary amylase,** the starch-splitting enzyme of saliva, requires chloride ion as an activator. Most, if not all, of the elements required by plants and animals in very small amounts — the so-called **trace elements,** manganese, copper, cobalt, zinc, iron, and others — serve as enzyme activators.

Enzymes may be present in the cell either dissolved in the liquid phase or bound to, and presumably an integral part of, one of the subcellular organelles. A water extract of ground liver contains all the 11 kinds of enzymes necessary to convert glucose to lactic acid. The respiratory enzymes, which catalyze the metabolism of lactic acid and the carbon chains of fatty acids and amino acids to carbon dioxide and water, are integral parts of the mitochondria.

The Mechanism of Enzyme Catalysis. Many years ago Emil Fischer, the German organic chemist, suggested that the specificity of the relationship of an enzyme to its substrate indicated that the two must fit together like a lock and key. The idea that an enzyme combines with its substrate to form a reactive intermediate enzyme-substrate complex, which subsequently decomposes to release the free enzyme and the reaction products, was formulated mathematically by Leonor Michaelis more than 50 years ago. By brilliant inductive reasoning, he assumed that such a complex does form, and then calculated what the relationships between enzyme concentration, substrate concentration, and the velocity of the reaction should be. Exactly these relationships are observed experimentally, which is strong evidence of the validity of Michaelis' assumption that an enzyme-substrate complex forms as an intermediate.

Direct evidence of the existence of enzyme-substrate complexes was obtained by David Keilin of Cambridge University and Britton Chance of the University of Pennsylvania. Chance isolated a brown-colored peroxidase from horseradish and found that when this was mixed with the substrate, hydrogen peroxide, a green-colored enzyme-substrate complex formed. This, in turn, changed to a second, pale red complex which finally split to give the original brown enzyme and the products of the reaction. By observing the changes of color, Chance was able to calculate the rates of formation and of dissociation of this complex.

It is clear that when it is part of an enzyme-substrate complex, the substrate is much more reactive than it is when free. It is not clear, however, *why* this should be true. One explanation postulates that the enzyme unites with the substrate at two or more places, and the substrate molecule is held in a position which strains its molecular bonds and renders them more likely to break.

One approach to the study of enzymic action is to investigate the detailed structure of the enzyme molecule itself. It is now possible to determine the kinds of amino acids present in an enzyme molecule, their relative numbers and the sequence of amino acids in the peptide chain or chains comprising the enzyme protein. Ribonuclease, which consists of 124 amino acids in a single chain that is looped on itself like a pretzel, was the first enzyme for which it was possible to give the exact sequence of amino acids for the entire molecule. Such analyses of the structure of the enzyme molecule may provide clues as to the mechanism of enzymic catalysis. It is probable that only a relatively small part of the enzyme molecule (termed the "active site") is involved in combining with the substrate. Studies of the sequence of amino acids and the three-dimensional structure of this part of the enzyme should be especially fruitful in throwing light on the mechanism of enzyme action.

4.4 FACTORS AFFECTING ENZYMIC ACTIVITY

Temperature. The velocity of most chemical reactions is approximately doubled by each 10 degree increase in temperature, and, over a moderate range of temperature, this is true of enzyme-catalyzed reactions as well. Enzymes, and proteins in general, are inactivated by high temperatures; the higher the temperature, the more rapidly the enzymic activity is lost. Native protein molecules exist at least in part as **spiral coils,** or helices, and the denaturation process appears to involve the unwinding of the helix. Enzyme inactivation is a reversible process if the temperature is not too high and has not been applied more than a short time.

Most organisms are killed by exposure to heat because their cellular enzymes are inactivated. The processes of protein denaturation and enzyme inactivation show a striking parallelism and this is one bit of substantiating evidence that enzymes are proteins. The enzymes of man and other warm-blooded animals operate most efficiently at a temperature of about 37° C.—body temperature—whereas those of plants and cold-blooded animals work optimally at about 25° C. Enzymes are generally not inactivated by freezing; their reactions continue slowly, or perhaps cease altogether at low temperatures, but their catalytic activity reappears when the temperature is again raised to normal.

Acidity. All enzymes are sensitive to changes in the acidity and alkalinity—the pH—of their environment and will be inactivated if subjected to strong acids or bases. Most enzymes exert their greatest catalytic effect only when the pH of their environment is within a certain rather narrow range. On either side of this optimum pH, as the pH is raised or lowered, enzymic activity rapidly decreases. The protein-digesting enzyme secreted by the stomach, pepsin, is remarkable in that its pH optimum is 2.0; it will work only in an extremely acid medium. The protein-digesting enzyme secreted by the pancreas, trypsin, in contrast, has a pH optimum of 8.5, well on the alkaline side of neutrality. Most intracellular enzymes have pH optima near neutrality, pH 7.0. This marked influence of pH on the activity of an enzyme is what would be predicted from the fact that enzymes are proteins. The number of positive and negative charges associated with a protein molecule, and the shape of the molecular surface, are determined in part by the pH. Probably only one particular state of the enzyme molecule, with a particular number of negative and positive charges, is active as a catalyst.

Concentration of Enzyme, Substrate and Cofactors. If the pH and temperature of an enzyme system are kept constant and if an excess of substrate is present, the rate of the reaction is directly proportional to the amount of *enzyme* present. This method is used, indeed, to measure the amount of some particular enzyme present in a tissue extract. If the pH, temperature and enzyme concentration of a reaction system are held constant, the initial reaction rate is proportional to the amount of *substrate* present, up to a limiting value. If the enzyme system requires a coenzyme or specific activator ion, the concentration of this substance may, under certain circumstances, determine the over-all rate of the enzyme system.

Enzyme Inhibitors. Enzymes can be inhibited by a variety of chemicals, some of which inhibit reversibly, others irreversibly. **Cytochrome oxidase,** one of the "respiratory enzymes," is inhibited by cyanide, which forms a complex with the atom of iron present in the enzyme molecule and prevents it from participating in the catalytic process. Cyanide is poisonous to man and other animals because of its action on the cytochrome enzymes. One of the enzymic steps in the conversion of glucose to lactic acid is inhibited by fluoride ion and another by iodoacetate. These substances, and others, have been used as tools by biochemists to investigate the properties and sequences of enzyme systems.

Enzymes themselves may act as poisons if they get into the wrong place. As little as 1 mg. of crystalline trypsin injected intravenously will kill a rat. Certain snake, bee and scorpion venoms contain enzymes that destroy blood cells or other tissues.

4.5 RESPIRATION AND CELLULAR ENERGY

All the phenomena of life—growth, movement, irritability, reproduction, and others—require the expenditure of energy by the cell. Living cells are not heat engines; they cannot use heat energy to drive these reactions but must use chemical energy, chiefly in the form of energy-rich phosphate bonds, abbreviated ~P. These bonds have a relatively high **free energy of hydrolysis,** ΔG (i.e., the difference in the energy content of the reactants and the products after the bond is split is relatively high). The free energy of hydrolysis is *not* localized in the covalent bond joining the phosphorus atom to the oxygen or nitrogen atom. Thus the term "energy-rich phosphate bond" is actually a misnomer, but it is so deeply ingrained by long usage that it is not likely to be changed.

The term **cellular respiration** refers to the

enzymatic processes within each cell by which molecules of carbohydrates, fatty acids and amino acids are metabolized ultimately to carbon dioxide and water with the conservation of biologically useful energy. Many of the enzymes catalyzing these reactions are located in the cristae and walls of the mitochondria (Fig. 3.12).

All living cells obtain biologically useful energy by enzymic reactions in which electrons flow from one energy level to another. For most organisms, oxygen is the ultimate electron acceptor; oxygen reacts with the electrons and with hydrogen ions to form a molecule of water. Electrons are transferred to oxygen by a system of enzymes, localized within the mitochondria, called the **electron transmitter system.**

Electrons are removed from a molecule of some foodstuff and transferred by the action of a specific enzyme to some primary electron acceptor. Other enzymes transfer the electrons from the primary acceptor through the several components of the electron transmitter system and eventually combine them with oxygen (Fig. 4.2). The chief source of energy-rich phosphate bonds, ~P, in the cell is from the flow of electrons through the acceptors and the electron transmitter system. This flow of electrons has been termed the "electron cascade," and we might picture a series of waterfalls over which electrons flow, each fall driving a water wheel, an enzymic reaction by which the energy of the electron is captured in a biologically useful form, that of the energy-rich phosphate bonds of ATP.

ATP is the "energy currency" of the cell. All the energy-requiring reactions of cellular metabolism utilize ATP to drive the reaction. Energy-rich molecules do not pass freely from one cell to another but are made at the site in which they are to be utilized. The energy-rich bonds of ATP that will drive the reactions of muscle contraction, for example, are produced in the muscle cells.

Processes in which electrons (e^-) are removed from an atom or molecule are termed

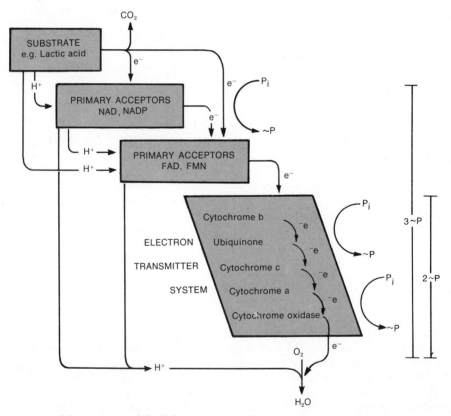

Figure 4.2 Diagram of the reactions of the "electron cascade," the succession of metabolic steps by which electrons are transferred from substrate to oxygen and the energy is trapped in a biologically useful form as energy-rich phosphate bonds. ~P. (Villee: Biology, 6th ed.)

oxidations; the reverse process, the addition of electrons to an atom or molecule, is termed **reduction.** An example of oxidation and reduction is the reversible reaction

$$Fe^{++} \rightleftharpoons Fe^{+++} + e^-$$

The reaction toward the right is an oxidation (the removal of an electron) and the reaction toward the left is a reduction (the addition of an electron). Each oxidation reaction, in which an electron is given off, must be accompanied by a reduction, a reaction in which the electron is accepted by another molecule, for electrons do not exist in the free state. The passage of electrons in the electron transmitter system is a series of oxidation and reduction reactions termed **biologic oxidation.** When the energy of this flow of electrons is captured in the form of ~P, the process is termed **oxidative phosphorylation.** In most biologic systems, two electrons and two protons (that is, two hydrogen atoms) are removed together and the process is known as **dehydrogenation.**

The specific compounds of the electron transmitter system that are alternately oxidized and reduced are known as **cytochromes.** Each cytochrome contains a heme group similar to the one present in hemoglobin. In the center of the heme group is an iron atom which is alternately oxidized and reduced—converted from Fe^{++} to Fe^{+++} and back—as it gives off and takes up an electron.

Another component of the electron transmitter system, **ubiquinone** (also called coenzyme Q), consists of a head, a six-membered carbon ring, which can take up and release electrons, and a long tail composed of 10 repeating units. These repeating units, called isoprenoid groups, each consisting of five carbon atoms, are the basic structural units of molecules of rubber.

Lactic acid, the acid of sour milk, is an important intermediate in metabolism. It undergoes a dehydrogenation to yield pyruvic acid. Its molecule contains the configuration $H-\overset{|}{\underset{|}{C}}-OH$ from which two hydrogens can be removed enzymatically by **lactic dehydrogenase** (LDH). In this reaction, and in all dehydrogenations, the electrons removed are transferred to a primary electron acceptor. The primary electron acceptor in this reaction is **nicotin-**

amide adenine dinucleotide, NAD. The product of the dehydrogenation reaction, pyruvic acid, cannot undergo dehydrogenation directly, for it does not have a structure suitable for attack by a dehydrogenase and must undergo further enzymic action to attain such a molecular configuration. The next reaction of pyruvic acid is a **decarboxylation,** the release of carbon dioxide. Carbon dioxide is derived in biologic systems only from carboxylic groups (COOH) by the process of decarboxylation. The product of the decarboxylation reaction, acetaldehyde, has two carbons instead of three but still is not in a form suitable for dehydrogenation. It must undergo a "make-ready" reaction, a reaction that will result in an $H-\overset{|}{\underset{|}{C}}-OH$ configuration suitable for dehydrogenation. The particular reaction that follows involves a complex organic molecule named **coenzyme A** which is abbreviated CoA—SH. In this abbreviation "—SH" represents the active end of the molecule, a sulfhydryl group composed of sulfur and hydrogen, and "CoA" represents the remainder of this complex molecule. The combination of coenzyme A with acetaldehyde results in a substance with an $H-\overset{|}{\underset{|}{C}}-OH$ group that can be dehydrogenated to give acetyl coenzyme A.

Another type of dehydrogenation, which involves a different molecular configuration, is exemplified by the oxidation of succinic acid. The dehydrogenation of molecules having a $-CH_2-CH_2-$ group involves a **flavin** as the hydrogen and electron acceptor. Succinic acid is oxidized to fumaric acid in the course of which two of its hydrogens are transferred to a flavin. The product, fumaric acid, cannot be dehydrogenated directly but undergoes a make-ready reaction in which a molecule of water is added to yield malic acid. Malic acid has a configuration of $H-\overset{|}{\underset{|}{C}}-OH$ which is suitable for dehydrogenation. The reaction catalyzed by malic dehydrogenase requires NAD as primary acceptor and yields oxaloacetic acid. Oxaloacetic acid may undergo several reactions, one of which is decarboxylation to yield pyruvic acid. The further metabolism of pyruvic acid to acetyl coenzyme A could occur by the reactions just discussed.

In the reactions by which carbohydrates, fats and proteins are oxidized, the cell utilizes these three simple types of reactions: dehydrogenation, decarboxylation and make-ready reactions. They may occur in different orders in different chains of reactions. All the dehydrogenation reactions are, by definition, oxidations — reactions in which electrons are removed from a molecule. The electrons cannot exist in the free state for any finite period of time and must be taken up immediately by other compounds, electron acceptors. Two of the primary electron acceptors of the cell are pyridine nucleotides — nicotinamide adenine dinucleotide (abbreviated NAD) and nicotinamide adenine dinucleotide phosphate (abbreviated NADP). The functional end of both pyridine nucleotides is the vitamin, **nicotinamide.** The nicotinamide ring accepts one hydrogen ion and two electrons from a molecule undergoing dehydrogenation (e.g., lactic acid) and becomes reduced nicotinamide adenine dinucleotide, NADH, releasing one proton.

NAD and NADP serve as primary electron and hydrogen acceptors in dehydrogenation reactions involving substrates with the

$$H—\overset{\displaystyle |}{\underset{\displaystyle |}{C}}—OH$$ configuration, such as the dehydrogenation of lactic or malic acid. The two pyridine nucleotides differ from one another in that NAD has two phosphate groups and NADP has three phosphate groups in the tail attached to the niacin ring. NAD may also be termed DPN (diphosphopyridine nucleotide) and NADP may be termed TPN (triphosphopyridine nucleotide). Most dehydrogenases specifically require either NAD or NADP as the hydrogen acceptor and will not work with the other; some enzymes are less specific and will work with either one, though they usually work more rapidly with one than the other.

Another primary hydrogen acceptor, **flavin adenine dinucleotide** (abbreviated **FAD**), serves in reactions involving the

$$—\overset{\displaystyle H}{\underset{\displaystyle H}{C}}—\overset{\displaystyle H}{\underset{\displaystyle H}{C}}—$$ configuration, as in the dehydro-

genation of succinic acid. In some reactions, **flavin mononucleotide (FMN)**, which consists of part of the FAD molecule, may serve in its place. The FAD of succinic dehydrogenase is bound very tightly to the protein part of the enzyme and cannot be removed easily. Such tightly bound cofactors are termed **prosthetic groups** of the enzyme. The pyridine nucleotide effective in the lactic dehydrogenase system, in contrast, is very loosely bound and is readily removed. Such loosely bound cofactors are termed **coenzymes.**

The reduced pyridine nucleotides, NADH or NADPH, cannot react with oxygen. Their electrons must be passed through the intermediate acceptors of the electron transmitter system before they can react with oxygen. The flavin primary acceptors usually pass their electrons to the electron transmitter system but some flavoproteins can react directly with oxygen. An enzyme that can mediate the transfer of electrons directly to oxygen is termed an **oxidase;** one that mediates the removal of electrons from a substrate to a primary or intermediate acceptor is termed a **dehydrogenase.**

4.6 THE CITRIC ACID CYCLE

The acetyl coenzyme A formed by the oxidation of lactic acid, or formed by the oxidation of fatty acids, undergoes a series of reactions involving dehydrogenation, decarboxylation and make-ready reactions which have been termed the **Krebs citric acid cycle.** Acetyl coenzyme A undergoes a make-ready reaction by combining with oxaloacetic acid, which contains four carbons, to yield citric acid, which contains six carbons (Fig. 4.3). Citric acid has neither an

$$H—\overset{\displaystyle |}{\underset{\displaystyle |}{C}}—OH$$ nor a $—CH_2—CH_2—$ group and

cannot undergo dehydrogenation. Two make-ready reactions involving the removal and addition of a molecule of water yield isocitric acid which can undergo dehy-

drogenation at its $H—\overset{\displaystyle |}{\underset{\displaystyle |}{C}}—OH$ group. The

hydrogen acceptor is a pyridine nucleotide, usually NAD, and the product is oxalosuccinic acid, which undergoes decarboxylation to yield α-ketoglutaric acid.

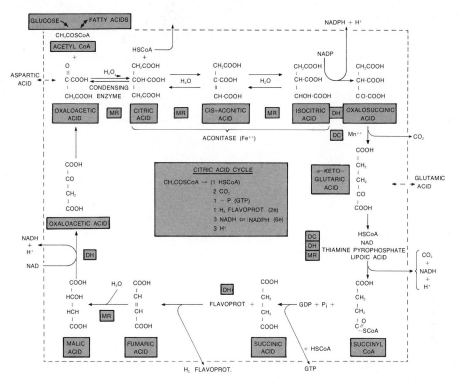

Figure 4.3 Diagram of the cyclic series of reactions, termed the Krebs citric acid cycle, by which the carbon chains of sugars, fatty acids and amino acids are metabolized to yield carbon dioxide. The reactions are designated as DH, dehydrogenations; DC, decarboxylations; and MR, make-ready. The over-all reaction effected by one "turn" of the cycle is summarized in the center. (Villee: Biology, 6th ed.)

Just as pyruvic acid is converted to the coenzyme A derivative of an acid with one less carbon atom, α-ketoglutaric acid is converted to succinyl coenzyme A by reactions requiring NAD, coenzyme A, thiamine pyrophosphate and lipoic acid as coenzymes. α-Ketoglutaric acid is metabolized by a dehydrogenation, a decarboxylation and a make-ready reaction using coenzyme A to yield succinyl coenzyme A.

The bond joining coenzyme A to succinic acid is an energy-rich one, $\sim S$, like the bond joining coenzyme A to acetic acid in acetyl coenzyme A. The energy of the bond of acetyl coenzyme A is utilized in bringing about the addition of the acetyl group to oxaloacetic acid. The energy in the $\sim S$ bond in succinyl coenzyme A can be converted to an energy-rich phosphate bond, $\sim P$, in the form of ATP. The reaction of succinyl coenzyme A with inorganic phosphate yields succinyl phosphate and coenzyme A. The phosphate group is then transferred to ADP to form ATP and free succinic acid. This is an example of an energy-rich bond synthesized at the **substrate level** by reactions not involving the electron transmitter system. Only a small fraction of the total energy-rich bonds made in metabolism are formed by reactions such as this. Normal cells metabolizing in a medium containing oxygen synthesize most of their ATP by oxidative phosphorylation in the electron transmitter system.

The further oxidation of succinic acid to fumaric, malic and oxaloacetic acid completes the cycle, for the oxaloacetic acid is then ready to combine with another molecule of acetyl coenzyme A to form a molecule of citric acid. In the course of this cycle, two molecules of CO_2 and eight hydrogen atoms are removed, and one molecule of $\sim P$ is synthesized at the substrate level. The Krebs citric acid cycle is the final common pathway by which the carbon chains of carbohydrates, fatty acids and amino acids are metabolized.

Fatty Acid Oxidation. The carbon chain

I. The activation reaction: $R \cdot CH_2 \cdot CH_2 \cdot COOH + ATP \longrightarrow R \cdot CH_2 \cdot CH_2 \cdot CO \cdot AMP + PyroPO_4$

$R \cdot CH_2 \cdot CH_2 \cdot CO \cdot AMP + HSCoA \longrightarrow R \cdot CH_2 \cdot CH_2 \cdot COSCoA + AMP$

2. Dehydrogenation: $CH_3(CH_2)_{10}CH_2 \cdot CH_2 \cdot \underline{CH_2 \cdot CH_2} \cdot COSCoA + Flavoprot. \rightleftharpoons CH_3(CH_2)_{10}CH_2 \cdot CH_2 \cdot CH =$

$CHCOSCoA + H_2 Flavoprot.$

3. Make–ready: $CH_3(CH_2)_{10}CH_2 \cdot CH_2 \cdot CH = CHCOSCoA + H_2O \rightleftharpoons CH_3(CH_2)_{10}CH_2 \cdot CH_2 \cdot CHOH \cdot CH_2 \cdot COSCoA$

4. Dehydrogenation: $CH_3(CH_2)_{10}CH_2 \cdot CH_2 \cdot CHOH \cdot CH_2 \cdot COSCoA + NAD \rightleftharpoons CH_3(CH_2)_{10} CH_2 \cdot CH_2 \cdot CO \cdot CH_2 \cdot$

$COSCoA + NADH + H^+$

5. Make–ready: $CH_3 (CH_2)_{10}CH_2 \cdot CH_2 \cdot COCH_2 \cdot COSCoA + HSCoA \rightleftharpoons CH_3 (CH_2)_{10}CH_2 \cdot \underline{CH_2 \cdot COSCoA} +$

$CH_3COSCoA$

Acetyl CoA   To Reaction 2 and Repeat Sequence

Sum: $C_{16}H_{32}O_2 \longrightarrow 8 CH_2 COSCoA + 7H\ Flavoprot. + 7 NADH + 7H^+$

28e

Figure 4.4 The series of reactions by which fatty acids are oxidized, two carbons at a time, to yield acetyl coenzyme A. (Villee: Biology, 6th ed.)

of a fatty acid is metabolized by a series of reactions which begins by a reaction with ATP and then with coenzyme A to yield the fatty acyl coenzyme A (Fig. 4.4). Palmitic acid, which contains 16 carbons, reacts to form palmityl coenzyme A. Palmityl coenzyme A undergoes a dehydrogenation between the second and third carbons of the chain. The hydrogen acceptor for this dehydrogenation is a flavin, for the configuration undergoing dehydrogenation is —CH_2—CH_2—. The product of this dehydrogenation undergoes a make-ready reaction, the addition of a molecule of water, to yield a molecule which has an H—C—OH configuration. This, in turn, undergoes a dehydrogenation utilizing NAD as the hydrogen acceptor (Fig. 4.4). The resulting molecule can undergo a make-ready reaction with coenzyme A to clip off a two-carbon unit, acetyl coenzyme A, and leave a carbon chain which is two carbons shorter. This molecule is in the activated state — it contains coenzyme A on its carboxyl group — and is ready to be dehydrogenated by the enzyme which uses flavin as a hydrogen acceptor. This repeating series of reactions, which includes dehydrogenations and make-ready reactions but not decarboxylations, cleaves a fatty acid chain two carbons at a time (Fig. 4.4). Seven such series of

reactions will split palmitic acid to eight molecules of acetyl coenzyme A.

Glycolysis. Glucose is also converted ultimately to acetyl coenzyme A. The series of glycolytic reactions begins, as with fatty acids, by one in which glucose is "activated." Glucose reacts with ATP to yield glucose-6-phosphate and ADP, a reaction catalyzed by the enzyme **hexokinase.** After this make-ready reaction, other make-ready reactions establish a configuration which can undergo dehydrogenation (Fig. 4.5). A rearrangement yields fructose-6-phosphate and the transfer of a second phosphate from ATP forms fructose-1,6-diphosphate. Fructose-1,6-diphosphate is split by an enzyme, **aldolase,** into two three-carbon sugars, glyceraldehyde-3-phosphate and dihydroxyacetone phosphate. These two are interconverted by a separate enzyme.

Glyceraldehyde-3-phosphate reacts with a compound containing an —SH group which is part of the enzyme molecule to yield an H—C—OH configuration that can undergo dehydrogenation with NAD as the hydrogen acceptor. The product of the reaction, phosphoglyceric acid bound to the SH group of the enzyme, then reacts with inorganic phosphate to yield 1,3-diphosphoglyceric acid and free enzyme —SH. The phosphate in carbon 1 is an en-

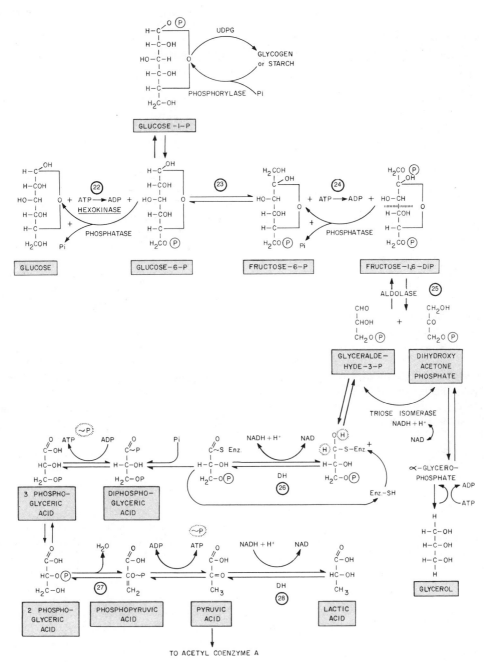

Figure 4.5 Diagram of the series of reactions by which glucose and other sugars are metabolized to pyruvic acid. Note that most of the steps are reversible, indicated by double arrows.

Other steps are essentially irreversible; glucose-6-phosphate is converted to glucose by a separate enzyme, a glucose-6-phosphatase, and not by the hexokinase that catalyzes the conversion of glucose to glucose-6-phosphate. (Villee: Biology, 6th ed.)

ergy-rich group which can react with ADP to form ATP. This, like the energy-rich phosphate of succinyl phosphate, is one made at the substrate level. The resulting 3-phosphoglyceric acid undergoes a make-ready rearrangement to yield 2-phosphoglyceric acid. Then, in an unusual reaction, an energy-rich phosphate is generated by the removal of water by a **dehydration,** rather than by the removal of two hydrogens, a **dehydrogenation.** The product, phosphopyruvic acid, can transfer its phosphate group to ADP to yield ATP and free pyruvic acid. This is the second energy-rich phosphate group generated at the substrate level in the metabolism of glucose to pyruvic acid.

Each glucose molecule yields two molecules of glyceraldehyde-3-phosphate and hence a total of four energy-rich phosphates is produced as glucose is metabolized to pyruvic acid. However, two energy-rich phosphates are utilized in the process, one to convert glucose to glucose-6-phosphate and the second to convert fructose-6-phosphate to fructose-1,6-diphosphate. The net yield in the process is 2 ~P (4 ~P produced minus 2 ~P used up in the reactions). Pyruvic acid is then metabolized to acetyl coenzyme A by the reactions described previously.

Amino Acid Oxidation. Amino acids are oxidized by reactions in which the amino group is first removed, a process called **deamination,** then the carbon chain is metabolized and eventually enters the Krebs citric acid cycle. The amino acid alanine, for example, yields pyruvic acid when deaminated, glutamic acid yields α-ketoglutaric acid, and aspartic acid yields oxaloacetic acid. These three amino acids can enter the Krebs citric acid cycle directly. Other amino acids may require several reactions in addition to deamination to yield a substance which is a member of the Krebs cycle but ultimately the carbon chains of all the amino acids are metabolized in just this way.

The Electron Transmitter System. The major reactions by which biologically useful energy is produced occur when the electrons are transferred from the primary acceptors, such as the pyridine nucleotides, through the electron transmitter system. The electrons entering the electron transmitter system from NADH have a relatively

high energy content. As they pass along the chain of enzymes they lose much of their energy, some of which is conserved in the form of ATP. The enzymes of the electron transmitter system are located within the substance of the mitochondrion in the mitochondrial membranes. They are very closely related spatially and it seems likely that the electrons flow through a solid phase rather than between enzymes that are in solution. The components of the electron transmitter system are given (Fig. 4.2) in the order of their oxidation-reduction potentials, which range from −0.32 volt for the pyridine nucleotides to +0.81 volt for oxygen. However, it is not known whether a particular electron in passing from pyridine nucleotide to oxygen must go through each and every one of the intermediates or whether it might skip some steps. It probably has to pass through at least three different steps to account for the 3 ~P which are made for each pair of electrons that pass from pyridine nucleotide to oxygen.

Oxidative phosphorylation is measured by the rate at which inorganic phosphate, P_i, is converted to ATP as NADH or some other substance undergoes oxidation. It is possible to prepare mitochondria by homogenizing the cells and separating the subcellular particles by centrifugation. Such mitochondria when removed carefully from the cell will carry out oxidative phosphorylation. It is even possible to disrupt mitochondria by ultrasonic vibration and obtain submitochondrial particles which will carry out oxidative phosphorylation. In these purified systems NADH will be oxidized and oxygen will be utilized only if ADP is present to accept the energy-rich phosphate produced by the flow of electrons. The flow of electrons is tightly coupled to the phosphorylation process and will not occur unless phosphorylation can occur also. This, in a sense, prevents waste, for electrons will not flow unless energy-rich phosphate can be formed.

The flow of electrons from pyridine nucleotides to oxygen, a total drop of 1.13 volts (from −0.32 to +0.81 volt), would yield 52,000 calories per pair of electrons if the process were 100 per cent efficient. This may be calculated from the formula $\Delta G' = -nF\Delta E$, where $\Delta G'$ is the change in free energy, n is the number of electrons involved (2), F is the Faraday (23,040 calories) and ΔE is the difference in the oxidation-reduction potentials of the reactants (1.13 volts). Under experimental conditions most cells will produce at most 3 ~P per pair of electrons as the electrons pass from pyridine nucleotide to oxygen. Each ~P is equivalent to about 7000 calories. Hence, the 3 ~P amount to about 21,000 calories. The efficiency of the electron transmitter system may then be calculated as 21,000/52,000 or about 40 per cent.

The fact that phosphorylation is tightly coupled to oxidation or electron flow provides a system of control which can regulate the rate of energy production and adjust it to the momentary rate of energy utilization. In a resting muscle cell, for example, oxidative phosphorylation will occur until all of the ADP has been converted to ATP. Then, since there are no more acceptors of ~P, phosphorylation must stop. Since oxidation is tightly coupled to phosphorylation, the flow of electrons will cease.

When the muscle contracts, the energy required is obtained from the splitting of the energy-rich terminal phosphate of ATP to yield ADP and inorganic phosphate plus energy. The ADP formed can then serve as an acceptor of ~P, phosphorylation begins, and the flow of electrons to oxygen occurs. Oxidative phosphorylation continues until all of the ADP has been converted to ATP. Electric generating systems have an analogous control device which adjusts the rate of production of electricity to the rate of utilization of electricity.

Some interesting calculations of the over-all energy changes involved in metabolism in the human body have been made by E. G. Ball of Harvard University. Since the conversion of oxygen to water involves the participation of hydrogen atoms and electrons, the total flow of electrons in the human body can be calculated and expressed in amperes. From the oxygen consumption of

an average 70 kg. man at rest—264 ml. per minute—and the fact that each oxygen atom requires two hydrogen atoms and two electrons to form a molecule of water, Dr. Ball calculated that 2.86×10^{22} electrons flow from foodstuff, via dehydrogenases, primary acceptors and the cytochromes, to oxygen each minute in all the cells of the body. Since an ampere equals 3.76×10^{20} electrons per minute, this flow of electrons amounts to 76 amperes. This is quite a bit of current, for an ordinary 100 watt light bulb uses just a little less than 1 ampere.

The flow of electrons from substrate to oxygen involves a potential difference of 1.13 volts (from -0.32 to $+0.81$ volt). Since volts $\times$ amperes = watts, $1.13 \times 76 = 85.9$ watts. The body, then, uses energy at about the same rate as a 100-watt light bulb, but differs from it in having a much larger flow of electrons passing through a much smaller voltage change.

The conversion of glucose to carbon dioxide and water yields about 4 Cal./gm. Let us now dissect the over-all equation

$$C_6H_{12}O_6 + 6\ O_2 \rightarrow 6\ CO_2 + 6\ H_2O + energy$$

and review where the energy is released (Table 4.1).

In glycolysis (reaction 1, Table 4.1), glucose is activated by the addition of 2 ~P and converted to 2 pyruvate + 2 NADH + 4 ~P. Then the two pyruvates are metabolized (reaction 2) to 2 CO_2 + 2 acetyl CoA + 2 NADH. Finally in the citric acid cycle (reaction 3), the two acetyl CoA are metabolized to 4 CO_2 + 6 NADH + 2 H_2FP + 2 ~P.

These reactions can be added together (reaction 4) by eliminating items that are present on both sides of the arrows. Then, since the oxidation of NADH in the electron transmitter system yields 3 ~P per mole, the 10 NADH = 30 ~P. The oxida-

TABLE 4.1 THE REACTIONS BY WHICH BIOLOGICALLY USEFUL ENERGY IS RELEASED

(1)	$C_6H_{12}O_6 + 2$ ~P $\longrightarrow$	2 pyruvate + 2 NADH + 4 ~P
(2)	2 pyruvate $\longrightarrow$	2 CO_2 + 2 acetyl CoA + 2 NADH
(3)	2 acetyl CoA $\longrightarrow$	4 CO_2 + 6 NADH + 2 H_2FP + 2 ~P
(4) Sum	$C_6H_{12}O_6 \longrightarrow$	6 CO_2 + 10 NADH + 2 H_2FP + 4 ~P
(5)	$C_6H_{12}O_6 + 6\ O_2 \longrightarrow$	6 CO_2 + 6 H_2O + 30 ~P + 4 ~P + 4 ~P

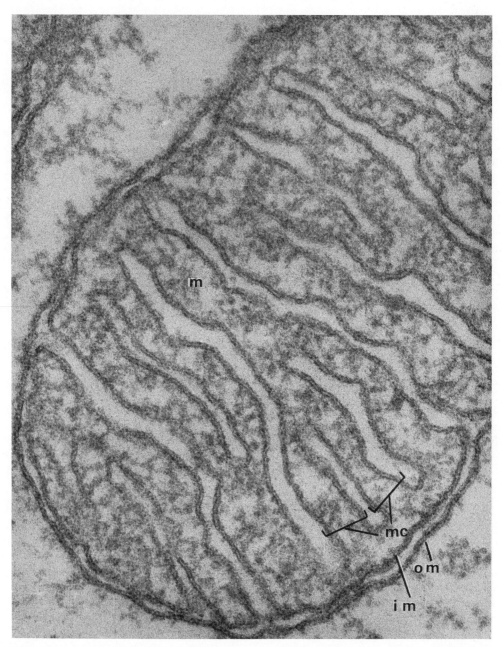

Figure 4.6 Electron micrograph of a mitochondrion from a pancreatic acinar cell. The double-layered unit membrane is evident in the smooth outer membrane (om) and in the inner membrane (im) which folds to form the cristae (mc). Magnified 207,000×. (Courtesy of G. E. Palade; from De Robertis, E. D., Nowinski, W. W., and Saez, F. A.: Cell Biology. Philadelphia, W. B. Saunders Company, 1970.)

tion of H_2FP yields 2 $\sim$P per mole and 2 $H_2FP = 4$ $\sim$P. Summing these we see that the complete aerobic metabolism of 1 mole (180 gm.) of glucose yields 38 $\sim$P (reaction 5). Each $\sim$P is equivalent to about 7000 calories and the 38 $\sim$P = 266,000 calories.

When a mole of glucose is burned in a calorimeter some 690,000 calories are released as heat. The metabolism of glucose in an animal cell releases 266,000/690,000 or about 40 per cent of the total energy as biologically useful energy, $\sim$P. The remainder of the energy is dissipated as heat.

4.7 THE MOLECULAR ORGANIZATION OF MITOCHONDRIA

Mitochondrial shapes range from spherical to elongated, sausage-like structures, but an average mitochondrion is an ellipsoid about 3μ long and a little less than 1μ in diameter. Some very generously endowed large cells, such as the giant ameba, *Chaos chaos*, have several hundred thousand mitochondria, and an average mammalian liver cell has about one thousand.

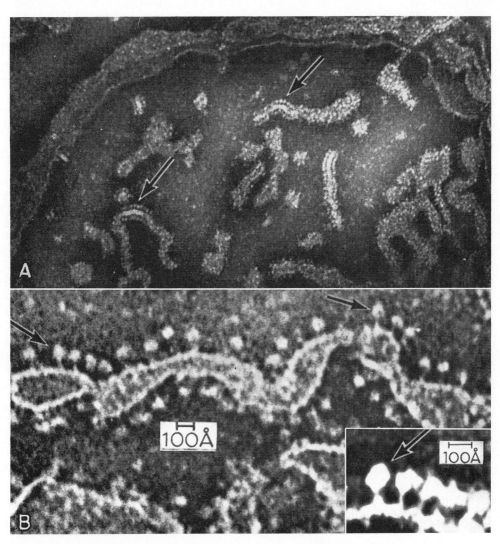

Figure 4.7 Electron micrograph of a mitochondrion swollen in a hypotonic solution and negatively stained with phosphotungstate. *A*, Isolated cristae 85,000×. *B*, At higher magnification, 500,000×, the elementary particles attached by a stalk to the surface of the cristae are evident. Insert: At 650,000× the polygonal shape of the elementary particle and its slender stalk are clearly visible. (Courtesy of H. Fernandez-Moran; from De Robertis, E. D., Nowinski, W. W., and Saez, F. A.: Cell Biology. Philadelphia, W. B. Saunders Company, 1970.)

The mitochondrial protein accounts for about 20 per cent of the total protein of the liver cell.

The inner membrane of the mitochondrion is folded repeatedly (Fig. 4.6) to form the shelflike cristae extending into, or all the way across, the central cavity. Both the inner and outer mitochondrial membranes are "unit membranes" composed of a core of a double layer of lipid molecules coated on both sides with a layer of protein molecules.

The enzymes of the Krebs cycle have been found in the soluble matrix within the mitochondrion. The enzymes of the electron transmitter system are tightly bound to the inner membrane. There is some evidence to suggest that successive enzymes in the chain are actually located adjacent to each other in the membrane. Each group of these electron transmitter enzymes, termed a **respiratory assembly,** is one of the fundamental units of cellular activity. It has been estimated that the mitochondrion of the liver cell contains some 15,000 respiratory assemblies and that they make up about one quarter of the mass of the mitochondrial membranes. Thus the mitochondrial membrane is not just a protective skin but an important functional part of the mitochondrion.

High resolution electron micrographs of mitochondria, made by Humberto Fernandez-Moran at the University of Chicago, have demonstrated the presence of small particles on the outer surface of the outer membrane and the inner surface of the inner membrane (Fig. 4.7). The particles on the inner membrane typically have a spherical knoblike head, a cylindrical stalk and a base plate. These particles may be sites of the enzyme reactions carried out by mitochondria. The particles of the inner membrane contain the respiratory assemblies, the electron transmitter system and the enzymes of oxidative phosphorylation.

It seems clear that much of the cell's biologically useful energy, ATP, is generated by enzyme systems located in the *inner* membrane of the mitochondrion, yet most of the energy utilized by the cell is required for processes that take place *outside of* the mitochondrion. ATP is used in the synthesis of proteins, fats, carbohydrates, nucleic acids and other complex molecules, in the transport of substances across the plasma membrane, in the conduction of nerve impulses, and in the contraction of muscle fibers, all of which are reactions occurring largely or completely outside of the mitochondrion in other parts of the cell. Several mechanisms, too complex to be considered here, have evolved to mediate the transfer of $\sim$P across the mitochondrial membrane. A single cell may have a thousand or more mitochondria and the function of each one must be controlled appropriately to generate the amount of energy required by the cell at any given moment. The rate at which $\sim$P is utilized by a cell may vary over a remarkably wide range as the cell becomes active or quiescent. This has been measured in the muscle cells of a frog. One hundred grams of frog muscle utilizes 1.6μ moles of $\sim$P per minute when quiescent and 3300μ moles of $\sim$P per minute in a state of tetanus (continuous contraction). The rate of production of ATP in the cell is controlled in large part by the rate of utilization of ATP by the cell. The flow of electrons in the electron transmitter system is tightly coupled to phosphorylation, and oxidative phosphorylation can occur only when there is ADP to be converted to ATP. These facts provide the basis for the system by which a cell's utilization of ATP, which produces ADP, could be used to regulate the rate at which ATP is produced. In addition, the structure and biological activity of some of the enzymes involved in glucose oxidation are affected by the concentration of ADP present. In this way an increased concentration of ADP can lead to increased activity of these enzymes and an increased production of ATP. The problem of the nature and interrelations of these various biological control systems is very much in the forefront in biology today.

4.8 THE DYNAMIC STATE OF CELLULAR CONSTITUENTS

The body of an animal or man appears to be unchanging as days and weeks go by and it would seem reasonable to infer that the component cells of the body, and even the component molecules of the cells, are equally unchanging. In the absence of any evidence to the contrary, it was generally held, until about 35 years ago, that the constituent molecules of animal and plant cells were relatively static and that, once formed, they remained intact for a long period of time. A corollary of this concept is that the molecules of food which are not used to increase the mass of protoplasm are rapidly metabolized to provide a source of energy. It followed from this that one could distinguish two kinds of molecules: relatively static ones that made up the cellular "machinery," and ones that were rapidly metabolized and thus correspond to cellular "fuel."

However, in 1938 Rudolf Schoenheimer and his colleagues at Columbia University began a series of experiments in which amino acids, fats, carbohydrates and water, each suitably labeled with some "heavy" or radioactive isotope, were fed to rats. Schoenheimer's experiments, which have been confirmed many times since, showed that the labeled amino acids fed to the rats were rapidly incorporated into body proteins. Similarly, labeled fatty acids were rapidly incorporated into the fat deposits of the body, even though in each case there was no increase in the *total* amount of protein or fat. Such experiments have demonstrated that the fats and proteins of the body cells — and even the substance of the bones — are constantly and rapidly being synthesized and broken down. In the adult the rates of synthesis and degradation are essentially equal so that there is little or no change in the total mass of the animal body. The distinction between "machinery" molecules and "fuel" molecules becomes much less sharp, for some of the machinery molecules are constantly being broken down and used as fuel.

The one exception to the rule of molecular flux is provided by the molecules of DNA that constitute the genes within the nucleus of the cell. Experiments with labeled atoms have shown that the molecules of DNA are remarkably stable and are broken down and resynthesized only very slowly, if at all. The amount of DNA per nucleus is constant, and new molecules of DNA are synthesized each time a cell divides. The stability of the DNA molecules may be important in ensuring that hereditary characters are transmitted to succeeding generations with as few chemical errors as possible. In contrast, the molecules of RNA undergo constant synthesis and degradation; indeed, the amount of RNA per cell may vary within wide limits.

From the rate at which labeled atoms are incorporated it has been calculated that one-half of all the tissue proteins of the human body are broken down and rebuilt every 80 days. The proteins of the liver and blood serum are replaced very rapidly, one-half of them being synthesized every 10 days. Some enzymes in the liver have half-times as short as two to four hours. The muscle proteins, in contrast, are replaced much more slowly, one-half of the total number of molecules being replaced every 180 days. The celebrated aphorism of Sir Frederick Gowland Hopkins, the late English biochemist, sums up this concept very succinctly: "Life is a dynamic equilibrium in a polyphasic system."

4.9 BIOSYNTHETIC PROCESSES

Our discussion thus far has dealt with processes that break down molecules of foodstuffs and conserve their energy in the biologically useful form of ~P. Cells have the ability to carry out an extensive array of biosynthetic processes utilizing the energy and, as raw materials, some of the five-, four-, three-, two- and one-carbon compounds that are intermediates in the metabolism of glucose, fatty acids, amino acids and other compounds.

This whole subject of intermediary metabolism, by which an enormous variety of compounds is synthesized, is much too complicated to be discussed in detail here. The enzyme controlling each reaction is genetically determined and the over-all maze of enzyme reactions by which any compound is synthesized includes a variety of self-adjusting control mechanisms to regulate and integrate the reactions. Several basic principles of cellular biosynthesis can be distinguished:

1. Each cell, in general, synthesizes its own proteins, nucleic acids, lipids, polysaccharides and other complex molecules and does not receive them preformed from other cells. Muscle glycogen, for example, is synthesized within the muscle cell and is not derived from liver glycogen.

2. Each step in the biosynthetic process is catalyzed by a separate enzyme.

3. Although certain steps in a biosyn-

thetic sequence will proceed without the use of energy-rich phosphate, the over-all synthesis of these complex molecules requires chemical energy.

4. The synthetic processes utilize as raw materials relatively few substances among which are acetyl coenzyme A, glycine, succinyl coenzyme A, ribose and pyruvate.

5. These synthetic processes are in general not simply the reverse of the processes by which the molecule is degraded but include one or more separate steps which differ from any step in the degradative process. These steps are controlled by different enzymes and this permits separate control mechanisms to govern the synthesis and the degradation of the complex molecule.

6. The biosynthetic process includes not only the formation of the macromolecular components from simple precursors but their assembly into the several kinds of membranes that comprise the outer boundary of the cell and the intracellular organelles. Each cell's constituent molecules are in a dynamic state and are constantly being degraded and synthesized. Thus, even a cell that is not growing, not increasing in mass, uses a considerable portion of its total energy for the chemical work of biosynthesis. A cell that is growing rapidly must allocate a correspondingly larger fraction of its total energy output to biosynthetic processes, especially the biosynthesis of protein. A rapidly growing bacterial cell may use as much as 90 per cent of its total biosynthetic energy for the synthesis of proteins.

The Formation of Anhydro Bonds. Although many of the steps in biosynthetic processes involve the formation of anhydro bonds — e.g., the peptide bonds of protein, the glycosidic bonds of polysaccharides, the ester bonds of lipids and nucleic acid — these bonds are *not* formed by reactions in which water is removed. The biosynthesis of sucrose in the cane sugar plant, for example, does not proceed via

$$glucose + fructose \rightleftharpoons sucrose + H_2O$$

This reaction would require energy, some 5500 calories per mole, to go to the right if all reactants were present in the concentration of 1 mole per liter. However, the concentration of glucose and fructose in the plant cell is probably less than 0.01 mole per liter, whereas the concentration

of water is very high, about 55 moles per liter. Thus, the equilibrium point of the reaction under these conditions would be very far to the left.

Instead, one or more of the reactants is activated by a reaction with ATP in which the terminal phosphate is enzymatically transferred to glucose with the conservation of some of the energy of ATP. The glucose phosphate, with a higher energy content than free glucose, can react with fructose via another enzyme-catalyzed reaction to yield sucrose and inorganic phosphate.

$$ATP + glucose \rightarrow$$
$$ADP + glucose\text{-}1\text{-}phosphate$$
$$glucose\text{-}1\text{-}phosphate + fructose \rightarrow$$
$$sucrose + phosphate$$
Sum: $$ATP + glucose + fructose \rightarrow$$
$$sucrose + ADP + P_i$$

This reaction proceeds to the right because there is a net decrease in free energy. The 7000 calories of the ~P bond are used to supply the 5500 calories needed to assemble the glucose and fructose into sucrose; the over-all decrease in free energy is 1500 calories per mole. Since water is not a product of this reaction, the high concentration of water in the cell does not inhibit it. The same principle applies to the synthesis of the peptide bonds of proteins, the ester bonds of lipids and nucleic acids, and the glycosidic bonds of polysaccharides, but in many of these not one but *two* terminal phosphates of ATP are transferred as a unit in the reaction by which one of the molecules is activated. The activated molecule may be X ~PP or X ~AMP, depending on which part of the ATP is transferred to the substrate molecule. In either case the reaction ultimately involves the hydrolysis of 2 ~P with an energy utilization of 14,000 calories per mole.

$$X + ATP \rightarrow X \sim PP + AMP$$
$$X \sim PP + Y \rightarrow XY + PP_i$$
$$PP_i + H_2O \rightarrow P_i + P_i$$
Sum: $$X + Y + ATP \rightarrow XY + AMP + 2P_i$$

4.10   BIOLUMINESCENCE

A number of animals, and some molds and bacteria as well, have an enzymic mechanism for the production of light. Luminescent animals are found among the

Figure 4.8 *Anomalops katoptron*, a luminescent fish from the waters of the Malay Archipelago. The crescent-shaped luminescent organs below the eyes are equipped with reflectors. (After Steche.) (Villee: Biology, 6th ed.)

protozoa, sponges, coelenterates, cteno-phores, nemerteans, annelids, crustaceans, centipedes, millipedes, beetles, flies, echi-noderms, mollusks, hemichordates, tuni-cates and fishes. From this wide and ir-regular distribution of the light-emitting ability it is clear that the enzymes for lu-minescence have appeared independently in a number of different evolutionary lines. It is sometimes difficult to establish that a given organism is itself luminescent; in a number of instances animals once believed to be luminescent have been shown in-stead to contain luminescent bacteria. When the bacteria are removed the animal is no longer able to emit light. Several different exotic East Indian fish have **light organs** under their eyes in which live luminous bacteria (Fig. 4.8). The light organs contain long cylindrical cells which are well pro-vided with blood vessels to supply an ade-quate amount of oxygen to the bacteria. The bacteria emit light continuously and the fish have a black membrane, somewhat similar to an eyelid, that can be drawn up over the light organ to turn off the light. How the bacteria come to collect in the fish's light organ, as they must in each newly hatched fish, is a complete mystery.

Some animals have accessory lenses, re-flectors and color filters with the light-pro-ducing organ, and the whole complex assembly is like a lantern. Certain shrimp have such complicated light-emitting or-gans.

The production of light is an enzyme-con-trolled reaction, the details of which differ in different organisms. Bacteria and fungi produce light continuously if oxygen is available. Most luminescent animals, in contrast, give out flashes of light only when

their luminescent organs are stimulated. The name **luciferin** has been given to the material which is oxidized to produce light and **luciferase** to the enzyme which cata-lyzes the reaction. The luciferin and luci-ferase from one species may be quite dif-ferent chemically from those in another. The oxidation of luciferin by luciferase can occur only in the presence of oxygen. It is possible to extract luciferin and luciferase from a firefly, mix the two in a test tube with added magnesium and adenosine triphos-phate, and demonstrate the emission of light in the test tube. The energy for the re-action is supplied by the ATP, and under certain conditions the amount of light emitted is proportional to the amount of ATP present. This system can be used to measure the amount of ATP in a tissue ex-tract.

The amount of light produced by certain luminescent animals is amazing. Many fire-flies produce as much light, in terms of lumens per square centimeter, as do modern fluorescent lamps. Different kinds of ani-mals may emit lights of different colors, red, green, yellow or blue. One of the more spectacular luminescent creatures is the "railroad worm" of Uruguay, the larva of a beetle, which has a row of green lights along each side of its body and a pair of red lights on its head. The light produced by lumines-cent organisms is entirely in the visible part of the spectrum; no ultraviolet or infrared light is produced. Since very little heat is given off in the process, bioluminescence has been called "cold light."

What advantage an animal derives from the emission of light can only be guessed at. For deep-sea animals, which live in perpetual darkness, light organs might be useful to enable members of a species to recognize one another, to serve as a lure for prey or as a warning for would-be predators. Experiments have shown that the light emitted by fireflies serves as a signal to bring the two sexes together for mating. The light emitted by bacteria and fungi probably serves no useful purpose to the organisms, but is simply a by-product of oxidative metabolism, just as heat is a by-product of metabolism in other plants and animals.

ANNOTATED REFERENCES

Baldwin, E. B.: Dynamic Aspects of Biochemistry. 4th ed. New York, Cambridge University Press, 1964. A detailed and technical but very well-written and interesting account of cellular metabolism.

Dixon, M., and E. C. Webb: The Enzymes. 2nd ed. New York, Academic Press, 1964. Presents, in considerable depth and detail, the structure and kinetic properties of enzymes together with a classification of enzymes by function.

Harvey, E. N.: Bioluminescence. New York, Academic Press, 1952. A classic presentation of the interesting phenomenon of "living light."

Henderson, L. J.: The Fitness of the Environment. New York, Macmillan, 1913. Advanced the thesis that the environment had to have certain physical and chemical characteristics for life to develop.

Lehninger, A. L.: Bioenergetics. New York, Benjamin, 1965. An excellent discussion of the principles of thermodynamics and their application to biologic systems.

Schoenheimer, R.: The Dynamic State of the Body Constituents. Cambridge, Harvard University Press, 1949. Describes the classic experiments that demonstrated the rapid renewal of the chemical constituents of tissues.

The standard college textbooks of biochemistry may be consulted for more detailed discussions of the topics presented in this chapter. These include:

McGilvery, R. W.: Biochemistry. Philadelphia, W. B. Saunders Co., 1970.

Mazur, A. and B. Harrow: Textbook of Biochemistry. 10th ed. Philadelphia, W. B. Saunders Co., 1970.

Karlson, P.: Introduction to Modern Biochemistry. (Translated by C. H. Derring.) 2nd ed. New York, Academic Press, 1970.

5. Structural and Functional Adaptations of Animals

All animals—vertebrate or invertebrate, multicellular or unicellular—have certain common problems of obtaining an adequate supply of nutrients and oxygen, of eliminating carbon dioxide and metabolic wastes, of responding suitably to stimuli from the environment, of moving to new areas and of reproducing their kind. To have survived, each species must have evolved some appropriate means of solving each of these problems. In later chapters, as we survey the Animal Kingdom, it will become apparent that an enormous variety of solutions to these problems has arisen in the course of evolution. As a basis for those discussions, however, we want first to emphasize the similar features in the structural and functional adaptations of animals.

5.1 TYPES OF NUTRITION

Each cell in every kind of organism needs a similar array of raw materials—sugars, amino acids, purines, pyrimidines, fatty acids and so on—to synthesize its own characteristic macromolecules such as glycogen, proteins, nucleic acids and lipids. Organisms that can synthesize their own raw materials are said to be **autotrophic** (self-nourishing). An autotroph needs only water, carbon dioxide, inorganic salts and a source of energy to survive. Green plants and the purple bacteria are photosynthetic autotrophs, obtaining energy from sunlight for the synthesis of organic molecules. Certain other bacteria are chemosynthetic autotrophs, obtaining energy for the synthesis of foodstuffs from the oxidation of certain inorganic substances—ammonia, nitrites or hydrogen sulfide. No animal is autotrophic; animals obtain their nutrients by eating autotrophs or by eating other animals which ate autotrophs. Ultimately, the foodstuff molecules of all animals are synthesized by energy obtained by these autotrophic organisms either from sunlight or from the oxidation of inorganic compounds.

The organisms which cannot synthesize their own raw materials from inorganic substances, and hence must live either by eating other organisms or by ingesting decaying matter, are called **heterotrophs.** The heterotrophs of the world, animals, fungi and most bacteria, cannot utilize light energy to synthesize sugars and other small organic molecules. But if supplied with these precursors, heterotrophs have the requisite enzymes to assemble their macromolecules.

Animals generally obtain their food as particles of some size which must be eaten and digested before it can be absorbed into the cell. In this type of nutrition, termed **holozoic,** the organism must find and catch other organisms; this has required the evolution of a variety of sensory, nervous and muscular structures to find and catch food and some sort of digestive system to convert the food into molecules small enough to be absorbed. Animals that feed chiefly upon plants are termed **herbivores,** those that eat other animals are called **carnivores,** and those that eat both plants and animals are known as **omnivores.** The structure and mode of functioning of the digestive system in different kinds of animals are correlated with the nature of food eaten, peculiarities of the manner of life, and so on. Carnivores, for example, characteristically secrete large amounts of proteolytic (protein-digesting) enzymes; whereas herbivores secrete less proteolytic but more carbohydrate-splitting, enzymes.

Although such familiar protozoa as amebae and paramecia take in food particles by **phagocytosis,** many protozoa, as well as yeasts, molds, most bacteria and parasites such as tapeworms, cannot ingest solid food. Instead, the required organic nutrients are absorbed through the cell membrane as dissolved molecules. The cells of molds and other fungi typically produce enzymes and secrete them to the exterior, and digestion occurs outside the cell. This type of nutrition is termed **saprophytic.** A tapeworm must depend upon predigested material obtained from its host and digested by its host's enzymes, a process termed **saprozoic** nutrition. Saprophytes can grow only in an environment which contains decomposing animal or plant bodies, or plant or animal by-products which will supply the necessary dissolved organic substances.

5.2 INGESTION, DIGESTION AND ABSORPTION

Some animals, such as sponges, clams, tunicates and certain whales, are **filter feeders** and make a living by filtering microscopic plants and animals from sea or pond water. The choanocytes of sponges take up these small particles by phagocytosis and digest them intracellularly. Tapeworms and certain other parasites live in the gut cavity of man and other vertebrates and absorb predigested nutrients from the host's gastrointestinal system. Certain protozoa take either living microscopic

plants or animals or dead bits of organic matter into their cells by phagocytosis. An ameba captures food by extruding cellular evaginations, **pseudopods,** which meet around the prey and form a **food vacuole** containing the particles to be eaten (Fig. 5.1). Hydrolytic enzymes synthesized in the cytoplasm are secreted through the vacuolar membrane into the cavity and digestion occurs within the vacuole as it circulates in the cell. Products of digestion are then absorbed through the membrane and utilized for the production of biologically useful energy or as substrates for the synthesis of macromolecules. Any undigested remnants are expelled and left behind as the animal moves on. Paramecia and other ciliates have a permanent **oral groove** lined with cilia. The beating of the cilia passes food particles into a cell mouth from which they are collected into food vacuoles. The choanocytes in the canals of sponges capture and ingest microscopic food particles in similar food vacuoles. Digestion in protozoa and sponges is **intracellular,** occurring in food vacuoles within the cytoplasm of the cell.

In coelenterates and platyhelminthes there has evolved a **gastrovascular cavity** lined with a gastrodermis. Small animals are taken in through the mouth and digestion begins within the gastrovascular cavity. The gastrodermis secretes digestive enzymes into the cavity and the prey is partially digested there. It is then absorbed into the gastrodermis cells where digestion is completed within food vacuoles. Digestion is partly extracellular in the gastrovascular cavity and partly intracellular within food vacuoles in the gastrodermis. The gastrovascular cavity of flatworms may be greatly branched, ramifying through most of the body and facilitating the distribution of digested food. Neither coelenterates nor flatworms have an anal aperture. Undigested wastes are excreted through the mouth.

In most of the remaining invertebrates and in all of the vertebrates, the digestive tract is a tube with two apertures; food enters the mouth and undigested residues leave by the anus. The advantages of such a system compared to the gastrovascular cavity of coelenterates are clear: food can pass in a single direction through the tube; the several sections of the tube can be

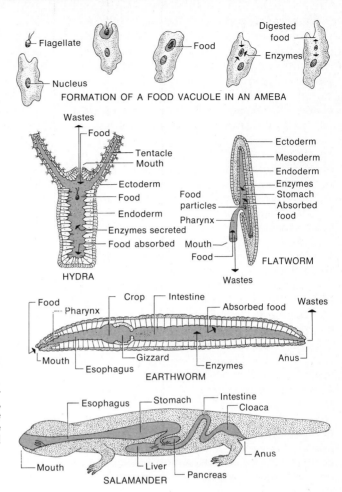

Figure 5.1 The structural basis of the process of digestion in ameba, hydra, flatworm, earthworm and salamander, illustrating the similarities and differences in the digestive systems of these widely different animal forms.

specialized to carry out certain functions in sequence, the simultaneous processing of several meals may occur, and there are no traffic problems as there are in coelenterates where incoming food and outgoing wastes must pass through the same opening.

The digestive tract may be short or long, straight or coiled, and may be subdivided into specialized organs. These organs, even though they may have similar names in different animals, may be quite different and may have different functions. The digestive system of the earthworm, for example, includes, in order, the **mouth;** a muscular **pharynx** which pumps food particles along and secretes a mucous material to lubricate them; an **esophagus;** a thin-walled **crop** where food is stored; a thick muscular **gizzard** where food is ground against small stones; and a long, straight **intestine** in which extracellular digestion occurs. The products of digestion are ab-

sorbed through the intestinal walls by diffusion, by facilitated diffusion or by active transport. The undigested residues pass out through the **anus.** Certain invertebrates—worms, squid, crustacea and sea urchins—have hard-toothed mouth parts which can tear off and chew bits of food.

The digestive systems of the vertebrates from fish to man are basically similar. The first portion of the intestine connected to the stomach is the small intestine, in which most foods are digested and through which most absorption occurs. This is followed by a large intestine in which digestion and absorption, especially the absorption of water, are completed. The **liver** and **pancreas** are large digestive glands, arising in development as outgrowths of the digestive tract, connected to the small intestine by ducts, and secreting bile and pancreatic juice respectively. These contain enzymes and other substances required for digestion.

Digestive Enzymes. Digestion, whether in ameba or man, involves the cleaving of complex macromolecules into simpler subunits by **hydrolysis** (the addition of water), a process catalyzed by enzymes called **hydrolases.** Specific hydrolases cleave the bonds within proteins, within fats and within polysaccharides. Polysaccharides such as starch and glycogen form an important part of the food ingested by most animals. The glucose units of these molecules are joined by glycosidic bonds, anhydro bonds linking carbon 4 or carbon 6 of one glucose molecule with carbon 1 of the adjacent one. These bonds are hydrolyzed by **amylases** which will split polysaccharides to the disaccharide maltose. Each kind of disaccharide is split by a specific enzyme—maltose by **maltase** present in saliva and intestinal juice, sucrose by **sucrase,** and lactose by **lactase** (β-galactosidase), both present in intestinal juice. The ultimate products of the digestion of carbohydrates are the hexoses such as glucose, fructose and galactose which are absorbed into the blood stream through the intestinal wall.

Several kinds of hydrolases attack the peptide bonds of proteins; each is specific for peptide bonds in a specific location in a polypeptide chain. **Exopeptidases** cleave the peptide bond joining the terminal amino acids to the peptide chain. **Carboxypeptidase** splits the peptide bond joining the amino acid with the free terminal carboxyl group to the chain and **aminopeptidase** removes the amino acid with a free terminal α-amino group. The **endopeptidases** cleave only peptide bonds within a peptide chain. **Pepsin** secreted by cells in the gastric mucosa, and **trypsin** and **chymotrypsin** secreted by the pancreas are endopeptidases but differ in their requirements for specific amino acids adjacent to the peptide bond to be cleaved. Pepsin requires tyrosine or phenylalanine adjacent to the bonds to be split, trypsin requires lysine or arginine and chymotrypsin requires tyrosine, phenylalanine, tryptophan, methionine or leucine. These endopeptidases split peptide chains into smaller fragments which are then cleaved further by exopeptidases. The combined action of endo- and exopeptidases results in splitting the protein molecules to free amino acids which are then absorbed through the intestinal wall into the blood stream by active transport. These powerful proteolytic enzymes would constitute a serious threat to any tissue secreting them. However, pepsin, trypsin and chymotrypsin are secreted as inactive precursors termed pepsinogen, trypsinogen and chymotrypsinogen. This prevents their digesting the proteins of the cell that produced them. In the gut, each is activated by the removal of a part of the precursor molecule to yield the active enzyme and one or more inactive fragments.

The digestion of fats is catalyzed by esterases that split the ester bond between glycerol and fatty acid. The principal mammalian esterase is **lipase** secreted by the pancreas. Like other proteins lipase is water-soluble, but its substrate, the fat, is not. The enzyme can attack only those molecules of fat at the surface of a fat droplet. The **bile salts** are surface active agents—detergents—that reduce the surface tension of fats. This breaks the large droplets of fat into many very small ones and greatly increases the surface area of fat exposed to the action of lipase, thereby increasing the rate of digestion of fat. Conditions in the intestine usually are not optimal for the complete hydrolysis of lipids to glycerol and fatty acids and the products absorbed include monoglycerides, diglycerides and some undigested fats, triglycerides.

The pancreas also secretes **ribonuclease** which splits the phosphate ester bonds linking adjacent nucleotides in ribonucleic acids. The phosphate ester bonds linking adjacent nucleotides in deoxyribonucleic acids are cleaved by another pancreatic enzyme, **deoxyribonuclease.** The cleavage of nucleic acids is completed by enzymes secreted by the intestinal mucosa. **Phosphodiesterase** removes nucleotides one at a time from the end of a polynucleotide chain. The nucleotides are attacked by **phosphatases** which remove the phosphate group and leave the nucleosides which are absorbed.

Phagocytosis and Pinocytosis. The plasma membrane has the remarkable property of capturing materials in the external environment by forming small, membrane-lined indentations which fold to enclose droplets of external medium. The resulting vacuoles then sink into the deeper layers of the cytoplasm where they are processed appropriately. In this way specific materials can be transferred from outside to inside the cell without any break in the continuity of the plasma membrane. This process of surface vacuolization, termed **phagocytosis** (cell eating) if the vacuoles are large and contain particulate matter, or **pinocytosis** (cell drinking) if the vacuoles are small and contain only dissolved substances, has been observed in many kinds of cells and studied especially in amebae and in white blood cells. The reverse process, in which membrane-lined vesicles fuse with the plasma membrane and release their contents to the exterior of the cell (seen especially in secretory cells), has been termed **emeiocytosis** (cell vomiting).

Pinocytosis can be induced in cells when proteins, acidic or basic amino acids, or certain other solutes are present in the surrounding medium. In amebae short pseudopodia form, each with a narrow, undulating channel extending from its tip to its base; vacuoles are formed at the inner end of the channel, separate from it, and pass deep into the cytoplasm. The plasma membrane of the cells of higher animals may exhibit undulations during pinocytosis but no channels are evident. Pinocytosis is an active, energy-requiring process and mitochondria are usually oriented near these active undulating surfaces. Pinocytosis can be inhibited by metabolic inhibitors and is varied by changes in the temperature, pH, and nutritional status of the cell. In a single cycle of pinocytosis an ameba takes up a volume of liquid equal to some 1 to 10 per cent of its total volume. The process requires the synthesis of an area of new membrane equivalent to as much as 6 per cent of the initial surface area of the ameba.

In amebae and in other cells, protein molecules are specifically bound to the membrane of the vacuole and concentrated there. The membrane then flows inward and forms a vesicle. The amount of protein taken up in pinocytosis is much greater than could be accounted for if the proteins were simply taken in as part of the solution being drunk by the cell. For example, an ameba can take up in five minutes an amount of protein equivalent to that present in 50 times its own volume of the protein-containing medium. Pinocytosis is thus a membrane activity which brings about the selective uptake of material from solutions. It can account for the rapid transport of large, water-soluble molecules such as proteins and nucleic acids across the plasma membrane. Human cells in tissue culture have been shown to take up proteins and nucleic acids from the medium. This is stimulated by the presence of polyamines such as polyornithine.

Amino acids not only serve as raw materials for the synthesis of specific cell proteins, but may undergo **deamination** and the carbon chains are then used to synthesize glycogen and other carbohydrates, to synthesize fatty acids, or are metabolized to carbon dioxide and water, yielding energy. The amino group may be excreted as ammonia or may be combined with carbon dioxide to form **urea** by a complex series of enzymatic reactions.

Sugars are absorbed and converted into glycogen for storage in liver and muscle. Liver glycogen can be converted enzymatically into glucose and secreted into the blood stream. One of the prime functions of the vertebrate liver is to maintain a constant concentration of glucose in the blood. It does this by absorbing glucose just after a meal when the blood coming from the intestine contains a great deal of glucose. The liver converts glycogen to glucose and secretes it into the blood stream between meals. The glycogen in muscle and other tissues is available only for local utilization; it cannot be converted to glucose since no glucose-6-phosphatase is present. Carbohydrates are rapidly converted to fats if more are taken in than can be used directly. These, plus the fats taken in as food, are stored for use between meals.

Nutritional Requirements. In addition to proteins, fats and carbohydrates, animals require water, minerals and vitamins to maintain health and to grow. The constant loss of minerals from the body in urine, sweat and feces must be balanced by the intake of equivalent amounts in the food. Most foods contain adequate amounts of minerals, and mineral deficiencies are comparatively rare. Certain human deficiency diseases may be traced to a lack of iron, copper, iodine, calcium or phosphorus. A disease which was resulting in the death of entire herds of sheep in Australia was finally shown to be due to a deficiency of cobalt. The soil in that region, and hence the grass eaten by the sheep, contained little or no cobalt which is required as a trace element for normal metabolism.

The extracellular fluids of vertebrates contain about 0.9 per cent salt, largely sodium chloride. The sodium and chloride ions play a major role in maintaining osmotic balance and acid-base balance in the body fluids. They are also present and play a functional role in the secretions of the digestive tract, the hydrochloric acid of the gastric juice and the slightly alkaline pancreatic and intestinal juices. Although large amounts of salts are secreted in the course of a day, they are reabsorbed almost completely and the loss via the digestive tract is negligible.

Minerals. Calcium and phosphorus are the chief constituents of bones and teeth and a childhood deficiency of either one — or of vitamin D, needed for their absorption and metabolism — will result in **rickets.** Phosphorus has an important role in metabolism as a constituent of nucleotides and nucleic acids. In addition, sugars and fatty acids must be phosphorylated before they can be utilized as sources of biologically useful energy. **Iodine** is a component of the hormone thyroxine, synthesized in the thyroid gland. **Iron** is a constituent of the red blood pigment hemoglobin and of the respiratory enzymes, the cytochromes, present in the electron transmitter system. This iron is used and reused; as long as there is no loss of blood, the amount of iron required in the daily diet is minimal. Because women lose a considerable amount of blood each month by menstruation, their iron reserves are typically very small and they are more likely than men to become anemic owing to an iron deficiency.

Traces of **copper** are needed as a component of certain enzyme systems and for the proper utilization of iron in hemoglobin. Traces of manganese, molybdenum, cobalt and zinc are required as components of certain enzyme systems; zinc, for example, is a component of carbonic anhydrase, alcohol dehydrogenase and a number of other enzymes. The role of **fluoride** ion in the prevention of dental caries is well known.

Water is required by every plant and animal and comprises some two thirds of the human body. Aquatic animals have no problem about obtaining water; indeed, their problem is to prevent the osmotic inflow of water and the consequent bursting of their cells. Most land animals drink water but certain desert animals obtain all they require from the food they eat and from the water formed when the foods are oxidized. Water plays a major role in temperature regulation.

Vitamins. Vitamins are relatively simple organic compounds, required in small amounts in the diet. They differ widely in their chemical structure but have in common the fact that they cannot be synthesized in adequate amounts by the animal and hence must be present in the diet. All plants and animals require these same substances to carry out specific metabolic functions, but organisms differ in their ability to synthesize them. Thus, what is a "vitamin" for one animal or plant is not necessarily one for another. Only man, monkeys and guinea pigs require vitamin C, **ascorbic acid,** in their diets; other animals can synthesize ascorbic acid from glucose. The mold *Neurospora* requires the vitamin biotin. Insects cannot synthesize cholesterol and for insects it might be argued that cholesterol is a vitamin.

The vitamins whose role in metabolism is known — niacin, thiamine, riboflavin, pyridoxine, pantothenic acid, biotin, folic acid and cobalamin — have been found to be constituent parts of one or more coenzymes. Two major groups of vitamins can be distinguished: those that are soluble in lipid solvents, the **fat-soluble vitamins** A, D, E and K; and those that are readily soluble in water, the **water-soluble vitamins** C and the B complex. The lack of any one of these vitamins produces a deficiency disease with characteristic symptoms, e.g., scurvy (lack of ascorbic acid), beriberi (lack of thiamine), pellagra (lack of niacin) and rickets (lack of vitamin D).

Thiamine pyrophosphate is the coenzyme for the oxidative decarboxylation of α-keto acids such as pyruvate and α-ketoglutarate. It is also the coenzyme for transketolase. **Riboflavin,** as riboflavin monophosphate and flavin adenine dinucleotide, is required for the reactions of electron transport in the mitochondria and for certain oxidations in the endoplasmic reticulum. Riboflavin occurs in most foods and is synthesized by the intestinal bacteria, so that deficiencies of riboflavin are quite rare. **Pyridoxine,** as pyridoxal phosphate, is the coenzyme for many different reactions involving amino acids—transamination, decarboxylation to amines and so forth—and for glycogen synthetase. **Biotin** is a coenzyme for the reactions in which carbon dioxide is added to an organic molecule, such as the conversion of pyruvate to oxaloacetate, and the carboxylation of acetyl coenzyme A to form malonyl coenzyme A (the first step in the biosynthesis of fatty acids). It is widely distributed in foods and only individuals who eat raw eggs in large quantities are likely to become deficient in biotin. Egg white contains a protein, **avidin,** which forms a tight complex with biotin and prevents its functioning.

Niacin is part of the coenzymes nicotinamide adenine dinucleotide (NAD) and nicotinamide adenine dinucleotide phosphate (NADP), which are cofactors for many dehydrogenases and serve as hydrogen acceptors or donors. The vitamin **folic acid** appears in the coenzyme tetrahydrofolic acid, a coenzyme for many reactions involving one-carbon transfers, and in biopterin, the coenzyme for the conversion of phenylalanine to tyrosine. **Cobalamin** (vitamin B_{12}) serves as the coenzyme in certain reactions involving one-carbon transfers such as the transfer of methyl groups, CH_3. Cobalamin is synthesized by bacteria but not by higher plants or animals—thus this is a "vitamin" for the green plants as well as for animals. Ascorbic acid is known to play a role in the hydroxylation of proline to form hydroxyproline during collagen formation, but the overall function of this vitamin is still unknown.

Vitamin A, or **retinol,** is converted in the retina to retinal, a component of the light-sensitive pigment, visual purple. A deficiency of retinol may lead to "night-blindness" and a severe deficiency will lead to xerophthalmia, a blindness due to the abnormal deposition of keratin as a film over the cornea. Keratin is also deposited in other mucous membranes of individuals suffering from xerophthalmia. Vitamin A undoubtedly has metabolic functions other than its conversion to retinal in the visual process but these are unknown. Plants synthesize β-carotene, an orange-yellow pigment, which can be split to give two molecules of retinol. Large amounts of retinol are stored in the liver, enough to supply a person for several years.

Vitamin D, **cholecalciferol,** plays a role in the movement of calcium ions through membranes, perhaps by stimulating the synthesis of a specific protein required in the transport process. Cholecalciferol can be formed in the skin from a precursor, cholesta-5,7-dienol, by the action of ultraviolet light, which cleaves the B ring of the precursor molecule. Thus cholecalciferol is a "vitamin" only if the person isn't exposed to an adequate amount of sunlight. Some ten structurally related compounds have varying amounts of vitamin D activity; one of the more potent ones is **calciferol,** the usual commercial form of the vitamin which is manufactured from a plant sterol, ergosterol, by irradiation. Excessive doses of vitamin D are toxic, causing hypercalcemia and the deposition of calcium in soft tissues. An excess of vitamin A is also toxic. Acute poisoning with vitamin A can occur by eating the livers of polar bears, a pound of which contains a 20-year supply of retinol for a man. Chronic toxicity is more common and results from prolonged overdosage with vitamin preparations.

A number of similar substances, referred to as **vitamin K,** play a role in the normal coagulation of blood by promoting the synthesis of prothrombin in the liver. These compounds are found in many kinds of food and are synthesized by intestinal bacteria; thus a deficiency of vitamin K is more often associated with some abnormality of its absorption than with a lack in the diet. Vitamin E, or α-**tocopherol,** plays some unknown role in metabolism. Male animals made deficient in tocopherol undergo degenerative changes in the testes and become sterile, and eggs from vitamin E-deficient hens fail to hatch. The skeletal muscles of vitamin E-deficient animals eventually undergo degeneration. Tocopherol may serve as an antioxidant, protecting certain labile cellular components from being oxidized by removing intermediate free radicals. Some investigators believe that the effects of tocopherol deficiency can be traced to the accumulation of fatty acid peroxides, which react with and destroy other cellular components. There is some evidence that vitamin E plays a role in the mitochondrial electron transport system, but the nature of that role is unclear.

5.3   THE EXCHANGE OF GASES BETWEEN CELLS AND THEIR ENVIRONMENT: RESPIRATION

The energy required for the myriad activities of cells is derived primarily from the reactions of biological oxidation (p. 66). The essential feature of these reactions is the transfer of hydrogen atoms or their electrons from donors to acceptors accompanied by the transfer of energy to phosphate ester bonds. The ultimate electron acceptor in the metabolism of most animal cells is oxygen, which is converted to water. Only small amounts of oxygen can be stored in blood or in tissues, and a continuous supply of oxygen is necessary for the continuation of cellular metabolism. **Cellular respiration** is defined as the sum of the processes in which oxygen is utilized, carbon dioxide is produced and energy is converted into biologically useful forms such as ATP. For these processes to continue, the supply of oxygen must be renewed constantly and the carbon dioxide produced must be removed. Thus, the survival of each cell requires the continuous exchange of gases with its environment.

Animals differ tremendously in their general levels of activity and hence in their requirements for energy and oxygen. Animal cells show corresponding differences in their susceptibility to oxygen deprivation. A mouse, using 2500 mm.^{3} of oxygen per gram and hour when resting, and as much as 20,000 mm.^{3} per gram and hour when active, quickly dies of suffocation when deprived of oxygen or when poisoned with carbon monoxide. An earthworm that utilizes 60 mm.^{3} of oxygen or a sea anemone that uses only 13 mm.^{3} of oxygen per gram per hour has a much lower rate of metabolism and does not readily suffocate. Life goes on in these lower animals at a much lower rate in general than in mammals and birds. There are some exceptions to this generalization, and some animals with a low rate of oxygen consumption are sensitive to oxygen deprivation.

The transfer of gases across the cell membrane to the surrounding body fluid or pond or sea water is also part of the respiratory process. In protists and small aquatic animals the exchange of gases is a simple process. Cellular respiration in the mitochondria of a paramecium swimming in a pond utilizes oxygen and lowers the concentration of oxygen in the cytoplasm. Oxygen diffuses into the paramecium from the pond water where the concentration of oxygen is greater, maintained by the diffusion of oxygen from the air into the water and by the production of oxygen in photosynthesis by green plants in the pond. Carbon dioxide produced in metabolic processes within the paramecium diffuses down a concentration gradient into the pond water. Some of the carbon dioxide diffuses out of the water into the air, some is used by plants in the pond in photosynthesis, and some is converted into bicarbonate and carbonate ions. In small animals the ratio of surface to volume is large enough so that the rate of diffusion of gases through the surface of the body is not the factor that limits the rate of respiration. In larger animals the ratio of surface to volume is smaller and cells located deep in the body cannot exchange oxygen and carbon dioxide with the environment rapidly enough by diffusion. Instead, cells exchange gases with the extracellular fluid that bathes them and this, in turn, exchanges gases with the external environment. Many animals have evolved specialized organs, gills, lungs or tracheal tubes for gas exchange with the environment (Fig. 5.2). The gas exchange organ must be thin-walled to facilitate diffusion and must be kept moist so that the gases are dissolved in water on both sides of the membrane. Some means of circulating the fluids bathing the cells must be provided. In most animals the systems for gas exchange and internal transport are closely interrelated in function.

Higher animals living on land and breathing air have ready access to oxygen, for it makes up about 21 per cent of the air. They may have a problem in keeping the respiratory surface moist, for the air to which it is exposed is usually not saturated with water vapor and thus tends to dry out the respiratory surface. Animals living in water and exchanging gases with it have no problem of desiccation, but even well-aerated water has only 10 ml. of oxygen per liter and the organism may require some mechanism to keep a continuous supply of fresh oxygen-rich water flowing over its gills or

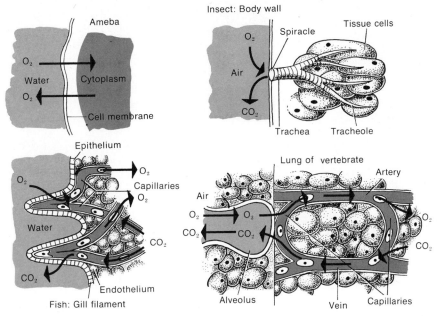

Figure 5.2   A diagram of some of the types of respiratory organs present in animals.

other respiratory surface. An earthworm in its burrow can obtain enough oxygen by diffusion through its moist skin and does not have to stir up the air. In contrast, a marine worm living in a burrow or tube must undulate its body constantly and provide a current of water if it is to obtain an adequate supply of oxygen from that dissolved in the sea water.

The **partial pressure** of a gas is the pressure due to that gas in a mixture of gases. It is calculated by multiplying the total pressure of the mixture of gases by the percentage of the gas in the mixture. Air has a pressure about 760 mm. Hg and is 21 per cent oxygen; hence the partial pressure of oxygen in air is 760 × 0.21 or 159 mm. Gas molecules dissolved in a liquid have a certain tendency to leave the liquid and enter the gaseous phase. This escaping tendency can be measured by the pressure of that gas in the gaseous phase in contact with the liquid which is required to prevent any net loss of the gas (i.e., to maintain equilibrium). When a liquid and gas are in contact, equilibrium is reached when the rate at which molecules pass from the liquid to the gas equals the rate at which they pass from gas to liquid. This escaping tendency, termed the "tension" of the gas, is numerically equal to the partial pressure of the gas with which it is in equilibrium. The gas tension is a measure of the tendency of the dissolved gas to diffuse out of the solution and is not a measure of the *quantity* of gas present.

The amount of gas in solution is a property of both the gas and the liquid and may vary considerably from one liquid to another. Water and blood in equilibrium with air would each have an oxygen tension of 159 mm. Hg, but the water would contain only 0.2 ml. of oxygen per 100 ml. whereas blood, because of the hemoglobin present, would contain 20 ml. of oxygen per 100 ml.

The Respiratory Surfaces. The protozoa and the simpler invertebrate animals, the sponges, coelenterates and flatworms, obtain oxygen from and give off carbon dioxide to the surrounding water by diffusion. This process is termed **direct respiration** since the cells exchange oxygen and carbon dioxide directly with the surrounding environment. The cells of the larger, more complex animals cannot exchange gases directly with the environment and some form of **indirect respiration** occurs. The cells exchange gases with the body fluids (**internal respiration**) and the body fluids exchange gases with the external environment via a specialized respiratory surface (**external respiration**) (Fig. 5.3). The respiratory surface

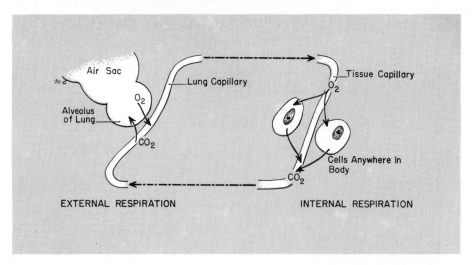

Figure 5.3   Diagram illustrating the exchanges of gases in external and internal respiration.

for many animals—such as the earthworm and frog—is simply the skin or perhaps the lining of the mouth. Fishes, many amphibia, mollusks, crustacea and some worms have developed **gills** as organs of gas exchange (Fig. 5.4). The fine filaments of tissue are covered with a thin epithelium and contain blood channels. Gases diffuse from the surrounding water through the thin, moist epithelium into the blood vessels. Each animal with gills has some arrangement to keep a current of water flowing over them. The fish opens its mouth, takes a gulp of water, then closes the mouth and forces the water out past the gills by contracting its mouth cavity.

The **spiracles** of insects and certain other arthropods represent a completely different solution to the problem of getting oxygen to the cells. Each segment of the body has a pair of openings (spiracles) through which air passes via a system of branched air ducts, termed tracheal tubes, to all of the internal organs (Fig. 5.4). The ducts terminate in microscopic, fluid-filled **tracheoles.** Oxygen and carbon dioxide diffuse through these to and from the adjacent cells. The larger insects can pump air through the tracheal tubes by contracting the muscles in their abdominal walls. The tracheal system is efficient; oxygen reaches the cells and carbon dioxide is removed by diffusion. No energy need be expended (as in the vertebrates) in maintaining a rapid flow of blood to keep the body cells supplied with oxygen.

The higher vertebrates have evolved **lungs** for external respiration. Lungs are hollow spaces, usually greatly subdivided into small hollow pockets (**alveoli**). The walls of the alveoli are very thin and supplied with a rich bed of capillaries. The network of elastic fibers between the alveoli supports them and makes the lung very pliable. The arrangement of the lung alveoli as pockets tends to minimize the loss of water and keeps the alveolar surface moist.

However different respiratory surfaces may appear structurally, they are essentially similar in consisting of a thin, moist membrane, richly supplied with blood vessels, separating body fluid and external environment. Oxygen molecules move from the air to the cells within the body along a steep diffusion gradient. The partial pressure of oxygen in air is about 150 mm. Hg, and that of the air in the lungs is about 105 mm. Hg. The oxygen tension of blood going from lungs to tissues is about 100 mm. Hg, and that of blood returning from tissues to lungs is about 40 mm. Hg. The oxygen tension in tissues may vary from 0 to 40 mm. Hg.

The transport of oxygen and carbon dioxide in most animals involves some sort of respiratory pigment present in the blood, either dissolved in the fluid or located within certain blood cells. Many of these pigments are heme proteins containing a porphyrin ring similar to that in chlorophyll and in the cytochromes (p. 67). The respiratory pigment contains a metal atom, typically iron or copper, to which oxygen atoms are bound. If human blood were simply water, it could carry only about 0.2 ml. of oxygen and 0.3 ml. of carbon dioxide in each 100 ml. The properties of the respiratory pigment, **hemoglobin,** enable whole blood to carry some 20 ml. of oxygen and 30 to 60 ml. of carbon dioxide per 100 ml. These properties also enable hemoglobin to serve simultaneously as an acid base buffer to minimize changes in the pH of the blood. The protein portion of hemoglobin is composed of four peptide chains, typically two α chains and two β chains, to which are attached four heme rings. An iron atom is bound in the center of each heme ring. Hemoglobin is the most commonly encountered respiratory pigment and occurs in all the major groups of animals above the flatworm. Mollusks, crustacea and certain other animals have other heme pigments such as the blue-green copper protein **hemocyanin.**

In the respiratory organ, the lung or gill, oxygen diffuses into the red cell from the plasma and combines with hemoglobin (Hb) to form oxyhemoglobin (HbO$_2$): $Hb + O_2 \rightleftharpoons HbO_2$.

The reaction is reversible and hemoglobin releases oxygen when it reaches a region where the oxygen tension is low, in the capillaries of the tissues. The combination of oxygen with hemoglobin and its release

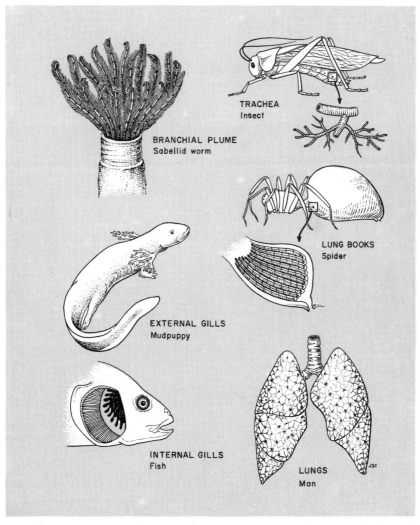

Figure 5.4   Diagrams of the types of respiratory organs found in animals.

from oxyhemoglobin are controlled by the concentration of oxygen and, to a lesser extent, by the concentration of carbon dioxide. Carbon dioxide reacts with water to form carbonic acid, H_2CO_3; hence an *increase* in the concentration of carbon dioxide results in an *increased* acidity of the blood. The oxygen-carrying capacity of hemoglobin *decreases* as blood becomes more acid; thus, the combination of hemoglobin with oxygen is controlled indirectly by the amount of carbon dioxide present. This results in an extremely efficient system. In the capillaries of the tissues the concentration of carbon dioxide is high and a large amount of oxygen is released from hemoglobin by the combined effects of the low oxygen tension and the high carbon dioxide tension. In the capillaries of the lung or gill carbon dioxide tension is lower and a large amount of oxygen is taken up by hemoglobin owing to the combined effects of the high oxygen tension and the low carbon dioxide tension.

The unique properties of hemoglobin also enable each liter of blood to transport some 50 ml. of carbon dioxide from tissue to lung while the pH of the blood changes only a few hundredths of a unit. Some carbon dioxide is carried in a loose chemical union with hemoglobin as **carbamino hemoglobin,** and a small amount is present as carbonic acid, but most of it is transported as bicarbonate ion, HCO_3^-. The CO_2 produced by cells dissolves in the tissue fluid to form H_2CO_3, a reaction catalyzed by **carbonic anhydrase,** and the carbonic acid is neutralized to bicarbonate by the sodium

and potassium ions released when oxyhemoglobin is converted to hemoglobin. Oxyhemoglobin is a stronger acid than reduced hemoglobin; hence some cations are released when HbO_2 is converted to Hb. In the course of evolution this one molecule has become endowed with all the properties needed for the transport of large amounts of oxygen and carbon dioxide with a minimal change in the pH of the blood.

The properties of the heme pigments are such that the amount of oxygen taken up by the pigment is not directly proportional to the oxygen tension; a graph of the relationship gives an S-shaped curve (Fig. 5.5). The blood is a more effective transporter of oxygen than it would be if the oxygen content were a simpler linear function of oxygen tension. The effect of carbon dioxide (really the effect of the change in pH brought about by the change in the carbon dioxide content) on the combination of oxygen with hemoglobin is shown in Figure 5.6. The oxygen dissociation curves for arterial blood, with low carbon dioxide tension, and for venous blood, with high carbon dioxide tension, illustrate how much more oxygen is delivered to the tissue by a given amount of blood as carbon dioxide is taken up in the tissue capillaries. The properties of the heme proteins in different species of animals are quite different and are in general adapted to the amount of carbon dioxide present in the environment. This is generally low in water-breathing animals and high in air-breathing animals. This points up the generalization that the evolution of

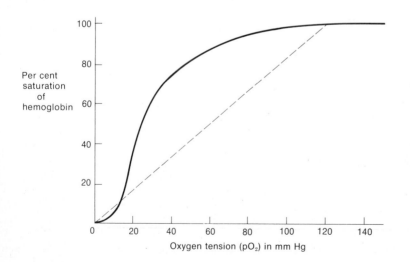

Figure 5.5 The combination of oxygen with hemoglobin as a function of oxygen tension. The dashed line shows the relation that would exist if hemoglobin bound oxygen as a linear function of oxygen tension. Because of the interactions between the four heme groups in a single hemoglobin molecule, the binding of oxygen to hemoglobin follows the S-shaped curve.

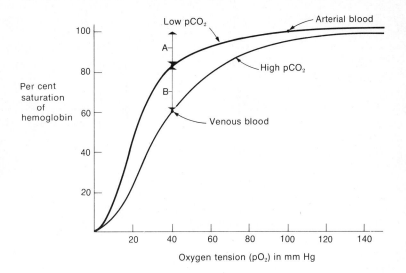

Figure 5.6 The effect of carbon dioxide tension ($_pCO_2$) on the delivery of oxygen to the tissues. The arrow A indicates the amount of oxygen released from hemoglobin as the $_pO_2$ falls from that of arterial blood (100 mm. Hg) to that of venous blood (40 mm. Hg). The arrow B indicates the additional amount of oxygen released because of the greater $_pCO_2$ in venous blood.

air-breathing animals from water-breathing ones involved marked changes not only in the structure of the respiratory organs, but also in the chemical properties of the heme proteins serving as blood pigments.

5.4 CIRCULATION

The metabolic processes of all cells require a constant supply of food and oxygen and constant removal of wastes, and all organisms have solved in one way or another the problem of transporting substances from one part of the body to another. The transport of hormones from the endocrine glands to their target organs and the transfer of heat to equalize body temperature are functions added in some of the higher animals.

In protozoa the transport of substances is effected by the diffusion of the molecules, aided generally by streaming movements of the cytoplasm itself. The flowing of the cytoplasm from rear to front as an ameba moves, and the circular movement of the cytoplasm in paramecia and other protozoa with a fixed shape are effective transport systems in these tiny animals (Fig. 5.7). Transport from cell to cell in simple multicellular animals such as sponges, coelenterates and flatworms occurs by diffusion. This is aided in some animals by the stirring of the body fluids brought about by the contraction of the muscles of the body wall. The rate of diffusion is directly proportional to the difference in concentration in

the two regions and inversely proportional to the distance separating them. From this we can see that an adequate supply of food and oxygen can be maintained by diffusion alone only if an animal is small and the distance over which diffusion occurs is short. In large animals the slower diffusion rate over the greater distance would not suffice. Such animals must develop some system of internal transport—some type of circulatory system. Not only absolute size, but also the shape and the activity of an animal determine the need for a circulatory system.

The **proboscis worms** or Nemertea are the simplest living animals to have a distinct circulatory system; it consists of a dorsal and two lateral blood vessels which extend the whole length of the body and are connected by transverse vessels. The earthworm has a more complex circulatory system: a **dorsal vessel** in which blood flows anteriorly, a **ventral vessel** and a **subneural vessel** in which blood flows posteriorly, and five pairs of pulsating tubes ("hearts") at the anterior end which drive blood from the dorsal to the ventral vessel (Fig. 5.8). In other segments of the body a network of vessels connecting dorsal and ventral vessels ramifies through the body wall and the wall of the intestine. The major advance of the annelid over the nemertean system is the addition of capillary networks in the body wall, intestine, and the glandular region of the metanephridia (kidneys).

The circulatory systems of the larger and more complex animals typically include

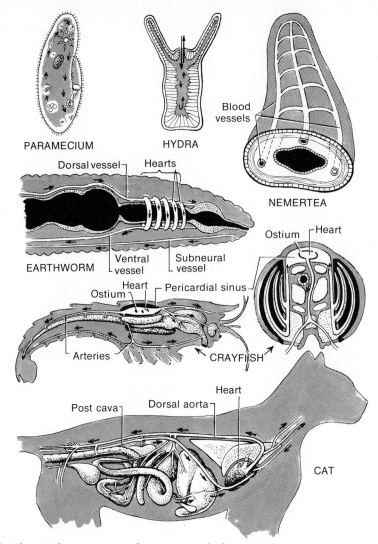

PARAMECIUM   HYDRA

Blood
vessels

NEMERTEA

Dorsal vessel   Hearts

Ostium   Heart

EARTHWORM   Ventral   Subneural
               vessel   vessel

Heart   Pericardial sinus
Ostium

Arteries   CRAYFISH

Heart

Post cava   Dorsal aorta

CAT

Figure 5.7   The circulatory systems of paramecium, hydra, nemertea, earthworm, crayfish and cat.

blood vessels and **heart** and the fluid within them—the **blood.** This, in turn, is composed of a fluid—**plasma**—and **blood cells** or corpuscles.

The circulatory system of some animals, such as annelid worms, squids and vertebrates is said to be "closed"; i.e., the blood in the course of circulation remains within blood vessels (Fig. 5.8). In contrast, the circulatory systems of arthropods and most mollusks are "open"; the blood vessels open to the body cavity, called a **hemocoel,** and blood circulates partly within blood vessels and partly through the cavity of the hemocoel in making a complete circuit. In the typical arthropod, the heart and other organs lie free in the hemocoel and are bathed in blood. In annelids and vertebrates, the internal organs lie in the coelomic cavity and are supplied by blood which reaches them in closed vessels. The arthropod heart is generally a single, elongate, muscular tube lying in the dorsal midline. Blood enters the heart from the pericardial sinus, which is part of the hemocoel, through paired openings (ostia) present in each segment and is moved forward by **peristaltic waves,** waves of contraction preceded by waves of relaxation along the tube. Blood is carried in vessels to the head and to other

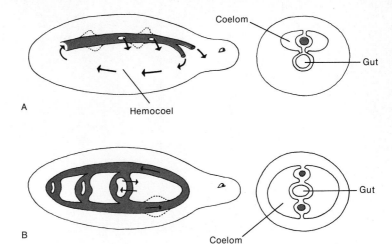

Figure 5.8 Diagrams illustrating an open (*A*) and a closed (*B*) circulatory system. The dotted lines indicate the location of pulsatile "hearts."

parts of the body, whence it returns to the heart through the hemocoel.

The hearts of most invertebrates can develop only very low pressures—a few millimeters of mercury—as they pump blood. In the closed circulatory systems of vertebrates, a higher pressure, as much as 100 to 200 mm. Hg, is required to drive the blood against the resistance of the tremendous number of narrow capillaries. This has led to the evolution of powerful, thick-walled, muscular hearts. The chamber of the vertebrate heart called the **ventricle** has quite thick walls. A certain amount of pressure is required to distend the muscular ventricle and cause the blood to flow in and fill it during the relaxation phase (**diastole**). The pressure in vertebrate veins is too low to do this. The vertebrate heart has a second chamber, the **atrium,** with walls thin enough to be filled by the low venous pressure yet strong enough to pump blood into the ventricle and distend it. The octopus, whose heart is similarly arranged with two different chambers, has the highest blood pressure, 35 to 45 mm. Hg, of any of the invertebrates.

The vertebrate heart is enclosed in a special cavity of the coelom, the **pericardial cavity,** separated from the rest of the body by a thin, strong sheet of connective tissue, the **pericardium.** This cavity provides space for the heart to change in volume as it beats.

The circulatory systems of all vertebrates are essentially similar: a closed system composed of heart, aorta, arteries, capillaries and veins arranged in a basically similar plan. Because of this similarity, students can learn much about the human circulatory system from the dissection of a dogfish or frog. **Arteries** carry blood away from the heart to the tissues, **veins** carry blood back to the heart from the tissues, and **capillaries** are minute, thin-walled vessels connecting the arteries to the veins and completing the circuit from heart to heart. Only capillaries have walls thin enough to permit the exchange of nutrients, gases and wastes between blood and tissues.

The principal evolutionary changes in the vertebrate circulatory system have been associated with the change from gills to lungs as respiratory organs. The changes in the pattern of circulation permit the delivery of oxygen-rich blood to the brain and muscles. The pattern of circulation in many lower vertebrates is such that some mixing of oxygen-rich and oxygen-poor blood occurs. Mammals and birds can be warm blooded because their circulatory systems supply enough oxygen to the tissues to support a metabolic rate high enough to maintain a constant high body temperature in cold surroundings.

5.5 THE REGULATION OF INTERNAL FLUIDS AND THE REMOVAL OF WASTES

Cells can survive and function only within a rather narrow range of conditions. A cell can function only as long as the compo-

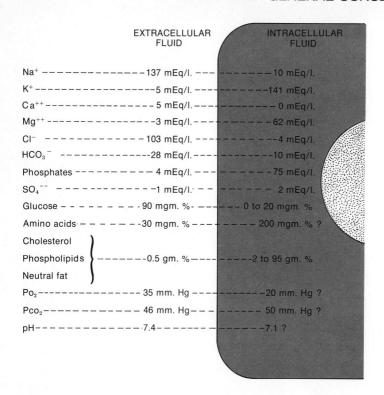

	EXTRACELLULAR FLUID	INTRACELLULAR FLUID
Na⁺	137 mEq/l.	10 mEq/l.
K⁺	5 mEq/l.	141 mEq/l.
Ca⁺⁺	5 mEq/l.	0 mEq/l.
Mg⁺⁺	3 mEq/l.	62 mEq/l.
Cl⁻	103 mEq/l.	4 mEq/l.
HCO₃⁻	28 mEq/l.	10 mEq/l.
Phosphates	4 mEq/l.	75 mEq/l.
SO₄⁻⁻	1 mEq/l.	2 mEq/l.
Glucose	90 mgm. %	0 to 20 mgm. %
Amino acids	30 mgm. %	200 mgm. % ?
Cholesterol Phospholipids Neutral fat	0.5 gm. %	2 to 95 gm. %
Po₂	35 mm. Hg	20 mm. Hg ?
Pco₂	46 mm. Hg	50 mm. Hg ?
pH	7.4	7.1 ?

Figure 5.9 Diagram illustrating the chemical composition of extracellular fluids and intracellular fluid. (Redrawn after Guyton, A. C.: Function of the Human Body, 3rd ed. Philadelphia, W. B. Saunders Company, 1969.)

sition of its internal fluid and the fluid bathing its plasma membrane are kept within certain bounds (Fig. 5.9). Evolution has been marked by the development of a host of mechanisms such as facilitated diffusion and active transport which, together with physical diffusion, maintain within each organism a fluid environment with a composition appropriate for the special needs of its cells. This principle of the need for constancy was concisely stated in the much-quoted aphorism of Claude Bernard, "La fixité du milieu intérieur est la condition de la vie libre"—the capacity to regulate the internal environment is essential for an organism to be able to live in a variety of environments. The term **homeostasis** was coined by Walter Cannon to refer to the tendency of organisms to maintain constant the conditions in their internal environment.

Maintaining the constancy of the body fluids can be complicated by the nature of the environment in which the animal lives, whether it is salt water, fresh water, or on the land. A limited homeostatic capacity may be one of the factors limiting an organism to one or another of these environments. The concentrations of salts in the cells of many marine animals—sponges,

starfish, crabs—are equal to those in sea water. These animals are "osmoconformers" (p. 100) and have little or no problem regarding salt and water balance. With most animals a considerable expenditure of energy is required to maintain constant the volume and composition of body fluids. Fish such as the salmon that migrate from fresh to salt water and back have an unusual capacity to adapt to changing environments. An animal living in fresh water has body fluids that are markedly hypertonic to the surrounding medium. Water tends to enter by osmosis through permeable membranes. As adaptations, many animals living in fresh water have developed skins covered with cuticle or scales that reduce their permeability to water. Others have developed water expulsion vesicles, flame cells and other devices for removing water molecules from the internal environment. A protozoan in fresh water is like a leaky boat that must be bailed constantly to stay afloat. Fresh-water sponges and coelenterates (e.g., *Hydra*) apparently have no contractile vacuoles and how they prevent the uptake of water or pump it out remains a mystery.

Animals living in salt water which is hypertonic to their internal medium (e.g., most marine vertebrates) have the opposite problem. They must conserve water and eliminate the salts that diffuse into the cell, down the concentration gradient. These animals have adapted in a variety of ways, such as the evolution of special glands to secrete the salts back into the sea water. Animals living on land have a comparable problem of preventing the loss of water through the skin and respiratory surfaces. This problem becomes acute in animals living in the desert. Some animals have behavioral adaptations and live in burrows or under stones; others have a scaly exoskeleton or a cuticle that minimizes the escape of water; still others have kidneys adapted to reabsorb a maximum amount of water from the urine.

In the course of the metabolic processes by which substances are utilized for the production of energy and the growth and maintenance of cells, wastes are produced which are toxic and must be removed from the body. The constant degradation of proteins, nucleic acids and other nitrogen-containing substances results in the formation of ammonia, urea, uric acid and creatinine. **Deamination,** the removal of the amino group from an amino acid, results in the formation of **ammonia** which is toxic, a potent inhibitor of many enzymes. Several energy-requiring enzymatic proc-

esses have evolved which convert ammonia to less toxic materials such as urea. Both urea and ammonia are water soluble and require the concurrent excretion of a certain amount of water. Some land organisms have evolved a sequence of enzymes by which the amino group is converted to uric acid and excreted in that form. Uric acid is poorly soluble in water; the water can be reabsorbed from the urine, and the uric acid can be excreted as a paste or dry powder. The enzymatic processes involved in the synthesis of urea and uric acid require ATP to drive them and constitute a constant drain upon the energy reserves of the body.

Small aquatic animals usually eliminate nitrogenous wastes as ammonia by physical diffusion through the body surface. The surface area of larger animals is too small for the elimination of wastes and special excretory organs have evolved. The simplest animals with specialized excretory organs are the flatworms and nemerteans which have **flame cells** (Fig. 5.10) equipped with flagella, and a branching system of excretory ducts from the flame cells to the outside. The flame cells lie in the fluid bathing the cells of the body and wastes diffuse into the flame cells and then into the excretory ducts. The beating of the flagella, which suggests a flickering flame when seen under the mi-

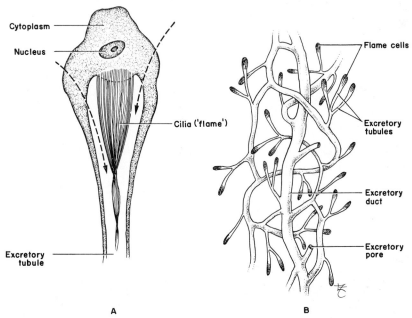

A
B

Figure 5.10   The excretory organs of the flatworm. *A,* A single flame cell. *B,* Flame cells are connected by excretory tubules and ducts to the pore which opens to the outside.

croscope, presumably moves fluid in the ducts out the excretory pores. Like the water expulsion vesicles of the protozoa, the chief role of the flame cells is to regulate the water content of the animals. The number of flame cells in a planarian is adjusted to the salinity of the environment. Planaria grown in slightly salty water develop few flame cells, but the number rapidly increases if the concentration of salt in the environment is decreased.

The excretory organs of higher animals are typically tubular in shape, open at one end to the exterior, and equipped with an intimate connection to the vascular system. In some, the inner end of the tubule is open to the coelomic cavity. The **nephridia** of the earthworm, paired excretory organs present in each segment of the worm, are excellent examples of such structures. Each nephridium is a long, coiled tubule opening at one end to the body cavity in a funnel-shaped structure lined with cilia, and at the other end to the outside of the body via an excretory pore. Around each tubule is a coil of capillaries, which permits the removal of wastes from the blood stream. As fluid passes through the nephridium, moved by the beating of the cilia in the funnel and the contraction of the muscles in the body wall, water and substances such as glucose are reabsorbed by the capillaries and the wastes are concentrated and pass out of the body. The earthworm excretes very dilute, copious urine, at a rate of about 60 per cent of its total body weight each day. Phagocytic chlorogen cells wander through the coelom, ingest waste particles and eventually deposit them in the skin as pigment.

The crustacean excretory organs are the **green glands,** paired structures located at the base of the antennae and supplied with blood vessels. Each gland consists of three parts: a coelomic sac, a greenish glandular

chamber with folded walls, and a canal which leads to a muscular bladder. Wastes pass from the blood to the coelomic sac and glandular chamber; the fluid in them is isotonic with the blood. Urine collects in the bladder and then is voided to the outside through a pore at the base of the second antenna.

The excretory organs of insects, the **malpighian tubules** (Fig. 5.11), are quite different from those of the crustaceans. They lie within the body cavity (hemocoel) and empty into the digestive tract. Each tubule has a muscular coat and its slow writhing assists the movement of the wastes in its lumen to the gut cavity. The tubules are bathed in blood in the hemocoel and their cells transfer wastes by diffusion or by active transport from the blood to the cavity of the tubule. The major waste product, uric acid, is slightly soluble in water and is excreted as a paste, an adaptation which conserves body water for the insect.

The kidneys of the vertebrates remove wastes from the blood and regulate its content of water, salts and organic substances. The structural and functional unit of the kidney is the **kidney tubule** (Fig. 5.12). This is in close contact with the blood stream, for a tuft of capillaries projects into the funnel-shaped **Bowman's capsule** at the end of the tubule. The tubule may be quite long and looped and in contact with additional capillaries along its length. It eventually opens to the outside of the body via collecting ducts and other intermediate tubes. Substances are filtered into the kidney tubules from the blood capillaries in the Bowman's capsules. Then, some substances are reabsorbed into the blood stream and others are secreted from the blood into the urine as the liquid flows through the tubules and past the additional capillaries there. The kidney must expend energy to move

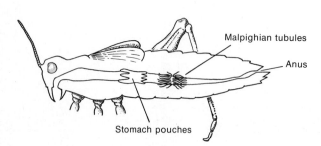

Malpighian tubules

Anus

Stomach pouches

Figure 5.11 Diagram of an insect, a grasshopper, showing its excretory organs, the Malpighian tubules.

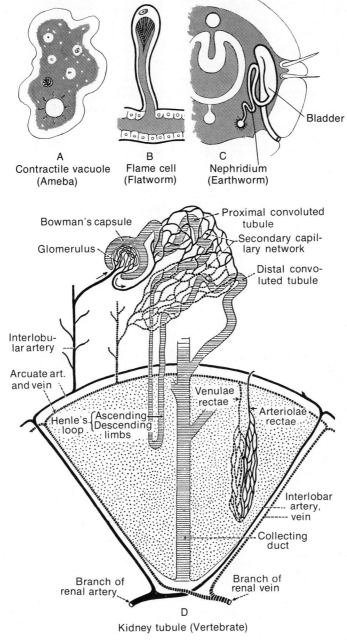

A
Contractile vacuole
(Ameba)

B
Flame cell
(Flatworm)

C
Nephridium
(Earthworm)

Bladder

Bowman's capsule

Glomerulus

Proximal convoluted tubule

Secondary capillary network

Distal convoluted tubule

Interlobular artery

Arcuate art. and vein

Venulae rectae

Arteriolae rectae

Henle's loop

Ascending Descending limbs

Interlobar artery, vein

Collecting duct

Branch of renal artery

Branch of renal vein

D
Kidney tubule (Vertebrate)

Figure 5.12 The excretory systems of (*A*) ameba, (*B*) flatworm, (*C*) earthworm and (*D*) vertebrate.

certain of the substances excreted or reabsorbed against a diffusion gradient. The excretory process would be very wasteful and inefficient if the urine leaving the body had the same composition as the fluid in the Bowman's capsule. However, as the urine passes down the kidney tubule, water, sugar, salts and many other substances are reabsorbed, whereas the waste products such as ammonia and urea are not. It is by this **selective reabsorption** of certain substances, and by the addition of others to the urine, that the kidney tubules regulate the composition of the blood and body fluids. In the higher animals such as man, the lungs, skin and digestive tract also remove certain wastes from the body.

We can distinguish "**osmoregulators**" and "**osmoconformers.**" Some animals, the osmoregulators, can maintain an internal osmotic concentration different from that of the external medium. This involves the expenditure of energy and a variety of mechanisms such as active transport and sodium pumps. The osmoconformers, by contrast, simply come into osmotic equilibrium with the environment, changing the intracellular fluid concentration as the external fluid changes. The osmoconformers are typically able to survive despite considerable changes in the osmotic pressure of the environment.

Some of the vertebrates living in or on the sea have evolved special means for dealing with salt. The blood and body fluids of marine bony fishes are hypotonic to sea water and the fish tends to lose water osmotically. The bony fishes drink salt water, retain the water and secrete the salt through their gills. Marine turtles and sea gulls can secrete the salt from the sea water they drink by special salt glands located in the head. The ducts from the salt glands empty either into the nasal cavities or onto the surface of the head. Marine mammals apparently eliminate their excess salt through their kidneys.

Fresh-water fish absorb salts by active transport across the gills, and excrete a copious, dilute urine to eliminate the water taken in osmotically through the gills and lining of the mouth (Fig. 5.13). Terrestrial animals have the dual problem of conserving their limited and precious supply of water while excreting their nitrogenous wastes. The complex kidney tubules of the higher vertebrates (Fig. 5.12) are admirably adapted to do this by the process of filtration, reabsorption and tubular secretion (Chapter 14).

5.6   PROTECTION

The complex physicochemical system that constitutes a cell requires protection against the many adverse effects of the surrounding environment. The ameba is an exception to the general rule that animals

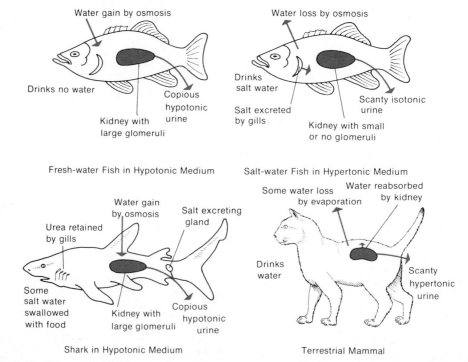

Figure 5.13   Water and salt balance in sharks, in fresh-water and salt-water fish and in a terrestrial mammal.

have some protective covering. The cellular contents of the ameba are separated from the surrounding environment only by the plasma membrane. Many other protozoa have a tough, flexible, complex **pellicle** surrounding the cell and some secrete hard, durable, calcareous or siliceous shells. All the multicellular animals have some protective covering or **skin** over the body. The skin may consist of one or many layers of cells and may be reinforced by scales, hair, feathers, shells or secretions of mucus or cutin. Hair, feathers and certain scales, such as those of reptiles, are composed of insoluble proteins called **keratins** derived from dead cells in the skin. The skin has several functions: it protects the underlying cells against mechanical and chemical injuries; it prevents the entrance of disease organisms; it prevents excessive loss of water from land animals and excessive uptake of water by fresh-water animals; and it protects underlying cells against the harmful effects of the ultraviolet rays in sunlight.

The skin is an effective thermostatically controlled radiator by which the body can eliminate the heat which is constantly produced in cellular metabolism. One of the factors controlling the rate of heat loss in higher vertebrates is the size of the blood vessels in the skin. To conserve heat in a cold environment, the blood vessels are constricted to decrease the rate of blood flow. The reverse occurs in a warm environment and the rate of heat loss can be increased by the evaporation of water, i.e., sweat, from the surface of the skin.

Many animals have some sort of framework or **skeleton** which protects and supports the body and against which muscles can act. In some animals blood or coelomic fluid acts as a hydrostatic or fluid skeleton. The engorgement of the foot of a clam and the erection of male copulatory organs are examples of fluid skeletons. The coelomic fluid of earthworms serves as a hydrostatic skeleton against which the muscles of the body wall can act to effect locomotion (p. 561). Some animals manage to survive without a skeleton but these are mostly aquatic forms.

Solid skeletons may be an **exoskeleton,** located on the outside of the body, or an **endoskeleton,** located within the body. The hard shells of lobsters, crabs and insects and the calcareous shells of oysters and clams are examples of exoskeletons. An exoskeleton provides excellent protection for the body, and muscles can be attached to its inner surface so as to move one part with respect to another. However, the presence of an exoskeleton usually interferes with growth. The arthropods have solved this problem by periodically shedding the shell. To do this the shell is first softened—that is, some of the calcium salts deposited in it are dissolved—the shell is split and the animal crawls out of the old shell. It then undergoes a period of rapid growth before the new shell, which formed under the old one, is hardened by the deposition of calcium salts. During this **molting** process the arthropod lacks protection and is weak and barely able to move. Hard, calcareous exoskeletons are present in most mollusks and arthropods, and in corals, bryozoa and a variety of lesser invertebrates. A clam or oyster secretes additional shell at the margin as it grows; the shell gets both larger and thicker as the animal grows. Many marine worms secrete calcareous tubes in which they live. Though these shells are not directly a part of the animal body, they are, in certain respects, the functional equivalent of an exoskeleton.

The vertebrate skeleton, lying within the soft tissues of the body and composed of many individual **bones** or **cartilages,** provides an excellent framework for their support and does not interfere with their growth. The arrangement of the parts of the skeleton is essentially the same in all the vertebrates. The details of this will be discussed in Chapter 11.

The region where two hard parts meet and move one on the other is known as a **joint.** The fundamental differences in the mechanics of vertebrate and arthropod joints are illustrated in Figure 5.14. The muscles of the vertebrate surround the bones; each is attached by one end to one bone and by its other end to another bone. Its contraction thus moves one bone with respect to the other. The muscles of the arthropod lie *within* the skeleton and are attached to its inner surface. In certain regions the arthropod exoskeleton is thin and flexible so that movements may occur. The muscle may stretch across the joint, so that its contraction will move one part on the next. Or, the muscle may be located en-

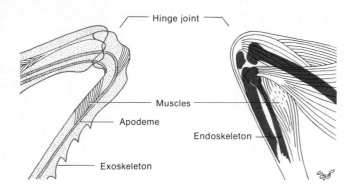

Figure 5.14 A comparison of the vertebrate endoskeleton and joint with the arthropod exoskeleton and joint. Note the difference in the arrangement of the muscles and skeletal elements at the two types of joints.

tirely within one section of the body or appendage and be attached at one end to a tough **apodeme,** a long, thin, firm part of the exoskeleton extending into that section from the adjoining one.

5.7 MOTION

We can correlate the movement so characteristic of animals with their heterotrophic nutrition and their need to reach a new part of the environment and a fresh supply of food. (Or to bring a new part of the environment and a fresh supply of food to them.)

The contraction of a cell involves the transformation of chemical energy into mechanical energy. The chemical energy of the energy-rich phosphate bonds synthesized in glycolysis and in biologic oxidation (p. 72) is converted into the mechanical energy of contractile protein molecules such as **actomyosin.** The basic process for the conversion of chemical to mechanical energy appears to be fundamentally similar in all cells, though the nature of the contracting protein molecule may differ somewhat. The mechanical behavior of a contracting muscle and the energy used can be measured and compared with the chemical, electrical and thermal changes coincident with contraction. These provide insights into the nature of the contractile mechanism.

Ameboid motion is the irregular flowing of the cellular contents seen in amebas, in the amebocytes of sponges, in the white blood cells of vertebrates and in the general process of cytoplasmic motion which occurs during cell division. The streaming cytoplasm bulges out in projections called **pseudopods** (false feet). Careful microscopic

study of a moving ameba reveals that not all of the cytoplasm streams simultaneously. A fibrous protein complex forms the semirigid nonmoving outer ectoplasmic tube (gel) within which is a less fibrous, more fluid mass of endoplasm (sol). At the posterior end of the moving ameba the gel contracts, causing the less fibrous endoplasm to flow forward. It is diverted laterally at the anterior end and added to the gel tube by conversion to gel. Fibrous gel is converted to fluid sol at the inner posterior gel margin and fluid sol is converted to fibrous gel at the anterior margin of the sol. The major unsettled questions at present concern the site of the major contraction and where and how the motive force is applied. Probably the ectoplasm contracts, pressing on the fluid endoplasm and driving it forward against the hyaline cap. By regulating the thickness of local areas in the ectoplasm the ameba can determine where a pseudopod forms and hence which way it will move. The animal has no permanent front and rear ends. Ameboid motion is a crawling motion, not a swimming one; the cell must be attached to some substratum in order to move.

Another type of motion involves the beating of slender, hair-like processes projecting from certain cells. These are termed **flagella** if each cell has one or a few long, whiplike processes and **cilia** if each cell has many short processes. Flagella are found on certain protozoa (the flagellates), on the collar cells of sponges and on certain cells lining the gastrovascular cavity of coelenterates. Cells equipped with cilia occur very widely: in certain protozoa (called ciliates), on the body surfaces of ctenophores, flatworms and rotifers, on the tentacles of bryozoa, certain worms and coelenterates, on the gills of

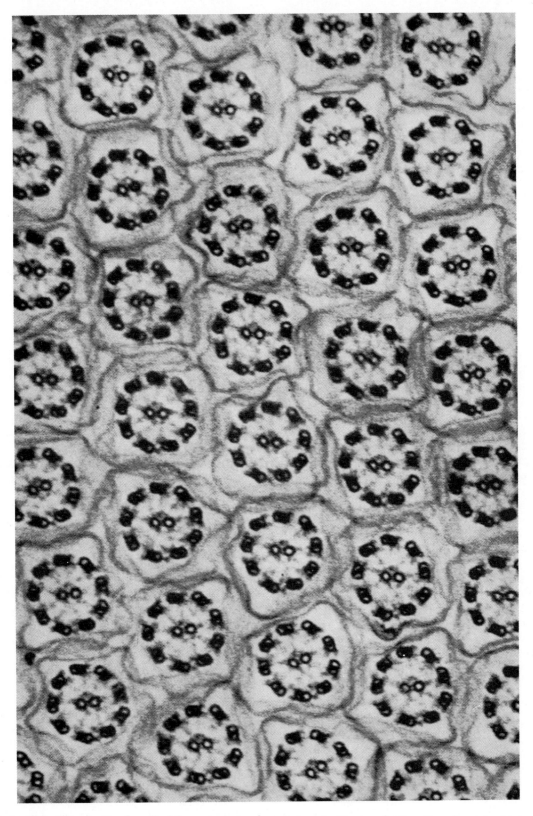

Figure 5.15 Lateral cilia of a gill of the mussel, *Anodonta cataracta*. Osmium fixation. Uranyl acetate staining. (From Gibbons, I. R. J. Biophys. Biochem. Cytol. *11*:179–205, 1961.) Magnification 110,000×.

clams and oysters, and lining certain ducts in the vertebrate body such as the bronchi and oviducts. The paramecium is a ciliate; some 2500 short cilia cover each single-celled animal. The cilia beat in a coordinated rhythm, not simultaneously but one after another, so that waves of ciliary movement pass along the body surface. The combined effort of the cilia beating backward moves the animal forward. The cilia beat somewhat obliquely so that the animal revolves on its long axis and moves in a spiral path.

With the aid of very high speed movie cameras (4000 frames per second) the nature of the ciliary beat has been analyzed. In swimming paramecia the cilia beat with a traveling helical wave that moves from the base to the tip of the cilia. Ciliary motion is not simply the back and forth movement of the cytoplasmic hair in a whip-like fashion. Instead the helical wave is composed of two traveling undulatory waves moving distally at right angles to each other and 90 degrees out of phase. When the animal changes direction, the cilia change their direction but continue to beat in a traveling helical wave. In stationary paramecia the ciliary beat is conicoidal, with the form of a three dimensional spiral.

The wave form of ciliary motion of protozoa appears to be simply one modification of the wave form of flagellar motion. Flagella may pull by a tip-to-base undulation or push by a base-to-tip undulation. It was known that cilia and flagella are basically similar in their ultrastructure. Each is composed of eleven fibrils extending the entire length of the cilium or flagellum. Two of the fibrils are central and the remaining nine are doublets arranged peripherally around the central two (Fig. 5.15). Cilia and flagella are similar in their biochemical constituents and in the nature of their movement. The beating of cilia and flagella results from the contraction of these fibrils but the nature of the process is unknown. At the base of each cilium is a basal body or kinetosome, similar in ultrastructure to the centriole.

The movements of paramecia indicate that they are capable of highly coordinated activity. When a paramecium strikes a solid object, the waves of ciliary beating reverse and the animal backs up a short distance. It turns slightly and goes ahead again, repeating the process if it strikes another obstruction. The rapidity with which the animal can change direction is astonishing.

Muscles. Motion in most animals is a function of the contraction of specialized muscle cells. The contractile material, **actomyosin,** is fundamentally similar in smooth, striated and cardiac muscles of vertebrates and in the muscles of invertebrates as well. Muscles that contract rapidly and briefly, such as the skeletal muscles of mammals, are striated, whereas those that contract slowly and remain contracted for a long time, such as those in the walls of the digestive tract or urinary bladder, are unstriated. This basic physiologic and histologic correlation is evident in the contractile cells of coelenterates, for those of jellyfish, which contract in twitches, have microscopic cross striations and those of sea anemones, which contract very slowly, are unstriated.

The muscles of animals all contract by basically similar biochemical and biophysical mechanisms, but the kind of movement resulting depends upon how the muscles are arranged and connected to each other and to the skeleton. Muscle contraction may result in deformation of the body or a part of it, in the pedal creeping of the gastropod, the peristaltic movement of the gut, or the undulatory motion of a swimming fish. The muscles may surround a cavity with a small aperture like the mantle cavity of the squid and their contraction may result in jet propulsion. Alternatively, muscles may be connected to bones or skeletal parts and their contraction, by a system of levers, moves one bone relative to another and results in locomotion.

Mollusks generally have slow, nonstriated muscles, but the scallop, which can swim actively by clapping its two shells together, has two muscles connecting the shells. One of these is nonstriated and contracts slowly, serving to keep the shells closed at rest, and the other is striated and twitches rapidly to power the swimming movements.

The arthropods have complex patterns of separate muscles rather than simple layers of muscles as in the worms. These muscles vary in size and attachment and provide for the movement of the segments of the body and their many-jointed appendages. A lobster or grasshopper has hundreds of separate muscles located within the exoskeleton and attached to its inner surface.

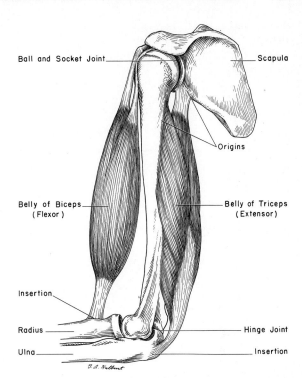

Ball and Socket Joint

Scapula

Origins

Belly of Biceps
(Flexor)

Belly of Triceps
(Extensor)

Insertion

Radius

Hinge Joint

Ulna

Insertion

Figure 5.16 Muscles and bones of the upper arm, showing origin, insertion and belly of a muscle, and the antagonistic arrangement of the biceps and triceps.

The muscles of vertebrates are generally attached to bones or cartilages as pairs which tend to pull in opposite directions (Fig. 5.16). Since muscles can pull but cannot push, this antagonistic arrangement allows for movement in both directions. The end of the muscle which remains relatively fixed when a muscle contracts is known as its **origin;** the end which moves is called the **insertion;** and the thick part between the two is called the **belly** of the muscle. Thus, the **biceps,** which bends or flexes the forearm, has its origin on the scapula and on the upper end of the humerus, and its insertion on the radius in the forearm. Its antagonist, the **triceps,** which straightens or extends the forearm, has its origin on the scapula and upper part of the humerus and its insertion on the ulna. The contraction of a muscle is stimulated by a nerve impulse reaching it via a motor nerve fiber from the central nervous system. The drug, **curare,** the chief ingredient of the arrow poison used by the South American Indians, blocks the junction between nerve and muscle so that impulses cannot pass and the muscle is paralyzed. A curare-paralyzed muscle can still be caused to contract by direct electric stimulation, a demonstration that muscle is independently irritable.

5.8 THE MECHANISM OF MUSCULAR CONTRACTION

The functional unit of vertebrate muscles, the **motor unit,** consists of a single motor neuron and the group of muscle cells innervated by its axon, all of which will contract when an impulse travels down the motor neuron. Each human has about 250 million muscle cells but only about 420,000 motor neurons in spinal nerves. Obviously, some motor neurons must innervate more than one muscle fiber. The degree of fine control of a muscle is inversely proportional to the number of muscle fibers in the motor unit. The muscles of the eyeball, for example, have as few as three to six fibers per motor unit, whereas the leg muscles have perhaps 650 fibers per unit.

If a single motor unit is isolated and stimulated with brief electric shocks of increasing intensity, beginning with shocks too weak to cause contraction, there will be no response until a certain intensity is reached, then the response is maximal. This phenomenon is known as the "all-or-none effect." In contrast, a whole muscle, made of many individual motor units, can respond in a graded fashion depending upon the number of motor units which are contracting

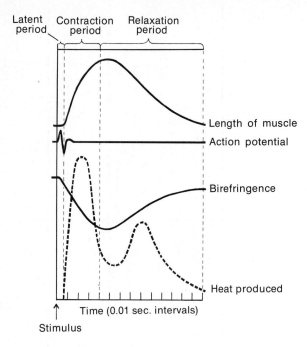

Latent period
Contraction period
Relaxation period

Length of muscle
Action potential
Birefringence
Heat produced

Time (0.01 sec. intervals)
↑ Stimulus

Figure 5.17 Diagram of the changes that occur in a muscle during a single muscle twitch. See text for discussion.

at any given time. Although an entire muscle cannot contract maximally, a single motor unit can contract *only* maximally or not at all. The strength of contraction of an entire muscle depends on what fraction of its constituent motor units are contracting and whether they are contracting simultaneously or alternately.

A muscle given a single stimulus, a single electric shock, responds with a single quick twitch. The changes which accompany a **single twitch** are shown in Figure 5.17. A twitch consists of (1) a very short **latent period,** the interval between the application of the stimulus and the beginning of the contraction; (2) a **contraction period,** during which the muscle shortens and does work; and (3) a **relaxation period,** longest of the three, during which the muscle returns to its original length. The latent period represents the interval between the conduction of the action current and the completion of the changes in the structure of the actomyosin which enables it to contract. The first event after the stimulation of a muscle is the initiation and propagation of an electrical response, the muscle **action potential,** followed by the changes in the structure of actomyosin observed as a change in the total birefringence of the muscle, by its shortening, and by the production of heat. Following a twitch there is

a **recovery period** during which the muscle is restored to its original condition. If a muscle is stimulated repeatedly at intervals short enough so that succeeding contractions occur before the muscle has fully recovered from the previous one, the muscle becomes fatigued and the twitches become feebler and finally cease. The fatigued muscle will regain its ability to contract if it is allowed to rest.

Muscles do not usually contract in individual twitches but in more sustained contractions evoked by a volley of nerve impulses reaching them in rapid succession. This state of sustained contraction is known as **tetanus;** the individual motor units are stimulated in rotation. Thus individual muscle fibers contract and relax, but do this in rotation so that the muscle as a whole remains partially contracted. The strength of the contraction depends on the fraction of the muscle fibers which contract at any given moment.

All normal skeletal muscles are in a state of sustained partial contraction, called **tonus,** as long as the nerves to the muscle are intact. Tonus, then, is a state of mild tetanus, maintained by a constant flow of nerve impulses to the muscle. A muscle under slight tension can react more rapidly and contract more strongly than one that is completely relaxed.

The problem of how a muscle can exert a

pull is far from solved, but it is now believed that the molecules of actin and myosin shorten by the sliding together of the two components. The energy for contraction is derived from the energy-rich phosphate bonds of adenosine triphosphate and phosphocreatine. These are renewed by the energy derived from the glycolysis of glycogen to lactic acid. This latter process can occur without utilizing oxygen and provides energy for the resynthesis of adenosine triphosphate and phosphocreatine.

Capturing food or evading enemies may call for prolonged bursts of muscular activity. Although both the rate of breathing and the rate of the heartbeat may increase markedly during prolonged exertion, these changes could not supply the muscles with enough oxygen to enable them to contract repeatedly if the contraction process itself required oxygen. That muscle contraction, and part of the recovery process, occur without the utilization of oxygen is clearly important for survival. During violent exercise glycogen is converted to lactic acid faster than the lactic acid can be oxidized. Lactic acid accumulates and the muscle is said to have incurred an "oxygen debt," which is repaid after the period of exertion by continued rapid breathing. This supplies enough extra oxygen to oxidize part of the accumulated lactic acid. Some of the energy released by the oxidation of the lactic acid in the Krebs cycle and the electron trans-

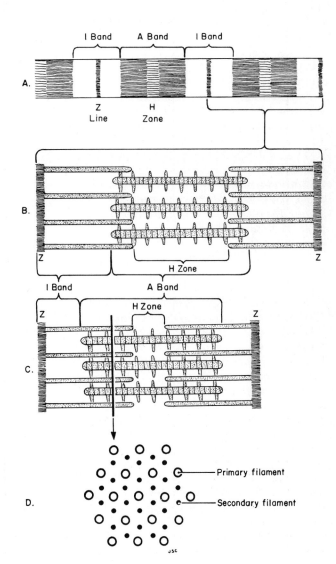

Figure 5.18 Diagrams illustrating the sliding filament hypothesis of the mechanism of muscle contraction. A, Diagram of part of a single myofibril showing the pattern of light (I) and dark (A) bands. B, Longitudinal view of the arrangement of thick and thin filaments within a myofibril in the relaxed state. C, Longitudinal view of the arrangement of thick and thin filaments in a contracted myofibril, showing that the I band decreases in thickness. Note that the two types of filaments appear to slide past one another during contraction. D, Transverse view through C at arrow, showing each thick primary filament surrounded by six thinner secondary filaments. (Adapted from H. E. Huxley.)

mitter system (Fig. 4.2) is used to resynthesize glycogen from the remainder of the lactic acid and to restore the energy-rich compounds, ATP and phosphocreatine, to their normal condition. A muscle that has contracted many times has depleted its stores of energy-rich phosphates and glycogen and has accumulated substantial amounts of lactic acid; it is unable to contract again and is said to be **fatigued.** A fatigued muscle will contract if stimulated directly and the nerve to the muscle can conduct impulses. The point of fatigue is the junction between nerve and muscle where the nerve impulse instigates muscle contraction.

Electron micrographs reveal that each muscle fiber is made of two kinds of longitudinal filaments, thicker **primary filaments** of myosin and thinner **secondary filaments** of actin. The alternating light and dark bands evident in the light microscope (see Fig. 3.18) consist of dense A bands and light I bands (Fig. 5.18). Each unit consists of an A band and half of each adjacent I band; it is separated from the next unit by a thin dense Z line through the middle of the I band. The central portion of the A band is the somewhat less dense H zone. The thick primary filaments occur only in the A band. The thinner secondary filaments are located in the I band but extend into the A band for some distance and interdigitate with the primary filaments. The secondary filaments appear to be smooth but the primary filaments appear to have minute spines every 60 to 70 Å along their length which project toward the adjacent secondary filament (Fig. 5.18B). These spines look like bridges connecting the two sets of filaments.

During contraction the length of the A band remains constant but the I band shortens and the length of the H zone within the A band decreases. Huxley and others have proposed that during contraction the filaments maintain their same length but the primary and secondary filaments slide past each other. During contraction the thin secondary filaments of actin are believed to extend farther into the A band, decreasing its central H zone and narrowing the I bands as the ends of the primary myosin filaments approach the Z band (Fig. 5.18C). The physicochemical mechanism by which this sliding may occur is not yet clear; perhaps the spiny bridges are broken and then re-formed farther along the filament. The energy of the ~P may be utilized in forming new bridges, new cross-linkages, between the primary and secondary filaments. If we assume that the energy of one ~P is required for the formation of each new crossbridge we can estimate, from the rate of contraction and the total number of crossbridges per fiber, how much ATP would be utilized per second by the fiber. The resultant figure is of the same order of magnitude as the experimentally determined rate of ATP utilization.

It was noted in Figure 5.17 that an action potential was associated with muscle contraction. In general, muscles are arranged with their fibers in parallel, so that the voltage difference in a large muscle is no greater than that of a single fiber. In the **electric organ** of the electric eel, however, the electric plates are modified muscle cells (motor end plates) arranged in series. Although each plate has a potential difference of about 0.1 volt, the discharge of the entire organ, made of several thousand plates, amounts to several hundred volts.

5.9   IRRITABILITY AND RESPONSE

The muscles just described, together with cilia, glands, nematocysts and so on, are **effectors** — they do things. To ensure that these effectors do the right things at the right time, animals are equipped with **receptors** — a variety of sense organs — and with nervous and endocrine systems to coordinate the activity of the effectors.

Irritability or excitability is a fundamental property of all kinds of cells. Waves of excitation are conducted, although very slowly, by eggs and by plant cells. Many of the ciliates have a network of **neurofibrils** which connect the bases of the cilia together with special fibrils to the gullet and other special structures of the body. It would appear that this net conducts impulses which coordinate the beating of the cilia and the functioning of the special organelles, for coordination is lost when the net is cut by a microneedle. There are no nerve cells in sponges, but waves of excitation can be conducted from cell to cell at about 1 cm. per minute. There are spindle-shaped contractile cells around the openings of the pores. These have been termed "independent effectors" because they respond to touch by contracting and thus combine sensory and motor functions.

The simplest special coordinating system is the **nerve net** found in coelenterates. The coelenterate nerve fibers are found all over the body in a diffuse network; a few sea anemones and medusae have rudimentary nerve trunks composed of aggregations of nerve fibers. Conduction in the nerve net progresses in all directions; the fibers are not actually fused together, but impulses pass from one fiber to an adjacent one in either direction.

5.10 THE NERVE IMPULSE

Galvani, in the eighteenth century, first showed that a muscle contracts when an electric shock is applied to the nerve leading to it. DuBois-Reymond in the nineteenth century showed that when a stimulus is applied to a sense organ electrical disturbances can be detected in the efferent nerves. With the development of improved instruments for detecting these weak currents, the electrical disturbances in nerve fibers were shown to have a potential of about 0.05 volt, to last for a very short time, about 0.0005 second, and to travel along the nerve at speeds as great as 120 meters per second.

The transmission of a nerve impulse is not simply an electrical phenomenon, like the passage of a current in a wire. It is a physicochemical process, which uses oxygen and produces carbon dioxide and heat. The transmission of a nerve impulse obeys the "all-or-none law": The conduction of the impulse is independent of the nature or strength of the stimulus starting it, provided that the stimulus is strong enough to start any impulse. The energy for the conduction of the impulse comes from the nerve, not from the stimulus, so that, although the speed of the conducted impulse is independent of the strength of the stimulus, it is affected by the state of the nerve fiber. Drugs or low temperature can retard or prevent the transmission of an impulse. The impulses transmitted by all types of neurons are believed to be essentially alike. That one impulse results in a sensation of light, another in a sensation of pain,

and a third in the contraction of a muscle is a function of the way the nerve fibers are connected and not of any special property of the impulses.

Our present knowledge of the nature of the nerve impulse has been derived in large part from experiments using the large axons found in squid, crayfish and certain worms. The giant axon of the squid is large enough, nearly 1 mm. in diameter, so that investigators can introduce microelectrodes and micropipettes into the substance of the nerve fiber and measure the electrical potential across the nerve membrane. According to the present **membrane theory** of nerve conduction, the electrical events in the nerve fiber are governed by the **differential permeability** of the plasma membrane of the neuron to sodium and potassium ions, and these permeabilities, in turn, are regulated by the **electric field** across the surface. The interaction of these two factors, differential permeability and electric field, leads to the requirement for a certain critical threshold of change for excitation to occur. Excitation is a regenerative release of electrical energy from the nerve membrane and the propagation of this change along the fiber is a brief, all-or-none electrical impulse called the **action potential.**

To understand what happens when a nerve impulse passes along the fiber, we must first have a clear picture of the state of the resting nerve fiber. The nerve fiber is, in essence, a long cylindrical tube whose surface membrane separates two solutions of different chemical composition, though they have the same total number of ions. In the external medium sodium and chloride ions predominate, whereas within the cell potassium and various organic anions predominate. The concentration of sodium in the external medium is some 10 times greater than that within the cell and the concentration of potassium in the internal medium is some 30 times greater than that in the surrounding medium. The neuronal membrane is about 50 Å thick, has high electrical resistance, low selective ionic permeability and high electrical capacity.

The concentrations of Na^+ and K^+ are kept relatively constant by a "sodium pump," by the active extrusion of sodium ions from the interior of the cell. The extrusion of sodium ions is accompanied by the entrance of potassium ions and there is a steady flux of ions in and out of the cell. As a result of this differential distribution of ions on the two sides of the membrane, there is a potential difference of 60 to 90 mV (the "resting" **membrane potential**) across the membrane, with the interior being negatively charged with respect to the outside (Fig. 5.19). The extrusion of sodium ions from the cell is an energy-requiring reaction, the energy for which is derived from metabolic processes within the nerve. When the nerve fiber is poisoned with a substance such as cyanide, the sodium pump is turned off.

The cell membrane has a relatively low ionic permeability so that, even when the sodium pump is turned off by a metabolic poison, it takes many hours before the concentration gradients of sodium and potassium across the membrane disappear. The cell membrane is differentially permeable to the two ions; it is much more permeable to potassium than to sodium. In the normal resting state of the nerve fiber,

there is an excess of positively charged ions on the outside of the membrane (Fig. 5.19).

Electrical studies of the cable properties of the nerve fiber have shown that the axon could hardly serve as a passive transmission line because its cable losses are enormous. When a weak signal is applied to the fiber, one too small to excite the usual relay mechanism of the fiber, it fades out within a few millimeters of its origin. The nerve impulse could not be propagated over the long distances in the nerve unless there were some process to boost the signal. This is the function of the excitatory process, which regenerates and reamplifies the signal all along the nerve fiber. The cable properties of the nerve fiber allow a change in electrical potential to spread along the nerve fiber for a short distance (even though it is rapidly attenuated) and thus stimulate the excitatory process in the adjacent portion of the nerve.

The excitation of a nerve, the generation of a nerve impulse, involves a momentary change in the permeability of the nerve membrane which permits sodium ions to enter. The entrance of the sodium ion leads to a **depolarization** of the nerve membrane; it becomes positively charged on the inside and negatively charged on the outside. Although the permeability of the membrane to sodium ions is very low at the usual resting potential, the permeability to sodium increases as the membrane potential decreases. This permits a leakage of sodium ions down the concentration gradient into the interior of the cell. This further decreases the membrane potential and further increases the permeability to sodium. The process is self-reinforcing and results in the upward spike of the action potential wave (Fig. 5.19). When the interior of the fiber becomes positively charged with respect to the exterior, the further net influx of sodium ions is prevented. The potential difference across the membrane may momentarily reach 0.04 volt, with the inside of the fiber now positively charged with respect to the outside. This is followed by a period of increased permeability to potassium, and potassium ions pass along their concentration gradient from inside the cell to outside.

The return of the membrane potential to its original state of negative inside and positive outside is not brought about by a

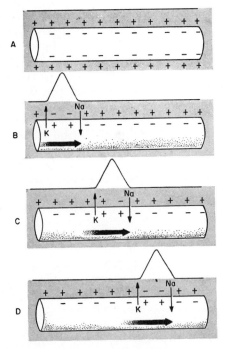

Figure 5.19 Diagram illustrating the membrane theory of nerve transmission. *A,* Resting nerve, showing the polarization of the membrane with positive charges on the outside and negative charges on the inside. *B–D,* Successive stages in the conduction of a nerve impulse, showing the wave of depolarization of the membrane and the accompanying action potential propagated along the nerve. (Villee: Biology, 6th ed.)

reversal of the movement of the ions. There is no expulsion of the amount of sodium ions that entered during the ascending phase of the action potential. Instead, an equivalent quantity of potassium ions leaks outward through the membrane. Both sodium and potassium ions are moving down their respective concentration gradients during the passage of an impulse. The actual quantities of ions involved, however, are so small that there is no detectable change in the concentration of either ion in the fiber during one impulse. The giant axon of the squid could conduct several hundred thousand impulses, even if its sodium pump were turned off, before its store of accumulated ions would be exhausted.

The propagation of each impulse is followed by a period of inexcitability, the **absolute refractory period,** during which the fiber cannot transmit a second impulse. Because of the changes in permeability that accompany the depolarization of the nerve membrane, the fiber cannot respond to a second stimulus. Excitability returns when the normal permeability relations have been restored.

The nerve impulse is thus a wave of "depolarization" that passes along the nerve fiber. The change in membrane potential in one region makes the adjacent region more permeable and, in this way, the wave of depolarization is transmitted along the fiber. The entire cycle of depolarization and repolarization requires only a few thousandths of a second.

Experiments by the English physiologist Adrian, published in 1926, provided the explanation as to how the nervous system transmits differences in intensity. By applying graded stimuli to an isolated sense organ and amplifying and measuring the impulses in its nerve, Adrian showed that variations in the intensity of a stimulus lead to variations in the *frequency* with which impulses are transmitted: the stronger the stimulus, the more impulses per second. This principle is true for both vertebrate and invertebrate sensory nerves. In contrast, a single impulse in a vertebrate motor nerve elicits a single twitch of all the muscle fibers in the motor unit. The differences in the strength of contraction of the muscle as a whole are due to variations in the total number of motor units actively contracting at any given moment. In the motor nerves of invertebrates, however, the frequency of the impulses does affect the strength of contraction of the muscle innervated. A single nerve impulse will, in general, not stimulate the muscle to contract. At least two successive impulses are required, and the strength of contraction is inversely proportional to the interval between the two. In many arthropods all the muscle fibers in a given muscle are innervated by branches of a single nerve fiber (axon). A single impulse in the axon will not produce contraction but repeated impulses will; the tension in the muscle increases with the frequency of the stimulation. It would appear that, although the arrival of a single impulse at the nerve-muscle junction is unable to bring about muscle contraction, it does affect the junction in such a way as to make it possible for a second impulse to do this if it arrives soon enough after the first. This phenomenon is known as **facilitation.**

The speed of propagation of the nerve impulse varies considerably from one nerve to another, and even more from one animal to another. Conduction is, in general, more rapid in those neurons with greater diameters. A number of animals—squid, lobsters and earthworms—have special **giant axons** which conduct impulses many times faster than the adjacent small fibers. Conduction is more rapid in those nerves surrounded by a thick myelin sheath. The speed of conduction is greater in those nerves in which the myelin sheath is interrupted periodically by nodes of Ranvier and the nerve impulses "jump" from one node to the next.

5.11 TRANSMISSION AT THE SYNAPSE

Where the tip of the axon of one nerve comes close to the tip of the dendrite of the adjoining nerve is a region called the **synapse.** Transmission of impulses across the synapse from one neuron to the next is slower than transmission along a nerve fiber. Synaptic transmission may be either electrical or chemical. Electrical transmission implies that, despite the apparent morphologic separation of the two neurons, an effective local circuit connection exists and permits enough current to pass from one to the other to generate an action

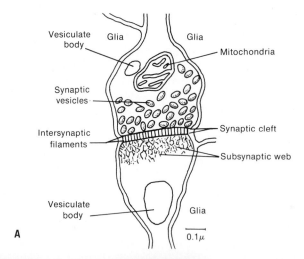

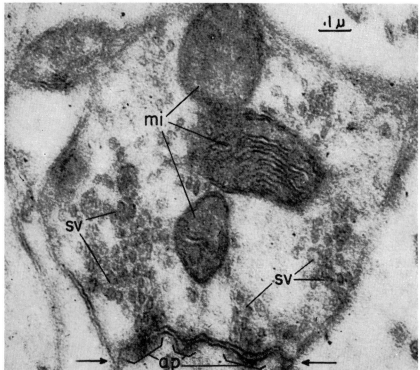

Figure 5.20 A, A common form of synapse in the mammalian brain, according to De Robertis (1962). The axonal (presynaptic) side is above; the dendritic (postsynaptic) side, below. B, Electron micrograph of a synaptic ending in the stimulated olfactory bulb of the rat. The synaptic ending contains three mitochondria (*mi*) and several synaptic vesicles (*sv*). The zone of contact between the two neurons is indicated by the two arrows. The nerve membranes appear to be thickened at "active points" (*ap*) in this zone of contact. (A from Brazier, M. A. B. (ed.): Brain and Behavior. Amer. Ass. Adv. Sci., 1962; B from De Robertis, E., and Pellegrino De Iraldi, A.: Anat. Rec., *139*:299, 1961).

potential. At most synapses, the two neuronal membranes are separated by a gap of some 200 Å and the impulse is transmitted across this gap by a special chemical transmitter. A specific chemical is synthesized by the neuron and released from the tip of the axon (**neurosecretion**) when a nerve impulse reaches it. This diffuses across the synaptic gap and attaches to specific molecular sites in the dendrite (an example of **chemoreception**), producing a change in the properties of the membrane of the dendrite and leading to the initiation of a new nerve impulse.

When a nerve impulse reaches the tip of certain vertebrate nerves it stimulates the secretion of packets of acetylcholine from synaptic vesicles at the tip of the axon (Fig. 5.20). This diffuses across the synaptic junction and attaches to a special chemoreceptor on the surface of the dendrite of the adjacent neuron. The combination of the chemical with the chemoreceptor leads to a depolarization of the membrane and sets up a new action potential which passes along that neuron. Tissues contain a powerful **cholinesterase,** an enzyme which specifically splits acetylcholine to its constituents, which are inactive, and thus the continued stimulation of the adjacent neuron is prevented. These synaptic vesicles provide a satisfactory explanation for the polarity of the synapse, for the fact that a nerve impulse will travel in one direction but not in the reverse between axon and dendrite.

The mechanism of synaptic transmission in other types of nerves is the subject of controversy. There is evidence that acetylcholine plays some role, perhaps the major one, in synaptic conduction in the central nervous system of vertebrates and certain invertebrates. Sympathin, a chemical similar to or identical with the hormone epinephrine, is the chemical mediator across certain synapses in the sympathetic nervous system. Synaptic transmission is greatly affected by the concentration of cations such as potassium and calcium, and these ions may play some direct role in transmission.

Much evidence has accumulated to show that chemical transmission at the synapse is a general phenomenon, and some investigators had argued that all synaptic transmission is chemical. However, the experiments of Furshpan and Potter in 1957 showed that transmission across the giant synapse in an abdominal ganglion of the nerve cord of the crayfish is by electrical means. The membranes at this special synapse are able to act as a rectifier and allow current to pass easily in one direction, from the axon of an interneuron to the dendrite of a motor neuron.

Synapses are important functionally because they are points at which the flow of impulses through the nervous system is regulated. Not every impulse reaching a synapse is transmitted to the next neuron. Information transmitted along the axon as a spike potential is coded and decoded at the synapse by processes involving graded thresholds and the facilitation and summation of inhibitory and excitatory impulses. The synapses, by regulating the route of nerve impulses through the nervous system, determine the response of the organism to specific stimuli.

The important details of the arrangement of the neurons to form the central nervous systems of the higher invertebrates and of the vertebrates will be discussed in later chapters. The invertebrate nervous system consists of one or more pairs of **ganglia**— collections of nerve cell bodies—at the anterior end of the body and one or more nerve cords extending posteriorly. The invertebrate nerve cord is solid and is typically located on the ventral side of the body; the vertebrate nerve cord is single, hollow, and located on the dorsal side of the body.

5.12 SENSE ORGANS

Nerve fibers can be stimulated directly by a variety of treatments—by electric shocks, by the application of chemicals or by mechanical cutting or crushing. In the intact organism, of course, sensory nerve fibers are activated by the **sense organs** to which they are connected.

The sense organs have dual functions: they *detect* environmental changes and *transmit* information concerning the change to the central nervous systems. In some sense organs the receptor cell is a primary neuron and it both detects and transmits. In other sense organs the receptor is an epithelial cell which detects, but the information is transmitted by an associated neuron. The sense organ receives a small amount of energy from the environment. Each is specialized to receive one particular form of energy very efficiently; it possesses some degree of intrinsic specificity. In addition to specificity and optimal sensitivity a sense organ must have a capacity for discrimination and for recording not merely "on" and "off" but also the rate, magnitude and direction of the change in the environment.

The unstimulated receptor neuron maintains a steady resting potential due to the difference in the ionic composition of the fluids on the two sides of the membrane. The stimulus changes the membrane's permeability to ions so that the potential difference either decreases (the cell is **depolarized**) or increases (the cell is **hyperpolarized**). The altered state of polarization, the **receptor potential,** spreads slowly down the dendrite, decaying exponentially. At special areas of the neuron, e.g., the axon hillock, depolarization leads to the generation of action potentials which travel along the neuron to the central nervous system.

Sense organs may be classified according to the type of stimulus to which they are sensitive. We can distinguish (1) **chemoreceptors**—smell and taste; (2) **mechanoreceptors**—touch, pressure, hearing and balance; (3) **photoreceptors**—sight; (4) **thermoreceptors**—heat and cold; and (5) undifferentiated nerve endings which serve the pain sense. Sense organs may also be classified by the location of the stimulus: thus **exteroceptors** supply information about the surface of the body (touch, pressure, taste, heat, cold); **proprioceptors** supply information about the position of the body (stretch receptors in muscles and joints, equilibrium organs which sense orientation in the field of gravity); **distance receptors** report on objects away from the body (sight, smell and hearing), and **interoceptors** provide sensations of pain, fullness, and so on, from internal organs.

When a sense organ is stimulated continuously it may either give off a continuous stream of nerve impulses or it may quickly cease responding to the stimulus. The proprioceptors of the body are generally of the first type, nonadaptive, whereas the exteroceptors are generally **adaptive,** and soon become nonresponsive to a continuing stimulus. The advantage of sense organ adaptation is clear: it prevents a continual train of nerve impulses impinging on the brain from all the body's sense organs, yet does not interfere with the body's responding to changes in the pattern of stimuli which are likely to be important for survival.

The actual excitation of the sensitive cells of the sense organ is either via mechanical stress, via chemical stimulation by contact of the molecules of some substance from the environment, or via some chemical process induced in the sense cell by the stimulus. An example of the latter is the chemical reaction induced by light falling on the sensitive cells of the retina of the eye.

The function of a sense organ in animals other than man can be deduced from its morphology and nerve connections. It can be investigated by connecting the efferent nerve to an amplifier and oscilloscope, applying stimuli to the sense organ, and measuring the resulting nerve impulses. It can also be investigated at the behavioral level, by training the animal to associate one situation with a given stimulus and a second situation with a different stimulus, and then observing its ability to distinguish between the first and second stimuli as they are gradually changed to resemble each other.

Chemoreceptors. Our own senses of taste and smell can be distinguished, for the taste buds are organs in the lining of the mouth which respond to substances in watery solution, whereas the olfactory epithelium is in the lining of the nose and responds to substances which enter as gases. In most lower animals, the distinction between taste and smell is blurred, for chemoreceptors are found over much of the surface of the head and part of the body in fish, and insects have chemoreceptors in their feet. Chemoreceptors are sensitive to remarkably small amounts of certain chemicals. Most people can detect ionone, synthetic violet odor, at a concentration of one part in 30 billion parts of air. Certain male insects can detect the pheromone given off by the female of the species over a distance of 3 km. (p. 402). Several thousand different odors can be recognized by man, but there is no clear correlation between the chemical composition of a substance and its smell.

Chemoreceptors are probably the most primitive of the distance receptors, and many kinds of animals depend solely upon them for finding food, avoiding predators and meeting mates.

Mechanoreceptors.   Mechanoreceptors are sensitive to stretch, compression or torque imparted to the cells by the weight of the body, the movement of parts one on another, the pressure of the substratum or the surrounding medium or by objects in the surroundings. Mechanoreceptors provide organisms with information about the primary attitude of the body with respect to gravity, the position of one part of the body with respect to another, and other information needed for locomotion and co-ordinated movements. Mechanoreceptors also provide information about the shape, weight, texture and topographical relationships of objects in the external environment.

The skin of man and other mammals contains several kinds of tactile mechanoreceptors. By making a survey of a small area of skin, point by point, and testing for regions sensitive to touch, pressure, temperature and pain, it has been found that receptors for each of these sensations are located in different spots. Then, by comparing the distribution of the types of sense organs and the types of sensations, it has been possible to identify the sense organ for each stimulus. In lower animals the sensory organs are less differentiated and the identification of a particular nerve ending with a given sensitivity is usually impossible.

The sense cells at the base of the bristles of insects are clearly mechanoreceptors, and indeed it has been possible to record impulses in the efferent nerves when the bristle is moved.

The mammalian ear is a remarkably complex organ which contains the senses of **hearing** and **equilibrium.** It can detect the direction of the force of gravity or of linear acceleration because it contains **otoliths,** masses of calcium carbonate, attached to slender processes of cells in such a way that the weight of the otolith will pull or push on these processes. Motion of the head about any of its axes is detected by the motion of the fluid in the semicircular canals, which moves clumps of hair-like processes attached to mechanoreceptors in the walls of the canals.

Organs of balance, called otocysts or **statocysts,** are found in most phyla of animals, even in coelenterates. These are usually hollow spheres of sense cells, in the middle of which is a statolith, a particle of sand or calcium carbonate, pressed by gravity against certain mechanoreceptors. As the animal's body changes position, the statolith is pressed against different sense cells and the animal is then stimulated to regain its orientation with respect to gravity.

The detection and analysis of sound waves involves the conversion of the sound waves to mechanical vibrations of the ear drum and middle ear bones, and then to waves of motion in the liquid filling the cochlea of the inner ear. The cochlea contains many sense cells with fibers of differing lengths which respond to sounds of different frequencies. The ear is basically a mechanoreceptor responding to the mechanical displacement of sense cells, or their fibers or hairs, produced by sound waves or by changes in position.

Many arthropods, especially insects, have sense organs which respond to sound waves; these organs consist of a fine membrane stretched in such a way that it is free to respond to the vibrations of sound waves. The nerve from the sound-sensitive organ of the locust has been tapped, and recordings of the nerve impulses from it show that it can respond to sound waves between 500 and 10,000 cycles per second. The human ear responds to frequencies between 20 and 20,000 c.p.s., dogs are sensitive to sounds as high as 40,000 c.p.s., and the sensitivity of the bat ear extends to high-pitched 80,000 c.p.s. noises.

Certain insects have balance organs which have evolved from the second pair of wings. These club-shaped structures, called **halteres,** beat up and down as the wings do and serve as "gyroscopes." When the direction of the beat is changed, sense organs in the base of the haltere are stimulated and give off nerve impulses. This has been shown by recording the nerve impulses passing through the nerves from the halteres.

Photoreceptors.   Almost all animals are sensitive to light and respond to variations in light intensity. Even protozoa which have no special light-sensitive organ show a generalized ability to respond to light. Many higher animals—usually the burrowing ones—have no recognizable "eyes" but have a general sensitivity to light over all or a large part of the body. Clams, for example, respond to sudden changes in light intensity by drawing in their siphon, and earthworms withdraw into their burrows when the light intensity is increased.

Most animals, even coelenterates, have some sort of specialized structure for the perception of light. A simple invertebrate eye usually consists of a cup-shaped layer of pigment cells which screen the light-sensitive cells from light coming from all directions but one. Light-sensitive cells are embedded between these pigment cells.

The cephalopods—the octopus, squid, and relatives—alone among the invertebrates have well-developed **camera eyes** which are superficially similar to vertebrate eyes, with retina, lens, iris, cornea and a mechanism for focusing for near and far vision. Although it is difficult to determine how well an octopus can see, we can infer from the structure of the eye that it should be the functional equivalent of the vertebrate eye.

The eyes of arthropods—insects and crabs—are **mosaic eyes,** composed of many visual units, perhaps thousands, called **ommatidia.** Each ommatidium has a clear outer cornea, under which is a lens which focuses the light on the end of the light-sensitive element made of eight or so retinal cells. These are believed to respond as a unit. Each ommatidium is separated from the adjacent ones by rings of pigment cells, so that it is a tube with light-sensitive elements at the base which can be reached only by light parallel to the axis of the tube. A mosaic eye presumably forms a very poor image composed of a series of rather large dots like a poor newspaper photograph. But a mosaic eye has a high flicker fusion rate and can detect flickers up to 265 per second. Because of this the insect eye is particularly sensitive to the motion of objects in its surroundings, for any movement would change the amount of light falling on one or more of the ommatidia.

Thermoreceptors.   Temperature-sensitive cells are found in a wide variety of animals, from the lowest to the highest. Ciliates such as paramecia will avoid warm or cold water and will collect in a region where the temperature is moderate. Some insects have thermoreceptors, either in the antennae or all over the body. Insects that suck blood from warm-blooded animals are attracted to their prey by the temperature gradients nearby. This has been shown experimentally, for blood-sucking bugs are much less able to find their prey after their antennae have been removed. Fish apparently have fairly sensitive thermoreceptors, for a change of only 0.5° C. will change the behavior of sharks and bony fish. Certain snakes have sensitive thermoreceptors on the sides of their heads by which they can detect the presence and location of warm-blooded prey.

All nerve impulses are qualitatively similar. The impulse set up by the ringing of a bell is exactly like the impulse initiated by the pressure of a pin against the skin, or the impulse in the optic nerve which results from light falling on the retina. The qualitative differentiation of stimuli must depend upon the pattern of connections between sense organ and brain. The ability to distinguish red from green, hot from cold, or red from cold is due to the fact that particular sense organs and their individual sensitive cells are connected to particular parts of the brain.

5.13 COORDINATION AND INTEGRATION

The activities of the several parts of a many-celled organism must be coordinated if that organism is to survive, and the greater the degree of complexity, the greater the specialization of the parts, the greater is the need for precise integration of their separate functions. Coordination of activity is achieved by two major systems, nervous and endocrine. The nerves and sense organs provide for rapid and precise adaptation to environmental factors. The **endocrine system,** the glands of internal secretion which secrete substances into the blood stream (or its equivalent in lower animals), provides for less rapid but longer lasting adaptations, such as general body growth, differentiation, development of sex organs and mating behavior, responses to stress, control of tissue metabolism and regulation of pigmentation. The reflexes and other nervous mechanisms by which coordination and integration are achieved will be discussed in Chapter 16.

The substances secreted by endocrine

glands, called **hormones,** cannot be defined as belonging to any particular class of chemicals; some are proteins, some are amino acids and some are steroids. Hormones may be defined as substances secreted by cells in one part of the body which are carried by the blood stream to some other part where they affect cell activities in a definite and characteristic fashion. Acetylcholine and sympathin fit this definition of a hormone and are sometimes referred to as **neurohormones** to emphasize this. Whether a hormone will affect a specific tissue, and the nature of the effect produced, is a function of the tissue; each tissue will respond only to certain hormones. Hormones produced in one animal will usually affect the cells of other animals in related species, orders and even, in some cases, classes. The endocrine glands of the vertebrates will be discussed in Chapter 17.

The processes under endocrine control in invertebrates include molting, pupation and metamorphosis in arthropods, pigmentation in mollusks and arthropods, and growth and differentiation of secondary sex characteristics in annelids and arthropods. The development of insects, by a series of molts and metamorphoses, is controlled by two hormones, ecdysone and juvenile hormone. Ecdysone is secreted by the **prothoracic glands** which, in turn, are under the control of a prothoracicotropic hormone secreted by the intercerebral gland in the brain. It induces molting accompanied by metamorphosis. Juvenile hormone is secreted by the corpora allata, paired glands in the posterior head region; it permits molting but inhibits metamorphosis. Transplantation of corpora allata into developing insects prevents metamorphosis for several successive molts so that giant adults eventually result.

The leaves of yew trees and certain weeds have been found to produce substances with ecdysone-like activity. It has been postulated that the capacity to secrete these substances has evolved as a protection to the plants against moths; moth larvae eating the leaves and ingesting these substances undergo premature molting and die. Other trees secrete substances with juvenile hormone-like activity. These, by preventing the larvae from becoming adults and reproducing, might protect the trees against certain insects. Field tests in which foliage has been sprayed with juvenile hormones give promise that these might be effective and relatively specific insecticides.

The molting of crabs and other crustaceans is a complex process involving many biochemical processes which must occur in proper sequence. This is under the control of at least two hormones, one secreted by the sinus glands in the eyestalk and the other secreted by the paired Y organs located beneath the external adductor muscles of the mandibles. The sinus glands are the expanded tips of neurosecretory cells whose nuclei constitute the X organ located on the surface of the brain. The hormones are actually produced in the X organ and released in the sinus gland. The removal of the eyestalk results in premature molting and in more frequent successive molts. If sinus glands from other crabs are transplanted to crabs without eyestalks, molting is delayed. Thus, the hormone of the X organ inhibits and delays molting. The hormone of the Y organ triggers the first stage of molting and removal of the Y organ prevents further molting.

The development of secondary sex characteristics in members of many invertebrate phyla is under hormonal control. When the gonads are removed surgically, or destroyed by parasites, the sex characteristics either fail to form or regress if present initially. The sinus gland of crustaceans and the corpora allata of insects secrete hormones which regulate the activity of the ovaries and thus are analogous to the gonadotropic hormones secreted by the vertebrate pituitary gland. The sinus gland hormone produces an increase in the concentration of blood sugar when injected into crabs, providing another interesting parallel between the secretions of the sinus gland and those of the vertebrate pituitary.

Hormones play a role in determining **pigmentation** in the octopus, squid, crabs, insects, fish, amphibia and reptiles. In most animals, color changes are produced by streaming movements of the pigment-laden cytoplasm of the color cells (**chromatophores**). The chromatophores of the cephalopod have smooth muscle fibers attached in such a way that their contraction spreads out the pigment-containing cytoplasm. Crustaceans can be separated into two major groups, those that darken and those that lighten in color when the eyestalk is removed. Injection of eyestalk extracts has diametrically opposite effects in the two types because of basic differences in the responses of the chromatophores. More recent experiments have shown that there are at least three different chromatophore-regulating hormones in crustaceans. Several endocrine organs are closely associated with the nervous system and undoubtedly evolved from such tissue; others evolved independently of the nervous system. It would seem useless to try to argue which is the more "primitive" coordinating system — nervous or endocrine. Both had their earliest traces in very primitive, single-celled animals and each type evolved independently of the other to its present state.

ANNOTATED REFERENCES

Baldwin, E. B.: An Introduction to Comparative Biochemistry. 4th ed. New York, Cambridge University Press, 1964. A fascinating presentation of principles and facts relative to the various physiologic and biochemical adaptations evolved by animals to enable them to survive.

Florey, E.: General and Comparative Animal Physiology. Philadelphia, W. B. Saunders Co., 1966. An excellent treatise of the functional aspects of a wide range of vertebrates and invertebrates.

Prosser, C. L. (Ed.): Comparative Animal Physiology. 3rd ed. Philadelphia, W. B. Saunders Co., 1973. Contains a wealth of information about the physiologic adaptations of vertebrates and invertebrates.

Schmidt-Nielsen, K.: Animal Physiology. 2nd ed. Englewood Cliffs, N.J., Prentice-Hall, Inc., 1964. A brief treatment in paperback form that covers the essentials of physiologic processes in animals.

Popular, relatively nontechnical accounts of many of the subjects discussed in this chapter are among the Scientific American articles published as offprints by the Wm. Freeman Co., San Francisco.

6. Reproduction

The processes needed for the day-to-day survival of the organism—nutrition, respiration, excretion, coordination, and the rest—were discussed in the preceding chapter. The survival of the species as a whole requires that its individual members multiply, that they produce new individuals to replace the ones killed by predators, parasites or old age. One of the fundamental tenets of biology, "omne vivum ex vivo" (all life comes only from living things), is an expression of this basic characteristic of all living things, their ability to reproduce their kind.

For centuries men believed that certain animals could arise from nonliving material by "spontaneous generation." For example, maggots and flies were thought to originate from dead animals, and frogs and rats to come from river mud. The classic experiments performed by Francesco Redi about 1670 disproved the theory of spontaneous generation. By the simple expedient of placing a piece of meat in each of three jars, leaving one uncovered, covering the second with fine gauze and the third with parchment, Redi demonstrated that although all three pieces of meat decayed, maggots appeared only on the uncovered meat. Maggots do not come from decaying meat, but hatch from eggs laid on the meat by blowflies. With the development of lenses and microscopes, and the subsequent increase in knowledge of eggs and larval forms, we now know that no animal arises by spontaneous generation.

The process of reproduction varies tremendously from one kind of animal to another, but we can distinguish two basic types: asexual and sexual. In **asexual reproduction** a single parent splits, buds or fragments to give rise to two or more offspring which have hereditary traits identical with those of the parent. **Sexual reproduction** involves two individuals; each supplies a specialized reproductive cell, a **gamete.** The male gamete, the **sperm,** subsequently fuses with the female gamete, the **egg,** to form the **zygote** or fertilized egg. The egg is typically large and nonmotile with a store of nutrients to support the development of the embryo which results if the egg is fertilized. The sperm is typically much smaller and motile, adapted to swim actively to the egg by the lashing movements of its long, filamentous tail. Sexual reproduction is advantageous biologically for it makes possible the recombination of the best inherited characteristics of the two parents and provides for the possibility that some of the offspring may be better adapted to survive than either parent was. Evolution can proceed much more rapidly and effectively with sexual than with asexual reproduction.

6.1 ASEXUAL REPRODUCTION

Asexual reproduction occurs commonly in plants, protozoa, coelenterates, bryozoa and tunicates but may occur even in the highest animals. The production of **identical twins** by the splitting of a single fertilized egg is a kind of asexual reproduction. The splitting of the body of the parent into two

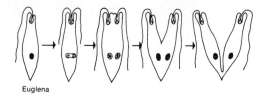

Euglena

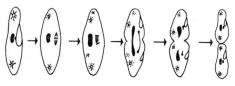

Paramecium

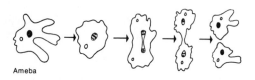

Ameba

Figure 6.1 Asexual reproduction in euglena, paramecium and ameba. In each, one cell divides mitotically to give rise to two cells.

more or less equal daughter parts, which become new whole organisms, is called **fission.** Fission occurs chiefly among the protists, the single-celled animals and plants; the cell division involved is mitotic (Fig. 6.1). Coelenterates typically reproduce by **budding;** a small part of the parent's body becomes differentiated and separate from the rest. It develops into a complete new individual and may take up independent existence, or the buds from a single

parent may remain attached as a colony of many individuals.

Salamanders, lizards, starfish and crabs can grow a new tail, leg or other organ if the original one is lost. When this ability to regenerate the whole from a part is extremely marked it becomes a method of reproduction. The body of the parent may break into several pieces and each piece then develops into a whole animal by regenerating the missing parts. A whole starfish can be regenerated from a single arm.

6.2 SEXUAL REPRODUCTION

Sexual reproduction is characterized by the development of a new individual from a zygote, or fertilized egg, produced in turn by the fusion of two sex cells, an egg and a sperm. Certain protozoa have a complicated process of sexual reproduction in which two individuals come together and fuse temporarily along their oral surfaces (Fig. 6.2). The nucleus of each one divides several times before one of the resulting daughter nuclei migrates across to the other animal and fuses with one of its nuclei. Following this the two animals separate and each reproduces asexually by fission. Paramecia are not differentiated morphologically into sexes, but T. M. Sonneborn has shown that there are as many as eight distinct, genetically determined **mating types.** A member of one mating group will mate only with some member of another group.

Before discussing sexual reproduction in metazoa, some generalizations about animal reproductive systems might be helpful.

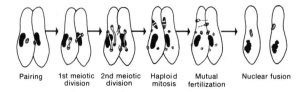

Pairing   1st meiotic   2nd meiotic   Haploid   Mutual   Nuclear fusion
           division   division   mitosis   fertilization

Figure 6.2 Sexual reproduction in paramecium. Two individuals with diploid micronuclei undergo conjugation (*left*). After meiosis (*second and third figures*) three of the nuclei degenerate and the fourth nucleus divides by mitosis (*fourth figure*). The organisms undergo mutual fertilization; one nucleus passes from each organism to the other. This is followed by fusion of the haploid nuclei to form a new diploid nucleus in each of the two organisms (*last figure*). The original macronuclei disappear. The two new diploid nuclei divide several times by mitosis and eventually establish both the new micronuclei and the new macronuclei.

The basic components of the male system are the male gonad, or **testis,** in which sperm are produced, and a **sperm duct** for the transport of sperm to the exterior. Various parts of the sperm duct, or areas adjacent to it, may be modified for other functions. (1) A part of the duct may be given over for sperm storage (often called a seminal vesicle). (2) There may be glandular areas for the production of seminal fluid, which serves as a vehicle, or carrier, for the sperm, and may also activate, nourish, or protect them. (3) The terminal part of the sperm duct may open onto, or into, a copulatory organ, or **penis,** for the transfer of sperm to the female.

The basic parts of the female system are the female gonad, or **ovary,** and a tube, the **oviduct,** for the transport of eggs to the exterior. The oviduct may be modified in the following ways. (1) There may be a glandular section for secretion of egg shells, cases, or cocoons. (2) The terminal portion may be adapted as a **vagina** for the reception of the male copulatory organ. (3) A section of the oviduct, the seminal receptacle, may be modified for the storage of sperm following their transfer from the male. (4) A section of the oviduct, the **uterus,** may be modified for egg storage or for the development of the fertilized eggs within the body of the parent.

The male and female systems of most animals do not possess all of these modifications and they cannot be considered as any kind of progressive sequence of changes within the Animal Kingdom. The presence or absence of specific adaptations of the reproductive system is correlated with a variety of different reproductive circumstances—whether the animal lives in the sea, fresh water, or on land; whether or not fertilization is external or internal; whether the eggs are liberated singly into the water to develop in the plankton, or whether they are deposited onto the bottom within envelopes; whether development is indirect (passes through a larval stage) or direct (no larval stage present).

Considering the great range of reproductive conditions that exist within any large diverse group of animals, it is to be expected that the reproductive systems would display a corresponding range of modifications, some of which may even have evolved independently a number of times.

6.3 MEIOSIS

The mitotic process is remarkably constant and ensures that the number of chromosomes per cell will remain unchanged through successive cell generations. The fusion of an egg and a sperm to form a fertilized egg would result in a doubling of the chromosome number in each successive generation if all cell divisions occurred by mitosis. However, at some point in the succession of cell divisions which constitute the life cycle of an individual—from the original fertilized egg through development, growth and maturation to the production of the fertilized egg in the next generation—there occurs a different type of cell division, called **meiosis.** In the higher animals, and in most of the lower ones, meiotic divisions occur during the formation of gametes. Meiosis is essentially a pair of cell divisions during which the chromosome number is reduced to half (Fig. 6.3). Thus the gametes contain only half as many chromosomes as the somatic cells, and when two gametes unite at fertilization, the fusion of their nuclei reconstitutes the normal number of chromosomes.

The reduction in chromosome number occurs in a very regular way. Chromosomes occur in *pairs* of similar chromosomes in somatic cells. As a result of meiosis, each gamete contains one and only one of each kind of chromosome, i.e., one complete set of chromosomes. This is accomplished by the **synapsis,** or longitudinal pairing, of like chromosomes and the subsequent separation of the members of the pair, one going to each pole. The like chromosomes which undergo synapsis during meiosis are called **homologous chromosomes.** They are identical in size and shape, have identical chromomeres along their length and contain similar hereditary factors. A set of one of each kind of chromosome is called the **haploid number** (n); a set of two of each kind is called the **diploid number** (2n). Gametes have the haploid number (e.g., 23 in man) and fertilized eggs and all the cells of the body have the diploid number (46 for man). A fertilized egg gets exactly half of its chromosomes (and half of its genes) from its mother and half from its father. Only the last two cell divisions which result in mature, functional eggs or sperm are meiotic; all other ones are mitotic.

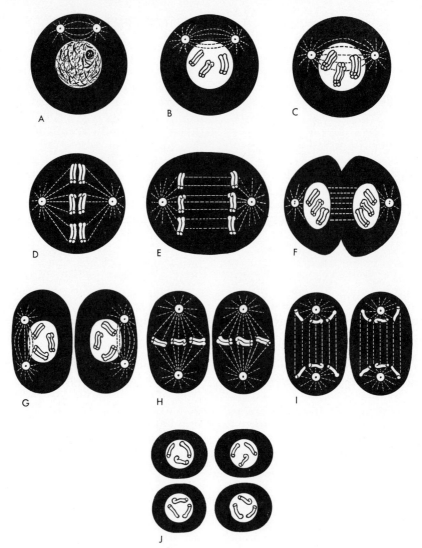

Figure 6.3   Meiosis in a hypothetical animal with a diploid chromosome number of six. It has three pairs of chromosomes, of which one is short, one is long with a hook at the end, and one is long and knobbed. *A*, Early prophase of the first meiotic division: chromosomes begin to appear. *B*, Synapsis: the pairing of the homologous chromosomes. *C*, Apparent doubling of the synapsed chromosomes to form groups of four identical chromatids, tetrads. *D*, Metaphase of the first meiotic division: the tetrads lined up across the equatorial plate. *E*, Anaphase: the chromatids migrating toward the poles. *F*, Telophase of the first meiotic division. *G*, Prophase of the second meiotic division. *H*, Metaphase of the second meiotic division. *I*, Anaphase of the second meiotic division. *J*, Mature gametes, each of which contains only one of each kind of chromosome.

Each of the meiotic divisions has the same four stages, prophase, metaphase, anaphase and telophase, found in mitosis. The chief differences between mitotic and meiotic divisions are seen in the prophase of the first meiotic division. Chromosomes appear as long thin threads which begin to contract and get thicker. The homologous chromosomes undergo synapsis, they pair longitudinally and come to lie side by side along their entire length, twisting around

each other. Each then becomes visibly double, as in mitosis, so that it consists of two threads. By synapsis and doubling, a bundle of four homologous chromatids, called a **tetrad,** is formed. Each pair of chromosomes gives rise to a bundle of four, so there are as many tetrads as the haploid number of chromosomes (23 in man). The centromeres have not divided and there are only two centromeres for the four chromatids.

While these events are occurring the centrioles go to opposite poles of the cell, a spindle appears between the centrioles, and the nuclear membrane dissolves. The tetrads then line up on the equatorial plate; this constitutes the metaphase of the first meiotic division. In the anaphase the daughter chromatids formed from each chromosome, still united by their single centromere, move as a unit to the poles. Thus the homologous chromosomes of each pair, but not the daughter chromatids of each chromosome, are separated in the first meiotic anaphase. In mitosis, in contrast, the centromeres do divide and the daughter chromatids pass to opposite poles. At telophase each pole has the haploid number of double chromosomes.

Typically, there is no interphase between first and second meiotic divisions, but the centrioles divide again, new spindles form (at right angles to the axis of the original spindle) and the haploid number of double chromosomes lines up on the equators of the spindles. Thus, the telophase of the first meiotic division and the prophase of the second are usually short and blurred together. The lining up of the chromosomes on the spindle constitutes the metaphase of the second division. The metaphases of the first and second meiotic divisions can be distinguished because in the first the chromosomes are arranged in bundles of four and in the second the chromosomes are arranged in bundles of two. There is no further doubling of the chromosomes; the centromeres divide and the daughter chromatids, now chromosomes, separate and pass to the poles.

In the anaphase of the second meiotic

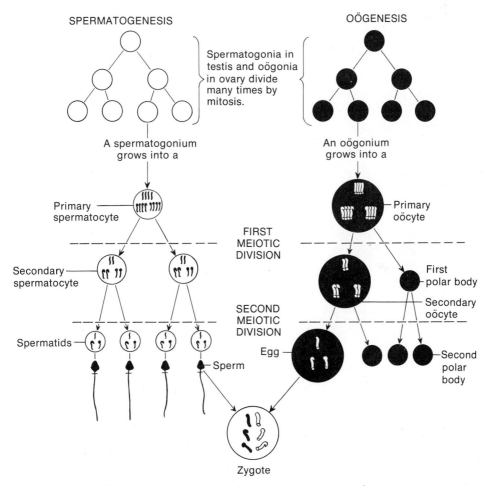

Figure 6.4 Comparison of the formation of sperm and eggs.

division a haploid set of single chromosomes passes to each pole. In the telophase,
the cytoplasm divides, the chromosomes
become longer, thinner and less easily
seen, and a nuclear membrane forms around
them. The net result of the two meiotic
divisions is a group of four nuclei, each of
which contains the haploid number of
chromosomes, that is, one and only one of
each kind of chromosome. These cells are
mature gametes and do not undergo any
further mitotic or meiotic divisions.

6.4 SPERMATOGENESIS

A typical testis consists of thousands of
cylindrical **sperm tubules,** in each of which
develop billions of sperm. The walls of the
sperm tubules are lined with unspecialized
germ cells called **spermatogonia.** Throughout development, the spermatogonia divide
by mitosis and give rise to additional spermatogonia to provide for the growth of the
testis. After sexual maturity, some spermatogonia begin to undergo **spermatogenesis,**

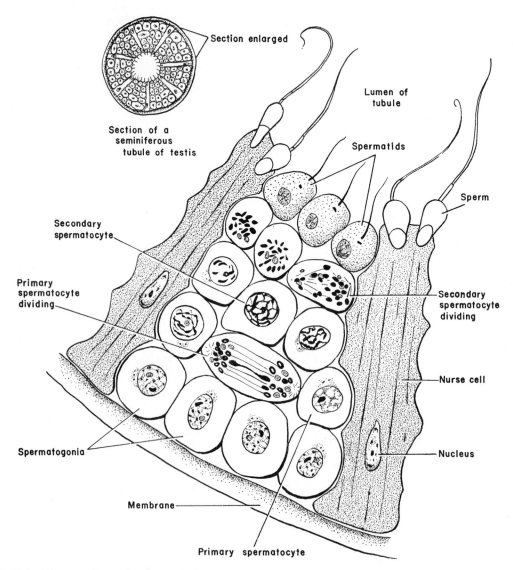

Figure 6.5 Diagram of part of a section of a human seminiferous tubule to show the stages in spermatogenesis and
in the transformation of a spermatid into a mature sperm. (Villee: Biology, 6th ed.)

which includes the two meiotic divisions followed by the cellular changes which result in mature sperm. Other spermatogonia continue to divide mitotically and produce additional spermatogonia for spermatogenesis at a later time. In most wild animals, there is a breeding season, either in spring or fall, during which the testis increases in size and spermatogenesis occurs. Between breeding seasons the testis is usually small and contains only spermatogonia. In man and most domestic animals, spermatogenesis continues throughout the year once sexual maturity has been attained.

The first step in spermatogenesis is the growth of the spermatogonia into larger cells, the **primary spermatocytes** (Fig. 6.4). Each primary spermatocyte divides (first meiotic division) into two cells of equal size, the **secondary spermatocytes.** These, in turn, divide (second meiotic division) to yield four equal-sized **spermatids.** The spermatid, a spherical cell with a generous amount of cytoplasm, is a mature gamete with the haploid number of chromosomes. Further changes (but no cell division) are required to convert it into a functional **spermatozoon.** The nucleus shrinks in size, becomes more dense, and forms the head of the sperm (Fig. 6.5). Most of the cytoplasm is shed, but some of the Golgi bodies aggregate at the anterior end of the sperm and form a cap (the acrosome). This contains enzymes which may play a role in puncturing the cell membrane of the egg.

The two centrioles of the spermatid move to a position just in back of the nucleus. A small depression appears on the surface of the nucleus and the proximal centriole takes up a position in the depression. The distal centriole gives rise to the axial filament of the sperm tail (Fig. 6.6). Like the axial filament of flagella, it consists of two longitudinal fibers in the center and a ring of nine pairs or doublets of longitudinal fibers surrounding the two (cf. Fig. 5.15).

The mitochondria move to the point

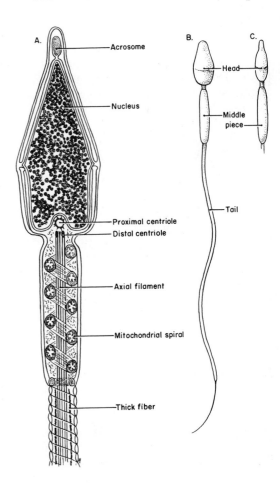

Figure 6.6 *A,* Diagram of the head and middle piece of a mammalian sperm, greatly enlarged, as seen in the electron microscope. *B* and *C,* Top and side views of a sperm seen by light microscopy. (Villee: Biology, 6th ed.)

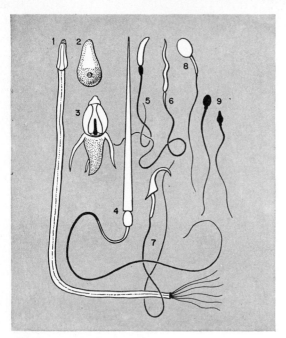

Figure 6.7 Spermatozoa from different species of vertebrates, illustrating the differences in size and shape. *1*, Gastropod. *2*, Ascaris. *3*, Hermit crab. *4*, Salamander. *5*, Frog. *6*, Chicken. *7*, Rat. *8*, Sheep. *9*, Man. (Villee: Biology, 6th ed.)

develop into **primary oöcytes.** When the human female is born, her two ovaries contain some 400,000 primary oöcytes, which have attained the prophase of the first meiotic division. These primary oöcytes remain in prophase for many years until the woman reaches sexual maturity. Then as each follicle matures the first meiotic division resumes and is completed at about the time of ovulation (Fig. 6.4). The two daughter cells, however, are not of equal size. One, the **secondary oöcyte,** receives essentially all the cytoplasm and yolk while the other, the **first polar body,** is essentially a bare nucleus.

The secondary oöcyte divides by the second meiotic division, again with an unequal division of cytoplasm, to yield a large **oötid,** with essentially all the yolk and cytoplasm, and a small **second polar body.** (The first polar body may divide at about the same time into two additional polar bodies.) The oötid undergoes further changes (but no cell division) and becomes a mature **ovum** (egg). The polar bodies disintegrate and disappear, so that each primary oöcyte forms a single ovum, in contrast to the four sperm derived from each primary spermatocyte. The formation of the polar bodies is a device to enable the maturing egg to get rid of its excess chromosomes, and the unequal division of the cytoplasm ensures the mature egg enough cytoplasm and yolk to survive and develop if it is fertilized.

The union of a haploid set of chromosomes from the sperm with another haploid set from the egg during fertilization re-establishes the diploid chromosome number. The fertilized egg, and all the body cells which develop from it, have the diploid number, two of each kind. Each individual gets half of his chromosomes (and half of his genes) from his father and half from his mother. Because of the nature of genic interaction, the offspring may resemble one parent much more than the other, but the two parents make equal contributions to its inheritance.

where head and tail join and form a middle piece. This provides energy for the beating of the tail which drives the sperm forward. Most of the cytoplasm of the spermatid is discarded; only a thin sheath remains surrounding the mitochondria in the middle piece and the axial filament of the tail.

The mature spermatozoa of different species exhibit a wide range of sizes and shapes (Fig. 6.7). The sperm of a few animals, such as the parasitic roundworm *Ascaris,* lack tails and crawl along by ameboid motion. Crabs and lobsters have curious tailless sperm with three pointed projections on the head. These hold the sperm in position on the surface of the egg while the middle piece uncoils like a spring and pushes the sperm nucleus into the egg cytoplasm, thereby accomplishing fertilization.

6.5 OÖGENESIS

The immature sex cells in the ovary, **oögonia,** undergo successive mitotic divisions to form additional oögonia. In the human female during the third month of fetal development the oögonia begin to

6.6 REPRODUCTIVE SYSTEMS

In some of the simpler invertebrates, such as the coelenterates, the testes and ovaries are the only sex structures present, and eggs and sperm are released directly from the gonads into the surrounding water. Most animals, however, have a system of ducts and glands which serve to carry gametes from the gonad to the exterior of the body and to protect and nourish them during the process.

Many of the lower animals are **hermaphroditic**; both ovaries and testes are present in the same individual and it produces both eggs and sperm. Some hermaphroditic animals, the parasitic tapeworms, for example, are capable of **self-fertilization.** (Does this violate the generalization that sexual reproduction involves two individuals?) Since a particular host animal may be infected with but one parasite, hermaphroditism is an important adaptation for the survival of the parasitic species. Most hermaphrodites, however, do not reproduce by self-fertilization; in the earthworm, for example, two animals copulate and each inseminates the other. In certain other species, e.g., the oyster, self-fertilization is impossible because the testes and ovaries produce gametes at different times.

The reproductive systems of different species have a fundamentally similar plan, but many variations on the theme are evident. The gonads and their ducts may be single, paired or multiple, perhaps present in several segments of the body.

Sperm, produced in the coiled seminiferous tubules of the testis, are transported in a series of ducts to the exterior, suspended in **seminal fluid,** secreted by glands associated with the reproductive tract. Seminal fluid contains glucose and fructose which the sperm metabolize, buffers which protect the sperm from the acids normally present in the urethra and female tract, and mucous materials which lubricate the passages through which the sperm travel.

In many vertebrates a number of accessory structures have developed to facilitate the transfer of sperm from the male to the female reproductive tract and to provide a place for the development of the fertilized egg. These structures have evolved either from or in close association with the urinary system, and the two together are frequently referred to as the **urogenital system.** In the male mammal, for example, the **vasa deferentia** empty into the **urethra,** which also carries urine from the bladder to the outside. The urethra of mammals is surrounded by the external reproductive organ, the **penis.** This consists of three columns of **erectile tissue** — spongy venous spaces which become filled with blood during sexual excitement to produce an erection of the penis.

Eggs are produced in the ovaries of the female and are typically surrounded and nourished by **nurse cells** during their development. At the time of ovulation, the eggs are released from the ovary into the abdominal cavity, whence they pass into the funnel-shaped end of the **oviduct.** Eggs are moved along the oviduct by the peristaltic contractions of its muscular wall or by the beating of cilia lining the lumen of the duct. The yolk of the bird's egg is formed while the egg is still within the ovary, but the egg white and shell are added by glands in the wall of the oviduct. The oviducts may open directly to the exterior or they may expand into a terminal duct, the **uterus,** which is a thick-walled muscular pouch in which the young develop. In mammals the uterus is connected with the exterior by the **vagina,** which is adapted to receive the penis of the male during copulation. Female mammals have a **clitoris,** the homologue of the male penis, just anterior to the opening of the vagina; it contains sense organs and erectile tissue which becomes engorged with blood during sexual excitement.

6.7 FERTILIZATION

The union of an egg and sperm is called **fertilization.** Most aquatic animals deliver their eggs and sperm directly into the surrounding water and the union of egg and sperm occurs there by chance meeting. This primitive and rather uncertain method of uniting the gametes is called **external fertilization.** Such animals usually have no accessory sex structures.

In other animals, fertilization occurs within the body of the female, usually in the oviduct, after the sperm have been transferred from the male to the female by copulation or by some other means. This method of **internal fertilization** requires some cooperation between the two sexes, and many species have evolved elaborate patterns of **mating behavior** to ensure that the two sexes are brought together, mate at the most appropriate time, and take care of the resulting offspring. A variety of secondary sex

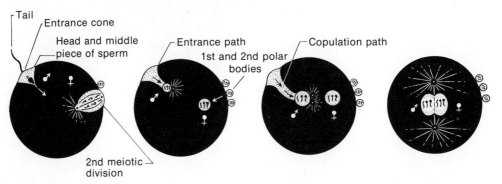

Figure 6.8 Diagram of the stages in the process of fertilization, the union of the egg and sperm.

characteristics have evolved and serve as stimuli to attract the opposite sex and elicit a mating response.

Fertilization involves not only the penetration of the egg by the sperm, but the union of the egg and sperm nuclei and the activation of the egg to undergo cleavage and development (Fig. 6.8). The egg may be in any stage from primary oöcyte to mature ovum at the time of sperm penetration, but the fusion of the sperm and egg nuclei occurs only after the egg has matured. There is experimental evidence that the eggs of some species secrete a substance, **fertilizin,** which is an important constituent of the jelly coat surrounding the egg. Fertilizin causes the sperm to clump together and stick to the surface of the egg. Other extracts of the egg jelly, which may be identical with fertilizin, stimulate sperm motility and respiration and prolong sperm viability.

After the entrance of one sperm, a **fertilization membrane** forms around the eggs of some species which prevents the entrance of other sperm. This prevents polyspermy and the possibility of the fusion of more than one sperm nucleus with the egg nucleus. It can be shown experimentally that such fusion of two or more sperm nuclei with one egg nucleus leads to abnormal development.

Eggs can be stimulated to cleave and develop without fertilization. The development of an unfertilized egg into an adult is known as **parthenogenesis** (virgin birth). Some species of arthropods have been found which apparently consist solely of females which reproduce parthenogenetically. In other species, parthenogenesis occurs for several generations, then some

males are produced which develop and mate with the females. The queen honeybee is fertilized by a male just once during her lifetime, in her "nuptial flight." The sperm are stored in a pouch connected with the genital tract and closed by a muscular valve. If sperm are released from the pouch as she lays eggs, fertilization occurs and the eggs develop into females—queens and workers. If the eggs are not fertilized they develop into males—drones.

Changes in temperature, in pH or in the salt content of the surrounding water, or chemical or mechanical stimulation of the egg itself will stimulate many eggs to parthenogenetic development. A variety of marine invertebrates, frogs, salamanders, and even rabbits have been produced parthenogenetically. The resulting adult animals are generally weaker and smaller than normal and are infertile.

The females of all birds, most insects, and many aquatic invertebrates lay eggs from which the young eventually hatch; such animals are said to be **oviparous** (egg-bearing). In contrast, mammals produce small eggs which are kept in the uterus and provided with nutrients from the mother's blood until development has proceeded to the stage where they can exist independently, to some extent at least. Such animals are said to be **viviparous** (live-bearing). In certain other forms—some insects, sharks, lizards and certain snakes—the female is **ovoviviparous;** she produces large, yolk-filled eggs which are retained within the female reproductive tract for a considerable period of development. The developing embryo forms no close connection with the wall of the oviduct or uterus and receives no nourishment from the mother.

The number of eggs produced by each female of a given species and the chance that any particular egg will survive to maturity are inversely related. In the evolution of the vertebrates from fish to mammals, the trend has been toward the production of fewer eggs, and the development of better parental care of the young. Fish such as the cod or salmon produce millions of eggs each year, but only a small number of these ever become adult fish; in contrast, mammals have few offspring but take good care of them so that the majority attain maturity. Fish and amphibia generally take no care of developing eggs, which are simply deposited in water and left to complete development unaided. The eggs of reptiles are usually laid in earth or sand and develop there without parental care, warmed by the sun. Birds, in contrast, have a complex behavior pattern for nest-building, incubating the eggs by sitting on them and caring for the newly hatched youngsters. The mammalian egg develops within the mother's uterus, where it is safe from predators and from harmful factors in the environment. Most mammals have a strong "maternal instinct" to take care of the newborn until they can shift for themselves. This pattern of behavior is, to a large extent, controlled by hormones secreted by the pituitary gland and ovary.

6.8 MATING BEHAVIOR

Since most animals breed only during relatively brief seasons of the year, the production and release of eggs and sperm must be synchronized if fertilization is to occur. Typically the males and females are triggered by some environmental cue such as a change in the photoperiod, a change in the ambient temperature, seasonal rainfall or by specific relations of tidal and lunar cycles (cf. the Palolo worm, p. 559). In some species the males and females must not only be brought to full sexual activity at the same time, but must be induced to move or migrate to specific mating and breeding grounds. The migration of salmon upstream to breed, the migration of eels to a specific breeding ground in the Central Atlantic, the migration of the gray whale to Baja California, the migration of turtles and birds are examples of such movements. Ani-

mals that live a solitary life during most of the year—seals, penguins, certain sea birds—come together for brief periods of mating at specific times and places.

The specific reproductive synchronization of a particular male for a particular female frequently involves some sort of **courtship behavior** (Fig. 35.16). The courtship, usually initiated by the male, may be a very brief ceremony, or in certain species of birds may last for many days. The courtship behavior serves two additional roles: it tends to decrease aggressive tendencies and it establishes species and sexual identification, i.e., it identifies a member of the same species but of the opposite sex. The members of many species normally fight whenever they meet and some special cue is needed if two animals are to avoid fighting long enough to mate. Special structural and functional adaptations have evolved in some species in which aggression seems to be especially difficult to control. The male praying mantis is smaller than the female and is usually attacked by her. Mating in this species can continue even after the male's head has been bitten off by the female, since the nervous activities controlling copulation are centered in an abdominal ganglion. The males of certain species of flies present their predatory females with little packages of food wrapped in silk thread to divert their attention. The male cockroach secretes a special substance on his back which is especially attractive to the female and diverts her attention during copulation. The males of other insects present the female with an empty silk balloon which diverts her attention and helps identify the species and sex of the male presenting it.

The song patterns of birds and the mating calls of certain fishes, frogs and insects provide effective cues for the discrimination of the species. Female frogs are attracted to calling males of their own species but not to calling males of related species.

6.9 EMBRYONIC DEVELOPMENT

The division, growth and differentiation of a fertilized egg into the remarkably complex and interdependent system of organs which is the adult animal is certainly one of the most fascinating of all biologic phenom-

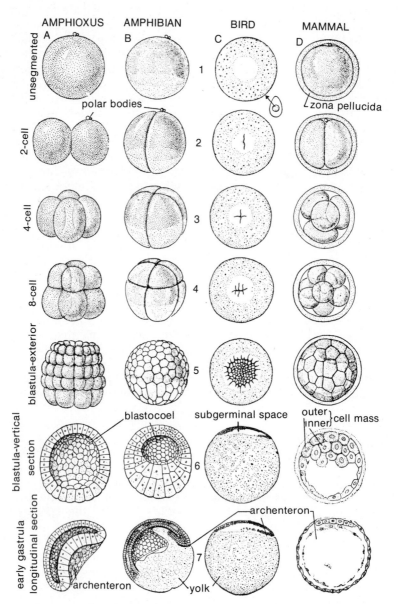

Figure 6.9 Stages in cleavage and early gastrulation in eggs of chordates. *A, Amphioxus* (holoblastic cleavage, isolecithal egg with little yolk). *B*, Frog (holoblastic cleavage, moderately telolecithal egg with much yolk). *C*, Bird (meroblastic discoidal cleavage, telolecithal egg with much yolk). *D*, Mammal (holoblastic cleavage, isolecithal egg with essentially no yolk). (From Storer and Usinger: General Zoology, 3rd ed. Copyright, 1957, by McGraw-Hill Book Co., Inc.)

ena. Not only are the organs complicated, and reproduced in each new individual with extreme fidelity of pattern, but many of these organs begin to function while they are still developing. The human heart begins to beat, for example, during the fourth week of gestation, long before its development is completed.

The early stages of development of practically all multicellular animals are fundamentally similar; differences in development become evident somewhat later.

When fertilization has been accomplished, the zygote divides repeatedly by a series of mitotic cell divisions termed **cleavage.** The pattern of cell division is determined largely by the amount of yolk present in the egg. An **isolecithal egg** has a relatively small amount of yolk distributed more or less evenly throughout the cytoplasm. **Telolecithal eggs** have a large amount of yolk which is more concentrated at the lower or **vegetal pole** of the

egg; the metabolically active cytoplasm is concentrated at the upper or **animal pole.** The insect egg is an example of a **centrolecithal** one; the yolk accumulates in the center of the egg and is surrounded by a thin layer of cytoplasm.

The plane of the first cleavage division of an isolecithal egg passes through the animal and vegetal poles of the egg yielding two equal cells, called **blastomeres** (Fig. 6.9). The second cleavage division passes through animal and vegetal poles at right angles to the first and divides the two cells into four. The third cleavage division is horizontal. Its plane is at right angles to the planes of the first two divisions, and the embryo is split into four cells above and four below this line of cleavage. Further divisions result in embryos containing 16, 32, 64, 128 cells and so on until a hollow ball of cells, the **blastula,** results. The wall of the blastula consists of a single layer of cells, and the fluid-filled cavity in the center of

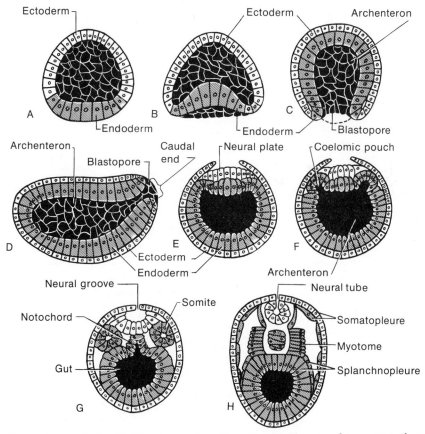

Figure 6.10 Stages in gastrulation (A–C) and mesoderm formation (F–H) in *Amphioxus.* Note that the mesoderm forms by the budding of pouches from the archenteron.

the sphere is called the **blastocoele.** Each of the cells in the blastula is small, and the total mass of the blastula is less than that of the original fertilized egg, for some of the stored food was used up in the cleavage process.

The single-layered blastula is soon converted into a double-layered sphere, a **gastrula,** by the process of gastrulation. In isolecithal eggs, gastrulation occurs by the pushing in (**invagination**) of a section of one wall of the blastula (Fig. 6.10). This pushed-in wall eventually meets the opposite wall and the original blastocoele is obliterated. The new cavity of the gastrula is the **archenteron** (primitive gut), the rudiment of the digestive system. The opening of the archenteron to the outside, the **blastopore,** marks the site of the invagination which produced gastrulation. The outer of the two walls of the gastrula, the **ectoderm,** eventually forms the skin and nervous system. The inner layer, lining the archenteron, is the **endoderm,** which will form the digestive tract, liver, pancreas and lungs.

Cleavage and gastrulation are markedly modified in telolecithal eggs by the presence of the large amount of yolk. In the frog egg, which may be called moderately telolecithal, the cleavage divisions in the lower part of the egg are slowed by the presence of the inert yolk (Fig. 6.11). The resulting blastula consists of many small cells at the animal pole and a few large cells at the vegetal pole. The lower wall of the blastula is much thicker than the upper one and the blastocoele is flattened and displaced upward. Only the small disc of cytoplasm at the animal pole of the hen's egg undergoes cleavage divisions; the lower, yolk-filled part of the egg never cleaves. The shallow cavity under the dividing cells is called the subgerminal space, since it is not homologous to the blastocoele of the frog egg. Gastrulation occurs in both frog and chick egg, and an archenteron is formed, but the process is greatly modified by the presence of the yolk. Gastrulation in the frog involves an invagination of the yolk-filled cells of the vegetal pole, a turning in of cells at the dorsal lip of the blastopore (**involution**), and a growth of ectoderm down and over the cells of the vegetal pole (**epiboly**) (Fig. 6.11).

In all multicellular animals, except sponges and coelenterates, which never develop beyond the gastrula stage, a third layer of cells, the **mesoderm,** develops between ectoderm and endoderm. In annelids, mollusks and certain other invertebrates, the mesoderm develops from special cells which are differentiated early in cleavage (p. 135). These migrate to the interior and come to lie between the ectoderm and

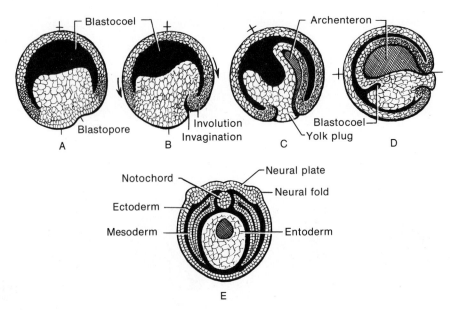

Figure 6.11　*A–D*, Successive stages in gastrulation and mesoderm formation in Amphibia. *E*, Transverse section of an early neurula stage.

endoderm. They then multiply to form two longitudinal cords of cells which develop into sheets of mesoderm between the ectoderm and endoderm. The coelomic cavity originates by the splitting of the sheets to form pockets and, hence, is called a **schizocoele.**

In primitive chordates the mesoderm arises as a series of bilateral pouches from the endoderm (Fig. 6.10). These lose their connection with the gut and fuse one with another to form a connected layer. The cavity of the pouches is retained as the coelom, which is called an **enterocoele** because it is derived indirectly from the archenteron. The mesoderm in amphibia is formed in part from the roof of the archenteron and in part from cells that roll in at the lips of the blastopore and move anteriorly as a sheet between ectoderm and endoderm (Fig. 6.11). In birds and mammals the **primitive streak** which develops on the surface of the developing embryo is homologous to the dorsal lip of the blastopore of lower forms. It is a thickened band of ectoderm and endoderm cells which marks the longitudinal axis of the embryo. At the primitive streak, cells migrate in from the surface, proliferate and form a sheet of mesoderm between ectoderm and endoderm. The primitive streak is a dynamic structure and persists even though the cells comprising it are constantly changing as they migrate.

However the mesoderm may originate,

it typically forms two sheets which grow laterally and anteriorly between the ectoderm and endoderm; one sheet becomes attached to the inner endoderm and the other to the outer ectoderm. The cavity between the two becomes the **coelom,** or body cavity. This splitting of the mesoderm permits the development of two independent sets of muscles, one in the body wall adapted for locomotion and the other in the gut wall adapted for churning and moving the contents of the gut.

The primitive skeleton of the chordates, the **notochord,** is a flexible, unsegmented, longitudinal rod found in the dorsal midline of all chordate embryos. It is formed at the same time and in a similar way as the mesoderm—as an outgrowth of the roof of the archenteron, from the dorsal lip of the blastopore, or from the primitive streak. Later in the development of vertebrates the notochord is replaced by the **vertebral column,** derived from part of the mesoderm.

The nervous system of chordates is derived from the ectoderm overlying the notochord. This first forms a thickened plate of cells, the **neural plate;** the center of the plate becomes depressed while the lateral edges rise as two longitudinal neural folds (Fig. 6.12). The folds eventually meet dorsally and form a hollow **neural tube.** The cavity of the tube becomes the central canal of the spinal cord and the ventricles of the brain. The ectodermal cells pinched off near the apex of each neural fold form

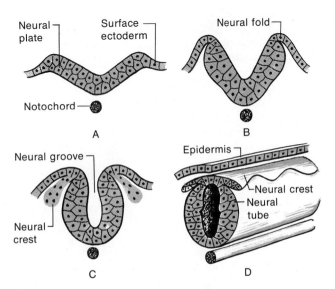

Figure 6.12 A series of cross sectional diagrams through the surface ectoderm to show the formation of the neural tube and neural crest. (After Arey.)

TABLE 6.1 DERIVATIVES OF GERM LAYERS AND
MAMMALIAN DEVELOPMENT

Ectoderm	Endoderm	Mesoderm
Epidermis of the skin	Lining of gut	Muscles—smooth, skeletal and cardiac
Hair and nails	Lining of trachea, bronchi and lungs	Dermis of skin
Sweat glands	Liver	Connective tissue, bone and cartilage
Brain, spinal cord, ganglia, nerves	Pancreas	Dentin of teeth
Receptor cells of sense organs	Lining of gallbladder	Blood and blood vessels
Lens of the eye	Thyroid, parathyroid and thymus glands	Mesenteries
Lining of mouth, nostrils and anus	Urinary bladder	Kidneys
Enamel of teeth	Lining of urethra	Testes and ovaries

a ridge, the **neural crest,** on each side of the neural tube. The cells of the neural crest become segmentally arranged and differentiate into afferent neurons of the cranial and spinal nerves. Other neural crest cells migrate out and become the postganglionic sympathetic fibers, the cells of the adrenal medulla, or the neurilemma sheath cells of the peripheral neurons.

The sheets of mesoderm grow ventrally and the ones from either side meet in the ventral midline; the coelomic cavities on the two sides then fuse into one. The mesoderm grows dorsally along each side of the notochord and neural tube and becomes differentiated into segmental blocks of tissue, the **somites,** from which the main muscles of the trunk develop. Other mesodermal cells become detached from the inner border of the somites, migrate inward, surround the notochord and neural tube, and develop into the **vertebrae.** The kidneys and their ducts, and the gonads and their ducts, are derived from the mesoderm originally located between the somites and the coelom.

The contributions of each germ layer to the development of a typical mammal are summarized in Table 6.1.

6.10 SPIRAL CLEAVAGE AND ITS EVOLUTIONARY IMPORTANCE

The acoelomate Platyhelminthes and Nemertea, and several eucoelomate phyla including the Mollusca and Annelida, share a pattern of early embryonic development called **spiral cleavage.** This pattern combines a precise system of cell division and a fixed, or predetermined, fate of the subsequent cells. Within any one species the pattern is constant, but the pattern may vary from one species to another. In a typical pattern of spiral cleavage, the first and second cell divisions are *meridional* and at right angles to each other, forming four cells of nearly equal size (Fig. 6.13).

The third division is *oblique*—the mitotic spindles are neither vertical nor horizontal, but inclined to one side. (One of them is indicated by the solid line in the eight-cell stage, Fig. 6.13.) As a result, the upper four cells are displaced circularly so that each upper cell touches *two* lower cells. The third division is also *unequal,* separating four upper, small **micromeres** from four lower, large **macromeres.** The four micromeres are called the **first quartette.**

The fourth division is also oblique but always in the opposite direction from the third (Fig. 6.13). The first quartette forms eight cells. The macromeres divide unequally, producing again an upper tier of small cells, the **second quartette,** and a lower tier of four macromeres.

The fifth division continues the pattern. The axes of division are oblique in the direction of those of the third division. The progeny of the first quartette become 16 cells, the second quartette becomes eight cells, and the macromeres divide unequally to form a **third quartette** of micromeres and a lower tier of four macromeres.

Divisions continue to be oblique, alternating to one side and then the other. If the

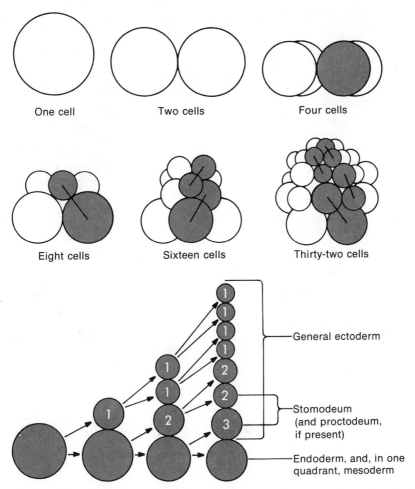

Figure 6.13 Spiral cleavage. One quadrant (the progeny of one cell of the four-cell stage) is shaded. Lines indicate the axes of the preceding mitoses. The lower diagram shows the fates of the cells of one quadrant. Numbers indicate the three quartettes or their progeny.

eight-, 16- and 32-cell stages are studied carefully, it should be evident that the embryo is twisted by this process first one way and then the other. This is particularly evident to an observer looking down on the top of the embryo when the macromeres are resting firmly on the bottom of a dish. It is this phenomenon that gives the pattern its name.

It should also be evident that the progeny of each cell of the four-cell stage remain together in one quadrant of the embryo.

Gastrulation usually begins after the sixth or seventh cleavage. In all the phyla showing spiral cleavage, the macromeres of the 32-cell stage or all their progeny pass into the interior. Usually, all the mesoderm develops from one of the macromeres, which

may be slightly larger than the others, while the other three become endoderm. The three quartettes form all of the ectoderm and its derivatives.

The occurrence of spiral cleavage in so many phyla suggests that it is a basic pattern of development in three-layered animals. The concept of a "main line" of evolution emerges, proceeding from the flatworms and nemerteans to the mollusks and annelids. The members of some phyla, such as the rotifers, gastrotrichs and nematodes, show further specialization and modification toward rigid embryonic fate and cell constancy, while others, including the chordates and echinoderms, show simplification and loss of rigidity in early development.

6.11   DEVELOPMENT AND IMPLANTATION OF THE MAMMALIAN BLASTOCYST

The first cleavage of the fertilized egg occurs about 30 hours after insemination, and the succeeding mitoses occur every 10 hours or so. By the time the developing ovum reaches the uterus, perhaps 3 to 7 days after fertilization, it is a tight ball of some 32 cells, called a **morula.** If the fertilized egg passes through the fallopian tube too rapidly and reaches the uterus prematurely it cannot be embedded in the wall. One type of contraceptive device, the **intrauterine coil,** may stimulate muscular contractions of the fallopian tube and uterus so that the fertilized ovum reaches the uterine cavity prematurely and dies before it can undergo implantation.

When the developing ovum reaches the uterine cavity it begins to differentiate into a **blastocyst** composed of an outer envelope of cells, the **trophoblast,** and an **inner cell mass,** a ball of cells at one pole within the trophoblast which is the precursor of the embryo. This stage implants in the endometrial lining of the uterus by secreting enzymes which erode the cells of the endometrium, permitting the embedded blastocyst to establish close contact with the maternal blood stream. The cells of the trophoblast grow and divide rapidly; they and the adjacent cells of the uterine lining, the **decidua,** form the placenta and fetal membranes. The cells of the endometrium heal over the site of entry of the blastocyst so that it lies wholly within the endometrium and out of the uterine lumen.

The embedding reaction of the endometrium can be elicited by pricking with a glass needle the uterine lining of a female suitably primed with estradiol and progesterone. This stimulus leads to "**pseudopregnancy**" and the uterus develops for a short period just as though an embryo were present.

The trophoblast initially consists of two layers of cells, an inner **cytotrophoblast** composed of individual cells and an outer **syncytiotrophoblast** composed of a multinucleate syncytium. The trophoblastic cells digest and phagocytize materials in the endometrium that were stored prior to implantation. The trophoblast soon is bathed in and nourished by the maternal blood. Menstruation normally occurs about 14 days after ovulation in the nonpregnant woman. To prevent this the embedded embryo must signal the maternal organism in some way. Because of the time required for traversing the oviduct, and because the fertilized egg remains in the lumen of the uterus for some days, about 11 days elapse between ovulation and implantation. Thus the embryo has only a very few days to provide the signal which will prevent menstruation. Fairly frequently the signal does not arrive in time and menstruation sweeps out the fertilized egg. The woman has been pregnant in the sense that she had a fertilized egg in her reproductive tract, but was never aware of it and menstruated at her usual time. One of the major contributions of the trophoblast is its secretion of **chorionic gonadotropin,** probably by the cells of the cytotrophoblast. Chorionic gonadotropin has properties similar to those of luteinizing hormone and luteotropic hormone of the pituitary; it prevents the corpus luteum from involuting. The secretion of chorionic gonadotropin begins the day the trophoblast is embedded in the endometrial lining.

The process of implantation has implicit in it two questions of general biological interest. Why does the trophoblast generally cease invading the endometrial lining once it has formed a connection with the maternal blood? Why doesn't it continue to invade as a group of cancer cells would? And why, since the cells of the trophoblast have the genotype of the developing fetus, a genotype different from that of the mother, do the maternal cells not react as though the trophoblast were a transplant and reject it, as an animal rejects a skin graft from another, genetically different member of the same species?

6.12   PROTECTION OF THE EMBRYO

The egg and the developing embryo are, in general, very susceptible to unfavorable environmental conditions, and a variety of adaptations have evolved in invertebrates and vertebrates to tide the embryo over this critical period.

The eggs of many parasitic worms are covered with **shells** which enable them to survive exposure to heat, cold, desiccation and digestive juices. The skate egg is covered by a tough leathery case that protects the developing embryo within. The eggs of most fish and amphibia are surrounded by a **jelly coat** which is of some value in protecting against mechanical shock. The eggs of reptiles and birds are protected by tough leathery or calcareous shells. The developing chick embryo takes in oxygen and gives off carbon dioxide through its shell.

The eggs of fish and amphibia are fairly large and contain yolk which supplies the nutrients for the developing embryo. These eggs are laid and typically develop in water, whence the oxygen, salts and water required for development are obtained. The embryos develop a pouch-like outgrowth of the digestive tract, the **yolk sac,** which grows around the yolk, elaborates enzymes to digest it, and transports the products in its blood vessels to the rest of the embryo.

The eggs of reptiles and birds develop on land rather than in water, and further adaptations were required to permit development in the absence of the large body of water. These forms have three additional membranes, the **amnion, chorion** and **allantois,** which are sheets of living tissue that grow out of the embryo itself. The amnion and chorion develop as folds of the body wall and surround the embryo; the allantois grows out of the digestive tract and functions along with the yolk sac in nutrition, excretion and respiration. Each of these membranes is composed of two germ layers in close apposition.

The formation of the amnion is a complex process and its details differ in different animals. A bilateral, double-walled outfolding of the body wall of the embryo grows upward and medially to surround the embryo and fuse above it, enclosing a space, the **amni-**

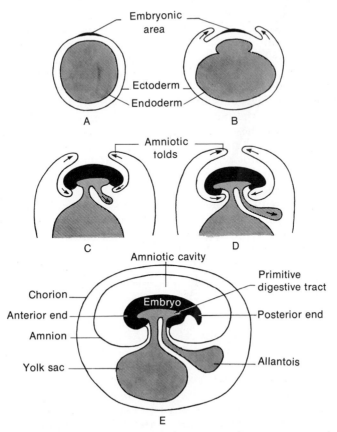

Figure 6.14 *A–E,* Steps in the formation of the extraembryonic membranes—amnion, chorion, yolk sac and allantois—in a typical mammal such as a pig. Arrows indicate direction of growth and folding.

otic cavity, between itself and the embryo (Fig. 6.14). This is filled with a clear watery fluid secreted in part by the embryo and in part by the amnion. The amnion develops from the inner part of the original fold; the outer part becomes the second fetal membrane, the chorion, which lies outside of and surrounds the amnion. The **chorionic cavity,** also known as the extraembryonic coelom (for the space is continuous with the coelomic cavity within the embryo), is the space between the amnion and chorion. The embryos of reptiles, birds and mammals develop in the liquid-filled amniotic cavity, their own private pond within the shell or uterus. This arrangement permits the embryo to move around to some extent but protects it from bumps and shocks.

The chorion of reptiles and birds comes to lie next to the shell and that of mammals is in contact with the maternal tissues of the uterus. The allantois, an outgrowth of the digestive tract, grows between the amnion and chorion and largely fills the chorionic cavity. The allantois of the bird and reptile typically fuses with the chorion to form a compound membrane, equipped with many blood vessels, by means of which the embryo takes in oxygen, gives off carbon dioxide and excretes certain wastes.

The mammalian allantois is usually small and has no function as a membrane, but supplies blood vessels to the **placenta,** the organ formed from chorion, allantois and maternal tissue. Finger-like projections, or villi, of the chorion grow into and become embedded in the lining of the uterus. These villi, their blood vessels and the uterine tissues with which they are in contact are called the placenta. This organ, in which fetal blood vessels come in close contact with maternal blood vessels, provides the developing mammalian fetus with nutrients and oxygen from the maternal blood, and eliminates carbon dioxide and waste products into the mother's blood. The two blood streams do not mix at all; they are always separated by one or more tissues. However, substances can diffuse, or be transported by some active process, from mother to fetus or the reverse. The form of the placenta, and the intimacy of the connection between maternal and fetal tissues, varies from one mammal to another.

The growth of the embryo and of the amnion brings the edges of the amniotic folds together to form a tube which encloses the yolk sac (which is usually small or vestigial), the allantois, the two umbilical arteries and the umbilical vein which pass to the placenta. This tube, the **umbilical cord,** is composed of a peculiar, jelly-like material which is unique to the cord.

The amnion, chorion and allantois, together with the eggshell or placenta, are adaptations which permit the embryos of the higher vertebrates to develop on land; they are a substitute for the pond or sea water in which the embryos of the lower vertebrates develop.

6.13 THE DIGESTIVE TRACT AND ITS DERIVATIVES

The neural tube and embryo elongate faster than the embryonic disc upon which the embryo is developing. As a result, the embryonic disc buckles at each end. The embryo continues to elongate, and the parts of the embryonic disc that originally lay anterior or posterior to the neural tube fold underneath the embryo (Fig. 6.15). Folds first separate the head and tail from surrounding structures. These folds deepen, and the folding process continues along each side until the embryo is more or less cylindrical in shape and remains connected to its surrounding membranes only by a narrow umbilical cord. The folding process is somewhat analogous to the gradual tightening of a pair of purse strings.

These folding processes gradually pinch off the dorsal part of the yolk sac and convert it into the primitive gut, or **archenteron,** of the embryo. The archenteron remains connected with the yolk sac by a narrow stalk that extends through the umbilical cord. The anterior part of the archenteron, the **foregut,** differentiates into the pharynx, esophagus, stomach and a small portion of the duodenum. The rest of the archenteron, the **hindgut,** forms most of the intestinal region and much of the embryonic cloaca. Only the linings of these organs are endodermal; the connective tissue and muscles in their wall are derived from mesoderm.

The pharyngeal pouches, thyroid gland, trachea and lungs develop as outgrowths from the pharynx. A ventral outgrowth from the posterior end of the foregut differenti-

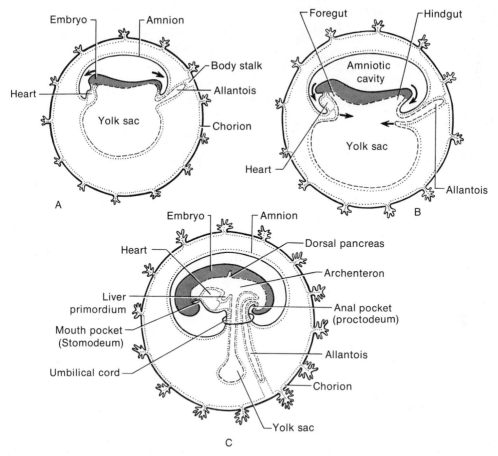

Figure 6.15 A series of diagrams of sagittal sections of embryos of different ages to show the folding processes that separate the embryo from its extraembryonic membranes. Solid lines represent ectoderm, broken lines endoderm, and stippled lines and shaded areas mesoderm. (Modified after Arey.)

ates into the liver and much of the pancreas, but part of the pancreas develops as a separate dorsal outgrowth (Figs. 6.15 and 6.16). This explains why the pancreas has two ducts, one entering the intestine with the bile duct and one independently.

The most anterior and posterior ends of the digestive tract develop from ectodermal pockets that invaginate and meet the archenteron. Initially, plates of tissue separate these pockets from the archenteron, but these plates eventually break down. The lining of the mouth, the enamel of the teeth and the secretory cells of the salivary glands are ectodermal in origin. The anterior and intermediate lobes of the pituitary gland develop as an ectodermal evagination from the roof of the mouth. The posterior lobe of the pituitary develops as an evagination from the floor of the diencephalic region of

the brain. Part of the embryonic cloaca is of ectodermal origin. A cloaca persists in the adults of most vertebrates but is divided in most mammals to form the dorsal rectum and the more ventral urogenital passages (part of the urethra in the male; part of the urethra and vagina in the female).

6.14 DIFFERENTIATION OF THE MESODERM

As the mesoderm spreads out from the primitive streak, its lateral portion splits into two layers (Fig. 6.16). This part of the mesoderm is known as the **lateral plate,** and the space between the two layers is the **embryonic coelom.** The embryonic coelom is continuous with the large **extraembryonic coelom,** or chorionic cavity, until the fold-

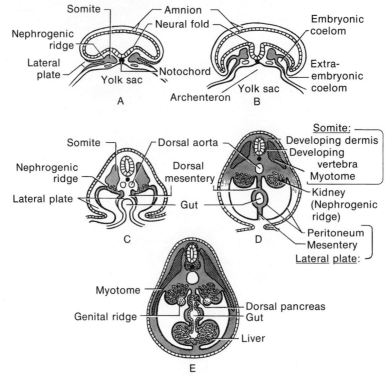

Figure 6.16 Diagrammatic cross sections through vertebrate embryos of different ages. The separation of the embryo from the yolk sac, the differentiation of the mesoderm, and the formation of the liver and dorsal pancreas are shown. (Modified after Patten.)

ing processes described above separate the embryo from surrounding structures. The inner layer of the lateral plate mesoderm, which lies next to the archenteron, forms the connective tissue and musculature (visceral muscles) of the digestive tract, the visceral peritoneum and the mesenteries. The outer layer forms the lateral wall of the coelom, that is, the parietal peritoneum, and may contribute to the musculature of the body wall.

Unlike the lateral plate, the mesoderm on each side of the neural tube and notochord becomes segmented and forms a series of paired **somites.** Some of the mesoderm of the somites spreads out beneath the surface ectoderm to form the dermis of the skin, some migrates around the neural tube and notochord and differentiates into the vertebral column and much of the skull, and the rest forms the segmented, embryonic skeletal muscle blocks, or **myotomes.** The myotomes extend out between the surface ectoderm and the lateral plate and develop into most of the musculature of the body wall and appendages (somatic mus-

cles). The segmentation of the muscles is retained in adult fishes, but muscle segmentation is largely lost during the later development of most higher vertebrates.

The resemblance of certain of the embryonic stages of the higher vertebrates to the adults of lower vertebrates, such as we see in the segmentation of the muscles, is regarded as strong evidence for evolution. In the late nineteenth century, Ernst Haeckel postulated that embryos pass through stages during their embryonic development (ontogeny) that resemble the adult stages of their evolutionary ancestors (phylogeny). In other words, "ontogeny recapitulates phylogeny, or the embryo climbs its own family tree." This generalization is no longer taken as literally as Haeckel intended. It is now clear that the embryos of higher vertebrates resemble the *embryos* and not necessarily the adults of lower vertebrates. Early vertebrates evolved a series of developmental stages that resulted in their characteristic organs. Higher vertebrates have certain differences, but these develop by introducing changes in the later stages of development rather than by altering the whole complex and intricately interrelated developmental sequence. Development, therefore, tends to be conservative, and the early embryos of different animals may bear marked resemblances to each other. However, the early development of an embryo may be altered and correlated with special conditions to which the embryo has become adapted. The extraembryonic membranes of mammals, for example, develop in advance of the main body of the embryo, and the placenta is formed very early. This is an adaptation of the embryo to intrauterine life. In reptiles the extraembryonic membranes develop only after the body of the embryo is well established.

A narrow band of mesoderm, known as the **nephrogenic ridge,** lies between the somites and the lateral plate. This part of the mesoderm differentiates into the kidney and contributes to the formation of the gonads.

The entire circulatory system develops from the mesoderm, and its development is rapid in all vertebrates. Transporting vessels are necessary for the embryo to obtain nutrients from the placenta, or yolk, as the case may be. The blood vessels differentiate by the hollowing out and coalescence of cords and knots of mesodermal cells that appear first in the mesodermal layer next to the yolk sac. A pair of vessels that are destined to become the heart develop in the anterior part of the embryonic disc before the neural tube is completely formed (Fig. 6.17). Subsequent foldings that give the embryo its shape carry these vessels beneath the front of the embryo. They fuse to form a single **cardiac tube,** and the cardiac tube differentiates into the series of chambers found in fish hearts (sinus venosus, atrium, ventricle, and conus arteriosus). Since the cardiac tube grows in length faster than the part of the coelom (the pericardial cavity) in which it lies, it folds and forms an S-shaped tube. The atrium, which originally lay posterior to the ventricle, thus comes to lie in front of the ventricle. Gradually, the cardiac tube differentiates into the adult heart. The atrium and ventricle become divided in mammals, the sinus venosus is incorporated into

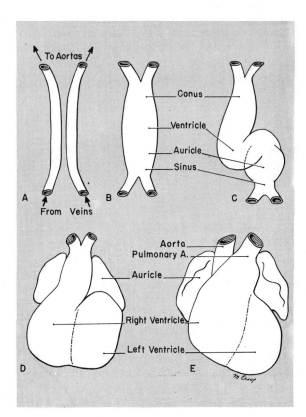

Figure 6.17 Stages in the development of a mammalian heart, seen in ventral view. (From Villee: Biology, 6th ed.)

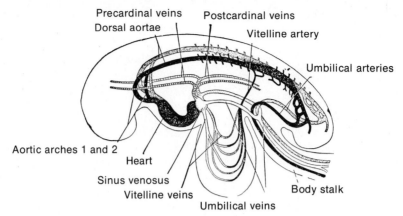

Figure 6.18   Lateral view of the major blood vessels of an early mammalian embryo. (From Arey after Felix.)

the right atrium, and the conus arteriosus forms part of the pulmonary artery and the arch of the aorta.

A series of paired **aortic arches,** which are similar in arrangement to those of a fish but are not interrupted by capillaries, carry blood from the heart up through the pharyngeal region to the **dorsal aorta** (Fig. 6.18). **Vitelline arteries** extend from the aorta to the yolk, and **umbilical arteries** follow the allantois to the chorion and developing placenta. Veins develop in a similar manner and return blood to the heart from the yolk sac, chorionic villi and the embryo itself. In the early mammalian embryo, the pattern of the veins resembles the pattern seen in fishes. **Cardinal veins** are present and the venae cavae do not develop until later.

6.15   GROWTH OF THE EMBRYO

The main morphologic changes in embryonic development take place with surprising speed. A human embryo four weeks old is only 5 mm. long, but it has already developed enough to be recognized as some sort of a vertebrate embryo (Fig. 6.19A). The development of all the organ systems is well under way, the heart has begun to beat, limb buds that will differentiate into arms and legs are protruding from the surface, and a small tail is present. Pregnancy may only be suspected at this time. At eight weeks (Fig. 6.19B), the embryo can be recognized as human. The face and distinct fingers and toes have developed. The organ systems are approaching their adult condi-

tion. Some of the bones are beginning to ossify and taste buds are developing on the tongue. The embryo is arbitrarily called a **fetus** from this age on.

Only relatively small changes occur in the organ systems during the remaining seven months of pregnancy, but a great increase in size takes place. An eight-week fetus has a crown-rump length of 30 mm. At term, its crown-rump length is about 35 cm. Among the morphologic changes that occur during this period are differentiation of the external genitilia, development of body hair, muscularization of the digestive tract, and myelinization of the neurons. Though the infant is well developed at the time of birth, development does not cease. Changes in the organ systems and in the relative size of body parts continue throughout infancy, childhood and adolescence. Human development is not really completed until the late teens.

6.16   TWINNING

Many offspring are born at the same time in pigs, rats and a number of other mammals. The number of piglets in a litter, for example, ranges from seven to 23. But many other mammals, including man and the other higher primates, whales and horses, normally have only one offspring at a time. Occasionally, multiple births occur in these mammals. Twins are produced about once in every 86 human births. Approximately three-fourths of these are **dizygotic,** or **fraternal twins.** Two eggs have been ovu-

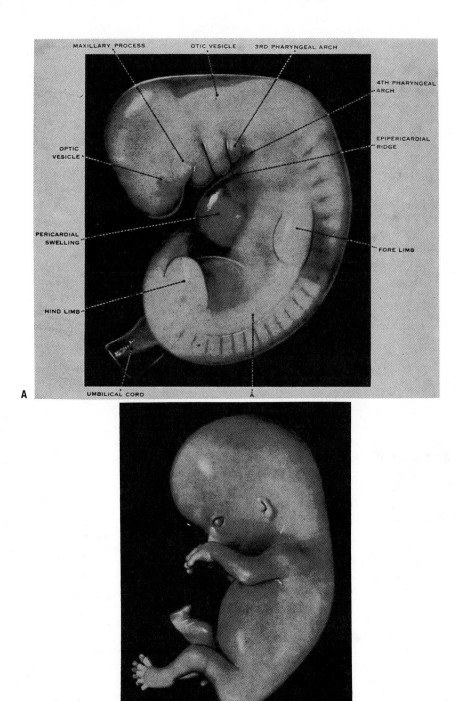

Figure 6.19 *A*, Side view of a human embryo about four weeks old; its crown-rump length is 5 mm. *B*, A human fetus about eight weeks old; its crown-rump length is 30 mm. (From Hamilton, Boyd and Mossman: Human Embryology, The Williams & Wilkins Co., Inc.)

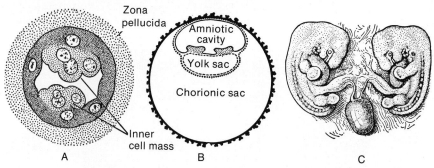

Figure 6.20 Two methods of monozygotic twinning: *A*, by division of inner cell mass; *B*, by the formation of two primitive streaks. In the latter case, the twins would be connected to a single yolk sac, as in *C*. (After Arey.)

lated and fertilized at about the same time. Such twins do not resemble each other any more closely than brothers or sisters born at different times, for they have somewhat different genetic constitutions. Fraternal twins occur more frequently in some families than in others, so it is possible that there are certain hereditary tendencies for the maturation and ovulation of more than one ovum during a single menstrual cycle.

More rarely, **monozygotic** or **identical twins** are formed. Only one egg is fertilized, but two embryos develop from it. Identical twins are always of the same sex and resemble each other closely, for they have identical genetic constitutions. Monozygotic twinning may occur in one of several ways. The two blastomeres produced by the first cleavage may separate and each become an embryo, the inner cell mass may subdivide, or two primitive streaks may develop upon a single embryonic disc (Fig. 6.20). Twins have been produced experimentally by the first method in lower vertebrates. This method is a possibility in mammals, but it is not so likely to occur as the others, for the mammalian egg and cleavage stages are surrounded by a strong membrane, the **zona pellucida,** that should prevent the blastomeres from separating. If embryos share the chorion, twinning occurs at the blastocyst stage or later.

A particularly interesting case of twinning is seen in the armadillo. This animal always has quadruplets, and the four individuals are always of the same sex. The fact that only one corpus luteum is found in the ovary, which means that only one follicle and egg matured and ovulated, indicates that all are identical twins. When the blastocyst is examined, it is discovered that the inner cell mass has subdivided into four parts.

If identical twins are produced by subdivision of the inner cell mass, or by the formation of two primitive streaks, one would expect to find occasional cases in which the separation is incomplete. Though fortunately rare, **conjoined twins** are born from time to time. All degrees of union have been found. Usually such individuals die in infancy, but the most famous pair, Chang and Eng, lived to be 63. Though Chinese, Chang and Eng were born in Siam. They worked for a circus, married and fathered 22 children! Their fame led to the popular term "Siamese twins" for such conjoined twins.

6.17 MORPHOGENETIC MOVEMENTS

Gastrulation in each type of embryo involves the movement or migration of cells which occur in specific ways and lead to specific arrangements of cells. These **morphogenetic movements** involve considerable parts of the embryo which stretch, fold, contract or expand. The movements are not, apparently, analogous to the contraction of muscles or to some sort of ameboid movement of the whole embryo. The forces driving these movements and the factors controlling their direction are unknown. A section of presumptive ectoderm from the early gastrula of the frog will expand actively, just as it does *in situ* in contributing to the movements of gastrulation, if removed from the embryo and grown in organ culture. A bit of the blastopore lip transplanted from one embryo into another embryo will invaginate and form an archenteric cavity independent of the archenteron of the host embryo. Invagination involves not only the movement of the cells, but changes in their shape, contractions at certain ends of the cells and expansions at others.

The movements of the cells and the positions they take up are guided, at least in part, by **selective affinities** of certain cells. This can be demonstrated if the cells of an early embryo are disaggregated and then incubated in various combinations. Epidermis cells become concentrated on the exterior of the cell mass and mesodermal cells take up a position between the epidermis and the endoderm. Neural plate cells form hollow vesicles resembling a neural tube or brain vesicle and mesodermal cells tend to arrange themselves around coelomic cavities. This sorting out of mixtures of cells is believed to result from their selective affinities. When cells touch as a result of random movement, they may remain in contact if held together or they may move apart if not bound strongly. You could postulate that different kinds of cells have qualitatively different specific affinities, or you could account for many of the experimental results by postulating quantitative differences in the degree of adhesiveness of different kinds of cells.

6.18 THE CONTROL OF DEVELOPMENT

Biologists have been interested for many years in the nature of the factors which regulate the complex, orderly processes leading to the production of a new adult from a fertilized egg. How can a single cell give rise to many different types of cells which differ widely in their morphologic, functional and chemical properties?

Early embryologists believed that the egg or the sperm contained a completely formed but minute germ which simply grew and expanded to give the adult. This **preformation theory** explained development by denying that it occurred! An extension of this theory postulated that each germ contained within it the germs for all succeeding generations,

each within the next. Some microscopists reported seeing this germ within the sperm or egg and described the "homunculus," a fully formed little man inside the egg or sperm! Others calculated the number of germs that were present in the ovaries of Eve, the mother of the human race, and suggested that when all of these were used up the human race would end.

The contrasting theory of **epigenesis,** first advanced by Wolff in 1759, stated that the unfertilized egg is not organized and that development involves progressive differentiation which is controlled by some outside force. We now know that development is not simply epigenetic, for there are certain potentialities localized in particular regions of the egg and the early embryo. The embryos of certain species, when separated into parts at an early stage, will develop normally; each part forms a complete, normal, though small, embryo. The embryos of other species show that certain potentialities are localized at an early stage, for neither of two parts can develop into a whole embryo. Each half develops only those structures it would have formed normally as part of the whole embryo. This localization of potentialities eventually occurs in the development of all eggs; it simply occurs at an earlier stage in some species than in others. It has been possible by experimental techniques to map out the areas of potentialities in the early amphibian gastrula and in the primitive streak stage of the chick (Fig. 6.21).

In the past, biologists have speculated that differentiation might occur (1) by some sort of segregation of genetic material during mitosis, (2) by the establishment of chemical gradients within the developing embryo, (3) by somatic mutations, (4) by the action of chemical organizers or (5) by the induction of specific enzymes.

Cellular differentiation might be explained if genetic material were parceled out differentially at cell division and the daughter cells received different kinds of genetic information. Although there are a few clear instances of differential nuclear divisions in animals such as *Ascaris* and *Sciara*, this does not appear to be a general mechanism of differentiation. The generalization that the mitotic process ensures the exact distribution of genes to each cell of the organism appears to be valid. Thus, the

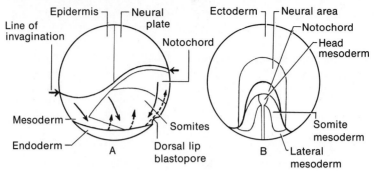

Figure 6.21   Embryo maps. *A*, Lateral view of a frog gastrula showing the presumptive fates of its several regions. *B*, Top view of a chick embryo showing location in the primitive streak stage of the cells which will form particular structures of the adult.

differences in enzymes and other proteins found in different cells must arise by differences in the *activity* of the same set of genes in different cells.

The induction of **adaptive enzymes** in bacteria has been used as a model system to provide an explanation for embryonic differentiation. Bacteria (and, to some extent, animals as well) can respond to the presence of some new substrate molecule by forming enzymes which will metabolize it. Jacques Monod, of the University of Paris, has suggested that extracellular or intracellular influences may initiate or suppress the synthesis of specific enzymes, thus affecting the chemical constitution of the cell and leading to differentiation. As an embryo develops, the gradients established as a result of growth and cell multiplication could result in quantitative and even qualitative differences in enzymes. As a result of the stimulation or inhibition of one enzyme, a chemical product could accumulate which would induce the synthesis of a new enzyme and thus confer a new functional activity on these cells.

Morphogenesis is probably too complicated a phenomenon to be explained in terms of a single phenomenon such as enzyme induction. Enzymes can indeed be induced in an embryo by the injection of a suitable substance. Adenosine deaminase, for example, has been induced by the injection of adenosine into a chick egg, but to date no enzyme has been induced which is not normally present to some extent in the embryo. Adult tissues show marked differences in their enzymatic activities, differences which might be the result of "adaptations" comparable to those seen in bacte-

ria. Adaptive changes in enzymes, however, are temporary and reversible, whereas differentiation is a permanent, essentially irreversible process. Cells may lose some of their morphologic characteristics but they retain all of their biochemical specificities.

Some interesting data bearing on the problem of morphogenesis were obtained by Briggs and King of the Lankenau Institute. They prepared a ripe frog's egg by removing its nucleus. A cell from an advanced stage of embryonic development is then separated from its neighbors and sucked up into a micropipette (Fig. 6.22). The plasma membrane of the cell is ruptured by the process and the nucleus plus some cytoplasmic debris is injected deep into the enucleated egg, after which the pipette is carefully withdrawn. The operated eggs begin cleaving and some develop normally and undergo metamorphosis. Nuclei obtained from late blastula or early gastrula stages, when there are as many as 16,000 cells, can, when transplanted in this fashion into an uncleaved egg, lead to the development of a normal embryo. Even nuclei from later stages of development, from the neural plate or from ciliated cells in the digestive tract of a swimming tadpole, may lead to the development of normal embryos when transplanted into enucleated eggs. The nuclei of cells from malignant tissue (cancer cells) but not from normal adult cells supported development when transplanted into enucleated eggs. It appears that the chromosomes of nuclei from cells in advanced stages of embryonic development or from adult cells are unable to divide rapidly enough to keep up with the rate of cytoplasmic division early in embryonic develop-

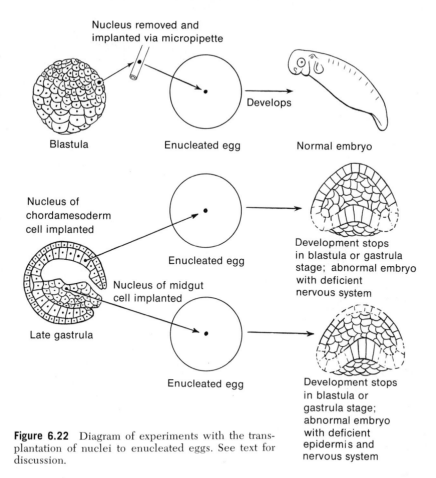

Figure 6.22 Diagram of experiments with the transplantation of nuclei to enucleated eggs. See text for discussion.

ment. Chromosomal duplication occurs too slowly and as a result the daughter cells receive incomplete sets of chromosomes and develop defectively. That an egg can develop normally when its nucleus is replaced with one from a highly differentiated cell is further proof that there is no segregation of genetic material during mitosis.

When a piece of the dorsal lip of the blastopore of a frog gastrula is implanted beneath the ectoderm of a second gastrula, the tissue heals in place and causes the development of a second brain, spinal cord and other parts at the site, so that a double embryo or closely joined Siamese twins results (Fig. 6.23). Many tissues show similar abilities to organize the development of an adjoining structure. The eyecup will initiate the formation of a lens from overlying ectoderm even if it is transplanted to the belly region, where the cells would normally form belly epidermis. Such experiments indicate that development is a coor-

dinated series of chemical stimuli and responses, each step regularly determining the succeeding one. The term "organizer" is applied to the region of the embryo with this property and also to the chemical substance given off by that region which passes to the adjoining tissue and directs its development.

It had been widely accepted that organizers can transmit their inductive stimuli only when in direct physical contact with the reactive cells. However, evidence from experiments by Niu and Twitty indicates that induction may be mediated by diffusible substances which can operate without direct physical contact of the two tissues. Niu and Twitty grew small groups of frog ectoderm, mesoderm and endoderm cells in tissue culture and found that ectoderm alone would never differentiate into nerve tissue. Ectoderm cells placed in a medium in which mesoderm cells had been grown for the previous week did differentiate into

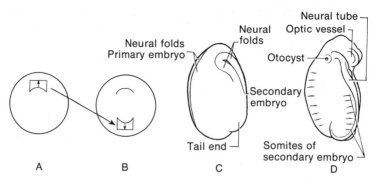

Figure 6.23   The induction of a second frog embryo by the implantation of the dorsal lip of the blastopore from embryo *A* onto the belly region of embryo *B*. Embryo *B* then develops through stage *C* to a double embryo, *D*.

chromatophores and nerve fibers. No comparable differentiation occurred when the ectoderm cells were placed in comparable cultures of endoderm cells. These experiments suggest that inductor tissues, such as chorda-mesoderm, contain and release diffusible substances which are capable of operating at a distance and inducing the differentiation of ectoderm. These substances appear to be nucleoproteins.

Additional evidence that morphogenetic substances are diffusible is provided by Clifford Grobstein's analysis of the differentiation of kidney tubules. Nephrogenic mesenchyme is normally induced to form renal tubules by the tip of the ureter. If the two rudiments are separated by treatment with trypsin and the dissociated cells are grown in tissue culture, they will reaggregate and form tubules. Certain tissues other than the ureter, and the ventral half of the spinal cord in particular, proved to be efficient inducers of renal tubules. Grobstein then separated the inducing and reacting tissues by cellophane membrane filters of varying thicknesses and porosities. The inducing substance can pass through membranes up to 60 microns thick with pores 0.4 micron in diameter. Electron micrographs revealed that induction occurred even when there was no cellular contact through the fine pores: the inducing principle is diffusible. It appears to be a large molecule, and is at least in part protein since it is inactivated by trypsin.

Evidence that steroids may play a morphogenetic role in development was obtained by Dorothy Price of the University of Chicago. When the reproductive tract of a fetal male rat is dissected out and grown in organ culture, development occurs normally if the testis is left in place. If both testes are removed, there is no development and differentiation of the accessory organs — vas deferens, seminal vesicles and prostate gland. However, if both testes are removed and a pellet of **testosterone,** the male sex hormone, is implanted, development proceeds normally. Thus testosterone can diffuse across a limited space and induce the development of male characters; it is a morphogenetic substance.

The bonds holding cells together can be broken and the cells can be made to dissociate by treating a tissue with dilute trypsin solution. In this way a suspension of individual, healthy cells can be obtained. When the dissociated cells are grown in tissue culture medium, the cells can reaggregate and continue to differentiate. This occurs in conformity with their previous pattern of differentiation. The cells reaggregate not in a chaotic mass but in an ordered fashion, forming recognizable morphogenetic units. The cells appear to have specific affinities, for epidermal cells join with each other to form a sheet, disaggregated kidney cells join to form kidney tubules, and so on. The question of how one cell recognizes another and joins with it to form a tubule, sheet or other morphologic unit is indeed a fascinating one.

Evidence that chemical differentiation precedes morphologic differentiation of a tissue has come from research using serologic and biochemical methods. The specific protein of the lens of the eye can be detected serologically in the chick embryo before the lens vesicle closes and before there is any evidence of morphologic differentiation. Cholinesterase is the enzyme which hydrolyzes acetylcholine and is believed to play an important role in the transmission of the nerve impulse. Edgar Boell showed by microchemical methods that the neural folds of the frog embryo, the parts which will form the central nervous system, have much more cholinesterase than the epidermis does. When epidermis is stimulated to form nervous system, by grafting a piece of chorda-mesoderm beneath it, the tissue becomes rich in cholinesterase.

In view of the extreme complexity of the developmental process it is indeed remarkable that it occurs so regularly and that so few malformations occur. About one child in 100 is born with some major defect, a cleft palate, clubfoot, spina bifida or the like. Some of these are inherited and others result from environmental factors. Experiments with fruit flies, frogs and mice have shown that x-rays, ultraviolet rays, temperature changes and a variety of chemical substances will induce alterations in development. The kind of defect produced is a function of the time in the course of development at which the environmental agent is applied and does not depend to any great extent on the kind of agent used. For example, x-rays, the administration of corti-

sone and the lack of oxygen will all produce similar defects in mice—harelip and cleft palate—if applied at comparable times in development. Such observations have led to the concept of **critical periods** in development, periods during which the development of a certain organ or organ system is occurring rapidly and hence is most susceptible to interference.

The property of reproduction, which we regard as one of the outstanding characteristics of living things, involves a great many complex and interdependent processes: the elaboration of hormones which regulate the development of gonads, secondary sex structures and the production of gametes in the parents; behavior patterns which bring the parents together and have them release their gametes at such a time and in such a place as to make their fusion probable; the union of male and female pronuclei followed by cleavage, gastrulation and morphogenesis; and devices for the care and protection of the developing young. Our descriptive knowledge of these phenomena is extensive but our understanding of the fundamental mechanisms involved in each of these processes is rudimentary. This is a fertile field for further investigation.

ANNOTATED REFERENCES

Arey, L. B.: Developmental Anatomy, 7th ed. Philadelphia, W. B. Saunders Co., 1965. A standard text on human development, emphasizing morphology.

Austin, C. R.: Fertilization. Englewood Cliffs, N.J., Prentice-Hall, Inc., 1965. A description of factors controlling the process of fertilization in various animals.

Balinsky, B. I.: An Introduction to Embryology, 3rd ed. Philadelphia, W. B. Saunders Co., 1970. A well-written and well-illustrated account of animal development with excellent accounts of gametogenesis, fertilization and early development.

Berrill, N. J.: Sex and the Nature of Things. New York, Dodd, Mead & Co., 1953. An interesting essay on the significance and nature of the sexual process in a variety of animals and plants.

Milne, L. J., and M. J. Milne: The Mating Instinct. Boston, Little, Brown and Co., 1954. A comparative account of the mating process in certain animals.

Page, E. W., C. A. Villee, and D. B. Villee: Human Reproduction. Philadelphia, W. B. Saunders Co., 1972. A general account of human reproduction ranging from the behavioral to the endocrine and biochemical aspects of the subject.

7.   Principles of Heredity

7.1   HISTORY OF GENETICS

It must have been thousands of years ago when man first made one of the fundamental observations of heredity—that "like tends to beget like." But his curiosity as to why this is true and how it is brought about remained unsatisfied until the beginning of the present century. A number of breeders, such as Kölreuter, who worked with tobacco plants about 1770, crossed different varieties of plants and produced hybrids. Kölreuter recognized that parental characters were transmitted by both the pollen and the ovule. Mendel's careful work with peas revealed the fundamental principles of heredity, but the report of his work, published in 1866, was far ahead of his time. It is clear that his work was known to a number of the leading contemporary biologists, such as the botanist Nägeli, but in the absence of our present knowledge of chromosomes and their behavior, its significance was unappreciated.

The chromosomal details of mitosis were described by Eduard Strasburger in 1876. Eduard van Beneden (1887) discovered the process of meiosis and understood its significance. Earlier that same year Weismann had pointed out, simply from theoretical considerations, that the chromosomal number in gametes must be half of that in somatic cells. It is conceivable that some brilliant theoretical biologist with these facts at hand might have postulated that, if hereditary factors were units located in the chromosomes, the mating of different parental types would yield offspring in predictable ratios. However, no such synthesis was made, and the existence of these definite ratios of the types of offspring resulting from a given mating remained to be demonstrated experimentally.

In 1900, three different biologists, working independently—de Vries in Holland, Correns in Germany and von Tschermak in Austria—rediscovered the phenomenon of regular, predictable ratios of the types of offspring produced by mating pure-bred parents. They then found Mendel's published report and, realizing his priority in these discoveries, gave him credit for his work by naming two of the fundamental principles of heredity **Mendel's laws.**

With the genetic and cytologic facts at hand, W. S. Sutton and C. E. McClung independently came to the conclusion (1902) that the hereditary factors are located in the chromosomes. They also pointed out that, since there are many more hereditary factors than chromosomes, there must be more than one hereditary factor per chromosome. By 1911, T. H. Morgan was able to postulate, from the regularity with which certain characters tended to be inherited together, that the hereditary factors (which he named "genes") were located in the chromosomes in linear order, "like the beads on a string."

7.2   MENDEL'S DISCOVERIES

Gregor Johann Mendel (1822–1884) was an Austrian abbot who spent some eight years breeding peas in the garden of his monastery at Brünn, now part of Czechoslovakia. He succeeded in reaching an understanding of the basic principles of heredity because (1) he studied the inheritance of single contrasting characters (such as green versus yellow seed color, wrinkled versus smooth seed coat), instead of attempting to study the complete inheritance of each organism; (2) his studies were quantitative; he counted the number of each type of offspring and kept accurate records of his crosses and results; and (3) by design or by good fortune, he chose a plant, and particular characters of that plant, that gave him clear ratios. If he had worked with other plants or with certain other characters of peas, he would have been unable to get these ratios. Now that the principles of heredity have been established, the explanation for these more complicated types of inheritance is clear.

Mendel established pure-breeding strains of peas with contrasting characters — yellow seed coat versus green seed coat, round seeds versus wrinkled ones — and then made crosses of the contrasting varieties. He found that the offspring of a cross of yellow and green all had yellow seed coats; the result was the same whether the male or the female parent had been the yellow one. Thus, the character of one parent can "dominate" over that of the other, but which of the contrasting characters is dominant depends upon the specific trait involved, not upon which parent contributes it. This observation, repeated for several different strains of peas, led Mendel to the generalization — the "law of dominance" — that, when two factors for the alternative expression of a character are brought together in one individual, one may be expressed completely and the other not at all. The character which appears in the first generation is said to be **dominant;** the contrasting character is said to be **recessive.**

Mendel then took the seeds produced by this first generation of the cross (called the **first filial generation,** abbreviated F_1), planted them and had the resulting plants fertilize themselves to produce the second filial generation, the F_2. He found that both the dominant and the recessive characters appeared in this generation, and upon counting the number of each type (Table 7.1) he found that, whatever set of characters he used, the ratio of plants with the dominant character to those with the recessive character was very close to 3:1. From such experiments Mendel concluded that (1) there must be discrete unit factors which determine the inherited characters, (2) these unit factors must exist in pairs, and (3) in the formation of gametes the members of these pairs separate from each other, with the result that each gamete receives only one member of the pair. The unit factor for green seed color is not affected by existing for a generation within a yellow seeded plant (e.g., the F_1 individuals). The two separate during gamete formation and, if a gamete bearing this factor for green seed coat fertilizes another gamete with this factor, the resulting seed has a green color. The generalization known as Mendel's first law, the **law of segregation,** may now be stated as: Genes exist in pairs in individuals, and in the formation of gametes each gene separates or segregates from the other member of the pair and passes into a different gamete, so that each gamete has one, and only one, of each kind of gene.

TABLE 7.1 AN ABSTRACT OF THE DATA OBTAINED BY MENDEL FROM HIS BREEDING EXPERIMENTS WITH GARDEN PEAS

Parental Characters	First Generation	Second Generation	Ratios
Yellow seeds × green seeds	All yellow	6022 yellow : 2001 green	3.01 : 1
Round seeds × wrinkled seeds	All round	5474 round : 1850 wrinkled	2.96 : 1
Green pods × yellow pods	All green	428 green : 152 yellow	2.82 : 1
Long stems × short stems	All long	787 long : 277 short	2.84 : 1
Axial flowers × terminal flowers	All axial	651 axial : 207 terminal	3.14 : 1
Inflated pods × constricted pods	All inflated	882 inflated : 299 constricted	2.95 : 1
Red flowers × white flowers	All red	705 red : 224 white	3.15 : 1

In other experiments Mendel observed the inheritance of two pairs of contrasting characters in a single cross. He mated a pure-breeding strain with round yellow seeds and one with wrinkled green seeds. The first filial generation all had round yellow seeds, but when these were self-fertilized he found in the F_2 generation all four possible combinations of seed color and shape. When he counted these he found 315 round yellow seeds, 108 round green seeds, 101 wrinkled yellow seeds, and 32 wrinkled green seeds. There is a close approximation of a 3:1 ratio for seed color (416 yellow to 140 green) and for seed shape (423 round to 133 wrinkled). Thus, the inheritance of seed color is independent of the inheritance of seed shape; neither one affects the other. When the two types of traits are considered together, it is clear that there is a ratio of 9 with two dominant traits (yellow and round): 3 with one dominant and one recessive (round and green): 3 with the other dominant and recessive (yellow and wrinkled): 1 with the two recessive traits (green and wrinkled). Mendel's second law, the **law of independent assortment,** may now be given as: The distribution of each pair of genes into gametes is independent of the distribution of any other pair.

7.3 CHROMOSOMAL BASIS OF THE LAWS OF HEREDITY

Each cell of every member of a given species of animal or plant contains a definite number of chromosomes; the constancy of the chromosome number is assured by the precise and regular events of mitotic division (p. 41). Many widely different species of animals and plants have the same number of chromosomes. It is not the number of chromosomes, but the nature of the hereditary factors within them, that differentiates species.

The constancy of the chromosome number in successive generations of the same species is assured by the precise separation of the members of the pairs of homologous chromosomes in the meiotic divisions leading to the formation of gametes. The normal number of chromosomes for somatic cells is reconstituted in fertilization, when the egg and sperm nuclei fuse.

The laws of heredity follow directly from the behavior of the chromosomes in mitosis, meiosis and fertilization. Within each chromosome are numerous hereditary factors, the **genes,** each of which controls the inheritance of one or more characteristics. Each gene is located at a particular point, called a **locus** (plural, loci), along the chromosome. Since the genes are located in the chromosomes, and each cell has two of each kind of chromosome, it follows that each cell has two of each kind of gene. The chromosomes separate in meiosis and recombine in fertilization and so, of course, do the genes within them. We currently believe that the genes are arranged in a linear order within the chromosomes; the **homologous chromosomes** have similar genes arranged in a similar order. When the chromosomes undergo synapsis during meiosis (p. 121) the homologous chromosomes become attached point by point and, presumably, gene by gene.

7.4 GENES AND ALLELES

Studies of inheritance are possible only when there are two alternate, contrasting conditions, such as Mendel's yellow and green peas or round and wrinkled ones. These contrasting conditions, inherited in such a way that an individual may have one or the other but not both, were originally termed allelomorphic traits. At present the terms allele and gene are used more or less interchangeably; both refer to the hereditary factor responsible for a given trait. The term allele emphasizes that there are two or more alternative kinds of genes at a specific locus in a specific chromosome.

Brown and black coat color are allelomorphic traits in guinea pigs. Each body cell of the guinea pig has a pair of chromosomes which contain genes for coat color; since there are two chromosomes, there are two genes per cell. A "pure" black guinea pig (one of a pedigreed strain of black guinea pigs) has two genes for black coat, one in each chromosome, and a "pure" brown guinea pig has two genes for brown coat. The genes themselves, of course, have no color; they are neither brown nor black. The brown gene controls the formation of an enzyme involved in one of the steps in the production of a brown pigment in the hair cells, whereas the black gene produces a different enzyme which alters the chemical reactions in the synthesis of pigment and results in the production of black pigment. In working genetic problems, letters are conventionally used as symbols for the genes. A pair of genes for black pigment is represented as **BB,** and a pair of genes for brown pigment by **bb.** A capital letter is used for one gene and the corresponding lower case letter is used to represent the gene for the contrasting trait, the allele.

7.5 A MONOHYBRID CROSS

The usage of genetic terms and some of the basic principles of genetics can be illustrated by considering a simple mono-hybrid cross, that is, a cross between two individuals that differ in a single pair of genes. The mating of a "pure" brown male guinea pig with a "pure" black female guinea pig is illustrated in Figure 7.1. During meiosis in the testes of the male the two **bb** genes separate so that each sperm has only one **b** gene. In the formation of ova in the female the **BB** genes separate so that each ovum has only one **B** gene. The fertilization of this egg by a **b** sperm results in an animal with the genetic formula **Bb.** These guinea pigs contain one gene for brown coat and one for black coat. What color would you expect them to be—dark brown, gray or perhaps spotted? In this instance they are just as black as the mother. The gene for black coat color is said to be **dominant** to the gene for brown coat color. It will produce black body color even when only one dose of the black gene is present. The brown gene is said to be **recessive** to the black one; it will produce brown coat color only when present in double dose (**bb**). We may define recessive genes as ones which will produce their effects only when an individual has two of them that are identical, and dominant genes as those which will produce their effects even when only one is present in an individual. The dominant gene is usually represented by a capital letter and the recessive gene by the corresponding lower case letter. The phenomenon of dominance supplies part of the explanation as to why an individual may resemble one of his parents more than the other despite the fact that both make equal

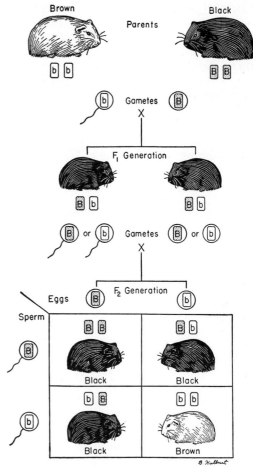

Figure 7.1 An example of a monohybrid cross: the mating of a brown with a black guinea pig. (Villee: Biology, 6th ed.)

contributions to his genetic constitution. In one species of animal black coat color may be dominant to brown while in another species brown might be dominant to black. The particular genetic relations that obtain in any given species must be determined by experiment.

An animal or plant with two genes exactly alike, two blacks (**BB**) or two browns (**bb**), is said to be **homozygous** or "pure" for the character. An organism with one dominant and one recessive gene (**Bb**) is said to be **heterozygous** or "hybrid." Thus, in the mating under consideration the black and brown parents were homozygous, **BB** and **bb,** respectively, and the offspring in the F₁ were all heterozygous, **Bb. Recessive genes** are those which will produce their effect only when homozygous; a **dominant gene** is one which will produce its effect whether it is homozygous or heterozygous.

In the process of gamete formation in these heterozygous black F_1 guinea pigs, the chromosome containing the **B** gene undergoes synapsis with, and then separates from, the homologous chromosome containing the **b** gene, so that each sperm or egg has a **B** gene or a **b** gene. No sperm or egg is without one or the other and none has both. Since there are two kinds of eggs and two kinds of sperm, the mating of two of these heterozygous black guinea pigs permits four different combinations of eggs and sperm. To see these possible combinations of eggs and sperm it is conventional to arrange them in a Punnett square (Fig. 7.1), devised by the English geneticist, R. C. Punnett. Gametes containing **B** genes and ones containing **b** genes are formed in equal numbers. There is no special attraction or repulsion between an egg and a sperm containing similar genes; an egg containing a **B** gene is just as likely to be fertilized by a **B** sperm as by a **b** sperm. The four possible combinations occur with equal frequency.

The possible types of eggs are written across the top of the Punnett square and the possible types of sperm are arranged down its left side, then the squares are filled in with the resulting zygote combinations (Fig. 7.1). Three-fourths of the offspring are either **BB** or **Bb,** and consequently have a black coat color, and one-fourth are **bb,** with a brown coat color. This 3:1 ratio is characteristically obtained in the second generation of a **monohybrid cross,** i.e., a mating of two individuals which differ in a single trait governed by a single pair of genes. The genetic mechanism responsible for the 3:1 ratios obtained by Mendel in his pea breeding experiments is now evident.

The appearance of an individual with respect to a certain inherited trait is known as its **phenotype.** The organism's genetic constitution, usually expressed in symbols, is called its **genotype.** In the cross we have been considering, the phenotypic ratio of the F_2 generation is 3 black coated guinea pigs: 1 brown coated guinea pig and the genotypic ratio is 1 **BB**:2 **Bb**:1 **bb.** The phenotype may be a morphologic characteristic—shape, size, color—or a physiologic characteristic such as the presence or absence of a specific enzyme required for the metabolism of a specific substrate. Guinea pigs with the genotypes **BB** and **Bb** are alike phenotypically; they both have black coats. One-third

of the black guinea pigs in the F_2 generation of the mating of black × brown are homozygous, **BB,** and the other two-thirds are heterozygous, **Bb.** The two types can be differentiated by making a **test cross,** by mating them with homozygous brown (**bb**) guinea pigs. If all of the offspring are black, what inference would you make about the genotype of the black parent? If any of the offspring are brown, what conclusion would you draw regarding the genotype of the black parent? Can you be more certain about one of these two inferences?

This sort of testing is of great importance in the commercial breeding of animals or plants where the breeder is trying to establish a strain that will breed true for a certain characteristic. When the individuals in a breeding stock are selected on the basis of their own phenotypes, the breeding program is not maximally effective because it does not differentiate between homozygous and heterozygous individuals. The method of **progeny selection** is one in which a breeder tests the genotypes of his breeding stock by making test matings and observing the offspring. If the offspring are superior with respect to the desired trait, the parents are thereafter used regularly for breeding. Two bulls, for example, may look equally healthy and vigorous, yet one will have daughters with qualities of milk production that are distinctly superior to those of the daughters of the other bull. By a method of progeny selection one breeding program raised the average annual egg production of a flock of hens from 114 to 200 eggs over a period of eight years.

7.6 LAWS OF PROBABILITY

It is important to realize that all genetic ratios are expressions of probability, based on the laws of chance or probability; they do not express certainties. One can state, perhaps more exactly, that in the mating of two individuals heterozygous for a given trait there are three chances out of four that any particular offspring will show the dominant trait and one chance out of four that it will show the recessive one. If two heterozygous black guinea pigs are mated and have exactly four offspring there is no guarantee that there will be exactly three black ones and one brown one. All might be black, or all might be brown. Since the probability that each offspring will be black is $3/4$, the probability that all four will be black is the product of the individual probabilities, $3/4 \times 3/4 \times 3/4 \times 3/4$ or $81/256$. The probability that all four will be brown is $1/4 \times 1/4 \times 1/4 \times 1/4$ or $1/256$. Any of the combinations of 3 black:1 brown, 2 black:2 brown, or 1 black:3 brown might appear. But if enough similar matings are made to produce a total of 400 offspring, the ratio of black to brown among the offspring will be very close to 300 to 100. The theoretical 3:1 ratio is approximated more and more exactly as the total number of individuals increases; this is predicted by the laws of probability and is actually found when genetic tests are made. Each mating, each union of an egg and a sperm, is an **independent event** which is not influenced by the results of previous matings. No matter how many black-coated offspring have been produced by the mating of two heterozygous black ones, the probability that the next offspring to be born will have a brown coat is one chance in four, and the probability that it will have a black coat is three chances in four.

7.7   INCOMPLETE DOMINANCE

In many different species and for a variety of traits it has been found that one gene is not completely dominant to the other. Heterozygous individuals have a phenotype which can be distinguished from that of the homozygous dominant; it may be intermediate between the phenotypes of the two parental strains. Such pairs of genes are said to be **codominant** or to show incomplete dominance. The mating of red short-horn cattle with white ones yields offspring which have an intermediate, roan-colored coat. The mating of two roan-colored cattle yields offspring in the ratio of 1 red:2 roan:1 white; thus the genotypic and phenotypic ratios are the same; each genotype has a recognizably different phenotype. This phenomenon of **incomplete dominance** is found with a number of traits in different animals and with some human characteristics. Studies of a number of human diseases inherited by recessive genes—sickle cell anemia, Mediterranean anemia, gout, epilepsy and many others—have shown that the individuals who are heterozygous for

the trait have slight but detectable differences from the homozygous normal individual.

7.8   A DIHYBRID CROSS

Frequently a geneticist must analyze the inheritance of two or more traits in the same group of individuals. A mating that involves individuals differing in two traits is called a **dihybrid cross.** The principles involved and the procedure of solving problems are exactly the same in monohybrid and in dihybrid or trihybrid crosses. In the latter the number of types of gametes is greater and the number of types of zygotes is correspondingly larger.

When two pairs of genes are located in different (nonhomologous) chromosomes, each pair is inherited independently of the other; that is, each pair separates during meiosis independently of the other. When a black, short-haired guinea pig (**BBSS**, short hair is dominant to long hair) and a brown, long-haired guinea pig (**bbss**) are mated, the **BBSS** individual produces gametes all of which are **BS**. The **bbss** guinea pig produces only **bs** gametes. Each gamete contains one and only one of each kind of gene. The union of **BS** gametes and **bs** gametes yields only individuals with the genotype **BbSs**. All of the offspring are heterozygous for hair color and for hair length and all are phenotypically black and short-haired.

When two of the F_1 individuals are mated, each produces four kinds of gametes in equal numbers—**BS, Bs, bS, bs;**—thus 16 combinations are possible among the zygotes (Fig. 7.2). There are 9 chances in 16 of obtaining a black, short-haired individual, 3 chances in 16 of obtaining a black, long-haired individual, 3 chances in 16 of obtaining a brown, short-haired individual and 1 chance in 16 of obtaining a brown long-haired individual.

Mendel's First Law, sometimes called the Law of the Purity of Gametes or the Law of Segregation, is illustrated by the mating of the black and brown guinea pigs described previously. Mendel's law may be stated as follows: Genes exist in individuals in pairs and in the formation of gametes each gene separates or segregates from the other member of the pair and passes into a different gamete so that each gamete has one and only

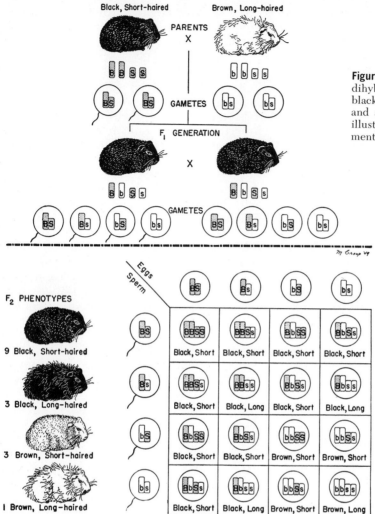

Figure 7.2 An example of a dihybrid cross: the mating of a black, short-haired guinea pig and a brown, long-haired one, illustrating independent assortment. (Villee: Biology, 6th ed.)

one of each kind of gene. Mendel's Second Law, the Law of Independent Segregation, is illustrated by this second mating. It states that the members of one pair of genes separate or segregate from each other in meiosis independently of the members of other pairs of genes and come to be assorted at random in the resulting gamete. The segregation of the **B-b** genes is independent of the segregation of the **S-s** genes. This law does not apply if the two pairs of genes are located in the same pair of chromosomes.

In a similar fashion problems involving three pairs of genes may be solved. An individual heterozygous for three pairs of genes will yield eight types of gametes in equal numbers. The union of these eight types of eggs and eight types of sperm will yield 64 possible zygotes in the F_2 generation. In peas, as Mendel demonstrated, yellow seed color (**Y**) is dominant to green (**y**), smooth seeds (**S**) are dominant to wrinkled (**s**), and tall plants (**T**) are dominant to dwarf (**t**). The mating of a homozygous yellow, smooth, tall plant (**YYSSTT**) with a homozygous green, wrinkled dwarf plant (**yysstt**) will produce offspring all of which are yellow, smooth and tall (**YySsTt**). When two of these F_1 plants are mated, F_2 offspring are produced in the ratio of 27 yellow, smooth, tall: 9 yellow, smooth, dwarf: 9 yellow, wrinkled, tall: 9 green, smooth, tall: 3 yellow, wrinkled, dwarf: 3 green, wrinkled, tall: 3 green, smooth, dwarf: 1 green, wrinkled, dwarf. Draw a Punnett square to verify these numbers.

7.9 DEDUCING GENOTYPES

The science of genetics resembles mathematics in that when the student has mastered the few basic principles involved he can solve a wide variety of problems. These basic principles include: (1) Inheritance is biparental; both parents contribute to the genetic constitution of the offspring. (2) Genes are not altered by existing together in a heterozygote. (3) Each individual has two of each kind of gene, but each gamete has only one of each kind. (4) Two pairs of genes located in different chromosomes are inherited independently. (5) Gametes unite at random; there is neither attraction nor repulsion between an egg and a sperm containing identical genes.

In working genetics problems, it is helpful to use the following procedure:

1. Write down the symbols used for each pair of genes.
2. Determine the genotypes of the parents, deducing them from the phenotypes of the parents and, if necessary, from the phenotypes of the offspring.
3. Derive all of the possible types of gametes each parent would produce.
4. Prepare the appropriate Punnett square and write the possible types of eggs across its top and the types of sperm along its side.
5. Fill in the squares with the appropriate genotypes and read off the genotypic and phenotypic ratios of the offspring.

As an example of the method of solving a problem in genetics, let us consider the following: The length of fur in cats is an inherited trait; the gene for long hair (l), as in Persian cats, is recessive to the gene for

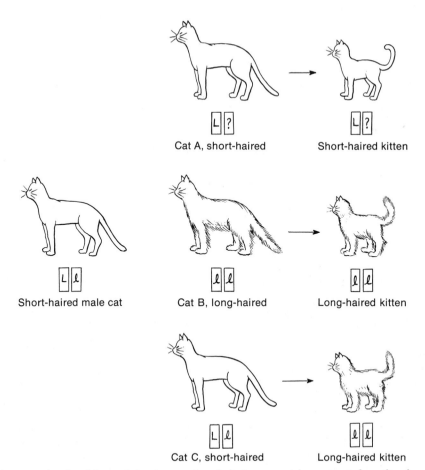

Figure 7.3 An example of problem-solving in genetics: deducing parental genotypes from the phenotypes of the offspring. See text for discussion.

short hair (**L**) of the common tabby cat. Let us suppose that a short-haired male is bred to three different females, two of which, A and C, are short-haired and one, B, is long-haired (Fig. 7.3). Cat A gives birth to a short-haired kitten, but cats B and C each produce a long-haired kitten. What offspring could be expected from further mating of this male with these three females?

Since the long-haired trait is recessive we know that all the long-haired cats must be homozygous. We can deduce, then, that cat B and the kittens produced by cats B and C have the genotype **ll**. All the short-haired cats have at least one **L** gene. The fact that any of the offspring of the male cat has long hair proves that he is heterozygous, with the genotype **Ll**. The kitten produced by cat B received one **l** gene from its mother but must have received the other from its father. The fact that cat C gave birth to a long-haired kitten proves that she, too, is heterozygous, and has the genotype **Ll**. It is impossible to decide, from the data at hand, whether the short-haired cat A is homozygous **LL** or heterozygous **Ll**. A test cross with a long-haired male would be helpful in deciding this. Further mating of the short-haired male with cat B would give half long-haired and half short-haired kittens, whereas further mating of the short-haired male with cat C would give three times as many short-haired kittens as long-haired ones.

7.10 THE GENETIC DETERMINATION OF SEX

The sex of an organism is a genetically determined trait. There is an exception to the general rule that all homologous pairs of chromosomes are identical in size and shape: the so-called **sex chromosomes.** In one sex of each species of animals there is either an unpaired chromosome or an odd pair of chromosomes, the two members of which differ in size and shape. In most species the females have two identical chromosomes, called X chromosomes, and males have either a single X chromosome or one X plus a generally somewhat smaller one called the Y chromosome. The existence of these unpaired chromosomes was discovered by C. E. McClung, in 1902, when he was studying the process of meiosis in

the testes of grasshoppers. He made the shrewd guess that these might play some role in sex determination. In a few animals, the butterflies and birds, the system is reversed and the male has two X chromosomes and the female one X and one Y. The Y chromosome usually contains few genes and in most species the X and Y chromosomes are distinguished by their different size and shape. Yet in meiosis the X and Y chromosomes act like homologous chromosomes; they undergo synapsis, separate, pass to opposite poles, and become incorporated into different gametes (Fig. 7.4). Human males have 22 pairs of ordinary chromosomes, called **autosomes,** one X and one Y chromosome, whereas females have 22 pairs of autosomes and two X chromosomes.

The experiments of C. B. Bridges revealed that the sex of fruit flies, *Drosophila*, is determined by the ratio of the number of X chromosomes to the number of haploid sets of autosomes. Males have one X and two haploid sets of autosomes, a ratio of 1:2, or 0.5. Females have two X and two haploid sets of autosomes, a ratio of 2:2. or 1.0. By genetic techniques possible in fruit flies, Bridges established abnormal flies with one X and three sets of autosomes. These flies, with a ratio of 0.33, had all their male characteristics exaggerated; Bridges called them "supermales." Other abnormal individuals, with three X and two sets of autosomes were "superfemales," with all the female characteristics exaggerated. Individuals with two X chromosomes and three sets of autosomes, a ratio of 0.67, were intersexes, with characters intermediate between those of normal males and normal females. All of these unusual flies, supermales, superfemales and intersexes, were sterile.

In man, and perhaps in other mammals, maleness is determined in large part by the presence of the Y chromosome. An individual with an XXY constitution is a nearly normal male in external appearance, though with underdeveloped gonads (Klinefelter's syndrome). An individual with one X but no Y chromosome has the appearance of an immature female (Turner's syndrome). It is possible to determine the "nuclear sex" of an individual by careful microscopic examination of some of his cells. Individuals with two X chromosomes have a "chromatin spot" at the edge of the nucleus which is

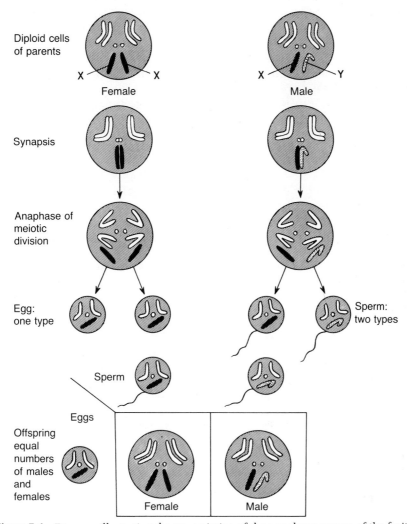

Figure 7.4 Diagram illustrating the transmission of the sex chromosomes of the fruit fly.

Figure 7.5 Sex chromatin in human fibroblasts cultured from skin of a female. The chromatin spot at the periphery of each nucleus is indicated by the arrow. (Feulgen, ×2200. Courtesy Dr. Ursula Mittwoch, Galton Laboratory, University College, London.) (Villee: Biology, 6th ed.)

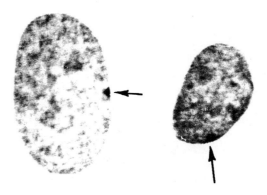

evident in cells from the skin or from the mucosal lining of the mouth (Fig. 7.5). Cells from male individuals with only one X chromosome do not show a chromatin spot.

The chromatin spot represents one of the two X chromosomes, which becomes condensed and dark-staining. The other X chromosome, like the autosomes, is a fully extended thread not evident by light microscopy. Mary Lyon has suggested that only one of the two X chromosomes in the female is active; the condensed chromatin spot represents the inactive X chromosome. Which of the two becomes inactive in any given cell is a matter of chance; thus the cells in the body of a female are of two types, according to which X chromosome is inactive. Since the two X chromosomes may have different genetic complements, the cells may differ in the effective genes present. Mice have several genes for coat color in the X chromosome, and females heterozygous for two such genes may show patches of one coat color in the midst of areas of the other, a phenomenon termed variegation. The inactivation of one X chromosome apparently occurs early in embryonic development and thereafter all the progeny of that cell have the same inactive X chromosome. Although one X chromosome appears to be inactive, there are marked abnormalities of development when one X chromosome is completely missing from the chromosomal complement of the cell, as in the XO condition of Turner's syndrome.

All the eggs produced by XX females have one X chromosome. Half of the sperm produced by XY males contain an X chromosome and half contain a Y chromosome. The fertilization of an X-bearing egg by an X-bearing sperm results in an XX, or female, zygote, and the fertilization of an X-bearing egg by a Y-bearing sperm results in an XY, or male, zygote. Since there are equal numbers of X- and Y-bearing sperm, there are equal numbers of male and female offspring. In human beings, there are approximately 107 males born for every 100 females, and the ratio at conception is said to be even higher, about 114 males to 100 females. One possible explanation of the numerical discrepancy is that the Y chromosome is smaller than the X chromosome, and a sperm containing a Y chromosome, being a little lighter and perhaps able to swim a little faster than a sperm containing an X chromosome, would win the race to the egg slightly more than half of the time. Both during the period of intrauterine development and after birth, the death rate among males is slightly greater than that among females, so that by the age of 10 or 12 there are equal numbers of males and females. In later life there are more females than males in each age group.

7.11 SEX-LINKED CHARACTERISTICS

The X chromosome contains many genes, and the traits controlled by these genes are said to be **sex-linked,** because their inheritance is linked with the inheritance of sex. The Y chromosome contains very few genes, so that the somatic cells of an XY male contain only one of each kind of gene in the X chromosome instead of two of each kind as in XX females. A male receives his single X chromosome, and thus all of his genes for sex-linked traits, from his mother. Females receive one X from the mother and one from the father. In writing the genotype of a sex-linked trait it is customary to write that of the male with the letter for the gene in the X chromosome plus the letter Y for the Y chromosome. Thus AY would represent the genotype of a male with a dominant gene for trait "A" in his X chromosome.

The phenomenon of sex-linked traits was discovered by T. H. Morgan and C. B. Bridges in the fruit fly, *Drosophila.* These flies normally have eyes with a dark red color, but Morgan and Bridges discovered a strain with white eyes. The gene for white eye, **w,** proved to be recessive to the gene for red eye, **W,** but in certain types of crosses the male offspring had eyes of one color and the female offspring had eyes of the other color. Morgan reasoned that the peculiarities of inheritance could be explained if the genes for eye color were located in the X chromosome; later work proved the correctness of this guess. Crossing a homozygous red-eyed female with a white-eyed male (**WW × wY**) produces offspring all of which have red eyes (**Ww** females and **WY** males). But crossing a homozygous white-eyed female with a red-eyed male (**ww × WY**) yields red-eyed females and white-eyed males (**Ww** and **wY**) (Fig. 7.6).

In man, **hemophilia** (bleeder's disease) and **color blindness** are sex-linked traits.

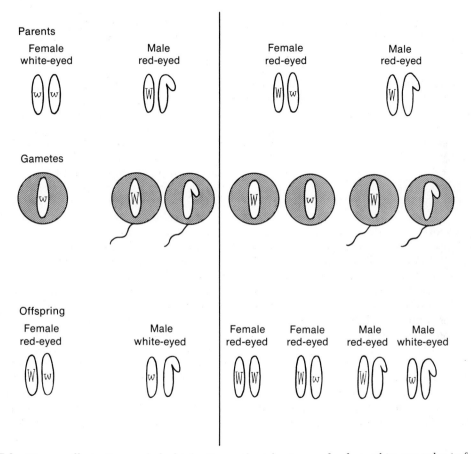

Figure 7.6 Diagram illustrating sex-linked inheritance, the inheritance of red vs. white eye color in fruit flies. See text for discussion.

About four men in every hundred are color-blind, but somewhat less than one per cent of all women are color-blind. Only one gene for color-blindness produces the trait in males, but two such genes (the trait is recessive) are necessary to produce a color-blind female.

Not all the characters which differ in the two sexes are sex-linked. Some, the **sex-influenced traits**, are inherited by genes located in autosomes rather than X chromosomes, but the expression of the trait, the action of the gene which produces the phenotype, is altered by the sex of the animal, presumably by the action of one of the sex hormones. The presence or absence of horns in sheep, mahogany-and-white spotted coat versus red-and-white spotted coat in Ayrshire cattle, and pattern baldness in man are examples of such sex-influenced traits.

7.12 LINKAGE AND CROSSING OVER

Each species of animal has many more pairs of genes than it has pairs of chromosomes. Obviously there must be many genes per chromosome. Man has 23 pairs of chromosomes, some large and some smaller, but thousands of pairs of genes. The chromosomes are inherited as units — they pair and separate during meiosis as units; thus, all the genes in any given chromosome tend to be inherited together and are said to be linked. If the chromosomal units never changed, the traits would always be inherited together and linkage would be absolute. However, during meiosis when the chromosomes are pairing and undergoing synapsis, homologous chromosomes may exchange entire segments of chromosomal material, a process called **crossing over** (Fig. 7.7). This exchanging of segments occurs at random along the length of the chromosome. Several exchanges may occur at different points along the same chromosome at a single meiotic division. It follows that the greater the distance is between any two genes in the chromosome, the greater will be the likelihood that an exchange of segments between them will occur.

In fruit flies the pair of genes **V** for normal wings and **v** for vestigial wings and the pair of genes **B** for gray body color and **b** for black body color are located in the same pair of chromosomes. They tend to be inherited together and are said to be linked. What would you predict the offspring would be like from a cross of a homozygous **VVBB** fly with a homozygous **vvbb** fly? They are all flies with gray bodies and normal wings and have the genotype **VvBb.** When one of these F_1 heterozygotes is crossed with a homozygous **vvbb** fly (Fig. 7.8), the offspring appear in a ratio which differs from that of the ordinary test cross for a dihybrid. If the two pairs of genes were not linked but were in different chromosomes, the offspring would appear in the ratio of $\frac{1}{4}$ gray-bodied, normal-winged: $\frac{1}{4}$ black-bodied, normal-winged: $\frac{1}{4}$ gray-bodied, vestigial-winged: $\frac{1}{4}$ black-bodied, vestigial-winged flies. If the genes were completely linked and no exchange of chromosomal segments occurred, then only the parental types — flies with gray bodies and normal wings and flies with black bodies and vestigial wings — would appear among the offspring, and these would be present in equal numbers (Fig. 7.8). However, there is an exchange of segments between the locus of gene **V** and the locus of gene **B**. Because of this crossing over of part of the chromosomes, some gray-bodied, vestigial-winged flies and some black-bodied, normal-winged flies (the crossover types) appear among the offspring (Fig. 7.8). Most of the offspring, of course, resemble the parents and are gray, normal or black, vestigial. In this particular instance, crossing over occurs between these two points in this chromosome in about one cell in every five or in 20 per cent of the total undergoing meiosis. In such crosses, about 40 per cent of the offspring are gray flies with normal wings. Another 40 per cent are black flies with vestigial wings. Ten per cent are gray flies with vestigial wings, and 10 per cent are black flies with normal wings. The distance between two genes in a chromosome is measured in "cross-over units" which represent the percentage of crossing over that occurs between them. Thus, **V** and **B** are said to be 20 cross-over units apart.

In a number of species the frequency of crossing over between specific genes has been measured. All of the experimental results are consistent with the hypothesis that genes are present in a linear order in the chromosomes. Thus, if the three genes A, B and C occur in a single chromosome, the amount of crossing over between A and

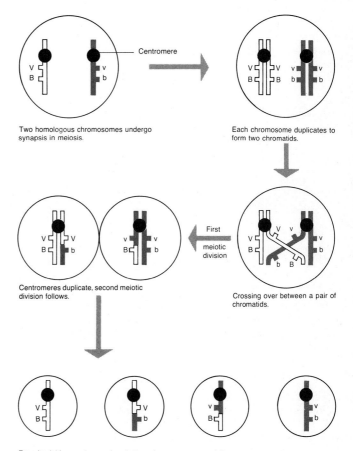

Two homologous chromosomes undergo synapsis in meiosis.

Centromere

Each chromosome duplicates to form two chromatids.

First meiotic division

Crossing over between a pair of chromatids.

Centromeres duplicate, second meiotic division follows.

Four haploid gametes produced; here two crossover and two noncrossover gametes.

Figure 7.7 Diagram illustrating crossing over, the exchange of segments between chromatids of homologous chromosomes. Crossing over permits recombination of genes (e.g., vB and Vb); the farther apart genes are located on a chromosome, the greater is the probability that crossing over between them will occur. (Villee: Biology, 6th ed.)

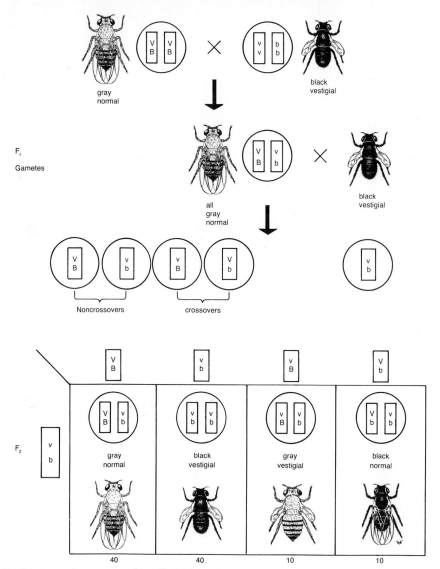

Figure 7.8 Diagram of a cross involving linkage and crossing over. The genes for vestigial versus normal wings and black versus gray body in fruit flies are linked; they are located in the same chromosome. (Villee: Biology, 6th ed.)

C is either the sum of, or the difference between, the amounts of crossing over between A and B and B and C. For example, if the crossing over between A and B is five units and between B and C is three units, the crossing over between A and C will be found to be either eight units (if C lies to the right of B) or two units (if C lies between A and B) (Fig. 7.9). By putting together the results of a great many such crosses, detailed maps of the location of specific genes on specific chromosomes have been made (Fig. 7.10). The mapping of human chromosomes is very difficult and has been proceeding

Figure 7.9 Diagram illustrating the principle of inferring the order of genes on a chromosome from their crossover values.

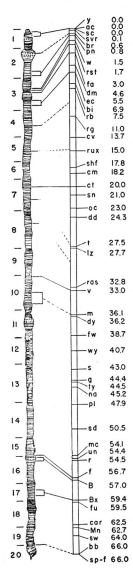

y	0.0
ac	0.0
sc	0.0
svr	0.1
br	0.6
pn	0.8
w	1.5
rst	1.7
fa	3.0
dm	4.6
ec	5.5
bi	6.9
rb	7.5
rg	11.0
cv	13.7
rux	15.0
shf	17.8
cm	18.2
ct	20.0
sn	21.0
oc	23.0
dd	24.3
t	27.5
lz	27.7
ras	32.8
v	33.0
m	36.1
dy	36.2
fw	38.7
wy	40.7
s	43.0
g	44.4
ty	44.5
na	45.2
pl	47.9
sd	50.5
mc	54.1
un	54.4
r	54.5
f	56.7
B	57.0
Bx	59.4
fu	59.5
car	62.5
Mn	62.7
sw	64.0
bb	66.0
sp-f	66.0

Figure 7.10 Diagram of the X chromosome of a fruit fly as seen in a cell of the salivary gland together with a map of the loci of the genes located in the X chromosome, with the distances between them as determined by frequency of crossing over. (Hunter and Hunter: College Zoology.)

slowly; several loci in the X chromosome have been mapped.

Since crossing over occurs at random, more than one crossover may occur in a single chromosome at a given time. One can observe among the offspring only the frequency of *recombination* of characters, not the actual frequency of crossovers. The frequency of crossing over will be somewhat larger than the observed frequency of recombination because the simultaneous

occurrence of two crossovers between two particular genes leads to the reconstitution of the original combination of genes (Fig. 7.11).

7.13   THE INTERACTIONS OF GENES

The relationship between the genes and their phenotypes discussed so far is simple and clear: each gene produces a single trait. Genetic research with many different kinds of animals and plants has revealed that the relationship between gene and trait may be quite complex. Several pairs of genes may interact to affect the production of a single trait; one pair of genes may inhibit or reverse the effect of another pair; or a given gene may produce different effects when the environment is altered in some way. The genes are inherited as units but may interact with one another in some complex fashion to produce the trait.

Complementary Genes. Two independent pairs of genes which interact to produce a trait in such a way that neither dominant can produce its effect unless the other is present too are called **complementary genes.** The presence of at least one dominant gene from each pair produces one character; the alternative condition results from the absence of either dominant or of both dominants.

Several different varieties of sweet peas with white flowers are known, and the mating of most white-flowered plants produces only white-flowered offspring. However, when plants from two particular white-flowered varieties were crossed, all the offspring had purple flowers! When two of these purple F_1 plants were crossed, or when they were self-fertilized, an F_2 generation was produced in the ratio of 9 purple to 7 white (Fig. 7.12). Subsequent analysis has shown that two pairs of genes located in different chromosomes are involved; one (**C**) regulates some essential step in the production of a raw material, and the other (**E**) controls the formation of an enzyme which converts the raw material into purple pigment. The homozygous recessive **cc** is unable to synthesize the raw material, and the homozygous recessive **ee** lacks the enzyme to convert the raw material

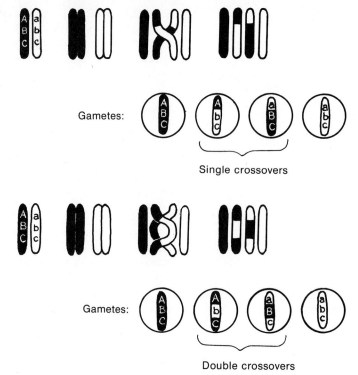

Figure 7.11   Diagram illustrating the results of single and double crossing over in a pair of homologous chromosomes. Note that the simultaneous occurrence of two crossovers in a given chromosome leads to a reconstitution of the original combination of genes, e.g., A and C.

into purple pigment. One of the white-flowered varieties was genotypically **ccEE** —lacked the gene for the synthesis of raw material—and the other was **CCee**, without the gene for the enzyme required for pigment synthesis. Crossing **CCee** and **ccEE** produces an F₁ generation all of which are **CcEe** and have purple flowers because they have both raw material and enzyme for the synthesis of the pigment. A pure-breeding variety of purple-flowered sweet peas could be established by self-fertilization of a plant with the genotype **CCEE**.

Supplementary Genes.   The term **supplementary genes** is applied to two independent pairs of genes which interact in the production of a trait in such a way that one dominant will produce its effect whether or not the second is present, but the second gene can produce its effect only in the presence of the first. In guinea pigs, in addition to the pair of genes for black versus brown coat color (**B** and **b**), the gene **C** controls the production of an enzyme which converts a colorless precursor into the pigment melanin and, hence, is required for

the production of any pigment at all in the coat. The homozygous recessive, **cc**, lacks the enzyme, no melanin is produced and the animal is a white-coated, pink-eyed **albino**, no matter what combination of **B** and **b** genes may be present. The eyes have no pigment in the iris and the pink color results from the color of the blood in the tissues of the eye. The mating of an albino, **ccBB**, with a brown guinea pig, **CCbb**, produces offspring all of which are genotypically **CcBb** and have black-colored coats! When two of these F₁ black guinea pigs are mated, offspring appear in the F₂ in the ratio of 9 black:3 brown:4 albino. Make a Punnett square to prove this.

Some combination of complementary and supplementary genes may be involved in the inheritance of a single trait. The dominant genes **C** and **R** are both necessary for the production of red kernels in maize, and the absence of either dominant results in white-colored kernels. There is, in addition, a **P** gene which produces purple-colored kernels if both **C** and **R** genes are present. The **P** gene is supplementary to the other

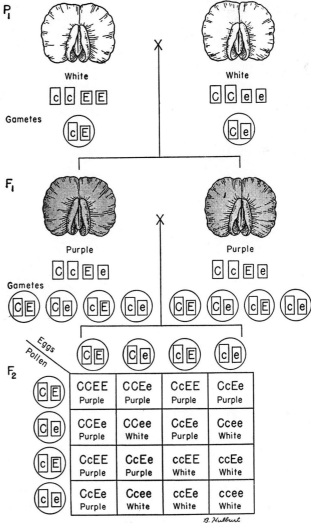

Figure 7.12 Diagram of a cross illustrating the action of complementary genes, the two pairs of genes which regulate flower color in sweet peas. At least one C gene and one E gene must be present to produce a colored flower. The absence of either one or both results in a white flower. (Villee: Biology, 6th ed.)

Phenotypes: 9 purple : 7 white

two pairs of genes and **C** and **R** are complementary.

The coat color of Duroc-Jersey pigs represents a slightly different type of gene interaction. Two independent pairs of genes (**R-r** and **S-s**) regulate coat color; at least one dominant of each pair must be present to give the full red-colored coat. Partial color, sandy, results when only one type of dominant is present, and an animal which is homozygous for both recessive (**rrss**) has a white-colored coat. The mating of two different strains of sandy-colored pigs, **RRss × rrSS**, yields offspring all of which are red, and the mating of two of these red F₁ individuals produces an F₂ generation in the ratio of 9 red : 6 sandy : 1 white (Fig. 7.13). Genes which interact in this fashion have been termed "mutually supplementary."

The inheritance of comb type in poultry provides an interesting example of genic interaction. Leghorns have single combs, Wyandottes have rose combs, and Brahmas have pea combs (Fig. 7.14). Each of these types is true breeding. Suitable crosses demonstrate that the gene for rose comb (**R**) is dominant to single (**r**) and that the gene for pea comb (**P**) is also dominant to its allele (**p**) for single comb. However, when a pea-combed fowl is mated with a rose-combed one, all of the offspring have a different type of comb, resembling half of

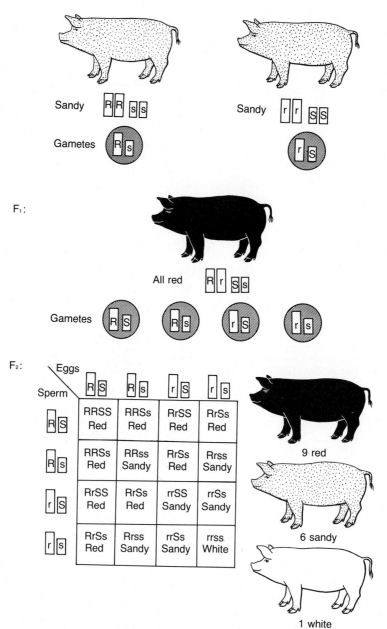

Figure 7.13 Diagram of the mode of inheritance of coat color in Duroc-Jersey pigs, illustrating inheritance by "mutually supplementary" genes.

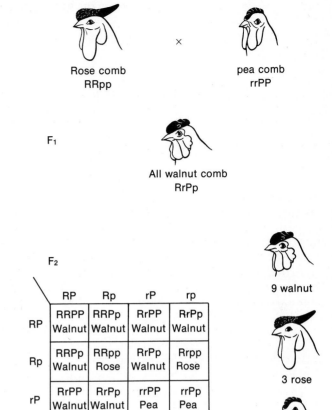

Figure 7.14 Diagram of the inheritance of comb types in chickens. See text for discussion.

a shelled walnut and called walnut. When two of these walnut-combed F_1 individuals are mated, offspring appear in the ratio of 9 walnut:3 pea:3 rose:1 single. We can deduce from this that the genotype of a single-combed fowl must be **rrpp;** a pea-combed fowl is either **PPrr** or **Pprr;** a rose-combed fowl is either **ppRR** or **ppRr,** and a walnut comb develops in animals with at least one **P** and one **R** gene. Thus, the genotypes **PPRR, PpRR, PPRr** and **PpRr** all yield walnut combs. Certain Malay varieties of chicken have walnut combs.

It is clear that there is nothing unusual about the method of inheritance of any of these genes; the phenotypic ratios observed are simply the result of some variation in the interaction of the genes in the production of the phenotype.

7.14 POLYGENIC INHERITANCE

Many human characteristics, height, body form, intelligence and skin color, and many commercially important characters of animals and plants, such as milk production in cows, egg production in hens, the size of

TABLE 7.2 POLYGENIC INHERITANCE OF SKIN COLOR IN MAN

Parents.. AaBb	AaBb	
(Mulatto)	(Mulatto)	
Gametes..AB Ab aB ab	AB Ab aB ab	

Offspring:

    1 with 4 dominants — AABB — phenotypically Negro
    4 with 3 dominants — 2 AaBB and 2 AABb — phenotypically "dark"
    6 with 2 dominants — 4 AaBb, 1 AAbb, 1 aaBB — phenotypically mulatto
    4 with 1 dominant — 2 Aabb, 2 aaBb — phenotypically "light"
    1 with no dominants — aabb — phenotypically white

fruits, and the like, are not separable into distinct alternate classes and are not inherited by single pairs of genes. Nonetheless these traits are governed by genetic factors; there are several, perhaps many, different pairs of genes which affect the same characteristic. The term **polygenic inheritance** is applied to two or more independent pairs of genes which affect the same character in the same way and in an additive fashion. When two varieties which differ in some trait controlled by polygenes are crossed, the F_1 are very similar to one another and are usually intermediate in the expression of this character between the two parental types. Crossing two F_1 individuals yields a widely variable F_2 generation, with a few members resembling one grandparent, a few resembling the other grandparent, and the rest showing a range of conditions intermediate between the two.

The inheritance of human skin color was carefully investigated by C. B. Davenport in Jamaica. He concluded that the inheritance of skin color in man is controlled by two pairs of genes, **A-a** and **B-b**, inherited independently. The genes for dark pigmentation, **A** and **B**, are incompletely dominant, and the darkness of the skin color is proportional to the sum of the dominant genes present. Thus, a full Negro has four dominant genes, **AABB**, and a white person has four recessive genes, **aabb**. The F_1 offspring of a mating of white and Negro are all **AaBb**, with two dominant genes and a skin color (mulatto) intermediate between white and Negro. The mating of two such mulattoes produces offspring with skin colors ranging from full Negro to white (Table 7.2). A mulatto with the genotype **AaBb** produces four kinds of eggs or sperm with respect to the genes for skin color: **AB, aB, Ab** and **ab**. From a Punnett square for the mating of two doubly

heterozygous mulattoes (**AaBb**) it will be evident that there are 16 possible zygote combinations: one with four dominants (black), four with three dominants (dark brown skin), six with two dominants (mulatto), four with one dominant (light brown skin) and one with no dominants (white skin). The genes **A** and **B** produce about the same amount of pigmentation and the genotypes **AaBb, AAbb** and **aaBB** produce the same phenotype, mulatto skin color.

This example of polygenic inheritance is fairly simple, for only two pairs of genes appear to be involved. With a larger number of pairs of genes, perhaps 10 or more, there are so many classes and the differences between them are so slight that the classes are not distinguishable; a continuous series is obtained. The inheritance of human stature is governed by a large number of pairs of genes, with shortness dominant to tallness. Since height is affected not only by these genes but also by a variety of environmental agents, there are adults of every height from perhaps 55 inches (140 cm.) up to 84 inches (203 cm.). If we measure the height of 1083 adult men selected at random and draw a graph of the number having each height, we will obtain a bell-shaped normal curve, or **curve of normal distribution** (Fig. 7.15). It is evident that there are few extremely tall or extremely short men, but many of intermediate height.

All living things show comparable variations in certain of their characteristics. If one were to measure the length of 1000 shells from the same species of clam, or the weight of 1000 hens' eggs, or the amount of milk produced per year by 1000 dairy cows, or the intelligence quotient (I.Q.) of 1000 grade school children, and make graphs of the number of individuals in each subclass, one would obtain a normal curve of distri-

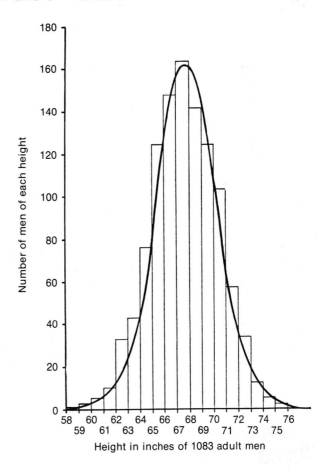

Figure 7.15 An example of a "normal curve," or curve of normal distribution: the heights of 1083 adult white males. The blocks indicate the actual number of men whose heights were within the unit range. For example, there were 163 men between 67 and 68 inches in height. The smooth curve is a normal curve based on the mean and standard deviation of the data. (Villee: Biology, 6th ed.)

bution in each instance. The variation is due in part to the action of polygenes and in part to the effects of a variety of environmental agents.

When a geneticist attempts to establish a new strain of hens that will lay more eggs per year, or a strain of turkeys with more breast meat, or a strain of sheep with longer, finer wool, he selects individuals which show the desired trait in greatest amount for further breeding. There is a limit, of course, to the effectiveness of selective breeding in increasing some desirable trait or in decreasing some undesirable one. When the strain becomes homozygous for all the genetic factors involved, further selective breeding will be ineffective.

The inheritance of certain traits depends not only on a single pair of genes which determines the presence or absence of the trait but also on a number of polygenes which determine the extent of the trait. For example, the presence or absence of spots in the coat of most mammals is determined by

a single pair of genes. The size and distribution of the spots, however, are determined by a series of polygenes. The term **modifying factors** has been suggested for polygenes which affect the degree of expression of another gene.

7.15   MULTIPLE ALLELES

In all the types of inheritance discussed so far, there have been only two possible alleles, one dominant and one recessive gene, which could be represented by capital and lower case letters respectively. In addition to a dominant and a recessive gene, there may be one or more additional kinds of gene found at that same location in the chromosome that affect the same trait in an alternative fashion. The term **multiple alleles** is applied to three or more different kinds of gene, three or more alternative conditions at a single locus in the chromosome, each of which produces a distinctive

TABLE 7.3 THE INHERITANCE OF THE HUMAN BLOOD GROUPS

Blood Group	Genotypes	Agglutinogen in Red Cells	Agglutinin in Plasma	Can Give Blood to Groups	Can Receive Blood from Groups
O	aa	none	a and b	O, A, B, AB	O
A	$a^A a^A$, $a^A a$	A	b	A, AB	O, A
B	$a^B a^B$, $a^B a$	B	a	B, AB	O, B
AB	$a^A a^B$	A and B	none	AB	O, A, B and AB

phenotype. In the population as a whole, there are distributed three or more different alleles, but each individual has any two, and no more than two, of the possible types of alleles, and any gamete has only one. The members of an allelic series are indicated by the same letter, with suitable distinguishing superscripts.

One series of multiple alleles which affects coat color in rabbits includes the dominant gene **C** for normal coat color, the recessive gene **c** which produces albino coat color when homozygous, and two other alleles, c^h and c^{ch}. The gene c^h, when homozygous, produces the "Himalayan" pattern of white coat over the body but with a dark color on the tips of the ears, nose, tail and legs. The gene c^{ch}, when homozygous, produces the "chinchilla" pattern of light gray fur all over the body. These alleles may be arranged in the series **C**, c^{ch}, c^h and **c**, in which each gene is dominant to the succeeding genes but recessive to the preceding ones. In other series of multiple alleles the genes may be incompletely dominant so that the heterozygote has a phenotype intermediate between those of its two parents, or one which is some combination of the two parental phenotypes.

Multiple alleles govern the inheritance of the human blood groups O, A, B and AB (p. 208). The three alleles of the series, a^A, a^B and **a**, regulate the kind of agglutinogen in the red blood cells (Table 7.3). Gene a^A produces agglutinogen A, gene a^B produces agglutinogen B, and gene **a** produces no agglutinogens. Gene **a** is recessive to the other two, but neither a^A nor a^B is dominant to the other; each produces its characteristic agglutinogen independently of the other. Transfusions of blood from one person to another are successful only when the two bloods are compatible, when the agglutinins in the plasma of the recipient do not react with the agglutinogens in the red cells of the donor to cause agglutination, clumping of the red cells. People with type O blood (no agglutinogens in their red cells) are known as "universal donors"; their blood can be transfused into the veins of persons with any of these blood groups. People with type AB blood are called "universal recipients"; they have no agglutinins in the plasma and hence their plasma will not cause agglutination of the red cells from any person.

Since blood types are inherited, and do not change in a person's lifetime, they are useful indicators of parentage. In cases of disputed parentage, genetic evidence can show only that a certain man or woman *could be* the parent of a particular child, and never that he *is* the parent. In certain circumstances, however, the genetic evidence can definitely exclude a particular man or woman as the parent of a given child. Thus, if a child of blood group A is born to a type O woman, no man with type O or type B blood could be its father (Table 7.4).

We now know of 11 different sets of blood groups, inherited by different pairs of genes, all of which are helpful in establishing paternity. The most important of these are the Rh alleles, which determine the presence or absence of a different agglutinogen, the **Rh factor**, first found in the blood of rhesus monkeys. There are actually several alleles at the **Rh** locus, but to simplify matters we shall consider just two: **Rh**, which produces the Rh positive antigen, and the recessive **rh**, which does not produce the antigen. Genotypes **RhRh** and **Rhrh** are phenotypically Rh positive and genotype **rhrh** is phenotypically Rh negative. An Rh negative woman married to an Rh positive man may have an Rh positive child. If some blood manages to pass across the placenta from the fetus to the mother it will stimulate the

TABLE 7.4 EXCLUSION OF PATERNITY
BASED ON BLOOD TYPES

Child	Mother	Father Must Be of Type	Father Cannot Be of Type
O	O	O, A, or B	AB
O	A	O, A, or B	AB
O	B	O, A, or B	AB
A	O	A or AB	O or B
A	A	A, B, AB, or O	————
B	B	A, B, AB, or O	————
A	B	A or AB	O or B
B	A	B or AB	O or A
B	O	B or AB	O or A
AB	A	B or AB	O or A
AB	B	A or AB	O or B
AB	AB	A, B, or AB	O

formation, in her blood, of antibodies to the Rh factor. Then, in a subsequent pregnancy, some of these Rh antibodies may pass through the placenta to the child's blood, and react with the Rh antigen in the child's red cells. The red cells are agglutinated and destroyed and a serious, often fatal, anemia, called **erythroblastosis fetalis,** ensues. This is now treated by massive blood transfusions in which essentially all of the blood of the newborn is replaced.

Extensive surveys have shown that approximately 41 per cent of native white Americans are type O, 45 per cent are type A, 10 per cent are type B, and 4 per cent are type AB. The frequency of the blood groups in other races may be quite different; American Indians, for example, have a low frequency of group A and a high frequency of group B. No one blood type is characteristic of a single race; the racial differences lie in the *relative frequency* of the several blood types. Studies of the relative frequencies of the blood groups found in different races living today and in mummies and prehistoric skeletons have provided valuable evidence as to the relationships of the present races of man.

7.16 INBREEDING AND OUTBREEDING

It is commonly believed that the mating of two closely related individuals—brother and sister or father and daughter—is harmful and leads to the production of monstrosities. The marriage of first cousins is forbidden by law in some states. Carefully controlled experiments with many different kinds of animals and plants have shown that there is nothing harmful in the process of inbreeding itself. It is, in fact, one of the standard procedures used by commercial breeders to improve strains of cattle, corn, cats and cantaloupes. It is not necessarily a bad practice in the human species. In all animals or plants it simply tends to make the strain homozygous.

All natural populations of individuals are heterozygous for many traits; some of the hidden recessive genes are for desirable traits, others are for undesirable ones. **Inbreeding** will simply permit these genes to become homozygous and lead to the unmasking of the good or bad traits. If a stock is good, inbreeding will improve it; but if a stock has many undesirable recessive traits, inbreeding will lead to their phenotypic expression.

The crossing of two completely unrelated strains, called **outbreeding,** is another widely used genetic maneuver. It is frequently found that the offspring of such a mating are much larger, stronger and healthier than either parent. Much of the corn grown in the United States is a special hybrid variety developed by the United States Department of Agriculture from a mating of four different inbred strains. Each year, the seed to grow this uniformly fine hybrid corn is obtained by mating the original inbred lines. If the hybrid corn were used in mating, it would give rise to many different kinds of corn since it is heterozygous for many different traits. The mule, the hybrid offspring of the mating of a horse and donkey, is a strong, sturdy animal, better adapted for many kinds of work than either of its parents. This phenomenon of **hybrid vigor,** or **heterosis,** does not result from the act of outbreeding itself, but from the heterozygous nature of the F_1 organisms which result from outbreeding. Each of the parental strains is homozygous for certain undesirable recessive traits, but the two strains are homozygous for *different* traits, and each one has dominant genes to mask the undesirable recessive genes of the other. As a concrete example, let us suppose that there are four pairs of genes, **A, B, C** and **D;** the capital letters represent the dominant gene for some desirable trait, and the lower case letters represent the recessive gene for its undesirable allele. If one parental strain is then **AAbbCCdd** and the other **aaBBccDD,** the offspring will all be **AaBbCcDd** and have all of the desirable and none of the undesirable traits. The actual situation in any given cross is undoubtedly much more complex and involves many pairs of genes.

7.17 PROBLEMS IN GENETICS

Since many students find that solving problems is helpful in learning the principles of genetics we have provided the following:

1. In peas, the gene for smooth seed coat is dominant to the one for wrinkled seeds. What would be the result of the following matings: heterozygous smooth × heterozygous smooth? Heterozygous smooth × wrinkled? Heterozygous smooth × homozygous smooth? Wrinkled × wrinkled?

2. In peas, the gene for red flowers is dominant to the one for white flowers. What would be the result of mating heterozygous red-flowered, smooth-seeded plants with white-flowered, wrinkled-seeded plants?

3. The mating of two black short-haired guinea pigs produced a litter which included some black long-haired and some white short-haired offspring. What are the genotypes of the parents and what is the probability of their having black short-haired offspring in subsequent matings?

4. Human color blindness is a sex-linked, recessive trait. What is the probability that a woman with normal vision whose husband is color-blind will have a color-blind son? a color-blind daughter? What is the probability that a woman with normal vision whose father was color-blind but whose husband has normal vision will have a color-blind son? a color-blind daughter?

5. The gene for white eye color (w) in fruit flies is sex-linked and recessive to normal red eye color (W). Give the results of mating (a) a heterozygous red-eyed female with a red-eyed male, (b) a white-eyed female with a red-eyed male and (c) a heterozygous red-eyed female with a white-eyed male.

6. A blue-eyed man, both of whose parents were brown-eyed, marries a brown-eyed woman whose father was blue-eyed and whose mother was brown-eyed. Their first child has blue eyes. Give the genotypes of all the individuals mentioned and give the probability that the second child will also have blue eyes.

7. Outline a breeding procedure whereby a true-breeding strain of red cattle could be established from a roan bull and a white cow.

8. Suppose you learned that shmoos may have long, oval or round bodies and that matings of shmoos gave the following results:

long × oval gave 52 long and 48 oval
long × round gave 99 oval
oval × round gave 51 oval and 50 round
oval × oval gave 24 long, 53 oval and 27 round

What hypothesis about the inheritance of shmoo shape would be consistent with these results?

9. A mating of an albino guinea pig and a black one gave six white (albino), three black and three brown offspring. What are the genotypes of the parents? What kinds of offspring, and in what proportions, would result from the mating of the black parent with another animal that has exactly the same genotype as it has?

10. Mating a red Duroc-Jersey hog to sow A (white) gave pigs in the ratio of 1 red:2 sandy:1 white. Mating this same hog to sow B (sandy) gave 3 red:4 sandy:1 white. When this hog was mated to sow C (sandy) the litter had equal numbers of red and sandy

piglets. Give the genotypes of the hog and the three sows.

11. A walnut-combed rooster is mated to three hens. Hen A (walnut-combed) has offspring in the ratio of 3 walnut:1 rose. Hen B (pea-combed) has offspring in the ratio of 3 walnut:3 pea:1 rose:1 single. Hen C (walnut-combed) has only walnut-combed offspring. What are the genotypes of the rooster and the three hens?

12. The size of egg laid by one variety of hens is determined by three pairs of genes; hens with the genotype **AABBCC** lay eggs weighing 90 gm. and hens with the genotype **aabbcc** lay eggs weighing 30 gm. Each dominant gene adds 10 gm. to the weight of the egg. When a hen from the 90 gm. strain is mated with a rooster from the 30 gm. strain, the hens in the F_1 generation lay eggs weighing 60 gm. If a hen and rooster from this F_1 generation are mated, what will be the weights of the eggs laid by the hens of the F_2 generation?

13. Mrs. Doe and Mrs. Roe had babies at the same hospital and at the same time. Mrs. Doe took home a girl and named her Nancy. Mrs. Roe received a boy and named him Harry. However, she was sure that she had had a girl and brought suit against the hospital. Blood tests showed that Mr. Roe was type O, Mrs. Roe was type AB, Mr. and Mrs. Doe were both type B, Nancy was type A and Harry was type O. Had an exchange occurred?

14. A woman who is type O and Rh negative is married to a man who is type AB and Rh positive. The man's father was type AB and Rh negative. What are the genotypes of the man and woman and what blood types may occur among their offspring? Is there any danger that any of their offspring may have erythroblastosis fetalis?

ANNOTATED REFERENCES

Carlson, E.: The Gene: A Critical History. Philadelphia, W. B. Saunders Co., 1966. Traces the development of the concept of the gene.

Herskowitz, I. H.: Genetics. 2nd ed. Boston, Little, Brown & Co., 1965. A standard, college-level text.

Strickburger, M. W.: Genetics. New York, Macmillan, 1968. A standard, college-level text.

Sturtevant, A. H.: History of Genetics. New York, Harper & Row, 1965. An authentic account of the rise of genetics by a man who played a prominent role in the development of this important science.

Whitehouse, H. L.: Towards an Understanding of the Mechanism of Heredity. 2nd ed. New York, St. Martin's Press, 1965. A discussion of genetics in terms of its hypotheses and how they were tested.

8.   Molecular and Mathematical Aspects of Genetics

Our modern concepts of genetics originated with the rediscovery of Mendel's Laws in 1900. Since that time, geneticists have been attempting to determine the physical structure and chemical composition of the hereditary units—the genes—and to discover the mechanisms by which they transfer biological information from one cell to another and control the development and maintenance of the organism. During the first half of this century, much was learned about the complexity of the molecular structure of proteins. As a result, nearly all biochemists assumed that any complex biological unit with such marked specificity as the gene must also be a protein. There was great difficulty, however, in explaining how protein molecules could be duplicated precisely, as genes must be with each cell division. The realization that genetic information is transferred by nucleic acid molecules rather than by protein molecules arose gradually. Despite the recent rapid advances in our understanding of many facets of biochemical genetics, our knowledge of the molecular organization of chromosomes, especially the complex chromosomes of higher plants and animals, is still very incomplete.

Our present belief that the nucleic acids, DNA and RNA, are the primary agents for the transfer of biological information arose gradually, culminating in 1953 in the proposal by James Watson and Francis Crick of a model of the DNA molecule that explained how it could transfer information and undergo replication. This proposal stimulated an enormous flood of research, and has led to the present "**central dogma**" of biology (Fig. 8.1): Genes are composed of DNA, and are located within the chromosomes. Each gene contains information coded in the form of a specific sequence of purine and pyrimidine nucleotides within its DNA molecule. The unit of genetic information, called a **codon,** is a group of three adjacent nucleotides that specify a single amino acid in a polypeptide chain. Thus the genetic code is a "**triplet**" code. The DNA molecule consists of two complementary chains of polynucleotides twisted about each other in a regular helix and joined by hydrogen bonds between specific pairs of purine and pyrimidine bases. The DNA molecule is replicated when the two strands of the helix separate, and each acts as a template for the formation of a new complementary strand. Each pair of strands, one old and one new, then twist together to form two daughter helices.

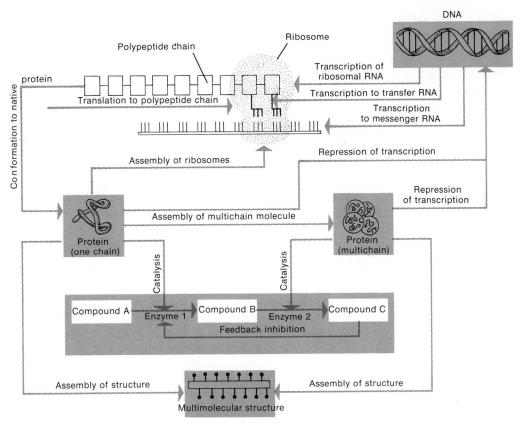

Figure 8.1 Overview of the process by which biological information is transferred from DNA via RNA to specific polypeptides. The peptide subunits are then assembled into multichain proteins. (Villee: Biology, 6th ed.)

8.1 THE CHEMISTRY OF CHROMOSOMES

Chromosomes have been shown to contain DNA, RNA, histones and other proteins. By homogenizing cells and isolating the nuclei by centrifugation, and then homogenizing the nuclei and isolating chromosomes by further centrifugation, A. E. Mirsky obtained pure chromosomal material for analysis. There are about 6×10^{-9} milligrams of DNA per nucleus in somatic cells and 3×10^{-9} milligrams of DNA per nucleus in egg or sperm cells (Table 8.1). In tissues

TABLE 8.1 AMOUNT OF DEOXYRIBONUCLEIC ACID (DNA) PER NUCLEUS IN ANIMAL TISSUES, EXPRESSED AS Mg. $\times 10^{-9}$

Species	Sperm	Red Cell	Liver	Heart	Kidney	Pancreas	Spleen
Shad	0.91	1.97	2.01				
Carp	1.64	3.49	3.33				
Brown trout	2.67	5.79					
Toad	3.70	7.33					
Frog		15.0	15.7				
Chicken	1.26	2.49	2.66	2.45	2.20	2.61	2.55
Dog			5.5		5.3		
Rat			9.47	6.50	6.74	7.33	6.55
Ox	3.42		7.05		6.63	7.15	7.26
Man	3.25	7.30	10.36		8.6		

known to be polyploid (having more than two sets of chromosomes per nucleus) the amount of DNA per nucleus corresponds to the multiple of chromosomes known to be present. Tetraploid cells, with four sets of chromosomes, have been found to have 12×10^{-9} milligrams of DNA per nucleus. The amounts of DNA and histone are essentially the same in all the cells of a given organism (Table 8.1). In marked contrast, the amounts of RNA and of several other proteins vary considerably from cell to cell. That the amount of DNA, like the number of genes, is constant in all the cells of the body, and the amount of DNA in germ cells is only half the amount in somatic cells, is evidence that DNA is an essential part of the gene. From the amount of DNA per cell, one can estimate the number of nucleotide pairs per cell and thus the amount of genetic information present (Table 8.2).

Treating chromosomes with **deoxyribonuclease** cleaves and removes the DNA, but a shadow of the chromosome structure remains. In contrast, a chromosome treated with proteolytic enzymes breaks into fragments. This suggests that the chromosome consists of a continuous protein core along which are local concentrations of DNA. Treating chromosomes with an agent (EDTA) that removes and binds divalent cations such as calcium, yields small particles the size of a chromosome band, which suggests that the DNA and protein of the chromosome are normally held together by divalent cations such as calcium and magnesium. When these cations are removed by EDTA the nucleic acids and proteins disaggregate.

The chromosomes of prokaryotic organisms appear to be enormous single molecules of DNA, with molecular weights of 1.2×10^8 (bacteriophage) or 2×10^9 (*Escherichia coli*). In these organisms the molecule of DNA appears to be circular, without any free ends. It has been suggested that this circularity prevents replication of the DNA and that the DNA molecule is converted from the circular to a linear form for only a brief moment during the replication process. The possibility remains that individual DNA molecules in the chromosomes of higher organisms are circular. However, electron micrographs published in 1966 by Dr. Margit Nass showed that the DNA in the nuclei of mouse fibroblasts is in linear fibrous strands 10 to 15 microns long. The mitochondria of many, perhaps all, cells contain a small amount of DNA (about 1 or 2 per cent of the total), and Dr. Nass's electron micrographs revealed that the mitochondrial DNA consists of circular strands some 5 microns long.

TABLE 8.2   THE AMOUNT OF DNA PER CELL IN ANIMAL AND PLANT CELLS AND IN VIRUS PARTICLES

	DNA (Mg. $\times 10^{-9}$ per Cell)	Nucleotide Pairs per Cell
Mammals	6	5.5×10^9
Birds	2	2×10^9
Reptiles	5	4.5×10^9
Amphibia	7	6.5×10^9
Fish	2	2×10^9
Insects	0.17–12	0.16×10^9
Crustacea	3	2.8×10^9
Mollusks	1.2	1.1×10^9
Echinoderms	1.8	1.7×10^9
Sponges	0.1	0.1×10^9
Higher plants	2.5–40	2.3×10^9
Fungi	0.02–0.17	0.02×10^9
Algae	3	2.8×10^9
Bacteria	0.002–0.06	2×10^6
T$_2$ bacteriophage	0.00024	2.2×10^5
λ bacteriophage	0.00008	7×10^4
Papilloma virus	—	6×10^3

8.2   THE ROLE OF DNA IN HEREDITY

The first direct evidence that DNA can transmit genetic information came from the experiments of Avery and his coworkers with the "transforming agents" isolated from pneumococci and certain other bacteria. A transforming agent isolated from strain III of pneumococcus (the bacteria causing pneumonia) will, when added to the culture medium, transform strain II organisms into strain III pneumococci. This is a "permanent" transformation; the strain II organisms so produced multiply to give only strain III offspring. Similar transforming agents have been found in other types of bacteria and, in each case, when they have been isolated and characterized, they have been found to be pure DNA.

DNA has been shown to be the carrier of genetic information in bacterial viruses that consist of a "head," made up of a protein membrane around a DNA "core," and a tail composed of protein. The virus becomes attached to the bacterial cell by the tip of its tail. The DNA present in the head of the virus is transferred inside the bacterial cell, but most of the tail and the head membrane remain outside. The viral protein on the outside of the bacterial cell can be sheared off by stirring the infected bacteria in a Waring Blendor. The bacteria containing the viral DNA will then produce a large number of virus particles identical with the one used for infection. Thus, most of the viral protein can be eliminated without interfering with the replication process, indicating that the protein is not important in the formation of new virus particles.

Experiments using labeled atoms have shown further that the atoms of the protein in the parent virus do not appear in the progeny, whereas more than 50 per cent of the atoms of the parental DNA do appear in the progeny. Within the bacterial cell the nucleic acid leads to the production both of additional nucleic acid cores and of protein coats, and many virus particles are formed and released when the infected bacterial cell bursts. In other experiments pure nucleic acid has been prepared from plant viruses and shown to have the viral activity characteristic of that species of virus.

Further evidence that DNA is the carrier of genetic information comes from experiments in which genetic recombination has been demonstrated between different strains of bacteria. On rare occasions two bacteria come together, a cytoplasmic bridge forms between them, and DNA passes across this bridge from the donor to the recipient cell. The amount of DNA transferred is proportional to the amount of genetic information transferred as measured on a genetic map. If the bacteria are left undisturbed, as much as one-third of the genetic material is transferred from donor to recipient. However, if bacteria are placed in a Waring Blendor at various times after the cytoplasmic bridge has formed and the cells are separated, varying amounts of DNA and of genetic material will be transferred. The propotionality between the amount of DNA and the length of the genetic map transferred indicates that genetic information is contained in DNA.

Nucleic acids absorb ultraviolet light very strongly with a maximum at 260 nanometers. The production of mutations by ultraviolet light is also maximal at 260 nanometers. When one compares the number of mutations produced per unit of energy delivered and the wave length at which the energy is delivered, an **action spectrum** for mutations is obtained. There is a close correlation between the action spectrum for the production of mutations and the absorption spectrum for nucleic acids. The simplest explanation of this phenomenon is that genes are composed of nucleic acids, that mutations are produced as nucleic acids absorb energy, and that the absorbed energy is effective in changing the nucleic acid molecule to yield a mutant gene.

8.3 THE WATSON-CRICK MODEL OF DNA

Highly purified DNA has been prepared from a wide variety of animals, plants and bacteria and found in each case to consist of a sugar — **deoxyribose,** phosphoric acid and nitrogenous bases. Four major kinds of bases are found in DNA: two purines, **adenine** and **guanine,** and two pyrimidines, **cytosine** and **thymine.** The ratios of these purine and pyrimidine bases are the same in the DNA from all the cells of a given species but may differ in samples of DNA taken from different species (Table 8.3). In all samples of DNA the total amount of purines equals the total amount of pyrimidines $(A + G = T + C)$, the amount of adenine equals the amount of thymine $(A = T)$, and the amount of guanine equals the amount of cytosine $(G = C)$. DNA from mammalian tissues is generally rich in adenine and thymine and relatively poor in guanine and cytosine, whereas nucleic acid from bacterial sources is, in general, rich in guanine and cytosine and relatively poor in adenine and thymine.

On the basis of these analytical results, and from a study of the x-ray diffraction patterns of a number of samples of DNA, Watson and Crick proposed in 1953 a model of the DNA molecule. The great usefulness of this model in providing a chemical explanation for the properties of DNA led to the awarding of Nobel prizes to Watson and Crick in 1962. In DNA the adjacent nucleo-

TABLE 8.3 RELATIVE AMOUNTS OF
PURINES AND PYRIMIDINES IN SAMPLES
OF DNA

Source	Adenine	Guanine	Cytosine	Thymine
Beef thymus	29.0	21.2	21.2	28.5
Beef liver	28.8	21.0	21.1	29.0
Beef sperm	28.7	22.2	22.0	27.2
Human thymus	30.9	19.9	19.8	29.4
Human liver	30.3	19.5	19.9	30.3
Human sperm	30.9	19.1	18.4	31.6
Hen red cells	28.8	20.5	21.5	29.2
Herring sperm	27.8	22.2	22.6	27.5
Wheat germ	26.5	23.5	23.0	27.0
Yeast	31.7	18.3	17.4	32.6
Vaccinia virus	29.5	20.6	20.0	29.9
Bacteriophage T_2	32.5	18.2	16.7	32.6

tides are joined together in a chain by phosphodiester bridges linking the 5′ carbon of the deoxyribose of one nucleotide with the 3′ carbon of the deoxyribose of the next nucleotide (Fig. 8.2). Watson and Crick suggested that the DNA molecule consists of two such polynucleotide chains wrapped helically around each other (a double helix), with the sugar-phosphate chain on the outside of the helix and the purines and pyrimidines on the inside. The two chains are held together by hydrogen bonds between specific pairs of purines and pyrimidines (Fig. 8.3).

The hydrogen bonds between purines and pyrimidines are such that adenine can bond to thymine and guanine can bond to cytosine but other combinations are not

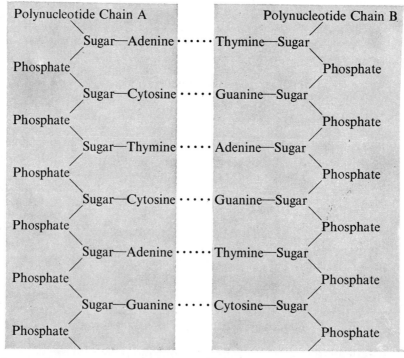

Figure 8.2   Schematic diagram of a portion of a DNA molecule, showing the two polynucleotide chains joined by hydrogen bonds (......). The chains are not flat, as represented here but are coiled around each other in helices (see Fig. 8.4). (Villee: Biology, 6th ed.)

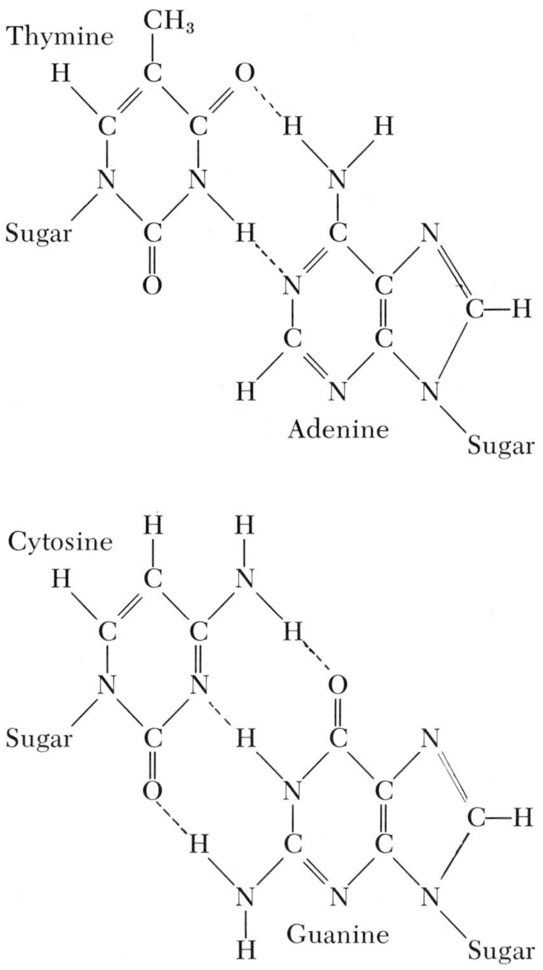

Figure 8.3 Diagram showing the specific pairing of adenine and thymine (above), joined by two hydrogen bonds (······), and of guanine and cytosine (below), linked by three hydrogen bonds. (Villee: Biology, 6th ed.)

possible; they would not fit into the repeating lattice formed by the two helical polynucleotide chains. Adenine could not form hydrogen bonds with guanine for there would not be enough room for these two relatively large molecules to be bonded and yet form parts of the helical chains. On the other hand, thymine could not be bonded to cytosine while both were parts of these helical chains because the distance between the two would be too great. Hydrogen bonds could not form between a thymine that was part of one chain and a cytosine that was part of another. *Two* hydrogen bonds form between adenine and thymine and *three* between guanine and cytosine (Fig. 8.3).

The specificity of the kind of hydrogen bonds that can be formed provides assurance that for every adenine in one chain there will be a thymine in the other, for every guanine in the first chain there will be a cytosine in the other, and so on. Thus, the two chains are *complementary* to each other; that is, the sequence of nucleotides in one chain dictates specifically and precisely the sequence of nucleotides in the other. The two strands run in opposite directions (have opposite polarities) and have their terminal phosphate groups at opposite ends of the double helix. When exactly scaled molecular models are made (Fig. 8.4), it can be seen that these combinations, adenine-thymine and guanine-cytosine, will fit into the space available whereas other combinations will not.

The most distinctive properties of the genetic material are that it carries information and undergoes replication. The Watson-Crick model explains how DNA molecules may carry out these two functions. When a DNA molecule undergoes replication, the two chains separate and each one brings about the formation of a new chain, which is complementary to it. Thus two new chains are established (Fig. 8.5). The nucleotides in the new chain are assembled in a specific order, because each purine or pyrimidine in the original chain forms hydrogen bonds with the complementary pyrimidine or purine nucleotide triphosphate from the surrounding medium and lines them up in a complementary order. Phosphate ester bonds are formed by the reaction catalyzed by **DNA polymerase** to join the adjacent nucleotides in the chain, and a new polynucleotide chain results (Fig. 8.6). The new and the original chains then wind around each other, and two new DNA molecules are formed. Each chain, in other words, serves as a template or a mold against which a new partner chain is synthesized. The end result is two complete double chain molecules, each identical to the original double-chain molecule. A second prime function of DNA, in addition to its role in replication, is that the information contained in its specific sequence of nucleotides must be transcribed some time between cell divisions. The product of the transcription process, now termed messenger RNA, then combines with ribosomes to carry out the synthesis of enzymes and other specific

(Text continued on page 185.)

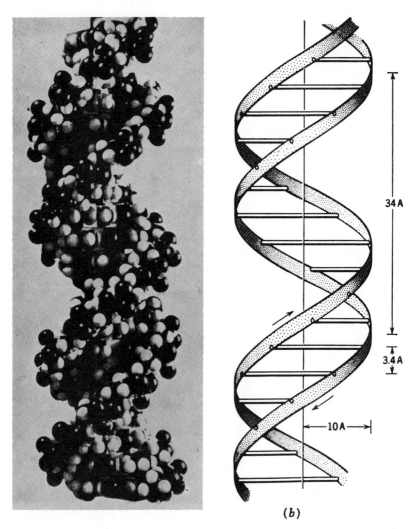

(b)

Figure 8.4 Photograph of a molecular model of deoxyribonucleic acid (through the courtesy of Dr. M. H. F. Wilkins). A schematic drawing of this two-stranded structure is shown on the right, together with certain of its dimensions in Ångstrom units. (Anfinsen: Molecular Basis of Evolution, John Wiley & Sons.)

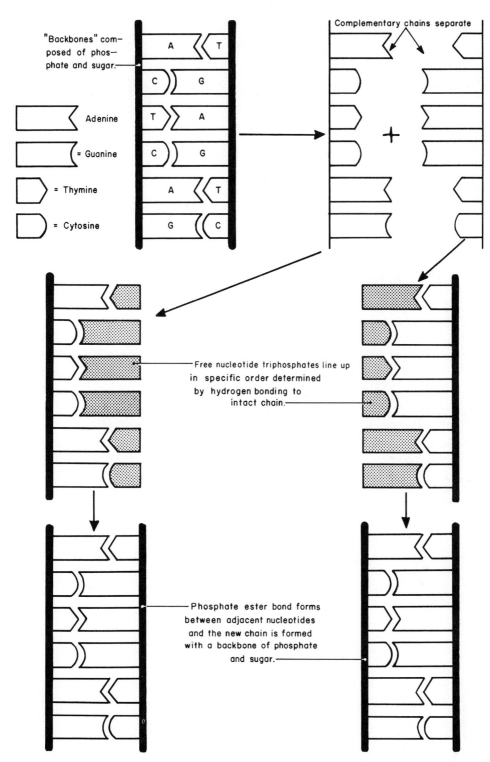

Figure 8.5 Diagrammatic scheme of how DNA molecules may undergo replication in the reaction catalyzed by the enzyme DNA polymerase. (Villee: Biology, 6th ed.)

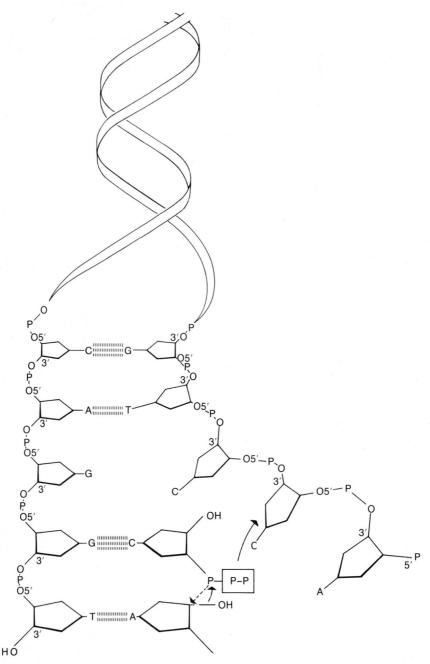

Figure 8.6 Mechanism of the replication of DNA by DNA polymerase. The two strands of the DNA double helix are shown separating. The left one, which runs from the 5′ phosphate to 3′ OH, is being copied starting from the bottom. The newly synthesized chain begins with a 5′ phosphate. In the new chain, as in the old, adenine forms base pairs with thymine and cytidine forms base pairs with guanine. The new phosphodiester bond is made between adjacent bases in the forming chain by an attack by the 3′ OH group of the deoxyribose on the bond between the inner phosphate and the outer two phosphates of the adjacent trinucleotide. The two outer phosphates are split off as inorganic pyrophosphate. (Villee: Biology, 6th ed.)

proteins. Thus we can visualize how each gene can lead to the production of a specific enzyme. We shall return to this subject in another section.

8.4 THE GENETIC CODE

The Watson-Crick model of the DNA molecule implied that genetic information is transmitted by some specific sequence of its constituent nucleotides. Since there are only four types of nucleotides—A, T, G and C—in the DNA and 20 or more kinds of amino acids in a peptide chain, it was obvious that there could not be a one-to-one correlation between nucleotide and amino acid in the coding process. If the code involved two nucleotides to specify an amino acid, the various combinations of four symbols taken two at a time would provide only 16 different combinations; again, this is not enough to account for the 20 or more different types of amino acids.

A triplet code of three nucleotides for each amino acid would permit 64 different combinations of four nucleotides taken three at a time. At first glance this would seem to provide for many more code symbols than are actually needed; however, experiments

have shown that the code is "degenerate"— as many as six different triplet codons may specify the same amino acid.

The fundamental characteristics of the genetic code are now well established: It is a triplet code with three adjacent nucleotide bases, termed a **codon**, specifying each amino acid (Table 8.4). Adjacent codons do not overlap; they do not share a given base. Each single base is part of only one codon. *The genetic code appears to be universal;* that is, the codons in the DNA and RNA specify the same amino acid in all the organisms that have been studied, from viruses to man.

Early in 1961 Crick postulated that three consecutive nucleotides in a strand of messenger RNA provide the code that determines the position of a single amino acid in a polypeptide chain. Experimental evidence to support this was quickly forthcoming from experiments of Nirenberg and Matthaei regarding the incorporation of specific labeled amino acids into protein by purified enzyme systems under the direction of artificial messenger RNA's of known composition.

Nirenberg used a synthetic polyuridylic acid (UUUUU . . .)—prepared by using the

TABLE 8.4 THE GENETIC CODE: THE SEQUENCE OF NUCLEOTIDES IN THE TRIPLET CODONS OF MESSENGER RNA WHICH SPECIFY A GIVEN AMINO ACID

First Position (5′ end)	Second Position	Third Position (3′ end)			
		U	C	A	G
U	U	Phe	Phe	Leu	Leu
	C	Ser	Ser	Ser	Ser
	A	Tyr	Tyr	Terminator	Terminator
	G	Cys	Cys	Terminator	Try
C	U	Leu	Leu	Leu	Leu
	C	Pro	Pro	Pro	Pro
	A	His	His	Glu·NH$_2$	Glu·NH$_2$
	G	Arg	Arg	Arg	Arg
A	U	Ileu	Ileu	Ileu	Met
	C	Thr	Thr	Thr	Thr
	A	Asp·NH$_2$	Asp·NH$_2$	Lys	Lys
	G	Ser	Ser	Arg	Arg
G	U	Val	Val	Val	Val
	C	Ala	Ala	Ala	Ala
	A	Asp	Asp	Glu	Glu
	G	Gly	Gly	Gly	Gly

enzyme polynucleotide phosphorylase — as messenger and found that phenylalanine was incorporated into protein. The addition of poly U to a ribosomal protein-synthesizing system led to the production of a polypeptide containing only phenylalanine. The inference that UUU is the code for phenylalanine was inescapable.

Comparable experiments by Nirenberg and by Severo Ochoa showed that polyadenylic acid provided the code for lysine and polycytidylic acid coded for proline. Further experiments with mixed nucleotide polymers (such as poly AC) as artificial messenger permitted the assignment of many other nucleotide combinations to specific amino acids.

These experiments did not reveal the *order* of the nucleotides within the triplets but this has been inferred from other kinds of experiments. Even when protein synthesis is not occurring, specific transfer RNA molecules will attach to ribosomes only when mRNA is present. Very fortunately this effect does not require a long molecule of mRNA (which would be difficult to synthesize); synthetic messenger RNA as short as a trinucleotide will suffice to promote specific binding of transfer RNA to the ribosome. Since it is possible to synthesize trinucleotides of known sequence, it has been possible to determine the coding assignments of all 64 possible triplets. Thus GUU but not UGU or UUG induces the binding of valine-tRNA to ribosomes. UUG induces the binding of leucine-transfer RNA. Since GUU and UUG code for different amino acids, it follows that the reading of the code in the messenger RNA strand makes sense only in one direction.

That messenger RNA is read three nucleotides at a time was firmly established by experiments carried out by Khorana. He and his colleagues synthesized a poly UC messenger containing the regularly alternating base sequence UCUCUCUC. When this was used in a protein-synthesizing system, the resulting polypeptide contained a regular alternation of serine and leucine. Mathematical analysis of this result shows that the coding unit must contain an odd number of bases. Khorana then synthesized the nucleotide sequence AAGAAGAAGAAG. When this nucleotide polymer was used as template in a protein-synthesizing system, the result was either polylysine (AAG), poly-

glutamate (GAA) or polyarginine (AGA). The type of peptide synthesized depends on which nucleotide in the polynucleotide chain happens to be read first. This "**frame shift**" **effect**, as in a movie camera, can be accounted for only if the chain is read in sequence, three nucleotides at a time, beginning from a fixed point.

It is apparent from Table 8.4 that the first two bases of the triplet have greater specificity than the third. Both AUU and AUC code for isoleucine; both bind isoleucine-transfer RNA to ribosomes. CCX and CGX code for proline and arginine respectively; the third base in the triplet, X, can be any of the four nucleotides.

It is now generally believed that the code is commaless; it has no "punctuation." No punctuation is necessary since the code is read out beginning at a fixed point and the entire strand is read, three nucleotides at a time, until the read-out mechanism comes to a specific "termination" code, which signals the end of the message. The nature of the signal, "begin reading here," at least in bacteria, also appears to be specified by a specific sequence of bases.

The specific properties of each protein depend in part on the sequence of the amino acids in the peptide chain or chains. A gene is a linear polynucleotide molecule and a protein is also a linear molecule with respect to its amino acid sequence. The essence of the genetic code is that the sequence of amino acids in the peptide chain is dictated by the order of the corresponding nucleotide bases in one of the two polynucleotide chains of the DNA molecule. Thus, the DNA molecule and the resulting polypeptide chain are said to be **colinear.**

This concept of colinearity was implicit in the original Watson-Crick model of the DNA molecule. Direct evidence of colinearity has come from analyses carried out by Charles Yanofsky of the genetic control of the enzyme tryptophan synthetase in bacteria. Yanofsky showed that the relative position of mutational sites within a cistron, as determined by genetic analysis of the mutants, corresponds to the relative position of the altered amino acids in the peptide chain of the enzyme molecule, as determined by direct chemical analysis of the peptide. Other examples of colinearity are derived from analyses of the genetic control of hemoglobin synthesis in man.

It is now clear that DNA does not control the production of a polypeptide by any direct interaction with the amino acid. Instead, it forms an intermediate template, an RNA molecule, which in turn directs the synthesis of the peptide chain. The DNA has been compared to a master model that is carefully preserved in the nucleus and used only to synthesize secondary working models which pass out to the ribosomal mechanism in the cytoplasm and are utilized for the actual synthesis of proteins.

The coding relationships between DNA, RNA and protein involve (1) the **replication** of DNA to form new DNA, (2) the **transcription** of DNA to form a messenger RNA template, and (3) the **translation** of the code of the messenger RNA template into the specific sequence of amino acids in a protein. The arrows in the formula indicate the direction of transfer of genetic information:

$$\text{DNA} \xrightarrow{\text{transcription}} \text{RNA} \xrightarrow{\text{translation}} \text{Protein}$$
$$\downarrow{\text{replication}}$$
$$\text{DNA}$$

8.5 THE SYNTHESIS OF DNA: REPLICATION

The synthesis of DNA is catalyzed by an enzyme system, **DNA polymerase,** first isolated from the cells of *Escherichia coli* by Kornberg and colleagues in 1957. DNA polymerase requires the triphosphates of all four deoxyribonucleosides (abbreviated dATP, dGTP, dCTP and dTTP) as substrates, magnesium ions, and a small amount of high molecular weight DNA polymer to serve as a primer and template for the reaction. The product of the reaction is more DNA polymer and a molecule of pyrophosphate for each molecule of deoxyribonucleotide incorporated.

$$\left.\begin{array}{l}\text{dATP}\\\text{dGTP}\\\text{dCTP}\\\text{dTTP}\end{array}\right\}_n \xrightarrow[\text{DNA polymerase}]{\substack{\text{DNA}\\\text{Mg}^{++}}} \text{DNA} + n\text{PPi}$$

The DNA polymerase from *Escherichia coli* can use template DNA prepared from a wide variety of sources — bacteria, viruses, mammals and plants — and will produce DNA with a nucleotide ratio comparable to that of the template used. Thus the sequence of nucleotides in the product is dictated by the sequence in the primer and not by the properties of the polymerase nor by the ratio of the substrate molecules present in the reaction mixture. Using a more highly purified enzyme, Kornberg was able in 1968 to synthesize biologically active viral DNA using viral DNA as primer. The DNA produced would infect bacteria just like "live" viruses.

Khorana and his colleagues prepared synthetic deoxyribonucleotide polymers containing adenylic and cytidylic acids and other polymers containing alternating thymidylic and guanylic acids. Neither one alone could serve as a template for DNA polymerase, but a mixture of the two, which forms a synthetic double-stranded helix with conventional pairing of the bases, can serve as a template.

The DNA template appears to have two functions in the DNA polymerase system: first, to provide 3'-OH groups which are free to serve as the growing end or primer of the polymer and, second, to provide coded information. A double-stranded molecule is required because each strand of the pair serves as a template for the extension of the complementary strand and as a primer for its own extension. The DNA-like polymer that is produced by the action of the DNA polymerase in the presence of double-stranded template is also double-stranded and has the same base composition as the template DNA. The ratios of the bases are those predicted by the Watson-Crick model.

In the cells of higher organisms, the synthesis of DNA occurs only during the interphase, when chromosomes are in their extended form and are not readily visible. Thus, if an enzyme similar to the Kornberg enzyme catalyzes the synthesis of DNA in vivo, there must be some sort of biological signal which initiates DNA synthesis at this time and turns it off at other times. It appears that both the enzyme DNA polymerase and the substrates, dATP, dGTP, dCTP and dTTP, are present all the time, and hence there must be some change in the DNA template which initiates DNA synthesis and then turns it off.

During DNA replication two new strands are formed, each complementary to one of the existing DNA strands in the double-

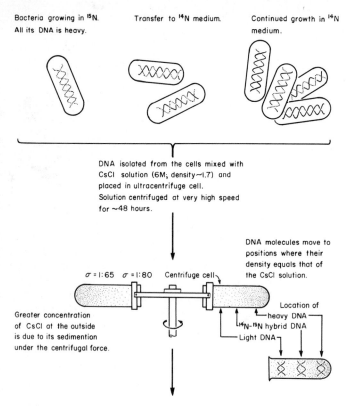

Bacteria growing in ^{15}N.
All its DNA is heavy.

Transfer to ^{14}N medium.

Continued growth in ^{14}N
medium.

DNA isolated from the cells mixed with
CsCl solution (6M; density~1.7) and
placed in ultracentrifuge cell.
Solution centrifuged at very high speed
for ~48 hours.

DNA molecules move to
positions where their
density equals that of
the CsCl solution.

$\sigma = 1.65$ $\sigma = 1.80$ Centrifuge cell

Location of
heavy DNA
^{14}N-^{15}N hybrid DNA
Light DNA

Greater concentration
of CsCl at the outside
is due to its sedimention
under the centrifugal force.

The location of DNA molecules within the centrifuge cell can be determined by ultraviolet
optics. DNA solutions absorb strongly at 2600 A.

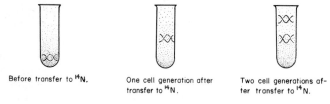

Before transfer to ^{14}N.

One cell generation after
transfer to ^{14}N.

Two cell generations af-
ter transfer to ^{14}N.

Figure 8.7 Diagram of the experiment of Meselson and Stahl which indicated that DNA is replicated by a semi-conservative mechanism: the two original strands of DNA are retained in the product, one in each daughter helix. (Villee: Biology, 6th ed.)

stranded helix. The double-stranded helix unwinds; one strand provides a template for one new one and the other original strand also provides a template for a second new strand. This is called a **semi-conservative** mechanism: the two original strands of DNA are retained in the product, one in each of the two daughter helices.

The experiment of Meselson and Stahl (Fig. 8.7) provided evidence that DNA replication is carried out by a semi-conservative mechanism, at least in bacteria. Bacteria were grown for several generations in a medium containing heavy nitrogen, ^{15}N. Thus, all of the nitrogen in the DNA of those bacteria was ^{15}N. When a sample of the DNA was isolated and centrifuged in a cesium chloride density gradient, the DNA

collected at a level which reflected the presence of the heavy nitrogen in the DNA molecules.

The bacteria were then removed from the ^{15}N medium, placed in a medium containing ordinary nitrogen, ^{14}N, and allowed to divide once in this medium. When some of the DNA from this generation was isolated and centrifuged, all the DNA was lighter and had a density corresponding to its being half labeled with ^{15}N and half labeled with ^{14}N. If the Watson-Crick theory is correct and replication is semi-conservative, this result would be expected because one strand of the double-stranded DNA in each organism is ^{15}N-labeled and the other is ^{14}N-labeled.

When the organisms were allowed to divide again in ^{14}N medium, each molecule of progeny DNA again received one parental strand and made one new strand containing ^{14}N. As a result, some double-stranded DNA containing only ^{14}N was formed and appeared as a light DNA on centrifugation. The parental ^{15}N-containing strands made complementary strands containing ^{14}N and appeared on centrifugation with a density characteristic of the half ^{15}N, half ^{14}N double-stranded state.

If the replication of the strands begins as the strands begin to untwist, Y-shaped molecules of DNA should be evident during the replication process. Such Y-shaped regions have been found by autoradiography of chromosomes of *Escherichia coli.*

8.6 TRANSCRIPTION OF THE CODE: THE SYNTHESIS OF MESSENGER RNA

Messenger RNA is synthesized by a **DNA-dependent RNA polymerase** found first in the nuclei of rat liver and subsequently in plants, bacteria and other animals. It requires DNA as a template and uses as substrate the four triphosphates of the ribonucleosides commonly found in RNA. The products are RNA and inorganic pyrophosphate. The reaction system can utilize single-stranded or native DNA or synthetic deoxyribopolynucleotides as template. These can be polymers of a single nucleotide or polynucleotides with some kind of repeating sequence.

Transfer RNA and ribosomal RNA are also produced in the nucleus by DNA-dependent RNA-synthesizing systems, and are transcribed from complementary deoxynucleotide sequences in DNA.

The RNA that is produced by the use of these carefully defined templates is exactly that predicted by the kinds of base pairing permitted in the Watson-Crick model. Although the DNA template is double-stranded and contains two different but complementary template sequences (which would have quite different genetic information), it appears that only one DNA strand is selected for transcription and only one kind of mRNA is produced. The molecular basis for this distinction between the two strands is unknown.

8.7 TYPES OF RNA: MESSENGER, RIBOSOMAL AND TRANSFER

Chemical analyses show that, in contrast to DNA, molecules of RNA do not usually have complementary base ratios and the inference is that RNA is not a double helix like DNA but is single-stranded. It is a long, unbranched molecule containing four kinds of ribonucleotides linked by 3′-5′ phosphodiester bonds. RNA contains ribose instead of deoxyribose and uracil rather than thymine.

Three kinds of RNA molecules are required for protein synthesis: **messenger RNA,** which transmits genetic information from the DNA molecule in the nucleus to the cytoplasm; **ribosomal RNA,** which serves some nonspecific function in connection with the cytoplasmic particles called ribosomes on which the protein synthesis occurs; and **transfer RNA,** which acts as an adaptor to bring amino acids into line in the growing polypeptide chain in the appropriate place.

Electron microscopy has established that most cells contain an extensive system of tubules with thin membranes termed the endoplasmic reticulum. Associated with the endoplasmic reticulum or floating freely in the cytoplasm are small particles termed ribosomes. Ribosomes are about half protein and half RNA. The RNA of ribosomes after extraction with phenol appears to consist of two components with molecular weights of about 600,000 and 1,300,000.

Molecules of transfer RNA are considerably smaller than the molecules of messenger or ribosomal RNA. Each functions as a specific adaptor in protein synthesis, binding to and identifying one specific amino acid; i.e., each of the 20 amino acids is attached to one or more specific kinds of transfer RNA. One portion of the nucleotide sequence in transfer RNA represents an **anticodon,** a nucleotide triplet complementary to the codon in messenger RNA that specifies that amino acid. Transfer RNA is unusual in containing not only the four usual ribonucleotides, adenylic, guanylic, cytidylic and uridylic, but also small amounts of unusual nucleotides such as **6-methylamino adenylic acid, dimethylguanylic acid** and **thymine ribotide.**

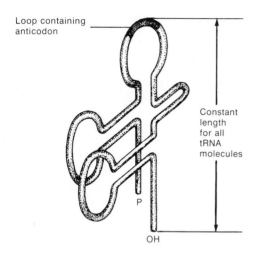

Loop containing anticodon

Constant length for all tRNA molecules

P

OH

Figure 8.8   A diagram of the three dimensional clover leaf structure of transfer RNA. One loop contains the triplet anticodon which forms specific base pairs with the mRNA codon. The amino acid is attached to the terminal ribose at the 3′ OH end, which has the sequence CCA of nucleotides. Each transfer RNA also has guanylic acid, G, at the 5′ end (P). The pattern of folding permits a constant distance between anticodon and amino acid in all transfer RNAs examined.

Transfer RNAs are polynucleotide chains of some 70 nucleotides. Each of the several kinds of transfer RNA has an identical sequence of nucleotides, CCA, at the 3′ end to which the amino acid is attached. Each in addition has guanylic acid at the opposite, 5′, end of the nucleotide chain. The chain is doubled back on itself, forming three or more loops of unpaired nucleotides; the folding is stabilized by hydrogen bonds between complementary bases in the intervening portions of the chain to form a double helix (Fig. 8.8). The loop nearest the amino acid acceptor (CCA) has seven nucleotides, with cytidine, pseudouridine and thymidine at positions 21, 22 and 23 from the CCA end. The triplet that is complementary to the codon is located in a seven-membered middle loop, and is preceded by uridine and followed by adenosine or a modified adenosine in the loop. Another unusual base, dimethylguanosine, is found eight positions before the anticodon, at the base of the larger (eight to twelve nucleotides) loop near the 5′ end. The folding results in a constant distance between the anticodon and the amino acid in all of the tRNAs examined so far. The triplet anticodons found in these tRNAs agree very well (Table 8.5) with the complements predicted to the known mRNA codons for those amino acids.

The first transfer RNA to be completely analyzed was the transfer RNA for alanine derived from yeast cells. This dramatic achievement of Robert Holley and his colleagues was recognized with the awarding of a Nobel Prize to Holley in 1968. The alanine tRNA has 77 nucleotides arranged in a unique sequence. Nine of these are unusual bases, with one or more methyl groups which are added enzymatically after the nucleotides have been linked by phosphodiester bonds. Certain of these unusual bases cannot form conventional base pairs, and may serve to disrupt the base pairing in other parts of the transfer RNA. This could expose specific chemical groups on the tRNA which form secondary bonds to messenger RNA or to the ribosome, or perhaps to the enzyme needed to attach the specific amino acid to its specific transfer RNA molecule. The exact sequence of nucleotides in several other transfer RNAs is now known. This implies that the sequence of the nucleotides in the genes that specify each of these particular kinds of transfer RNA is also known. The sequence in the genic DNA can be inferred from the Watson-Crick rules of specific base pairing. The gene specifying the transfer RNA for alanine would also be 77 nucleotides long. A complete, double-stranded DNA molecule with this sequence of 77 nucleotides was synthe-

TABLE 8.5 AGREEMENT BETWEEN THE ANTI-CODONS FOUND IN ISOLATED tRNAs AND THE ANTICODONS PREDICTED FROM THE GENETIC CODE

Amino Acid	mRNA Codons →	Complements ←	Observed Anticodons [°] ←
Alanine	GpCpA	UpGpC	
	GpCpG	CpGpC	
	GpCpC	GpGpC	IpGpC
	GpCpU	ApGpC	
Phenylalanine	UpUpC	GpApA	OMGpApA
	UpUpU	ApApA	
Tyrosine	UpApU	ApUpA	
	UpApC	GpUpA	Gpψpa
Serine	ApGpC	GpCpU	
	ApGpU	ApCpU	
	UpCpC	GpGpA	IpGpA
	UpCpU	ApGpA	
	UpCpA	UpGpA	
	UpCpG	CpGpA	
Valine	GpUpU	ApApC	
	GpUpC	GpApC	IpApC
	GpUpA	UpApC	
	GpUpG	CpApC	

[°] The anticodons contain some "unusual" bases, including I (inosine), OMG (2′-O-methylguanosine) and ψ (pseudouridine).

sized in 1970 by Khorana and was found to be transcribed to give alanine tRNA. This was the first gene to be synthesized.

8.8 THE SYNTHESIS OF A SPECIFIC POLYPEPTIDE CHAIN

The first step in the synthesis of a peptide requires the activation of the amino acid by an enzyme system in the cytoplasm. There is a separate specific amino acid activating enzyme for each amino acid. The enzyme first catalyzes the reaction of the amino acid (aa) and ATP to form the amino acid adenylic acid compound (aa-AMP) and release pyrophosphate. The same enzyme then catalyzes the transfer of the amino acid to the specific transfer RNA for the amino acid, yielding the transfer RNA-amino acid compound and free adenylic acid. The amino acid is attached at the end of the transfer RNA which contains cytidylic, cytidylic and adenylic acid; the amino acid is attached to the ribose of the terminal adenylic acid. If these three nucleotides are removed the transfer RNA is unable to function.

Francis Crick had predicted on theoretical grounds that some sort of nucleic acid molecule must serve as an adaptor in the course of protein synthesis. He argued that since there is no simple steric correspondence between a polynucleotide chain and a polypeptide chain which would enable the nucleotide sequence to specify the amino acid sequence directly, the amino acids might be lined up in appropriate register by means of small RNA adaptor molecules. These adaptor molecules could assemble at specific places on the nucleic acid template by their complementary sequences of nucleotides. This hypothesis has now been validated by a wide variety of experiments. For example, Chapevelle prepared a com-

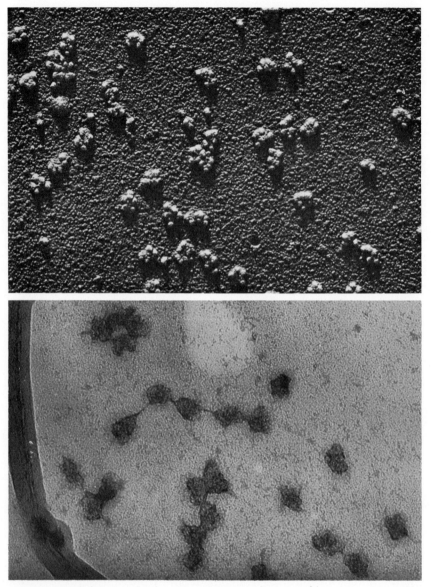

Figure 8.9 Electron micrographs of polyribosomes isolated from the reticulocytes of a rabbit. The upper preparation was shadowed with gold and the lower one was stained with uranyl nitrate. The electron micrographs show that polyribosomes tend to occur in clusters of four, five or six and the clusters are connected by a thin strand of mRNA. (Photographs courtesy of Dr. Alexander Rich.)

plex of cysteine with its specific transfer RNA (cysteinyl-tRNA^{cys}), and then converted the cysteine to alanine while it was still bound to the transfer RNA (alanyl-tRNA^{cys}). When this was added to a protein-synthesizing system, a polypeptide was made which contained *alanine* at the sites in the peptide chain where cysteine should have been. Thus these experiments provided direct proof that the ordering of specific amino acids into their appropriate place is dictated by the specific transfer RNAs and not by the amino acid that is bound to it.

The amino acid bound to its specific transfer RNA is transferred to the ribosomes. The role of the ribosome is to provide the proper orientation of the amino acid-transfer RNA precursor, the messenger or template RNA, and the growing polypeptide chain, so that the genetic code on the template or messenger RNA can be read accurately. There are some 15,000 ribosomes in a rapidly growing cell of *E. coli*, each with a molecular weight of nearly 3,000,000. These ribosomes account for nearly one-third of the total mass of the cell. Only one polypeptide chain can be formed at a time on any given ribosome.

The template for the synthesis of a specific protein is supplied by the messenger RNA formed on one strand of the double helix of DNA. The messenger RNA leaves the nucleus and passes to the cytoplasm, where it becomes associated with the ribosome. Ribosomes from different kinds of cells may differ somewhat in their mass, in the composition of their RNA and in the ratio of RNA to protein, but there is a general similarity in their structures. Bacterial ribosomes have a molecular weight of about 2,600,000, and are composed of two portions—a larger one, molecular weight 1,800,000, and a smaller one, molecular weight 800,000. The ribosome can be separated into these two subunits if they are placed in a solution with a low concentration of magnesium ion. When observed in the electron microscope, the subunit structure becomes apparent. The smaller subunit seems to sit on the flat surface of the larger subunit like a cap. The ribosomes of higher organisms tend to be somewhat larger than the bacterial ones, and are composed of two or four subunits. Each ribosome contains several kinds of proteins bound to the RNA. Ribosomes have GTPase activity, which appears to play an important but undetermined role in the transfer of amino acids from transfer RNA to the forming peptide chain. Protein-synthesizing particles somewhat similar to ribosomes are also found in the nucleus, in the chloroplasts of plant cells and in mitochondria.

During protein synthesis the messenger RNA appears to move across the site of a ribosome at which protein synthesis occurs.

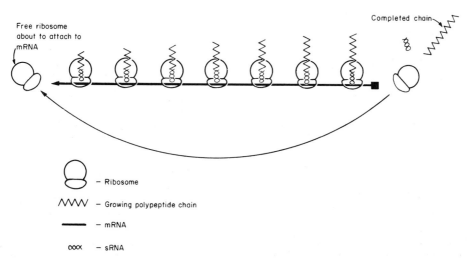

Figure 8.10 The postulated mechanism by which a polypeptide chain is synthesized. Each ribosome "rides" along the messenger RNA, reading and translating the genetic message. Amino acids are added as dictated by the specific base-pairing of the mRNA codon and the tRNA anticodon. (After Watson, J. D.: The Molecular Biology of the Gene. New York, W. A. Benjamin, 1965.)

In doing this it brings successive codons into a position on which the appropriate amino acid transfer RNA anticodons can be arranged. The transfer of the amino acid to the growing peptide chain of a ribosome requires guanosine triphosphate, one or more specific enzymes, and glutathione.

Protein synthesis has been studied intensively in preparations of rabbit reticulocytes, which are engaged primarily in making just one protein—hemoglobin. Alexander Rich and his coworkers showed that the ribosomes most active in protein synthesis are those that interact in clusters of five or more (Fig. 8.9). These clusters, termed **polyribosomes**, are held together by the strand of messenger RNA. Peptide chains are formed by the sequential addition of amino acids beginning at the N-terminal end, the end having a free amino group. Studies of ribosomes in the electron micro-

scope suggest that individual ribosomes become attached to one end of a polyribosome cluster, and gradually move along the messenger RNA strand as the polypeptide chain attached to it increases in length by the sequential addition of amino acids (Fig. 8.10). Thus each ribosome appears to ride along the extended messenger RNA molecule, "reading the message" as it goes. The ribosome is believed to play a part in bringing the transfer RNA molecule into line at the right position. After it completes the reading of one molecule of messenger RNA and releases the polypeptide that it has synthesized, the ribosome appears to jump off the end of one messenger RNA chain, move to a new messenger RNA and begin reading it (Fig. 8.11). The growing peptide chain always remains attached to its original ribosome; there is no transfer of a peptide chain from one ribosome to another. Several

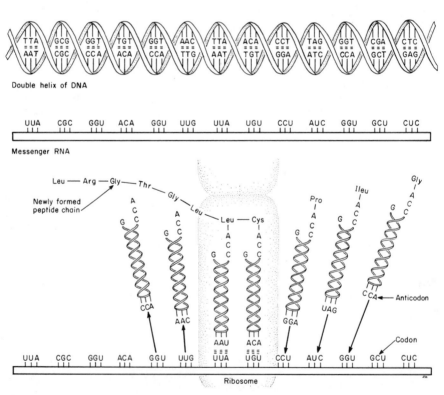

Figure 8.11 Diagram of the postulated mechanism of protein synthesis on the ribosome, illustrating the relationship between the triplet code of the DNA helix, the complementary triplet code of messenger RNA and the complementary triplet code (anticodon) of transfer RNA. Molecules of tRNA charged with specific amino acids are depicted coming from the right, assuming their proper place on mRNA at the ribosome, transferring the amino acid to the growing peptide chain, and then (*left*) leaving the ribosome to be recharged with amino acids for further reactions. The growing polypeptide chain remains attached to its original ribosome. (Villee: Biology, 6th ed.)

ribosomes may be working simultaneously on a single strand of messenger RNA, each reading a different part of the message.

All the processes of gene replication, gene transcription and protein synthesis depend upon the formation of specific, though relatively weak, hydrogen bonds between specific pairs of purine and pyrimidine bases. The specificity of these bonds ensures the remarkable accuracy of the process; mistakes in base pairing occur less than one time in a thousand.

8.9 CHANGES IN GENES: MUTATIONS

Although genes are remarkably stable and are transmitted to succeeding generations with great fidelity, they do from time to time undergo changes called **mutations.** After a gene has mutated to a new form this new form is stable and usually has no greater tendency to mutate again than the original gene. A mutation has been defined as any inherited change not due to segregation or to the normal recombination of unchanged genetic material. Mutations provide the diversity of genetic material which makes possible a study of the process of inheritance, and investigations of the nature of the mutation process have provided important clues as to the nature of the genetic material itself.

Some mutations, termed **chromosomal mutations,** are accompanied by a visible change in the structure of the chromosome.

A small segment of the chromosome may be missing (a **deletion**) or be represented twice in the chromosome (a **duplication**) (Fig. 8.12). A segment of one chromosome may be translocated to a new position on a new chromosome (a **translocation**), or a segment may be turned end for end and attached to its usual chromosome (an **inversion**). Point mutations or gene mutations produce no visible change in chromosome structure, and it is assumed that these involve such small changes at the molecular level that they are not evident under the microscope. From the current theory that the genic material is DNA arranged in a double helix, and that genic specificity resides in a particular sequence of nucleotides, it would follow that a gene mutation is some change in the sequence of nucleotides within a particular region of the DNA.

From your knowledge of the DNA molecule, you might predict that replacing one of the purine or pyrimidine nucleotides by an analogue such as azaguanine or bromouracil would result in mutation. In several experiments in which such analogues were incorporated into bacteriophage, no mutations were evident. Because of the degeneracy of the genetic code (Table 8.4), a number of changes in base pairs could occur without changing the amino acid specified. In other experiments, the incorporation of bromouracil into DNA did lead to an increased rate of mutation. Other chemicals known to be mutagenic include nitrogen mustards, epoxides, nitrous acid and alkylating agents. These are all chemicals which can react

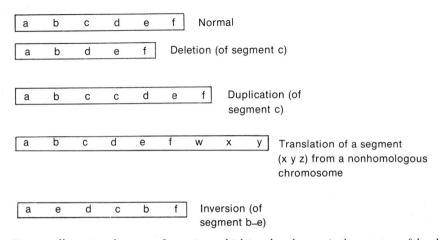

Figure 8.12 Diagram illustrating the types of mutations which involve changes in the structure of the chromosome.

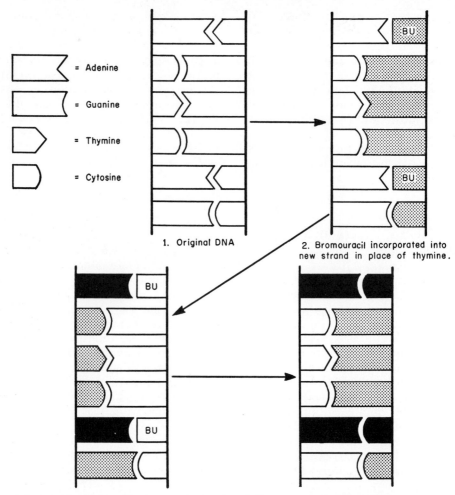

1. Original DNA

= Adenine

= Guanine

= Thymine

= Cytosine

2. Bromouracil incorporated into new strand in place of thymine.

3. Strand with bromouracil leads to production of new strand with guanine paired to the bromouracil.

4. New, mutant DNA which contains no analogue bases, but has nucleotide sequence different from original, with GC pairs in place of AT.

Figure 8.13 Diagrammatic scheme of how an analogue of a purine or pyrimidine might interfere with the replication process and cause a mutation, an altered sequence of nucleotides in the DNA, indicated in black. The nucleotides of the new chain at each replication are indicated by the dotted blocks. In this instance, two new GC pairs are indicated. Probably a single substitution of a GC pair for an AT pair would be sufficient to cause a mutation if it occurred in the triplet code at a point that changed the amino acid specified, i.e., in the first or second base of the triplet, or in the third member of certain triplets (see Table 8.4). (Villee: Biology, 6th ed.)

with specific nucleotide bases in the DNA and change their nature. When an analogue is incorporated into DNA it may lead to mistakes in the pairing of nucleotides during subsequent replication processes. For example, when bromouracil is incorporated into DNA in place of thymine it will pair with guanine rather than with adenine, the normal pairing partner of thymine. This would lead to the substitution of a GC pair of nucleotides at the point in the double helix previously occupied by an AT pair of nucleotides (Fig. 8.13).

The presence of mutagenic materials may

increase the frequency of mistakes in nucleotide base pairing, which would result in the production of DNA molecules containing only natural bases, but with an altered base order. The change in normal base order is significant, because during subsequent replications, the altered base sequence will be reproduced by the normal process of DNA synthesis. The definition of mutation includes the restriction that the change that has been introduced into the DNA molecule must be propagated subsequently for an indefinite number of times.

"Spontaneous" gene mutations may result

from errors in base pairing during the replication process; thus the A-T normally present at a given site may be replaced by G-C, C-G or T-A. The altered DNA will be transcribed to give an altered mRNA, and this will be translated into a peptide chain with one amino acid different from the normal sequence. If the altered amino acid is located at or near the active site of the enzyme, the altered protein may have markedly decreased or altered catalytic properties. If the altered amino acid is elsewhere in the protein, it may have little or no effect on the properties of the enzyme and may thus go undetected. The true number of gene mutations may be much greater than the number observed.

If a single nucleotide pair were inserted into or deleted from the DNA molecule, it would shift the reading of the genetic message, alter all the codons lying beyond that point, and change completely the nature of the resulting peptide chain and its biological activity. Thus if the normal sequence is CAG TTC ATG, read (CAG)(TTC)(ATG), the insertion of a G between the two T's results in CAG TGT CAT G, read (CAG)(TGT) (CAT) (G . . .).

Gene mutations can be induced not only by exposing the cell to certain chemicals but by a variety of types of radiation — x-rays, gamma rays, cosmic rays, ultraviolet rays, and the several types of radiation that are by-products of atomic power. Mutations occur spontaneously at low, but measurable, rates which are characteristic of the species and of the gene; some genes are much more prone to undergo mutation than others. Natural radiations, such as cosmic rays, probably play some role in causing "spontaneous mutations," but other factors, such as errors in base pairing, play a role. The rates of spontaneous mutation of different human genes range from 10^{-3} to 10^{-5} mutations per gene per generation. Since man has a total of some 2.3×10^4 genes, this means that the total mutation rate is on the order of one mutation per person per generation. Each one of us, in other words, has some mutant gene that was not present in either of our parents.

8.10 GENE-ENZYME RELATIONS

If we assume that a specific gene leads to the production of a specific enzyme by the method outlined above, we must next inquire how the presence or absence of a specific enzyme may affect the development of a specific trait. The expression of any trait is the result of a number, perhaps a large number, of chemical reactions which occur in series, with the product of each reaction serving as the substrate for the next: A→B→C→D. The dark color of most mammalian skin or hair is due to the pigment **melanin** (D), produced from dihydroxyphenylalanine (dopa) (C), produced in turn from tyrosine (B) and phenylalanine (A). Each of these reactions is controlled by a particular enzyme; the conversion of dopa to melanin is mediated by **tyrosinase. Albinism,** characterized by the absence of melanin, results from the absence of tyrosinase. The gene for albinism (a) does not produce the enzyme tyrosinase but its normal allele (A) does.

The earliest attempts to connect the action of a particular gene with a particular reaction were studies of the inheritance of flower colors, in which the specific flower pigments could be extracted and analyzed. Other studies of the inheritance of coat color in mammals and eye color in insects were also able to relate specific genes with specific enzymic reactions in the synthesis of these pigments.

A major advance was made when George Beadle and Edward Tatum conceived the idea of looking for mutations in the mold *Neurospora* which interfere with the reactions by which chemicals essential for its growth are produced. The wild type *Neurospora* requires as nutrients only sugar, salts, inorganic nitrogen and biotin. A mixture of these comprises the so-called "minimal medium" for the growth of the wild type *Neurospora* (Fig. 8.14). Exposure of the conidia (haploid asexual spores) to x-rays or ultraviolet rays produces many mutations. After irradiation the mold is supplied with a "complete" medium, an extract of yeast which contains all the known amino acids, vitamins, purines, pyrimidines, and so on. Any nutritional mutant produced by the irradiation is thus able to survive and reproduce to be tested subsequently.

Some of the cells from the irradiated mold are then placed on minimal medium. If these are unable to grow we know that a mutant has been produced which is unable to produce some compound essential for

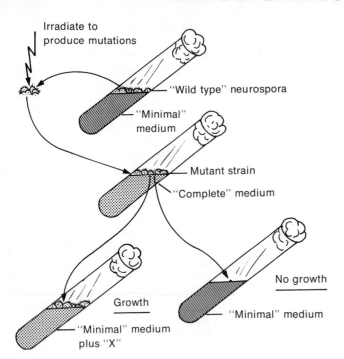

Figure 8.14 The method of producing and testing biochemical mutants in *Neurospora*. See text for discussion.

growth. By trial and error, by adding substances to the minimal medium in groups or singly, the required substance is identified. In each instance genetic tests show that the mutant strain produced by irradiation differs from the normal wild type by a single gene, and chemical tests show that the addition of a single chemical substance to the minimal medium will enable the mutant strain to grow normally. The obvious inference is that each normal gene produces a single enzyme which regulates one step in the biosynthesis of this particular chemical; the mutant gene does not produce this enzyme. It has been possible in certain instances to show that the particular enzyme can be extracted from the cells of normal *Neurospora* but not from cells of the mutant strain. The synthesis of each of these nutritional substances involves a number of different steps, each mediated by a separate, gene-controlled enzyme. An estimate of the minimal number of steps involved can be obtained from the number of different mutants that will interfere with the production of that substance.

Similar one-to-one relationships of gene, enzyme and biochemical reaction in man were first described by the English physician A. E. Garrod in 1908. **Alkaptonuria** is a trait, inherited by a recessive gene, in which

the patient's urine turns black on exposure to air. The urine contains homogentisic acid, a normal intermediate in the metabolism of phenylalanine and tyrosine. The tissues of normal people have an enzyme which oxidizes homogentisic acid so that it is eventually excreted as carbon dioxide and water (Fig. 8.15). Alkaptonuric patients lack this enzyme because they lack the gene which controls its production. As a result, homogentisic acid accumulates in the tissues and blood and spills over into the urine. Garrod used the term "inborn errors of metabolism" to describe alkaptonuria and comparable conditions such as phenylketonuria and albinism.

The question of whether the genes normally operate so as to lead to the production of the *maximum* number of enzymes all the time has been given consideration in recent years. From a variety of experimental evidence it would appear that this is not the case. Each gene is probably "repressed" to a greater or lesser extent under normal conditions and then, in response to some sort of environmental demand for that particular enzyme, the gene becomes "derepressed" and leads to an increased production of the enzyme. When a single gene is fully derepressed it can lead to the synthesis of fantastically high amounts of enzymes—one en-

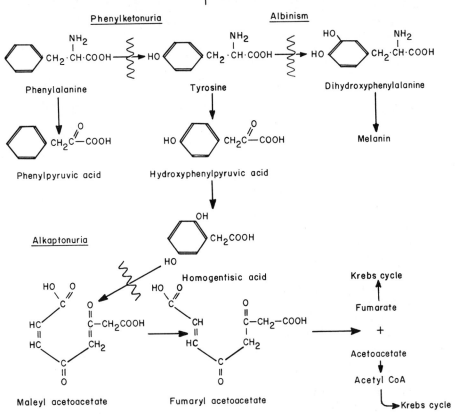

Figure 8.15 Pathway by which phenylalanine and tyrosine are metabolized. Mutants which interfere with the production of the enzymes that catalyze each of these steps may result in "inborn errors of metabolism" such as phenylketonuria, albinism and alkaptonuria. (Villee: Biology, 6th ed.)

zyme may comprise 5 to 8 per cent of the total protein of the cell! If all enzymes were produced at this same fantastically high rate, metabolic chaos would result. Thus, the phenomena of gene repression and derepression would appear to be necessary to prevent this chaos and to provide a means for increasing or decreasing the rate of synthesis of one particular enzyme in response to variations in environmental requirements.

The control mechanism postulated by Jacob and Monod from their studies of β-galactosidase in *E. coli* states that in addition to the structural genes that provide the code for the synthesis of specific proteins there are "regulator genes" that code for the synthesis of "repressors"; that the repressors may be active or inactive depending on whether they are bound to small molecules, the inducers or corepressors; and that active repressors bind to operator sites in the DNA and turn off the transcription of the adjacent structural genes. The

structural genes under the control of one repressor are termed an operon.

There is clear evidence that the synthesis of many proteins in *E. coli* and in other organisms is not influenced by substances such as corepressors and inducers in the external environment; either their operator sites are turned on all the time or their genes are not associated with an operator site.

8.11 WHAT IS A GENE?

In our previous discussions, the gene was defined simply as the unit of inheritance, a unit that can undergo mutation and subsequently be detected by the change it produces in the phenotype of the organism. With the techniques of classical genetics, two genes can be defined as being alleles or non-alleles—the same gene or different genes—on the basis of whether or not they can undergo recombination or whether they affect the same biochemical function.

For example, in the cross of gray, normal-winged fruit flies with black, vestigial-winged flies discussed in the previous chapter the offspring all had gray bodies and normal wings. When these offspring were crossed with black, vestigial-winged flies, four types of offspring appeared in the second generation: two similar to the parental types, which occurred in greater frequency, and two crossover types, gray-bodied, vestigial-winged flies and black-bodied, normal-winged flies which were less frequent among the offspring.

In this example the two genes were twenty crossover units apart and therefore the probability of finding some crossovers among the offspring was quite high. However, if the two genes were located very close together in the chromosome, we might find only two kinds of offspring—gray-bodied, normal-winged ones and black-bodied, vestigial-winged ones—simply because recombination was such a rare event that it was not detected in the particular experiment. If this were found, the geneticist would conclude that both body color and wing shape were determined by the same gene.

To put it another way, a geneticist can be sure that two characters are inherited by different genes only if recombination between the genes can be observed. When two genes affect the same character in similar fashion it may be quite difficult to determine whether the two genes are in fact alleles.

The genetic analysis of the lozenge genes in *Drosophila* made by Green and Green in 1949 revealed three different loci very close together in one chromosome. All of these produce the same phenotype, a change in pigmentation and shape of the eye of the fruit fly termed lozenge. These lozenge genes are recessive; heterozygous flies have normal eyes. However, flies that are heterozygous for two of these mutant alleles will have different phenotypes depending upon whether the two mutant alleles are in the same chromosome strand or are in opposite members of the pair of homologous chromosomes (Fig. 8.16).

Two mutant alleles in the same chromosome resulted in normal eyes, whereas two mutant alleles in different members of the homologous pair of chromosomes produced the mutant phenotype, lozenge eyes. It was

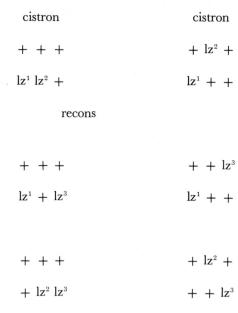

Figure 8.16 Diagrams of the possible arrangements of the three lozenge genes, lz^1, lz^2 and lz^3, within the lozenge cistron. In the *cis* configurations, one of the members of the homologous pair of chromosomes contains only wild type alleles (+++) and thus can produce the wild type (normal eye). In the *trans* configurations, neither of the two homologous chromosomes has a complete set of normal alleles, a complete wild type cistron, and the eyes are the lozenge phenotype. (Villee: Biology, 6th ed.)

inferred that this entire region is one functional genetic unit and that the separate loci are substructures of the parent gene. A mutation at any one of the three loci might lead to an impairment of gene function so that it produces a lozenge instead of a normal eye.

When both mutant alleles are present in one chromosomal strand (the so-called *cis* position), the opposite chromosome strand is composed entirely of wild-type alleles and functions to produce a wild-type eye (Fig. 8.16). When the two lozenge genes are in opposite members of the chromosome pair, there is no complete normal gene able to function and thus the lozenge phenotype is produced. When two such genes are present in opposite members of the homologous pair of chromosomes, they are said to be in the *trans* position.

Seymour Benzer suggested the terms cistron, muton and recon to sharpen the definitions of the different aspects of genetic function. He defined **cistron** as the genetic unit of biochemical function. Two genes are different in a biochemical sense, they belong to different cistrons, if they lead to the production of different enzymes. The cistron may be a very complex structure containing many subunits distinguishable by genetic tests. It is defined as that part of a chromosome within which recessive mutants determine a mutant phenotype when in combinations of two in the *trans* position but not when in the *cis* position. Thus, the term cistron indicates the smallest unit of genetic material that must be intact for normal function.

A cistron can be subdivided by appropriate genetic tests into the ultimate units of recombination termed **recons** by Benzer. The recon is "the smallest element in a one-dimensional array that is interchangeable but not divisible by recombination." Thus, the smallest segment of a chromosome that is indivisible by recombination is the recon, and it may be as small as a single nucleotide pair.

The **muton** is defined as the smallest portion of a chromosome which when altered gives rise to a mutant form of the organism. There are clear examples of mutations of a single base pair which give rise to the mutant form of an organism and therefore the muton may be as small as one single pair of nucleotide bases. Although both mutons and recons may be as small as one base pair, the cistron is a much larger unit containing perhaps thousands of nucleotide pairs.

A gene could be defined as that part of the hereditary material which specifies the amino acid sequence of a specific protein. Although this definition of a gene is difficult to apply in many cases, it remains the best theoretical definition — "a gene is that section of DNA involved in the determination of the amino acid sequence of a single polypeptide chain."

8.12 GENES AND DIFFERENTIATION

We now have a detailed working hypothesis as to how biological information is transferred from one generation of cells to the next and how this information may be transcribed and translated in each cell so that specific enzymes and other proteins are synthesized. The operation of this system would produce a multicellular organism in which each cell would have the same assortment of enzymes as every other cell. Additional hypotheses are needed to account for: (1) the means by which the amount of any given enzyme produced in a cell is regulated, (2) the control of the time in the course of development when each kind of enzyme appears, and (3) the mechanism by which unique patterns of proteins are established in each of the several kinds of cells in a multicellular organism despite the fact that they all contain identical quotas of genetic information.

Any satisfactory model of the developmental process must be able to explain how genetic and nongenetic factors can interact to control the differentiation of cells and tissues.

Cellular differentiation might be explained if genetic material were parceled out differentially at cell division and the daughter cells received different kinds of genetic information. Although there are a very few clear instances of differential nuclear divisions in animals such as *Ascaris* and *Sciara*, this does not appear to be a general mechanism of differentiation. By and large, genes are neither lost nor gained during developmental processes. The generalization that the mitotic process ensures the exact distribution of genes to each cell of the organism is a valid one. Thus the differences in the kinds of enzymes and other proteins found in different cells of the same organism must arise by differences in the *activity* of the same set of genes in different cells. The experiments of Briggs and King (p. 146) showed that even the nucleus from a differentiated cell taken from an advanced stage of embryonic development can, when placed in an enucleated egg, lead to the development of a normal embryo. Thus it clearly had retained a full set of genetic information.

Some striking evidence regarding the differential activity of genes comes from cytologic studies of insect tissues. In certain insect tissues the chromosomes undergo repeated duplication, and the daughter strands line up exactly in register, locus by locus, so that characteristic bands appear along the length of the giant chromosome. When these bands are examined carefully, either in the same tissue at different times

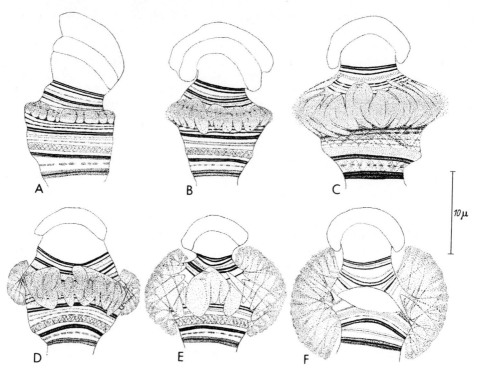

Figure 8.17 Diagrams illustrating changing appearance of a polytene chromosome of a salivary gland of *Chironomus tentans* as a chromosome puff gradually appears. The material comprising the puff has been shown by histochemical tests and by autoradiography with tritium labeled uridine to be largely ribonucleic acid. (From Beermann: Chromosoma, 5:139–198, 1952.)

or in different tissues at the same time, certain differences in appearance become evident. A particular section of a chromosome may have the appearance of a diffuse puff (Fig. 8.17). Histochemical tests and autoradiographic evidence from experiments using tritium-labeled precursors have shown that the puff consists of RNA. It has been inferred that genes show this puffing phenomenon when they become active and that the puff represents the messenger RNA produced by the active gene in that band. It has been possible to correlate the appearance of puffs at certain regions in the chromosome with specific cellular events such as the initiation of molting and pupation.

The turning on and off of the synthesis of specific protein—differentiation at the molecular level—could occur by some process involving the genic DNA, the transcription of the DNA to form messenger RNA, the combination of the messenger RNA with the ribosome during protein synthesis or even some transformation of the ultimate protein product. In view of the tremendous number of kinds of DNA that

are represented by the genic complement of a cell in a multicellular plant or animal, we might ask what prevents that cell from producing continuously all the tremendous variety of messenger RNAs and their corresponding protein products that are possible. What determines which molecules of DNA are to be transcribed at any given moment in a given cell?

A mechanism which controls the transcription of DNA to regulate the production of messenger RNA would probably be the most economical one biologically, for it would clearly be to the cell's advantage not to have its ribosomes encumbered with nonfunctional molecules of messenger RNA. A cell optimally should produce only those kinds of messenger RNA that will code for the specific proteins required at that moment.

The continued synthesis of any protein requires the continued synthesis of its corresponding messenger RNA. Each kind of messenger RNA has a half-life ranging from a few minutes in certain microorganisms to 12 or 24 hours or longer in man and other mammals. Although each molecule of RNA template can serve to direct the synthesis of many molecules of its protein, the RNA is eventually degraded and must be replaced. This provides a mechanism by which a cell can alter the kind of protein being synthesized as new types of messenger RNA replace the previous ones. Thus the cell can respond to exogenous stimuli with the production of new types of enzymes.

The induction of enzymes by environmental stimuli has been cited as a model for embryonic differentiation. Bacteria and, to some extent, animal cells, respond to the presence of certain substrate molecules by forming enzymes to metabolize it. Jacques Monod has suggested that extracellular or intracellular influences may initiate or suppress the synthesis of specific enzymes, thus affecting the chemical constitution of the cell and leading to differentiation. As the embryo develops, the gradients established as a result of growth and cell multiplication could result in quantitative and even qualitative differences in enzymes. The induction or inhibition of one enzyme could lead to the accumulation of another chemical product which would induce the synthesis of a new enzyme and confer a new functional activity on these cells.

Morphogenesis appears to be too complex a phenomenon to be explained in terms of any single phenomenon such as enzyme induction. Enzymes can be induced in an embryo by the injection of an appropriate substance. Adenosine deaminase, for example, has been induced in the chick embryo by the injection of adenosine; however, no enzyme has been induced which is not normally present to some extent in the embryo. Adult tissues such as the liver may show marked differences in their enzymatic activities, differences which might be the result of adaptations comparable to those seen in bacteria. Adaptive changes in enzymes, however, are temporary and reversible, whereas differentiation is a permanent, essentially irreversible process.

Most biologists believe that an important factor in differentiation is the production of new kinds of messenger RNA which code for new, specific proteins characteristic of the new kind of tissue. The new proteins, with specific enzymatic properties, catalyze specific biochemical processes which are reflected in the differentiation of the tissue. The DNA of the genes that are not being

transcribed at any given moment may be bound to a histone or to some other kind of protein which makes the DNA unavailable for the transcription system. There is some evidence suggesting that in the nucleus some genes are free and can be transcribed and others are bound and not transcribable.

Studies of the control of the production of a specific globulin by the pea seed provide an example of such regulation of gene activity. This specific protein is synthesized in the seed but not in any other part of the pea plant. James Bonner has provided evidence that the DNA which codes for the synthesis of **pea seed globulin** is bound to histone in the cells elsewhere in the plant but within the pea seed the histone is removed; that particular segment of the DNA becomes free and can be transcribed, and ultimately forms messenger RNA which leads to the production of the globulin. The question of what controls the binding and release of a specific segment of DNA remains to be answered.

An alternative hypothesis suggests that DNA may be transcribed to form messenger RNA but the mRNA is "masked" and inactive as a template for protein synthesis until it is subsequently unmasked by a separate process. There is some evidence that this form of control may be operative especially during early embryonic development. The mRNA may be synthesized in the egg but be inactive until activated by the fertilization process and freed to undergo translation.

Studies of the synthesis of RNA in the nucleus have shown that a large fraction of it, perhaps 80 per cent of the total RNA synthesized, is destroyed without leaving the nucleus; only about 20 per cent of the nuclear RNA is identical with RNA present in the cytoplasm. Some have suggested that this "heterogeneous" RNA serves in some way to regulate genic action. There appears to be far more DNA in the cells of multicellular organisms than is necessary to serve as template for the mRNA used in directing the synthesis of protein. Is the rest of the DNA part of some enormous, complex regulatory system which controls the activity of the DNA that does produce mRNA? Others interpret the rapid turnover of nuclear RNA as indicating that all genic DNA is transcribed into RNA all the time, but much of the RNA produced is rapidly destroyed before leaving the nucleus. Only the RNA that is stabilized in some way survives and passes to the cytoplasm. This second interpretation implies that gene action is regulated not during transcription but subsequently, between transcription and translation, by some process which selectively stabilizes certain kinds of RNA.

There may be marked qualitative and quantitative differences in the kinds of enzymes present in different cells and tissues of the same organism. Mammalian liver cells, for example, have a glucose-6-phosphatase and can convert glycogen and other precursors to free glucose, whereas skeletal muscle cells lack this enzyme. Enzymes that catalyze the same reaction in different tissues may differ in their molecular size, their amino acid composition, their immunologic properties and their responses to hormones and other control mechanisms.

Even within a single tissue or a single cell, multiple molecular forms of an enzyme, termed **isozymes,** may be found. All of these proteins catalyze the same general reaction but have distinct chemical and physical properties. The different molecular forms may bear a different net charge and thus be separable by the process of electrophoresis.

Studies of the lactic dehydrogenases indicate that each enzyme molecule is composed of four subunits bound together. There appear to be two kinds of subunits, A and B, each of which is a polypeptide chain with a specific sequence of amino acids. The entire molecule is analogous to a hemoglobin molecule which is composed of two α and two β polypeptides. However, in the lactic dehydrogenase molecule any combination of the two types of subunits is permissible. The combinations of two kinds of subunits taken four at a time (A_4, A_3B, A_2B_2, AB_3, B_4) add up to the five kinds of lactic dehydrogenases that are typically observed in different tissues.

It appears that there are two genes, one for each of the subunits, but the question of why different types of tissues have characteristically different ratios of the different chains in the tetramers remains unanswered. A very curious observation, one that requires explanation, is that the lactic dehydrogenase in the breast muscle of chickens changes during embryonic development from a pure B_4 isozyme through a series of intermediates to a pure A_4 type in the adult.

The experiments of Clifford Grobstein and others emphasize that complex cellular interactions may be involved in regulating differentiation and that extrinsic factors as well as nuclear factors may play a role in this process. Grobstein found that embryonic pancreatic epithelium will continue to differentiate in organ culture only in the presence of mesenchyme cells. This requirement can be met not only by pancreatic mesenchyme from the mouse but by mesenchyme from a variety of other sources, even embryonic chick mesenchyme.

The mesenchyme can be replaced by a chick embryo juice, and the active principle of the juice appears to be a protein, for it is inactivated by trypsin but not by ribonuclease or deoxyribonuclease. It can be sedimented by high speed centrifugation, and appears to be a large protein molecule. This factor is only weakly effective in causing differentiation of salivary gland epithelium in culture and is ineffective in inducing the formation of kidney tubules or of cartilage from their respective mesenchymes. From these and other experiments there appears to be a spectrum of protein factors, each of which is more or less specific for the differentiation of one kind of cell.

Embryonic tissues have differential sensitivities to changes in nutrients, to the presence of inhibitors and antimetabolites, and to various environmental agents. These factors may change the course of differentiation and mimic the phenotype of a mutant gene, producing what is termed a **phenocopy.**

One of the clearest demonstrations that the same genes operating in dissimilar environments may have different morphologic effects was provided by experiments with races of frogs found in nature in Florida, Pennsylvania and Vermont. Each of these races normally develops at a speed which is adapted to the normal length of the spring and summer seasons. Southern frogs develop slowly and northern frogs develop more rapidly. When northern frogs are raised under southern environmental conditions their development is overaccelerated, whereas when southern frogs are raised under northern conditions their development is over-retarded.

It is possible to fertilize an egg with sperm from a different race and remove the original egg nucleus before the sperm nucleus can unite with it. In this way it was possible to set up a cell with "northern" genes in "southern" cytoplasm and the reverse. Northern genes operating in southern cytoplasm resulted in a poorly regulated development. The animal's head grew more rapidly than the posterior region and was disproportionately large. When southern genes were introduced into northern cytoplasm, development was again poorly regulated but the head rather than the posterior region was retarded and was disproportionately small.

Genes from the Pennsylvania race of frog acted as "northern" with Florida cytoplasm and as "southern" with Vermont cytoplasm. Thus the same set of genes produced quite opposite morphologic effects when operating in different cytoplasmic environments.

Cellular differentiation may involve the differential activation of specific genetic sites in different tissues, and it may involve mechanisms operating at the ribosomal level or even at the cell surface where the transport of substances into and out of the cell is regulated.

Differentiation may be controlled at least in part by influences originating outside the cell, from neighboring cells in early differentiation or from distant cells by materials such as hormones or the mesenchymal proteins studied by Grobstein. Such systemic influences participate in the integration of the differentiation of individual cells into the larger pattern of differentiation of the tissues of the whole organism. Eventually, it should be possible to bridge the gap between studies of development at the level of the whole organism and studies at the molecular level and trace in detail the sequence of events from the initial action of the gene to the final expression of its phenotype.

8.13 LETHAL GENES

Certain genes produce such a tremendous deviation from the normal development of an organism that it is unable to survive. The presence of these **lethal genes** can be detected by certain upsets in the expected genetic ratios. For example, some mice in a certain strain had yellow coat color, but experimenters found it impossible to establish a true-breeding strain with yellow coat. Instead, when two yellow mice were bred,

offspring were produced in the ratio of 2 yellow : 1 nonyellow. A yellow mouse bred to a black mouse gave half yellow mice and half black mice among the offspring. Then investigators noticed that the litters of yellow × yellow matings were somewhat smaller than other litters of mice, being only about three-quarters as large. They reasoned that one-quarter of the embryos, those homozygous for yellow, did not develop. When the uterus of the mother was opened early in pregnancy the abnormal embryos, those homozygous for the yellow trait, were found. Embryos homozygous for yellow color begin development, then cease developing, die and are resorbed.

"Creeper" fowl have short legs and short wings and, when two "creeper" fowl are bred, the offspring are in the ratio of two "creepers" to one normal. One-quarter of the embryos—those homozygous for "creeper"—have marked abnormalities of the vertebrae and spinal cord and die without hatching.

These lethal genes, yellow and "creeper," produce a phenotypic effect when heterozygous and hence are said to be dominant. Many, perhaps most, of the lethal genes appear to have no effect when heterozygous but cause death when homozygous and are called recessive lethal genes. These can be detected only by special genetic techniques. When wild populations of fruit flies and other organisms are analyzed, the presence of many recessive lethals is revealed. In the light of our present theory about the relations between genes and development we can suppose that a lethal gene is a mutant which causes the absence of some enzyme of primary importance in intermediary metabolism. The absence of this enzyme prevents the proper development of the organism.

Each recessive gene described so far produces its trait when it is homozygous, and each dominant gene produces its effect when it is homozygous or heterozygous, but other genes are known which do not always produce their expected phenotypes. Genes which always produce the expected phenotype are said to have complete penetrance. If only 70 per cent of the individuals of a stock homozygous for a certain recessive gene show the character phenotypically, the gene is said to have 70 per cent penetrance. The term **penetrance** refers to the statistical regularity with which a gene produces its effect when present in the requisite homozygous (or heterozygous) state. The percentage of penetrance of a given gene may be altered by changing the conditions of temperature, moisture, nutrition, and so forth, under which the organism develops.

Some stocks which are homozygous for a recessive gene may show wide variations in the appearance of the character. Fruit flies homozygous for a recessive gene which produces shortening and scalloping of the wings exhibit wide variations in the *degree* of shortening and scalloping. Such differences are known as variations in the **expressivity** or expression of the gene. The expressivity of the gene may also be altered by changing the environmental conditions during the organism's development. In view of the long and sometimes tenuous connection between the gene in the nucleus of the cell and the final production of the trait, it is easy to understand why the expression of the trait might vary or why the mutant trait might be completely absent.

8.14　THE MATHEMATICAL BASIS OF GENETICS: THE LAWS OF PROBABILITY

The discussions of heredity in Chapter 7 were concerned with inheritance in individuals, with the appearance of the offspring of two individuals with specific traits. Geneticists may also be concerned with the genetic characteristics of a population as a whole. It is possible, with the aid of simple mathematical methods, to make inferences about the mode of inheritance of a trait from its distribution in a population. All genetic events are governed by the laws of probability and, although the outcome of any single event is highly uncertain, in a large number of events the laws of probability provide a reasonable prediction of the fraction of those events which will be of one type or the other. In tossing a coin, where the probability, p, of obtaining a "heads" is one chance in two, or $1/2$, one cannot predict the outcome of any *single* toss of the coin. But in 100 tosses about 50 will come up "heads" and 50 will come up "tails." Probabilities are usually expressed as the fraction obtained by dividing the number of "favorable" events by the total number of possible events.

When expressed in this fashion, the limits of probability are from 0 to 1. A probability of 0 indicates that the event is impossible; there is no favorable possibility. A probability of 1 represents a certainty; that is, all the possible events are favorable ones. If you are engaged in a game of chance in which the favorable event involves turning up a "three" on a die, the probability of this favorable event is one in six. If a bag contains ten red, 40 black and 50 white marbles the chance of picking a single red marble out of the bag is ten out of one hundred or $1/10$.

Three types of probabilities can be distinguished. **A priori probabilities** are those which can be specified in advance from the nature of the event. For example, in flipping a coin the probability of obtaining a head is one in two and in casting a die the probability of obtaining a six is one in six; these probabilities are independent of whether the event actually occurs. In contrast, **empiric probabilities** are obtained by counting the number of times a given event occurs in a certain number of trials. For example, if a surgeon performs a certain type of operation on 500 people and 40 of them succumb, then the probability of death in this type of operation is 40 in 500 or 0.08. Such empiric probabilities have to be used in many fields of research where there is no theoretical basis, no *a priori* basis, for predicting the outcome. This type of probability is used in setting up the "risk tables" widely used by insurance firms. If a scientist collects data about the number of individuals with a certain trait in a given population and wants to know the probability that these numbers agree with the ratio expected on the basis of some genetic theory (1:1 or 3:1, etc.), he uses the methods of **sampling probability.**

If two events are independent the probability of their coinciding is the *product* of their individual probabilities. For example, the probability of obtaining a head on the first toss of a coin is $1/2$ and the probability of obtaining a head on the second toss of a coin (an independent event) is $1/2$. The probability of obtaining two heads on successive tosses of the coin is the product of their probabilities, $1/2 \times 1/2$, or $1/4$. There is one chance in four of obtaining two heads on

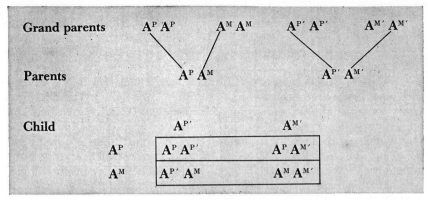

Figure 8.18 An example of the application of the laws of probability to genetics, illustrating both the "product law" of independent events and the "sum law" of mutually exclusive events. See text for discussion. (Villee: Biology, 6th ed.)

two successive tosses of a coin. This "product rule" of probability also holds for three or more independent events. For example, the probability of choosing at random an individual who is male, has blood group A and was born in June is $0.5 \times 0.45 \times 0.084 = 0.0189$.

The probability that one or another of two mutually exclusive events will occur is the sum of their separate probabilities. For example, in rolling a die the probability that the die will come up *either* two or five is $\frac{1}{6} + \frac{1}{6} = \frac{1}{3}$.

The application of these considerations to genetics is illustrated in Figure 8.18. Let us consider the probability that a child will inherit from his father the particular allele that the father inherited in turn from his father, the child's grandfather. The father has two alleles, A^P and A^M, obtained from the grandfather and grandmother respectively. He will pass onto his son one of these two. The "favorable" event is the transmission of the A^P gene and the total number of events is two; thus, the probability that the son will receive from his father the same allele that the father received from his grandfather is one in two. The probability that the child will receive from his father the allele that the father obtained from the grandfather (A^P) and also will obtain from his mother the allele that she received from the grandmother ($A^{M'}$) is the product of their independent occurrences, $\frac{1}{2} \times \frac{1}{2} = \frac{1}{4}$. The probability that the child will obtain either the two alleles from the two grandfathers, $A^P A^{P'}$, or the two alleles from the two grandmothers, $A^M A^{M'}$, is the sum of their independent probabilities, $\frac{1}{4} + \frac{1}{4} = \frac{1}{2}$.

8.15 POPULATION GENETICS

The question that sometimes puzzles beginning geneticists is: Why, if brown eye genes are dominant to blue eye genes, haven't all the blue eye genes disappeared? The answer lies partly in the fact that a recessive gene, such as the one for blue eyes, is not changed by having existed for a generation next to a brown eye gene in a heterozygous individual, **Bb.** The remainder of the explanation lies in the fact that as long as there is no selection for either eye color, that is, as long as people with blue eyes are just as likely to marry and have as many children as people with brown eyes, successive generations will have the same proportion of blue- and brown-eyed people as the initial one.

A brief excursion in mathematics will show why this is true. If we consider the distribution of a single pair of genes, **A** and **a,** any member of the population will have the genotype **AA, Aa** or **aa.** No other possibilities exist. Now let us suppose that these genotypes are present in the population in the ratio of $\frac{1}{4}$**AA**: $\frac{1}{2}$**Aa**: $\frac{1}{4}$**aa.** If all the members of the population select their mates at random, without regard to whether they are **AA, Aa** or **aa,** and if all of the types of pairs produce, on the average, comparable numbers of offspring, the succeeding generations will also have genotypes in the ratio $\frac{1}{4}$**AA**: $\frac{1}{2}$**Aa**: $\frac{1}{4}$**aa.** This can be demonstrated by putting down all the possible types of matings, the frequency of their occurrence at random, and the kinds and proportions of offspring produced by each type of mating. When all the types of

TABLE 8.6 THE OFFSPRING OF THE RANDOM MATING OF A POPULATION COMPOSED OF ¼**AA**, ½**Aa** AND ¼**aa** INDIVIDUALS (Villee: Biology, 6th ed.)

Mating Male Female	Frequency	Offspring
AA × AA	1/4 × 1/4	1/16 AA
AA × Aa	1/4 × 1/2	1/16 AA + 1/16 Aa
AA × aa	1/4 × 1/4	1/16 Aa
Aa × AA	1/2 × 1/4	1/16 AA + 1/16 Aa
Aa × Aa	1/2 × 1/2	1/16 AA + 1/8 Aa + 1/16 aa
Aa × aa	1/2 × 1/4	1/16 Aa + 1/16 aa
aa × AA	1/4 × 1/4	1/16 Aa
aa × Aa	1/4 × 1/2	1/16 Aa + 1/16 aa
aa × aa	1/4 × 1/4	1/16 aa
		Sum: 4/16 AA + 8/16 Aa + 4/16 aa

offspring are summed, it will be found that the next generation will also have genotypes in the ratio ¼AA : ½Aa : ¼aa (Table 8.6).

G. H. Hardy, an English mathematician, and G. Weinberg, a German physician, independently observed in 1908 that the frequencies of the members of a pair of allelic genes in a population are described by the expansion of a binomial equation. If we let p be the proportion of **A** genes in the population and q be the proportion of **a** genes in the population, since a gene must be either **A** or **a**, $p + q = 1$. Thus, if we know the value of either p or q, we can calculate the value of the other.

When we consider all of the matings in any given generation, a p number of **A**-containing eggs and a q number of **a**-containing eggs are fertilized by a p number of **A**-containing sperm and a q number of **a**-containing sperm: $(p\mathbf{A} + q\mathbf{a}) \times (p\mathbf{A} + q\mathbf{a})$. The proportion of the types of offspring of all these matings is described by the algebraic product: $p^2\mathbf{AA} + 2pq\mathbf{Aa} + q^2\mathbf{aa}$. If p, the frequency of gene **A**, equals ½, then q, the frequency of gene **a**, equals $1 - p$ or $1 - ½$ or ½. From the formula, the frequency of genotype **AA**, p^2, equals $(½)^2$ or ¼; the frequency of **Aa**, $2pq$, equals $2 \times ½ \times ½$, or ½, and the frequency of **aa**, q^2, equals $(½)^2$ or ¼. Any population in which the distribution of alleles **A** and **a** conforms to the relation $p^2\mathbf{AA} + 2pq\mathbf{Aa} + q^2\mathbf{aa}$ is in genetic equilibrium. The proportions of these alleles in successive generations will be the same (unless altered by selection or mutation).

8.16 GENE POOLS AND GENOTYPES

The genetic constitution of a population of a given organism is termed the **gene pool.** Stated differently, all the genes of all the individuals in a population make up the gene pool. This may be contrasted to the **genotype,** which is the genetic constitution of a single *individual*. Any individual may have only two alleles of any given gene. In contrast, the gene pool of the population may contain any number of different alleles of a specific gene. The A B O blood groups (p. 172) are inherited by three alleles, $\mathbf{a^A}$, $\mathbf{a^B}$, and **a**. In the population there are three alleles, but any given individual can have no more than two of the three.

The gene pools of different populations may differ in the ratios or proportions of the specific alleles. One population may have the alleles **A** and **a** in a ratio of 0.5 to 0.5. Another population of the same species may have the two alleles in the ratio 0.7A : 0.3a. If all the individuals in this second population have equal chances of surviving to adulthood and equal chances of producing gametes, then 70 per cent of the sperm produced by the entire population of males would have gene **A** and 30 per cent would have gene **a**. Similarly 70 per cent of the eggs produced by the entire population of females would contain gene **A** and 30 per cent would have gene **a**. The random union of the eggs and sperm would result in offspring in the ratio of 0.49**AA** + 0.42**Aa** + 0.09**aa**. Notice that the gene pool in the offspring is identical to the gene pool of the parents!

	sperm	
	.7A	.3a
.7A	.49AA	.21Aa
.3a	.21Aa	.09aa

eggs

In a more general sense this can be represented as $(p\mathbf{A} + q\mathbf{a})^2$, where p represents the frequency of one allele (0.7 for gene **A** in this example) and q represents the frequency of the other allele (0.3 for gene **a** in this example). Multiplying $(p\mathbf{A} + q\mathbf{a}) \times (p\mathbf{A} + q\mathbf{a})$ gives $p^2\mathbf{AA} + 2pq\mathbf{Aa} + q^2\mathbf{aa}$. Since $p = 0.7$, $p^2 = 0.7 \times 0.7$ or 0.49, the frequency of **AA**. The value of $q = 0.3$ and $q^2 = 0.3 \times 0.3$ or 0.09, the frequency of **aa**. Finally, the value of $2pq = 2 \times 0.7 \times 0.3$ or 0.42, the frequency of **Aa**.

By similar calculations you can show that the next generation, and each succeeding generation, will contain an identical gene pool, 0.7**A** and 0.3**a**. The three kinds of genotypes will be presented in the ratio 0.49**AA** : 0.42**Aa** : 0.09**aa** in succeeding generations provided that: (1) There are no mutations for **A** or **a**; (2) the three kinds of genotypic individuals have equal probabilities of surviving, mating and producing offspring, and there is no selection of mates according to these genotypes; and (3) the population of individuals is large enough so that chance cannot play a role in determining gene frequencies.

This principle that a population is genetically stable in succeeding generations is termed the **Hardy-Weinberg Principle**. Since the principle was established, J. B. S. Haldane, R. A. Fisher and Sewall Wright have developed mathematical methods for analyzing the inheritance of a given trait in a population. Subsequently it has become clear that the process of evolution, stated in the simplest terms, represents departure from the Hardy-Weinberg principle of genetic stability. Evolution involves changes in the gene pool of a population that result from mutations and selection. Thus an understanding of the Hardy-Weinberg Prin-

ciple is of prime importance in understanding the mechanism of evolutionary change.

The values of p and q, which are **gene frequencies**, cannot be measured directly. However, since the recessive phenotype can be distinguished you can determine q^2, the frequency of genotype **aa**. From this you can calculate the gene frequencies q (which is the square root of q^2) and p (which is $1 - q$). Finally you can calculate the frequencies of the other genotypes, $p^2\mathbf{AA}$ and $2pq\mathbf{Aa}$. To calculate the number of individuals in a population that are genetic carriers for a given trait (i.e., are heterozygotes, **Aa**) you need to know only that it is inherited by a single pair of genes and the frequency with which the homozygous recessive individuals appear in the population.

Albinos are individuals with no pigment at all in their skin or hair. **Albinism** is an inherited trait in which the individual lacks a specific enzyme, **tyrosinase**. Tyrosinase catalyzes one of the reactions involved in the production of the dark pigment **melanin**. Albinism is inherited by a single pair of genes, and albinos, the homozygous recessive individuals, occur about once in 20,000 births. From this fact you can calculate that the frequency of **aa** individuals (q^2) is $1/20{,}000$. From the value of q^2 you can determine q by taking the square root of q^2. The square root of $1/20{,}000$ is about $1/141$. Since $p = 1 - q$ or $1 - 1/141$, $p = {}^{140}/141$. You now have values for both p and q, and can calculate the value of $2pq$, which represents the frequency of the genetic "carrier" **Aa** individuals: $2 \times {}^{140}/141 \times 1/141 = 1/70$. Surprising as it may seem, one person in 70 is a **carrier** of albinism, although only one person in 20,000 is homozygous and displays the trait. At first glance it may seem odd that there are so many carriers in a population that contains so few homozygous recessives, yet reflecting on the mathematical relations involved should lead you to realize that this must be true. When q is small (such as $1/141$), then q^2 will be very small, but $2pq$ will be much larger.

The ability or inability to taste **phenylthiocarbamide** (PTC) and related compounds with the N—C=S group is an

inherited trait. Some people find that this substance has a bitter taste; others report

TABLE 8.7 THE INHERITANCE OF THE ABILITY TO TASTE
PHENYLTHIOCARBAMIDE

Marriage	No. of Families	Offspring		Proportion of Nontasters	
		Tasters	Nontasters	Observed	Calculated [*]
taster × taster	425	929	130	0.123	0.124
taster × nontaster	289	483	278	0.336	0.354
nontaster × nontaster	86	5	218	0.979	1.0

[*] From the Hardy-Weinberg Law, it follows that among marriages of parents with unlike traits (i.e., taster × nontaster), the fraction of the offspring with the recessive trait is $q/1 + q$. Among marriages of parents with like traits (i.e., taster × taster), the fraction of offspring with the recessive trait, nontaster, is $(q/1 + q)^2$. Since 29.8 per cent of the population were nontasters, $q^2 = 0.298$, $q = 0.545$, $q/1 + q = 0.354$ and $(q/1 + q)^2 = 0.124$. (Villee: Biology, 6th ed.)

it to be completely tasteless. L. H. Snyder examined 3643 people and found that 70.2 per cent were "tasters" and 29.8 per cent were "nontasters" for PTC. If this trait is inherited by a single pair of genes, with "tasting" dominant to "nontasting," then the mathematics underlying population genetics would predict that in a population of marriages of tasters with tasters, 12.4 per cent of the children will be nontasters (Table 8.7). It similarly predicts that in a population of marriages of tasters with nontasters, 35.4 per cent of the children will be nontasters. In Snyder's study the percentages actually found were 12.3 per cent and 33.6 per cent, respectively. From this close agreement, we may conclude that the original assumption, that the tasting-nontasting trait is inherited by a single pair of genes, is correct. Although some 30 per cent of whites are nontasters, they are very rare in populations of Negroes, Eskimos and American Indians. Thus the **Tt** gene pools in these populations are quite different.

As another example of the use of the Hardy-Weinberg Principle, let us consider the inheritance of the blood groups M, MN and N. In the United States, among the white population 29.16 per cent have blood type M, 49.58 per cent have blood type MN, and 21.26 per cent have blood type N. Applying the Hardy-Weinberg Principle to these data, $q^2 = 0.2126$, from which we calculate that $q = 0.46$. To estimate the value for p, we subtract 0.46 from 1 and arrive at 0.54. The square of this estimate of p, $(0.54)^2$, is equal to 0.29 and $2pq = 2 \times 0.54 \times 0.46$ or 0.49. The excellent agreement between the values observed and the theoretical values is evidence that M and N blood types are inherited by a single pair of genes, with neither gene being dominant to the other (they may be termed **codominants**) so that the heterozygote shows the two kinds of blood antigens.

8.17 HUMAN CYTOGENETICS

Many of the basic principles of genetics were discovered by experiments in lower organisms in which it was possible to relate genetic data with cytologic events. This involved making smears of cells and examining them under the microscope to see the number and the structure of the chromosomes present. Some of the organisms used in genetics, such as the fruit fly *Drosophila,* have few chromosomes (four pairs) and, at least in certain cells of the body (for example, the salivary glands), the chromosomes are quite large and the details of their structure are readily evident. Not only are the chromosomes of mammals more numerous and rather small, but mammalian cells are difficult to fix for cytologic examination. Advances in recent years in the techniques of fixing and staining cells, however, have made possible cytologic examination of human cells as well as those of other mammals. This led in 1956 to the finding that the chromosome number for man is 46 rather than 48, the number which had been quoted for the previous 30 years or more.

Although the normal chromosome number in man is 46, some rare instances of abnormal chromosome numbers have been reported. These usually are associated with some change in the phenotype of the individual. Changes in chromosome number are termed *ploidy*. An individual may be polyploid, having one or more complete extra sets of chromosomes, or he may have one or two extra chromosomes with the total chromosome number of 47 or 48. A severely defective male child was found to be a *triploid* individual with a total of 69 chromosomes. He had 66 autosomes, two X chromosomes and one Y chromosome. It seems likely that this zygote was formed from a normal haploid egg which was fertilized by an unusual diploid sperm or from an exceptional diploid egg fertilized by a normal haploid sperm.

A number of cases have now been reported of individuals who show a gain or a loss of a single chromosome, presumably caused by "nondisjunction." **Nondisjunction** refers to the failure of a pair of homologous chromosomes to separate normally during the reduction division. Two X chromosomes, for example, might fail to separate and both might enter the egg nucleus, leaving the polar bodies with no X chromosome. Alternatively, the two joined X chromosomes might go into the polar body, leaving the female pronucleus with no X chromosome. Nondisjunction of the XY chromosomes in the male might lead to the formation of sperm which have both an X and a Y chromosome or to sperm with neither an X nor a Y chromosome. Chromosomal nondisjunction may occur during either the first or second meiotic division; it may also occur during mitotic divisions and lead to the establishing of a group of abnormal cells in an otherwise normal individual.

Cytogenetic studies have clarified the origin of one of the more distressing abnormal conditions in man, that of **mongolism** or Down's syndrome. Individuals suffering from this have abnormalities of the face, eyelids, tongue and other parts of the body, and are greatly retarded in both their physical and mental development. The term "mongolism" was originally applied to this condition because affected individuals often show a fold of the eyelid similar to that typical of members of the Mongolian race. Mongolism is a relatively common congenital malformation, occurring in 0.15 per cent of all births. It had been known for

some time that the appearance of mongolism is related to the age of the mother and that it increases greatly with maternal age. For example, mongolism is a hundredfold more likely in the offspring of women 45 years or older than in the offspring of mothers under 19. The occurrence of mongolism, however, is independent of the age of the father, and it is also independent of the number of preceding pregnancies in the woman. Cytogenic studies revealed that mongols have one extra chromosome, a total of 47. The presence of this extra small chromosome is believed to arise by nondisjunction in the maternal oöcyte.

Mongolism should be inherited as though it were a dominant gene, since a mongoloid would form gametes half of which have the normal complement of 23 chromosomes and half of which have 24 chromosomes. In the rare cases where mongols have had offspring they have produced normal and mongol children in about equal proportions.

Another condition caused by an upset in chromosome number is that of individuals who are outwardly nearly normal males but have small testes. They produce few sperm, they have seminiferous tubules which are very aberrant in appearance, and they usually have gynecomastia (a tendency for formation of female-like breasts). This condition, called **Klinefelter's syndrome,** usually becomes apparent only after puberty, when the small testes and gynecomastia may bring the individual to the attention of his physician. The cells of these individuals show a chromatin spot, and at one time they were thought to be XX individuals, i.e., genetic females. However, when their chromosomes were examined cytologically and counted it was found that they have 47 chromosomes; their cells have *two* X and one Y chromosome. The fact that they are nearly normal males in their external appearance emphasizes the male-determining effect of the Y chromosome in man.

Another condition resulting from changes in chromosome number is **Turner's syndrome,** in which the external genitalia, though feminine, are those of an immature female. The internal reproductive tract is present and resembles that of an immature but perfectly formed female. The uterus is present but small, and the gonads may be absent. The cells of these individuals are "chromatin negative," which suggests that they are males. However, they have only one X chromosome but no Y chromosome. This type of individual again emphasizes the importance of the Y chromosome in determining the male characteristics.

An individual with an extra chromosome, with three of one kind, is said to be **trisomic,** and an individual lacking one of a pair is said to be **monosomic.** Thus mongols are trisomic for chromosome 21, and individuals with Turner's syndrome are monosomic for the X chromosome.

In summary, the primary genetic material in all living systems is nucleic acid. The genetic material is DNA in all organisms except a few of the viruses in which only RNA is present. Genetic information is transferred from one generation to the next in the form of specific sequences of nucleo-

tides in the nucleic acid chain. It would appear that the number of nucleotides that form a unit in this informational code, a codon, is three. A gene is a localized sequence of nucleotides in a nucleic acid which carries the information for the synthesis of a peptide chain with a specific sequence of amino acids. The function of the gene is to transfer this specificity to some nongenic macromolecule, presumably messenger RNA. The model of DNA proposed by Watson and Crick provides for gene specificity, gene replication and gene mutation. A gene mutation is believed to be an alteration of the specific sequence of nucleotides in the nucleic acid molecule which leads to the formation of a different type of protein or perhaps prevents the formation of that protein. Each gene is believed to control the production of a single specific protein.

ANNOTATED REFERENCES

Frisch, L. (Ed.): The Genetic Code. Cold Spring Harbor Symposium on Quantitative Biology. Cold Spring Harbor, N.Y. Vol. 31, 1966. A rich source of facts and theories detailing the state of our knowledge of biochemical genetics as of June 1966.

Hartman, P. E., and S. R. Suskind: Gene Action. Englewood Cliffs, N.J. Prentice-Hall, Inc., 1964. A brief paperback presentation of molecular genetics.

Ingram, V.: The Biosynthesis of Macromolecules. New York, W. A. Benjamin, 1965. A clear, logically presented discussion of the genetic control of protein synthesis.

Markert, C. L.: Developmental Genetics. Englewood Cliffs, N.J., Prentice-Hall, Inc., 1964. A summary of the application of current genetic theories to the problem of differentiation.

Turpin, R., and J. Lejeune: Human Afflictions and Chromosomal Aberrations. London, Pergamon Press, 1969. An excellent source of information on human genetics.

Watson, J. D.: Molecular Biology of the Gene. 2nd ed. New York, W. A. Benjamin, 1971. A masterful summary of biochemical genetics based on experiments with bacteria and viruses.

Wolstenholme, G. E. W., and C. M. O'Connor: The Biochemistry of Human Genetics. Boston, Little, Brown and Company, 1959. The proceedings of a Ciba symposium on the chemical aspects of human inheritance.

9. The Concept of Evolution

An immense variety of forms of life inhabit every conceivable place on land and in the water and exhibit tremendous variations in size, shape, degree of complexity, and methods of obtaining food, of evading predators and of reproducing their kind. How all these species came into existence, how they came to have the particular adaptations which make them peculiarly fitted for survival in a particular environment, and why there are orderly degrees of resemblance between forms which permit their classification in genera, orders, classes and phyla, are fundamental problems of zoology. From detailed comparisons of the structures of living and fossil forms, from the sequence of the appearance and extinction of species in times past, from the physiologic and biochemical similarities and differences between species, and from the analyses of heredity and variation in many different animals and plants has come one of the great unifying concepts of biology, that of **evolution.**

The concept of evolution can emerge logically and naturally from our understanding of the principles of genetics. This is not, however, how it arose historically. It came as a result of countless observations of similarities and differences in structures and functions of the various kinds of animals and plants in different parts of the world. It came from Charles Darwin's profound insight into the arrangement of the pieces of the puzzle of how these similarities and differences may have arisen.

9.1 THE PRINCIPLE OF ORGANIC EVOLUTION

The term evolution means an unfolding, or unrolling, a gradual, orderly change from one state to the next. The planets and stars, the topography of the earth, and the chemical compounds of the universe have undergone gradual, orderly changes sometimes called **inorganic evolution.** The principle of **organic evolution,** now universally accepted by biologists, simply applies this concept to living things: all the various plants and animals living today have descended from simpler organisms by gradual modifications which have accumulated in successive generations.

Evolution is continuing to occur; indeed, it is occurring more rapidly today than in many of the past ages. In the last few hundred thousand years, hundreds of species of animals and plants have become extinct and other hundreds have arisen. The process is usually too gradual to be observed, but there are some remarkable examples of evolutionary changes which have taken place within historic times. For example, some rabbits were released early in the fifteenth century on a small island near Madeira called Porto Santo. There were no other rabbits and no carnivorous enemies on the island and the rabbits multiplied at an amazing rate. In 400 years they became quite different from the ancestral European stock; they were only half as large, had a different color pattern, and were more nocturnal animals. Most important, they could not produce offspring when bred with members of the European species. They were, in fact, a new species of rabbit.

9.2   DEVELOPMENT OF IDEAS ABOUT EVOLUTION

The idea that the present forms of life have arisen from earlier, simpler ones was far from new when Charles Darwin published *The Origin of Species* in 1859. The oldest speculations about evolution are found in the writings of certain Greek philosophers, Thales (624–548 B.C.), Anaximander (588–524 B.C.), Empedocles (495–435 B.C.) and Epicurus (341–270 B.C.). The spirit of this age of Greek philosophy was somewhat similar to that of our own age, for simple, natural causes were sought to explain all phenomena. Since they knew very little biology, however, their ideas about evolution were extremely vague and can scarcely be said to foreshadow our present theory of organic evolution. Aristotle (384–322 B.C.), who was a great biologist as well as a philosopher, knew a great deal about animals and plants and wrote detailed, accurate descriptions of many of them. He observed that organisms could be arranged in graded series from lower to higher and drew the correct inference that one evolved from the other. However, he had the metaphysical belief that the gradual evolution of living things occurred because nature strives to change from the simple and imperfect to the more complex and perfect. An evolutionary explanation of the origin of plants and animals was given by the Roman poet Lucretius (99–55 B.C.) in his poem *De Rerum Natura.*

With the Renaissance, interest in the natural sciences quickened and the increasing knowledge of the many kinds of animals led more and more scientists to consider the concept of evolution favorably. Among these were Hooke (1635–1703), Ray (1627–1705), Buffon (1707–1788), Erasmus Darwin (1731–1802) and Lamarck (1744–1829). Even before the Renaissance men had discovered shells, teeth, bones and other parts of animals buried in the ground. Some of these corresponded to parts of familiar living animals, but others were strangely unlike any known form. Many of the objects found in rocks high in the mountains, far from the sea, resembled parts of marine animals. In the fifteenth century, the versatile artist and scientist, Leonardo da Vinci, gave the correct explanation of these curious finds, and gradually his conclusion, that they were the remains of animals that had existed at one time but had become extinct, was accepted. This evidence of former life suggested to some people the theory of **catastrophism** – the idea that a succession of catastrophes, fires and floods, have periodically destroyed all living things, followed each time by the origin of new and higher types by acts of special creation.

Three Englishmen in the eighteenth and early nineteenth centuries laid the foundations of modern geology, and by their careful, cogent arguments advanced the theory of **uniformitarianism** to replace the concept of catastrophism. In 1785 James Hutton developed the concept that the geologic forces at work in the past were the same as those operating now. He arrived at this conclusion after a careful study of the erosion of valleys by rivers and the formation of sedimentary deposits at the mouths of rivers. He demonstrated that the processes of erosion, sedimentation, disruption and uplift, carried on over long periods of time, could account for the formation of fossil-bearing rock strata. The publication of John Playfair's *Illustrations of the Huttonian Theory of the Earth* in 1802 gave further explanation and examples of the idea of uniformitarianism in geologic processes. Sir Charles Lyell, one of the most influential geologists of his time, finally converted most of the contemporary geologists to the theory of uniformitarianism by the publication of his *Principles of Geology* (1832). A necessary corollary of the idea that slowly acting geologic forces have worn away mountains and filled up seas is that geologic time has been immensely long. This idea, completely revolutionary at the time, paved the way for the acceptance of the theory of organic evolution, for the process of evolution requires an extremely long time.

The earliest theory of organic evolution to be logically developed was that of Jean Baptiste de Lamarck, the great French zoologist whose *Philosophie Zoologique* was published in 1809. Lamarck, like most biologists of his time, believed that all living things are endowed with a vital force that controls the development and functioning of their parts and enables them to overcome handicaps in the environment. He believed that any trait acquired by an organism during its lifetime was passed on to succeeding generations—that acquired characters are inherited. Developing the notion that new organs arise in response to the demands of the environment, he postulated that the size of the organ is proportional to its use or disuse. The changes produced by the use or disuse of an organ are transmitted to the offspring and this process, repeated for many generations, would result in marked alterations of form and function. Lamarck explained the evolution of the giraffe's long neck by suggesting that some short-necked ancestor of the giraffe took to browsing on the leaves of trees, instead of on grass, and that, in reaching up, it stretched and elongated its neck. The offspring, inheriting the longer neck, stretched still farther, and the process was repeated until the present long neck was achieved.

Both Buffon and Erasmus Darwin had similar ideas about the role in evolution of the direct response of the organism to its environment but had not expressed them so clearly. This theory, called **Lamarckism,** provides an explanation for the remarkable adaptation of many plants and animals to their environment but is completely unacceptable because of the overwhelming genetic evidence that acquired characteristics cannot be inherited. The theoretical distinction between somatoplasm and germ plasm made by Weismann (1887) refuted all theories of evolution based on the inheritance of acquired characters. Acquired characters are present only in the body cells (somatoplasm) and not in the germ cells (germ plasm), and only traits present in the germ plasm are transmitted to the next generation.

9.3 BACKGROUND FOR THE ORIGIN OF SPECIES

Charles Darwin made two great contributions to the body of scientific knowledge: he presented a wealth of detailed evidence and cogent arguments to show that organic evolution had occurred, and he formulated a theory, that of **natural selection,** to explain the mechanism of evolution.

Darwin was born in 1809 and was sent at the age of 15 to study medicine at the University of Edinburgh. Finding the lectures intolerably dull, he transferred, after two years, to Christ's College, Cambridge University, to study theology. Many of Darwin's friends at Edinburgh were interested in geology and zoology, and at Cambridge he joined a circle of friends interested in collecting beetles. Through them he came to know Professor Henslow, the naturalist. Shortly after leaving college, and upon the recommendation of Professor Henslow, Darwin was appointed naturalist on the ship *Beagle,* which was to make a five-year cruise around the world preparing navigation charts for the British Navy. The *Beagle* left Plymouth in 1831 and cruised slowly down the east coast and up the west coast of South America. While the rest of the company mapped the coasts and harbors, Darwin studied the animals, plants and geologic formations of both coastal and inland regions. He made extensive collections of specimens and copious notes of his observations. The *Beagle* then spent some time at the Galápagos Islands, west of Ecuador, where Darwin continued his observations of the flora and fauna, comparing them to those on the South American mainland. These observations convinced Darwin that the theory of special creation was inadequate and set him to thinking about alternative explanations.

Upon his return to England in 1836, Darwin spent his time assembling the notes of his observations for publication and searching for some reasonable explanation for the diversity of organisms and the peculiarities of their distribution. As Darwin wrote in his notebook:

On my return home in the autumn of 1836 I immediately began to prepare my journal for publication, and then saw how many facts indicated the common descent of species. . . . In

July (1837) I opened my first notebook for facts in relation to the origin of species, about which I had long reflected, and never ceased working for the next twenty years. . . . Had been greatly struck from the month of March on character of South American fossils, and species on Galápagos Archipelago. These facts (especially latter) origin of all my views. . . .

In October (1838), that is fifteen months after I had begun my systematic inquiry, I happened to read for amusement *Malthus on Population,* and being well prepared to appreciate the struggle for existence which everywhere goes on, from long-continued observation of the habits of animals and plants, it at once struck me that under these circumstances favorable variations would tend to be preserved, and unfavorable ones to be destroyed. The result of this would be the origin of new species. Here then I had at last got a theory by which to work.

Darwin spent the next 20 years accumulating data from many fields of biology, examining it critically, and building up a tremendous body of facts that demonstrated that evolution had occurred and formulating his arguments for natural selection. In 1857 he submitted a draft of his theory to a number of scientific friends for comment and criticism. Alfred Russel Wallace, a naturalist and explorer who was studying the flora and fauna of Malaya and the East Indies, was similarly struck by the diversity of living things and the peculiarities of their distribution. Like Darwin, he happened to read Malthus' treatise and came independently to the same conclusion, that evolution occurred by natural selection. In 1858 Wallace sent a manuscript to Darwin, and asked him, if he thought it of sufficient interest, to present it to the Linnaean Society. Darwin's friends persuaded him to present an abstract of his own work along with Wallace's paper and this was done at a meeting of the Linnaean Society in July, 1858. Darwin's monumental *On the Origin of Species by Means of Natural Selection* was published in November, 1859.

The time was ripe for the formulation and acceptance of the theory of organic evolution. The publication of Lyell's *Principles of Geology* and the subsequent acceptance of the idea of geologic evolution, the publication of Malthus' ideas on population growth and pressure and the struggle for existence, together with the vast accumulation of information about the distribution of living and fossil forms of life, and studies of comparative anatomy and embryology, all showed the inadequacy of the theory of special creation. Because the time was ripe, Darwin's theory rapidly gained acceptance.

9.4 THE DARWIN-WALLACE THEORY OF NATURAL SELECTION

Darwin's explanation of the way in which evolution occurs can be summarized as follows:

1. Variation is characteristic of every group of animals and plants, and there are many ways in which organisms may differ. (Darwin and Wallace did not understand the cause of variation, and assumed it was one of the innate properties of living things. We now know that inherited variations are caused by mutations.)

2. More organisms of each kind are born than can possibly obtain food and survive. Since the number of each species remains fairly constant under natural conditions, it must be assumed that most of the offspring in each generation perish. If all the offspring of any species remained alive and reproduced, they would soon crowd all other species from the earth.

3. Since more individuals are born than can survive, there is a struggle for survival, a competition for food and space. This contest may be an active kill-or-be-killed struggle, or one less immediately apparent but no less real, such as the struggle of plants or animals to survive drought or cold. This idea of competition for survival in an overpopulated world was derived from Malthus.

4. Some of the variations exhibited by living things make it easier for them to survive; others are handicaps which bring about the elimination of their possessors. This idea of "the survival of the fittest" is the core of the theory of natural selection.

5. The surviving individuals will give rise to the next generation and, in this way, the "successful" variations are transmitted to the succeeding generations. The less fit will tend to be eliminated before they have reproduced.

Successive generations in this way tend to become better adapted to their environment; as the environment changes, further adaptations occur. The operation of natural selection over many generations may produce descendants which are quite different from their ancestors, different enough to be separate species. Furthermore, certain members of a population with one group of variations may become adapted to the environment in one way, while others, with a different set of variations, become adapted in a different way, or become adapted to a different environment. In this way two or more species may arise from a single ancestral stock.

Animals and plants exhibit many variations which are neither a help nor a hindrance to them in their struggle for survival. These are not affected directly by natural selection but are transmitted to succeeding generations.

Darwin's theory of natural selection was so reasonable and well documented that most biologists soon accepted it. One of the early, serious objections to the theory was that it did not explain the appearance of many apparently useless structures in an organism. We now know that many of the visible differences between species are not important for survival but are simply incidental effects of genes that have other physiologic effects of great survival value. Other nonadaptive differences may be controlled by genes that are closely linked in the chromosomes to genes for traits which are important for survival. Still other nonadaptive characteristics may become fixed in a population by chance, by the phenomenon of "genetic drift."

Another of the early objections to the theory was that new variations would be lost by "dilution" as the individuals possessing them bred with others without them. We now know that although the phenotypic expression of a gene may be altered when the gene exists in combination with certain other genes, the gene itself is not altered and is transmitted unchanged to succeeding generations.

9.5 POPULATIONS AND GENE POOLS

The concepts of the "struggle for survival" and "survival of the fittest" were key points in the Darwin-Wallace theory of natural selection, but it is now realized that the actual physical struggle between animals for survival or the competition between plants for space, sun and water is much less important as an evolutionary force than Darwin believed. The evolution of any given kind of organism occurs over many generations during which individuals are born and die, but the population has a certain continuity. Thus, the unit in evolution is not the individual but rather a *population* of individuals.

A population of similar individuals living within a circumscribed area and interbreeding is termed a **deme** or a **genetic population.** The territorial limits of any given deme may be quite vague and difficult to define and the number of individuals in the deme may fluctuate widely with time. A deme commonly overlaps with one or more adjacent demes to some extent. The next larger unit of population in nature is the **species,** composed of a series of intergrading demes.

The relative frequencies of the genes in a population will remain constant from one generation to the next (1) if the population is large, (2) if there is no selection for or against any specific gene or allele, i.e., if mating occurs at random, (3) if no mutations occur, and (4) if there is no migration of individuals into or out of the population. The operation of the Hardy-Weinberg Principle will result in maintaining a given gene frequency in a population. The essential feature of the process of evolution is a gradual change in the gene frequencies of a population when this Hardy-Weinberg equilibrium is upset either because mutations occur, because reproduction is nonrandom, or because the population is small, so that the gene frequencies in successive generations will be determined by chance events.

The demes and species occurring in nature tend to continue unchanged for many generations or centuries. This implies that there has been no change in the genetic constitution of the deme and no change in the environmental factors. When a change in a population does occur, this reflects either a change in the genetic factors brought about by mutation or a change in the environmental factors, which lead to selective survival of one or another phenotype.

One of the basic concepts of population genetics and of evolution is that each population is characterized by a certain **gene** pool. Each individual in the population is genetically unique and has a specific genotype. However, if we count all the alleles of a given gene (A_1, A_2, A_3 . . .) in a population (or in a statistically valid sample of the population) we could then calculate the fraction of the total represented by allele A_1, allele A_2 and so on. A population in genetic equilibrium has a gene pool which is constant from one generation to the next; i.e., the frequency of each allele in the population remains unchanged.

In contrast, a population undergoing evolution is one in which the gene pool is changing from generation to generation. The gene pool of a population may be changed (1) by mutation, (2) by hybridization, that is, by the introduction into the population of genes from some outside population, or (3) by natural selection. Recombination brought about by crossing over and by the assortment of chromosomes in meiosis may also lead to new combinations of genes and phenotypes with some specific advantage or disadvantage for survival that would be reflected in a change in the gene pool.

Evolution by natural selection simply means that those individuals with certain traits have more surviving offspring in the next generation and so contribute a proportionately greater percentage of genes to the gene pool of the next generation than organisms with other traits. New inherited variations arise primarily by mutation, and if the organisms with a new mutation survive and have, on the average, more offspring that survive than the organisms without that mutation, then in successive generations the gene pool of the population will gradually change.

This process, termed **nonrandom reproduction** or **differential reproduction**, implies that the conditions of the Hardy-Weinberg equilibrium do not apply to that population. The individuals that produce more surviving offspring in the next generation are usually, but not necessarily, those that are best adapted to the given environment. Well adapted individuals may be healthier, better able to obtain food and mates and better able to care for their offspring, but the primary factor in evolution is how many of their offspring survive to be parents of the next generation.

The ultimate raw material of evolution is a mutation which establishes an alternative allele at a given locus and makes possible an alternative phenotype. Evolutionary changes are possible only when there are alternative phenotypes that may survive or perish. However, the process of selection does not in general operate gene by gene but rather individual by individual and on the basis of the effects of the individual's entire genetic system.

When a mutation first appears, only one or a few organisms in the population will bear the mutant gene, and these will breed with other members of the population from which the mutant arose. The change in the gene pool so that the mutant gene appears with greater and greater frequency in the population is a gradual process which may occur only over many generations. When one is considering evolutionary changes in a larger population, the success or lack of success of some new mutant gene will depend largely on its ability to confer on its possessors the capacity to leave a larger number of surviving individuals in the next generation.

Most, perhaps all, genes have many different effects on the phenotype (they are said to be **pleiotropic**). Some of the effects of a given gene may be advantageous for survival (termed **positive selection pressure**) and others may be disadvantageous (termed **negative selection pressure**). Whether the frequency of a given allele increases or decreases will depend on whether the sum of the positive selection pressures due to its advantageous effects are greater or lesser than the sum of the negative selection pressures due to its harmful effects.

In the reproduction of small populations chance alone may play a considerable role in determining the composition of the succeeding generation. The equilibrium of the genetic pool of the population can be changed by chance processes rather than by natural selection. This role of chance in the evolution of small breeding populations has been described by Sewall Wright as **genetic drift.** Within small interbreeding populations, heterozygous gene pairs tend to become homozygous for one allele or the other by chance rather than by selection. This may lead to the accumulation of certain disadvantageous characters and the subsequent elimination of the group possessing those characters.

The role that genetic drift actually plays

in the evolution of organisms in nature has been the subject for debate among biologists, but there seems little doubt that it does play at least a minor role. Certainly, many animal and plant populations in nature are divided into subgroups small enough to be affected by the chance events underlying genetic drift.

Genetic drift represents an exception to the Hardy-Weinberg Law underlying the tendency for a population to maintain its proportion of homozygous and heterozygous individuals. The Hardy-Weinberg Law is based on statistical events and, like all statistical laws, holds only when the number of individuals involved is large enough. Genetic drift may explain the common observation that closely related species in different parts of the world frequently differ in curious, even bizarre, ways which appear to have no particular adaptive value.

The role of chance in evolution is particularly evident when a species moves into a new area, for the number of individuals moving into that area is usually small. These first colonizers from which the entire new population develops rarely constitute a representative sample of the gene pool of the original population, but instead differ from the parent population in the frequencies of specific genes. These differences may be quite marked but the new colonizing generation differs from the parent population in ways which are random rather than selective.

This effect is most apparent on islands and other areas of geographic isolation, and helps to account for the differences evident in island populations as compared to their mainland relatives. When a species is expanding continuously, the populations at the edge of the range, invading new areas, are likely to be small and differ genetically from the main body of the population. In all these situations, when the breeding population is small, chance rather than selection may play a large role in determining the evolution of a particular group.

9.6 DIFFERENTIAL REPRODUCTION

The evolutionary forces of mutation, genetic drift and the migration of genes from one population to another by hybridization can lead to the establishment of diversity among living things, but these processes operate by and large at random. In contrast, a key feature of evolution is the tendency of organisms to become adapted so as to survive and reproduce in the particular environment that they inhabit. The evolutionary process, then, is not random with respect to the establishing of adaptive features of organisms.

The process of differential reproduction may be nonrandom with respect to (1) the union of male and female gametes, (2) the production of viable zygotes, or (3) the development and survival of the zygotes until they are adults and capable of producing their own offspring. Nonrandom reproduction tends to produce nonrandom, directional changes in the genetic pool which lead to nonrandom, directional evolution. Random mating implies that any given male must be as likely to mate with any one female in the population as with any other, or that the gametes of any two individuals must be equally likely to unite. However, mating in nature is seldom completely at random.

Differential reproduction, i.e., a change in the gene pool in successive generations, usually occurs in some way that is correlated with the genotypes of the individuals concerned and tends to produce directional changes in evolution. Differential reproduction may result from nonrandom mating — that is, the union of male and female gametes by the mating of specific individuals — from the differential viability of the resulting zygotes or from differences in the development and survival of the offspring until they are in turn the parents of the next generation.

Behavior patterns of courtship and mating within a given species which lead to the acceptance or refusal of one individual by another in mating comprise one kind of force directing differential reproduction through **nonrandom mating.** Any gene mutation that causes a change from the usual pattern of courtship will generally have a negative selective advantage and tend to be eliminated. In many fishes and birds some brightly colored part on the male serves as a stimulus to the opposite sex which is necessary before copulation can be begun. Mutations that lead to the formation of bigger, brighter spots may confer a selective advantage on their possessors. Darwin recognized this evolutionary force and termed it **sexual selection.** However, it is clear that this is just one kind of natural selection, one factor that may result in differential reproduction.

A second aspect of differential reproduction is **differential fecundity,** which refers to the number of viable zygotes that are produced by a given mating. There may be genetic differences in the number of gametes produced by individuals and in the proportion of those gametes that can unite with others to form viable zygotes. In organisms with a low probability of individual survival, high fecundity will be an evolutionary advantage. However, in other organisms where the chances of the survival of an individual are high, an extremely high fecundity may actually reduce the chances of survival of any given individual by reducing the opportunity for parental care and feeding of the offspring.

Even after two gametes have united to produce a viable zygote, the resulting organism must go through a fairly long period of development and growth before becoming sexually mature and able to contribute to the next generation. It is this aspect of differential reproduction that has played perhaps the largest role in determining the course of evolution. It is the differential success of organisms in a population in surviving to the reproductive age and contributing to the next generation's gene pool that has been responsible for the more obvious features of organic adaptation.

The survival of an organism to sexual maturity and its reproductive performance obviously require its competence to withstand a variety of elements in the physical and biological environment. The physical factors of sunlight, moisture, temperature, gravity, light and darkness may be of prime importance in determining the survival or elimination of certain genotypes. Each organism has to live amid other organisms, which leads to competition to eat and avoid being eaten.

Plants must compete for room in the soil and for sunlight as well as for water and inorganic salts. Each is constantly threatened by animals that may eat it before it has the opportunity to reach sexual maturity and release the spores or seeds for the next generation. Animals are under similar pressure to avoid being eaten or killed and to find food for themselves. Any adaptation that improves an organism's ability to find food and avoid being eaten will, of course, play a role in its differential reproduction.

Thus, evolutionary changes resulting from differential reproduction are of such a nature as to maintain or improve the average ability of that population to produce successive generations under the conditions of the environment. Since there are a great many different ways by which the process of differential reproduction can be facilitated, there are a great many different ways in which natural selection may operate.

Natural selection, in general, does not operate upon the phenotypes of single genes but upon the phenotypic effects of the entire genetic system or **genome.** One group of organisms may survive despite some obviously disadvantageous character while another may be eliminated despite some traits that are highly advantageous for its getting along in life. The plants and animals that ultimately survive and are the parents of the next generation are those with qualities whose sum total renders them a little better able than their competitors to survive and reproduce their kind. The environment itself may change from time to time; thus, a characteristic of adaptive value at one time may be useless or even deleterious at another.

9.7 MUTATIONS, THE RAW MATERIAL OF EVOLUTION

The Dutch botanist Hugo de Vries, one of the three rediscoverers of Mendel's laws, was the first to emphasize the importance in evolution of sudden large changes rather than the gradual accumulation of many small changes postulated by Darwin. In his experiments with plants, such as the evening primrose, de Vries found that many unusual forms, which differed markedly from the ancestral wild plant, appeared and bred true thereafter. He applied the term **mutations** to these sudden changes in the characteristics of an organism (earlier breeders had called them "sports"). Darwin had observed such changes but thought they occurred too rarely to be of importance in evolution. Darwin believed that these sudden changes would upset the harmoni-

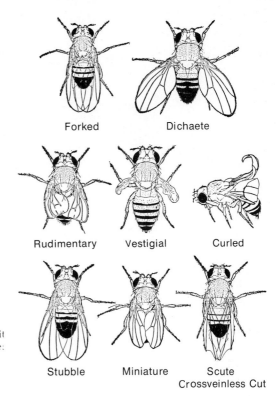

Forked   Dichaete

Rudimentary   Vestigial   Curled

Stubble   Miniature   Scute
                  Crossveinless Cut

Figure 9.1 Some wing and bristle mutants in the fruit fly, *Drosophila melanogaster*. (Drawn by E. M. Wallace; Sturtevant and Beadle: An Introduction to Genetics.)

ous relations between the various parts of an organism and its adaptation to the environment. Thousands of breeding experiments with plants and animals since the turn of the century have shown that such mutations do occur constantly and that their effects may be of adaptive value. With the development of the gene theory, the term mutation has come to refer to sudden, discontinuous, random changes in the genes and chromosomes, although it is still used to some extent to refer to the new type of plant or animal.

In the plants and animals most widely used in breeding experiments—corn and fruit flies—some 400 to 600 mutations, respectively, have been detected. The fruit fly mutations are tremendously varied and include all shades of body color from yellow through brown and gray to black; red, white, brown or purple eyes; crumpled, curled, shortened, and peculiarly shaped wings—even the complete absence of wings; oddly shaped legs and bristles; and such extraordinary changes as a pair of legs growing on the forehead in place of the antennae (Fig. 9.1). Mutations are found in

domestic animals; the six-toed cats of Cape Cod and the short-legged Ancon sheep are two of many examples of the persistence of a single mutation.

Early in the present century there was a heated discussion as to whether evolution was the result of natural selection or of mutation. As more was learned about heredity, it became clear that natural selection can operate only when there is something to be selected, that is, when mutations present alternative ways of coping with the environment. The evolution of new species, then, involves both mutation and natural selection.

One of the current controversies in evolutionary theory concerns the possible role of small and large mutations in the origin of new species. The Neo-Darwinists argue that new species (and all the higher categories) evolve by the gradual accumulation of many small mutations; thus, there should exist many forms intermediate between the original species and the new one. Other biologists believe that new species and genera arise in a single step by a **macromutation,** a major change in the genetic system which

Figure 9.2   A comparison of the structure of the tail of the primitive bird Archaeopteryx (*right*) and the tail of a modern bird (*left*). (Villee: Biology, 6th ed.)

produces a major change in the pattern of development. This results in an adult that is morphologically and physiologically quite different from its parents. The macromutationists would hold that one should not expect to find forms which are intermediate between the original species and the new one. Many macromutations result only in "monsters" which would be unable to survive. (The term monster simply means any form which is markedly different from the usual type of the species and does not necessarily imply that it is ugly.) Other macromutations may give rise to what Richard Goldschmidt called "hopeful monsters," organisms which are enabled by their mutation to occupy some new environment. The evolution of the extinct ancestral bird, *Archaeopteryx*, into modern birds, Goldschmidt believed, may have occurred by a macromutation. *Archaeopteryx* (Fig. 9.2) had a long reptile-like tail covered with feathers; a macromutation which altered development so that the tail was greatly shortened would result in a "hopeful monster" with the fan-shaped arrangement of tail feathers seen in modern birds. This new shape of the tail, which is better suited for flying than the long tail of *Archaeopteryx*, gave its possessors an advantage in the struggle for existence. There is, of course, no proof that modern birds evolved in this way, but there is ample evidence that similar marked skeletal changes may result from a single mutation. The stubby tail of the Manx cat is the result of a mutation which causes the tail vertebrae to shorten and fuse. Professor Goldschmidt did not deny that small mutations may occur and accumulate but held that they can lead only to varieties or geographic races and not to species, genera and the higher taxonomic divisions.

The causes of natural or spontaneous mutations are unknown. Both gene and chromosome mutations can be produced artificially by a variety of agents: x-rays and alpha, beta and gamma rays emitted by radioactive elements, neutrons, ultraviolet rays, chemicals such as the war gas known as nitrogen mustard, even heat and cold are slightly effective. Cosmic rays and other particles bombarding the earth may account for some of the spontaneous mutations. Errors in the process of gene replication could lead to the production of mutant genes.

Both spontaneous and artificially induced mutations occur at random; the appearance of a mutation bears no relationship to the kind of inducing agent or to the particular need of the organism at that time. There is no way of producing to order a particular kind of mutation — a particular kind of biochemical mutant in *Neurospora*, for example. An investigator who wants to use some particular mutant has no choice but to irradiate many organisms, produce hundreds or even thousands of mutations, and then select the particular one he wants.

Whatever the causes of mutations may be, their central role in evolution as the raw material for natural selection is now generally accepted. Some evolutionists have in the past objected that the spontaneous or induced mutations observed in the laboratory could not be the basis for evolution because almost all of them are deleterious, and because the differences between species are usually slight variations, affecting many different parts of the organism and inherited by means of multiple factors, whereas the mutations observed in the laboratory are usually large variations, involving a single organ and inherited by single gene differences. Studies in the genetics of wild populations have shown that mutations that occur in the wild, like the ones observed in the laboratory, are usually for detrimental traits. We must keep clearly in mind that the animals and plants living today are the result of a long and vigorous process of natural selection. In the course of their evolution, most of the possible mutations have occurred, and the beneficial ones have been selected and preserved. The organisms are well adapted to their surroundings and further mutations are much more likely to be harmful than helpful. However, a few of the mutations seen in the laboratory and in wild populations are beneficial and have survival value. Mutations may produce traits which are deleterious in one environment but advantageous in another. Sickle cell anemia, for example, is generally disadvantageous, but the resistance to malaria it confers is advantageous in regions such as Central Africa where malaria is widespread.

Closer study of populations has shown that the sort of variations which differentiate a species do appear in stocks bred in the laboratory. However, being somewhat more difficult to detect and study, they were missed in some of the earlier work. More recent experiments indicate that such mutations occur at an even greater rate than the larger, more obvious ones.

9.8 BALANCED POLYMORPHISM

It might be assumed that selection would inevitably result in a population completely homozygous for whichever pair of alleles at a given locus provides for a trait of the greatest adaptive value. Although this indeed may happen, it is not the only possibility in differential reproduction. Variation itself may be adaptive for a population, because a completely homozygous population would have no genetic substratum on which natural selection could act. A population with a good prognosis for survival is one that has maintained enough variation to permit further adaptive changes. Observations on wild populations of flies and other organisms have shown that their genetic pools do change adaptively in response to such changes in the environment as the alternations of the seasons.

There are instances in which the heterozygote is fitter for reproductive efficiency and survival than either of the corresponding homozygous individuals; however, the heterozygous state cannot be maintained in the population unless a certain number of the comparatively less fit homozygous individuals are also produced. Depending on the relative selective values of the heterozygous and homozygous states, there is a particular ratio of alleles in the genetic pool that will result in the optimal proportion of heterozygotes and homozygotes. The preservation of variation by selection in this way is termed a **balanced polymorphism.** More and more evidence is coming to light that the populations that are most effective reproductively do maintain a fairly high degree of heterozygosity.

9.9 ADAPTIVE RADIATION

The phenomenon termed adaptive radiation is a general feature of the evolution of most plants and animals. Whenever a group of organisms has been provided with the opportunity to spread into a number of new ecological habitats to which it has physical access and in which it has the possibility of surviving, these radiations have occurred. Because of the competition for food and living space, there is a tendency for each group of organisms to spread out and occupy as many different habitats as they can reach

Figure 9.3   Adaptive radiation. All the various mammals shown have evolved from the common ancestor depicted in the center. Each of the descendants has become adapted to a different type of habitat. (Villee: Biology, 6th ed.)

and which will support them. This evolution, from a single ancestral species, of a variety of forms which occupy different habitats is called **adaptive radiation.** It is obviously advantageous in enabling organisms to tap new sources of food and to escape from some of their enemies.

The placental mammals provide a classic illustration of the process, for from a primitive, insect-eating, five-toed, short-legged creature that walked with the soles of its feet flat on the ground have evolved all the present-day types of placental mammals (Fig. 9.3). There are dogs and deer, adapted for terrestrial life in which running rapidly is important for survival; squirrels and primates, adapted for life in the trees; bats, equipped for flying; beavers and seals, which maintain an amphibious existence; the completely aquatic whales, porpoises and sea cows; and the burrowing animals, moles, gophers and shrews. The number and shape of the teeth, the length and number of leg bones, the number and attachment sites of muscles and the thickness and color of the fur are some of the structures that are involved in adaptation.

There was a comparable major adaptive radiation of the reptiles in earlier geologic ages. Adaptive radiation may take place on a much smaller scale, as represented by the variety of ground finches found today on the Galápagos Islands west of Ecuador. Some of these are ground birds that feed mainly on seeds, others feed mainly on cactus, and still others have taken to living in trees and eating insects. There have been evolutionary changes in the size of the beak and in its structure. The essence of adaptive radiation, then, is the evolution of a variety of different forms from a single ancestral form with each of the descendants being adaptively specialized in a unique way to survive in a particular habitat.

Adaptive radiation in which a single ancestral type gives rise to several descendant lines adapted in different ways to different environments may be termed **divergent evolution** (Fig. 9.4). The opposite phenomenon is also fairly frequent in evolution — that is, two or more unrelated groups become adapted to similar environments and tend to develop features that are at least superficially similar. The evolution of similar sets of characteristics in groups of quite different evolutionary ancestry is termed **convergent evolution.** An often quoted example of convergent evolution is the development of wings in flying reptiles, in birds and in mammals as well as in insects.

The dolphins and porpoises (which are mammals), the extinct ichthyosaurs (which were reptiles) and both bony and cartilaginous fishes have all evolved streamlined shapes, dorsal fins, tail fins and flipper-like fore and hind limbs which make them look much alike (Fig. 9.5). The moles and gophers, in adapting to a burrowing life, have evolved similar fore and hind leg structures adapted for digging, but the mole is an insectivore and the gopher is a rodent. The eye of the squid and the eye of the simpler vertebrates such as the fish are also very similar in structure and provide yet another example of convergent evolution.

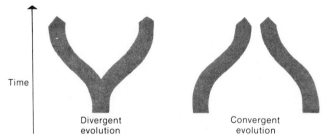

Figure 9.4 Diagram illustrating the difference between divergent and convergent evolution. A single stock may branch to give two diverging stocks which become more and more different as evolution proceeds. In convergent evolution two stocks originally quite different come to resemble each other more and more as time passes, probably because they occupy a comparable habitat and become adapted to similar conditions. (Villee: Biology, 6th ed.)

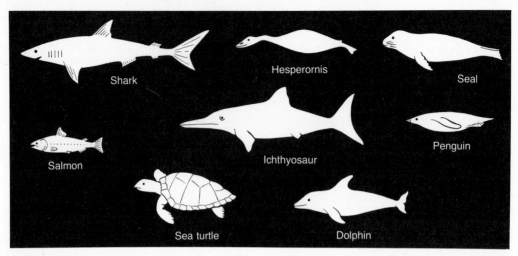

Figure 9.5 Convergent evolution. All of these aquatic vertebrates have a marked superficial similarity despite their distant relationship, because of their adaptations to similar environments.

9.10 SPECIATION

The unit of classification for both plants and animals is the species. It is difficult to give a definition of this term which can be applied uniformly throughout the animal and plant kingdoms, but a **species** may be defined as a population of individuals with similar structural and functional characteristics, which have a common ancestry and in nature breed only with each other. A species is a collection of demes or populations within which interbreeding may occur —a group of populations with a common gene pool. It is implicit in this definition that there is no free flow of genes between two different species. One species is isolated from the next by reproductive barriers. These barriers are not absolute and the occasional hybrid between species that may occur does not alter their status as separate species.

Closely related species are grouped together in the next higher unit of classification, the **genus** (plural, genera). The scientific names of plants and animals consist of two words, the genus and the species, given in Latin. This **binomial system** of naming organisms was first used consistently by the Swedish naturalist, Carl Linnaeus. Linnaeus, like many biologists before him, recognized that living things occur in discontinuous groups so that plants and animals can be assigned to separate, distinct kinds or species. Linnaeus catalogued and described plants in his *Species Plantarum*

(1753) and animals in *Systema Naturae* (1758). This system of classification was based on structural similarities; but since these structural similarities are determined by the specific relationships of different gene pools, the modern classification of organisms based on evolutionary relationships is similar in many respects to the one set up by Linnaeus. Using the binomial system, the scientific name of the domestic cat, *Felis domestica,* applies to all the varieties of tame cats—Persian, Siamese, Manx, Abyssinian and so on. All of them belong to the same species, and they all are capable of interbreeding. Related species in the same genus are the lion, *Felis leo,* the tiger, *Felis tigris* and the leopard, *Felis pardus.* The dog, *Canis familiaris,* belongs to a different genus. The name of the genus is written first and is capitalized, and the name of the species is written second and is not capitalized.

Just as several species may be grouped to form a genus, a number of related genera may be grouped to constitute a **family.** Similarly, families may be grouped into **orders,** orders into **classes** and classes into **phyla** (singular, phylum). The phyla are the large major divisions of the plant and animal kingdoms just as the species are the basic small units of this evolutionary classification. The complete classification of man is phylum Chordata, subphylum Vertebrata, class Mammalia, subclass Eutheria, order Primates, family Hominidae, genus *Homo,* species *sapiens.*

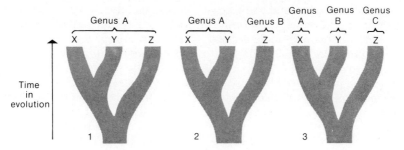

Figure 9.6 A single original stock evolves in the course of time into three different groups of organisms—X, Y and Z. Different biologists looking at these three groups may either assign them all to the same genus (1) or to two genera (2) or to three different genera (3). Some biologists prefer taxonomic systems with a single large classification and many subdivisions. Others prefer separate subdivisions for each group that can be recognized. There is no single "correct" way of classifying organisms in any subdivision of the plant or animal kingdom. (Villee: Biology, 6th ed.)

Many plants and animals fall into easily recognizable, natural groups and their classification presents no difficulty. Others seem to lie on the borderline between two groups, with some characteristics in common with each, and are difficult to assign to one or the other. The number of the principal groups and the organisms included within each vary according to the basis of classification used and the judgment of the scientist making the classification (Fig. 9.6). Some taxonomists like to group things together in units that exist already. Others prefer to establish separate categories for forms that do not fall

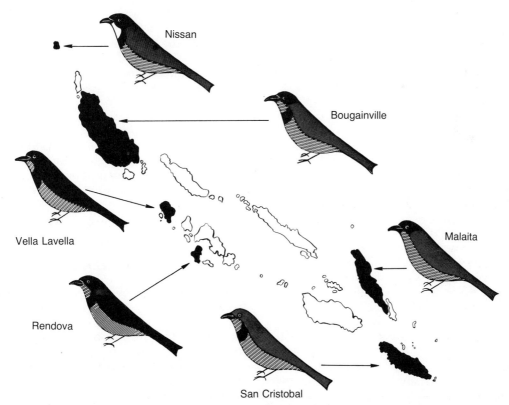

Figure 9.7 The distribution of the subspecies of the golden whistler (*Pachycephala pectoralis*) in the Solomon Islands. A major factor in the evolution of these subspecies has been their geographic isolation on separate islands. Green-colored plumage is indicated by cross-hatching; yellow by light gray tone. (Modified from Dobzhansky.)

readily into one of the recognized classifications. For example, different taxonomists consider that there are from 10 to 36 animal phyla and from four to 12 plant phyla.

In providing an explanation for the origin of a new species, we must describe how the summation of unit evolutionary changes in a population may culminate eventually in the establishment of new species, genera, families and orders. This requires that **reproductive barriers** arise between the incipient species, as they are becoming established. When interbreeding between subgroups of a population becomes progressively less frequent and the resulting hybrids become progressively less fertile, the several groups eventually become different species. Any factor that decreases the amount of interbreeding between groups of organisms is termed an "**isolating mechanism.**"

A very common type of isolation is **geographic,** the separation of groups of related organisms by some physical barrier, such as a river, desert, glacier, mountain or ocean (Fig. 9.7). In a mountainous region the individual mountain ranges afford effective barriers between the valleys. Valleys only a short distance apart, but separated by ridges always covered with snow, typically have species of plants and animals that are peculiar to those valleys. Thus there are usually more different species in a given area in a mountainous region than on open plains. For example, in the mountains of the western United States there are 23 species and subspecies of rabbits, whereas in the much larger plains area of the Midwest and the East there are only eight species of rabbits.

The isthmus of Panama provides another striking example of geographic isolation. On either side of the isthmus the phyla and classes of marine invertebrates are made up of different but closely related species. For some 16,000,000 years during the Tertiary period, there was no connection between North and South America, and marine animals could migrate freely between what is now the Gulf of Mexico and the Pacific Ocean. When the isthmus of Panama re-emerged, the closely related groups of animals were isolated and the differences between the fauna in the two regions represent the subsequent accumulation of hereditary differences.

Geographic isolation is usually not permanent, and the two previously isolated groups may come into contact again and resume interbreeding unless **genetic isolation** or interspecific sterility has arisen in the meantime. The various races of man have resulted from geographic isolation and the accumulation of chance mutations in different stocks. Since interracial sterility has not developed, the differences tend to disappear when geographic isolation is removed.

Genetic isolation results from one or more mutations that occur independently of mutations for structural or functional features. Genetic isolation may appear only after a long period of geographic isolation has produced striking differences between two groups of organisms, or it may originate within a single, otherwise homogeneous group of organisms. For example, in a species of fruit fly, *Drosophila pseudoobscura*, a mutation for genetic isolation has produced two groups of flies which, though externally indistinguishable, are completely sterile when cross-mated. These two groups are isolated genetically as effectively as if they lived on different continents. As generations pass and different mutations accumulate in each group by chance and by selection, the two groups will undoubtedly become visibly different. Two groups of organisms living in the same geographic area may be **ecologically isolated** if they occupy different habitats. Marine animals living in the intertidal zone are effectively isolated from other organisms living only a few feet away below the low-tide mark. Ecologic isolation might result from the simple fact that two groups or organisms breed at somewhat different times of the year.

9.11 THE ORIGIN OF SPECIES BY HYBRIDIZATION

The crossing of two different varieties or species, called **hybridization,** provides another way in which new species may originate. The new species may combine the best characters of each of the parental species, thereby becoming better able to survive than either of its parents. Hybridization is used routinely by animal and plant breeders to establish new combinations of desirable characters.

When two different species are crossed,

and especially ones with different chromosome numbers, the offspring are usually sterile. The unlike chromosomes cannot pair properly, cannot undergo synapsis in the process of meiosis, and the resulting eggs and sperm do not receive the proper assortment of chromosomes. However, if one of these interspecific hybrids undergoes a doubling of the chromosome number, meiosis can then occur normally and fertile eggs and sperm are produced. The hybrid will breed true thereafter and generally will not produce fertile offspring when bred with either of the parental species. It is widely believed that this process has been important in the evolution of the higher plants; more than half of the higher plants appear to be polyploids. There are species of wheat with 14, 28 and 42 chromosomes, species of roses with 14, 28, 42 and 56 chromosomes, and species of violets with every multiple of 6 from 12 to 54. The fact that similar series of plants with related numbers of chromosomes can be established by experimental breeding lends credence to the idea that these natural series arose by successive hybridization and chromosome doubling.

Although there are many examples of species of plants produced by hybridization and chromosome doubling, this process appears to have played a negligible role in the evolution of animals. Two explanations of this have been advanced: the gametes of animals are more sensitive to imbalances of chromosomes and are nonviable unless a normal haploid set is present; since the sexes are separate in most animals, the random segregation of several pairs of sex chromosomes in a polyploid animal might lead to the formation of sterile combinations.

9.12 PHYLOGENY

The evolutionary history of any group of organisms is termed its **phylogeny.** It is basic to many aspects of biological research to know which organisms are most closely related—i.e., which ones have common ancestors in the recent past, and which ones have common ancestors only in the more distant past. To establish the phylogenetic relationships of a group of organisms, each investigator must examine as many charac-

teristics of each type as possible, looking for patterns of similarities and dissimilarities that may provide clues. Phylogeneticists originally were restricted largely to comparing morphologic characters—patterns of bones, muscles and nerves—but now a host of physiologic, biochemical, immunologic and cytologic characters can be examined and used to test the validity of the relationships inferred on the basis of morphologic characters. It is reassuring to find that the evolutionary relationships inferred on the basis of the newest, most sophisticated biochemical analyses of the types of proteins found in different species agree remarkably well with the evolutionary relationships established a century ago on the basis of gross morphologic similarities.

Many of the earlier paleontologists and other students of evolution were led to the conclusion that evolution tends to progress in a straight line. Fuller examination of the accumulating fossil data, however, has led to the rejection of this concept. The horse was said to have evolved in a straight line from the primitive *Hyracotherium* (a small animal, the size of a fox, with four toes on the front feet and three toes on the hind feet) to the modern *Equus*. The complete fossil record shows that there were many side branches in horse evolution (Fig. 9.8). The evolution of the present-day horse is not at all the simple progression along a single straight line of evolution that it was once thought to be. The evolution of the horse was said to show the following "trends": an increase in size, a lengthening of the legs, enlargement of the third digit and reduction of the others, an increase in the size of the molar teeth and in the complexity of the patterns of ridges on their crowns, and increases in the size of the lower jaw and the skull. There are now so many exceptions to each of these that the concept of a straight-line evolution of the horse has been abandoned.

Our increasing knowledge of how genes act in controlling development enables us to explain whatever straight-line trends in evolution may be real in terms of conventional evolution by mutation and selection. Many different types of developmental patterns may arise by random mutation, yet most of them will result in unharmonious processes, ones which will not interdigitate properly and will lead to the death of the organism.

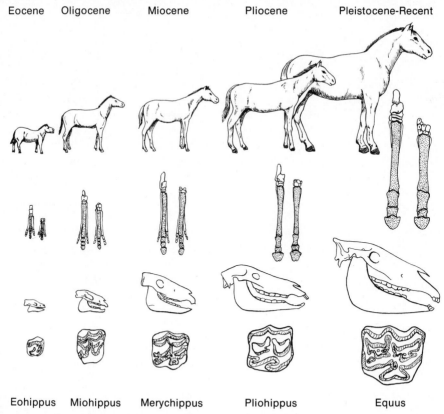

Eocene   Oligocene   Miocene   Pliocene   Pleistocene-Recent

Eohippus   Miohippus   Merychippus   Pliohippus   Equus

Figure 9.8   Stages in the evolution of the horse, illustrating (top) the changes in size and shape, (second row) the bones of the fore and hind feet, (third row) the skull, and (bottom) the grinding surfaces of the second upper molar tooth. *Eohippus* is a synonym of *Hyracotherium*.

Others, with no particular value for survival, will remain or be eliminated by chance. The ones most likely to survive, perhaps, are those which provide for further improvement in some peculiar adaptive structure already present. Thus, straight-line evolutionary series can be explained as the result of random mutation and selection occurring along one of the few possible lines of development.

An explanation for the overdevelopment of parts is now possible as well: genes do not function independently but must operate against the background of many other genes also present. Those controlling larger horns, for example, might cause the horns to be proportionately larger than the rest of the body, and if other genes cause an increase in total body size, the horns might become lethal to their possessors, as they did in the now extinct Irish deer, whose antlers had a spread of 3.5 meters!

9.13   THE ORIGIN OF LIFE

The modern theories of mutation, natural selection and population dynamics provide us with a satisfactory explanation of how the present-day animals and plants evolved from previous forms by descent with modification. The question of the ultimate origin of life on this planet has been given serious consideration by many different biologists. Some have postulated that some kind of spores or germs may have been carried through space from another planet to this one. This is unsatisfactory, not only because it begs the question of the ultimate source of these spores, but because it is extremely unlikely that any sort of living thing could survive the extreme cold and intense irradiation of interplanetary travel. Evidence for life in other parts of the cosmos has come from the discovery in 1961 of what appear to be fossils of microscopic organisms, somewhat like algae, in meteorites. This does not provide evidence, however, that living organisms could be transported through space in a meteorite.

The spontaneous origin of living things at the present time is believed to be extremely improbable. Francesco Redi's experiments showed, about 1680, that maggots do not arise *de novo* from decaying meat and laid to rest the old superstition that animals could appear by spontaneous generation. Some 200 years later Louis Pasteur showed conclusively that microorganisms such as bacteria do not arise by spontaneous generation but come only from previously existing bacteria. Other investigators have shown since then that even the smallest organisms, but filterable viruses, do not come from non-viral material by spontaneous generation. The multiplication of viruses requires the presence of previously existing viruses. Although the spontaneous generation of life at present is unlikely, it is most probable that billions of years ago, when chemical and physical conditions on the earth's surface were quite different from those at present, the first living things did arise from non-living material.

This concept, that the first things did evolve from nonliving things, and suggestions as to what the sequence of events may have been have been put forward by Pflüger, J. B. S. Haldane, R. Beutner, and particularly by the Russian biochemist A. I. Oparin, in his book, *The Origin of Life* (1938). The earth originated some five billion years ago, either as a part broken off from the sun, or by the gradual condensation of interstellar dust. Most authorities now agree that the earth was very hot and molten when it was first formed and that conditions consistent with life appeared on the earth only perhaps three billion years ago. Twenty-two different amino acids were isolated recently from Precambrian rocks from South Africa that are at least 3.1 billion years old. At that time the earth's atmosphere contained essentially no free oxygen—all the oxygen atoms were combined as water, or as oxides. The primitive atmosphere was strongly reducing, composed of methane, ammonia and water originating by "outgassing" from the earth's interior.

A number of reactions are known by which organic substances can be synthesized from inorganic ones. Originally the carbon atoms in the earth's crust were present mainly as metallic carbides. These could react with water to form acetylene, which would subsequently polymerize to form compounds with long chains of carbon atoms. High-energy radiation such as cosmic rays can catalyze the synthesis of organic compounds. This was shown by Melvin Calvin's experiments in which solutions of carbon dioxide and water were irradiated in a cyclotron and formic, oxalic and succinic acids, which contain 1, 2 and 4 carbons respectively, were obtained. These, as you know, are intermediates in certain metabolic pathways of living organisms. Irradiating solutions of inorganic compounds with ultraviolet light, or passing electric charges through the solution to simulate lightning also produce organic compounds. Harold Urey and Stanley Miller in 1953 exposed a mixture of water vapor, methane, ammonia and hydrogen gases to electric discharges for a week and demonstrated the production of amino acids such as glycine and alanine, together with other complex organic compounds. The earth's atmosphere in prebiotic times probably contained water vapor, methane, ammonia and hydrogen gas from which irradiation could produce a tremendous variety of organic materials. Amino acids and other compounds could be produced in nature at the present time by lightning discharges or ultraviolet radiations; however, any organic compound produced in this way might undergo spontaneous oxidation or it would be taken up and degraded by molds, bacteria and other organisms.

The details of the chemical reactions that could give rise, without the intervention of living things, to carbohydrates, fats and amino acids have been worked out by Oparin and extended by Calvin and others. Most, if not all, of the reactions by which the more complex organic substances were formed probably occurred in the sea in which the inorganic precursors and the organic products of the reaction were dissolved and mixed. The sea became a sort of dilute broth in which these molecules collided, reacted and aggregated to form new molecules of increasing size and complexity (this might be called the "chicken soup" theory of evolution). As more has been learned of the role of specific hydrogen bonding and other weak intermolecular forces in the pairing of specific nucleotide bases and the effectiveness of these processes in the transfer of biological information, it has become clear that similar forces could have operated early in evolution before "living" organisms appeared.

The known forces of intermolecular attraction, and the tendency for certain molecules to form liquid crystals, provide us with means by which large, complex, specific molecules can form spontaneously. Oparin suggested that natural selection can operate at the level of these complex molecules before anything recognizable as life is present. As the molecules came together to form colloidal aggregates, these aggregates began to compete with one another for raw materials. Some of the aggregates, which had some particularly favorable internal arrangement, would acquire new molecules more rapidly than others and would eventually become the dominant types.

Once some protein molecules had formed and had achieved the ability to catalyze reactions, the rate of formation of additional molecules would be greatly stepped up. Next, in combination with molecules of nucleic acid, these complex protein molecules probably acquired the ability to catalyze the synthesis of molecules like themselves. These hypothetic, autocatalytic particles composed of nucleic acids and proteins would have some of the properties of a virus, or perhaps of a free-living gene. The next step in the development of a living thing was the addition of the ability of the autocatalytic particle to undergo inherited changes—to mutate. Then, if a number of these free genes had joined to form a single larger unit, the resulting organism would have been similar to certain present-day viruses. A major step in early evolution was the development of a protein-lipid membrane around this aggregate which permitted the accumulation of some and the exclusion of other molecules. All the known viruses are parasites that can live only within the cells of higher animals and plants. However, a little reflection will suggest that free-living viruses, ones which do not produce a disease, would be very difficult to detect; such organisms may indeed exist.

The first living organisms, having arisen in a sea of organic molecules and in contact with an atmosphere free of oxygen, presumably obtained energy by the fermentation of certain of these organic substances. These heterotrophs could survive only as long as the supply of organic molecules in the sea broth, accumulated from the past, lasted. Before the supply was exhausted, however, the heterotrophs evolved further and became autotrophs, able to make their own organic molecules by chemosynthesis or photosynthesis. One of the by-products of photosynthesis is gaseous oxygen, and it is likely that all the oxygen in the atmosphere was produced and is still produced in this way. It is estimated that all the oxygen of our atmosphere is renewed by photosynthesis every 2000 years and all the carbon dioxide molecules pass through the photosynthetic process every 300 years. All the oxygen and carbon dioxide in the earth's atmosphere are the products of living organisms and have passed through living organisms over and over again during the course of evolution.

An explanation of how an autotroph may have evolved from one of these primitive, fermenting heterotrophs was presented by N. H. Horowitz in 1945. According to Horowitz' hypothesis, an organism might acquire, by successive mutations, the enzymes needed to synthesize complex from simple substances, in the reverse order to the sequence in which they are used in normal metabolism. Let us suppose that our first primitive heterotroph required organic compound Z for its growth. Substance Z, and a variety of other organic compounds, Y, X, W, V, U, etc., were present in the organic sea broth which was the environment of this heterotroph. They had been synthesized previously by the action of nonliving factors of the environment. The heterotroph would survive nicely as long as the supply of compound Z lasted. If a mutation occurred which enabled the heterotroph to synthesize substance Z from substance Y, the strain of heterotroph with this mutation would be able to survive when the supply of substance Z was exhausted. A second mutation, which established an enzyme catalyzing a reaction by which substance Y could be made from the simpler substance X, would again have great survival value when the supply of Y was exhausted. Similar mutations, setting up enzymes enabling the organism to use successively simpler substances, W, V, U, . . . and finally some inorganic substance, A, would eventually result in an organism able to make substance Z, which it needs for growth, out of substance A by way of all the intermediate compounds. When, by other series of mutations, the organism was able to synthesize all of its requirements from simple inorganic compounds, as the green plants can, it would have become an autotroph. Once the first simple autotrophs had evolved, the way was clear for the further evolution of the vast variety of green plants, bacteria, molds and animals that inhabit the world today.

These considerations lead us to the conclusion that the origin of life, as an orderly natural event on this planet, was not only possible, it was almost inevitable. Furthermore, with the vast number of planets in all the known galaxies of the universe, many of them must have conditions which permit the origin of life. It is probable, then, that there are many other planets on which life as we know it exists. Wherever life is possible, it should, if given enough time, appear and ramify into a wide variety of types. Some of these may be quite dissimilar from the ones on this planet, but others may be quite like those found here; some may, perhaps, be like ourselves. Living things on other planets might have a completely different kind of genetic code, or might be composed of elements other than carbon, hydrogen, oxygen and nitrogen.

It seems unlikely that we will ever know how life originated, whether it happened only once or many times, or whether it might happen again. The theory (1) that organic substances were formed from inorganic substances by the action of physical factors in the environment; (2) that they interacted to form more and more complex substances, finally enzymes, and then self-reproducing systems ("free genes"); (3) that these "free genes" diversified and united to form primitive, perhaps virus-like heterotrophs; and (4) that autotrophs then evolved from these heterotrophs, has the virtue of being quite plausible. Many of the parts of this theory have been subjected to experimental verification.

9.14 PRINCIPLES OF EVOLUTION

However much students of evolution may disagree as to the nature of mutations, the kind of mutations involved in evolution, and the degree to which such factors as natural selection, isolation, genetic recombination and population dynamics may affect the evolution of some particular organism, there are several fundamental principles upon which they are agreed: changes within the genes and chromosomes are the raw material of evolution, some sort of isolation is necessary for the establishment of a new species, and natural selection by differential reproduction is involved in the survival of some, but not all, of the mutations which occur. In addition, there are five principles of evolution to which nearly all biologists would subscribe.

1. Evolution occurs more rapidly at some times than at others. At the present time it is occurring rapidly, with many new forms appearing and many old ones becoming extinct.

2. Evolution does not proceed at the same rate among different types of organisms. At one extreme are the brachiopods, some species of which have been exactly the same for the last 500,000,000 years at least, for fossil shells found in rocks deposited at that time are identical with those of animals living today. In contrast, several species of man have appeared and become extinct in the past few hundred thousand years. In general, evolution tends to occur rapidly when a new species first appears, and then gradually slows down as the group becomes established and adapted to its particular environment.

3. New species do not evolve from the most advanced and specialized forms already living, but from relatively simple, unspecialized forms. The mammals, for example, did not evolve from the large, specialized dinosaurs, but from a group of rather small and unspecialized reptiles.

4. Evolution is not always from the simple to the complex. There are many examples of "regressive" evolution, in which a complex organism has given rise to simpler ones. Most parasites have evolved from free-living ancestors which were more complex than they; wingless birds, such as

the cassowary and emu, have descended from birds that could fly; many wingless insects have evolved from winged ones; the legless snakes came from reptiles with appendages; the whale, which has no hind legs, evolved from a mammal that had the customary two pairs of legs and so on. These are all reflections of the fact that mutations occur at random and not necessarily from the simple to the complex or from the imperfect to the perfect. If there is some ad-

vantage to a species in having a simpler structure, or in doing without some structure altogether, any mutations which happen to occur for such conditions will tend to accumulate by natural selection.

5. Evolution occurs by *populations*, not by individuals; evolutionary processes are brought about by the processes of mutation, nonrandom reproduction, natural selection and genetic drift.

ANNOTATED REFERENCES

Blum, H. F.: Time's Arrow and Evolution. Revised ed. Gloucester, Mass., Peter Smith, 1969. A discussion of theories of the origin of life.

Bryson, V., and H. J. Vogel (Eds.): Evolving Genes and Proteins. New York, Academic Press, 1965. Proceedings of a symposium relating biochemistry, genetics and evolution, especially the evolutionary implications of our expanding knowledge of the structure and function of macromolecules.

Carter, G. S.: Animal Evolution: A Study of Recent Views on Its Causes. London, Sidgwick & Jackson, 1951. A detailed, technical account of the evolution of animals.

Darwin, C.: The Origin of Species. 1859. Available in a number of recent reprint editions. This classic is well worth sampling for its clear, logical arguments and for its wealth of examples.

Dobzhansky, T.: Genetics and the Origin of Species. 3rd ed. New York, Columbia University Press, 1951. A detailed, technical presentation of the neo-Darwinian viewpoint of the role of small mutations and natural selection in evolution.

Goldschmidt, R. B.: The Material Basis of Evolution. New Haven, Yale University Press, 1940. Presents in detail Goldschmidt's views on the importance of large mutations in the evolution of species.

Irvine, W.: Apes, Angels and Victorians. New York, McGraw-Hill Book Co., 1955. Presents a clear picture of the impact of the theory of evolution on Victorian England and a vivid portrayal of Thomas Huxley's championing of Darwin's theory.

Jukes, T. H.: Molecules and Evolution. New York, Columbia University Press, 1966. An account of certain biochemical reactions responsible for the structure, function and survival of organisms and their bearing on evolution.

Keosian, J.: The Origin of Life. 2nd ed. New York, Van Nostrand-Reinhold Publishing Co., 1968. An interesting updating of Oparin's and Calvin's theories.

Mayr, E.: Animal Species and Evolution. Cambridge, Harvard University Press, 1963. A scholarly account of the problem of speciation.

Oparin, A. I.: The Origin of Life. New York, Macmillan, 1938. A translation of Oparin's classic arguments as to how life may have evolved from nonliving systems.

Osborn, H. F.: From the Greeks to Darwin. New York, Macmillan, 1913. An interesting and classic account of the history of ideas on evolution.

Ross, H. H.: A Synthesis of Evolutionary Theory. Englewood Cliffs, N.J., Prentice-Hall, Inc., 1964. A short book summarizing current views of the theory of evolution.

Savage, J. M.: Evolution. New York, Holt, Rinehart and Winston, 1963. A readable, relatively brief presentation of the evidence for and theories of evolution.

Simpson, G. G.: The Meaning of Evolution. New Haven, Yale University Press, 1950. An excellent, nontechnical presentation of evolutionary concepts.

Stebbins, G. L., Jr.: Processes of Organic Evolution. Englewood Cliffs, N.J., Prentice-Hall, 1966. A clear, literate paper-back summary of evolution and natural selection.

Williams-Ellis, A.: Darwin's Moon. London and Edinburgh, Blackie & Son, 1966. An excellent biography of Alfred Russel Wallace with a vivid portrayal of Wallace's life and times which reemphasizes his solid contributions to the advancement of the theory of evolution.

10. The Evidence for Evolution

The evidence that organic evolution has occurred is so overwhelming that no one who is acquainted with it has any doubt that new species are derived from previously existing ones by descent with modification. The fossil record provides direct evidence of organic evolution and gives the details of the evolutionary relationships of many lines of descent. Most of the facts underlying the several subdivisions of biology acquire significance and make sense only when viewed against the background of evolution.

10.1 THE FOSSIL EVIDENCE

The evidence of life in former times is now both abundant and diverse. The science of **paleontology,** which deals with the finding, cataloguing and interpretation of **fossils,** has aided immensely in our understanding of the lines of descent of many vertebrate and invertebrate stocks. The term "fossil" (Latin *fossilium,* something dug up) refers not only to the bones, shells, teeth and other hard parts of an animal's body which may survive, but to any impression or trace left by previous organisms. In view of the large number of fossils that have been found, it is sobering to realize that only a small fraction of animals died under conditions that favored their preservation as fossils and

that only a small fraction of these fossils have been dug up and studied to date.

Footprints or trails made in soft mud, which subsequently hardened, are a common type of fossil. For example, the tracks of an amphibian from the Pennsylvanian period, discovered in 1948 near Pittsburgh, revealed that the animal moved by hopping rather than by walking, for the footprints lay opposite each other in pairs.

The commonest vertebrate fossils are skeletal parts. From the shape of bones, and the position of the bone scars which indicate points of muscle attachment, paleontologists can make inferences about an animal's posture and style of walking, the position and size of its muscles, and hence the contours of its body. Careful study of fossil remains has enabled paleontologists to make reconstructions of what the animal must have looked like in life (Figs. 9.2 and 10.1).

In some fossils, the original hard parts, or more rarely the soft tissues of the body, have been replaced by minerals, a process called **petrifaction.** Iron pyrites, silica and calcium carbonate are some of the common petrifying minerals. The petrified muscle of a shark more than 300,000,000 years old was so well preserved by petrifaction that not only individual muscle fibers, but even their cross striations, could be observed in thin sections under the microscope. A

Figure 10.1   An example of a fossil, the remains of *Archaeopteryx*, a tailed, toothed bird from the Jurassic period. (Courtesy of the American Museum of Natural History.)

famous example of the process of petrifaction is the Petrified Forest in Arizona.

Molds and casts are superficially similar to petrified fossils but are produced in a different way. **Molds** are formed by the hardening of the material surrounding a buried organism, followed by the decay and removal of the body of the organism. The mold may subsequently be filled by minerals which harden to form **casts** which are exact replicas of the original structures. Some animal remains have been exceptionally well preserved by being embedded in tar, amber, ice or volcanic ash. The remains of woolly mammoths, deep frozen in Siberian ice for more than 25,000 years, were so well preserved that the meat was edible!

The formation and preservation of a fossil require that some structure be buried. This may take place at the bottom of a body of water, or on land by the accumulation of wind-blown sand, soil or volcanic ash. Many of the men and animals living in Pompeii were preserved almost perfectly by the volcanic ash from the eruption of Vesuvius. Sometimes animals were trapped and entombed in a bog, quicksand or asphalt pit, such as the famous La Brea tar pits in Los Angeles which have provided superb fossils of Pleistocene animals.

10.2   THE GEOLOGIC TIME TABLE

Studies of the earth's crust have shown that it consists of sheets of rock lying one on top of the next. There are five major rock strata and each of these is subdivided into minor strata. These layers were generally formed by the accumulation of sediment— sand or mud—at the bottom of oceans, seas or lakes. Each rock stratum contains certain characteristic kinds of fossils which can now be used to identify deposits made at the same time in different parts of the world. Geologic time has been divided, according to the succession of these rock strata, into eras, periods and epochs (Table 10.1). The duration of each period or epoch can be estimated from the thickness of the sedimentary deposits, although, of course, the rate of deposition was not exactly the same in different places and at different times.

The layers of sedimentary rock should occur in the sequence of their deposition, with the newer strata on top of the older ones, but subsequent geologic events may have changed the relationship of the layers. Not all of the expected strata may occur in some particular region, for that land may have been exposed rather than submerged during one or more geologic ages. In some regions the strata formed previously have subsequently emerged, been washed away, and then relatively recent strata have been deposited directly on very ancient ones. Certain sections of the earth's crust, in addition, have undergone massive foldings and splittings, so that early layers have come to lie on top of later ones. The age of a rock stratum may be determined by a study of its fossils, for some kinds of fossils were deposited in only one era or period.

Rock deposits are now dated largely by taking advantage of the fact that certain radioactive elements are transformed into other elements at rates which are slow and essentially unaffected by the pressures and temperatures to which the rock has been subjected. Half of a given sample of uranium will be converted to a special isotope of lead in 4.5 billion years. Hence, by measuring the proportion of uranium and lead in a bit of crystalline rock, its age can be measured. In this way, the oldest rocks of the earliest geologic period are calculated to be about 3,500,000,000 years old and the latest Cambrian rocks to be 500,000,000 years old. These dates have been confirmed by newer methods in which the radioactive decay of rubidium 87 (half-life 47 billion years!) and potassium 40 has been utilized to measure the ages of micas and feldspars. Events in more recent times can be dated quite accurately by the decay of carbon 14, which has a half-life of 5568 years.[*] Relatively short periods of geologic time are estimated by measuring the rate at which waterfalls recede upstream as they wear away the rocks over which they tumble or by counting the annual deposits of clay on the bottoms of ponds and lakes.

Between the major eras there were widespread geologic disturbances, called **revolutions,** which raised or lowered vast regions of the earth's surface and created or eliminated shallow inland seas. These revolutions produced great changes in the distribution of sea and land organisms and wiped out many of the previous forms of life. The Paleozoic era ended with the revolution that raised the Appalachian mountains and, it is believed, killed all but 3 per cent of the forms of life existing then. The Rocky Mountain revolution (which raised the Andes, Alps and Himalayas as well as the Rockies) annihilated most reptiles of the Mesozoic.

10.3 THE GEOLOGIC ERAS

Archeozoic Era. The rocks of the oldest geologic era are very deeply buried in most parts of the world but are exposed at the bottom of the Grand Canyon and along the shores of Lake Superior. The oldest geologic era, the Archeozoic, begins not with the origin of the earth but with the formation of the earth's crust, when rocks and mountains were in existence and the processes of erosion and sedimentation had begun. The Archeozoic era lasted about two billion years, about as long as all the succeeding eras combined. It was characterized by widespread volcanic activity and large upheavals which resulted in the raising of mountains. The heat, pressure and churning associated with the movements of the earth's crust probably destroyed most of whatever fossil remains there may have been, but a few traces of life remain. Scattered through the Archeozoic rocks are flakes of graphite, pure carbon, which are probably the transformed remains of plants and animal bodies. Although graphite can originate inorganically, its distribution in the rocks suggests that it was formed organically. If the amount of graphite in these rocks can be taken as a measure of the amount of living things in the Archeozoic, and there are reasons for believing that this is justified, then life must have been abundant in the Archeozoic seas, for there is more carbon in these rocks than in the coal beds of the Appalachians.

Proterozoic Era. The second geologic era, which lasted about one billion years, was characterized by the deposition of large quantities of sediment and by at least one great period of glaciation during which ice sheets stretched to within 20 degrees of the equator. There was less volcanic activity in this than in the preceding era, and the rocks are better preserved. Only a few fossils have been found in Proterozoic rocks but they show not only that life was present but

---

[*] Organic carbon is derived by CO_2 fixation from atmospheric CO_2 and the ratio of ^{12}C to ^{14}C in living organisms is the same as that in the atmosphere. No exchange of carbon atoms with the atmosphere occurs after death and the ^{14}C in the body is slowly transformed into ^{14}N. The age of organic remains can be estimated from their $^{12}C/^{14}C$ ratio and the half-life of ^{14}C.

TABLE 10.1   GEOLOGIC TIME TABLE

Era	Period	Epoch	Dura-tion in Mil-lions of Years	Time from Be-ginning of Period to Pres-ent (Mil-lions of Years)	Geologic Conditions	Plant Life	Animal Life
Cenozoic (Age of Mammals)	Quaternary	Recent	0.011	0.011	End of last ice age; climate warmer	Decline of woody plants; rise of her-baceous ones	Age of man
		Pleistocene	1	1	Repeated glaciation; four ice ages	Great extinction of species	Extinction of great mammals; first hu-man social life
	Tertiary	Pliocene	12	13	Continued rise of mountains of west-ern North Amer-ica; volcanic ac-tivity	Decline of forests; spread of grass-lands; flowering plants, monocoty-ledons developed	Man evolved from manlike apes; ele-phants, horses, camels almost like modern species
		Miocene	12	25	Sierra and Cascade mountains formed; volcanic activity in north-west U.S.; climate cooler		Mammals at height of evolution; first manlike apes
		Oligocene	11	36	Lands lower; cli-mate warmer	Maximum spread of forests; rise of monocotyledons, flowering plants	Archaic mammals extinct; rise of anthropoids; fore-runners of most living genera of mammals
		Eocene	22	58	Mountains eroded; no continental seas; climate warmer		Placental mammals diversified and specialized; hoofed mammals and carnivores es-tablished
		Paleocene	5	63			Spread of archaic mammals

Rocky Mountain Revolution (Little Destruction of Fossils)

Era	Period	Epoch	Dura-tion in Mil-lions of Years	Time from Be-ginning of Period to Pres-ent (Mil-lions of Years)	Geologic Conditions	Plant Life	Animal Life
Mesozoic (Age of Reptiles)	Cretaceous		72	135	Andes, Alps, Hima-layas, Rockies formed late; ear-lier, inland seas and swamps; chalk, shale de-posited	First monocotyle-dons; first oak and maple forests; gymnosperms de-clined	Dinosaurs reached peak, became ex-tinct; toothed birds became ex-tinct; first modern birds; archaic mammals com-mon
	Jurassic		46	181	Continents fairly high; shallow seas over some of Eu-rope and Western U.S.	Increase of dicoty-ledons; cycads and conifers com-mon	First toothed birds; dinosaurs larger and specialized; primitive mam-mals
	Triassic		49	230	Continents exposed; widespread des-ert conditions; many land depos-its	Gymnosperms dom-inant, declining toward end; ex-tinction of seed ferns	First dinosaurs, pter-osaurs and egg-laying mammals; extinction of prim-itive amphibians

Table 10.1 continued on opposite page.

TABLE 10.1 GEOLOGIC TIME TABLE (*Continued*)

Era	Period	Epoch	Duration in Millions of Years	Time from Beginning of Period to Present (Millions of Years)	Geologic Conditions	Plant Life	Animal Life
Appalachian Revolution (Some Loss of Fossils)							
Paleozoic (Age of Ancient Life)	Permian		50	280	Continents rose; Appalachians formed; increasing glaciation and aridity	Decline of lycopods and horsetails	Many ancient animals died out; mammal-like reptiles, modern insects arose
	Pennsylvanian		40	320	Lands at first low; great coal swamps	Great forests of seed ferns and gymnosperms	First reptiles; insects common; spread of ancient amphibians
	Mississippian		25	345	Climate warm and humid at first, cooler later as land rose	Lycopods and horsetails dominant; gymnosperms increasingly widespread	Sea lilies at height; spread of ancient sharks
	Devonian		60	405	Smaller inland seas; land higher, more arid; glaciation	First forests; land plants well established; first gymnosperms	First amphibians; lungfishes, sharks abundant
	Silurian		20	425	Extensive continental seas; lowlands increasingly arid as land rose	First definite evidence of land plants; algae dominant	Marine arachnids dominant; first (wingless) insects; rise of fishes
	Ordovician		75	500	Great submergence of land; warm climates even in Arctic	Land plants probably first appeared; marine algae abundant	First fishes, probably fresh-water; corals, trilobites abundant; diversified mollusks
	Cambrian		100	600	Lands low, climate mild; earliest rocks with abundant fossils	Marine algae	Trilobites, brachiopods dominant; most modern phyla established
Second Great Revolution (Considerable Loss of Fossils)							
Proterozoic			1000	1600	Great sedimentation; volcanic activity later; extensive erosion, repeated glaciations	Primitive aquatic plants—algae, fungi	Various marine protozoa; towards end, mollusks, worms, other marine invertebrates
First Great Revolution (Considerable Loss of Fossils)							
Archeozoic			2000	3600	Great volcanic activity; some sedimentary deposition; extensive erosion	No recognizable fossils; indirect evidence of living things from deposits of organic material in rock	

that evolution had proceeded quite far before the end of the era. Plants and animals were differentiated, multicellular forms had evolved from unicellular ones and some of the major groups of plants and animals had appeared. Sponge spicules, jellyfish, and the remains of fungi, algae, brachiopods and annelid worm tubes have been found in Proterozoic rocks. Several rich deposits of Pre-Cambrian fossils have been found in South Australia. These include jellyfish, corals, segmented worms and two animals with no resemblance to any known fossil or living form. The bodies of these animals were soft and were strengthened only by spicules of calcium carbonate.

Paleozoic Era. A second great revolution ended the Proterozoic era. During the ensuing 370,000,000 years of the Paleozoic every phylum and class of animals except birds and mammals appeared. Some of these animals appeared and became extinct in a short time (geologically speaking), and their fossils provide convenient markers by which rocks of the same era in different localities can be correlated.

The fossil deposits of the first three periods of the Paleozoic era, the Cambrian, Ordovician and Silurian, were mostly laid down in the seas. Large shallow seas covered most of the continents during these three periods and they teemed with life. Many of these forms had hard skeletons or armor coverings which left a good fossil record. The organisms living in the Cambrian were so varied and complex that they must have evolved from ancestors dating back to the Proterozoic era. Apparently, both plants and animals lived in the sea, and the land was a curious lifeless waste until the Ordovician, when plants became established on land. The Cambrian seas contained small floating plants and animals that were eaten by primitive shrimplike crustaceans and swimming annelid worms. The sea floor was covered with simple sponges, corals, echinoderms growing on stalks, snails, pelecypods and primitive cephalopods. An exceptionally well-preserved collection of Cambrian fossils was found in the mountains of British Columbia; it included annelids, crustaceans, and a connecting link similar to peripatus. The most numerous animals were brachiopods and trilobites. Brachiopods, sessile, bivalved plankton feeders, flourished in the Cambrian and the rest

of the Paleozoic. One of the present-day brachiopods, *Lingula*, is the oldest known genus of animals and is almost identical with its Cambrian ancestors. The trilobites (see Fig. 26.9) were primitive arthropods, with flattened, elongated bodies covered dorsally by a hard shell. The shell had two longitudinal grooves that divided the body into three lobes. On the ventral side of the body was a pair of legs on each somite but the last, and each leg was biramous, had an outer gill branch and an inner walking or swimming branch. Most trilobites were only 5 to 8 cm. long but the largest was about 60 cm. They reached their peak of importance in the late Cambrian and then dwindled and became extinct in the Permian.

Evolution since the Cambrian has been characterized by the elaboration and ramification of the lines already present rather than by the establishment of entirely new forms. The original, primitive members of most lines were replaced by more complex, better adapted ones. The Ordovician seas contained, among other forms, giant cephalopods, squidlike animals with straight shells 5 to 7 meters long and 30 cm. in diameter. The Ordovician seas were apparently quite warm, for corals, which grow only in warm waters, lived as far north as Ontario and Greenland. The first vertebrates, the jawless, limbless, armored, bottom-dwelling fishes called **ostracoderms** (p. 679), appeared in the Ordovician. These lived in fresh water and their bony armor may have served as a defense against their chief predator, the carnivorous giant arachnids called **eurypterids** (p. 581). Two important events of the Silurian were the evolution of land plants and of the first air-breathing animals, primitive scorpions.

The evolution of the vertebrates, from ostracoderms to placoderms, cartilaginous and bony fishes, amphibians, reptiles, birds and mammals will be traced in Chapters 29 to 34. The Devonian seas contained corals, sea lilies and brachiopods in addition to a great variety of fishes. Trilobites were still present but were declining in numbers and importance. The first land vertebrates, the amphibians called **labyrinthodonts**, appeared in the latter part of the Devonian; this period also saw the first true forests of ferns, "seed ferns," club mosses and horsetails and the first wingless insects and millipedes.

Figure 10.2 Texas in the Permian period, about 280,000,000 years ago. Various pelycosaurs are shown. Some had large fins, others were essentially like lizards. In the lower illustration is a salamander-like amphibian with a flat, triangular skull. (Copyright, Chicago Natural History Museum, from the painting by Charles R. Knight.)

The Mississippian and Pennsylvanian periods are frequently grouped together as the Carboniferous, for during this time there flourished the great swamp forests whose remains gave rise to the major coal deposits of the world. The earliest stem reptiles appeared in the Pennsylvanian and from these there evolved in the succeeding Permian period a group of early, mammal-like reptiles, the **pelycosaurs,** from which the mammals eventually evolved (Fig. 10.2). Two important groups of winged insects, the ancestors of the cockroaches and the ancestors of the dragonflies, evolved during the Carboniferous.

The Permian period was characterized by widespread changes in topography and climate. The land began to rise early in the period, so that the swamps and shallow seas were drained, and the Appalachian revolution that ended the period, together with widespread glaciation, killed off a great many kinds of animals. The trilobites finally disappeared and the brachiopods, stalked echinoderms, cephalopods, and many other kinds of invertebrates were reduced to small, unimportant, relict groups.

Mesozoic Era. The Mesozoic era began some 230,000,000 years ago and lasted some 167,000,000 years. It is subdivided into the Triassic, Jurassic and Cretaceous periods.

During the Triassic and Jurassic most of the continental area was above water, warm and fairly dry. During the Cretaceous the Gulf of Mexico expanded into Texas and New Mexico, and the sea once again overspread large parts of the continents. There were great swamps from Colorado to British

Figure 10.3 Western Canada in the Cretaceous period, about 110,000,000 years ago. The land was low, well watered, and covered with numerous swamps. Most of the dinosaurs were harmless, plant-eating Ornithischians, reptiles with bird-like pelvic bones. Two types of duck-billed dinosaurs can be seen—three large, uncrested ones in the upper portion, and two kinds of crested ones in the lower portion. In the upper right foreground is a heavily armored, four-footed dinosaur covered with bony plates and spines. In the upper right and lower left background are ostrich dinosaurs—tall slender animals, with the general proportions of an ostrich, but with short forelegs and a long, slender tail. (Copyright, Chicago Natural History Museum, from the painting by Charles R. Knight.)

Columbia (Fig. 10.3). In the latter part of the Cretaceous the interior of the North American continent was further submerged and cut in two by the union of a bay from the Gulf of Mexico and one from the Arctic Sea. The Rocky Mountain revolution ended the Cretaceous with the upheaval of the Rockies, Alps, Himalayas and Andes mountains. The Mesozoic is characterized by the tremendous evolution, diversification and specialization of the reptiles and is commonly called the Age of Reptiles. Mammals originated in the Triassic and birds in the Jurassic. Most of the modern orders of insects appeared in the Triassic, and snails, bivalve mollusks and sea urchins underwent important evolutionary advances.

At the end of the Cretaceous a great many reptiles became extinct; they were apparently unable to adapt to the marked changes brought about by the Rocky Mountain revolution. As the climate became colder and drier, many of the plants which served as food for the herbivorous reptiles disappeared. Some of the herbivorous reptiles were too large to walk about on land when the swamps dried up. The smaller, warm-blooded mammals which appeared were better able to compete for food, and many of these ate reptilian eggs. The demise of the many kinds of reptiles was probably the result of a combination of a whole host of factors rather than any single one.

Cenozoic Era. The Cenozoic era, extending from the Rocky Mountain revolution to the present, is subdivided into the earlier Tertiary period, which lasted some 62,000,000 years, and the present Quaternary period, which includes the last million or million and one-half years.

The Tertiary is subdivided into five epochs, the Paleocene, Eocene, Oligocene, Miocene and Pliocene. The Rockies, formed at the beginning of the Tertiary, were considerably eroded by the Oligocene, and the North American continent had a gently rolling topography. Another series of uplifts in the Miocene raised the Sierra Nevadas and a new set of Rockies, and resulted in the formation of the western deserts. The climate of the Oligocene was rather mild, and palm trees grew as far north as Wyoming. The uplifts of the Miocene and Pliocene, and the successive ice ages of the Pleisto-

cene, killed off many of the mammals that had evolved.

The last elevation of the Colorado Plateau, which initiated the cutting of the Grand Canyon, occurred almost entirely in the short Pleistocene and Recent epochs, the two subdivisions of the Quaternary period. Four periods of glaciation occurred in the Pleistocene, between which the sheets of ice retreated. At their greatest extent, these ice sheets extended as far south as the Missouri and Ohio rivers and covered 4,000,000 square miles of North America. The Great Lakes, which were carved out by the advancing glaciers, changed their outlines and connections several times. It is estimated that at one time, when the Mississippi river drained lakes as far west as Duluth and as far east as Buffalo, its volume was more than 60 times as great as at present. During the Pleistocene glaciations enough water was removed from the oceans and locked in the vast sheets of ice to lower the level of the oceans as much as 100 meters. This created land connections, highways for the dispersal of many land forms, between Siberia and Alaska at Bering Strait and between England and the continent of Europe. Many mammals, including the saber-toothed tiger, the mammoth and the giant ground sloth, became extinct in the Pleistocene after primitive man had appeared.

The fossil record available today makes it impossible to doubt that the present species arose from previously existing, different ones. For many lines of evolution the individual steps are well known; other lines have some gaps which remain to be filled by future paleontologists.

Even if there were no fossil record at all, the results of the detailed studies of the morphology, physiology and biochemistry of present-day animals and plants, of their mode of development, of the transmission of inherited characteristics, and of their distribution over the earth's surface would provide overwhelming proof that organic evolution has occurred.

10.4   THE EVIDENCE FROM TAXONOMY

The science of taxonomy began long before the doctrine of evolution was accepted; indeed the founders of scientific taxonomy, Ray and Linnaeus, were firm believers in the fixity, the unchangingness, of species. Present-day taxonomists are concerned with the naming and describing of species primarily as a means of discovering or clarifying evolutionary relationships. This is based upon the assumption that the degree of resemblance in homologous structures is a measure of the degree of relationship. The characteristics of living things are such that they can be fitted into a hierarchical scheme of categories, each more inclusive than the previous one — species, genera, families, orders, classes and phyla. This can best be interpreted as proof of evolutionary relationship. If the kinds of animals and plants were not related by evolutionary descent, their characteristics would be present in a confused, random pattern and no such hierarchy of forms could be established.

The basic unit of taxonomy is the **species,** a population of closely similar individuals, which are alike in their morphologic, embryologic and physiologic characters, which in nature breed only with each other and which have a common ancestry. It is difficult to give a definition of species that is universally applicable. The definition must be modified slightly to include species whose life cycle includes two or more quite different forms (many coelenterates, parasitic worms, larval and adult insects and amphibians, for example). A population that is spread over a wide territory may show local or regional differences which may be called subspecies. Many instances are known in which a species is broken up into a chain of subspecies, each of which differs slightly from its neighbors but interbreeds with them. The subspecies at the two ends of the chain, however, may be so different that they cannot interbreed. Such a series of geographically distributed subspecies is called a *Rassenkreis* (German, race-circle).

The classification of living organisms into well-defined groups is possible because most of the intermediate forms have become extinct. If representatives of every type of animal and plant that have ever lived were still living today, there would be many series of intergrading forms and the division of these into neat taxonomic categories would be difficult indeed. The present-day species have been compared to the terminal twigs of a tree whose main branches and

trunk have disappeared. The fascinating puzzle for the taxonomist is to reconstruct the missing branches and put each twig on the proper branch.

10.5 THE EVIDENCE FROM MORPHOLOGY

A study of the details of the structure of any particular organ system in the diverse members of a given phylum reveals a basic similarity of form which is varied to some extent from one class to another. The skeletal, muscular, circulatory and excretory systems of the vertebrates provide especially clear illustrations of this principle, but this is generally true of all systems in all phyla. Only those similarities based on homologous organs (p. 258) can be used in classification. Homologous organs are basically similar in their structure, in their relationship to adjacent structures, in their embryonic development, and in their nerve and blood supply. A seal's front flipper, a bat's wing, a cat's paw, a horse's front leg and a human hand, though superficially dissimilar and adapted for quite different functions, nevertheless are homologous organs. Each consists of almost the same number of bones, muscles, nerves and blood vessels arranged in the same pattern, and their mode of development is very similar. The existence of such homologous organs implies a common evolutionary origin.

Many species of animals have organs or parts of organs which are useless and often small or lacking some essential part; in related organisms, the organ is full-sized, complete and functional. There are more than 100 such **vestigial organs** in the human body, including the appendix, the coccyx (fused tail vertebrae), the wisdom teeth, the nictitating membrane of the eye, body hair, and the muscles that move the ears (Fig. 10.4). Such organs are the remnants of ones which were functional in the ancestral forms, but when some change in the environment rendered the organ no longer necessary for survival it gradually became reduced to a vestige. This appears at first glance to be an application of Lamarck's idea of the role of "use and disuse" of an organ in evolution, but the

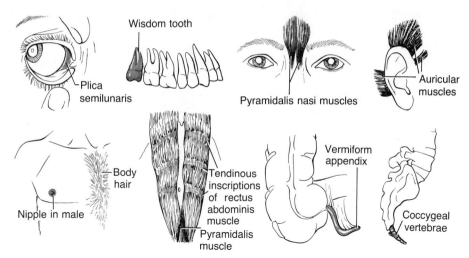

Figure 10.4 Diagrams of some of the vestigial organs of the human body.

underlying mechanism is quite different. Mutations for the decrease in the size and functional importance of an organ are occurring constantly; as long as the organ is necessary for survival, such mutations are lethal and eliminate their possessors. But if the organ is no longer needed for survival, such mutations will not be lethal, and they may accumulate and lead to the reduction of the organ.

10.6 THE EVIDENCE FROM COMPARATIVE PHYSIOLOGY AND BIOCHEMISTRY

The study of the physiologic and biochemical traits of organisms generally requires complex apparatus and is more difficult than the direct observation of morphologic characteristics. Yet, as such studies have been made using a wide variety of animal types, it has become clear that there are functional similarities and differences which parallel closely the morphologic ones. Indeed, if one were to establish taxonomic relationships based on physiologic and biochemical characters instead of on the usual structural ones, the end result would be much the same.

The fundamental similarity of the chemical constituents and patterns of enzymes present in cells of different animals was presented in Chapter 4. There are, however, certain chemical constituents, certain enzymes and certain hormones that are found in some animals and not in others. The distribution of these biochemical characters strongly parallels the evolutionary relationships inferred from other evidence.

The blood serum of each species of animal contains certain specific proteins. The degree of similarity of these serum proteins can be determined by **antigen-antibody** reactions. To perform the test, an experimental animal, usually a rabbit, is injected with a small amount of the serum—for example, a sample of human serum. The proteins of the injected serum are foreign to the rabbit's blood and hence act as antigens. The plasma cells of the rabbit respond by producing antibodies specific for human blood-protein antigens. These antibodies are then obtained by withdrawing blood from the rabbit and allowing it to clot; the antibodies are in the serum. When a dilute sample of this serum is mixed with a drop of human serum, the antibody for human serum reacts with the human serum antigen and produces a visible precipitation. The strength of the reaction can be measured by making successive dilutions of the human serum, mixing each dilution with a fresh sample of the antibody solution (the rabbit serum), and observing at what point the precipitation no longer occurs. When serum from an animal other than man is mixed with rabbit serum containing antibodies for human serum proteins, there is either no precipitation at all, or else a precipitation occurs only with concentrated antigen solutions. By testing, in turn, the sera of a variety of animals with rabbit serum containing antibodies for human serum proteins, the degree of similarity between the proteins can be determined. If the serum of another animal contains proteins which are similar to those of man, a precipitation will occur. In this way, man's closest "blood relations" have been found to be the great apes, and then, in order, the Old World monkeys, the New World monkeys, and finally the tarsioids.

The serum of the lemur gives the smallest amount of precipitation when mixed with antibodies specific for human serum.

The biochemical relationships of a variety of forms, tested in this way, correlate with and complement the relationships determined by other means. Cats, dogs and bears are closely related, as determined by this test; cows, sheep, goats, deer and antelopes constitute another closely related group. This test reveals that there is a closer relationship among the modern birds than among the mammals, for all of the several

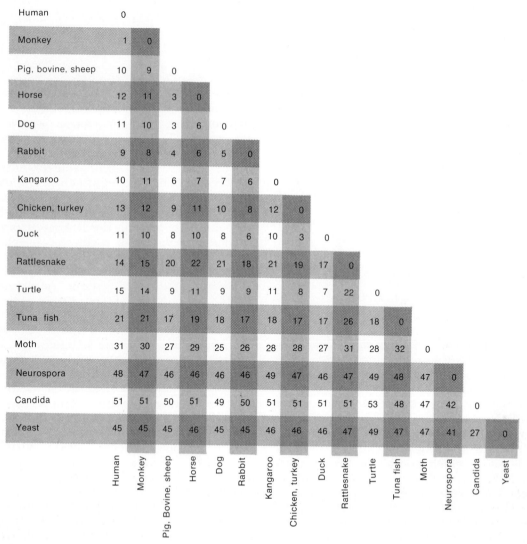

Figure 10.5 A diagram illustrating the differences in amino acid sequences in the cytochrome c's from a variety of different species of animals and plants. The numbers refer to the number of different amino acids in the cytochrome c's of the species compared. (From Dayoff, M. O., and R. V. Eck: Atlas of Protein Sequence and Structure. Silver Springs, Maryland, National Biomedical Research Foundation, 1968.)

hundred species of birds tested give strong and immediate reactions with serum containing antibodies for chicken serum. From other tests it was concluded that birds are more closely related to the crocodile line of reptiles than to the snake-lizard line, which corroborates the paleontologic evidence. Similar tests of the sera of crustaceans, insects and mollusks have shown that those forms regarded as being closely related from morphologic or paleontologic evidence also show similarities in their serum proteins.

Investigations of the sequence of amino acids in the α and β chains of hemoglobins from different species have revealed great similarities, of course, and specific differences, the pattern of which demonstrates the order in which the underlying mutations, the changes in nucleotide base pairs, must have occurred in evolution. The evolutionary relationships inferred from these studies agree completely with those based on morphologic studies. Analyses of the amino acid sequence in the protein portion of the cytochromes provide further concurring evidence of evolutionary relationships (Fig. 10.5). Thus evidence of evolutionary relationships can be adduced from similarities and differences in *molecular* structure as well as from gross morphology.

The properties of specific enzymes, which reflect their molecular structure, also provide evidence of evolutionary relationships. The pattern and rates of reaction of lactate dehydrogenase and certain other enzymes with the normal pyridine nucleotide coenzyme (NAD) and with analogues of NAD can be used to demonstrate evolutionary relationships.

It might seem unlikely that an analysis of the urinary wastes of different species would provide evidence of evolutionary relationship, yet this is true. The kind of waste excreted depends upon the particular kinds of enzymes present, and the enzymes are determined by genes, which have been selected in the course of evolution. The waste products of the metabolism of purines, adenine and guanine, are excreted by man and other primates as uric acid, by other mammals as allantoin, by amphibians and most fishes as urea, and by most invertebrates as ammonia. Vertebrate evolution has been marked by the successive loss of enzymes required for the stepwise degrada-

tion of uric acid. Joseph Needham made the interesting observation that the chick embryo in the early stages of development excretes ammonia, later it excretes urea, and finally it excretes uric acid. The enzyme uricase, which catalyzes the first step in the degradation of uric acid, is present in the early chick embryo but disappears in the later stages of development. The adult frog excretes urea, but the tadpole excretes ammonia. These are biochemical examples of the principle of recapitulation.

10.7 THE EVIDENCE FROM EMBRYOLOGY

The importance of the embryologic evidence for evolution was emphasized by Darwin and brought into even greater prominence by Ernst Haeckel in 1866, when he developed his theory that embryos, in the course of development, repeat the evolutionary history of their ancestors in some abbreviated form. This idea, succinctly stated as "Ontogeny recapitulates phylogeny," stimulated research in embryology and focused attention on the general resemblance between embryonic development and the evolutionary process, but it now seems clear that the embryos of the higher animals resemble the *embryos* of lower forms, not the adults, as Haeckel had believed. The early stages of all vertebrate embryos, for example, are remarkably similar, and it is not easy to differentiate a human embryo from the embryo of a fish, frog, chick or pig (Fig. 10.6).

In recapitulating its evolutionary history in a few days, weeks or months the embryo must eliminate some steps and alter and distort others. In addition, some new characters have evolved which are adaptive and enable the embryo to survive to later stages. For example, mammalian embryos, which have many early characteristics in common with those of fish, amphibia and reptiles, have other structures which enable them to survive and to develop within the mother's uterus rather than within an egg shell. Such secondary traits may alter the original characters common to all vertebrates so that the basic resemblances are blurred. The concept of recapitulation must be used with due caution, rather than rigorously, but it

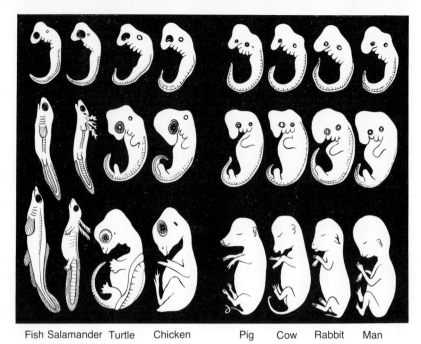

Fish Salamander Turtle   Chicken   Pig   Cow   Rabbit   Man

Figure 10.6 Comparison of early and later stages in the development of vertebrate embryos. Note the similarity of the earliest stages of each.

does provide an explanation for many otherwise inexplicable events in development.

Studies of the embryonic forms may provide the only means for identifying the relationships of certain organisms. *Sacculina*, for example, is an extremely aberrant barnacle which parasitizes crabs. The adult form is a saclike structure which sends processes into the tissues of the host to absorb nourishment. It resembles no other organism and its relationship became clear only when it was found that its larva is like that of other barnacles until it becomes attached to the abdomen of the host. Then it loses its appendages and other structures and becomes the adult, saclike creature.

The concept of recapitulation is very helpful in understanding the curious and complex development of the vertebrate circulatory and excretory systems. It is also useful, when not taken too literally, in getting a broad picture of the whole of development. Thus, the fertilized egg can be compared to the putative single-celled flagellate ancestor of all animals, and the blastula can be compared to a colonial protozoan or to some hypothetic blastula-like animal which has been postulated to be the ancestor of all Metazoa. Haeckel believed that the ancestor of coelenterates

and all the higher animals was a gastrula-like organism with two layers of cells and a central cavity connected by a blastopore to the outside. After gastrulation, development follows one of two main lines. In the echinoderms and chordates the blastopore becomes the anus, or comes to lie near the anus. In the annelid-mollusk-arthropod line the blastopore becomes the mouth or comes to lie near the mouth. In both lines the mesoderm develops between the ectoderm and endoderm. In the chordate-echinoderm line the mesoderm develops, at least in part, as pouches from the primitive digestive tract, whereas in the annelid-mollusk line the mesoderm usually originates from special cells differentiated early in development.

All chordate embryos develop, shortly after the mesoderm begins to appear, a dorsal hollow nerve cord, a notochord and pharyngeal pouches. The early human embryo at this stage resembles a fish embryo, with gill pouches, pairs of aortic arches, a fishlike heart with a single atrium and ventricle, a primitive fish kidney, and a well-differentiated tail complete with muscles for wagging it (Fig. 10.7). At a slightly later stage the human embryo resembles a reptilian embryo. Its gill

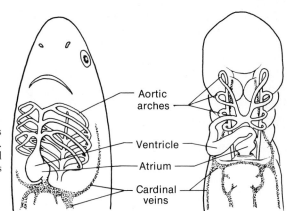

Figure 10.7 Ventral views of the heart and aortic arches of a human embryo (right) and an adult shark (left). Both have a single atrium and single ventricle, several aortic arches, and anterior and posterior cardinal veins emptying into the heart.

pouches regress; the bones which make up each vertebra, and which had been separate as in the most primitive fishes, fuse; a new kidney, the mesonephros, forms and the pronephros disappears or becomes incorporated into other structures; and the atrium becomes divided into right and left chambers. Still later in development the human embryo develops a mammalian, four-chambered heart and a third kidney, the metanephros. During the seventh month of intrauterine development the human embryo, with its coat of embryonic hair and in the relative size of body and limbs, resembles a baby ape more than it resembles an adult human.

Our increasing understanding of physiologic genetics provides us with an explanation of the phenomenon of recapitulation. All chordates have in common a certain number of genes which regulate the processes of early development. As our ancestors evolved from fish, through amphibian and reptilian stages, they accumulated mutations for new characteristics but kept some of the original "fish" genes, which still control early development. Later in development the genes which the human shares with amphibians influence the course of development so that the embryo resembles a frog embryo. Subsequently, some of the genes which we have in common with reptiles come into control. Only after this do most of the peculiarly mammalian genes exert their influence, and these are followed by the action of genes we have in common with other primates. The anthropoid apes, which have the most immediate ancestors in common with us, have the most genes in common with us and their development is identical with ours except for some fine details. A pig or rat, whose ancestors are the same as ours only up to the stage of the primitive placental mammals, has fewer genes in common and has developmental processes that diverge at an earlier time. In general, during development, the general characteristics that distinguish phyla and classes appear before the special characteristics that distinguish genera and, finally, species. Within each phylum, the higher forms pass through a sequence of developmental stages which are similar to those of lower forms but achieve a different final form by adding changes at the end of the original sequence and by altering certain of the earlier embryonic stages they share with the lower forms.

10.8 THE EVIDENCE FROM GENETICS AND CYTOLOGY

For the past several thousand years man has been selecting and breeding animals and plants for his own uses, and a great many varieties, adapted for different purposes, have been established. These results of artificial selection provide striking models of what may be accomplished by natural selection. All of our breeds of dogs have descended from one, or perhaps a very few, species of wild dog or wolf, yet they vary so much in color, size and body proportions that if they occurred in the wild they would undoubtedly be considered separate species. They are all interfertile and are known to come from common ancestors, so they are regarded as varieties of a single species. A comparable range of varieties has been produced by artificial selection in cats, chickens, sheep, cattle and horses. Plant breeders have established by selective breeding a tremendous variety of plants. From the cliff cabbage, which still grows wild in Europe, have come cultivated cabbage, cauliflower, kohlrabi, Brussels sprouts, broccoli and kale.

Geneticists have been able to trace the ancestry of certain modern plants by a combination of cytologic techniques in which the morphology of the chromosomes is compared and by breeding techniques which compare the kinds of genes and their order in particular chromosomes in a series of plants. In this way, the present cultivated tobacco plant, *Nicotiana tabacum,* was shown to have arisen from two species of wild tobacco, and corn was traced to teosinte, a grasslike plant which grows wild in the Andes and Mexico. The cytologic details of the structure of the giant chromosomes of the salivary glands of fruit flies have been of prime importance in unraveling the evolutionary history of the many species of *Drosophila.*

10.9 THE EVIDENCE FROM THE GEOGRAPHIC DISTRIBUTION OF ORGANISMS

In the course of the voyage of the *Beagle,* Darwin was greatly impressed by his observations that the plants and animals of South America and the Galápagos Islands were *not* found everywhere that they could exist if climate and topography were the only factors determining their distribution. Central Africa, for example, has elephants, gorillas, chimpanzees, lions and antelopes, while Brazil, with a similar climate and other environmental conditions, has none of these, but does have prehensile-tailed monkeys, sloths and tapirs, animals not found in Africa. The facts of biogeography, the geographic distribution of plants and animals, were of prime importance in leading both Darwin and Alfred Russel Wallace to the conclusion that organic evolution had occurred by natural selection. The present distribution of organisms, and the sites at which their fossil remains are found, are understandable only on the basis of the evolutionary history of each species.

The **range** of each species is that particular portion of the earth in which it is found. The range of a species may be restricted to a few square miles or less or, as with man, may include almost the entire earth. In general, the ranges of closely related species or subspecies are not identical, nor are they widely separated, but are adjacent and separated by a barrier of some sort. This generalization was stated by David Starr Jordan and is known as **Jordan's rule.** The explanation for this should be clear from the discussion of the role of isolation in species formation. A single species cannot be subdivided as long as interbreeding can occur throughout the whole population. But when some barrier is interposed between two parts of the population so that interbreeding is prevented, the two populations will, in the subsequent course of time, accumulate different gene mutations.

One of the fundamental assumptions of biogeography is that each species of animal or plant originated only once. The place where this occurred is known as its **center of origin.** The center of origin is not a single point, but the range of the population when the new species was formed. From this center of origin each species spreads out, under the pressure of an increasing population, until it is halted by a barrier of some kind: a physical one such as an ocean, mountain or desert, an environmental one such as unfavorable climate, or a biologic barrier such as the absence of food or the presence of other species which prey upon it or compete with it for food or shelter.

As one might expect, regions which have been separated from the rest of the world for a long time, such as South America and Australia, have a unique assemblage of animals and plants. Australia has a population of monotremes and marsupials that is found nowhere else. Australia became separated from Malaya during the Mesozoic, before placental mammals evolved, and its primitive mammals were not eliminated, as were the monotremes and most of the marsupials in the other parts of the world, by the competition of the better adapted placental mammals. The Australian marsupials evolved into a wide variety of forms, each adapted to some particular combination of environmental factors.

The kinds of animals and plants found on oceanic islands resemble, in general, those of the nearest mainland, yet include species found nowhere else. Darwin studied the flora and fauna of the Cape Verde Islands, some 400 miles west of Dakar in Africa, and of the Galápagos Islands, a comparable distance west of Ecuador. On each archipelago the plants and the nonflying animals were indigenous, but those of Cape Verde resembled African species and those of the Galápagos resembled South American ones. Organisms from the neighboring continent migrated or were carried to the island and subsequently evolved into new species. The animals and plants found on oceanic islands are only those that could survive the trip there. There are, for example, no frogs or toads on the Galápagos and no terrestrial mammals, even though conditions would favor their survival.

There are many facts of the present-day distribution of animals and plants which can be explained only by knowledge of their history. Alligators, for example, are found only in the rivers of southeastern United States and in the Yangtse River in China. Sassafras, tulip trees and magnolias are found only in the eastern United States, Japan, and eastern China. The explanation for these curious patterns of distribution lies in the fact that early in the Cenozoic era the northern hemisphere was much flatter than at present and the North American continent was connected with eastern Asia by a land bridge at what is now Bering Strait. The climate of the whole region was much warmer than at present, and fossil evidence shows that alligators, magnolia trees and sassafras were distributed over the entire region. Later in the Cenozoic, as the Rockies increased in height, the western part of North America became much colder and drier. During the Pleistocene the ice sheets moving down from the north met the desert and mountain regions of western North America, and the animals and plants that had lived in that region either became extinct or migrated. In southeastern United States and in eastern China were regions untouched by the glaciations and here the alligators and magnolia trees survived. Because the alligators and magnolias of the two regions have been separated for several million years, they have had the opportunity to accumulate different random mutations. They are, thus, slightly different but closely related species of the same genera.

10.10 THE BIOGEOGRAPHIC REALMS

Careful studies of the distribution of plants and animals over the earth have revealed the existence of six major biogeographic realms, each characterized by the presence of certain unique organisms (Fig. 10.8). These realms were originally defined on the basis of the distribution of mammals, but they have proved to be valid for many other kinds of animals and plants as well. The various parts of each realm may be widely separated and have quite different conditions of climate and topography, but it has been possible, during most geologic eras, for organisms to pass more or less freely from one part to another. In contrast, the six realms are separated from each other by major barriers of sea, desert or mountains.

The **Palearctic** realm includes Europe, Africa north of the Sahara desert, and Asia north of the Himalaya and Nan-Ling mountains, plus Japan, Iceland and the Azores and Cape Verde Islands. The animals indigenous to the Palearctic are moles, deer, oxen, sheep, goats, robins and magpies.

The **Nearctic** realm includes Greenland and North America north of the northern plateau of Mexico. This contains many of the same animals as the Palearctic, plus species of mountain goats, prairie dogs, opossums, skunks, raccoons, bluejays, turkey buzzards and wren-tits found nowhere else. The land bridge connecting North America and Asia at Bering Strait in former geologic times permitted the migration back and forth of many kinds of animals and plants. The flora and fauna of the Palearctic and Nearctic realms are similar in many respects and the two are sometimes combined as the **Holarctic** region.

The **Neotropical** realm consists of South America, Central America, southern Mexico and the islands of the West Indies. Its distinctive fauna includes alpacas, llamas, prehensile-tailed monkeys, blood-sucking bats, sloths, tapirs, anteaters, and a host of bird species—toucans, puff birds, tinamous and others—found nowhere else in the world.

The part of Africa south of the Sahara, plus the island of Madagascar, comprises the **Ethiopian** realm. The gorilla, chimpanzee,

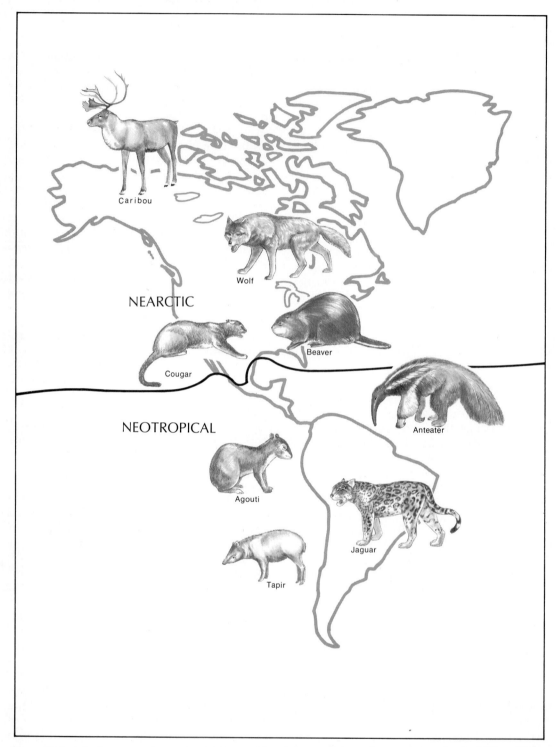

Figure 10.8   The biogeographic realms of the world with some of their characteristic animals. (Villee: Biology, 6th ed.)

(*Continued on opposite page*)

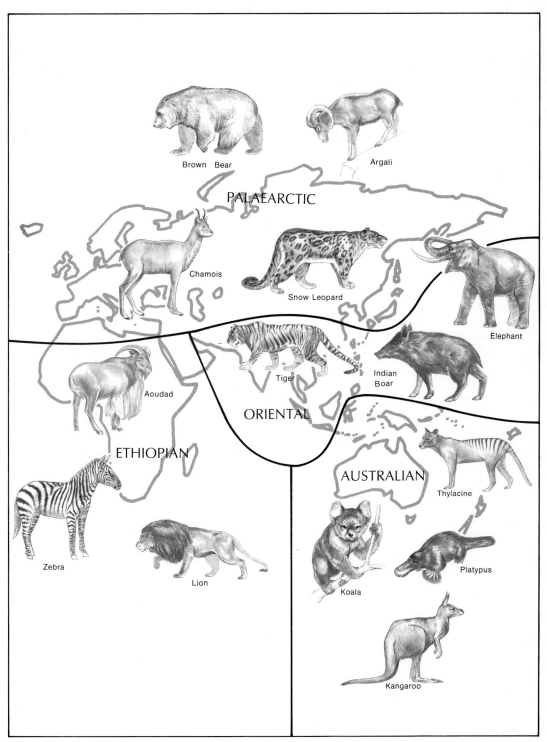

Figure 10.8 *(continued)*

zebra, rhinoceros, hippopotamus, giraffe, aardvark, and many birds, reptiles and fishes live only in this realm.

The **Oriental** realm includes India, Ceylon, southeast Asia, southern China, the Malay peninsula and some of the islands of the East Indies — the Philippines, Borneo, Java and Bali. Some of the animals peculiar to it are the orangutan, black panther, Indian elephant, gibbon and tarsier.

Australia, New Zealand, New Guinea, and the remaining islands of the East Indies, those east of Celebes and Lombok, make up the **Australian** realm. The line separating the Oriental and Australian realms, known as Wallace's Line, separates Bali and Lombok, goes through the straits of Macassar between Borneo and Celebes, and passes east of the Philippines. Although the islands of Bali and Lombok are separated by a channel only 30 kilometers wide, their respective animals and plants are more unlike than are those of England and Japan, almost on the opposite sides of the world from each other. Native to the Australian realm are the duck-billed platypus, echidna, kangaroo, wombat, koala bear, and other marsupials. Its assortment of curious birds includes the cassowary and emu, the lyrebird, cockatoo and bird-of-paradise.

Why certain animals appear in one region yet are excluded from another in which they are well adapted to survive (and in which they flourish when introduced by man) can be explained only by their evolutionary history.

ANNOTATED REFERENCES

Baldwin, E. B.: An Introduction to Comparative Biochemistry. 4th ed. Cambridge, Cambridge University Press, 1964. A classic exposition of some of the biochemical similarities among animals that aid in defining evolutionary relationships.

Berrill, N. J.: The Origin of Vertebrates. London, Oxford University Press, 1955. Contains excellent descriptions of fossil vertebrates and presents the theory that vertebrates evolved from ascidians.

Bryson, V., and H. J. Vogel (Eds.): Evolving Genes and Proteins. New York, Academic Press, 1965. The proceedings of a symposium held at Rutgers University, containing a series of articles which compare the structures of specific proteins and nucleic acids in different species and show how this evidence may be used to deduce evolutionary relationships.

Colbert, E. H.: Evolution of the Vertebrates. New York, John Wiley & Sons, Inc., 1955. A text recommended for its presentation of information regarding vertebrate fossils.

Dodson, E. O.: Evolution: Process and Product. New York, Reinhold Publishing Corp., 1960. A good introductory text describing the general field of evolutionary phenomena.

Jukes, T. H.: Molecules and Evolution. New York, Columbia University Press, 1966. A readable introductory text that relates modern genetics and the evolution of specific molecular structures.

Moore, R. C., C. G. Lalicker and A. G. Fischer: Invertebrate Fossils. New York, McGraw-Hill Book Co., 1952. An excellent source of information regarding extinct invertebrate animals.

Raymond, P. E.: Prehistoric Life. Cambridge, Harvard University Press, 1939. A very readable description of the important vertebrate fossils.

Romer, A. S.: Vertebrate Paleontology. 3rd ed. Chicago, University of Chicago Press, 1966. A comprehensive, well-written text, highly recommended as a source book for further reading.

Shrock, R. R., and W. H. Twenhofel: Principles of Invertebrate Paleontology. New York, McGraw-Hill Book Co., 1953. An excellent text describing the evolution of the invertebrates.

PART TWO

VERTEBRATE FORM AND FUNCTION

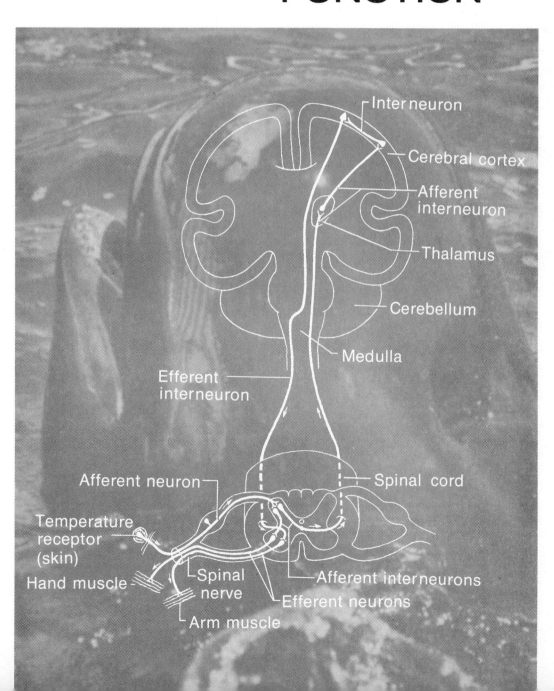

11. Protection, Support and Movement

11.1 THE VERTEBRATES

The vertebrates, or backboned animals, include man and for this reason they have been studied more intensively than other animals. A consideration of their form and function will give you an understanding of many basic aspects of animal organization that can be applied later when you study the different groups of animals.

The ancestral vertebrates were fishes well adapted to life in the water. Three major groups survive today. Most primitive of contemporary species are the jawless lampreys and other cyclostomes belonging to the class **Agnatha** (Fig. 11.1). Cyclostomes are the direct descendants of an ancient and now extinct group of armored fishes known as the **ostracoderms.** Two other lines of evolution lead from ostracoderms to carti- laginous and bony fishes. Fishes in each line of evolution became more active and aggressive, evolving jaws and improved methods of locomotion. Sharks, skates and similar fishes have an internal skeleton composed of cartilage, and hence are placed in the class **Chondrichthyes,** which means cartilaginous fishes. Remaining fishes, in- cluding the sturgeon, trout, minnows, and lungfishes, have at least a partially ossified internal skeleton and are placed in the class

Osteichthyes, i.e., bony fishes. It is from certain primitive, fresh-water members of the latter group that terrestrial vertebrates, or **tetrapods,** evolved.

Members of the class **Amphibia,** which includes our present day frogs and sala- manders, were the first tetrapods. They evolved legs and certain other adaptations for land life, but they must live in wet or damp habitats because they retain many fish- like attributes, including only a rudimentary ability to conserve body water. Turtles, lizards, snakes, crocodiles, and other mem- bers of the class **Reptilia** can conserve body water, and they were the first group of vertebrates to penetrate extensively the terrestrial environment. A few of them have subsequently readapted to an aquatic mode of life. Reptiles resemble more primitive vertebrates in being cold-blooded, or **ecto- thermic;** their body temperature is linked closely to the ambient temperature, and in cold weather they are sluggish or go into a dormant state. Birds and mammals (classes **Aves** and **Mammalia**) evolved from reptiles independently of one another, and each group became warm-blooded, or **endo- thermic.** Most of their body heat is gen- erated internally by metabolic processes. They can maintain their body temperature at a rather high and constant level so that

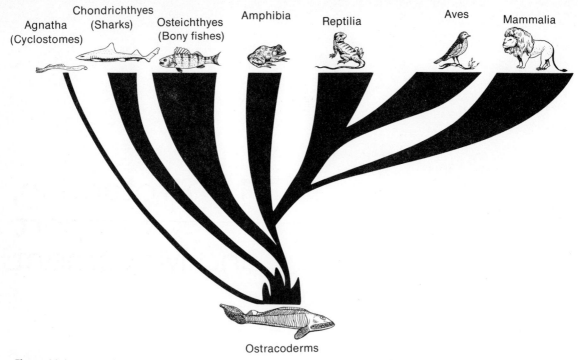

Figure 11.1 An evolutionary tree to show the interrelationships between the classes of living vertebrates. Ostracoderms, now extinct, belong to the class Agnatha.

they can be active under a wide range of environmental conditions.

Mammals include man, rats, cats, pigs and other endothermic vertebrates having hair and nourishing their young with milk produced by mammary glands. In Chapters 11 through 17, emphasis will be placed on the mammalian organ systems, and on those evolutionary transformations among lower vertebrates that are helpful in understanding the mammalian condition.

11.2 HOMOLOGY AND ANALOGY

In order to determine the transformations that have occurred during evolution, one must distinguish between various types of resemblances and have a clear understanding of the implications of each. There are two broad categories of resemblance: homology and analogy. **Homologous organs** are organs in different animals which share a structural similarity. By definition this similarity is due to an inheritance from a common ancestor. Homologous organs may resemble each other closely, as the bones of the forelimb of a man and a monkey, or

they may be superficially quite unlike, as the scales of a snake and the feathers of a bird. Whether superficially alike or unlike, homologous organs share certain deep-seated, basic similarities in structure that derive from their common ancestry. This includes such features as positional relationships to other parts of the body, similarity in vascular and nerve supply, and a common mode of embryonic development. The scales of a snake and the feathers of a bird resemble each other closely in their early stages of embryonic development. It should also be clear from these examples that homologous organs may, or may not, have similar functions.

Analogous organs are organs in different animals with a functional similarity that is in no way related to a common ancestry; that is to say, analogous organs are non-homologous organs with a similar function. Often analogous organs have little gross structural resemblance to each other; for example, the gills of a crayfish and the lungs of a man. Other analogous organs, such as the wings of insects and vertebrates, may have a superficial structural resemblance (Fig. 11.2). Their common function, flight,

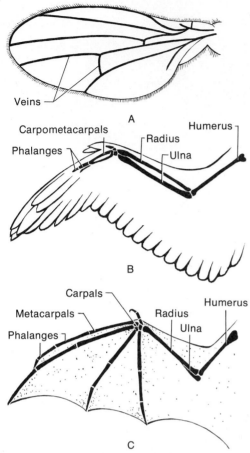

Figure 11.2 Wings of a (A) fly, (B) bird and (C) bat. The bones of the wings of the bird and bat are homologous to each other. The fly's wing is similar to the vertebrate wings in function, but it has only a superficial structural resemblance to them. It is, therefore, analogous and homoplastic to them.

has only one structural solution, the evolution of some sort of an airfoil. Analogous organs with a structural resemblance are said to be **homoplastic.**

Evolutionary relationships must be traced through a careful identification of homologous organs, for, by definition, they imply descent from a common ancestor. A potential source of error, however, is to confuse homoplastic organs, which have evolved independently, with homologous ones. Usually, homoplastic resemblances are superficial and not basic. In the wing example, it is clear that the bones supporting the wings of the bird and the bat have a basic resemblance, i.e., are homologous, whereas the veins supporting the insect wing are an entirely different sort of structure. The

implication is that the bird and the bat had a common ancestor in which the pattern for the bones in the pectoral appendage had already been established; the insect evolved a wing quite independently and from different materials. Another source of confusion is that a pair of organs may be homologous at one level of comparison and homoplastic at another. Although supported by homologous bones, the flying surface of the bird wing is composed of feathers, whereas the bat's wing is composed of a skin membrane. At this level of comparison the two wings are homoplastic. All of this suggests that birds and bats had a remote common ancestor with a certain arrangement of bones in the pectoral appendage, but that the wings of birds and bats evolved independently in the two groups; bats did not evolve from birds.

11.3 THE INTEGUMENT

The vertebrate skin, or **integument,** is the outermost layer of the body and separates the organism from its external environment. It helps to maintain a constant internal environment and protects the body against a variety of mechanical and chemical injuries. Yet the skin does not completely isolate the organism from its environment, since many sensory stimuli are received by the skin, and some exchange of gases, water and excretory products may occur through it. In addition, a variety of bony plates, scales, feathers, hair, pigment cells and glands develop from the skin and serve a variety of purposes. The skin is truly a "jack-of-all-trades."

In all vertebrates, the skin is made up of an outer stratified epithelium (the **epidermis**) and a deeper, rather thick layer of dense connective tissue (the **dermis**). New cells are continually produced by divisions in the basal layer of the epidermis, the **stratum germinativum.** As they are pushed toward the surface they flatten, accumulate a horny protein known as **keratin,** and gradually die. Since keratin is insoluble in water, these layers of dead cells form a horny **stratum corneum** on the skin surface that reduces water exchanges. The stratum corneum is not very thick in fishes or amphibians (Fig. 11.3), but it becomes much thicker and more effective in reptiles and mammals. The outer cells of the stratum

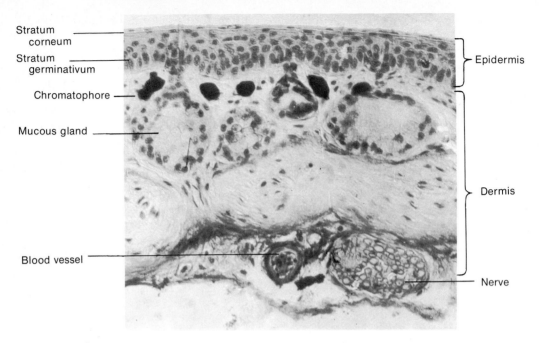

Stratum corneum

Stratum germinativum

Chromatophore

Mucous gland

Blood vessel

Epidermis

Dermis

Nerve

Figure 11.3   A photomicrograph of a vertical section through the skin of a frog.

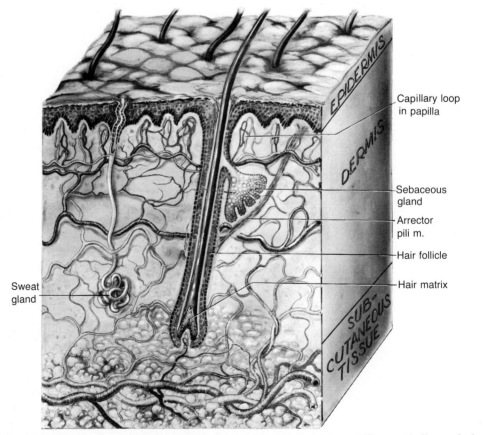

EPIDERMIS

DERMIS

SUB-CUTANEOUS TISSUE

Capillary loop in papilla

Sebaceous gland

Arrector pili m.

Hair follicle

Hair matrix

Sweat gland

Figure 11.4   Diagrammatic vertical section through human skin. (Modified from Pillsbury, Shelley and Kligman; Manual of Cutaneous Medicine, 1961.)

corneum are eventually lost, sometimes as an extensive sheet but often as small groups of cells. Dandruff is a familiar example. Well-defined, intermediate layers can also be recognized where the epidermis is especially thick, as on the palm of the hand and the sole of the foot.

The dermis is composed of fibrous connective tissue; bone may develop in it in certain regions. The dermis is richly supplied with blood vessels, some of which lie close to the surface and enter papilla-like projections of the dermis that extend into the base of the epidermis (Fig. 11.4). In addition to their nutritive function, these vessels in mammals play an important role in thermoregulation. An increased blood flow through them dissipates heat; a reduced flow conserves heat. Nerves and microscopic sense organs that receive stimuli of touch, pressure, temperature and pain are abundant in the dermis, but only a few naked nerve endings penetrate the epidermis. Fat may accumulate in special fat cells, adipocytes, in the deeper parts of the dermis and in the subcutaneous tissue. The fat serves as a reserve supply of food, as a thermal insulator, and as a cushion against mechanical injury. The blubber of whales serves as a good insulation in the aquatic environment. Hair is not an efficient insulator in aquatic animals, for its thermal insulation depends on air trapped within it, and it has been lost on most of the body surface of adult whales.

Though the skin itself is relatively simple, its derivatives are numerous and complex. These may be grouped into bony structures, horny structures, glands and pigment. The bony structures develop within the dermis, though parts of them may become exposed if the overlying epidermis wears off. Thick **bony scales** and plates were prominent in certain extinct fishes and have been retained in reduced form in most groups of living fishes (Fig. 11.5A). Certain of the dermal plates in the head region early in evolution became associated with the skull and pectoral girdle, and these have been retained by later vertebrates as integral parts of the skeleton. Most of the primitive bony scales have been lost in tetrapods, but the dermis retains the ability to form bone and becomes heavily ossified in certain species. The shell of a turtle is composed of dermal plates covered by large horny laminae; a comparable condition is found in the skin of certain lizards and crocodiles and in the shell of the armadillo. The **antlers** of deer (Fig. 11.6) are also composed of dermal bone. During its development, the antler is covered by skin, the **velvet,** but this sloughs off when the antler is fully formed. Antlers branch, are shed annually and, with the exception of the reindeer and caribou, are found only on males. The **horns** of sheep and cattle, in

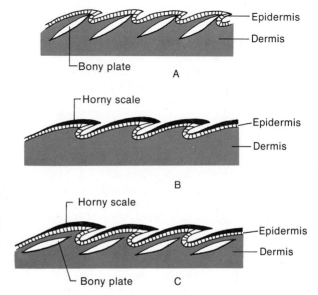

Figure 11.5 Vertical sections through the skin of vertebrates to show the relationship of the various types of scales. *A,* Bony scales of a fish; *B,* horny scales of a snake; *C,* horny scales and bony plates as in the skin of certain lizards, crocodiles and turtles.

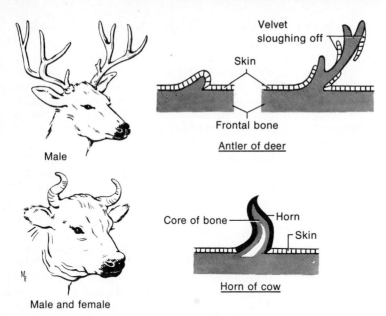

Figure 11.6   A diagram to show the differences between antlers (deer) and horns (cow). Antlers are annual growths that are shed in the winter; horns are permanent outgrowths.

contrast, do not branch, are not shed and occur in both sexes. These horns have a core of bone covered by a highly cornified skin.

Horny skin derivatives develop by the accumulation of keratin in the cells of the epidermis. Reptiles have a covering of **horny scales** that reduce water loss through the skin (Fig. 11.5B). As the animal increases in size, the horny scales are periodically shed and newly formed ones are exposed beneath them. Bony scales, in contrast, are not shed but increase in size by the addition of new bone. Except for their retention in such regions as the feet of birds and the tails of certain mammals, horny scales are not present in most birds and mammals, though a prominent stratum corneum persists.

Feathers are believed to be modified horny scales, but the **hairs** of mammals are regarded as a different kind of horny skin derivative. A hair lies within a **hair follicle** (Fig. 11.4), which is composed of a tubular invagination of the epidermis supported by surrounding fibers of the dermis. A **hair papilla,** containing blood vessels and nerves, protrudes into the base of the follicle and nourishes the adjacent epithelial cells. These proliferate rapidly and add to the base of the hair, which extends up through the follicle as a column of keratinized cells. A small smooth muscle, the **arrector pili,** is associated with each follicle. When temperatures fall, these muscles are stimulated to contract and pull the hair into a more erect

position, thereby increasing the depth of the hair layer and its effectiveness in insulation. They also depress the skin between the hairs, leaving little hillocks where the hairs emerge. We are familiar with this as "goose flesh." In addition to their thermal function, hairs are effective mechanical amplifiers of the sense of touch; a slight movement of a hair can stimulate nerve ends wrapped around its base. This property is extremely well developed in the tactile whiskers of cats and many other mammals. Other specializations of hair include the quills of a porcupine and even the "horn" of a rhinoceros, which lacks a core of bone, appears to be a clump of specialized hairs.

Other horny derivatives of the integument include **claws,** which first appear in reptiles and may be modified as **nails** or **hoofs** in certain mammals, the **whalebone plates** with which toothless whales filter plankton from the water flowing through their mouths, and the covering of the horns of sheep and cattle.

Individual mucus-secreting cells are common in the epidermis of fishes, and multicellular **mucous glands** are abundant in amphibian skin (Fig. 11.3). Many fishes and amphibians also have a few cutaneous **poison glands** that produce a secretion that is irritating and can be dangerous to their predators. Reptiles have lost the mucous and poison glands, and only a few glands, chiefly scent glands, are present in their dry, horny skin. This paucity persists in birds, but glands have again become abundant in mammalian skin. Alveolar-shaped **sebaceous glands,** epithelial outgrowths from the hair follicles (Fig. 11.4), discharge their oily secretions onto the hairs. Coiled, tubular **sweat glands** are also abundant in parts of the skin of most mammals. A little urea and some salts are eliminated in the sweat, but sweat glands are particularly important in secreting water whose evaporation cools the body surface. Contractions of myoepithelial cells, which are located peripheral to the secretory cells, aid in expelling sweat. The **mammary glands** of female mammals are integumentary glands that produce milk during the period of lactation; myoepithelial cells are also present in them. Musk and other **scent glands,** serving for sexual recognition, are common in many mammals, although they do not occur in man.

Pigment within the skin, the degree of vascularity, and sometimes the presence of refractive granules or surface striations interact to produce the brilliant colors of many vertebrates. Some color patterns conceal the animal from predators, or enable an animal to lie unseen as it awaits its prey, but others are used for species recognition or as warning signals to frighten another animal or establish territory. Lower vertebrates, such as the frog (Fig. 11.3), may have several kinds of pigment contained in stellate-shaped cells known as **chromatophores.** Changes in general color tone are effected by the migration of a dark pigment known as **melanin** within the cells. When the skin of a frog darkens, melanin streams into the processes of the cells; when it becomes pale, the pigment concentrates near the center of the cell (Fig. 17.8). Pigment migration is controlled by nerves in some vertebrates and by hormones in others. Chromatophores are rare in mammals, but melanin is present within and between the cells of the epidermis. Some melanin is present in the skin of all human beings (except albinos, p. 197), but it is especially abundant in the skin of Blacks.

11.4 THE SKELETON

Nature and Parts of the Vertebrate Skeleton. Organisms must remain small and

slow moving unless they have a skeleton for support and to serve as levers on which muscles can act. All vertebrates have a skeleton that provides for this and that encloses and protects some of the more delicate internal organs. Certain of the central cavities of the bones of higher vertebrates contain red bone marrow in which red blood cells and certain of the white cells are produced. The vertebrate skeleton is basically an internal skeleton, for it develops within the skin or in deeper body tissues. None of it is a secretion on the body surface, as is the exoskeleton of certain invertebrates.

The skeleton can be subdivided into a **dermal skeleton,** consisting of the bony scales and plates mentioned earlier in this chapter, and an **endoskeleton,** situated more deeply. During early embryonic development the endoskeleton is composed of cartilage, but the cartilage is replaced by bone in most adults. This bone is called **cartilage replacement bone** to distinguish it from the **dermal bone** that develops in more superficial parts of the body without any cartilaginous precursor. These types of bone differ only in their mode of development; they are the same histologically.

The endoskeleton and its associated dermal bones can be further subdivided into somatic and visceral skeletons:

Somatic skeleton (skeleton of the body wall)
    Axial skeleton (vertebral column, ribs, sternum
      and most of the skull)
    Appendicular skeleton (girdles and limb
      bones)
Visceral skeleton (skeleton of the pharyngeal
    wall, primitively associated with the gills)

The Fish Skeleton. The parts of the skeleton can be seen more clearly in a dogfish (a small shark) than in terrestrial vertebrates (Fig. 11.7). The dogfish skeleton is typical of the skeleton of primitive vertebrates, except that the skeleton is entirely cartilaginous. The failure of the dogfish's skeleton to ossify is believed to represent the retention of an embryonic condition rather than a primitive adult condition. The vertebral column is composed of vertebrae, each of which has a biconcave **centrum** that develops around and largely replaces the embryonic notochord. Dorsal to each centrum is a **vertebral arch** surrounding the **vertebral canal** in which the spinal cord lies, short **ribs** attach to the vertebrae, a sternum

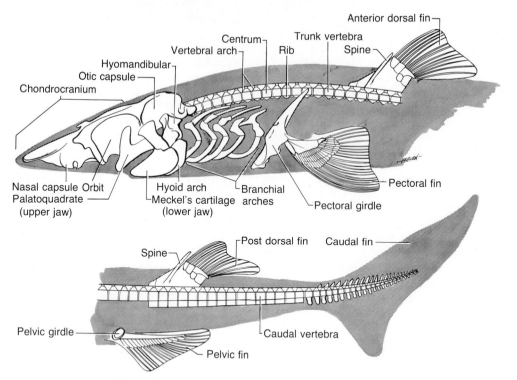

Figure 11.7   A lateral view of the skeleton of a dogfish.

is absent, and the individual vertebrae are rather loosely held together. A strong vertebral support is not necessary in the aquatic environment.

Most of the skull of the dogfish is an odd-shaped box of cartilage encasing the brain and major sense organs. This belongs to the axial skeleton and is known as the **chondrocranium.** It forms the core of the skull of all vertebrates. Other basic components of a vertebrate skull include the anterior arches of the visceral skeleton and dermal bones that encase the chondrocranium and anterior visceral arches. These dermal bones have been lost during the evolution of cartilaginous fishes, but they were present in the fishes ancestral to tetrapods.

The visceral skeleton consists of seven pairs of >-shaped visceral arches. The arches are hinged at the apex of the >; they are interconnected ventrally, but are free dorsally. Each arch lies in the wall of the pharynx and supports gills in very primitive vertebrates. In jawed vertebrates the first or **mandibular arch** becomes enlarged and, together with associated dermal bones, forms the upper and lower jaws. The entire jaw

in the dogfish is derived from the mandibular arch; there are no surrounding dermal bones. The second or **hyoid arch** has moved forward in the dogfish and helps to support the jaws. Its dorsal portion, known as the **hyomandibular,** extends as a prop from the **otic capsule** (the part of the chondrocranium housing the inner ear) to the angle of the jaw. The gill slit that in primitive fish lay between the mandibular and hyoid arches is reduced to a spiracle. The third to seventh visceral arches are known as **branchial arches;** they support the gills and complete gill slits lie between them.

The appendicular skeleton is very simple in the dogfish. A U-shaped bar of cartilage, the **pectoral girdle,** lies in the body wall posterior to the gill region and supports the **pectoral fins.** The **pelvic girdle** is a transverse bar of cartilage in the ventral body wall anterior to the termination of the digestive tract. It supports the **pelvic fins** but is not connected with the vertebral column.

The Tetrapod Skeleton. The numerous changes that have occurred in the evolution of the skeleton from primitive fishes are correlated with the shift from water to land, and with the greater activity of the tetrapod. They can be understood by a study of the frog and human skeletons (Figs. 11.8 and 11.9). The vertebral column must support the weight of the body in all tetrapods, and it has become much stronger. It is thoroughly ossified, and the individual vertebrae are strongly united by overlapping **articular processes** borne on the vertebral arches. A well-developed **vertebral spine** extending dorsally from the arch, and lateral **transverse processes** serve for the attachment of ligaments and muscles. Ribs articulate on the centrum and transverse processes.

Correlated with changes in the method of locomotion and the independent movement of various parts of the body, there is more regional differentiation of the vertebral column. All tetrapods have at least one neck or **cervical vertebra,** the **atlas,** that articulates with the skull; a series of **trunk vertebrae;** one or more **sacral vertebrae** that articulate with the pelvic girdle; and tail or **caudal vertebrae.** Amphibians have a single cervical vertebra and only one sacral vertebra. Frogs, which have a jumping method of locomotion, have a short trunk in which short ribs have fused onto the transverse processes, and two caudal vertebrae have fused together to form a solid **urostyle** on which certain jumping muscles attach. Man and other mammals have seven cervical vertebrae of which the first two,

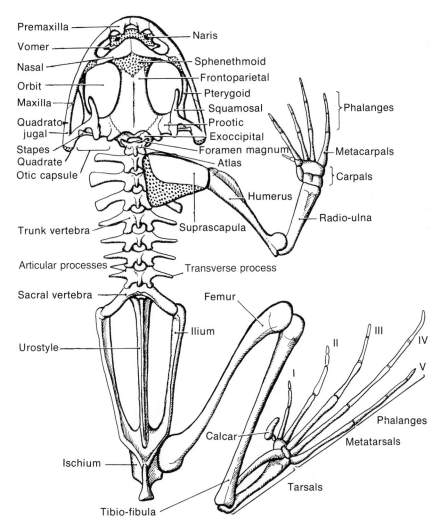

Figure 11.8 A dorsal view of the frog's skeleton. Major cartilaginous areas are stippled. Roman numerals refer to digit numbers. (After Parker and Haswell.)

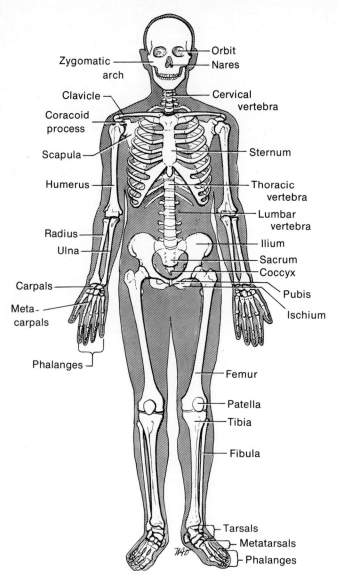

Figure 11.9 A ventral, or anterior, view of the human skeleton.

the atlas and **axis,** are modified to permit extensive movement of the head. The skull rocks up and down at the joint between the skull and atlas; turning movements occur between the atlas and axis. The vertebral column of the mammalian trunk is differentiated into thoracic and lumbar regions. Man has 12 **thoracic** and six **lumbar vertebrae.** Only the thoracic vertebrae bear distinct **ribs,** most of which connect, via the costal cartilages, with the ventral breast bone, or **sternum.** Rudimentary ribs, present in the other regions during embryonic development, fuse onto the transverse processes. The number of sacral vertebrae has increased as the tetrapods have evolved more effective terrestrial locomotion. Typically, amphibians have one, reptiles two and mammals three. The greater number in man (five) is correlated with the additional problems of support inherent in a bipedal gait. Man has no external tail, and the four caudal vertebrae that remain are fused together to form an internal **coccyx** to which certain anal muscles attach. Representative human vertebrae are shown in Figure 11.10.

The skull of a frog or man (Figs. 11.8 and 11.9) can be divided into a portion housing the brain and encasing the inner ear, the **cranium,** and the bones forming the jaws and encasing the eyes and nose, the **facial skeleton.** The brain and cranium are very large in mammals. The eyes are lodged in sockets known as **orbits.** A **temporal fossa**

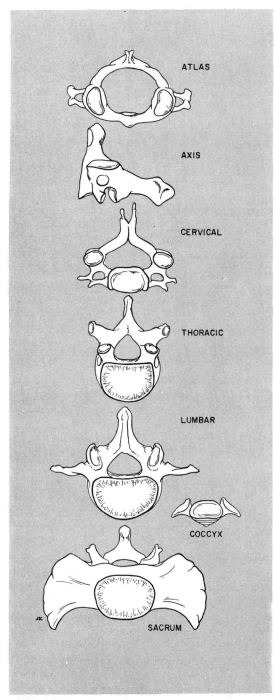

ATLAS

AXIS

CERVICAL

THORACIC

LUMBAR

COCCYX

SACRUM

Figure 11.10 Types of mammalian vertebrae as represented by those of man. The atlas is seen in lateral view; the others are viewed from the anterior end. (From Villee: Biology, 6th ed.)

lies posterior to each orbit in mammals and contains certain of the powerful jaw muscles. It is bounded laterally and ventrally by a handle-like bar of bone, the **zygomatic arch,** from which other muscles extend to the lower jaw. External nostrils, or **nares,** lead to the paired nasal cavities and internal nostrils, or **choanae,** lead from these cavities. The choanae of frogs open into the roof of the front part of the mouth cavity, but in mammals a bony **hard palate** extends the nasal cavities more posteriorly so the choanae open into the pharynx. This permits a mammal to chew and manipulate food in the mouth while continuing to breathe; breathing need be stopped only momentarily when the food is swallowed. An endotherm needs more food and oxygen than an ectotherm, since it must be able to sustain a high rate of metabolism and generate enough heat to maintain a constant body temperature in a cold environment. An **external acoustic meatus** leads to the eardrum, which is sunk beneath the body surface in mammals. There are many foramina for blood vessels and nerves; most conspicuous is the large **foramen magnum** through which the spinal cord enters the skull.

Many of the individual bones forming the skull of the frog and man are shown in Figures 11.8 and 11.11. As the brain grew larger during the course of mammalian evolution, the cartilage replacement bones of the chondrocranium no longer completely encased it. They form a ring of bone around the foramen magnum (the **occipital bone**), encase the inner ear (part of the **temporal bone**), and form the floor of the cranium. The sides and roof of the cranium are completed by dermal bones such as the **frontal** and **parietals** and by a portion of the mandibular arch known as the **alisphenoid.** The last is a cartilage replacement bone.

Of particular importance are changes in the jaws and in the small bones of the inner ear that transmit pressure waves from the eardrum to the inner ear. The jaws of frogs and most vertebrates are formed of dermal bones that encase the cartilaginous mandibular arch, but the jaw joint lies between the posterior ends of the rods forming the arch, that is between the dorsal palatoquadrate and ventral Meckel's cartilage. Usually a **quadrate bone** ossifies from the posterior end of the palatoquadrate

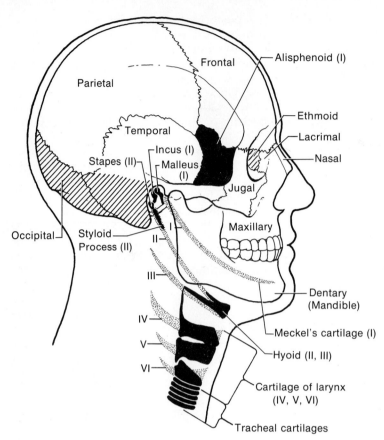

Figure 11.11 Components of the human skull, hyoid and larynx. Dermal bones have been left plain, chondrocranial derivatives are hatched, those parts of the embryonic visceral skeleton that disappear are stippled, parts of the visceral arches that persist are shown in black. Roman numerals refer to visceral arches and their derivatives. (Modified after Neal and Rand.)

and an **articular bone** from the end of Meckel's cartilage. Recall that the dorsal part of the hyoid arch (the hyomandibular) helps to suspend the palatoquadrate in the dogfish (Fig. 11.7). In most tetrapods this element has been transformed into the **stapes,** a slender rod of bone extending from the eardrum to the part of the skull housing the inner ear (Fig. 11.8), but it still retains a connection with the quadrate. During the evolution of mammals the jaw mechanism became stronger and the bite more powerful. The dentary, a dermal bone of the lower jaw, enlarged and contacted the squamosal bone (part of the mammalian temporal bone) of the upper part of the skull (Fig. 11.11). A new joint evolved between these bones, and the original jaw joint, located just posterior to the new one, became redundant. The articular and quadrate, which had become quite small, were overspread by the eardrum and incorporated in the middle ear as the outer two auditory ossicles, **malleus** and **incus** respectively. The incus of mammals, like the quadrate

from which it evolved, connects with the stapes. Mammals have three auditory ossicles which form a pressure amplifying leverage system (Chap. 15). It is of interest to observe that the auditory ossicles have the same relationship to each other as their homologues in fish.

The ventral part of the hyoid arch, together with the remains of the third visceral arch, form the **hyoid bone** (a sling for the support of the tongue) and the **styloid process** of the skull, to which the hyoid is connected by a ligament. With the loss of gills in tetrapods, the remaining visceral arches have become greatly reduced, but parts of them form the cartilages of the larynx.

Although the appendicular skeleton of the dogfish is quite different from that of terrestrial vertebrates, there is a close resemblance between the appendicular skeleton of crossopterygian fishes that gave rise to terrestrial vertebrates and the skeleton of terrestrial vertebrates (Fig. 11.12). The humerus of the frog or human arm and the

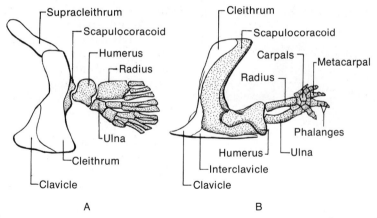

Figure 11.12   Lateral views of the appendicular skeleton of a crossopterygian (*A*) and ancestral amphibian (*B*) to show the changes that occurred in the transition from water to land. Dermal bones have been left plain, cartilage replacement bones are stippled. (*A* modified after Gregory; *B* after Romer.)

femur of the leg (Figs. 11.8 and 11.9) represent the single proximal bone of the crossopterygian fin; the **radius** and **ulna,** or **tibia** and **fibula,** the next two bones. These bones are fused in frogs as an adaptation for jumping. The **carpals** or **tarsals, metatarsals** or **metacarpals,** and **phalanges** of the hand or foot are homologous with the more peripheral elements of the crossopterygian fin. We tetrapods have a single bone in the proximal part of the appendage followed by two bones in the second part because this pattern was established by our piscine ancestors.

The girdles of tetrapods are necessarily stronger than those of fish. The pectoral girdle is bound onto the body by muscles, but the pelvic girdle extends dorsally and is firmly attached to the vertebral column. A **pubis, ischium** and **ilium** are present on each side of the pelvic girdle, though all have fused together in the adult. Our pectoral girdle includes a **scapula,** a **coracoid process,** which is a distinct bone in most lower tetrapods, and a **clavicle.** The clavicle is the only remnant of a series of dermal bones that are primitively associated with the girdle. All other girdle bones are cartilage replacement bones.

11.5   JOINTS

Where bones come together they do not simply abut one upon the other but are united by **joints** of varying degrees of complexity. The structure of joints is correlated with the functions the bones perform and the degree of movement between them. Three broad categories can be recognized: immovable joints, slightly movable joints and freely movable joints.

Bones of the cranium (e.g., parietal and frontal, Fig. 11.11) provide a solid protective housing for the brain and are united by immovable joints known as **sutures.** The connective tissue **periosteum,** which covers the surfaces of all bones, forms a septum that dips between the bones. The bones can grow at their periphery, but they remain firmly articulated with each other at all times.

The vertebrae must form a firm supporting beam, yet a beam that is capable of limited bending. Most vertebrae are united by slightly movable **symphyses.** An elastic pad of fibrocartilage separates them and they are held together by **ligaments**—tough bands of connective tissue that extend across the articulation.

Since the appendages are used as flexible struts in locomotion, a great deal of motion occurs between most of their bones. We find freely movable articulations of several kinds: ball and socket joints between the appendages and the girdles, a hinge joint at the knee, a combination of a hinge and pivot joint at the elbow, etc. All are often called **synovial joints,** for they contain a synovial cavity that permits free movement. Structurally they are quite complex (Fig. 11.13). The adjacent bones are usually shaped so that a protuberance of one fits into a depression in the other. This, of

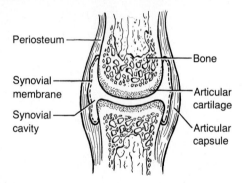

Figure 11.13 A diagrammatic section through a synovial joint. Ligaments, not shown, cross the joint peripheral to the articular capsule and provide additional strength to the joint in certain areas.

course, makes the bones less likely to disarticulate during movement. **Articular cartilages** cover the ends of the bones and provide a smooth, somewhat elastic surface. A **synovial membrane** lines the synovial cavity and produces, probably by filtration from blood vessels within it, a lymphlike synovial fluid which lubricates the articulation. Stiffness of joints in old age results in part from a reduction in the amount of this fluid. The periosteum covering the bones extends across the joint as an **articular capsule**, ligaments strengthen the joint, and it is also supported by the tonus of surrounding muscles.

11.6 MUSCLES

The movement of the vertebrate body and its parts depends upon the contraction of muscles, and muscles play an important role in supporting the body of terrestrial vertebrates. The nature of muscle contraction and

the source of the energy required have been considered earlier. At this time we will be concerned with certain aspects of the evolution of the muscular system and the biomechanics of muscle action.

Histologically, muscles may be classified as smooth, cardiac and skeletal (Fig. 3.18). In tracing their evolution it is more convenient to divide them into **somatic muscles,** associated with the body wall and appendages, and **visceral muscles,** associated with the pharynx and other parts of the gut tube. This grouping parallels the major subdivisions of the skeletal system. Somatic muscles are striated and under voluntary control. Most of the visceral muscles are smooth and involuntary; however, the visceral muscles associated with the visceral arches, called branchial muscles, are striated and under voluntary control.

Evolution of Somatic Muscles. Most of the somatic musculature of fishes consists of segmental **myomeres** (Fig. 11.14). This is an effective arrangement for bringing about the lateral undulations of the trunk and tail that are responsible for locomotion. The muscles of the paired fins are very simple and consist of little more than a single dorsal **extensor,** or **abductor,** that pulls the fin up and caudally, and a ventral **flexor,** or **adductor,** that pulls the fin down and anteriorly.

The transition from water to land entailed major changes in the somatic muscles. The appendages became increasingly important in locomotion, and movements of the trunk and tail less important. The segmental nature of the trunk muscles is largely lost, although traces can be seen in the **rectus abdominis** that extends longitudinally on each side of the midventral line (Figs. 11.15

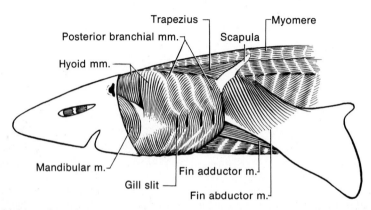

Figure 11.14 A lateral view of the anterior muscles of a dogfish. (Modified after Howell.)

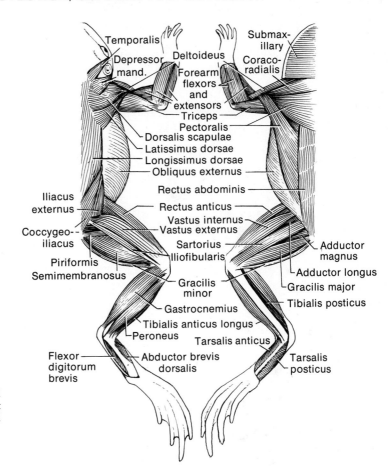

Temporalis
Depressor mand.
Deltoideus
Submaxillary
Coraco-radialis
Forearm flexors and extensors
Triceps
Pectoralis
Dorsalis scapulae
Latissimus dorsae
Longissimus dorsae
Obliquus externus
Rectus abdominis
Iliacus externus
Rectus anticus
Vastus internus
Vastus externus
Coccygeo-iliacus
Sartorius
Iliofibularis
Adductor magnus
Piriformis
Semimembranosus
Adductor longus
Gracilis major
Gracilis minor
Tibialis posticus
Gastrocnemius
Tibialis anticus longus
Peroneus
Tarsalis anticus
Flexor digitorum brevis
Abductor brevis dorsalis
Tarsalis posticus

Figure 11.15 Superficial skeletal muscles of the frog in a dorsal (left side of figure) view and a ventral (right side) view.

and 11.16). Back muscles remain powerful for they play an important role in supporting the vertebral column and body, but trunk muscles on the flanks form thin sheets, such as the **external oblique,** that support the abdominal viscera and assist in breathing movements. Some trunk muscles, including the **serratus anterior,** attach onto the pectoral girdle and in a quadruped help transfer body weight to the girdle and appendage. The primitive single fin extensor and flexor became divided into many components, and these became larger and more powerful. Despite the complexity of tetrapod appendicular muscles, it is possible to divide them into a dorsal group that evolved from the fish extensor and a ventral group derived from the flexor. Our **latissimus dorsi** and **triceps** (Fig. 11.16), for example, are dorsal appendicular muscles, whereas the **pectoralis** and **biceps** are ventral appendicular muscles.

Evolution of Branchial Muscles. Branchial muscles are well developed in fishes

and are grouped according to the visceral arches with which they are associated (Fig. 11.14): **mandibular muscles** and certain of the **hyoid muscles** are concerned with jaw movements; most of the rest, with respiratory movements of the gill apparatus. Branchial muscles obviously become less important in tetrapods, for the gills are lost and the visceral arches are reduced. Nevertheless, certain ones are retained. Those of the mandibular arch remain as the **temporalis, masseter** and other jaw muscles (Figs. 11.15 and 11.16). Most of those of the hyoid arch move to a superficial position and become the **facial muscles** that are responsible for facial expressions. Those of the remaining arches are associated with the pharynx and larynx and some, e.g., the **sternocleidomastoid** and **trapezius,** become important muscles associated with the pectoral girdle.

Biomechanics of Muscles. As explained in Section 5.7, muscles **arise** from one bone and **insert** upon another, which is moved

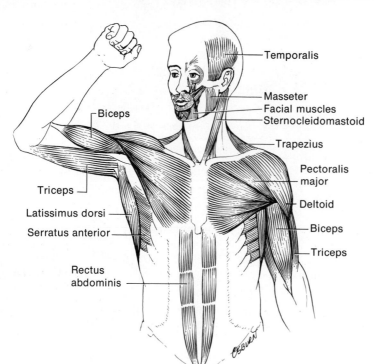

Figure 11.16 An anterior view of certain of the superficial muscles of man.

as the muscle contracts. They are arranged in antagonistic sets so that contraction of one muscle pulls the bone in one direction and contraction of the other pulls it in the opposite direction. **Flexion** describes the bending of the trunk toward the midventral line, and the movement of a distal segment of an appendage toward a more proximal one with a consequent diminishing of the angle between them, e.g., movement of the forearm toward the upper arm. **Extension** is the opposite movement. **Abduction** describes the movement of a part toward a point of reference; **adduction**, movement away from the reference. The midventral line of the body is taken as the reference for limb abduction and adduction. The forward movement of the entire limb at the shoulder or hip is called **protraction**; the opposite movement is **retraction.** Origins, insertions, and actions of the major muscles of the frog are presented in Table 11.1 as an example of the complexity of the muscular system of a tetrapod. Most of these muscles are illustrated in Figure 11.15.

Many parts of the skeleton represent levers acted on by muscles. The muscles usually attach to the bones in such a way

that the fulcrum is at one end of the lever, and the muscle attachment is nearer the fulcrum than the point at which the lever exerts its force (Fig. 11.17). Consequently the power arm of the lever (the perpendicular distance from the fulcrum to the line of action of the muscle) is less than the force arm (the perpendicular distance from the fulcrum to the point of application of the force). Muscles must contract with power greater than the force to be overcome because muscle power × the length of the power arm = the force generated × the length of the force arm. Such a lever is mechanically inefficient from the point of view of power-force relationships, but it provides for compactness and relatively high speed of movement because a short contraction of a muscle brings about a more extensive movement at the end of the lever. Changes in the relative lengths of the two lever arms can adapt an organism for more power or greater speed. The front leg of an armadillo, for example, is adapted for digging, and it has a relatively long humerus and short lower leg. This gets the insertions of muscles acting on the humerus farther away from the fulcrum, thereby increasing the length of the power arm relative to the length of the entire limb, which represents the force arm. The front leg of a horse is adapted for running, and speed is increased by a short humerus, which brings the insertion of humeral muscles close to the fulcrum, and by a long forearm and hand.

TABLE 11.1 MAJOR ATTACHMENTS AND ACTIONS OF IMPORTANT FROG MUSCLES

Muscle	Origin	Insertion	Action
Jaw Muscles			
Temporalis	Dorsal surface of skull	Mandible anterior to jaw joint	Closes jaws
Depressor mandibulae	Dorsal surface of skull, connective tissue of back	Mandible posterior to jaw joint	Opens jaws
Submaxillary	Medial surface of mandible	Longitudinal connective tissue septum between submaxillaries of opposite sides	Raises floor of mouth and pharynx during breathing and swallowing
Trunk Muscles			
Obliquus externus	Connective tissue of back	Longitudinal midventral connective tissue septum	Supports abdomen, helps compress lungs during breathing
Rectus abdominis	Sternum	Anteroventral surface of pelvic girdle	Supports abdomen, flexes back
Longissimus dorsi	Posterior end of skull, dorsal surface of vertebrae	Urostyle and ilium	Extends back
Coccygeoiliacus	Urostyle	Ilium	Helps extend back
Foreleg Muscles			
Dorsalis scapulae	Surface of scapula and suprascapula	Proximal end of humerus	Abducts arm, helps protract arm
Latissimus dorsi	Dorsolateral surface of trunk	Proximal end of humerus	Abducts arm, helps retract arm
Triceps	Base of scapula, shaft of humerus	Proximal end of radioulna	Extends forearm
Deltoideus	Base of scapula, clavicle	Proximal end of humerus	Adducts arm, protracts arm
Pectoralis	Ventral surface of trunk and sternum	Proximal end of humerus	Adducts arm, retracts arm
Coracoradialis	Ventral surface of coracoid	Proximal end of radioulna	Flexes forearm
Hindleg Muscles: Thigh			
Triceps femoris (rectus anticus, vastus internus and externus)	Ilium and capsule of hip joint	Proximal end of tibiofibula	Extends shank
Iliofibularis Semimembranosus Gracilis major and minor	Posterodorsal part of pelvic girdle	Proximal end of tibiofibula	Flexes shank, during shank extension helps retract and abduct thigh
Sartorius	Anteroventral part of pelvic girdle	Proximal end of tibiofibula	Flexes shank, helps adduct and protract thigh
Adductor magnus Adductor longus	Ventral part of pelvic girdle	Distal portion of femur	Adducts thigh
Hindleg Muscles: Shank			
Gastrocnemius	Distal end of femur	Plantar aponeurosis and terminal segment of each toe	Extends foot
Tibialis posticus	Shaft of tibiofibula	Proximal end of tarsals	Helps extend foot
Peroneus Tibialis anticus longus	Distal end of femur	Proximal end of tarsals	Flexes foot, helps extend shank

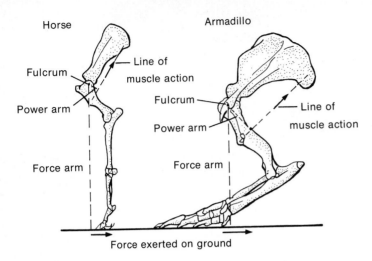

Horse

Armadillo

Line of
muscle action

Fulcrum

Power arm

Fulcrum

Line of
muscle action

Power arm

Force arm

Force arm

Force exerted on ground

Figure 11.17 Forelimbs of a horse and an armadillo drawn to the same size to show how changes in power and force arm relationships adapt the limb lever system for speed or power. (After J. Maynard Smith, from Young: Life of Mammals. Oxford University Press, 1957.)

ANNOTATED REFERENCES

General

The following references contain a great deal of information on the anatomy and physiology of vertebrate organ systems.

Beck, W. S.: Human Design. New York, Harcourt Brace Jovanovich, 1971. A carefully integrated account of form and function in man from the molecular to the organ level.

Bloom, W., and D. W. Fawcett: A Textbook of Histology. 9th ed. Philadelphia, W. B. Saunders Co., 1968. The microscopic and ultrastructure of the cells and tissues of mammalian organ systems are thoroughly described.

Cannon, W. B.: The Wisdom of the Body. New York, W. W. Norton & Co., 1939. A classic and very readable account of mammalian physiology.

Gordon, M. S. et al.: Animal Function: Principles and Adaptations. New York, Macmillan, 1968. A useful textbook of comparative physiology.

Guyton, A. C.: Textbook of Medical Physiology. 4th ed. Philadelphia, W. B. Saunders Co., 1971. A detailed consideration of mammalian physiology is presented in this standard textbook.

Marshall, P. T., and G. M. Hughes: The Physiology of Mammals and Other Vertebrates. Cambridge, Cambridge University Press, 1965. An excellent account which gives the reader a good understanding of the structural and functional unity of the organ systems of vertebrates.

Prosser, C. L., and F. A. Brown, Jr.: Comparative Animal Physiology. 2nd ed. Philadelphia, W. B. Saunders Co., 1961. A very valuable source book on the physiology of both invertebrates and vertebrates.

Romer, A. S.: The Shorter Version of the Vertebrate Body. 4th ed. Philadelphia, W. B. Saunders Co., 1971. The morphological aspects of the evolution of vertebrates are thoroughly considered in this widely used textbook of comparative anatomy.

Ruch, T. C., and H. D. Patton: Physiology and Biophysics. 19th ed. Philadelphia, W. B. Saunders Co., 1965. An advanced text and source book on mammalian physiology.

Schmidt-Nielsen, K.: How Animals Work. Cambridge, Cambridge University Press, 1972. Water and heat balance, respiration, countercurrent mechanisms, and body size are explored in this collection of essays originally presented as the Hans Gadow Lectures in 1970–71.

Wessels, N. K. (Ed.): Vertebrate Adaptations. San Francisco, W. H. Freeman and Company, 1969. A collection of articles from the Scientific American, including many cited in subsequent chapters. Each group of articles on an adaptive topic is preceded by a general introduction by the editor.

Young, J. Z.: The Life of Mammals. New York, Oxford University Press, 1957. Structure and function are carefully considered in this thorough account of the gross and microscopic anatomy and the physiology of mammals.

Young, J. Z.: The Life of Vertebrates. 2nd ed. Oxford, Clarendon Press, 1962. One or more chapters are devoted to the anatomy, physiology and evolution of each of the classes of vertebrates in this excellent text.

Protection, Support and Movement

Gray, J.: Animal Locomotion. New York, W. W. Norton and Co., 1968. A thorough analysis of the principles and types of animal locomotion.

Hill, A. V.: Muscular Movement in Man. New York, McGraw-Hill Book Co., 1927. A classic account by one of the pioneer investigators of muscle physiology.

Montagna, W.: The Structure and Function of Skin. 2nd ed. New York, Academic Press, 1962. A thorough study of the skin and its regional variations is presented in this book.

Thompson, D'Arcy, W.: On Growth and Form. Revised ed. New York, Macmillan, 1942. Biomechanical principles are used to explain the form of skeletons and other structures.

12. Digestion and Respiration

A fundamental characteristic of living organisms is their ability to take in materials quite unlike themselves and to synthesize their own unique cellular constituents from them. Grass becomes beef and beef becomes human flesh by the alchemy of living organisms. Animals must take into their bodies many different substances to provide the raw materials and energy necessary for the synthesis and maintenance of the wide variety of substances present in cells, for reproduction and for the various activities of the body. These include energy-rich organic compounds, vitamins, oxygen, water and mineral salts. The carbohydrates, fats, proteins and vitamins are synthesized by plants and other animals.

In vertebrates oxygen enters through the respiratory system—gills or lungs—and through the skin in certain animals; the other materials enter through the digestive system. These are the intake systems of the body, but they also serve to some extent in the removal of waste products. Some toxins are removed by the digestive system, and most of the carbon dioxide produced in cellular respiration is eliminated by the respiratory system along with some water and, in fishes, some nitrogenous wastes from the metabolism of proteins and nucleic acids.

The vertebrate digestive tract is a tube passing through the body with openings at either end. Food is taken into this tract, where most of it is digested and absorbed. The undigested and unabsorbed residues are eliminated as **feces** from the posterior end of the tract. The process of elimination, known as **defecation,** should not be confused with excretion, which is the discharge of the by-products of metabolism. Excretion is primarily a function of the excretory and respiratory systems and the skin. Most of the material in the feces has, in fact, neither entered the tissues of the body nor taken part in metabolism.

12.1 THE MOUTH

The basic pattern of the digestive system is similar in all vertebrates. In very primitive vertebrates the mouth is unsupported by jaws, but most vertebrates have jaws and a good complement of teeth that aid in food-getting.

Teeth are similar in structure to the scales of sharks, which are composed of enamel- and dentin-like materials, and are believed to have evolved from bony scales. A representative mammalian tooth (Fig. 12.1) consists of a **crown** projecting above the gum, a **neck** surrounded by the gum, and one or more **roots** embedded in sockets in the jaws. The crown is covered by a layer of **enamel.** Enamel is the hardest substance

276

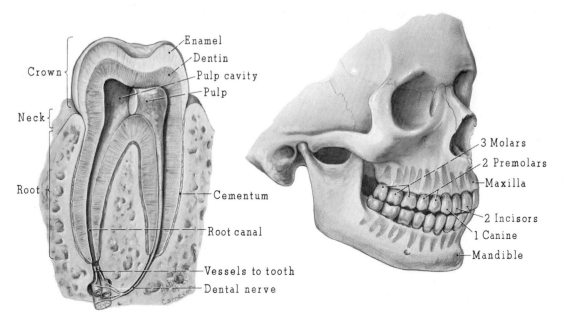

Figure 12.1   The teeth of man. *Left*, a vertical section through a molar tooth; *right*, types of teeth. (From Jacob and Francone: Structure and Function in Man, 2nd ed., W. B. Saunders Co., 1970.)

in the body and consists almost entirely of crystals of calcium salts. Calcium, phosphate and fluoride are important constituents of enamel and all must be present in the diet in suitable amounts for proper tooth development and maintenance. The rest of the tooth is composed of **dentin,** a substance very similar to bone. In the center of the tooth is a **pulp cavity** containing blood vessels and nerves. A layer of **cement** covers much of the root and holds the tooth firmly in place in the jaw.

The teeth of most vertebrates are cone-shaped structures used primarily for seizing and holding the prey. In mammals, the teeth are differentiated into several types that are used not only for seizing food but often for its mechanical breakdown (p. 751). Mammalian teeth, unlike those of lower vertebrates, are not continuously replaced by new sets. Man, for example, first has a set of deciduous or **milk teeth** —two incisors, one canine and two premolars on each side of each jaw. These are later replaced by **permanent teeth;** in addition, three molars develop on each side of each jaw behind the premolars. The molars last throughout life and are not replaced (Fig. 12.1).

Once food is in the mouth, a fish easily manipulates and swallows it, for the flow of water aids in carrying it back into the pharynx. Oral glands and a tongue are poorly developed in fishes. The evolution of these structures accompanied the transition from water to land, and they became more elaborate in the higher tetrapods. The **tongue** of frogs and anteaters is specialized as a food gathering device (Fig. 12.2), but its chief function in most mammals is to manipulate food in the mouth and to aid in swallowing. In many mammals, the tongue pushes the food between the teeth so that the food is thoroughly masticated and mixed with saliva. Then the food is shaped into a ball, a **bolus,** and moved into the pharynx by raising the tongue (Fig. 12.4). The tongue also bears numerous microscopic taste buds, and the human tongue is of great importance in speech.

In addition to a liberal sprinkling of simple glands in the lining of the mouth cavity, mammals have evolved several pairs of conspicuous **salivary glands** that are connected to the mouth by ducts. The location of the **parotid, submandibular** and **sublingual glands** of man is shown in Figure 12.3. In primitive terrestrial vertebrates, oral glands simply secrete a mucous and watery fluid that lubricates the food, and this is still the major function of our saliva. The saliva of most mammals and of a few other tetrapods contains **salivary amylase**

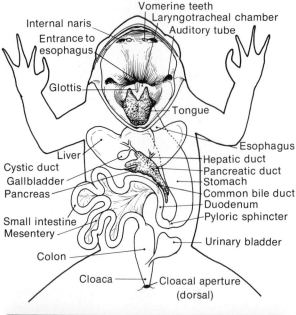

Vomerine teeth
Laryngotracheal chamber
Auditory tube
Internal naris
Entrance to esophagus
Glottis
Tongue
Esophagus
Liver
Hepatic duct
Cystic duct
Pancreatic duct
Gallbladder
Stomach
Pancreas
Common bile duct
Duodenum
Pyloric sphincter
Small intestine
Mesentery
Urinary bladder
Colon
Cloaca
Cloacal aperture (dorsal)

Figure 12.2   A ventral view of the frog's digestive system. The liver lobes have been turned forward to show the gallbladder. Tongue action is shown in the inset. Notice that the tongue is stretched to over half the length of the body. (Inset from Van Riper in Natural History, Vol. LXVI, No. 6)

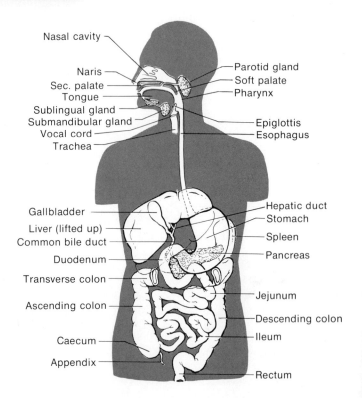

Nasal cavity
Naris
Parotid gland
Sec. palate
Soft palate
Tongue
Pharynx
Sublingual gland
Submandibular gland
Epiglottis
Vocal cord
Esophagus
Trachea

Gallbladder
Hepatic duct
Liver (lifted up)
Stomach
Common bile duct
Spleen
Duodenum
Pancreas
Transverse colon
Ascending colon
Jejunum
Descending colon
Caecum
Ileum
Appendix
Rectum

Figure 12.3   The digestive system of man.

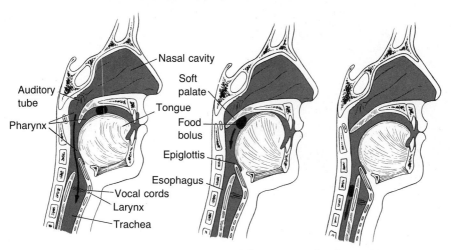

Figure 12.4   Diagrams showing the shifts in the position of the tongue, soft palate and epiglottis when a bolus of food is swallowed. (Slightly modified after Villee: Biology, 6th ed.)

which hydrolyzes polysaccharides such as starch. The poison glands of reptiles and the glands of vampire bats that secrete an anticoagulant are other specialized oral glands.

12.2   THE PHARYNX AND ESOPHAGUS

Part of the **pharynx** of man lies above the **soft palate** (Fig. 12.4) and receives the internal nostrils, or **choanae,** and the openings of the pair of eustachian or **auditory tubes** from the middle ear cavities. Another part lies beneath the soft palate just posterior to the mouth cavity. The rest of the pharynx lies posterior to these parts and leads to the esophagus and larynx. Passage of the food into the pharynx initiates a series of reflexes: The muscular soft palate rises and prevents food from entering the nasal cavities, breathing momentarily stops, the larynx is elevated and the epiglottis swings over the glottis, preventing food from entering the larynx; the tongue prevents food from returning to the mouth, and muscular contractions of the pharynx move the bolus into the esophagus.

The pharynx of tetrapods is a rather short region in which the food and air passages cross, but in fishes it is a more extensive area associated with the gill slits. Gill pouches are present in the embryos of mammals, and some of them give rise to glandular structures such as the **thymus** and **parathyroids,** but only parts of the first two remain in adults. The middle ear cavity and the auditory tube develop from the first pouch (the spiracle of fishes), and part of the second forms the fossa in which the palatine tonsil lies. The **thyroid gland** and the lungs are outgrowths from the floor of the pharynx. Glands derived from the pharynx are endocrine in nature and will be considered in Chapter 17.

Successive waves of contraction and relaxation of the muscles, known as **peristalsis,** propel the bolus down the esophagus to the stomach. The muscles relax in front of the food and contract behind it. When the food reaches the end of the esophagus the cardiac sphincter, which closes off the entrance to the stomach, relaxes and allows it to enter. The esophagus is generally a simple conducting tube, but in some animals its structure has been modified for storage. The crop of the pigeon, for example, is a modified part of the esophagus.

12.3   THE COELOM AND STOMACH

The stomach and intestines of vertebrates lie within a large, mesodermal body cavity, or **coelom.** Coelomic epithelium lines the body cavity and extends as epithelial sheets known as **mesenteries** to cover the visceral organs (Fig. 12.5). Blood vessels and nerves reach the viscera through the mesenteries. The presence of a coelom, together with a small amount of watery coelomic fluid,

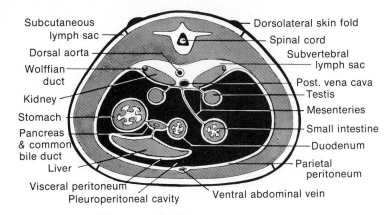

Figure 12.5 A diagrammatic cross section through the trunk of a frog viewed from behind. At a more anterior level, a mesentery would pass to the stomach rather than to the intestine.

facilitates the expansion, contraction and other functional movements of the viscera. The part of the coelom in the abdominal region of the frog and most vertebrates is known as the **pleuroperitoneal cavity.** In mammals the portion of it surrounding the lungs is a separate pair of **pleural cavities,** and the part containing the digestive organs is the **peritoneal cavity.** Coelomic epithelium in the latter cavity is called **peritoneum.** The heart of all vertebrates lies in another section of the coelom known as the **pericardial cavity.**

The **stomach** is usually a **J**-shaped pouch (Figs. 12.2 and 12.3) whose chief functions are the storage and mechanical churning of food and the initiation of the enzymatic hydrolysis of proteins. Lampreys, lungfishes and some other primitive fishes do not have a stomach, and the absence of this organ is thought to have been a characteristic of the ancestral vertebrates. The early vertebrates were probably filter-feeders that fed more or less continuously on minute food particles that could be digested by the intestine alone. Presumably, the evolution of jaws and the habit of feeding less frequently and on larger pieces of food required an organ for the storage and initial conversion of this food into a state in which it could be digested further in the intestine. In most vertebrates both mechanical and chemical breakdown of food begins in the stomach.

After food enters the stomach, the **cardiac sphincter** at its anterior end and the **pyloric sphincter** at the posterior end close. Muscular contractions of the stomach churn the food, breaking it up mechanically and mixing it with the gastric juice secreted by tubular-shaped **gastric glands.** The gastric

juice contains **hydrochloric acid** and the proteolytic enzyme **pepsin.** In addition, **rennin** is particularly abundant in the stomach of young mammals and causes the milk protein casein to coagulate and remain in the stomach long enough to be acted on by pepsin. Rennin has been extracted for centuries from the stomachs of calves and used to curdle milk; this is an important step in the manufacture of cheese.

When the food has been reduced to a creamy material known as **chyme,** and most of the microorganisms that entered the stomach with it have been killed by the action of the gastric juices, the pyloric sphincter opens and the food passes into the small intestine. The most fluid food passes first. Indeed, upon entering the stomach, water passes almost immediately into the intestine. The acid food enters the intestine in spurts and is quickly neutralized by the alkaline secretions flowing into the intestine from the liver and pancreas.

One striking modification of the stomach is seen in the cow and other ruminants. A cow's stomach consists of a series of four chambers (Fig. 12.6). Food passes first into the **rumen,** where it is temporarily stored and from which it is regurgitated from time to time as the animal ruminates, or chews its cud. The rumen, which has a capacity up to about 200 liters, contains a large colony of bacteria and other microorganisms which play a twofold role in the animal's nutrition. They produce **cellulases,** which split the β-glucosidic bonds of cellulose, and other enzymes that convert the glucose to smaller units, chiefly acetic acid, that are absorbed directly from the rumen. No vertebrate can synthesize cellulases, so herbivores cannot

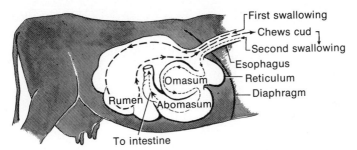

First swallowing
Chews cud
Second swallowing
Esophagus
Reticulum
Diaphragm
Omasum
Rumen
Abomasum
To intestine

Figure 12.6   Course of food through the stomach of a cow.

digest cellulose without the aid of micro-organisms contained in some part of their digestive tract. Secondly, as the micro-organisms multiply, they synthesize amino acids and proteins which are "harvested" and digested by the cow. Eventually, material from the rumen passes through the reticulum and omasum, where much of the water is pressed out of the food mass, and into the abomasum. The abomasum alone contains gastric glands.

12.4   THE LIVER AND PANCREAS

The liver and pancreas are large glandular outgrowths from the anterior part of the intestine. The **liver,** in fact, is the largest organ of the body. Its cells continually secrete **bile,** which passes through hepatic ducts into the **common bile duct** and then up the cystic duct into the **gallbladder.** Bile does not enter the intestine immediately, for a sphincter at the intestinal end of the bile duct is closed until food enters the intestine. Contraction of the wall of the gallbladder forces bile out. The bile that is finally poured into the intestine is concentrated, for a considerable amount of water and some salts are absorbed from the bile in the gallbladder.

Although bile contains no digestive enzymes, it nevertheless has a twofold digestive role. Its alkalinity, along with that of the pancreatic secretions, neutralizes the acid food entering the intestine and creates a pH favorable for the action of pancreatic and intestinal enzymes. Its **bile salts** emulsify fats, breaking them up into smaller globules and thereby providing more surfaces on which fat-splitting enzymes can

act. These salts are also essential for the absorption of fats and fat-soluble vitamins (A, D, E, K). Most of the bile salts are not eliminated with the feces but are absorbed in the intestine along with the fats and are carried back to the liver by the blood stream to be used again.

The color of bile (green, yellow, orange or red in different species) results from the presence of **bile pigments,** derived from the breakdown of hemoglobin in the liver. The bile pigments are converted by enzymes of the intestinal bacteria to the brown pigments responsible for the color of the feces. If their excretion is prevented by a gallstone or some other obstruction of the bile duct, they are reabsorbed by the liver and gallbladder, the feces are pale, and the pigments accumulate in the skin, giving it the yellowish tinge characteristic of jaundice.

All of the blood returning from the intestine, where it has absorbed a variety of materials, passes through the liver before entering the general circulation of the body. In the minute capillary-like spaces of the liver the blood comes into intimate contact with the hepatic cells, which take up, store, interconvert, and alter in many ways the absorbed food molecules. The liver cells also detoxify certain poisonous substances and excrete some of them in the bile.

The **pancreas** is an important digestive gland, producing quantities of enzymes that act upon carbohydrates, proteins, fats and nucleic acids. These enzymes enter the intestine by way of a pancreatic duct that joins the common bile duct. An accessory pancreatic duct is present in some vertebrates and empties directly into the intestine. The pancreas contains patches of endocrine tissue, the **islets of Langerhans,** which will be considered in Chapter 17.

12.5   THE INTESTINE

Most of the digestive processes, and virtually all of the absorption of the end products of digestion, occur in the intestine. Enzymes acting in the intestine are produced by the pancreas, by many of the epithelial cells lining the intestine, and in birds and mammals by small, tubular-shaped **intestinal glands.** Adequate surface area for absorption is made available by the length of the intestine and by outgrowths and foldings of the lining.

The structural details of the intestine vary considerably among vertebrates. Primitive fishes have a short, straight intestine extending posteriorly from the stomach. Since its internal surface is increased by a helical fold known as the spiral valve, it is called a **valvular intestine.** Tetrapods have lost the spiral valve and make up for this by an increase in the length of the intestine, which becomes more or less coiled. The tetrapod intestine has become further differentiated into an anterior **small intestine** and a posterior **large intestine** (Figs. 12.2 and 12.3). The first part of the small intestine is known as the **duodenum** and, in mammals, the two succeeding parts are the **jejunum** and **ileum.** The large intestine, or **colon,** of the frog and most vertebrates leads to a posterior chamber known as the **cloaca.** The cloaca also receives the products of the urinary and reproductive systems, and opens on the body surface by the **cloacal aperture.** In mammals, the cloaca has become divided into a ventral part which receives the urogenital products, and a dorsal **rectum** which opens on the body surface at the **anus.** A blind pouch called the **caecum** is present at the junction of small and large intestines of mammals. This is very long in such herbivores as the rabbit and horse and contains colonies of bacteria that digest cellulose. Man has a small caecum with a vestigial **vermiform appendix** on its end. An **ileocaecal** valve located between the small and large intestine prevents bacteria in the colon from moving back up into the small intestine.

A transverse section of the small intestine of a mammal illustrates the microscopic structure of the digestive tract (Fig. 12.7). There is an outer covering of **visceral peritoneum** (the serous coat), a layer of **smooth muscle,** a layer of vascular connective tissue, the **submucosa** and, finally, the innermost layer, the **mucosa** or mucous membrane. The outer fibers of the muscular coat are usually described as longitudinal; the inner, as circular. Actually, both layers are spiral; the outer is an open spiral and

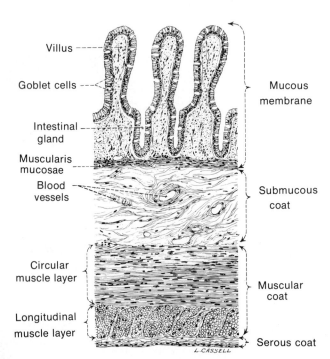

Villus

Goblet cells

Intestinal gland

Muscularis mucosae

Blood vessels

Circular muscle layer

Longitudinal muscle layer

Mucous membrane

Submucous coat

Muscular coat

Serous coat

L. CASSELL

Figure 12.7 Diagrammatic cross section of a portion of the small intestine. Numerous capillaries and a lymphatic vessel enter each villus. (From King and Showers: Human Anatomy and Physiology, 5th ed.)

the inner, a tight spiral. The relaxations and contractions of these layers are responsible for the peristaltic and churning movements that mix the food with digestive enzymes and move it along the digestive tract. The mucosa consists of a layer of smooth muscle, connective tissue and, finally, the simple columnar epithelium next to the lumen. Many mucus-producing **goblet cells** are present in the lining epithelium and their secretion helps to lubricate the food and to protect the lining of the intestine. In the small intestine of mammals and birds, the mucosa bears numerous minute, finger-shaped **villi** containing blood capillaries and small lymphatic vessels. The villi protrude into the lumen and increase the intestinal surface manyfold. They are moved about by the muscle layer in the mucosa, the **muscularis mucosae,** and by strands of smooth muscle that extend into them. Tubular-shaped intestinal glands known as the **crypts of Lieberkühn** lie at the base of the villi. Mitotic division of epithelial cells in the bottom of the crypts continually produce new cells that migrate outward and over the villi from where they are sloughed off. There is a rather rapid turnover of epithelial cells for an individual one lasts only about 48 hours. The intestinal glands produce many digestive enzymes and a watery intestinal fluid.

Although most of the water ingested and in the gastrointestinal secretions is absorbed in the small intestine, the material that enters the large intestine is still quite fluid. Most of the remaining water and many salts are absorbed as the residue passes through the colon. If the residue passes through very slowly so that too much water is absorbed, the feces become very dry and hard and **constipation** may result; if it goes through very rapidly and little water is absorbed, **diarrhea** results. Many bacteria reside in the colon and synthesize a variety of vitamins which are absorbed from the colon. The bacteria reproduce very rapidly, and many are eliminated. As much as 25 per cent of the feces may consist of bacteria.

12.6 DIGESTION OF FOODS

As we explained in section 2.5, carbohydrates, proteins and fats are all composed of smaller molecules that are held together by **anhydro bonds,** i.e., bonds formed essentially by the removal of a molecule of water. The chemical breakdown, or **digestion,** of these foods is a process of **hydrolytic cleavage,** in which the anhydro bond is split by the addition of water. It requires, however, a great many different enzymes to effect the splitting because the various foodstuffs contain a great variety of molecular species, and enzymes are very specific with respect to the types of molecules they can cleave. Although digestion is a continuous process involving the simultaneous breakdown of a variety of materials by a host of enzymes, it is convenient to discuss it by considering each major food category in turn.

The building units of most carbohydrates are six carbon sugars (single sugars or **monosaccharides**) including **glucose, galactose** and **fructose.** These are united by pairs into larger double sugars (**disaccharides**) and these in turn into complex **polysaccharides** such as **starch, glycogen** and **cellulose.** Enzymes known as amylases first attack the polysaccharides and split them into disaccharides. Vertebrates do not synthesize an enzyme that can attack the bonds in cellulose, but starch and glycogen are first split by **salivary amylase** into the disaccharide, **maltose.** Chloride ions present in the saliva are necessary to activate amylase. Its pH optimum is close to neutrality, so its action is eventually stopped by the acidic gastric juice of the stomach. But, since it takes one-half hour or longer for the food and gastric juice to become thoroughly mixed, 40 per cent or more of the starches are split before the amylase is inactivated. The starches not broken down in the stomach are soon converted to maltose in the neutral environment of the small intestine by **pancreatic amylase.** There may also be traces of amylase in the intestinal juice. Maltose and the disaccharides sucrose and lactose are acted on by specific enzymes: **maltase, sucrase,** and **lactase** (Table 12.1). These and other enzymes produced by the intestinal cells are not completely free in the main part of the intestinal lumen, for cell free extracts of intestinal juice contain very little enzyme. Rather, the enzymes appear to be localized among the numerous microvilli that comprise the striated border of the intestinal epithelium (Fig. 12.8). The disaccharides are hydrolyzed as they enter this border; all

TABLE 12.1 MAJOR DIGESTIVE ENZYMES

Produced by	Enzyme	Substrate Acted upon	Product
Salivary glands	Amylase	Starch	Maltose (double sugar)
Gastric glands	Pepsinogen, converted to pepsin	Proteins	Peptides
	Rennin	Casein	Coagulated casein
Pancreas	Amylase	Starch	Maltose
	Trypsinogen, converted to trypsin	Proteins	Peptides
		Chymotrypsinogen	Chymotrypsin
	Chymotrypsinogen, converted to chymotrypsin	Proteins	Peptides
	Exopeptidases	Peptides	Amino acids
	Lipase	Emulsified fat	Fatty acids and glycerol
	Nucleases	Nucleic acids	Nucleotides
Intestinal glands	Amylase	Starch	Maltose
	Maltase	Maltose	Glucose (single sugar)
	Sucrase	Sucrose (double sugar)	Glucose and fructose
	Lactase	Lactose (double sugar)	Glucose and galactose
	Enterokinase	Trypsinogen	Trypsin
	Exopeptidases	Peptides	Amino acids
	Lipase	Emulsified fats	Fatty acids and glycerol
	Nucleases	Nucleic acids	Nucleotides

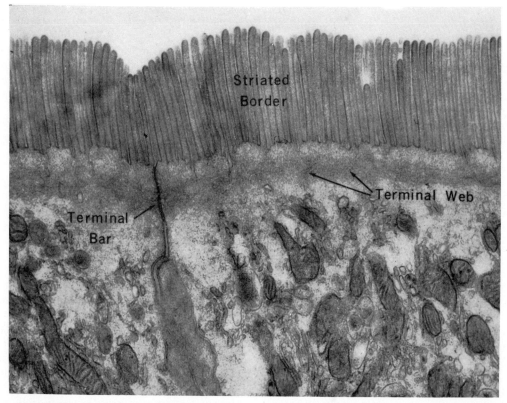

Figure 12.8 Electron micrograph of part of the surface of two epithelial cells from a villus to show the striated border composed of numerous microvilli. The terminal bar is a specialization of the plasma membrane of adjacent cells that unites them tightly and closes the intercellular space. Original photograph, ×30,000. (From Bloom and Fawcett: A Textbook of Histology, 9th ed.)

yield glucose; sucrose and lactose also yield fructose and galactose respectively. It is in the form of these monosaccharides that most carbohydrates are absorbed.

You will recall that proteins are composed of **amino acids** linked together by a type of anhydro bond known as a peptide bond. Proteolytic enzymes are secreted by the cells that produce them as inactive precursor substances. They are activated, usually by cleaving off a part of their molecules, only when in the lumen of the digestive tract and in the presence of protein to be digested. This, together with a protective coating of mucopolysaccharide over the exposed surface of the lining epithelial cells, normally prevents these powerful enzymes from injuring the tissues of the digestive tract. If these control mechanisms break down, an ulcer may result, or a tissue rich in proteolytic enzymes, such as the pancreas, may start to digest itself. Proteolytic enzymes are either **endopeptidases,** capable of attacking protein molecules at many points along their amino acid chains, except for the terminal groups, or **exopeptidases,** which strip off terminal amino acids from chain fragments. The first of the endopeptidases is **pepsin.** Pepsin is secreted as the enzyme precursor, **pepsinogen,** by the **chief cells** of the gastric glands. The conversion of pepsinogen to pepsin is initiated by hydrochloric acid produced by the **parietal cells** of these glands but, once formed, pepsin helps to continue the activation of additional pepsinogen. The acid also brings the stomach contents close to pH 2.0, the optimum for pepsin action. Pepsin action stops when the chyme enters the duodenum, but other endopeptidases, **trypsin** and **chymotrypsin,** then go to work. Both of these are secreted by the pancreas as inactive precursors, **trypsinogen** and **chymotrypsinogen.** Their activation is initiated by **enterokinase,** an intestinal enzyme that converts trypsinogen to trypsin. Once formed, trypsin helps to continue the activation of trypsinogen, and it alone activates chymotrypsinogen. Since the various endopeptidases split bonds adjacent to specific amino acids, their concerted action is necessary for the fragmentation of protein molecules. Pepsin only splits bonds adjacent to phenylalanine or tyrosine; trypsin, bonds next to arginine or lysine; and chymotrypsin, bonds next to leucine, methionine, phenylalanine, tyro-

sine, or tryptophan. Exopeptidases secreted by the pancreas and small intestine attack the terminal peptide bonds, stripping off one amino acid after another. Exopeptidases known as **carboxypeptidase** split off the terminal amino acid with the free carboxyl group (COOH, Fig. 2.6); others known as **aminopeptidases** separate the terminal amino acid with the free amino group (NH_2). The free amino acids resulting from the digestion of proteins are then absorbed into the blood stream.

Fats emulsified by bile salts are attacked by esterases which cleave the ester bonds joining fatty acids to glycerol. **Lipase,** most of which is secreted by the pancreas although some comes from the intestinal glands, is the principal esterase of vertebrates. The mixture of bile salts, fatty acids and partly digested fats collectively emulsifies fats further into particles, many of which are small enough to be absorbed directly.

Carbohydrates, proteins and fats constitute the bulk of the food digested, but many other complex materials are degraded by less familiar enzymes. For example, nucleic acids are hydrolyzed by pancreatic and intestinal enzymes. Esterases known as **ribonuclease** and **deoxyribonuclease** split the phosphate ester bonds linking the nucleotides in RNA and DNA respectively. Other enzymes separate each of these into their constituent pentoses, phosphoric acid and nitrogenous bases. In man, as much as 100 grams of enzymes are secreted each day by the digestive tract. These enzymes are, of course, protein, and are themselves digested, inactivated and absorbed as digestion is completed.

12.7 THE CONTROL OF DIGESTIVE SECRETIONS

Each of the various enzymes is secreted at an appropriate time: We salivate when we eat, and gastric juice is produced when food reaches the stomach. The control of these digestive secretions is partly nervous and partly endocrine. The smell of food or its presence in the mouth stimulates sensory nerves that carry impulses to a salivating center in the medulla of the brain. From there the impulses are relayed along motor nerves to the salivary glands, which then secrete.

The control of gastric secretion is more complex. Years ago the famous Russian physiologist Pavlov performed an experiment in which he brought the esophagus of a dog to the surface of the neck and severed it. When the dog ate, the food did not reach the stomach, yet some gastric juice was secreted provided that the **vagus nerve,** which carries motor fibers to the stomach and other internal organs, was left intact. If the vagus nerve was cut, this secretion did not occur. This experiment proved that the control of gastric secretion was at least partly nervous. Subsequently, it was discovered that if the vagus was cut but food was permitted to reach the stomach, a considerable flow of gastric juice was produced. Obviously, the vagus nerve is not the only means of stimulating the gastric glands. Further investigation revealed that, when partly digested food reaches the pyloric region of the stomach, certain of the mucosal cells produce the hormone **gastrin,** which is absorbed into the blood through the stomach wall and ultimately reaches the gastric glands, stimulating them to secrete.

Other factors are also involved. Certain foods, particularly partly digested meats, release substances of unknown composition that stimulate gastric secretion. The absorption of certain products from the small intestine, among them alcohol, further stimulates gastric secretion. These are among the reasons for restricting meats and alcohol from the diets of ulcer patients.

Gastric secretion is reduced and finally stopped as the stomach empties and food enters the duodenum. When food, especially fats, enters the duodenum, the duodenal mucosa produces the hormone **enterogastrone** which, on reaching the stomach, inhibits the secretion of the gastric glands and slows down the churning action of the stomach. This not only helps to prevent the stomach from digesting its own lining but also enables fatty foods to stay for a longer period in the duodenum where they can be acted on by bile salts and lipase.

One of the first hormones to be discovered was **secretin,** which initiates pancreatic secretion. In 1902 Bayliss and Starling were investigating the current belief that the secretion of pancreatic juice was under nervous control. They found that the pancreas secreted its juice when acid food entered the small intestine even though the nerves to and from the intestine were cut. A stimulant of some sort apparently traveled in the blood. The injection of acids into the blood stream had no effect, so they reasoned that some stimulating principle must be produced by the intestinal mucosa upon exposure to acid foods. When they injected extracts of such a mucosa into the circulatory system the pancreas secreted. It is now recognized that secretin itself only stimulates the pancreas to produce a copious flow of a bicarbonate-rich alkaline liquid which neutralizes the acid material entering from the stomach. A second hormone produced by the intestinal mucosa, **pancreozymin,** is necessary to stimulate the release of pancreatic enzymes.

Secretin has a side effect on the liver, for it increases slightly the rate of bile secretion. However, another hormone, **cholecystokinin,** which is also produced by the duodenal mucosa when acid food is in the duodenum, is largely responsible for causing the gallbladder to contract and release the bile. Vagal stimulation also plays a role in the release of bile.

12.8 ABSORPTION AND UTILIZATION OF MATERIALS

Circular folds in the intestinal mucosa, the villi, and microvilli form an enormous surface area for absorption; the surface area of the human small intestine may be as much as 550 square meters. Some larger food molecules can be engulfed by intestinal cells and taken into vacuoles by a process known as **pinocytosis.** Many suckling mammals can absorb intact protein, including antibodies, by this method. Adults lose this ability, and most foods, apart from some fats, are absorbed as small molecules by diffusion and active transport. Little is known about the mechanism for the active transport of food products into the mucosal cells, but several lines of evidence make it clear that one operates. Absorption continues after most of the food in the intestinal lumen has been taken up and any concentration gradient that would have favored diffusion has been reversed. There are some surprising variations in the rates of absorption of similar sized molecules; if glucose, for example, is given an arbitrary absorption rate of 100, galactose has a rate of 115, and fructose one of 44. Poisons that interfere with the metabolism of mucosal cells greatly reduce the rate and amount of absorption.

The absorption of fats and fatty acids presents a special problem for, unlike the other end products of digestion, they are not water soluble. Apparently, their uptake is facilitated by their combining with bile salts, for this makes a soluble complex. Bile salts are freed within the mucosal cells, and the fatty acids and glycerol are combined with phosphate to form water soluble **phospholipids.** Unhydrolyzed fat and most of the phospholipids enter the lymphatic capillaries, but the other absorbed materials enter the capillaries of the blood vessels.

Energy Requirements. The absorbed products may be used as raw materials for the synthesis of cellular components, as a source of energy, or they may be stored for later use. The minimum energy requirements for a young man are about 1600 Calories per day; they are slightly less for a young woman. This is known as the **basal metabolic rate,** for it represents the amount of energy needed to keep alive when no muscular work is being done and no food is being digested. With heavy physical work, energy requirements may go as high as 6000 Calories per day. A person leading a rather sedentary life requires 2500 to 3000 Calories a day. Most of us establish a balance between the amount of energy expended and the amount taken in with our food. If we consume more food than is required to balance our expenditure of energy, we of course gain weight.

All kinds of food yield energy when metabolized, but not to the same extent. When burned completely in a calorimeter, 1 gm. of carbohydrate or protein yields about 4 Calories, and 1 gm. of fat 9.5 Calories. Though carbohydrates do not contain so many Calories per gram as fats, they constitute the major body fuel for most people. Normally, our diet contains more carbohydrates than fats or proteins.

Carbohydrate Metabolism. The various kinds of monosaccharides that are absorbed are carried to the liver, where most are first converted to glucose-6-phosphate and then to glycogen (animal starch) for storage.

When energy is needed by body cells, liver glycogen is reconverted to glucose-6-phosphate, and this is transformed to glucose, which is released into the blood stream. Muscle cells can also store glucose as glycogen, but they lack the enzyme (glucose-6-phosphatase) to hydrolyze glucose-6-phosphate to glucose, so muscle glycogen can serve as a fuel for the muscle cells but not as a general reserve for the other cells of the body.

The role of the liver in maintaining a constant level of glucose in the blood was discussed in section 5.2. The glucose molecules are carried to all the cells of the body, where they are oxidized via the Krebs citric acid cycle to carbon dioxide and water and some of the energy is conserved as energy-rich phosphate groups, $\sim P$. Not all sugars are converted to glycogen and released as glucose. Some are converted by the liver and other cells to fat and stored in adipose tissue. It is a common observation that an excessive intake of carbohydrates or proteins is just as fattening as an excessive intake of fats.

Protein Metabolism. Studies on nitrogen intake and output in animals show that body proteins are continually being broken down and resynthesized. Most of the amino acids in the diet, therefore, are used as raw material for the synthesis of protein needed in the growth, repair and maintenance of tissue; to maintain an adequate supply of enzymes; and for the production of certain hormones including adrenalin, insulin and thyroxin. Many amino acids can be synthesized from other substances, but certain **essential amino acids** can be obtained only from the diet. Most plant proteins lack one or more essential amino acids, but all are present in a diet that includes meat, eggs and dairy products.

Amino acids are not stored in any appreciable quantity. Excess acids are **deaminated,** the amino group being stripped off; the rest of the molecule can then enter the citric acid cycle to be used immediately as a source of energy or can be converted to glycogen or fat. Carnivores obtain most of their glycogen from amino acids. Deamination occurs principally in the cells of the liver, but it can take place in any of the cells of the body. After deamination, the amino group is converted to ammonia, a toxic substance that would be injurious if it accumulated in the cells. In mammals, ammonia is combined with carbon dioxide to form the less toxic **urea,** which is excreted by the kidney. Urea synthesis takes place in the liver and kidney cells and involves a number of intermediate steps, including ornithine, citrulline and arginine, in a series of reactions known as the **urea cycle.**

Fat Metabolism. As we have pointed out, most fatty acids and glycerol are resynthesized into fat during absorption and the fat is transported as small globules, the **chylomicrons,** by the lymphatic system to the blood vessels. Within two or three hours after absorbing a fatty meal, the chylomicrons disappear from the blood; some are taken up by liver cells; others are digested within the blood stream by **lipoprotein lipase.** Lipoprotein lipase is produced in great quantities by the fat depots of the body (mesenteries, subcutaneous connective tissue, intermuscular connective tissue), and it is believed that most of the hydrolyzed fat is quickly absorbed and resynthesized by these tissues. In addition to being stored for future use, the fat in certain of these tissues helps to insulate the body against heat loss. Fat absorbed by the liver may be stored or undergo various conversions to other lipid materials. As the need arises for lipid material or more energy, fatty acids and glycerol are mobilized from the adipose tissue and liver and transported chiefly in combination with protein or as phospholipids to the various tissues.

Lipids are constituents of the plasma membrane and the membranes of many cell organelles. Fats are also a very important fuel reserve, for they can be metabolized in the citric acid cycle. They yield about twice as much metabolic water as other foodstuffs. Camels, for example, derive a great deal of their water from the metabolism of fat stored in their humps, and much of the water required by the embryos of egg-laying terrestrial vertebrates is derived from fat metabolism.

Vitamins. Other absorbed materials include minerals, steroids, the various building blocks of nucleotides, water and vitamins. By definition, the vitamins are organic substances that an animal needs in minute amounts and must obtain from its food, for they cannot be synthesized by the animal in question. Organisms differ in their capacities to synthesize these important molecules. All animals need ascorbic acid, for example, but most can synthesize it; only in primates and guinea pigs is it a vitamin needed in the diet. If a vitamin is lacking, metabolic processes dependent on it are impaired and a **deficiency disease** results. A list of the more common vitamins needed by man and their characteristics is presented in Table 12.2. Since their chemical composition was unknown when they were first discovered, they were designated by letters; we now know the chemical structure for most of them, and many can be synthesized in the laboratory.

Vitamins of the B complex and ascorbic acid are water soluble and are absorbed easily, but the others are lipid soluble and require bile salts to be absorbed in adequate quantities. Many are stored in body cells, particularly in liver cells. Most vitamins of the lipid soluble group are present in sufficient quantity to provide for body needs for a long time. A normal, adult person has enough vitamin D to last for several months, and enough vitamin A for two or three years. On the other hand, vitamin K and most of the water soluble ones (except for B_{12}) are stored only in small amounts; metabolic processes dependent upon them are soon interrupted if there is not an adequate intake.

Vitamins were discovered through an analysis of deficiency diseases, some of which have long plagued man. **Beriberi** has been common for centuries among Orientals and other peoples who subsist largely on polished rice. Rice hulls, which contain thiamine, prevent the disease when added to the diet. **Pellagra** used to be common in our southern states, for cornmeal, which formerly made up a large part of the diet, contains very little niacin. **Scurvy** was long the scourge of sailors, explorers and others who could not get fresh fruits and vegetables and the ascorbic acid they contain. Many Civil War prisoners such as the ones in Andersonville prison were victims of this disease. Captain James Cook was among the first to notice that feeding his crew such unusual foods (to sailors at least) as sauerkraut reduced the incidence of scurvy. He reported his findings to the Royal Society in 1776, and about two decades later, when more was known about the disease, the British Navy periodically enforced a ration

TABLE 12.2 COMMON VITAMINS

Vitamins	Common Sources	Function	Disease and Symptom if Deficient in Diet
A, Retinal	Butter, eggs, fish liver oils. Carotene in plants can be converted to A	Maintenance of epithelial cells, component of visual pigments	Scaly skin, easy infection, night blindness, xerophthalmia
B Complex [*] B_1, Thiamine	Yeast, meat, whole grain, eggs, milk, green vegetables	Coenzyme in decarboxylation, carbohydrate metabolism	Beriberi: nerve and muscle degeneration
B_2, Riboflavin	Same as B_1, colon bacteria	Coenzyme in cellular oxidations, electron transport	Stunted growth, cracked skin
B_6, Pyridoxine	Same as B_1	Coenzyme in amino acid metabolism	Dermatitis
G, Niacin	Same as B_1	Coenzyme in cellular oxidations	Pellagra: inflammation of skin, nervous disorders
B_{12}, Cobalamin, and Folic Acid	Same as B_1	Coenzymes in DNA synthesis, cell division	Anemia (Red cell precursors are most rapidly dividing cells.)
C, Ascorbic Acid	Citrus fruits, fresh vegetables. Destroyed on cooking	Collagen formation, maintenance of connective tissues and capillary walls	Scurvy: bleeding gums, swollen joints, general weakness
D, Cholecalciferol	Eggs, meat, liver oil. Cholesta-5,7-dienol in skin converted to D on exposure to ultraviolet light	Absorption and utilization of calcium and phosphorus	Rickets: weak bones, defective teeth
E, Alpha-tocopherol	Green vegetables, wheat germ, vegetable oils	Maintenance of reproductive cells	Sterility in poultry, rats, and possibly man
K, Several Naphthoquinone Compounds	Green vegetables, colon bacteria	Synthesis of prothrombin in liver, hence normal blood clotting	Bleeding

[*] Other B complex vitamins (pantothenic acid, biotin) are seldom deficient in human diets.

of lime juice on members of all crews. British sailors have been called "limeys" ever since. **Rickets** is a disease of children who have not received sufficient vitamin D; it is characterized by marked malformation of the skeleton (Fig. 12.9).

The mechanism of vitamin action is an important area of biochemical research. A derivative of vitamin A (retinal) is a component of the visual pigments, and plays an essential role in the chemistry of vision (p. 347). The B complex vitamins are converted to coenzymes which serve as intermediary acceptors and donors of vari-

ous atoms and groups in many metabolic pathways. Thiamine (B_1), for example, participates in the decarboxylation of pyruvate. In its absence pyruvate is not converted to acetyl coenzyme A and cannot enter the citric acid cycle (Fig. 4.3, p. 69); carbohydrate metabolism is reduced and lipid metabolism increased. Since the nervous system is highly dependent on carbohydrates as a source of energy, thiamine deficiency leads to the neuromuscular disorders characteristic of beriberi. The other B vitamins function as coenzymes in different metabolic reactions. The bio-

Figure 12.9 Illustration of hypophosphatemic vitamin D-refractory rickets of simple type. Mother and four-year-old daughter show typical deformities. (Courtesy of D. Fraser, J.A.M.A., *176*:281, 1961.)

chemical functions of ascorbic acid and of most of the lipid soluble vitamins are not yet clear. They are not coenzymes functioning in acceptor-donor reactions.

Although vitamins are essential in the diet, an excess of certain ones can be toxic. Excess vitamin D causes an abnormal deposition of calcium in soft tissues. Most of our vitamin D does not come from our diet as such, but is manufactured in the skin by the action of ultraviolet light on cholesta-5,7-dienol.

12.9 RESPIRATORY MEMBRANES

Cellular respiration is an oxidative process in which the energy in the absorbed food molecules is made available for the various cellular activities. To maintain it, oxygen must be continuously supplied and the by-products, carbon dioxide and water, must be continuously removed. In vertebrates this involves the uptake of oxygen and the release of carbon dioxide in the respiratory organ, the transport of these gases by the blood and their exchange between the blood and cells. Gas transport by the blood, together with exchanges between the blood and cells, were considered in Chapter 5. Here we are concerned with the structure and function of the vertebrate respiratory organs, in which gas exchange with the environment occurs.

All respiratory surfaces, whether in a worm, a fish or a man, consist of a moist, semipermeable, vascular membrane exposed to the external environment so that gas exchange by diffusion can take place between the blood and the environment. The entire body surface of primitive organisms may serve as a respiratory membrane, but the respiratory surface in the higher animals is generally confined to a limited region and protected in various ways. This reduces the chance of mechanical injury and the amount of body water lost or gained by osmosis via this route, but restricting the size of this membrane poses the problem of providing enough surface for gas exchange. Each kind of vertebrate has had to solve the dilemma of how to expose these delicate membranes to the environment while protecting them adequately.

12.10 THE RESPIRATORY SYSTEM OF FISHES

The respiratory organs of fish are **gills** located in the gill slits and attached to the visceral arches. Since water contains only about 1 per cent dissolved oxygen in contrast to 21 per cent in air, fish must move a large volume of water across their gills. The mechanics of this in a typical bony fish has recently been worked out by Dr. George Hughes. During inspiration both the pharynx and **opercular chambers** expand (Fig. 12.10), causing a decrease in pressure within them relative to the surrounding water. Water is drawn in only through the mouth, for a thin membrane on the free edge of the operculum acts as a valve, preventing entry

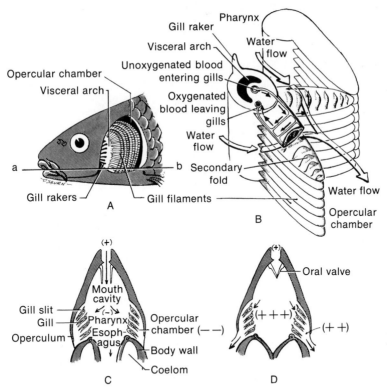

Figure 12.10 External respiration in fishes. *A*, The operculum has been cut away to show the gills in the gill chamber. *C* and *D*, Frontal sections through the mouth and pharynx in the plane of line *a–b* in *A*. Inspiration occurs in *C* and expiration in *D*. Relative water pressures in the various parts of the system are shown by + and −. *B*, An enlargement of several gill filaments. (*B* modified after Hughes.)

in this direction. Once in the pharynx, the water passes across the gills into the opercular chamber as a result of the lower pressure there. The opercular chamber acts as a suction pump. Expiration begins with a closure of the mouth, or of **oral valves,** and a contraction of the pharynx and opercular chambers so that the pressure in them exceeds that of the surrounding water. During both inspiration and expiration there is a pressure gradient moving water across the gills. Water does not go down the esophagus, for this is collapsed except when swallowing. Food and other particles are prevented from clogging the gills by **gill rakers,** which act as a strainer.

The gills themselves consist of numerous filaments bearing secondary folds perpendicular to them. Capillary beds are located in the latter, and it is here that gas exchange with the water occurs. This exchange is particularly efficient because the direction of the blood flow in the capillary beds is

counter to that of the water current. This ensures that for the duration of the contact there is a greater oxygen tension in the water than in the adjacent capillary bed (Fig. 12.11). The blood can continue to pick up oxygen from the water as long as it is not fully saturated and any oxygen remains in the water. If blood and water were flowing in the same direction, some point of equilibrium would soon be reached at which the blood was not fully saturated, yet considerable oxygen remained in the water.

In addition to gas exchange, the body gains or loses water through the gills and, except for cartilaginous fishes, some nitrogenous wastes are excreted here. The saltwater teleosts also excrete salts through the gills (p. 333).

A number of fishes live in water which has a low oxygen content and they supplement gill respiration by occasionally gulping a bubble of air. The oxygen in the air can be extracted by gills so long as they

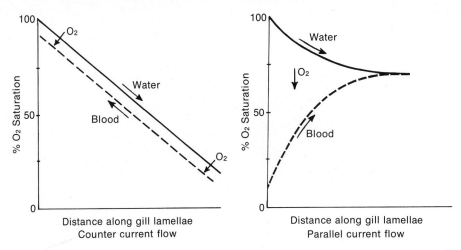

Figure 12.11   Theoretical diagrams to illustrate the effects of counter current or parallel flow on the exchange of oxygen between water and blood. The equilibrium attained in parallel flow would be somewhere above 50 per cent saturation of the blood because of the presence of hemoglobin, yet the blood becomes less fully saturated than it would during counter current flow. (Modified after Hughes.)

remain moist and are not collapsed. Closing the opercular chamber enables the mud-skipper of southeastern Asia to keep its gills moist for a while, and even to come out onto the land. The European loach swallows air and extracts the oxygen in a special chamber of its intestine! Other fish have vascular outgrowths from various parts of the pharynx or opercular chamber that serve as accessory respiratory organs. Seemingly, the development of lungs, which are ventral outgrowths from the pharynx, by early bony fishes was just one of many adaptations which have evolved to supplement aquatic respiration.

Extracting oxygen from swamp water, which is probably the environment in which animals with lungs evolved, poses the problem of saturating the blood with oxygen in an environment with a low oxygen and high carbon dioxide content. As we explained in Chapter 5, the presence of carbon dioxide reduces the oxygen-carrying capacity of respiratory pigments such as hemoglobin. The chemical properties of the hemoglobin of contemporary swamp fish have changed in such a way that it can take up more oxygen in the presence of a given amount of carbon dioxide. This change must also have occurred during the evolution of ter-restrial vertebrates, for the carbon dioxide content of the lungs is always higher than that of the external environment, though, of course, not so high as that in the tissues.

12.11   THE RESPIRATORY SYSTEM OF TERRESTRIAL VERTEBRATES

The lungs of early bony fishes evolved into hydrostatic swim bladders in most of their descendants, but they were retained in some that remained in fresh water, and it is from certain of these fishes that tetra-pods evolved. Larval amphibians resemble many larval fishes in having external gills protruding from the body surface beside the gill slits. **Lungs** develop later and begin to function before the aquatic larva meta-morphoses to a terrestrial adult. The internal surface of the sac-like amphibian lung is increased somewhat by pocket-like folds (Fig. 12.12), but the surface area is far smaller than in higher tetrapods. Amphib-ians also have a relatively inefficient method of ventilating the lungs. A frog, for example, moves air in and out of its lungs by a three phase **buccopharyngeal pump.** During the first phase, the floor of the pharynx is lowered and fresh air entering through the nostrils goes into the distended postero-ventral portion of the buccopharyngeal cavity (Fig. 12.13). The **glottis,** which is a slit-like opening from the pharynx to the laryngotracheal cavity and lungs, remains closed during this phase, but opens in the second one. A slight contraction of the flanks then drives stale air from the lungs,

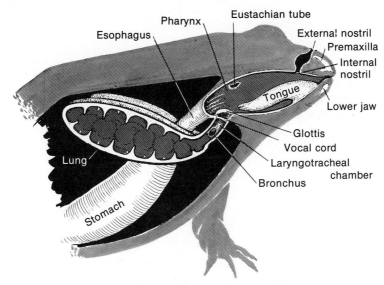

Figure 12.12   A diagrammatic longitudinal section of the respiratory system of the frog.

across the dorsal portion of the buccopharyngeal cavity, and out the nostrils. There is little mixing of stale and fresh air. Finally in phase three the nostrils are closed by the lower jaw pushing against the moveable premaxillary bones (Fig. 12.12), the glottis remains open, and a contraction of the pharynx drives the fresh air into the lungs. This type of pump, which utilizes fish-like pharyngeal movements, may provide the animal with enough oxygen, for air is about 21 per cent oxygen. However, there is not

enough exchange of air to remove enough carbon dioxide. Modern amphibians supplement pulmonary ventilation by gas exchange through the skin, which is thin, moist, and vascular. About 75 per cent of the oxygen needs of a frog are obtained through the lungs and the rest from the skin, but 66 per cent of the carbon dioxide is eliminated through the skin. The need to supplement pulmonary respiration by cutaneous respiration, and the loss of body water that this entails, is a factor that has

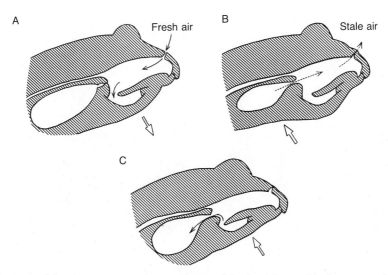

Figure 12.13   Diagram of the three stages in the ventilation of the frog's lungs. Black arrows indicate the course of fresh and stale air; white arrows the movements of the floor of the pharynx and flanks.

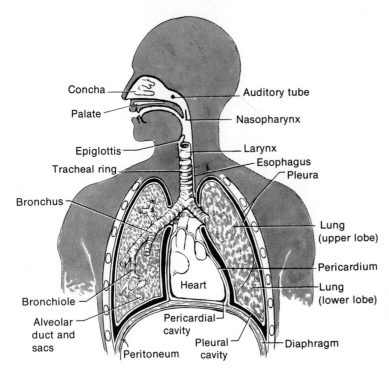

Concha — Auditory tube
Palate — Nasopharynx
Epiglottis — Larynx
Tracheal ring — Esophagus
Pleura
Bronchus — Lung (upper lobe)
Pericardium
Heart — Lung (lower lobe)
Bronchiole — Pericardial cavity
Alveolar duct and sacs — Pleural cavity — Diaphragm
Peritoneum

Figure 12.14 Respiratory system of man. Details of the alveolar sacs, here drawn enlarged, are shown in 12.16.

prevented amphibians from fully exploiting the terrestrial environment. Cutaneous respiration also places a limit on the maximum size of amphibians; small organisms have more body surface relative to mass than large ones. Higher terrestrial vertebrates have evolved lungs with greater internal surface areas, and more efficient ways of ventilating the lungs. They have thus been able to dispense with cutaneous respiration and the constraints that it imposes.

In mammals (Fig. 12.14), air is drawn into the paired **nasal cavities** through the **external nostrils,** or **nares.** These cavities are separated from the mouth cavity by a bony palate, and the animal can breathe while food is in its mouth. The surface area of the cavities is increased by a series of ridges

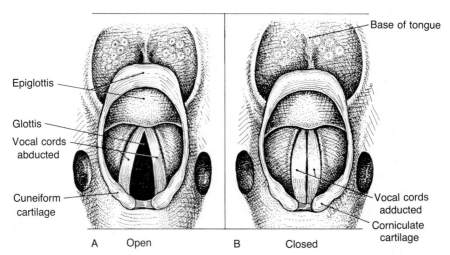

Base of tongue
Epiglottis
Glottis
Vocal cords abducted
Cuneiform cartilage
Vocal cords adducted
Corniculate cartilage

A   Open   B   Closed

Figure 12.15   A view of the vocal cords looking into the larynx from above. *A,* Normal position of vocal cords; *B,* position of cords during speech. (From Jacob and Francone: Structure and Function in Man, 2nd ed., W. B. Saunders Co., 1970.)

known as **conchae,** and the nasal mucosa (in addition to having receptors for smell) is vascular and ciliated and contains many mucous glands. In the nasal cavities the air is warmed and moistened and minute foreign particles are entrapped in a sheet of mucus, which is carried by ciliary action into the pharynx where it is swallowed or expectorated. Inspired air is moistened in primitive tetrapods such as the frog, but cold-blooded tetrapods do not need so much conditioning of the air as birds and mammals.

Air continues through the **internal nostrils,** or **choanae,** passes through the **pharynx** and enters the **larynx,** which is open except when food is swallowed. The raising of the larynx during swallowing can be demonstrated by placing your hand on the Adam's apple, the external protrusion of the larynx. The **epiglottis** flips back over the entrance of the larynx when it is raised.

The larynx is composed of cartilages derived from certain of the visceral arches and serves both to guard the entrance to the windpipe, or **trachea,** and to house the **vocal cords** (Fig. 12.15). The vocal cords are a pair of folds in the lateral walls of the larynx. They can be brought close together, or be moved apart, by the pivoting of laryngeal cartilages connected to their dorsal ends. The **glottis** of mammals is the opening between the vocal cords. When the cords are close together, air expelled from the lungs vibrates them and they, in turn, vibrate the column of air in the upper respiratory passages just as the reed in an organ pipe vibrates the air in the pipe. This phenomenon is called **phonation,** or the production of a vocal sound. **Speech** is the shaping of the vocal sounds into patterns that have meaning for us; it is accomplished by the pharynx, mouth, tongue and lips. Muscle fibers in the vocal cords and larynx control the tension of the cords and the pitch of our voice. We normally speak when air is expired. Ventriloquists have trained themselves to speak during inspiration.

The **trachea** extends down the neck and finally divides into **bronchi** that lead to the pair of **lungs.** Unlike the esophagus, which is collapsed except when a ball of food is passing through, the trachea is held open by C-shaped cartilaginous rings, and air can move freely back and forth. Its mucosa continues to condition the air.

The lungs of amphibians lie in the anterodorsal part of the pleuroperitoneal cavity. The pleural cavities of mammals lie within the chest, or **thorax,** and are separated from the peritoneal cavity by a muscular **diaphragm.** A coelomic epithelium, the **pleura,** lines the pleural cavities and covers the lungs. Each bronchus enters a lung, accompanied by blood vessels and nerves, in a mesentery-like fold of pleura (Fig. 12.14).

The bronchi branch profusely within the lungs, and the walls of the respiratory passages become progressively thinner (Fig. 12.14). Each passage eventually terminates in an **alveolar sac** whose walls are so puckered by pocket-shaped **alveoli** that it resembles a cluster of miniature grapes (Fig. 12.16). The alveolar walls are extremely thin and could not be detected with certainty until they were studied with the electron microscope. A network of capillaries, which is so dense that there is little space left between the individual vessels, covers the alveoli. Gas exchange by diffusion occurs between the air in the alveoli and the blood. Endotherms need a large surface for gas exchange to sustain their high level of metabolic activity. The alveolar surface of human lungs is estimated to be 55 square meters.

Lipoprotein secreted by alveolar cells is present in the thin film of liquid that lines the alveoli. It acts as a detergent to reduce surface tension within the alveoli, and this is important in several ways. It helps in water conservation, for the less the surface tension, the less water will be drawn from the capillaries into the lumina of the alveoli. A reduced surface tension also reduces the amount of muscular effort needed to expand the alveoli as air is inspired. Finally, the lipoprotein helps to equalize forces acting in the alveoli during the different phases of respiration. Surface tension is great in a sphere of small radius and decreases as the radius increases. The alveoli are small at the end of expiration and surface tension is therefore high, but the tension is reduced by the concomitant concentration of the lipoprotein in the reduced surface area of the alveoli. Surface tension decreases when the alveoli expand as air is inspired and their radii increase, but this is compensated for by the attenuation of the lipoprotein in the expanding surface area of the alveoli.

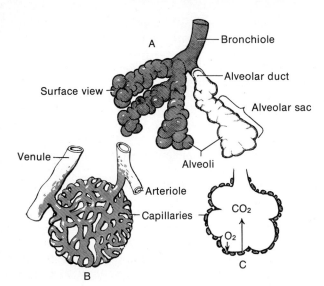

Figure 12.16 *A*, Termination of the respiratory passages in the mammalian lung; *B*, a further enlargement to show the dense capillary network covering a single alveolus; *C*, an alveolus in section. Alveoli have a diameter of 0.2 to 0.3 mm.

12.12 THE MECHANICS AND CONTROL OF BREATHING

Mammalian lungs are ventilated by changing the size of the thoracic cavity and, consequently, the pressure within the lungs. The lungs follow the movements of the chest wall, for they are at once separated from it and united to it by the adhesive force of a thin layer of fluid that lies within the pleural cavities. During normal, quiet inspiration, the size of the thorax is increased slightly, intrapulmonary pressure falls to about 3 mm. Hg below atmospheric pres-

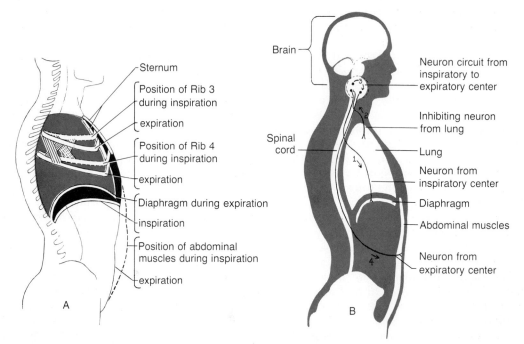

Figure 12.17 Mechanics and control of breathing. *A*, The elevation of the ribs and depression of the diaphragm during inspiration increases the size of the chest cavity, indicated by the black area. *B*, A diagram of the nervous mechanism for controlling the rhythm of breathing. See text for explanation.

sure, and air passes into the lungs until intrapulmonary and atmospheric pressures are equal. During normal **expiration,** the size of the thorax is decreased, intrapulmonary pressure is raised to about 3 mm. Hg above atmospheric pressure, and air is driven out of the lungs until equilibrium is again reached. During inspiration, the thorax is enlarged by the contraction of the dome-shaped **diaphragm** and the **external intercostal muscles.** The diaphragm pushes the abdominal viscera posteriorly and increases the length of the chest cavity (Fig. 12.17A); the external intercostals raise the sternal ends of the ribs and expand the dorsoventral diameter of the chest. Expiration results primarily from the relaxation of the inspiratory muscles and the elastic recoil of the lungs and chest wall, which are stretched during inspiration. But during heavy breathing, antagonistic expiratory muscles can decrease the size of the thoracic cavity. Contraction of **abdominal muscles** forces the abdominal viscera against the diaphragm and pushes them anteriorly; **internal intercostals** pull the sternal ends of the ribs posteriorly.

The lungs of an adult man can hold about 6 liters of air, but in quiet breathing they contain only about half this amount, of which 0.5 liter is exchanged in any one cycle of inspiration and expiration. This half liter of **tidal air** is mixed with the 2.5 liters of air already in the lungs. Vigorous respiratory movements can lower and raise the intrapulmonary pressure 60 mm. Hg below and above atmospheric pressure, and under these conditions 4 to 5 liters of air can be exchanged. This maximum is known as the **vital capacity.** There is always, however, at least a liter of **residual air** left in the lungs to mix with the tidal air, for the strongest respiratory movements cannot collapse all the alveoli and respiratory passages. Since the inspired air always mixes with a certain amount of stale air already in the lungs, alveolar air always has a lower oxygen content and a higher content of carbon dioxide than atmospheric air. Alveolar air is also saturated with water vapor.

Respiratory movements are cyclic and are controlled by inspiratory and expiratory centers (collectively called the **respiratory center**) in the medulla of the brain. The respiratory center has a rhythmic activity alternately sending out impulses to the inspiratory and expiratory muscles. The basis for this rhythmicity is not completely understood, but a neuronal feed-back mechanism does play an important role. The inspiratory center sends out impulses along the nerves to the inspiratory muscles (neuron #1, Fig. 12.17B), and we breathe in. The alveoli fill with air, become stretched, and the resultant sensory impulses traveling to the respiratory center inhibit inspiration (#2). At the same time, impulses that were initiated in the inspiratory center and took a rather circuitous route within the brain reach the expiratory center (#3) and stimulate it to send impulses out to the expiratory muscles (#4). As we exhale, inhibiting impulses stop coming from the stretch receptors in the lungs, and the inspiratory center becomes active again.

The respiratory center receives many other signals indicating important changes in the body and modifies the depth and rate of breathing accordingly. Increased metabolism during exercise, for example, results in an increased carbon dioxide content of the blood. This, in turn, increases the excitability of the respiratory center and we breathe more rapidly and deeply. The same thing happens when we voluntarily hold our breath. Since the lungs are not being ventilated, carbon dioxide accumulates in the alveolar air and blood and eventually reaches a level that activates the respiratory center, and we breathe again involuntarily. One cannot suffocate by holding one's breath. The accumulation of carbon dioxide in the blood of a newborn baby normally initiates its breathing.

The oxygen content of the blood is less important than the carbon dioxide content in stimulating the respiratory center. However, chemoreceptive cells in the carotid bodies, associated with the carotid arteries supplying the head, are activated by any marked drop in oxygen content. Impulses initiated by them stimulate the inspiratory center.

Receptors in the larynx and trachea can also affect respiration. If food inadvertently enters these passages, these receptors are stimulated and a very vigorous expiration, i.e., a cough, results. The cough reflex is one of many safeguards in the body that are activated if something goes wrong with the primary control mechanism, in this case the swallowing reflex.

ANNOTATED REFERENCES

Additional information on digestion and respiration can be found in the general references on vertebrate organ systems cited at the end of Chapter 11.

Andersen, H. T.: Physiological adaptations in diving vertebrates. Physiological Review 46:212, 1966. An excellent summary of the respiratory, vascular and other adaptations made by terrestrial vertebrates that have readapted to an aquatic mode of life.

Baldwin, E. B.: Dynamic Aspects of Biochemistry. 3rd ed. Cambridge, Cambridge University Press, 1957. The chemistry of digestion and many aspects of metabolism are lucidly described.

Clements, J. A.: Surface tension in the lungs. Scientific American 207:120, (Dec.) 1962. An account of the discovery and great importance of the surface active agent in the alveoli of mammalian lungs.

Comroe, J. H., Jr.: The lung. Scientific American 214:57, (Feb.) 1966. A fascinating summary of pulmonary anatomy and physiology.

Conference for National Cooperation in Aquatics: The New Science of Skin and Scuba Diving. New York, Association Press, 1962. This standard and widely used manual includes excellent discussions on the physiology of diving.

Fulton, J. F.: Selected Readings in the History of Physiology. Springfield, Ill., Charles C Thomas, 1930. Interesting accounts of the discovery of digestive and respiratory processes are included in this very readable book.

Hughes, G. M.: Comparative Physiology of Vertebrate Respiration. Cambridge, Harvard University Press, 1963. A review of breathing mechanisms, gas transport and cellular respiration from fish to mammals.

13. Blood and Circulation

All animals, from the simplest protozoan to the most complex vertebrate, must have some means for distributing nutrients and other materials throughout their bodies. As we pointed out in Chapter 5, the simple diffusion of molecules is adequate to transport substances in the smaller and less active organisms. But the vertebrates and many of the higher invertebrates are so large and active that diffusion alone cannot suffice. Complex circulatory systems are necessary for the rapid transport of nutrients absorbed from the alimentary tract, and of oxygen from the lungs, to all the tissues, and for carrying carbon dioxide and other metabolic wastes to the sites where they are discharged from the body.

13.1 THE VERTEBRATE CIRCULATORY SYSTEM

The vertebrate circulatory system not only transports gases, nutrients and waste products but conveys hormones to supplement the nervous system in integrating body activities. It plays an important role in maintaining **homeostasis,** i.e., the constancy of the internal environment. It maintains the pH, salt content and fluid volume of the body fluids within narrow limits. The rate of its circulation through the skin is a factor in the control of body temperature in birds and mammals. Special cells in the blood function in wound healing and in protecting the body from the invasion of viruses and bacteria.

The circulatory system includes not only the complex system of vessels but also the fluids within them. About one-third of the 15 liters of extracellular fluid in the body of an adult man is **blood.** The remainder includes the **tissue fluid** that lies between and bathes the cells of the body, the **lymph** that moves slowly in the lymph vessels, the **cerebrospinal fluid** within and around the central nervous system, the aqueous and vitreous humors of the eye, and the fluids in the coelom. The chief difference between blood and tissue fluid or lymph is the presence in the blood of red blood cells and abundant soluble proteins.

The fundamental pattern of the vessels in a mammal is shown in Figure 13.1. A muscular **heart** propels blood through **arteries** to **capillaries** in the tissues. Exchanges between the circulatory system and the tissues can occur only through the walls of the capillaries. Molecules of nutrients, wastes, oxygen, carbon dioxide and water, but not most of the large protein molecules or the red blood cells, pass readily through the capillary walls. The tissues are drained by the veins, which return blood to the heart, and by a separate system of **lymph capillaries.** Lymph capillaries lead to **lymph vessels,** which pass through **lymph nodes,**

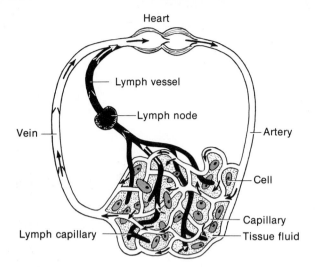

Figure 13.1   The fundamental structure of the mammalian circulatory system. Arrows indicate the direction of the flow of blood and tissue fluid.

and finally empty into the veins where the venous pressure is lowest, a short distance from the heart. The lymph nodes are an important link in the body's defense mechanisms. They produce one kind of white blood cell (**lymphocytes**) and contain cells that engulf foreign particles.

13.2   BLOOD PLASMA

Blood is one of the tissues of the body. It consists of a liquid component, the **plasma**, and several types of formed elements—red blood cells, white blood cells and platelets (Fig. 13.2)—which flow along in it. Plasma is about 90 per cent water, 7 to 8 per cent soluble proteins, 1 per cent electrolytes, and the remaining 1 to 2 per cent includes (1) nutrients such as glucose, amino acids, lipids, and vitamins; (2) intermediates of metabolic processes such as pyruvic and lactic acid; (3) waste products such as urea and uric acid; (4) dissolved gases; and (5) hormones. Many of these molecules are free in solution; others are bound to certain

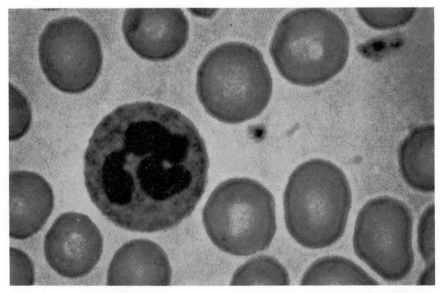

13.2   Photomicrograph of human blood cells. Many erythrocytes and one large leukocyte (a neutrophil) can be seen. The single dark dot and the clump of several dots are platelets. (From Zucker, Scientific American, Vol. 204, No. 2.)

plasma proteins. Some of the trace metals (copper, for example) are bound to specific transport proteins. Clearly plasma is a complex liquid in dynamic equilibrium with other body fluids; many substances move in and out of it, but its over-all composition remains remarkably constant. A constant level of nutrients is maintained by a careful balance between absorption, storage, removal from storage, utilization, and excretion. We have considered the important role of the liver in maintaining the concentration of blood glucose.

The chief plasma proteins are albumins, globulins and fibrinogen. Other components of the plasma can easily pass through the semipermeable capillary walls, but the proteins are rather large molecules and most remain in the blood in the capillary bed. They exert an osmotic pressure that is responsible for the return of water from the tissue fluids. Hydrostatic pressure, i.e., blood pressure, forces the water out of the capillaries into the tissue fluid. These two forces normally just balance and keep the blood volume constant. Many of the proteins have additional functions. Fibrinogen is essential for blood clotting, and the numerous globulins have functions in blood clotting (cf. section 13.4), as transport proteins, and as immunoglobulins, or antibodies that provide immunity to many infectious diseases. Most of the plasma proteins are synthesized in the liver, but the immunoglobulins are formed in the lymphoid tissues and plasma cells (cf. section 13.6).

The plasma proteins, hemoglobin in the red blood cells and certain of the inorganic salts in the blood are also important buffers. A **buffer** is a mixture of a weak acid and its salt, or of a weak base and its salt. A buffer tends to prevent a change in the pH of a solution when an acid or base is added. Complex animals such as mammals cannot tolerate wide fluctuations in pH, and the pH of the blood is held remarkably constant, at about 7.4. Buffers combine reversibly with the hydrogen ions (H^+) released by the dissociation of acids into their constituent ions. Acidic substances are constantly produced as by-products of cell metabolism and enter the blood. The carbon dioxide produced in cellular respiration tends to increase the acidity of the blood for it combines with water to form carbonic acid, H_2CO_3. Basic substances, which release hydroxyl ions (OH^-), are much less common by-products of metabolism. Buffers neutralize their effects by releasing hydrogen ions, which combine with the hydroxyl ions to form water (H_2O). Eventually, the acidic or basic substances are removed from the body, carbon dioxide by the lungs and the others by the kidneys.

13.3 RED BLOOD CELLS

The red blood cells, or **erythrocytes,** are the most numerous of the formed elements of the blood, there being about 5,000,000 of them in each cubic millimeter of blood in an adult human. Mammalian red cells lose their nuclei as they develop, and become biconcave discs. Such a shape provides more surface area than a sphere of equal volume, and the increased surface area, in turn, facilitates the passage of materials through the plasma membrane.

Erythrocytes contain the respiratory pigment **hemoglobin,** which acts as a buffer and is essential for the transport of oxygen and carbon dioxide. As we explained in section 5.3, hemoglobin (Hb) combines with oxygen in the capillaries of the lungs, where the oxygen tension is high, to form **oxyhemoglobin** (HbO_2), and oxyhemoglobin releases oxygen in the tissue capillaries, where the oxygen tension is low. It has been estimated that we would need a volume of blood 75 times as great or the blood would have to circulate very much faster than it does if all of the oxygen were carried in physical solution instead of in combination with hemoglobin.

The uptake and release of oxygen by hemoglobin is intimately related to the transport of carbon dioxide. Tissues have a high carbon dioxide tension; this causes CO_2 to diffuse into the blood. Only a small portion is carried in physical solution. Up to 20 per cent combines directly with oxyhemoglobin to form **carbaminohemoglobin** ($HbCO_2$), and in so doing tends to drive off the oxygen as shown in the first line in the diagrammatic red cell in Figure 13.3A.

This reaction does not have any marked effect upon the pH of the blood. The rest of the carbon dioxide unites with water to form carbonic acid. This reaction occurs much more rapidly in the erythrocytes than

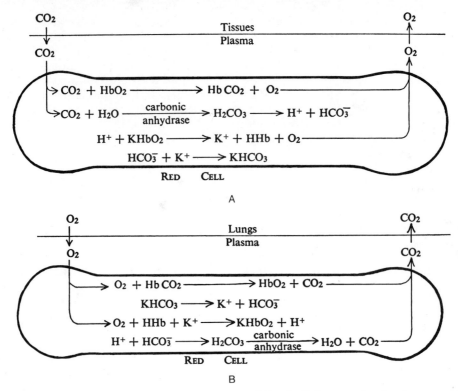

Figure 13.3 Diagrams of the major chemical reactions in a red blood cell that are related to the transport of carbon dioxide and oxygen. *A*, Carbon dioxide enters a red cell in a tissue capillary and oxygen is released; *B*, the opposite reactions occur within a cell in a lung capillary.

in the plasma because they contain the enzyme **carbonic anhydrase,** which speeds up the reaction some 1500 times. Most of the carbonic acid, in turn, dissociates into its constituent hydrogen and bicarbonate ions (second line, Fig. 13.3*A*). The hydrogen ions produced combine with oxyhemoglobin, and this facilitates the dissociation of oxyhemoglobin and the release of oxygen to the tissues. Potassium ions (K^+), which are loosely associated with oxyhemoglobin, are also released as oxyhemoglobin takes up hydrogen ions to become reduced hemoglobin (third line, Fig. 13.3*A*). Most of the bicarbonate ions diffuse out of the cell into the plasma, where they combine with sodium. As the cell loses these negative ions, chloride ions (Cl^-) diffuse in, so the electrochemical balance of the cell is maintained. Thus most of the carbon dioxide is carried as carbonic acid and bicarbonate ions.

These reactions associated with the transport of carbon dioxide also maintain the pH

of the cell close to neutrality. Oxyhemoglobin is a stronger acid than hemoglobin, and the conversion of oxyhemoglobin (HbO_2) to hemoglobin (Hb) would tend to raise the pH within the red cell (make it more alkaline). The formation and dissociation of carbonic acid would tend to lower the pH within the red cell (make it more acid). These two opposing phenomena tend to balance each other and the pH of the erythrocyte is maintained essentially unchanged.

All these reactions are reversible, and their direction is determined by the relative levels of carbon dioxide and oxygen. In the tissues, carbon dioxide is continually produced, oxygen continually consumed. There is a relatively high level of carbon dioxide, so the reactions move in the direction of binding carbon dioxide and releasing oxygen. As blood passes through the capillaries in the lungs, oxygen diffuses in and carbon dioxide diffuses out. There is a relatively high oxygen tension

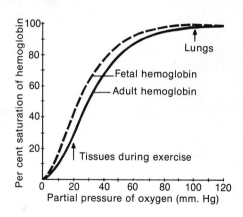

Figure 13.4 Oxygen dissociation curves for fetal and adult hemoglobin. At a partial pressure of 100 mm. Hg in the lungs, adult hemoglobin becomes about 98 per cent saturated. It can give up nearly 70 per cent of its oxygen when it reaches the tissues, where the partial pressure of oxygen may be as low as 20 mm. Hg.

so the reactions move in the direction of binding oxygen and releasing carbon dioxide as shown in Figure 13.3*B*.

These reactions enable the blood to carry a great deal more oxygen and carbon dioxide than it could in simple physical solution; they prevent the pH of the blood from changing greatly; and they facilitate the release of oxygen in the tissues of the body and the release of carbon dioxide in the lungs. The amount of oxygen that can be delivered to the tissues is shown in the oxygen dissociation curve for hemoglobin in Figure 13.4. (Review pp. 88 to 93 for an interpretation of such a curve.)

The oxygen-carrying capacity of hemoglobin is determined by its structure. Each molecule of hemoglobin is composed of a protein (a **globin**) which in turn is made up of four polypeptide chains. One prosthetic group (a **heme**), containing an atom of ferrous iron, is associated with each chain, hence there are four iron atoms in each molecule of hemoglobin. During the formation of oxyhemoglobin, one molecule of molecular oxygen (O_2) may form a very loose bond with each iron atom. Since a single erythrocyte contains as many as 265,000,000 molecules of hemoglobin, a tremendous amount of oxygen can be carried.

The reversible association of oxygen with iron is influenced by many factors. Oxygenation of one or more hemes affects the ease of oxygenation of neighboring ones, and these interactions influence the shape of the oxygen dissociation curve for hemoglobin (Fig. 13.4). The nature of the globin also affects the oxygen binding properties of the hemes. Heme alone does not bind

reversibly with oxygen, and differences in the peptide chains of the globin give hemoglobins with different properties. Fetal hemoglobin, for example, has a greater affinity for oxygen than adult hemoglobin. At any given partial pressure of oxygen it has a higher per cent saturation than maternal hemoglobin (Fig. 13.4), which, of course, enables it to take oxygen away from the mother's blood. Fetal hemoglobin is gradually replaced by the adult type within a year of birth.

Mature mammalian erythrocytes do not survive indefinitely. Experiments which involve tagging them with radioactive iron show that they have an average life span of 120 days. Cells lining the blood spaces of the spleen and liver eventually engulf or **phagocytize** the red cells and digest them. The iron of the heme is salvaged by the liver and is reused, but the heme is metabolized to bile pigments and excreted. Under normal circumstances, the erythrocyte population is in a steady state with new ones being synthesized as rapidly as the old are destroyed. But if delivery of oxygen to the tissues is reduced, as during hemorrhage or when ascending to a high altitude, more cells are made available so that delivery of oxygen to the tissues is restored to normal levels. Additional erythrocytes can be released immediately into the circulating blood from reserve stores in the spleen, but the reduced oxygen tension also triggers a set of reactions that leads to an increased rate of synthesis of erythrocytes. Cells in the kidney in particular respond to a lower oxygen tension by synthesizing a hormone known as **erythropoietin.** This passes in the blood to the red bone marrow where it stimulates the first stage in red cell production, the formation of hemocytoblasts from primordial stem cells. In lower vertebrates, red cells are produced in vascularized connective tissues of the kidney, liver and spleen. These sites are important during the embryonic development of mammals, but the red bone marrow is the primary source of erythrocytes in the adult.

Erythrocyte destruction and production are surprisingly rapid. From the total number of red cells in the body and their average life span, one can calculate that about 2,500,000 are made and destroyed each second of the day and night. If the rate of production of cells or of hemoglobin decreases, some type of **anemia** results. Anemia is characterized by a decrease in the number of red cells per cubic millimeter of blood, by a decrease in the amount of hemoglobin per red cell, or both. In pernicious anemia the number of erythrocytes steadily decreases. Eating large quantities of liver increases the rate of red cell formation, for liver is rich in vitamin B_{12}, which is necessary for normal erythrocyte development. A person with pernicious anemia cannot absorb enough B_{12}, even though the requisite amount may be present in the diet, because the lining of his stomach does not secrete enough "intrinsic factor," necessary for the absorption of B_{12}. If an excess is made available by giving foods especially rich in B_{12}, enough can be absorbed. It is now known that vitamin B_{12} is a necessary coenzyme in DNA synthesis, which precedes cell division. Since erythrocyte precursors are among the most rapidly dividing cells in the body, a deficiency in vitamin B_{12} is first manifested by anemia.

13.4 PLATELETS AND HEMOSTASIS

Platelets are non-nucleated blobs of cytoplasm that continually bud off from giant cells in the bone marrow. They number about 300,000 per cubic millimeter of blood in man and live for eight to ten days. They, and the thrombocytes of lower vertebrates, play an important role in **hemostasis,** that is stopping blood flow after an injury. Hemostasis involves a short term

constriction and plugging of small blood vessels followed by the formation of a blood clot which, by its contraction, holds the injured tissues together until healing occurs. Nerve reflexes cause a vasoconstriction of small blood vessels, but platelets assist in the early stages of hemostasis for they swell at the site of the injury and release various substances. Released serotonin prolongs vasoconstriction, and released adenosine diphosphate aggregates the platelets, forming a temporary plug in the vessels.

The platelets, as well as the injured tissues, also release **thromboplastin** and other substances that initiate a "cascade of proenzyme-enzyme transformations" in the plasma that result in the formation of a clot. The conversion of one proenzyme to an active enzyme triggers the conversion of a second one and so on through at least five steps, several of which require the presence of calcium ions. When whole blood is removed from the body, clotting can be prevented by adding substances such as sodium citrate that bind calcium ions. One advantage of this complexity is the amplification that results. The few molecules involved in the first step affect more and more at each successive step until the reaction finally involves countless billions. The final enzyme in the initial phase of the clotting reaction, **prothrombinase,** in the presence of phospholipids released by the platelets and calcium ions, then converts a plasma globulin, **prothrombin,** into **thrombin** by splitting off certain peptide fragments. Thrombin, in turn, acts as an enzyme and catalyzes the conversion of the soluble protein **fibrinogen** into insoluble **fibrin.** This involves the removal of two peptides from the fibrinogen molecule to form activated **fibrin monomer.** Fibrin monomer polymerizes rapidly into long fibrin threads which form a delicate network within which the formed elements of the blood are trapped. A **clot** has formed. The clot subsequently contracts and squeezes out the liquid phase, called **serum.** Serum lacks fibrinogen and cannot clot. Vitamin K, necessary for clotting, does not enter into this series of reactions directly, but is essential for the production of prothrombin in the liver.

The clotting reactions can be summarized in the following scheme:

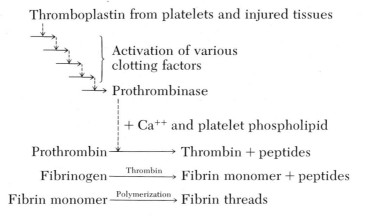

Thromboplastin from platelets and injured tissues

Activation of various clotting factors

Prothrombinase

+ Ca^{++} and platelet phospholipid

Prothrombin $\longrightarrow$ Thrombin + peptides

Fibrinogen $\xrightarrow{\text{Thrombin}}$ Fibrin monomer + peptides

Fibrin monomer $\xrightarrow{\text{Polymerization}}$ Fibrin threads

Many safeguards reduce the chance of clots occurring accidentally in the blood vessels. Specific inhibitors in the blood prevent activation of each of the numerous enzymes unless a critical concentration of the activating substances is achieved. A chance accumulation of activating substances is unlikely because these are removed by the liver. Finally, a fibrinolysin system, which dissolves fibrin in clots that may start to form, can be activated. However, despite these mechanisms, a clot, known as a **thrombus,** may develop in a vessel, and can be very serious if it plugs a vessel that supplies a vital area.

In the hereditary disease **hemophilia** there is a deficiency of one of the factors necessary for the formation of prothrombinase, so that clots do not form and the slightest scratch may lead to fatal bleeding. This disease attracted special attention because it appeared in several different European royal families and was apparently inherited from Queen Victoria of England.

13.5 WHITE BLOOD CELLS

Five types of white blood cells, or **leukocytes,** can be recognized—**lymphocytes, monocytes, neutrophils, eosinophils** and **basophils** (Fig. 3.19). They differ in the size and shape of the nucleus, and in the size and nature of the granules present in the cytoplasm. Collectively, there are only about 7000 leukocytes per cubic millimeter in human blood. They are produced in the lymph nodes, the spleen and red bone marrow, and live from one to four days.

Although they are passively carried by the blood, most leukocytes can also creep about by ameboid motion. This enables them to squeeze between the cells of the capillary walls and enter the tissues where their physiological functions are performed. Many are lost from the body through the lungs, digestive tract and kidneys.

These cells play important roles in protecting the body against disease organisms. Bacteria penetrating through a wound produce toxins which, together with factors in injured tissue, cause blood vessels in the infected area to dilate and become more permeable to liquids and to white blood cells. The area becomes inflamed and swollen; vast numbers of leukocytes migrate into the region and begin to engulf and digest bacteria and destroyed cells. Neutrophils are among the first cells to become active, but if the infection is severe they break down and contribute, along with the bacteria and tissue debris, to the **pus** that forms. As time goes on, monocytes and lymphocytes transform into large phagocytic cells known as **macrophages,** and they clean up the area. Other lymphocytes transform into fibroblasts which, along with fibroblasts in the connective tissue, repair the injury by producing fibers and forming **scar tissue.**

13.6 IMMUNITY

An additional and important defense is the body's capacity to produce **antibodies** that combat foreign substances (**antigens**) that enter the body. Antigens are usually proteins associated with invading bacteria or viruses, but any foreign protein and certain foreign polysaccharides and DNA may induce antibody formation. Antibodies are immunoglobulins that combine with the antigens and neutralize them; they may cause the invading microorganisms to clump, or **agglutinate,** thereby effectively preventing a further penetration of the body; they may cause the invading microorganisms to break up and dissolve (a phenomenon known as **lysis**); or they may make the invaders more susceptible to phagocytosis. The antigen-antibody reaction is generally very specific. Antibodies that have developed in response to mumps viruses, for example, will not combine with other antigens. The configuration of the antigen and antibody molecules is such that only antibodies that have developed in response to a given antigen can fit on the surface of the antigen and react with it.

Once the body has produced antibodies in response to an antigen, it retains the capacity to produce similar antibodies for many years. If a subsequent invasion of the same type of antigen occurs during this period, antibodies specific for it may be present, and more can be synthesized very quickly by sensitized body cells. The body has developed an **immunity** to this particular antigen. The immunity that is acquired as a result of having once had mumps, smallpox and certain other infectious diseases lasts a very long time, generally for life. The immunity to certain other diseases lasts for a much shorter time and, after it is lost, one can get the disease again.

However, it is necessary to contract a disease to develop an immunity to it. During the late eighteenth century, Edward Jenner observed that milkmaids and others who handled the udders of cows infected with cowpox never got smallpox. In 1796, he took a bit of the material from the pustules of an infected cow and scratched it into the skin of a person. Individuals so treated acquired a mild disease but thereafter were immune to smallpox. Cowpox is caused by the **vaccinia** virus; smallpox, by a different but related **variola** virus. Vaccinia does not cause serious disease in man, but is similar enough to variola so that antibodies that develop in response to it are effective in combating variola. Jenner's experiments were the beginning of the **vaccination technique.** Since then, many kinds of vaccines have been developed. Usually a related and less virulent microorganism, which could serve as the basis of a vaccine, is not available, but vaccines can be produced by rendering the disease organisms harmless by appropriate treatment, and injecting them. Although the organisms are incapable of causing the disease, they are still capable of inducing antibody formation. One of man's recent triumphs over disease has been the development by this method of a vaccine for rubella, or German measles.

The production of antibodies in response to an antigen that enters the body, either through disease or vaccination, is known as an **active immunity.** Active immunities are typically long-lasting. Temporary protection can be given a person who has been exposed to a new and dangerous antigen by injecting an antiserum containing immunoglobulins from an organism that has already produced the appropriate antibodies. A person bitten by a poisonous snake, for example, can be given antiserum obtained from a horse that has developed antibodies specific to this snake venom. The horse has developed an active immunity in response to successive injections of small amounts of the snake venom, but the immunity conferred on the person receiving the antiserum is a **passive immunity.** A newborn infant receives a passive immunity to certain diseases by the passage of maternal antibodies through the placenta, or in the early milk. Passive immunities last only a few days or weeks.

The cells that produce antibodies have been identified only in recent years, although it has long been known that lymph nodes, spleen and other lymphoid tissues are somehow involved. The lymphoid tissues contain phagocytic cells that engulf bacteria, viruses and other foreign material entering the body, but they do not themselves synthesize the antibodies. Rather they are believed to transfer moderate quantities of antigen to certain lymphocytes circulating through these tissues. Lymphocytes competent to respond to the antigenic stimulus multiply, develop an elaborate granular endoplasmic reticulum capable of protein synthesis (Fig. 13.5), lodge in the tissues and, now known as **plasma cells,** synthesize the appropriate antibodies. The few days required for this multiplication and transformation explains the latent

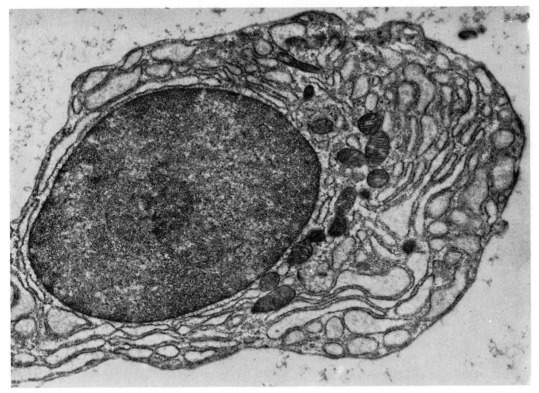

Figure 13.5 Electron micrograph of a human plasma cell showing the hypertrophied granular endoplasmic reticulum, which is associated with protein synthesis. (From Bloom and Fawcett after Van Breeman.)

period between the body's exposure to an antigen and the synthesis of antibodies.

Our understanding of the mechanism whereby the lymphocytes that respond to the antigenic stimulus produce the appropriate antibody is far from complete. The **instructive theory** postulates that antigen can enter any lymphocyte not previously exposed to antigen, act as a template, and instruct the cell's protein-synthesizing machinery to make antibodies that will combine properly with the antigen. The **clonal selection theory** has become more widely accepted in recent years. It postulates that the various lymphocytes in the body have the genetic machinery needed to synthesize antibodies in response to any conceivable antigen. There is a "dictionary" of lymphocytes, each precoded to make one or more antibodies; the lymphocytes with the appropriate genetic machinery respond to (or are selected by) the antigen that enters the body, and then multiply to form a clone of responsive cells. The clonal selection theory requires that there be thousands of dif-

ferent types of lymphocytes with different genetic codes. However, the number of possible antigens is finite, and many share certain groups of amino acids to which the antibodies respond. It has been estimated that as few as 10,000 types of lymphocytes might be sufficient to meet the body's needs.

The clonal selection theory helps us to understand how the body may distinguish between "nonself" and "self." The plasma cells synthesize antibodies in response to a foreign protein ("nonself") but normally do not synthesize antibodies to their own ("self") proteins. The thousands of types of lymphocytes presumably develop by mutation during embryonic development. Those mutants that would induce unfavorable antigen-antibody reactions with the embryo's own tissues would be eliminated quickly so that by birth, or shortly thereafter, an individual should have a wide assortment of lymphocytes capable of responding only to foreign antigens.

Among the many lines of evidence that support the clonal selection theory are skin

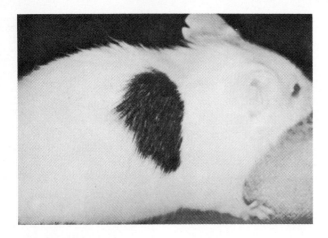

Figure 13.6   A white mouse of strain *A*, that early in life developed a tolerance to lymphoid tissue from strain *B*, accepts a skin graft from a black mouse of strain *B*. (After P. R. Russell, from Beck: Human Design. Harcourt Brace Jovanovich.)

grafting experiments. Normally an antigen-antibody reaction causes a mature mouse (or any other vertebrate) in a strain, A, to reject a skin graft from a donor in a different genetic strain, B. But if lymphoid tissue from a strain B mouse is implanted into the A-strain mouse about the time of birth, the A individual will later accept a graft from a B individual (Fig. 13.6). Apparently the B cells are introduced into the A mouse early enough so that it accepts them as "self," and any lymphocytes that would produce antibodies in response to the B cells are eliminated from the "dictionary" of responsive cells.

13.7   THE ABO BLOOD GROUPS

When the practice of transfusing blood from one person to another was begun, it was found that the transfusions were sometimes successful, but more often they were not and erythrocytes in the blood of the recipient would clump (agglutinate), with fatal results. Careful analysis by Landsteiner at the beginning of this century showed that specific antigenic proteins, called A and B, might be present in the plasma membranes of the erythrocytes. These antigens are called **agglutinogens**, since they may cause agglutination of the red cells. Some individuals have protein A, some B, some both A and B, and some neither. Naturally occurring antibodies (**agglutinins**) specific for these agglutinogens, and designated *a* and *b*, may be present in the plasma. An individual with red cells containing a certain agglutinogen does not, of course, have

the agglutinin specific for it in his plasma. If he did, his own red cells would be agglutinated.

Four groups of persons, O, A, B and AB, can be recognized according to the presence or absence of these agglutinogens and agglutinins (Fig. 13.7). Before a transfusion is made, both donor's and recipient's cells are typed by adding them to serum containing an *a* agglutinin (anti-A serum) and to serum containing a *b* agglutinin (anti-B serum), and observing which, if either, causes the cells to clump. Transfusions between members of the same group are perfectly safe, and transfusions between different groups are also safe provided that the donor's erythrocytes do not contain an agglutinogen that will react with the recipient's agglutinins. The agglutinins in the donor's plasma become so diluted in the recipient that they have no effect and they may be disregarded unless an unusually large transfusion is given. Members of Group O, who have neither of the agglutinogens, can give blood to members of any group and are "universal donors." But since their plasma contains both of the agglutinins, they can receive blood only from members of their own group. Members of Group AB, in contrast, have neither agglutinin, and can receive blood from members of any group. Since they have both agglutinogens, they can give blood only to members of their own group. They are "universal recipients." Members of Groups A and B can give blood to members of Group AB and receive from members of Group O. The inheritance of these blood groups was considered in section 7.15.

Blood group	Agglutinogen in erythrocyte	Agglutinins in plasma	Reaction with	
			Anti-A serum	Anti-B serum·
O Universal donor	none	a and b	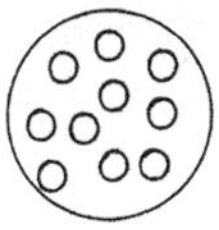	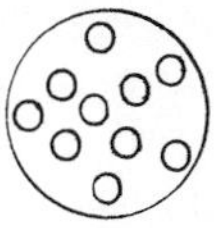
A	A	b	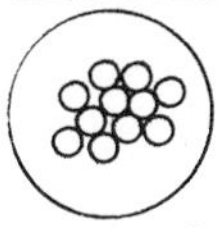 agglutination	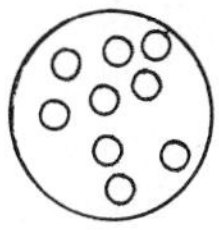
B	B	a	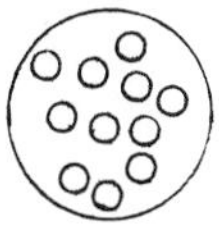	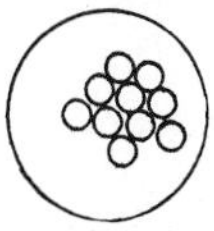 agglutination
AB Universal recipient	A and B	none	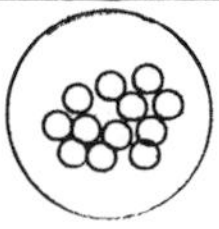 agglutination	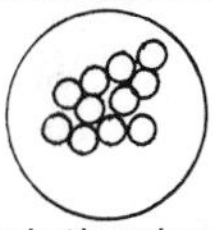 agglutination

Figure 13.7 The ABO blood groups. The two right-hand columns show the appearance of the red cells of each group when exposed to anti-A and anti-B serum. The blood group to which an individual belongs is identified by the antiserum in which its red cells agglutinate; if there is no agglutination with either antiserum, the individual belongs to group O.

13.8 THE Rh ANTIGEN

A number of other inherited antigenic proteins may be present in the blood. Most are rare or are not apt to be involved in transfusions, but one that is common is the **Rh antigen**, so called because it was first discovered in the rhesus monkey. The Rh system is very complex for there are six or more inherited factors that interact to produce one or more of several variants of Rh antigen in an Rh positive person, or none at all in an Rh negative person. The frequency of Rh positive and Rh negative individuals varies considerably among different groups of people. About 60 per cent of most European groups are Rh+ and 40 per cent are Rh—; Nigerians are 86 per cent Rh+ and 14 per cent Rh—; Eskimos are all Rh+. Rh antibodies do not occur normally in human blood, but are formed in response to foreign Rh antigens.

An Rh negative woman will develop Rh antibodies if she receives Rh positive blood either by transfusion or during pregnancy if she receives blood cells from an Rh positive fetus. There is normally no mixing of maternal and fetal bloods but fetal red cells may make their way into the maternal blood stream through breaks in the placenta or when the placenta is pulled away during delivery or during an abortion. During a subsequent pregnancy some of the maternal Rh antibodies may pass into the fetus and cause agglutination and hemolysis of its red cells. This condition, **erythroblastosis fetalis,** may be fatal, or may result in injury to the brain from the bile pigment (bilirubin) formed from the breakdown of hemoglobin released by the hemolysis of the red cells. A newborn infant showing symptoms of it can be saved by extensive transfusions. Ordinarily, not enough Rh positive blood enters the mother to cause any harm, but her blood contains certain antibodies, and if she subsequently needs a transfusion for any reason, blood from an Rh negative person must be used.

13.9 PATTERNS OF CIRCULATION

Heart, arteries, capillaries and veins constitute the **cardiovascular system;** the lymphatic vessels and nodes comprise the **lymphatic system.** Most vertebrates have both, but primitive vertebrates such as cyclostomes and cartilaginous fishes have no lymphatic system.

Primitive Fishes. The cardiovascular system has undergone some striking changes during the evolution of vertebrates. Most of these are correlated with the shift from gills to lungs as the site of external respiration that occurred during the transition from water to land and with the development of the efficient, high pressure circulatory system necessary for an active terrestrial vertebrate.

In a primitive lungless fish (Fig. 13.8), all of the blood entering the heart from the veins has a low oxygen and a high carbon dioxide content; i.e., it is venous blood. The heart consists of a **sinus venosus,** a single **atrium,** a single **ventricle** and a **conus arteriosus** arranged in linear sequence. The muscular contraction of the heart increases the blood pressure, which is very low in the veins, and sends the blood out through an artery, the **ventral aorta,** to five or six pairs of **aortic arches** that extend dorsally through capillaries in the gills to the **dorsal aorta.** Carbon dioxide is removed and oxygen is added as the blood flows through the gills; i.e., it changes to arterial blood. The dorsal aorta distributes this through its various branches to all parts of the body.

Blood pressure decreases as blood flows along because of the friction between the blood and the lining of the vessels. Blood pressure is reduced considerably as the blood passes through the capillaries of the gills, for friction is greatest in vessels of small diameter. Mean blood pressure in the ventral aorta of a dogfish during heart contraction, for example, is about 20 mm. Hg; that in the dorsal aorta is about 15 mm. Hg. This relatively low blood pressure in the aorta becomes even lower when the blood reaches the capillaries in the tissues. Circulation in primitive fishes is rather sluggish and not conducive to great activity.

Veins drain the capillaries of the body (where blood pressure is further reduced) and lead to the heart, but not all veins go directly to the heart. In primitive fish, blood returning from the tail first passes through capillaries in the kidneys before entering veins leading to the heart. Veins that drain one capillary bed and lead to another are called portal veins, and these particular veins are known as the **renal portal system.** Another group, known as the **hepatic portal system,** drains the digestive tract and leads to capillary-like passages in the liver. Since much of the blood returning to the heart has passed through one or the other of these portal systems in addition to the capillaries in

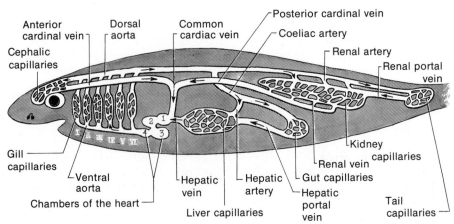

Figure 13.8 The major parts of the cardiovascular system of a primitive fish. *1*, Sinus venosus; *2*, atrium; *3*, ventricle; *4*, conus arteriosus of the heart. The aortic arches are numbered with Roman numerals. Only traces of the first aortic arch remain in the adults of most fishes.

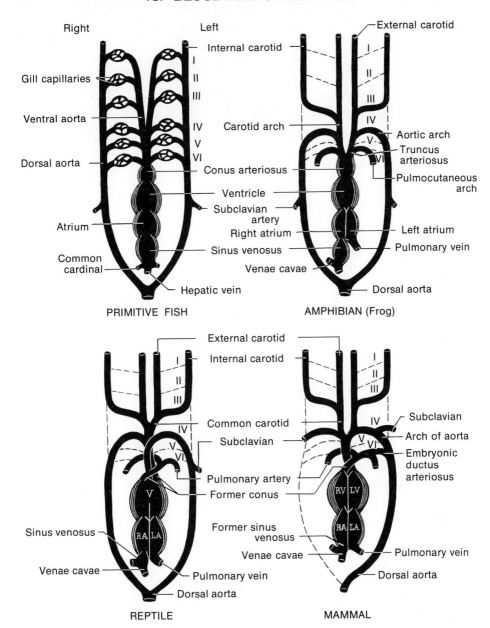

Figure 13.9 Diagrams of the heart and aortic arches to show the changes that occurred in the evolution from primitive fishes to mammals. All are ventral views. The heart tube has been straightened so that the atrium lies posterior to the ventricle.

the gills and tissues, its pressure is near 0 mm. Hg.

It is not difficult to appreciate the significance of a hepatic portal system, since the liver plays such an important role in the metabolism of foods, but the adaptive significance of a renal portal system in primitive vertebrates is less clear.

Amphibians. Correlated with the shift from gills to lungs, many changes occurred

in the heart and aortic arches (Fig. 13.9). The aortic arches were reduced in number, the first two and the fifth being lost. Those that remain are no longer interrupted by gill capillaries. In a primitive tetrapod, such as the frog, the third pair of aortic arches forms part of the **internal carotid** arteries supplying the head; the fourth, the **systemic arches** leading to the dorsal aorta; and the sixth, the **pulmocutaneous**

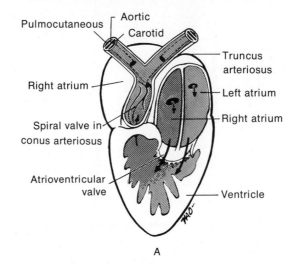

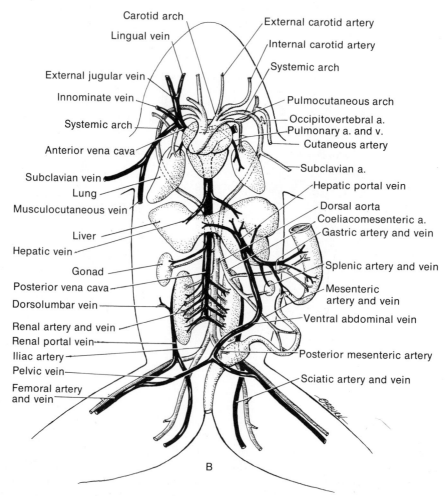

Figure 13.10 *A*, ventral view of a dissection of the heart of a frog showing the partial separation that occurs in the ventricle and conus arteriosus between blood high in oxygen content entering from the lungs and left atrium and less highly oxygenated blood entering from the skin, body and left atrium. *B*, ventral view of the major arteries and veins of the frog. Veins are shown in black; arteries are white. Certain of the anterior veins have been omitted from the right side of the drawing and certain of the anterior arteries from the left side.

arches leading to the lungs and skin. New veins, the **pulmonary veins,** return aerated blood from the lungs to the heart. The heart now receives blood from both the body and lungs. Though blood streams from the body and lungs are separated in the frog by a divided atrium, the two converge in a single ventricle (Fig. 13.10A). Recent experiments using indicators to identify the streams have shown that, although there is some mixing, a high degree of separation is achieved. The slight differences in the times the streams enter the ventricles, a spongy ventricular muscle that helps hold them apart, the deflective action of a spiral valve within the conus arteriosus, and the internal partitioning of the arterial trunks that lead from the conus to the vessels supplying the body and lungs all play a role in separating the two streams. The pulmocutaneous arch receives primarily right atrial blood relatively low in oxygen content, the systemic arch receives a rather complete mixture of blood, and the carotid arch receives primarily left atrial blood with a high oxygen content. The mixture of left and right atrial blood going to the body is not as detrimental as may at first appear, because some of the right atrial blood comes from the skin where aeration also occurs. When the frog is under water the lungs are not used, and the lack of a divided ventricle is probably an advantage because it allows aerated blood returning to the right atrium from the skin to be distributed directly to the body. The lack of a divided ventricle in modern amphibians may be an adaptation related to the evolution of cutaneous respiration. Living lungfish have a partial interventricular septum that provides a more complete separation of the blood from the body and that from the lungs.

The loss of gills and the evolution of a double circulation through the heart result in a much higher blood pressure in the arteries of a primitive tetrapod than in a fish. Blood in the dorsal aorta of a frog has a mean pressure of about 30 mm. Hg, twice that of the dogfish. This increases the circulatory rate, but this benefit is somewhat offset by the fact that the blood delivered to the tissues is mixed to some extent and does not contain so much oxygen, relatively, as it did in a fish. The pattern of the major arteries and veins of a frog is shown in Figure 13.10B.

Reptiles and Birds. Higher tetrapods depend upon their lungs for external respiration. Since no respiration occurs in the skin, there is no mixing of aerated blood from the skin with blood from the body. The mixing in the heart of arterial blood from the lungs with venous blood from the body is lessened in reptiles by a partial division of the ventricle and by a complex, tripartite division of the conus (Fig. 13.9). In a lizard, for example, the venous blood from the body enters the right side of the heart and most of it is sent to the lungs via the pulmonary arteries. After returning to the left side of the heart, this blood, now rich in oxygen content, is distributed to the head and also to the trunk via the right fourth aortic arch. Venous blood on the right side of the heart that did not go to the lungs mixes with oxygen-rich blood and is sent to the trunk via the left fourth aortic arch. It has been suggested that this is a mechanism that permits some venous blood to by-pass the lungs at an evolutionary stage when the lungs may not be able to handle the volume of blood returning from the body to the heart. The head and brain receive oxygen-rich blood; the trunk, whose tissues are less sensitive to some oxygen deprivation, receives a mixed blood. Birds have no mixing of blood at all, for their ventricle is completely divided and they have lost the left fourth aortic arch of their reptilian ancestors.

Mammals. Mammals too have a completely divided ventricle, so there is no mixing of oxygen-rich and oxygen-depleted bloods (Fig. 13.9). Venous blood from the body enters the **right atrium,** into which the primitive sinus venosus has become incorporated. Arterial blood from the lungs enters the **left atrium.** The atria pass the blood on to the **right** and **left ventricles,** respectively. The ventricular walls of mammals are more muscular than those of lower vertebrates and can increase the blood pressure considerably. The primitive conus arteriosus has become completely divided, part contributing to the pulmonary artery leading from the right ventricle to the lungs and the rest to the arch of the aorta leading from the left ventricle to the body.

The sixth pair of aortic arches form the major part of the mammalian **pulmonary arteries,** and the third pair contribute to the **internal carotid arteries.** But it will be

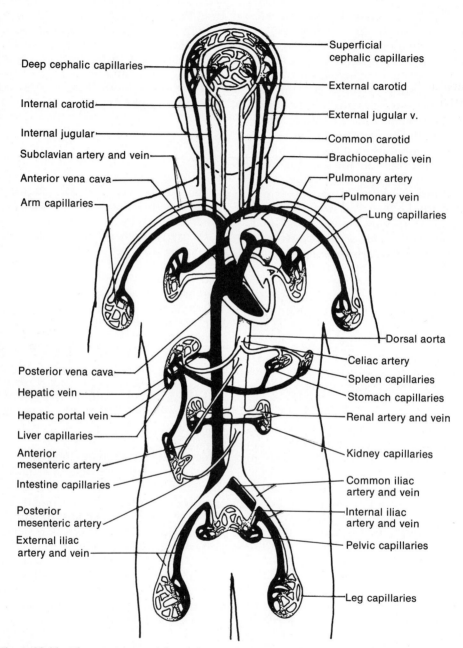

Figure 13.11 The major parts of the cardiovascular system of man as seen in an anterior view.

observed in Figure 13.9 that only the left side of the fourth arch, known as the **arch of the aorta,** leads to the dorsal aorta. The right fourth arch contributes to the right **subclavian artery** to the shoulder and arm but does not connect with the aorta.

The major change in the veins is the complete loss of a renal portal system. Blood from the tail and posterior appendages enters a **posterior vena cava,** which continues forward to the heart. It receives blood from the kidneys but does not carry blood to them. An **anterior vena cava** drains the head and arms. The hepatic portal system is still present. The pattern of the major arteries and veins of man is shown in Figure 13.11.

These evolutionary changes have resulted in a very efficient cardiovascular system. Mammals have relatively more blood than lower vertebrates, it is dis-

tributed under greater pressure, and there is no mixing of arterial and venous blood. Man, for example, has 7.6 ml. of blood per 100 gm. of body weight compared with 2 ml. per 100 gm. in a fish. The mean pressure in the dorsal aorta of man is about 100 mm. Hg.

As blood pressures have increased during the evolution of the cardiovascular system, more liquids and plasma proteins have escaped from the capillaries into the tissue fluid than have been returned by the veins. A separate **lymphatic system** evolved; this plays a role in returning fluid and plasma proteins from the tissues to the cardiovascular system. Lymphatic vessels arise as outgrowths from the veins and, in general, tend to parallel the veins and ultimately empty into them.

The system reaches its greatest development in mammals (Fig. 13.12). **Lymphatic**

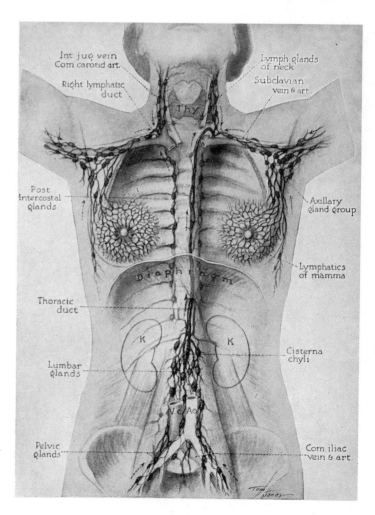

Figure 13.12 Major lymphatic vessels and nodes in the trunk of a human female. (Courtesy of S. H. Camp Co.)

capillaries occur in most of the tissues of the body. They are more permeable than cardiovascular capillaries, and pressures within them are exceedingly low. Their high permeability makes them the most likely route for the spread of microorganisms, or cancer cells, within the body. **Lymph nodes** lie at many points where small lymphatic vessels converge. They are major sites for the production of lymphocytes, and cells within them can phagocytize invading bacteria or respond to them by producing antibodies. Lymphatic vessels draining the hind limbs and the pelvic and abdominal regions finally converge upon a lymph sac, the **cisterna chyli,** located posterior to the diaphragm and dorsal to the aorta. The largest lymphatic vessel of the body, the **thoracic duct,** extends anteriorly from this point, receives the drainage from the left side of the anterior parts of the body, and finally enters the left brachiocephalic vein. A smaller **right lymphatic duct** drains the right anterior parts of the body and enters the right brachiocephalic vein.

13.10 THE FETAL CIRCULATION

The placenta of the mammalian fetus, rather than the digestive tract, lungs and kidneys, is the site for exchange of materials. This, together with the fact that the vessels in the lungs of the fetus are not developed enough to handle the total volume of blood that is circulating through the body, requires certain differences in the fetal circulation (Fig. 13.13). Blood rich in oxygen returns from the placenta in an **umbilical vein,** passes rather directly through the liver via the **ductus venosus,** and enters the posterior vena cava, where it is mixed with blood returning from the posterior half of the fetus. The posterior vena cava empties into the right atrium, which also receives venous blood from the head by way of the anterior vena cava.

The lungs cannot accommodate all of this blood early in development, yet a large volume of blood must pass through all chambers of the heart to ensure their

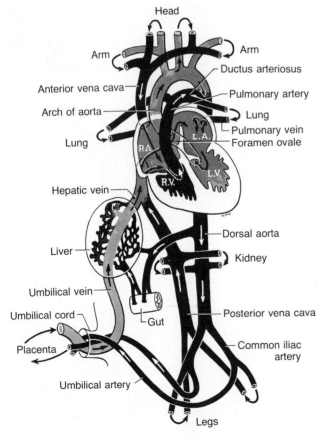

Figure 13.13 Circulation in a fetal mammal. The shading gives some indication of the mixing of the blood, though there is more mixing than can be indicated diagrammatically. The lightest shading represents blood with the highest oxygen content; the darkest shading, blood with the lowest oxygen content. (Modified after Patten.)

normal development. The entrance of the posterior vena cava is directed toward an opening, the **foramen ovale,** in the partition separating the two atria. Most of the blood from the posterior vena cava tends to go through this into the left atrium, thence to the left ventricle and out to the body through the arch of the aorta. This blood by-passes the lungs yet permits the left side of the heart, which otherwise would receive little blood from the collapsed lungs, to function and develop normally. The rest of the blood from the posterior vena cava enters the right ventricle along with the blood from the anterior vena cava and starts out the pulmonary artery toward the lungs. However, only a fraction of this blood passes through the lungs to return to the left atrium and mix with blood from the posterior vena cava. Most of the blood in the pulmonary artery goes through another by-pass, the **ductus arteriosus,** to the dorsal aorta. The ductus arteriosus represents the dorsal part of the left sixth aortic arch (Fig. 13.9). Since the ductus arteriosus enters the aorta after the arteries to the head have been given off, the head receives the blood with the highest oxygen content. After the entrance of the ductus arteriosus, the blood in the aorta is highly mixed. This is the blood that is distributed to the rest of the body and, by way of umbilical arteries, to the placenta.

13.11 CHANGES AT BIRTH

Throughout fetal life the lungs and most of the vessels within them are collapsed. The resistance to blood flow through the lungs (pulmonary resistance) is greater than the resistance to flow through the body and placenta (systemic resistance). As a consequence there is a large flow of blood through the placenta and a small flow through the lungs. The volume of blood returning to the left atrium is less than the volume entering the right atrium. This facilitates the opening of the valve in the foramen ovale and the continued by-passing of the lungs (Fig. 13.14A).

These conditions are immediately reversed at birth. The placenta is expelled and blood volume in the systemic circuit is reduced. Carbon dioxide accumulates in the fetal blood, activating the respiratory system; the lungs fill with air and the pulmonary vessels that were collapsed open up. Resistance to blood flow in the pulmonary circulation is now less than that in the systemic circulation. More blood flows through the lungs and returns to the left atrium than during fetal life. The valve in the foramen ovale is pushed against the interatrial septum and soon adheres to it by the growth of tissue. This by-pass of the lungs is cut off.

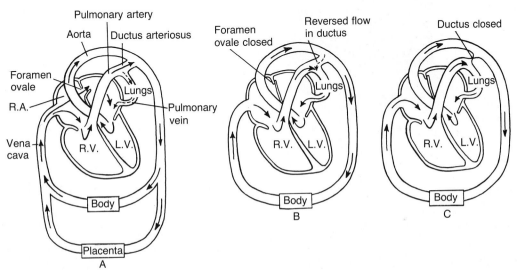

Figure 13.14 Circulatory changes at birth: *A,* fetal; *B,* neonatal; *C,* adult circulation. Solid arrows represent major blood circuits; broken arrows, lesser circuits.

The circulatory pattern is now very close to the adult condition except that the ductus arteriosus remains open (Fig. 13.14*B*). Since pulmonary resistance is less than systemic resistance, the direction of flow in the ductus arteriosus is reversed. Some of the blood which has already been through the lungs and is leaving the heart in the arch of the aorta flows back to the lungs through the ductus arteriosus. This pattern of circulation, which lasts from several hours to a day or two in the human infant, is known as the **neonatal circulation.** Experiments on newborn lambs show that this reversal of flow, and the consequent double aeration of some of the blood at a time when the lungs are not functioning at maximal efficiency, is of great significance. If the ductus arteriosus is experimentally tied off during this period, the hemoglobin is 10 to 20 per cent less saturated with oxygen than it normally is.

Muscles in the wall of the ductus arteriosus eventually contract and stop all blood flow through it, and the adult pattern is established (Fig. 13.14*C*). As time goes on the duct becomes permanently occluded by the growth of fibrous tissue into its lumen. The stimulus for these changes is unknown, but if the duct remains open, an undue strain is placed upon the heart, for it must pump this extra amount of blood that is recirculating through the lungs in addition to the normal amount of blood to the tissues.

13.12   FLOW OF BLOOD AND LYMPH

The Heart.   The heart (Fig. 13.15) is the pump that provides the pressure gradient necessary for the blood and lymph to flow. It lies within a division of the coelom, the **pericardial cavity,** which contains some lymph-like fluid that lubricates it and facilitates its movements. It is covered with a smooth coelomic epithelium, the **visceral pericardium,** and is lined by the simple squamous epithelium, the **endothelium,**

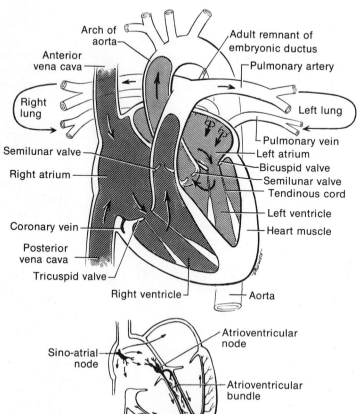

Figure 13.15 The adult mammalian heart. *Upper,* course of blood through the heart; *lower,* distribution of the specialized cardiac muscle that forms the conducting system of the heart.

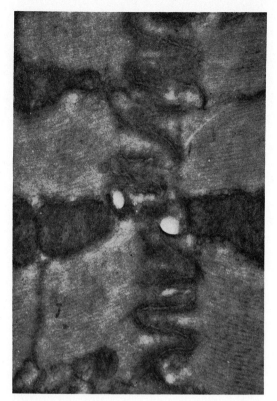

Figure 13.16 Electron micrograph of a portion of the end-to-end union (intercalated disc) of two cardiac muscle cells or fibers to show the interdigitation of their plasma membranes. Parts of three myofibrils can be seen in each muscle fiber. (From Bloom and Fawcett.)

which lines all parts of the circulatory system. The rest of its wall is composed of **cardiac muscle,** which is unique in that its fibers branch and anastomose profusely (see Fig. 3.18). Prominent cross bands occur periodically. For a long time their nature was a puzzle, but electron micrographs (Fig. 13.16) reveal that cells uniting end-to-end interdigitate complexly with each other in this region. This specialization presumably joins the cellular units more tightly and enables them better to withstand the tensions that are continually developed as the heart beats. The musculature of the atria is separated from that of the ventricles, but the individual muscle cells or fibers in each are more intimately united than in other types of muscle. Atria or ventricles respond as a unit. Any stimulus that is strong enough to elicit a response will elicit a total response. Thus, the atria and ventricles follow the "all-or-none" law that applies to individual motor units of skeletal muscle.

During a heart cycle, the atria and ventricles contract and relax in succession. Contraction of these chambers is known as **systole;** relaxation, as **diastole.** Ventricular systole is very powerful and drives the blood out into the pulmonary artery and arch of the aorta under high pressure. Since the muscle fibers of the ventricles are arranged in a spiral, the blood is not just pushed out but is virtually wrung out of them. When the ventricles relax, their elastic recoil reduces the pressure within them, and blood enters from the atria. Atrial contraction does not occur until the ventricles are nearly filled with blood. The atria are primarily antechambers that accumulate blood during ventricular systole.

Blood being pumped by the heart is prevented from moving backward by the closure of a system of valves (Fig. 13.15). One with three cusps, known as the **tricuspid valve,** lies between the right atrium and ventricle; one with two cusps, the **bicuspid valve,** between the left chambers. These valves operate automatically as pressures change, opening when atrial pressure is greater than ventricular, closing when ventricular pressure is greater. **Tendinous cords** extend from the free margins of the cusps to the ventricular wall and prevent them from turning into the atria during the powerful ventricular contractions. When the ventricles relax, blood in the pulmonary artery and aorta, which is under pressure, tends to back up into them. This closes the pocket-shaped **semilunar valves** at the base of these vessels that prevent blood from returning to the ventricles. Abnormalities in the structure of the valves occurring congenitally or produced by disease organisms may prevent their closing properly. Blood then leaks back during diastole; the leaking blood is heard as a "heart murmur."

Cardiac muscle has an inherent capacity for beating, and the hearts of vertebrates, if properly cultured, will continue to beat rhythmically when excised from the body. Each contraction is initiated in the **sino-atrial node,** or "pacemaker"—a node of specialized cardiac muscle (**Purkinje fibers**) located in that part of the wall of the right atrium into which the primitive sinus venosus is incorporated (Fig. 13.15). The impulse spreads to all parts of the atria and to another node of Purkinje fibers, the **atrio-**

ventricular node, from which it continues along pathways of Purkinje fibers to all parts of the ventricles. Since there is no muscular connection between atria and ventricles, an impulse can reach the ventricular muscles only through the Purkinje fibers. It does so very rapidly, so that ventricular contraction begins at the apex of the heart and spreads quickly toward the origin of the great arteries leaving the heart.

The rhythmicity of the sino-atrial node results from the leakage of positively charged sodium ions, which are abundant outside all cells, through the plasma membranes of the cells of the node. The resting electrical potential of the membrane is reversed and the cells become active. A similar phenomenon occurs when nerve cells and other muscle fibers become active. As sodium ions are pumped out of the cell, the original resting potential of the membrane is restored. The unique feature of the cells of the sino-atrial node is that sodium ions begin to leak in again after the resting potential has been re-established without the specific stimulus that is needed to activate nerve or other muscle cells. The cycle of inward leakage and outward pumping of sodium ions accounts for the rhythmicity of the sino-atrial node.

Control of Cardiac Output. Though the heart has an inherent rhythm, its rate of contraction and the volume of blood pumped per stroke can be regulated by a number of extrinsic factors so as to adjust the cardiac output to body requirements. Nervous pathways are present for many cardiac reflexes. Motor nerves that increase or decrease the heart rate go to the heart from the vasomotor center in the medulla of the brain, and sensory impulses from many parts of the body reach this center. For example, receptors in the right atrium are stimulated by the increase in the pressure of the venous blood returning to the heart which occurs during exercise. They travel to the vasomotor center (*1*, Fig. 13.17) and, by inhibiting the depressor nerve fibers and activating accelerator fibers, cause an increased number of nerve impulses to go out on the accelerator nerve (*2*). The rate of heart beat and strength of contraction increases to accommodate the greater return of venous blood. Blood pressure also rises and, if it exceeds a certain threshold, pressure receptors in the aortic arch and other vessels near the heart are activated. Impulses travel to the vasomotor center (*3*) and, again by an appropriate combination of inhibition and excitation, the rate of impulses traveling down the accelerator nerve is decreased

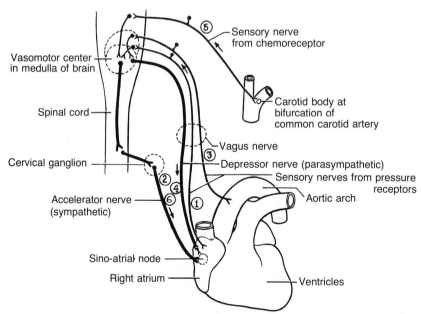

Figure 13.17 Diagram of major nerve reflexes controlling heart output and blood pressure. Motor nerve fibers have been drawn more heavily than sensory fibers. The motor nerves to the heart are parts of the autonomic nervous system, which is discussed more fully in Chapter 16.

and an increased number go out on the depressor nerve (4). Heart rate and force of contraction are reduced to appropriate levels. Other reflexes are initiated by chemoreceptors. A reduction in the level of oxygen in the arterial blood is detected by receptors in the carotid body (5), and this can lead to an increase in heart rate (6). Yet other extrinsic factors such as temperature, carbon dioxide content of the blood, and one's emotional state can influence heart rate by acting on the vasomotor center.

In addition to these extrinsic factors, heart muscle itself can make certain adjustments. For example, the increased pressure and more rapid return of venous blood during exercise stretches the heart musculature. This causes it to contract with greater force and to send out the greater volume of blood received during each period of atrial diastole. Within physiologic limits, the greater the tension on cardiac (or any other) muscle, the more powerful will be its contraction. This capacity of the heart to adjust its output per stroke to the volume of blood delivered to it is known as Starling's **law of the heart.**

The heart of a normal adult man who is not exercising sends about 70 ml. of blood per beat out into the aorta. At the normal rate of 72 beats per minute, this is a total output of 5 liters per minute, which is approximately equivalent to the total amount of blood in the body. A similar observation made in 1628 by William Harvey helped to lead him to the conclusion that the blood recirculates. Until that time it was believed that blood was continually produced in the liver, pumped to the tissues and consumed. Harvey's calculations showed that the amount of blood pumped by the heart each hour was much more than could possibly be produced and consumed. He made the correct inference that the blood must recirculate, even though he could not see the microscopic capillaries that connect arteries and veins.

Although a large volume of blood flows through the cavities of the heart, this blood does not provide for the metabolic needs of the heart musculature. A pair of **coronary arteries** arise from the base of the arch of the aorta and supply capillaries in the heart wall. This capillary bed is drained ultimately by a **coronary vein** that empties into the right atrium. Obviously, any damage to the coronary vessels, the plugging of one of the larger arteries by a thrombus, for example, could have serious consequences, for the heart muscles cannot function without a continuing supply of oxygen and food.

The Arteries.   Arteries are lined with endothelium and have a relatively thick wall containing elastic connective tissue and smooth muscles. The walls of the larger arteries are richly supplied with elastic

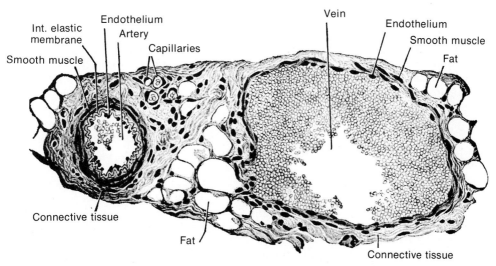

Figure 13.18   A drawing of a cross section through a small artery and an accompanying vein. Several capillaries lie between them. (Modified from Bloom and Fawcett.)

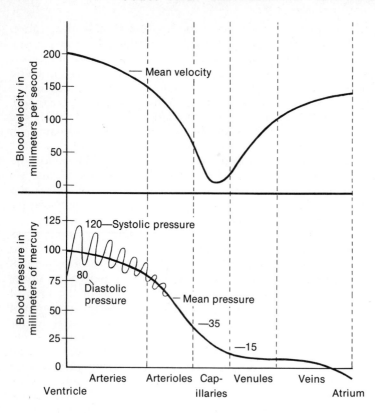

Figure 13.19 Variation in blood velocity and pressure in different parts of the cardiovascular system. The velocity does not return to its original value, for the cross sectional area of the veins is greater than the cross sectional area of the arteries. The blood pressure in the veins near the heart is less than atmospheric pressure because of the negative pressure within the thorax.

tissue. The force of each ventricular systole forces blood into the arteries and stretches them to accommodate it. During diastole, the elastic recoil of the first part of the artery to expand helps to push the blood into the adjacent part of the artery, which in turn expands. If the arteries were rigid pipes, they would deliver blood to the tissues in spurts that coincided with ventricular systole. The blood would pound like steam rushing into empty radiator pipes. The elasticity of the large arteries transforms what would otherwise be an intermittent flow into a steady flow. The alternate stretching and contracting of the arteries travels peripherally very rapidly (7.5 meters per second) and constitutes the **pulse**, but the blood itself does not move as fast.

The smaller arteries (Fig. 13.18), and especially the **arterioles** preceding the capillaries, contain a relatively large amount of smooth muscle, and they regulate the supply of blood to the various organs. **Vasodilator** and **vasoconstrictor nerves** supply these muscles, causing them to relax or contract. If a region of the body becomes very active,

its small arteries enlarge, and the blood flow through them is increased. If an area is not particularly active, its small arteries constrict, and blood flow is reduced. In this way, maximum use is made of the volume of blood available. The body does not contain enough blood to supply fully all tissues and organs at the same time.

As the arteries extend to the tissues, they branch and rebranch. Each time the lumen becomes smaller, but the total cross sectional area of all these branches increases greatly. The velocity of blood flow, therefore, decreases, for the blood, like a river widening out and flowing into a lake, is moving into an area that grows larger and larger. The mean blood pressure is also decreased continually because of the friction of the blood moving in the vessels (Fig. 13.19). Blood pressure continues to decrease as the blood flows through the capillaries and veins. The rate of flow, however, increases as the blood passes from the capillaries to the venules, and as these smaller veins lead into fewer larger ones. The blood is now moving into a smaller and smaller area and, like water flowing out of a lake into a narrowing river, moves faster and faster.

Capillary Exchange. Capillaries are small and exceedingly thin-walled vessels. Their diameter is about that of the blood cells, and their walls consist of little more than an endothelial lining, which is continuous with that of the larger vessels. The capillary wall is a semipermeable membrane, and molecules that are small enough can easily pass back and forth between the blood and the surrounding tissue fluid (Fig. 13.20). Some large molecules, including a fraction of the plasma proteins, also leak out. Most substances are exchanged by simple diffusion following concentration gradients. There is more glucose and oxygen in the blood than in the tissue fluid, so their net movement is out of the capillaries. There are more wastes and carbon dioxide in the tissue fluid, so their net movement is into the capillaries.

The exchange of water is more complicated than the exchange of solutes, for its movement depends upon two opposing forces. The blood pressure tends to force water out of the capillaries, whereas the osmotic pressure exerted by the plasma protein molecules tends to draw water back in. The osmotic pressure of the blood, which depends to a large extent on the plasma proteins, drops only slightly from the arterial to the venous ends of the capillary bed, but blood pressure decreases sharply. At the arterial end of the capillary bed, blood pressure is greater than osmotic pressure and some water is driven out of the capillaries. At the venous end, osmotic pressure is greater than blood pressure, and water is drawn back into the capillaries. Excess residual liquid and the few proteins that leak out are drained by the lymphatics.

Not all capillaries function concurrently. Within a capillary bed there are certain thoroughfare channels through which some blood flows from arterioles to venules all of the time. But tiny muscular **precapillary sphincters** are situated at the beginning of the capillaries leading from these channels, so other parts of the bed may be open or closed according to the needs of the tissues. Total blood flow through a capillary bed is regulated both by the arterioles and precapillary sphincters.

Venous and Lymphatic Return. The structure of the veins is fundamentally similar to that of arteries, though a vein is larger and has a much thinner and more flaccid wall than its companion artery (Fig. 13.18). Since they are larger, the veins hold more blood than the arteries and are an important reservoir for blood. Lymphatic vessels have even thinner walls. Valves present in both veins and lymphatics permit the blood and lymph to flow only toward the heart. It is sometimes easy to demonstrate the valves in the veins on the back of your hand. Push your finger on a vein at the point where several join on the back of your wrist and move your finger distally along the vessel. This will force the blood out of the vein, and you will notice that blood does not re-enter this vein from the others at the wrist, for valves prevent it from doing so. Remove your finger

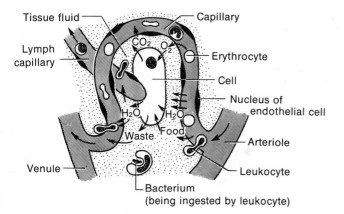

Tissue fluid — Capillary

Lymph capillary

CO_2 O_2

Erythrocyte

Cell

Nucleus of endothelial cell

H_2O H_2O

Waste Food

Arteriole

Venule

Leukocyte

Bacterium (being ingested by leukocyte)

Figure 13.20 Exchange of materials in a capillary bed. Solutes enter and leave all parts of a capillary. Most of the water leaves at the arterial end and re-enters at the venous end. Less than 1 per cent of the water that leaves the capillaries is returned by the lymphatic system.

from the vein and it immediately fills with blood from the periphery. William Harvey in the seventeenth century demonstrated the presence of valves in the veins in a similar way (see Fig. 1.5).

Though blood pressure is low in the veins (Fig. 13.19), and lowest in the large veins near the heart, it is still the major factor in the return of blood. Two other factors assist it. One is the fact that the elastic lungs are always stretched to some extent and tend to contract and pull away from the walls of the pleural cavities. This creates a slight subatmospheric or negative pressure within the thoracic cavity, which is greatest during inspiration. The larger veins, of course, pass through the thorax, and the reduction of pressure around them decreases the pressure within them and increases the pressure gradient. The other factor is that the contraction and relaxation of body muscles exert a "milking" action on the veins. When the muscles contract, their bulging squeezes the veins and forces the blood toward the heart, for the valves in the veins prevent the blood from moving in any other direction. All these factors increase during exercise, which makes for a more rapid return of blood and an increased cardiac output.

The return of lymph is dependent upon similar forces. The tissue fluid itself has a certain pressure derived from the flow of liquid out of the capillaries. This establishes a pressure gradient in the lymphatics that is made steeper by the negative intrathoracic pressure. The "milking" action of surrounding muscles and, for lymphatics returning from the intestine, the movement of the villi, help considerably. Some lower vertebrates have lymph "hearts" – specialized pulsating segments of lymphatic vessels.

ANNOTATED REFERENCES

Attention is again directed to the general references on vertebrate organ systems cited at the end of Chapter 11.

Burnet, F. M.: The mechanism of immunity. Scientific American *204*:58, (Jan.) 1961. Theories of immunity are discussed by this investigator, who shares a Nobel prize in medicine for his work in this field.

Harvey, W.: Anatomical Studies on the Motion of the Heart and Blood. Springfield, Ill., Charles C Thomas, 1931. This is one of a number of available translations of Harvey's important studies in 1628 which established the circulation of the blood and introduced the experimental method into biology.

Krogh, A.: The Anatomy and Physiology of Capillaries. New Haven, Yale University Press, 1922. The classic study on this subject written by one of the pioneer investigators of capillary circulation.

Mayerson, H. S.: The lymphatic system. Scientific American *208*:80, (June) 1963. The importance of this second drainage system of the tissues is thoroughly discussed.

Wiener, A. S.: Blood Groups and Blood Transfusions. 3rd ed. Springfield, Ill., Charles C Thomas, 1943. An old but classic treatise which traces the history of the discovery of blood groups and their applications to problems of transfusion, anthropology, disputed paternity and forensic medicine.

Wiggers, C. J.: The heart. Scientific American *196*:87, (May) 1957. A fine discussion of the activities of the heart and the safety factors that enable it to continue operating even though partially impaired by coronary disease.

Zuckerkandle, D.: The evolution of hemoglobin. Scientific American *212*:110, (May) 1965. An analysis of the various types of mammalian hemoglobin and the evolutionary information which they can provide.

Zweifach, B. W.: The microcirculation of the blood. Scientific American *200*:54, (Jan.) 1959. A discussion of the factors that control circulation in capillary beds.

14. The Urogenital System — Excretion and Reproduction

Functionally, the kidneys have nothing in common with the reproductive organs. Kidneys help excrete wastes and regulate the composition of body fluids; reproductive organs perpetuate the species. But the two systems are morphologically interrelated in vertebrates because certain excretory ducts are used for discharging gametes, and it is convenient to treat them together as the **urogenital system.** First, we shall consider the excretory portion of the system, and then relate the reproductive organs to it.

Although the kidneys come to mind when one thinks of excretion in vertebrates, they do not have a monopoly on the removal of the waste products of metabolism. Special salt-secreting glands are present in most marine vertebrates, and the gills and lungs, the skin and, to some extent, the digestive tract play a role in excretion. Gills eliminate carbon dioxide and some nitrogenous wastes; lungs, carbon dioxide; the skin (especially in amphibians), a certain amount of carbon dioxide and traces of salts and nitrogenous wastes; the digestive tract, bile pigments and certain metal ions. The kidneys remove most of the nitrogenous wastes in terrestrial vertebrates, but this is not their only function. By removing, or conserving, water, salts, acids, bases and various organic substances as the situation requires, they play a vital role in regulating the composition of the blood and the internal environment of the body.

14.1 EVOLUTION OF THE KIDNEYS AND THEIR DUCTS

The **kidneys** of vertebrates are paired organs that lie dorsal to the coelom on each side of the dorsal aorta. All vertebrate kidneys are composed of units called kidney tubules, or **nephrons,** which remove materials from the blood, but the number and arrangement of the nephrons differ in the various groups of vertebrates. Comparative studies have led to the conclusion that each kidney in ancestral vertebrates contained one nephron for each of those body segments that lay between the anterior and posterior ends of the coelom (Fig. 14.1A). These nephrons drained into a **wolffian,** or **archinephric, duct** which continued posteriorly to the cloaca. Such a kidney may be regarded as a complete kidney, or **holonephros,** for it extends the entire length of the coelom. A holonephros is found today in the larvae of certain cyclostomes but not in any adult vertebrate.

In the kidney of adult fishes and amphib-

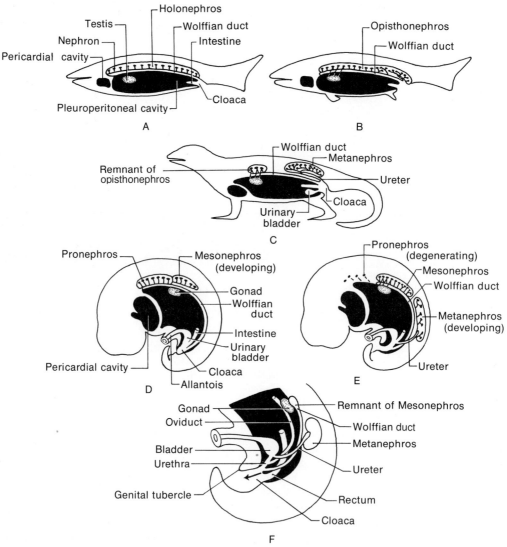

Figure 14.1 A comparison of the evolution and embryonic development of the kidney and its ducts. *A, B* and *C:* The evolutionary sequence of kidneys. *A*, Hypothetical ancestral vertebrate with a holonephros; *B*, a fish with an opisthonephros; *C*, a reptile with a metanephros. *D* and *E*, The developmental sequence of kidneys in a reptile. *F*, A mammalian embryo in which the cloaca is becoming divided by the growth of the fold indicated by the arrow. The ventral part of the cloaca contributes to the urethra in the male. It becomes further subdivided in the female and contributes to both urethra and vagina. In both sexes, the dorsal part of the cloaca forms the rectum.

ians (see Fig. 14.1*B*), the most anterior tubules have been lost, some of the middle tubules are associated with the testis, and there is a concentration and multiplication of tubules posteriorly. Such a kidney is known as a posterior kidney or **opisthonephros.**

Reptiles, birds and mammals (Fig. 14.1*C*) have lost all the middle tubules not associated with the testis and have an even greater multiplication and posterior concentration of tubules. The number of nephrons is particularly large in birds and mammals; their high rate of metabolism yields a large amount of wastes to be removed. It is estimated that man has about 1,000,000 nephrons per kidney, whereas certain salamanders have less than 100. The tubules producing urine drain into a **ureter,** which evolved as an outgrowth from the wolffian duct. The wolffian duct itself has been taken over completely by the male genital system. The kidney of the higher vertebrates is known as a **metanephros.**

The evolutionary sequence of kidneys is holonephros, opisthonephros and metanephros. In the embryonic development of vertebrates, we find a slightly different sequence, but one that also involves a posterior concentration of kidney functions (Fig. 14.1D and E). In an early embryo of a reptile, for example, segmentally arranged tubules appear dorsal to the anterior end of the coelom, form the wolffian duct, and disappear. These transitory tubules constitute a **pronephros.** Then a middle group of tubules, known as the **mesonephros,** appear and connect with the wolffian duct. These function during much of embryonic life, but when the metanephric tubules develop, all the mesonephric tubules are lost except for those associated with the testes. The embryonic sequence of kidneys in the development of a higher vertebrate is pronephros, mesonephros and metanephros.

Most tetrapods have a **urinary bladder** that develops as a ventral outgrowth from the cloaca. Generally, the excretory ducts from the kidneys lead to the dorsal part of the cloaca, and urine must flow across it to enter the bladder; however, in mammals (Figs. 14.1F and 14.2) the ureters lead directly to the bladder, and the bladder opens to the body surface through a short tube, the **urethra.** The cloaca becomes divided and disappears as such in all but the most primitive mammals. The dorsal part of the cloaca forms the rectum and the ventral part contributes to the urethra of higher mammals (Fig. 14.1F).

Urine is produced continually by the kidneys and is carried down the ureters by peristaltic contractions. It accumulates in the bladder, for a smooth muscle sphincter at the entrance of the urethra and a striated muscle sphincter located more distally along the urethra are closed. Urine is prevented from backing up into the ureters by valve-like folds of mucous membrane within the

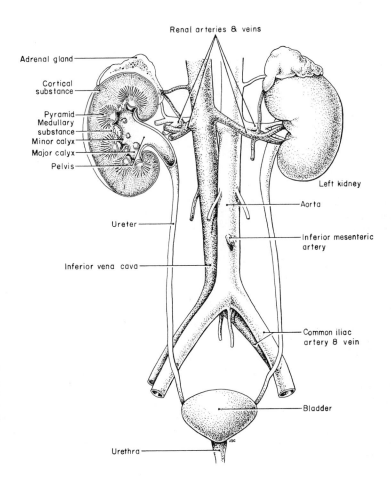

Renal arteries & veins
Adrenal gland
Cortical substance
Pyramid
Medullary substance
Minor calyx
Major calyx
Pelvis
Left kidney
Aorta
Ureter
Inferior mesenteric artery
Inferior vena cava
Common iliac artery & vein
Bladder
Urethra

Figure 14.2 The human excretory system as seen in a ventral view. The right kidney has been sectioned to show the internal structures. (From Villee: Biology, 6th ed.)

bladder. When the bladder becomes filled, stretch receptors are stimulated and a reflex is initiated which leads to the contraction of the smooth muscles in the bladder wall and the relaxation of the smooth muscle sphincter. Relaxation of the striated muscle sphincter is a voluntary act.

14.2   THE NEPHRON AND ITS FUNCTION

Nephron Structure.   The excretory ducts and the urinary bladder are important adjuncts to the kidneys, but the essential work of the system, the selective removal of materials from the blood, is performed by the individual kidney tubules. The general nature and function of these tubules was described in Chapter 5. The mammalian nephron may be taken as an example. The proximal end of each nephron (Fig. 14.3) is known as **Bowman's capsule.** It is a hollow ball of squamous epithelial cells, one end of which has been pushed in by a knot of

capillaries called a **glomerulus.** Bowman's capsule and the glomerulus constitute a **renal corpuscle.** The rest of the nephron is a tubule largely composed of cuboidal epithelial cells and subdivided in mammals into a **proximal convoluted tubule,** a **loop of Henle** and a **distal convoluted tubule.** A **collecting tubule** receives the drainage of several nephrons and leads to the **renal pelvis,** an expansion within the kidney of the proximal end of the ureter (Fig. 14.2). The location of the different parts of a nephron within the kidney and their relationship to blood vessels have important functional consequences. As shown in Figure 14.3, the renal corpuscles and convoluted tubules lie in the outer part, or **cortex,** of the kidney and a dense capillary network surrounds the convoluted tubules; the loops of Henle extend toward the center, or **medulla,** of the kidney. Most of the human nephrons extend only a short distance into the medulla, but about one-fifth of them (the **juxtamedullary nephrons**) have long loops of Henle that extend, along

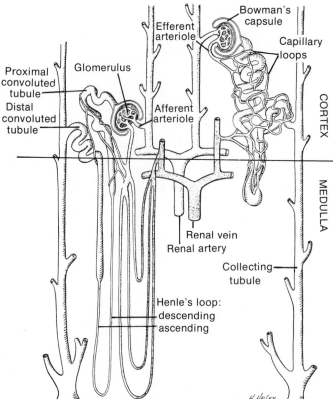

Figure 14.3 A diagram of mammalian nephrons and associated blood vessels. A juxtamedullary nephron is shown on the left; a cortical nephron on the right. (After Smith.)

with the collecting tubules and capillary loops, far into the medulla. It is the convergence of these structures into subdivisions of the renal pelvis (the **calyces**) that forms the renal **pyramids** (Fig. 14.2).

Urine Formation. The kidney tubules produce **urine,** a watery solution containing waste products of metabolism removed from the blood, but not plasma substances needed by the organism. The most abundant nitrogenous waste in man and other mammals is urea, but lesser amounts of ammonia, uric acid and creatinine are present. The yellowish color of urine is due to **urochrome,** a pigment derived from the breakdown of hemoglobin and, hence, related to the bile pigments.

That the first step in urine formation is **glomerular filtration** was demonstrated in the 1920s by Dr. A. N. Richards who developed a micropipette technique for removing and analyzing minute samples of fluid from the lumen of Bowman's capsule. Water, various ions, and small organic molecules, including simple sugars and amino acids as well as nitrogenous wastes, pass easily through the semipermeable membranes of the renal corpuscle. Filtration permits all small molecules that are free in the plasma to pass through, and they appear in the filtrate in the same concentration as in the plasma. Blood cells and large molecules, including fats and plasma proteins, stay in the blood. The only small molecules to be held in the blood are those bound to plasma proteins; among these are iron and other trace minerals and certain vitamins.

Estimates of the filtration rate are obtained by the use of **inulin,** a polysaccharide from artichokes that is filtered easily and is not removed or added to the filtrate as the fluid continues down the kidney tubules. For example, when inulin is injected intravenously until the plasma concentration reaches 1 mg. per ml., 130 mg. of inulin appears in the urine per minute. For this to occur, 130 ml. of plasma must be filtered each minute. This adds up to 188 liters of filtrate per day! Such a large volume is not surprising when one realizes that the kidneys have an unusually rich blood supply. They receive about one quarter of the cardiac output in a mammal, which enables them to process the total blood volume every four or five minutes.

Materials are filtered from the glomeruli

as they are in other capillary beds, but the arrangement of the blood vessels is such that more material passes out of the glomerulus than returns to it. A glomerulus lies between an **afferent arteriole,** a branch of a renal artery, and an **efferent arteriole,** which leads to capillaries on other parts of the tubule (Fig. 14.3). The efferent arteriole is the smaller of the two, and this ensures a high **filtration pressure** in the glomerular capillaries that drives water and many solutes from the blood. The filtration pressure exceeds the pressures within the kidney tubule and the osmotic pressure of the blood (both of which promote the return of fluid to the blood), by 20 mm. of mercury. As a consequence there is a substantial net production of glomerular filtrate.

The volume of urine in man is only about 1 per cent of the glomerular filtrate, and nearly all of the glucose, amino acids and ions present in the glomerular filtrate are taken back into the blood by **tubular reabsorption** as the filtrate passes down the tubules. Other substances may be added to the filtrate by **tubular secretion.** Two important techniques have helped to elucidate these processes. One is a study of **clearance rates.** For example, each minute the kidneys of man excrete an amount of urea equal to that in 75 ml. of plasma. It can be said that 75 ml. of plasma has been cleared of urea, or that urea has a clearance rate of 75 ml. per minute. Other substances have different clearance rates. From inulin studies we know that the filtration rate is 130 ml. per minute. If the clearance rate of a freely filterable substance (such as urea) is less than this, some of this substance must be reabsorbed in the tubules; if the clearance rate exceeds 130 ml. per minute, then some of the substance must be added by tubular secretion.

Sites of reabsorption and secretion have been determined by the **stop-flow technique.** A catheter is inserted into the renal pelvis and clamped. As liquid in the tubules backs up, pressure increases and filtration stops, but the different parts of the tubules continue to act on the now stationary column of liquid. After a suitable interval, the catheter is opened and a series of samples are taken and analyzed. The first samples come from the collecting tubules, and successively later ones from more and more proximal parts of the tubules. From such studies it is concluded that in mammals virtually all of the glucose and amino acids that may have escaped into the filtrate, and some urea are reabsorbed in the proximal convoluted tubule. Sodium, chloride and bicarbonate ions are reabsorbed in both the proximal and distal convoluted tubules. More than 99 per cent of the water in the glomerular filtrate is reabsorbed; this occurs at many levels in the tubules. Creatinine, ammonia, hydrogen and potassium ions and various drugs (penicillin) are among the few substances added to the filtrate by tubular secretion in mammals, and most of this occurs in distal parts of the tubule. But in certain teleosts, which have lost their renal corpuscles, this is an essential way of eliminating waste products.

Reabsorption, which plays such an important role in urine formation, involves both the passive diffusion of materials back into the capillaries surrounding the tubules and the active uptake of materials by the tubular cells and their secretion into the blood against a concentration gradient. This, of course, requires the expenditure of energy by the tubular cells. Apart from urea, most of the reabsorption of solutes that takes place in the proximal convoluted tubule is an active process because the concentration of these substances in the filtrate in this region is the same as their concentration in the blood. Those materials that can be actively reabsorbed are taken back in varying amounts, depending upon their concentration in the blood. If the concentration of one of these materials in the blood and glomerular filtrate rises above a certain level, known as the **renal threshold,** not all of it will be reabsorbed into the blood from the tubule, and the amount present in excess of the renal threshold is excreted. The quantitative value of the renal threshold differs for different substances. In **diabetes mellitus** the impaired cellular utilization of glucose leads to a high concentration of glucose in the blood; the renal threshold for glucose (about 150 mg. of glucose per 100 ml. blood) is exceeded; and sugar appears in the urine in large amounts. The osmotic pressure of the body fluids is controlled by the amount of salts that are taken back into the blood from the glomerular filtrate and the pH of the fluids by the elimination or retention of basic and acidic substances.

Water is reabsorbed passively. As solutes are actively taken out of the filtrate in the proximal tubule, the water in the filtrate would tend to become more concentrated than it is in the blood. However, water molecules passively diffuse out as fast as solutes are pumped out; hence the concentration of the filtrate remains the same as that of the blood. About 80 per cent of all the water taken back is reabsorbed in this way in the proximal tubule, but the tubular filtrate cannot be made more concentrated (i.e., to contain less water) than the blood by this mechanism.

In most of the lower vertebrates the urine is not more concentrated than the blood; however, the higher vertebrates, and especially mammals, do produce a hypertonic urine. The unique feature of the mammalian nephron is the loop of Henle, and for a long time it has been implicated in the ability of these vertebrates to produce a very concentrated urine, but the mechanism of its operation has only recently been discovered. One important factor is that the descending and ascending limbs of the loop of Henle lie parallel to each other so that the direction of flow of fluid in one is opposite to that in the other (Fig. 14.4). Another factor is the active transport of sodium ions out of the ascending limb into the interstitial fluid, and the passive diffusion of these ions back into the descending limb. Together these factors result in a countercurrent multiplying mechanism. Sodium pumped out of the ascending limb goes right back into the descending limb. The recycling of sodium in the loop of Henle, together with additional sodium being continually brought to the loop in the glomerular filtrate, results in an accumulation of sodium in the loop of Henle and the surrounding interstitial fluid of the medulla. A concentration gradient is established as shown in Figure 14.4. The degree of accumulation depends upon the length of the loop; there is more opportunity to pump out sodium in a long loop. The importance of the capillary loops associated with the juxtamedullary nephrons is that here too we have a countercurrent mechanism that permits most of the sodium that starts out of the medulla in the blood to diffuse back into the blood entering the medulla. This, combined with the rather sluggish rate of

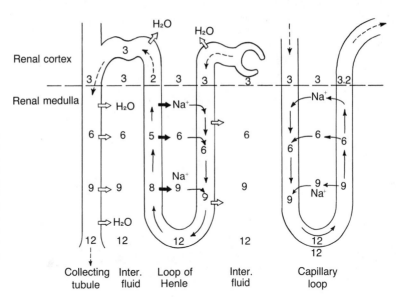

Figure 14.4 Diagram of the countercurrent multiplying mechanism of the mammalian kidney. The general direction of fluid movement is shown by broken arrows; active sodium transport, by heavy black arrows; passive sodium transport, by small black arrows; water movements, by white arrows. Numerals show the relative concentration of osmotically active solutes. In hundreds (add 00), they express the concentrations in milliosmols per liter.

blood flow through these vascular loops, means that little sodium is carried away from the medulla by this route.

The osmotic gradient established by these mechanisms makes it possible for additional water to be passively reabsorbed and for a hypertonic urine to be produced. Water simply follows the osmotic gradient, moving from an area of low osmotic pressure (high concentration of water) to one of high osmotic pressure (low concentration of water). The glomerular filtrate, which was isotonic to the blood in the proximal tubule, loses water as it passes down into the loop of Henle. It does not regain this water as it ascends the loop of Henle, for the cells of the ascending limb of the loop have a low permeability to water. However, the filtrate becomes more dilute because of the large amount of sodium pumped out. By the time the filtrate reaches the distal tubule it is again isotonic, or in some cases hypotonic, to the blood. The filtrate now descends through the medulla again, this time in the collecting tubule, loses additional water, and becomes very hypertonic.

Kidney Regulation. The amount of material reabsorbed or excreted by the kidney depends to some extent on the rate of glomerular filtration. If this is too high, certain essential substances are flushed through the tubules before reabsorption is completed; if the filtration rate is too low, some products normally excreted may be reab-

sorbed. Filtration pressure, and hence rate, can be varied within limits by the constriction or relaxation of the afferent arterioles leading to the glomeruli. In recent years a **juxtaglomerular apparatus** has been discovered that may enable the kidney to monitor the nature of its output and modify filtration pressure accordingly. Shortly before it empties into the collecting tubule, a distal convoluted tubule passes between the afferent and efferent arterioles (Fig. 14.5), and comes into an intimate relationship with modified muscular cells (juxtaglomerular cells) around the afferent arteriole. These cells contain granules identified as a vasoconstrictor substance. The degree of constriction of the juxtaglomerular cells appears to be influenced by the concentration of materials in the distal tubule.

The amount of water excreted and hence the volume of body fluids are also very much affected by an **antidiuretic hormone.** This hormone, which is released by the neural lobe of the pituitary (p. 389), increases the permeability of the cells of the collecting tubule to water. If an excess of water is present in the body fluids, the blood volume and pressure increase. This raises the glomerular filtration pressure, and more filtrate is produced. An increase in the amount of water in the tissue fluid inhibits the release of the antidiuretic hormone, the permeability of collecting tubule cells is lowered, and less water is reab-

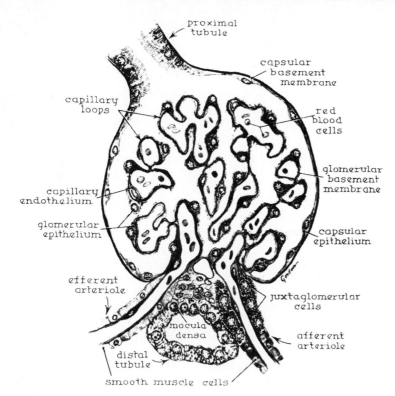

Figure 14.5   A diagram of the juxta-glomerular apparatus. (From Ham: Histology. J. B. Lippincott Co.)

sorbed. Increased production of filtrate and decreased reabsorption of water rapidly bring the volume of body fluids down to normal. If the volume of body fluids falls below normal, as in a severe hemorrhage, these factors work in the opposite direction: Less glomerular filtrate is produced, more water is reabsorbed, and the volume of body fluid is soon raised to normal.

The osmotic pressure of the tubular contents also affects the amount of water removed. If a large amount of salts or sugars is being eliminated, the osmotic pressure of the tubular contents is increased and less water can be reabsorbed. The urine volume is greater when there is a large amount of osmotically active substances in the urine, as after a large intake of salt or in diabetes mellitus.

Water and Salt Balance.   The nephrons of other vertebrates are essentially similar in structure and function to mammalian nephrons, although there are differences in detail. In addition to being associated with the glomerulus, some of the nephrons of primitive vertebrates are connected with the coelom via a **nephrostome** and can remove materials from the coelomic fluid.

This is analogous to the nephridia of the earthworm (p. 98). It may have been the primitive condition in vertebrates, for in the tubules of still more primitive vertebrates the glomerulus protrudes into the coelom, instead of into the beginning of the tubule, and the glomerular filtrate is discharged into the coelom.

The size of the renal corpuscles and other details of the nephron vary with the environment in which the animal lives. Primitive fresh-water fishes have large renal corpuscles that produce copious amounts of filtrate and do not have special water-reabsorbing mechanisms (Fig. 14.6). The concentration of salts within their bodies is greater than that in the surrounding medium, and water moves by osmosis into their bodies. Their problem is to pump out the excess water, yet retain the needed salts. The type of tubule found in these fishes is well adapted for this. The primary function of this primitive tubule may have been water regulation, for much of the nitrogenous waste is eliminated by diffusion through the gills.

Most salt-water fishes have the opposite problem, for the concentration of salt in the sea is greater than in their bodies; they lose water by osmosis. Small glomeruli, or even aglomerular tubules in some species, reduce the loss, and the deficit is made up by drinking salt water. Excess salt is then excreted by specialized cells in the gills. Marine cartilaginous fishes are an exception. A considerable amount of urea is retained in their tissues since not so much urea is eliminated by the tubules and, unlike other fishes, little is lost through the gills. The retained urea raises their internal osmotic pressure sufficiently for water to diffuse in. Their nephrons resemble those of fresh-water fishes. Excess salt taken in with their food is eliminated by a digitiform gland that discharges into the terminal part of the intestine.

Amphibians retain the primitive fresh-water type of tubule and have little control over the loss of water. Frogs can lose each day through their skin and urine an amount of water equivalent to one-third of their body weight. The need to soak up water and to keep the skin moist for gas exchange is a factor that compels frogs to stay near water. Water is conserved in reptiles by their horny skin and by the small size of their glomeruli. Less water is removed

from the blood by these glomeruli than by the large ones of primitive fresh-water fishes and amphibians. Birds and mammals have glomeruli of moderate size but have evolved loops of Henle that make it possible for a hypertonic urine to be produced. Mammals that live in deserts have exceptionally long loops of Henle and can remove more water from the urine than other mammals. Some terrestrial vertebrates (toads and many reptiles) also reabsorb water from the urinary bladder, although ordinarily urine is not changed after it leaves the collecting tubules.

Some terrestrial vertebrates have secondarily adapted to a marine environment and many of these resemble typical marine fishes in that they drink sea water and excrete excess salt by way of specialized head glands. These glands open beside the eyes in sea turtles and into the nasal cavities in most marine birds (Fig. 32.12, p. 739). Marine mammals apparently eliminate excess salt through their kidneys.

As we pointed out in Chapter 5, animals can also save water by converting ammonia into nitrogenous wastes that require less water for their removal. Ammonia, which is produced by the deamination of amino

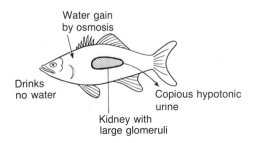

FRESH-WATER FISH IN HYPOTONIC MEDIUM

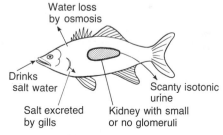

SALT-WATER FISH IN HYPERTONIC MEDIUM

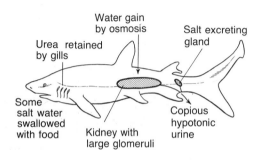

SHARK IN HYPOTONIC MEDIUM

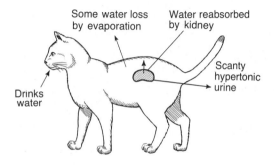

TERRESTRIAL MAMMAL

Figure 14.6 Water and salt balance in representative vertebrates.

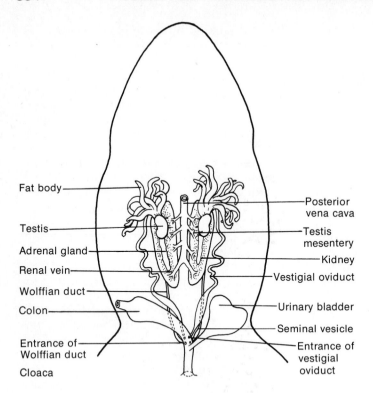

Fat body

Testis

Adrenal gland

Renal vein

Wolffian duct

Colon

Entrance of
Wolffian duct

Cloaca

Posterior
vena cava

Testis
mesentery

Kidney

Vestigial oviduct

Urinary bladder

Seminal vesicle

Entrance of
vestigial
oviduct

Figure 14.7 Ventral view of the urogenital system of a male frog. The
vestigial oviduct shown in this figure
tends to be absent in many male frogs.

acids, is a very toxic compound, but it is
highly soluble in water and can be excreted
rapidly if ample water is available to carry
it away. If an animal converts its ammonia to
urea, some water can be conserved, for each
molecule of urea is formed from two molecules of ammonia. If ammonia is converted
to uric acid, more water can be saved, for
uric acid has a low toxicity, is relatively
insoluble and can be excreted as an in-

soluble paste. Ammonia is the primary
nitrogenous waste of fresh-water fishes,
whereas urea and uric acid are excreted
by terrestrial vertebrates.

14.3   THE GONADS

From a biologic point of view, all the
structures and processes that permit an

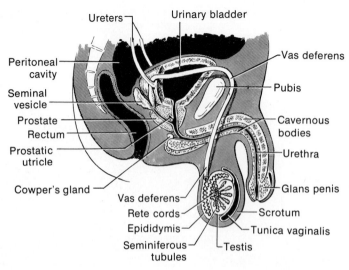

Ureters

Urinary bladder

Peritoneal
cavity

Seminal
vesicle

Prostate

Rectum

Prostatic
utricle

Cowper's gland

Vas deferens

Rete cords

Epididymis

Seminiferous
tubules

Vas deferens

Pubis

Cavernous
bodies

Urethra

Glans penis

Scrotum

Tunica vaginalis

Testis

Figure 14.8 A diagrammatic sagittal section through the pelvic region of a man to
show the genital organs. The prostatic
utricle is a vestige of the oviduct that is
present in the sexually indifferent stage
of the embryo. (Modified after Turner.)

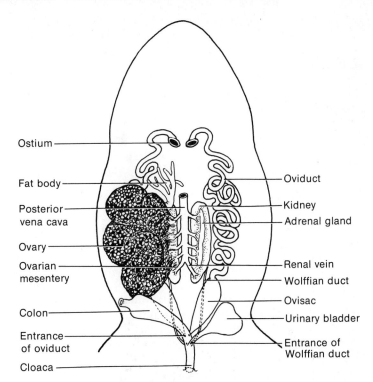

Figure 14.9 Ventral view of the urogenital system of a female frog. The left ovary has been removed.

Labels: Ostium, Fat body, Posterior vena cava, Ovary, Ovarian mesentery, Colon, Entrance of oviduct, Cloaca, Oviduct, Kidney, Adrenal gland, Renal vein, Wolffian duct, Ovisac, Urinary bladder, Entrance of Wolffian duct

animal to survive are of no avail unless the species can reproduce its kind. The general aspects of reproduction, including the production of gametes in the gonads, fertilization, and the early development of the embryo, were considered in Chapter 6. At this time we shall deal more specifically with the reproductive organs of vertebrates and their role in reproduction.

Reproduction is sexual in vertebrates, and the sexes are separate in all but a few bony fishes. The **testes** (Figs. 14.7 and 14.8) are paired organs of modest size, each con-

sisting of numerous, highly coiled **seminiferous tubules,** whose total length in man has been estimated at 250 meters! This provides an area large enough for the production of billions of sperm. As the sperm mature, they enter the lumen of the tubule and move toward the genital ducts. Finger-shaped **fat bodies** are associated with the gonads of frogs. They contain food reserves utilized during hibernation and in the spring breeding season when the animals become very active.

The ovaries of fishes or amphibians that

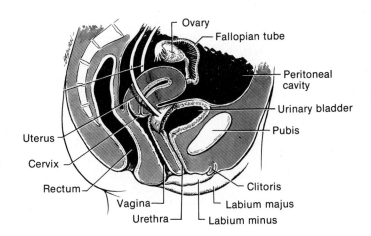

Figure 14.10 A diagrammatic sagittal section through the pelvic region of a woman to show the genital organs. (Modified after Turner.)

Labels: Ovary, Fallopian tube, Peritoneal cavity, Urinary bladder, Pubis, Uterus, Cervix, Rectum, Vagina, Urethra, Clitoris, Labium majus, Labium minus produce thousands or hundreds of eggs fill much of the body cavity (Fig. 14.9). Higher vertebrates produce fewer eggs, for fertilization is internal and the eggs are deposited in situations where there is a greater chance for survival of the developing embryos. The ovaries of reptiles and birds are still large, and the eggs contain much yolk. Mammalian eggs contain very little yolk and the ovaries are quite small, 2.5 cm. in length in a human being (Fig. 14.10). The eggs are not free within the ovary, for each one is surrounded by a **follicle** of epithelial and connective tissue cells. When the egg is ripe, the follicle bursts and the egg is discharged into the coelom, a process known as **ovulation** (Fig. 14.13).

In nearly all vertebrates, the gonads are suspended by mesenteries in the abdominal cavity, and they remain there throughout life. But in the males of most mammals the testes undergo a posterior migration, or descent, and move out of the main part of the abdominal cavity into a sac of skin known as the **scrotum** (Fig. 14.8). As they move into the scrotum, they carry a coelomic sac, the **tunica vaginalis,** down with them so that, despite their superficial position, they lie within a portion of the coelom.

In most mammals spermatogenesis does not go to completion unless the testes are descended. They remain descended in the majority of species, but in rabbits and rodents they are migratory—descending into the scrotum during the breeding season, withdrawing into the abdominal cavity at other times. Spermatogenesis, like other vital processes, can only occur within a limited temperature range. This range is exceeded by the temperature in the abdominal cavity but not by the temperature in the scrotum, which is approximately 4° C. lower. In order to test this hypothesis, Dr. Carl R. Moore of the University of Chicago confined the testes of rats to the abdominal cavity and found that spermatogenesis did not occur. Indeed, the seminiferous tubules underwent regression. He also insulated the scrotum of a ram in which the testes were descended. This raised the temperature, and again spermatogenesis did not occur. Apparently, during the evolution of endothermism in mammals, spermatogenesis did not become adapted to the higher body temperatures.

14.4 REPRODUCTIVE PASSAGES

Once the sperm and eggs have been produced, they must be removed from the body and be brought together to form a zygote. This is a simple procedure in primitive vertebrates such as cyclostomes. No reproductive ducts are present, and both eggs and sperm simply break out of the gonad into the coelom. Ciliary currents carry them to the posterior end of the coelom, where they are discharged through a pore into the cloaca. Fertilization and development occur outside the body.

Embryonic Formation of Reproductive Ducts. Other vertebrates have a system of ducts for the removal of the gametes, and some of them are excretory ducts. In order to understand this relationship, it is necessary to go back to a period in embryonic development when the embryo is **sexually indifferent** (Fig. 14.11). Its sex is determined genetically at the time of fertilization (p. 158), but early in development the embryo has the morphologic potentiality of differentiating into either a male or a female, for the primordia of both male and female duct systems are present. A pair of **oviducts** are present, each one opening anteriorly into the coelom through a funnel-shaped ostium and connecting posteriorly with the

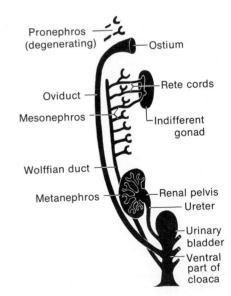

Figure 14.11 A diagrammatic ventral view of the urogenital organs of the sexually indifferent stage of an embryonic mammal.

Figure 14.12   Leopard frogs in amplexus.

cloaca. The developing gonad, which is not recognizable as an ovary or a testis at first, is adjacent to each mesonephros, and **rete cords** develop and connect the gonad with some of the mesonephric tubules. Gametes could thus pass through the rete cords, the mesonephric tubules and the wolffian duct. If the embryo differentiates into a male, the route through the mesonephros materializes, and the embryonic oviduct disappears, leaving at most a few traces. If the embryo differentiates into a female, the route through the coelom and oviducts is used, the oviducts develop further, and those parts of the male system not necessary for excretion largely disappear.

Male Vertebrates. In the males of most fishes and amphibians (frog, Fig. 14.7), the rete cords become the **vasa efferentia,** which extend through the mesentery, carrying sperm from the seminiferous tubules in the testis to the anterior part of the kidney. The frog's kidney is an opisthonephros, but its anterior portion develops from the embryonic mesonephros. Sperm pass through kidney tubules into the wolffian duct, which carries both sperm and urine to the cloaca, though not at the same time.

Reptiles, birds and mammals (man, Fig. 14.8) have metanephric kidneys, and sperm pass from each testis to an **epididymis,** thence out a **vas deferens** to the urethra.

This, seemingly, is a different pattern, but it is not so different as it first appears. Rete cords connect the seminiferous tubules with the epididymis, and the epididymis represents those mesonephric tubules that were associated embryonically with the testis, together with a highly convoluted portion of the wolffian duct. The vas deferens represents the rest of the wolffian duct, and most of the urethra represents the ventral part of a divided cloaca. Man thus utilizes passages homologous to those of a frog.

Other differences between the male reproductive organs of lower and higher vertebrates are correlated with differences in mode of reproduction. Frogs mate in the water. The male grasps the female in an embrace termed amplexus (Fig. 14.12) and sprays sperm over the eggs as they are discharged. Fertilization is external. This mating procedure is perfectly satisfactory for species that mate in water, but the gametes are too delicate for external fertilization in the terrestrial environment. To accomplish internal fertilization, male mammals have a **penis** with which sperm are deposited in the female reproductive tract, and a series of **accessory sex glands** that secrete a fluid in which the sperm are carried. The penis develops around the urethra and contains three **cavernous bodies** composed of spongy **erectile tissue.** Arterial dilatation coupled with a restriction of venous return causes the vascular spaces within the erectile tissue to become filled with blood during sexual excitement, making the penis turgid and effective as a copulatory organ. The accessory sex glands are a pair of **seminal vesicles,** which connect with the distal end of the vasa deferentia; a **prostate gland** surrounding the urethra at the point of entrance of the vasa deferentia; and a pair of **Cowper's glands** located more distally along the urethra.

Female Vertebrates. Eggs are removed from the coelom in most female vertebrates by a pair of oviducts, but the oviducts are modified for various modes of reproduction. Fishes and nearly all amphibians reproduce in the water. Most are oviparous, fertilization is external, and the eggs develop into larvae that can care for themselves. In the frog (Fig. 14.9), each oviduct is a simple, coiled tube that extends from the anterior end of the coelom to the cloaca. The oviducts contain glandular cells that secrete layers of jelly about the eggs, and their lower ends are expanded for temporary storage of the eggs, but they are not otherwise specialized.

Fertilization is internal in reptiles, birds and mammals. They reproduce on the land, and the free larval stage has been replaced by the evolution of a cleidoic egg (p. 712). Most reptiles and all birds are oviparous and the eggs develop externally. The oviducal glands, which secrete the albumin and a shell around the egg, are more complex in the oviducts of reptiles than in those of amphibians and fishes, but in other respects the oviducts of reptiles have not changed greatly. Birds have lost the right oviduct along with the right ovary, but the remaining left oviduct is essentially similar to a reptilian one.

Most mammals and a few fishes and reptiles have become viviparous; they retain the fertilized egg within the reproductive tract until embryonic development is complete and the embryo receives its nutrients from the mother. The oviducts are modified accordingly. In the human female (Figs. 14.10 and 14.13), the **ostium** lies adjacent to the ovary and may even partially surround it. When ovulation occurs, the discharged eggs are close enough to the ostium to be easily carried into it by ciliary currents. The anterior portion of each oviduct is a narrow tube known as the **fallopian tube,** and eggs are carried down it by ciliary action and muscular contractions. The remainder of the primitive oviducts have fused with each other to form a thick-walled, muscular **uterus** and part of the **vagina.** The terminal portions of the vagina and urethra develop from a further subdivision of the ventral part of the cloaca. The vagina is a tube specialized for the reception of the penis, and it is lined with stratified squamous epithelium. It is separated from the main body of the uterus, in which the embryo develops, by a sphincter-like neck of the uterus known as the **cervix.** The orifices of the vagina and urethra are flanked by paired folds of skin, the **labia minora** and

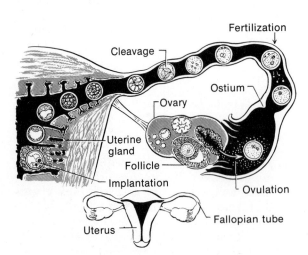

Figure 14.13 A diagram to show the path of an egg from the ovary to the uterus and the changes that occur en route. The last stage is about a week and one-half old. (Modified after Dickinson.)

labia majora. A small bundle of sensitive erectile tissue, the **clitoris,** lies just in front of the labia minora. Structures comparable to these are present in the sexually indifferent stage of the embryo and develop into more conspicuous organs in the male. The labia majora are comparable to the scrotum; the labia minora and clitoris, to the penis. A pair of glands, homologous to Cowper's glands in the male, discharge a mucous secretion near the orifice of the vagina. A fold of skin, the **hymen,** partially occludes the opening of the vagina but is ruptured during the first intercourse.

14.5 MAMMALIAN REPRODUCTION

Fertilization. During copulation, the sperm that have been stored primarily in the epididymis are ejaculated by the sudden contraction of muscles in and around the male ducts, and the accessory sex glands concurrently discharge their secretions. The seminal fluid that is deposited may contain as many as 400,000,000 sperm. Mucus in the seminal fluid serves as a conveyance for the sperm; proteolytic enzymes break the mucus down into a more watery fluid after the semen has been deposited in the vagina and permit the sperm to become highly motile. Fructose provides a source of energy, alkaline materials prevent the sperm from being killed by acids normally in the vagina, and certain fatty acids (prostaglandins) promote the contraction of the smooth muscle in the walls of the uterus and fallopian tubes.

Sperm move from the vagina through the uterus and up the fallopian tube in a little over one hour. How they do this is not entirely understood. They can swim, tadpole fashion, by the beating of the tail, but muscular contractions of the uterus and fallopian tubes, and ciliary currents in the tubes must help considerably. Fertilization occurs in the upper part of the fallopian tube (Fig. 14.13), but the arrival of an egg and the sperm in this region need not coincide exactly. Sperm retain their fertilizing powers for a day or two, and the egg moves slowly down the fallopian tube, retaining its ability to be fertilized for about a day. The chance of fertilization is further increased

in many species of mammals (but not in human beings) by the female coming into "heat" and receiving the male only near the time of ovulation. Ovulation, "heat," and changes in the uterine lining in preparation for the reception of a fertilized egg are controlled by an intricate endocrine mechanism that will be considered in Chapter 17.

Only one sperm fertilizes each egg, yet unless millions are discharged, fertilization does not occur. One reason for this is that only a fraction of the sperm deposited in the vagina reach the upper part of the fallopian tube. The others are lost or destroyed along the way. When the egg enters the fallopian tube, it is still surrounded by a few of the follicle cells that encased the egg within the ovary (Fig. 3.21), and a sperm cannot penetrate the egg until these are dispersed. This requires an enzyme, **hyaluronidase,** which can break down **hyaluronic acid,** a component of the intercellular cement. Hyaluronidase is believed to be produced by the sperm themselves, and large numbers are apparently necessary to produce enough of it. After the follicle cells are dispersed and one sperm has fertilized the egg, a **fertilization membrane** is raised from the surface of the egg and additional sperm penetration is not possible.

Establishment of the Embryo in the Uterine Lining. The fertilized egg passes down the fallopian tube into the uterus, undergoing cleavage along the way. Energy for early development is supplied by the small amount of food within the egg and by secretions from glands in the uterine lining. About a week after fertilization the embryo of most mammals penetrates the uterine lining, apparently by secreting digestive enzymes, and the lining folds over it. As explained earlier (p. 138), a placenta, which provides the metabolic requirements of the embryo, is formed by the union of the chorioallantoic membrane of the embryo with the uterine lining. The degree of union of fetal and maternal tissues differs among the various groups of mammals. In some, including the pig, the fetal chorion simply rests against the uterine lining. There is no breakdown of fetal or maternal tissue. In human beings the union is more intimate, for microscopic **chorionic villi,** which contain the fetal capillaries, penetrate the uterine lining, break down mater-

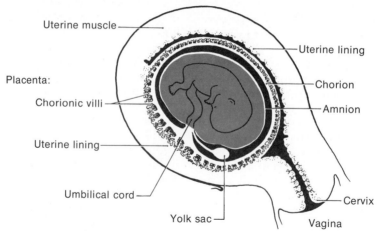

Uterine muscle

Uterine lining

Placenta:

Chorion

Chorionic villi

Amnion

Uterine lining

Umbilical cord

Cervix

Yolk sac

Vagina

Figure 14.14   A young human embryo surrounded by its extraembryonic membranes and lying within the uterus. Notice that the whole complex of embryo and membranes is embedded in the uterine lining. Villi are present all over the surface of the chorion at this stage, but only those on the side toward the uterine wall enlarge and contribute to the definitive placenta. (Modified after Patten.)

nal tissue, and become bathed in maternal blood (Fig. 14.14). Since maternal blood and fetal chorion are in contact, this type of union is called a **hemo-chorial placenta.** It should be emphasized that, except for occasional breaks in the placental membrane, no blood is exchanged between fetus and mother. Most gases and waste products simply diffuse across the membrane. However, active transport may play a role in the transfer of certain substances; the fetal blood contains a higher concentration of amino acids, calcium and ascorbic acid than the maternal blood.

Birth.   As the embryo develops, the uterus

enlarges considerably to accommodate it. At the time of conception, the human uterus does not protrude far above the pubic symphysis (Fig. 14.10), but nine months later, when embryonic development has been completed, it extends up in the abdominal cavity nearly to the level of the breasts. During this enlargement, the individual muscle fibers in its wall increase in size, and additional muscle develops from undifferentiated cells in the uterine wall. The uterus becomes a powerful muscular organ ready to assume its role in childbirth, or **parturition.**

The factors that initiate birth are not en-

Figure 14.15   Photographs of two models from the Dickinson-Belski series on human birth. (From the *Birth Atlas*, published by Maternity Center Association, New York.)

tirely clear, but hormones secreted by the ovary and pituitary and perhaps by the fetal adrenal gland may play a role. Hormones produced by the pituitary, ovary and also the placenta have prepared the mother's body for the birth. The mammary glands have enlarged and are ready for milk production, the uterine musculature has increased, and the pubic and other pelvic ligaments have relaxed so that the pelvic canal can enlarge slightly. Birth begins by a series of involuntary uterine contractions, "labor," that gradually increase in intensity and push the fetus, generally head first, against the cervix (Fig. 14.15). The cervix gradually dilates, but in human beings as much as 18 hours or more may be required to open the cervical canal completely at the first birth. The sac of amniotic fluid that surrounds the fetus acts as a wedge and also helps to open the cervix. The amnion normally ruptures during this process, and the amniotic fluid is discharged. When the head begins to move down the vagina, particu-larly strong uterine contractions set in, and the baby is born within a few minutes. A few more contractions of the uterus force most of the fetal blood from the placenta to the baby, and the umbilical cord can be cut and tied, although tying is unnecessary since contraction of the umbilical arteries would prevent excessive bleeding of the infant. Other mammals simply bite through the cord. Within a week, the stump of the cord shrivels, drops off, and leaves a scar known as the **navel.**

Uterine contractions continue for a while after birth, and the placenta and remaining extraembryonic membranes are expelled as the "after-birth." Much of the uterine lining is lost at birth, for the human placenta is an intimate union of fetal membranes and maternal tissue. Uterine contractions prevent excessive bleeding at this time. Following the birth, the uterine lining is gradually reconstituted and the uterus decreases in size, though it does not become as small as it was originally.

ANNOTATED REFERENCES

Attention is again called to the general references on vertebrate
organ systems cited at the end of Chapter 11.

Asdell, S. A.: Patterns of Mammalian Reproduction. 2nd ed. Ithaca, N.Y., Comstock Publishing Co., 1964. An important source book on differences in reproduction and reproductive cycles that occur in the various kinds of mammals, from the aardvark to the zebu.

Baldwin, E. B.: An Introduction to Comparative Biochemistry. 4th ed. Cambridge, Cambridge University Press, 1964. Contains an interesting account of the osmotic and excretory problems of vertebrates living in different environments.

Masters, W. H. and V. F. Johnson: Human Sexual Response. Boston, Little, Brown and Co., 1966. A careful analysis of the physiological aspects of human mating.

Pincus, G.: Control of conception by hormonal steroids. Science 153:493, 1966. A review of the benefits and potential dangers of oral contraceptives.

Schmidt-Nielsen, K.: Desert Animals. Oxford, Clarendon Press, 1964. The ways vertebrates meet the problems of water conservation in arid environments are thoroughly discussed.

Schmidt-Nielsen, K.: Salt glands. Scientific American 200:109, (Jan.) 1959. The reasons that many marine vertebrates can drink salt water and man cannot are discussed.

Smith, H. W.: From Fish to Philosopher. Boston, Little, Brown and Co., 1953. A very entertaining and readable account of vertebrate evolution with emphasis on kidney structure and function.

Smith, H. W.: The kidney. Scientific American 188:40, (Jan.) 1953. An excellent summary of the evolution of the kidney and its relation to the environment of vertebrates.

Turner, C. D. and J. T. Bagnara: General Endocrinology, 5th ed. Philadelphia, W. B. Saunders Co., 1971. Contains an excellent chapter on the biology of sex and reproduction.

Villee, C. A. (Ed.): The Control of Ovulation. London, Pergamon Press, 1961. Summarizes studies on the process of ovulation and its control in a series of papers prepared by leading students of the subject.

15. Sense Organs

If an organism is to be successful and survive in the complex world in which it lives, the activities of all its organs must be integrated so that the organism will make appropriate responses to changes in its external and internal environment. In the higher animals, integration is accomplished by special **receptors,** or sense organs, which detect changes in the environment, and by the **nervous system,** which receives and integrates information from all the sense organs and sends impulses to appropriate **effectors** (muscles, glands), whose activity brings about the appropriate response. Many vertebrate and invertebrate effectors are regulated, in part, by hormones that are secreted by endocrine glands and transported in the blood stream. Endocrine integration tends to be general rather than specific in its action; that is to say, one hormone may affect more than one organ. Endocrine integration is generally slower but longer lasting than nervous integration; it is especially effective in controlling continuing processes such as metabolism and growth. In a few instances, e.g., in the control of pancreatic secretion, endocrine integration is specific and rapid, but most of the specific and rapid adjustments are achieved by the sense organs and the nervous system. Nervous integration is highly specific; the neurons carry impulses from specific receptors to the spinal cord or brain, from which impulses go out along other neurons to specific effectors. It is rapid because the nerve impulse can travel very fast—as fast as 140 meters per second in the larger, myelinated mammalian neurons—and a second impulse can follow after a brief recovery period that lasts at most only several milliseconds.

It will be recalled from Chapter 5 that our ability to perceive different kinds of stimuli (touch, light, sound, etc.) is a function of the specificity of the receptors, which are attuned to specific stimuli, and of their specific connections within the nervous system. The nerve impulse that is initiated is not specific and is fundamentally the same regardless of where it originates. Differences in intensity are coded in the number of nerve fibers activated and in the number of impulses passing along a given fiber. Awareness of the sensation depends on the precise part of the brain the impulse reaches. This can be demonstrated by bypassing the receptor and stimulating its neurons directly. The subject then feels the same sort of sensation as if the receptors themselves had been stimulated.

15.1 RECEPTOR MECHANISMS

Organisms have evolved the capability of sensing those changes in the external and internal environments that are of survival value. We, in common with other vertebrates, are able to detect changes in many features of the environment, but our knowledge of the world is not complete. We cannot detect cosmic and radio waves without

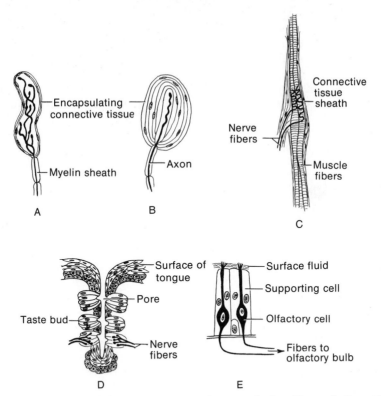

Figure 15.1 A group of mammalian receptors. *A*, Meissner's corpuscle found beneath the epidermis, assumed to be sensitive to touch; *B*, Pacinian corpuscle found in the dermis and many internal organs, sensitive to pressure; *C*, neuromuscle spindle, sensitive to muscle tension (proprioception); *D*, taste buds between papillae on the surface of the tongue; *E*, olfactory cells in the nasal mucosa. (*A* modified after Ranson; *B* and *E* after Gardner; *C* after Maximow and Bloom.)

instruments, and there may be other perturbations in the environment of which we are completely unaware. Receptors can be grouped into four broad categories according to the nature of the environmental change that activates them: **Chemoreceptors** in the mouth and nose detect chemical changes in the environment; various kinds of **mechanoreceptors** in many parts of the body are affected by touch, pressure, muscle stretch, vibrations and balance; **photoreceptors** respond to changes in light; and **thermoreceptors** in the skin and mouth respond to changes in temperature (Fig. 15.1).

All of these receptors consist of one or more specialized cells or of free nerve endings that may or may not be encapsulated in connective tissues. Physically, they respond as miniature transducers for, like the piezoelectric crystal in a phonograph pickup, they convert one form of energy into another. Each receptor is easily affected by the type of energy to which it is attuned and converts this to an electric current, **the generator potential**, which, in turn, initiates a nerve impulse. The generator potential has been studied particularly in certain mechanoreceptors. It depends, like the action potential of muscle and nerve, on changes in membrane permeability and the flow of ions. But unlike muscle and nerve, the activity of a receptor cell is not an all-or-none phenomenon; it varies with the strength of the stimulus. The generator potential must attain a certain threshold to initiate a nerve impulse. If the intensity of the potential exceeds the threshold, it initates additional impulses.

15.2 SOME MICROSCOPIC RECEPTORS

All receptor cells are microscopic, and most are widely scattered throughout the body. After briefly examining certain of these, we will consider more thoroughly the eye and the ear, two sense organs which are aggregations of receptive cells along with tissues that help to gather and amplify the stimuli.

The chemoreceptors of taste are small groups of receptive cells, the **taste buds,** found in mammals in association with certain of the tongue papillae. In lower vertebrates, they are found in many parts of the mouth and pharynx and, in some fishes, even on the skin of the head. Most investigators recognize four primary taste qualities localized to some extent on different parts of the tongue. Thus the tip of the tongue is particularly sensitive to sweet and salty stimuli; the sides, to sour or acid stimuli; the back of the tongue, to bitter stimuli. It is believed that other tastes are based on a combination of these four primary sensations.

The olfactory epithelium within the nose contains **olfactory cells,** the chemoreceptors of smell. Unlike other receptors, processes of these cells lead directly to the brain, hence they are considered to be **neurosensory cells.** Less is known about the kinds of qualities of odor, but there are doubtless many. It has been proposed that man can distinguish seven primary odors (camphoraceous, musky, floral, pepperminty, ethereal, pungent and putrid). Substances having these odors have molecules of different size and shape, and there is evidence that these fit into sites of specific size and shape on different receptor cells. Our final sensation of a substance is usually a complex mixture of taste, odor, texture and sight, and it is sometimes difficult to resolve the specific contribution of each particular sensation.

Man's chemical senses, especially olfaction, are somewhat rudimentary. In most vertebrate groups, apart from birds, they provide vital sensory clues to the external world: clues to food, predators, mates, and even the way home. It has been demonstrated that salmon with plugged nasal sacs are unable to find their home tributary in which to spawn, but normal salmon readily recognize their own river, apparently by recalling its distinctive odor!

Particularly important mechanoreceptors are the **muscle spindles** (Fig. 15.1C) that occur in specialized parts of most muscles, and similar organs in tendons. They mediate our sense of **proprioception,** or **kinesthesis,** by detecting changes in the tensions that are developed in the muscles. You can, for example, put your hands behind your back and without touching anything detect their movements by tensions developed in the muscles. It is this sort of feedback from the muscles that is essential for coordinated muscle activity, for example, for the appropriate amount of relaxation of one group of muscles as another group contracts.

Stimuli of touch, pressure, cold, warmth and pain are received by specialized, microscopic receptors and free nerve endings in the skin (Fig. 15.1A and B). There is no doubt that there is a physiologic specificity to the reception and transmission of these stimuli; that is, the excitation of certain neurons results in sensations of cold, of touch, and so forth. This principle was clearly set forth by the famous German physiologist Johannes Müller near the middle of the last century and is known as the doctrine of specific nerve energies. At one time, it was assumed that this physiologic specificity was mirrored by a morphologic specificity in the end organ. To some extent this is true, but there are not enough specialized endings in most parts of the skin to account for all of the physiologic specificity. Considerable sorting out and interpretation of stimuli must occur in the brain.

15.3   THE EYE

Ancestral vertebrates had eyes of two types—a median eye on the top of the head, which probably distinguished only between light and dark, and a pair of image-forming eyes on the sides of the head. Cyclostomes and a few reptiles retain a functional median eye but, in most groups it has become a small organ, the **pineal body,** attached to the top of the brain. The mammalian pineal body is a small glandlike organ with an endocrine role (p. 399).

Structure of the Mammalian Eye. Although the lateral, image-forming eyes of different groups of vertebrates vary in their adaptation for seeing beneath water, in the air and under varying light intensities, all are alike in their major features. Those of mammals may be taken as an example. Each eyeball is an oval-shaped organ constructed

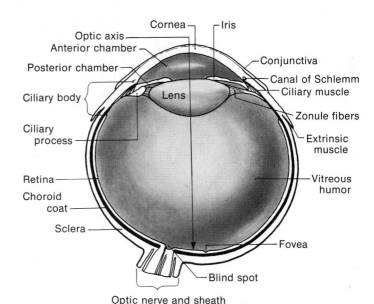

Figure 15.2 Diagram of a section through a mammalian eye.

on the principles of a simple camera (Fig. 15.2). It has a small opening at the front, the **pupil**, through which light enters; a **lens**, which brings the images of objects into sharp focus; and a light-sensitive **retina**, which is analogous to the film.

The wall of the eyeball is composed of three layers of tissue. The outermost one is a dense, fibrous connective tissue that gives strength to the wall. Most of this layer is opaque and is known as the **sclera**, but its anterior portion, through which light passes, is clear and is called the **cornea**. The surface of the cornea is covered with a layer of stratified epithelium, the **conjunctiva**, which is continuous with the epidermis.

The next layer of the eyeball wall is a darkly pigmented and very vascular **choroid coat**. Its pigmentation absorbs light rays, thereby reducing internal reflections that might blur the image, and its vessels nourish the retina. The anterior portion of the choroid coat, together with a nonsensitive portion of the retina, extends in front of the lens and forms the **iris** — an opaque disc with the pupil in its center. The iris prevents the light from entering the eye except through the center of the lens, which is optically the most efficient part. The amount of light entering the eye is controlled by circularly and radially arranged smooth muscles in the iris that constrict or dilate the pupil. In this respect, the iris is analogous to the iris diaphragm of a camera or microscope. The

thickened portion of the choroid around the base of the iris is the **ciliary body**. A number of **zonule fibers** extend from it to the lens and help to hold it in place. Muscles within the ciliary body affect the lens through the zonule fibers and focus an image on the retina.

The retina is the innermost layer of the eyeball. It consists of a **pigmented layer**, intimately associated with the choroid, and a **nervous layer**, which contains millions of receptor cells, the **rods** and **cones**, and neurons that initiate the processing of light signals and relay the information to the brain by way of the **optic nerve**. The rods and cones lie in the surface of the nervous layer that faces the choroid, and light must pass through most of the retina before it can stimulate them. This apparently illogical arrangement is explained by the mode of development of the eye. The retina develops from an outgrowth of the brain which, in turn, develops from an infolding of the surface ectoderm (Fig. 15.3). What was the outer surface of the ectodermal cells becomes the inner surface of the nervous layer of the retina. The polarity of the cells is retained during their various developmental gymnastics. The fact that the retina and optic nerves are developmentally parts of the brain also explains why at least two afferent neurons (**bipolar** and **ganglion cells**) are involved in transmitting impulses from the rods and cones. Chains of neurons are

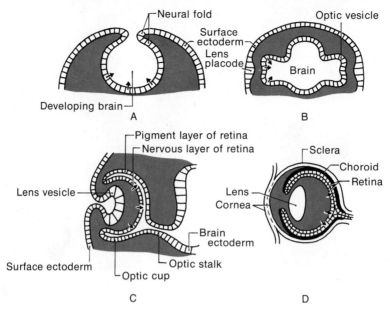

Figure 15.3   The development of the eye. *A*, Cross section through an embryo in which the anterior portions of the neural folds are closing to form the brain; *B*, the optic vesicles evaginate from the sides of the forebrain; *C*, an optic cup develops from each optic vesicle and the lens forms from adjacent surface ectoderm; *D*, the choroid, sclera and part of the cornea develop from surrounding mesoderm. Arrows indicate the original polarity of the ectoderm cells. (*D* from Romer.)

common in brain tracts, but in most nerves only one neuron extends from a receptor cell to the brain or spinal cord.

The cavities within the eye are filled with liquid. A gelatinous **vitreous humor** occupies the large chamber that lies between the lens and the retina and helps to hold the lens in place. A watery **aqueous humor** fills the **posterior chamber** between the iris and the lens and the **anterior chamber** between the iris and the cornea. The aqueous humor is secreted continually by the ciliary body and drained through the **canal of Schlemm** at the base of the cornea. This humor performs a twofold function. Firstly, it helps to nourish the cornea and lens, which are devoid of blood vessels; secondly, by maintaining the intraocular pressure, it helps to maintain the turgidity and shape of the eyeball. An imbalance between production and drainage of aqueous humor can lead to an increased intraocular pressure and the disease **glaucoma,** in which the pressure flattens and eventually injures the retina.

The eyeball lies in the orbit of the skull, and six **extrinsic ocular muscles,** which move the entire eyeball, extend from it to the walls of the orbit. A pair of movable

eyelids cover the eyeball, and the cornea is kept moist, cleansed, and possibly nourished, by the secretion of tears from several **tear glands.** Tears are drained from the median corner of the eye by a **lacrimal duct,** which leads into the nasal cavity. Pigs, cats and many other mammals have a third lid, the **nictitating membrane,** in the median corner of the eye. It is moved passively over the cornea when the eyeball is retracted slightly, and aids in cleaning and protecting the eye. This membrane is reduced to a vestigial **semilunar fold** in man.

Refraction of Light.   Light that enters the eye is bent toward the optic axis in such a way that it forms a sharp, though inverted, image upon the retina (Fig. 15.4*A*). The lens is important in bending the light rays but the cornea, humors and the retina itself are also involved. The cornea is the major refractive agent in terrestrial vertebrates, for the difference between the refractive index of air and the cornea is greater than that between any of the other refractive media. The action of the cornea places the image approximately on the retina; the lens brings it into sharp focus.

When the eye is at rest, distant objects are in focus. The refractive power of the eye

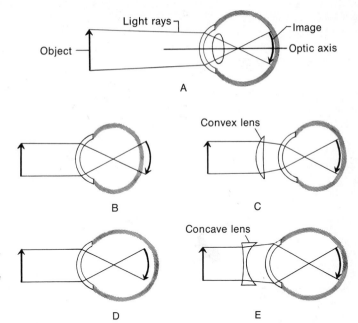

Figure 15.4 Image formation by the eye. A, Normal eye; B, far-sighted eye; C, far-sighted eye corrected by a convex lens; D, near-sighted eye; E, near-sighted eye corrected by a concave lens.

must be increased in viewing a near object or its image would be blurred, for the image would come into sharp focus theoretically at a point behind the retina. Accommodation for near vision is accomplished in mammals by the contraction of muscles within the ciliary body. This brings the point of origin of the zonule fibers a bit closer to the lens and releases the tension of these fibers. The front of the elastic lens bulges out slightly, and its refractive powers are increased accordingly. When the ciliary muscles are relaxed, intraocular pressure pushes the wall of the eyeball outward, increases the tension of the zonule fibers, and the lens is flattened a bit. The lens becomes less elastic with age, and our ability to focus on near objects decreases.

The refractive parts of the eye form a sharp image of an object on the retina only in an eyeball of appropriate length. If the eyeball is shorter than normal, as it is in far-sighted people, the image of an object in theory falls behind the retina. Accommodation is necessary to bring the image into focus, and the power of accommodation may not be great enough to focus on a near object. This can be corrected by placing a convex lens in front of the eye (Fig. 15.4B and C). Near-sighted people have eyeballs that are longer than normal and the image falls short of the retina. This can be corrected by a concave lens (Fig. 15.4D and E).

Chemistry of Vision. Light that strikes the rods and cones activates them and they, in turn, initiate nerve impulses. The outer segment of each rod contains a great elaboration of the membrane system of the cell, and a great deal of the pigment visual purple, or **rhodopsin**, is associated with these membranes.

In common with other visual pigments, rhodopsin consists of a color-bearing component, **cis-retinal**, bound to a protein known as **opsin**. Upon absorbing light energy, cis-retinal isomerizes, that is, it undergoes a change in its geometrical configuration, to **trans-retinal** (Fig. 15.5). Trans-retinal and opsin constitute a compound called **lumirhodopsin**, but trans-retinal and opsin do not "fit" well together as a result of the changed geometry. Lumirhodopsin is unstable; it decays through **metarhodopsin** and the trans-retinal and opsin separate. It should be emphasized that light energy is needed only in the cis-trans isomerization; the other changes do not require light. Some of the changes affect, in some way not yet clear, the permeability of the membranes with which rhodopsin is associated and lead to the development of the generator potential. Many believe that photoisomerization is the activating process, partly

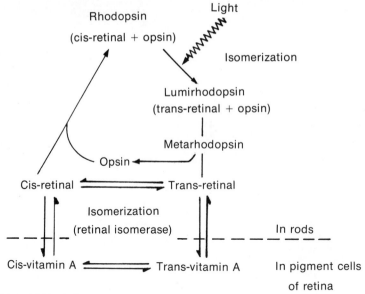

Figure 15.5   A diagram of the chemical changes that occur during rod vision.

because of its great speed and the speed with which light affects the rods. Rhodopsin is extremely sensitive to light. It has been calculated that a single photon of light can activate a molecule of rhodopsin and lead to the activation of an entire rod, but it does not follow, as we shall see, that a single active rod can initiate an impulse in a neuron going to the brain.

Resynthesis of rhodopsin first requires the isomerization of trans-retinal to cis-retinal. This process requires energy and the enzyme **retinal isomerase**. Cis-retinal readily combines with opsin to form rhodopsin.

A cycle of breakdown and resynthesis of rhodopsin goes on continuously if the eyes are exposed to any light, but shifts in the equilibrium point of the cycle can adapt the eye for seeing in very dim or in bright light. Low levels of rhodopsin are present in the cones of an eye adapted to bright light, for much has been broken down into trans-retinal, and some of the trans-retinal, which is an aldehyde of retinol (vitamin A), has been gradually converted to vitamin A. Most of the retinal involved in these reactions is contained in the pigment cells of the retina. In an eye adapted for seeing in dim light, a maximum amount of rhodopsin is present in the rods,

and some of the reserves of retinal have been used in its synthesis. Shifts between fully light and dark adapted retinas are gradual processes requiring some minutes. On coming out of a dark cinema to a bright day, one is dazzled until the level of rhodopsin is reduced. Conversely, one cannot see well in going from a bright to a dark room until the level of rhodopsin has increased.

Vitamin A is a precursor of retinal and normal vision requires that there be an adequate dietary intake of the vitamin. However, a great deal of vitamin A is stored in liver cells, and a dietary deficiency must continue for weeks before the ability to see in dim light is impaired, a condition referred to as **night blindness.**

Cones are activated only by light of greater intensity, and they are responsible for color vision. It has long been assumed that there are probably three types of cones sensitive to wave lengths of light in the blue, green and red parts of the spectrum. Presumably these interactions are responsible for our range of color vision. This hypothesis is consistent with the three known types of color blindness, but only recently has it been possible to get more direct evidence for this trichromatic theory. By exploring isolated retinas from individuals who have just died with a very small beam of light, it has been possible to find three types of cones with absorption spectra maxima in the blue, green or red parts of the spectrum. The light-sensitive pigment of the cones is **iodopsin,** which consists of retinal and an opsin different from that in rhodopsin. Chemical differences between the three cone types are uncertain.

Organization of the Retina. As we have seen, the retina develops embryonically as an outgrowth of the brain, so it is not surprising that its organization is very complex, nor that the retina itself plays a role in the processing of visual information. There are about seven million cones in the retina of each human eye. These are particularly dense in a small part of the retina opposite the optic axis which is known as the **fovea** (Fig. 15.2). Only cones are present in this region, and they become less dense toward the periphery of the retina, where they finally disappear. Each retina contains about 120 million rods, and they are most dense near the periphery of the eye. Rods and cones synapse with **bipolar cells** (Fig. 15.6) which in turn synapse with **ganglion cells** whose axons cross the retina, congregate at the blind spot, and go to the brain as the **optic nerve.**

The synaptic relationships of rods and cones differ in an important way. Many rods, sometimes several hundred of them in the peripheral part of the retina, converge on a single bipolar cell, and a number of bipolar cells may converge on one ganglion cell. Far fewer cones converge on a single bipolar cell, and in the fovea many of the cones synapse with a single bipolar and ganglion cell. This difference in neuron pathways, together with a difference in threshold of illumination, explains why rods are more efficient than cones in light of low intensity. Stimuli from a number of rods, each of which is below the threshold of a bipolar cell, converge upon a single cell, have an additive effect upon it, and may activate it. Less summation, or none, of this type occurs in the more "private line" system of the cones. Since the greatest density of rods and the greatest degree of convergence occur in the periphery of the retina, one can see best in dim light by looking out of the side of the eye so that the image falls to the side of the fovea. Rods are particularly abundant in the eyes of nocturnal vertebrates.

Although rod vision is more sensitive to light than cone vision, it also follows from their different pathways that rod vision is less acute. The eye cannot distinguish which rods are receiving light among a number of rods that converge on a single bipolar cell. Cone vision in the fovea is extremely acute, because most of the cones here have a "private line" to the brain. Dis-

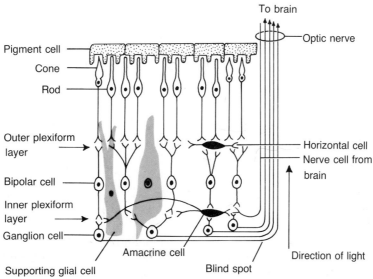

Figure 15.6 A diagrammatic vertical section through the retina to illustrate major types of interconnections among the component cells.

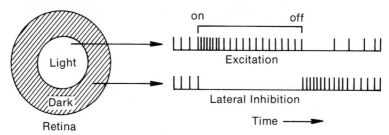

Figure 15.7   Differences in the rate of discharge of impulses of a ganglion cell receiving from an illuminated part of the retina (Excitation), and one receiving from an adjacent nonilluminated part (Lateral Inhibition). Impulses are represented by the vertical lines. (Modified from Guyton, Textbook of Medical Physiology, 4th ed., W. B. Saunders Co.)

tinct pathways to the brain from cones sensitive to different waves lengths of light are also necessary for an animal to distinguish and interpret colors. Rod vision is analogous to a high-speed, coarse-grained, black and white photographic film; cone vision, to a slower, fine-grained, color film.

Other complexities occur within the retina (Fig. 15.6). A network of **horizontal cells** interconnects the rods and cones and the bipolar cells at the level of the synapses between receptors and bipolar cells, and a second network of **amacrine cells** interconnects the bipolar cells at the level of their synapses with the ganglion cells. These interconnections, of which some are inhibitory and others facilitory, appear to be the morphological basis for phenomena one can identify in ganglion cells. Ganglion cells have a certain "background" rate of discharge of impulses, even in the dark. If a beam of light impinges on the retina, ganglion cells receiving impulses from the activated rods and cones first increase their rate of discharge; then their rate tapers off to nearly the background level (Fig. 15.7). When the light is turned off, the rate of discharge may stop momentarily before the background rate resumes. Ganglion cells receiving from the area of the retina immediately adjacent to the illuminated area respond in the opposite way. They are inhibited and cease their background discharge during the period of illumination of their neighbor's field; when the light goes off, they briefly increase their rate of discharge. These mechanisms would appear to increase the contrast between areas of the retina receiving different intensities of illumination. There are other mechanisms, including one in a frog that enables

the retina to be particularly responsive to the movement of a small object.

Although most cells in the retina carry impulses to the brain, there are a few nerve cells that carry impulses from the brain to the retina. These fibers probably enhance or suppress retinal activity, but we know little about them.

Eyes of Other Vertebrates.   The eyes of all vertebrates are essentially alike, but those of primitive vertebrates differ from mammalian eyes in several important respects, for the problems associated with sight beneath water are not identical with those in the air. For one thing, the water itself cleans and moistens the eye, and fishes have not evolved movable eyelids or tear glands. Secondly, the refractive index of water is nearly the same as that of the cornea, so the cornea of a fish's eye does not bend light rays. Most refraction is accomplished by the lens, which is nearly spherical and, hence, has a greater refractive power than the oval lens of tetrapods. It is interesting in this connection that the lens of a frog's eye flattens a bit during metamorphosis, when a change in environment occurs. Finally, the method of accommodation differs, for the lens is moved back and forth in camera fashion in fishes and amphibians and does not change shape.

15.4   THE LATERAL LINE AND THE EAR

Equilibrium.   All vertebrates have the ability to perceive differences in the orientation of their bodies with respect to their surroundings and to maintain their equilibrium. Although vision and proprioceptive impulses from the muscles play a part, this ability is primarily a function of the inner ear. The inner ear is embedded within the otic capsule of the skull and consists of a complex of membranous walled sacs and canals, the **membranous labyrinth,** which are filled with a liquid **endolymph** and surrounded by a protective liquid cushion, the **perilymph** (Fig.15.8). The dorsal part of the membranous labyrinth consists of three **semicircular canals,** each of which is perpendicular to the other two. Two lie in the vertical plane, but at right angles to each other, and one is in the horizontal plane at right angles to the other two. Each has a round swelling, an **ampulla,** at one of its ends in which there is a patch of **hair cells**—receptor cells bearing hairlike processes. The three semicircular canals connect with a chamber known as a **utriculus** and this, in turn, connects with a more ventral chamber known as a **sacculus.** Both of these chambers contain patches of hair cells. Calcareous **otoliths** are in contact with these cells. Different parts of the membranous labyrinth are concerned with different aspects of equilibrium. Differences in the position of the head and body (**static equilibrium**) affect the way in which gravity pulls the otoliths upon the underlying hair cells. Rapid forward movement (**linear acceleration**) causes the otoliths, which have more inertia than the surrounding endolymph and hence lag, to push back upon certain hair cells. During sudden turns of the head in various planes (**angular acceleration**), the endolymph in the semi-

circular canals, because of its inertia, does not move as fast as the head and the hair cells in the ampullae. This differential in rate of movement stimulates the hair cells.

Phonoreception in Fishes. The part of the ear concerned with equilibrium is essentially the same in all vertebrates, but the part concerned with phonoreception or hearing differs considerably among vertebrates. By hearing we mean the detection of pressure waves resulting from a mechanical disturbance some distance away. The human ear can detect waves with frequencies of 20 to about 20,000 cycles per second, and some animals, e.g., bats, can detect frequencies of over 100,000 cycles per second. Frequencies lower than 20 cycles per second are usually perceived as vibrations and not as sound.

Mammals, birds and some reptiles have a **cochlear duct,** an elongated cul-de-sac extending from the sacculus that is clearly concerned with phonoreception. Fishes have a homologous but very small diverticulum known as the **lagena.** The rudimentary nature of this structure, together with early experiments in which fishes were shown to be unresponsive to sounds made in the air, led to the conclusion that they could not hear. Later this conclusion was shown to be wrong, for it was realized that most air-borne sound waves are reflected by the air-water surface, but sounds generated in the water travel rapidly and far. By using underwater listening and sound-generating devices, investigators have discovered that aquatic organisms produce

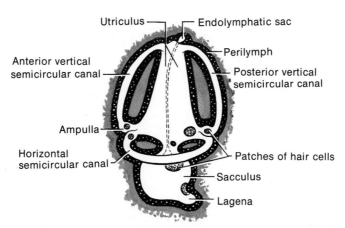

Figure 15.8 The left ear of a fish seen in a lateral view. Only an inner ear is present, embedded within spaces in the otic capsule of the skull. (Modified after Kingsley.)

and respond to a great many sounds. But a fish must overcome one problem. Since their ears and tissues are mostly water, they are essentially transparent to a sound generated in the water; the sound waves pass right through the fish without interruption. For a fish to detect the waves, there must be some tissue that will respond to the waves differently than the body as a whole. In most fishes this is simply a large calcareous **otolith** in the sacculus. Catfish, suckers and minnows also utilize an air-filled **swim bladder** as a hydrophone. The swim bladder, which has evolved as a modification of a lung-like structure, is located in the dorsal part of the body cavity. Pressure waves passing through the body change the volume of the swim bladder, and these movements are transmitted via a chain of small bones derived from the vertebrae (**weberian ossicles,** Fig. 30.20) to the sacculus. This mechanism is particularly effective, for members of this group of fishes can respond to a wider range of frequencies (60 to 6000 cycles per second) than other fishes, and they can also discriminate better between frequencies.

Lateral Line. Clearly, fishes can detect underwater sounds by means of a part of the membranous labyrinth. In addition, fishes have a **lateral line system** that is sensitive to currents, to changes in pressure and to vibrations of low frequency. It consists of groups of hair cells similar to those in the ear. Most of these are arranged in a longitudinal canal extending the length of the trunk and tail and in a series of canals that ramify over the head, but often other groups

of hair cells are scattered over the surface (Fig. 15.9A). The canals are embedded in the skin, but connect with the surface through a series of pores. The system is sometimes described as one of "distant touch," because the fish can detect its approach to an object, or an object's approach to it (Fig. 15.9B), by the resulting deflection of certain of the groups of hair cells. Neurons from these receptors enter an **acoustico-lateralis area** of the brain along with neurons from the ear, which suggests that there is a close relationship between the ear and lateral line. The inner ear develops embryonically in close association with certain lateral line canals, and it may have evolved in the same way. Larval amphibians have a lateral line system, but it is lost during metamorphosis. Higher vertebrates never have this system at all.

Phonoreception in Tetrapods. In all tetrapods, a part of the membranous labyrinth, generally the lagena or cochlear duct, is specialized for phonoreception, and various devices have evolved which transmit either ground- or air-borne vibrations to it. Frogs have an external tympanic membrane (Fig. 15.10), which responds to vibrations in the air, and a stapes, which transmits the vibrations across the middle ear cavity to a **fenestra ovalis** in the otic capsule. The fenestra ovalis communicates with the inner ear.

The hearing apparatus of mammals is basically similar but much more elaborate (Fig. 15.11). Most mammals have a well-developed external ear consisting of a canal, the **external auditory meatus,** and an external flap, the **auricle.** Together they act

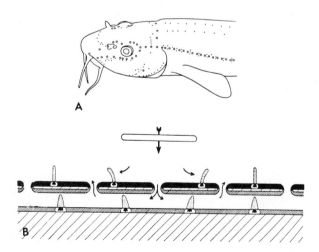

Figure 15.9 Lateral line organs of a fish. *A*, Surface view of the anterior end of the catfish, *Nemachilus*. Circles indicate the position of the lateral line; small dots, isolated groups of hair cells. *B*, Diagrammatic vertical section through a lateral line canal and several isolated groups of hair cells. Water currents (small arrows) and the deflection of groups of hair cells caused by an approaching object are shown. (From Dijkgraaf, S.: Bau und Funktionen der Seitenorgane und des Ohrlabyrinths bei Fischen. Experientia 8:205–216, 1952.)

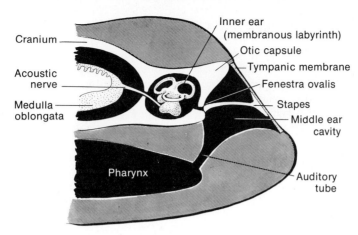

Figure 15.10 A diagrammatic cross section through the head of a frog to show the ear and its relation to surrounding parts.

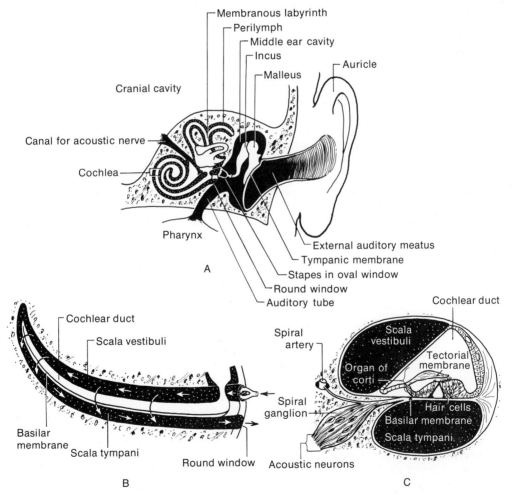

Figure 15.11 The mammalian ear. *A*, Schematic drawing of the outer, middle and inner ear of a human being. *B*, Diagram of the cochlea as though it were uncoiled. *C*, An enlarged cross section through the cochlea. The cochlear duct and other parts of the membranous labyrinth are filled with endolymph and are shown in white; perilymph is black with white stipples.

as an ear trumpet, concentrating and increasing slightly the pressure of the sound waves by the time they reach the delicate **tympanic membrane,** which is situated in a protected site at the internal end of the meatus. Acoustic pressure variations cause slight movements of the tympanic membrane. Three auditory ossicles (the hammer-shaped **malleus,** the anvil-shape **incus** and the stirrup-shaped **stapes,** arranged in sequence) transmit these vibrations across the middle ear cavity to the **fenestra ovalis,** or oval window. The stapes evolved from a part of the hyoid arch of fishes, and the malleus and incus were derived from the posterior part of the mandibular arch when a new jaw joint evolved in mammals anterior to the former one. These three ossicles form a lever system that reduces the displacement amplitude but increases the pressure amplitude of the sound waves. The movement of the foot plate of the stapes against the membrane within the oval window is only about one-half as extensive as the movement of the tympanic membrane, but the force of the movement is slightly greater. Two small muscles, derived from visceral arch muscles, attach onto the ossicles and reflexly dampen their oscillation to sounds of very high intensity. They form a protective mechanism somewhat analogous to the iris of the eye. The major pressure amplification results from the fact that the tympanic membrane has nearly 20 times the surface area of the membrane in the oval window. Virtually all of the force that impinges on the tympanic membrane reaches the membrane in the oval window and, since this membrane is smaller, the force per unit area is increased. Pressure amplification is essential for an efficient transfer of energy from the light, compressible external air to the dense, incompressible liquid of the inner ear.

The **middle ear cavity,** in which the ossicles lie, evolved from the first gill slit, homologous to the spiracle of many fishes. It connects with the pharynx via the **auditory (eustachian) tube** and, hence, indirectly with the outside of the body. The pharyngeal opening of the auditory tube is normally closed, but if pressures become unequal on the two sides of the tympanic membrane, the auditory tubes usually reflexly open and pressures are equalized. If pressure differences on the two sides of the membrane are great, a situation that often occurs during SCUBA diving, it is sometimes necessary to swallow in order to open the tubes.

A long **cochlear duct** has evolved from the lagena of fishes, and it contains the actual receptive structure, the **organ of Corti** (Fig. 15.11*B* and *C*). The cochlear duct is filled with endolymph and is a part of the membranous labyrinth. Pressure waves reach the cochlear duct via specialized perilymphatic channels. A **scala vestibuli** begins at the oval window, extends along the cochlear duct, curves around its apex and returns as the **scala tympani** to a **fenestra rotunda,** or round window, that is separated by a delicate membrane from the middle ear cavity. The scala vestibuli and scala tympani have a different origin from the cochlear duct, but all three are in intimate association and collectively form the spiral-shaped **cochlea.**

Pressure waves induced by the stapes at the oval window pass through the scala vestibuli, cross the cochlear duct, and escape through the scala tympani and round window. The **basilar membrane,** which supports the organ of Corti, is set in vibration. Since the basilar membrane and the **tectorial membrane** of the organ of Corti are hinged at different places, a pressure wave causes a slight differential movement between them and develops a shearing force that stimulates the intervening hair cells. The ear is extremely sensitive for, at certain frequencies, it can detect any displacement of these membranes considerably less than the diameter of a hydrogen atom! Sensory neurons of the acoustic nerve extend from the hair cells to the brain.

Sound analysis by the cochlea is very complex and not completely understood. It is well established that pitches or tones of different frequencies are detected in different regions of the cochlea. According to the current hypothesis, a movement of the stapes against the liquid in the scala vestibuli causes the proximal part of the basilar membrane to move toward the scala tympani. This sets up a tension on the basilar membrane and initiates a wave that travels along the membrane. The mechanism is analogous to a wave sent along a whip by a jerk at the proximal end. The wave travels along the entire basilar membrane (Fig. 15.12), but physical properties of the mem-

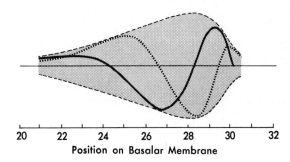

Figure 15.12 Displacement of the basilar membrane resulting from a traveling wave of a specific frequency. The shape of the wave at two instances in time is shown. Dashes outline the entire envelope of the wave. For any one frequency, maximum displacement of the membrane occurs only at one point. (From Case after Von Békésy.) (From G. Von Békésy: J. Acoust. Soc. Amer., *19:*452, 1947.)

brane, including width and elasticity, change progressively so the length and amplitude of the wave varies. Waves of different frequency reach maximal amplitude at different points. Long wave lengths (low notes) cause a maximal displacement near the apex of the cochlea; short wave lengths (high notes), a maximal displacement at the proximal end of the basilar membrane. Von Békésy has demonstrated the essential truth of this view by direct observations on the ears of fresh cadavers. More intense sounds, in addition to stimulating hair cells to move vigorously, cause a longer length of basilar membrane to vibrate, but the displacement peak remains the same for any given frequency. Two different musical instruments playing the same note have different qualities because of differences in the number and kinds of overtones or harmonics also present. Both instruments stimulate the same part of the basilar membrane, but they differ with respect to the other parts of the membrane stimulated.

In an organ as elaborate as the ear, many things can go wrong. Infections may enter the middle ear via the auditory tube and affect the auditory ossicles. The stapes may become locked in the oval window by an abnormal growth of bone, or the individual ossicles may fuse together. Conduction deafness of these types can be corrected by a hearing aid that amplifies vibrations enough to be transmitted directly through the skull bones to the cochlea. More rarely, the acoustic nerve or the cochlea may be damaged. Deafness of this sort cannot be corrected. If only a part of the cochlea is injured, one may become deaf only to sounds of certain frequencies. The continuing, loud, high-pitched noises to which boilermakers are subjected sometimes destroy a part of the cochlea, and the men become deaf to sounds of this frequency.

ANNOTATED REFERENCES

Attention is again called to the general references on vertebrate organ systems cited at the end of Chapter 11.

Amoore, J. E., J. W. Johnson, Jr., and M. Rubin: The stereochemical theory of odor. Scientific American *205:*74, (Feb.) 1964. An analysis of the theory that smell is based on the shapes of odoriferous molecules and receptor cells.

Case, J.: Sensory Mechanisms. New York, Macmillan, 1966. Prepared for Macmillan's Current Concepts in Biology series, this is a thorough account of the sense organs of invertebrates and vertebrates. Much recent work is included.

Davson, H. (Ed.): The Eye. New York, Academic Press, 1962. A four-volume reference book on the eye.

Geldard, F.: The Human Senses. New York, John Wiley & Sons, Inc., 1953. A valuable and detailed account of the human senses.

Lowenstein, W. R.: Biological transducers. Scientific American *203:*98, (Aug.) 1960. A discussion of the way receptor cells convert the energy they receive into electric pulses.

Michael, C. R.: Retinal processing of visual images. Scientific American 220:105, (May) 1969. A discussion of the interactions of retinal cells that increase an organism's ability to interpret important parameters of the visual field, including contrast and movement.

Miller, W. H., F. Ratliff, and H. K. Hartline: How cells receive stimuli. Scientific American 205:222, (Sept.) 1961. A general analysis of the problem of reception.

Polyak, S.: The Vertebrate Visual System. Chicago, University of Chicago Press, 1957. A very valuable source book.

Van Bergeijk, W. A., J. R. Pierce, and E. E. David, Jr.: Waves and the Ear. Garden City, Doubleday & Co., 1960. An excellent and authoritative account of all aspects of hearing.

Von Békésy, G.: The ear. Scientific American 197:66, (Aug.) 1957. An excellent account of the ear by the man who made many of the basic discoveries in the physiology of hearing.

Walls, G. L.: The Vertebrate Eye and Its Adaptive Radiation. Bloomfield Hills, Mich., Cranbrook Institute of Science, 1942. An old but still very valuable source book on the evolution of the eye and its adaptations to different environments.

Young, R. W.: Visual cells. Scientific American 223:81, (Oct.) 1970. A review of the ultrastructure and function of the rods and cones and their method of regeneration.

16.   Nervous Coordination

The nervous system provides for the co-ordination and integration of the body's many activities. It receives information about the current external and internal environments from the receptors, it integrates different sensory modalities and compares them with stored information about past events, and finally it computes a response and activates the appropriate effectors. The nervous system is composed of nerve cells or **neurons,** which conduct the impulses, and of supporting cells known as **neuroglia.** We previously considered the morphology and many aspects of the physiology of these cells (pp. 108–113).

16.1   ORGANIZATION OF THE NERVOUS SYSTEM

The neurons can be assigned either to the **central nervous system,** consisting of the brain and spinal cord, or to the **peripheral nervous system,** which includes the nerves that extend between the central nervous system and the receptors and effectors. The

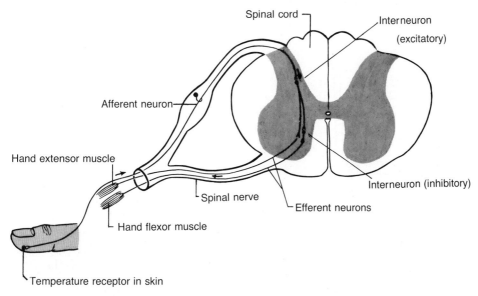

Figure 16.1   Diagram of the types of neurons involved in a withdrawal reflex in the spinal cord. The pathways shown here are superimposed upon neural mechanisms (not shown) that maintain muscle tone.

neurons themselves can be grouped into three broad categories: (1) sensory or **afferent neurons,** which transmit impulses from the sense organs to the brain or cord; (2) motor or **efferent neurons,** which transmit impulses from the brain or cord to the muscles and other effectors of the body; and (3) **interneurons,** which lie entirely within the central nervous system interposed between the other two.

The interrelations of the various kinds of neurons can be seen by considering a specific example. When you touch a hot stove, for example (Fig. 16.1), receptors in the skin are stimulated and initiate impulses in afferent neurons. These neurons are part of a spinal nerve and extend into the spinal cord, where they end in synapses with interneurons. The interneurons, in turn, transmit impulses to efferent neurons, which extend from the cord and carry impulses back through the spinal nerve to a group of extensor muscles of the hand. Their contraction withdraws your hand from the stove. For the movement to be effective, however, the antagonistic flexor muscles should relax. This relaxation involves the inhibition of impulses going to these muscles. Normally, some impulses go out to all of the muscles of the body continually and cause a

partial contraction, a condition called muscle **tonus.** There is now evidence for the existence of inhibitory neurons in the spinal cord which, upon activation, release an inhibitory transmitter substance at their synapses with the motor neurons. According to the chemical theory of synaptic transmission, inhibitory substances act by making the cell membrane of the postsynaptic neuron more permeable to certain ions. Chloride (Cl^-) flows into the postsynaptic cell and potassium (K^+) flows out; the cell membrane becomes hyperpolarized and hence less subject to stimulation by excitatory transmitter substances. Excitatory substances depolarize the membrane and thereby initiate a nerve impulse.

The stimulus and response just described is a simple **spinal reflex,** and the neuronal pathway along which the impulse travels is called a **reflex arc.** Reflexes are fixed patterns of response to stimuli and they need not involve an awareness of the stimulus. The impulse need not pass through any of the higher centers in the brain in order that the response occur. An impulse may be carried to the cerebral cortex of the brain by other connector neurons, **afferent interneurons** (Fig. 16.2). You then become aware of the stimulus and may voluntarily

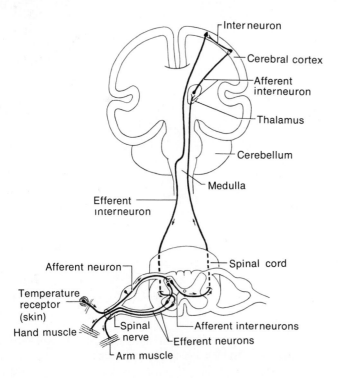

Figure 16.2 Diagram of the neurons involved in the passage of impulses to and from the brain, and the relationship of these to the neurons in a reflex arc (shown simplified).

decide to do something about it, perhaps withdraw your whole arm or turn off the stove. If so, impulses will pass out from the brain along **efferent interneurons** to the appropriate efferent neurons.

The **withdrawal reflex** described above requires at least two interneurons. Many other kinds of reflexes occur in the spinal cord and brain. A **stretch reflex,** the automatic contraction of a muscle when it or its tendon is unduly stretched, involves fewer neurons. In the familiar knee jerk, afferent neurons from proprioceptor cells in the tendon of the large muscle in the front of the thigh synapse directly with motor neurons extending to the same muscle; as the tendon is stretched, the muscle contracts. But even here, inhibitory neurons are connected to the antagonistic muscles. More complex reflexes may involve several regions of the body. If a drop of acid is placed on the flank skin of a frog, both hind legs will converge on this spot and alternately flex and extend in an attempt to scrape off the acid. This will happen even if the entire brain has been destroyed. Complex, coordinated reflexes of this type are possible because interneurons extend from the afferent neurons through the cord to many different efferent neurons.

Reflexes of the types described are present in all individuals as soon as the neuronal pathways have developed. These are inherited or **inborn reflexes,** and they are not dependent upon the training that the individual receives. Other reflexes, known as **conditioned reflexes,** develop as a result of specific training. Conditioned reflexes were first demonstrated in the early part of the twentieth century by Pavlov, the Rus-sian physiologist who also performed experiments on the control of gastric secretion. In a classic experiment, Pavlov fed a dog and simultaneously rang a bell. The bell, of course, had nothing to do with salivation, and at the beginning of the experiment would not induce salivation by itself. Salivation was reflexly stimulated by the sight or smell of food. The bell was rung each time the dog was fed, and the dog gradually learned to associate the sound of the bell with food. Eventually, ringing the bell without presenting food would initiate salivation. A classic conditioned reflex is established when a new sensory clue (the bell) becomes associated with an inborn reflex (salivation). The neuronal mechanism involved is not known. More complex conditioned reflexes can be developed. Once the dog has been conditioned to the bell, it can be secondarily conditioned to a flash of light preceding the bell. Elaborate reflex responses can be built up in this way as a result of specific training.

Reflexes in the spinal cord and brain form the basis of a great many of our responses, but other neuronal interrelations play important roles in integrating and correlating the body's activities. Most pathways within the nervous system involve many neurons, not just two or three as in the simpler reflexes, and this permits a variety of complex interrelations. Some pathways are **divergent** (Fig. 16.3A). The axon of a neuron may branch many times, synapse with a number of different neurons, and these, in turn, may branch further. Such an arrangement permits a single impulse to exert an effect over a wide area; a single impulse may ultimately activate a thousand or more neurons.

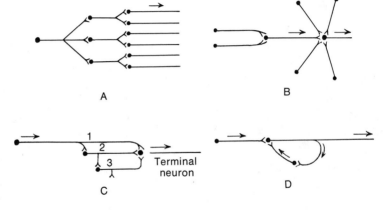

Figure 16.3 Diagrams of important types of neuronal interrelationships. *A,* A divergent pathway; *B,* a convergent pathway; *C,* a multiple chain circuit; *D,* a closed chain circuit. A given neuron may be involved in more than one of these pathways.

Other pathways are **convergent** (Fig. 16.3*B*); neurons coming from many different areas converge upon a single neuron or group of neurons. The convergence of neurons upon centers in the brain and upon the cell bodies of efferent neurons are examples of this type of pathway. Efferent neurons may receive impulses originating from 15 or 20 different sources. The response of the last neuron in a convergent pathway is the result of the interaction of a variety of excitatory and inhibitory influences. Convergent pathways are important in forming the structural basis for the integrative activity of the nervous system.

Many neuronal circuits, including those diagrammed in Figure 16.3A and B, involve the passage of impulses only as long as the first neuron continues to be stimulated. When the stimulation stops, the passage of impulses stops. Other arrangements in the nervous system ensure the continuation of the impulse for a period of time after the stimulus has stopped. One of these is the **multiple chain circuit** (Fig. 16.3*C*). The first neuron is stimulated momentarily, an impulse travels rapidly to the terminal neuron and also, via a branch, to a second neuron. The second neuron is stimulated and a moment later sends a second impulse to the same terminal neuron and also, via a branch, to a third neuron, which is stimulated and sends yet a third impulse to the same terminal neuron. If a great many neurons are involved, the terminal neuron will receive a whole series of impulses, and receive them for some time after the initial stimulus has stopped. In another arrangement, the **closed chain circuit** (Fig. 16.3*D*), one or more branches of the neurons in the circuit transmit impulses back to a point near the beginning of the circuit. Once such a circuit is activated, impulses could continue indefinitely unless the neurons became fatigued or were inhibited. Possibly, such circuits form the basis for the spontaneous activity of the inspiratory center and similar centers in the brain.

16.2 PERIPHERAL NERVOUS SYSTEM

Spinal Nerves. The vertebrate body is segmented (although segmentation is obscure in the head region), and there is a pair of peripheral nerves for each body segment: those arising from the spinal cord are known as **spinal nerves;** those from the brain, as **cranial nerves.** Afferent and efferent neurons lie together in most of a spinal nerve, but near the cord the nerve splits into a dorsal and a ventral root, and the neurons are segregated (Fig. 16.4). The **dorsal root** contains the afferent neurons and bears an enlargement, the dorsal root ganglion, which contains their cell bodies. The cell bodies of afferent neurons are nearly always located in ganglia on both spinal and cranial nerves. The afferent neurons enter the spi-

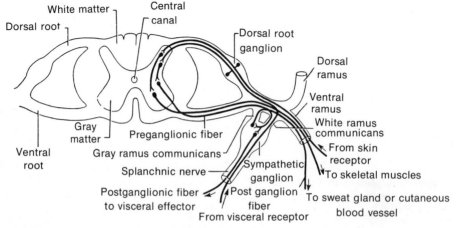

Figure 16.4 A diagrammatic cross section through the spinal cord and a spinal nerve. Each spinal nerve is formed by the union of dorsal and ventral roots and divides laterally into several branches (rami) going to different parts of the body. The dorsal ramus contains the same types of neurons as the ventral ramus.

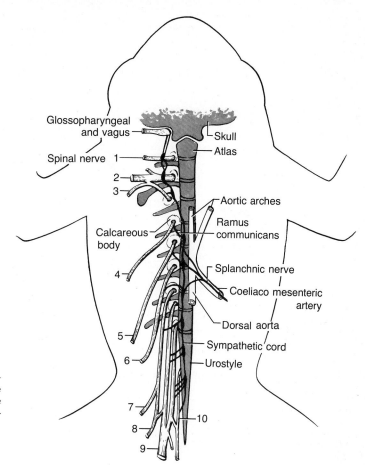

Glossopharyngeal and vagus

Spinal nerve 1

2

3

Calcareous body

4

5

6

7

8

9

Skull

Atlas

Aortic arches

Ramus communicans

Splanchnic nerve

Coeliaco mesenteric artery

Dorsal aorta

Sympathetic cord

Urostyle

10

Figure 16.5 A ventral view of the ventral rami of the spinal nerves and the sympathetic cord lying on the right side of the vertebral column of a frog. (Modified after Gaupp.)

nal cord and generally terminate in synapses with the dendrites or cell bodies of interneurons. Most of these cell bodies are located in the dorsal portion of the gray matter of the cord. The **ventral root** contains the efferent neurons, and their cell bodies nearly always lie in the ventral portion of the gray matter of the cord.

The spinal nerves of all vertebrates are essentially alike, although in the most primitive vertebrates the roots do not unite peripherally, and some of the efferent neurons leave the cord in the dorsal root. In most vertebrates, the roots unite to form a spinal nerve that divides into a dorsal branch, or **dorsal ramus,** which supplies the skin and muscles in the dorsal part of the body, a **ventral ramus,** which innervates the lateroventral parts of the body, and frequently one or more **communicating rami** to the visceral organs. Afferent and efferent neurons are found in each ramus. Man has

31 pairs of spinal nerves. Those supplying the receptors and effectors of the limbs are larger than the others, and their ventral rami are interlaced to form a complex network, or **plexus,** from which nerves extend to the limbs (Fig. 16.5).

Cranial Nerves. The nerves from the nose, the eye and the ear have evolved along with the organs of special sense. They are composed entirely of afferent fibers, except for a few efferent neurons in the optic and vestibulocochlear nerve that extend to the sense organs and may modulate their activity. The other cranial nerves contain large numbers of both afferent and efferent fibers, and they are considered to be serially homologous with the separate roots of the spinal nerves of primitive vertebrates. Some of them are essentially the cephalic counterparts of dorsal roots; others, the counterparts of ventral roots. The location of the cell bodies of the neurons of cranial

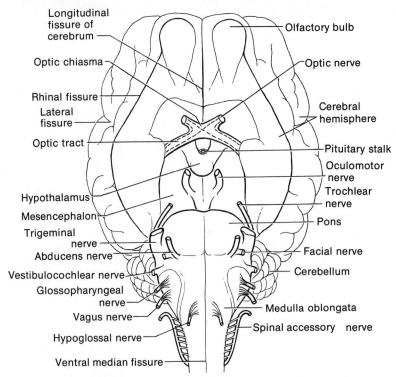

Figure 16.6   A ventral view of the brain of a sheep. The stumps of all but the first pair of cranial nerves are visible. The olfactory nerves consist of the processes of olfactory cells (cf. Fig. 15.1E), which enter the olfactory bulbs in many small groups that cannot be seen with the unaided eye. The rhinal fissure separates the ventral, olfactory portion of each cerebral hemisphere from the rest of the hemisphere. The paths of the optic fibers in the optic chiasma have been indicated by broken lines.

nerves, and of their endings within the brain, follows the pattern described for spinal neurons.

Reptiles, birds and mammals have 12 pairs of cranial nerves, if we omit the minute and poorly understood nervus terminalis. Though distributed to the nasal mucosa, this nerve is not olfactory. The other cranial nerves and their distribution are shown in Table 16.1, and their stumps can be seen in a figure of a sheep brain (Fig. 16.6).

Fishes and amphibians lack discrete spinal accessory and hypoglossal nerves. The homologues of neurons that are segregated in the spinal accessory of higher vertebrates are included in the vagus of fishes and amphibians, and the homologues of neurons in the hypoglossal are included in several minute nerves emerging from the occipital region of the skull. The trigeminal, facial, glossopharyngeal and vagus nerves of fishes are primarily associated with the muscles of the visceral arches and, as shown in Table 16.1, they supply the derivatives of this musculature in the higher vertebrates. Muscles change in shape and function during the course of evolution, but their innervation remains remarkably constant.

Autonomic Nervous System.   Most of the efferent fibers in the spinal and cranial nerves supply somatic muscles of the body and visceral muscles associated with the gill region. But in addition to these, certain of the cranial and spinal nerves contain other efferent fibers going to muscles in the walls of the gut, heart, blood vessels and other internal organs; to the small muscles associated with the hairs; to the ciliary and iris muscles in the eye; and to many of the glands of the body (Fig. 16.7). These efferent fibers constitute the **autonomic nervous system.** The organs supplied by these fibers function automatically, requiring no thought on our part. Indeed, they cannot be controlled voluntarily. It should

TABLE 16.1 CRANIAL NERVES OF MAN

Nerve	Origin of Afferent Neurons	Distribution of Efferent Neurons
I, Olfactory	Olfactory portion of nasal mucosa (smell)	
II, Optic	Retina (sight)	A few to retina
III, Oculomotor	A few fibers from proprioceptors in extrinsic muscles of eyeball (muscle sense)	Most fibers to four of the six extrinsic muscles of eyeball, a few to muscles in ciliary body and pupil
IV, Trochlear	Proprioceptors in extrinsic muscles of eyeball	Another extrinsic muscle of eyeball
V, Trigeminal	Teeth, and skin receptors of the head (touch, pressure, temperature, pain); proprioceptors in jaw muscles	Muscles derived from musculature of first visceral arch, i.e., jaw muscles
VI, Abducens	Proprioceptors in extrinsic muscles of eyeball	One other extrinsic muscle of eyeball
VII, Facial	Taste buds of anterior two-thirds of tongue (taste)	Muscles derived from musculature of second visceral arch, i.e., facial muscles; salivary glands; tear glands
VIII, Vestibulocochlear	Semicircular canals, utriculus, sacculus (sense of balance); cochlea (hearing)	A few to cochlea
IX, Glossopharyngeal	Taste buds of posterior third of tongue; lining of pharynx	Muscles derived from musculature of third visceral arch, i.e., pharyngeal muscles concerned in swallowing; salivary glands
X, Vagus	Receptors in many internal organs: larynx, lungs, heart, aorta, stomach	Muscles derived from musculature of remaining visceral arches (excepting those of pectoral girdle), i.e., muscles of pharynx (swallowing) and larynx (speech); muscles of gut, heart; gastric glands
XI, Spinal Accessory	Proprioceptors in certain shoulder muscles	Visceral arch muscles associated with pectoral girdle, i.e., sternocleidomastoid and trapezius
XII, Hypoglossal	Proprioceptors in tongue	Muscles of tongue

be emphasized that the autonomic nervous system is usually defined as a motor system, and the afferent fibers that return from internal organs are not a part of this system, even though they may be in nerves composed largely of autonomic fibers.

The autonomic nervous system is morphologically unique in that the autonomic neurons that emerge from the central nervous system do not extend all the way to the effectors, as do other efferent neurons. They go only to a **peripheral ganglion** in which there is a relay, and a second set of autonomic fibers continues from the ganglion to the organ. Autonomic fibers having their cell bodies in the central nervous system and extending to a peripheral ganglion are known as **preganglionic fibers;** those having their cell bodies in the ganglia and extend-ing to the organs are the **postganglionic fibers.**

The autonomic nervous system is subdivided into **sympathetic** and **parasympathetic systems.** Most organs innervated by the autonomic nervous system receive fibers of both types. The preganglionic sympathetic fibers leave the central nervous system through the ventral roots of spinal nerves in the thoracic and anterior lumbar regions (Figs. 16.4 and 16.7), and pass through the ramus communicans to a **sympathetic cord,** one of which lies on each side of the vertebral column. These fibers may synapse with the postganglionic fibers in the **sympathetic ganglia** along the sympathetic cord, or they may continue from the sympathetic cord through **splanchnic nerves** to **collateral ganglia** located at the

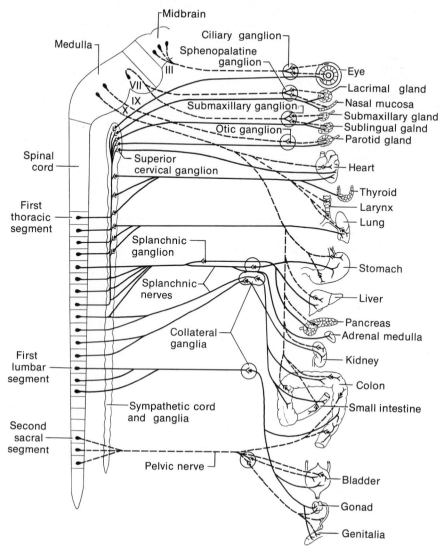

Figure 16.7 The human autonomic nervous system. Sympathetic fibers are drawn in solid lines; parasympathetic fibers in broken lines. The sympathetic fibers that go to the skin are not shown. (After Howell.)

base of the coeliac and mesenteric arteries. Postganglionic sympathetic fibers continue from the ganglia to the organs they supply. Those to the skin re-enter the spinal nerves (Fig. 16.4), but the others tend to follow along the arteries to the organs (Fig. 16.7). Preganglionic parasympathetic fibers are distributed to the organs through the oculomotor, facial, glossopharyngeal and vagus nerves and through a pelvic nerve derived from certain spinal nerves in the sacral region. Preganglionic parasympathetic fibers are longer than those of the sympathetic system, for they end in ganglia that are very

near the organs they supply or are in the walls of the organs. Relatively short postganglionic parasympathetic fibers continue to the muscle and gland cells.

Sympathetic and parasympathetic systems usually have opposite effects upon the organs innervated. Sympathetic stimulation speeds up the rate and increases the force of the heart beat (Fig. 13.17), causes most peripheral arteries to constrict, thereby increasing the blood pressure; increases the glucose content of the blood; and, in general, has effects that enable the body to adjust to conditions of stress. It inhibits the

TABLE 16.2 SELECTED EFFECTS OF AUTONOMIC STIMULATION *

Organ	Sympathetic Stimulation	Parasympathetic Stimulation
Skin		
Hair muscles	Contraction	....
Sweat glands	Secretion	....
Eye		
Iris sphincter	....	Contraction
Iris dilator	Contraction	....
Circulatory system		
Heart (rate and force)	Increased	Decreased
Coronary arteries	Dilation	Constriction
Most other arteries	Constriction	Dilated
Lung bronchi	Dilation	Constriction
Digestive organs		
Muscles of stomach and intestine	Decreased peristalsis	Increased peristalsis
Salivary glands	Some mucus secretion	Secretion
Gastric and intestinal glands	....	Secretion
Pancreas	....	Secretion
Liver	Bile flow inhibited Glucose released	Bile flow stimulated
Urinary organs		
Urinary bladder muscles	Relaxation	Contraction
Bladder sphincter	Contraction	Relaxation
Adrenal medulla	Secretion	....

* Dotted line indicates the absence of innervation.

activity of the digestive tract generally. Parasympathetic stimulation, on the other hand, speeds peristalsis of the digestive tract and similar vegetative processes, but it slows down the heart and decreases blood pressure. Sometimes the effects of the two systems complement each other. The salivary glands are activated primarily by the parasympathetic fibers, but sympathetic stimulation does increase their activity. A summary of the major effects of the two systems is presented in Table 16.2.

Ingenious experiments performed by Loewi in 1921 demonstrated the basis of the different effects of sympathetic and parasympathetic fibers. He removed the heart of a frog, leaving only its nerve supply intact, then perfused a salt solution through it and into another completely isolated heart. Both hearts continued to beat. When the vagus nerve (parasympathetic fibers) going to the first heart was stimu-lated, the rates of both hearts slowed down; when the sympathetic fibers were stimu-lated, the rates of both hearts increased. Apparently, some substance secreted by the nerves going to the first heart entered the salt solution and reached the second heart. Further work revealed that two **neurohumors** are produced. **Acetylcho-line** is secreted by the parasympathetic and **sympathin** by the sympathetic fibers. Acetylcholine may also be involved in the transmission of the nerve impulse across the synapses in other parts of the nervous system and across the junction between neuron and muscle. It may also play a role in the transmission of the nerve impulse along the neuron. Sympathin has been found only in connection with post-ganglionic sympathetic fibers, but it appears to be identical to norephinephrine, one of the hormones secreted by the medullary cells of the adrenal glands (p. 386). There is fairly clear evidence that these cells are themselves modified postganglionic sympathetic fibers.

16.3 CENTRAL NERVOUS SYSTEM

Spinal Cord. A small **central canal** (Fig. 16.4) extends through the center of the spinal cord, gray matter surrounds the central canal, and white matter lies peripheral to the gray. The **gray matter** is dark in color,

for it is composed of the cell bodies of neurons and of unmyelinated fibers; the **white matter** is light because it is composed of fibers surrounded by fatty myelin sheaths. The gray matter forms continuous longitudinal columns, which are H-shaped in cross section. There are a pair of **dorsal columns,** a pair of **ventral columns** and a **gray commissure** connecting the columns of opposite sides. The dorsal column has sensory functions, for it contains the dendrites and cell bodies of afferent interneurons, with which many afferent neurons synapse (Fig. 16.4). The ventral column contains the dendrites

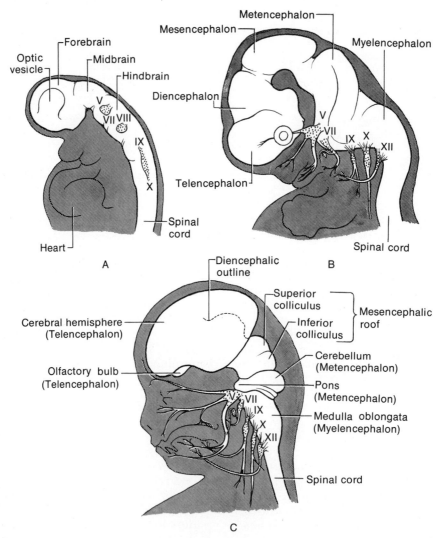

Figure 16.8 Three stages in the development of the human brain. *A*, The three primary brain regions can be recognized in an embryo that is about three and one-half weeks old. *B*, All five brain regions are evident in an embryo seven weeks old. *C*, The various structures found in a fully developed brain are beginning to differentiate in an embryo eleven weeks old. (After Patten.)

and cell bodies of the efferent neurons. The gray commissure is composed of fibers crossing from one side of the spinal cord to the other.

Much of the white matter consists of the fibers of afferent neurons, some of which extend some distance in the central nervous system before entering the gray matter, and of afferent interneurons, which end in the brain. The rest of the white matter consists of the processes of efferent interneurons extending from the brain to the efferent neurons. Most afferent impulses that enter the spinal cord cross to the opposite side before they reach the brain, and efferent impulses coming from the brain cross within the brain. Thus, afferent impulses initiated on the left side of the body reach the right side of the brain, and efferent impulses initiated in the right side of the brain reach the left side of the body.

Though all of the white matter looks the same, careful experimentation has enabled neuroanatomists to localize the various groups of fibers that comprise it. Impulses initiated by temperature receptors on the left side of the body, for example, are carried to the brain by fibers located in the lateral portion of the white matter on the right side of the cord (Fig. 16.2). A lesion in this part of the cord would prevent one from being conscious of temperature changes on the opposite side of the body posterior to the lesion, though one would still respond reflexly to such changes.

The Brain. *Major Parts of the Brain.* A brief consideration of the embryonic development of the brain makes it easier to understand its major divisions and parts. The brain develops as a series of enlargements of the anterior portion of the embryonic neural tube (Fig. 16.8). In an early embryo there are only three swellings (forebrain, midbrain and hindbrain), but the forebrain and hindbrain are later subdivided, so that five regions are present in an adult. The forebrain divides into a **telencephalon** and a **diencephalon.** The telencephalon differentiates into a pair of **olfactory bulbs,** which receive the endings of olfactory cells, and a pair of **cerebral hemispheres,** which become the major brain center in the higher vertebrates. The lateral walls of the diencephalon become the **thalamus,** its roof the **epithalamus,** and its floor the **hypothalamus.** Fibers in the optic nerves cross below the hypothalamus and form an **optic chiasma** (Fig. 16.6). All the optic fibers cross and go to the opposite side of the brain in most vertebrates, but only half of them cross in mammals. The pituitary gland is attached to the hypothalamus just posterior to the chiasma, and the pineal body is attached to the epithalamus. No further division occurs in the midbrain, or **mesencephalon,** but its roof differentiates into a pair of optic lobes in all vertebrates. In addition to the optic lobes, or **superior colliculi,** the mesencephalic roof of mammals bears a pair of **inferior colliculi.** The hindbrain divides into a **metencephalon,** the dorsal portion of which forms the **cerebellum,** and a **myelencephalon,** which becomes the **medulla oblongata.** The regions of the brain and their major derivatives are summarized in Table 16.3.

TABLE 16.3 BRAIN REGIONS AND DERIVATIVES

Brain Region	Major Derivatives
Forebrain (Prosencephalon)	
Telencephalon	Olfactory bulbs
	Cerebral hemispheres
Diencephalon	Epithalamus, Pineal body
	Thalamus
	Hypothalamus, Pituitary gland (part)
Midbrain (Mesencephalon)	Superior colliculi
	Inferior colliculi
Hindbrain (Rhombencephalon)	
Metencephalon	Cerebellum
	Pons
Myelencephalon	Medulla oblongata

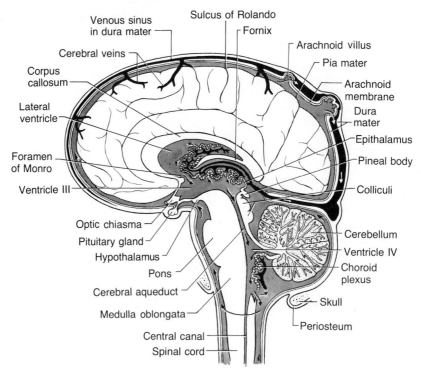

Figure 16.9 A sagittal section of the human brain and its surrounding meninges. Cerebrospinal fluid is produced by the choroid plexuses, circulates as indicated by the arrows, and finally enters a venous sinus in the dura mater. (Modified after Rasmussen.)

Ventricular System. The central canal of the spinal cord extends into the brain and is continuous with several large, interconnected chambers known as **ventricles** (Fig. 16.9). A lateral ventricle lies in each cerebral hemisphere and each is connected with the third ventricle in the diencephalon by a **foramen of Monro.** The **cerebral aqueduct of Sylvius** extends from the third ventricle through the mesencephalon to a fourth ventricle in the metencephalon and medulla oblongata. All of these passages are filled with a lymphlike **cerebrospinal fluid,** much of which is produced by vascular **choroid plexuses.** Choroid plexuses develop in the thin roof of the diencephalon and medulla and are also present in the lateral ventricles of mammals. Cerebrospinal fluid escapes from the brain through foramina in the roof of the medulla and slowly circulates in the spaces between the layers of connective tissue, the **meninges,** that encase the brain and spinal cord. The innermost meninx, the **pia mater,** is a vascular membrane that is closely applied to the surface of the brain and spinal cord. Certain parts of it help to form the choroid plexuses. A delicate **arachnoid membrane** lies peripheral to the pia, and a very tough **dura mater** forms a protective envelope around the entire central nervous system. The cerebrospinal fluid lies in the subarachnoid space between the arachnoid and pia, and it extends far into the nervous tissue in **perivascular spaces** that surround the blood vessels penetrating the brain. It is produced continuously and reenters the circulatory system by filtering into certain venous sinuses located in the dura mater covering the brain. The cerebrospinal fluid forms a protective liquid cushion about the brain and spinal cord and it may help to nourish the tissue of the central nervous system. The perivascular spaces probably serve some of the functions of a lymphatic system by returning any plasma protein that may have escaped from capillaries in the brain. Nervous tissue does not have a lymphatic drainage.

Medulla Oblongata. Brain functions are

exceedingly complex and far from completely understood. The medulla oblongata (Figs. 16.10 and 16.11) lies between the spinal cord and the rest of the brain and is fundamentally the same in all vertebrates. The gray columns of the spinal cord extend into the medulla, but within the brain they become discontinuous, breaking up into discrete islands of cell bodies known as **nuclei.** The dorsal nuclei receive the afferent neurons from cranial nerves that are attached to this region and contain the cell bodies of afferent interneurons. These are sensory nuclei, just as the dorsal columns of the cord are sensory columns. The ventral nuclei contain the cell bodies of the efferent neurons of the cranial nerves and, hence, are motor nuclei. In mammals, reflexes that regulate the rate of heart beat, the diameter of arterioles, respiratory movements, salivary secretion, swallowing and many other processes are mediated by these nuclei. Afferent impulses come into the sensory nuclei, are relayed by the interneurons to the moter nuclei, and efferent impulses go out to the effectors. Motor and sensory nuclei associated with other cranial nerves are also found in the metencephalon and mesencephalon.

Reticular Formation. Between the motor and sensory nuclei throughout the brain stem there is a network of thousands of cell bodies and their processes that is known as the **reticular formation.** Dendrites and axons of these neurons branch profusely and have numerous connections with each other and with many sensory and motor pathways passing to and from the brain (Fig. 16.11). Certain of these neurons receive an input from as many as 4000 other neurons and feed out to 25,000 other neurons! The reticular formation is considered to be a very primitive part of the brain from which the long ascending and descending fiber tracts and the various centers evolved. Certain of its specific functions in mammals are now being clarified. A convergence and interaction of neurons from many of the sense organs, the cerebellum, the cerebrum, the thalamus, the hypothalamus and other areas occurs here, and the region appears to be able to suppress, enhance, or otherwise modify impulses passing through it. For example, branches from cells in the sensory nuclei, in addition to synapsing with motor cells and ascending to higher centers, feed into this system. If the sensory input is the type the individual has learned is important (an unusal noise, the cry of a baby at night, etc.), the reticular formation, in turn, sends nonspecific impulses to the cerebral cortex. These fan out widely in the cortex and rouse the animal if it is asleep, or help to keep it alert if it is awake. Without this source of stimulation, the cortex does not function and cannot interpret the specific

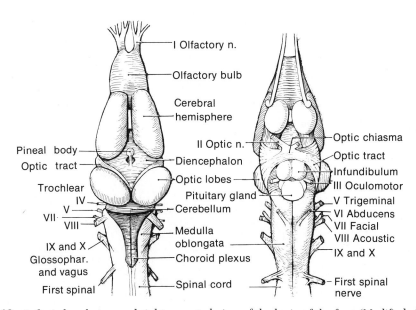

Figure 16.10 *Left,* A dorsal view and *right,* a ventral view of the brain of the frog. (Modified after Gaupp.)

sensory information brought to it. In short, the reticular formation appears to be very much involved in what we call consciousness, that is, an awareness of one's self and one's environment. Descending fibers in the reticular formation form a part of a motor pathway from the cerebrum known as the **extrapyramidal pathway,** and impulses passing down to motor neurons of cranial and spinal nerves may also be facilitated or inhibited.

Cerebellum and Pons. All vertebrates have a **cerebellum,** which develops in the dorsal part of the metencephalon, and is a center for balance and motor coordination. It is small in many of the lower vertebrates such as the frog (Fig. 16.10), in which muscular movements are not complex, but it is very large in birds and mammals. Impulses enter it from most of the sense organs, but those from the proprioceptive organs in the muscles and the parts of the ear concerned with equilibrium play a particularly prominent role (Fig. 16.11). The cerebellum keeps track of the orientation of the body in space and the degree of contraction of the skeletal muscles. In mammals, this informa-

tion is sent, by way of part of the reticular formation and thalamus, to the cerebrum, where voluntary movements are initiated. Copies, so to speak, of the motor directives sent out by the cerebrum to the muscles are also sent to the cerebellum. It monitors the responses of the body and returns corrective signals to the cerebrum or, in some cases, directly to the muscles. Much of the gray matter of the mammalian cerebellum lies on the surface, where there is more room for the increased number of cell bodies. The surface is also complexly folded, which further increases the surface area available for cell bodies.

The floor of the metencephalon is unspecialized in lower vertebrates and simply contributes to the medulla oblongata, but this region differentiates into a **pons** in mammals (Figs. 16.6 and 16.11). Evolution of the pons is correlated with the elaboration of the cerebellum. It contains nuclei that relay cerebral impulses into the cerebellum, and transverse fibers that interconnect the two sides of the cerebellum.

Optic Lobes. In fishes and amphibians, the optic lobes (Figs. 16.10 and 16.12) re-

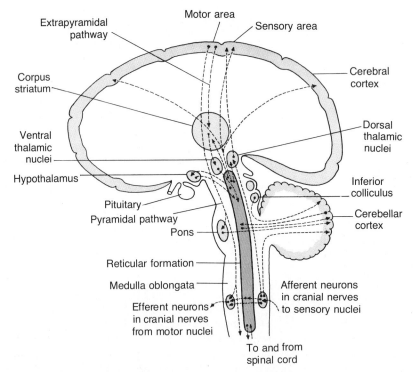

Figure 16.11   Diagrammatic lateral view of the human brain on which some of the major masses of gray matter, the reticular formation and certain important pathways have been projected.

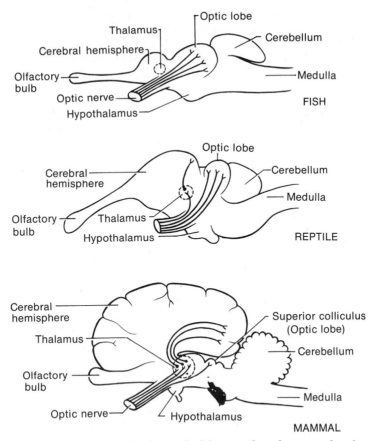

Figure 16.12 Diagrammatic lateral views of the brain of a fish, a reptile and a mammal to show the increasing importance of the cerebral hemispheres, and the decreasing importance of the optic lobes, as integration centers. The shift in optic pathways is shown; a similar shift occurs for other sensory modalities.

ceive the termination of all the fibers in the optic nerve, and they also receive projections from most of the other sense organs. This sensory information is integrated, and motor impulses are sent to the appropriate efferent neurons. The optic lobes are the master integrating center of the brain, insofar as these vertebrates have such a center. The cerebral hemispheres of the lower vertebrates are rather small and are concerned almost exclusively with integrating olfactory impulses. In reptiles, some of the optic and other types of sensory data are sent to the cerebral hemispheres, and the cerebrum begins to assume certain of the functions of the optic lobes (Fig. 16.12). Still more sensory information is sent to the cerebral hemispheres of birds and mammals, and the hemispheres of mammals have taken over most of the functions of the optic lobes. The optic lobes (**superior colliculi**) of mammals

remain as relatively small centers that regulate the movements of the eyeballs and pupillary and accommodation reflexes. A pair of **inferior colliculi** are present posterior to them, and they are a center for certain auditory reflexes.

Thalamus and Hypothalamus. Much of the thalamus is a relay center to and from the cerebral hemispheres, and it has become enlarged during the course of evolution as the cerebral hemispheres have assumed a dominant role in integrating the activities of the body. All the sensory impulses that go to the cerebrum, except those from the olfactory organ, are relayed in nuclei in the dorsal part of the thalamus (Figs. 16.2 and 16.11). Many motor impulses descending from the cerebrum go directly to the motor nuclei and columns, but some of these are relayed in nuclei in the ventral part of the thalamus.

The thalamus is, however, more than just a relay station in the higher vertebrates. Considerable processing of the sensory input occurs here, for the thalamic nuclei have many interconnections with each other, with the hypothalamus, with the reticular formation and with many parts of the cerebral hemispheres. Pain, temperature and certain other sensory modalities reach the level of conscious awareness in the thalamus. Other sensations are to some extent sorted out in this region, and a determination is made as to which of these we will concentrate on among the hundreds of stimuli continually impinging upon us. The thalamus also influences the manner in which we view many stimuli, that is, the different degree of agreeableness or disagreeableness that we may place upon the same type and intensity of stimuli at different times. Many of our emotions, such as pleasure and fear of punishment, are influenced by the thalamus. Experimental rats with an electrode implanted in a "pleasure center" will push a lever that stimulates the center thousands of times per hour. Other areas appear to be "punishing centers," and experimental animals will avoid activity that activates an electrode implanted in one of them.

The rest of the diencephalon has not changed very much during vertebrate evolution. The hypothalamus is an important center for the control of many autonomic functions. Body temperature, water balance, appetite, carbohydrate and fat metabolism and sleep are among the processes regulated by the hypothalamus in mammals. The hypothalamus exerts its control by neuronal connections with the motor nuclei and columns and by the production of neurosecretions. Some of these secretions travel along axons to the posterior lobe of the pituitary gland where they are released as hormones into the blood; others enter blood vessels that lead to the anterior lobe of the pituitary and influence the release of its secretions (p. 393). The parts of the hypothalamus that produce neurosecretions are, in a sense, neuroendocrine transducers, for they convert nervous signals into endocrine signals. Damage to the hypothalamus may be fatal, for so many vital processes are disturbed.

Cerebral Hemispheres. As the cerebral hemispheres assumed the dominant role in nervous integration during the course of evolution, they enlarged and grew posteriorly over the diencephalon and mesencephalon (Fig. 16.12). A layer of gray matter has developed on the surface of the cerebrum and has formed a **gray cortex** which provides more area for the increased number of cell bodies. The cortex is also complexly folded, forming numerous ridges (**gyri**) with furrows (**sulci**) between them, and this further increases the surface area. Over 12 billion neurons, and even more neuroglial cells, are present.

Parts of the cerebral hemispheres are still concerned with their primitive function of olfactory integration, but their great

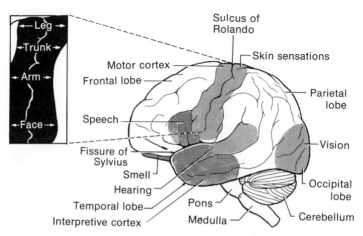

Figure 16.13 Important cortical areas of the human brain as seen in a lateral view. Most association areas of the cortex have not been hatched.

enlargement is correlated with the evolution of other integration centers (Fig. 16.13). Afferent impulses from the eyes, ears, skin and many other parts of the body are carried to the cerebral cortex by afferent interneurons after being relayed in the thalamus. The impulses terminate in specific parts of the cerebral cortex; their locations have been determined by correlating brain injuries with loss of sensation and also by electrical stimulation during brain operations. Many human brain operations are performed under local anesthesia, and the patient can describe the sensations that are felt when particular regions are stimulated. Impulses from the skin terminate in the gyrus that is located just posterior to the **central sulcus of Rolando,** a prominent sulcus extending down the side of each hemisphere and dividing the hemisphere into an anterior **frontal** and a posterior **parietal lobe.** The sensory areas of the skin are projected upside down. Impulses from the head are conducted to the lower part of the gyrus, whereas those from the feet reach the upper part. The extent of the area receiving impulses from any part of the body is proportional to the number of sense organs in that

part of the body, not to the surface area of the body. Thus, the area receiving impulses from the hands is as extensive as the area receiving impulses from the trunk (Fig. 16.14A).

Impulses from the ear are carried to part of the **temporal lobe,** which is separated from the frontal and parietal lobes by the **lateral fissure of Sylvius.** Impulses from the eye are received in the **occipital lobe,** which lies just posterior to the parietal lobe. The path of the optic fibers of mammals is an exception to the generalization that most afferent impulses cross at some point during their ascent to the brain. Half of the fibers in each optic nerve cross in the optic chiasma and end up on the opposite side of the brain, but the other half do not. Thus, destruction of one occipital lobe results in inability to perceive images that fall on half of each retina rather than complete loss of vision in one eye (Fig. 16.6).

Appropriate motor impulses to the striated muscles are initiated in response to all the sensory data that enter the cerebrum. The cell bodies of the efferent interneurons are contained in the motor cortex, which lies just anterior to the sulcus of Rolando. The

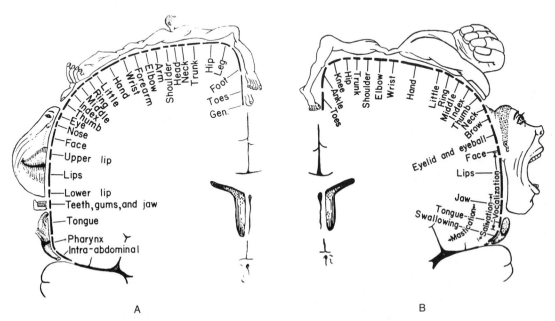

A B

Figure 16.14 Vertical sections through (A) the sensory cortex receiving from the skin and (B) the motor cortex. The proportions of the cortex related to the various body regions have been emphasized by drawing body parts in similar proportions. (From Penfield and Rasmussen: The Cerebral Cortex, published by Macmillan.)

motor cortex is subdivided, in the manner of the adjacent sensory cortex, into areas associated with the different parts of the body. Fibers to the hand occupy a large portion of it, for the muscles that control finger movements contain more motor units than do most muscles (Fig. 16.14*B*). This is correlated with the intricacy of our finger movements. Many efferent interneurons pass directly to the motor nuclei of the brain and to the motor columns of the spinal cord, crossing to the opposite side along the way (Fig. 16.11). This direct pathway to the lower motor neurons is called the **pyramidal pathway.** It is phylogenetically a new pathway, being present only in mammals and reaching its greatest development in primates. Skilled and learned movements are for the most part mediated by this pathway. Many other motor impulses leave the cortex by way of the **extrapyramidal pathway,** which involves the relay of motor impulses in a mass of gray matter (the **corpus striatum**) situated deep within each cerebral hemisphere, and additional relays in the thalamus and reticular formation of the brain stem. This is phylogenetically an older system, and it is associated with grosser movements, automatic postural adjustments and stereotyped responses.

Billions of **association neurons** lie between the areas where sensory impulses terminate and motor pathways begin. In most parts of the cortex, the neurons are arranged in six vertical layers with numerous interconnections between them and with the afferent neurons bringing impulses into the cortex and efferent ones carrying impulses out. Such an organization makes possible exceedingly complex functional interrelationships between different parts of the cortex and between the cortex and subcortical regions. Information in one hemisphere can also affect the other by crossing on **commissural fibers** that extend between them. A particularly large commissure, known as the **corpus callosum** and found only in placental mammals, can be seen in a sagittal section of the brain (Fig. 16.9). All these interconnections permit the integration of many different sensory modalities, the comparison of current sensory input with information stored in the brain, and the composition of a motor response appropriate to the present requirements of the organism.

In man and other higher mammals, large parts of the cortex known as association areas are composed primarily of association neurons and their interconnections (Fig. 16.13). Presumably such complex mental processes as thought, learning and memory occur here.

Regardless of what memory is, or of how or where information is stored, one part of the cortex does appear to be involved in recall and in relating present experience with relevant past events. This has been called the **interpretive** or **psychical cortex,** and it occupies a large part of each temporal lobe. Electrical stimulation of the various sensory areas produces somewhat vague sensations of light, sound, or touch, but stimulation of the interpretive cortex sometimes causes a vivid and detailed recall of past events – of sights and sounds and feelings long forgotten. At other times, its stimulation alters the interpretation one places on present events. Objects may seem suddenly strange, distant or near. Dr. Penfield of the Montreal Neurological Institute considers that the interpretive cortex makes some contribution to the reflex comparison of the present with the past and thus contributes to the interpretation of the present.

Brain research is a challenging field. Many of the brain's functions are beyond our present comprehension, but we are beginning to understand this complex organ.

ANNOTATED REFERENCES

Attention is again called to the general references on vertebrate organ systems cited at the end of Chapter 11

Bitterman, M. E.: The evolution of intelligence. Scientific American *212*:92 (Jan.) 1965. An interesting report on the qualitatively different sorts of intelligence found in different classes of vertebrates.

Gardner, E.: Fundamentals of Neurology. 5th ed. Philadelphia, W. B. Saunders Co.,

1968. A good and concise account of the morphology and physiology of the human nervous system.

Hydén, H.: Satellite cells in the nervous system. Scientific American *205*:62 (Dec.), 1961. A discussion of the author's theory that changes in RNA metabolism in neurons and neuroglia are associated with learning.

Katz, B.: Nerve, Muscle, and Synapse. New York, McGraw-Hill Book Co., 1966. An excellent review of neuron physiology and synaptic transmission written for the McGraw-Hill series in the New Biology.

Noback, C. R.: The Human Nervous System. New York, McGraw-Hill Book Co., 1967. An excellent and superbly illustrated discussion of the nervous system and sense organs. Structural and functional aspects of the subject are closely integrated.

Sherrington, C. S.: Integrative Action of the Nervous System. New Haven, Yale University Press, 1948. A classic on the subject, written by one of the pioneers in neurophysiology.

Walter, W. G.: The Living Brain. New York, W. W. Norton & Co., 1953. An authoritative and nontechnical account of the evolution of the brain and its great elaboration in man.

Wilson, V. J.: Inhibition in the central nervous system. Scientific American *214*:102 (May), 1966. A careful discussion of the evidence for inhibition in spinal reflexes.

17. Hormonal Integration

The activities of the several parts of the higher, more complex animals are integrated by two major coordinating systems: the nervous system, discussed in the previous chapter, and the endocrine system. Nerves and sense organs enable an animal to adapt very rapidly—with responses measured in milliseconds—to changes in the environment. The swift responses of muscles and glands are typically under nervous control. The glands of the endocrine system secrete substances called **hormones,** which diffuse or are transported by the blood stream to other parts of the body and coordinate their activities. The responses under endocrine control are generally somewhat slower—measured in minutes, hours or weeks—but longer lasting than those under nervous control. The long-range adjustments of metabolism, growth and reproduction are typically under endocrine control. Hormones play a key role in maintaining constant the concentrations of glucose, sodium, potassium, calcium, phosphate and water in the blood and extracellular fluids of man and other vertebrates. Hormonal regulation and integration of cellular processes occur in insects, crustaceans, annelids, mollusks, and certain other invertebrates as well as in the vertebrates.

17.1 ENDOCRINE GLANDS

Endocrine glands secrete their products into the blood stream rather than into a duct leading to the exterior of the body or to one of the internal organs as do exocrine glands and, hence, are called ductless glands or glands of internal secretion. The pancreas has both endocrine and exocrine functions; it secretes enzymes which pass via the pancreatic duct to the duodenum and it secretes hormones which are transported to other parts of the body in the blood stream. In the toadfish the exocrine and endocrine parts of the pancreas are anatomically separate.

The term "hormone" was originated in 1905 by the British physiologist E. H. Starling, who was studying the control of the exocrine function of the pancreas by **secretin,** a substance produced in the duodenal mucosa. Starling defined a hormone as "any substance normally produced in the cells in some part of the body and carried by the blood stream to distant parts, which it affects for the good of the body as a whole." In current usage, the term hormone refers to special chemical substances produced by some restricted region of an organism which diffuse, or are transported by the blood stream, to another region of the organism, where they are effective in very low concentrations in regulating and coordinating the activities of the cells.

The hormones isolated and characterized to date have proved to be proteins, peptides, amino acids, fatty acids or steroids; thus, we cannot define a hormone as a member of some particular class of organic compound. All the hormones are required for normal body function and each must be present in a certain optimal amount. Either a hyposecretion (deficiency) or hypersecretion (excess) of any one may result in a characteristic pathologic condition.

Some practical aspects of endocrinology, such as the results of the castration of men and animals, have been known for several thousand years. However, it was not until 1849 that Berthold, from the results of experiments in which testes were transplanted from one bird to another, postulated that these male sex glands secrete some blood-borne substance essential for the differentiation of the male secondary sex characters. In 1855 the British physician Thomas Addison described the signs and symptoms of the human disease which now bears his name; he realized that this was associated with the deterioration of the cortex of the adrenal. The first attempt at endocrine therapy was made in 1889, when the French physiologist Brown-Séquard injected himself with testicular extracts and claimed that they had a rejuvenating effect. Epinephrine was the first hormone to be isolated and chemically identified (1902).

A few hormones, such as thyroxine or growth hormone, affect the metabolic conditions of every cell in the body; every cell responds to the presence of the hormone and every cell shows an altered metabolic state when deprived of the hormone. Most hormones, however, affect only certain cells in the body, despite the fact that the blood stream carries them to all parts of the organism. Only the pancreas responds to the secretin circulating in the blood. The cells that respond to a given hormone are called the "target organ" of that hormone. The thyroid gland is the target organ of thyrotropin or TSH secreted by the pituitary and the ovary or testis is the target organ of the gonadotropins FSH and LH from the pituitary. Some hormones, such as estradiol, have marked effects on their primary target organs, the uterus, vagina and mammary glands, lesser effects on other characters, the voice, distribution of hair on the body, bone growth and so forth and still smaller effects on other tissues.

It is now clear that the ability of a tissue to respond to estradiol is correlated with the presence in the tissue of a protein that specifically takes up and binds estradiol. The uterus, vagina, pituitary and hypothalamus contain this protein and can accumulate estradiol multifold from the blood. It is assumed that this protein plays a role in transporting the estradiol from outside the cell into the nucleus where it produces its effect. There is evidence for comparable protein receptors in the oviduct which are specific for progesterone and receptors in the prostate specific for testosterone. The specificity of the effects of the peptide and protein hormones is also a property of the cells responding to them, although the mechanism involved may be something other than a "receptor" molecule.

17.2 METHODS OF INVESTIGATING THE ENDOCRINE GLANDS

The complete understanding of the role of an endocrine gland requires information about (1) the number and kinds of hormones it secretes, (2) what chemical and physical properties each of these hormones has, (3) where and how they are made within the endocrine organ, (4) what factors control their production and secretion, (5) how they are transported to the target organ, (6) how the target organ recognizes and takes up its specific hormones, (7) how the hormone acts to alter the metabolism of the target organ, (8) how hormones are broken down and eliminated from the body, (9) how they may be produced synthetically and (10) what use they may have in the treatment of disease.

To determine whether a gland suspected of producing a hormone is in fact an endocrine gland, an investigator usually begins by removing the gland surgically and observing the effect on the animal. Next he replaces the gland with one transplanted from another animal and determines whether the changes induced by removing the gland can be reversed by replacing it. In replacing the gland he is careful to ensure that the new gland becomes connected only with the vascular system of the recipient and not a duct of some sort. He might next try feeding an animal from which the gland has been removed with dried glands to determine whether the active substance, the hormone, can be replaced in this way. Finally he extracts the gland with a variety of solvents to determine the solubility characteristics of the active material and gain a clue as to its chemical nature. By making an extract of the gland, or perhaps by making an extract of blood or urine, and purifying the extract by a sequence of appropriate chemical and physical methods, he finally obtains the pure compound and can determine its exact chemical structure. At each step of his extraction and purification procedure he tests his material by injecting it into a test animal from which the gland has been removed. In this way he finally determines what specific chemical in the original extract will reverse the changes caused by removing the gland.

Hormones are remarkably effective substances and a very small quantity produces a marked effect in the structure and function of one or another part of the body. Only small amounts of a hormone are secreted at any one time by the endocrine gland and the amount circulating in the blood is very small. Even the amount excreted each day in the urine is not very large. Because of this the isolation of a pure hormone can be a difficult job indeed. To obtain a few milligrams of pure estradiol, one of the female sex hormones, more than two tons of pig ovaries were extracted! When the hormone secreted by a given gland has been extracted and its chemical structure has been determined, it can then be prepared synthetically to be used in treating diseases resulting from a deficiency of that hormone.

Much has been learned about endocrine function by careful observation of the symptoms of human diseases resulting from the hypo- or hypersecretion of hormones. Further information has been derived from the careful study of strains of rats, mice and other animals with particular endocrine abnormalities—dwarf mice, obese mice, diabetic mice, and so on.

The location of the human endocrine glands is shown in Figure 17.1. Their relative position in the body is much the same in all the vertebrates. The source and physiologic effects of the principal mammalian hormones are listed in Table 17.1.

17.3　THE THYROID GLAND

All vertebrates have a bilobed thyroid gland located in the neck. In mammals the two lobes are located on either side of the larynx and are joined by a narrow isthmus of tissue extending across the ventral surface of the trachea near its junction with the larynx. The thyroid has an exceptionally

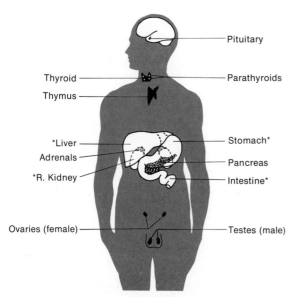

Thyroid

Thymus

*Liver

Adrenals

*R. Kidney

Ovaries (female)

Pituitary

Parathyroids

Stomach*

Pancreas

Intestine*

Testes (male)

Figure 17.1 The human body, showing the location of the endocrine glands. The starred organs, though not primary endocrine glands, do secrete one or more hormones.

TABLE 17.1 VERTEBRATE HORMONES AND THEIR PHYSIOLOGIC EFFECTS

Hormone	Source	Physiologic Effect
Thyroxine	Thyroid gland	Increases basal metabolic rate
Parathormone	Parathyroid glands	Regulates calcium and phosphorus metabolism
Calcitonin	Ultimobranchial bodies	Regulates calcium and phosphorus
Insulin	Beta cells of islets in pancreas	Increases glucose utilization by muscle and other cells, decreases blood sugar concentration, increases glycogen storage and metabolism of glucose
Glucagon	Alpha cells of islets in pancreas	Stimulates conversion of liver glycogen to blood glucose
Secretin	Duodenal mucosa	Stimulates secretion of pancreatic juice
Cholecystokinin	Duodenal mucosa	Stimulates release of bile by gallbladder
Epinephrine	Adrenal medulla	Reinforces action of sympathetic nerves; stimulates breakdown of liver and muscle glycogen
Norepinephrine	Adrenal medulla	Constricts blood vessels
Cortisol	Adrenal cortex	Stimulates conversion of proteins to carbohydrates
Aldosterone	Adrenal cortex	Regulates metabolism of sodium and potassium
Dehydroepiandrosterone	Adrenal cortex	Androgen, stimulates development of male sex characters
Growth hormone	Anterior pituitary	Controls bone growth and general body growth; affects protein, fat and carbohydrate metabolism
Thyrotropin	Anterior pituitary	Stimulates growth of thyroid and production of thyroxine
Adrenocorticotropin (ACTH)	Anterior pituitary	Stimulates adrenal cortex to grow and produce cortical hormones
Follicle-stimulating hormone (FSH)	Anterior pituitary	Stimulates growth of graafian follicles in ovary and of seminiferous tubules in testis
Luteinizing hormone (LH)	Anterior pituitary	Controls production and release of estrogens and progesterone by ovary and of testosterone by testis
Prolactin (LTH)	Anterior pituitary	Maintains secretion of estrogens and progesterone by ovary; stimulates milk production by breast; controls "maternal instinct"
Oxytocin	Hypothalamus, via posterior pituitary	Stimulates contraction of uterine muscles and secretion of milk
Vasopressin	Hypothalamus, via posterior pituitary	Stimulates contraction of smooth muscles; antidiuretic action on kidney tubules
Melanocyte stimulating hormone	Anterior lobe of pituitary	Stimulates dispersal of pigment in chromatophores
Testosterone	Interstitial cells of testis	Androgen; stimulates development and maintenance of male sex characters
Estradiol	Cells lining follicle of ovary	Estrogen; stimulates development and maintenance of female sex characters
Progesterone	Corpus luteum of ovary	Acts with estradiol to regulate estrous and menstrual cycles
Prostaglandins	Seminal vesicle	Stimulate uterine contractions
Chorionic gonadotropin	Placenta	Acts together with other hormones to maintain pregnancy
Placental lactogen	Placenta	Has effects like prolactin and growth hormone
Relaxin	Ovary and placenta	Relaxes pelvic ligaments

rich blood supply, which reflects its function as an endocrine gland. The thyroid develops as a ventral outgrowth of the floor of the pharynx, but the connection with the pharynx is usually lost early in development. The thyroid consists of cuboidal epithelial cells arranged in hollow spheres one cell thick. The cavity of each follicle is filled with a gelatinous colloid secreted by the epithelial cells lining it (Fig. 17.2).

The follicle cells have a remarkable ability to accumulate iodide from the blood. This is used in the synthesis of the protein thyroglobulin, which is secreted into the colloid and stored. Proteolytic enzymes in the colloid hydrolyze thyroglobulin to its constituent amino acids, one of which is **thyroxine,** a derivative of the amino acid tyrosine containing 65 per cent iodine. Thyroxine passes into the blood stream, where it is transported loosely bound to certain plasma proteins. In tissues, thyroxine, which contains four atoms of iodine, may be converted to triiodothyronine, which contains one less atom of iodine and is several times more active than

thyroxine. It is not yet clear whether the hormone active at the cellular level is thyroxine itself, triiodothyronine, or some closely related derivative.

The first clue to thyroid function came from observations on human disease in 1874 by the British physician Sir William Gull, who noted the association of spontaneous decreased function of the thyroid and puffy dry skin, dry brittle hair, and mental and physical lassitude. The Swiss surgeon Kocher removed the thyroid from a series of patients and then noted that they developed the same symptoms as Gull's patients. In 1895, using a calorimeter to measure the rate of metabolism in patients by the amount of heat they produced, Magnus-Levy found that persons with **myxedema** (Gull's disease) had notably lower than normal metabolic rates. When these patients were fed thyroid tissue, their metabolic rate was raised toward normal. This led to the idea that the thyroid secretes a hormone which regulates the metabolic rate of all body cells.

The role of thyroid hormone in all verte-

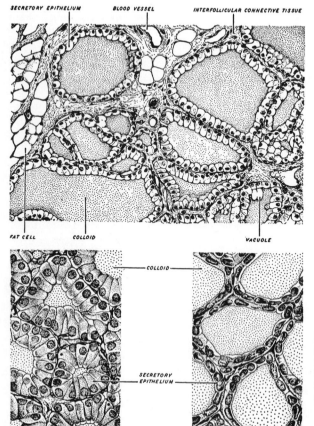

Figure 17.2 *Upper*, Cells of the normal thyroid gland of the rat. *Lower left*, Thyroid from a normal rat which had received ten daily injections of thyrotropin. *Lower right*, Thyroid from a rat six months after complete removal of the pituitary gland. (Turner: General Endocrinology, 2nd ed.)

brates is to increase the rate of the oxidative, energy-releasing processes in all body cells. The amount of energy released by an organism under standard conditions at rest, measured in a calorimeter by the amount of heat given off, or calculated from the amount of oxygen consumed, is decreased in thyroid deficiency and increased when thyroid is administered or when the gland is overactive. Thyroxine added to a suspension of mitochondria alters the permeability of the mitochondrial membrane, causes the mitochondria to swell, and perhaps in this way uncouples oxidative phosphorylation from electron transport. Whether thyroxine has a similar effect on mitochondria within an intact cell is not yet clear. Complete removal of the thyroid gland from a mammal reduces the animal's metabolic rate to half of the normal value, and the body temperature decreases slightly. Since foods are metabolized at a lower rate, they tend to be stored and the animal becomes obese. Not only is the metabolic rate of the intact animal decreased by thyroid deficiency, but individual bits of tissue removed from the animal and incubated in vitro show a decreased oxygen consumption and decreased utilization of carbohydrates, fats and proteins.

By its action on metabolic processes, thyroxine has a marked influence on growth and differentiation. Extirpating the thyroid of young animals causes decreased body growth, retarded mental development, and delayed or decreased differentiation of gonads and external genitalia. All of these changes are reversed by the administration of thyroxine. The metamorphosis of frog and salamander tadpoles into adults is controlled by the thyroid. Removal of the larval thyroid completely prevents metamorphosis, and administering thyroxine to tadpoles causes them to metamorphose prematurely into miniature adults (Fig. 17.3). The effect of thyroxine on amphibian metamorphosis appears not to be simply a secondary result of its effect on metabolism, for tadpole metabolism can be increased by dinitrophenol but premature metamorphosis does not occur. Some specific effect of thyroxine on metamorphosis appears to be involved.

The production and discharge of thyroxine is regulated by the hormone **thyrotropin** secreted by the anterior lobe of the pituitary gland. In 1916, P. E. Smith found

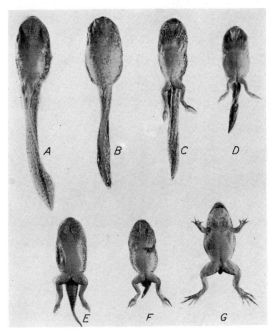

Figure 17.3 The effect of thyroid feeding upon the tadpoles of *Rana catesbiana*. A, The untreated control, which was killed at the end of the experiment. The metamorphosed animal at the lower right (G) was killed two weeks after starting the feeding of thyroid gland. The remaining animals (B to F) were removed from the experiment at intervals during this period. Note the effect of thyroid substances on the metamorphosis of the mouth, tail and paired appendages. (Turner: General Endocrinology, 3rd ed.)

that the removal of the pituitary of frog tadpoles produced deterioration of the thyroid and prevented metamorphosis. A similar pituitary control of thyroid function has been found in rats, man and other mammals. The secretion of thyrotropin by the pituitary is regulated in part by the amount of thyroxine in the blood. Thus, a decreased production of thyroxine by the thyroid leads to less thyroxine in the blood stream and this stimulates the pituitary to release thyrotropin, which passes to the thyroid gland and raises its output of thyroxine. When the blood level of thyroxine is brought back to normal, the release of thyrotropin is decreased. By this "feedback" mechanism the output of thyroxine is kept relatively constant, and the basal metabolic rate is kept within the normal range. Since iodine is an essential atom in thyroxine, a deficiency of this element leads to decreased synthesis of thyroxine. The decreased thyroxine production in iodine deficiency results in in-

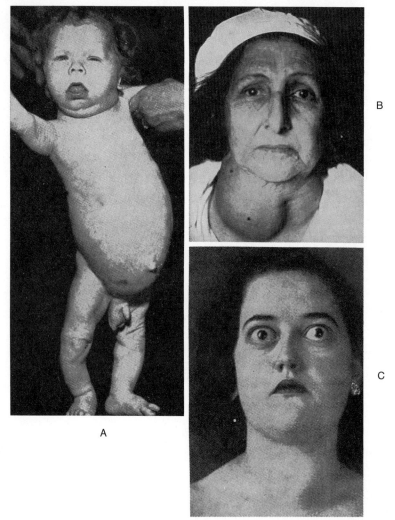

Figure 17.4 *A*, A cretin. *B*, Simple goiter. *C*, Exophthalmic goiter. (*A* and *B* from Selye: Textbook of Endocrinology, published by Acta Endocrinologia, Inc.; *C* from Houssay: Human Physiology, published by McGraw-Hill Book Company, Inc.)

creased secretion of thyrotropin, which stimulates the thyroid follicle cells to enlarge and to increase in number. Enlargement of the thyroid is known as **goiter.** Thiouracil and related compounds are goitrogenic. They block the reactions by which iodide is oxidized and fixed onto the tyrosine molecule, thus decreasing the production of thyroxine and secondarily increasing the production and release of thyrotropin from the pituitary.

Thyroid deficiency in infancy produces a dwarfed, mentally retarded child known as a **cretin** (Fig. 17.4*A*). A cretin has an enlarged tongue, coarse features, malformed bones, distended belly and wrinkled, cold skin. If thyroid therapy is begun early enough, normal development of the brain and body can be induced. Thyroid deficiency in adults results in **myxedema,** characterized by decreased metabolic rate, mental deterioration, obesity, loss of hair and cold, rough skin. Simple goiter, or enlarged thyroid, results usually from a deficiency of iodine, with a secondary increase in the size of the thyroid due to its stimulation by thyrotropin (Fig. 17.4*B*). Iodine is deficient in the soil and water of certain parts of the world and, hence, deficient in plants grown there and in the animals eating these plants. The prevalence of human goiter has been greatly decreased by the practice of adding iodide to table salt.

The overproduction of thyroid hormone produces Graves' disease, or **exophthalmic goiter** (Fig. 17.4C). The thyroid may be enlarged, or may be of nearly normal size, but it produces excessive amounts of hormone, with a resulting increased basal metabolic rate, increased production of heat, loss of weight, increased heart rate and blood pressure, nervousness, and exophthalmos, or protrusion of the eyeballs. Hyperthyroidism can be treated by surgical removal of part of the thyroid or by its destruction with x-rays or with radioactive iodine.

Calcitonin, secreted by the ultimobranchial bodies embedded in the thyroid, is a peptide hormone containing 32 amino acids. Its molecular weight is about 4000. It plays a role, together with parathyroid hormone, in regulating the metabolism of calcium and phosphorus. It decreases the concentration of calcium in the blood, decreases the excretion of calcium in the urine and stimulates the deposition of calcium phosphate in the bones.

17.4 THE PARATHYROID GLANDS

Embedded in the thyroid glands of terrestrial vertebrates are small masses of tissue called **parathyroid glands.** There are usually two pairs of parathyroids, which develop as outgrowths of the third and fourth pairs of pharyngeal pouches. Each gland consists of solid masses and cords of epithelial cells, rather than of spherical follicles as in the thyroid. The hormone secreted by the parathyroids, **parathormone,** regulates the concentrations of calcium and phosphorus in the blood and body fluids. It is essential for life; the complete removal of the parathyroids results in death in a few days. Parathormone is a single peptide chain containing 84 amino acids; its molecular weight is 8500. It is inactivated by proteolytic enzymes and cannot be administered orally. Parathormone promotes the absorption of calcium from the lumen of the intestine, the release of calcium from the bones and the reabsorption of calcium from the glomerular filtrate in the kidney tubules. Parathyroidectomy produces a decreased

concentration of calcium in the serum, a decreased excretion of phosphorus, and a resulting increase in the amount of phosphorus in the serum. The animal is subject to muscular tremors, cramps and convulsions, a condition known as **tetany,** which results from the low level of calcium in the body fluids. An injection of a solution of calcium stops the tetanic convulsions, and further convulsions can be prevented by repeated administration of calcium.

Parathyroid deficiencies are rare, occurring occasionally when the glands are removed inadvertently during an operation on the thyroid, or when degeneration results from an infection. The administration of parathormone cannot be used for the long-term treatment of parathyroid deficiencies, for the patient becomes refractory to repeated injections of the extract. The deficiency can be treated successfully by a diet rich in calcium and vitamin D and low in phosphorus.

Hyperfunction of the parathyroid, induced by a tumor of the gland, is characterized by high calcium and low phosphorus content of the blood and by increased urinary excretion of both calcium and phosphorus. The calcium comes, at least in part, from the bones and, as a result, the bones become soft and are easily broken. The increased level of calcium in the body fluids eventually leads to deposits of calcium in abnormal places — the kidneys, intestinal wall, heart and lungs.

17.5 THE ISLET CELLS OF THE PANCREAS

Scattered among the acinar cells which secrete the digestive enzymes are clusters of endocrine cells, the **islets of Langerhans,** which are quite different in appearance and staining properties. They have a richer supply of blood vessels than the acinar cells and have no associated ducts. The islet cells can be differentiated into two or more types by the staining reactions of their cytoplasmic granules. The pancreas develops as two outgrowths from the duodenum which grow together and fuse in most vertebrates. The islet cells develop as buds from the pancreatic ducts and eventually lose all connection with the ducts. In some bony fishes the acinar and islet tissues form spatially separate organs. The pancreas of the cyclostomes is ductless and located in the wall of the duodenum or in the liver.

The human disease **diabetes** had been recognized for many centuries but its cause and cure were equally unknown. A similar condition was produced experimentally in dogs by von Mering and Minkowski in 1889 when they surgically removed the pancreas while studying its role in digestion. Many attempts were subsequently made to feed pancreas or to prepare an extract for injection into diabetics, but all were unsuccessful. The proteolytic enzymes made by the pancreas destroyed the protein hormone before it could be extracted. Finally, in 1922, Banting and Best prepared an extract of fetal pancreas which had antidiabetic potency. The endocrine cells of the pancreas become active before the exocrine ones do. The first preparation of pure crystalline **insulin** was made in 1927 by Abel. The present commercial insulin is extracted from beef, sheep or hog pancreas by an acid alcohol method which rapidly inactivates the proteolytic enzymes. Insulin is a protein with a molecular weight of 12,000. The brilliant work of F. Sanger in England made known the exact sequence of the amino acids in each of the two peptide chains making up the insulin molecule. One chain contains 21 amino acids and the other contains 30. Insulin is produced in the β cells of the pancreas as a single peptide composed of 84 amino acids. This peptide, termed **proinsulin,** undergoes folding, three disulfide bonds are formed, and a peptide chain containing 33 amino acids is removed from the center of the chain, leaving two peptide chains joined by the disulfide bridges.

Most commercial preparations of insulin contain a second hormone, which *increases* blood sugar concentration instead of decreasing it as insulin does. This hormone, now christened **glucagon,** has been separated from insulin, crystallized, and found to be a single peptide chain composed of 27 amino acids. Glucagon is secreted by the alpha cells of the islets, and insulin by the beta cells.

Insulin and glucagon both take part in the regulation of carbohydrate metabolism, along with certain hormones secreted by the pituitary, adrenal medulla and adrenal cortex. Glucagon activates, via cyclic AMP,

the enzyme **phosphorylase,** which is involved in the conversion of liver glycogen to blood glucose, and thus raises the concentration of glucose in the blood. Insulin increases the rate of conversion of blood glucose to intracellular glucose-phosphate, thereby decreasing the blood glucose level, increasing the storage of glycogen in skeletal muscle, and increasing the metabolism of glucose to carbon dioxide and water. A deficiency of insulin decreases the utilization of sugar, and the alterations in carbohydrate metabolism which result secondarily produce many other changes in the metabolism of proteins, fats and other substances.

The surgical removal of the pancreas, or its hypofunction in diabetes mellitus, produces impaired glucose utilization, which results in high concentration of glucose in the blood (**hyperglycemia**) and the excretion of large amounts of glucose in the urine (**glycosuria**) because the concentration of sugar in the blood exceeds the renal threshold (p. 330). Extra water is required to excrete this sugar, the urine volume increases, and the patient tends to become dehydrated and thirsty. Because the tissues are unable to get enough glucose from the blood, they break down protein and convert the carbon chains of the amino acids into glucose. Much of this is excreted and there is a steady loss of weight. The fat deposits are also mobilized and broken down, and the concentration of fat in the blood may increase to the point where the blood has a milky appearance. The fatty acids are not metabolized completely but tend to accumulate as partially oxidized **ketone bodies** such as acetoacetic acid. These acidic substances accumulate in the blood and are excreted in the urine, causing an acidosis (loss of base) which finally results in coma and death. The injection of insulin alleviates all of these symptoms; with the utilization of glucose made normal by insulin, all the other metabolic conditions return to normal.

The effect of an injection of insulin lasts for only a short time, a day at most, for the insulin is gradually destroyed in the tissues. Long-lasting insulins, such as protamine zinc insulin and globin insulin, have been developed, and they reduce the number of injections needed by most diabetics to one a day.

The administration of a large dose of insulin to a normal or a diabetic person causes a marked decrease in the blood sugar level. The nerve cells, which require a certain amount of glucose for normal function, become hyperirritable and then fail to respond as the glucose level decreases. The patient becomes bewildered, incoherent, and comatose and may die unless some glucose is administered. There are rare cases of pancreatic tumors that, by hypersecretion of insulin, cause recurring attacks of convulsions and unconsciousness by reducing the blood glucose level.

The secretion of insulin is controlled by the concentration of glucose in the blood. When the amount of blood glucose rises, e.g., after a meal, the secretion of insulin is stimulated and it acts to restore the glucose level to normal. A rise in the concentration of insulin in the blood can be detected within two minutes after an experimentally induced rise in blood glucose. A major effect of insulin is a dramatic increase in the rate of transport of glucose into skeletal muscle and adipose tissue. This leads to a decrease in the concentration of glucose in the blood and removes the stimulus for the secretion of insulin. The secretion of glucagon is stimulated by low concentrations of glucose. As the glucose content of the blood falls below the optimal range, the release of glucagon from the α cells of the pancreas is stimulated, the glycogen phosphorylase system of the liver is stimulated by the glucagon and the glucose released by the liver returns the concentration of glucose in the blood to normal.

17.6   THE ADRENAL GLANDS

The small, paired **adrenal glands** of mammals are located at the anterior end of each kidney. The two human glands weigh only about 12 gm. but have a richer supply of blood vessels per mass of tissue than any other organ of the body. Each adrenal consists of two parts, an outer, pale, yellowish pink **cortex** and a dark, reddish brown inner **medulla.** In cyclostomes and fishes the two parts are spatially separate; in amphibians,

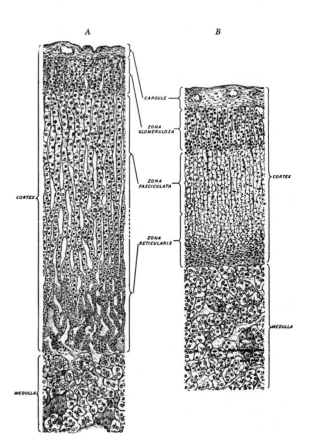

Figure 17.5 Sections through the adrenal cortex and medulla of (A) normal and (B) hypophysectomized rats. Since the functional capacity of the adrenal cortex is conditioned by the release of ACTH, hypophysectomy results in tremendous shrinkage of the cortex. The medulla is not influenced by hypophysectomy. Both sections are drawn to scale. (Turner: General Endocrinology, 3rd ed.)

reptiles and birds their anatomic relations are quite variable and the two parts are interspersed. Cortical tissue develops from coelomic mesoderm near the mesonephric kidneys, whereas the medullary tissue is ectodermal, derived from the neural crest cells which also form the sympathetic ganglia.

The cells of the medulla are arranged in irregular cords and masses around the blood vessels (Fig. 17.5). The medulla secretes two closely related hormones, **epinephrine** (also called adrenalin) and **norepinephrine.** These are comparatively simple chemicals derived from the amino acid tyrosine. Epinephrine produces an increase in heart rate, a rise in blood pressure, a decrease in the glycogen content of liver and muscle, an increase in blood glucose, and an increase in the rate at which blood coagulates. It causes dilation of the pupils of the eye, goose flesh, and dilation of most blood vessels but constriction of those of the skin, so that the skin becomes pale. Norepinephrine has much weaker effects on blood sugar and heart rate but is a more powerful vasoconstrictor.

The secretion of the adrenal medulla functions during emergencies to reinforce and prolong the action of the sympathetic nervous system. Epinephrine secretion is greatly increased by stresses such as cold, pain, trauma, emotional states and certain drugs. The changes resulting from the action of the sympathetic nerves and epinephrine would prepare an animal to attack its prey, defend itself against enemies or run away: (1) The efficiency of the circulatory system is increased by increased blood pressure and heart rate and the dilation of the large blood vessels. (2) The increase in the ability of blood to coagulate and the constriction of the vessels in the skin tend to minimize the loss of blood if the animal is wounded. (3) The intake of oxygen is increased by the increased rate of breathing and dilation of the respiratory passages. (4) The mobilization of the glycogen stores of the liver and muscle makes glucose available for energy. (5) The release of ACTH from the pituitary is stimulated. ACTH, in turn, stimulates the release from the adrenal cortex of glucocorticoids which increase the breakdown of protein and make further carbohydrate available.

Epinephrine is widely used clinically in treating asthma (it dilates respiratory passages), in increasing blood pressure and in stimulating a heart that has stopped beating.

The adrenal cortex is more complex than the medulla both structurally (for it is composed of three layers of cells) and functionally (for it secretes a number of hormones with different types of activity). The cortex is composed of three zones: an outer glomerulosa, a middle fasciculata and an inner reticularis (Fig. 17.5). Cells are formed by mitosis in the outer layer and are pushed inward to the reticularis, where they degenerate and disappear. The cells of the fasciculata are believed to be most active in hormone production. The embryos of man and other primates have adrenals as large as the kidneys, due to the presence of a large mass of cells, the **fetal zone,** interposed between the cortex and medulla. The fetal zone regresses and disappears after birth.

The hormones of the adrenal cortex are all steroids and include (1) **glucocorticoids,** which stimulate the conversion of proteins to carbohydrates, (2) **mineralocorticoids,** which regulate sodium and potassium metabolism, and (3) **androgens,** which have male sex hormone activity. The most potent glucocorticoid is **cortisol** (Compound F). The most potent mineralocorticoid is **aldosterone; deoxycorticosterone** is an effective regulator of salt and water metabolism and is widely used clinically.

The adrenal cortex of both males and females produces dehydroepiandrosterone and adrenosterone, steroids with slight male sex hormone activity. In addition the cortex produces small amounts of the much more potent male sex hormone, testosterone, the principal hormone secreted by the testis. Hyperfunction of the adrenal cortex in male children may increase the production of androgens and cause precocious sexual maturity. The child has the muscular development, hair distribution and voice of a man. Cortical hyperfunction in females may cause masculinization—growth of a beard, deep voice, regression of the ovaries, uterus and vagina, and development of the clitoris to resemble a penis.

Steroids are produced from cholesterol, which in turn is made by the union of two-carbon acetyl coenzyme A units. The cholesterol content of the adrenal cortex exceeds that of any other organ; as much as 5 per cent of the wet weight of the gland may

TABLE 17.2 STEROID HORMONES

Adrenal Cortex	Ovary	Testis	Placenta
Cortisol	Estradiol	Testosterone	Progesterone
Deoxycorticosterone	Progesterone	Androstenedione	Estradiol
Aldosterone	Androgens	Estradiol	
Androsterone		Estrone	
Dehydroepiandrosterone		Corticoids	
Estradiol			

be cholesterol. Steroids are synthesized from cholesterol not only in the adrenal cortex but in the testis, ovary and placenta as well (Table 17.2).

The complete removal of the adrenal cortex, or its hypofunction in **Addison's disease,** results in an increased excretion of sodium in the urine and a corresponding excretion of chloride, bicarbonate and water. The loss of sodium produces an acidosis, and the loss of body fluid leads to lowered blood pressure and a decreased rate of blood flow. The concentration of potassium in the body fluids increases. There is a marked decline in blood sugar concentration and in the glycogen content of liver, muscle and other tissues. It is clear from experimental evidence that the animal's ability to produce carbohydrates from proteins is greatly impaired.

The appetite for food and water decreases, and there is loss of weight. There are marked upsets in the digestive tract, with diarrhea, vomiting and pain. Muscles are more readily fatigued and less able to do work. The basal metabolic rate decreases, and the animal is less able to withstand exposure to cold and other stresses. Death ensues within a few days after complete adrenalectomy. The skin of a patient with Addison's disease develops a peculiar bronzing in patches, owing to the deposition of melanin.

The development and function of the adrenal cortex is regulated by adrenocorticotropic hormone, ACTH, secreted by the anterior lobe of the pituitary. ACTH increases the size of the adrenal cortex by stimulating the synthesis of RNA and protein. In addition, it stimulates the production of corticoids by increasing the activity of one or more of the enzymes involved in the conversion of cholesterol to cortisol. Stimulation of the adrenal cortex by ACTH leads to an increased production of cortisol and an increased concentration of that hormone in the blood. This in turn decreases, by a negative feedback control, the secretion of ACTH by the pituitary. Cortisol may do this directly by inhibiting the synthesis of ACTH in the pituitary or indirectly by decreasing the production of corticotropin-releasing factor (CRF) by the hypothalamus.

The primary stimulus that elicits the secretion of cortisol is some sort of physical stress—an injury, a burn, a painful disease, exposure to heat or cold—that sends impulses to the brain that are forwarded to the hypothalamus. The hypothalamus secretes the corticotropin-releasing factor which passes along a special portal system of blood vessels directly to the anterior lobe of the pituitary. The CRF stimulates the appropriate cells in the pituitary to secrete ACTH, and this is carried in the blood to the adrenal cortex where it stimulates the production and release of cortisol. The mobilization of amino acids and lipids from peripheral tissues and the gluconeogenesis in the liver provide substrates for the repair of the damage and decrease the stimulus which led to the production of the releasing factor. The increased concentration of cortisol in the blood acts in the hypothalamus or pituitary or both to decrease the production and release of ACTH.

An inherited defect of any one of the several enzymes involved in the synthesis of cortisol from pregnenolone and cholesterol may lead to enlargement of the adrenal cortex. The commonest defect is a deficiency of the enzyme adding a hydroxyl group at carbon 21 of the steroid. Such defects lead to an accumulation of intermediates in the biosynthetic pathway. Some of these intermediates cannot be converted to cortisol, but yield androgens such as androstenedione. This may be converted either in the adrenal or elsewhere in the body into testosterone, the most potent androgen. The failure of the adrenal cortex to secrete cortisol results in the oversecretion of ACTH, since there is nothing to shut it off. The adrenal grows larger and secretes even more androgens and the individual becomes virilized. A female fetus lacking the enzyme has external genitalia that are masculinized to varying degrees but a male fetus with the enzymatic defect may show no abnormality at birth. After birth virilization progresses in both males and females with enlargement of the phallus, early development of pubic and axillary hair, lowering of the voice and other effects of androgens. Patients with this condition, adrenal cortical hyperplasia, can be treated by injecting cortisol to turn off the production of ACTH by the pituitary.

17.7 THE PITUITARY GLAND

The pituitary gland, or hypophysis cerebri, is an unpaired endocrine gland lying in a small depression on the floor of the skull, just below the hypothalamus of the brain, to which it is attached by a narrow stalk. Its only known function is the secretion of hormones. The pituitary has a double origin: a dorsal outgrowth (**Rathke's pouch**) from the roof of the mouth grows up and surrounds a ventral evagination (the **infundibulum**) from the hypothalamus (Fig. 17.6). Both parts are of ectodermal origin. Rathke's pouch soon loses its connection to the mouth, but the connection to the brain, the infundibular stalk, remains. The hypophysis has three lobes: anterior and intermediate lobes derived from Rathke's pouch and a posterior lobe from the infundibulum. The pituitary, like the adrenal, is a double gland whose parts have quite different functions. The anterior lobe has no nerve fibers and is stimulated to release its hormones by hormonal factors reaching it through its blood vessels. The anterior lobe receives a double blood supply, arterial and portal. Some branches of the internal carotid artery pass directly to the pituitary, others serve a capillary bed around the infundibular stalk and the median eminence of the hypo-

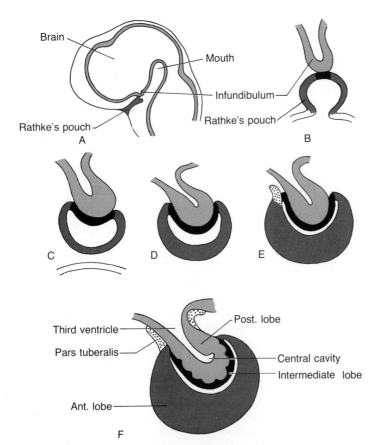

Figure 17.6 The development of the pituitary gland. *A*, Sagittal section through head of young embryo. *B–F*, Sagittal sections of successive stages of developing pituitary gland.

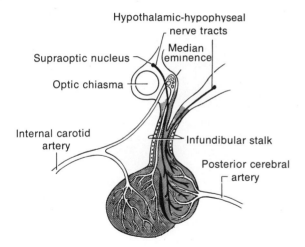

Supraoptic nucleus

Optic chiasma

Hypothalamic-hypophyseal nerve tracts

Median eminence

Internal carotid artery

Infundibular stalk

Posterior cerebral artery

Figure 17.7 Blood supply of the pituitary gland.

thalamus (Fig. 17.7). Portal veins from these capillaries then pass down the infundibular stalk and empty into the capillaries surrounding the secretory cells of the anterior lobe. There is, thus, a direct route for releasing factors secreted by the hypothalamus to pass directly to the anterior lobe and control the secretion of its hormones.

The anterior lobe contains at least five different types of cells that differ in their shape, size, staining properties and the kind of granules present in the cytoplasm. It seems likely that each type produces and secretes a different kind of hormone. The cell secreting growth hormone has been identified as a rounded cell whose cytoplasm is packed with dense, round, acidophilic granules. These have been isolated and shown to have a high content of growth hormone. The cells secreting prolactin stain deeply with carmine stain and contain granules that are larger and more ovoid than the granules in the growth hormone-secreting cells.

The intermediate lobe contains basophil cells smaller than those of the anterior lobe, some with and some without granules. The posterior lobe is composed of many nonmyelinated nerve fibers and branching cells (pituicytes) which contain brownish cytoplasmic granules.

The posterior lobe contains two hormones, **oxytocin** and **vasopressin.** The latter is also known as **antidiuretic hormone,** or ADH. The brilliant work of Vincent du Vigneaud, for which he was awarded the Nobel Prize in 1955, led to the isolation of these two hormones, the determination of their molecular structure, and their synthesis. Each is a peptide containing nine amino acids, seven of which are identical in the two. It is of considerable interest that these two substances, with quite different physiologic properties, differ in only two amino acids. Oxytocin stimulates the contraction of the uterine muscles and is sometimes injected during childbirth to contract the uterus. Vasopressin causes a contraction of smooth muscles; its contraction of the muscles in the wall of arterioles causes a general increase in blood pressure. It also regulates the reabsorption of water by the cells of the distal convoluted tubules and Henle's loop in the kidney (p. 331).

These two hormones are not produced in the posterior lobe but are secreted by neurosecretory cells in the supraoptic and paraventricular nuclei of the brain. They then pass along the axons of the hypothalamic-hypophyseal tract and are stored and released by the posterior lobe. An injury of these brain nuclei, of the posterior lobe or of the connecting nerve tracts, may lead to a deficiency of ADH and the condition known as **diabetes insipidus.** In this disease the patient's kidneys have a lessened ability to reabsorb water and his urine volume increases from the normal 1 or 2 liters to 10 to 25 liters per day. He suffers from excessive thirst and drinks copiously. Injection of ADH relieves all the symptoms, but the injections must be repeated every few days. A comparable condition can be produced in experimental animals by severing the

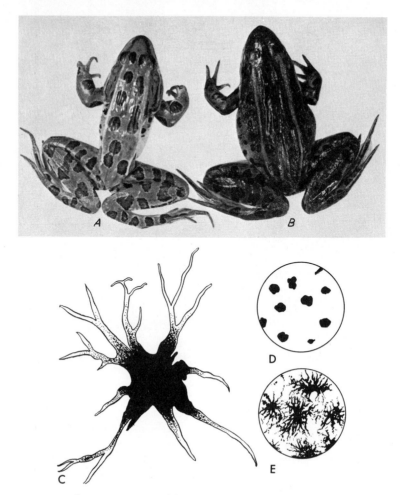

Figure 17.8   Integumentary adaptations in normal frogs (*Rana pipiens*). *A*, Light-adapted animal; *B*, dark-adapted animal. (Turner: General Endocrinology, 2nd ed.) *C*, A chromatophore, greatly magnified, showing the pigment. *D*, A section of skin of frog adapted to a warm, light environment. *E*, Skin adapted to a cool, dark environment.

hypothalamic-hypophyseal tract by electrolytic lesions accurately placed with a microelectrode.

The intermediate lobe of the pituitary secretes a hormone, **intermedin,** which darkens the skin of fishes, amphibians and reptiles by dispersing the pigment granules in the chromatophores. The skin of a frog becomes darkened in a cool, dark environment and light-colored in a warm, light place (Fig. 17.8). Hypophysectomy produces a permanent blanching of the skin, and injection of intermedin causes darkening. The location of the pigment in the chromatophore is controlled directly by the amount of intermedin present, not by nerves. The pituitaries of birds and mammals are rich in intermedin, but there is no known function for this hormone in these animals; it does not affect their pigmentation.

The anterior lobe of the pituitary secretes growth hormone (somatotropin), thyrotropin, adrenocorticotropin (ACTH), follicle-stimulating hormone (FSH), luteinizing hormone (LH) and prolactin (lactogenic hormone). The importance of these hormones is demonstrated by the marked abnormalities which follow hypophysectomy: cessation of growth in young animals, regression of gonads and reproductive organs, and atrophy of the thyroid and adrenal cortex (Fig. 17.9).

Growth hormone was the first pituitary

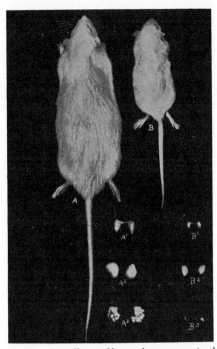

Figure 17.9 The effects of hypophysectomy in the rat. *A*, Normal littermate control; *B*, littermate hypophysectomized when 36 days of age. These photographs were made at 144 days of age, when the control animal weighed 264 gm. and the hypophysectomized rat weighed 80 gm. A^1, A^2 and A^3 are thyroid, adrenals and ovaries from normal animal; B^2, B^2 and B^3 are thyroids, adrenals and ovaries from hypophysectomized animal. Note marked differences in size. (Turner: General Endocrinology, 2nd ed.)

hormone to be described. As early as 1860 it was recognized that gigantism was correlated with an enlargement of the pituitary. A growth-promoting extract of beef pituitaries was prepared by Evans and Long in 1921 and pure growth hormone was isolated in 1944. Human growth hormone is a single peptide chain of 188 amino acids. It controls general body growth and bone growth and leads to an increase in the amount of cellular protein (Fig. 17.10). Overactivity of the pituitary during the growth period leads to very tall, but well-proportioned persons, and underactivity leads to small persons of normal body proportions, called midgets. After normal growth has been completed, hypersecretion of growth hormone produces **acromegaly,** characterized by the thickening of the skin, tongue, lips, nose and ears and by growth of the bones of the hands, feet, jaw and skull. Other bones have lost their ability to respond to growth hormone.

A race of hereditary dwarf mice is known whose pituitaries apparently lack the type of cell which secretes growth hormone. These animals can be induced to grow to normal size by implanting a pituitary from a normal mouse. Growth hormones from different species have been found to differ slightly in their amino acid composition

Figure 17.10 The effect of growth hormone on the dachshund. Top, normal dog. Bottom, dog injected with growth extract for a period of six months. (From Evans, Simpson, Meyer and Reichert.)

and in their effectiveness. Thus, beef growth hormone will cause growth in rats but not in man or monkeys. Growth hormone prepared from human or monkey pituitaries will stimulate growth in man and monkeys. In addition to its general effects on growth, this hormone affects protein, lipid and carbohydrate metabolism, leading to increased protein synthesis, a mobilization of lipid from tissues and an increased lipid concentration in the blood, and an increased deposition of glycogen in liver and muscle but an increased concentration of glucose in the blood.

Adrenocorticotropic hormone, ACTH, is a peptide containing 39 amino acids in a single chain. ACTH became famous because of the remarkable results it sometimes gives in the treatment of allergies and arthritis. However, the prime, and perhaps the only, physiologic function of ACTH is to stimulate the adrenal cortex to grow and to release cortical steroids. The adrenal cortex undergoes a prompt atrophy after the removal of the pituitary and can be returned to normal by the injection of ACTH.

The extirpation of the pituitary also causes atrophy of the thyroid. The gland decreases in size and the follicle cells become flattened. The thyroid is returned to normal by the implantation of a pituitary gland or by the administration of an extract containing **thyrotropin.** The injection of thyrotropin in a normal animal causes growth of the thyroid and thickening of the follicle cells so that they become columnar rather than cuboidal epithelium (Fig. 17.2). Thyrotropin is a glycoprotein with a molecular weight of about 25,000. Its carbohydrate components include mannose, fucose and N-acetyl galactosamine.

The gonadotropins, **follicle-stimulating hormone,** FSH, and **luteinizing hormone,** LH, secreted by the anterior lobe of the pituitary, are both glycoproteins. FSH, with a molecular weight of about 31,000, contains about 8 per cent carbohydrate, including some **sialic acid.** FSH loses its biological activity when the sialic acid is split by the enzyme **neuraminidase** or when the other carbohydrate components are removed by treating with amylase. LH is a slightly smaller protein, with a molecular weight of about 26,000. LH's from different species have carbohydrate contents ranging from 4.5 per cent in human LH to 11 per cent in sheep LH. A third gonadotropin, **prolactin** or **luteotropic hormone,** is a protein composed of a single peptide chain of 205 amino acids, and has a molecular weight of about 25,000.

The three glycoprotein pituitary hormones—TSH, FSH and LH—are each composed of two subunits, termed α and β. The α subunits of all three are very similar, if not identical, whereas the β subunits are distinctive and appear to be responsible for the biological specificity of the hormone. The hormones can be separated into their subunits, which have little or no biological activity, and then recombined to give full activity. One can combine a TSH α subunit with an FSH β subunit and get a protein with FSH activity.

The ovaries or testes of a hypophysectomized young animal never become mature; they neither produce gametes nor secrete enough sex hormones to develop the secondary sex characters. Hypophysectomy of an adult results in involution and atrophy of the gonads. Both FSH and LH are necessary for achieving sexual maturity and for the regulation of the estrous cycle. The effect of FSH is primarily on the development of graafian follicles in the ovaries: it does not produce any significant release of estrogen. Luteinizing hormone controls the release of ripe eggs from the follicle, the formation of corpora lutea, and the production and release of estrogens and progesterone. **Prolactin,** or lactogenic hormone, maintains the secretion of estrogens and progesterone and stimulates the secretion of milk by the breast. It is effective, however, only after the breast has been appropriately stimulated by estrogen and progesterone. Prolactin induces behavior patterns leading to the care of the young (the "maternal instinct") in mammals and in other vertebrates as well. Roosters treated with prolactin will take care of chicks, taking them to food and water, sheltering them under their wings, and protecting them from predators.

The development and functioning of the testis is also controlled by FSH and LH. FSH increases the size of the seminiferous tubules, and both FSH and LH are needed for normal spermatogenesis. LH, but not FSH, stimulates the interstitial cells of the testis to produce male sex hormone.

The control of pituitary function, which

ensures that the proper amount of each of these hormones will be released at the proper moment in response to the demands of the organism, is indeed complex. The release of each tropic hormone is controlled in part by the level of the target hormone in the circulating blood, e.g., the release of ACTH is inhibited by cortisol. This provides for a cutoff mechanism so that in a normal animal the secretions of the pituitary and its target organs are kept in balance.

The hypothalamus provides an important control of pituitary function. Axons from certain centers in the hypothalamus end in the median eminence (Fig. 17.7). The tips of these axons secrete specific releasing factors which are carried by the portal veins to the hypophysis, where they stimulate the release of specific pituitary hormones. For example, evidence of ACTH secretion is obtained when the median eminence is stimulated electrically but not when the stimulus is applied to the supraoptic nuclei whose axons pass to the posterior lobe of the pituitary. The electrical stimulus is ineffective if the blood vessels between the hypothalamus and pituitary are cut. If the nerve fibers to the median eminence are destroyed, ACTH is no longer released in response to stresses.

All the living vertebrates have pituitaries which are basically similar, and they all appear to secrete the same battery of hormones. The intermediate lobes of birds and mammals secrete intermedin, although these forms have no chromatophores; birds secrete luteinizing hormone but have no corpora lutea; and all vertebrates secrete prolactin, but only mammals have its target organ, the mammary glands.

17.8 THE TESTIS

In between the seminiferous tubules of the testes are hormone-secreting cells, the interstitial cells of Leydig which produce and secrete the male sex hormones (androgens) such as **testosterone.** Although Berthold concluded in 1849 that the testis produces a blood-borne substance needed for the development of male sex characters, the major androgen, testosterone, was not identified until 1935.

Testosterone has a general effect on metabolism, inducing growth by stimulating the formation of cell proteins. The administration of androgens leads to an increase in body weight owing to the synthesis of protein in muscle and to a lesser extent in the liver and kidney.

Testosterone and other androgens stimulate the development and maintenance of the **secondary male characters:** the enlargement of the external genitals, the growth of the accessory glands such as the prostate and seminal vesicles, the growth of the beard and body hair, and the deepening of the voice. The secondary sex characters of other animals, the antlers of deer and the combs, wattles and plumage of birds, are controlled by androgens. Male sex hormones are responsible, in part, for increased libido in both sexes and for the development of mating behavior.

The removal of the testis (castration) of an immature male prevents the development of the secondary sex characters. A castrated man, a **eunuch,** has a high-pitched voice, beardless face, and small genitals and accessory glands. Castration was practiced in the past to provide guardians for harems and sopranos for choirs. Many domestic animals are castrated to make them more placid. The injection of testosterone into a castrated animal restores all the sex characters to normal. The anal fin of the male mosquito fish, *Gambusia*, is differentiated into a penis-like organ used to transfer sperm to the female. This fails to develop if the fish is castrated but appears if the castrate male or the female is treated with testosterone.

It should be clearly understood that males produce female sex hormones (estrogens) and that females produce androgens in considerable amounts. One of the richest sources of female sex hormones is the urine of stallions. The normal differentiation of the sex characters is a function of a balance between the two.

The failure of the testes to descend normally from the body cavity to the scrotal sac, called **cryptorchidism,** produces sterility but has little or no effect on the production of testosterone. Microscopic examination of an undescended testis shows that the cells in the seminiferous tubules regress, but the interstitial cells are normal. The cells of the seminiferous tubules are particularly susceptible to heat, and the temperature of the body cavity, 3 or 4 degrees higher than that of the scrotal sac, destroys them. It is probable that the elevated temperature during a prolonged fever makes a man sterile for some time. In many wild animals the testes remain in the body cavity except during the breeding season, when they descend into the scrotal sac.

The removal of the pituitary causes regression of both interstitial cells and seminiferous tubules of the testis. Androgen secretion is decreased, and the secondary sex characters regress. Normal development of the seminiferous tubules and spermatogenesis apparently require the combined action of FSH, LH and testosterone. The administration of excessive amounts of testosterone or estrogen may produce regression of the testes, presumably by inhibiting the release of FSH and LH from the pituitary.

The cyclic growth and regression of the testes in animals with periodic breeding seasons appears to be mediated via the pituitary. Such animals have very low amounts of gonadotropin in the nonbreeding season. Changes in the ambient temperature or in the amount of daily illumination produce stimuli which are mediated

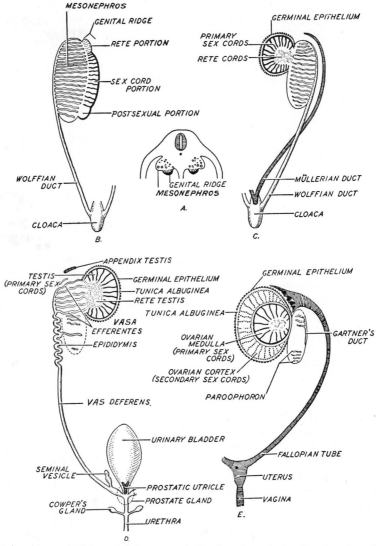

Figure 17.11 The development of the genital system. A, Section through the dorsal region of an early embryo. B, The wolffian body and genital ridge in frontal section. C, The indifferent stage. D, Differentiation of the male genitalia. E, Differentiation of the female genitalia. (Modified from Turner, 2nd ed.)

by the brain and hypothalamus to induce gonadotropin secretion by the pituitary, the consequent growth and functional state of the testes and the appearance of the secondary sex characters.

17.9 THE OVARIES

The ovaries of vertebrates are endocrine organs as well as the source of eggs; they produce the steroid hormones **estradiol** and **progesterone.** Some mammalian ovaries produce a third hormone, the protein **relaxin.**

Both ovaries and testes develop from mesoderm, from the **genital ridge** on the ventral side of the mesonephros (Fig. 17.11). Each ovary consists of closely packed cells covered by a thickened mesothelium called the germinal epithelium. During embryonic development the **primordial germ cells** originate in the eipthelium of the yolk sac and migrate to their final position in the gonad.

As each oöcyte develops, it becomes surrounded by other cells derived from the germinal epithelium which form a spheri-cal follicle about it (Fig. 17.12). These cells proliferate and form a thick layer, the **stratum granulosum,** around the egg. A cavity, the **antrum,** filled with liquid appears in the mass of follicle cells. The connective tissue of the ovary forms a sheath, the **theca,** around the follicle. As the follicle enlarges and its antrum becomes dilated with follicular fluid, it is pushed near the surface of the ovary. It finally bursts and releases the egg into the peritoneal cavity, whence it passes into the oviduct. The release of the egg is known as **ovulation.** If the egg is fertilized in the oviduct, it will subsequently become embedded in the lining of the uterus and begin development.

The follicular cells remaining after the rupture of the follicle multiply and increase in size, filling the cavity left by the follicle. Cells from the theca grow in along with the granulosa cells and the two form the **corpus luteum.** This yellowish structure, a solid mass of cells about the size of a pea, projects from the surface of the ovary. If the egg is fertilized the human corpus luteum persists for months, but if no fertilization

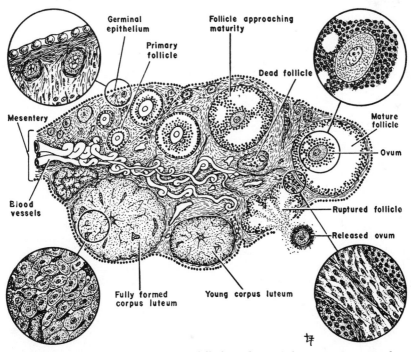

Figure 17.12 Stages in the development of an egg, follicle and corpus luteum in a mammalian ovary. Successive stages are depicted clockwise, beginning at the mesentery. Insets show the cellular structure of the successive stages. (Villee: Biology, 6th ed.)

takes place it regresses after about two weeks to a small patch of whitish scar tissue, the corpus albicans.

Histochemical evidence indicates that the thecal cells are the source of estrogen and that these plus the granulosa cells of the corpus luteum are the source of progesterone. The primary estrogen is 17-β-estradiol; other estrogens such as estrone and estriol may be metabolites of estradiol. Estradiol stimulates the changes which occur at sexual maturity: the growth of the accessory sex organs, uterus and vagina, the development of the breasts, changes in skeletal structure such as the broadening of the pelvis, the change in voice quality, the growth of pubic hair and the onset of the menstrual cycle. Both progesterone and estradiol are required for the growth of the uterine lining in each menstrual cycle to the stage at which implantation of the fertilized egg is possible. Progesterone is required for the maintenance of the developing embryo in the uterus and, together with estradiol, regulates the development of the breasts during pregnancy.

It is a curious fact that the primary female sex hormone **estradiol,** with 18 carbons, is synthesized from the 19-carbon male sex hormone **testosterone.** This in turn is synthesized from the second type of female sex hormone, **progesterone,** which has 21 carbons. Progesterone is also a precursor in the adrenal cortex of both glucocorticoids and mineralocorticoids.

17.10   THE ESTROUS AND MENSTRUAL CYCLES

The females of most mammalian species show cyclic periods of the sex urge and will permit copulation only at certain times, known as periods of **estrus** or "heat," when conditions are optimal for the union of egg and sperm. Most wild animals have one estrous period a year, the dog and cat have two or three and rats and mice have estrous periods every five days. Estrus is characterized by heightened sex urge, ovulation and changes in the lining of the uterus and vagina. Following estrus, the uterine lining thickens, and its glands and blood vessels develop to provide optimal conditions for implantation.

The menstrual cycle of the primates is characterized not by periods of mating urge but by periods of bleeding caused by the degeneration and sloughing of the uterine lining. Ovulation occurs about midway between two successive menstruations, or periods of bleeding. Most primates show little or no cyclic change in the sex urge and permit copulation at any time in the menstrual cycle.

The menstrual cycle is controlled by the interaction of hormones secreted by the hypothalamus, pituitary and ovary and includes events in the ovary, uterus and vagina. One menstrual cycle, from the beginning of one period of bleeding to the next, lasts 28 to 30 days in the human female (Fig. 17.13).

The lining of the uterus is almost completely sloughed off at each menstruation and thus is thinnest just after the menstrual flow. At that time, under the influence of FSH from the pituitary, one or more of the follicles in the ovary begin to grow rapidly. The follicular cells produce estradiol, which stimulates the growth of the uterine lining (the endometrium) and some growth of the uterine glands and blood vessels. The rupture of the follicle in ovulation does not occur automatically when a certain size is reached but is induced by a sharply defined surge of LH from the pituitary. Ovulation occurs about 14 days before the beginning of the next menstrual period. The corpus luteum develops and, under the stimulation of LH and prolactin, secretes progesterone and estradiol. These promote further growth of the endometrium. The endometrial glands grow further and become secretory, and the blood vessels become long and coiled. Progesterone decreases the activity of the uterine muscles and brings the uterus into a condition such that the developing embryo formed from the fertilized egg can become implanted and develop. Progesterone inhibits the development of other follicles. If fertilization and implantation do not occur the corpus luteum begins to regress, it secretes less progesterone, and the endometrium, no longer provided with sufficient progesterone to be maintained, begins to slough. Thus menstruation ensues, completing the cycle.

If the egg has been fertilized and implants in the endometrium, the cells of the

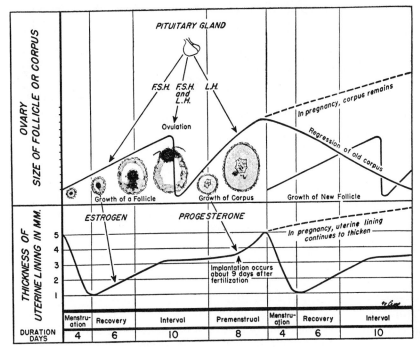

Figure 17.13 The menstrual cycle in the human female. The solid lines indicate the course of events if the egg is not fertilized; the dotted lines indicate the course of events when pregnancy occurs. The actions of the hormones of the pituitary and ovary in regulating the cycle are indicated by arrows. (Villee: Biology, 6th ed.)

trophoblast in the developing placenta secrete **chorionic gonadotropin.** Its strong luteinizing and luteotropic activities maintain the corpus luteum and stimulate the continued secretion of progesterone. By the sixteenth week or so of pregnancy the placenta itself produces enough progesterone so that the corpus luteum is no longer needed and undergoes involution.

Progesterone blocks ovulation not by a direct effect on the ovary but by preventing the secretion of the releasing factors FRF and LRF by the hypothalamus. The **oral contraceptives** contain synthetic estrogens and progestins which block ovulation in a similar manner. The natural hormones, estradiol and progesterone, are rapidly metabolized in the body but the synthetic hormones have slight changes in the molecular structure that markedly decrease the rate at which they are destroyed. The oral contraceptives, like the natural hormones, inhibit the release of LH by the pituitary and thus prevent ovulation. A woman taking "the pill" has no midcycle surge of LH and FSH and does not ovulate. Some of the synthetic progestins alter the character of the mucus secreted by the cer-

vix of the uterus and make it less readily penetrated by sperm, thus decreasing the probability of fertilization.

Progesterone also stimulates the growth of the glands and ducts of the breasts during the latter months of pregnancy and prepares them for the action of prolactin secreted by the pituitary which, together with oxytocin, stimulates the flow of milk.

17.11 THE HORMONES OF PREGNANCY

The **placenta,** which develops in part from the extraembryonic membranes of the fetus and in part from the lining of the uterus, is primarily an organ for the support and nourishment of the fetus. It is also an endocrine organ that produces hormones similar to those of the ovary and pituitary. These placental hormones, together with those of the maternal endocrine glands, control the many adaptations necessary for the continuation and successful termination of pregnancy.

The placenta secretes **chorionic gonadotropin,** produced by the cells of the chori-

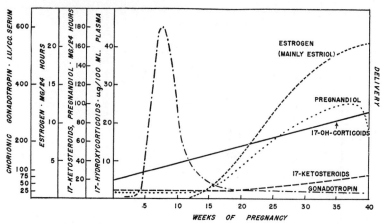

Figure 17.14 Hormone levels in blood and urine during pregnancy.

onic villi. Its effects are similar to, yet distinct from, those of the pituitary gonadotropins. The placenta does not merely accumulate a hormone made elsewhere, for bits of placenta grown in tissue culture produce chorionic gonadotropin. One of the earliest signs of pregnancy is the appearance of this hormone in the blood and urine. The peak of chorionic gonadotropin production is reached in the second month of pregnancy, after which the amount in blood and urine decreases to low levels (Fig. 17.14). Several pregnancy tests involve the effect of this gonadotropin, obtained from a sample of urine from the woman suspected of being pregnant, on sperm release in the frog or African toad or on the production of corpora lutea in rats or rabbits. These tests are quite accurate and make possible a diagnosis of pregnancy within a few weeks of conception. More sensitive immunoassays and radioimmunoassays for HCG can diagnose pregnancy just a few days after implantation and have now largely replaced the earlier bioassays. Chorionic gonadotropin stimulates the corpus luteum to remain functional and not regress as it would in the absence of pregnancy.

The human placenta (and probably the placentas of other mammals, too) secretes another protein hormone, placental lactogen, with properties somewhat similar to those of pituitary lactogenic hormone. This hormone has marked lactogenic properties and also serves as a synergist to pituitary growth hormone. The placenta secretes both estrogens and progesterone which reinforce the ovarian hormones in the maintenance of pregnancy.

In some animals, such as the rabbit, the placenta is a significant source of relaxin. This protein hormone, also produced by the ovary, functions to relax the ligaments of the pelvis to facilitate the birth of the young. Relaxin is effective only after the connective tissue of the pubic symphysis has been sensitized by the action of estradiol. Relaxin also inhibits the motility of the uterine muscles.

The production of estrogens and progesterone, as reflected by the amount present in blood and urine, increases gradually throughout pregnancy, reaches a peak just before, or at the time of, parturition, and then abruptly declines after birth (Fig. 17.14). The factors which determine the onset of labor, the expulsion of the fetus from the uterus, remain a mystery. There are many hormonal changes which occur at about the time of parturition—decreases in estrogen and progesterone, and an increase in chorionic gonadotropin—but whether these are causes, effects, or unrelated phenomena remains to be determined.

17.12 OTHER ENDOCRINE GLANDS

The thymus lies in the upper part of the chest, just above the heart. Its cells closely resemble lymph tissue. The thymus is large

TABLE 17.3 HORMONES OF THE DIGESTIVE TRACT

Hormone	Secreted by	Stimulus for Secretion	Target Organ	Response of Target Organ
Gastrin	Pyloric mucosa	Presence of food in stomach	Mucosa of stomach fundus	Secretion of gastric juice
Secretin	Duodenal mucosa	Presence of acid food in duodenum	Pancreas	Secretion of pancreatic juice
Enterogastrone	Duodenal mucosa	Neutral fat	Stomach	Decreased motility and secretion of HCl
Cholecystokinin	Duodenal mucosa	Acid food	Gallbladder	Contraction of gallbladder

during the years of rapid growth but begins to regress after puberty. It has been postulated that it affects growth or sexual maturity, but extirpation of the gland or the administration of extracts fails to reveal any endocrine function.

The **pineal gland,** a small round structure on the upper surface of the thalamus, lying between the cerebral hemispheres, is derived embryologically as an outgrowth of the brain. It secretes **melatonin,** a methoxy indole synthesized from tryptophan, which is converted to 5-hydroxy tryptamine (serotonin). Serotonin is acetylated at the amine nitrogen and a methyl group is added to the 5-hydroxy group by **hydroxy indole-O-methyl transferase,** an enzyme found uniquely in the pineal. The enzyme is stimulated by norepinephrine, released by the tips of the sympathetic nerves that extend to the pineal from a cervical sympathetic ganglion. Light falling on the retina of the eye increases the synthesis of melatonin by the pineal. A small nerve, the **inferior accessory optic tract,** passes from the optic nerve through the medial forebrain and connects with the sympathetic nervous system. The pineal secretion inhibits ovarian functions either directly or via an effect on the pituitary. Girls blind from birth undergo puberty earlier than normal, apparently because they lack the inhibitory effect of melatonin on ovarian function.

The cells of certain parts of the digestive tract are known to secrete hormones in response to the presence of certain kinds of food which stimulate the production and release of digestive juices. These are summarized in Table 17.3.

17.13 ENDOCRINE INTERRELATIONSHIPS

It is now becoming clear that each gland affects the functioning of almost every other one and that together they constitute an interrelated and interdependent system which coordinates body activities. When the role of the pituitary in regulating the activity of the thyroid, adrenals and gonads was first discovered, the pituitary was described as a "master controlling gland." But in view of the reciprocal effects of the hormones of these glands of the pituitary, and of the further control of the pituitary imposed by the hypothalamus, it is probably unwarranted to regard the pituitary as a special master gland.

The interplay of estradiol, progesterone, FSH and LH in regulating the menstrual cycle, and of estrogen, progesterone and prolactin in producing the development and functioning of the breasts, is now well established. The rate of cell metabolism and the relative rates of utilization of carbohydrates, fats and proteins are under the complex control of thyroxine, insulin, epinephrine, glucagon, growth hormone, cortisol, estradiol and testosterone. Normal growth requires not only growth hormone and thyroxine but also insulin, androgens and others.

Stresses such as trauma, burns, cold, starvation, hemorrhage, intense sound or light and anoxia provoke a pattern of adaptation which tends to resist damage from the stress. The stress stimulates the release of epinephrine from the adrenal medulla which, in turn, leads to the release of ACTH by the anterior lobe of the pituitary. The adrenal cortical hormones released by the action of the ACTH produce changes in mineral and carbohydrate metabolism and in tissue reactivity which adapt the animal to resist the effects of the stress. Long-continued stresses eventually overcome the body's adaptive ability and produce exhaustion and shock. In the absence of either the hypophysis or the adrenal cortex, the body's ability to tolerate stress is greatly decreased.

17.14   MECHANISMS OF HORMONE ACTION

Any theory of the molecular mechanism by which a given hormone produces its specific effects in specific tissues must account for the high degree of **specificity** of many hormones and for the remarkable biological **amplification** inherent in hormonal processes. Hormones circulate in the blood in very low concentrations—steroid hormones at concentrations of 10^{-9} M or less and peptide and protein hormones at concentrations of 10^{-12} M. The several current theories regarding the mechanism of hormone action are alike in suggesting that the hormone goes to or into a cell and combines with some specific **receptor.** Many cells do not recognize certain hormones, presumably because they have no receptor for that hormone and because the receptors are highly specific. The hypotheses differ as to what the receptor is, where it is located and what happens after the hormone is bound to its receptor. Appropriate experiments with highly labeled steroid hormones have shown the presence of specific protein receptors for estradiol in the uterus, for progesterone in the oviduct, for aldosterone in the kidney and for testosterone in the prostate gland. They indicate that these receptors are in the nuclei of the cells.

The effects of hormones in facilitating the entrance of certain substances in the cell, such as the uptake of glucose by muscle cells stimulated by insulin, has suggested that the hormone combines with a protein or some other substance in the cell membrane. This leads to a change in the molecular architecture of the membrane and hence in its permeability to specific substrates.

Another hypothesis states that the receptor is a protein whose enzymatic activity is

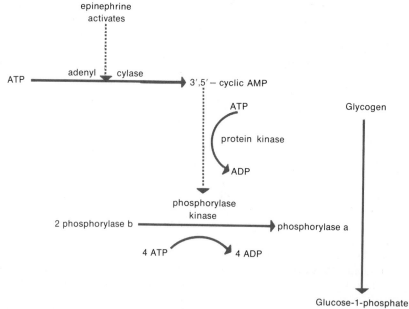

Figure 17.15   The sequence of enzymatic events by which epinephrine or glucagon stimulates adenyl cyclase and brings about the synthesis of 3′,5′-adenosine monophosphate (cyclic AMP). This in turn activates a protein kinase that phosphorylates phosphorylase kinase which in turn phosphorylates and activates phosphorylase. This finally brings about the cleavage of glycogen and the secretion of glucose. (Villee: Biology, 6th ed.)

altered by the combination. One variation of this hypothesis currently popular is that the hormone combines with a specific **adenyl cyclase** and stimulates the production of **cyclic 3′,5′-adenylic acid** from ATP. The cyclic AMP is regarded as a "second messenger" that mediates the effect of the hormone. Epinephrine, for example, has been shown to stimulate the adenyl cyclase of liver cells (Fig. 17.15). The resulting cyclic AMP is an activator of another enzyme, a protein kinase, that transfers a phosphate group from ATP to a third enzyme, phosphorylase kinase, and activates it so that it can convert an inactive fourth enzyme, phosphorylase b, to active phosphorylase a. The latter then catalyzes the production of glucose-1-phosphate from glycogen. At each of these successive steps there is an amplification of 10- to 100-fold, so that a very small amount of epinephrine will lead to the production of a very large amount of glucose-1-phosphate.

A third general hypothesis suggests that the hormone enters the nucleus and activates specific genes that were previously repressed so that they are transcribed (Fig. 17.16). This leads to the production of new kinds of messenger RNA which code for the synthesis of new specific proteins. This theory also accounts for the marked amplification of hormonal effects, for a very small amount of hormone, by turning on the transcription of a specific gene, could result in many molecules of messenger RNA and many more molecules of protein.

Whether hormones are typically used up as they regulate metabolism in their target tissues is not clear. Estradiol is not used up or changed chemically as it stimulates the growth of the uterus. The hormones bound to their receptors appear to be relatively stable, but hormones circulating in the blood have relatively short biological half-lives. They are inactivated and eliminated

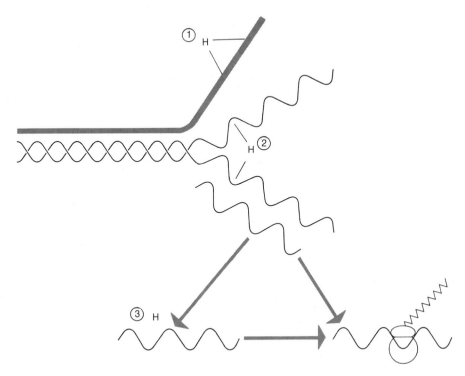

Figure 17.16 A diagram illustrating some of the possible sites at which a hormone may regulate genetic transcription, either by combining with a histone (*1*) or other protein which is bound to the DNA or by interacting with the DNA itself (*2*). The hormone might act by combining with a specific messenger RNA (*3*) and protecting it from being inactivated. (Villee: Biology, 6th ed.)

from the body and must be replaced by new hormone molecules synthesized in the appropriate endocrine gland.

It seems unlikely that all hormones have a common molecular mechanism by which their effects are produced. Indeed there is evidence that certain hormones produce their effects not by a single mechanism but by several different mechanisms acting in parallel. The theories current at any given moment tend to reflect our general knowledge of cellular and molecular biology. Theories implicating effects of hormones on cell membranes have given way to theories implicating an effect on altering the activity of an enzyme and these in turn have been replaced by theories involving effects of the hormone on the genetic mechanism which result in the synthesis of specific kinds of RNA and proteins.

17.15 PHEROMONES

In recent years it has been appreciated that the behavior of animals may be influenced not only by hormones — chemicals released into the internal environment by endocrine glands and which regulate and coordinate the activities of other tissues — but also by **pheromones** — substances secreted by *exocrine* glands, released into the *external* environment and which influence the behavior of other *animals* of the same species. We are used to thinking that information can be transferred from one animal to another by sight or sound; pheromones represent a means of communication, a means of transferring information, by smell.

Some pheromones act in some way on the recipient's central nervous system and produce an immediate effect on its behavior. Among these are the sex attractants of moths and the trail pheromones and alarm substances secreted by ants. Other pheromones act more slowly and trigger a chain of physiological events in the recipient which affect its growth and differentiation. These include the regulation of the growth of locusts and control of the numbers of reproductives and soldiers in termite colonies.

The sex attractants of moths provide one of the more spectacular examples of phero-

mones. Among the ones that have been isolated and identified are **bombykol**, a 16-carbon alcohol with two double bonds, secreted by female silkworms, and **gyplure**, 10-acetoxy-Δ^7-hexadecenol, secreted by female gypsy moths. The male has an extremely sensitive device in his antennas for sensing the attractant and responds by flying upwind to the source. He cannot determine the direction of the source by flying up a concentration gradient because the molecules are nearly uniformly dispersed except within a few meters of the source. With a gentle wind the attractant given off by a single female moth covers an area several thousand meters long and as much as 200 meters wide. An average silkworm contains some 0.01 mg. of sex attractant. It can be shown experimentally that when as little as 10,000 molecules of attractant are allowed to diffuse from a source 1 cm. from a male he responds appropriately. He can have received only a few hundred of these molecules, perhaps less. Thus the amount of attractant in one female could stimulate more than one billion males! The attractants, generally hydrocarbons, contain 10 to 17 carbons in the chain, which provides for the specificity of the several kinds of attractants.

The sex attractant of the American cockroach is not a long chain alcohol like bombykol or gyplure but has a central three-carbon ring to which methyl groups and a propanoxy group are attached. Sex attractants have been tested as possible specific insecticides. By putting sex attractant on stakes placed every ten meters in a large field, investigators could blanket the air with sex attractant, thus confusing the males and greatly decreasing the probability of their finding females and mating with them.

The fire ants, when returning to the nest after finding food, secrete a "trail pheromone" which marks the trail so that other ants can find their way to the food. The trail pheromone is volatile and evaporates within two minutes, so that there is little danger of ants being misled by old trails. Ants also release alarm substances when disturbed and this (rather like ringing the bell in a firehouse) in turn transmits the alarm to ants in the vicinity. These alarm substances have a lower molecular weight than the sex attractants and are less specific, so that members of several different species respond to the same alarm substance.

Worker bees, on finding food, secrete **geraniol,** a 10-carbon, branched chain alcohol, to attract other worker bees to the food. This supplements the information conveyed by their wagging dance (p. 800). Queen bees secrete 9-ketodecanoic acid which, when ingested by worker bees, inhibits the development of their ovaries and their ability to make royal cells in which new queens might be reared. This substance also serves as a sex attractant to male bees during the queen's nuptial flight.

In colonial insects, such as ants, bees and termites, pheromones play an important role in regulating and coordinating the composition and activities of the population. A termite colony includes morphologically distinct queen, king, soldiers and nymphs or workers. All develop from fertilized eggs; however, queens, kings and soldiers each secrete inhibitory substances, pheromones, that act on the corpus allatum of the nymphs and prevent their developing into the more specialized types. If the queen dies there is no longer any "antiqueen" pheromone released and one or more of the nymphs develop into queens. The members of each colony will permit only one queen to survive and will eat up any excess ones. Similarly the loss of the king termite or a reduction in the number of soldiers permits other nymphs to develop into the specialized castes to replace them. Males of migratory locusts secrete a substance from the surfaces of their skin which accelerates the growth of young locusts.

There are examples of pheromones in mammals as well as in insects. When female mice are placed four or more per cage there is a greatly increased frequency of pseudopregnancy. If their olfactory bulbs are removed this effect disappears. When more females are placed together in a cage their estrous cycles become very erratic. However, if one male mouse is placed in the cage his odor can initiate and synchronize the estrous cycles of all the females (the "Whitten effect") and reduce the frequency of reproductive abnormalities. Even more curious is the finding (the "Bruce effect") that the odor of a strange male will block pregnancy in a newly impregnated female mouse. The nerve impulses from the nose pass to the hypothalamus and block the output of prolactin-releasing factor. The subsequent lack of prolactin leads to regression of the corpora lutea and the failure of the fertilized ova to implant.

The question of whether there are human pheromones remains unanswered but of interest in this respect is the observation of the French biologist J. LeMagnen that the odor of 15-hydroxypentadecanoic acid is perceived clearly only by sexually mature females and that it is perceived most sharply at about the time of ovulation! Males and young girls are relatively insensitive to this substance but male subjects become more sensitive to it after an injection of estrogen.

Analyses of the menstrual cycles of the students at an American women's college showed a statistically significant tendency for the increasing synchronization of menstrual cycles among room-mates and close friends. The study ruled out several possible explanations for the phenomenon but suggested some pheromonal effect between girls who were together much of the time. The study also demonstrated a significant shortening of the menstrual cycles of women who were with male companions three or more times per week; those who dated twice a week or less had longer and more irregular cycles. This appears to be analogous to the Whitten effect, but whether it involves a pheromone is unknown.

ANNOTATED REFERENCES

Barrington, E. J. W.: Introduction to General and Comparative Endocrinology. Oxford, Clarendon Press, 1963. A comparative treatment of endocrine principles.
Gorbman, A., and H. A. Bern: A Textbook of Comparative Endocrinology. New York, John Wiley & Sons, Inc., 1962. An excellent presentation of the broad, evolutionary aspects of endocrine systems in vertebrate and invertebrate animals.
Prosser, C. L.: Comparative Animal Physiology. 3rd ed. Philadelphia, W. B. Saunders

Company, 1973. Contains several chapters concerned with comparative endocrinology.

Turner, C., and Bagnara, J.: General Endocrinology. 5th ed. Philadelphia, W. B. Saunders Company, 1971. An excellent, up-to-date treatment of the basic biological aspects of endocrinology.

Williams, R. H. (Ed.): Textbook of Endocrinology. 4th ed. Philadelphia, W. B. Saunders Company, 1968. A standard reference text by many specialists dealing primarily with medical endocrinology.

Young, W. C. (Ed.): Sex and Internal Secretions. 3rd ed. (2 vols.) Baltimore, Williams & Wilkins Company, 1961. A collection of essays by 28 experts dealing with many phases of reproductive physiology.

PART THREE

THE
ANIMAL
GROUPS

18. The Classification of Animals

Over 1,500,000 different kinds of organisms have been described, about two-thirds of which are multicellular, heterotrophic, and motile and are termed animals, or **metazoans.** Some of these, such as sponges, corals, and barnacles, cannot move about; such attached animals, however, have clearly evolved from motile ancestors and still retain the ability to move parts of the body. The various members of the Animal Kingdom are all related through some degree of common ancestry, and have varying degrees of similarity in structure and function.

The great number and diversity of animals require some framework of order within which they can be viewed. That framework is a system of classification, and the branch of zoology concerned with classification is called **taxonomy** or **systematic zoology.**

18.1 THE SPECIES CONCEPT

The basic unit in the classification of animals and other organisms is the **species.** A species is a population of organisms which is distributed over certain geographical areas, its **range,** and which occurs in certain kinds of **habitats** within that geographical area. For example, the giraffe, *Giraffa camelopardalis,* inhabits the open bush and savannah country of East Africa

from the Sudan to South Africa, and up to Angola in the west. All of the giraffes constitute the species population.

The science of genetics provides the theoretical basis of the species concept. All members of the same species population have a common gene pool and are capable of breeding with one another. The genetic constitution determines the ways in which one species differs from another. There will be structural, biochemical, physiological and behavioral differences. Those differences which can be used for easy recognition are said to be **diagnostic.** For example, it is easier to recognize a giraffe by its long neck than by its carbohydrate metabolism. However, it is important to remember that species differ from one another in all the ways expressed by their genotypes and not just by a few conspicuous features.

Although the members of a species can breed with one another, they normally cannot breed with members of other species. Structural differences eliminate the possibility of mating between most species of animals. Closely related species with similar structure may be separated by different ranges, or if they occur within the same geographical area they may be isolated by different habitats, different breeding times, different behavior or by genetic incompatibility, which may result in lethal or sterile offspring. Species that do not interbreed in the wild because of such isolating mechanisms occasionally interbreed in captivity. The lion-tiger hybrids that occasionally appear in zoos are an example.

18.2 THE GROUPING OF SPECIES

A species population may be thought to have not only a dimension in space (range) but also a dimension in time. The population extends backward in time, merging with other species populations much like the branches of a tree (Fig. 18.1). Thus, species have varying degrees of evolutionary relationships with one another depending upon the distance to the point of population mergence or common ancestry. The evolutionary relationship of organisms is known as **phylogeny** and this relationship provides a theoretical basis for the grouping, or classification, of organisms. Closely related species are grouped together within a **genus.** Related genera are grouped together within a **family;** related families within **orders;** orders within **classes;** and classes within **phyla** (Fig. 18.2). Theoretically at least, any category in this hierarchy of classification should constitute a **monophyletic**

grouping, i.e., all the species or subgroups within any group should have a common ancestry.

The evolutionary history of only a very few species is known directly from fossil records, and species relationships must therefore be determined indirectly. Related species should have similar gene pools and thus similar phenotypes. The degree of similarity between species, reflected by their similar gene pools, can be used as a measure of the closeness of their relationship. Although the degree of similarity can be studied at the biochemical level, morphological similarities are more commonly used to determine relationships because morphological characteristics are generally simpler to work with.

As a category, the species is relatively easy to recognize and delimit. If the individuals sampled from a population of organisms are fairly similar and their diagnostic characteristics do not intergrade with those of other related species, the population constitutes a species. Higher categories, however, are much more subjective and difficult to define. For example, in Figure 18.1, should species A–F be placed within a single genus or do they represent three genera, A–B, C–E, and F? (All of the groupings

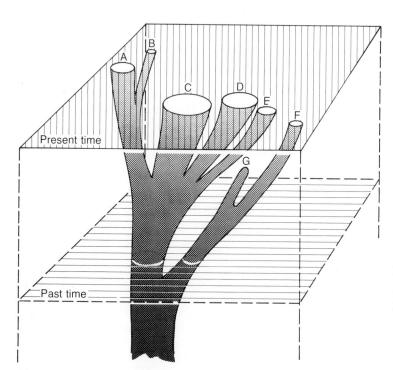

Figure 18.1 Diagrammatic representation of the evolutionary relationships of six (hypothetical) monophyletic species. Circular cross sections of the branches represent the species at the present time; junctions of branches represent points of common ancestry. Species G is extinct.

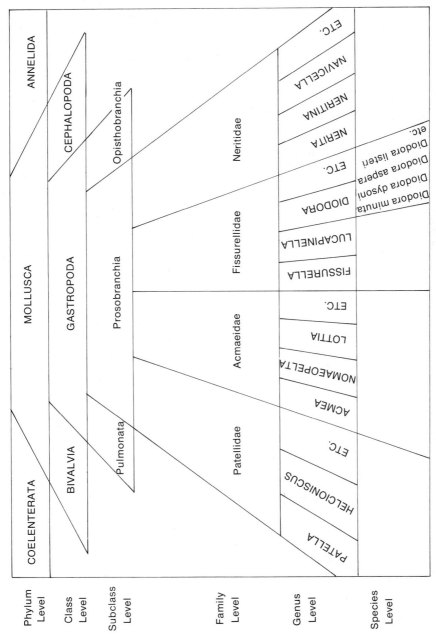

Figure 18.2   Part of the classification of mollusks to show the hierarchy of categories. (Adapted from Blackwelder.)

would be monophyletic.) The same type of problem exists in determining the limits of family, order, class, and phylum. In general, the greater the gap, i.e., the extent of dissimilarity, between species, the more likely a separate grouping is justified.

The approach to classification just described assumes that species are actual biological entities that can be determined and studied. This is known as a *phyletic* viewpoint and is the approach taken by many zoologists. Other taxonomists, however, hold a *phenetic* view of classification, meaning that they believe that species are only artificial abstractions devised by man. They claim that a classification based on phylogeny will never be attained and that artificiality is unavoidable and should be simply accepted. It is important to note that phyleticists and pheneticists differ only in their interpretation of what classification means, not in the method of classification. Both would agree that the best classification is one which is based upon all discernible characteristics.

18.3 TAXONOMIC NOMENCLATURE

The naming of species and higher taxonomic categories is another aspect of taxonomy. The scientific name of a species is composed of words usually derived from Latin or Greek, of which the first is the **generic** name and the second is the **trivial** name. Together the generic and trivial names constitute the species name, which is therefore a binomial, i.e., composed of two words. All species in the same genus have the same generic name. As an illustration, the scientific names of the categories to which the familiar box turtles belong are listed below. Note that the family name ends in *idae* and that the species name is always italicized.

Phylum Chordata
  Subphylum Vertebrata
    Class Reptilia
      Order Chelonia — Turtles and tortoises
        Family Emydidae — Fresh-water and marsh turtles
          Genus *Terrapene* — Box turtles

Terrapene carolina — Common box turtle
Terrapene ornata — Ornate box turtle

Scientific names are derived from many sources — descriptive adjectives, mythological figures, names of persons and geographical place names. However, there is no necessity to know the derivation of a name. The name has significance only as an identifying label or tag, similar to a person's name.

Law of Priority. All taxonomic names must be unique, i.e., the same name cannot be applied to more than one species or higher taxonomic category. When duplication occurs, the first species or category to which the name was applied takes priority in the claim to the name. The application of the **Law of Priority** for animals begins with the tenth edition of the *Systema Naturae*, a catalog of the plants and animals known to the Swedish biologist Carolus Linnaeus in 1758. The *Systema Naturae* represents the first uniform application of the present system of binomial nomenclature, although the system was not invented by Linnaeus. Detailed rules governing the construction of scientific names and the application of the Law of Priority have been formulated in an international code and are continually reviewed by an International Commission on Zoological Nomenclature.

Role of the Taxonomist. No systematic zoologist could possibly know all of the million species of animals, so most workers are specialists in a particular group. Such a specialist can recognize a new species in his group when one is encountered. A description of the new species along with its name must be published before it can be officially numbered among the known species. A number of specimens of the new species are designated as **type specimens** and are deposited in the collections of a museum for future reference.

An important task of many taxonomists is the identification of animal species. Identification, mistakenly termed classification by many people, is the determination of the species to which specimens belong. Ecological and other types of research may require the collection of large numbers of animals. If the collector is unfamiliar with the groups involved, he may ask a specialist in that group for help with identification.

The largest collections of animals for taxonomic purposes are located in museums, which may have educational as well as research facilities. Research collections are never on public display, but they are available to zoologists who visit the museum or who wish to borrow specimens. Museum specimens are usually obtained from scientific collecting expeditions or they are given to the museum by zoologists following the completion of a research project. The collections are catalogued and maintained by a staff of museum taxonomists, called curators, who normally are also engaged in research. The size of a museum's collection often varies greatly from one group of animals to another and from the collections of other museums. The differences in holdings simply reflect the sources of material and interests of the curators in the past history of the museum.

Museums may be a part of a university, for example, the Museum of Comparative Zoology at Harvard and the Peabody Museum at Yale; they may be private institutions supported by endowments, like the American Museum of Natural History in New York and the Philadelphia Academy of Natural Sciences; or they may be supported by state or federal government, as is true of the National Museum of Natural History of the Smithsonian Institution.

18.4 ADAPTIVE DIVERSITY

The Animal Kingdom displays a tremendous adaptive diversity. Every conceivable habitat within the ancestral oceanic environment has been exploited, and there have been numerous invasions of organisms into fresh water and onto land. The entry of animals into new habitats posed new problems of existence, which were met with modification of old structures and new systems for new functions. Since different animal groups invaded the same habitats, the evolutionary history of animals is filled with convergent adaptations. The varied success of the invading groups is reflected in part by the number of individuals and the number of species.

Adaptation to new conditions rarely has resulted in the loss of all of a species' ancestral characteristics. Some are modified to the demands of the new environment; some remain unchanged and are said to be **primitive.** Thus every species possesses some primitive and some specialized features. A primitive species is one which has retained many ancestral characteristics. But the term primitive is useful only in a comparative sense. One can speak of a primitive animal, a primitive vertebrate, a primitive mammal or a primitive man. But a primitive man is not a primitive mammal, nor is a primitive mammal a primitive vertebrate. Animals may be primitive because they lack certain structures, but they do not lack the structure because they are primitive. Hydras do not lack an anus, gills, and blood vessels because they are primitive, but because the hydra's small size and mode of existence makes them unnecessary. The absence of such structures probably reflects a primitive condition, since it is unlikely that hydras evolved from animals which possessed them. However, structures may be lost as well as gained through evolution.

The evolutionary history of the Animal Kingdom can be depicted as having a tree-like form (a **phylogenetic tree**), the diverging lines dividing trunks and branches representing diverging evolutionary lines. Those groups of animals which diverged from the main line early in the history of the Animal Kingdom, i.e., from the lower part of the phylogenetic tree, are often described as being "lower animals." Others, which arose later at a higher level of the tree, are often said to be "higher animals." However, these terms are often carelessly used and seem to imply that all animals and animal structures fit into a hierarchy of simple to complex. Even the "lowest" animals, such as sponges, may display certain specialized and unique features in addition to their primitive characteristics.

18.5 HOW TO STUDY ANIMAL GROUPS

The study of animal groups, or any group of organisms for that matter, can be a tedious and painful task if it amounts to nothing more than committing to memory a list of characteristics of one group after another. A much more useful and interesting approach is one which emphasizes relationships and adaptations. This is the approach that will be utilized in the following chapters. Two questions should be continually asked as these chapters are studied: (1) How is the structure and physiology of the group correlated with its mode of existence or the environmental demands of the habitat in which it lives? (2) How does the structure and physiology of the group reflect the group's evolutionary origin and its relationship to other groups? These questions can provide a meaningful framework into which facts about animal groups can be fitted. Moreover, with a little knowledge of the structural ground plan of animal phyla and an understanding of the problems posed by different modes of existence and different environments, it is possible to make intelligent guesses about possible adaptations that might be encountered.

There are about thirty animal phyla, more than we can possibly discuss in detail in this book. Our discussion of animal groups will be limited to the major phyla. Minor phyla—those with relatively few species—will be mentioned only in the following synopsis of the Animal Kingdom. Detailed accounts of these groups can be found in the references cited at the end of the chapter.

Protozoa are often treated as animals, for many are able to move and are heterotrophic. But protozoa are unicellular organisms, excluded by definition from the Metazoa, or multicellular animals. They have had a separate evolutionary history and their complexity has developed in a different way from that of metazoans. Protozoa will be discussed in the next chapter.

18.6 SYNOPSIS OF THE PHYLA OF METAZOAN ANIMALS

The following diagnoses are limited to distinguishing characteristics. The approximate number of species described to date is indicated in parentheses.

Parazoa. Many-celled animals with poorly defined tissues and no organs.

PHYLUM PORIFERA (10,000). Sponges. Sessile; no anterior end; some primitively radially symmetrical, but most are irregular. Mouth and digestive cavity absent; body organized about a system of water canals and chambers. Marine, but a few found in fresh water.

Eumetazoa. Many-celled animals with organs, mouth, and digestive cavity.

Radiata. Tentaculate radiate animals with few organs. Digestive cavity, with mouth the principal opening to the exterior.

PHYLUM CNIDARIA, or COELENTERATA (9000). Hydras, hydroids, jellyfish, sea anemones, corals. Free-swimming, or sessile, with tentacles surrounding mouth. Specialized cells bearing stinging organoids called nematocysts. Solitary or colonial. Marine, with a few in fresh water.

PHYLUM CTENOPHORA (90). Comb jellies. Free-swimming; biradiate, with two tentacles and eight longitudinal rows of ciliary combs (membranelles). Marine.

Bilateria. Bilateral animals.

Protostomes. Cleavage spiral and determinate; mouth arising from or near blastopore.

Acoelomates. Area between body wall and internal organs filled with mesenchyme.

PHYLUM PLATYHELMINTHES (12,700). Flatworms. Body dorsoventrally flattened; digestive cavity (when present) with a single opening, the mouth. The turbellarians are free-living, a few terrestrial. Trematodes and cestodes are parasitic.

PHYLUM MESOZOA (50). An enigmatic group of minute parasites of marine invertebrates. No organs, and body composed of few cells.

PHYLUM RHYNCHOCOELA, or NEMERTINA (650). Nemerteans. Long, dorsoventrally flattened body. Digestive cavity with mouth and anus. Marine, with a few terrestrial and in fresh water.

PHYLUM GNATHOSTOMULIDA (43). Gnathostomulids. Minute wormlike animals. Body covered by a single layer of epithelial cells, each of which bears a single cilium. Anterior end with bristle-like sensory cilia. Mouth cavity with a pair of cuticular jaws. Marine.

Pseudocoelomates. Animals in which the blastocoel persists, forming a body cavity. Digestive tract with mouth and anus. Body usually covered with a cuticle.

PHYLUM ROTIFERA (1500). Rotifers. Anterior end bearing a ciliated crown; posterior end tapering to a foot. Pharynx containing movable cuticular pieces. Microscopic. Largely fresh-water, some marine, some inhabitants of mosses.

PHYLUM GASTROTRICHA (175). Gastrotrichs. Slightly elongated body with flattened ciliated ventral surface. Few to many adhesive tubes present; cuticle commonly ornamented. Microscopic. Marine and fresh-water.

PHYLUM KINORHYNCHA (64). Kinorhynchs. Slightly elongated body. Cuticle segmented and bearing recurved spines. Spiny, retractile anterior end. Less than 1 mm. in length. Marine.

PHYLUM NEMATODA (10,000). Roundworms. Slender cylindrical worms with tapered anterior and posterior ends. Cuticle often ornamented. Free-living species usually only a few millimeters or less in length. Marine, fresh-water, terrestrial, and parasitic forms.

PHYLUM NEMATOMORPHA (230). Hairworms. Extremely long threadlike bodies. Adults free-living in damp soil, in fresh water and a few marine. Juveniles parasitic.

PHYLUM ACANTHOCEPHALA (500). Acanthocephalans. Small wormlike endoparasites of arthropods. Anterior retractile proboscis bearing recurved spines.

PHYLUM ENTOPROCTA (60). Entoprocts. Body attached by a stalk. Mouth and anus surrounded by a tentacular crown. Mostly marine.

Schizocoelous Coelomates. Body cavity a coelom; or, if a body cavity is absent, the coelom has been lost. Digestive tract with mouth and anus.

PHYLUM PRIAPULIDA (8). Priapulids. Cucumber-shaped marine animals, with a large anterior proboscis. Body surface covered with spines and tubercles. Peritoneum of coelom greatly reduced.

PHYLUM SIPUNCULIDA (250). Sipunculids. Cylindrical marine worms. Retractable anterior end, bearing lobes or tentacles around mouth.

PHYLUM MOLLUSCA (128,000). Snails, chitons, clams, squids, octopods. Ventral surface modified in the form of a muscular foot, having various shapes; dorsal and lateral surfaces of body modified as a shell-secreting mantle, although shell may be reduced or absent. Marine, fresh-water, and terrestrial.

PHYLUM ECHIURIDA (60). Echiurids. Cylindrical marine worms, with a flattened nonretractile proboscis. Trunk with a large pair of ventral setae.

PHYLUM ANNELIDA (8700). Segmented worms—polychaetes, earthworms, leeches. Body wormlike and metameric. A large longitudinal ventral nerve cord. Marine, fresh-water and terrestrial.

PHYLUM TARDIGRADA (350). Water bears. Microscopic segmented animals. Short cylindrical body bearing four pairs of stubby legs terminating in claws. Fresh-water and terrestrial in lichens and mosses; few marine.

PHYLUM ONYCHOPHORA (65). Onychophorans. Terrestrial, segmented, wormlike animals, with an anterior pair of antennae and many pairs of short conical legs terminating in claws. Body covered by a thin cuticle.

PHYLUM ARTHROPODA (923,000). Crabs, shrimp, mites, ticks, scorpions, spiders, insects. Body metameric with jointed appendages and encased within a chitinous exoskeleton. Vestigial coelom. Marine, fresh-water, terrestrial, or parasitic.

PHYLUM PENTASTOMIDA (70). Wormlike endoparasites of vertebrates. Anterior end of body with two pairs of leglike projections terminating in claws and a median snoutlike projection bearing the mouth.

Lophophorate Coelomates. Mouth surrounded or partially surrounded by a crown of hollow tentacles (a lophophore).

PHYLUM PHORONIDA (70). Phoronids. Marine, wormlike animals with the body housed within a chitinous tube.

PHYLUM BRYOZOA (4000). Bryozoans. Colonial, sessile; the body usually housed within a chitinous or chitinous-calcareous exoskeleton. Mostly marine, a few fresh-water.

PHYLUM BRACHIOPODA (260). Lamp shells. Body attached by a stalk and enclosed within two unequal dorsoventrally oriented calcareous shells. Marine.

Deuterostomes, or Enterocoelous Coelomates. Cleavage radial and indeterminate; mouth arising some distance (anteriorly) from blastopore. Mesoderm and coelom develop primitively by enterocoelic pouching of the primitive gut.

PHYLUM CHAETOGNATHA (50). Arrow worms. Marine planktonic animals with dart-shaped bodies bearing fins. Anterior end with grasping spines flanking a ventral preoral chamber.

PHYLUM ECHINODERMATA (5300). Starfish, sea urchins, sand dollars, sea cucumbers. Secondarily pentamerous radial symmetry. Most existing forms free-moving. Body wall containing calcareous ossicles usually bearing projecting spines. A part of the coelom modified into a system of water canals; external projections used in feeding and locomotion. Marine.

PHYLUM POGONOPHORA (80). Marine, deepwater animals, with a long body housed within a chitinous tube. Anterior end of body bearing from one to many long tentacles. Digestive tract absent.

PHYLUM HEMICHORDATA (80). Acorn worms. Body divided into proboscis, collar and trunk. Anterior part of trunk perforated with varying number of pairs of pharyngeal clefts. Marine.

PHYLUM CHORDATA (39,000). Pharyngeal clefts, notochord and dorsal hollow nerve cord present at some time in life history. Marine, fresh-water and terrestrial.

Subphylum Urochordata (1600). Sea squirts, or tunicates. Sessile, non-metameric invertebrate chordates enclosed within a cellulose tunic. Notochord and nerve cord present only in larva. Solitary and colonial. Marine.

Subphylum Cephalochordata (1). *Amphioxus.* Fishlike metameric invertebrate chordates.

Subphylum Vertebrata (37,-000). The vertebrates—fish, amphibians, reptiles, birds, mammals. Metameric. Trunk supported by a series of cartilaginous or bony skeletal pieces (vertebrae) surrounding or replacing notochord in the adult.

ANNOTATED REFERENCES

Blackwelder, R. E.: Taxonomy. New York, John Wiley and Sons, Inc., 1967. A general account of the theories and methods of zoological taxonomy.

Mayr, E.: Principles of Systematic Zoology. New York, McGraw-Hill Book Co., 1969. A general account of zoological taxonomy including the text of the International Code of Zoological Nomenclature.

The following invertebrate zoology texts and reference works contain general discussions of all of the invertebrate phyla, including the minor phyla omitted from this text. References which deal solely with particular groups will be listed at the end of each chapter of the following survey.

Barnes, R. D.: Invertebrate Zoology. 2nd ed. Philadelphia, W. B. Saunders Co., 1968.

Bayer, F. M., and H. B. Owre: The Free-Living Lower Invertebrates. New York, Macmillan, 1968.

Hickman, C. P.: Biology of the Invertebrates. St. Louis, The C. V. Mosby Co., 1967.

Hyman, L. H.: The Invertebrates. Six volumes. New York, McGraw-Hill Book Co., 1940–1967. Vol. I (1940): Protozoa through Ctenophora; Vol. II (1951): Platyhelminthes and Rhynchocoela; Vol. III (1951): Acanthocephala, Aschelminthes, and Entoprocta; Vol. IV (1955): Echinodermata; Vol. V (1959): Smaller Coelomate Groups; Vol. VI (1967): Mollusca, Part 1.

Kaestner, A.: Invertebrate Zoology. Three volumes. New York, Wiley—Interscience, 1967–69.

Meglitsch, P. A.: Invertebrate Zoology, 2nd ed. Oxford, Oxford University Press, 1972.

19. The Protozoans

Protozoans are unicellular, **eukaryotic** (nucleated) organisms. The animals which are the subject of the greater part of this book are **metazoans**–multicellular, motile **heterotrophs.** But motility and heterotrophic nutrition are also characteristic of many protozoans. For these reasons they are commonly considered to be animals and traditionally are the starting point for surveys of the Animal Kingdom.

Whether one considers protozoans to be "animals" is simply a matter of definition, and there are many biologists who would prefer to restrict the concept of animals to metazoans. However, there are other reasons which make it useful to examine protozoans before surveying the more typical animals which comprise the Metazoa. Metazoans undoubtedly evolved from some group of protozoans, most probably the flagellated protozoans, and the metazoan characteristics of motility and heterotrophic nutrition are probably a protozoan inheritance. Despite these similarities, however, protozoans and metazoans are *very* different from each other, especially in the way they have become complex.

The evolution of complexity in multicellular plants and animals has occurred through a division of labor among the cells, certain cells becoming specialized for certain functions. Complexity in unicellular organisms has evolved through the specialization of different parts of the cell, for although protozoans are single cells, they are also complete organisms. The cell must perform not just certain functions but must retain the ability to perform all of the functions demanded of an organism.

Specialized parts of the cytoplasm are called **organelles.** Organelles are found in the cells of all eukaryotic organisms, unicellular or multicellular, but are developed most elaborately in unicellular organisms. In many groups the cells are fantastically complex. *Paramecium,* for example, far exceeds in complexity any cell found within the human body.

19.1 CLASSIFICATION

The classification of protozoans poses a number of problems. The principal subgroups are distinctive and easily defined, but they share few characteristics with each other. They are similar chiefly in being **unicellular.** Traditionally the protozoans have been considered a phylum, but many biologists today believe that their unicellular condition alone is too general a feature with which to define a phylum. They advocate raising the various subphyla to the level of phyla so that what was a single phylum becomes a group of phyla. The name Protozoa refers to the group and can be used like Metazoa to designate a subkingdom of the Animal Kingdom, if one wishes to consider protozoans to be animals.

Many modern biologists favor classifications which divide living organisms into a number of kingdoms rather than the tradi-

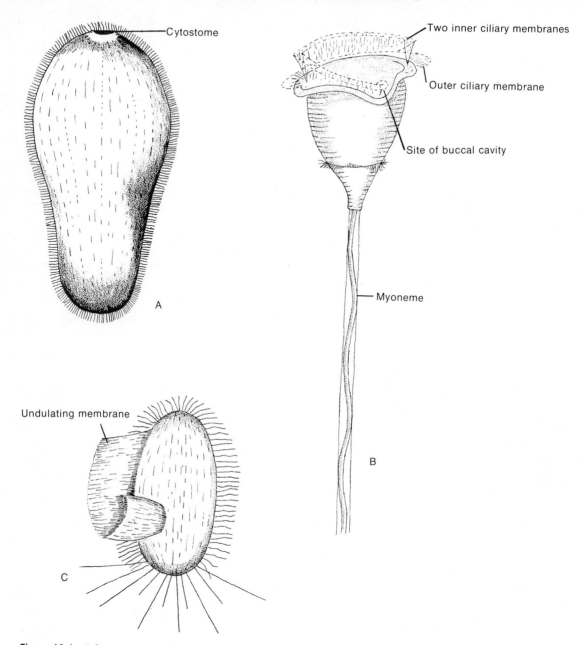

Figure 19.1   Ciliates. *A, Prorodon,* a primitive ciliate; *B, Vorticella; C, Pleuronema; D, Stentor; E, Tintinnopsis,* a marine ciliate with a test composed of foreign particles. (*A* and *E* after Faure-Fremiet from Corliss; *B* after Pierson from Kudo; *C* after Noland from Corliss; *D* after Tartar from Manwell.)   (*Continued on opposite page*)

tional two kingdoms embracing plants and animals. Of the many schemes proposed, the popular protistan concept raises the old protozoan classes to the rank of phyla and places them within the kingdom Protista. The members of this kingdom, which includes the protozoans, algae, and fungi, are characterized by unicellular or multi-cellular bodies composed of nucleated cells. Gametes are produced by single cells, not by multicellular gonads, and development never involves an embryo. The **Metazoa** (animals), **Embryophyta** (plants), and **Monera** (bacteria and blue-green algae) constitute separate kingdoms.

We will consider the Protozoa to be a

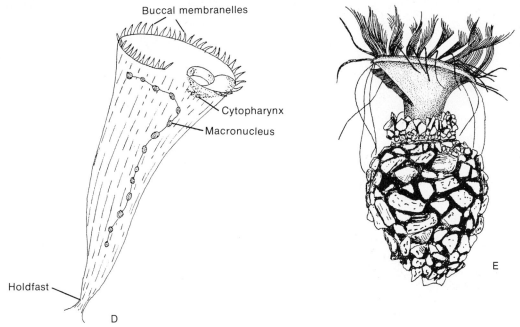

Buccal membranelles

Cytopharynx

Macronucleus

Holdfast

D

E

Figure 19.1 *(continued)*

group of phyla, and for the sake of simplicity and convenience we also will consider them to be members of the Animal Kingdom.

The flagellated protozoans are generally considered to be primitive and ancestral to the other protozoans, but let's break with tradition and start our examination of protozoan organisms with the ciliates. The ciliates are the most homogeneous, the most specialized, and the most animal-like protozoans and will serve especially well to contrast with the metazoans we will be studying.

19.2 PHYLUM CILIATA

Members of the phylum Ciliata are characterized by possessing at some time in their life history ciliary organelles for locomotion and feeding. Most ciliates are the size of the familiar *Paramecium*, although a few species are large enough to be barely visible to the naked eye. Ciliates are widespread in the sea and in fresh water, and a few species are commensals within the gut of vertebrates.

There is almost always a distinct anterior end and primitive species are radially symmetrical (Fig. 19.1). However, most ciliates are asymmetrical. The constant shape is maintained by a **pellicle,** a living outer layer of denser cytoplasm containing the peripheral and surface organelles. Studies with electron microscopy have revealed that this region is extremely complex (Fig. 19.2). The ciliate peripheral organelles, like many other cellular organelles, are largely derived from folding of the unit membrane.

The cilia, the most conspicuous surface organelles, arise from subsurface basal granules, or **kinetosomes** (Fig. 19.3). The kinetosomes are connected together in longitudinal rows by fibrils, and all the fibrils and kinetosomes of a row make up a **kinety.** The kineties constitute the subsurface, or infraciliature.

The layer of cilia covering the general body surface is called the **somatic** ciliature. Primitively longitudinal rows of somatic cilia cover the entire surface of the body, but in many species the somatic ciliature is reduced to girdles, tufts, and bristles, or is lacking altogether (Fig. 29.1). However, even those species with no somatic cilia possess an infraciliature persisting from cilia of earlier developmental stages.

The pellicle of many ciliates, including *Paramecium,* contains large vesicles (alveoli) around the cilia (Fig. 19.3). Although lying beneath the cell membrane, they contribute to the surface configuration charac-

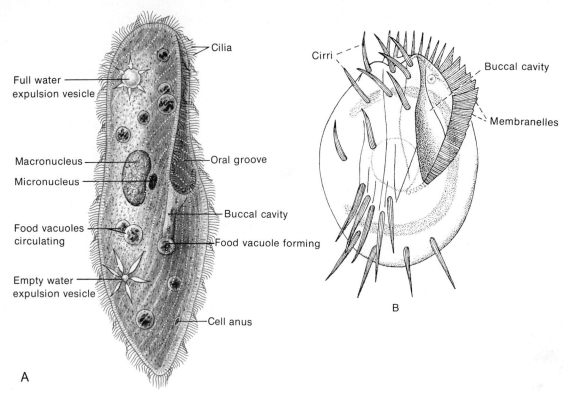

Figure 19.2 *A, Paramecium* (generalized drawing combining features of several species); *B, Euplotes.* (*B* after Pierson from Kudo.)

teristic of the species. Other common pellicular organelles of ciliates are bottle- or rod-shaped bodies, termed **trichocysts** (Fig. 19.3). In some species these can be discharged and transformed into fine threads. Their functional significance varies according to species. They may serve in anchoring the organism during feeding, in defense, or in prey capture. Other pellicular organelles found in many ciliates include vesicles which secrete mucoid substances, perhaps used in the formation of protective encasements (cysts) during adverse environmental conditions.

Locomotion. The majority of ciliates swim by ciliary propulsion. In its beat each cilium undulates in two planes, one at right angles to the other, and describes a helix (Fig. 19.4). The propulsive force is a helical wave which travels from the base to the tip of the cilium, driving the organism in the opposite direction. The cilia thus function like boat propellers. When the cilia are directed backward, the organism moves forward. To reverse, the cilia change position, driving water in the opposite direction, but the

nature of the beat of the cilium itself remains the same. Frequently ciliates are stationary with the cilia still beating. High speed cinematography has shown that at such times the cilia are oriented at right angles to the body surface and the undulations of the beat describe a spiral instead of a helix (Fig. 19.4).

In normal forward movement the entire body of the animal spirals, probably because the cilia are directed backward but somewhat obliquely to the long axis of the body. It is presumed that the subsurface kinety system is in some way responsible for the synchronization and direction of the beat.

Some of the most specialized ciliates, members of the order **Hypotricha,** have the somatic ciliature restricted to isolated tufts of fused cilia located in rows or groups on one side of the body. Fused cilia have more limited movements than do separate cilia and commonly the undulations are restricted to one plane, producing a simple lashing motion forward and backward.

Some ciliates are sessile. *Stentor,* a trum-

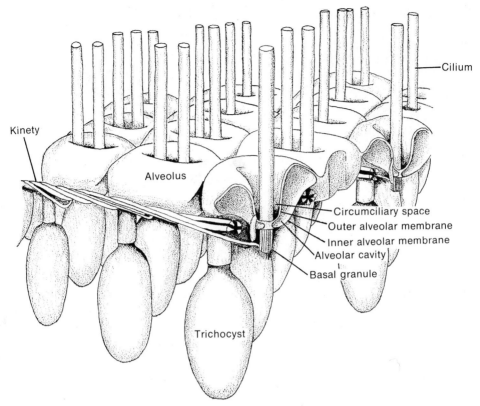

Figure 19.3 Pellicular system in *Paramecium*. (After Ehret and Powers from Corliss.)

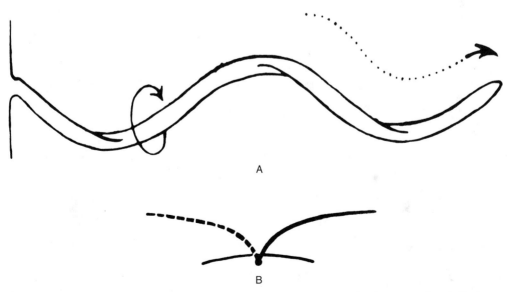

Figure 19.4 Ciliary motion. *A*, A single cilium of a swimming *Paramecium*, showing a traveling helical wave progressing from base to tip (dotted arrow). The motion gives the illusion that the cilium is rotating (as shown by solid arrow) but it actually is not. *B*, Lateral view of spiral beat of a cilium of a stationary *Paramecium*. (After Kuznicki.)

pet-shaped form, is attached by the tapered end and shortens by the contraction of pellicular fibrils similar to muscle myofibrils. *Stentor* can also release itself and swim about. *Vorticella* has a bell-shaped body connected to the substratum by a long stalk which contains a contractile fibril (Fig. 19.1). The ciliate retracts, coiling the stalk like a spring, and extends by a sudden release and popping movement. Some related forms are colonial and the individuals of the colony are connected together by a common stalk. Some other sessile ciliates are "tubicolous," living in shells composed of foreign material cemented together (Fig. 19.1).

Nutrition. Ciliates possess a mouth and it is this structure, along with motility, that gives them such an animal character. However, the ciliate mouth, or **cytostome,** is an organelle, an aperture within a single cell. It opens into a short canal, the **cytopharynx,** which leads into the interior, more fluid cytoplasm, or endoplasm. Primitively, the mouth is located at the anterior end (Fig. 19.1A), but in most ciliates it has been dis-

placed posteriorly to varying degrees (Fig. 19.2). Species in which the oral apparatus is located at the anterior end are usually carnivores, and this is probably the primitive mode of feeding in ciliates. They prey upon rotifers, gastrotrichs, other protozoans, including other ciliates, and the mouth can be opened to a great diameter to ingest prey (Fig. 19.5). The prey or contents of the prey pass into a food vacuole, which forms within the endoplasm at the end of the cytopharynx.

The higher ciliates are suspension feeders and possess a more complex buccal apparatus. Typically, the mouth lies at the bottom of a preoral chamber, the buccal cavity, which contains compound ciliary organelles (Figs. 19.2 and 19.5). These organelles, which constitute the buccal as opposed to the somatic ciliature, consist of two types: undulating membranes and membranelles. An undulating membrane is a long row of fused cilia, which forms a movable membrane (Fig. 19.1C). A membranelle is a short row of fused cilia, which forms a plate (Fig. 19.2B). Membranelles are typically arranged

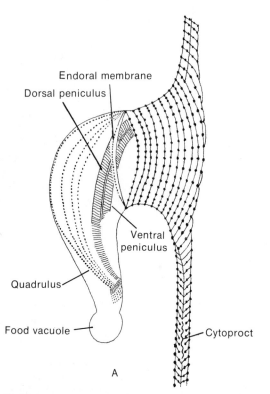

Endoral membrane

Dorsal peniculus

Ventral peniculus

Quadrulus

Food vacuole

Cytoproct

A

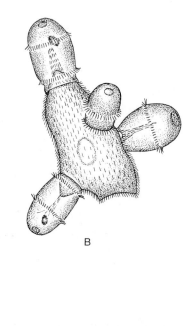

B

Figure 19.5 *A*, Buccal organelles of *Paramecium; B, Didinium,* a raptorial ciliate, attacking a *Paramecium.* (*A* after Yusa from Manwell; *B* after Mast from Dogiel.)

in fairly large numbers one behind another. The function of the buccal ciliature is to produce a feeding current and drive suspended food particles – detritus, bacteria, etc. – into the cytopharynx.

In the sessile *Vorticella* and the trumpet-shaped *Stentor* the ciliary organelles wind around the distal end of the animal and spiral down into a pit on one side (Fig. 19.1*B* and *D*). In the familiar *Paramecium* the buccal cavity, cytostome, and cytopharynx are located at the posterior end of a lateral oral groove. The somatic cilia within the oral groove produce a feeding current which sweeps from front to back, and the compound ciliary organelles of the funnel drive food particles down into the mouth (Figs. 19.2 and 19.5).

In these suspension feeders the food particles collect at the end of the cytopharynx within a food vacuole, which gradually increases in size like a soap bubble at the end of a pipe (Figs. 19.2 and 19.5). When the vacuole reaches a certain size it breaks free from the cytopharynx and circulates within the fluid endoplasm. Like other organelles the food vacuoles are composed of unit membranes.

Studies on digestion in ciliates have demonstrated that the contents of the vacuole first become increasingly acid. The acid phase is followed by an alkaline phase, which results from the secretion of enzymes within an alkaline medium into the vacuole. This phase coincides with an increase in the size of the vacuole. The digestive enzymes include all of the major classes – **amylases, lipases, nucleases** and **proteases.** Products of digestion are absorbed from the vacuole into the surrounding cytoplasm. Electron microscopy has revealed that during digestion and absorption there is extensive pinocytosis of the vacuole membrane into the exterior surrounding cytoplasm. The small food vacuole containing residual undigestible waste eventually connects with the cytoproct, a fixed cell anus in the pellicle (Fig. 19.2*A*). The contents of the vacuole are discharged and the vacuole itself disappears.

Although only a few ciliates are parasitic, many are commensals. Among the most interesting commensals are the highly specialized species living within the stomach of some hoofed mammals. These are capable of digesting cellulose to acetate.

Water Balance. Because of their microscopic size, ciliates have no special structures for gas exchange or excretion. They do have one or more organelles for water balance, called **contractile vacuoles** or **water expulsion vesicles,** which have fixed positions within the body. *Paramecium,* for example, has a water expulsion vesicle at each end (Fig. 19.2). In most ciliates the organelle is composed of a ring of radiating tubules which deliver droplets of water centrally (Fig. 19.2). These droplets coalesce into a single vesicle which gradually increases in size. On reaching a definitive size, the vacuole rapidly discharges through a canal in the pellicle. Then a new vacuole forms.

One might expect that only fresh-water ciliates would have water expulsion vesicles, but they are also found in marine species. They apparently serve in marine forms to rid the body of water taken in during feeding. That this is true is indicated by the fact that the water expulsion vesicles of marine ciliates pulsate at a slower rate than do those in fresh water and that in forms such as *Paramecium,* which have one anterior and one posterior vesicle, the posterior organelle, which is near the site of food intake, pulsates at a more rapid rate than does the anterior vesicle.

Reproduction. Ciliates are distinguished from other protozoans in possessing two types of nuclei. One type, the **macronucleus,** is large and governs the non-reproductive functions of the cell. It is highly polyploid and its shape varies greatly: in *Paramecium* it is bean-shaped (Fig. 19.2); in *Vorticella* it is U-shaped; and in *Stentor* it has the form of a string of beads (Fig. 19.1).

The **micronuclei,** which range in number from one to 80, are small, round, and typically located in the vicinity of the macronucleus. They are diploid and function in reproduction.

Asexual. Ciliates reproduce asexually by means of transverse fission, the fission plane cutting across the longitudinal rows of kinetosomes (Fig. 19.6). Mitotic spindle fibers form for each micronucleus, but the macronucleus usually divides by constriction. Regeneration of organelles lost in fission is a complex process and depends largely upon replication of existing structures. Of great importance in the reformation process are the kinetosomes. In many

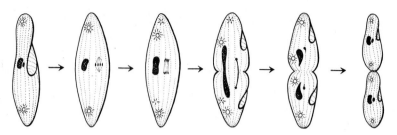

Figure 19.6 Transverse fission in *Paramecium*.

species these bodies replicate early in the division process and each half receives an equal number. In some highly specialized ciliates, such as the hypotrichs, all of the organelles are resorbed during division and new organelles are formed from a small number of persisting "germinal" kinetosomes.

Sexual. Sexual reproduction involves a process of **conjugation** and an exchange of nuclear material. Two individuals, called **conjugants,** meet, probably by random contact, adhere and their cytoplasm fuses in the region of adhesion (Fig. 19.7). The macronucleus is not involved in conjugation and is resorbed during the course of the process. All of the micronuclei undergo two meiotic divisions. Then all but one of these haploid nuclei disappear. The remaining nucleus divides mitotically to form two haploid nuclei: a stationary and a wandering nucleus (Fig. 19.7). The wandering nucleus of each conjugant migrates to the opposite conjugant and fuses with the stationary nucleus to form a synkaryon, or "**zygote,**" nucleus. The conjugants separate and the zygote nucleus undergoes a number of mitotic divisions to restore the number of micronuclei characteristic of the species. The macronucleus develops from one of these micronuclei.

Restoration of nuclear number is commonly associated with cytosomal divisions. The process is highly variable and is best illustrated with an example. The adult *Paramecium caudatum* has one macronucleus and one micronucleus. Following conjugation the "zygote" nucleus divides three times to produce eight nuclei (Fig. 19.7). Four become macronuclei and four micronuclei. However, three of the micronuclei degenerate. The remaining micronucleus undergoes two mitotic divisions accompanying two cytosomal divisions (Fig. 19.7). Each of the resulting four daughter individuals obtains one micronucleus from these divisions and inherits one of the four macronuclei formed prior to fission.

In some species of ciliates, including *Paramecium,* the individuals of the population belong to genetically determined mating types. In *P. bursaria,* for example, there are a number of mating types within each variety and conjugation can only occur between individuals of different mating types (Table 19.1). Apparently adhesion of the cytoplasm will not take place between individuals of the same mating type.

Note that conjugation itself does not result in any increase in the number of individuals but does provide for an exchange of genetic material within the population. All of the individuals that result from the asexual reproduction of an original parent without intervening conjugation are genetically identical and constitute a **clone.** There have been numerous studies to determine how frequently conjugation must occur. Some species can undergo several

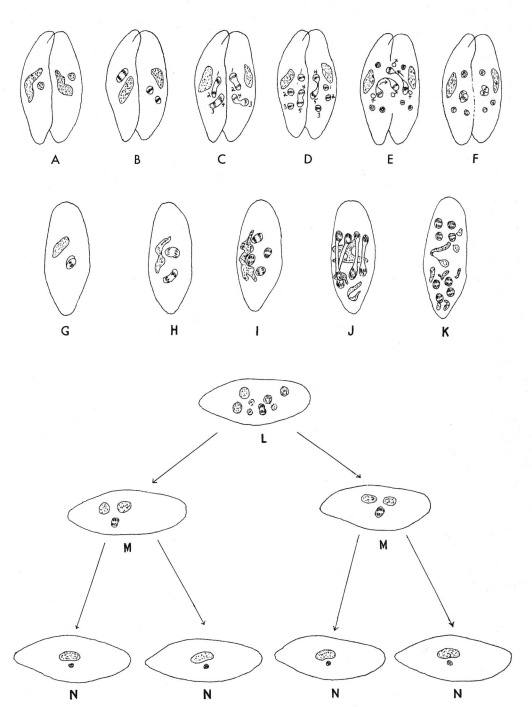

Figure 19.7 Sexual reproduction in *Paramecium caudatum*. A to F, Conjugation. B to D, Micronuclei undergo three divisions, the first two of which are meiotic. E, "Male" micronuclei are exchanged. F, They fuse with the stationary micronucleus of the opposite conjugant. G, Exconjugant with macronucleus and zygote micronucleus; other micronuclei have been resorbed. H to K, Three divisions of zygote, forming either micronuclei; old macronucleus is resorbed. L, Four micronuclei form macronuclei; three are resorbed. L and M, Remaining micronucleus divides twice in course of two cytoplasmic divisions. Resulting daughter cells each receive one of the four macronuclei in L. N, Normal nuclear condition restored. (Modified after Calkins from Wichterman.)

TABLE 19.1 BREEDING RELATIONS IN *PARAMECIUM BURSARIA*

Variety	I				II								III			
Mating Type	I	II	III	IV	I	II	III	IV	V	VI	VII	VIII	I	II	III	IV
I I	−	+	+	+	−	−	−	−	−	−	−	−	−	−	−	−
II	+	−	+	+	−	−	−	−	−	−	−	−	−	−	−	−
III	+	+	−	+	−	−	−	−	−	−	−	−	−	−	−	−
IV	+	+	+	−	−	−	−	−	−	−	−	−	−	−	−	−
II I	−	−	−	−	−	+	+	+	+	+	+	+	−	−	−	−
II	−	−	−	−	+	−	+	+	+	+	+	+	−	−	−	−
III	−	−	−	−	+	+	−	+	+	+	+	+	−	−	−	−
IV	−	−	−	−	+	+	+	−	+	+	+	+	−	−	−	−
V	−	−	−	−	+	+	+	+	−	+	+	+	−	−	−	−
VI	−	−	−	−	+	+	+	+	+	−	+	+	−	−	−	−
VII	−	−	−	−	+	+	+	+	+	+	−	+	−	−	−	−
VIII	−	−	−	−	+	+	+	+	+	+	+	−	−	−	−	−
III I	−	−	−	−	−	−	−	−	−	−	−	−	−	+	+	+
II	−	−	−	−	−	−	−	−	−	−	−	−	+	−	+	+
III	−	−	−	−	−	−	−	−	−	−	−	−	+	+	−	+
IV	−	−	−	−	−	−	−	−	−	−	−	−	+	+	+	−

A plus sign indicates the possibility of mating, and a minus sign indicates that mating does not occur. No mating type will mate with another individual of its own type, but it will mate with any other mating type of its variety. Three varieties are found in the United States, and no mating type of one variety will mate with any mating type of any other variety. (From Sonneborn, 1947.)

hundred asexual divisions and perhaps can continue this indefinitely. In other species there does seem to be an absolute requirement for conjugation after a given period of time or the clone will perish.

Most ciliates are capable of forming cysts under adverse environmental conditions. The body becomes encased within a protective secreted covering; there is some loss of water; and the metabolic rate is sharply reduced. Encystment is very important for the survival of ciliates when pools and ditches dry up, and for dispersion by wind and on the muddy feet of aquatic birds and mammals.

19.3   PHYLUM MASTIGOPHORA

The phylum Mastigophora, which literally means *whip-bearer,* includes those protozoan organisms which possess one to many flagella as locomotor organelles. However, other than possession of flagella, the members of the phylum are very heterogeneous, which probably indicates that the phylum is an assemblage of only distantly related groups.

The flagellates are generally believed to be the oldest of the eukaryotic organisms and the ancestors, directly or indirectly, of the other major groups of organisms — ciliates, algae, fungi, plants, metazoan animals, and others. The ciliates, for example, probably derive from some group of multinucleated mastigophorans that possessed many flagella.

A flagellum has an ultrastructure similar to that of a cilium but is longer (Fig. 19.8). It also functions essentially like a cilium. The flagella of most flagellates are located at the anterior end of the body and pull the body forward like an airplane propeller. As in ciliates, the entire organism tends to move in a spiral pathway (Fig. 19.8).

Phytoflagellates. The phylum is composed of two groups, the phytoflagellates and the zooflagellates. The phytoflagellates are plant-like marine and fresh-water forms. Most possess chlorophyll and exhibit autotrophic nutrition; the ten orders are classified with different groups of algae by algologists. The body is commonly asymmetrical

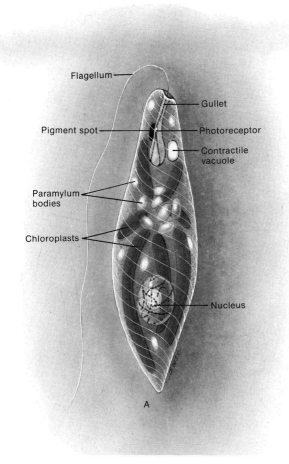

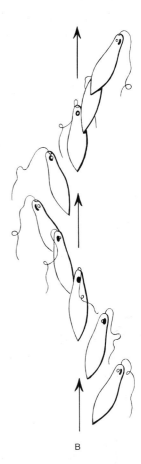

Figure 19.8 *Euglena.* *A,* A lateral view of *Euglena viridis.* *B,* A diagram showing successive positions in the spiral swimming pattern of *Euglena.* The position of the pigment spot shows the rotation that occurs.

and the structure varies greatly. They possess one or two flagella, which are generally carried at the anterior end. A non-living cell wall or envelope is often present. Descriptions of a few of the more common groups will illustrate the diversity of the phytoflagellates.

The mostly fresh-water **euglenids** have rather spindle-shaped bodies covered by a living pellicle (Fig. 19.8). The pellicle is flexible and many euglenids are capable of bending or ameboid movement. One or two flagella arise from a deep recess of the anterior end and in the same vicinity there is a light sensitive "eye" spot containing a red carotene pigment. One or two water expulsion vesicles are also present.

Euglenids are typically green but there are some colorless species, such as *Peranema*. This common little flagellate swims against the bottom, and its large, conspicu-

ous, forward-projecting flagellum undergoes lateral undulations in propelling the organism forward. Another smaller flagellum trails. *Peranema* is predaceous and feeds on other protozoans, including *Euglena.* The prey is ingested through an anterior cytostome, which can be greatly distended in swallowing (Fig. 19.9).

The members of the Volvocida are small green phytoflagellates possessing two flagella and a cellulose wall. Although there are some species, such as *Chlamydomonas*, which are solitary (Fig. 19.10), the group as a whole tends to be colonial, forming plate-like or spherical aggregations. The most highly developed colonies are the large hollow spheres of *Volvox* (Fig. 19.10). The entire colony moves, rotating through the water, by the synchronized beating of the flagella.

The brownish or yellowish dinoflagel-

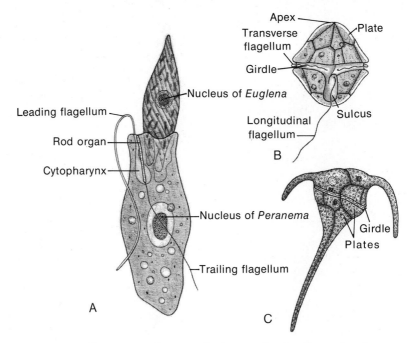

Apex
Plate
Transverse flagellum
Girdle
Nucleus of *Euglena*
Leading flagellum
Rod organ
Cytopharynx
Sulcus
Longitudinal flagellum
B
Nucleus of *Peranema*
Girdle
Plates
Trailing flagellum
A
C

Figure 19.9 *A*, A *Peranema* swallowing a *Euglena; B*, a fresh-water dinoflagellate, *Glenodinium cinctum; C, Ceratium*, a genus of common marine dinoflagellates. (*A* after Chen; *B* after Pennak; *C* after Jorgenson from Hyman.)

lates are common members of marine plankton, although there are also fresh-water species. The various species of dinoflagellates have a variety of shapes, but most are more or less oval or shaped like a top (Fig. 19.9*B*). Many are encased within a nonliving cellulose wall, which is often divided into plates and sculptured (Fig. 19.9*A* and *B*). One flagellum is located within a transverse groove which rings the body, and the

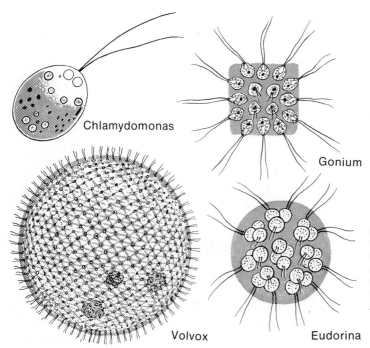

Chlamydomonas

Gonium

Volvox

Eudorina

Figure 19.10 Examples of the order Volvocida showing solitary form (*Chlamydomonas*), simple colonies (*Gonium, Eudorina*), and a complex colony (*Volvox*). Colonial forms are embedded in a matrix of transparent jelly.

other is located posteriorly in a longitudinal groove (Fig. 19.9*B*).

Many species of dinoflagellates, as well as some other phytoflagellates, lose their flagella and pass into a non-motile vegetative, or palmella, phase. The symbiotic algae of many corals, scyphomedusae, and other invertebrates are dinoflagellates in a palmella state.

The red tides that periodically plague the Gulf coast and other parts of the world are commonly caused by species of dinoflagellates. Optimum environmental conditions result in tremendous population growths, or blooms, of certain species. They occur in such great densities that their metabolic wastes poison fish and other marine animals.

Zooflagellates. The zooflagellates lack chlorophyll and are heterotrophic. Most species are either commensal or parasitic, and they constitute a much smaller part of the flagellate fauna than the phytoflagellates. They are undoubtedly a polyphyletic assemblage which evolved from different groups of phytoflagellates through loss of chlorophyll and autotrophic nutrition. The following examples will serve to illustrate their diversity.

The choanoflagellates are a strange little group of free-living fresh-water forms with bodies resembling the collar cell of sponges (Fig. 19.11). Most are colonial and attached, but one species has the collar cells embedded in a gelatinous mass (Fig. 19.11). Some zoologists have suggested that the sponges may have evolved from the choanoflagellates.

Trypanosoma is the most notable parasitic zooflagellate. They live in the blood stream of vertebrate hosts, particularly mammals, and are transmitted by an insect host, usually tsetse flies. In the blood stream the organism has an elongate, flattened body. A single flagellum arises posteriorly and extends anteriorly and laterally along the side of the body as an undulating membrane (Fig. 19.12). When ingested by the bloodsucking insect host, the trypanosome loses its flagellum, curls up, and becomes an intracellular parasite of the insect gut cells. Some get into the buccal and salivary glands and are passed back into the vertebrate host during another feeding.

Although most species of *Trypanosoma* are probably not pathogenic, a few African species cause serious diseases in domestic mammals and man. In man the flagellate invades the fluid of the cerebro-spinal canal,

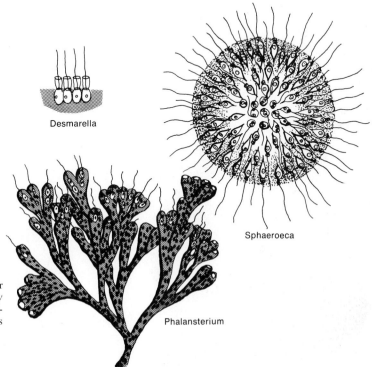

Desmarella

Sphaeroeca

Phalansterium

Figure 19.11 Examples of the order Choanoflagellida showing solitary and colonial forms. The matrix holding colonial individuals together is very thick.

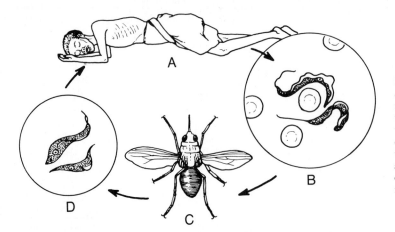

Figure 19.12 Life cycle of *Tryp-anosoma*. A, African sleeping sickness victim. Active trypanosomes in the blood (*B*) are sucked up by the tsetse fly (*C*). The protozoa reproduce in the digestive tract, migrate to the salivary glands where they attach to the walls and finally become infective (*D*), passing into a new host during salivary secretion.

causing the fatal disease known as sleeping sickness. The development of drugs which attack the flagellate, and of pesticides to control the tsetse fly, has reduced but not eradicated the disease.

The most complex flagellates, indeed among the most complex protozoans, are the gut symbionts of termites and wood roaches. The body is commonly sac-like and the anterior end bears a cap and a kinetosome complex from which spring the large number of flagella (Fig. 19.13). These flagellates engulf at the posterior end bits of wood ingested by the host. The flagellates have enzymes, β-glucosidases, to transform the cellulose to glucose within their food vacuoles. The product, glucose, is shared with the host, which is incapable of digesting cellulose. The insect obtains a new gut fauna following each molt by licking other individuals, by rectal feeding, or by eating fecal cysts.

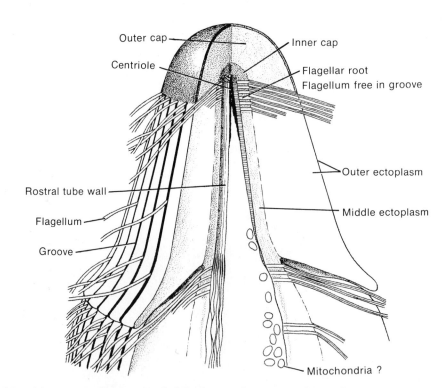

Figure 19.13 Ultrastructure of anterior end of *Trichonympha*, a gut symbiont of termites. The numerous flagella emerge from deep longitudinal grooves on the body surface. (After Pitelka and Schooley.)

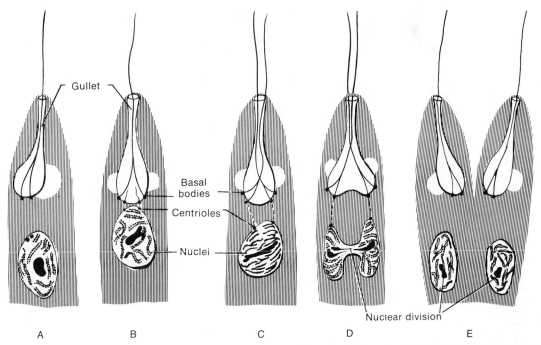

Gullet

Basal bodies

Centrioles

Nuclei

Nuclear division

A   B   C   D   E

Figure 19.14 Details of asexual reproduction in *Euglena*. *A*, The centriole has already divided. *B*, Each centriole produces a new basal body and flagellum. The nucleus is in prophase and the contractile vacuole is double. *C*, The old pair of flagellar roots separate and fuse with the new roots. *D*, Mitosis proceeds and the gullet begins to divide. *E*, Anterior end dividing following duplication of organelles. (Redrawn from Ratcliffe, 1927.)

Reproduction. Asexual reproduction is typically by binary fission, but in contrast to ciliates, flagellates divide longitudinally to produce two more or less equal halves (Fig. 19.14). The kinetosomes and other organelles duplicate or re-form before or after actual division.

Not much is known about sexual reproduction in most flagellates. The volvocids are an exception. This group beautifully illustrates the probable origin of **gametes**, an evolution that may have occurred more than once among flagellates. In *Chlamydomonas* the motile **isogametes** are identical and only a little smaller than the adults. In *Platydorina* the gametes are flagellate and motile, but one is much larger than the other. *Volvox* possesses typical **heterogametes.** The sperm are small, flagellate, and motile, and the egg is large and nonmotile. As in most algae and plants, meiosis in the volvocids is post-zygotic; i.e., it occurs *after* fertilization, not before, as in metazoans and ciliates.

19.4 PHYLUM SARCODINA

The phylum Sarcodina contains the amebas and other protozoans which possess flowing extensions of the body known as **pseudopodia.** Pseudopodia are used in feeding in all Sarcodina; in some, the pseudopodia have a locomotor function as well. Sarcodina possess fewer organelles than ciliates and flagellates and are therefore relatively simple in structure. However, skeletal structures have reached a degree of development that is equaled by few other protozoa.

Amebas

The Sarcodina include four major groups. The most familiar are the fresh-water and marine amebas, some of which are naked and some of which are enclosed within a shell (Fig. 19.16). Amebas have straplike or large, blunt pseudopodia, used in locomotion and in feeding (Fig. 19.15).

In the shelled species, the shell is secreted or composed of mineral particles cemented together. For example, the fresh-water *Arcella* possesses a brownish dome-shaped shell composed of a secreted keratinoid material (Fig. 19.16). *Difflugia*, another genus of fresh-water amebas, has a shell shaped like an egg and composed of mineral particles embedded within a secreted ce-

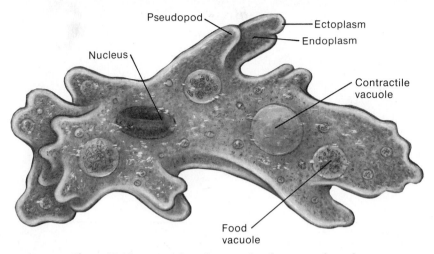

Figure 19.15 An ameba. The animal is flowing to the right.

menting material (Fig. 19.16). A large opening in the shell permits extension of the body and the pseudopodia.

Locomotion. The pseudopodia and other parts of the body are bounded by a thick layer of gelatinous ectoplasm. Ectoplasm and the more fluid interior endoplasm are different colloidal states of cytoplasm, and ameboid flow involves a rapid change from sol to gel at the pseudopodial tip. The addition of ectoplasm at the point of conversion builds up the wall of the pseudopodial tube into which the endoplasm flows. At the posterior end, the reverse is taking place.

No theory yet provides a completely satisfactory explanation of ameboid motion. Several theories postulate that it depends upon a contraction process. The "front contraction" theory holds that protein chains are extended in the endoplasm and that at the pseudopodial tip, where conversion from endoplasm to ectoplasm is occurring, the chains are folded or contracted. Thus, according to this view, the cell body is literally pulled or hauled along at the front end. The "rear contraction" theory contends that contraction occurs at the posterior end and that the body is pushed or squeezed forward.

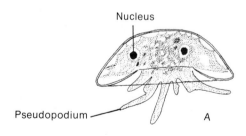

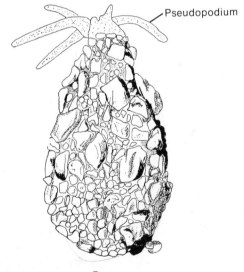

Figure 19.16 Shelled amebas. A, *Arcella vulgaris*, side view; B, *Difflugia oblonga* with test of mineral particles. (Both after Deflandre.)

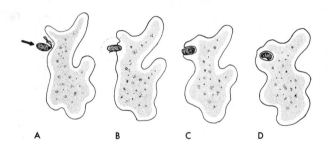

Figure 19.17 An ameba capturing a large flagellate. The flagellate hits the side of the ameba and slips into the crevasse at the base of a pseudopod. The ectoplasm of this region erupts and rapidly engulfs the prey. Stages shown are at about one second intervals.

A   B   C   D

The factors initiating flow are still uncertain. Several pseudopodia may form simultaneously, but one eventually dominates. Also, pseudopodial movement occurs in more than the single plane usually depicted.

Nutrition. Amebas feed on diatoms, algae, rotifers, and other protozoans. The food is surrounded by pseudopodia and eventually enclosed within a food vacuole along with some water (Fig. 19.17). Digestion proceeds as in ciliates, passing through an acid and then an alkaline phase. The vacuole containing undigestible residue ruptures at the posterior end.

A number of commensal and parasitic amebas inhabit the gut of different animals, including man. The commensal species, such as *Entamoeba coli* of man, feed on bacteria and intestinal debris. The parasitic species invade the intestinal tissue. *Entamoeba histolytica* parasitizes the large intestine of man and is the cause of amebic dysentery, or amebiasis (Fig. 19.18). Both commensal and parasitic species leave the host as cysts in the feces and reinfestation occurs through the mouth.

A water expulsion vesicle is present in fresh-water amebas but is absent from marine forms. This is true of other Sarcodina besides the amebas. Unlike that of ciliates, the vesicle does not form from radiating tubules nor does it have a fixed position within the body.

Foraminiferans

Foraminiferans are marine Sarcodina that secrete a chambered calcareous shell (Fig. 19.19). The shell is perforated by many small openings through which the pseudopodia extend. Foram pseudopodia are delicate anastomosing strands that form a food-trapping net over the shell surface. The pseudopodial cytoplasm is adhesive and any small organism swimming into the net is held and slowly surrounded by cyto-

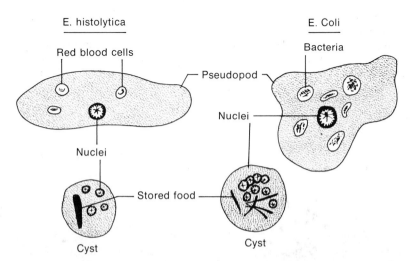

Figure 19.18 A parasite, *Entamoeba histolytica* (left), and a commensal, *E. coli* (right), of the human large intestine. Active amebas above, cysts below, that are passed in the feces and can infect new individuals.

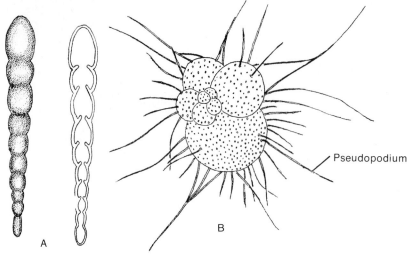

Figure 19.19 Foraminiferans. *A*, Shell of *Rheophax nodulosa* (entire, and in section); *B*, living *Globigerina*. (After Cushman from Calvez.)

plasm. Digestion occurs in vacuoles outside of the shell, and the products of digestion diffuse to the interior through the shell perforations.

Although there are some planktonic forams, such as *Globigerina* (Fig. 19.19), the majority are benthic and move slowly over the bottom by means of pseudopodia.

In all forams the shell is composed at first of a single chamber with an opening at one end. When the foram outgrows the first chamber, it secretes another one. This process continues, producing a multichambered shell that is occupied entirely by one individual. The shape and arrangement of the chambers vary according to species (Fig. 19.19).

Accumulations of foram shells are an important constituent of ocean bottom sediments and are sometimes termed "globigerina ooze" because of the large number of individuals of the genus *Globigerina*. Foram ooze is absent below 3700 meters because the great pressure causes the shells to dissolve. Forams have an extensive fossil record that begins in the Cambrian, and there are great limestone deposits composed largely of foram shells. The paleontology of these organisms is especially important in the search for oil deposits.

Heliozoans

Heliozoans are mostly fresh-water Sarcodina. Their spherical bodies are free or attached by a stalk (Fig. 19.20). The body is composed of a central cytoplasmic core, or medulla, which contains one to many nuclei, and an outer cortex of highly vacuolated cytoplasm (Fig. 19.20). Radiating from the cortex are many needle-like pseudopodia, called axopods, from which the name Heliozoa, sun animals, is derived. The axopods contain a central cytoplasmic rod which extends from the medulla. Many heliozoans possess a skeleton of siliceous scales, tubes, spheres, or needles embedded in the cortex (Fig. 19.20). Where the skeleton is composed of needles, they radiate out of the cortex like the axopods. Some species even have sand grains or living diatoms embedded within the cortex.

The axopods function solely as food-trapping organelles. On contact, small organisms adhere to the axopods, which then liquefy and withdraw. The prey is covered by cytoplasm and is gradually withdrawn into the cortex to be digested within a food vacuole.

Radiolarians

The radiolarians are another group of marine Sarcodina in which skeletal structures are highly developed. They are planktonic and reach a fairly large size, several millimeters in some species. The body is somewhat like that of heliozoans in being more or less spherical with a central nucleated core of cytoplasm and a broad outer

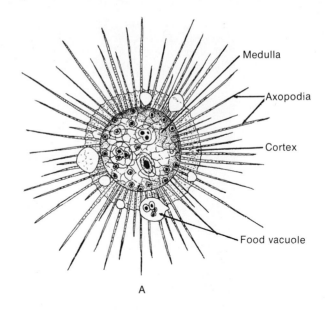

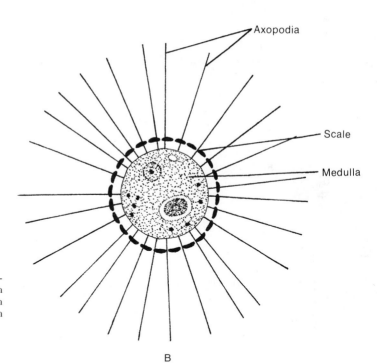

Figure 19.20 Heliozoans. *A, Actino-sphaerium; B, Pinaciophora*, with skeleton of scales. (*A* after Doflein from Tregouboff; *B* after Penard from Hall.)

vacuolated cortex (Fig. 19.21). The cortical cytoplasm of many species contains symbiotic dinoflagellates in a palmella state. The pseudopodia are axopods, or net-like as in forams. Radiolarians differ from heliozoans in that the central cytoplasm is encased within a chitin-like capsule, which is perforated to permit communication with the outer cortex (Fig. 19.21).

Some radiolarians have a skeleton of radiating strontium sulfate needles, but most have a skeleton of silicate arranged as concentric spherical lattices within and outside of the cortical cytoplasm (Fig. 19.21). The beautiful skeletons are very elaborate and not always symmetrical.

The pseudopodia project through the skeletal openings and function as food-

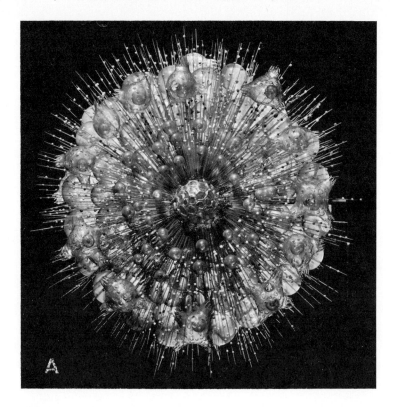

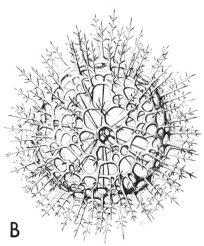

Figure 19.21 *A*, Glass model of a radiolarian, *Trypanosphaera regina*; *B*, the internal, siliceous skeleton of a radiolarian. (*A*, courtesy of the American Museum of Natural History; *B*, courtesy of E. Giltsch, Jena.)

trapping structures in the same way as in heliozoans and forams.

Plankton samples taken at different depths reveal a distinct vertical stratification of radiolarians, which are capable of some depth regulation through the retraction or extension of the radiating skeletal rods or the cortical cytoplasm. Some are found only at great depths (4600 m.); some species undergo seasonal movements, dropping from the surface to lower depths in the summer. This could be passive movement following a density gradient, since the density of sea water increases with decreasing temperature. At great depths where calcium carbonate shells of forams dissolve, radiolarian skeletons often accumulate to predominate the ooze.

The fossil record of radiolarians extends back into the pre-Cambrian and they have contributed to great sedimentary deposits.

Reproduction in Sarcodina. Asexual re-

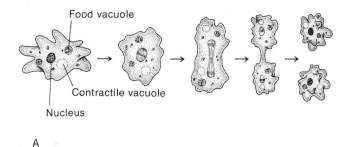

Food vacuole

Contractile vacuole

Nucleus

A

B

C

Figure 19.22 *A*, Fission in a naked ameba. *B* and *C*, two stages in the division of *Euglypha*, a shelled ameba: *B*, formation of skeletal plates on cytoplasmic mass protruding from aperture; *C*, division of nucleus. (*B* and *C* after Sevajakov from Dogiel.)

production is generally by binary fission (Fig. 19.22). In shelled amebas, heliozoans, and radiolarians the skeleton is divided, or else one daughter cell gets the skeleton and the other secretes a new one (Fig. 19.22). Sexual reproduction usually involves the fusion of isogametes, which in radiolarians and foraminiferans are flagellated. Forams have a complex life cycle in which a sexual generation alternates with an asexual one. Both stages spawn numerous individuals, each of which produces a shell. Old shells never divide.

Considerable evidence suggests that the Sarcodina evolved from flagellate ancestors. Radiolarians and forams have flagellated developmental stages, and some flagellates lose their flagella and undergo ameboid stages. An interesting little group of mastigophorans, called the Rhizomastigida, look like naked amebas but have a long flagellum (Fig. 19.23).

19.5 SPOROZOANS

Two phyla of parasitic organisms, the Sporozoa and the Cnidospora, were once placed together within the old class Sporozoa because of the presence of a spore-like infective stage in the life cycle of many species. The similarities, however, appear to represent convergence associated with

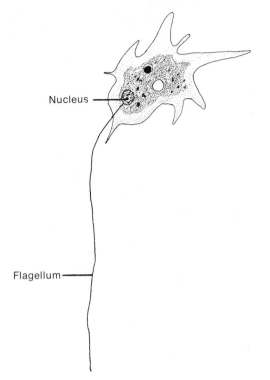

Nucleus

Flagellum

Figure 19.23   *Mastigameba.* (After Calkins, modified from Hyman.)

parasitism, and the members of the phylum Sporozoa are probably more closely related to the Sarcodina and the phylum Cnidospora to the flagellates than they are to each other.

Most sporozoan species are intracellular parasites, infecting both invertebrate and vertebrate hosts. The life cycles typically involve (1) a stage in which spores are formed, (2) a stage of asexual multiplication by fission, and (3) a sexual stage leading to gamete formation and a zygote. Some sporozoans require only one host to complete the life cycle; others require two.

The nature and life cycle of sporozoans can be illustrated by the **coccidians,** which include the parasites causing malaria in man and coccidiosis in domesticated animals.

Although in decline today, malaria was once widespread throughout the world and was one of the worst scourges of mankind. The untreated disease can be long-lasting and terribly debilitating. As late as 1957 it was estimated that 250 million persons were suffering from malaria, of which 2.5 million cases would be fatal. Malaria has been a continual and important factor in the shaping of human history.

Malaria is caused by species of the genus *Plasmodium.* There are more than 50 species of *Plasmodium,* all of which require a mosquito as one host and various vertebrates, mostly birds, for the other. Man is the vertebrate host for four species.

The introduction of the parasite into a human host is brought about by the bite of certain species of mosquitoes, which inject the spore stages (sporozoites) along with their salivary secretions into the capillaries of the skin. The parasite is carried by the blood stream to the liver where it invades a liver cell. Here further development results in asexual reproduction through multiple fission. These daughter cells (cryptozoites) invade other liver cells and continue to reproduce. After a week or so there is an invasion of red blood cells by parasites produced in the liver. Within the red cell the parasite increases in size and undergoes multiple fission. These individuals (merozoites), produced by fission within the red cell, escape and invade other red cells. The liberation and re-invasion does not occur continually but occurs simultaneously from all infected red blood cells. The timing of the event depends upon the period of time required to complete the developmental cycle within the host's cells. The release causes chills and fever, the typical symptoms of malaria. A specific periodicity is characteristic of different species. For example, in *Plasmodium vivax* 48 hours are required to complete the development of erythrocytic stages, i.e., the period from invasion of the red cell to release of daughter cells. The chills and fever produced by this species thus recur at approximately 48-hour intervals.

Eventually some of the parasites invading red cells do not undergo fission but become transformed into **gametocytes.** The gametocyte remains within the red blood cell. If such a cell is ingested by a mosquito, the gametocyte is liberated within the new host's gut. After some further development, a male gametocyte (microgametocyte) fuses with a female gametocyte (macrogametocyte) to form a zygote. The zygote enters the stomach wall and gives rise to a large number of spore stages (sporozoites). It is these stages that are introduced into the human host by mosquitoes.

Species of *Eimeria* cause coccidiosis in domesticated animals, especially chickens.

This parasite has a life cycle that involves stages similar to those of malaria, but all of which take place in the cells of the gut wall of a single host. The intestinal tissue may be so badly damaged that sloughing and hemorrhaging occur and may lead to death. Spore stages are passed in the feces and the next host is invaded through the mouth.

CLASSIFICATION OF PROTOZOA

Subkingdom Protozoa

Unicellular eukaryotic organisms. Most are heterotrophic and move by means of flagella, cilia, or pseudopodia.

PHYLUM MASTIGOPHORA, or FLAGELLATA. Unicellular organisms possessing one or more flagella.

    Subphylum Phytomastigina. Plant-like flagellates. Autotrophic. (These organisms might also be placed within the algal phyla.)

    Subphylum Zoomastigina. Animal-like flagellates. Heterotrophic.

PHYLUM SARCODINA. Unicellular organisms possessing pseudopodia.

    CLASS AMEBIDA. Naked amebas. Large pseudopodia; no shell.

    CLASS TESTACIDA. Shelled amebas. Strap-like pseudopodia; shell present.

    CLASS FORAMINIFERA. Forams. Strand-like interconnecting pseudopodia; calcareous perforated shell which is usually chambered.

    CLASS HELIOZOA. Heliozoans. Needle-like pseudopodia, or axopods. Sessile or free spherical bodies. Siliceous skeletons of separate pieces.

    CLASS RADIOLARIA. Radiolarians. Axopods or strand-like pseudopodia. Cell with central capsule surrounded by a vacuolated cortex. Skeleton usually siliceous; composed of strontium sulfate in some.

PHYLUM SPOROZOA. Intracellular parasitic organisms, with non-flagellated spore-like stages. *Plasmodium, Eimeria.*

PHYLUM CNIDOSPORA. Mostly intracellular parasitic organisms with flagellated spore-like stages.

PHYLUM CILIATA. Ciliates. Ciliated unicellular organisms possessing two types of nuclei.

    Subphylum Holotrichia. With simple and uniform somatic cilia. Buccal cilia absent or inconspicuous. *Didinium, Paramecium.*

    Subphylum Peritrichia. Adults usually lack body cilia, but the apical end of the body typically bears a conspicuous buccal ciliature. Mostly attached ciliates. *Vorticella.*

    Subphylum Spirotrichia. With reduced body cilia and well-developed conspicuous buccal ciliature. *Stentor, Euplotes* (hypotrichs).

    Subphylum Suctoria. Sessile, stalked tentaculate ciliates lacking cilia as adults.

ANNOTATED REFERENCES

Accounts of the protozoans can be found in the references cited at the end of Chapter 18. The parasitology texts listed at the end of Chapter 21 cover the parasitic protozoans. The references cited below are devoted exclusively to protozoans.

Corliss, J. O.: The Ciliated Protozoa: Characterization, Classification, and Guide to the Literature. New York, Pergamon Press, 1961.

Dodson, E. O.: The kingdoms of organisms. Systematic Zool. *20:*265–281, 1971. This paper and Whittaker's review present various schemes for the classification of living organisms.

Dogiel, V. A.: General Protozoology. 2nd ed. New York, Oxford University Press, 1965. A good introductory protozoology text. Emphasis on biology of protozoans rather than taxonomy.

Honisberg, B. M. (Chairman): A revised classification of the phylum Protozoa. J. Protozoology, *11:*7–20, 1964. A classification of the protozoans proposed by the Committee on Taxonomy and Taxonomic Problems of the Society of Protozoologists.

Manwell, R. D.: Introduction to Protozoology. New York, St. Martin's Press, 1961. An introductory protozoology text. Emphasis on general biology of protozoans rather than taxonomy.

Poindexter, J. S.: Microbiology: An Introduction to Protists. New York, Macmillan, 1971.

Tartar, V.: The Biology of Stentor. New York, Pergamon Press, 1961. A complete study of every aspect of this genus of ciliates.

Whittaker, R. H.: New concepts of the kingdoms of organisms. Science, *163*:150–159, 1971.

20. Sponges

Sponges, which compose the phylum **Porifera,** lack most of the typical animal features. There is no head or anterior end; there is no mouth or gut cavity; and the body is immobile. In fact, some zoologists believe that sponges had an evolutionary origin separate from other animals. Certainly they departed early from the main line of animal evolution, and their structure reflects a combination of primitive and specialized characteristics.

With the exception of one fresh-water family, the 5000 species of sponges are marine. They live attached to the bottom, most commonly to rock, shell, coral, pilings and other hard surfaces. Bottom-dwelling animals, both marine and fresh-water, are called **benthic,** and all organisms which are attached to the substratum are said to be **sessile.** Many of the peculiarities of sponges are correlated with their sessile mode of existence.

20.1 STRUCTURE OF SPONGES

The simplest sponges are shaped like little vases (Fig. 20.1). The interior cavity, the **spongocoel,** opens to the outside through a large opening at the top, the **osculum.** The body of the sponge surrounding the spongocoel is perforated by pores; the phylum name, Porifera, means pore-bearer. The outer surface of the sponge body is covered by flattened epidermal cells, **pinacocytes,** and

the pores are formed by **porocytes,** cells which are perforated like a ring. The spongocoel is lined with flagellated **collar cells,** so called because of a collar-like extension

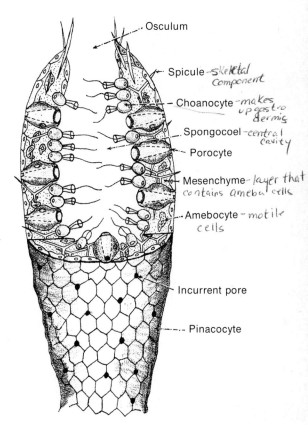

Osculum

Spicule – *skeletal component*

Choanocyte – *makes up gastro dermis*

Spongocoel – *central cavity*

Porocyte

Mesenchyme – *layer that contains ameba cells*

Amebocyte – *motile cells*

Incurrent pore

Pinacocyte

Figure 20.1   Structure of an asconoid sponge.

of the cell membrane around the base of the flagellum. Between the epidermis and the flagellated cells lies a layer of **mesenchyme,** containing ameboid cells of different types and skeletal pieces embedded within a gelatinous protein matrix. The skeleton of most sponges consists of **spicules** of calcium carbonate or silicon dioxide secreted by the amebocytes (Fig. 20.2). Each spicule may be a single needle or several needles fused at certain angles to each other. They are generally microscopic and unconnected, but in some sponges the spicules are fused together to form a complex skeleton. The shape of the spicule and the number of rays present are characteristic for each species of sponge, and spicule structure is an important characteristic in the classification of sponges. A small number of sponges possess an organic skeleton of spongin fibers (Fig. 20.3) instead of, or in addition to, spicules and it is from such species that commercial sponges are obtained. In preparing commercial sponges the living tissue is allowed to decompose and is then washed away from the spongin skeleton (Fig. 20.4). The natural sponge industry has greatly declined as a result of the development of synthetic substitutes.

Many sponges are brightly colored red, orange, yellow, or purple. The color is derived from pigment located within the amebocytes. Many of the fresh-water species are green from symbiotic algae living within the amebocytes.

20.2 PHYSIOLOGY OF SPONGES

The flagellum of the collar cells describes a spiral in its beat, the motion characteristic of most flagellar movement. As a result, water is driven from a collar cell in the same manner as air is blown from a fan (Fig. 20.5). The beating of all the flagella lining the spongocoel causes water to be sucked in through the pores and driven up and out of the spongocoel through the osculum. The rate of water flow can be regulated to some extent by changes in the diameter of the osculum. Cells with contractile fibrils are located around the osculum, and their contraction brings about a reduction in the size of the opening.

Many aspects of sponge physiology depend upon the water current passing through the body. Sponges are filter feeders, removing bacteria and other fine suspended organic matter from the water. An initial screening is provided by the pores, which permit only very small particles to enter. When such particles pass over the collar cells, they may become trapped on the collar surface. Electron microscopy has revealed that the collar is actually com-

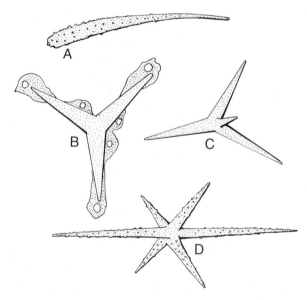

Figure 20.2 Sponge spicules. *A* and *C,* Three types of calcium carbonate spicules found in members of the class Calcispongiae. Similar types, made of silicate, occur in members of the class Demospongiae. *B,* A three-pronged calcareous spicule being secreted by amebocytes. *D,* A six-pronged spicule characteristic of the members of the class Hexactinellida. (*B* modified from Minchin.)

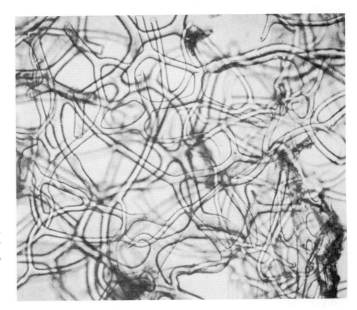

Figure 20.3 Spongin fibers (they appear translucent in photograph). (Courtesy of the General Biological Supply House, Inc.)

Figure 20.4 *Haliclona*, an irregular leuconoid sponge. Note the many oscula which open over the surface. (By Betty M. Barnes.)

posed of fine parallel fibrils between which water passes (Fig. 20.5). The trapped particle passes down to the base of the collar where it is engulfed by the cell. Digestion occurs within the collar cell (Fig. 20.5), or if the collar cell is small, the food particle is transferred to an amebocyte for intracellular digestion. In either case the products of digestion are passed by diffusion to the other cells of the body.

Filter feeding has evolved in many animals and often represents, as in sponges, an adaptation to a sessile or slow-moving existence. Filter feeding usually involves (1) a water stream from which suspended food particles can be removed (the animal may actively create the flow of water or it may utilize a natural current); (2) some part of the body functioning as a filter for the removal of particles; and (3) some mechanism for the transfer of trapped particles to the mouth (a food vacuole in sponges). In sponges, the collar cells lining the spongocoel provide for all of these filter feeding requirements.

The current of water passing through the sponge body not only provides a source of food material but also serves for gas exchange, for the removal of wastes, and for the transfer of gametes.

20.3 HYDRODYNAMICS AND BODY FORM

Sponges with the simple vase-like structure described earlier are called **asconoid** sponges and are small, not more than a few centimeters high. The asconoid structure imposes limitations in size, for as the volume of the spongocoel increases, the flagellated surface area does not increase proportionally (volume increases by the cube; surface area only by the square). Consequently, a large asconoid sponge would contain more water than its collar cells could efficiently move. In the evolution of sponges this problem was solved by repeated folding of the flagellated layer to increase its surface area. The first stage of folding is exhibited by **syconoid** sponges, in which the flagellated layer is evaginated outward into finger-like projections (Fig. 20.6). The evaginations are called **flagel-**

lated canals and the corresponding invaginations of the external surface are called **incurrent canals.** Pores are located between the incurrent and flagellated canals. The flagellated canals of syconoid sponges open into a central spongocoel devoid of collar cells.

In the great majority of sponges the surface area of the flagellated layer has been further increased by the formation of many small chambers within which the collar cells are located. In these so-called **leuconoid** sponges, water enters dermal pores on the body surface and passes through a system of incurrent canals, eventually reaching the flagellated chambers (Fig. 20.6). Most leuconoid sponges have no spongocoel and water leaves the body through converging excurrent canals opening to the exterior through an osculum. The development of flagellated chambers greatly increases the water-moving ability of a sponge. One leuconoid sponge 10 cm. × 3 cm. × 3 cm. was found to possess some 2,250,000 flagellated chambers and could pass 22.5 liters of water through its body in 24 hours.

Most leuconoid sponges are irregular in shape, with many oscula located over the body surface (Fig. 20.4). A far greater size is attained than in asconoid and in syconoid forms. The largest leuconoid sponge would more than fill a bushel basket. Most of the sponges commonly encountered in shallow water are leuconoid.

The glass sponges, members of the class **Hexactinellida,** have retained both radial symmetry and a spongocoel, although the body structure is more or less leuconoid. The glass spicules of these sponges are often highly organized and fused together, forming a beautiful lattice skeleton (Fig. 20.7). Most glass sponges occur in water deeper than 50 meters and some are adapted for living on soft bottoms by means of a basal tuft of spicules that anchors the animal into the substratum.

20.4 REGENERATION AND REPRODUCTION

The truly remarkable regenerative powers of sponges are well illustrated by the classical experiment of forcing a small bit of

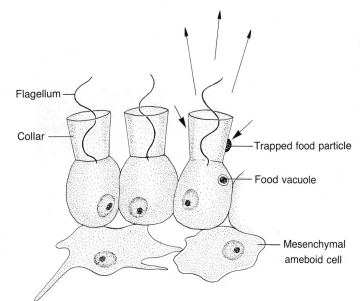

Figure 20.5 Section of flagellated layer and underlying mesenchyme, showing three choanocytes, or collar cells. Arrows indicate direction of water current.

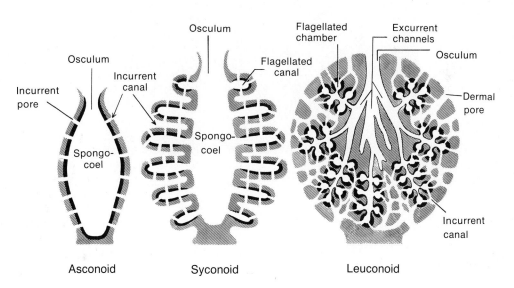

Figure 20.6 The three structural types of sponges. In each the choanocytes are shown in black.

Figure 20.7   Photograph of the skeleton of the glass sponge, *Euplectella*, Venus's-flower-basket. The spicules are fused to form intersecting girders. (Courtesy of the American Museum of Natural History.)

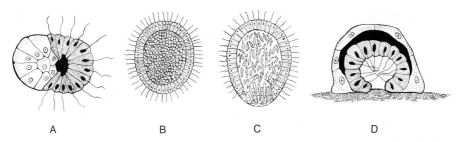

Figure 20.8 Sponge development. *A*, An amphiblastula larva. *B* and *C*, Parenchymella larvae. *D*, Gastrulation of an amphiblastula larva following settling. (*C* and *D* redrawn from Hyman.)

sponge through bolting silk and dissociating the cells. Within a short period of time, the dissociated cells reaggregate in the proper relationship. Ability to regenerate is closely correlated with asexual reproduction, which is also highly developed in sponges. A bud or small fragment of sponge breaks free of the parent and gives rise to a new sponge. Some sponges produce special asexual reproductive bodies consisting of an aggregate of essential cells, especially amebocytes. Certain amebocytes are **totipotent,** i.e., capable of giving rise to any other type of cell, and these play an important role in sponge regeneration and growth. Commercial sponges are sometimes propagated by attaching a piece of sponge to a bit of rock or concrete. A marketable sponge is obtained after several years' growth.

The majority of sponges are hermaphroditic. Hermaphroditism is another common adaptation to a stationary existence in animals. Any individual which settles near another can provide sperm or eggs. Cross fertilization can occur—this is typical of most hermaphroditic animals—but there is no requirement to be near an individual of the opposite sex.

Sperm and eggs originate from amebocytes and develop within the mesenchyme; they are not located within a gonad. Mature sperm are shed into the water canals and exit through the osculum. Some are carried into the water canals of neighboring sponges. These sperm are trapped and engulfed by collar cells, or amebocytes, which then move to an adjacent ripe egg. The carrier cell plasters itself against the egg and transfers the sperm. Fertilized eggs

may be released into the water canals, carried out with the water stream, and undergo development in the sea water; or the eggs may be brooded and develop within the parent mesenchyme.

In either case, embryonic development leads to a free-swimming larva, a stage which is important for species dispersion in sessile animals. In most species the larval condition is reached at the blastula stage. In some species, the larva (an **amphiblastula** larva) is hollow and one hemisphere is flagellated (Fig. 20.8). In other sponges, the larva (a **parenchymella** larva) is solid and covered with flagellated cells (Fig. 20.8). After a free-swimming existence among the plankton, the larva settles to the bottom, undergoes gastrulation (Fig. 20.8), and develops into an adult sponge. During development of the adult form following settling, some leuconoid sponges pass through asconoid and syconoid stages, reflecting perhaps the evolutionary sequence of sponge structure.

Although sponges are primitive in that they lack organs, including a gut, and have only a small number of different kinds of cells, they are highly specialized animals in other respects. The specializations are largely adaptations to a stationary mode of existence—the absence of a head or anterior end, the circulation of water through the body for filter feeding, gas exchange and water removal, and the condition of hermaphroditism. Certainly sponges evolved early in the evolution of the Animal Kingdom, and it seems highly unlikely that sponges gave rise to any other groups of animals.

CLASSIFICATION OF SPONGES

Phylum Porifera

Sponges. Sessile aquatic animals with no anterior end and lacking a mouth and digestive cavity. Body organized about a system of water canals and chambers.

CLASS CALCISPONGIAE. Calcareous sponges. Spicules are usually separate and composed of calcium carbonate. Body form is asconoid, syconoid, or leuconoid. Calcareous sponges are mostly drab in color and small, not exceeding 10 cm. in height.

CLASS HEXACTINELLIDA. Glass sponges. Spicules are siliceous, six-pointed and often fused together to form a highly organized skeleton. Although leuconoid, many species are cup-, vase-, or urn-like in shape, reaching a height of 10 to 30 cm. Most glass sponges occur in deeper water (450 to 900 m.) than do other sponges and are thus less frequently encountered.

CLASS DEMOSPONGIAE. This class contains the greatest number of sponge species and includes most of the commonly encountered North American sponges. The skeleton of Demospongiae is composed of separate siliceous spicules. But some species, the commercial sponges, possess a skeleton of spongin fibers and some possess both spongin fibers and siliceous spicules. The body structure is always leuconoid with many oscula, and an irregular symmetry prevails. Many Demospongiae are brightly colored. The small number of fresh-water sponges belong to this class.

Detailed accounts of sponges may be found in the references listed at the end of Chapter 18.

21. Coelenterates

The phylum **Cnidaria,** or **Coelenterata,** contains many familiar animals—hydras, hydroids, jellyfish, sea anemones, and corals. Almost all are marine; only the hydras and a few species of small jellyfish occur in fresh water.

21.1 COELENTERATE STRUCTURE

Coelenterates are radially symmetrical, similar parts being arranged around the central axis of the body. The oral end of the axis terminates in the mouth and a circle of tentacles (Fig. 21.1). The mouth opens into a blind **gastrovascular cavity.** A layer of cells, the **epidermis,** covers the outer surface of the body; another layer, the **gastrodermis,** lines the gastrovascular cavity (Fig. 21.1). The **mesoglea,** located between these two layers, varies from a thin noncellular membrane, as in hydras, to a thick jelly layer with or without cells, as in jellyfish. The epidermis and gastrodermis contain several kinds of cells, and each

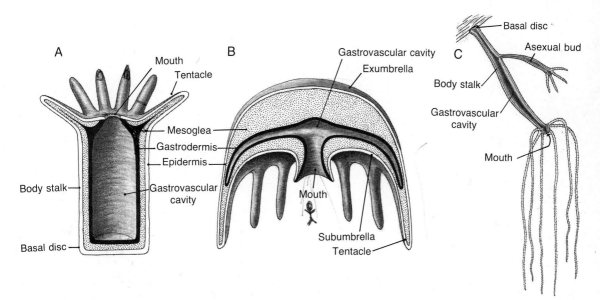

Figure 21.1 *A*, Polypoid body form; *B*, medusoid body form; *C*, an attached hydra with tentacles hanging in water. (*C* after Hyman.)

447

would be functionally equivalent to several different cell layers in higher animals. Coelenterates have no organs.

Two types of coelenterate body form can be distinguished. **Polypoid** coelenterates, such as hydras and sea anemones, have a cylindrical body with the oral end (bearing the mouth and tentacles) directed upward and the aboral end attached to the substratum (Figs. 21.1 and 21.20). **Medusoid** coelenterates, such as jellyfish (Figs. 21.1 and 21.15), have bell- or saucer-shaped bodies with the aboral end convex and directed upward, and the oral end concave and directed downward. Medusae are usually unattached and free-swimming.

21.2   MOVEMENT

The body and tentacles of coelenterates can be extended or contracted and bent to one side or the other. Movement is brought about by the contraction of longitudinal and circular muscle fibers. However, the fibers are located not in true muscle cells but in basal extensions of epidermal and gastrodermal cells (Fig. 21.2). These **epitheliomuscle cells** and **nutritive muscle cells** have characteristics of the epithelial and muscle tissues of most other animals. The contractile layers are variously developed in different coelenterates. The gastrodermal fibrils of hydras are poorly devel-

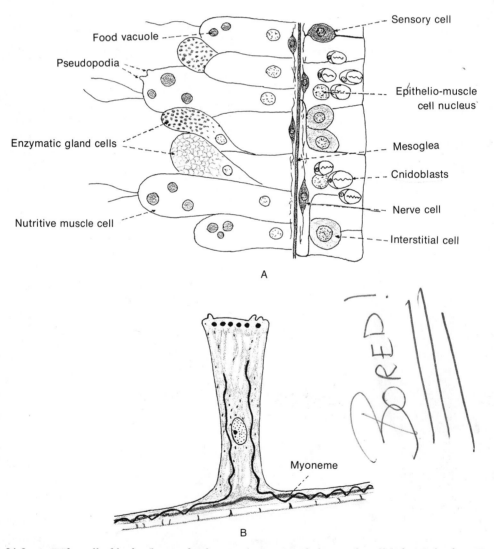

Figure 21.2   *A,* Body wall of hydra (longitudinal section); *B,* an epitheliomuscle cell (After Gelei from Hyman.)

21.3 NUTRITION AND NEMATOCYSTS

Coelenterates are carnivorous. Small forms such as hydras and corals feed upon **planktonic** animals, especially crustaceans, but the larger jellyfish and sea anemones can consume small fish. Gland cells lining the gut secrete proteolytic enzymes, which rapidly digest the tissues of the prey. Mixing of the gut contents is aided by the beating of the flagella of the gastrodermal cells. Small fragments of tissue are then engulfed by another type of gastrodermal cell, and digestion is completed intracellularly (Fig. 21.2). Undigestible waste materials are ejected through the mouth when the animal contracts.

Coelenterates capture their prey with the aid of special stinging cells, called **cnidocytes,** located largely in the epidermal layer (Fig. 21.2). A cnidocyte contains the **nematocyst,** the actual stinging element. The undischarged nematocyst is composed of a bulb and a long thread coiled within the bulb (Fig. 21.3). When discharged, the nematocyst is expelled from the cnidocyte, and the thread is everted out of the bulb in the process. The mechanism of discharge is not completely understood, but it is believed that stimulation by contact or chemical substances produced by the prey changes the permeability of the bulb wall and fluid rushes into the interior. The elevated fluid pressure both everts the thread and hurls the nematocyst from the cnidocyte. Some types of nematocysts function by entanglement, the thread wrapping around the prey (**volvent nematocysts**). Other nematocysts (**penetrant nematocysts**) are driven into the body of the prey and may inject a toxin (Fig. 21.3). It is these toxic penetrants that produce the sting of jellyfish and other coelenterates. Hydras produce nematocysts with sticky threads (**glutinant nematocysts**) which function in adhesion to the bottom when the animal is moving upon its tentacles. Nematocysts vary in detail in different groups of coelenterates and sometimes are of importance in identification.

Although nematocysts may be located throughout the epidermis, they are especially prevalent on the tentacles, where they may be concentrated in ring or wart-like batteries (Fig. 21.10). In hydras, contact with prey may result in the discharge of 25 per cent of the tentacular nematocysts. Following discharge the old cnidocytes are resorbed and new cnidocytes differentiate from **interstitial cells** (Fig. 21.2), totipotent cells similar to certain amebocytes of sponges.

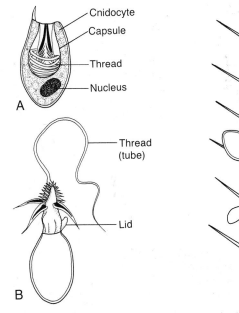

Figure 21.3 *A,* Undischarged penetrant nematocyst within cnidocyte of hydra; *B,* a discharged penetrant nematocyst; *C,* discharged penetrant and volvent nematocysts on a crustacean bristle. (After Hyman.)

After the prey has been quelled by the nematocysts, it is pulled by the tentacles toward the mouth, which often dilates enormously in swallowing. Mucus secreted by glands in the mouth region facilitates the process of swallowing.

There are no special systems for internal transport, gas exchange, or secretion. The size, surface area, and structure of coelenterates permit all of these processes to take place by diffusion.

21.4   NERVOUS SYSTEM

The coelenterate nervous system displays a number of primitive features. The neurons, located at the base of the epidermis and gastrodermis, are usually arranged as **nerve nets** rather than as nerve bundles (Fig. 21.4). The association of nerve cells to form conducting chains between receptors and effectors shows varying degrees of complexity. Sensory neurons may connect directly with motor neurons, or sensory and motor neurons may be separated by interneurons. There are even neurons with both sensory and motor processes. The neurons may possess from two to many processes; the axons and dendrites of inter-

neurons are not differentiated. Conduction occurs in a radiating manner with interneurons capable of transmitting impulses in both directions. Neuronal junctions are synaptic in all coelenterates, including hydras, where the neuronal processes were once thought to be in direct contact with one another.

21.5   GROWTH AND REPRODUCTION

Polypoid coelenterates exhibit a high level of regenerative ability. In fact, some polypoid species appear to undergo continuous cell replacement. In hydras, for example, old cells at the basal disc and ends of the tentacles are continually being replaced by new cells arising by mitosis within the body stalk. When a major part of the body, such as the oral end, is lost, the remaining part undergoes reorganization to form a new mouth region and tentacles. The original oral-aboral polarity, however, is retained, and tentacles and mouth will form only at the reorganized oral end.

Asexual reproduction is very common in coelenterates, especially in polypoid species. New individuals are usually formed

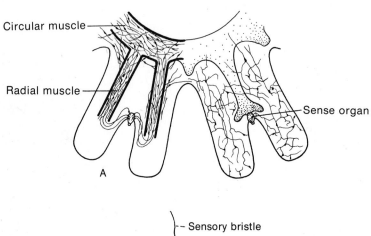

Circular muscle

Radial muscle

Sense organ

A

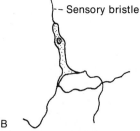

Sensory bristle

B

Figure 21.4   *A*, Section of the bell margin of the scyphozoan medusa *Aurelia*, showing the nerve net system; *B*, sensory nerve cell. (*A* after Horridge; *B* after the Hertwigs from Hyman.)

Figure 21.5 *Gonionemus:* planula larva that develops from the egg. (After Hyman.)

by **budding.** A bud arises as an outpocketing of the body wall and thus contains an extension of the gastrovascular cavity and all of the body wall layers (Fig. 21.1). The distal, or outer, end acquires the oral end polarity and forms a mouth and tentacles. The body separates from the parent, or in colonial species may remain attached as a new individual of the colony.

The sexes of most coelenterates are separate, but some are hermaphroditic. The gametes develop from interstitial cells and form aggregations in specific locations in the epidermis or gastrodermis. Since there is no surrounding wall of sterile cells, the coelenterate gonads are not comparable to the gonads of higher animals.

Fertilization is commonly external with development occurring in the plankton. At the completion of gastrulation, a characteristic larval stage, termed a **planula,** is attained. The planula is slightly elongate and radially symmetrical (Fig. 21.5), composed of a solid interior mass of cells surrounded by an outer layer of flagellated cells. The course of development from the planula stage varies with different groups of coelenterates.

21.6 CLASS HYDROZOA

The class Hydrozoa includes the hydras and many colonial species called **hydroids.**

While hydrozoan coelenterates are very abundant, they are small and not as conspicuous as the larger jellyfish, sea anemones, and corals. Most hydrozoans are marine, but the few species of fresh-water coelenterates—the hydras and a few jellyfish—are members of this class.

The members of the class Hydrozoa may exhibit a medusoid body form, a polypoid body form or both during the life history. The mesoglea is never cellular, nematocysts are limited to the epidermis, and the gametes develop within the epidermis.

The hydrozoans are characterized by colony formation, skeleton formation, polymorphism, and reduction of the medusoid stage. All degrees of each condition occur and apparently each evolved independently a number of times within the class. An understanding of these conditions will be helpful in gaining an appreciation of hydrozoan diversity.

Colony Formation. Some hydrozoans—the hydras and the small hydrozoan jellyfish—are solitary, but many species are colonial. The colonial forms are mostly polypoid and are called hydroids (Fig. 21.6). If one can imagine hydras budding but the bud remaining attached to the parent, some notion of a hydroid colony can be attained. The individuals, or polyps, of a hydroid colony are usually attached to a main stalk, which is in turn anchored to the substratum (algae, rock, shell, wharf piling) by a root-like stolon. The arrangement of polyps on the stalk and the branching of the stalk vary with the species (Fig. 21.6). In some hydroids, separate polyps spring directly from the stolon (Fig. 21.6). The tissue layers of the stalks and stolons are continuous with the tissue layers of the polyps. Thus all of the polyps are interconnected, and there is a common gastrovascular cavity for the entire colony.

Skeleton Formation. The solitary hydrozoans have no skeletons and some hydroid colonies in which the polyps arise directly from the substratum have only anchoring skeletons. Most hydroid colonies, however, are three to 10 cm. high and support is provided by an external chitinous skeleton secreted by the epidermis. In some species the skeletal envelope is limited to the stalk (Fig. 21.11); in others the skeleton extends around the body of the polyps but is open at the oral end for the emergence of the ten-

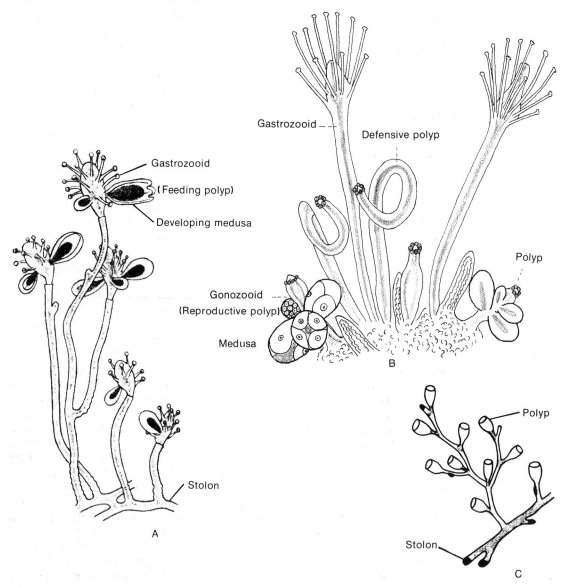

Figure 21.6   Hydroid colonies. *A, Coryne tubulosa,* a dimorphic form in which medusoids are formed directly on gastrozooids. *B, Hydractinia.* Polyps arise separately from mat of stolons and skeleton is limited to stolons. *C,* Skeleton of a branching type of hydroid colony. (*A* modified after Naumov; *B* after Hyman; *C* after Kuhn from Naumov.)

tacles and mouth (Fig. 21.7). In a few species the opening at the oral end is closed by a lid.

Polymorphism.   The existence of two or more structurally and functionally different kinds of individuals within the same species is termed polymorphism. This division of labor has evolved in many colonial or social animals, including the hydrozoans. The unspecialized, commonest type of individual in hydroid colonies is the feeding polyp (gastrozooid), resembling a hydra

(Figs. 21.6 and 21.7). These polyps capture food and carry out the initial extracellular phase of digestion. Intracellular digestion and absorption occur throughout the colony since the gastrovascular cavities are continuous.

Some hydroid colonies include defensive individuals, highly modified polyps with club-like bodies bearing great numbers of nematocysts (Fig. 21.6). Tentacles and mouth have disappeared.

Gametes are produced by reproductive

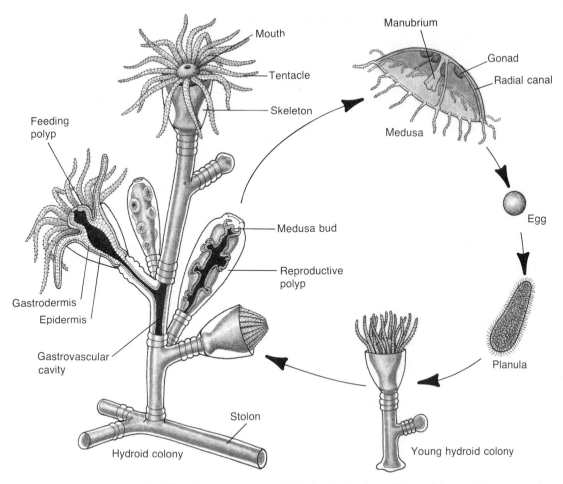

Figure 21.7 Life cycle of *Obelia*, showing structure of the hydroid colony. (Adapted from various sources.)

individuals. These are not polyps but medusae, or jellyfish. The medusae are produced from buds formed by polypoid individuals. In some hydroids, medusae are budded from the body of feeding polyps (Fig. 21.6). In many hydroids, however, there has been a further division of labor and medusae are budded from special reproductive polyps (gonozooids) no longer involved in feeding (Figs. 21.6 and 21.7).

Hydroid colonies with feeding polyps that produce medusoids are said to be dimorphic, i.e., having two kinds of individuals. Those with feeding polyps plus special polyps for the budding of medusoids are termed trimorphic. And those which in addition have defensive polyps are tetramorphic.

Most colonial hydrozoans are hydroids,

i.e., polypoid. However, one group of colonial species, the **Siphonophora**, includes large, pelagic colonies composed of both medusoid and polypoid individuals. There are individuals adapted as floats and pulsating swimming bells, from which hang feeding polypoid individuals with long tentacles. The siphonophoran **Portuguese man-of-war** has a single gas-filled float from which the tentacles of the polyps may hang down several meters (Fig. 21.8). The nematocysts can produce painful stings, and entanglement with the tentacles can be a dangerous encounter for a swimmer in deep water.

Medusa Reduction. Sexual individuals are medusae. Hydrozoan jellyfish (**hydromedusae**) are small, usually not more than a few centimeters in diameter (Figs. 21.9

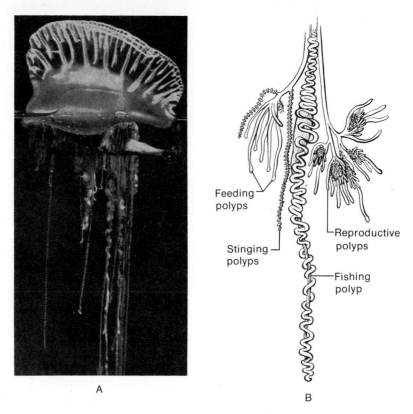

Feeding polyps

Stinging polyps

Reproductive polyps

Fishing polyp

A

B

Figure 21.8   *A, Physalia*, the Portuguese man-of-war; *B*, a cluster of polyps from *Physalia*, showing the various modifications of the individual polyps. (*A*, courtesy of the New York Zoological Society; *B* after Hyman.)

and 21.10). The large jellyfish often encountered at beaches belong to another class of coelenterates. A varying number of tentacles hang from the bell margin of the hydromedusa, and the **manubrium,** a fold of body wall surrounding the mouth, hangs from the center of the convex undersurface. The contraction of a ring of muscle fibers within the bell margin reduces the diameter of the bell and drives water from beneath the animal. These pulsations propel the animal mostly in an upward direction and maintain the animal at a specific depth.

Hydromedusae feed on other small planktonic organisms. The gastrovascular cavity is not a single sac as in polyps. The mouth opens into a central stomach from which extend four radial canals (Figs. 21.9 and 21.10). The radial canals connect in turn with a ring canal that extends around the bell margin.

The gametes of hydromedusae develop within the epidermis beneath the radial canals (Figs. 21.9 and 21.10). The eggs are fertilized when shed into the sea water and development to the planula larva occurs in the plankton. In some species, fertilization and early development occur within the epidermis.

Hydrozoa exhibit great variations in their life cycles. Some species have no polypoid individuals; the planulae from the free-swimming medusoid adults develop directly into hydromedusae (Fig. 21.13). In the course of that development, there is a stage called the **actinula** which looks like either a polyp or a medusa, depending upon whether the mouth is turned upward or downward (Fig. 21.13).

Some hydroids produce free-swimming medusae by budding from polypoid individuals (Fig. 21.7). Even more common in hydroids is the formation of medusae which remain attached to the parent polyp (Fig. 21.11).

Finally there are polypoid species such as hydra in which there is no medusa at all.

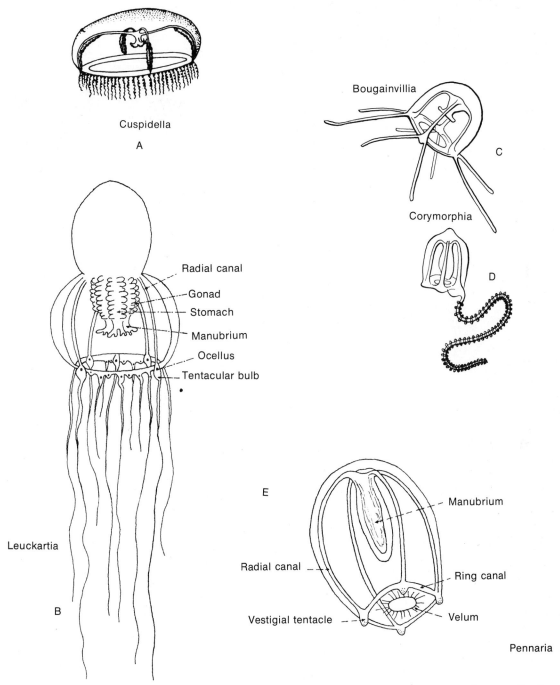

Cuspidella

A

Bougainvillia

C

Corymorphia

D

Radial canal

Gonad

Stomach

Manubrium

Ocellus

Tentacular bulb

Leuckartia

B

E

Manubrium

Radial canal

Ring canal

Vestigial tentacle

Velum

Pennaria

Figure 21.9 Medusae of five genera of hydrozoans. All are only a few centimeters or less in diameter. (*A* after Mayer, from Naumov; *B* after Hyman; *C* and *D* after Mayer; *E* after Mayer from Hyman.)

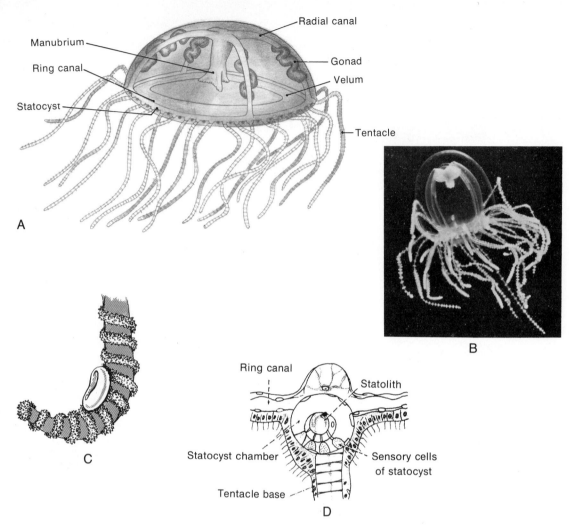

Figure 21.10 *A*, Lateral view of the hydromedusa *Gonionemus*. *B, Gonionemus vertens* actively swimming. *C*, A portion of a tentacle from *Gonionemus*, showing rings of nematocysts. *D*, Statocyst of hydromedusa. (*B*, courtesy of D. P. Wilson; *C* after Hyman; *D* after the Hertwigs from Hyman.)

Eggs and sperm are produced in the epidermis of the body column (Fig. 21.12).

We can account for the diversity of hydrozoan life cycles by postulating that the coelenterate polyp represents a persistent larval stage, and that in the evolution of the Hydrozoa there has been a tendency for the medusa to become reduced. Living hydrozoans reflect varying degrees of polypoid development and medusoid reduction.

According to this view, the medusa is the more primitive coelenterate body form. The hydrozoan life cycle initially lacked a polypoid stage, and the planula developed first into an actinula and then into an adult medusa (Fig. 21.13). Such a life cycle is exhibited by some living hydrozoans. In other hydrozoans, the actinula stage perhaps became attached to the bottom by the aboral end and produced more actinulae by budding. The latter developed directly into medusae (Fig. 21.13). This attached actinula became the first polyp and secondary budding would represent the formation of medusa buds. Such a life cycle has the advantage of permitting a long life for an actinula, which can give rise to not one but many medusae. This type of life cycle occurs in such hydrozoans as the marine *Gonionemus* and the little fresh-water jellyfish *Craspedacuspa* (Fig. 21.12).

With the evolution of the polyp, there

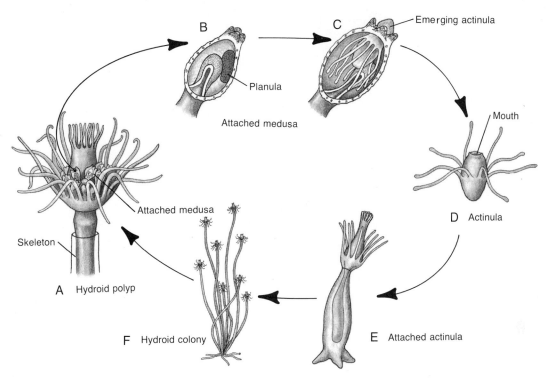

Figure 21.11 Life cycle of *Tubularia*. *A*, Skeleton is restricted to stalk of polyp; medusae are formed on gastrozooid and remain attached. *B*, Egg develops into planula within attached parent medusa. *C*, An actinula larva is released from medusa (*D*) and eventually settles to the bottom and develops into a new hydroid colony (*E* and *F*). (After Allman from Bayer and Owre.)

now occurred a progressive reduction of the medusa. The medusa developed by budding but was not free or detached. There was a gradual reduction of medusoid tissues until the polyp epidermis itself produced gametes.

Present evidence indicates that colonies evolved independently in different lines of hydrozoans and this occurred independent of the evolutionary trend toward medusa reduction. A solitary polyp would produce more polyps by budding during certain times of the year and then produce medusa buds at another season. If the first type of buds remained attached to each other, this could have led to the evolution of a hydroid colony. Colony formation did not evolve in polyps of all hydrozoans. The hydras, for example, probably evolved from some line of solitary polyps which underwent reduction of the medusoid form.

21.7 CLASS SCYPHOZOA

The medusae of the class Scyphozoa are the coelenterates to which the name jelly-

fish is usually applied; they are considerably larger and more conspicuous than hydrozoan medusae (Fig. 21.14). The medusoid body form is dominant in the Scyphozoa, the polypoid form being strictly a larval stage. In contrast to hydromedusae, the mesoglea of scyphomedusae may be **cellular**, nematocysts are present in the gastrodermis, and the gametes develop within the gastrodermis rather than within the epidermis.

Scyphomedusae are about the size of a saucer and the translucent body is often tinted with orange, pink, purple, and other colors. Tentacles of varying numbers and length hang from the margin of the bell. Four long divisions of the manubrium, called oral arms, hang from the center of the underside of the bell and are often more conspicuous than the tentacles (Figs. 21.14 and 21.15). The nematocysts, especially those in the tentacles and oral arms, can produce painful stings. The sea nettles, such as the Chesapeake *Crysaora*, can be a painful nuisance to swimmers at certain times of the year when they occur in large

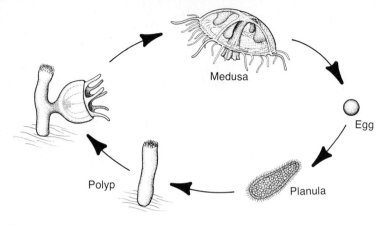

A Craspedacusta

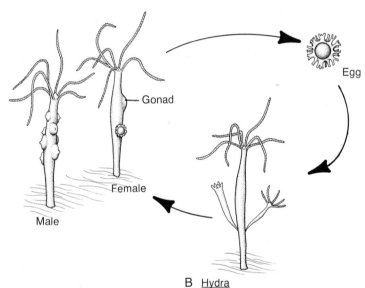

B Hydra

Figure 21.12   Life cycles of the fresh-water *Craspedacusta* (A) and *Hydra* (B). Both have solitary polypoid stages; the medusoid stage is absent in *Hydra.*

numbers (Fig. 21.14). The sea wasps of the Indopacific have such virulent nematocysts that their sting causes severe lesions and even death.

The epidermis of the scyphozoan bell margin contains two types of sensory structures. Clusters of photoreceptor cells form simple eyes, or **ocelli,** which detect general light intensity. Balancing organs, called **statocysts,** are composed of a vesicle with an interior body (**statolith**) in contact with hair-like endings of receptor cells located in the vesicle wall (Fig. 21.10). The gravitational responses of the statolith stimulate different receptor cells depending upon the orientation of the animal.

Ocelli and statocysts occur in hydromedusae, but in the scyphomedusae they are associated at specific sites between the scallop-like folds of the bell margin to form a **rhopalium** (Figs. 21.15 and 21.16). The scyphomedusae are larger than hydromedusae but their swimming movements are similar. They are primarily vertical and horizontal movement depends largely upon water currents. If the bell becomes tilted, stimuli from the statocysts bring about asymmetrical contractions and the bell is righted.

Scyphozoans feed on animals of various sizes, including fish, which come in contact with the tentacles and oral arms. Some species, such as *Aurelia,* feed on plankton

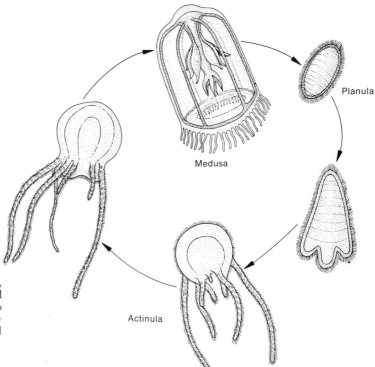

Medusa

Planula

Actinula

Figure 21.13 Life cycle of *Aglaura*, a hydrozoan which has no polypoid stage. Planula larva develops into an actinula, which develops directly into a medusa. (From Bayer and Owre.)

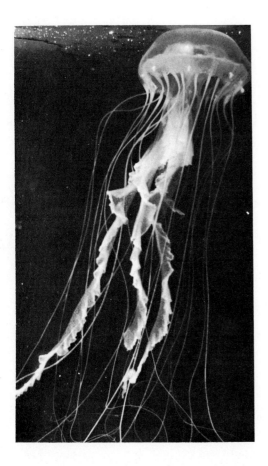

Figure 21.14 The sea nettle, *Crysaora quinquecirrha*, a common scyphozoan along the Atlantic coast. (Courtesy of William H. Amos.)

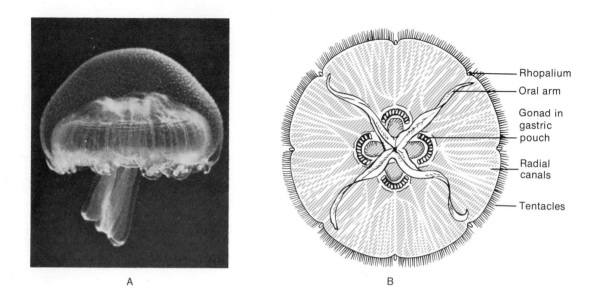

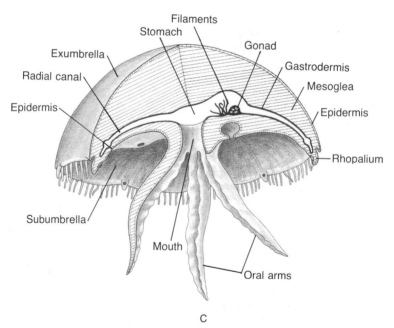

Figure 21.15 *Aurelia*, a scyphozoan medusa. *A*, Young specimen with bell contraction; *B*, ventral view; *C*, side view in section. (*A*, courtesy of D. P. Wilson.)

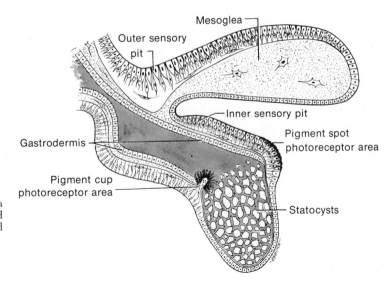

Figure 21.16 Section through a rhopalium showing the hood and the various sensory areas. (Modified from Hyman after Schewiakoff.)

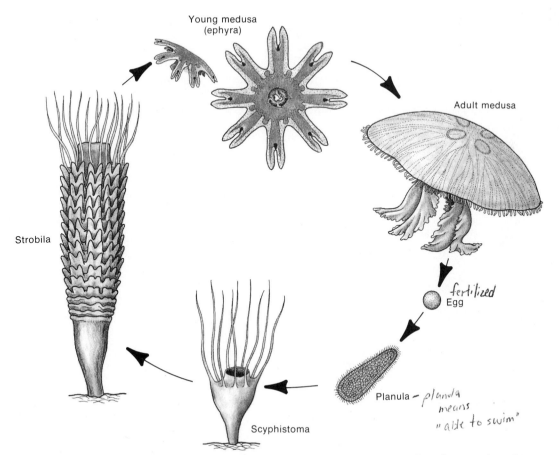

Figure 21.17 Life cycle of the scyphozoan *Aurelia*. The polypoid stage is a larva and produces medusae by transverse budding.

that adheres to the undersurface of the bell. The plankton is then swept by flagellated surface cells to the bell margin which is wiped by the oral arms. There is a central stomach as in the hydromedusae, but the stomach floor bears filaments containing nematocysts (Fig. 21.15). These gastrodermal nematocysts perhaps function to quell any prey which is still alive when it enters the stomach. In most of the commonly encountered north temperate jellyfish, numerous canals extend radially to the bell margin. A ring canal may or may not be present.

Gametes develop in the gastrodermis of the stomach floor, and the "gonads" are often conspicuous within the transparent body (Fig. 21.15). The eggs and sperm are shed through the mouth. Fertilization and early development may occur in the sea water, or the eggs may be brooded on the oral arms. In either case, a free-swimming planula larva is attained. The planula develops into a polypoid larval stage, called a **scyphistoma,** which is about the size of a hydra and lives attached to the bottom (Fig. 21.17). The scyphistoma has a life span of one to several years during which it feeds and produces more scyphistomae by budding. At certain seasons it ceases feeding and undergoes a special

form of budding to produce young jellyfish. The buds are produced at the oral end of the body and in the course of formation are stacked up like plates (**strobila**). When detached, the young jellyfish (**ephyra**) is tiny and displays only a rudimentary medusoid form (Fig. 21.17). There are some species in which the entire scyphistoma transforms into a little medusa.

In describing the evolution of hydrozoans, we theorized that the polyps evolved from a larval stage. Note that the scyphozoan life cycle displays just such a polypoid larva, which by budding produces both more polyps and medusae. Thus this theory of the origin of the polypoid stage is supported by evidence from comparative development of scyphozoans as well as hydrozoans.

21.8   CLASS ANTHOZOA

The 6000 species of anthozoans constitute the largest class of coelenterates and include the familiar sea anemones and various types of corals. The class is entirely polypoid, and the cellular mesoglea, gastrodermal nematocysts, and gastrodermal gametes suggest that the anthozoans may have evolved from the polypoid larva of some ancient group of scyphozoans.

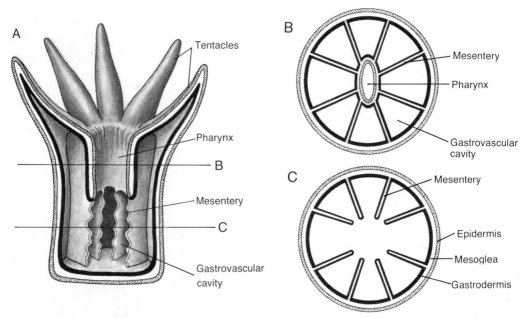

Figure 21.18   Structure of an anthozoan polyp. *A*, Longitudinal section; *B*, cross section taken at level of pharynx; *C*, cross section taken below level of pharynx.

The distinctive feature of anthozoans is the presence of a **pharynx** and **mesenteries.** The pharynx is a tube which hangs from the mouth into the gastrovascular cavity like a sleeve (Fig. 21.18). Since the pharynx is derived from an infolding of the body wall around the mouth, it possesses the same layers as the outer body wall but reversed; the lumen of the pharynx is lined with epidermis. The mesenteries are sheetlike partitions which extend into the gastrovascular cavity from the outer body wall toward the pharynx (Fig. 21.18). Each is composed of two layers of gastrodermis separated by a layer of mesoglea. At least eight of the mesenteries, called complete mesenteries, connect with the mesoglea and gastrodermis of the pharynx. In many anthozoans, there are additional incomplete mesenteries extending partway into the gastrovascular cavity (Fig. 21.20).

The functional significance of the mesenteries is difficult to understand. One might expect that they would provide greater surface area for digestion and absorption, but studies have shown that only the free margin is involved in digestion and absorption. The mesenteries may serve for internal gas exchange and may limit the diameter of the body, but these factors alone could not account for the evolution of mesenteries.

Sea Anemones. The largest of the anthozoans are the solitary sea anemones (Fig. 21.19). The majority live attached to hard substrates, although there are some forms which burrow in sand or mud. Most are one to several centimeters in diameter and are often brightly colored. A species on the Great Barrier Reef of Australia attains a diameter of one meter.

The body of sea anemones is heavier than that of hydrozoan polyps, and the oral end, bearing tentacles peripherally and a central slit-shaped mouth, is disc-like. When sea anemones contract, the upper rim of the body is pulled over the oral disc-like collar (Fig. 21.19). The muscle fibers are entirely gastrodermal. Circular fibers are located

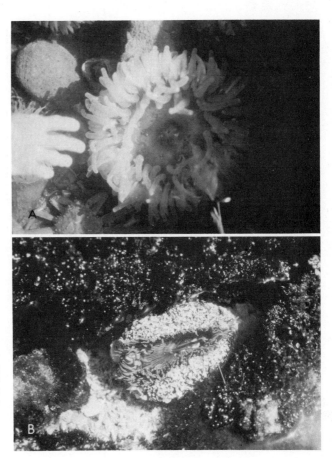

Figure 21.19 Sea anemones. *A,* View of expanded oral end. *B,* A contracted sea anemone in which the oral surface is partially covered by the upper surface of column. This species is typically clothed with shell fragments. (By Betty M. Barnes.)

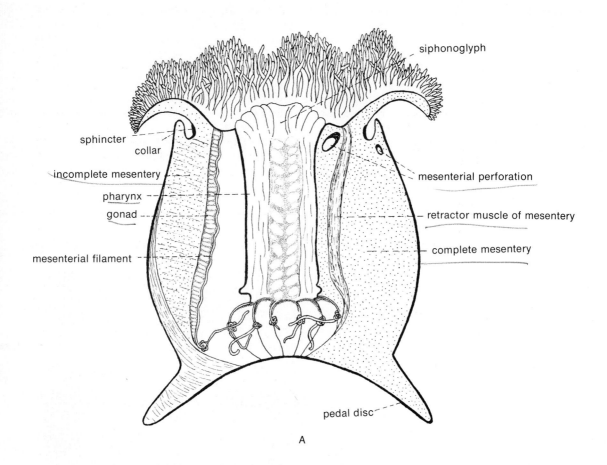

siphonoglyph

sphincter

collar

incomplete mesentery

pharynx

gonad

mesenterial filament

mesenterial perforation

retractor muscle of mesentery

complete mesentery

pedal disc

A

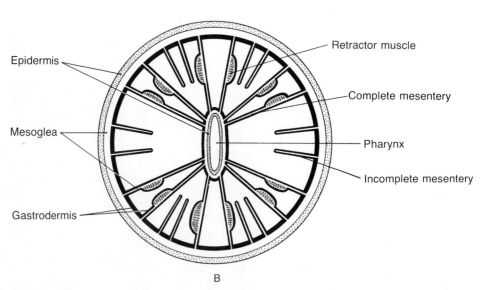

Epidermis

Mesoglea

Gastrodermis

Retractor muscle

Complete mesentery

Pharynx

Incomplete mesentery

B

Figure 21.20 Structure of a sea anemone. *A*, Longitudinal section; *B*, cross section at the level of the pharynx. (*A* after Hyman.)

in the gastrodermis of the body wall, and longitudinal retractor fibers are located in the mesenteries.

Sea anemones feed upon small fish and other invertebrates. The prey is passed down the pharynx and into the center of the gastrovascular cavity below the pharynx. If the prey is large, partial digestion may be necessary before fragments can pass into some section of the gastrovascular cavity between complete mesenteries. Sea anemones typically possess six or more pairs of incomplete mesenteries which provide additional digestive surfaces (Fig. 21.20). A perforation in the upper part of each complete mesentery permits circulation of the gastrovascular contents.

Some sea anemones reproduce asexually by splitting of the body or by the regeneration of fragments of tissues separated from the base of the body. The sexes are usually separate, and the gametes develop in gastrodermal bands just behind the free edge of the mesenteries (Fig. 21.20). Fertilization and early development may occur externally in the sea water or within the gastrovascular cavity. The planula larva develops into a ciliated planktonic polypoid larva, in which mesenteries and pharynx make their appearance. The polyp soon settles and becomes attached as a young sea anemone.

Madreporarian Corals. Most other anthozoans are various types of corals. The spe-

Figure 21.21 *A*, Polyps of the star coral; *B*, a coral polyp within its skeletal cup. (*A*, courtesy of the American Museum of Natural History; *B* after Hyman.)

A

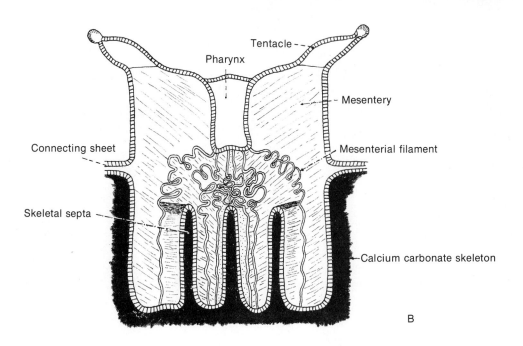

B

cies with which the name coral is most generally associated are the **madreporarian,** or stony corals. The polyps of madreporarian corals are similar to those of sea anemones but are usually smaller and almost always are connected together in colonies (Fig. 21.21). The connection is by means of a lateral fold of the body wall, and the entire colony has the form of a sheet. A calcium carbonate skeleton is secreted by the epidermis of the undersurface of the connecting sheet and the lower part of the polyp. The living colony is thus sitting on top of a skeleton which is actually external. The lower part of the polyp is situated within a skeletal cup, the bottom of which contains radiating septa that project up into folds in the base of the polyp (Figs. 21.21 and 21.22).

Species of corals display various growth forms. Some are low and encrusting; others are upright and branching (Fig. 21.22). The surface configuration of the skeleton depends upon the relative positions of the polyps. The coral appears pockmarked when the polyps are well separated. In brain coral, which has a surface configuration of troughs and ridges, the polyps are joined together in rows (Fig. 21.22). No skeletal material is deposited between the polyps within a row, but skeleton is deposited between the rows.

The gastrodermal cells of most madreporarian corals contain symbiotic algae. The algae utilize the nitrogenous and carbon dioxide wastes of the coral and facilitate the deposition of calcium carbonate ($CaCO_3$). The coral obtains glycerol from the algae. Phosphate is cycled between the two. When either light or algae are absent, $CaCO_3$ deposition is greatly reduced. Coral with symbiotic algae does not occur at depths below the penetration of light.

Octocorallian Corals. The remaining anthozoans are the tropical octocorallian corals, which include the sea fans, sea whips, sea pens, sea pansies, and others. There are only eight mesenteries, all complete, within the body of the polyp and only eight tentacles, which are **pinnate,** i.e.,

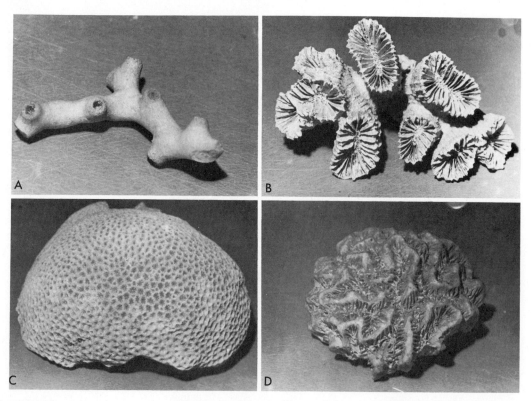

Figure 21.22 Skeletons of madreporarian, or stony, corals. The cups in which the polyps are located are widely separated in *A* and *B*. *D* is a brain coral, in which the polyps are arranged in rows. (By Betty M. Barnes.)

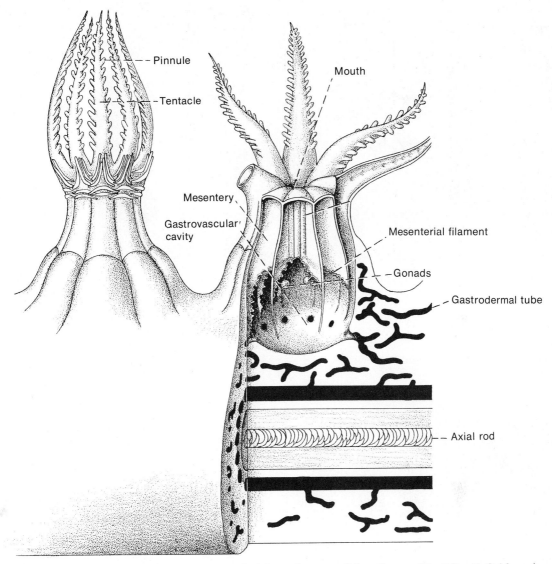

Figure 21.23 Structure of a gorgonian coral of the anthozoan subclass Octocorallia. (After R. C. Moore.)

they have little side branches (Fig. 21.23). Octocorallian polyps are usually very small but are organized into colonies which may attain considerable size. The interconnection of the polyps of a colony is quite different from that of madreporarian coral. A common mass of tissue, composed of mesoglea and gastrodermal tubes, unites the lower half of the colony. A skeleton of separate microscopic pieces and sometimes also an organic horny material may be present *within* the mesoglea (Fig. 21.23).

A skeleton supports the colony and is especially important in large upright forms. A central horny skeletal core extends through the long rod-like colonies of sea whip (Fig. 21.24). The sea fan is similar to the sea whip but its branches are all in one plane with lattice-like cross connections (Fig. 21.24).

Coral Reefs. The contemporary coral reefs, located in the Caribbean and in the tropical parts of the Indo-Pacific oceans, have been built chiefly by madreporarian corals. Over 200 species of living corals contribute to the Great Barrier Reef along the coast of Australia and 35 species to the Caribbean reefs. The coralline foundation creates an environment for a vast assem-

Figure 21.24 Octocorallian corals. A part of the lattice-like colony of a sea fan can be seen in the lower part of the figure. To either side of the sea fan are branching erect rod-like forms. The hundreds of tiny polyps which compose a colony are visible in the specimen on the left. (By Betty M. Barnes.)

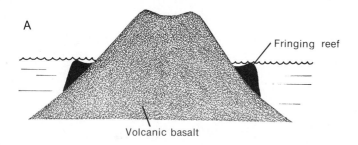

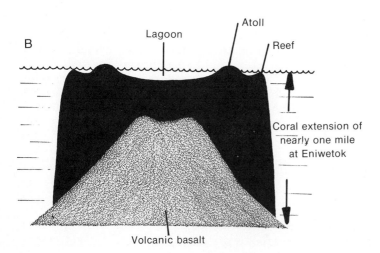

Figure 21.25 Formation of an atoll. *A,* Fringing reef around an emergent volcano. *B,* Continuous deposition of coral as volcanic cone subsides leads to the formation of a great coralline cap; emergent part of cap is atoll.

blage of animals. Only part of the exposed coral is living. Old skeletal remains form attachment and boring surfaces for many sessile animals, including octocorallian corals. The labyrinths of crevices, holes, and grottos are populated with a great number of different invertebrates and vertebrates.

As the reef increases in thickness, there is a gradual compaction of the lower deposits. Core drillings of coral reefs have disclosed coralline deposits of great depths. On the Pacific atoll of Eniwetok coral limestone extends downward for almost a mile before reaching basaltic rock. Since the deposition of coral by living colonies occurs only in the upper lighted zone, thick deposits of coral can only be explained by fluctuations in sea level and by subsidence of the bedrock upon which the reef is resting. Both have occurred at Eniwetok. The geologic history of Eniwetok began during the early Cenozoic Period, when a fringing reef developed around an emergent volcanic cone (Fig. 21.25). This oceanic peak gradually subsided to below sea level. Coral deposition occurred at a rate equaling subsidence, and the original fringing reef was transformed into an atoll resting on vertical walls almost one mile in height (Fig. 21.25). Fluctuating sea level periodically caused emergence and submergence. The last drop of 600 feet in sea level gave Eniwetok its present emergent form.

Subsidence of the Gulf Basin and fluctuating sea level have also contributed to reef formation in the Caribbean. The Bahamas represent an old coral reef, and submarine caverns with stalactites indicate a period of past emergence and subsequent submergence.

Although existing coral reefs have been formed by madreporarian coral, reefs in the past have been produced by other organisms. Two extinct groups of anthozoans, the tabulate and rugose corals, formed great reefs during the Paleozoic. Reefs have also been formed in the past by algae which secrete calcium carbonate. Coralline algae are still prevalent today, and are common members of reef communities, but there are no existing reefs composed primarily of living coralline algae.

CLASSIFICATION OF COELENTERATES

Phylum Coelenterata

CLASS HYDROZOA. Common but inconspicuous forms, usually with both polypoid and medusa stages. Hydroid colony formed by attached buds. Most colonies surrounded by a chitinous perisac.

Order Trachylina. Medusoid hydrozoans lacking or having a poorly developed polypoid generation. This order contains perhaps the most primitive members of the class.

Order Hydroida. Hydrozoans with a well-developed polypoid generation. Medusoid stage present or absent. The majority of hydrozoans belong to this order.

Order Siphonophora. Pelagic hydrozoan colonies of polypoid and medusoid individuals. Colonies with float or large swimming bells. Largely in warm seas. *Physalia* (Portuguese man-of-war). *Stephalia, Nectalia.*

Order Milleporina. Colonial polypoid forms having a heavy calcareous skeleton. Free, minute medusa. Single genus *Millepora* common in tropical waters.

Order Stylasterina. Colonial polypoid hydrozoans provided with a calcareous skeleton. Usually with an upright, branching growth form. Medusa absent. Mostly tropical. *Stylaster, Stylantheca, Spinipora, Allophophora.*

CLASS SCYPHOZOA. Medusa stage is most conspicuous, with bell diameters commonly from 2 to 40 cm. Widespread marine forms, mostly free-swimming.

Order Stauromedusae, or *Lucernariida.* Sessile scyphozoans attached by a stalk on the aboral side of the trumpet-shaped body. Chiefly in cold littoral waters.

Order Cubomedusae, or *Carybdeida.* Scyphomedusae with bells having four flattened sides and simple margins. Tropical and subtropical oceans.

Order Coronatae. Bell of medusa with a deep groove or constriction, the coronal groove, extending around the exumbrella. Many deep sea species.

Order Semaeostomae. Scyphomedusae with bowl-shaped or saucer-shaped bells having scalloped margins. Gastrovascular cavity with radial canals or channels extending from central stomach to bell margin. Occur throughout the oceans of the world.

Order Rhizostomae. Bell of medusa lacking tentacles. Oral arms of manubrium branched and bearing secondary mouth openings which lead into arm canals. Arm canals of manubrium pass into stomach. Original mouth lost through fusion of oral arm. Tropical and subtropical littoral scyphozoans.

CLASS ANTHOZOA. Either solitary or colonial, with medusoid stage completely absent. Includes sea anemones and corals, plus many heterogeneous forms.

Subclass Octocorallia. Polyps with eight mesenteries and eight pinnate tentacles. All colonial.

Order Telestacea. Lateral polyps on simple or branched stems. Skeleton of calcareous spicules.

Order Alcyonacea. Soft corals. Coenenchyme forming a rubbery mass and colony having a massive, mushroom, or a variously lobate growth form. Skeleton of separate calcareous spicules. Largely tropical.

Order Coenothecalia. Contains only the Indo-Pacific blue coral, *Heliopora*, having a massive calcareous skeleton.

Order Gorgonacea. Horny corals or gorgonian corals. Common tropical and subtropical octocorallian coelenterates having a largely upright plantlike growth form and a skeleton of a hornlike organic material. Separate or fused calcareous spicules may be present.

Order Pennatulacea. Sea pens. Colony having a fleshy, flattened or elongate body, or rachis. Skeleton of calcareous spicules.

Subclass Zooantharia. Polyps with more than eight tentacles, and tentacles rarely pinnate. Solitary or colonial.

Order Zoanthidea. Small colonial anemone-like anthozoans having one siphonoglyph and no skeleton.

Order Actiniaria. Sea anemones. Solitary anthozoans with no skeleton, with mesenteries in hexamerous cycles, and usually with two siphonoglyphs.

Order Madreporaria, or *Scleractinia.* Stony corals. Mostly colonial anthozoans secreting a heavy external calcareous skeleton. Sclerosepta arranged in hexamerous cycles. Many fossil species.

Order Rugosa, or *Tetracoralla.* An extinct order of mostly solitary corals possessing a system of major and minor radiating sclerosepta. Cambrian to Permian.

Order Corallimorpharia. Tentacles radially arranged. Resemble true corals but lack skeletons.

Order Ceriantharia. Anemone-like anthozoans with greatly elongate bodies adapted for living in sand burrows. One siphonoglyph; mesenteries all complete.

Order Antipatharia. Black or thorny corals. Gorgonian-like species with upright, plantlike colonies. Polyps arranged around an axial skeleton composed of a black thorny material and bearing thorns. Largely in deep water in tropics.

Subclass Tabulata. Extinct colonial anthozoans with heavy calcareous skeletal tubes containing horizontal platforms, or tabulae, on which the polyps rested. Sclerosepta absent or poorly developed.

Detailed accounts of coelenterates may be found in the references listed at the end of Chapter 18.

22. The Flatworms

All the remaining members of the Animal Kingdom are either bilaterally symmetrical or, if radially symmetrical, have clearly evolved from bilateral ancestors. The body can be cut along only one plane — anterior to posterior, dorsal to ventral — to produce two mirror image halves. Bilaterality is a widespread animal feature, correlated with motility. **Cephalization** (development of the head) is another consequence of motility. It is advantageous for a motile animal to

have many sensory receptors concentrated at the end of the body that has first contact with the environment. In evolution, nervous tissues also became concentrated at the front end and assumed a functional dominance over other parts of the body. The earliest bilateral animals were probably marine bottom dwellers. One surface of the body, perhaps that bearing the mouth, was kept toward the bottom. This led to a differentiation between dorsal and ventral sur-

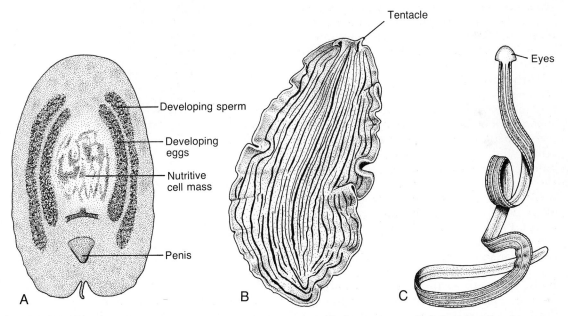

Figure 22.1 Turbellarian flatworms. *A, Polychoerus*, an acoel; *B, Prostheceraeus*, a polyclad; *C, Bipalium kewense*, a cosmopolitan land planarian found in Florida, Louisiana, California, and in greenhouses throughout the United States. (*A* from Barnes: Invertebrate Zoology, 2nd ed.; *B* redrawn from a photograph by D. P. Wilson; *C* after Hyman.)

faces, and along with the differentiation of anterior and posterior ends, established a bilateral symmetry.

Most zoologists believe that the flatworms, members of the phylum **Platyhelminthes,** are probably the most primitive of all the bilateral animals. The phylum is composed of three classes. The class **Turbellaria** contains the **free-living flatworms.** The class **Trematoda,** the **flukes,** and the class **Cestoda,** the **tapeworms,** are both entirely parasitic and probably evolved independently from turbellarians.

22.1 CLASS TURBELLARIA

Members of the class Turbellaria, the free-living flatworms, occur both in the sea and in fresh water, and a few species are terrestrial in humid forests. The aquatic turbellarians are bottom dwellers, living within algal masses and beneath stones and other objects. The common and familiar fresh-water **planarians** inhabit the undersurface of stones in springs, streams, and lakes. There are also marine flatworms which live with other animals. *Bdelloura,* for example, lives on the gills of horseshoe crabs.

Many tiny marine turbellarians live in the spaces between sand grains. Animals which live in these interstitial spaces compose the **interstitial fauna.** Virtually every phylum of animals has some representatives adapted for this habitat.

Turbellarians range in size from microscopic species to ones more than 60 cm. in length (land planarians), but most are less than 10 mm. long. The body tends to be dorso-ventrally flattened, the condition to which the name Platyhelminthes (*platy:* flat; *helminth:* worm) refers, and the larger the species, the more pronounced is the flattened shape (Fig. 22.1*B*). The anterior end often bears eyes and in some species tentacles. Planarians commonly possess lateral projections which give the anterior end a triangular shape (Figs. 22.1, 22.2 and 22.6).

A ciliated epidermal layer covers the body, although in the planarians only the ventral surface is ciliated. Beneath the epidermis is a muscle layer of circular, diagonal, and longitudinal fibers (Fig. 22.3). A network of loosely connected cells, called **mesenchyme,** surrounds the gut and other organs,

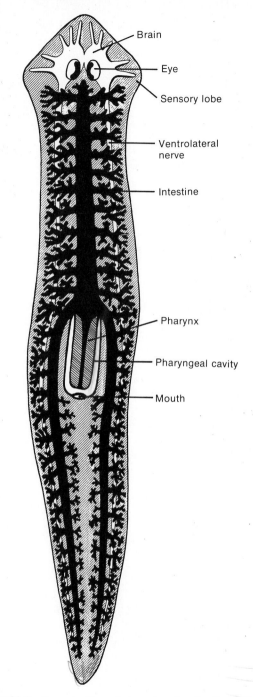

Figure 22.2 *Dugesia.* Dorsal view, showing the digestive and nervous systems. The mouth opens ventrally. (After Hyman).

and fills the interior of the body (Fig. 22.3). Flatworms are therefore said to possess a solid, or **acoelomate,** body structure, there being no cavity between the body wall and the internal organs.

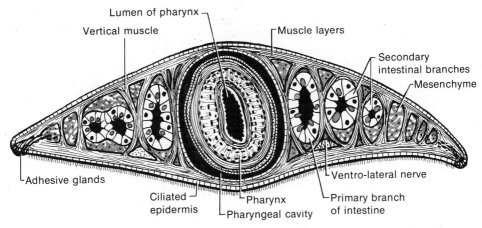

Lumen of pharynx
Vertical muscle
Muscle layers
Secondary intestinal branches
Mesenchyme
Adhesive glands
Ciliated epidermis
Pharynx
Pharyngeal cavity
Ventro-lateral nerve
Primary branch of intestine

Figure 22.3   Diagrammatic cross section of the planarian *Dugesia* at the level of the pharynx.

22.2   LOCOMOTION

Very small flatworms are not greatly flattened and swim or crawl about bottom debris by ciliary propulsion. Contractions of the muscle layer permit turning, twisting, and folding of the body (Fig. 22.4). The movement of turbellarians over 3 mm. in length involves delicate undulatory waves of muscle contraction in addition to cilia. The dorso-ventral flattening of the body is probably in part an adaptation for locomotion. With increased size, a flattened shape preserves a large surface area upon which the body can be carried. Glands in the epidermis and underlying tissue secrete a mucous film over which the animal glides.

22.3   NUTRITION

Turbellarians are **carnivorous,** feeding upon other small invertebrates. The mouth is located along the mid-ventral line. In most species the gut is differentiated into a pharynx adapted for ingestion, and an intestinal sac which functions in digestion and absorption. No anus is present and wastes are ejected through the mouth as in coelenterates.

The simplest turbellarians have no gut cavity. The mouth opens into a short nonmuscular ciliated pharynx, which ends in a large internal mass of nutritive cells (Fig. 22.5). The mass envelops the prey after it is swallowed. These small turbellarians belong to the marine order **Acoela** and display a number of characteristics which may be primitive. The name Acoela in this case refers to the absence of a gut cavity and not to the lack of a body cavity.

The pharynx of most turbellarians is muscular and functions in feeding. In some species the pharynx is a muscular bulb which can be projected out of the mouth. These are small turbellarians, and the gut has the form of a simple unbranched sac (Fig. 22.5). Most of the larger flatworms, including the fresh-water planarians, have

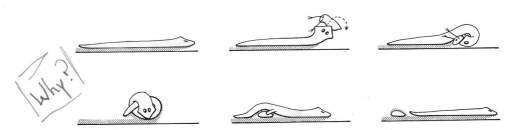

Figure 22.4   Hunting and feeding in *Dugesia*. A small crustacean (*Daphnia*) is captured and eaten, its tough exoskeleton remaining as an empty shell.

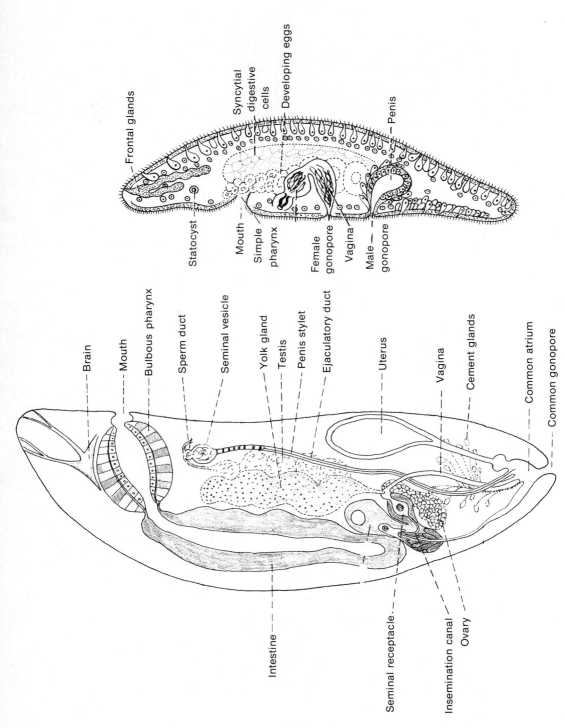

Figure 22.5 Sagittal sections of two turbellarians. *A*, The acoel *Convoluta*; *B*, the rhabdocoel *Anoplodiera voluta*, which is an endocommensal within the gut of sea cucumbers. (Both after Westblad from Hyman.)

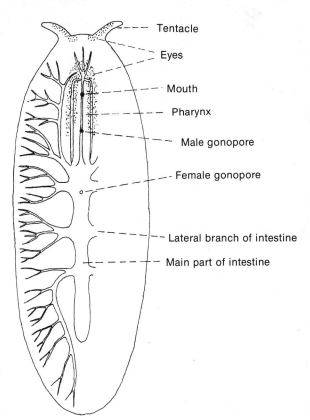

- - - Tentacle

- - Eyes

- - Mouth

- - Pharynx

- - Male gonopore

- Female gonopore

- - Lateral branch of intestine

- - Main part of intestine

Figure 22.6 Digestive system of a polyclad. (After Hyman.)

a tubular pharynx and a highly branched intestinal sac (Fig. 22.2). The tubular pharynx is muscular and, except for the end attached to the intestinal sac, lies free within a pharyngeal cavity. The cavity opens to the exterior through the mouth. When the animal feeds, the pharyngeal tube is projected out of the mouth (Fig. 22.7). In some flatworms, the pharynx is directed **anteriorly** (Fig. 22.6); in others, including the

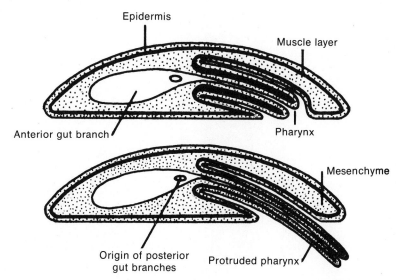

Epidermis

Muscle layer

Anterior gut branch

Pharynx

Mesenchyme

Origin of posterior gut branches

Protruded pharynx

Figure 22.7 Longitudinal section of the fresh-water triclad, *Polycelis*, showing retracted and protruded tubular pharynx. (After Jennings.)

planarians, the pharynx is directed **poste-riorly.** In flatworms with the latter condition, the mouth is located behind the middle of the body (Fig. 22.2).

During feeding, flatworms crawl upon their prey, pinning them down and enveloping them with mucus (Fig. 22.4). The prey is often swallowed whole. However, planarians, as well as many other species with a tubular pharynx, do not swallow the prey entirely. Rather, the extended pharynx is inserted into the body of the prey or into dead animal matter with the aid of **proteolytic enzymes** secreted by pharyngeal glands. Fragments of the body of the prey are then pumped into the intestine. The many divisions of the gut, which extend throughout much of the body, greatly increase the surface area available for digestion and absorption.

Digestion occurs in much the same way as in coelenterates. An initial extracellular digestion within the lumen of the intestinal sac is followed by intracellular digestion within the cells of the intestinal wall. The same processes occur in the acoel flatworms, the mass of nutritive cells secreting enzymes around the enveloped prey.

22.4 GAS EXCHANGE, INTERNAL TRANSPORT AND WATER BALANCE

There are no special organs for gas exchange or internal transport. Gas exchange takes place across the general body surface, which, because of its flattened shape, is sufficiently great to meet oxygen demands. Moreover, the small vertical distances resulting from the flattened shape greatly facilitate internal transport by diffusion. Diffusion is also sufficient for the movement of food materials, for in the larger flatworms the gut branches are so extensive that no tissue is very far from some intestinal cells.

Most flatworms possess a system of tubules, called **protonephridia** (Fig. 22.8). These are sometimes said to be excretory organs, and indeed there are some animals which possess excretory protonephridia, but the nitrogenous wastes of flatworms, which are chiefly ammonia, are removed by general diffusion across the body surface, and the protonephridial system appears to function chiefly in water balance. A protonephridium consists of a single tubule or

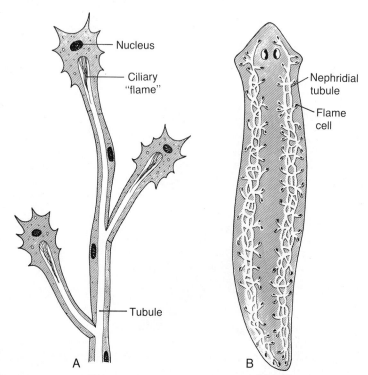

Nucleus

Ciliary "flame"

Nephridial tubule

Flame cell

Tubule

A

B

Figure 22.8 Protonephridial system of the planarian *Dugesia. A,* Detail showing flame cells and tubules; *B,* the tubule network.

a branching tubule that opens to the outside by a nephridiopore. The internal end of the tubule is embedded in mesenchyme and its blind end contains either a flagellum or a tuft of cilia. The beating of the terminal flagellum or cilia has been likened to a flame and the terminal cells are sometimes called flame cells. Excess water is passed through the terminal tubule wall and then driven through the lumen to the outside. One to four pairs of protonephridia may be present. They are absent or poorly developed in marine flatworms, in which the interstitial fluid is isosmotic with the surrounding sea water (**osmoconformers**). The fresh-water flatworms live in a hyposmotic environment and must continually pump out excess water; they have many highly developed protonephridia.

22.5 NERVOUS SYSTEM AND SENSE ORGANS

The simplest nervous system, found in some acoels, is a **nerve net** system located in the epidermal layer, with processes supplying the underlying muscle fibers. Some acoels, like coelenterates, also possess epitheliomuscle cells.

An epidermal nerve net appears to be the primitive location and organization of the animal nervous system. Such a system provides the minimum necessary integration between epidermal receptors and adjacent underlying muscle effectors. An epidermal nerve net is found not only in coelenterates and some flatworms, but also in some higher groups, such as echinoderms. Moreover, the nervous systems of all animals arise in the outer ectoderm, perhaps recapitulating their original epidermal position.

In the majority of flatworms, however, the neurons are organized in bundles, or nerves, which lie just below the epidermis. The neuronal processes are not well differentiated into dendrites or axons, and there are few ganglia, that is, the cell bodies of the neurons are not clustered together. The most primitive arrangement of nerves in flatworms appears to be one in which there are four or five pairs of longitudinal cords radially arranged around the body—dorsal, dorso-lateral, lateral, ventro-lateral, and ventral (Fig. 22.9). The cords merge at the

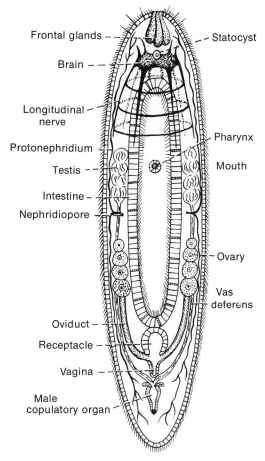

Figure 22.9 Dorsal view of a generalized rhabdocoel turbellarian. (Modified from Ax.)

anterior end, and the slight enlargement is sometimes called a brain.

In the majority of turbellarians, the radial arrangement of nerve cords has been lost through the predominance of certain pairs and the loss of others. Planarians have a very large ventral pair and a reduced lateral pair (Fig. 22.3). The other pairs are absent.

A statocyst is present in many flatworms (Fig. 22.9). Some acoels with a nerve net have an anterior statocyst surrounded by a heavy concentration of neurons. This concentration has been suggested as the forerunner of a brain.

Flatworms usually have two or more eyes (Fig. 22.2), but they are never highly developed. They are generally located at the anterior end, sometimes on or around tentacles, but in some flatworms the eyes are located along the sides of the body. The

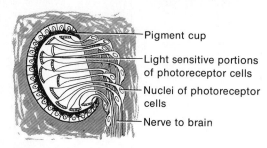

Figure 22.10 Diagrammatic section through the eye of a planarian. Light reaches the sensitive elements from the right.

- Pigment cup
- Light sensitive portions of photoreceptor cells
- Nuclei of photoreceptor cells
- Nerve to brain

simplest eyes, found in acoels, are merely patches of photoreceptors and pigment cells in the epidermis. Usually, however, the eye has a cup-like form resulting from the invagination of the pigment cells (Fig. 22.10). In some animals the opening of the retinal cup becomes filled in and covered with a lens-cornea, but lenses are not present in most flatworm eyes.

The eyes of invertebrates are usually specializations of just the skin. Only in vertebrates is part of the eye (the retina) derived from the embryonic brain and part from the skin. Most eyes have a cup-like form and arise through the invagination of the embryonic rudiment of the ectodermal eye.

The function of the eye of the great majority of animals, and undoubtedly its original function, is to provide information about the general light intensity of the surrounding environment. Such information may have important implications for an animal. A low light intensity might mean that the animal is safely under cover or that it is dusk or night, at which time activity is increased or decreased. A point of high light intensity, such as that produced by the sun or at the opening of a burrow or at the water's surface might provide a cue for orientation. A sudden change in light intensity might indicate the passing of a predator and initiate a withdrawal or escape reflex.

Flatworms avoid bright light and the response is probably an adaptation to remain under cover.

22.6 REGENERATION AND REPRODUCTION

Many flatworms are capable of asexual reproduction, and this ability is closely correlated with the ability to regenerate. The most common method of asexual reproduction is by **transverse fission.** In many fresh-water planarians, for example, a fission plane forms behind the pharynx, and during movement the animal breaks in two. Following fission, the two halves regenerate the missing parts. In a number of small flatworms, such as *Stenostomum* and *Microstomum,* fission planes and regeneration proceed more rapidly than the separation of the parts, and chains of individuals are formed as a result (Fig. 22.11).

Mesenchyme is the principal source of new cells for regeneration, which involves both reorganization of remaining tissues and addition of new cells. As in coelenterates, the regenerative parts retain their original polarity (Fig. 22.12). A piece taken from the middle of the body always regenerates a new anterior end at the severed anterior surface and a new posterior end at the posterior surface. There is also a lateral polarity; thus, a two-headed planarian can be produced by cutting the anterior end longitudinally along the midline (Fig. 21.12). The rate of regeneration of pieces taken at different levels reflects a distinct metabolic gradient along the anterior-posterior axis.

Figure 22.11 Chain of attached daughter individuals resulting from incomplete transverse fission in *Stenostomum.* (After Child from Hyman.)

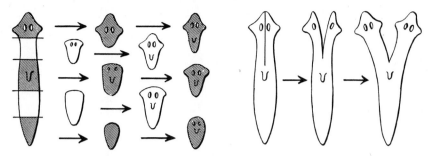

Figure 22.12 Polarity and regeneration in *Dugesia*. Left, each of five pieces regenerates, but the rapidity with which the head develops depends upon the level of the piece. Right, a two-headed form produced by repeated splitting of the anterior end.

Anterior pieces regenerate more rapidly than do posterior pieces of equal size. Or, reflecting the same phenomenon, posterior pieces must be larger than anterior pieces for regeneration to be completed in the same length of time.

Flatworms are hermaphroditic, but most exhibit cross-fertilization (a condition true of hermaphroditic animals in general; self-fertilization is rare). There is usually simultaneous and mutual copulation (Fig. 22.13), the penis of each animal being received by the female system of the other animal, and the sperm are stored for a period of time. The eggs are fertilized internally and deposited on the bottom in jelly masses or in cocoons.

The reproductive systems of most turbellarians are complex and highly variable but reflect adaptations for the reproductive conditions just described. The male system (Figs. 22.5, 22.9 and 22.14) may contain one

testis or many pairs of testes. A small duct connects each testis with a main sperm duct that extends along each side of the body. The sperm ducts join together to form an ejaculation duct, which exits through a penis. The penis is muscular and often armed with a stylet (but not in planarians). Parts of the sperm ducts are frequently modified as seminal vesicles. Glands are present, most commonly associated with the ejaculation duct. Commonly the penis is located within a chamber, the genital atrium, which may also contain the terminal part of the female system. The atrium opens to the outside through a gonopore.

The female system (Figs. 22.5, 22.9 and 22.14) contains either a single ovary or one to many pairs of ovaries, but only a single pair of oviducts is present. In many turbellarians, including the planarians, there are yolk glands located along the length of the oviduct. Yolk cells are released as the eggs

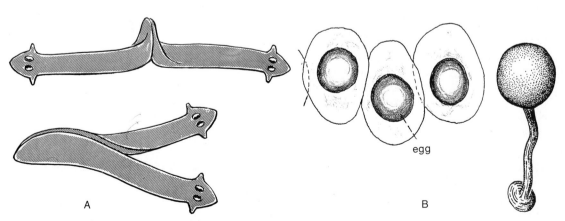

egg

A   B

Figure 22.13 *A*, The planarian *Dugesia* copulating; *B*, an egg cocoon of a fresh-water planarian. (Both after Hyman.)

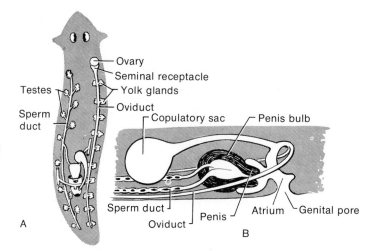

Figure 22.14 Reproductive system of the fresh-water planarian *Dugesia*. *A*, A dorsal view showing the male organs on the left and the female organs on the right. *B*, A side view of the copulatory organs. (After Hyman.)

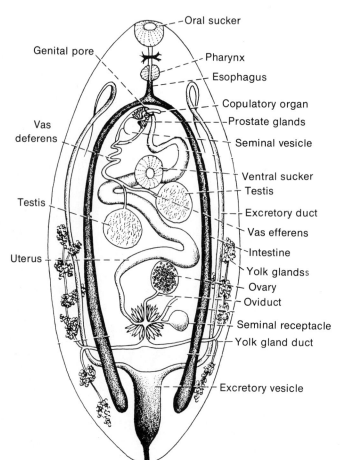

Figure 22.15 Structure of a generalized trematode. (After Chandler and Read.)

travel down the oviduct, and the eggs, when deposited, are surrounded by yolk. The deposition of yolk outside an egg cell (**ectolecithal**) is an unusual condition. In most animals the eggs are **entolecithal,** that is, the yolk is contained within the egg cytoplasm. A seminal receptacle, a vagina or atrium, and glands for the production of egg envelopes are typically present. There may be a separate female gonopore, or there may be a common genital atrium and gonopore for both systems.

Acoel turbellarians exhibit a number of interesting conditions which may be primitive for the phylum. Acoels have no true gonads, but as in coelenterates, gametes develop as aggregates within the mesenchyme (Fig. 22.15). The female system either is completely lacking or consists of only a gonopore and a vagina. There are no oviducts; the eggs reach the exterior by rupture of the body wall.

The eggs of some turbellarians are deposited in gelatinous masses. Many species, especially fresh-water forms, lay their eggs within capsules, or cocoons, which in the planarians are attached to stones or other objects by little stalks (Fig. 22.13).

Some fresh-water turbellarians produce two kinds of eggs. Summer eggs have a thin shell, and hatch within a short period of time; winter or dormant eggs have a thick resistant shell and are capable of withstanding cold and desiccation. Having two types of eggs is an adaptation shared with many other fresh-water animals, especially those living in pools and small streams.

Spiral cleavage is present in turbellarians with entolecithal eggs, but has been lost in the many flatworms with ectolecithal eggs. Some marine flatworms have larvae.

22.7 Symbiosis and Parasitism

Before describing the two parasitic classes of flatworms, we need to consider briefly the general nature of parasitism. Many organisms depend for their existence upon an intimate physical relationship with other organisms. Such a relationship is known as **symbiosis.** The larger organism is termed the host and the smaller, the **symbiote.** The symbiote always derives some benefit from the relationship but the conse-quence to the host varies. When both the symbiote and the host benefit, the relationship is called **mutualism.** Mutualism is the least common type of symbiosis, but we have already encountered examples among the sponges and coelenterates. Algae live within the amebocytes of fresh-water sponges and within the gastrodermal cells of madreporarian corals. The metabolism of each species is enhanced by the presence of the symbiotic partner.

Commensalism, a more common symbiotic relationship, has evolved many times within the Animal Kingdom. In commensalism the host is neither benefited nor harmed. The symbiote, or commensal, is protected by the host or utilizes excess food of the host. The shrimp living within the spongocoel of glass sponges and the clown fish living among the tentacles of sea anemones are commensals. The sponge and sea anemones can exist without their symbiotes, but the commensal shrimp and fish cannot exist alone.

The most common symbiotic relationship is **parasitism.** Here the host is harmed by the presence of the symbiote, or parasite. It is important to recognize that it is the host individual that is harmed. The relationship may not necessarily be seriously detrimental to the host species. Indeed, if the host species suffers too greatly from its parasites and becomes extinct, the parasite species is adversely affected.

Parasites which live on the outside of their host are said to be **ectoparasitic;** those on the inside, **endoparasitic.** Most parasites can utilize several closely related or ecologically similar species as hosts. Some parasites require two hosts to complete their life cycles, in which case the host for the larval or developmental stage is termed the **intermediate host;** the host for the adult stage is termed the **primary host.**

The principal benefit derived by a parasite from his relationship with the host is nutritive. The parasite feeds upon the host's tissues or body fluids or utilizes the food ingested by the host. When the parasite is enclosed by some part of the host's body, there may be secondary advantages, such as protection.

Although the symbiote gains an easy source of food material from a parasitic existence, the relationship poses problems and is not without cost. The primary problem of parasites is that of reaching and penetrating new hosts.

From the few generalizations which have been made thus far about the nature of a parasitic existence, many of the adaptations that are encountered in parasites can be predicted.

The anterior end of the body, the mouth region, and the anterior parts of the gut of parasites are commonly adapted for attaching, piercing, and sucking. Ectoparasites, which feed infrequently, may have a part of the gut modified for storage. On the other hand, some endoparasites utilize digested food of the host and have lost the gut entirely.

The problems of penetration and attachment to the host have resulted in the evolution of a variety of structures, such as suckers, hooks, and teeth. Enzymes may facilitate penetration in some species. The most common point of entrance for endoparasites is through the mouth of the host. The parasite may remain in the gut of the host or, in some species, break through the gut wall to reach other organs.

The problem of reaching new hosts has been met in most parasites through the production of enormous numbers of eggs or other developmental stages. The species is perpetuated if only a few individuals encounter the proper host and reach adulthood. Needless to say, the reproductive system of parasites is highly developed for the production of great numbers of gametes, and a larger proportion of body tissue is assigned to this function than is true of free-living animals. In fact, there are some parasites in which most of the body is concerned with reproduction.

The high reproductive output of parasites demands a high expenditure of energy. The energy costs are met by the host which is providing food for the parasite.

The structure of endoparasites usually reflects the less rigorous demands of the limited and uniform environment in which they live. Locomotor processes are reduced and the sense organs found in free-living species are absent. The nervous system in turn is greatly reduced. Gas exchange organs are never present.

A "good" parasite doesn't kill its host, for a dead host results in a dead parasite. Moreover, a host may carry a small population of parasites without any serious consequences.

Where the stress of a parasitic infection is manifest in the host, the condition is recognized as a disease. Parasitic disease may be caused by a number of processes of the parasite. Cells and tissues may be destroyed when the parasite feeds upon, penetrates into, or moves within the host. Blood vessels, ducts, or the gut may be clogged by parasites. The host may be robbed of food. The parasite may produce a substance which has a toxic or allergic effect on the host.

Some parasites cause the host's tissues to undergo abnormal mitotic activity and consequent growth. Plant galls, for example, represent the response of the surrounding plant tissues to the deposition of an egg by certain insects.

Not only has symbiosis evolved many times in the Animal Kingdom, but it probably originated in different ways. Some symbiotes probably evolved from ancestral species which used the host simply as a substratum on which they attached or moved about; others, from species that used the host as one of a number of protective retreats, or habitations. Many parasites undoubtedly evolved from predatory ancestors, but some may have originated from forms which lived in decaying organic matter. For example, the bot flies, which lay their eggs in the wounds of cattle, probably evolved from flies which deposited their eggs in carrion.

Once having shifted over to a dependent symbiosis, many species, especially parasites, exploited the relationship and underwent adaptive radiation resulting in thousands of species.

The parasitic flatworms comprise two classes: the Trematoda, or flukes, and the Cestoda, or tapeworms. These two groups of flatworm parasites, along with the parasitic roundworm, include the great majority of parasitic worms of great economic and medical significance. The Trematoda and Cestoda evolved independently from the free-living turbellarians.

22.8 CLASS TREMATODA

The class Trematoda, to which the flukes belong, contains over 6000 species of ectoparasites and endoparasites. The majority

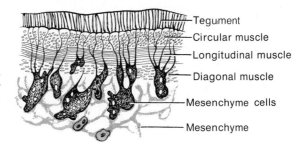

Figure 22.16 Part of the body wall of a trematode. Note that an epidermis is missing, and that the covering tegument lies directly on tissue of mesodermal origin.

are parasites of vertebrates, especially fish, but immature stages are harbored by invertebrates.

Trematodes are flattened and oval in shape (Fig. 22.15), the majority being not more than a few centimeters long. The body is covered by a nonciliated cytoplasmic syncytium, the tegument, the nuclei of which are sunken into the mesenchyme (Fig. 22.16). The mouth is located at the anterior end. Adhesive suckers are usually present around the mouth and may also be present mid-ventrally and at the posterior end. The digestive tract is composed of a muscular pumping pharynx behind the mouth, a short esophagus, and usually two long blind intestinal sacs (Fig. 22.15). Trematodes feed on cell fragments, mucus, tissue fluids, and blood of the host. Protonephridia are present and the nervous system is similar to that of turbellarians, except that sensory structures are absent in endoparasitic forms.

Mutual copulation and cross fertilization is the rule and the reproductive system is adapted for the production of a tremendous number of eggs, which are ectolecithal and surrounded by separate shells. Egg production has been estimated to be 10,000 to 100,000 times greater in trematodes than in the free-living turbellarians.

Life Cycles. Eggs are passed out of the host and hatch as mobile aquatic ciliated miracidium larvae (Fig. 22.17), which provide the means of reaching a new host. In many trematodes, development requires an intermediate host. The miracidium enters an invertebrate, commonly a snail, and develops further into a tailed cercaria larva (Fig. 21.17). The cercaria passes from the first intermediate to a second intermediate

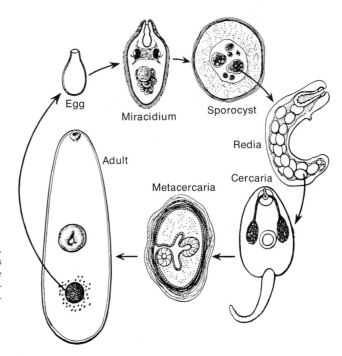

Figure 22.17 Life cycle of a digenetic fluke. The stages shown here belong to various species. The arrows indicate whether one stage becomes the next or whether it produces the next by reproduction. (After Hyman.)

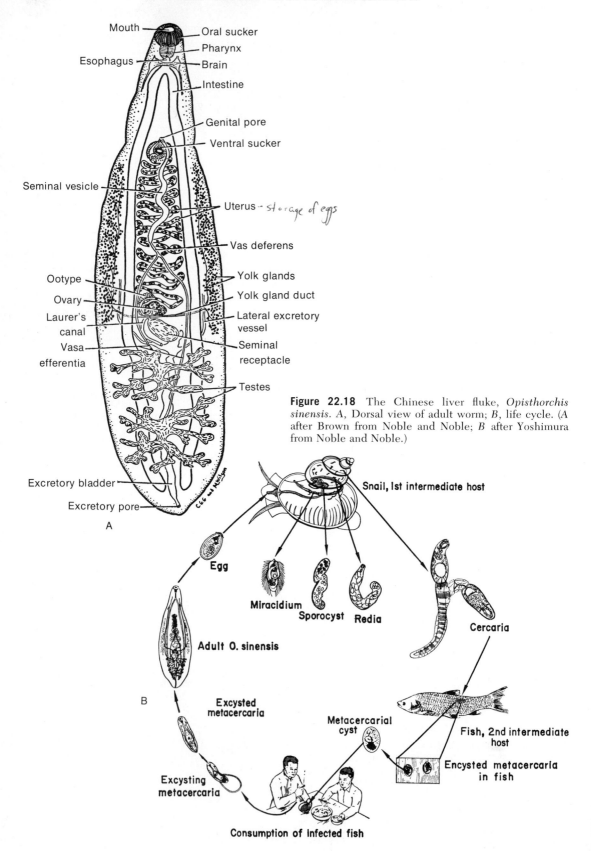

Mouth — Oral sucker
— Pharynx
Esophagus — — Brain
— Intestine

— Genital pore
— Ventral sucker

Seminal vesicle —

— Uterus - *storage of eggs*

— Vas deferens

Ootype — — Yolk glands
Ovary — — Yolk gland duct
Laurer's canal — — Lateral excretory vessel
Vasa efferentia — — Seminal receptacle

— Testes

Excretory bladder —
Excretory pore —
A

Snail, lst intermediate host

Egg

Miracidium
Sporocyst   Redia

Cercaria

Adult O. sinensis

Fish, 2nd intermediate host

Excysted metacercaria

Metacercarial cyst

Encysted metacercaria in fish

Excysting metacercaria

Consumption of infected fish

B

Figure 22.18 The Chinese liver fluke, *Opisthorchis sinensis. A,* Dorsal view of adult worm; *B,* life cycle. (*A* after Brown from Noble and Noble; *B* after Yoshimura from Noble and Noble.)

host, where it develops into a developmental stage, called a metacercaria. If the second intermediate host is eaten by the primary host, the metacercaria is freed from the tissues of the intermediate host and develops into an adult. There are many variations of this generalized outline of the life cycles and space only permits a few examples.

Bendenia melleni. This ectoparasite lives on the skin and eyes of a number of different marine fish. The worm can cause blindness and serious damage to the host's integument. The eggs, when shed, fall to the bottom. A larva, similar to a miracidium and having a precocious adhesive organ, escapes from the egg, and if a host fish is encountered, attaches to the skin. Following attachment and while feeding upon the host, the miracidium develops into an adult parasite.

Echinostoma revolutum. A great many trematodes infect the gut or gut derivatives of their primary host. Lungs, bile ducts, pancreatic ducts, or intestines are common sites. The life cycle of *Echinostoma revolutum*, which infects the intestine of a variety of aquatic birds and mammals, is a good illustration of the generalized life cycle described earlier. The eggs pass out with the host's feces and hatch in water. The free-swimming miracidium can penetrate and develop within species of three different genera of snails. A cercaria larva leaves this snail and penetrates another snail, where it undergoes further development. If the second intermediate host is ingested by the primary host, the immature worm is liberated and develops into an adult.

Opisthorchis sinensis. This species, commonly called the Chinese liver fluke, lives in the bile ducts of man, dogs, cats, foxes, and other fish-eating mammals (Fig. 22.18). The intermediate hosts for the miracidium and cercaria are a snail and a fish respectively. Human infestation has been common in the Orient because human feces are used to fertilize ponds and raw fish from these ponds are eaten. A few worms cause no disease symptoms, but several hundred can cause destruction of liver tissue, clogging of ducts, formation of bile stones, and hypertrophy of the liver.

Fasciola hepatica. Known as the sheep liver fluke, this species is also an inhabitant of its host's bile ducts, and man, cattle, pigs, rodents, and other mammals may be hosts in addition to sheep (Fig. 22.19). The life cycle is similar to that of the Chinese liver fluke,

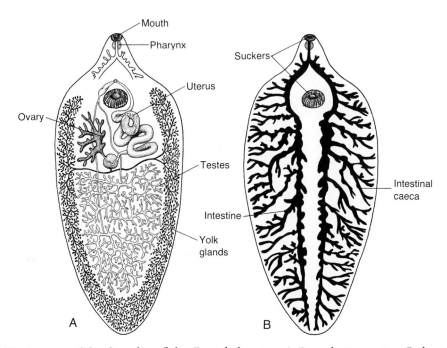

Figure 22.19   Structure of the sheep liver fluke, *Fasciola hepatica. A,* Reproductive system; *B,* digestive system.

but a snail is the only intermediate host. The cercariae leave the snail and encyst as metacercariae on vegetation along the edges of ponds and streams.

22.9 CLASS CESTODA

About 3400 species of cestodes, or tapeworms, have been described. All are endoparasites and the majority are adapted for living in the gut of vertebrates.

The body form of adult tapeworms is quite different from that of turbellarians and trematodes. The anterior end consists of a knob-like **scolex,** containing suckers and hooks, which anchors the worm to the host (Fig. 22.20). A narrow neck region connects the scolex to the **strobila,** which makes up the greater part of the tapeworm body. The strobila is composed of flattened sections, called **proglottids,** arranged in a linear series. New proglottids are continually being formed in the neck region and old ones detached at the end of the strobila. Tape-

worms are generally long, and some species, with hundreds of proglottids, reach lengths of 40 feet.

As in trematodes, the body is covered by a tegument. However, a digestive system is completely absent and digested food of the host is absorbed through the tegument. Longitudinal nerve cords and excretory ducts run the length of the strobila. Much of the body tissue is given over to the reproductive system, which is complete within each proglottid (Fig. 22.20). A common gonopore is present on one edge of each proglottid. Mutual copulation between the proglottids of two different worms generally occurs, but copulation between proglottids on the same strobila and self-fertilization within one proglottid are known.

The eggs, which are surrounded by a shell as in trematodes, may be continually shed through the gonopore into the host's intestine, or they may be stored in a blind sac, called the uterus. In the latter case, terminal proglottids packed with eggs break

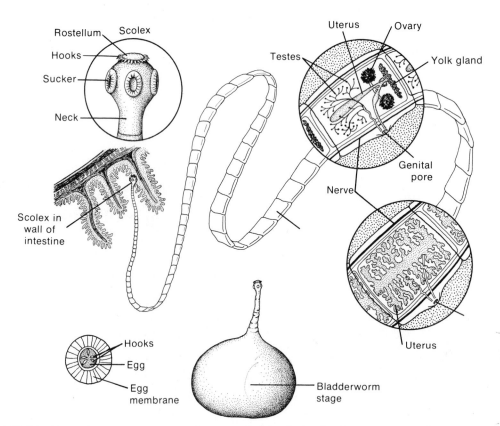

Figure 22.20   The pork tapeworm, *Taenia solium.* Insets show the head and an immature and a mature section of the body. (Villee: Biology, 6th ed.)

away from the strobila and may rupture within the host's intestine or they may leave with the feces and rupture later.

Life Cycles. One or more arthropod or vertebrate intermediate hosts are required to complete the life cycle, which involves an **oncosphere** and a **cystocercus** larva. The following examples illustrate the basic pattern of the life cycle, but there are many variations.

Species of the genus *Taenia* are among the best known tapeworms. *Taenia solium,* the pork tapeworm, lives in the intestine of man and reaches a length of up to 10 feet. Man is the only primary host. The proglottids are passed into the host's feces and the eggs may lie dormant in moist soil for long periods of time. The eggs do not hatch unless swallowed by a pig, dog, or certain other mammals, including man. On hatching, a spherical oncosphere larva bearing a number of hooks bores into the intestinal wall, where it is picked up by the circulatory system and transported to striated muscle. Here the larva leaves the blood stream and develops into a cystocercus stage (Fig. 22.20). The cystocercus, sometimes called a bladder worm, is an oval stage about 10 mm. in length, with the scolex invaginated into the interior. If raw or insufficiently cooked pork is ingested by man, the cystocercus is freed, the scolex evaginates, and the larva develops into an adult worm within the gut.

A severe infection of adult tapeworms may cause diarrhea, weight loss, and occasionally perforation of the intestinal wall. The worms may be eliminated with drugs. Much more serious is cystocercus infection, and man may be the intermediate host for the cystocercus of other tapeworms besides that of *Taenia solium.* Sometimes cystocerci lodge in the brain or in tissues other than striated muscle. Cystocerci can be removed only by surgery.

Soil mites are the intermediate hosts of sheep tapeworms, and copepods—tiny aquatic crustaceans—are the intermediate hosts for the tapeworms of many fish. Insects are intermediate hosts of many tapeworms.

CLASSIFICATION OF FLATWORMS

Phylum Platyhelminthes

Flatworms. Body dorsoventrally flattened; digestive cavity (when present) with a single opening, the mouth.

CLASS TURBELLARIA. Free-living ciliated flatworms.

Order Acoela. Small marine flatworms, usually measuring less than 2 mm. in length. Mouth and sometimes a simple pharynx present, but no digestive cavity. Protonephridia absent. No distinct gonads. Oviducts and yolk glands absent.

Order Rhabdocoela. Marine and fresh-water flatworms, usually less than 3 mm. in length. Enteron in the form of a simple elongated sac without lateral diverticula and with a simple or bulbous pharynx. The rhabdocoels were formerly considered a single order, but modern authorities have now raised the suborders to ordinal rank.

Order Alloeocoela. Flatworms of moderate size, averaging between 1 and 10 mm. Enteron may be a simple or branched sac with a simple, bulbous, or tubular pharynx. Majority are marine. Like the rhabdocoels, the former suborders of the order Alloeocoela are now given the rank of orders.

Order Tricladida. Relatively large marine, fresh-water and terrestrial flatworms, known as planarians, ranging from 2 to more than 60 cm. in length in the terrestrial forms. Digestive system with a tubular muscular pharynx and a three-branched intestinal sac. Both mouth and pharynx located in the middle of the body. Eyes present in most species.

Order Polycladida. Marine flatworms of moderate size, averaging from 2 to 20 mm. in length with a greatly flattened and more or less oval shape. Intestinal sac elongate and centrally located with many highly branched diverticula. Eyes numerous. Yolk glands absent.

CLASS TREMATODA. Parasitic flatworms known as flukes. Body oval. Ciliated epidermis absent; a living cuticle overlying muscle layer of body wall. Digestive tract present; mouth at anterior end.

> *Order Monogenea.* Mostly ectoparasites of fish. Anterior end with adhesive glands; posterior end with adhesive organ-bearing suckers. Life cycle does not require an intermediate host.
>
> *Order Digenea.* This order of endoparasites contains the majority of flukes. Mouth within an anterior sucker; ventral or posterior sucker also present. Life cycle requires one or more intermediate hosts.

CLASS CESTODA. Tapeworms. Endoparasitic flatworms lacking a digestive system. Ciliated epidermis lacking; body covered by a cuticle.

> **Subclass Eucestoda.** Body composed of a scolex and strobila. Larva possesses six hooks. Includes almost all tapeworms.
>
> **Subclass Cestodaria.** Fifteen species of fish parasites. Body leaf-shaped without a scolex and strobila.

ANNOTATED REFERENCES

Detailed accounts of flatworms may be found in the references listed at the end of Chapter 18. The works listed below deal specifically with the biology of animal parasites and commensals and the general biology of symbiosis.

Cheng, T. C.: The Biology of Animal Parasites. Philadelphia, W. B. Saunders Co., 1964.

Gotto, R. V.: Marine Animals: Partnerships and Other Associations. New York, American Elsevier Publishing Co., 1969.

Noble, E. R. and Noble, G. A.: Parasitology. 3rd ed. Philadelphia, Lea and Febiger, 1971.

Read, C. P.: Parasitism and Symbiology. New York, Ronald Press, 1970.

23. Pseudocoelomates

Flatworms are **acoelomates.** Almost all other bilateral animals possess a body cavity or at least are derived from forms which had a body cavity. The body cavity of most animals is a **coelom,** but in others the body cavity, termed a **pseudocoel,** has a different embryonic origin from the coelom of other animals. When gastrulation occurs, the blastocoel is not obliterated but remains to become the adult body cavity (Fig. 23.1). A pseudocoel therefore does not have a peritoneal lining as does a coelom.

Pseudocoelomates are a diverse assemblage of free-living and parasitic animals, but they share a number of features in addition to the pseudocoel. They are all poorly cephalized; i.e., they lack a distinct head with well-developed sensory structures. Free-living species are all essentially aquatic. Even the nematodes which live in soil are dependent upon water films around soil particles. Most pseudocoelomates are tiny, usually microscopic, and they possess neither a blood vascular system nor organs for gas exchange.

The organs, such as the intestine and gonads, of many pseudocoelomates have fixed numbers of cells, which are constant for a particular species. For example, the intestine of the roundworm *Rhabditis longicauda* is always composed of 18 cells. Nuclear divisions are completed at hatching and subsequent growth involves an increase in the size of the cells but no cell division occurs. This strange condition of cell constancy is possibly correlated with the small

size of pseudocoelomates. The cells of these tiny animals are perhaps near the lower limit of effective size for metazoans, and cell constancy may be an adaptation to preserve an adequate cell size.

The body of pseudocoelomates is covered by a flexible non-living **scleroprotein cuticle,** which is often sculptured (Fig. 23.9). A varying number of adhesive glands is usually present and these open through the cuticle individually as small tubes (Fig. 23.6). The adhesive tubes are used for temporary anchorage to the substratum.

Water balance is achieved by a pair of **protonephridia** or by structures probably derived from protonephridia. The nervous system is relatively simple, composed of a varying number of longitudinal nerves.

The digestive tract of pseudocoelomates, like that of higher animals, is a tube with two openings, a **mouth** and an **anus.** Different sections of the gut tube become specialized for certain parts of the digestive process. The specializations present in each animal are closely correlated with diet and feeding habits. In pseudocoelomates, a muscular pharynx and a stomach or intestine are the principal specializations of the gut tube.

Reproduction in most pseudocoelomates is entirely sexual, and with few exceptions, the sexes are separate. Cleavage is determinate, but the spiral condition is often lost. Development is typically direct in free-living species.

There are seven groups of pseudocoelo-

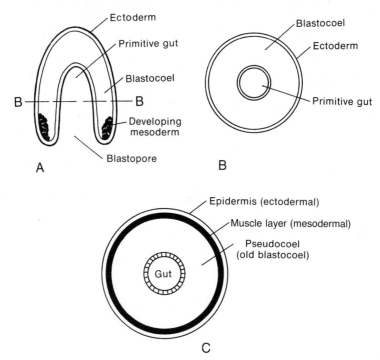

Figure 23.1 Embryonic origin of the pseudocoel. *A*, Frontal section of the gastrula, showing the remains of the blastocoel and the developing mesoderm; *B*, cross section of *A* taken at level B–B; *C*, diagrammatic cross section of adult pseudocoelomate.

mate animals. Some zoologists believe that many of these groups are closely enough related to constitute classes of a single phylum, called the **Aschelminthes.** Other zoologists object to the phylum Aschelminthes, pointing out that although pseudocoelomates share a number of features, there are no distinctive unifying characteristics. These objections seem justified to us and we will treat each of the pseudocoelomate groups as a separate phylum. Limitations of space permit us to describe only the three largest phyla of pseudocoelomates.

23.1 PHYLUM ROTIFERA

The phylum Rotifera consists of some 1500 species of microscopic animals, which are very common in fresh water. There are some marine rotifers and some which live in the water films of mosses.

The body of a rotifer is composed of a somewhat cylindrical trunk and foot (Figs. 23.2 and 23.3). The anterior end of the trunk bears a crown of **cilia,** a distinguish-

ing characteristic of the phylum. The distribution of crown cilia varies in different rotifers (Figs. 22.2 and 22.3). A common type of crown is one in which the cilia are arranged in the form of two discs (Fig. 23.2). The two discs of cilia beat in a circular manner, one clockwise and one counterclockwise, and look like two wheels spinning, from which the name *Rotifera*—wheel bearer—is derived.

The foot is a narrow posterior extension of the trunk and commonly terminates in a pair of adhesive glands, which open to the exterior through tubes, called toes (Fig. 23.3*A*). In some rotifers, the foot is divided into ringlike sections, which can telescope into one another (Fig. 23.2).

Rotifers swim by means of the ciliated crown and crawl in a leech-like fashion, using the terminal adhesive glands as a means of attachment. Most rotifers are found on algae and bottom debris and swim and crawl intermittently. Some rotifers are planktonic and are never found on the bottom (Fig. 23.3*C*), and a few are sessile and live in tubes (Fig. 23.3*B*).

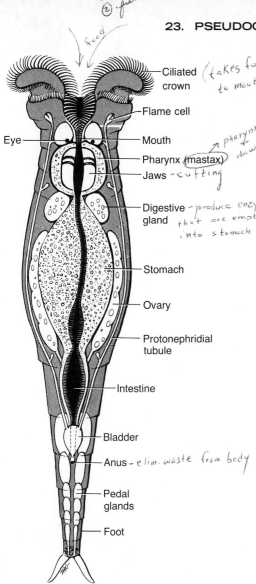

① small
② fresh water

food

Ciliated crown — (takes food to mouth)

Flame cell

Eye

Mouth

Pharynx (mastax) — pharynx + jaws

Jaws — cutting

Digestive gland — produce enzyme that are emptied into stomach

Stomach

Ovary

Protonephridial tubule

Intestine

Bladder

Anus — elim. waste from body

Pedal glands

Foot

Figure 23.2 Ventral view of a rotifer, *Philodina roseola*, showing many of the internal structures. (Redrawn from Hyman.)

A second distinguishing feature of rotifers is the pharyngeal apparatus, called the **mastax.** The mastax differs from the pharynx of other pseudocoelomates in possessing seven cuticular teeth-like pieces projecting into the lumen (Fig. 23.2). The teeth vary in size and shape depending upon the feeding habit. Predatory rotifers have a mastax which can project from the mouth and grasp and seize prey, mostly protozoans and other rotifers (Fig. 23.4*B*). Rotifers with disc-like ciliated crowns are filter feeders. The beating cilia drive small particles into the mouth. The mastax of these

rotifers contains broad flattened pieces adapted for grinding (Fig. 23.4*A*).

Digestion is probably extracellular within the large stomach. A short intestine passes from the stomach to the anus, which is located on the ventral side at the end of the trunk.

Reproduction is entirely sexual. The male, which is usually smaller than the female, stabs the female on any part of her body with his penis. This type of copulation, called **hypodermic impregnation,** occurs in other animals, including some flatworms. The sperm then migrate to the eggs for fertilization. During the life of a female, 10 to 20 shelled eggs, one for each ovarian nucleus, are deposited singly on the bottom or are attached to the body of the female. When the females hatch, they have all of the adult features and attain sexual maturity after a growth period of a few days. The smaller males are sexually mature on hatching. Sessile rotifers are free-swimming for a brief period after hatching.

Many fresh-water species produce both rapidly hatching thin shelled eggs and dormant thick shelled eggs. **Parthenogenesis** is common. In some species, both parthenogenesis and development from fertilized eggs occur. But in other species there are no known males, and all individuals are females produced parthenogenetically. Parthenogenesis is perhaps an adaptation for a rapid expansion in number of a population. Populations of few individuals would be common in fresh-water pools and streams which are subject to desiccation and other extreme environmental conditions.

As might be expected, the rotifers living on mosses are capable of withstanding extreme environmental conditions. During the short periods of time when these plants are filled with water, the animals swim about in the water films on the leaves and stems. They are capable of undergoing **desiccation,** usually without the formation of protective covering, and can remain in a dormant state for as long as three to four years (Fig.23.5). The resistance to both low temperatures and lack of moisture in such a dormant state is remarkable. Some species have been placed in liquid helium ($-272°C$) and in vacuum desiccators without suffering damage. The few other kinds of animals which live on mosses have similar abilities.

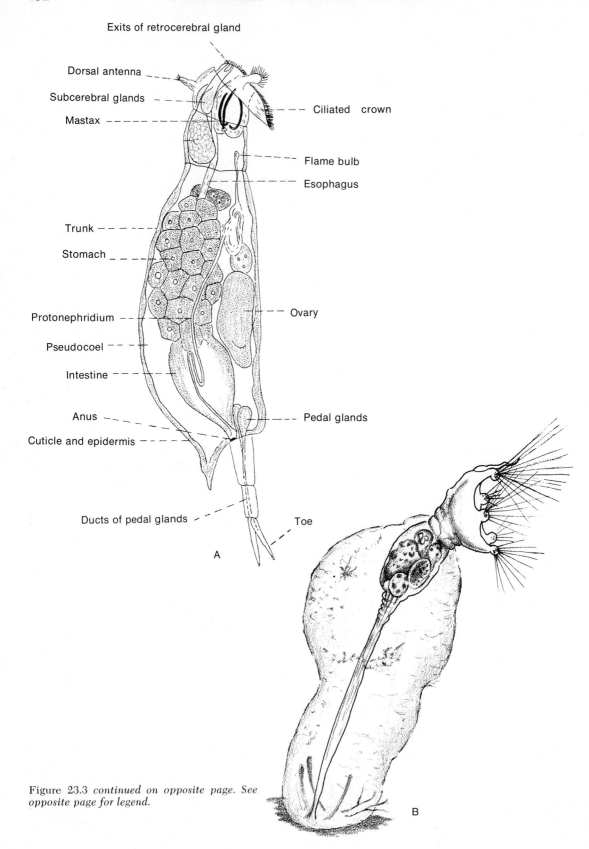

Exits of retrocerebral gland

Dorsal antenna

Subcerebral glands

Mastax

Ciliated crown

Flame bulb

Esophagus

Trunk

Stomach

Protonephridium

Ovary

Pseudocoel

Intestine

Anus

Pedal glands

Cuticle and epidermis

Ducts of pedal glands

Toe

A

B

Figure 23.3 *continued on opposite page. See opposite page for legend.*

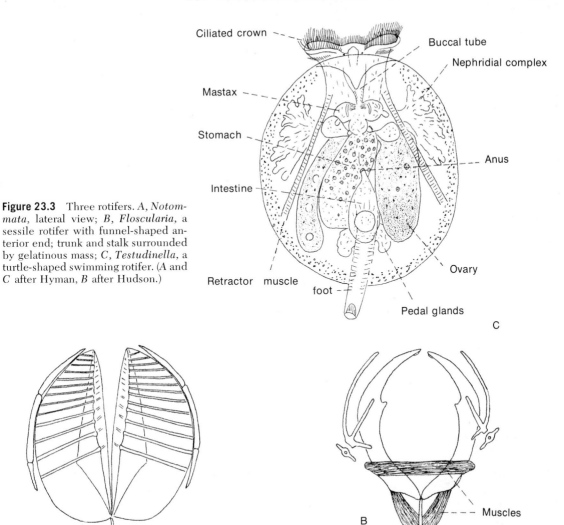

Ciliated crown

Mastax

Stomach

Intestine

Buccal tube

Nephridial complex

Anus

Ovary

Retractor muscle

foot

Pedal glands

C

Figure 23.3 Three rotifers. *A, Notommata,* lateral view; *B, Floscularia,* a sessile rotifer with funnel-shaped anterior end; trunk and stalk surrounded by gelatinous mass; *C, Testudinella,* a turtle-shaped swimming rotifer. (*A* and *C* after Hyman, *B* after Hudson.)

A

B

Muscles

Figure 23.4 Two types of mastax teeth. *A,* Mastax with teeth adapted for grinding. Note the two large ridged plates that provide grinding surfaces. *B,* Mastax with grasping, forceps-like teeth. (*A* after Beauchamp from Hyman; *B* after Hyman.)

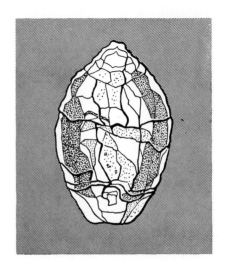

Figure 23.5 A desiccated rotifer from a dried-up puddle. It can absorb water and become active again in a few minutes.

23.2 PHYLUM GASTROTRICHA

Gastrotricha constitutes a smaller phylum of pseudocoelomates. Its members are found in the sea and in fresh water but are less common than rotifers. Gastrotrichs are about the same size as rotifers. The body is elongate and the ventral surface is flattened and ciliated, from which the name *Gastrotricha*—stomach hairs—is derived. The anterior end may bear bristles or tufts of cilia. The posterior end is sometimes forked (Fig. 23.6). The **cuticle** of gastrotrichs is commonly in the form of scales, which are sometimes ornamented with spines (Fig. 23.6A). Adhesive tubes are present, often in rows along the sides of the body.

Gastrotrichs glide over the bottom propelled by the ventral cilia and may temporarily attach with the adhesive tubes. They feed on bacteria, small protozoa, algae, and detritus. Food is swept into the anterior mouth by cilia or is pumped in by the muscular pharynx.

In contrast to most other pseudocoelomates, gastrotrichs are **hermaphroditic.** However, in most fresh-water species, the

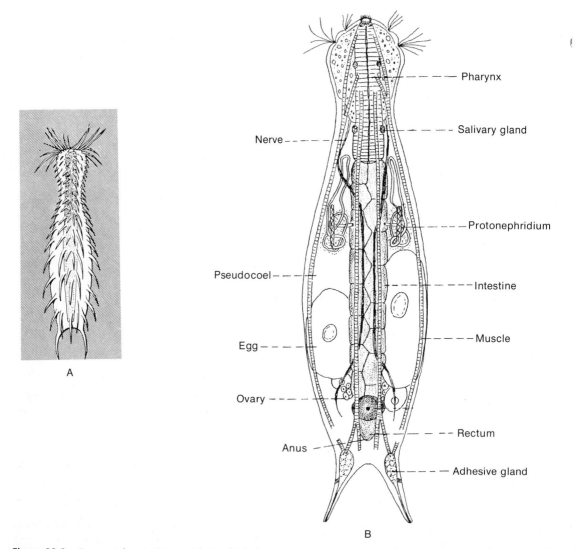

Figure 23.6 Gastrotrichs. *A*, Gastrotrich in which the cuticle is ornamented with spines; *B*, Internal structure of a gastrotrich, ventral view. (*B* modified after Zelinka from Pennak.)

male system is degenerate, and all individuals are thus parthenogenetic females. As in fresh-water rotifers, both dormant and non-dormant eggs are produced.

23.3 PHYLUM NEMATODA

The more than 10,000 species of nematodes, or roundworms, constitute the largest phylum of pseudocoelomates and one of the most widespread and abundant groups of animals. They occur in enormous numbers in marine and fresh-water bottom sediments and in the water films around soil particles. One square meter of bottom mud off the Dutch coast has been reported to contain as many as 4,420,000 nematodes. Several hundred billion nematodes may be present in an acre of good farmland. Decomposing plant and animal bodies often contain large populations of nematodes. For example, examination of *one* decomposing apple revealed some 90,000 nematodes belonging to several different species.

The phylum includes many parasitic species, exhibiting all degrees of parasitism on both plant and animal hosts. Food crops, domestic animals, and man himself are parasitized by many species and consequently roundworms are of tremendous economic and medical importance.

Most free-living nematodes are less than a millimeter in length. While parasitic nematodes may also be minute, species parasitic in the gut of vertebrates may attain much larger sizes. Many are about the length of a pencil, and the horse nematode, *Parascaris equorum,* reaches a length of 35 cm.

The body of nematodes is long, cylindrical, and tapered at both ends (Fig. 23.7). Lips, small sensory bristles, papillae, and slits encircle the mouth at the anterior end (Fig. 23.8*A*). The anus is located a short distance in front of the posterior end, which in free-living species usually terminates in an adhesive gland. The complex nematode cuticle is composed of several layers, often with a sculptured outer layer. The cuticle is periodically shed, but unlike the exoskeleton of insects and crustaceans, the nematode cuticle grows between molts. Beneath the cuticle is an epidermis composed of cells whose nuclei are restricted to the mid-dorsal, mid-ventral and mid-lateral lines (Fig.

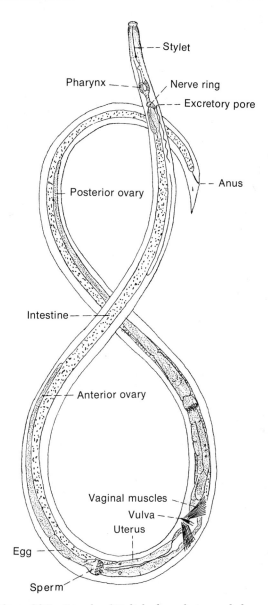

Figure 23.7   Female of *Dolichodorus heterocephalus,* an aquatic nematode. (After Cobb from Chitwood and Allen.)

23.9). The subepidermal muscle layer is composed only of longitudinal fibers. Contractions of the muscle fibers produce undulatory or thrashing locomotor movements.

The mouth opens into a **buccal cavity,** which may be provided with teeth or a **stylet** (Fig. 23.8). The buccal cavity is connected to a tubular muscular **pharynx,** which functions as a pump. The remainder of the gut, a long straight intestine, is the site of digestion and absorption (Fig. 23.7).

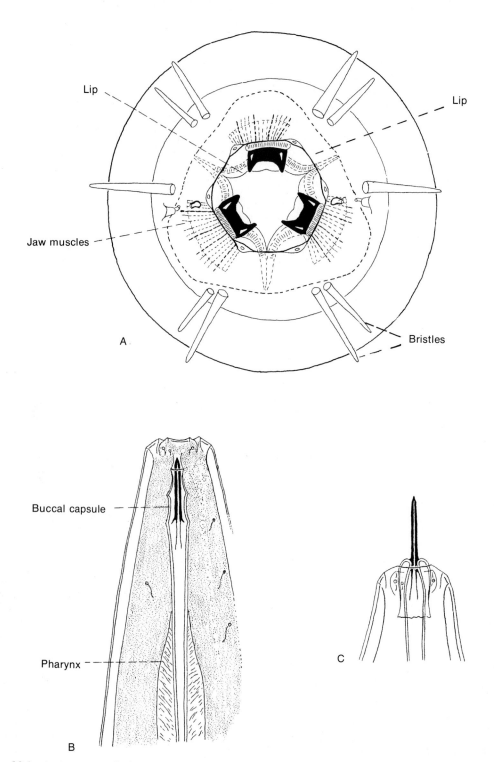

Figure 23.8 *A*, Anterior end of a marine nematode; *B*, anterior end of a predaceous nematode, with a retracted, spear-like stylet; *C*, stylet projected. (*A* after de Coninck from Hyman; *B* and *C* after Thorne from Hyman.)

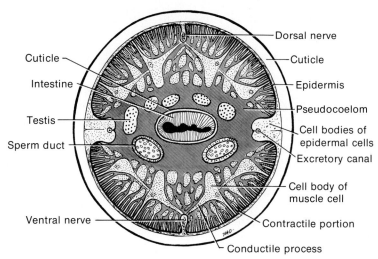

Figure 23.9 Cross section through the middle region of a male *Ascaris lumbricoides*. The testes are sectioned several times because they lie folded in the body.

Many free-living nematodes are predaceous, while others feed on the contents of plant cells. Some consume dead organic matter. Carnivorous nematodes use the teeth or stylet to puncture the prey. For example, *Mononchus papillatus*, a soil nema-tode, consumes as many as 1000 other nematodes during its life span of approximately 18 weeks by attaching its lips to its prey, making an incision with one of the buccal teeth, and then pumping out the contents with the pharynx. Many herbivorous nema-

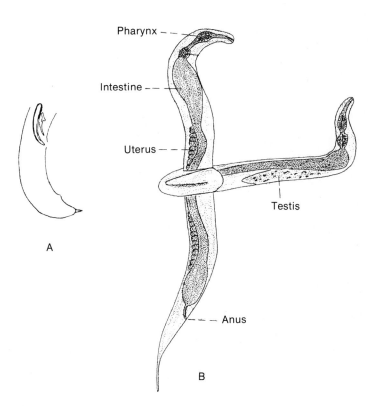

Figure 23.10 *A,* Posterior of a male *Aphelenchoides,* showing right penile spicule. *B,* Termite parasite, *Pristionchus aerivora,* in copulation. (*A* after Cobb from Chitwood and Allen; *B* after Merrill and Ford from Hyman.)

todes use a stylet to puncture plant cells, the contents of which are then pumped out with the pharynx (Fig. 23.8*B*).

The female reproductive system is paired and tubular, and includes two ovaries (Fig. 23.7). Each oviduct empties into a uterus and then into a common vagina which opens to the exterior in the midregion of the body. Male nematodes are usually smaller than females, and the posterior end of the male is curled like a hook (Fig. 23.10*B*). The male reproductive system is a long coiled tube composed of a testis, sperm duct, seminal vesicle, and muscular ejaculatory duct. The latter opens into the rectum so that the anus functions as a gonopore as well as for egestion of wastes. The rectum contains two short, curved spicules, which can be projected from the anus (Fig. 23.10*A*). During copulation, the posterior end of the male is curled around the body of the female in the region of the gonopore. The spicules are used to hold open the gonopore of the female during sperm transmission. The sperm of nematodes are peculiar in lacking a flagellum. The eggs possess a thick shell, which may be sculptured, and the sculpturing is often important in identifying the particular kind of infection present in a fecal smear (Fig. 23.11). Free-living nematodes deposit their eggs in the bottom de-

bris and soil in which they live. The young have most of the adult features on hatching and undergo several molts before attaining maturity. Adults do not molt but continue to grow.

Parasitic Nematodes. Parasitic nematodes attack both plants and animals and exhibit all degrees of complexity in their relationship with the host and in their life cycle. Some species are parasitic only during the juvenile stages; others are parasitic only during adulthood; still others are parasitic throughout their life cycle. The life cycle may be completed within a single host, or one or two intermediate hosts may be required.

Ascarids. The **ascaroid** nematodes, which are intestinal parasites of man, dogs, cats, pigs, cattle, horses, chickens and other vertebrates, are entirely parasitic within a single host. The life cycle typically involves transmission by the ingestion of eggs passed in the feces of another host. The juvenile stages, usually called larvae, are not confined to the intestinal lumen and frequently involve temporary invasion of other tissues of the host.

A typical ascarid life cycle is that of *Toxicara canis*, a parasite of dogs. The adult male of this species reaches 10 cm. in length and the female 18 cm. Eggs begin develop-

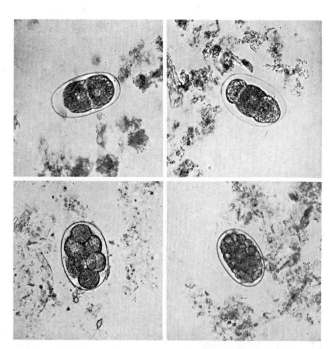

Figure 23.11 Human fecal smear containing eggs of the hookworm *Necator americanus*. (From Davidsohn and Henry: Todd-Sanford Clinical Diagnosis by Laboratory Methods, 14th ed. W. B. Saunders Company.)

ment in the host's feces and become infective in about five days. If ingested by another dog, the eggs hatch in the small intestine. The young worms penetrate the intestinal walls, enter the venous system and eventually are carried by the blood to the lungs. Here they break into the alveoli and migrate up the air passageways. They are swallowed when they reach the pharynx and finally lodge in the intestine.

The human ascarid, *Ascaris lumbricoides,* is one of the best known parasitic nematodes. Physiological studies suggest that *Ascaris* produces enzyme inhibitors that protect the worm from the host's digestive enzymes. The ascarids feed on the host's intestinal contents.

Hookworms. The **hookworms** are another group of parasites of the digestive tract of vertebrates. Most members of this group feed on the host's blood. The mouth region is usually provided with cutting plates, hooks, teeth or combinations of these structures for attaching and lacerating the gut wall (Fig. 23.12). A heavy infestation of hookworms can produce serious danger to the host through loss of blood and tissue damage. Hookworms are one of the parasites causing serious infections in humans. It is estimated that millions of people in

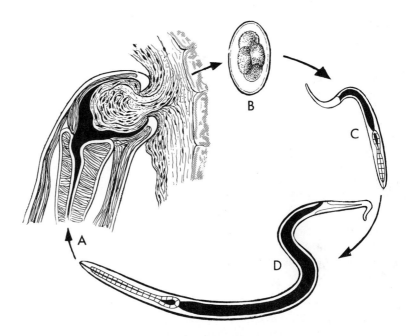

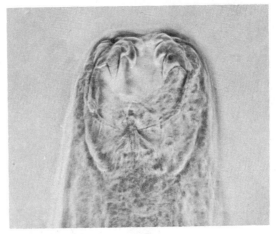

Figure 23.12 Hookworm. A, Longitudinal section through head of adult showing mouthful of intestinal wall being sucked. Eggs (*B*) pass out in the host feces, hatch in the soil (*C*) and grow to the infective stage (*D*). These penetrate the host skin and migrate by way of the blood, lungs, and throat to the small intestine. *E*, Anterior end of a dog hookworm, showing buccal region and teeth. (*A* after Ash and Spitz; *B, C,* and *D* after Chandler; *E* from Barnes: Invertebrate Zoology, 2nd ed.)

E

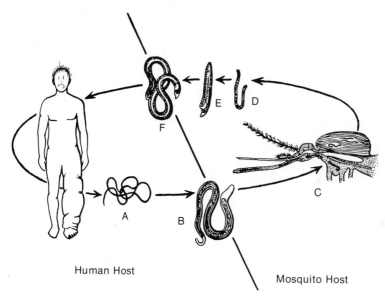

Figure 23.13 *Wuchereria bancrofti.* Adult worms in human lymphatic tissue (*A*) release microscopic larvae into the blood (*B*). If these are taken up by a mosquito (*C*) they migrate to the thoracic muscle, where they metamorphose and grow (*D, E* and *F*). The infective stage (*F*) migrates to the proboscis where it can penetrate into man while the insect is feeding.

India, Pakistan and Indonesia are infected with the Old World species, *Ancylostoma duodenale.*

The life cycle of hookworms involves an indirect migratory pathway by the juveniles, as in ascaroids. The fertilized eggs leave the host in its feces and hatch outside the host's body on the ground. The larva gains re-entry by penetrating the host's skin (feet in man) and is carried in the blood to the lungs. From the lungs the larva migrates to the pharynx where it is swallowed and passes to the intestine. Not all species of this group gain re-entry by skin penetration. Some, such as dog hookworms, have a pattern similar to the dog ascarid.

Filarioids. The **filarioid** nematodes have life cycles requiring an intermediate host.

The filarioids are threadlike worms inhabiting the lymphatic glands, coelom, and some other sites in the vertebrate host, especially birds and mammals. The female is **ovoviviparous** and the larvae are called **microfilariae.** Bloodsucking insects, such as fleas, certain flies, and especially mosquitoes, are the intermediate hosts. A number of species parasitize man, producing **filariasis.**

The chiefly Asian *Wuchereria bancrofti* illustrates the life cycle. The male is 40 mm. by 0.1 mm. and the female is about 90 mm. by 0.24 mm. (Fig. 23.13). Adults live in the lymph glands of man. The microfilariae are found in the blood and are present in the peripheral blood stream. When certain species of mosquitoes bite the host, the mi-

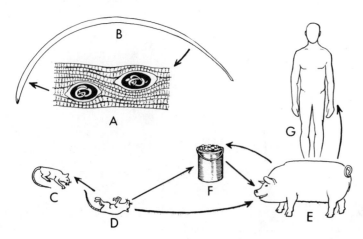

Figure 23.14 *Trichinella spiralis.* Larvae encysted in muscle (*A*) mature into intestinal worms when eaten (*B*). These give birth to larvae that burrow into the host, encysting in muscle. The natural reservoir is rodents (*C*) and similar animals which eat their dead (*D*). Pigs (*E*) will also eat dead rodents. Furthermore, killed rodents and pig scraps are fed to them in garbage (*F*). Man can become infected by eating insufficiently cooked meat containing larvae.

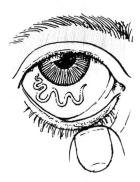

Figure 23.15   Adult of the African eye worm, *Loa loa*, visible in the white of the eye. (After Fülleborn.)

crofilariae enter the mosquito with the host's blood. Development within the intermediate host involves a migration through the gut to the **thoracic** muscles and after a certain period into the **proboscis.** From the proboscis the microfilariae are introduced back into the primary host when the mosquito feeds (Fig. 23.13). In severe filariasis, the blocking of the lymph vessels by large numbers of worms results in extreme edema of host tissues, particularly in the legs, breast, and scrotum. Such enlargement is called **elephantiasis** (Fig. 23.13).

Some other important nematode parasites of man are *Enterobius vermicularis*, intestinal pinworms of children; *Trichinella spiralis*, the juvenile stages of which encyst in the striated muscles, causing trichinosis (Fig. 23.14); *Loa loa*, the African eye worm (Fig. 23.15); and the 120 cm. long guinea worm, *Dracunculus medinesis*, which in the final part of its life cycle migrates to the subcutaneous tissue and produces an ulcerated opening to the exterior.

CLASSIFICATION OF PSEUDOCOELOMATES

(Classification of the pseudocoelomate phyla is technically difficult; therefore, we will supply no explicit listing of classification here. The interested student can consult the references below and at the end of Chapter 18 for further information.)

ANNOTATED REFERENCES

Detailed accounts of the pseudocoelomate phyla may be found in the references listed at the end of Chapter 18. Parasitic forms are described in the parasitology texts at the end of Chapter 22. The following are a few works devoted to nematodes.

Goodey, T.: Soil and Freshwater Nematodes. New York, John Wiley & Sons, Inc., 1951.
Lee, D. L.: The Physiology of Nematodes. San Franscisco, W. H. Freeman & Co., 1965.
Levine, N. D.: Nematode Parasites of Domestic Animals and of Man. Minneapolis, Burgess Publishing Co., 1968.

24. The Mollusks

24.1 COELOM AND COELOMATES

All the remaining members of the Animal Kingdom either have a coelom or have evolved from coelomate ancestors. The coelom appears during development as a cavity within the mesoderm (Fig. 24.1), and has a mesodermal lining, the **peritoneum.** Internal organs may bulge into the coelom, but they are still retroperitoneal, i.e., covered by peritoneum and not literally "in" the coelomic cavity.

Neither the manner in which the coelom evolved nor its original adaptive significance is known. Given the varieties of development discussed in Chapter 6, it can

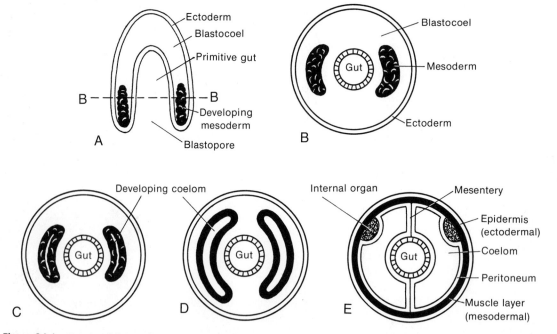

Figure 24.1   Origin of the coelom in protostomes. *A*, Frontal section of a late gastrula. *B*, Cross section of gastrula at level indicated in *A*. *C*, Coelom developing as a split (schizocoel) in the mesodermal mass. *D*, Further development of the mesoderm and coelom. Note that the old blastocoel is gradually obliterated. *E*, Cross section through an adult coelomate protostome. Muscle layer develops from mesoderm. Internal organs lie behind peritoneum (retroperitoneal), which lines coelom.

be seen that the coeloms of living animals arise in a number of different ways. They also serve a variety of functions, among them: (1) as a medium for transport of substances; (2) as a site for the deposition of nitrogenous wastes before the removal of these wastes by excretory organs; (3) as a space in which gametes may undergo development; or (4) as a space for growth and development of organs during embryogeny.

Yet another evolutionary puzzle in coelomates involves the development of circulatory systems. Metabolically active animals generally have closed circulatory systems, but animals with *closed* systems do not always have evolutionary ancestors with *open* systems; nor do all animals with low metabolic rates have open systems. Apparently, closed systems evolved independently a number of different times in the evolutionary history of animals.

Comparative studies of the development of coelomates led to the recognition of two major groups of metazoan phyla, the **protostomes** and the **deuterostomes.** These groups probably represent two principal lines of evolution within the Animal Kingdom. The protostome phyla—mollusks, annelids, arthropods, and certain lesser phyla—are characterized by **determinate spiral cleavage;** the **mesoderm** can be traced back to a single cell of the **blastula.** The **coelom** is a schizocoele, formed by an internal splitting of the mesodermal masses, a process called schizocoely (Fig. 24.1*C*). The **mouth** originates from the blastopore, hence the name protostome—first mouth. Protostomes commonly possess a **trochophore** larva (Fig. 24.6) described on page 506. Although flatworms lack a coelom, their development by spiral cleavage justifies classifying them as acoelomate protostomes (page 412).

The features just described do not characterize the embryogeny of all protostomes; rather, they represent the primitive or basic plan of protostome development and are most commonly encountered among marine species. They have been partly or completely lost in the many species whose reproductive circumstances have led to marked changes in the amount and distribution of yolk material (in turn, the amount and distribution of yolk greatly influence the cleavage pattern) and to the suppression of larval stages, a characteristic of most fresh-water and terrestrial animals and even of some marine species.

In contrast to protostomes, the deuterostome coelomates are characterized by **indeterminate radial cleavage.** The mouth appears during development as a new opening at the opposite end of the embryo from the blastopore, hence the name deuterostome—second mouth. The formation of the mesoderm and coelom is quite different from that of protostomes (p. 641).

24.2 GENERAL FEATURES OF THE PHYLUM MOLLUSCA

The phylum **Mollusca,** the second largest phylum of animals, includes more than 128,000 living and 35,000 fossil species. Mollusks are about as well known taxonomically as mammals and birds. This reflects, at least in part, the long popularity of shell collecting.

Two distinguishing characteristics of the mollusks are the adaptation of the ventral surface of the body as a muscular foot for locomotion, and the modification of the integument of the dorsal surface, the **mantle,** for the secretion of a protective calcareous shell. The shell has been lost in some mollusks, but the mantle almost always remains.

24.3 THE ANCESTRAL MOLLUSKS

We have no direct evidence of the first mollusks. The phylum evolved in the Archeozoic seas and by the time of the Paleozoic, when the fossil record first becomes clear, the different molluscan classes were already defined. Nevertheless, we can make some inferences about the nature of the ancestral mollusks on the basis of what can be observed in living forms. The characteristics shared by some or all of the seven classes of mollusks probably were present in the ancestral form. Although it is not particularly important that we know the nature of the first mollusk, such a hypothetical ancestor can serve as an introduction to the general features of the phylum.

Biologists postulate that the ancestral mollusk was small, perhaps about two centimeters in length, and adapted for living on hard rocky bottoms in the Archeozoic seas (Fig. 24.2). By piecing together bits of evidence we can infer that the animal's ventral surface formed a broad, flat, muscular creeping **foot.** The **head** was well developed, bearing a pair of tentacles with a simple **eye** at the base of each. The dorsal surface of the body was covered by a low shield-shaped **shell** secreted by the underlying mantle and composed of one to several layers of calcium carbonate covered on the outside by an organic layer, the **periostracum.** Much of the shell of living mollusks is

secreted by the mantle margin so that the shell grows in diameter as well as in thickness. The shell of the ancestral mollusks provided protection as long as the animal was attached to the rocky substratum; pairs of retractor muscles extending from the foot to the shell permitted the shell to be pulled down over the body.

The overhanging shell and mantle at the posterior end created a large mantle cavity containing two gills and two excretory openings (Fig. 24.2). The **anus** was located at the dorsal side of the cavity opening. The bipec-

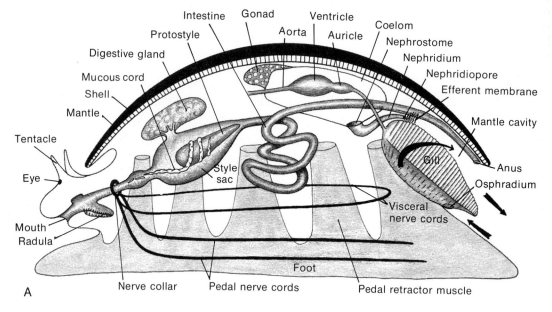

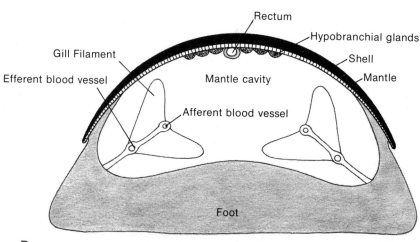

Figure 24.2 *A,* Hypothetical ancestral mollusk (lateral view). Arrows indicate path of water current through mantle cavity. (Adapted from various authors.) *B,* Transverse section through body of ancestral mollusk at level of mantle cavity.

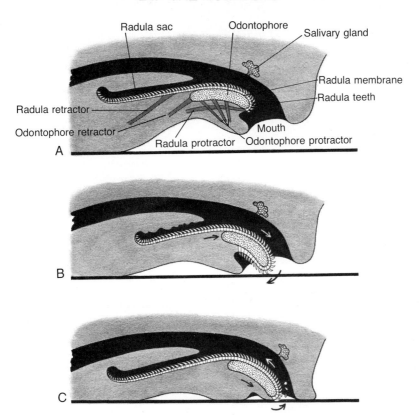

Radula sac   Odontophore   Salivary gland

Radula membrane
Radula teeth

Radula retractor

Odontophore retractor

Mouth
Radula protractor   Odontophore protractor
A

B

C

Figure 24.3 Molluscan radula. *A*, Mouth cavity, showing radula apparatus (lateral view). *B*, Protraction of the radula against the substratum. *C*, Forward retracting movement, during which substratum is scraped by radula teeth.

tinate **gill** structure postulated for the ancestral mollusk is similar to that found in many living species. Each gill was composed of a longitudinal axis to which was attached on either side flattened gill filaments (Fig. 24.2). Blood circulated through the filaments, flowing from a supply to a drainage vessel in the gill axis. The filaments were ciliated and produced a ventilating current of water that entered the lower part of the mantle cavity, made a U-turn across the gills, and then flowed out of the cavity dorsally and posteriorly (Fig. 24.2). Wastes discharged by the **nephridia** and anus were removed in the exhalant stream of water.

As in most living mollusks, the mouth cavity of the ancestral mollusk probably contained a **radula,** a unique rasping organ consisting of a belt-like membrane bearing a large number of teeth (Fig. 24.3). The radula rested upon a cartilaginous supporting skeleton, the **odontophore,** supplied with protractor and retractor muscles. The ancestral mollusks were probably **microphagous**—they fed upon small particles of

algae scraped with the radula from rocks.

In living mollusks the end of the radula and odontophore can project from the mouth and lick the adjacent surface, like the tongue of a mammal. New teeth are secreted at the posterior end of the radula as old teeth are worn away anteriorly. Particles of algae brought into the mouth are enveloped by mucus secreted by the **salivary glands.** A rope of mucus containing the algal particles then passes posteriorly through an **esophagus** to the **stomach.** The posterior end of the stomach, the **style sac,** is ciliated and rotates the mucous mass, winding in the string of mucus like a windlass winds in a rope (Figs. 24.2 and 24.4). Particles dislodged from the mucus are sorted by a ridged ciliated area at the anterior end of the stomach (Fig. 24.4). Large undigestible particles are conducted posteriorly to the **intestine,** and fine digestible particles are conveyed to the ducts of a pair of **digestive glands** located to either side of the stomach. Digestion occurs intracellularly within the digestive

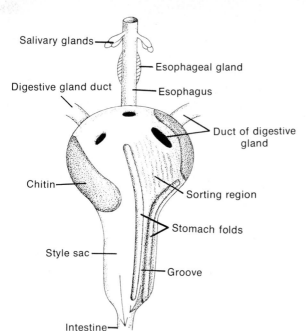

Salivary glands

Esophageal gland

Digestive gland duct

Esophagus

Duct of digestive gland

Chitin

Sorting region

Stomach folds

Style sac

Groove

Intestine

Figure 24.4 Diagram of a primitive molluscan stomach. (Modified from Owen.)

gland. The long intestine functions only in the formation of feces.

An open blood vascular system provides for internal transport in all mollusks and was undoubtedly characteristic of the ancestral form. The heart, located within the coelomic cavity, is composed of a **ventricle** and two lateral **auricles,** which receive blood from each gill (Fig. 24.2). The contractile ventricle forces blood via an anterior aorta to the tissue spaces of the body—head, foot, visceral mass, and mantle. Blood then collects within larger sinuses from which it is returned to the heart by way of the gills (Fig. 24.5A).

The pair of excretory organs, often called **kidneys,** are **metanephridia.** In contrast to the blind protonephridia of flatworms and pseudocoelomates, these tubes open at the inner end into the coelom by way of a ciliated **nephrostome** (Fig. 24.5B). The outer end opens into the mantle cavity. Like most excretory organs, the molluscan nephridia function to concentrate metabolic wastes in urine. Special glandular tissue lining the coelom secretes waste into the coelomic fluid. Fluid containing wastes collected by the blood from body tissues filters from the heart into the coelomic fluid. Coelomic fluid containing wastes passes into the inner end of the nephridium. As the fluid passes

through the nephridial tube, some substances are selectively reabsorbed into the body and others are secreted by the cells into fluid. The resulting mixture, urine, is expelled into the mantle cavity and flushed out by the exhalant ventilating current.

The nervous system of the ancestral mollusk was probably composed of a nerve ring around the esophagus, from the underside of which extended a ventral pair of pedal cords innervating the foot, and a more dorsal pair of visceral cords innervating the mouth and the organs of the visceral mass (Fig. 24.2). Sense organs included the **tentacles,** a pair of simple eyes at the base of the tentacles, a pair of **statocysts** in the foot, and the **osphradia.** An osphradium is a patch of sensory epithelium near each gill, which monitors chemical substances in the water current passing through the mantle cavity (Fig. 24.2).

A pair of **gonads** were located to the front and sides of the coelom (Fig. 24.2). Eggs and sperm were released from the gonads into the coelom and were then carried to the mantle cavity by way of the nephridia. In the ancestral mollusks fertilization was probably external either within the mantle cavity or in the surrounding sea water.

In living mollusks cleavage is typically spiral and the resulting gastrula develops

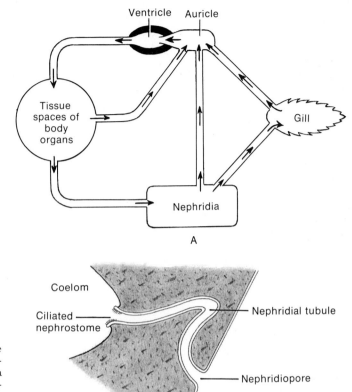

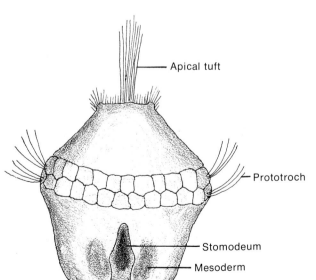

Figure 24.5 *A,* Diagram showing the principal features of the molluscan circulatory system. Auricles, gills and nephridia are usually paired. *B,* Diagram of a metanephridium.

Figure 24.6 The trochophore larva of *Patella,* a marine snail. (After Patten.)

into a free-swimming trochophore larva (Fig. 24.6). The larval body is ringed about the middle with a girdle of **cilia,** the **prototroch,** and the anterior pole bears an apical tuft of cilia. A digestive tract is present, and the mouth is located below the prototroch. The anus opens at the posterior pole. The beating cilia of the prototroch provide for locomotion and may also serve to collect fine plankton for food.

In the ancestral mollusks the trochophore probably developed directly into the adult body; the body lengthened behind the prototroch, the foot formed behind and below the mouth, and mantle and shell developed on the dorsal side. The young animal settled to the bottom to take up the adult mode of existence.

Using this hypothetical ancestral form as a basis for comparison, we will examine five of the seven classes of mollusks, omitting two small classes, the **Scaphopoda,** or tooth shells, and the **Aplacophora.**

24.4 CLASS GASTROPODA

Among the 35,000 living species of gastropods, the largest and most diverse class of mollusks, are snails, whelks, conchs, limpets, and sea slugs. Most gastropods are marine, but some live in fresh water and others have become adapted for life on land.

Torsion and Shell Spiraling.   Gastropods, like the hypothetical ancestral mollusk, possess a well developed **head,** a broad, flat, creeping **foot,** and a **shell** composed of a single piece. However, gastropods are distinguished from all other mollusks by the curious twisting of the visceral mass which occurs during development (Fig. 24.7). This condition, called **torsion,** involves a 180° counterclockwise (viewed from above) twist which results in the **mantle cavity** and **anus** being located at the anterior end of the body. The gut, nervous system and blood vascular system are correspondingly twisted. During embryonic development

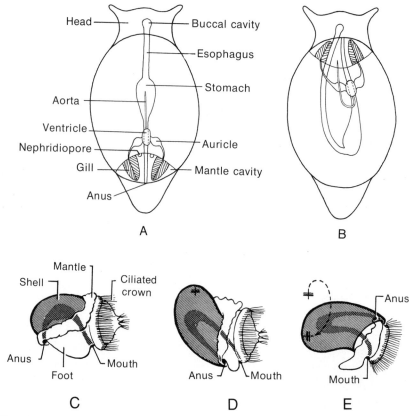

Figure 24.7   Dorsal view of hypothetical ancestral gastropod. *A,* Prior to torsion; *B,* after torsion; *C* to *E,* embryonic torsion in the gastropod *Acmaea* (a limpet). (*A* and *B* after Graham; *C, D* and *E* after Boutan, 1899.)

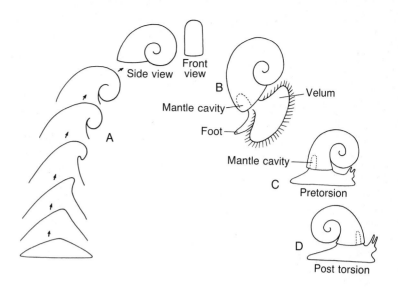

Figure 24.8 *A,* Evolution of a plano-spiral shell. Height of the shield-like shell of hypothetical ancestral mollusk increases and a peak forms. Peak is pulled forward and coiled under. Aperture is reduced and animal can withdraw into spiral shell, which is more compact and less awkward to carry than would be a straight conical shell. Note that shell is bilaterally symmetrical. *B,* Diagram of a gastropod larva before torsion. Velum is the ciliated larval locomotor organ. *C,* Hypothetical pretorsion "gastropod" with a plano-spiral shell. *D,* Post-torsion gastropod. Torsion does not affect plano-spiral shell except to place coils of shell posteriorly. Mantle cavity is now anterior.

the mantle cavity forms first at the posterior end of the body but shifts to an anterior position (Fig. 24.7, *C* to *E*). The evolutionary significance of torsion is uncertain, although several theories have been put forward to account for it.

The spiral shell of gastropods is *not* a consequence of torsion; indeed, the spiral shell is formed during development before torsion occurs. Paleontological evidence also indicates that shell spiraling is not a direct consequence of torsion. The shield-like shell of the ancestral mollusk protected only its dorsal surface, and the animal would have been vulnerable when dislodged from its substratum. In the course of evolution the gastropod shell became higher and more cone-like, with a reduced aperture (Fig. 24.8). Such a shell provided a protective retreat into which the entire body could be withdrawn. The spiral condition appears to have been an adaptation for making the shell more compact and less awkward to carry than if it were a long cone.

The gastropod shell is balanced obliquely

upon the body with the spire directed posteriorly and upward on the right side (Figs. 24.9 and 24.12). The large first whorl of the shell bulges into the right side of the mantle cavity (Fig. 24.9). The partial occlusion of the mantle cavity has resulted in the reduction and loss of the right gill and in turn the reduction and loss of the right auricle and right kidney. Some primitive gastropods still have two gills (Fig. 24.10), two auricles and two kidneys, but most have only those on the left side.

Evolution of Water Circulation and Gas Exchange. Whatever the evolutionary significance of torsion may have been, the anterior position of the anus imposed a sanitation problem on the early gastropods, for waste would have been dumped on top of the head. The problem has been solved in most aquatic gastropods by the evolution of structures to generate an oblique water current. Except in certain primitive species which have the ancestral bipectinate gills (Fig. 24.10), the single left gill lacks the filaments on one side and is anchored by

Figure 24.9 Position in which shell is carried in gastropods. Slot (*A*) which is found in some primitive gastropods indicates the original mid-dorsal line of the mantle cavity. Note that the mantle cavity is restricted largely to the left side. (*B* after Yonge.)

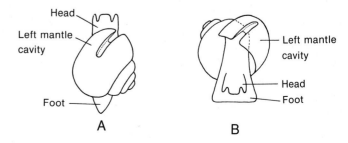

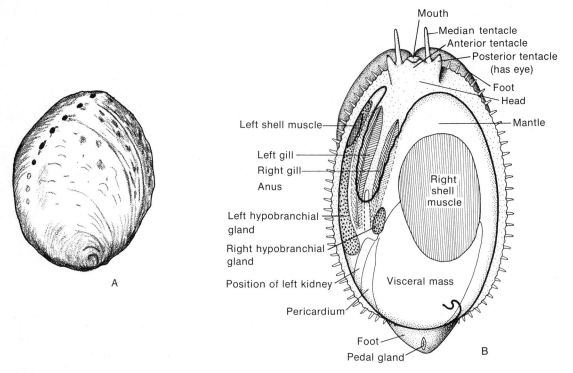

Figure 24.10   The abalone, *Haliotis*. A, Dorsal view of shell, which consists largely of a single expanded whorl. B, Ventral view, showing two bipectinate gills. Note that the right gill, which is lost in most gastropods, is much smaller than the left. (After Bullough.)

the axis to the mantle wall. Water enters the mantle cavity on the left side of the head, makes a turn across the gills, and exits on the right side of the head (Fig. 24.11). The anus is located at the right edge of the mantle cavity and wastes are removed by the exhalant water current. Commonly the edge of the mantle on the left side is drawn out to form a siphon into which the inhalant water stream passes (Fig. 24.12).

Most marine snails, including the common whelks, conchs, slipper shells, and limpets, comprise the subclass **Prosobranchia.** The name Prosobranchia—"front gills"—refers to the position of the gills in the anterior mantle cavity. From the prosobranchs two other large subclasses of gastropods evolved. Members of the subclass **Opisthobranchia**—the bubble shells, sea butterflies, sea hares, and sea slugs—are characterized by a degree of detorsion, or untwisting of the visceral mass, from which the name Opisthobranchia (back gill) is derived. The significance of detorsion is as uncertain as torsion, but it has been accompanied by reduction of the shell and mantle cavity. Primitive opisthobranchs, such as the bubble shells, possess a shell and look

like prosobranchs but the mantle cavity and gill are located on the right side as a result of detorsion. Other opisthobranchs exhibit such a degree of shell reduction or loss that they have secondarily become bilaterally symmetrical. Bilateral opisthobranchs include sea hares (Fig. 24.13) and the beautiful sea slugs. The shell and mantle cavity of sea slugs have disappeared and gas exchange occurs either through secondary gills located around the now posterior anus or across the external mantle surface. In the latter group the mantle surface is greatly increased by a large number of club-shaped projections.

The term *secondary* refers to the reattainment of a habit, condition, or structure that is similar to some ancestral condition. Most mollusks exhibit a primary bilateral symmetry inherited from the ancestral form. As a result of torsion, gastropods became asymmetrical and lost the primary bilateral symmetry. With detorsion, opisthobranchs such as sea slugs reacquired bilateral symmetry, but since the bilateral symmetry is a new feature evolved from asymmetrical ancestors, it is said to be secondary.

The Prosobranchia gave rise to a third

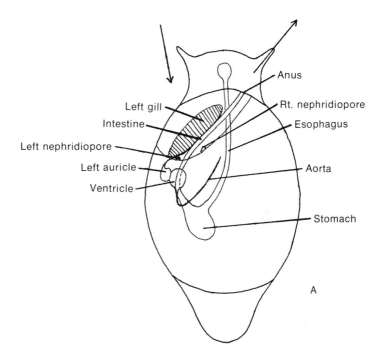

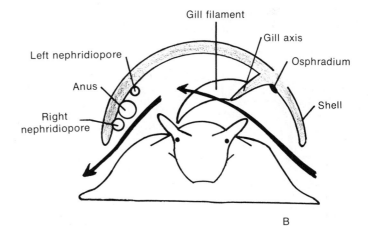

Figure 24.11 Typical prosobranch gastropod, with a single gill and mantle cavity on left side of animal. Arrows indicate path of water current through mantle cavity. *A*, Dorsal view; *B*, diagrammatic transverse section through head and mantle cavity. (After Graham.)

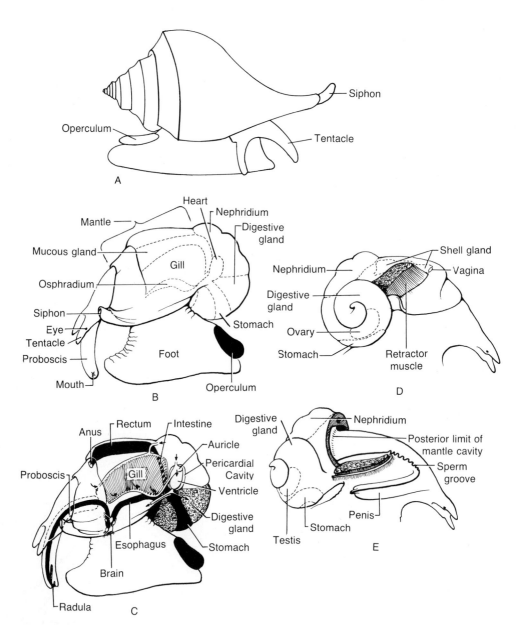

Figure 24.12   Anatomy of *Busycon canaliculatum* (shell removed). *A*, Position of shell when foot is extended. *B*, Left side, showing external organs and internal organs visible through the integument. *C*, Same view with digestive, respiratory, circulatory and nervous systems indicated. *D*, Female, showing portion of the right side. *E*, Male, portion of the right side with mantle and retractor muscle cut short. In *D* and *E* the proboscis is withdrawn.

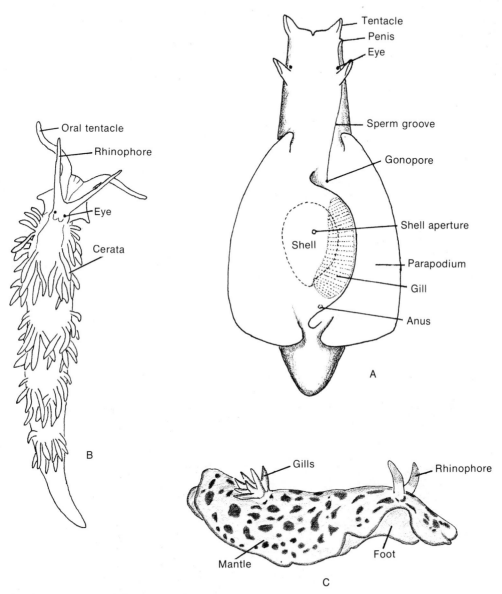

Figure 24.13 Opisthobranch gastropods. *A*, The sea hare, *Aplysia* (dorsal view). *B*, *Eolidia papillosa*, a sea slug (nudibranch). *C*, *Glossodoris*, a sea slug with secondary anal gills. Color in life is yellow with red spots. (*A* after Bullough; *B* after Pierce; *C* after a photograph by Zahl.)

subclass, the **Pulmonata,** adapted for living on land. The mantle cavity of the pulmonates evolved into a lung (Fig. 24.14).

Pulmonates illustrate the general animal adaptations for gas exchange on land. The gills have disappeared, and the mantle wall has become highly vascular and functions as the gas exchange surface. Desiccation is minimized by reducing the opening into the mantle cavity to a very small pore (Fig. 24.14). The animal ventilates its mantle cavity by raising and lowering the floor of the mantle cavity, thereby moving air out of and into the chamber.

Many pulmonates have secondarily returned to an aquatic existence. Some of these fresh-water pulmonates must come to the surface to obtain air; others have abandoned the lung altogether and acquired secondary gills outside of the mantle cavity.

Shell. The shells of gastropods exhibit great variations. The spire may be high or low; the surface may be smooth or highly sculptured; and the coloration may be dull

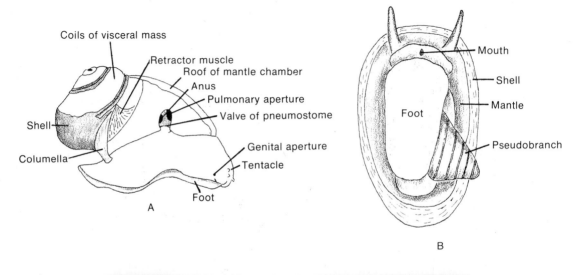

Figure 24.14   Pulmonates. *A,* The land snail, *Helix,* partially dissected to show pore-like opening (pneumostome) into lung, or mantle cavity. *B,* The fresh-water pulmonate *Ferrissia,* a limpet, with secondary gill (ventral view). *C,* A terrestrial slug. Opening into lung seen at the lower edge of the saddle-like mantle. (*A* after Rowett: Histology and Embryology; *B* after Pennak; *C* by Betty M. Barnes.)

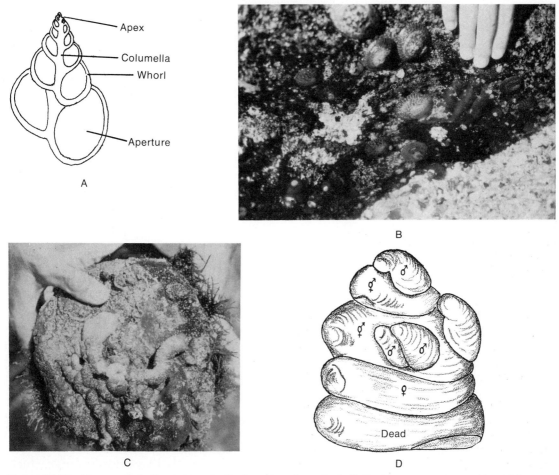

Figure 24.15 Gastropod shells. *A*, Longitudinal section through shell. *B*, *Acmaea*, a common intertidal proso-branch limpet on rocky shores. *C*, *Vermetus*, a worm shell, attached to a rock. *D*, Slipper shell, *Crepidula*. A semi-sessile prosobranch in which the shell represents a single expanded whorl. These commonly occur in clusters, attaching to each other. (*B* and *C* by Betty M. Barnes; *D* after Coe.)

or brightly colored and patterned. The ventral margin of the aperture of the shell of many prosobranchs is drawn out to house the siphon (Fig. 24.12). Another common modification is a greatly expanded last whorl. The spiral apex develops during juvenile stages. Then the last whorl becomes extremely large and covers most of the body of the animal as a low convex plate. Abalones, slipper shells and limpets have shells of this type. The shell of limpets has become secondarily bilateral.

The central axis of the asymmetrical spiral shell is the **columella,** to which the single large rectractor muscle of the body is anchored (Fig. 24.15A). The gastropod withdraws into the shell by folding the middle of the foot. The anterior half of the foot

and the head are withdrawn first, followed by the posterior half of the foot. Many prosobranchs have a round plate, the **operculum,** on the back of the foot, which plugs the aperture when the animal is withdrawn (Fig. 24.12).

Locomotion. Gastropods creep upon the broad, flat ventral foot by a process of body deformation. Small transverse waves of contraction sweep along the foot from back to front, one wave closely following another (Fig. 24.16). The area of the foot in the contracted region is lifted; at relaxation it is replaced on the substratum a little in front of the point from which it was raised (Fig. 24.16B). During each wave of contraction a small section of the foot performs a little step; the summation of all of these little

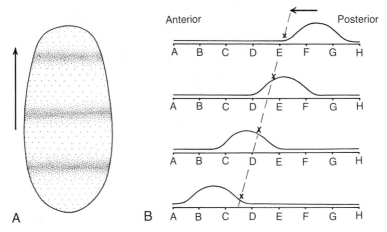

Figure 24.16   Pedal creeping in gastropods. *A*, Ventral view of foot of a gastropod showing three transverse waves of contractions moving from back to front of foot. *B*, Diagrammatic sections of a region of the foot during the course of movement of a wave of contraction (arrow). In region of contraction the foot is lifted from substratum. In the course of contraction a region of the foot (x) is placed back upon the substratum in advance of the point from which it was lifted. Letters indicate fixed positions of substratum. Diagonal dashed line shows forward movement of contracted region of foot.

steps gives the appearance of a gliding motion. The substratum is lubricated by large amounts of mucus produced by a large gland in the foot. Gastropods living on soft substrata such as sand and mud move by means of the cilia which cover the foot surface. Many species living on soft bottoms can burrow into the substratum.

The foot of abalones, limpets, slipper shells and other gastropods living on hard substrata functions as an adhesive organ (Fig. 24.15). The sucker-like foot and the low plate-like shell enable the animal to live attached to rocks that are swept by currents. Some gastropods are semi-sessile or sessile. Slipper shells, for example, move about very little (Fig. 24.15). Worm shells have lost the foot, and the shell, which has uncoiled and become tube-like, is attached to rocks and other objects (Fig. 24.15).

The ability to swim has evolved in a number of different gastropod groups, all of which have the foot modified in some way as a fin. The best swimmers are the prosobranch heteropods and the opisthobranch sea butterflies (Fig. 24.17). The sea hares swim intermittently, moving by beautiful undulations of the lateral fin folds of the foot.

Nutrition.   Primitive gastropods inhabiting rocks are **microphagous,** feeding on fine algae scraped up with the radula. Most gastropods are **macrophagous,** feeding on large food masses, with the radula retained as the feeding organ. Some species, especially pulmonates, feed on algae or terrestrial vegetation; other gastropods are carnivorous. The radula of carnivores usually has large heavy teeth and may be located at the end of an extensible proboscis (Figs. 24.12 and 24.18). The prey—other invertebrates, especially other mollusks—is held with the foot, and the tissues of the victim are torn and devoured with the radula. Some carnivores use the radula as a drill to penetrate the shells of bivale prey, which have been previously softened with secretions from a gland on the foot or proboscis (Fig. 24.19A). Another group of carnivores, the tropical cone shells, kill prey by hurling a long, detached, harpoon-like radula tooth, carrying a poison, into the victim (Fig. 24.18B).

Many aquatic gastropods are scavengers or detritus feeders, and some, such as the Atlantic coast mud snail *Nassarius obsoletus,* may occur in enormous numbers on intertidal beaches (Fig. 24.19). A few filter-feeding gastropods, like the semi-sessile slipper shells, use the gills to filter plankton from the ventilating current.

Some gastropods are ectoparasites of bivalves (Fig. 24.20); others are endoparasites in the body wall of echinoderms such as starfish, or in the coelom of sea cucumbers. The foot of endoparasites is greatly reduced, and the shell is lacking in coelomic parasites. In all, the buccal region is modified for piercing and sucking.

The stomach of macrophagous gastropods

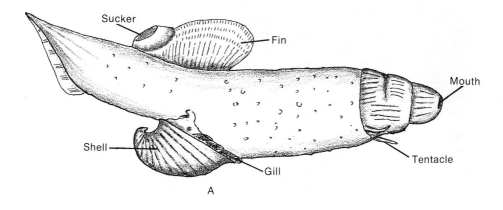

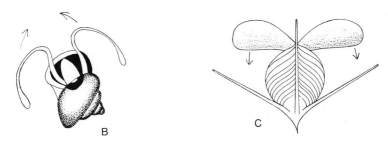

Figure 24.17 Swimming gastropods. *A*, *Carinaria*, a pelagic prosobranch. *B* and *C*, Two opisthobranch sea butter-flies, in which the feet are modified as fins. (*A* after Abbott; *B* and *C* after Morton.)

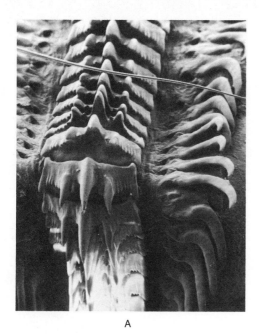

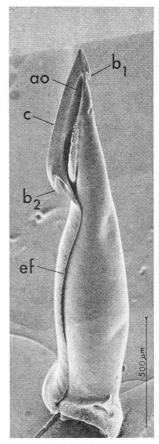

Figure 24.18   Scanning electron photomicrographs of radula teeth. *A*, Radula of a carnivorous drilling gastropod. There are only three teeth to a transverse row but each tooth bears several cusps. *B*, Harpoon radula tooth of the poisonous gastropod *Conus imperialis*. (*A* from Carriker, M. R.: Excavation of boreholes by the gastropod, *Urosalpinx*. Amer. Zool., 9:917–933, 1969; *B* from Kuhn, A. J., et al.: Radula tooth structure of the gastropod *Conus imperialis*. Science, *176*:49–51. Copyright 1972 by the American Association for the Advancement of Science.)

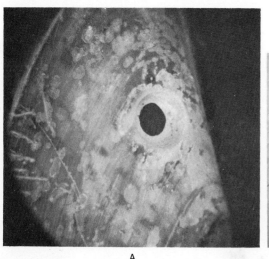

Figure 24.19   *A* A bivalve shell which has been drilled by a gastropod. Note the beveled edge of the hole. *B*, *Nassarius obsoletus* feeding at low tide. This intertidal prosobranch scavenger and deposit feeder occurs in enormous numbers on protected beaches on the east coast of the United States. (By Betty M. Barnes.)

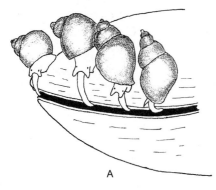 

Figure 24.20 *A, Brachystomia,* an ectoparasitic snail, feeding on the body fluids of a clam. *B, Stylifer,* an endoparasitic snail in the body wall of a starfish. (Both after Abbott.)

has undergone profound changes. The style sac and sorting region have disappeared, and the stomach has the form of a simple sac in which extracellular digestion occurs. Enzymes are provided by the salivary glands, glands along the esophagus, or by the digestive glands. Absorption occurs within the digestive glands and the intestine functions only in the formation of feces.

Circulation, Gas Exchange, Excretion and Water Balance. Except for the asymmetry produced by torsion and the loss of the right gill, the circulatory system of gastropods is like that described for the ancestral mollusks. The blood of most gastropods contains **hemocyanin,** a respiratory pigment.

Ammonia is the excretory waste of aquatic gastropods, but terrestrial pulmonates excrete uric acid. By adding a **ureter** formed from the mantle wall, the excretory opening of land pulmonates is located outside of the mantle cavity. The anus also opens outside the mantle cavity; thus the lung is not fouled with waste. Pulmonates are not especially adapted for avoiding water loss through desiccation, and large amounts of water are lost in the mucus secreted when crawling. Most pulmonates are therefore restricted to humid environments or must be nocturnal. During the winter or during very dry periods the animals hide in leaf mold, beneath wood or stone, or attach the shell to vegetation by mucous cords. The aperture of the shell is then covered with a mucous film, which dries to form a protective diaphragm. The animal is inactive during this time and the metabolic rate drops to a very low level. Such a period of dormancy (termed **estivation**) is a common adaptation of semi-tropical and tropical

animals living in regions with long dry seasons. Nocturnal activity and estivation permit some pulmonates to inhabit deserts and other arid regions.

Sense Organs. The head of gastropods bears one or two pairs of sensory tentacles. In limpets the mantle edge may be fringed with tentacles.

The two eyes located at the base of the tentacles (Figs. 24.12 and 24.21) are not highly developed; indeed, in most species they are not capable of object discrimination. The **osphradium** is an important sense organ, especially in carnivorous gastropods. The mobile siphon moves about, drawing in water from different areas, and the chemoreceptors of the osphradium enable the animal to determine the location of prey or other food.

Reproduction. Most species of prosobranchs have separate sexes. Pulmonates and opisthobranchs are hermaphroditic. In primitive prosobranchs the gametes are discharged through the nephridia. In most gastropods, however, a complex reproductive tract has developed to which the origi-

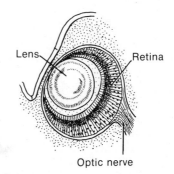

Figure 24.21 Section through the eye of a whelk. (After Helger.)

nal right nephridium contributes only a small part. Copulation followed by internal fertilization is the usual mode of reproduction. The eggs of aquatic species are deposited in strings, masses, or within special cases molded by the foot (Fig. 24.22). The large yolky eggs of pulmonates are laid singly and deposited in small clusters in the soil and leaf mold or beneath bark, logs or stones.

Cleavage is typically spiral. In primitive marine prosobranchs the initial trochophore larva develops further into a **veliger** larva. Most marine prosobranchs pass through the trochophore stage within the egg membranes and hatch as veligers. The veliger larva displays many gastropod features such as a spiral shell and foot (Fig. 24.23). Torsion occurs during the veliger stage (Fig. 24.27). Like the trochophore, the veliger is planktonic and swims by means of the **velum,** a large ciliated organ derived from the prototroch of the trochophore. The velum also collects fine suspended particles upon which the larva feeds. In many marine and in all fresh-water and terrestrial species, development is direct, i.e., no larval stage is present. Little snails emerge from the eggs or cases. In all of these forms with direct development, the eggs contain more yolk than the eggs of species with a feeding larval stage.

24.5 POLYPLACOPHORA AND MONOPLACOPHORA

The **Polyplacophora** and the **Monoplacophora** have a broad, flat, creeping foot. The class Polyplacophora (or Amphineura) includes about 600 species of **chitons** ranging from a few centimeters to over 35 cm. in length and adapted for living on rocks and other hard substrata. In many ways chitons parallel the prosobranch limpets. The head is reduced and much of the ventral surface is occupied by the broad foot (Figs. 24.24 and 24.25). Chitons, however, are distinguished from all other mollusks in possessing a shell composed of eight plates, hence the name Polyplacophora—bearer of many plates. The plates are arranged linearly from anterior to posterior and overlap one another (Fig. 24.24). The lateral margins of the plates are overgrown by the mantle to varying degrees; in some chitons the plates are completely buried in the mantle (Fig. 24.25).

Beyond the plates the mantle forms a heavy girdle surrounding the body (Fig. 24.24). The girdle may be smooth or covered with scales or calcareous spines; it is an important area of sensory reception.

Chitons crawl about much like snails but may be immobile for long periods of time. When the chiton is clinging to rocks and

Figure 24.22   The egg case of *Busycon*. (Photo by Hugh Spencer.)

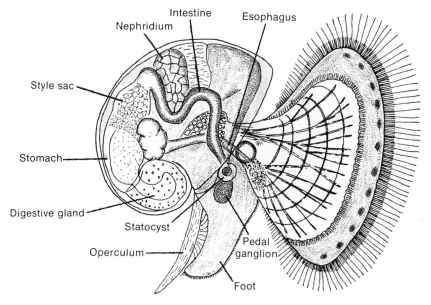

Figure 24.23   Lateral view of the veliger larva of gastropod (slipper shell). (After Werner from Raven.)

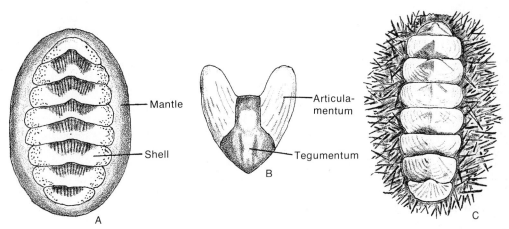

Figure 24.24   *A*, Common Atlantic coast chiton, *Chaetopleura apiculata. B*, Single shell plate of *Katharina. C*, *Chiton.* (*A* after Pierce; *C* after Borradaile and others.)

Figure 24.25   Photograph of two species of chitons from the northwest Pacific. The shell plates of the larger species are completely covered by the mantle. (By Betty M. Barnes.)

other objects, the mantle margin is held tightly against the substratum. The inner edge of the mantle is then lifted to create a partial vacuum. This vacuum, combined with the adhesion of the foot, enables the chiton to hold tenaciously to the substratum. A knife is often required to remove the animals from rocks or shells.

The mantle cavity of chitons is limited to two lateral troughs, or grooves, one on each side of the body between the foot and mantle edge (Fig. 24.26). Within each groove are many gills. If, as we speculated, the first mollusks had only a single pair of gills in a posterior mantle cavity, the gills and mantle cavity of chitons are secondary adaptations, perhaps associated with the flattened body. Indeed, the structure of chiton gills is quite unlike that of primitive gastropods.

The ventilating current, propelled by cilia on the gills, flows through the mantle grooves from anterior to posterior (Fig. 24.26). The current enters each groove through an opening created by the raised mantle margin. After passing the length of the groove and over the gills, the two currents exit as a single stream at the posterior end of the body, where the mantle edge is raised.

Chitons feed largely on algae, scraped from rock surfaces with a radula. The stomach is unlike that described for the ancestral mollusks or gastropods, but a large digestive gland is present.

A single pair of auricles collects blood from all of the gills on each side. A single pair of nephridia opens into the posterior half of the mantle groove (Fig. 24.26).

Chitons have separate sexes and each of the two gonads is provided with a gonoduct that opens near the nephridiopore in the mantle trough (Fig. 24.26). The nephridia do not serve as gonoducts in chitons. The animals do not copulate but sperm released by a male fertilize the eggs in the sea, or fertilization occurs in the mantle trough of the female. Eggs are shed singly or in strings, or the eggs may be brooded within the mantle cavity. Spiral cleavage results in a trochophore larva, which develops directly into the adult without an intervening veliger (Fig. 24.27).

The members of the class Monoplacophora were first known from fossils and were thought to be extinct until 10 living specimens were dredged up from a great depth in the Pacific, off the coast of Central America in 1952. Since that time additional specimens have been collected from depths of 2000–7000 m. in other parts of the world.

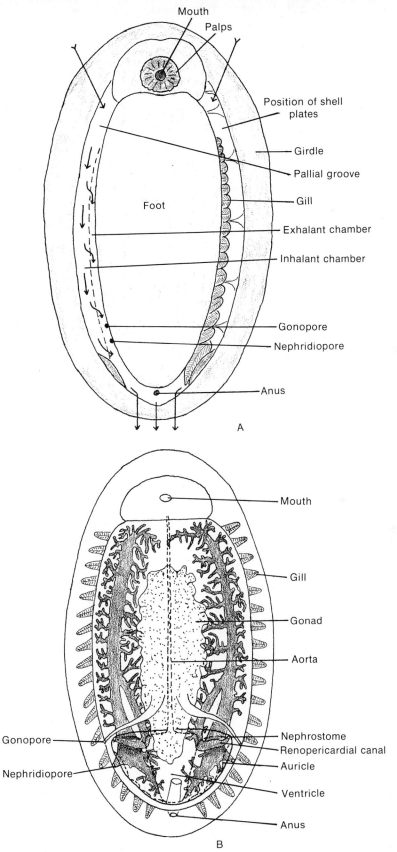

Figure 24.26 Structure of a chiton, ventral views. *A*, External structure, showing direction of water currents; *B*, internal structure. (*A* after Yonge; *B* after Lang and Haller.)

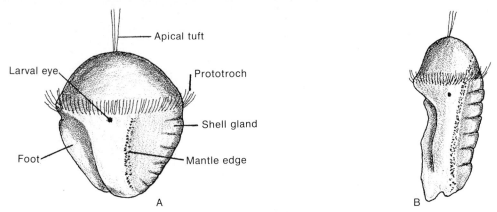

Figure 24.27   Development of chitons. *A*, Trochophore; *B*, metamorphosing trochophore. (Both after Heath from Dawydoff.)

The specimens belong to two species of *Neopilina,* relics of a class that once contained more widely distributed species.

The dorsal surface of monoplacophorans is covered by a symmetrical shield-shaped shell, the apex of which is a little peaked and directed anteriorly (Fig. 24.28). The ventral surface is similar to that of chitons, with the mantle cavity in the form of two grooves located to either side of the foot. The unusual feature of monoplacophorans, and the one that evoked special interest, is the replication of parts (Fig. 24.28). There are eight pairs of pedal retractor muscles and six pairs of nephridia. The mantle groove contains five pairs of unipectinate gills drained by two pairs of auricles. The ventilating current is probably similar to that of chitons. A radula is present, and the stomach is similar to that described for the ancestral mollusk. Two pairs of nephridia function as gonoducts for two pairs of gonads.

The significance of replication of structures in monoplacophorans is not clear. Some zoologists consider the replication to be primitive and an indication that ancestral mollusks were segmented animals. However, the lack of any other evidence of segmentation elsewhere within the phylum makes such a view questionable.

24.6   CLASS BIVALVIA

The class **Bivalvia,** or Pelecypoda, contains 20,000 species of marine and freshwater mollusks commonly called clams or bivalves; it includes the familiar mussels, cockles, oysters, and scallops. The distinguishing characteristics of the class represent adaptations for burrowing in a soft substratum. The body is greatly compressed laterally (Fig. 24.29). The head is very much reduced, and the shell is composed of two lateral pieces, or valves, hinged together dorsally. Lateral compression has resulted in a great overhang of the mantle and shell and the large mantle cavity extends to both sides of the body. The anteriorly directed foot is laterally compressed and somewhat blade-like, hence the name Pelecypoda— hatchet foot.

Mantle and Shell.   The mantle margin possesses three folds (Fig. 24.30). The muscular inner fold of the mantle edge of one side is pressed against that of the other side when the valves are closed. The middle fold is sensory. The outer fold secretes the **periostracum** and the outer part of the calcareous material of the shell. The remainder of the calcareous material is secreted by the general mantle surface (Fig. 24.30). The calcareous portion of the shell consists of two to four layers laid down as thin sheets (nacre) or prisms, or in more complex ways.

Pearls are formed by the deposition of concentric layers of calcareous material around a parasite, a sand grain, or some other foreign object which becomes lodged between the mantle and shell. Most bivalves can produce pearls, but the finest pearls are formed by species of the pearl oyster, *Pinctata,* which inhabits the warmer parts of the Pacific. Cultured pearls are produced by introducing a tiny foreign particle beneath the mantle. Bead pearls are pro-

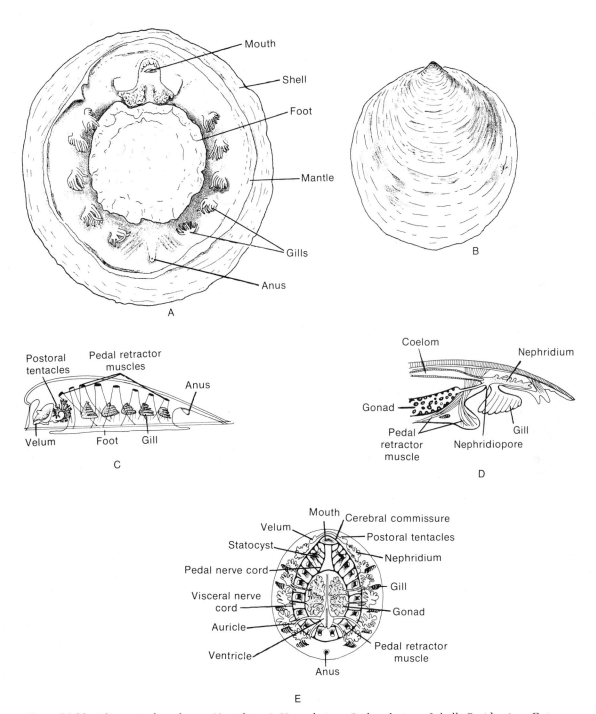

Figure 24.28 The monoplacophoran, *Neopilina*. *A*, Ventral view; *B*, dorsal view of shell; *C*, side view; *D*, transverse section through one half of body; *E*, internal anatomy. (All adapted from Lemche and Wingstrand.)

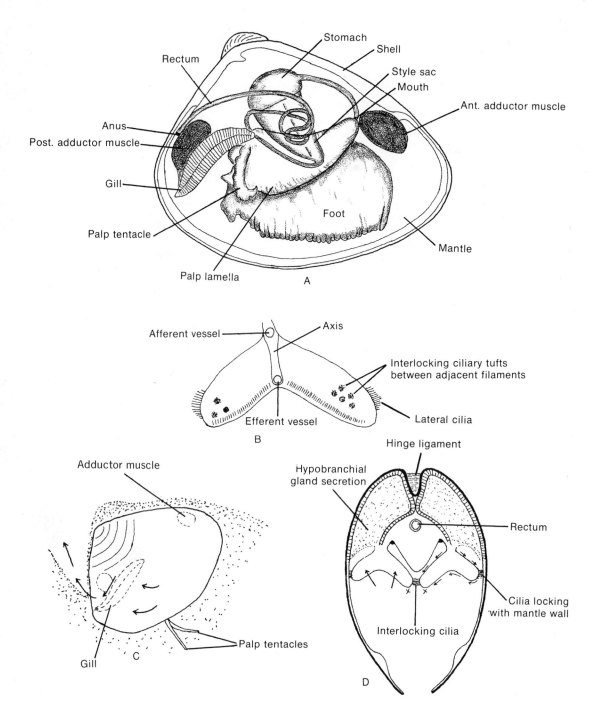

Figure 24.29 The protobranch, *Nucula*. *A*, Body with right valve and mantle removed (lateral view). *B*, Gill, showing lateral filaments (transverse section). Note similarity to gill of the ancestral mollusk shown in Figure 24.2. *C*, Feeding position. Arrows indicate direction of water current through mantle cavity. *D*, Position of gills in mantle cavity (transverse section). (All after Yonge.)

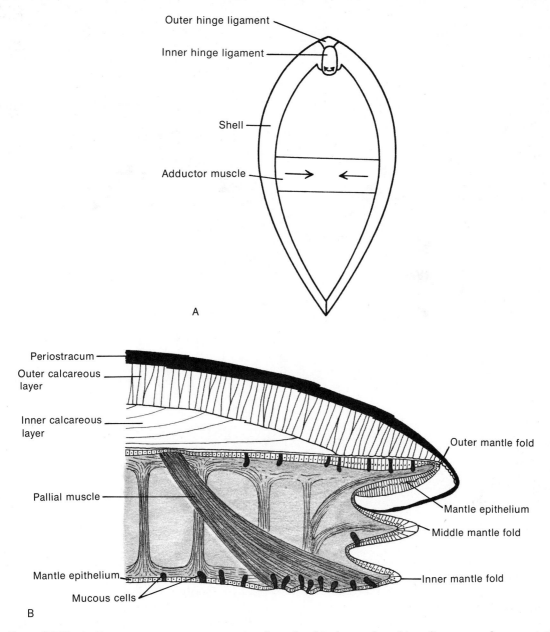

Outer hinge ligament

Inner hinge ligament

Shell

Adductor muscle

A

Periostracum

Outer calcareous layer

Inner calcareous layer

Outer mantle fold

Pallial muscle

Mantle epithelium

Middle mantle fold

Mantle epithelium

Mucous cells

Inner mantle fold

B

Figure 24.30 *A*, Diagrammatic transverse section through a bivalve to show hinge ligament and antagonistic adductor muscle. *B*, Transverse section through margin of shell and mantle of a bivalve, showing mantle lobes. (After Kennedy.)

duced by using a shell bead core, around which the oyster secretes a thin layer of mother-of-pearl.

The hinge ligament which connects the two valves dorsally is composed of the same organic material that forms the periostracum. The valves are opened by tension resulting from compression and stretching of the elastic ligament (Fig. 24.30) and closed by anterior and posterior adductor muscles extending transversely between the valves. The attachment points of the muscles are visible as scars on the inner surface of the valves. Commonly, this inner surface in the region of the hinge possesses teeth or ridges, which interlock with sockets and grooves on the opposite valve.

Protobranch Bivalves. It is generally

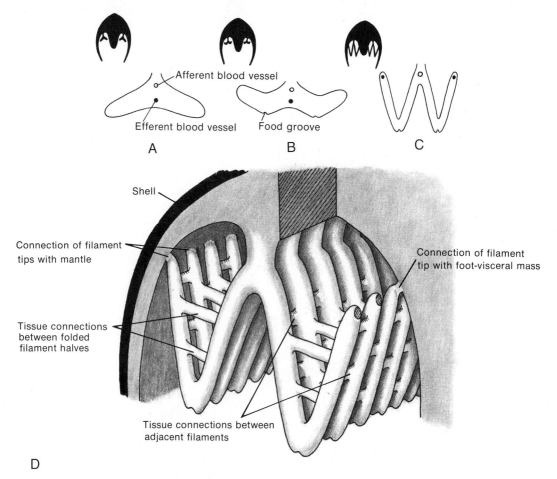

Afferent blood vessel

Efferent blood vessel   Food groove

A   B   C

Shell

Connection of filament
tips with mantle

Connection of filament
tip with foot-visceral mass

Tissue connections
between folded
filament halves

Tissue connections between
adjacent filaments

D

Figure 24.31 Evolution of lamellibranch gills. *A*, Primitive protobranch gill (position relative to foot-visceral mass and mantle indicated in cross section; see Figure 24.29). *B*, Development of food groove in hypothetical intermediate condition. *C*, Folding of filaments at food groove to produce the lamellibranch condition. *D*, Tissue connections which provide support for the folded lamellibranch filaments.   *(Continued on opposite page)*

agreed that the early bivalves were shallow burrowers in soft substrata and belonged to the subclass Protobranchia, some species of which are still living today. The protobranchs possess a single pair of posterior-lateral bipectinate gills like those of primitive gastropods (Fig. 24.29). The ventilating current enters the mantle cavity between the posterior and ventral gape of the valves, passes up through the gills, and exits posteriorly and dorsally.

The early protobranch bivalves, like some living species, were probably indirect deposit feeders. Deposit feeding is the consumption of dead organic particles—fragments of plant and animal bodies—which have settled (deposited) to the bottom and become mixed with sand particles. Direct deposit feeding refers to the ingestion of

deposit material obtained by direct contact of the mouth with the substratum. Indirect deposit feeding involves an appendage of some sort to reach the substratum and transport deposit material back to the mouth. The appendages in protobranchs are a pair of long tentacles. Each tentacle is associated with two large flap-like folds, called **labial palps,** located to either side of the mouth (Fig. 24.29). During feeding the tentacles are extended into the bottom sediments. Deposit material adheres to the mucous-covered surface of the tentacle and then is transported by cilia back to the palps. Each pair of palps functions as a sorting device. The inner opposing surfaces are ridged and ciliated (similar to Fig. 24.23*C*). Light particles are carried by certain cilia to the mouth; heavy particles are carried by other

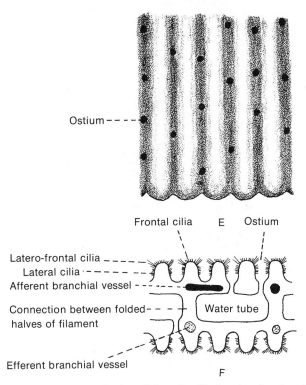

Ostium - - - -

Frontal cilia   E   Ostium

Latero-frontal cilia - - - - -
Lateral cilia · - - - - -
Afferent branchial vessel - - - - -
Connection between folded - - -
halves of filament

Water tube

Efferent branchial vessel

F

Figure 24.31 (*continued*) *E*, Surface view of a lamellibranch gill showing five fused adjacent filaments. The extensive tissue connections between filaments leave only small openings (ostia) for the entrance of water. *F*, Frontal section of *E*.

cilia to the palp margins, where they are ejected to the mantle cavity.

The evolution of indirect deposit feeding in protobranch bivalves was correlated with the invasion of a soft bottom environment and the shift of the mouth away from the substratum as the body became laterally compressed for burrowing. The radula disappeared.

Evolution of Lamellibranchs. In some group of early protobranch bivalves filter feeding evolved. An explosive evolution followed and the filter feeders, called **lamellibranchs,** came to dominate the bivalve fauna. The gills and ventilating current of protobranchs preadapted them for filter feeding. The term preadaptation implies that the basic features of a new condition existed in the ancestral stock but served some other function. As the lamellibranchs evolved, plankton in the ventilating current came to be utilized as a source of food, the gills became the filters, and the gill cilia (which originally served for the transport of sediment) became adapted for the transport of the trapped plankton from the filter to the mouth.

The principal modification of the gills for filtering was the lengthening and folding of the gill filaments, which greatly increased their surface area (Fig. 24.31). The long folded filaments are supported by the development of cross connections between the two halves, by connections between adjacent filaments, and by connection of the tips of the filaments to the foot or mantle wall (Fig. 24.31D). The folding converted the original single gill into two gills, each new gill formed by one series of folded filaments. The lengthened filaments and their attachment to one another give the gills a sheet-like form, hence the name Lamellibranchia—sheet gills. Four large, broad, filtering surfaces are present, two on each new gill (Figs. 24.31 and 24.32).

Although adjacent filaments are connected together, the points of attachment are behind the original ciliated margin, which remains distinct, and openings (**ostia**) remain for the passage of water between the filaments (Fig. 24.31E and F). The interior space between the two folded halves of the filaments forms water tubes, which connect with the **suprabranchial cavity,** the

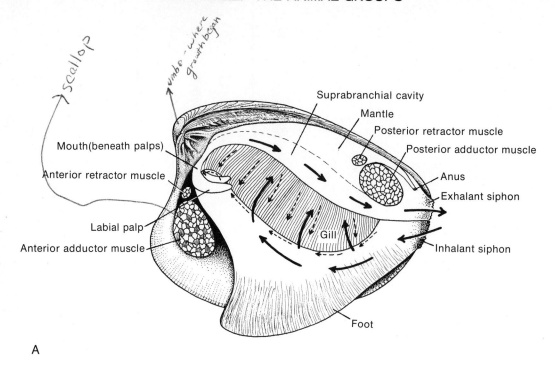

Scallop

umbo – where growth began

Mouth(beneath palps)

Anterior retractor muscle

Labial palp

Anterior adductor muscle

Suprabranchial cavity

Mantle

Posterior retractor muscle

Posterior adductor muscle

Anus

Exhalant siphon

Inhalant siphon

Gill

Foot

A

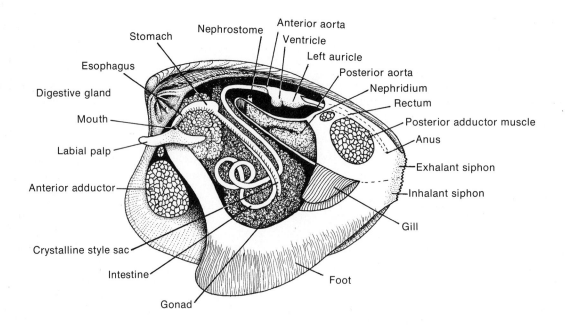

Stomach

Esophagus

Digestive gland

Mouth

Labial palp

Anterior adductor

Crystalline style sac

Intestine

Gonad

Nephrostome

Anterior aorta

Ventricle

Left auricle

Posterior aorta

Nephridium

Rectum

Posterior adductor muscle

Anus

Exhalant siphon

Inhalant siphon

Gill

Foot

B

Figure 24.32 Anatomy of *Mercenaria mercenaria. A,* Interior of the right valve. *B,* Partial dissection, showing some of the internal organs.

portion of the mantle cavity above the gills (Figs. 24.31D and 24.32). In lamellibranchs the ventilating current, now also the feeding current, enters posteriorly and ventrally as in protobranchs. On reaching the gills the water, propelled by the lateral gill cilia, enters openings between the filaments on all four of the gill surfaces. Within the interior water tubes, the water stream flows upward to the suprabranchial cavity, where it turns posteriorly and flows outward through the shell gape (Fig. 24.32).

Suspended particles are filtered out by special latero-frontal cilia as the ventilating current passes between the filaments (Fig. 24.31F). The filtered particles are passed to short frontal cilia, become covered with mucus, and are transported downward or upward to food grooves. The food grooves are located ventrally along the gill margins and are dorsally adjacent to the points where the gill is attached (Fig. 24.33). Few lamellibranchs possess the five food grooves that could be present, but the number and distribution varies from group to group (Fig. 24.33B).

The food grooves carry the collected particles posteriorly to the labial palps, which retain their original sorting function (Fig. 24.33C). Fine particles, mostly phytoplankton, are conveyed to the mouth.

A cord of mucus filled with phytoplankton is carried down the esophagus and wound into the stomach, which is much like that of protobranchs and the ancestral mollusk (they are all microphagous). However, the mass of mucus within the style sac has become compacted into a stiff rod, the **crystalline style** (Fig. 24.34). The rod is secreted by the style sac and rotated by the style sac cilia. In addition to mucus, the style contains carbohydrate-splitting enzymes. The rotating end of the style is abraded against a chitinous piece, the **gastric shield,** located at the anterior end of the stomach. The rotation of the crystalline style not only winds in the mucous rope from the esophagus and stirs the contents of the stomach, but the abrasion of the end liberates enzymes and initiates the digestion of carbohydrates within the stomach. The churning of the stomach mass throws particles against the sorting region, which separates and conveys fine particles to the digestive glands. Here digestion is completed intracellularly. The rejected coarse particles are carried along a groove to the intestine, where they are compacted into fecal pellets and eventually ejected.

Adaptive Groups of Bivalves. The evolution of filter feeding enabled lamellibranchs to exploit habitats other than shallow depths with soft bottoms. Although bivalves are not nearly as diverse as gastropods, several groups have evolved, each adapted for survival in a specific environment.

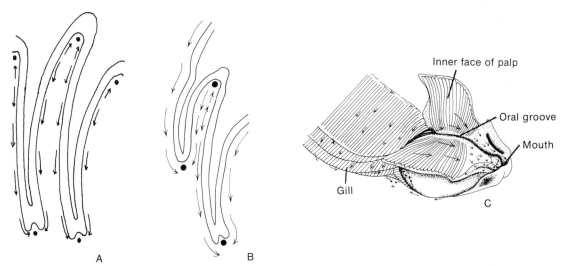

Figure 24.33 Transport and sorting of filtered particles by lamellibranch gills. A and B, Transverse sections of gills on one side (inner gill on right) showing direction of frontal cilia beat and position of anteriorly moving food tracts (black dots). A shows primitive condition with five tracts; B shows condition in lamellibranchs. C, One pair of palps spread apart and anterior section of gills of oyster. Arrows indicate ciliary tracts. (A and B after Atkins; C after Yonge.)

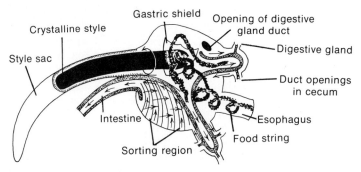

Figure 24.34 Diagram of the stomach and digestive gland duct of a lamellibranch, showing rotation of crystalline style and winding in of mucous food string. Arrows indicate ciliary pathways. (After Morton.)

Soft-Bottom Burrowers. Most lamellibranchs have continued to inhabit sand and mud bottoms but the ability to utilize the ventilating current in feeding has enabled many species to survive while burrowing deeper into the substratum.

Burrowing is accomplished with the foot, which is extended anteriorly between the gape of the valves into the substratum (Fig. 24.35). Initial extension of the foot is provided by a pair of protractor muscles, which extends from the foot to a point on the shell below the anterior adductor muscle. The substratum is simultaneously softened by water ejected from the mantle cavity as the valves are closed. The closing valves exert pressure on the water remaining within the

mantle cavity, which in turn exerts pressure on the visceral mass, driving blood into the foot (Fig. 24.35). The elevated blood pressure further extends the foot and dilates the distal end, anchoring it in the substratum. Two pairs of pedal retractors, one anchored on the shell near the posterior adductor muscle and one near the anterior adductor muscle, now pull the valves down upon the anchored foot. The pedal retractors may contract somewhat alternately, causing the valves to rock and facilitating their movement through the substratum.

Some bivalves can burrow much more rapidly than others. Species of *Donax* which live on surf beaches may be washed up by one wave and burrow back into the sand as

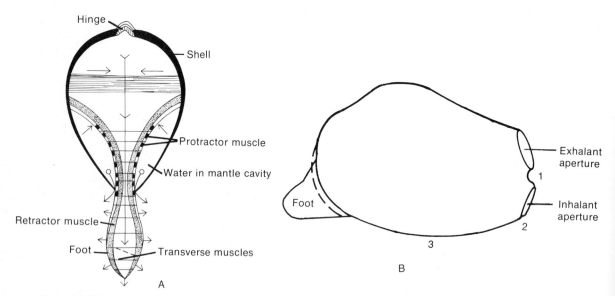

Figure 24.35 A, Diagrammatic cross section of a bivalve, showing hydrostatic forces that produce dilation of foot. Central vertical arrow indicates flow of blood into foot. B, Areas of mantle fusion in bivalves: (1) between inhalant and exhalant apertures or siphons, the most common point of fusion; (2) fusion below inhalant aperture or siphon; and (3) fusion between inhalant aperture and foot aperture. (A modified after Trueman.)

the wave recedes. Some bivalves burrow quite deeply; the common razor clam *Tagelus* burrows to depths of one meter and can anchor tenaciously with the engorged dilated foot (Fig. 24.36).

Burrowers in soft bottoms must cope with the tendency of the surrounding sediment to enter with the ventilating and feeding water stream. The problem becomes greater as the animal burrows deeper. Various adap-

tations have evolved to reduce the entrance of sediment into the mantle cavity. Relatively permanent burrows with mucus-compacted walls are formed by many species, especially those that burrow deeply. There has also been an evolutionary tendency for the opposing mantle edges to fuse together. The most common fusion points are around the apertures for inhalant and exhalant water currents, but fusion may

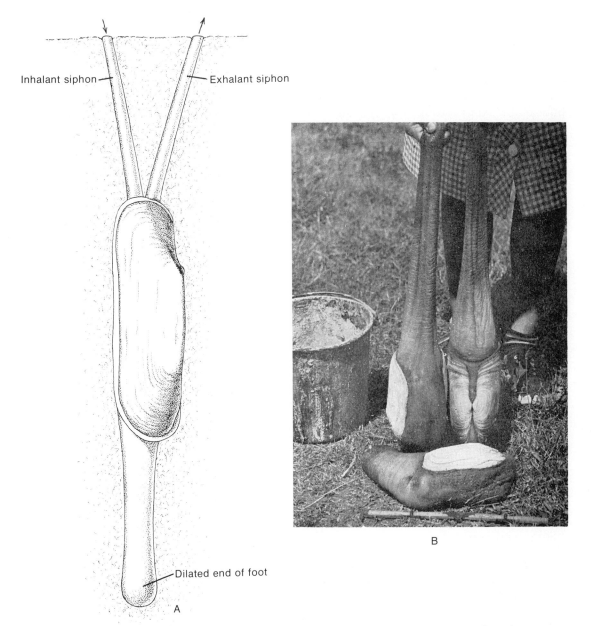

Figure 24.36 A, The razor clam *Tagelus*, a common deep-burrowing bivalve in shallow water along the east coast of the United States. B, The geoduck, *Panope generosa*, a giant California bivalve in which body and siphons cannot be enclosed within valves; mantle fused. (From Milne and Milne: Animal Life. Prentice-Hall, Inc., 1959.)

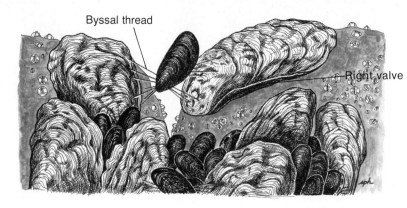

Byssal thread

Right valve

Figure 24.37 Sessile bivalves: mussels attached by byssal threads; oysters attached by right valve.

also occur ventrally, leaving only an opening for the protrusion of the foot (Figs. 24.35 and 24.36). **Siphons,** fused, tube-like extensions of the mantle around the inhalant and exhalant openings, are yet another adaptation to reduce the intake of sediment (Fig. 24.36). The siphons are extended upward to the surface and permit the animal to obtain water relatively free of sediment.

Attached Surface Dwellers. Many bivalves have abandoned burrowing and live on the surface of the sea bottom, especially on hard surfaces such as rock and coral. The animal attaches itself in one of two ways. It may, like the oyster, lie on its side with one valve fused to the substratum (Fig. 24.37). The foot is absent and the anterior adductor muscle is reduced or absent. Other bivalves, such as the common mussels, are attached by means of horny **byssal threads,** secreted by a gland in the foot (Fig. 24.37). The foot is small and finger-like and contains a groove for the placement and molding of the thread. A large number of byssal threads may be secreted and radiate out from the ventral gape of the valves like guy wires.

Sediment is much less a problem for bivalves living attached to the surface of hard substrata. It is not surprising therefore that these species exhibit no mantle fusion and do not possess siphons.

Some surface-dwelling bivalves, including both mussels and oysters, inhabit the intertidal zone and may be found attached to rocks, pilings, and sea walls which are exposed at low tide. During this period the animals are inactive and keep the valves closed to reduce desiccation.

Unattached Surface Dwellers. A few bivalves, such as scallops and file shells, rest unattached on the bottom or are attached only temporarily (Fig. 24.38). These bivalves can swim in a jerky manner for short distances by rapidly clapping the valves and driving a jet of water from the mantle cavity. Such unattached surface dwellers exhibit many adaptations similar to those of attached species. The foot is reduced. Only the posterior adductor muscle is present; it is located in the middle of the valves. (This is the part of the scallop that is eaten.) Correlated with their mobility, the sensory lobe of the mantle margin is highly developed and may bear eyes and tentacles (Fig. 24.38).

Hard-Bottom Burrowers. Several groups of bivalves have evolved the ability to burrow into peat, clay, sandstone, coral and even limestone rock. The invasion of this type of habitat occurred independently within a number of different lamellibranch groups. The ancestors of some were probably soft-bottom burrowers and acquired the ability to burrow in rock through adaptations to successively harder substrates. The ancestors of others were probably surface dwellers attached to the substrate by byssus threads.

Boring bivalves use the anterior margins of the valves to drill. Cutting occurs when the two valves are opened, pulled down, or rocked against the head of the burrow. The drilling force is provided by pulling the body against the attachment of the sucker-like foot or against the attachment of byssal threads.

Shipworms are highly modified to drill into wood. The valves, which function as the drill, are very tiny and no longer cover the greatly elongated body and siphons (Fig. 24.39). The animal occupies the entire

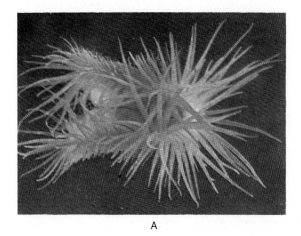

A

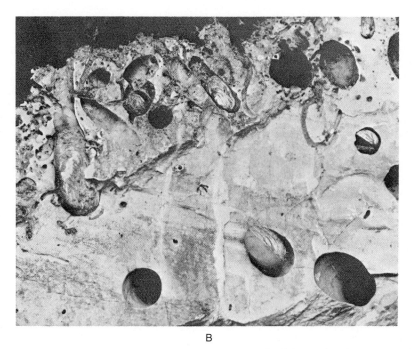

B

Figure 24.38 *A*, File shell, *Lima*, a swimming bivalve. *B*, Burrows of rock-boring bivalves excavated in limestone. (Both courtesy of D. P. Wilson.)

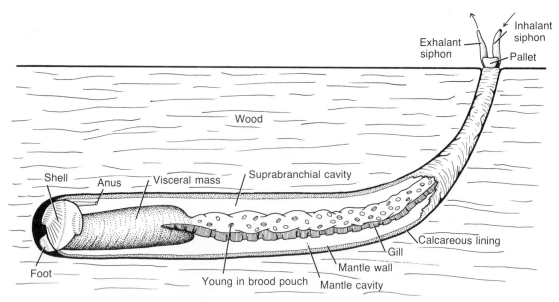

Figure 24.39 A shipworm, a wood-boring bivalve.

burrow, which is lined with calcium carbonate secreted by the mantle, and the ends of the siphons are located at the burrow opening. Shipworms are no longer filter feeders but ingest the sawdust that they excavate in drilling and digest it with cellulase. Shipworms can do extensive damage to marine timbers.

Reproduction and Development in Bivalves. Most bivalves have separate sexes. The small number of hermaphroditic forms includes some oysters and scallops. The two nephridia serve as gonoducts in the protobranchs, but separate gonoducts are present in most lamellibranchs (Fig. 24.32). The gametes are shed in the exhalant water current, and fertilization occurs in the surrounding sea water. Some bivalves, including most fresh-water species, brood their eggs within the water tubes of the gills.

Successive trochophore and veliger larvae are typical of the development of marine bivalves. A number of characteristics of the class appear in the veliger (Fig. 24.40). There is some lateral compression of the body and a bivalve shell. The finger-like foot provides the larva with considerable mobility when it first settles.

Many fresh-water bivalves have a highly specialized developmental history. There is no veliger, but a modified larval stage,

a **glochidium**, is released from the brood chambers of the female. The glochidium settles to the bottom of the stream or lake. When certain species of fish swim over the bottom, the glochidia become attached to the fins or gills by means of an attachment thread or a hook on the ventral margin of each valve. The tissues of the host then overgrow the glochidium, which now becomes a parasite for the remainder of larval development. The fish serves for the dispersion of the bivalves. When development is complete, the young clam breaks free of the host, drops to the bottom and becomes a free-living adult.

24.7 CLASS CEPHALOPODA

The class **Cephalopoda** includes *Nautilus*, **squids, cuttlefish** and **octopods.** There are only some 200 living species, but the rich fossil record indicates that in past geologic periods the class contained thousands of members.

The cephalopod characteristics represent adaptations for a swimming, carnivorous mode of existence, although there are many species that have secondarily adopted other life styles. The dorsoventral axis has become greatly lengthened (Fig. 24.41). The foot has become divided into **tentacles,** or

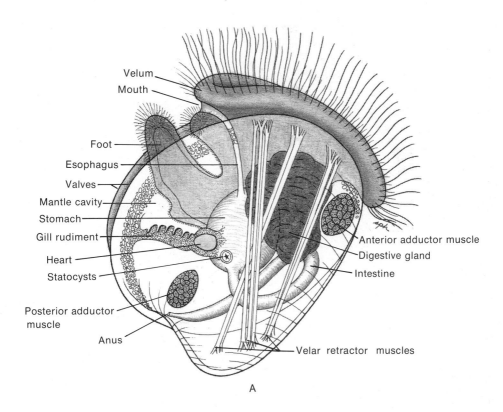

Velum

Mouth

Foot

Esophagus

Valves

Mantle cavity

Stomach

Gill rudiment

Heart

Statocysts

Posterior adductor
muscle

Anus

Anterior adductor muscle

Digestive gland

Intestine

Velar retractor muscles

A

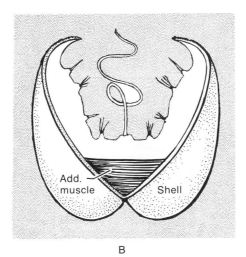

Add.
muscle

Shell

B

Figure 24.40 *A*, A fully developed veliger larva of an oyster. *B*, Glochidium, the larva of a fresh-water bivalve. (*A* after Galtsoff.)

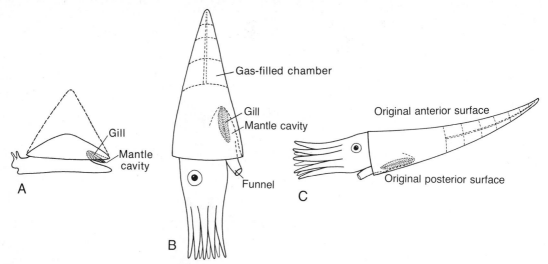

Figure 24.41   Evolution of a cephalopod. *A*, Lateral view of hypothetical ancestral mollusk. Dashed line indicates increase in length of dorso-ventral axis and a change in the shape of the shell from a shield to a cone. *B*, An early cephalopod oriented in a position that is comparable to the ancestral mollusk. *C*, Actual swimming position of an early cephalopod.

arms, and has shifted somewhat anteriorly around the mouth, hence the name cephalopod—"head foot." Cephalopods swim by means of jet propulsion. The force is generated by the contraction of the **mantle,** which becomes locked about the **head,** and water is expelled from the mantle cavity through a short funnel derived from a part of the foot.

The great increase in the dorsoventral axis and the assumption of a swimming habit have led to a shift in the orientation of the body. The tentacles (which represent the original ventral surface) are directed forward, and the **visceral mass** (which represents the original dorsal surface) is now directed posteriorly.

Cephalopod Shells. Only *Nautilus* among living cephalopods possesses a well developed **shell,** but a shell was characteristic of thousands of fossil species (Figs. 24.41 and 24.42). The cephalopod shell, unlike that of other mollusks, is divided by transverse septa into interior chambers. The living animal occupies only the outer chamber, which opens to the exterior. The posterior chambers are filled with gas which provides buoyancy for swimming. The gas is secreted by a cord of mantle tissue which extends back through the septa. *Nautilus* possesses a plano-spiral shell; that is, the shell is coiled but still bilaterally symmetrical. Many fossil species had plano-spiral shells, but others had straight tusk-like coni-

cal shells, which is probably the more primitive condition. These early cephalopods swam with the straight conical shell more or less horizontal.

In squids and cuttlefish the shell has become greatly reduced and completely overgrown by the mantle (Figs. 24.43 and 24.44). In octopods the shell has disappeared completely.

The largest living cephalopods are giant squids of the genus *Architeuthis*, which may reach 16 meters in length and are bottom dwellers at 200 to 400 meters. There are no giant octopods. The largest shelled fossil species were ones with straight conical shells up to five meters long and ones with coiled shells that were two meters in diameter.

Locomotion.   The squids are the most powerful cephalopod swimmers. Flying squids of the genus *Onchyteuthis,* which can shoot out of water and glide for short distances, have been collected from the decks of ships. The body of squids is torpedo-shaped; its posterior lateral fins are used as stabilizers. The arms are held together and function as a rudder. The siphon can be directed anteriorly or posteriorly to permit backward or forward swimming.

In very rapid swimming, contractions of the mantle muscles are synchronized by a system of **giant neurons.** The greater the diameter of a neuronal process, the faster it will conduct an impulse. The motor neu-

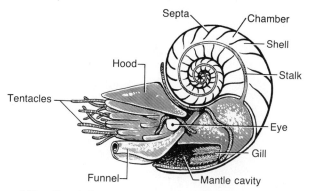

Figure 24.42 Lateral view of *Nautilus*. A diagrammatic section of the chambered shell is shown. The mantle of the left side is cut away to show the mantle cavity and two of the gills. When the animal retracts, the leathery hood protects the shell opening. (Combined from several sources.)

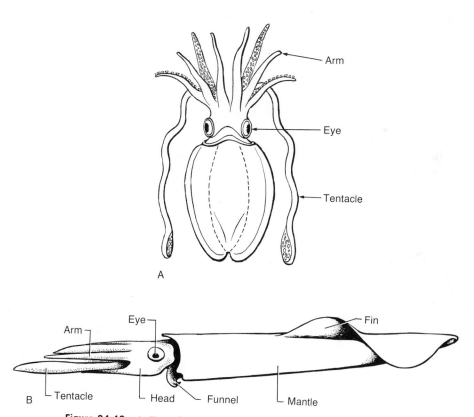

Figure 24.43 A, Dorsal view of a cuttlefish; B, lateral view of a squid.

Figure 24.44   *A*, A "small" relative of the giant squid, the oceanic squid, *Ommastrephes caroli*. This remarkably intact specimen was stranded. A meter ruler gives the scale. *B*, Octopus pursuing a crab. (*A*, courtesy of Douglas P. Wilson; *B* by Fritz Goro — courtesy of LIFE Magazine, © 1955, Time Inc.)

rons radiating out of the mantle ganglia are of different diameters depending upon the distance that the impulse must be transmitted. Long neurons have larger diameters than do short neurons. As a result of these structural differences, an impulse originating in the mantle ganglion reaches all of the muscle fibers at the same time, ensuring simultaneous contraction.

Cuttlefish have shorter bodies than squids and are agile but not powerful swimmers. Their fins undulate and contribute to propulsion. Many species live in shallow water, hovering and darting over the bottom. Some lie on the bottom, covered with sand, during the day or while waiting for prey.

The body of octopods is rather globular (Fig. 24.44B). They are largely bottom dwellers and crawl about with the arms, swimming only to escape. They are most commonly found on rocky and coralline bottoms. Some deep-sea octopods have become pelagic. The arms are connected together by webbing and the animals swim like jellyfish (Fig. 24.45).

Nutrition. Cephalopods are highly adapted for raptorial feeding and a carnivorous diet. The prey is seized by the many tentacles. Squids and cuttlefish have eight arms and two long prehensile tentacles; octopods possess only the eight arms. Except in *Nautilus*, the arms and tentacles are provided with suckers.

The principal ingestive organ is a pair of powerful, highly mobile, beak-like, horny jaws, which can cut and tear prey (Fig. 24.46B). The radula functions as a tongue for pulling in pieces of flesh bitten off by the jaws. As a further aid to dispatching the victim, a pair of salivary glands have been modified as poison glands (Fig. 24.46).

Fish, shrimp, crabs, and other mollusks serve as food. The two prehensile tentacles of squids and cuttlefish can be shot out to seize prey. Some squids can dart backward into a school of fish, seize one, and quickly bite out a chunk of flesh behind the head. The fish is rapidly eaten until the only remains are the intestines, which are dropped to the bottom. Octopods feed on crabs and snails and may remove the occupant by injecting poison through a hole drilled in the shell with the radula.

Internal Structure and Physiology. Many features of their internal structure and physiology reflect the active, raptorial mode of life of cephalopods. The stomach is quite different from that of other mollusks. Digestion is extracellular, and the digestive glands produce large amounts of powerful proteolytic enzymes. Cilia are no longer needed in the gills and mantle cavity, since muscle contraction provides for the propulsion of the ventilating current, now also the locomotor current. A pair of **secondary hearts** elevates the blood pressure on entrance into the gills (Fig. 24.45), and blood passes through the gill filaments within **capillaries**, increasing the rate of transport, and reflecting a higher metabolic rate. Within the blood, oxygen is carried by **hemocyanin.** A single pair of **nephridia** provides for excretion.

The nervous system and sense organs of cephalopods are among the most highly developed of invertebrates. The ganglia, which in many other mollusks are located at different points along the nerve cords, are concentrated in cephalopods at the anterior end to form a complex **brain,** housed within a cartilaginous capsule (Fig. 24.47). Functional centers of the brain have been identified, and the behavior of the octopus and other cephalopods has been studied extensively.

The **eye** is highly developed and parallels the eye of vertebrates to a striking degree. A **retina,** a movable **lens,** an **iris diaphragm** and a **cornea** are present (Fig. 24.47), but the photoreceptors in the retina are directed *toward* the source of light (*direct eye*) in-

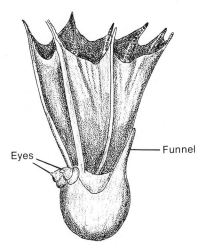

Figure 24.45 *Amphitretus*, a pelagic octopod with webbed arms for swimming. (After Hoyle from Parker and Haswell.)

Eyes

Funnel

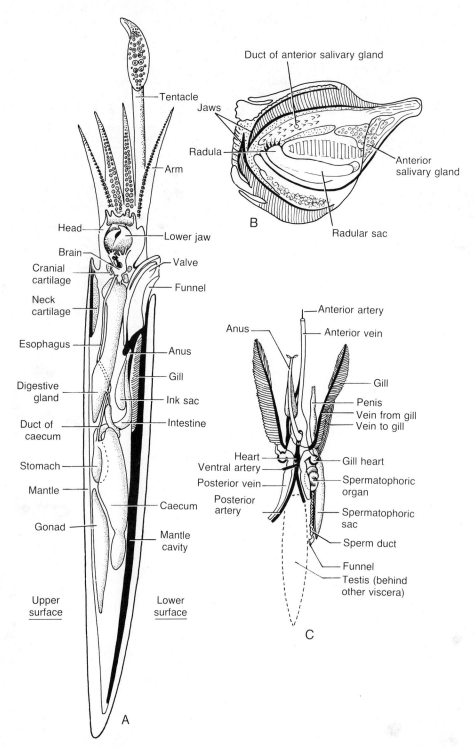

Figure 24.46   Anatomy of *Loligo. A*, Lateral view with body wall removed, showing digestive and nervous systems. *B*, Sagittal section through buccal mass of *Loligo. C*, Ventral view of part of the circulatory system and of the male reproductive system (drawn to the same scale as *A*, so that it can be turned on edge and fitted into *A*). (*A* and *C* after Williams; *B* after Owen.)

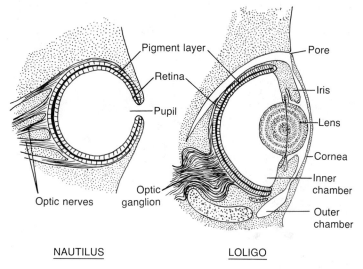

Figure 24.47 Cephalopod eyes. *Left*, The pinhole camera type in *Nautilus*. (After Borradaile et al.) *Right*, The lens type complete with shutter (iris) in *Loligo*. (After Williams.)

stead of *away* from the source of light as in vertebrates (*indirect eye*). The large number of photoreceptor cells present suggests that object discrimination is possible. Object discrimination requires that there be enough photoreceptors present so that the "neuron light points" can form an object when put together. The response of a photoreceptor or a group of photoreceptors would represent the equivalent of a point of light. The mechanism is something like image formation on a television screen. Behavioral studies suggest that although many cephalopods possess eyes which would permit object discrimination, they are very nearsighted animals.

The arms are provided with many tactile cells and chemoreceptors and play an important role in the structural and chemical discrimination of surfaces. A **statocyst** is located on either side of the brain.

An **ink gland** opens into the intestine and is provided with an ejector mechanism, which enables the animal to discharge a cloud of dark ink. It has been suggested that the cloud functions not as a smoke screen, but as a dummy, confusing the predator and enabling the animal to escape. The alkaloids in the ink may narcotize the chemoreceptors of predators such as fish.

Chromatophores are large, pigment-laden cells to which muscle fibers are attached. Yellow, orange, red, blue, and black chromatophores may be present, depending upon the species. The contraction of the muscle cells expands the chromatophores, and the color changes produced appear to function as background simulation in the courting behavior of males, in aggressive responses between individuals, and in alarm reactions to intruders.

Deep-sea cephalopods do not have chromatophores but many are **bioluminescent.** The light is produced in special organs called **photophores** located on different parts of the body, often in patterns and composed of different colors. Luminescent ink may also be present but in deep-sea forms the ink gland is absent.

Reproduction and Development. The sexes are separate in cephalopods. The gonad in each sex is provided with a relatively complex gonoduct (Fig. 24.46). Copulation occurs during swimming and involves the transfer of sperm in packets, termed **spermatophores.** Masses of sperm are cemented to one another or encased within special secretions in certain regions of the sperm duct. The shape of the spermatophore is often determined by the region of the sperm duct which molds it. The spermatophores of many cephalopods look like baseball bats (Fig. 24.48).

In copulation, the male seizes the female head on and interlocks his arms with hers. One especially modified arm of the male reaches into his mantle cavity and plucks a mass of spermatophores from the gono-

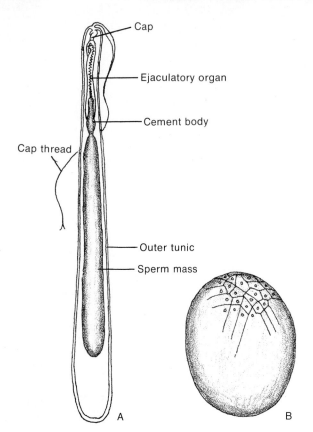

Figure 24.48 *A*, Spermatophore of the squid *Loligo; B*, incomplete cleavage in the yolky egg of *Loligo*. (After Watase from Dawydoff.)

duct. The copulatory arm of the male then deposits the spermatophores in the mantle cavity of the female, or (in some species) within a special pouch on the membrane around the mouth. The spermatophores discharge their sperm shortly after being deposited within the mantle cavity, and the eggs are fertilized on leaving the female gonoduct. Where the sperm are deposited on the buccal membrane, the eggs are fertilized as they leave the female siphon for deposition.

In some pelagic (deep-sea) cephalopods the eggs are planktonic. In most, however, the eggs are deposited in strings on the bottom or are attached to stones or other objects. Octopods remain with their eggs, guarding and cleaning them of sediment.

Cephalopod eggs are telolecithal with large amounts of yolk material. Cleavage is therefore meroblastic and discoidal as in the eggs of reptiles and birds. Development is direct, and the young have the adult form on hatching.

24.8 THE ANCESTRAL MOLLUSK: A REVIEW

Having studied members of five of the seven molluscan classes, you are now in a better position to appreciate the way a hypothetical ancestor can be derived and to determine whether or not the description of the ancestral mollusks as given in the beginning of this chapter is justified. Is there evidence for describing the first mollusks differently? For some features, there would seem to be little basis for question. Four of the five classes studied possess a **radula.** Three of the five have a flat **foot.** Three of the five have a **shell** composed of a single piece, and the multiple **valves** of bivalves and chitons appear to be later adaptations to permit articulation.

The most debatable derivations are the position of the **mantle cavity** and the number and nature of the **gills.** One might argue, using the chitons and monoplacophorans as evidence, that the ancestral mollusks had many gills within a pair of lateral mantle grooves, and this is the hypothesis held by some zoologists. However, two pairs of gills are found in three of the five classes. Especially convincing is the close similarity of the structure of the gills of primitive gastropods and primitive bivalves, two classes which are quite different in other respects. The mantle cavity in gastropods is clearly posterior before torsion, and in bivalves there is still a posterior ventilating current.

The origin of mollusks is obscure, but the most likely ancestor is some ancient metazoan group stemming from flatworms, which are similar to mollusks in having a broad ventral surface upon which they creep. The presence of spiral cleavage and a trochophore larva clearly links the mollusks to another large phylum, the Annelida.

CLASSIFICATION OF THE MOLLUSCA

CLASS GASTROPODA. Snails. Mantle and visceral mass exhibit some degree of torsion. Shell typically spiraled. Foot flattened and head well developed.

    Subclass Prosobranchia. Marine and fresh-water, gill-bearing species in which the mantle cavity and contained organs are located anteriorly. Shell present. Most marine snails with well-developed shells belong to this subclass.

    Subclass Opisthobranchia. Visceral mass has undergone detorsion and some degree of reduction of shell and mantle cavity is exhibited. Hermaphroditic. Entirely marine. Bubble shells, sea hares and sea slugs.

    Subclass Pulmonata. Gills absent and anterior mantle cavity converted into a lung. Shell usually present but never an operculum. Hermaphroditic. Terrestrial and fresh-water. Most land snails of temperate regions are pulmonates.

CLASS POLYPLACOPHORA, or AMPHINEURA. Chitons. Body greatly flattened dorso-ventrally. Head reduced. Shell composed of eight linearly arranged, overlapping plates. Marine.

CLASS APLACOPHORA. Small group of aberrant marine mollusks with worm-like bodies and no shells.

CLASS MONOPLACOPHORA. Small group of marine deep-water species. Body flattened dorso-ventrally. Shield-like shell. Various organs replicated—eight pairs of retractor muscles, five pairs of gills, six pairs of nephridia and two pairs of auricles.

CLASS BIVALVIA, or PELECYPODA. Clams, or bivalves. Body greatly flattened laterally. Shell composed of two lateral valves hinged dorsally. Head reduced.

    Subclass Protobranchia. Primitive marine bivalves with one pair of unfolded gills.

    Subclass Lamellibranchia. Marine and fresh-water filter feeding bivalves with folded gills. Most bivalves belong to this subclass.

    Subclass Septibranchia. Small group of marine species having the gills modified as a pumping septum.

CLASS SCAPHOPODA. Tusk or tooth shells. Burrowing marine mollusks having a tusk-like shell open at each end.

CLASS CEPHALOPODA. *Nautilus*, cuttlefish, squids, and octopods. Marine, mostly swimming mollusks having the foot divided into tentacles, or arms. Shell, when present, divided into chambers, of which only the most recent is occupied by the living animal.

    Subclass Nautiloidea. Mostly fossil cephalopods with straight or coiled chambered shells. *Nautilus* is only living species.

    Subclass Ammonoidea. Fossil cephalopods with coiled shells, having wrinkled septa.

    Subclass Coleoidea. Shell internal or absent. Cuttlefish, squids, and octopods.

ANNOTATED REFERENCES

The references listed below are devoted solely to mollusks. Accounts of mollusks may also be found in the general references listed at the end of Chapter 18.

Abbott, R. T.: American Sea Shells. Princeton, New Jersey, D. Van Nostrand Co., 1964. A semitechnical guide to the marine mollusks of the Atlantic and Pacific coasts of North America.

Fretter, V., and Graham, A.: British Prosobranch Molluscs. London, The Ray Society, 1962.

Morton, J. E.: Molluscs. 4th ed. London, Hutchinson University Library, 1967. A general account of the mollusks.

Purchon, R. D.: The Biology of the Mollusca. Oxford, Pergamon Press, 1968. A general account of the mollusks.

Wilbur, K. M., and Yonge, C. M. (Eds.): Physiology of Mollusca. New York, Academic Press, Vol. I, 1964; Vol. II, 1966. A detailed account of all aspects of molluscan physiology.

25. The Annelids

The phylum **Annelida,** the segmented worms, includes the familiar earthworms and leeches as well as many marine and fresh-water worms. The most striking annelid characteristic is the division of the body into similar parts, or **segments,** arranged in a linear series along the anterior-posterior axis. Each segment is termed a **metamere** (Fig. 25.1) and this condition is called **metamerism.** Only the trunk is segmented. Neither the **head,** or **prostomium,** containing the brain and bearing the mouth, nor the terminal part of the body, the **pygidium,** which carries the **anus,** is a segment. New segments arise in front of the pygidium; the oldest segments lie just behind the head. Some longitudinal structures, such as the gut and the principal blood vessels and nerves, extend the length of the body, passing through successive segments; other structures are repeated in each segment, reflecting the metameric organization of the body.

25.1 METAMERISM AND LOCOMOTION

Metamerism appears to have evolved twice in the evolutionary history of the Animal Kingdom, once in the evolution of annelids and arthropods and once in the evolution of chordates. In both instances the condition appears to have evolved as

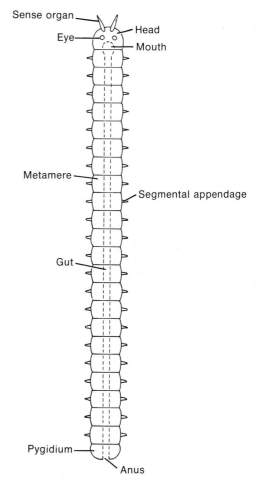

Figure 25.1   Diagram of a metameric animal, such as an annelid worm.

547

an adaptation for locomotion, but for different kinds of locomotion. In chordates metamerism probably represented an adaptation for undulatory swimming (p. 678). Annelid metamerism probably arose as an adaptation for burrowing.

The ancestors of the annelids were in all probability elongate coelomate animals which burrowed in marine sand and mud. Peristaltic contraction of the muscles of the body wall provides an efficient means of locomotion in such an environment. Con-

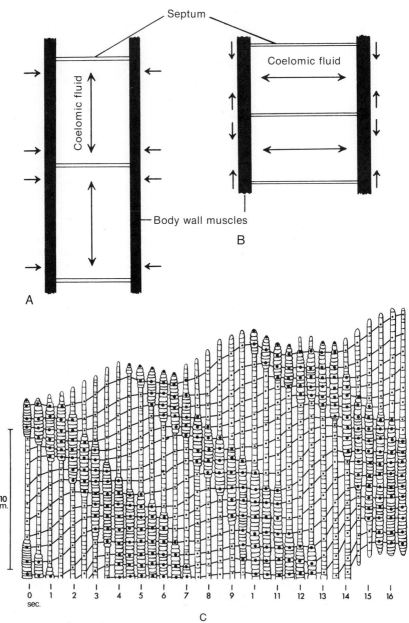

Figure 25.2 *A* and *B*, Diagrammatic frontal section through two annelid segments. External arrows indicate direction of force exerted by body wall muscles. Internal arrows indicate direction of force exerted by coelomic fluid pressure: (*A*) during circular muscle contraction, and (*B*) during longitudinal muscle contraction. *C*, Diagram showing mode of locomotion of an earthworm. Segments undergoing longitudinal muscle contraction are marked with larger dot and drawn twice as wide as those undergoing circular muscle contraction. The forward progression of a segment during the course of several waves of circular muscle contraction is indicated by the horizontal lines connecting the same segments. (After Gray and Lissmann.)

traction of the circular muscles elongates the anterior end of the body and thrusts the head forward into the substratum. The head is then anchored to the substratum and contractions of the longitudinal muscles of the body wall pull forward the more posterior parts of the body. In an unsegmented, elongate coelomate, the **coelomic fluid** would form a continuous skeletal bar through the body; alternate contraction of the circular and longitudinal muscles, by exerting pressure on the coelomic fluid, would result in lengthening and shortening of the entire body. The locomotion movements can be smoother, more continuous, and better controlled if the action of the body wall muscles is localized, i.e., restricted to sections of the body (Fig. 25.2). Metamerism permits such localization and this probably accounts for the evolution of this condition in annelids. The **coelom** is compartmented by transverse **septa,** and the longitudinal and circular muscles are organized into segmental blocks corresponding to the coelomic compartmentation. A coelomic compartment and its surrounding muscle blocks constitute a segment, and the body is composed of a series of such segments. Each segment is a functional as well as a structural unit, and the elongation and shortening resulting from the interaction of the fluid skeleton and the body wall muscles is restricted to a small part of the body (Fig. 25.2). In locomotion the segments function in groups, and waves of muscle contraction sweep down the length of the body, affecting a number of segments simultaneously. The

wave of *longitudinal* contraction is followed by a wave of *circular* contraction. Each section of the body is thus alternately elongated and shortened.

It is important to note that annelidan metamerism is basically a modification of the body wall muscles and the coelom. The **nervous, circulatory** and **excretory** systems are also metameric, i.e., parts are repeated in each segment. But the metamerism of these structures probably reflects an adjustment of these supply systems to the primary metamerism of the muscles and coelom (Fig. 25.3).

The typical annelidan has a **nervous system** with a dorsal **brain** and a ventral longitudinal **nerve cord,** a straight tubular **gut,** and a closed circulatory system (Fig. 25.3). The excretory organs are usually **metanephridia** (see p. 506), typically one pair per segment (Fig. 25.3).

25.2 CLASSIFICATION

The phylum is divided into three classes. The **Polychaeta,** the largest of the three classes, contains the marine annelids. The **Oligochaeta** includes the fresh-water annelids and earthworms, and the **Hirudinea** comprises the leeches. The latter clearly evolved from the oligochaetes, but the origin of the oligochaetes is more obscure. They may have evolved from some group of estuarine polychaetes, or polychaetes and oligochaetes may have arisen independently from some common ancestor.

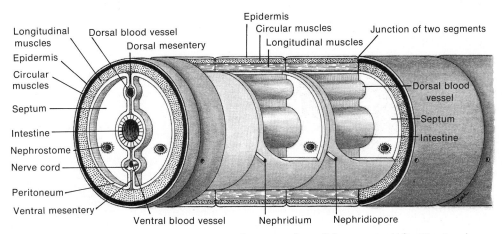

Figure 25.3 Diagrammatic lateral view of a series of annelid segments. (After Kaestner.)

25.3   CLASS POLYCHAETA

The members of the class Polychaeta are distinguished from other annelids by the presence of paired, segmented appendages, **parapodia,** which carry projecting bundles of chitinous rods, the **setae** (Fig. 25.4). The setae grip the substratum when the animal is crawling.

The parapodia vary greatly in shape and size but when well developed, each is composed of an upper **notopodium,** and a lower **neuropodium.** Each contains setae, a small, tentacle-like process at the base called a **cirrus,** and an internal heavy chitinous rod (**aciculum**) to which the parapodial muscles are attached (Fig. 25.5).

The head, or prostomium, of polychaetes is often highly developed and may bear various sensory and feeding structures (Fig. 25.5). The mouth is located beneath the prostomium, just in front of the first one or two modified trunk segments (the **peristomium**).

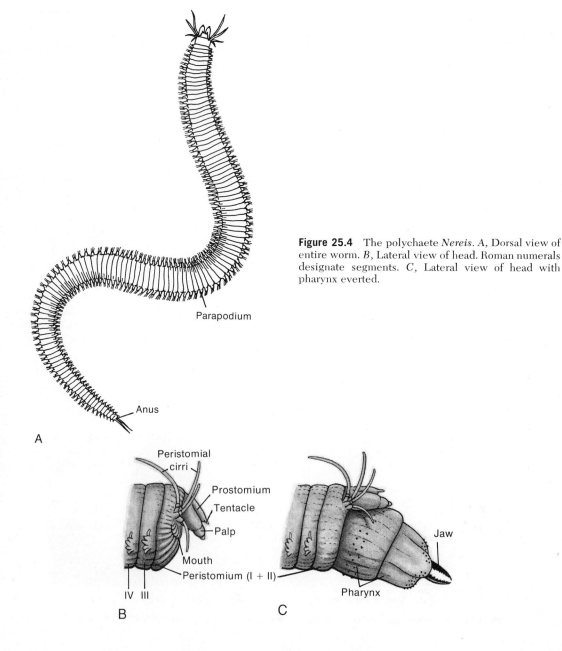

Figure 25.4   The polychaete *Nereis. A,* Dorsal view of entire worm. *B,* Lateral view of head. Roman numerals designate segments. *C,* Lateral view of head with pharynx everted.

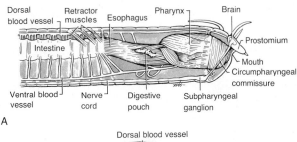

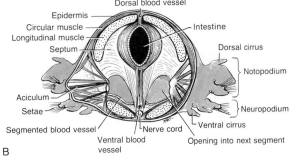

Figure 25.5 Structure of the polychaete *Nereis*. *A*, Longitudinal section through anterior end. *B*, Transverse section at level of intestine.

The digestive tract is usually composed of an eversible muscular **pharynx** and a long straight **intestine.** The pharynx is everted (turned inside out) through the mouth by special protractor muscles or by elevated coelomic fluid pressure (Fig. 25.4).

The more than 5000 species of marine polychaetes exhibit great diversity both in form and in habit. All types of marine habitats contain polychaetes; sandy and muddy bottoms harbor an especially rich fauna. An appreciation of polychaete diversity can be gained by examining the adaptive groups which comprise the class, for variations in structure are closely correlated with different modes of existence and feeding habits.

Surface Dwellers. Many polychaetes live on the bottom of the sea, but never on exposed surfaces. They are found beneath stones and shells, in rock and coral crevices, and in algae and sessile animals. The heads of surface-dwelling polychaetes are well developed and carry several kinds of sensory structures—one or two pairs of **eyes,** up to five **antennae,** and a pair of **palps** (Fig. 25.6).

The parapodia are large and function like legs in crawling. At the beginning of the power stroke, the parapodium is moved forward and placed against the substratum. The parapodial muscles then pull the parapodium backward. The parapodium does not lose contact with the substratum during the back sweep; consequently the body is driven forward. Each stroke constitutes a small step, and the movements of the parapodium occur in sequence as a result of **waves of contraction** passing along the length of the body (Fig. 25.7). The waves of contraction on one side of the body alternate with those on the opposite side. In rapid movement, alternating contractions of the longitudinal muscles of the body wall throw the body into undulatory waves which generate additional thrust against the substratum.

The crawling mode of locomotion of surface-dwelling polychaetes is reflected in the structure of the body wall and coelom. The longitudinal muscles are more highly developed than the circular muscles (Fig. 25.5), and the septa are somewhat reduced, since the coelom does not function as a localized hydrostatic skeleton as it does in peristaltic movement.

Many surface dwellers are carnivores and feed on other small invertebrates, including polychaetes. Others are algae eaters, scavengers, or detritus feeders. Almost all have a muscular pharynx provided with one or more chitinous jaws used for seizing prey or tearing off pieces of algae.

Pelagic Polychaetes. Several families of polychaetes are adapted for living in the open ocean. These worms never come to the bottom but resemble the surface-dwelling polychaetes in some ways. They have well-developed heads and the large para-

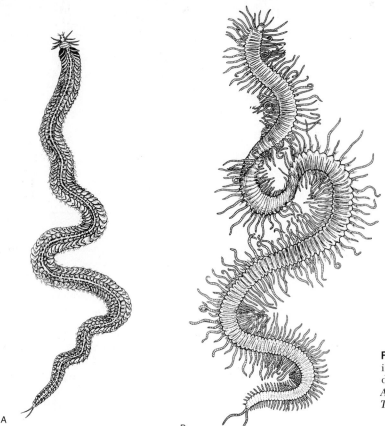

A

B

Figure 25.6 Two surface-dwelling polychaetes. Note the well-developed head and parapodia. *A, Phyllodoce maculata; B, Trypanosyllis zebra.*

podia are used as paddles in swimming (Fig. 25.8). They are commonly carnivorous. Like many other small pelagic or planktonic animals, these polychaetes tend to be pale or transparent.

Gallery Dwellers. Many polychaetes are adapted for burrowing in sand or mud. Some excavate extensive burrow systems, or galleries, which open to the surface at numerous points (Fig. 25.9). The wall of the burrow is kept from caving in by its lining of mucus. The body of gallery dwell-

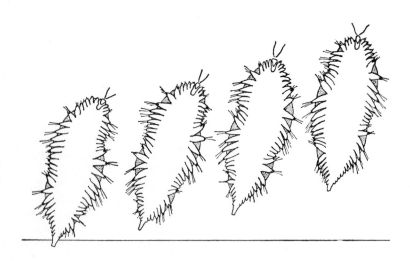

Figure 25.7 Parapodial movement in polychaete crawling. Waves of movement on one side of body alternate with the opposite side. (After Mettam.)

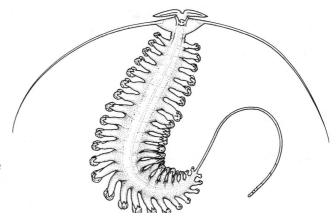

Figure 25.8 *Tomopteris renata*, a pelagic polychaete.

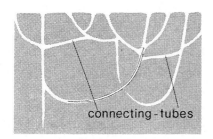

connecting-tubes

Figure 25.9 Burrow system of *Glycera alba*, showing worm lying in wait for prey. (After Oekelmann and Vahl.)

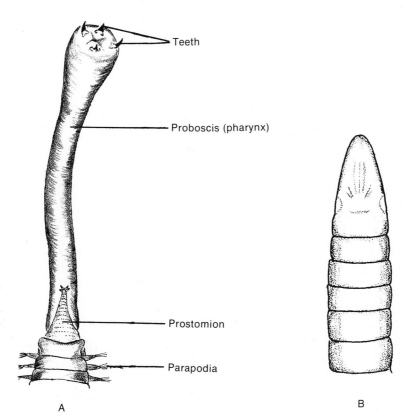

— Teeth

— Proboscis (pharynx)

— Prostomion

— Parapodia

A   B

Figure 25.10 Gallery dwellers. *A*, Anterior end of *Glycera* with proboscis everted. *B*, Anterior end of *Drilonereis*.

ers is usually long and the prostomium is commonly conical, or a simple lobe devoid of eyes and other sensory structures (Fig. 25.10).

Gallery dwellers may use the parapodia to crawl through the burrow system. However, many crawl by peristaltic contractions; the parapodia of these forms are somewhat reduced and function primarily to anchor the segments against the burrow wall. Septa and circular muscles are well developed.

Some gallery dwellers are carnivores. The bloodworm *Glycera*, for example, feeds on small surface-dwelling invertebrates. The worm lies in wait below the surface and can detect slight changes in water pressure created by the movement of an animal above (Fig. 25.9). When prey is in the vicinity, *Glycera* slowly creeps toward the opening and then everts the long proboscis-like pharynx with great force. The victim is seized with four teeth at the proboscis tip (Fig. 25.10).

Other gallery dwellers are direct deposit feeders and consume the substratum through which the gallery system penetrates. The organic material is then digested and the mineral material is defecated.

Sedentary Burrowers. Some burrowing polychaetes construct simple vertical burrows with only one or two openings to the surface (Fig. 25.11). These sedentary burrowers, in contrast to gallery dwellers, move about relatively little. Peristaltic crawling is the rule, and the parapodia are reduced to ridges which bear special hook-like setae that aid in gripping the burrow wall. The prostomium lacks sensory structures, although special feeding appendages may be present.

Some sedentary burrowers are direct deposit feeders. *Arenicola*, the lugworm, for example, lives head down within its **L**-shaped burrow and eats the sand at the end (Fig. 25.11). Its soft eversible pharynx lacks jaws. With rhythmic periodicity, the animal backs up to the surface and defecates castings of sand. An intertidal sand flat may display thousands of castings reflecting the large population of worms living below the surface (Fig. 25.12).

Indirect deposit feeding is a common adaptation of sedentary burrowers, and such worms have typically evolved tentacle-like head appendages. The head of *Amphitrite*, for example, is provided with a great mass of long tentacles. These spread over the surface from the opening of the burrow (Fig. 25.11). Detritus material adheres to

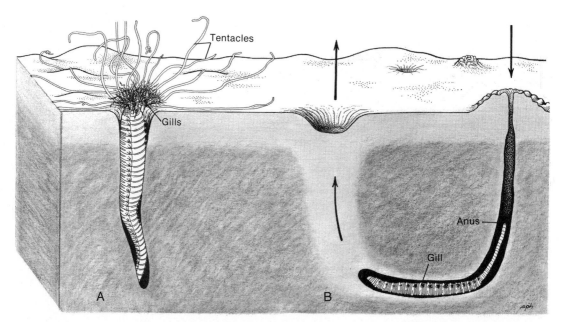

Figure 25.11 Polychaetes: sedentary burrowers. *A, Amphitrite,* an indirect deposite feeder with tentacles outstretched over the substratum. *B, Arenicola,* the lugworm. This worm, a direct deposit feeder, eats the substratum at the end of its burrow. Arrows indicate direction of ventilating current. (*A* modified from Wells.)

Figure 25.12 Lugworm (*Arenicola*) and castings on a mud flat exposed at low tide in Puget Sound. (By Betty M. Barnes.)

the mucus on the tentacles and then is conveyed to the mouth in a ciliated tentacular gutter.

Tubiculous Polychaetes. Many polychaetes live in tubes. A tube-dwelling, or tubiculous, habit is found in other ani-

mals besides polychaetes, but the polychaetes have exploited this particular mode of existence to a greater degree than any other group of animals.

The tube may function as a protective retreat, or as a lair from which the worm may pounce upon passing prey. In soft bottoms, the opening of the tube projects above the sand surface and thus provides the worm with access to water free of sediment. Tubes may also permit worms to live on hard exposed substrates, such as rock, coral, or shell.

The tube may be composed entirely of hardened material secreted by the worm or it may be composed of foreign material cemented together. Secreted tubes are commonly horny or parchment-like, but one family of fanworms secretes tubes of calcium carbonate (Fig. 25.14). Sand grains are the usual building materials of cemented tube (Fig. 25.14), although some worms may use pieces of shell. The tubes

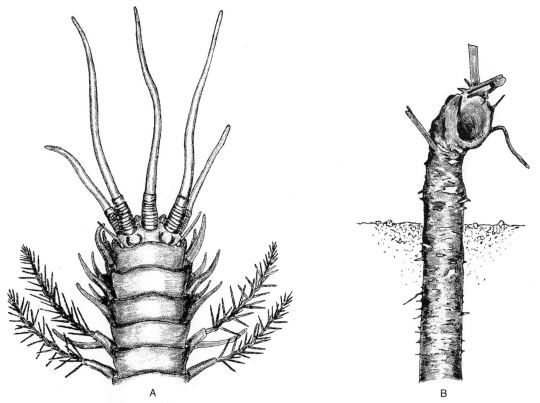

A B

Figure 25.13 *A*, Head and first two gill-bearing segments of *Diopatra*, a carnivorous tube dweller (dorsal view). *B*, Funnel-shaped parchment tube of *Diopatra*, to which pieces of shell and algae are attached.

and the cementing material are commonly secreted by glands on the ventral surface of a number of segments. The construction methods have been studied in many species and have been found to vary greatly.

Carnivorous and Sedentary Tube Dwellers. Tube-dwelling polychaetes may be divided into **carnivorous** tube dwellers and **sedentary** tube dwellers. The carnivorous group contains predatory species which feed on other polychaetes and small invertebrates. They lie in wait at the mouth of the tube and seize the victim as it passes by (Fig. 25.13). As might be expected, these worms are very agile and crawl within the tube using the parapodia. The body may be projected some distance out of the tube opening. Their adaptations—sensory structures on the head and well-developed para-

podia—are essentially like those of surface dwellers.

The sedentary tubiculous polychaetes move about within the tube less actively than do the carnivorous tube-dwellers and only the extreme ends of the body project from the tube opening. This group is similar in many respects to sedentary burrowers. The head usually lacks special sensory structures, although feeding appendages may be present. The animal moves within the tube by peristaltic contractions; hence the parapodia tend to be reduced to ridges with hooked setae.

Nutrition. The structural diversity of tube-dwelling polychaetes is in large part correlated with their different modes of feeding. A few, such as the bamboo worms, are **direct deposit feeders.** Like *Arenicola*,

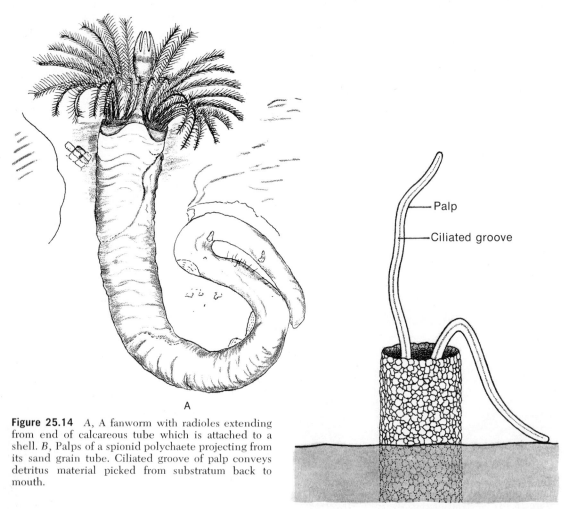

Palp

Ciliated groove

A

B

Figure 25.14 *A,* A fanworm with radioles extending from end of calcareous tube which is attached to a shell. *B,* Palps of a spionid polychaete projecting from its sand grain tube. Ciliated groove of palp conveys detritus material picked from substratum back to mouth.

the bamboo worm lives head down in sand grain tubes and eats the substratum at the bottom of the tube. Periodically the worm backs to the surface and defecates the mineral material. The eversible pharynx lacks jaws.

Some sedentary tube-dwellers are **indirect deposit feeders.** There are species which feed in the same manner as described for *Amphitrite*, a sedentary burrower (Fig. 25.11). Members of the family Spionidae are common tubiculous polychaetes with two long, tentacle-like palps (Fig. 25.14). The palps of these tiny worms are projected from the opening of the tube and pick up deposit material from the surrounding substratum. The collected particles are then conveyed back to the mouth along the palp within a ciliated groove.

Filter feeding has evolved in several families of sedentary, tube-dwelling polychaetes. The beautiful fanworms are a notable example. The members of the two fanworm families (one of which secretes a calcareous tube and the other a parchment tube or a tube of cemented materials), have feather-like head structures, called **radioles.** When feeding, the radioles project from the opening of the tube, often in the form of a funnel (Fig. 25.14). The cilia on the radioles create a water current, and plankton and other suspended particles passing through the radioles are filtered out. The trapped particles are then transported by cilia down the radiole into the mouth.

The members of the family **Chaetopteridae** use a mucous bag to filter the water pumped through the U-shaped or straight tube (Fig. 25.15). This remarkable filter is secreted by certain parapodia held out from the body like a basketball hoop (Fig. 25.15). Water passes through the mucous film leaving suspended particles behind. Mucus is continuously spun out as the end of the bag is gathered and rolled up within a ciliated **cupule.** Periodically the animal halts the secretion, and the remainder of the mucous bag is rolled up within the cupule. The resulting ball containing trapped food particles is transported forward along a ciliated groove to the mouth (Fig. 25.15).

Internal Transport, Gas Exchange, and Excretion in Polychaetes. The closed blood vascular system of polychaetes has a relatively simple basic plan (Fig. 25.16). A

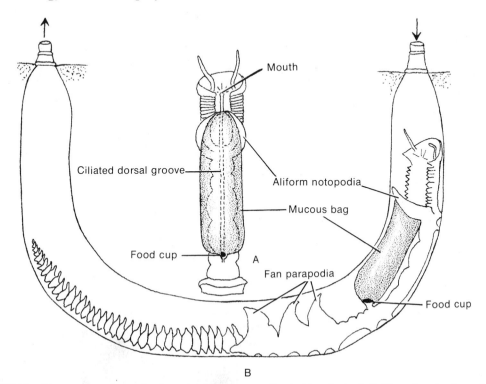

Figure 25.15 *Chaetopterus* during feeding. *A*, Anterior part of body (dorsal view). *B*, Worm in tube (lateral view). (After MacGinitie.)

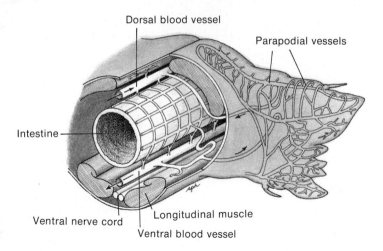

Dorsal blood vessel

Parapodial vessels

Intestine

Ventral nerve cord

Longitudinal muscle

Ventral blood vessel

Figure 25.16 Vascular system within a segment of *Nereis*. Arrows indicate direction of blood flow; anterior is to the right. (Modified from Nicoll.)

contractile, longitudinal vessel located above the gut functions as the heart and conveys blood anteriorly. At the anterior end of the body, paired vessels carry the blood around the gut to a ventral longitudinal vessel in which blood passively flows posteriorly. In each segment the ventral vessel gives rise to a vessel supplying the gut and to paired vessels supplying the body wall, parapodia and nephridia. These segmental vessels eventually return blood to the dorsal vessel.

Very small polychaetes and those with long, thread-like bodies have no gills. The varying structure and the distribution of the gills present in the larger species indicate that they have evolved independently in different groups. Most commonly the gills are modifications of the parapodia (Fig. 25.13) or outgrowths of some part of the parapodium. However, gills are found in other parts of the body in some polychaetes. The gills of the sedentary burrower, *Amphitrite,* for example, are located dorsally at the anterior end of the body (Fig. 25.11). The radioles of fanworms serve as gills in addition to functioning as filters.

The ventilating current may be produced by cilia or by contractions of the gills. Gallery dwellers, sedentary burrowers, and tube dwellers drive a ventilating current through the burrow or tube, usually by undulations of the body or by peristaltic contractions (Fig. 25.11*B*).

Polychaetes with gills always possess a respiratory pigment, usually **hemoglobin.** Polychaete hemoglobin occurs either dis-

solved in the blood plasma or within corpuscles in the coelomic fluid. When present in solution in the plasma, it is always a very large molecule. The hemoglobin of *Arenicola,* for example, is composed of 96 heme units, whereas vertebrate hemoglobin contains only four heme units. In contrast, the hemoglobin in the coelomic corpuscles of some polychaetes is a small molecule. The coelomic hemoglobin of the carnivorous gallery dweller, *Glycera,* for example, consists of a single heme unit. The presence of many small hemoglobin molecules dissolved in the blood plasma would exert considerable osmotic pressure and would seriously disturb the fluid balance between the tissues and the blood. This problem is solved in polychaetes by having a small number of very large hemoglobin molecules in solution; it is solved in vertebrates by having the hemoglobin located within corpuscles. Vertebrate blood contains about four times as much hemoglobin per ml. as that in polychaetes, reflecting the greater requirement for gas transport in vertebrates. The location of coelomic hemoglobin of polychaetes in corpuscles may represent a means of preventing the loss of the pigment through the nephridia.

A second respiratory pigment found in some polychaetes, especially fanworms, is **chlorocruorin.** Chlorocruorin is also an iron porphyrin but differs from heme in one of the side chains. This difference is sufficient to make this pigment green rather than red in color.

It seems likely that the gas-transporting function of hemoglobin and chlorocruorin evolved a number of times within annelids. Indeed, the widespread but sporadic distribution of hemoglobin throughout the Animal Kingdom suggests an independent evolution of this function within other phyla as well.

A third kind of respiratory pigment, **hemerythrin,** is found in one genus of polychaetes. Hemerythrin is not a porphyrin. Rather it is similar to hemocyanin in that the oxygen molecule is bound between two iron atoms. This pigment gives the blood a pink color.

Polychaete excretory organs are metanephridia (p. 506). A pair of nephridia is present in each segment, but the nephrostome of each nephridium opens through the septum into the segment anterior to the one containing the tubule (Fig. 25.3). The nephridiopore opens near the neuropodium.

Reproduction and Development. The sexes of polychaetes are separate. The **gametes** are produced by the coelomic **peritoneum,** and mature within the coelomic fluid. Primitively most segments produce gametes, but

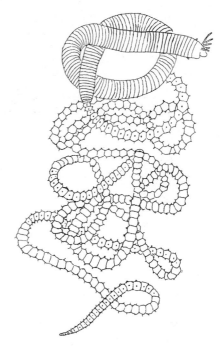

Figure 25.18 *Eunice viridis,* the Samoan palolo worm, with posterior region of gamete bearing segments. (After Woodworth from Fauvel.)

in many polychaetes gamete production is restricted to the segments in certain regions of the body. The **nephridia** commonly function as the **gonoducts,** but some species have separate gonoducts.

The animals do not copulate. Instead, the gametes are shed into the sea water where fertilization occurs. The likelihood of fertilization in some polychaetes is increased by swarming behavior. Males and females with ripe gametes swim to the surface at the same time in large numbers, shedding their eggs and sperm simultaneously (Fig. 25.17). Swarming usually occurs at specific times of the year, determined by annual, lunar, and tidal periodicities. The times at which swarming will occur in certain species can be predicted. The most notable example is the Samoan palolo worm. At the beginning of the last lunar quarter of October or November the worms release the posterior gamete-bearing regions of the body (Fig. 25.18), which come to the surface in enormous numbers and rupture.

Some polychaetes undergo structural modification for the swarming reproductive period, and the head and parapodia take on a different appearance from the non-repro-

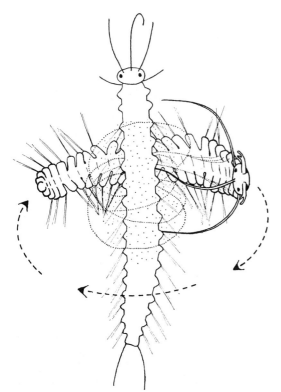

Figure 25.17 Syllid polychaetes during swarming. Male is swimming around female and releasing sperm. (After Gidholm.)

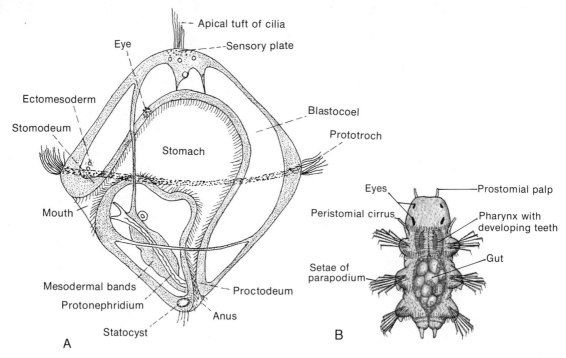

Figure 25.19   *A*, An annelid trochophore; *B*, a metamorphosing polychaete. (*A* after Shearer from Hyman; *B* from life.)

ductive individuals. This condition is called **epitoky.**

The eggs, especially those of swarming polychaetes, are often planktonic. However, there are species that deposit their eggs within jelly masses attached to the tube or burrow.

Cleavage is typically spiral and leads to a trochophore larva, which is essentially like that of mollusks (Fig. 25.19). During the course of larval life, the trochophore lengthens, and segments with parapodia and setae appear behind the prototroch and

mouth (Fig. 25.19). As development proceeds, the young worm settles to the bottom and takes up the adult mode of existence.

25.4   CLASS OLIGOCHAETA

In fresh water and on land, the annelids are represented by some 3000 species of oligochaetes. These are far less diverse than the polychaetes and in many respects parallel the burrowing polychaete gallery dwellers. The simple conical or lobe-like **prostomium** lacks sensory structures (Fig.

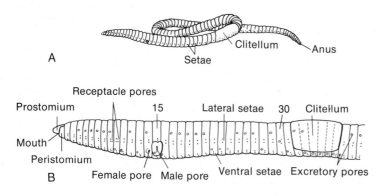

Figure 25.20   Earthworm. *A*, Entire worm. *B*, Lateral view of the anterior 40 segments of *Lumbricus*. Reproductive openings are found on segments 9, 10, 14 and 15. On each segment the excretory pore is either ventral, near the ventral setae, or lateral, above the lateral setae, with much variability between worms.

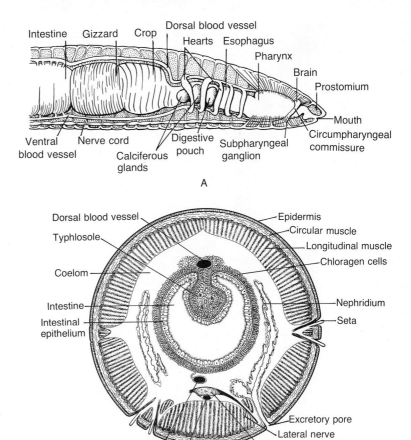

Figure 25.21 Structure of the earthworm *Lumbricus*. *A*, Lateral view of anterior section of body; *B*, cross section taken at the level of the intestine.

25.20). Oligochaetes crawl by peristaltic contractions. Parapodia are completely absent, but **setae** serve to anchor segments in crawling (Figs. 25.20 and 25.21). There are fewer setae per segment than in polychaetes, hence the name Oligochaeta— "few setae." Correlated with their peristaltic locomotion, the **coelom, septa,** and circular and longitudinal **muscles** are well developed and reflect a high order of metamerism.

Aquatic oligochaetes are usually less than three cm. in length and a few are almost microscopic. They live in algae, and in the bottom debris and mud of ponds, lakes, and streams. Some, such as the members of the family Tubificidae, are tubiculous (Fig. 25.22), but this habit is much less widespread than in polychaetes. Setae are generally short, but species living in algae and loose debris may have long setae (Fig. 25.22). Most aquatic oligochaetes live in shallow water, but some Tubificidae occur in enormous numbers in the lower anaerobic regions of deep lakes and have been reported from deep ocean bottoms.

Many oligochaetes live in wet boggy soil and intergrade with the more terrestrial earthworms, of which there are four major families. The largest annelids are Australian earthworms of the genus *Megascolides*, which may attain a length of three meters.

Earthworms are gallery dwellers, with short, heavy setae. They display few structural modifications for a terrestrial existence. Earthworms, like most other oligochaetes, lack special organs for gas exchange. The high concentration of oxygen in air and the highly vascularized body wall permit adequate gas exchange across the general body surface despite the large size that earthworms may attain. The surface is kept moist for gas exchange by epidermal mucous glands and by coelomic fluid, which is released through a dorsal pore between each segment.

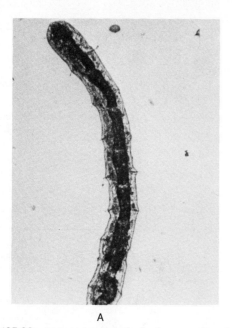
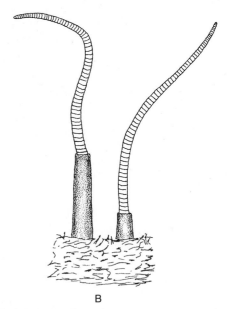

A   B

Figure 25.22   *A,* Anterior end of a fresh-water oligochaete. Dark interior tube is the gut. Setae appear as dark lines at right angles to body wall. *B, Tubifex,* a fresh-water, tube-dwelling oligochaete. Posterior end of body is projected from tube and waved about in water, facilitating gas exchange. (*A* by Betty M. Barnes; *B* after Pennak.)

Desiccation is minimized by behavioral adaptations. Earthworms come to the surface only at night or during rains. During dry periods or during freezing winter weather, earthworms burrow deeper in the soil. Some tropical Asian earthworms migrate down to depths of more than three meters. Estivation and hibernation are characteristic of many species.

Well-drained soils with a large amount of organic matter provide the best environment for earthworms. Within such soils the population is vertically stratified. Small species and small individuals live in or near the upper humus layer; larger species, such as *Lumbricus,* tunnel down to depths of one to two meters. Wet soils may lack adequate oxygen, and soils which are too acidic lack the free calcium ions necessary for the removal of excess carbon dioxide.

Nutrition.   Decomposing plant tissue is the principal food of most oligochaetes, although some fresh-water species feed upon algae, and a few are raptorial on other small invertebrates. Earthworms feed on organic matter, especially dead vegetation, in the humus layer of the soil. They are **direct deposit feeders,** cycling soil through the **gut** and digesting the organic matter. The castings may be deposited at the surface.

Food is ingested by means of a muscular **pharynx.** In many aquatic species the pharynx is everted as a mucus-laden pad; in earthworms it functions as a pump (Fig. 25.21). From the pharynx food passes into the **esophagus,** which is often modified at different levels as a **crop** and one or more **gizzards** (Fig. 25.21). The thin-walled crop functions in storage and the muscular gizzard grinds the food to smaller particles. *Lumbricus* has one crop and one gizzard at the beginning of the intestine (Fig. 25.21).

The remainder of the gut is a long straight **intestine,** the site of extracellular digestion and absorption. The surface area of the earthworm's intestine is nearly doubled by a dorsal longitudinal fold, the typhlosole (Fig. 25.21).

Surrounding the intestine is a layer of chloragen cells. These, like the vertebrate liver, are an important site of intermediary metabolism. The synthesis and storage of glycogen and fat, the deamination and synthesis of urea, and the removal of silicates (in terrestrial species) occur in chloragen cells.

The circulatory system, nephridia, and nervous system of oligochaetes are essentially like those of polychaetes. The contractile dorsal blood vessel provides the principal force for blood propulsion. In some oligochaetes, certain anterior commissural vessels (five pairs around the esophagus in *Lumbricus*) are conspicuously contractile (Fig. 25.21). These "hearts" function as accessory pumps.

Hemoglobin is present in solution in the plasma of the larger oligochaetes, including most of the earthworms.

The **nephridia** play a role in salt and water balance, for in fresh-water species and in earthworms with a plentiful water supply the urine is hypotonic to the coelomic fluid. Like other fresh-water animals, oligochaetes must eliminate the water entering by osmosis.

Giant fibers are present in the ventral nerve cord and earthworms have three especially large giant fibers grouped together in the dorsal part of the cord. The giant fibers transmit impulses controlling rapid contraction of the body.

Sense organs are lacking in most species, but the integument is richly supplied with sensory receptors, including photoreceptors.

Reproduction. In contrast to polychaetes, all oligochaetes are **hermaphroditic.** Moreover, there are distinct **gonads,** restricted to a few segments in the anterior third of the body.

The paired **ovaries** are located within a single segment (Fig. 25.23). The eggs mature within the coelomic fluid of that segment and exit through a pair of simple **oviducts.** These have ciliated funnels which penetrate the posterior septum and open

onto the ventral surface of the next posterior segment (Fig. 25.23).

Seminal receptacles are present, but in oligochaetes these structures are inpocketings of the ventral body wall and are not connected to the other parts of the female system (Fig. 25.23).

There are one or two testicular segments, each containing a pair of **testes** (Fig. 25.23). In some earthworms a transverse partition separates the lower part of the coelomic cavity containing the testes, and the sperm mature within this special chamber. The **septum** of the testicular segment is usually greatly evaginated to form a pouch-like **seminal vesicle.** The paired sperm ducts penetrate the posterior septum and may pass through several segments before opening onto the ventral surface (Figs. 25.20 and 25.23). In animals with two pairs of testes, the sperm ducts from the two testes of one side unite. The exact number of reproductive structures and their location in specific segments varies greatly.

The epidermis of certain adjacent segments in the anterior third of the body contains many glands which produce secretions for various parts of the reproductive process. These segments, collectively called the **clitellum,** are characteristic of oligochaetes. The position, number of segments included, and degree of glandular development of the clitellum vary considerably in different oligochaetes. The clitellum of aquatic species is usually composed of

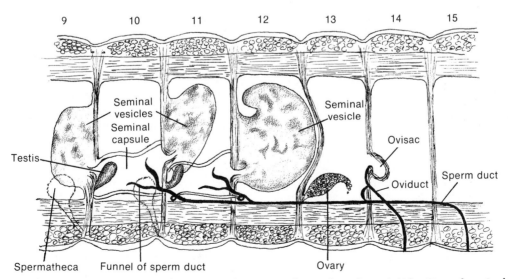

Figure 25.23 Reproductive segments of the earthworm *Lumbricus* (lateral view). (After Hesse from Avel.)

Figure 25.24 A, Two earthworms copulating. (Photograph of living animals made at night, courtesy of General Biological Supply, Chicago, Ill.) B, Movement of sperm during copulation in lumbricid earthworms. C, Earthworm cocoon. (B after Grove and Cowley from Avel; C after Avel.)

only two segments, those that contain the genital pores. The clitellum may be visible only during the reproductive period. The earthworm *Lumbricus*, in contrast, has a conspicuous, permanent, girdle-like clitellum of six to seven segments located behind the genital pores (Fig. 25.20). The clitellum of some earthworms includes as many as 60 segments.

Copulation with mutual transfer of sperm is characteristic of oligochaetes. Two worms come together with ventral surfaces opposed and with their anterior ends facing in opposite directions. The worms are held together by mucus secreted by the clitellum and sometimes by special genital setae. In most oligochaetes sperm are passed directly from the male gonopores into the seminal receptacles. However, in *Lumbricus* and

other members of the same family, the seminal receptacles of one worm are located some distance anterior to the male gonopores of the copulating partner (Fig. 25.24). The sperm on being released pass down a pair of grooves on the ventral surface and then cross over at the level of the seminal receptacles. The grooves are arched over by mucus and thus separated from the grooves of the opposite worm. Copulation in *Lumbricus* is a process requiring two to three hours.

A few days after copulation the clitellum secretes a dense material which will form the **cocoon**. Albumin is secreted inside the cocoon. The secretions form an encircling band, which slips forward over the body. Eggs from the female gonopores and sperm from the seminal receptacles are collected en route. The cocoon eventually slips off the anterior end of the worm, the two ends sealing in the process (Fig. 25.24). Fertilization takes place within the cocoon which is left in the bottom mud and debris or in the soil. Development is direct and young worms emerge from the end of the cocoon. Only one egg completes development within the cocoons of *Lumbricus terrestris.*

25.5 CLASS HIRUDINEA

The members of the class Hirudinea, the **leeches,** constitute the smallest and the most aberrant of the annelid classes. Only some 300 species have been described. The body is dorsoventrally flattened, with the anterior segments modified as a small **sucker** surrounding the mouth and the posterior segments forming a larger sucker behind the anus (Fig. 25.25). The head is greatly reduced, setae are absent, and there is little external evidence of metamerism. The body is ringed with many annuli, but they do not correspond to the 34 segments that compose the leech body (Fig. 25.25). Only the nephridia and nervous system indicate the original segmentation.

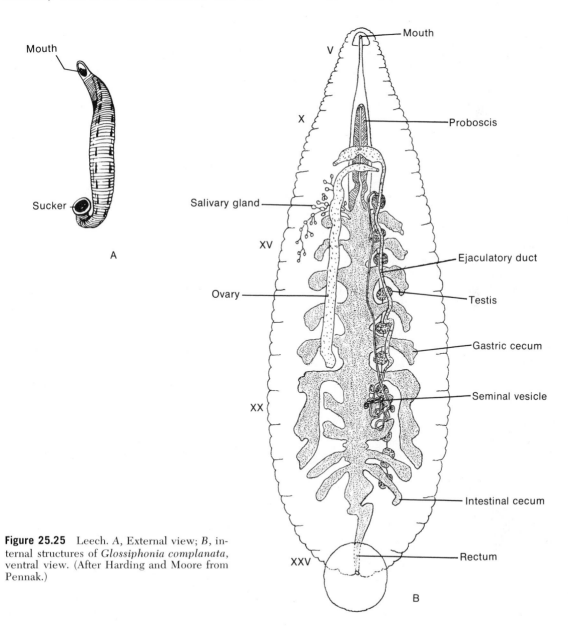

Figure 25.25 Leech. *A*, External view; *B*, internal structures of *Glossiphonia complanata*, ventral view. (After Harding and Moore from Pennak.)

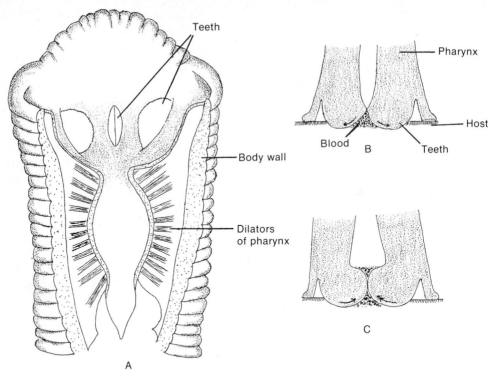

Figure 25.26   *A*, Anterior end of *Hirudo*, the medicinal leech (ventral dissection). *B* and *C*, Ingestion by *Hirudo*. Outward movement of jaws (*B*); followed by medial movement of teeth and dilation of pharynx (*C*). (*A* modified after Pfurtscheller from Pennak; *B* and *C* after Herter from Harant and Grassé.)

Most leeches are two to five cm. in length and are never as small as many polychaetes and other annelids. *Hirudo medicinalis*, the medicinal leech (Fig. 25.26), may reach a length of 20 cm. Leeches are commonly olive green and brown in color.

Locomotion.   Leeches crawl in a looping manner. The body is extended and the an-terior sucker attached. Then the posterior sucker is released, pulled forward, and re-attached. Many leeches can swim by dorso-ventral undulations of the flattened body. Correlated with these modes of movement, the longitudinal muscles of the body wall are powerfully developed. The septa have been lost and the coelom is reduced by the

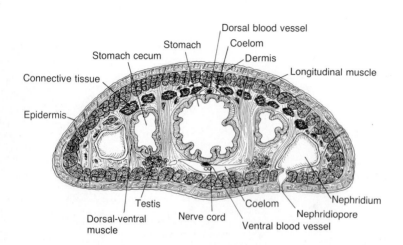

Figure 25.27   Cross section of a leech showing the reduction of the coelom by connective tissue.

invasion of connective tissue to a system of interconnecting spaces, or sinuses (Fig. 25.27).

Nutrition. Contrary to popular notion, not all leeches are bloodsuckers. Many are predaceous and feed upon other small invertebrates, especially annelids, snails, and insect larvae. The bloodsucking leeches are ectoparasitic on invertebrates such as snails and crustaceans, and on vertebrates. All classes of vertebrates may be hosts but fish and other aquatic forms, including water fowl, are the most common victims. In humid areas of the tropics, members of the family Haemadipsidae attack terrestrial birds and mammals, attaching to the moist thin membranes of the nose and mouth. They climb upon the host when it brushes through vegetation on which the leeches are located. Some ectoparasitic leeches, such as the fish leeches of the family Piscicolidae, may remain on the host throughout most of their lives; others are located on the host only when feeding.

Many leeches, both predatory and parasitic, ingest food through a tubular jawless proboscis, which can be extended through the mouth from a proboscis cavity (Fig. 25.25). The method by which the proboscis penetrates the prey is not understood, although it is thought that enzymes may be utilized. Predatory species usually swallow the prey whole.

The **pharynx** of other leeches, mostly the bloodsuckers, serves as a pump but cannot be everted. The mouth cavity in those leeches contains three blade-like **teeth** (Fig. 25.26). When the anterior sucker is attached, the teeth slice through the skin of the host. A salivary secretion, called **hirudin**, prevents the coagulation of blood, which is ingested by the pumping action of the pharynx.

An **esophagus** connects the pharynx with a large stomach, or **crop**, which commonly possesses lateral pouches, or **caeca** (Fig. 25.25). An **intestine** comprises the posterior third of the digestive tract and opens through the **anus** just above the posterior sucker. The stomach functions both in digestion and as a storage center. Many leeches may consume two to ten times their weight in blood at a single feeding. Water is removed from the ingested blood, and the remaining material may require as long as six months to be digested. Symbiotic bacteria provide the enzymes for a part of the digestive process and produce some substance which prevents decomposition of the blood cells. Thus a meal may be stored for a long time between infrequent feedings. Leeches have survived without feeding for as long as a year and a half.

Some leeches have a blood vascular system like that of their oligochaete ancestors. Others have lost the original blood vessels and the interconnecting coelomic channels

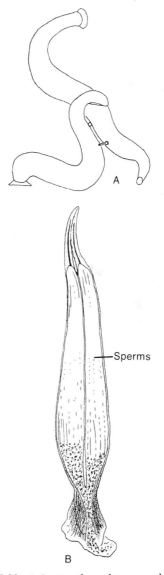

Figure 25.28 A, Section through two copulating *Erpobdella*. Upper individual is injecting spermatophores into lower individual. B, Spermatophore of the rhynchobdellid, *Haementeria*. (A after Brumpt; B after Pavlovsky from Harant and Grassé.)

have been converted into a blood vascular system. Hemoglobin provides for gas transport, but gills are present in only a few leeches. Although the head is reduced, some leeches possess eyes. Special sensory papillae are arranged in rings on the annuli.

Reproduction. The presence of a **clitellum** and a **hermaphroditic** reproductive system is taken as evidence that leeches evolved from oligochaetes. Copulation involves a mutual exchange of **spermatophores** (Fig. 25.28), which in some leeches are transferred by means of hypodermic impregnation. The spermatophores are injected into the ventral surface of the opposite leech (Fig. 25.28), and the injected sperm make their way to the **ovaries** of the female system, where fertilization occurs.

The clitellum produces a **cocoon** as in oligochaetes. Many fish leeches attach their cocoon to the host. Other leeches deposit the cocoon in water, beneath stones and other objects, and in soil. The members of one family of leeches brood their eggs within a modified cocoon attached to the ventral surface of the body.

CLASSIFICATION OF THE PHYLUM ANNELIDA

CLASS POLYCHAETA. Marine annelids in which setae are carried on lateral segmental parapodia. Metamerism usually well developed. Prostomium (head) variable, but commonly bears sensory or feeding structures. Sexes separate and gametes produced by the peritoneum of numerous segments.

CLASS OLIGOCHAETA. Fresh-water annelids and the terrestrial species known as earthworms. Parapodia absent but setae present. Metamerism well developed. Prostomium a simple lobe without sensory or feeding structures. Hermaphroditic, with gonads present in a few specific segments. Epidermis of certain segments modified as a clitellum for the secretion of a cocoon.

CLASS HIRUDINEA. Marine, fresh-water, and terrestrial annelids known as leeches. No parapodia or setae. Body more or less dorsoventrally flattened with anterior and posterior segments modified as suckers. Metamerism greatly reduced. Some ectoparasitic. Hermaphroditic, with a clitellum.

ANNOTATED REFERENCES

Detailed accounts of the annelids may be found in many of the references listed at the end of Chapter 18. The following works are devoted solely to annelids.

Dales, R. P.: Annelids. London, Hutchinson University Library, 1963. A brief, general account of the annelids.

Laverack, M. S.: The Physiology of Earthworms. New York, Pergamon Press, 1963.

Mann, K. H.: Leeches (Hirudinea). New York, Pergamon Press, 1962. A general biology of leeches.

26. The Arthropods

The phylum **Arthropoda** is a vast assemblage of animals. At least three quarters of a million species have been described; this is more than three times the number of all other animal species combined. The tremendous adaptive diversity of arthropods has enabled them to survive in virtually every habitat; they are perhaps the most successful of all the invaders of the terrestrial environment.

A number of features indicate that arthropods evolved from annelids, perhaps from some group of surface-dwelling polychaetes. Arthropods are **metameric** and the segments bear lateral appendages, which may be homologous to the parapodia of polychaetes. The nervous system, with its large ventral **nerve cord,** is essentially like that of annelids, and the dorsal tubular **heart** may be the homologue of the dorsal contractile blood vessel of annelids.

26.1 SUPPORT AND MOVEMENT

The distinguishing feature of arthropods, one which has been an important factor in the evolutionary success of the group, is the **chitinous exoskeleton** that covers the entire body, including the appendages. The exoskeleton is divided into plates on the trunk and head and forms cylinders on the appendages (Fig. 26.1). At the junction points between skeletal pieces the chitin is thin and folded, permitting movement (Fig. 26.2).

Distinct **muscle bundles** are anchored to the inner side of the skeleton and extend or flex parts at a joint, depending upon the point of attachment (Fig. 26.1). Thus the skeletal muscle system of arthropods operates using a system of levers in much the same way as the skeletal muscle system of vertebrates. The principal difference is that arthropod muscles are attached to the *inner* side of an *external* skeleton whereas vertebrate muscles are attached to the *outer* side of an *internal* skeleton.

An exoskeleton is a highly efficient supporting and locomotor framework for animals such as arthropods which are no more than a few centimeters long. For large animals, such as vertebrates, an internal skeleton is more efficient; an external skeleton would be very heavy and unwieldy.

Despite the locomotor and supporting advantages of an external skeleton, it poses problems for a growing animal. The solution to this problem evolved by the arthropods has been the periodic shedding of the skeleton, a process called **molting** or **ecdysis.**

The arthropod skeleton is composed of a thin outer **epicuticle** and a much thicker **procuticle.** The fully developed procuticle consists of an outer **exocuticle** and an inner **endocuticle** (Fig. 26.2). Both layers are composed of chitin and protein bound together to form a complex glycoprotein, but the exocuticle in addition has been tanned; i.e., with the participation of phenols, its molecular structure has been further stabilized by the formation of additional cross linkages.

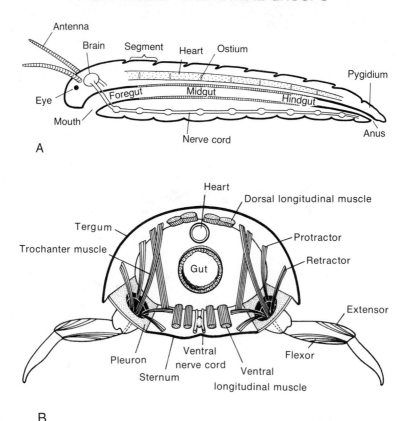

Figure 26.1 Structure of a generalized arthropod. *A*, Sagittal section; *B*, cross section.

Before the old skeleton is shed, the epidermal layer (**hypodermis**) becomes detached from the skeleton and secretes enzymes—chitinase and protease—which erode away the untanned endocuticle (Fig. 26.3). Muscle attachments and nerve connections are unaffected, and the animal can continue to move. Following the secretion of enzymes, the epidermis produces a new epicuticle, the protein-lipid composition of which is unaffected by the enzymes produced earlier. Beneath the new epicuticle a new procuticle is laid down. At this point the animal is encased within both an old and a new skeleton (Fig. 26.3). The old skeleton now splits along certain predetermined lines and the animal pulls out of the old encasement. Immediately after shedding, the new skeleton is soft and pliable and is stretched to accommodate the increased size of the animal.

The periods between molts are known as **instars,** and the length of the instars becomes longer as the animal becomes older. Some arthropods, such as lobsters, continue to molt throughout their life, although the event becomes less and less frequent. Other arthropods, such as insects and spiders, have fixed numbers of instars, the last being attained with sexual maturity.

Only vestiges of the coelom remain in adult arthropods, although it appears during the course of embryonic development. The loss is probably related to the shift from a fluid internal skeleton to a solid external skeleton.

The primitive arthropod trunk was composed of a large number of segments, each bearing a pair of similar appendages (Fig. 26.9). The evolution of different modes of existence with different types of locomotion —swimming, crawling, running, pushing, climbing—and with different types of feeding behavior has led to a reduction in the number of segments. Segments dropped out, fused together, or become specialized. Appendages became specialized for many functions other than locomotion—prey capture, filter feeding, food handling, gas exchange, ventilation, copulation, egg brood-

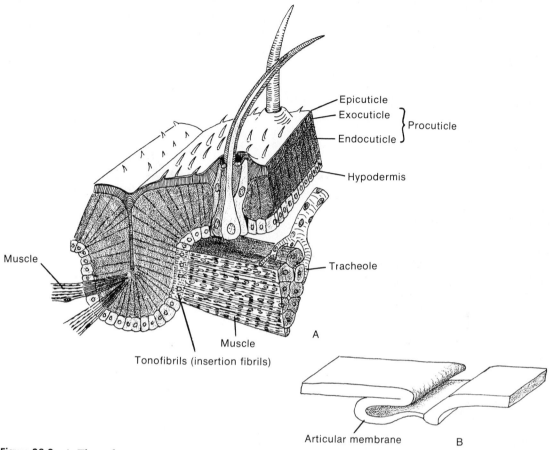

Figure 26.2 *A*, Three dimensional section of an arthropod integument (insect). *B*, An intersegmental joint. Articular membrane is folded beneath segmental plate. (*A* after Weber from Kaestner; *B* after Weber from Vandel.)

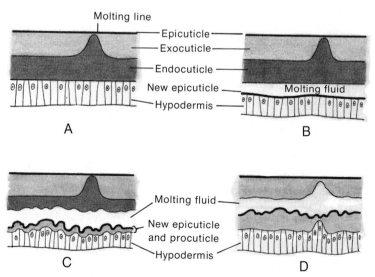

Figure 26.3 Molting in an arthropod. *A*, The fully formed exoskeleton and underlying epidermis between molts. *B*, Separation of the epidermis and secretion of molting fluid and the new epicuticle. *C*, Digestion of the old endocuticle and secretion of the new procuticle. *D*, The animal just before molting, encased within both new and old skeleton.

ing, etc. In the crayfish, for example, the oral appendages are adapted for feeding, the large claws are adapted for grasping, the legs for crawling, the abdominal appendages for egg brooding, and the last appendages form a flipper with the terminal section of the trunk (Fig. 26.24). Although these appendages are structurally quite different, they are all derived from originally similar segmental appendages. They are therefore said to be **serially homologous.** The fusion, loss, and differentiation of segments and appendages has been repeated many times in the history of different arthropod groups.

The legs of arthropods are much longer and more slender than the parapodia of polychaetes and the body sags down between them. However, there are some similarities to the crawling gait of polychaetes. Each leg performs an effective and a recovery stroke in which the leg is lifted, swung forward, and placed down upon the substratum. The legs on one side of the body carry out these movements in sequence, comprising a locomotor wave; the locomotor waves on the two sides of the body alternate with each other as in crawling polychaetes.

Arthropod muscle bundles are composed of striated fibers attached to the skeleton by acellular fibrils that extend up to the epicuticle. In sharp contrast to the condition in vertebrates (p. 105), each arthropod muscle contains relatively few fibers and is innervated by only a small number of neurons. One neuron may supply more than one muscle. Moreover, several types of motor neurons—fast neurons, slow neurons, and inhibitory neurons—may supply a single muscle. The terms "fast" and "slow" do not refer to the speed of transmission but to the nature of the response. The impulses of fast neurons produce a postsynaptic potential that results in a rapid, powerful contraction. With repetitive firing of the neuron the muscle fiber does not relax between impulses and contraction is sustained (tetanic). The impulses of slow fibers produce a postsynaptic potential that results in only a small twitch, but a volley of impulses, by summation and facilitation, can cause the muscle to contract to varying degrees. The impulses of inhibitory neurons block contraction.

In vertebrates, graded responses are in large part dependent upon the *number* of motor units contracting. Arthropod muscles are not organized as motor units, and graded responses are dependent upon slow fibers and the interaction of the various kinds of neurons supplying the muscle.

26.2 DIGESTIVE SYSTEM

The arthropod gut differs from that of most other animals in having large stomodeal and proctodeal regions (Fig. 26.1). The derivatives of these ectodermal portions are lined with chitin and constitute the **foregut** and **hindgut.** The intervening region, derived from endoderm, forms the **midgut.** The foregut is chiefly concerned with ingestion, trituration, and storage of food; its parts are variously modified for these functions depending upon the diet and mode of feeding. The midgut is the site of enzyme production, digestion, and absorption; however, in some arthropods enzymes are passed forward and digestion begins in the foregut. Very commonly the surface area of the midgut is increased by outpocketings forming pouches or large digestive glands. The hindgut functions in the absorption of water and the formation of feces.

26.3 INTERNAL TRANSPORT, GAS EXCHANGE, AND EXCRETION

The circulatory system of arthropods is open; a large **pericardial sinus** surrounds the dorsal **heart** (Fig. 26.1). Blood flows from the pericardial sinus into the heart through small lateral openings called **ostia.** In primitive tubular hearts there is one pair of ostia per body segment. When the heart contracts, the ostia close, and blood is propelled anteriorly or posteriorly out of the heart through arteries. The arterial system varies greatly in the various kinds of arthropods but generally is more extensive in large species. The arteries deliver the blood to tissue spaces where various exchanges take place. From the tissue spaces (collectively termed the **hemocoel**) blood drains into a system of larger sinuses and eventually returns to the pericardial sinus around the heart.

Most arthropods have organs for gas exchange. Their **gills** are usually modifications of appendages or outgrowths of the integument associated with an appendage (Fig. 26.4). The surface area of the gill is

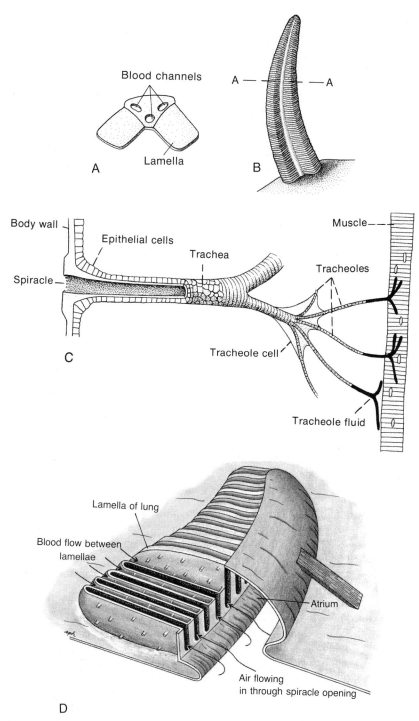

Figure 26.4 Arthropod gas exchange organs. *A* and *B*, Crab gill. Gill consists of a series of flat plates (**lamellae**) stacked upon one another and attached to a central axis (*B*). Cross section shown in *A* taken at the level indicated in *B*. *C*, Diagrammatic section of an insect tracheal system showing relationship of spiracle and tracheoles to tracheae. *D*, Section through a book lung. (*A* and *B* modified from Calman; *C* modified from Ross.)

increased by flattening, folding, or branching in various ways. The chitin over the exchange surface is thin and offers no obstacle to the passage of gases.

The gas exchange organs of terrestrial arthropods are typically internal. However, they are derived from invaginations of the integument and thus are lined with chitin. The commonest gas exchange organ in terrestrial arthropods, and one which has evolved independently in many different groups, is a system of air-conducting tubes called **tracheae.** The system opens to the exterior through **spiracles;** in many species these can be closed by valves. The inner end of the tubes are blind and terminate in very fine fluid-filled tubules, called **tracheoles,** which directly penetrate the tissue being supplied with oxygen (Fig. 26.4). There is usually no special ventilating system. Gases move through the air and fluid-filled sections of the tube down a diffusion gradient.

Book lungs, present in spiders and scorpions, are terrestrial gas exchange organs resembling internal gills. A slit-like spiracle opens into an internal chitin-lined chamber (Fig. 26.4). One wall of the chamber is thrown into a large number of leaf-like folds, and gases pass between the blood circulating *within* the folds and the air moving *between* the folds.

The arthropods with a respiratory pigment most commonly have **hemocyanin.**

Hemoglobin occurs only sporadically. Respiratory pigments are absent from the blood of arthropods with tracheal systems; the blood in these animals plays only a small role in gas transport.

The loss of the coelom probably accounts for the disappearance of nephridia and the evolution of the diverse excretory organs of arthropods. Most of the excretory organs are blood-bathed sacs opening by a duct to the body surface. The excretory organs of insects and some arachnids, such as spiders, consist of a few to many blind **tubules** lying free in the **hemocoel** and opening into the posterior section of the **gut** (Fig. 26.54). Waste materials are removed from the blood and secreted into the lumen of the excretory organ. Aquatic arthropods excrete ammonia. Some terrestrial species, such as insects, excrete uric acid; others, such as spiders, excrete guanine.

26.4 NERVOUS SYSTEM AND SENSE ORGANS

The nervous system of arthropods exhibits the same basic design as that of annelids—a dorsal anterior **brain** and a ventral **nerve cord** (Figs. 26.1 and 26.54). However, the fusion and loss of segments is reflected in a corresponding forward migration and fusion of the ventral **ganglia** of many arthropods. For example, in a crab all of the

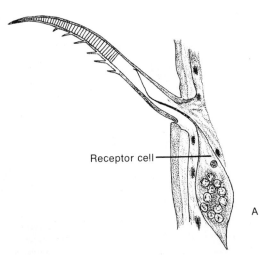
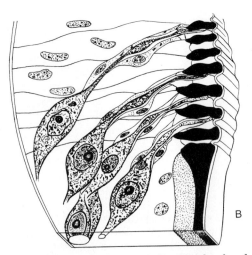

Figure 26.5   *A,* Sensory hair of a mite. *B,* Section through part of a lyriform organ of a spider. The distal end of the sensory neuron is shown attached to the incurved cuticular membrane of the slit. (*A* after Gossel from Millot; *B* modified from Kaston and others.)

Receptor cell

ganglia of the ventral nerve cord have fused together anteriorly. This ventral mass of ganglia plus the circumesophageal connectives and the brain form a large collar that encircles the esophagus.

The **sensory receptors** of arthropods are usually associated with some modification of the chitinous exoskeleton, which otherwise would act as a barrier to the detection of external stimuli. An important and very common type of receptor is one connected with **hairs, bristles,** or **setae.** When the bristle is moved, the receptor ending in the shaft or base is stimulated (Fig. 26.5). Some arthropods, such as spiders, have hairs so delicate and sensitive that they can respond to air-borne vibrations. Other common modifications for receptors are slits, pits or other openings in the exoskeleton. The opening is covered by a thin membrane, to the underside of which is attached a nerve ending (Fig. 26.5). Such sense organs detect vibrations or other forces which change the tension of the skeleton.

Chemoreceptors are especially concentrated on the appendages or mouthparts, where they may be lodged in setae or the general skeletal surface.

The appendages of arthropods contain many kinds of receptors and are of great importance in receiving information about the surrounding environment. A leg, for example, may serve not only for movement but also for touch, taste, and smell. One or more appendages (e.g., the **antennae**) may be given over entirely to the function of monitoring environmental signals.

Most arthropods have **eyes,** which can vary greatly in complexity. Some are simple and have only a few photoreceptors. Others are large with thousands of retinal cells and can form a crude image. A transparent **lens-cornea** is the usual skeletal contribution to the eye, and the focus is always fixed since the immovable lens is continuous with the surrounding skeleton (Fig. 26.6).

Insects and many crustaceans, such as crabs and shrimp, have a **compound eye,** so called because it is composed of many long cylindrical units possessing all of the elements for light reception. Each unit, called an **ommatidium,** contains an outer **cornea,** a middle **crystalline cone,** and an inner **retinula** (Fig. 26.7). On the external surface, the cornea of each ommatidium is

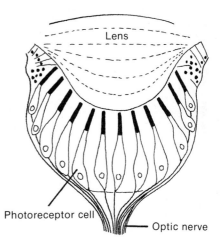

Figure 26.6 Section through the eye of a scorpion.

distinct and forms one facet of the compound eye. The cornea functions as the lens and the crystalline cone funnels light down to the retinula, which is equivalent to the retina and is composed of a rosette of seven cells. Their inner photosensitive surfaces (**rhabdomeres**) together form a central **rhabdome.** Moveable screening pigment is commonly present between ommatidia.

The term "compound eye" is misleading, for it implies that each ommatidium forms a separate image. This is not true. The seven retinula cells function as a single photoreceptor unit and together transmit a single signal. The image formed by the entire eye, although sometimes called mosaic, depends upon all of the signals transmitted by the ommatidia, and is thus really no different from that of other types of eyes (Fig. 26.7).

Compound eyes may function differently in bright light than they do in weak light. In bright light, the screening pigment is *extended* between the ommatidia and each retinula responds only to the light received by its facet and crystalline cone (Fig. 26.7A). This type of image, called an apposition image, is believed to be especially effective in detecting movement, for slight changes in the position of the moving object will stimulate different ommatidia.

In weak light the screening pigment is *contracted* and light received by one ommatidium can cross over to adjacent units. Thus a retinula may be fired by the light received through several ommatidia (Fig. 26.7C). Under these conditions the eye is

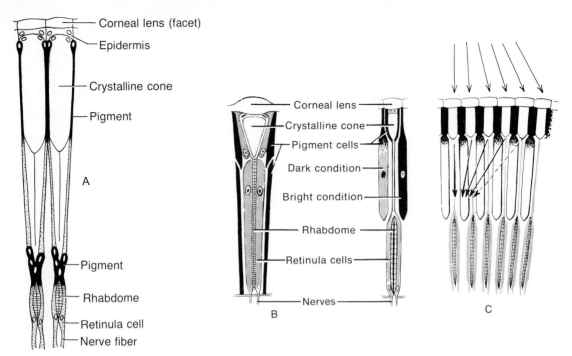

Figure 26.7   Compound eye. *A*, Two ommatidia from the compound eye of a crayfish. In each, light passing through the two lenses is focused on the outer end of the rhabdome, which is made of seven fused rods, or rhabdomeres, striated thickenings along the inner edges of seven retinula cells. Pigment screens out stray light. *B*, Insect ommatidia, showing a diurnal type (*left*) and a nocturnal type (*right*). In the nocturnal type, the pigment is shown in two positions, adapted for very dark conditions on the left side, and for relatively bright conditions on the right. *C*, Nocturnal type of eye adapted for dark conditions, showing how light can be concentrated upon one rhabdome from several lenses. If the pigment moved downward, light from peripheral lenses would be screened out.

said to form a "superposition image," but it is unlikely that there is any object discrimination at all. Rather, the eye is detecting changes in general light intensity or the position of a bright light source or shadow.

Although compound eyes are commonly described as being capable of forming both an apposition and a superposition image, in actual fact they tend to be especially adapted for functioning either in bright light or in weak light, but not for both conditions. Most species are either diurnal or nocturnal or live in habitats in which there is reduced light. Hence the eyes usually function in only limited conditions. Eyes especially adapted for functioning in bright light may have a large amount of screening pigment and the long crystalline cone is very close to the retinula. Eyes adapted for functioning in dim light may have little screening pigment and the short crystalline cone is some distance from the retinula, permitting light rays to cross over from one ommatidium to another. (Fig. 26.7C).

Visual acuity varies greatly and depends upon the number of photoreceptors present, regardless of the type of eye. The eyes of some hunting spiders and the compound eyes of many insects and some crabs appear to be capable of detecting objects. In all of these species the number of photoreceptors is very great. The eye of the dragon fly, for example, has 10,000 ommatidia, and the eye of a wolf spider contains 4500 photoreceptors. However, when one compares these numbers with the 80,000,000 photoreceptors of the human eye, it is evident that the images formed in even the most highly developed arthropod eyes must be very crude.

Color discrimination has been demonstrated in several arthropods and is an important adaptation of bees and other insects which depend on flowers as a food source.

26.5   REPRODUCTION AND DEVELOPMENT

Arthropods have separate sexes, each with a single pair of **gonads,** but the nature

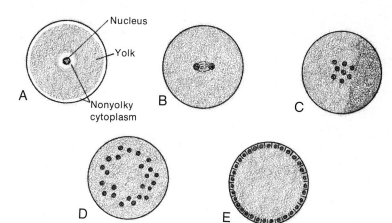

Figure 26.8 Superficial cleavage. A, Centrolecithal egg; B and C, nuclear divisions; D, migration of nuclei toward periphery of egg; E, blastula.

of the reproductive system and the location of reproductive openings are too variable to permit generalization. The egg contains a relatively large amount of **yolk** arranged concentrically about a central **nucleus.** The non-yolky cytoplasm is limited to a thin layer immediately about the nucleus and at the egg surface (Fig. 26.8). Such an egg is called **centrolecithal** and is characterized by incomplete superficial cleavage. The nucleus divides initially without any cleavage of the cytoplasm (Fig. 26.8). After a number of such nuclear divisions, the daughter nuclei migrate to the surface, where cell membranes begin to form but do not extend far into the interior yolk mass between nuclei. This stage is a solid **blastula.**

26.6 ARTHROPOD CLASSIFICATION

Living arthropods may be divided into two subphyla. Those that lack antennae are commonly placed in the subphylum **Chelicerata,** which takes its name from the fact that the first post-oral appendages are feeding appendages called chelicerae. To this group belong the horseshoe crabs, scorpions, spiders, mites, and ticks. Arthropods with antennae are placed within the subphylum **Mandibulata;** the first post-oral appendages in the members of this group are feeding appendages called mandibles. This group includes all of the insects as well as shrimp, crabs, millipedes and centipedes.

26.7 TRILOBITES

The most primitive known arthropods are the extinct trilobites, which inhabited the Paleozoic oceans from the Cambrian through the Permian. They are usually placed within a separate subphylum, the **Trilobitomorpha.**

The flattened oval body of most trilobites was about three to 10 cm. in length and composed of an anterior **cephalon,** a middle **thorax,** and a posterior **pygidium** (Fig. 26.9). The cephalon, originating from the fusion of the head with four trunk segments, formed a solid shield-shaped **carapace,** which bore a pair of eyes on the dorsal surface. The middle section of the body, the thorax, was composed of many unfused segments, and the posterior pygidium contained a number of fused segments. Two dorsal longitudinal furrows extended the length of the body and divided it into three transverse sections, from which the name trilobite is derived.

A pair of antennae was located on the undersurface of the body just in front of the mouth. Behind the mouth, each segment, whether fused with other segments or separate, carried a pair of identical appendages (Fig. 26.9). In all other arthropods, at least some of the segmental appendages are different from others. Each appendage contained two branches—one leg-like, on which the animal crawled, and one which contained parallel filaments and was believed to have functioned as a gill (Fig. 26.9).

Most trilobites were bottom dwellers; however, variations in their body form and size suggest that some species bur-

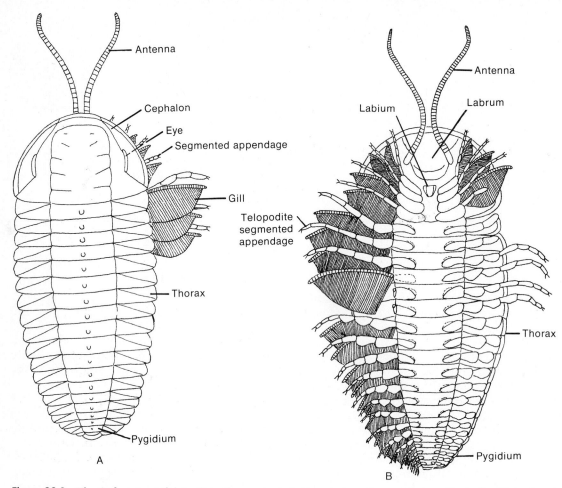

Figure 26.9 The Ordovician trilobite, *Triarthrus eatoni*. A, Dorsal view; B, ventral view. (Both after Walcott and Raymond from Störmer.)

rowed through sand, some were swimmers, and some were planktonic. Surprisingly, we know a great deal about the development of trilobites, for complete series of larval stages of numerous species have been preserved in the fossil record.

26.8 THE CHELICERATES

Members of the subphylum Chelicerata have no antennae and the first post-oral appendages are feeding appendages called **chelicerae.** Most chelicerates are further distinguished from other arthropods in having the body composed of a **cephalothorax** and an **abdomen** (Fig. 26.10). The cephalothorax contains a number of trunk segments which have fused with the head, and the fused region is covered dorsally by a single skeletal piece, the **carapace.** Behind the chelicerae the cephalothorax carries a pair of appendages, called **pedipalps,** and four pairs of legs. There are no distinct abdominal appendages.

Class Merostomata

The members of the class Merostomata are marine gill-bearing chelicerates, called horseshoe crabs, which inhabit sandy bottoms in shallow water. The class has been known from the Ordovician, but there are only five species living today. Four are Asian and one, *Limulus polyphemus,* is found along the Atlantic coast of the United States and in the Gulf of Mexico. They are among the largest living chelicerates, reaching a length of 60 cm.

The name horseshoe crab comes from the large, convex horseshoe-shaped carapace

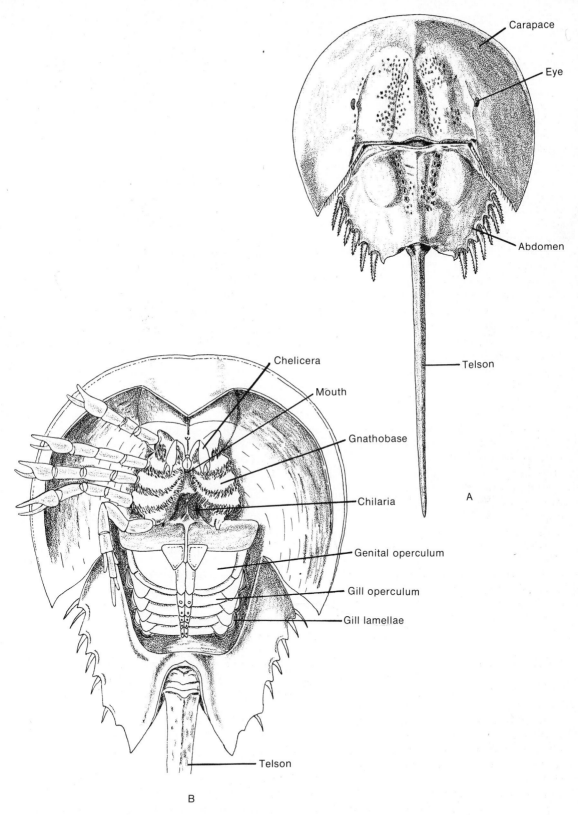

Figure 26.10 *A*, A horseshoe crab (dorsal view). *B, Limulus polyphemus,* showing appendages on one side (ventral view). (*A* after Van der Hoeven from Fage; *B* after Versluys and Demoll from Kaestner.)

which covers the cephalothorax (Fig. 26.10). Dorsally the carapace carries a pair of eyes, and the underside of the cephalothorax bears the six pairs of appendages typical of most chelicerates—a pair of chelicerae, a pair of pedipalps, and four pairs of legs (Fig. 26.10). However, the pedipalps are not markedly different from the posterior appendages and should properly be called legs. All except the last pair of appendages are **chelate,** that is, they possess pincers formed by the two terminal segments of the appendage (Fig. 26.10). The last pair of appendages is used for sweeping away sand and mud.

The abdominal segments are fused together and terminate in a long spike-like **telson** (Fig. 26.10). The underside of the abdomen carries five pairs of **book gills,** each composed of a large number of thin plates (**lamellae**) protected by a flap-like cover (Fig. 26.10).

Horseshoe crabs are harmless animals. They crawl over or push through sand in shallow water, and small individuals swim upside-down using the gills as paddles. The telson is used to right the animal if flipped over.

Horseshoe crabs are omnivores and scavengers, and their diet includes the soft-bodied invertebrates and algae they encounter as they plow through the bottom.

Adults congregate in shallow water during the reproductive period. Fertilization occurs at the time of egg deposition while the male is clasped to the back of the female abdomen with his modified hook-like pedipalps. The eggs are deposited in the sand, and a tiny swimming "trilobite" larva is released at hatching. The superficial trilobite appearance is gradually lost with the attainment of adult form.

Class Arachnida

The great majority of chelicerates, some 60,000 species, are members of the class **Arachnida.** In contrast to merostomes, arachnids are terrestrial and lack gills. They are widely distributed, living in vegetation, in leaf mold, and beneath bark, logs and stones.

The class is divided into eight orders, five of which contain the most common and familiar species. Scorpions (order **Scor-**

pionida) are large arachnids with big chelate pedipalps and a long segmented abdomen terminating in a **sting** (Fig. 26.11). Paired eyes are mounted on tubercles in the middle of the carapace, and two to five additional pairs of eyes may be present along the anterior lateral margins. Scorpions are secretive, largely nocturnal animals of the tropics and semitropics. In the United States they are common only in the Gulf and southwestern states.

Scorpions are the most ancient arachnids, known from the Silurian, and probably were the first terrestrial arthropods. Their ancestors may have been the extinct, giant eurypterids, a group of Paleozoic merostomes (Fig. 26.11). Although most eurypterids were marine, some invaded brackish and fresh water. Fossil evidence suggests that the first scorpions were also aquatic.

Pseudoscorpions (order **Pseudoscorpionida**) are only a few millimeters in length and are common inhabitants of leaf mold in both tropical and temperate regions. These tiny arachnids have large chelipeds like scorpions, but the segmented abdomen is short and lacks a terminal sting (Fig. 26.12).

Spiders (order **Araneae**) comprise the largest of the arachnid orders. Over 25,000 species have been described and they frequently occur in far greater numbers than most people are aware. An ungrazed meadow, for example, may support as many as 2,250,000 spiders per acre. The abdomen is unsegmented and connected to the cephalothorax by a narrow waist (Figs. 26.13 and 26.16). The pedipalps are small and leg-like. Eight eyes are arranged in two rows of four each across the front of the carapace.

Daddy-long-legs, or harvestmen (order **Opiliones,** or **Phalangida**), are distinguished from other arachnids by their very long legs and segmented abdomen broadly joined to the cephalothorax (Fig. 26.12). A tubercle on the center of the carapace bears a single pair of eyes. The members of this order are common arachnids in both temperate and tropical regions.

The order containing the mites and ticks (order **Acarina**) is the second largest and most diverse group of arachnids. Over 20,000 species have been described to date, and the order is still poorly known taxonomically.

Mites are usually less than a millimeter in length, and their adaptive diversity may in

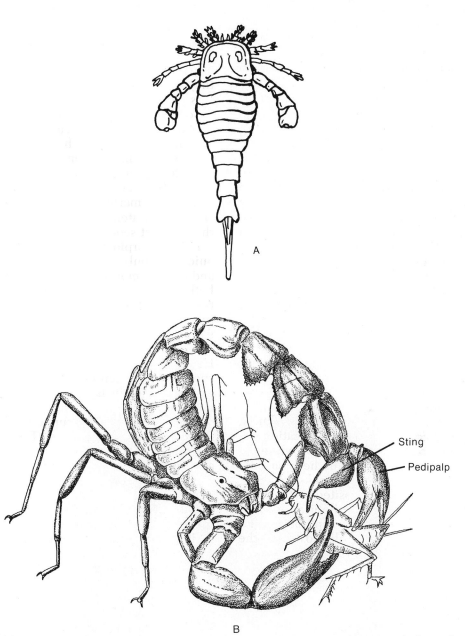

Sting

Pedipalp

B

Figure 26.11 A, Dorsal view of a eurypterid, an extinct group of Paleozoic merostomes, which may have been ancestral to the scorpions. B, The North African scorpion, *Androctonus australis,* capturing a grasshopper. (B after Vachon from Kaestner.)

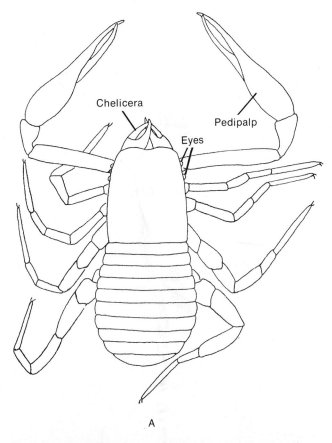

A

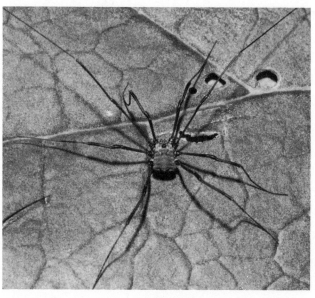

B

Figure 26.12 *A*, Pseudoscorpion (dorsal view); *B*, dorsal view of a harvestman. (*A* after Hoff; *B* by Betty M. Barnes.)

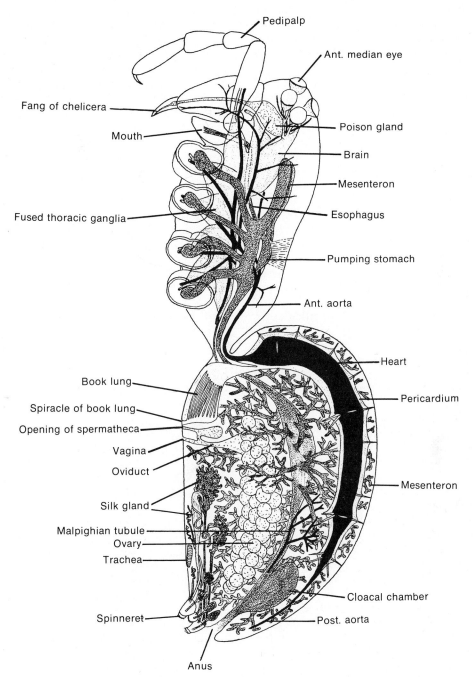

Figure 26.13 Internal anatomy of a two-lunged spider. (After Comstock.)

part be attributed to their small size, which has enabled them to exploit many types of microhabitats. However, such habitats must contain adequate water vapor, for the small size of mites also makes them vulnerable to desiccation. Where conditions are favorable, as in leaf mold, mites occur in enormous numbers.

The abdomen of mites is unsegmented and broadly fused with the cephalothorax. The entire body is thus covered dorsally by a single skeletal piece (Fig. 26.14). The pedipalps are small and usually leg-like.

Silk. Two groups of arachnids evolved the means of producing and utilizing silk. The tiny pseudoscorpions have silk glands in the chelicerae and use the silk to build retreats or nests, especially during the time of reproduction.

The **silk glands** of spiders are located in the abdomen and open through conical **spinnerets** at the end of the abdomen (Fig. 26.13), each spinneret bearing numerous spigots. A particular species may possess from two to six different kinds of silk glands. Silk is a protein and hardens not on exposure to the air but by polymerization during the process of being drawn out.

Spiders utilize silk in many ways, but contrary to popular notion, only some spiders use silk to construct webs for trapping prey. Silk is used to build nests, which are used as retreats, for reproduction, or for overwintering; in all spiders the eggs are encased within a silken **cocoon.**

Most spiders lay down a **dragline** behind them, anchoring it at intervals to the substratum. The dragline not only functions as a safety line for the spider, but is also an important means of communication between members of the species. A spider may determine from another dragline whether its owner is male or female and whether it is immature or adult.

Small spiders and newly hatched spiders use the silk as a means of dispersal. They climb to favorable take-off points, tilt the abdomen upward, and release a strand of silk. When air currents produce sufficient pull, the spider lets go and sails out to whatever new habitat and fate the wind will take it. The wide distribution of many species of spiders is undoubtedly correlated with this ballooning phenomenon. The larger, more primitive tarantulas, which do not balloon, have a relatively limited distribution.

Feeding. Most arachnids are predatory animals and other arthropods are their usual prey. As an aid to dispatching prey, three orders of arachnids have independently evolved **poison glands.** The poison glands of scorpions are located in the terminal sting at the end of the abdomen (Fig. 26.11). The prey is caught with the large pedipalps and then stabbed by the poison barb with a forward thrust of the abdomen, which is curled back over the body. Although the sting of scorpions may be painful, few species have a poison dangerous to man. Members of the genus *Centruroides*, which occurs in Mexico, New Mexico, and Arizona, are dangerous. The poison is neurotoxic but would be fatal only to children.

Pseudoscorpions have poison glands in their large scorpion-like pedipalps. The glands open at the tips of the fingers, and the poison is injected as the prey is held. The pedipalps are too tiny to be of any danger to man.

Spiders have a poison gland associated with each chelicera, which consists of a terminal **fang** that folds down against a larger basal piece (Fig. 26.13). The gland, which may extend back into the head, opens by a duct through the end of the fang and the poison is injected through a bite.

The poison of a very small number of spiders is dangerous to man. Few tarantulas have a toxic bite, despite popular mythology. The members of the cosmopolitan genus *Latrodectus*, which includes the black widows, are perhaps the most notorious of the poisonous species. *Latrodectus mactans* is widely distributed in the United States. The poison is neurotoxic, and the symptoms, which include pain in the legs and abdomen, are somewhat like those for appendicitis. However, the consequences of a bite, although severe and painful, are not likely to be fatal except in children.

The poison of the brown recluse spider, *Loxosceles reclusa*, which occurs in midwestern and southeastern United States, causes necrosis of the tissues around the wound, and healing is difficult.

The method of catching prey can be a basis for dividing spiders into two adaptive groups, **hunters** and **web builders.** Hunting spiders include the tarantula, wolf spider, crab spider, jumping spider, and others. They spin silk draglines, nests and cocoons,

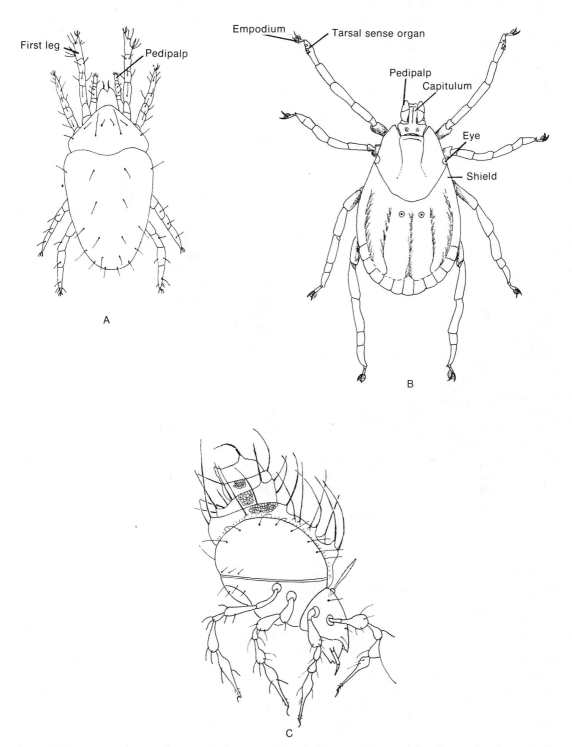

Figure 26.14 *A*, Dorsal view of a mite, *Tydeus starri*. *B*, tick, *Dermacentor variabilis; C*, an oribatid mite, *Belba jacoti*, carrying five shed skins. (*A* after Baker and Wharton; *B* and *C* after Wilson from Baker and Wharton.)

but they do not use silk to capture prey. Rather, the prey is stalked, pounced upon, and bitten. Hunting spiders generally have heavier legs and more highly developed eyes than do web builders (Fig. 26.15). Wolf spiders and jumping spiders have the largest eyes of any arachnids, and some species appear to be capable of object discrimination.

Web-building spiders construct webs to trap prey. Sheet webs, tangle webs, funnel webs and orb webs are common types, and each is characteristic of a different family. Note that only one major family builds the orb web that is so commonly associated with spiders. Web-building spiders are aerialists and have long slender legs for climbing about the silken lines (Fig. 26.15). The thread is hooked beneath a small middle claw that lies between the two main terminal claws characteristic of the legs of all spiders. Eyesight is poor, but web builders are able to detect and interpret the various vibrations of the web with great facility. Web vibrations inform an orb weaver, for example, about the size of the struggling prey and whether it is securely caught. The spider approaches the prey, throws out additional silk, and then gives it a fatal bite.

The web-building habit of spiders may have evolved from random draglines laid down in close quarters by some ancestral species. An occasional insect caught in such

A

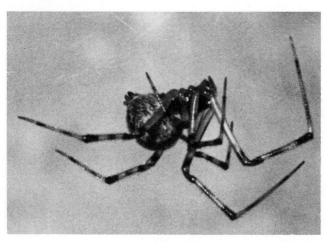

B

Figure 26.15 A, Wolf spider, a member of a large family of hunting spiders. Note the somewhat elongate body and heavy legs. B, House spider, a very common member of a large family of spiders which catch their prey in webs composed of irregular arranged threads. The body of this spider is somewhat globose and the legs are long and slender. (Both by Betty M. Barnes.)

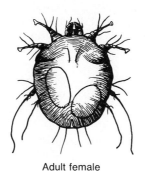

Mange mite burrowing in skin

Figure 26.16 The mange mite, *Sarcoptes scabiei*. These pass their entire life cycle on the host. Eggs laid in the burrows hatch into young mites that begin burrows of their own. Note the suckers on the anterior legs. (After Craig and Faust.)

Adult female

a primitive snare could have provided the selective advantage that led to the evolution of the more complex traps of modern spiders.

Arachnids are unusual in that they begin digestion of their prey outside their bodies. While the tissues are torn by the chelicerae, enzymes are secreted by the midgut, passed forward through the foregut, and poured out of the mouth into the prey. The partly digested tissues are then sucked in by the pumping action of a part of the foregut (Fig. 26.13). Digestion is completed and absorption occurs in the midgut, which may be greatly evaginated and ramify into various parts of the body.

There are some exceptions to the predatory feeding habit of most arachnids. Harvestmen are omnivores and feed upon vegetable material and dead animal remains in addition to live invertebrates. The greatest diversity in feeding is displayed by mites. Some are herbivorous and have mouthparts adapted for piercing the cells of plants and sucking out the contents. Such plant-feeding mites are often considered to be parasites.

Members of several groups of mites are scavengers. The oribatids, called beetle mites because of their convex and highly sclerotized dorsal surface (Fig. 26.14C), occur in large numbers in leaf mold and eat decaying vegetation and animal remains. Hair and feather mites spend their lives on the skin of mammals and birds, where they feed upon sloughed skin cells, gland secretions, and fragments of hair and feathers.

Parasitic mites are largely ectoparasites although some species attack the respiratory membranes of vertebrates. The animal parasites feed upon the blood of the host or upon skin or lymph. Ticks are blood-sucking parasites of mammals and birds,

and the chelicerae are adapted for penetrating and anchoring into the skin of the host (Fig. 26.14B). A tick feeds until it is engorged. Then it drops off the host and does not feed again until the next instar.

Chiggers, or redbugs, are ectoparasites on the skin of mammals during their post-hatching instar. In feeding, the minute mite secretes enzymes that produce a deep well from which it sucks out the digested contents. The mite secretions produce an irritating reaction resembling a mosquito bite but lasting much longer. Following feeding the chigger falls off the host and is predaceous as an adult.

Mange and itch mites (Fig. 26.16) spend their entire lives tunneling through the skin of their mammalian host. Eggs are deposited in the burrows and the young upon hatching follow the same existence as the adults.

Gas Exchange. Four pairs of book lungs provide for gas exchange in scorpions. The slit-like openings are located on the ventral surface of the anterior abdominal segments. Primitive spiders, such as tarantulas, have two pairs of book lungs, but most spiders have one pair of book lungs and one pair of tracheae. The openings are on the ventral surface of the abdomen (Fig. 26.13).

All other arachnids utilize tracheae for gas exchange, but the different locations of the spiracles in different groups suggest that tracheae evolved independently a number of times within the class.

Reproduction and Development. The paired gonads are located in the abdomen, and in both sexes a median **gonopore** opens onto the anterior ventral surface.

Indirect transfer of **spermatophores** appears to be a primitive condition in arachnids and perhaps represents the early arthropod solution to the problem of sperm transmission on land. The male deposits a stalked spermatophore on the ground (Fig. 26.17). The female then moves over the spermatophore and takes it up within her gonopore. Chemical, tactile, and sometimes visual signals are important in attracting and orienting the female to the spermatophore.

Among the orders of arachnids we are examining, scorpions, pseudoscorpions, and some mites produce spermatophores, and the peculiar mode of sperm transfer in spiders probably evolved from indirect sperm transfer by spermatophores. Spermatophore transfer in scorpions is preceded by a "courtship" dance, during which the large pedipalps of the male are locked with those of the female. In the course of the dance, the male deposits a spermatophore and then maneuvers the female so that it is taken up into her gonopore (Fig. 26.17).

The primitive pattern in pseudoscorpions appears to be one in which the male deposits a spermatophore in the absence of a female. Later a passing female is attracted chemotactically to the spermatophore. Most pseudoscorpions have evolved various mechanisms to increase the likelihood of the female encountering the spermatophore. The males of some species, on encountering a female, lay down a long silk signal line from the spermatophore. The female follows the chemical signals of the line to the spermatophore. Other species have a courtship pattern in which the spermatophore is not deposited until the male encounters the female. He then entices her over the spermatophore with a courtship dance.

Our knowledge of reproductive behavior in mites is still very poor. Some species transfer sperm indirectly by spermatophores; others transfer sperm directly, utilizing a penis.

The process of sperm transfer in spiders is remarkable, and is paralleled in few other animals. The copulatory organs of the male are the ends of the pedipalps, which resemble a pair of boxing gloves (Fig. 26.18). Although complex in structure, the basic element is a blind coiled tube, the inner end of which functions as a storage chamber; the distal end serving as an ejaculatory duct that opens to the exterior through a long projecting process (Fig. 26.18).

Prior to mating, the male spins a tiny sperm web, on which a droplet of semen is secreted. The two pedipalps are then rapidly dipped into the droplet until the semen is taken up within the reservoir of the palpal organ. The male now seeks a female.

The male is frequently smaller than the female, and the predatory nature of spiders makes it important for the male to ensure that the female does not mistake him for potential prey. The precopulatory behavior of any given species thus displays features that are correlated with its feeding adaptations. For example, hunting spiders depend upon visual cues in addition to chemical and tactile stimuli in their sexual encounters. Elaborate courtship dances, involving

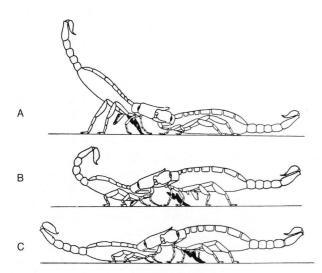

Figure 26.17 Sperm transfer in scorpions. A, While holding female's pedipalps with his own, male on left deposits spermatophore on ground. B, Female is pulled over spermatophore. C, Spermatophore taken up into female gonopore. (After Angermann.)

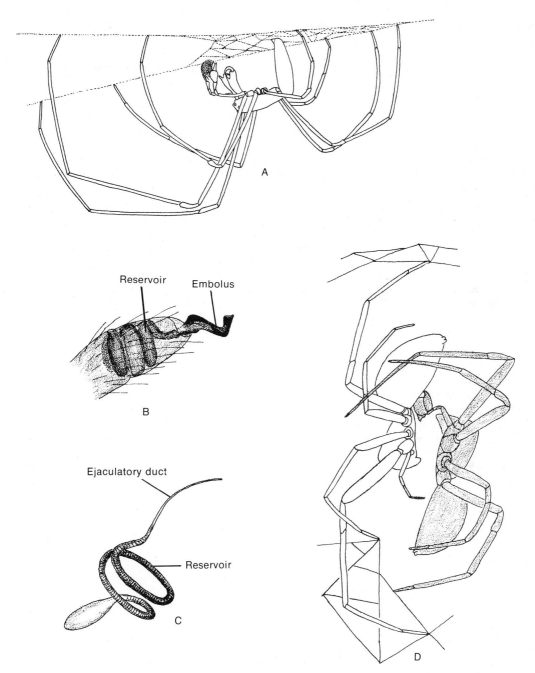

Figure 26.18 *A*, Male tetragnath spider in sperm web, filling palps from globule of semen. *B*, Simple palp of *Filistata hibernalis*. *C*, Receptaculum seminis, the semen-containing part of the male palp. *D*, Mating position of *Chiracanthium* (male shaded). (*A* after Gerhardt from Millot; *B* and *C* after Comstock; *D* after Gerhardt from Kaston.)

posturing by the male, occur in some wolf spiders and jumping spiders which have highly developed eyesight. Male web-building spiders use the vibrations of the web to determine the presence of the female and to advertise his presence and desire to enter.

Chemical and tactile signals are important in locating the female and in the precopulatory behavior of all spiders. An amputated leg of a female will evoke courting postures in the male. The same leg washed in ether and dried initiates no responses, but the male will posture before the apparently empty watchglass containing the evaporated washings. Such chemical substances produced by one individual that trigger behavioral responses in another, called **pheromones** (p. 402), are important in evoking and directing behavioral patterns in many arthropods.

Following final precopulatory stroking and palpating by the male, the palpal organ is locked onto the chitinous plate containing the female reproductive openings by means of special palpal projections, and the ejaculatory process is inserted into the seminal receptacles (Fig. 26.18). Sperm is transferred from one palp at a time. This unusual mode of sperm transfer in spiders probably had its origins in transfer by spermatophore. The male of some arachnid ancestral to the spiders may have used the palp to place a spermatophore into the female gonopore. The claw of the palp (the palpal organ of the male is homologous to the claw) might then have become adapted for the transfer process. The deposition of sperm within a spermatophore disappeared and the palpal claw became further modified to handle semen.

Harvestmen and many mites transfer sperm directly without spermatophores. These arachnids deposit their fertilized eggs in soil, leaf mold, or beneath bark, but spiders place their eggs in silk cocoons, which are then usually left beneath stones, bark, or leaf mold, or are attached to vegetation.

Brooding is common. Wolf spiders and fisher spiders carry their cocoons about with them. After hatching, wolf spiderlings are carried on the back of their mother. Pseudoscorpion eggs develop within a **brood sac** attached to the female reproductive openings. The eggs of scorpions develop within the body of the female. Following birth, the young are carried about on the back of the mother (Fig. 26.19).

Most arachnids have direct development, and the young at hatching or at birth resemble the adult form. Mites, however, hatch as a six-legged "larva," which acquires the fourth pair of legs at the first post-hatching instar (Fig. 26.20).

26.9 THE MANDIBULATES

The remaining arthropods, which includes the great majority, are **antennate** and belong to the subphylum Mandibulata. The first post-oral appendages are jaw-like mandibles. Behind the mandibles are typically one or two pairs of small food-handling appendages called **maxillae** (Fig. 26.21).

Figure 26.19 Female scorpion carrying young. (Courtesy of H. L. Stahnke.)

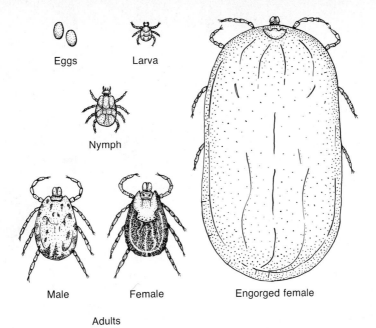

Eggs   Larva

Nymph

Male   Female   Engorged female

Adults

Figure 26.20 The common tick, *Dermacentor andersoni*. Eggs laid on the ground hatch into six-legged larvae that feed on small mammals. These drop off, molt into eight-legged nymphs that return to small mammals. After molting on the ground again the adults attack large mammals. The females become enormous after mating and eventually fall to the ground to lay a thousand or more eggs. (After Chandler.)

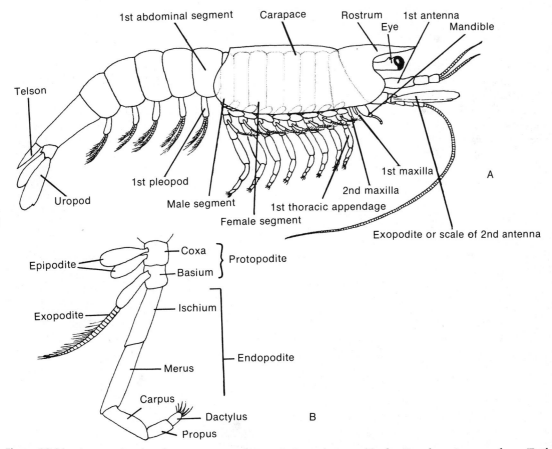

1st abdominal segment   Carapace   Rostrum   1st antenna
Eye   Mandible

Telson

1st pleopod
Male segment
Female segment
1st thoracic appendage
2nd maxilla
1st maxilla
Uropod
A

Exopodite or scale of 2nd antenna

Epipodite
Coxa
Basium
Protopodite

Exopodite
Ischium
Endopodite

Merus

Carpus
Dactylus
Propus
B

Figure 26.21 A generalized malacostracan crustacean. *A*, Lateral view of body; *B*, a thoracic appendage. (Both after Calman.)

Class Crustacea

The class Crustacea contains the shrimp, lobsters, crayfish, and crabs; it is the only large group of aquatic arthropods. Most are marine, but there are many species found in fresh water, and some, notably the sow bugs, are adapted for a terrestrial existence.

Crustaceans differ from all other mandibulates in possessing two pairs of antennae. The first pair is probably homologous to the antennae of insects, centipedes, millipedes, and other mandibulates, but the second pair is unique to crustaceans. Behind the mandibles are two pairs of maxillae. Thus the head appendages, which are constant for all members of the class, are two pairs of antennae, one pair of mandibles, and two pairs of maxillae (Fig. 26.21).

The trunk varies too greatly to permit many generalizations. A **carapace** formed from the fusion of the dorsal exoskeleton terga of a varying number of segments is commonly present. It may cover only a small part of the dorsal surface of the trunk or it may greatly overhang the sides of the body (Fig. 26.21). In some crustaceans, the entire body is enclosed within a carapace. The terminology of crustacean appendages is based on the assumption that, primitively, each appendage was a simple two-branched (**biramous**) structure. The outer branch (**exopodite**) and inner branch (**endopodite**) were attached to a basal piece (**protopodite**) (Fig. 26.21). The trunk appendages have been adapted for a wide range of functions, and there is little uniformity in the way differentiation has occurred.

Crustaceans possess either compound eyes or a simple median eye, but not both. The **median**, or **nauplius**, eye is a little cluster of three or four pigment cups containing photoreceptors. It is characteristic of the nauplius larva (Fig. 26.22), but in some groups the larval eye is retained in the adult.

The larger crustaceans are often brightly colored, the pigment being located in the outer layer of the exoskeleton. A blue pigment, astaxanthin, is denatured when heated and is responsible for the red color of boiled shrimp, lobsters, and crabs. The larger crustaceans also possess **chromatophores**, which are visible through the exoskeleton. These are cells with branching processes, and the pigment granules flow into and out of the processes. Many colors of pigments may be apparent; some chromatophores contain pigments of more than one color. The chromatophores enable the animal to adapt to the color of its background; the most common change is simple darkening or lightening.

The ancestral crustaceans were probably small swimming animals, and many living groups are planktonic. The larger species are generally benthic.

Although crustaceans have a wide range of diets and feeding habits, filter feeding is very common, especially among small crustaceans. The **filter** is formed by closely spaced **setae** on certain appendages, usually head or anterior trunk appendages (Fig. 26.23).

A number of generalizations can be made about crustacean reproduction. Copulation is the rule, and certain appendages are adapted for clasping the female and transferring sperm. Fertilization may be internal or it may occur externally at the time of egg deposition, and the eggs are usually brooded on certain parts of the body (Fig. 25.31). The earliest hatching stage, the **nauplius larva**, possesses three pairs of appendages—first antennae, second antennae, and mandibles (Fig. 26.22). The acquisition and differentiation of additional segments and appendages occurs in subsequent developmental instars, and these later larval stages have been given various names. Bottom-dwelling species settle upon attaining the adult form.

The structural diversity of crustaceans makes them difficult to describe without

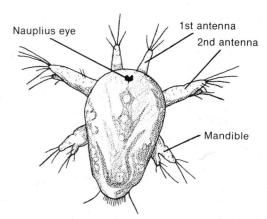

Figure 26.22 Nauplius larva of *Cyclops fuscus*, a copepod. (After Green.)

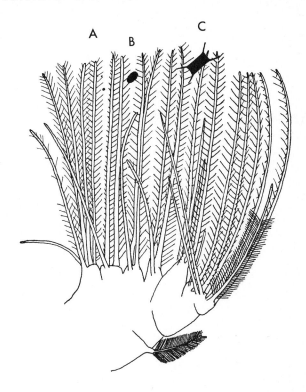

Figure 26.23 Filtering setae on the left maxilla of the copepod, *Calanus*. Three diatoms, representing species of different sizes, are drawn to scale to indicate the filtering ability of setae. Diatom *B* is approximately 25 microns long. (After Dennell from Marshall and Orr.)

considering some of the different subgroups which comprise the class.

Decapods. The **Decapoda**, the largest of the crustacean orders, contains over 8500 species. It also contains the biggest and most familiar crustaceans—shrimp, lobsters, crayfish, and crabs. The name Decapoda refers to the five pairs of legs, including the first pair, which are frequently modified as large claws. In front of the legs are three pairs of small appendages called **maxillipeds**, which, like the maxillae, function in food handling (Figs. 26.24 and 26.25). The three pairs of maxillipeds and the five pairs of legs are appendages of the anterior trunk region, called the **thorax.** The thorax is covered by a carapace and the head-thorax region together is usually called the **cephalothorax** (Fig. 26.24).

Primitively, the **abdomen** is large, as in shrimp, lobsters, and crayfish, and carries six pairs of biramous appendages. The last pair is flattened and forms a **tail fan** with the terminal **telson**. The first five pairs are called **swimmerets,** or **pleopods** (Fig. 26.24). In crabs the abdomen is greatly reduced and flexed beneath the thorax (Fig. 26.25). This change to a short body form shifts the center of gravity forward beneath the cepha-

lothorax and represents an adaptation for greater mobility in crawling.

The abdomen of hermit crabs is housed within an empty gastropod shell; the soft, twisted abdomen bears reduced appendages (Fig. 26.26). The hermit crab carries the gastropod shell by clasping the columella with a modified uropod and jamming the fourth and fifth pairs of legs against the inside of the shell. When the hermit crab becomes too large, it finds another shell and leaves the old one. Most hermit crabs can use the shells of many different gastropod species. Only empty shells are used; the original occupant is never killed. Hermit crabs probably evolved from forms which backed the body into crevices and other retreats.

Most decapods are bottom dwellers, but many shrimp can swim, using the pleopods for propulsion. The best swimmers, however, are the portunid crabs, which have the fifth pair of legs adapted as paddles (Fig. 26.25A).

Decapods possess **gills** that project upward from near the base of the thoracic appendages and are enclosed within a protective branchial chamber created by the great lateral overhang of the carapace

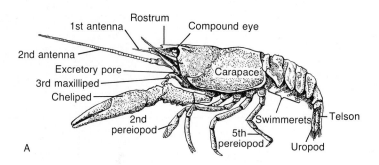

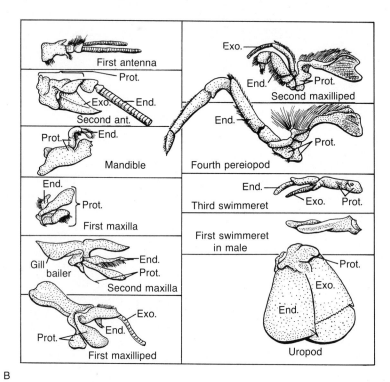

Figure 26.24 *A,* Lateral view of a crayfish. *B,* Appendages of the crayfish: prot. = protopodite; exo. = exopodite. Those on the left are drawn to a larger scale than those on the right. First antenna is not considered to be a segmental appendage. (Both after Howes.)

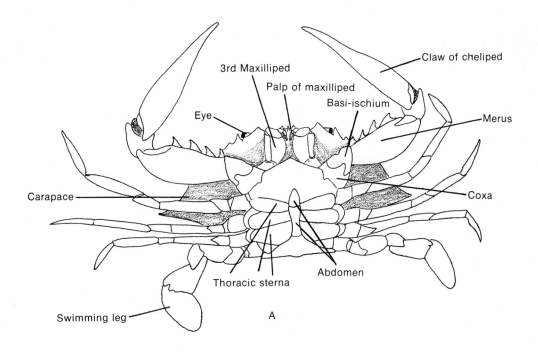

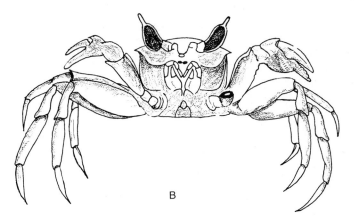

Figure 26.25 *A*, Ventral view of a brachyuran crab of the family Portunidae. The fifth pair of legs is adapted for swimming. *B*, A ghost crab, *Ocypode*. (*A* after Schmitt from Rathbun; *B* after Schmitt.)

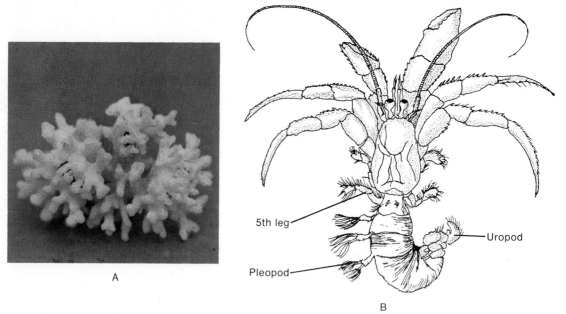

A

B

5th leg

Pleopod

Uropod

Figure 26.26   Hermit crab. *A,* In shell; *B,* out of shell. (*A* courtesy of the American Museum of Natural History; *B* after Calman from Kaestner.)

(Fig. 26.27). Each gill consists of a central axis, to which are attached either **flat plates** or **filaments,** depending on the species. The ventilating current is produced by the rapid sculling motion of the **gill bailer,** a semilunar process of the second maxilla (Fig. 26.24*B*). The gill bailers drive two exhalant water streams out of the branchial chamber to either side of the mouth. Water enters the branchial chamber at the back of the carapace or between the legs. In crabs, where the carapace is tightly sealed along the ventral margin, the inhalant aperture is at the base of the **claw,** or **cheliped** (Fig. 26.28).

Several groups of crabs have invaded the land with varying degrees of success. Amphibious fiddler crabs (*Uca*) burrow in the intertidal zone (Fig. 26.29). At high tide the burrow is flooded but at low tide the crabs emerge from the burrow to feed, utilizing water contained within the branchial chambers for gas exchange.

The related ghost crabs (*Ocypode*) are more terrestrial. These crabs live above the high tide mark and in many areas are common inhabitants of dunes (Fig. 26.25*B*). In the tropics, land crabs live in forests and thickets well back from the sea, as much as hundreds of miles inland. Almost all terrestrial species continue to use the gills as organs of gas exchange, which may explain the restriction of land invasions to crabs, for

only in these decapods is the branchial chamber sufficiently closed to make possible the retention of the moisture needed for gas exchange. Most land crabs are nocturnal and remain in burrows during the day.

Although the majority of decapods are predators or scavengers, there are many exceptions. Recently fallen fruits and leaves are an important food source for many land crabs. Some decapods are filter feeders. Certain crabs living within the U-shaped tubes of *Chaetopterus* filter the worm's water current with outstretched maxillipeds. Mole crabs are also filter feeders. These animals burrow backward in sand and project their plumose antennae as filters. Some species, including the Atlantic coast mole crab, live on wave-swept beaches and filter the current of the receding waves with their second antennae.

The amphibious fiddler crabs feed on fine organic matter deposited onto the surface of the intertidal zone of protected beaches, estuaries and marshes. A mass of sand or mud is placed in the mouthparts, which are then flooded with water pumped forward from the gill chambers. Organic material is floated off and separated by fine setae on the feeding appendages. The remaining mineral material is ejected as a small spitball (Fig. 26.29).

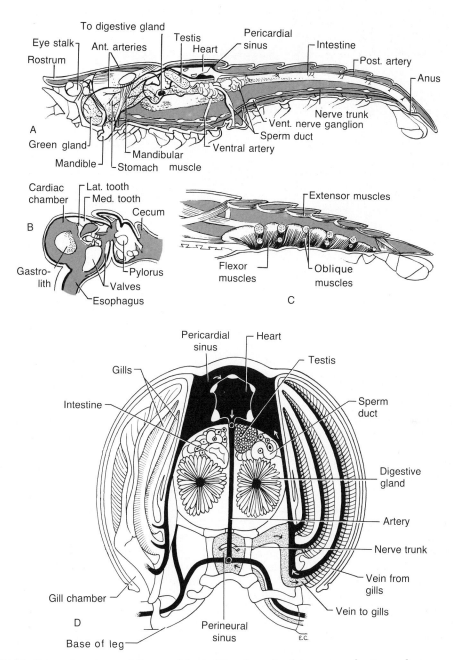

Figure 26.27 Internal anatomy of the crayfish. *A,* Digestive, circulatory, reproductive and nervous systems; *B,* stomach (enlarged); *C,* musculature of the abdomen; *D,* cross section just behind the third pair of legs. (All after Howes.)

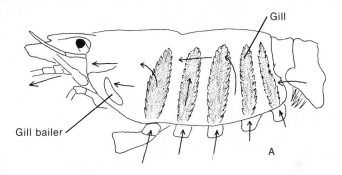

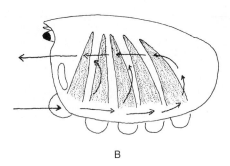

Figure 26.28 Path of ventilating current through gill chamber of (A) a crayfish, and (B) a crab.

The decapod digestive tract is complex. In lobsters, crayfish and crabs, the foregut includes a very large chitin-lined cardiac **stomach,** containing one dorsal and two lateral **teeth** that form a **gastric mill** (Fig. 26.27). A narrow constriction separates the **cardiac** stomach from the **pyloric** stomach, part of which is derived from the foregut and part from the midgut. Associated with the midgut section and derived from it is a pair of large **digestive glands** (Fig. 26.27).

The cardiac stomach functions both as a **crop** and a **gizzard.** The grinding action of the gastric mill plus the action of **enzymes** passed forward from the digestive glands reduces the food masses to fluid and fine particulate matter, which is then conducted along channels through the pyloric stomach to the digestive glands. Here absorption takes place.

The excretory organs of decapods are **antennal glands,** also called green glands. The organ consists of a blood-bathed sac in the head and a duct which opens to the outside by a pore at the base of the second antennae.

Most decapods are marine and are **osmoconformers,** i.e., the salt content of the blood is determined by that of the surround-ing sea water. When such crustaceans are placed in water of decreasing salinity, there is a corresponding decrease in the concentration of salts in their blood. When this salt concentration in the blood falls below a certain value, the animal dies. The ability to "osmoregulate," to maintain a certain concentration of blood salts despite the dilution of the surrounding medium, has evolved in a number of decapods which have invaded fresh water. Estuarine and fresh-water species include some shrimp, the widely distributed crayfish, and some crabs. In all of these, the green glands function to remove excess water, but only in the crayfish do the excretory organs also function in salt balance and produce a hyposmotic urine. In other fresh-water crustaceans the urine is isotonic to the blood and the gills are the osmoregulatory organs. Salts lost in the urine are replaced by the absorption of salts from the water of the ventilating current (Fig. 26.30). This mechanism of salt balance is strikingly parallel to that of fresh-water fish (p. 100) which use the gills in the same way.

Decapods have **compound eyes** and some species, such as crabs, have a high degree of visual acuity. The eyes are located on

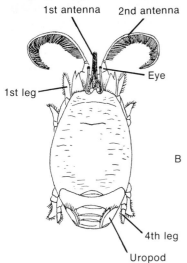

1st antenna 2nd antenna

Eye

1st leg

B

4th leg

Uropod

Figure 26.29 *A*, Fiddler crab by burrow. The large balls of sand were excavated from the burrow; the small pellets are composed of sand that has been filtered of organic material and ejected by the mouth parts. *B*, Mole crab. (After Verill, Smith, and Harger, from Waterman and Chase.)

Figure 26.30 The osmoregulatory ability of three crabs. The osmotic concentration of blood changes with the concentration of salts in the surrounding sea water. The spider crab cannot osmoregulate, and the osmotic concentration of its blood conforms to the salt concentration of the sea water. Below and above certain concentrations, the crab dies. The blood of both the mud crab and the woolly-handed crab conforms to an upper range of external salt concentration, but at the lower concentrations the crabs maintain higher osmotic concentrations in the blood by absorbing salts through the gills. These two crabs can live in brackish estuarine environments where the spider crab can not. The woolly-handed crab can also excrete salt at high blood concentrations. (Adapted from Schmidt-Nielsen.)

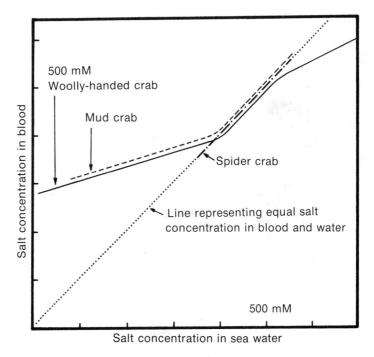

stalks and have a wide visual arc (Fig. 26.25).

In the base of each first antenna is a **statocyst** which opens by a **pore** to the exterior. The animal uses sand grains as **statoliths,** and must replace them following each molt. A newly molted crayfish given iron filings instead of sand grains will place them within the statocyst chamber and then will orient itself to a magnetic field rather than to gravity.

The **hormones** and **neurosecretions** of crustaceans are perhaps better known than those of any other invertebrates. There is experimental evidence for hormones which regulate chromatophores, molting, growth, sexual development, and reproduction. A number of hormones, such as those controlling chromatophores, are released from a gland in the eye stalk. These are easily studied since the eye stalk can be removed and extracts of the eye stalk can be prepared and tested.

The single pair of **testes** or **ovaries** is located in the dorsal part of the thorax. The male **sperm duct** descends to the ventral side and opens at the base of the fifth pair of legs (Fig. 26.31). The **oviduct** opens to the exterior at the base of the third pair of legs (Fig. 26.31). During copulation the male assumes various positions astride the female, and, using the greatly modified anterior two pairs of pleopods, transfers sperm to the female gonopores or to a median seminal receptacle between the fourth or fifth pair of legs (Figs. 26.24 and 26.31).

The eggs are fertilized internally or on release from the oviduct. The fertilized eggs pass to the ventral surface of the abdomen, where they are attached to the pleopods by an adhesive material on the egg surface. Here they are brooded, often forming a conspicuous mass. Even in crabs the reduced folded abdomen retains its brooding function, and the eggs form a large mass, or sponge, between the folded abdomen and the undersurface of the thorax (Fig. 26.31).

Some shrimp hatch as a nauplius larva; but in most decapods the hatching stage is a later planktonic larval stage, called a **zoea,** in which all of the thoracic appendages are present (Fig. 26.31). The adult form is gradually attained through a series of instars, and the young shrimp, lobster,

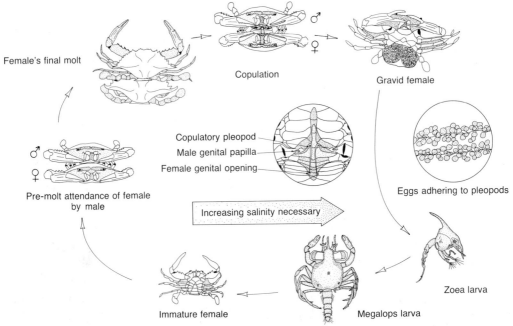

Figure 26.31 Life cycle of the commercial blue crab, *Callinectes sapidus*. Adults are common off shore as well as in very brackish water. Females living in brackish water must return to water of high salinity to permit hatching of the eggs. The hatching stage is a zoea larva, which lives in the plankton for a period of time. Successive molts transform the zoea to a megalops larva, possessing many crab features but with a still unfolded abdomen. Further development leads to the adult body form. Prior to the female's last molt she is carried about beneath a male, and copulation occurs after molting. Eggs are laid some time later and are attached to pleopod setae beneath the folded abdomen. (Courtesy of C. Piling; adapted from various sources.)

or crab settles to the bottom and takes up the adult mode of existence.

Copulation and brooding occur on land in terrestrial crabs, but most must go back to the sea to permit hatching of the eggs and a planktonic larval life.

Direct development takes place in most fresh-water decapods, including the crayfish. In these species the eggs are brooded on the abdomen throughout development.

Many species of shrimp and crabs are eaten by man throughout the world. In the United States the blue crab, *Calinectes sapidus*, a swimming portunid (Fig. 26.31), is the principal commercial species on the Atlantic and Gulf coasts. On the Pacific coast, *Cancer*, a member of another family, is commercially important. The Alaskan King crab is a large decapod that is crablike in form, but is more closely related to the hermit crabs than to the true crabs.

Species of *Pennaeus* are important commercial shrimp throughout the world, including the United States. The Japanese have developed the art of shrimp farming, raising the animals from egg to a marketable size of 20 to 25 cm. The shrimp are packed live and shipped in seaweed. The cost of shrimp farming is very high and is profitable in Japan only because of the high prices the Japanese will pay for live shrimp. Declining catches and rising demand will eventually make shrimp farming in the United States economically feasible.

The class Crustacea is divided into subclasses, and the decapods belong to the largest subclass, the **Malacostraca.** In these crustaceans the trunk is composed of a thorax of eight segments, which bear the legs, and an abdomen of six segments, which usually carry five pairs of biramous **pleopods** and a terminal pair of flattened **uropods** (Fig. 26.21).

Amphipods and Isopods. The two largest orders of malacostracans other than decapods are the **Amphipoda** and **Isopoda.** These malacostracans, which are only about a centimeter in length, share a number of features. The **compound eyes** are on the sides of the head and not on stalks as in decapods. No carapace is present, and the thoracic and abdominal regions are not sharply demarcated on the dorsal side (Figs. 26.30 and 26.31). As in decapods, there are eight pairs of **thoracic appendages,** but only

the first pair are food-handling maxillipeds; the other seven are legs. In both orders development is direct, and the young are brooded beneath the thorax in a chamber, the **marsupium,** formed by inward shelf-like projections from the bases of the legs (Fig. 26.32).

Isopods tend to be dorsoventrally flattened and the pleopods are modified for gas exchange (Fig. 26.32). Amphipods, in contrast, are laterally flattened and look somewhat like shrimp (Fig. 26.33). The gills of amphipods are simple processes from the bases of the legs.

The majority of isopods and amphipods are marine and most are bottom dwellers, swimming only intermittently with the pleopods. They live in bottom debris, in algae, and among sessile animals. The common skeleton shrimp are very aberrant amphipods, found climbing about hydroids and bryozoans (Fig. 26.34). Many amphipods burrow in sand or mud and some are tube dwellers. The tubes are composed of mud, sand grains, shell, or plant fragments bound together with cement from glands on certain appendages.

Both orders include fresh-water species; in favorable habitats amphipods may occur in great numbers. Both groups have also invaded land, but the beach fleas, which occur beneath drift at the high tide mark, represent the limits of the amphipod incursion. Beach fleas get their name from their ability to jump by a rapid backward extension of the abdomen.

The pill bugs, sow bugs, and woodlice, all isopods, represent the only truly successful crustacean invasion of land. These isopods are widely distributed in both temperate and tropical regions, where they are commonly found beneath stones, wood, and in leaf mold. They reduce desiccation by living in protected habitats, by avoiding light, and by their ability to roll up into a ball, covering the less highly sclerotized ventral surface. The gills remain as gas exchange organs, and the marsupium is filled with water when developing eggs are carried.

Most isopods and amphipods are omnivores and scavengers, but among members of both orders there are predaceous species, such as some skeleton shrimp. Filter feeding has evolved in some amphipods, with the various anterior appendages used

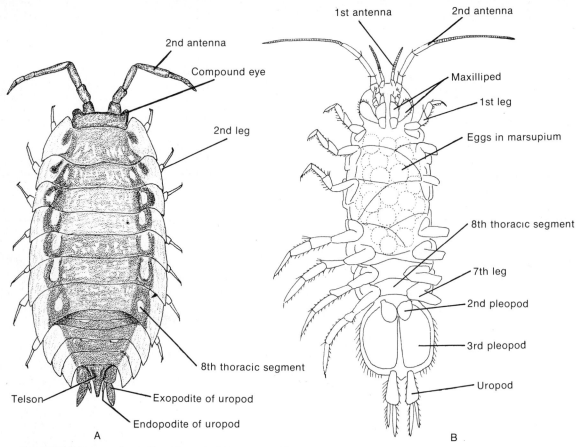

Figure 26.32   Isopods. *A*, Dorsal view of *Oniscus asellus*, a terrestrial isopod. *B*, Ventral view of the fresh-water isopod *Asellus*. (*A* after Paulmier from Van Name; *B* after Van Name.)

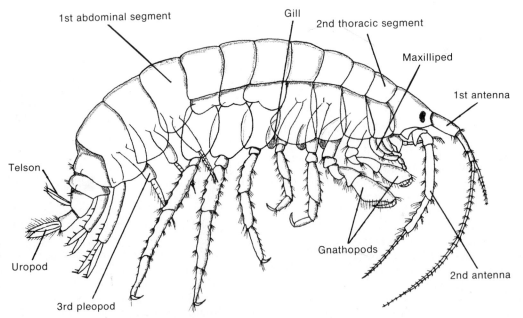

Figure 26.33   Male of the amphipod, *Gammarus*. (After Sars from Calman.)

Figure 26.34 The skeleton shrimp, *Caprella equilibra,* clinging to seaweed. (By D. P. Wilson.)

as filters. The terrestrial sow bugs feed chiefly on dead plant and animal remains.

Some isopods that burrow at the high tide mark and certain species of amphipod beach fleas are able to return to their habitat when they move down to the intertidal zone to feed or are for other reasons displaced. Orientation is achieved by sensing the slope of the beach and the angle of the sun. The latter ability, however, requires a "map sense," i.e., knowledge of the orientation of the beach with regard to compass points, and an internal clock to compensate for the changing sun angles as the day passes. Animals can be misdirected experimentally by altering the sun's apparent position with a mirror. A comparable orientation ability, using the angle of the sun, has been demonstrated in certain wolf spiders living on the banks of lakes and streams, and in bees,

which return to particular flowers to obtain pollen and nectar (p. 630).

Branchiopods. Most of the non-malacostracan crustaceans are very small, averaging only a few millimeters in length. They are found in all sorts of habitats, but many species are planktonic. Most are filter feeders, and the filter feeding planktonic species constitute an important link in the food chain between the photosynthetic level and the larger predators.

The majority of non-malacostracan crustaceans belong to one of four subclasses. The subclass **Branchiopoda** contains the water fleas, fairy shrimp, and brine shrimp (Figs. 26.35 and 26.36). All possess flattened **leaf-like trunk appendages** bordered by fine setae but the body form is quite diverse. Fairy shrimp have an elongate body without a carapace, and there are many seg-

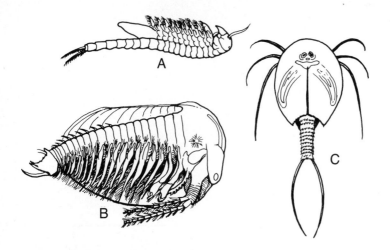

Figure 26.35 Branchiopods. *A*, A fairy shrimp (they swim upside down). *B*, A clam shrimp. One of the valves of the carapace has been removed. *C*, A tadpole shrimp. (All after Borradaile et al.)

ments and appendages. Water fleas look like plump little birds. The trunk, but not the head, is enclosed within a bivalved carapace which greatly overhangs the body. The head bears large biramous second antennae and a single median compound eye.

The flattened foliaceous appendages of branchiopods function in gas exchange, from which the name branchiopoda—**gill foot**—is derived. The appendages also serve in filter feeding, the fine setae bordering the appendages functioning as filters. Fairy shrimp also use the appendages for swimming, but water fleas swim with the antennae.

In contrast to most of the other subclasses, branchiopods are chiefly inhabitants of fresh water. Only the water fleas are found in the sea, where they may occur in large numbers in the plankton. Water fleas are the only branchiopods inhabiting lakes. Most other branchiopods are restricted to ponds and temporary pools and are often distributed sporadically. The brine shrimp *Artemia,* which belong to the same group as fairy shrimp, are an exception. They are adapted for living in salt lakes and can tolerate salinities near the saturation point of salt.

The internal structure of water fleas can be easily observed through the transparent carapace when these animals are placed on a depression slide and studied under a microscope. Most of the structures illustrated in Figure 26.36 can be seen in a specimen of *Daphnia.* The ease with which water fleas can be collected, maintained, and observed has made them the subjects of

many experimental studies of their orientation to light, their heart beat and their reactions to various environmental factors.

Water fleas brood their eggs in the space beneath the carapace above the back of the trunk (Fig. 26.36). Fairy shrimp brood their eggs within **ovisacs** attached to the female gonopore and secreted at the time of egg deposition.

Water fleas have direct development, but in other branchiopods the egg hatches as a nauplius larva. The retention of larval stages in fresh-water branchiopods is an exception to the general rule of suppression of larval stages in fresh water.

Branchiopods exhibit a number of reproductive and developmental adaptations to the environmental stresses common in fresh-water lakes and ponds. Parthenogenesis is common. Many species produce thin-shelled eggs which hatch rapidly and thick-shelled eggs which remain dormant during periods of drought or freezing. Note that these adaptations are the same as those exhibited by fresh-water flatworms and rotifers.

Copepods.   The subclass **Copepoda** is the largest of the non-malacostracan subclasses. The 4500 species are mostly marine, but there are also many fresh-water forms. The copepod body, usually less than three mm. in length, is cylindrical and tapered (Fig. 26.37). The head bears a single **median nauplius eye** but no compound eyes. The first pair of antennae are very large. The trunk is composed of an anterior **thorax** bearing biramous appendages, and a narrower posterior **abdomen,** which lacks ap-

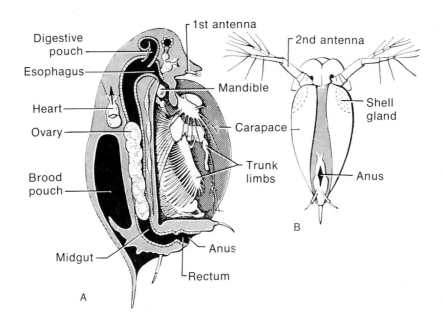

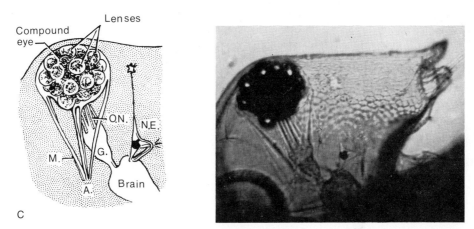

Figure 26.36 *Daphnia*, the water flea. *A*, Side view, with one side of the carapace removed to show enclosed body and organs. *B*, Ventral view, with trunk appendages omitted. *C*, Part of the head of *Daphnia* showing compound eye with protruding lenses and muscles (*M.*) of the right side attached to the side of the head (at *A.*). Also shown are the optic nerves (*O.N.*), optic ganglion (*G.*), brain, and nauplius eyes (*N.E.*).

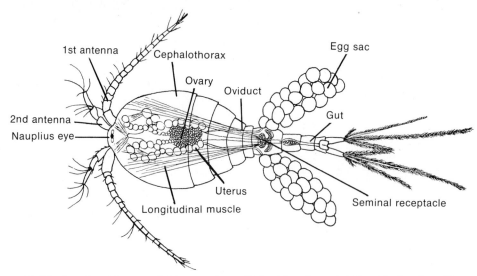

Figure 26.37 A cyclopoid copepod, *Macrocyclops albidus* (dorsal view). (After Matthes from Kaestner.)

pendages but usually carries a pair of terminal processes.

Copepods may be planktonic; they may live near the bottom, where they swim about over debris; they may be interstitial (live in the spaces between sand particles); and some are parasitic. Each swims or crawls with his thoracic appendages. The antennae of swimming copepods serve to decrease the tendency to sink when the thoracic appendages are not beating. Planktonic species have many setae on especially long antennae. Interstitial copepods are tiny and have short antennae.

Although some copepods are grazers or are predatory, most planktonic forms are

Figure 26.38 Parasitic copepods, *Penella exocoeti*, on a flying fish. The copepods are in turn carrying the barnacle, *Conchoderma virgatum* (striped body). (Modified after Schmitt.)

filter feeders. Certain of the post-oral head appendages produce the water current and function as the filter. Diatoms are the principal source of food, and copepods play an important role in the food chain. Studies on the marine planktonic copepod *Calanus finmerchious,* which is 5 mm. long, have shown that as many as 373,000 diatoms may be filtered out and digested every 24 hours.

About one quarter of the copepod species are parasitic, attacking the skin, fins, and gills of marine and fresh-water fish. These are commonly referred to as fish lice (Fig. 26.38). All degrees of modification from the typical copepod body form exist, and some species are so aberrant that they no longer look like crustaceans.

Copepods have no gills, and many species also lack a heart and blood vessels.

Some copepods shed their eggs singly into the water, but many brood their eggs within secreted ovisacs attached to the female gonopore. A nauplius larva is the hatching stage in both marine and fresh-water forms.

Ostracods. Members of the subclass Ostracoda are tiny marine and fresh-water crustaceans with the body entirely enclosed within a bivalved carapace (Fig. 26.39). Ostracods are sometimes called mussel shrimp and indeed parallel the bivalve mollusks in many ways. The two **valves,** usually two mm. or less in length, may have interlocking teeth and are held together dorsally by an

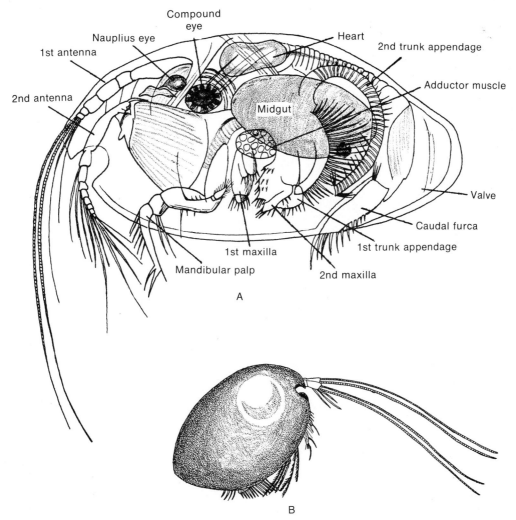

Figure 26.39 Ostracods. *A.* Lateral view of a female marine ostracod, *Cypridina,* with left valve removed. *B,* Lateral view of a marine ostracod swimming. Valves contain an antennal notch. (*A* after Claus from Calman; *B* after Müller from Schmitt.)

elastic hinge. The valves are closed by a bundle of **adductor muscle fibers** extending transversely between the two valves near the middle of the body (Fig. 26.39). The chitinous skeleton is impregnated with calcium carbonate. As a consequence, ostracods have an extensive fossil record dating from the Cambrian. Sculpturing of the valves and the location of the adductor muscle scars are the principal taxonomic characteristics of fossil species.

Within the two valves the ostracod body is mostly head, for the trunk is greatly reduced (Fig. 26.39). The two antennae are large and the other head appendages are well developed. However, the number of trunk appendages is reduced to two, one, or zero.

A nauplius eye is characteristic of all ostracods; sessile compound eyes are present in the members of one order.

There are some planktonic ostracods, in-

cluding the 23 mm. giant *Gigantocypris,* but most species are benthic and scurry over the bottom, crawling and swimming with the large antennae. Along with copepods and water fleas, ostracods are very common crustaceans in fresh-water ponds and small pools.

Many ostracods are filter feeders utilizing their different post-oral appendages as filters, but some species are predators or scavengers.

There are no gills. Only in one order is there a heart or blood vessels, but blood is propelled through tissue spaces, including the valves of the carapace, by the movements of the body.

Eggs are released singly or brooded within the carapace and a nauplius with a bivalved carapace is the hatching stage.

Barnacles. The subclass **Cirripedia** contains the barnacles. These marine animals differ from other crustaceans and most other

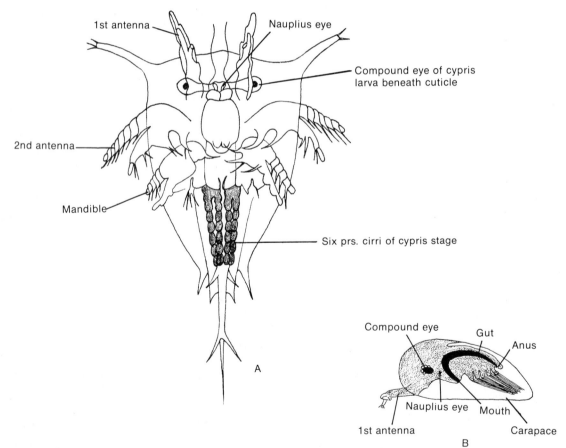

Figure 26.40 *A,* Nauplius larva of *Balanus* with structures of cypris larva visible beneath exoskeleton. *B,* Free-swimming cypris larva. (*A* after Claus from Calman; *B* after Korschelt from Kaestner.)

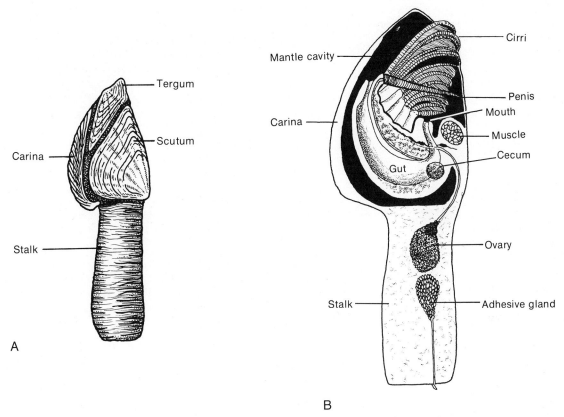

Figure 26.41 *Lepas*, a stalked barnacle. *A*, Entire animal showing calcareous plates. *B*, Internal structure. (After Broch from Kaestner.)

arthropods in being attached to the substratum. Correlated with the sessile habit, the body is enclosed within a bivalved carapace covered with protective **calcareous plates** (Figs. 26.40 and 26.41). Their attached condition and calcareous covering makes them so unlike other members of the class that it was not until 1830, when the larvae were discovered, that barnacles were recognized as crustaceans.

If one can imagine an ostracod, which is also enclosed within a bivalved carapace, attached to the substratum by the antennae, and the valves covered by calcareous plates, one has some idea of the structure of barnacles. The late larval stage of barnacles is enclosed within a bivalved carapace and looks much like an ostracod. It settles to the bottom and attaches by means of a **cement gland** in the base of the first antennae.

All barnacles are attached, but the group may be divided into **stalked** and **sessile** forms. The body of stalked barnacles is composed of a fleshy **peduncle**, which represents the pre-oral part of the body, and a **capitulum,** which represents the post-oral part of the body (Fig. 26.41). The capitulum carries the bivalve carapace, often called the **mantle,** and the **gape** is directed upward. The plates covering the carapace vary in number, but there are always two pairs of plates, **scuta** and **terga,** which flank either side of the gape.

The body of sessile barnacles consists mostly of the capitulum, for the pre-oral part is reduced to a platform on which the capitulum rests and is attached to the substratum (Fig. 26.42). The platform of sessile barnacles and the peduncle of stalked species both contain the antennal cement gland, indicating that the two structures are homologous. In sessile barnacles the basal plates of the capitulum form a rigid circular wall surrounding the upper movable lid-like terga and scuta.

Although some stalked barnacles are found on rocks, many species live attached to floating objects such as timbers, or on the

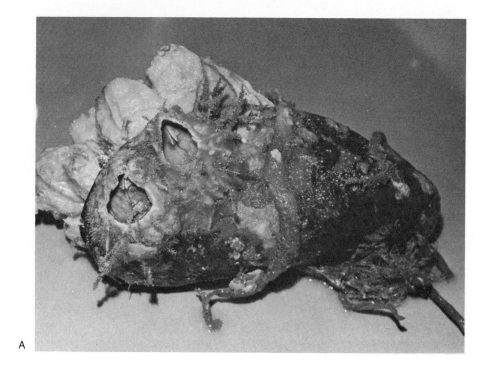

A

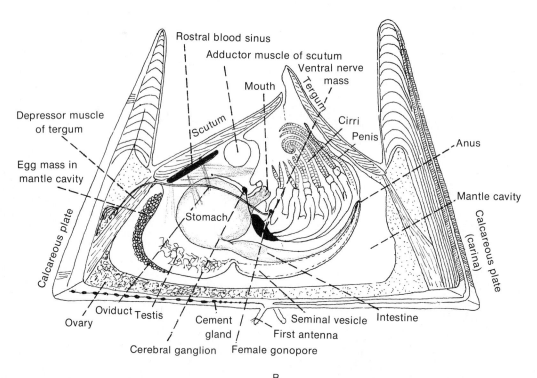

B

Figure 26.42 *Balanus*, a sessile barnacle. *A*, A cluster of animals attached to a mussel shell. Two are viewed from above and show the movable terga and scuta within the surrounding immovable wall of plates. *B*, Lateral view of animal in section. (*A* by Betty M. Barnes; *B* after Gruval from Calman.)

bodies of larger swimming animals. Whales, porpoises, sea turtles, sea snakes, crabs, and even jellyfish are utilized as substrata by different species. Correlated with the utilization of other animals as attachment surfaces, there has been a tendency toward a reduction of the protective calcareous covering so that some barnacles have become almost naked.

Most sessile barnacles have become adapted for a life on rock or other hard substrata, and the heavy, somewhat fused, ring of wall plates probably represents an adaptation for protection against currents, pounding waves, and browsing fish. However, like stalked barnacles, some sessile barnacles have become adapted for living on other animals, such as crabs and whales.

The tendency of barnacles to utilize other animal bodies as substrata probably led to parasitism. Approximately one-third of the members of the class are parasitic and are so highly modified that they are recognizable as barnacles only by their larval stages. Other crustaceans are the principal host.

Both stalked and sessile barnacles are filter feeders, and the trunk bears six pairs of long, coiled, biramous appendages, called **cirri,** from which the name of the class is derived. In feeding, the cirri unroll and project through the gape of the carapace as a large basket (Fig. 26.43). They perform a rhythmic scooping motion and the many fine setae of the appendages filter out plankton.

Neither eyes, gills, heart, nor blood vessels are present. Correlated with their attached existence, barnacles, unlike most other arthropods, are **hermaphroditic.** At copulation, a long, extensible, tubular penis is projected from the gape of one individual into the mantle cavity of a neighboring barnacle. Fertilization and brooding occur within the mantle cavity, and a nauplius is the usual hatching stage. The nauplius develops into a bivalved **cypris larva** which settles and attaches to the bottom. The many changes involved in metamorphosis include the loss of antennae and the development of calcareous plates. In the molts that accompany growth the plates are not shed; only the inner chitinous lining of the mantle is molted.

Barnacles are major fouling organisms on pilings, sea walls, buoys, and ship bottoms. Much research has been expended to develop anti-fouling measures, such as antifouling paint, for a badly fouled ship may have its speed reduced by as much as 35 per cent. Recently, barnacle cement has been the subject of a number of investigations in the hope that there might be practical applications for dentistry.

The Myriapodous Arthropods

The myriapodous arthropods comprise four classes of mandibulates which were once placed within a single class, the **Myriapoda.** They are the **Diplopoda** (which contains the millipedes), the **Chilopoda** (which contains the centipedes), and two small classes, the **Symphyla** and the **Pauropoda.** Although the myriapodous arthropods belong to separate classes, they share a number of characteristics. All are terrestrial and secretive, living in soil and leaf mold and beneath stones, logs, and bark. The body is composed of a **head** and a long **trunk** with many segments and legs (Fig. 26.44A). The head bears a pair of **antennae,** a pair of **mandibles,** and one or two pairs of **maxillae.** The eyes are not compound.

Gas exchange organs are **tracheae,** and a separate system of tubules and spiracles is present in each segment. Excretory organs are **Malpighian tubules,** and the heart is a long dorsal tube with ostia in each segment.

Reproduction parallels the arachnids in that sperm transfer is indirect and involves a spermatophore.

Adaptations for locomotion have been a primary theme in the evolutionary history

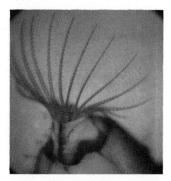

Figure 26.43 A feeding *Chelonebia*, showing the outstretched cirri. *Chelonebia* is a sessile barnacle, which lives attached to the carapaces of swimming crabs. The conspicuous plates at the base of the cirri are the two terga. One scutum can be seen on the right side. (By E. Ryan.)

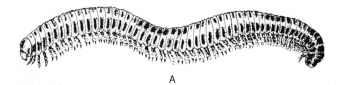

A

B

Figure 26.44   Millipedes. *A*, A juliform millipede; *B*, a flat-back millipede and two snails
on a piece of wood.

of centipedes and millipedes. These animals have become adapted for running,
climbing, pushing, and wedging beneath
objects. Yet, in contrast to most other arthropods where increased motility is correlated
with a reduction in number of legs and compaction of the body, myriapods have retained a long trunk with many appendages.

Millipedes.   Millipedes, members of the
class Diplopoda, have the body adapted for
pushing, the force being generated by the
large number of legs. As many as 12 to 52
legs may be involved in one wave of movement sweeping down the length of the
body. The large number of legs requires
a large number of segments, but a large
number of segments results in a weakened
trunk column. This problem was solved by
the formation of double segments, or **diplosegments,** i.e., each trunk section of a millipede represents two fused segments. Each
double segment contains two pairs of legs,
two sets of spiracles, two pairs of heart ostia
and two ventral nerve ganglia (Fig. 26.45).

The common cylindrical (juliform) millipedes are best adapted for pushing through
leaf mold and other loose debris (Fig. 26.44).
The head is rounded and the many legs are

attached near the mid-ventral line of the
cylindrical smooth body. Some species may
become as large as pencils in size.

Flatbacked millipedes are adapted for
wedging into confining places, such as beneath stones or bark. The body is dorsoventrally flattened and the laterally projecting
terga create a protective working space for
the legs (Figs. 26.44 and 26.45).

Both flatbacked and juliform millipedes
possess **repugnatorial (stink) glands** on
the diplosegments. These glands secrete
compounds that contain iodine, quinone, or
hydrocyanic acid and are believed to serve
a defensive function.

Millipedes are chiefly scavengers, feeding on living and decaying vegetation and
on dead animal remains.

Eggs are deposited freely in the soil, or in
protective mud capsules, or in a nest composed of excrement or soil. The young on
hatching possess only a small number of
segments and legs.

Centipedes.   Centipedes, members of the
class Chilopoda, are predaceous and most
are adapted from running. Behind the
mandibles and the two pairs of maxillae
are a pair of large poison claws used for

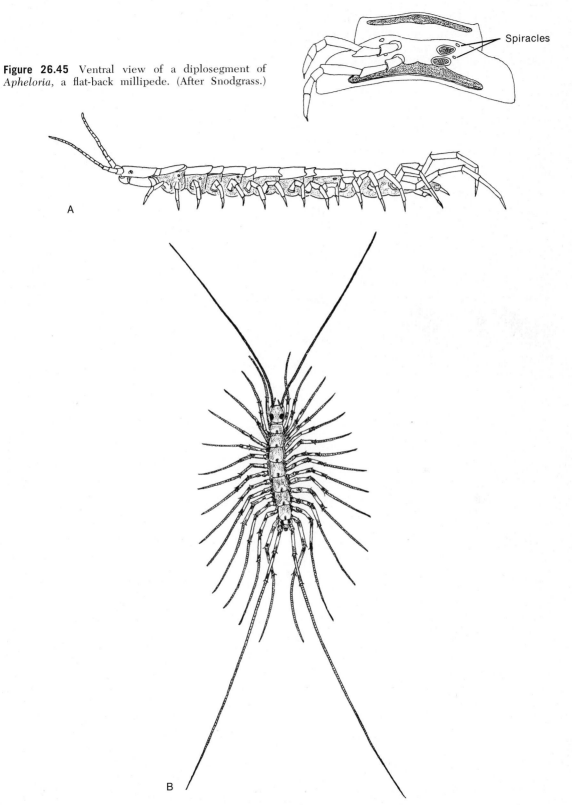

Figure 26.45 Ventral view of a diplosegment of *Apheloria*, a flat-back millipede. (After Snodgrass.)

Spiracles

A

B

Figure 26.46 Centipedes. *A*, A lithobiomorph centipede with unequal and overlapping tergal plates. *B*, *Scutigera*, the common house centipede. (Both after Snodgrass.)

seizing and killing prey (Fig. 26.47). Centipedes feed mostly on other arthropods, and although large specimens can inflict a painful bite, only a few tropical species are actually dangerous.

Only one pair of legs is present per trunk segment, for the segments are not doubled as in millipedes. To reduce its tendency to undulate or wobble when running, the trunk of many species is strengthened by having overlapping terga of different lengths (Fig. 26.46).

The centipedes most highly adapted for running are members of the order **Scutigeromorpha** (Fig. 26.46). They inhabit rocky, open areas and can catch moving insects. Some species occur in human habitations, commonly in bathtubs and sinks. The scutigeromorphs have overlapping trunk terga; the legs are very long to provide for a greater step; the leg length increases along the trunk from anterior to posterior, preventing interference; and the distal part of the leg is composed of many small segments, providing greater traction

through increased contact with the ground.

Although some centipedes deposit their eggs in the soil, most brood their eggs, wrapping the body around the developing young.

The class **Symphyla** contains 45 species of tiny myriapods which inhabit leaf mold. They are somewhat like centipedes except that there are no poison claws (Fig. 26.48). The mouthparts are strikingly like those of insects, and some zoologists believe that perhaps the symphylans are closely related to the myriapodous ancestors of insects.

Class Insecta

The insects are not only the largest class of arthropods but also the largest group of animals, including more than three quarters of a million species. The great adaptive radiation that the class has undergone has led to the occupation of virtually every type of terrestrial habitat, and some groups have even invaded fresh water. The great success of insects over other terrestrial arthropods

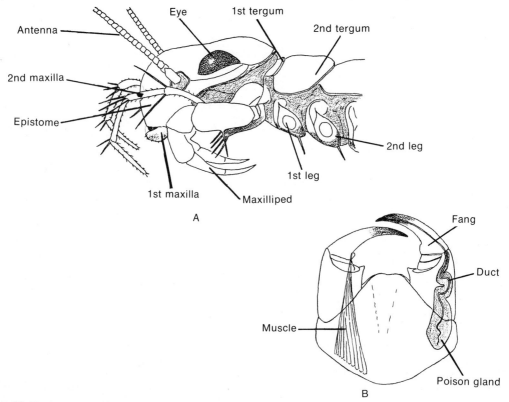

Figure 26.47 *A*, The head of the centipede *Scutigera coleoptrata* (lateral view). *B*, Maxillipeds, or poison claws, of *Otocryptops* (ventral view). (Both after Snodgrass.)

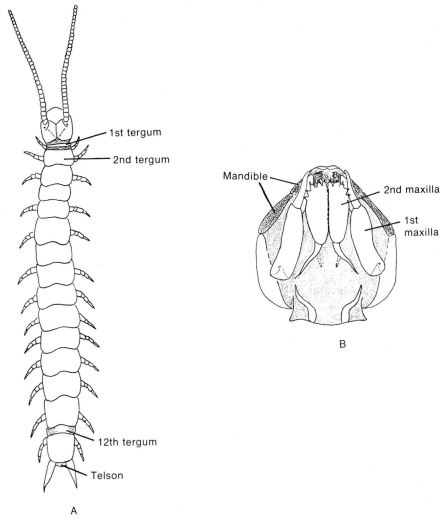

Figure 26.48   Symphyla. *A, Scutigerella immaculata* (dorsal view). *B,* Ventral view of head. (Both after Snodgrass.)

can be attributed in part to the evolution of flight, which provided advantages for dispersal, escape, access to food, or more favorable environmental conditions. The ability to fly evolved in reptiles, birds, and mammals, but the *first* flying animals were insects.

Insects are of great ecological significance in the terrestrial environment. Two-thirds of all flowering plants are dependent upon insects for pollination. The principal pollinators are bees, wasps, butterflies, moths, and flies, and the three orders represented by these insects have an evolutionary history that is closely tied to that of the flowering plants, which underwent an explosive evolution in the Cretaceous.

Insects are of enormous importance for man. Mosquitoes, lice, fleas, bedbugs and a host of flies can contribute directly to human misery. Some contribute indirectly as vectors of human diseases or of diseases of man's domesticated animals: mosquitoes (malaria, elephantiasis, yellow fever); tsetse fly (sleeping sickness); lice (typhus and relapsing fever); fleas (bubonic plague); and the housefly (typhoid fever and dysentery). Our domesticated plants are dependent upon some insects for pollination but are destroyed by others. Vast sums are expended to control insect pests, which can greatly reduce the high agricultural yields necessary to support large human populations. But the overzealous use of pesticides can in turn be hazardous to the environment.

External Structure. Despite the great diversity of insects, the general structure is relatively uniform. The body is composed of **head, thorax,** and **abdomen** (Fig. 26.49). A large pair of **compound eyes** occupies the lateral surface of the head. Between the compound eyes are three small **simple eyes,** or **ocelli,** and a pair of **antennae** (Fig. 26.50).

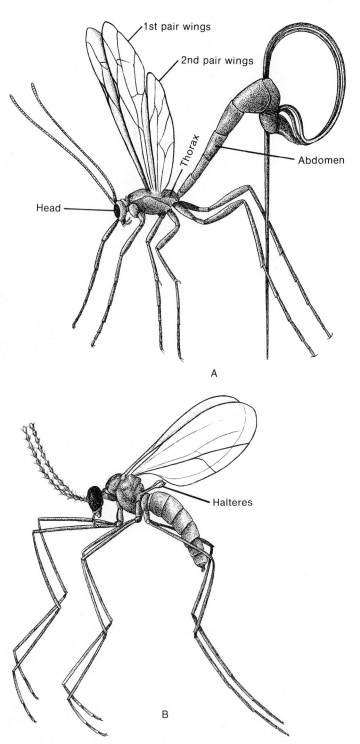

Figure 26.49 *A*, A parasitic female ichneumon fly (order Hymenoptera). *B*, A gall gnat (Order Diptera). (*A* after Lutz; *B* after Usda from Borror and Delong.)

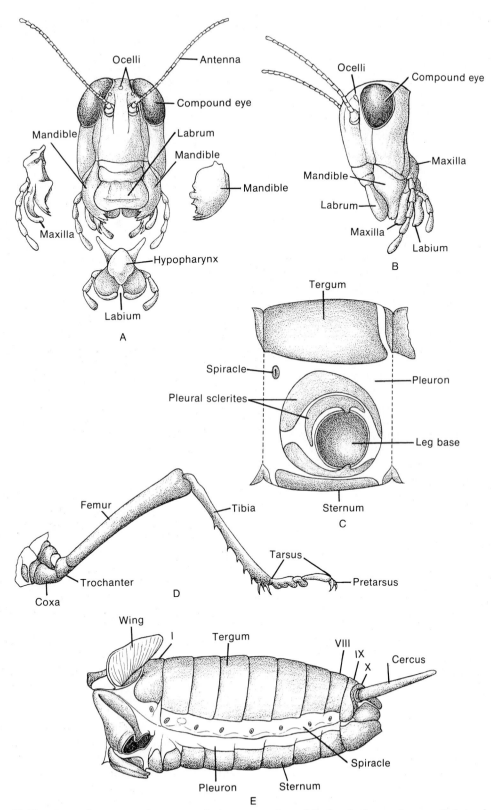

Figure 26.50 External structure of an insect. *A*, Anterior surface of the head of a grasshopper. *B*, Lateral view of the head of a grasshopper. *C*, Lateral view of a wingless thoracic segment. *D*, Leg of a grasshopper. *E*, Lateral view of the abdomen of a male cricket. (*C–E* after Snodgrass.)

The feeding appendages consist of a pair of **mandibles** and two pairs of **maxillae.** The second pair of maxillae are fused together and called the lower lip, or **labium.** An upper lip, or **labrum,** formed by a shelf-like projection of the head, covers the mandibles anteriorly. Near the base of the labium, a median process, the **hypopharynx,** projects from the floor of the oral cavity.

The thorax is composed of three segments: a **prothorax, mesothorax,** and **metathorax.** Each segment bears a pair of legs, and the last two segments may each carry a pair of wings (Fig. 26.49).

The abdomen is composed of nine to 11 segments, which lack appendages; however, the terminal reproductive structures are believed by some entomologists to be derived from segmental appendages.

Flight. Most insects have wings and comprise the subclass **Pterygota.** Primitive insects, such as proturans, thysanurans, and collembolans (members of the subclass **Apterygota**), are wingless and apparently diverged from the main stream of insect evolution before the origin of flight (Fig. 26.51). Some insects—ants, termites, fleas, and lice—are secondarily wingless.

The wings develop as folds of the body wall and are thus composed of two layers of skeletal material applied together. Support is provided by a net-like or strut-like arrangement of tubular thickenings, called **veins.**

Each wing articulates with the edge of the **tergum,** but its inner end rests on a dorsal pleural process, which acts as a fulcrum (Fig. 26.52). The wing is thus somewhat analogous to a seesaw off center. Upward movement of the wings results *indirectly* from the contraction of vertical muscles within the thorax, depressing the tergum. Downward movement of the wings is produced either *directly*, by contraction of muscles attached to the wing base (dragonflies and roaches); or *indirectly*, by the contraction of transverse horizontal muscles raising the tergum (bees, wasps, and flies).

Up and down movement alone is not sufficient for flight. The wings must at the same time be moved forward and backward. A complete cycle of a single wing beat describes an ellipse (grasshoppers) or a figure eight (bees and flies), during which the wings are held at different angles to provide both lift and forward thrust. In many insects the two pairs of wings are locked together in flight.

The raising or lowering of the wings resulting from the contraction of one set of flight muscles stretches the antagonistic muscles, which then also contract. Insect wing beat thus involves the alternate contraction of the antagonistic elastic systems. The beat frequency varies greatly—4 to 20 beats/sec. in butterflies and grasshoppers; 190 beats/sec. in the honeybee and housefly; and 1000 beats/sec. in certain gnats. At low frequencies (30 beats/sec. or less) there is usually one nerve impulse to one muscle contraction. At higher frequencies, however, the contraction is **myogenic,** originating from the stretching caused by the contraction of the antagonistic muscles, and there are a number of beats, or oscillations, between each nerve impulse.

Rapid contraction is facilitated by the nature of the muscle insertion. A very slight decrease in muscle length during contraction can bring about a large movement of the wing (as a seesaw with the fulcrum near one end). A large part of insect flight muscles is occupied by giant mitochondria, which provide for the high rate of respiration in these cells.

Flying ability varies greatly. Many butterflies and damselflies have a relatively slow wing beat and limited maneuverability.

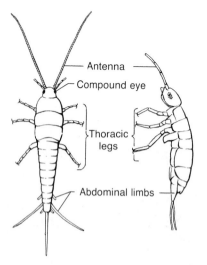

Figure 26.51   Primitive wingless insects (Apterygota), showing a silverfish (*left*) and a springtail (*right*). (After Lubbock [*left*] and Carpenter and Folsom [*right*].

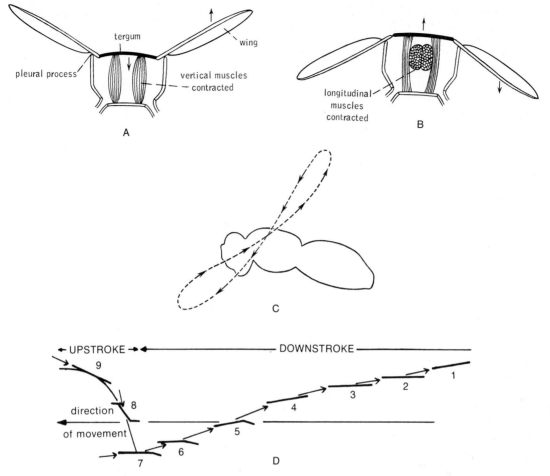

Figure 26.52 Diagrams showing relationship of wings to tergum and pleura, and the mechanism of the basic wing strokes in an insect. *A*, Upstroke resulting from the depression of the tergum through the contraction of vertical muscles. *B*, Downstroke resulting from the arching of the tergum through the contraction of longitudinal muscles. *C*, An insect in flight, showing the figure 8 described by the wing during an upstroke and a downstroke. *D*, Changes in the position of the forewing of a grasshopper during the course of a single beat. Short arrows indicate direction of wind flowing over wing and numbers indicate consecutive wing positions. (*A* and *B* after Ross; *C* after Magnan from Fox and Fox; *D* after Jensen from Chapman: The Insects: Structure and Function. London, University of London Press Ltd., 1971. Courtesy of the American Elsevier Publishing Co., Inc.)

At the other extreme, some flies, bees, and moths can hover and dart. The fastest flying insects are hummingbird moths and horseflies, which have been clocked at over 33 miles per hour. Gliding, an important form of flight in birds, occurs in only a few large insects.

There is no flight control center in the insect nervous system, but the eyes and the sensory receptors on the antennae, wings, and in the wing muscles themselves provide continual feedback information for flight control. Horizontal stability is maintained in part by a dorsal light reaction: the insect keeps the dorsal ommatidia of the eyes under maximum illumination from above. Deviation because of rolling is corrected by slight changes in wing position to bring the dorsal part of the eyes back to maximum illumination.

Members of the order **Diptera** (flies, gnats, and mosquitoes) have the second pair of wings reduced to knobs, called **halteres.** The halteres beat with the same frequency as the forewings and function as gyroscopes for the control of flight instability (pitching, rolling, and yawing).

Flight speed is probably determined by air flow over receptors on the antennae and movement of objects from front to back across the eyes.

Flight is inhibited by contact of the **tarsi**

with the ground. At rest the wings are either held outstretched, directed back over the abdomen, or folded up. The anterior pair of wings of beetles (order **Coleoptera**) are heavy shield-like covers for the membranous second pair of wings, which are folded over the abdomen at rest.

Nutrition. Primitive insects were herbivores that chewed plants with heavy jaw-like **mandibles.** In the evolution of various groups, the mouthparts have become adapted for many different modes of feeding and now a wide range of diets is utilized. Sucking, piercing and sucking, and chewing and sucking insects have the mouthparts elongated to form a heavy **beak,** but since this feeding habit evolved independently a number of times, the beak is formed in different ways.

The primitive herbivorous chewing habit has been retained in many members of a number of major orders, such as grasshoppers, crickets, and beetles. The raptorial dragonflies and some beetles, on the other hand, use their chewing mouthparts for seizing and tearing prey. A sucking proboscis which rolls up when not in use is characteristic of butterflies and moths (Fig. 26.53); bees have both chewing and sucking mouthparts. Bugs, aphids, cicadas, and many mosquitoes, gnats and flies have piercing and sucking mouthparts to pierce plant tissue or to penetrate the bodies of other animals.

Salivary secretions play an important role in feeding. Lubrication of food, digestion of sugar and pectins, anticoagulation, and production of venom are provided by the salivary glands of different groups.

The **foregut** of insects is variously modified to suit the diet and mode of feeding, but a **pharynx, crop,** and **proventriculus** are most commonly encountered (Fig. 26.54). The proventriculus may function as a gizzard or as a valve into the midgut.

The insect **midgut** (the **ventriculus,** or **stomach**) is the site of enzyme production, digestion and absorption, as in other arthropods (Fig. 26.54). In those species which ingest solid foods, the foregut-midgut junction secretes a thin cuticle, the **peritrophic membrane,** which surrounds the food mass as it passes through the midgut. Supposedly the peritrophic membrane protects the delicate midgut walls from abrasion by the food mass. The membrane is permeable to en-

zymes and the products of digestion. Outpocketings of the midgut, called **gastric caecae,** are characteristic of many insects, but their function is still uncertain.

The **hindgut,** composed of an **intestine** and **rectum,** opens to the exterior at the end of the abdomen. In many insects the hindgut is an important site of water absorption, and some species have special rectal structures to facilitate the process.

Excretion, Gas Exchange, and Internal Transport. The excretory organs are **Malpighian tubules.** Two to several hundred tubes arise from the hindgut-midgut junction. Sometimes they are long and coiled. They lie more or less freely in the **hemocoel,** and wastes picked up from the surrounding blood are deposited as uric acid in the tubule lumen. The sludge-like sediment passes into the intestine where more water may be removed. The fecal matter eliminated from the anus thus consists of undigested residues and excretory wastes.

Tracheae are the gas exchange organs of insects; they have been studied more extensively in this group of animals than in any others. A **spiracle** is located on each side of the last two thoracic segments and the first seven or eight abdominal segments (Fig. 26.55). Usually the spiracle is provided with a closing device and is located at the bottom of a pit. Various filtering devices reduce clogging from dust. Internally, the tracheal system exhibits a more or less ladder-like plan, with longitudinal and transverse trunks. The larger tubes may have the walls strengthened by ring-like thickenings. The fluid-filled **tracheoles** are minute, usually less than a micron in diameter, and a number of tracheoles may arise from a single tracheal cell. The tracheoles are not shed at molting, but the new cuticle fuses with the old cuticle of the persisting tracheole. The distribution of tracheoles reflects the oxygen demands of the tissues served. In the wing muscles, for instance, the tracheoles actually push into the muscle cells.

Diffusion along concentration and pressure gradients is the principal means by which oxygen reaches the tracheoles and carbon dioxide passes to the outside. Insects weighing more than a gram and extremely active insects possess additional ventilating mechanisms. The movement of body muscles may compress the larger

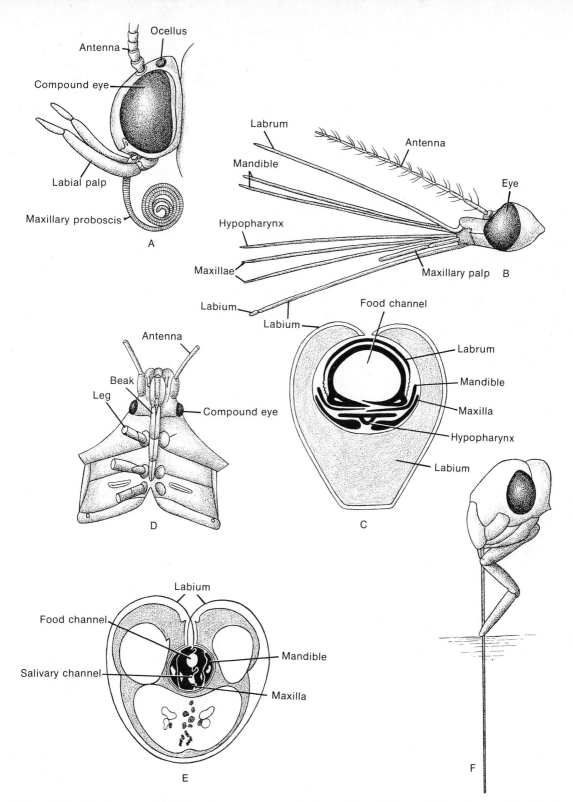

Figure 26.53 Mouth parts of sucking and piercing insects. *A*, Lateral view of the head of a moth, a sucking insect. *B*, Lateral view of the head of a mosquito, showing separated mouth parts. *C*, Cross section of mouth parts of a mosquito in their normal functional position. *D*, Ventral view of anterior half of a hemipteran, showing beak. *E*, Cross section through a hemipteran beak, showing the food and the salivary channels enclosed within the stylet-like maxillae and the mandibles. *F*, A hemipteran penetrating plant tissue with its stylets. (*A* after Snodgrass; *B* and *C* after Waldbauer from Ross; *D* after Hickmann; *E* after Poisson; *F* after Kullenberg.)

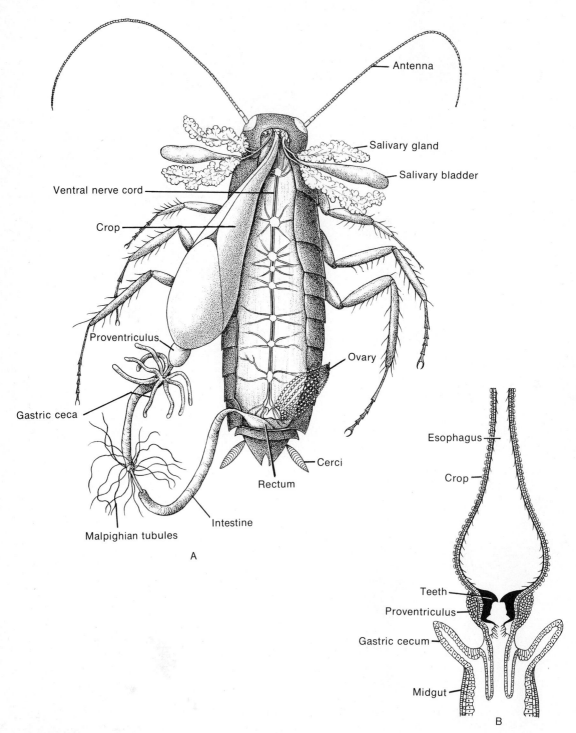

Figure 26.54 *A*, Internal structure of a cockroach. *B*, Longitudinal section through the foregut and anterior part of the midgut of a cockroach. (*A* after Rolleston; *B* after Snodgrass.)

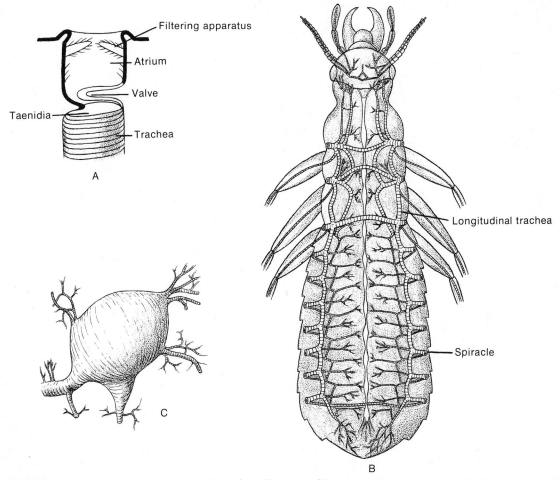

Figure 26.55 Insect tracheal system. *A*, Spiracle with atrium, filtering apparatus, and valve. *B*, Typical plan of tracheal system. *C*, An air sac. (*A* after Snodgrass; *B* after Ross; *C* after Snodgrass.)

tracheae or special dilated regions called **air sacs.** Compression coupled with a pattern of opening and closing of spiracles moves air from one part of the system to another.

Adult aquatic insects continue to use tracheae and obtain oxygen by contact with the surface or from bubbles or films of air trapped on the surface of the body. Aquatic larvae may have gills, but even here, tracheae provide for internal gas transport. Oxygen dissolved in the water diffuses across the gill cuticle into underlying tracheae. Within the tracheal system it follows a diffusion gradient to the body tissues.

The **heart** is usually a long abdominal tube with nine pairs of ostia, and the only vessel is an anterior **aorta** leading into the thorax and head. Blood flows from posterior to anterior, although reversal of flow is known to occur in some forms.

Insect blood is unusual in containing large amounts of organic compounds, especially amino acids. High concentrations of uric acid, organic phosphates, and a nonreducing sugar, trehalose, are also present.

Reproduction and Development. The single pair of gonads is located in the abdomen and the gonoducts from each side unite posteriorly before opening at the end of the abdomen through a short median **vagina** in the female or an **ejaculatory duct** in the male. In addition, the female system usually includes a **seminal receptacle** and **accessory glands** associated with the vagina, and the male system includes paired **seminal vesicles** and **accessory glands** (Figs. 26.56 and 26.57).

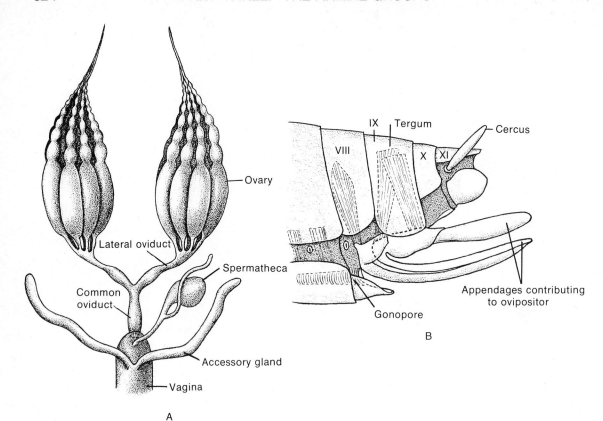

Figure 26.56 *A*, Reproductive system in a female insect. *B*, Lateral view of the posterior end of the abdomen, showing reproductive opening and appendages, forming ovipositor. (Both after Snodgrass.)

The sperm of many insects are transferred within spermatophores. In a few primitive groups the spermatophores are transferred indirectly. Like arachnids and myriapods the spermatophores are deposited on the ground and then picked up by the female. In most insects, however, transfer is direct, and a tubular **penis** is inserted by the male into the vagina of the female. The posterior abdominal segments of the male of many moths, butterflies, true flies, and other insects bear clasping structures which are used to hold the abdomen of the female during copulation (Fig. 26.57).

Sperm are stored in the seminal receptacle, and fertilization occurs internally as the eggs pass through the oviduct. The egg leaves the ovary encased within a hard shell but a tiny opening, or **micropyle,** at one end of the egg permits entrance of the sperm.

Certain parts of the terminal segments of the female are modified as an **ovipositor,** through which the eggs pass upon leaving the gonopore. By means of the ovipositor, eggs can be buried in soil, excrement, or carrion, injected into plant tissue, or ap-

plied to twig, leaf, soil, water, or other surfaces. The eggs are generally deposited in batches, cemented to each other and to the substratum by secretions from the accessory glands.

Parthenogenesis occurs in a number of insect groups. The condition in aphids closely parallels that of the crustacean water fleas, where there are successive generations of parthenogenetic females followed by the appearance of males. In bees, unfertilized eggs produce males; fertilized eggs produce females.

The degree of development at hatching is quite variable. The newly hatched young of primitive wingless insects are similar to the adults, except in size and sexual maturity. In contrast, the young of the members of winged orders—grasshoppers, crickets, dragonflies, leaf hoppers, bugs, and many others—are similar to the adults but lack the wings and reproductive system, which gradually develop during the course of subsequent instars (Fig. 26.58). This type of development is called **gradual,** or **incomplete, metamorphosis (hemime-**

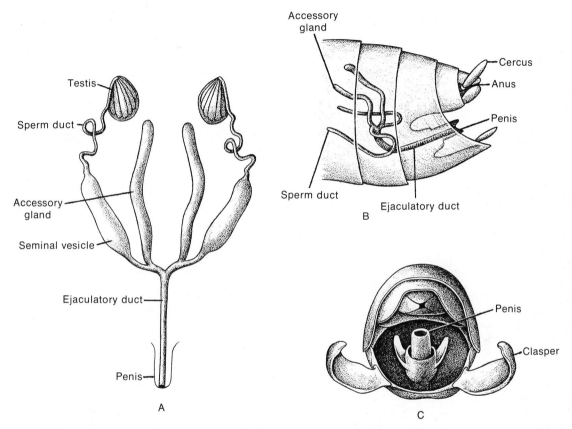

Figure 26.57 Reproductive system in a male insect. *A*, General plan of system. *B*, Lateral view of the posterior end of the abdomen, showing reproductive opening and other structures. *C*, Posterior view of the abdomen, showing penis and claspers. (All after Snodgrass.)

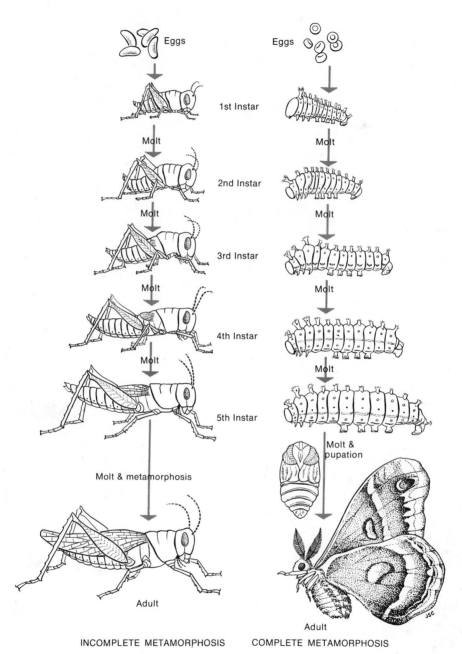

Figure 26.58　Comparison of gradual, or incomplete, metamorphosis of a grasshopper and complete metamorphosis of a cecropia moth. (Villee: Biology, 6th ed.)

tabolous development) and the immature stages are called **nymphs** or, when aquatic, **naiads.** Aquatic development with incomplete metamorphosis is characteristic of dragonflies, stoneflies, and mayflies.

The higher orders of insects, such as beetles, butterflies, moths, bees, wasps, and flies, undergo **complete metamorphosis (holometabolous development)** involving more radical changes between the immature and adult forms. The hatching stage is an active feeding worm-like **larva,** called a caterpillar (butterflies and moths), grub (beetles), or maggot (flies). The larva increases in size through successive instars, and at the end of the larval period, passes into a quiescent **pupal** stage, in which a radical transformation takes place. In many groups most of the larval structures break down and are resorbed, and residual germinal areas (**imaginal discs**) give rise to adult structures. A completely formed adult emerges at the end of pupation, which typically occurs within a protective location (soil) or within some sort of encasement (cocoon of some moths).

The transformation of immature insects into reproducing adults is known to be under endocrine control. A hormonal secretion from the brain stimulates a gland in the prothorax, the **prothoracic gland,** which produces **ecdysone A** and **B,** hormones that stimulate growth and molting. During the larval stages, juvenile hormone is secreted by the **corpora allata,** glands associated with the brain. This hormone is responsible for the maintenance of larval structures; it inhibits metamorphosis. The **juvenile hormone** can exert its effect only after the molting process has been initiated. It thus must act in conjunction with the prothoracic hormones. When a relatively high level of juvenile hormone is present in the blood, the result is a larva-to-larva molt. When the level of the juvenile hormone is less, the molt is larva-to-pupa, and in the absence of the juvenile hormone, there is a pupa-to-adult molt.

Insect larvae are a specialized developmental condition, and as we have seen, are absent in primitive groups. Larvae are able to utilize food sources and habitats which would be unavailable to the adults and vice versa. For example, the caterpillars of butterflies are chewing herbivores, in contrast to the nectar-feeding habit of the adults. In some insects, certain moths for instance, the short-lived adults have lost the ability to feed at all.

An unusual form of "twinning," called **polyembryony,** occurs in some small wasps. The eggs develop parasitically within the body of caterpillars. At an early stage of development the embryonic mass fractures into a number of smaller masses, each of which develops into a complete individual. In *Litomastix* as many as 3000 larvae may develop from one egg!

Parasitism. There are many parasitic insects and the condition has undoubtedly evolved several times within the class.

Insect parasitism represents an adaptation to meet the habitat-nutrition needs of different stages in the life cycle. For some insects, parasitism provides a new food source and habitat for the adults; for others, a new food source for the larvae.

Adult fleas and lice (Fig. 26.59) are bloodsucking ectoparasites on the skin of birds and mammals (there is one group of chewing lice). The eggs and immature stages of fleas may develop on the host or in the host's nest or habitation.

Many species of wasps and flies illustrate

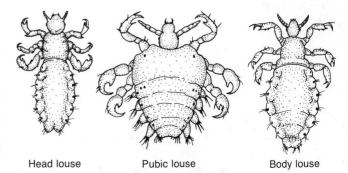

Figure 26.59 Anoplura. The three varieties of human lice. The head louse, *Pediculus humanus* var. *capitis,* and body louse, *P. h.* var. *corporis,* are interfertile varieties of one species that rarely interbreed because one lives on the head, laying eggs on the hairs, while the other lives on the clothed portion of the body, laying eggs in the clothing. The pubic louse, *Phthirus pubis,* lives in the pubic region and occasionally in the armpits. (After Patton and Evans.)

Head louse   Pubic louse   Body louse

larval parasitism. Certain small wasps insert an egg into the leaves and stems of plants. The surrounding plant tissue reacts to form a large mass, or gall. The egg develops within the gall and the larval stage feeds upon the surrounding plant tissue. Other wasps deposit their eggs in the bodies of insects, especially insect larvae.

The screw-worm fly, a species of blowfly and a pest of domestic animals, lays its eggs in the wounds and nostrils of mammals and the larvae feed on living tissue. The parasitic condition was probably preceded by the deposition of eggs in carrion, for this is the habit of many non-parasitic species of blowflies.

Communication. Both social and non-social insects utilize chemical, tactile, visual, and auditory signals as methods of communication. Many examples of chemical communication by pheromones (p. 402) are now known and the following are just a few of the many striking examples in insects. The males of some moths can locate females by means of air-borne substances detected from a distance as great as 4.5 km. Substances deposited on the ground by ants returning from a foraging trip serve as a trail marker for other ants. This type of communication is especially important in the complex movements of tropical army ants. Substances produced by the death and decomposition of the body of an ant within the colony stimulate other workers to remove the body. If a live ant is painted with an extract from a decomposing body, the painted ant will be carried live and struggling from the nest.

Among the more unusual visual signals are the luminescent flashings of fireflies. The significance of these signals is still not understood but they apparently play some role in sexual attraction.

Sound production is especially notable in grasshoppers, crickets, and cicadas. The chirping sounds of the first two are produced by rasping. The front margin of the forewing or the hind leg acts as a scraper and is rubbed over a file formed by veins of the forewing. Where scraper and file are both located on the forewings, as in crickets, the wings cross over, and one forewing functions as a scraper and the other as a file. Each species of cricket produces a number of songs which differ from the songs of the other species. Cricket songs function in

sexual attraction, and aggression. The static-like sounds of cicadas, which serve to aggregate individuals, are produced by vibrations of special chitinous abdominal membranes.

The remarkable mechanisms of communication in bees are described in the next section.

Social Insects. Colonial organization has evolved in a number of animal phyla, but only among some insects and vertebrates are individuals functionally interdependent, yet morphologically separate. The condition is therefore usually described as being a social organization.

Social organizations have evolved in two orders of insects, the **Isoptera,** which contains the termites, and the **Hymenoptera,** which includes the ants, bees, and wasps. In all social insects, no individual can exist outside of the colony nor can it be a member of any colony but the one in which it developed. There is cooperative brood care and an overlap of generations. All social insects exhibit some degree of **polymorphism** and the different types of individuals of a colony are termed **castes.** The principal castes are **male, female** (or **queen**) and **worker.** Males function only for the insemination of the queen, which produces new individuals for the colony. The workers provide for the support and maintenance of the colony. Worker castes are always sterile females, and caste determination of female is a developmental phenomenon regulated by the presence or absence of certain substances provided in the immature stages by other members of the colony. However, only in bees is much known about the actual determination processes.

Termites live in galleries constructed in wood or soil, and in some species the colony may be huge and structurally very complex. The colony is built and maintained by workers and soldiers (Fig. 26.60). The soldiers are sterile females with large heads and mandibles and serve for the defense of the colony. Workers and soldiers are wingless; wings are present in the males and queens only during a brief nuptial flight, during which copulation and dispersion occur.

Ant colonies resemble those of termites and are housed within a gallery system in soil, wood, or beneath stones. There may

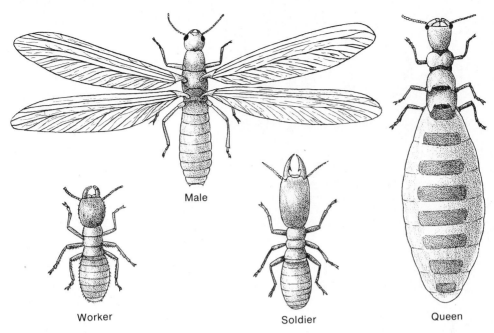

Male

Worker

Soldier

Queen

Figure 26.60   Castes of the common North American termite, *Termes flavipes*. (After Lutz.)

be a soldier caste in addition to workers. Wings are present only during the nuptial period.

Polymorphism is less highly developed in wasps and bees. There is no soldier caste and workers are winged, but many of these insects exhibit remarkable adaptations for a social organization.

The honeybee, *Apis mellifera*, is the best known social insect. This species is believed to have originated in Africa and to be a recent invader of temperate regions, for unlike other social bees and wasps of temperate regions, the honeybee colony survives the winter and multiplication occurs by the division of the colony, a process called swarming. Stimulated at least in part by the crowding of workers (20,000 to 80,000 in a single colony), the mother queen leaves the hive along with part of the workers (a **swarm**) to found a new colony. The old colony is left with developing queens. On hatching, a new queen takes several nuptial flights during which copulation with males (**drones**) occurs, and she accumulates enough sperm to last her lifetime. The male dies following copulation, when his reproductive organs are literally exploded into the female. A new queen may also depart with some of the workers as an after-swarm, leaving the remaining workers to yet an-

other developing queen. Eventually the old colony will consist of about one-third of the original number of workers and their new queen.

Honeybee colonies are large. The workers' life span is not long, and a queen may lay one thousand eggs per day. The diet provided these larvae by the nursing workers results in their developing into sterile females, i.e., additional workers. The nursing behavior of the workers is a response to a pheromone ("queen substance") produced by the queen's mandibular glands. At the advent of the swarming or when the vitality of the queen diminishes, the production of this pheromone declines. In the absence of the inhibiting effect of the pheromone, the nursing workers construct royal cells, into which eggs and royal jelly are placed. The exact composition of this complex food is still unknown, but those larvae fed upon it develop into queens in about 16 days. At the same time that queens are being produced, unfertilized eggs are deposited into cells similar to those for workers. These haploid eggs develop into drones.

A remarkable feature of honeybee social organization is the temporal division of the workers. As can be seen from the graph on Figure 26.61, the first activities of the

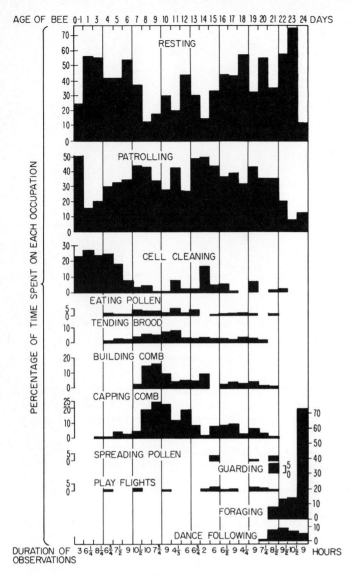

Figure 26.61 The activities of a single worker honeybee during the first 24 days of her adult life (Redrawn from Ribbands, 1953; based on data of Lindauer, 1952.)

worker are maintenance tasks within the hive. During this period there is secretion by wax, mandibular, and other glands involved in comb construction, food storage, and larval care. After about three weeks, such glandular activity declines, and the bee begins a period of foraging outside of the hive, its final service to the colony. The many functions performed in the lifetime of a worker are not strictly sequential but a worker shifts from one task to another (Fig. 26.61). A large amount of time is spent by the older worker bees in resting and patrolling. Patrolling, or "determination" of hive needs, plus the ability of workers to change tasks, enables a colony to adjust

to changing environmental conditions and is believed to have contributed to the success of the species.

Communication between members of a honeybee colony is highly evolved; some aspects, such as the tail-wagging dance, set the honeybees apart from all other social insects. A successful foraging scout returns to the hive and communicates to other workers the nature, direction and distance of a food source. The nectar and pollen on the scout's body provides the information about the kind of food that has been found. The scout bee also executes an excited dance which is a ritualization of the flight path. The dancing bee circles to the right and to the left, with a straight-line run between the two semi-circles (Fig. 26.62). During the straight-line run, the bee wags her abdomen and emits audible pulsations. Von Frisch, a pioneer in the study of communication in bees, discovered that the orientation of the circular movements shows the direction of the food, and that the frequency of the tail-wagging runs indicates the distance. The closer the food source is to the hive, the greater the frequency of tail-wagging runs. Bees use the angle of the sun and light polarization as a means of orientation, and the dance of the scout bee indicates the location of the food in reference to the sun's position. If the tail-wagging run is directed upward, the food is located toward the sun; if the tail-wagging run is directed downward, the food is located away from the sun. The inclination of the run to the right or to the left of vertical indicates the angle of the sun to the right or to the left of the food source. An internal "clock" compensates for the passage of time between discovery of the food and the start of the dance, so that the information is correct even though the sun has moved during the interval. On cloudy days, the polarization of the light rays and ultra-violet light act as indirect references in the absence of the sun. If the food source is closer than 80 meters, the clues provided by chemoreception are sufficient for finding the food, and the tail-wagging dance is not performed by the scout bee.

Although the tail-wagging dance has been decoded, the sensory modality by which it is transmitted to other bees is still uncertain. The hive is dark so that the dance cannot be easily detected visually. The surrounding bees must receive the dancer's vibrations through their antennae or legs. The sound pulsations of the dancer apparently also indicate the distance of the food source from the hive. The average number of vibrations is proportional to the distance of the food from the hive.

26.10 ARTHROPOD PHYLOGENY

Most zoologists agree that there is convincing evidence to support the belief that arthropods evolved from annelids or at least from some common ancestor. However, there is much less agreement as to how the major arthropod groups might be related to each other. The difficulty centers

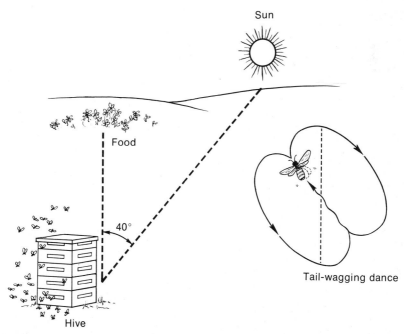

Figure 26.62 Diagram illustrating the inclination of the straight tail-wagging run by a scout bee to indicate the location of a food source by reference to the sun. The food source is located at an angle 40 degrees to the left of the sun. The tail-wagging run of the scout bee is therefore upward (indicating that food is toward sun) and inclined 40 degrees to the left (indicating the angle of the food source to the sun). (After Von Frisch.)

on the mandibulate arthropods. The terrestrial myriapodous classes (millipedes and centipedes) and the insects are generally thought to be related, but do they share a common ancestry with the crustaceans? Current views are divergent and conflicting. The traditional view holds that the subphylum Mandibulata is a natural group. The supporters of this position consider the presence of antennae, mandibles, and maxillae among the mandibulates as evidence of common ancestry, but living or fossil intermediate forms have not been discovered.

Another view of arthropod phylogeny takes the opposite position and considers the phylum to be polyphyletic. The supporters of this view believe that "arthropodization" evolved twice, once as an adaptation for a terrestrial existence in a line that led to myriapods and insects and again as an adaptation for a marine bottom existence. The latter evolution produced the trilobites, crustaceans, and chelicerates. According to this polyphyletic view of arthropod origins, the similar head appendages and compound eyes of insects and crustaceans had independent origins and represent evolutionary convergence. In both theories the crustaceans and chelicerates are considered to be culminations of divergent lines from a remote ancestral form, to which the trilobites are closest of known arthropods.

Much more evidence than is presently available is needed to raise either of these ideas regarding arthropod evolution to something beyond pure speculation.

CLASSIFICATION OF THE PHYLUM ARTHROPODA

Subphylum Trilobitomorpha

The fossil trilobites. A single pair of antennae present; the many post-oral trunk appendages are all similar.

Subphylum Chelicerata

Antennae absent; the first post-oral appendages are a pair of chelicerae.

CLASS MEROSTOMATA. Body composed of a cephalothorax and abdomen. Marine, with abdominal gills.

　Subclass Xiphosura. Horseshoe crabs. Abdominal segments fused.

　Subclass Eurypterida. Extinct merostomes. Abdominal segments not fused.

CLASS ARACHNIDA.* Body composed of a cephalothorax and abdomen. Terrestrial; gills absent.

　　　　Order Scorpionida. Scorpions. Large chelate pedipalps and long segmented abdomen with terminal sting.

　　　　Order Pseudoscorpionida. Tiny arachnids with large pedipalps and short segmented abdomens.

　　　　Order Araneae. Spiders. Abdomen unsegmented and attached to cephalothorax by a narrow waist.

　　　　Order Opiliones. Harvestmen, or daddy-long-legs. Legs usually very long; the short, segmented abdomen is broadly joined to the cephalothorax.

　　　　Order Acarina. Mites and ticks. Very small arachnids with unsegmented abdomen broadly fused with cephalothorax.

CLASS PYCNOGONIDA. Sea spiders. Marine chelicerates(?) having a very narrow trunk and four pairs of long legs.

Subphylum Mandibulata

At least one pair of antennae is present and first post-oral appendages are a pair of mandibles.

CLASS CRUSTACEA.* Head with two pairs of antennae.

　Subclass Branchiopoda. Fairy shrimp, tadpole shrimp, clam shrimp, and water fleas. Fresh-water and marine crustaceans with leaf-like setose appendages.

---

* Abbreviated.

Subclass Ostracoda. Mussel, or seed shrimp. Tiny marine and fresh-water crustaceans in which the entire body is enclosed within a bivalved carapace.

Subclass Copepoda. Copepods. Very small marine and fresh-water crustaceans having a cylindrical tapered body with long first antennae.

Subclass Cirripedia. Barnacles. Marine, sessile crustaceans in which the body is enclosed within a bivalved carapace that is typically covered with calcareous plates.

Subclass Malacostraca.* Trunk composed of an eight-segmented thorax, on which the legs are located, and a six-segmented abdomen.

Order Amphipoda. Amphipods. Small marine and fresh-water crustaceans in which the body is laterally compressed and the eggs are brooded within a ventral thoracic marsupium.

Order Isopoda. Isopods. Marine, fresh-water and terrestrial (wood lice, pill bugs) crustaceans in which the body is dorsoventrally flattened and the eggs are brooded within a ventral thoracic marsupium.

Order Decapoda. Shrimp, crayfish, lobsters, and crabs. Thorax covered by a carapace; thoracic appendages consist of three pairs of maxillipeds and five pairs of legs.

CLASS DIPLOPODA. Millipedes. Elongate trunk composed of many similar doubled segments, each bearing two pairs of legs.

CLASS PAUROPODA. Minute grublike animals which inhabit leaf mold. Elongate trunk of eleven segments, nine of which bear legs.

CLASS SYMPHYLA. Small centipede-like mandibulates which inhabit leaf mold. Trunk contains 12 leg-bearing segments.

CLASS CHILOPODA. Centipedes. Elongate trunk of many leg-bearing segments. First trunk segment carries a pair of large poison claws.

CLASS INSECTA. Insects. Body composed of head, thorax and abdomen. Thorax bears three pairs of legs and usually two pairs of wings.

Subclass Apterygota. Primary wingless insects. The five orders include the silverfish and springtails.

Subclass Pterygota.† Winged insects, or if lacking wings, the wingless condition is secondary.

Order Ephemeroptera. Mayflies.

Order Odonata. Dragonflies and damselflies.

Order Orthoptera. Grasshoppers and crickets.

Order Isoptera. Termites.

Order Plecoptera. Stoneflies.

Order Dermaptera. Earwigs.

Order Mallophaga. Chewing lice, bird lice.

Order Anoplura. Sucking lice.

Order Thysanoptera. Thrips.

Order Hemiptera. True bugs.

Order Homoptera. Cicadas, leaf hoppers, aphids.

Order Neuroptera. Lacewings, ant lions, snakeflies, dobsonflies.

Order Coleoptera. Beetles, weevils.

Order Trichoptera. Caddisflies.

Order Lepidoptera. Butterflies and moths.

Order Diptera. True flies, midges, gnats, mosquitoes.

Order Hymenoptera. Ants, bees, wasps, sawflies.

Order Siphonaptera. Fleas.

ANNOTATED REFERENCES

Detailed accounts of arthropods may be found in the references listed at the end of Chapter 18. The works listed below deal exclusively with specific arthropod groups.

Borror, D. J., and De Long, D. M.: An Introduction to the Study of Insects. Revised ed.

---

* Greatly abbreviated.

† Abbreviated.

New York, Holt, Rinehart and Winston, 1964. A general entomology text. This work and that of Ross (see below) are good starting points for a more detailed study of insects, especially for students interested in insect taxonomy.

Chapman, R. F.: The Insects: Structure and Function. New York, American Elsevier Publishing Co., 1969. A detailed account of the anatomy and physiology of insects.

Cloudsley-Thompson, J. L.: Spiders, Scorpions, Centipedes, and Mites. New York, Pergamon Press, 1958. A discussion of the natural history and ecology of terrestrial arthropods other than insects.

Englemann, F.: Physiology of Insect Reproduction. New York, Pergamon Press, 1970.

Gertsch, W. J.: American Spiders. New York, Van Nostrand, 1949. An old, but still valuable account of the natural history of spiders.

Krishna, K., and F. M. Weesner (Eds.): Biology of Termites, Vol. I, 1969; Vol. II, 1970. New York, Academic Press.

Lockwood, A. P. M.: Aspects of the Physiology of Crustacea. San Francisco, W. H. Freeman, 1967.

Ross, H. H.: A Textbook of Entomology. 3rd ed. New York, John Wiley and Sons, 1965.

Savory, T.: Arachnida. New York, Academic Press, 1964. A general treatment of the biology of arachnids.

Schaller, F.: Soil Animals. Ann Arbor, Mich., University of Michigan Press, 1962. An account of the various invertebrates found in soil and leaf mold.

Schmitt, W. L.: Crustaceans. Ann Arbor, Mich., University of Michigan Press, 1965. A natural history of the crustaceans.

Weygoldt, P.: The Biology of Pseudoscorpions. Cambridge, Mass., Harvard University Press, 1969.

Wilson, E. O.: The Insect Societies. New York, Academic Press, 1971. A superb general biology of the social insects.

27. Bryozoans

The members of the phylum Bryozoa, sometimes called moss animals, are common marine organisms, but their small size and atypical form make them virtually unknown to the layman. They live attached to rocks, shells, pilings, jetties, and ship bottoms, and a few species occur in fresh water.

Bryozoans can be used to illustrate the predictive value of some of the information we have developed about animals thus far. The members of this phylum are **sessile, colonial,** and **coelomate,** and individuals are usually less than 0.5 mm. in length. On the basis of these four facts, what other features —structural or functional—would you expect to find in bryozoans? Think a few minutes and make some educated guesses before you continue reading.

The following bryozoan features are correlated with their sessile life, colonial organization, and minute size, and are therefore not unexpected:

Correlated with their colonial organization:

1. Bryozoans are **polymorphic.**

Correlated with their sessile condition:

2. Bryozoans are encased within an **exoskeleton.** A protective and supporting skeleton is characteristic of many sessile animals, such as sponges, hydrozoans, corals, and barnacles.

3. Bryozoans are **filter feeders.** Filter feeding, a common adaptation for a slow-moving or immobile life, is illustrated by such animals as sponges, sedentary tubiculous polychaetes, worm shells, slipper shells, bivalves, and barnacles.

4. Bryozoans are **hermaphroditic.** Hermaphroditism is a common adaptation of animals which cannot move about, such as many sponges, some sea anemones, some corals, some bivalves, slipper shells, and barnacles.

5. A **larval stage** is present in the development of bryozoans. Larvae are the principal means of dispersion for most sessile animals.

Correlated with their very small size:

6. Bryozoans have no special internal transport system, for internal distances are short enough to permit transport solely by diffusion.

7. Bryozoans have no gas exchange organs. The ratio of surface area to volume is sufficiently favorable to permit gas exchange across the general body surface.

8. Bryozoans have no excretory system. The excretion of highly soluble ammonia and the large ratio of surface area to volume permit elimination by diffusion across the general body surface.

27.1 STRUCTURE OF A BRYO-ZOAN INDIVIDUAL

Individuals (**zooids**) of a bryozoan colony are shaped something like a little rectangu-

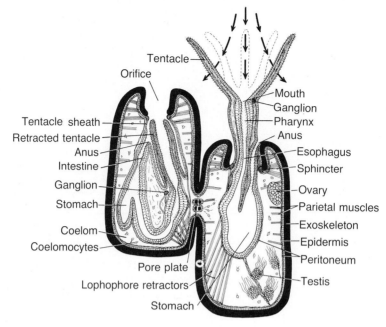

Figure 27.1   Structure of two zooids of a marine bryozoan. (After Marcus from Hyman.)

lar box or coffin, in which each of the sides represents one of the usual morphological surfaces—anterior, posterior, ventral, etc. (Fig. 27.1). The body is covered with an exoskeleton commonly composed of **chitin.** A layer of calcium carbonate is typically found just beneath the chitin; both are secreted by a single layer of epidermal cells. Immediately beneath the epidermis is a layer of **peritoneum;** the rigidity of the exoskeleton would make ordinary body wall muscles useless. The interior of the body is occupied by a spacious **coelom.**

The principal organ in the coelom is the **gut,** which is U-shaped to avoid the obstruction resulting from the attachment of the posterior end to other members of the colony. At the anterior end the skeletal housing is perforated by a circular orifice through which the two ends of the gut project.

The feeding organ of bryozoans is a crown of ciliated tentacles, the **lophophore.** When the lophophore is protruded, the outstretched tentacles usually form a funnel with the **mouth** in the center of the base and the **anus** projecting outside of the base (Figs. 27.1, 27.2 and 27.4). The hollow tentacles contain an extension of the coelom. When the lophophore is protruded, a sheath

of body wall extends from the rim of the orifice up to the base of the lophophore. On retraction of the lophophore into the orifice, this tentacular sheath is reversed and surrounds the bunched tentacles within the skeletal housing (Fig. 27.1).

The lophophore is protruded by increased pressure in the coelomic fluid. The pressure is elevated by the inbowing of the body wall on contraction of certain transverse muscle bands. In many bryozoans the muscle bands attach to the inner side (frontal membrane) of the ventral surface, which has become thin and easily depressed (Fig. 27.2). Withdrawal of the lophophore back into the orifice is brought about by a special retractor muscle.

During feeding, the **cilia** on the outstretched tentacles drive water down into and out the sides of the lophophore funnel, and food particles suspended in the incoming water stream are driven by cilia downward into the mouth (Fig. 27.1).

The nervous system consists of a **ganglion** and **nerve ring** around the anterior end of the gut. Fibers extend from the nerve ring to the tentacles and other parts of the body. No special systems for internal transport, gas exchange, or excretion are present.

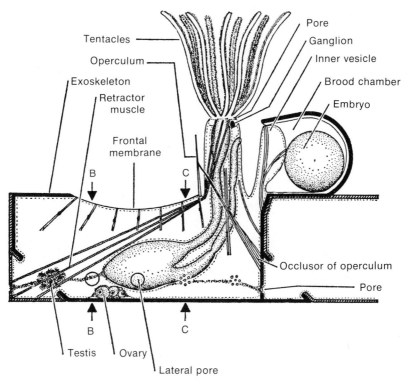

Figure 27.2 Lateral section through an encrusting bryozoan. This individual possesses a brood chamber containing an embryo. The orifice of this species is closed by a lid, or operculum. (After Ryland.)

27.2 ORGANIZATION OF COLONIES

In most bryozoans the individuals are so attached to each other that the resulting colonies form erect plant-like growths or encrusting sheets over rocks and shells (Fig. 27.3). In either case the ventral, or frontal, surface is exposed and the dorsal surface is attached. To visualize an encrusting colony, imagine a large number of rectangular box-like individuals lying on their backs on a rock. The dorsal surface is attached to the substratum, the lateral, anterior, and posterior surfaces are attached to surrounding individuals, and the ventral surface is exposed (Fig. 27.4). Since the anterior end is generally blocked by adjacent individuals, the orifice is commonly shifted over onto the ventral (exposed) sur-

Figure 27.3 *Amathia*, a stoloniferous bryozoan. The zooids are very tiny and arranged around the cylindrical stolons, which compose the conspicuous part of the colony.

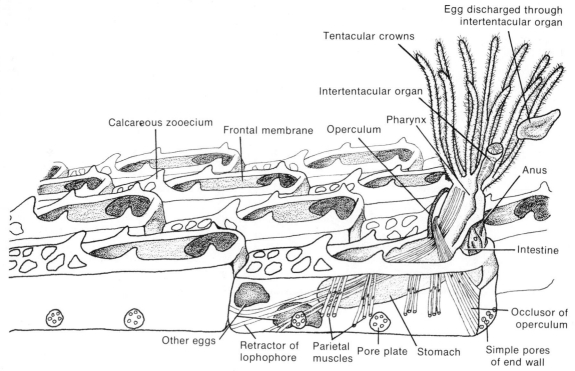

Figure 27.4   Lateral view of a portion of a colony of an encrusting bryozoan, *Electra*. (Modified from Marcus.)

face. Such an encrusting colony follows the contours of the shell or rock on which it is growing and may reach a diameter of many centimeters. The colonies are commonly white or pale hues of red or other colors.

Erect colonies may be foliaceous, dendritic, or strap-like, but the ventral surface is still the one that is exposed; individuals are attached to one another by the dorsal surface. *Bugula,* one of the most common and widely distributed bryozoans and an important marine fouling organism, illustrates the erect growth form. The brownish colonies are branching and dendritic, and the individuals spiral around an axis formed by their attaching dorsal surfaces (Fig. 27.5). An overlapping arrangement keeps the anterior end bearing the orifice clear of other individuals.

Members of a colony are connected to each other more intimately than by the fusion of their exoskeletons. Pores in the walls permit diffusion of coelomic fluid from one individual to another, and food material can be distributed to non-feeding members of the colony (Figs. 27.1 and 27.4).

Bryozoans are **polymorphic,** but the degree and nature of polymorphism vary. The most common members of the colony are the feeding individuals already described. The colonies of some species include highly modified defensive individuals which protect the colony against other settling organisms (Fig. 27.5B). Some species have vegetative stoloniferous individuals which creep over and anchor to the substratum (Fig. 27.5A). Reproductive individuals with special brooding structures are still another type found in many bryozoan colonies (Fig. 27.2).

27.3   REPRODUCTION

Most bryozoans are **hermaphroditic.** The gonads bulge into the coelom, with one or two ovaries at the anterior end and testes at the posterior end. There are no gonoducts; **gametes** are shed into the coelom and exit through a pore in one or two of the tentacles or through a special opening in the region of the lophophore. In many bryozoans the lat-

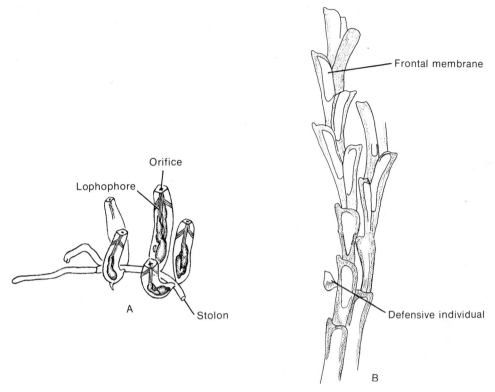

Figure 27.5 *A, Bowerbankia,* a marine bryozoan in which the feeding individuals arise from a creeping stolon, formed by highly modified members of the colony. *B, Bugula,* a genus of common marine bryozoans which form erect branching colonies. (*A* after Mature; *B* after Hyman.)

ter opening is mounted on an elongate process called the **intertentacular organ,** and it is here that the eggs are released (Fig. 27.4).

In the very few species in which fertilization has been observed, sperm swept in with the feeding current fertilized the eggs as they were released, or else the sperm entered and fertilized the eggs within the intertentacular organ.

Most bryozoans brood their few eggs during the early stages of development, commonly in special brooding chambers. For example, *Bugula* and other related forms brood their eggs within a large hood-like chamber that develops at the anterior end of reproductive individuals (Fig. 27.2).

The developing eggs are released from the brood chambers as larvae which are somewhat like the trochophore of annelids and mollusks. Following a planktonic existence of varying length, the larva settles to the bottom to become the parent member of a new colony. The body wall evaginates and separates as a new individual, the particular way in which subsequent budding occurs accounting for the growth form of the colony.

During their life span the individuals of a colony undergo phases of degeneration and regeneration, and a large colony typically exhibits zones of individuals in various stages of development. At the outer perimeter of an encrusting colony, for example, budding and development of new individuals occur. Further inward is a zone of fully developed individuals, which are feeding and reproducing. Still further inward is a zone of degenerating members. The gut and lophophore are phagocytized, and the remains of these organs are lodged in the coelom as dark balls. A new gut and lophophore form from the persistent body wall in regenerating zooids, which form still another zone to the inside of the degenerating one.

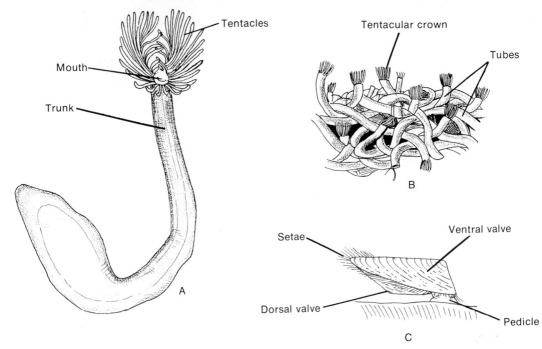

Figure 27.6 *A,* The phoronid, *Phoronis architecta,* removed from tube. *B,* Part of a cluster of *Phoronis hippocrepia.* *C,* Lateral view of the brachiopod *Discinisca* attached to substratum. (*A* after Wilson from Hyman; *B* after Shipley from Hyman; *C* after Morse from Hyman.)

27.4 OTHER LOPHOPHORATES

Bryozoans are not the only animals having a lophophore. Two other small phyla, the **Phoronida** and the **Brachiopoda,** are lophophorates. The phoronids are tubiculous, worm-like, marine animals (Fig. 27.6*A* and *B*). The brachiopods, although represented today by only a small number of species, were very abundant in the past, and their rich fossil record dates back to the Cambrian. Brachiopods superficially resemble bivalve mollusks in having the body enclosed within two calcareous valves (Fig. 27.6*C*). However, in brachiopods the valves are dorsoventrally oriented and the ventral valve is larger than the dorsal one. A fleshy stalk, which emerges from a hole in the back of the ventral shell, attaches these animals upside down to the bottom.

ANNOTATED REFERENCES

Detailed accounts of bryozoans, phoronids, and brachiopods can be found in many of the references listed at the end of Chapter 18. The works listed below deal exclusively with the biology of two of these lophophorate phyla.

Rudwick, M. J. S.: Living and Fossil Brachiopods. London, Hutchinson University Library, 1970.
Ryland, J. S.: Bryozoans. London, Hutchinson University Library, 1970.

28. Echinoderms

Echinoderms, hemichordates and chordates constitute the **deuterostomes,** the second great evolutionary line of the Animal Kingdom. In contrast to the protostomes, the mouth arises as a new opening located opposite the blastopore, which forms the anus. Cleavage is radial, not spiral (Fig. 6.13), and the fate of the blastomeres is fixed much later in development than in protostomes. Primitively, the mesoderm and coelom arise as paired outpocketings of the embryonic gut; the coelom is an **enterocoel** (Fig. 28.1), in contrast to the schizocoelous mode of coelom formation in many protostomes (p. 502).

The more than 5000 species of the phylum Echinodermata represent the largest group of invertebrate deuterostomes, for most of the other deuterostomes are vertebrates. Echinoderms are entirely marine and include the familiar starfish, which are per-

haps the invertebrate symbols of the sea, as well as brittle stars, sea urchins, sand dollars, sea cucumbers, and sea lilies.

However, the most striking feature of echinoderms is their **pentamerous,** or five part, radial symmetry. The mouth is in the center of one side of the body and forms one end of the oral-aboral axis, around which the five similar sections are arranged (Fig. 28.2). In contrast to coelenterates, the radial symmetry of echinoderms is clearly secondary, i.e., the echinoderms evolved from bilateral ancestors. Such an ancestry is indicated by the echinoderm larva, which is distinctly bilateral; radial symmetry develops in the course of a complex larval metamorphosis (Fig. 28.9). Moreover, there still remain irregular features within the adult radial symmetry.

The body wall of echinoderms contains a skeleton of small calcareous pieces, or

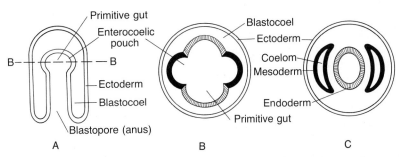

Figure 28.1   Mesoderm and coelom formation in deuterostomes. *A,* Lateral view of a gastrula. *B,* Section of gastrula (*A*) taken at level B–B and showing developing enterocoelic pouches, outpocketings of the primitive gut. *C,* Cross section of a later gastrula in which pouches have separated as mesodermal vesicles each containing a coelomic cavity.

Figure 28.2 Aboral view of four starfish, showing variation in body form. *A, Linkia*, a common genus with long arms; *B, Culcita*, a genus with such short arms that body is pentagonal; *C*, the sun star, *Crossaster; D*, the coral-eating *Acanthaster*. (*A, B*, and *D*, courtesy of the U.S. National Museum; *C*, courtesy of D. P. Wilson.)

ossicles. Associated with the skeleton are **calcareous surface spines,** from which the name Echinodermata, or "spiny skin," is derived.

A third characteristic distinguishing echinoderms from other animals is a unique system of tube-like body wall appendages, the **tube feet,** and a system of canals derived from the coelom. Together they make up the so-called **water vascular system.**

Echinoderms are almost entirely bottom dwellers, and hard substrata of rock, shell, and coral were the habitats of both ancestral and many contemporary echinoderms. But within each class of echinoderms are species which invaded soft bottoms and be-

came adapted for life in sand. The adaptations for soft substrata are quite diverse and reflect the different prior adaptations of their hard-bottom ancestors.

28.1   CLASS ASTEROIDEA

The class **Asteroidea** contains the most familiar echinoderms, the starfish, or sea stars. In the members of this group the body is drawn out into arms that are not sharply set off from a central disc. There typically are five arms, which vary in length from species to species, but the sun stars have many arms (Fig. 28.2). The body surface may ap-

Figure 28.3 Oral view of *Coscinasterias*, a starfish with six to eight arms. The many tube feet with suckers can be seen within the ambulacral grooves.

pear smooth or granular, or may bear conspicuous spines, as in the coral-eating crown-of-thorns, *Acanthaster.* Starfish are usually brightly colored, with hues of red, orange and blue predominating. Most are 12 to 24 cm. in diameter.

The mouth is located in the center of the oral surface, which is directed downward. An **ambulacral groove** radiates from the mouth to the tip of each arm, and contains the tube feet, the locomotor structures of sea stars (Fig. 28.3).

Body Wall. The body surface of asteroids is covered by a ciliated epithelium. Beneath the epithelium is a thick layer of connective tissue, which secretes the skeleton of small calcareous ossicles. The ossicles of all echinoderms are somewhat similar to vertebrate bone, for they are perforated by irregular canals filled by cells. In starfish the ossicles form a movable skeletal lattice (Fig. 28.4).

A muscle layer below the dermis enables the arms to bend. Ciliated peritoneum forms the innermost layer of the body wall and lines the very large coelom that fills the interior of the arms and disc (Fig. 28.6).

A number of special structures are associated with the body wall. All asteroids bear calcareous spines which may be projections

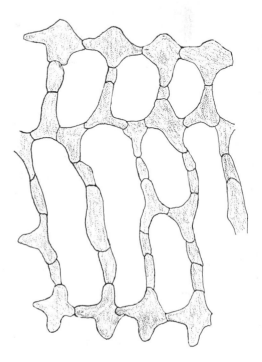

Figure 28.4 Lattice-like arrangement of the skeletal ossicles (enlarged) of a starfish. Surface view after overlying tissue had been removed. (After Fisher from Hyman.)

of the deeper dermal skeleton, or special ossicles resting on top of the deeper skeleton (Figs. 28.2 and 28.5). The spines vary greatly in size and prominence. Even those species which seem to be smooth and spineless in fact have the body surface covered with many tiny, closely fitting spines.

Located between the spines are small finger-like projections of the body wall, called **papulae,** which function in gas exchange and excretion (Figs. 28.5 and 28.6). The thin wall of each papula is composed of an inner peritoneum and outer epidermis, and coelomic fluid circulates within the interior.

Other structures associated with the body wall of many starfish are minute jaw-like appendages called **pedicellariae.** A pedicellaria contains two ossicles which form the jaws; these are embedded in epithelium

and provided with muscles for opening and closing the gape. The pedicellariae are believed to function in the killing of small organisms which might settle on the bodies of starfish. Such organisms, as well as sediment, are swept clear by the ciliated surface epithelium.

The Water Vascular System. The water vascular system, unique to echinoderms, is composed of tubular outpocketings of the body wall—the tube feet, or podia—and an internal system of canals derived from the coelom. The system opens to the exterior through the button-like aboral **madreporite,** which is perforated by tiny canals. The canals from the madreporite converge into a vertical **stone canal** that extends orally to a **ring canal** embedded in the ossicles around the mouth (Fig. 28.7). From the ring canal

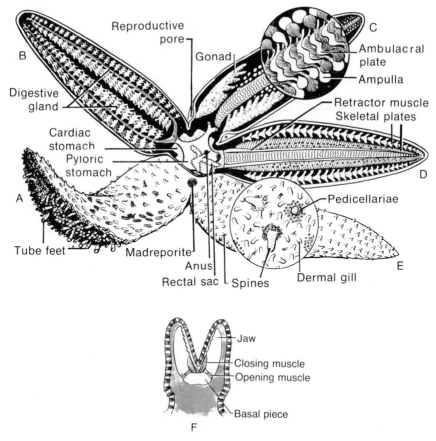

Figure 28.5 *Asterias* viewed from above with the arms in various stages of dissection. *A*, Arm turned to show lower side. *B*, Upper body wall removed. *C*, Upper body wall and digestive glands removed, with a magnified detail of the ampullae and ambulacral plates. *D*, All internal organs removed except the retractor muscles, showing the inner surface of the lower body wall. *E*, Upper surface, with a magnified detail showing surface features. *F*, Structure of a pedicellaria of the starfish *Asterias.*

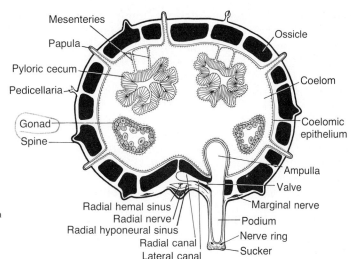

Figure 28.6 Diagrammatic cross section through the arm of a sea star.

a **radial canal** extends into each arm, passing between the ossicles at the top of the ambulacral groove. At frequent intervals, **lateral canals** leave the radial canals, those on one side alternating with those on the other (Figs. 28.6 and 28.7). Each lateral canal terminates in an aborally directed bulb-like **ampulla,** which bulges into the coelom, and an orally directed tube foot, which projects into the ambulacral groove (Fig. 28.6).

Fluid pressure plays an important role in the functioning of the tube feet. Muscular contraction of the ampulla forces fluid into the podium and at the same time closes a valve in the lateral canal, preventing backflow. Hydraulic pressure extends the podium, and the sucker at its tip is brought into contact with the substratum. Following adhesion, longitudinal muscles of the podium contract, shortening the tube foot, forcing water back into the ampulla, and pulling the body forward. When a podium is extended, it is also swung forward. Thus each podium performs a little step and the combined activity of all the podia enables a starfish to grip objects tenaciously and to crawl about. The podia do not move synchronously, but they are coordinated to the extent that they all step in the same direction. One arm acts as the leading arm and exerts a temporary dominance over the other arms.

Nervous System, Gas Exchange, and Excretion. The nervous system can be regarded as primitive in that it is intimately

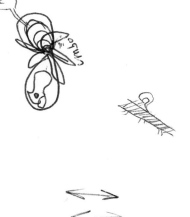

Figure 28.7 Diagram of the asteroid water vascular system.

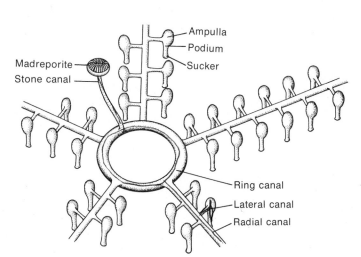

associated with the epidermal layer. The epidermis surrounding the mouth contains a pentagonal **nerve ring,** which at each angle supplies a **radial nerve** to an arm (Fig. 28.6). The radial nerve is located in the epidermis in the middle of the ambulacral groove. Fibers in the nerve ring and the radial nerve make connection with neurons of a general epidermal nerve plexus.

An intact nervous system is necessary for the coordination of the tube feet. If a radial nerve is cut, the tube feet distal to the cut will continue to move but may not step in synchrony with those in other arms. If the nerve ring is cut, the podia in all of the arms become uncoordinated. It is believed that temporary arm dominance during movement resides at the junction point of the radial nerve with the nerve ring.

An **eye spot** at the tip of each arm, composed of a cluster of photoreceptor and pigment cells, constitutes the only sense organ, but individual receptor cells are present in the general body epidermis and are especially concentrated in the epidermis of the podia and the margins of the ambulacral groove.

Gas exchange and excretion in starfish occur through the surface of the papulae and podia, and internal transport is provided by the coelomic fluid. In almost all echinoderms, gas exchange organs are infoldings or outpocketings of the body wall and coelomic fluid is the medium for internal transport. These gas exchange organs must have evolved independently in different echinoderm groups, for there is little consistency in their structure or location.

Soft-Bottom Starfish.   Podia with suckers represent an adaptation for life on hard substrata such as rock, shell, and coral — habitats in which sea stars often abound. Many sea stars are adapted for living on sandy bottoms and can even burrow to some degree. Soft-bottom starfish have doubled ampullae which provide greater pressure for thrusting the suckerless, pointed tube feet into the sand.

The papulae of soft-bottom sea stars are protected from the surrounding sand and sediment by special table-like spines, which create a protective cover under which the papulae are lodged and a ventilating current flows (Fig. 28.8). It is these covering spines which give a smooth appearance to the surface of many soft-bottom sea stars.

Figure 28.8   Diagrammatic section through the surface structures of a soft-bottom asteroid. The large table-like spines bear smaller spines above and protect the papulae below.

Nutrition.   The digestive system of sea stars reflects the radial plan of body structure. The mouth opens into a large thick-walled **cardiac stomach** * which fills most of the central disc. The cardiac stomach opens in turn into a smaller aboral **pyloric stomach** (Fig. 28.5). A pair of digestive glands, located in each arm, discharges separately or by a common duct into the periphery of the pyloric stomach, giving it a star-shaped appearance. A short intestine extends from the top of the pyloric stomach to the inconspicuous anus at the center of the aboral surface. Associated with the intestine are rectal ceca, outpocketings of unknown function (Fig. 28.5).

Most sea stars are carnivores and scavengers. Crustaceans, mollusks, and other echinoderms are common prey. Species with short arms swallow their prey whole; those with longer arms wrap them about the body of the prey and by increased pressure of the coelomic fluid evert the cardiac stomach over the victim.

Asterias and other species of the same family feed largely on bivalve mollusks, penetrating their prey in a remarkable way. The starfish humps over the clam, the mouth directed over some part of the gape. The pull exerted by the arms produces a very slight opening between the valves, and through this opening the stomach of the starfish is slithered. These animals can pass the stomach through a gape of no more than 0.1 mm. It is this ability rather than the ability to pull which enables a starfish to prey upon clams. Many bivalves cannot close the valves tightly enough to prevent entrance of the starfish's stomach. The Pa-

---

* The unfortunate use of the names cardiac and pyloric stomach in asteroids, and in decapod crustaceans, reflects the attempts by early zoologists to apply the established terminology of human anatomy to the lesser known invertebrates.

cific sea star *Pisaster* can penetrate the gape of bivalves which have been wired closed!

Asteroids can be a serious pest of oyster beds and are sometimes removed by dragging a large mop across the bottom. The pedicellaria attach to the threads, and the starfish can be brought to the surface and destroyed.

Enzymes produced by the digestive glands are passed to the pyloric and cardiac stomachs. In those species which evert the stomach the enzymes flow out onto the everted surface and initiate digestion outside the body. Thus, when a starfish attacks a bivalve, the digestion of the clam's adductor muscle causes the valves to gape widely. The stomach is later retracted, bringing with it the partially digested prey. The digestive glands appear to be the principal site of absorption.

Reproduction and Development. Like most other echinoderms, starfish have separate sexes; there are usually two gonads to an arm, with a simple gonoduct from each leading to an inconspicuous gonopore at the base of the arm (Fig. 28.5). Correlated with the simple nature of the reproductive system, there is no copulation, and the eggs are shed freely into the sea water where fertilization takes place. Development occurs in the plankton. The eggs contain little yolk material, and radial equal cleavage results in a hollow blastula. Gastrulation occurs by the invagination of cells of the vegetal hemisphere, resulting in a narrow finger-like archenteron. From the tip of the archenteron arise two lateral outpocketings, the enterocoelic pouches, which eventually separate as coelomic vesicles (Fig. 28.1). These primary vesicles later divide to form three pairs of vesicles. Their walls give rise to all of the mesodermal derivatives of the body. The remainder of the archenteron contributes to the gut proper and connects anteriorly with a stomodeal invagination from the ectoderm.

The appearance of ciliated bands on the body surface indicates that development has reached a larval stage, called a **bipinnaria** (Fig. 28.9). The bipinnaria larvae and the larvae of all other echinoderms are distinctly bilateral and are believed to represent recapitulations of the symmetry of ancestral echinoderms. The course of further development is marked by the appearance of **larval arms,** extensions of the body over which the ciliated locomotor bands run. The larval arms bear no relationship to the arms of the adult.

With the formation of an attachment device—three small anterior arms and a sucker—the bipinnaria larva becomes a **brachiolaria** larva. On encountering a suitable substratum, the brachiolaria anchors by the anterior adhesive structures, and a radical and complex metamorphosis ensues. The larval arms and much of the larval gut degenerate, and the adult structures, including the arms, arise from the posterior part of the larval body. After its metamorphosis the young starfish becomes detached from the substratum and takes up the adult benthic existence.

Some arctic and antarctic starfish brood

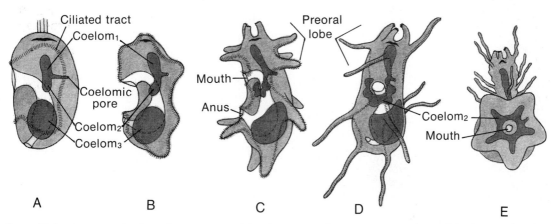

Figure 28.9 Diagrammatic lateral views of larval development and metamorphosis of a starfish, showing development of the coelom and water vascular system. *A* and *B*, Early bipinnaria larva. *C*, Brachiolaria larva. *D*, Attached metamorphosing larva. *E*, Young starfish developing from posterior part of old larva. (Adapted from various sources.)

their eggs beneath or on various parts of the body. There are only a few large yolky eggs and development is direct. Such brooding behavior is found in many cold-water species of other echinoderm classes. Why brooding is largely limited to cold-water species is not clearly understood.

The regenerative ability of starfish is well known. Any fragment of the body which contains a portion of the central disc is capable of regeneration. But the process is slow and may take as long as a year. Only in some starfish is regeneration coupled with asexual reproduction, when the central disc is cleaved along certain predetermined lines and each part grows into a new starfish.

28.2 CLASS OPHIUROIDEA

Closely related to the Asteroidea is the largest class of echinoderms, the **Ophiuroidea,** which contains the basket stars and brittle stars or serpent stars. The ophiuroid body, like that of the asteroids, is composed of arms and a central disc, but the arms are long, slender, and sharply set off from the disc (Fig. 28.10). The arms are highly mobile and easily broken, characteristics which led to the names serpent star and brittle star. In basket stars the arms are branched.

Although the central disc of basket stars may reach a diameter of 12 cm., most brittle stars are small and have a central disc no larger than a dime. They are commonly mottled or banded but are usually not as brilliantly colored as starfish.

The arms of brittle stars are largely composed of a string of large ossicles, called **vertebrae** because of their superficial similarity to the bones in the vertebrate backbone (Fig. 28.11). The articulation of the arm ossicles and their musculature give the arms great mobility: brittle stars and basket stars move about by flexing their arms. The body is propelled forward by the backward movements of whichever arms are located laterally. Spines along the sides of the arms increase traction. An arm can be broken at any point if seized by a predator. This ability to undergo self-amputation, or **autotomy,** is an escape adaptation also found in some arthropods, such as crabs.

Ophiuroids are especially common in rock and coral habitats with rough uneven surfaces studded with various objects, including the bodies of sessile organisms. Brittle stars and basket stars can climb, and some species clamber about sponges and branched gorgonian corals.

Some echinologists consider the ophiuroids the most successful of modern echinoderms and attribute their success in part to their motility, small size, and ability to utilize the protective cover of crevices, holes, and other natural retreats.

The **madreporite** of ophiuroids is in an ossicle located on the oral rather than the aboral surface. Correlated with the evolution of a different mode of locomotion and the reduction of the arm coelom, ampullae are absent, and the suckerless podia are extended by water pressure generated by the contraction of a section of the podial canal (Fig. 28.11).

Ophiuroids employ a variety of feeding methods; in fact, several methods may be used by a single species. They may be scavengers, using the looping motion of the arms and the five jaw-like ossicles which frame the mouth to ingest the bodies of dead animals (Fig. 28.11). They may feed on detritus raked into the mouth with the arms. Suspension feeding is also common. The arms are often lifted up into the water current and suspended plankton and detritus are trapped on the arm surfaces or on strands of mucus draped between spines. The collected material is pushed as balls by the podia along the length of the arms to the mouth.

Ophiuroids have no papulae; their gas exchange organs are pouches, called **bursae,** which are infoldings of the body wall of the oral disc to either side of the base of each arm (Fig. 28.11). By means of cilia or by pressure changes within the disc itself, a ventilating current passes into and out of the slit opening into a bursa. Coelomic fluid is again the medium for internal transport.

Most brittle stars have separate sexes but it is within this class that some hermaphroditic echinoderms are encountered. There are from one to many gonads attached to the inner wall of each of the 10 bursal sacs. Gametes reach the exterior by rupturing into the sac and passing out with the ventilating current. Brooding species use the bursae as brood chambers.

Fertilization occurs in the sea water and

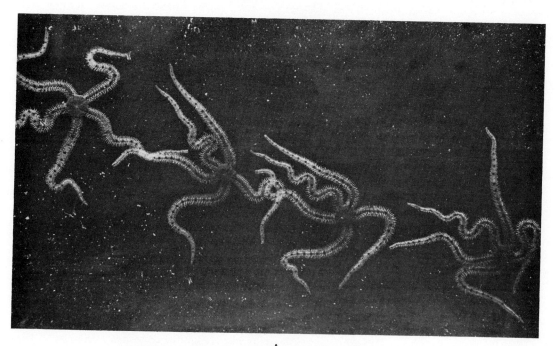

A

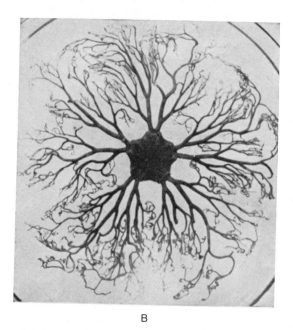

B

Figure 28.10 *A*, A Caribbean brittle star, shown in repetitive flash photographs, pulls itself along with its two anterior arms and shoves with the other three. Ophiuroids are far more agile and flexible than starfish. *B*, The basket star, *Gorgonocephalus*, showing the branched arms. (*A* by Fritz Goro, courtesy of LIFE Magazine, © 1955, Time, Inc.; *B*, courtesy of the American Museum of Natural History.)

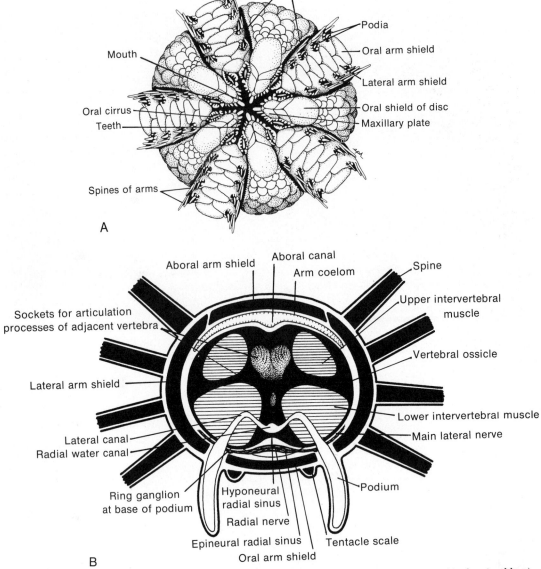

Figure 28.11   *A*, Oral view of disc of a brittle star *B*, Cross section of a brittle star arm. (*A* after Streklov.)

a bilateral larva develops as in asteroids. However, the number and position of the arms is different, and the larva can be recognized as an ophiuroid (Fig. 28.12). There is no attachment stage and the entire course of development occurs in the plankton. A tiny brittle star settles to the bottom on completion of metamorphosis.

28.3   CLASS ECHINOIDEA

The class **Echinoidea** contains the sea urchins, heart urchins, sand dollars and sea

biscuits. The members of this class have a body which is not drawn out into arms but is spherical or discoidal (Fig. 28.13). The skeletal ossicles are flattened plates fused together to form a rigid internal shell, or **test,** and the body surface is covered with movable **spines** mounted upon tubercles on the test (Fig. 28.15).

The class contains two adaptive groups, the sea urchins, most of which are adapted for life on hard bottoms, and the sand dollars and heart urchins, which are adapted for burrowing in sand.

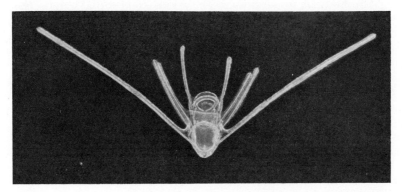

Figure 28.12 Pluteus larva of the brittle star, *Ophiothrix fragilis*. (Courtesy of Douglas P. Wilson.)

Figure 28.13 Echinoids. *A*, Side view of the common Atlantic sea urchin, *Arbacia punctulata*, showing long spines and podia. *B*, *Heterocentrotus*, a sea urchin with large spines and many small short spines. *C*, A heart urchin viewed from the anterior, or oral end. Mouth can be seen on the lower side. *D*, The Atlantic five-slotted sand dollar, *Mellita quinquiesperforata*. (By Betty M. Barnes.)

Sea Urchins. The body of sea urchins is spherical with the oral pole containing the mouth directed downward, and the aboral pole containing the anus directed upward. Although there are no arms, five ambulacral areas bearing the tube feet radiate out and upward over the test to the aboral pole (Fig. 28.14). Thus the body is divided around the equator into alternating ambulacral and interambulacral meridians.

The long movable spines are covered at least at the base by the general surface epithelium. Pedicellariae are located between the spines but, unlike those of asteroids, the echinoid pedicellariae have three jaws and are mounted at the end of a long stalk (Fig. 28.15). Some species have pedicellariae with poison glands.

The water vascular system differs from that of asteroids in a few respects. One of the aboral plates is modified as a madre-porite. The radial canals run from the ring canal encircling the mouth at the oral pole up beneath the inside of the test to the aboral pole. Each ampulla connects with the podium by a pair of little canals, which actually perforate the ossicle (Fig. 28.15). Thus an ambulacral area can be recognized on a dried test by the two lines of paired holes.

Sea urchins use the podia to move about in the same manner as asteroids. The podia are long and can reach beyond the spines. Locomotion may also be aided by the pushing movement of the spines. Some sea urchins tend to settle themselves in rock depressions and some actually excavate burrows in rock. Some kinds of sea urchins cover themselves with shells, pebbles, and other debris.

When a sea urchin is alarmed, the spines become erect and immobile. Bending and

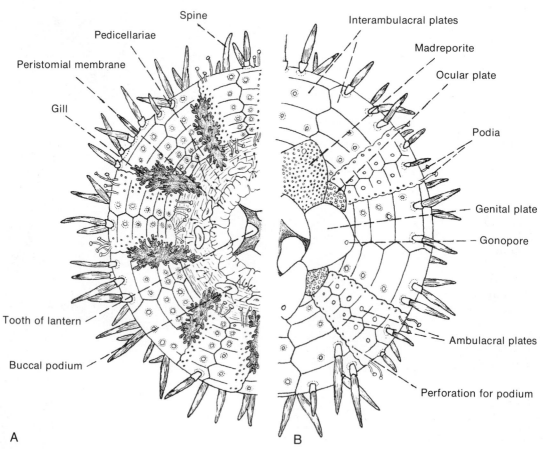

Figure 28.14 The common Atlantic coast sea urchin, *Arbacia punctulata*. A, Oral view; B, aboral view. (Modified from Petrunkevitch.)

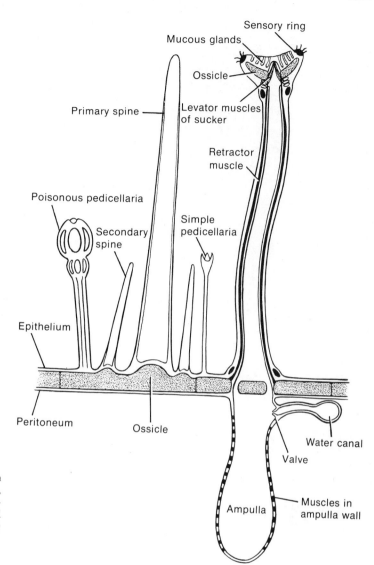

Figure 28.15 Diagrammatic section through the body wall of a sea urchin, showing one ambulacral and one interambulacral ossicle and associated structures. (After Nichols, in part.)

erection are operated by two different sets of muscle fibers and motor innervations around the socket of the spine.

Sea urchins are scavengers on plant and animal material, which is ingested with a complex movable jaw apparatus (known as **Aristotle's lantern**) containing five projecting teeth (Fig. 28.16). The gut is tubular and loops about inside the test.

Five pairs of gills, which are highly dissected outpocketings of the body wall located to either side of the ambulacral areas at the oral pole, provide for gas exchange (Fig. 28.14). Coelomic fluid is pumped into and out of the gills.

Sand Dollars and Heart Urchins. Sand dol-

lars and heart urchins, the soft-bottom echinoids, burrow in sand. These animals, when moving, always keep the same meridian forward and thus have a definitive anterior end. Shifts in the position of the oral center or anus or both have led to a degree of bilateral symmetry, so that sand dollars and heart urchins are sometimes called **irregular echinoids**. In sand dollars, for example, the oral-aboral axis is so depressed that the body is flattened (Fig. 28.13). The mouth is still in the center of the oral surface, but the anus has shifted out of the aboral center and is located eccentrically. Many sand dollars have the test perforated by slots, and in the Atlantic Coast five-

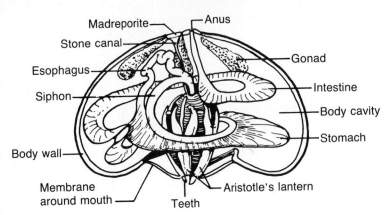

Madreporite — Anus
Stone canal —
Esophagus —
Siphon —
Body wall —
Membrane around mouth —
Teeth
Aristotle's lantern
— Gonad
— Intestine
— Body cavity
— Stomach

Figure 28.16 A sea urchin, *Arbacia punctulata*, with one side of the body wall removed and only some of the structures shown. The teeth protrude from Aristotle's lantern, of which only the outer structures are indicated. The digestive tract circles twice around the body, once in each direction. A second tube, the siphon, by-passes the esophagus and stomach. Each of the five gonads opens above, near the anus.

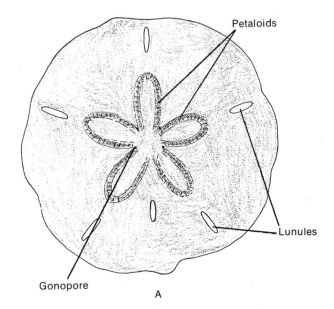

Petaloids
Lunules
Gonopore

A

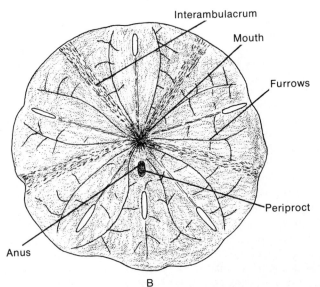

Interambulacrum
Mouth
Furrows
Periproct
Anus

B

Figure 28.17 The West Indian six-slotted sand dollar, *Mellita sexiesperforata. A*, Aboral view; *B*, oral view. (After Hyman.)

slotted sand dollar the anus is located at the inner margin of the posterior slot.

The spines of irregular echinoids are tiny and literally clothe the body; it is the movement of the spines which enables these animals to crawl slowly through the sand. Sand dollars live below the intertidal zone and burrow at a slight angle into the surface layer of sand. Heart urchins construct semipermanent burrows with a small opening to the surface (Fig. 28.18).

On the aboral surface, the podia of irregular echinoids are very wide and flattened as a modification for gas exchange. They form five petal-like areas, which are especially conspicuous in sand dollars. The dense clothing of spines extends just beyond the podial gills and prevents the gills from being smothered by sand.

Soft-bottom echinoids are **indirect deposit feeders.** In sand dollars, fine particulate organic matter, but not sand grains, can drop down between the dense clothing of spines. The particles are then driven by cilia to the oral surface, where they pass into a branching system of food grooves (Fig. 28.17). The food grooves converge into five principal grooves corresponding to the ambulacral areas. The food grooves are lined by podia which push the collected food masses to the mouth.

Four or five gonads are suspended radially from the inside of the test; each opens by a short duct through a conspicuous gono-

pore located in one of five large aboral genital plates (Fig. 28.14). Fertilization is external, and the larva looks much like that of ophiuroids. Metamorphosis occurs in the plankton, and a tiny sea urchin or sand dollar settles to the bottom.

28.4 CLASS HOLOTHUROIDEA

The **Holothuroids,** or sea cucumbers, are similar to sea urchins in lacking arms, but in holothuroids the oral-aboral axis is greatly lengthened, so that the animal has a worm-like or cucumber-like shape and lies on its side (Fig. 28.19). Unlike other echinoderms the skeleton is reduced to microscopic ossicles, and the body wall has a leathery texture. In many sea cucumbers the surface bears numerous tubercles or wart-like projections. Sea cucumbers are rather drab animals, with greenish and brownish colors predominating. Most are from six to 30 cm. in length.

Many sea cucumbers live on hard substrata, lodging themselves beneath and between stones, or within coral crevices. Some attach to sea weeds or hydroids. These hard-bottom sea cucumbers move by means of tube feet; in some species, three ambulacra are kept against the substratum as a sole and the two upper ambulacra have reduced podia (Fig. 28.19).

Soft-bottom sea cucumbers bury themselves in sand. Many construct temporary

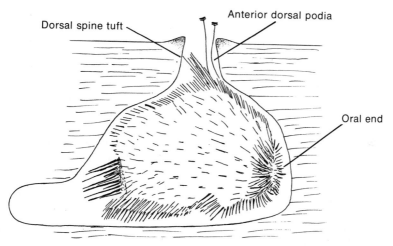

Figure 28.18 Heart urchin, *Echinocardium flavescens,* in its sand burrow (lateral view). (Modified after Gandolfi-Hornyold.)

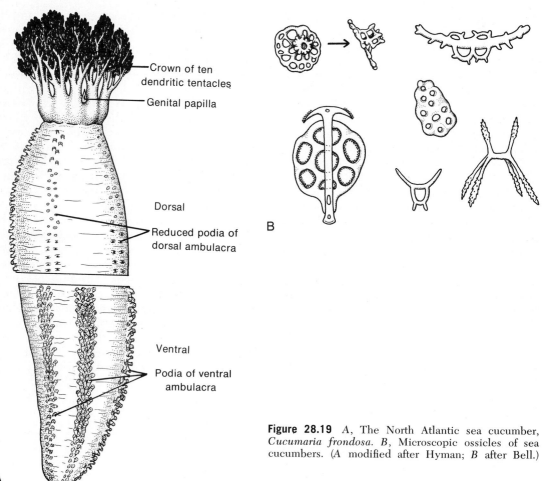

Crown of ten
dendritic tentacles

Genital papilla

Dorsal

Reduced podia of
dorsal ambulacra

B

Ventral

Podia of ventral
ambulacra

Figure 28.19 *A*, The North Atlantic sea cucumber, *Cucumaria frondosa*. *B*, Microscopic ossicles of sea cucumbers. (*A* modified after Hyman; *B* after Bell.)

A

Anus

Figure 28.20 A burrowing wormlike synaptid sea cucumber.

burrows, from which the oral end of the animal projects to the surface. The very worm-like synaptid sea cucumbers burrow beneath the surface by peristaltic contractions (Fig. 28.20). The podia have completely disappeared.

At the oral end of the body a circle of **tentacles** representing modified podia surrounds the mouth (Figs. 28.19 and 28.21). The tentacles are outstretched and collect plankton or deposit material from the surrounding sea bottom. They are then retracted, and stuffed one at a time into the mouth. The worm-like synaptid sea cucumbers are direct feeders; a column of sand is often visible through the transparent body wall.

The gut of sea cucumbers is tubular and terminates before the anus in a muscular **cloaca** which is involved in gas exchange. The gas exchange organs are unusual, tubular branching structures, called **respiratory trees,** that arise as evaginations of the cloacal wall and extend up into the coelom (Fig. 28.21). The pumping action of the cloaca moves a ventilating current of sea water into and out of the respiratory trees.

In the Indo-Pacific, a little commensal pearlfish (*Carapus*) uses the base of the respiratory tree as a home. To enter its home the fish nudges the anus of the sea cucumber with its snout and then backs in, tail first, sometimes twisting against the pressure of the closing anus. The fish has reduced scales and no pelvic fins, and the anus has shifted far forward.

Some sea cucumbers possess a cluster of tubular evaginations from the base of the respiratory trees. These tubules, which look like thin spaghetti, are called **tubules of Cuvier** and can be shot out of the anus. They elongate in the process and are very

Figure 28.21 The sea cucumber, *Thyone briareus,* cut open along one side. The digestive tract has been moved to one side to show the respiratory trees, retractor muscles of the anterior end, and the internal surface of the body wall with its five ambulacra. In holothurians the madreporite lies in the body cavity, so that the water vascular system is not filled with sea water but with coelomic fluid.

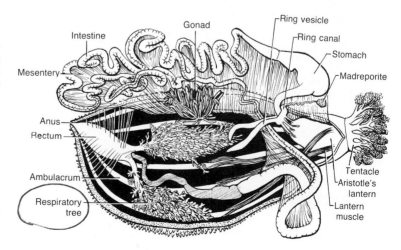

Intestine · Gonad · Ring vesicle · Ring canal · Stomach · Madreporite · Mesentery · Anus · Rectum · Ambulacrum · Respiratory tree · Tentacle · Aristotle's lantern · Lantern muscle

adhesive. An intruder or predator which disturbs a sea cucumber enough to evoke the discharge of the tubules can become enmeshed in a death trap of adhesive threads.

Another strange process exhibited by many sea cucumbers is the spewing out of internal organs, a process called **evisceration.** The respiratory trees and gut may be discharged through a ruptured cloaca or through the mouth. The loss is followed by regeneration. Evisceration has been most commonly observed in crowded laboratory conditions; its significance and the extent to which it is a normal event are still poorly understood.

There is only one gonad in sea cucumbers, and the gonopore opens between two of the tentacles (Fig. 28.19). The bilateral larva possesses ciliated locomotor bands but never develops the long larval arms characteristic of other echinoderm larvae (Fig. 28.22). Transformation to a young sea cucumber occurs before settling.

28.5   CLASS CRINOIDEA

The echinoderm classes we have examined thus far have the mouth directed downward and the oral surface placed

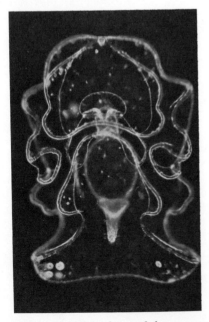

Figure 28.22   Auricularia larva of the sea cucumber *Labidoplax.* (Courtesy of Douglas P. Wilson.)

against the substratum. However, in one living class of echinoderms, the **Crinoidea,** the reverse is true: the oral surface is directed upward. Moreover, many members of this class are attached to the substratum by a **stalk.** The attached crinoids are believed to be the most primitive of the living echinoderms and probably illustrate the ancestral mode of existence of the phylum and the original function of the water vascular system.

Crinoids consist of two groups, the sessile sea lilies, in which the aboral surface is connected to the substratum by a stalk (Fig. 28.23), and the free-moving feather stars, in which a stalk is absent. Sea lilies are usually less than 70 cm. in length and most occur in relatively deep water. The stalk, which is composed of ossicles and can bend, is attached to the aboral surface of the body proper, or **crown.** The pentamerous crown bears arms which fork repeatedly, so that there may be many multiples of five, depending upon the species. All along the length of the arms there are side branches, called **pinnules.** A ciliated ambulacral groove with flanking suckerless podia runs the length of the arms and up onto the pinnules (Fig. 28.24). The grooves from all of the arms converge to the mouth in the center of the upward directed oral surface.

The arms are composed mostly of ossicles but the articulation and musculature permit considerable movement, especially vertical movement.

Feather stars are similar to sea lilies, from which they clearly have evolved. They are stalked and attached in the last stages of development. Then the crown breaks free and swims away as a tiny feather star. Although the animal is unattached, its oral surface is still directed upward (see Figure 28.25). They perch for long periods of time by means of an aboral ring of claw-like projections, called **cirri,** on rocks, coral, or even on soft substrata. Feather stars swim intermittently. The arms are raised and lowered in cycles of five, and the swimming motion is very graceful.

The largest number of living crinoid species are feather stars, and the greatest number of species is found in the Pacific. There are many shallow-water species in the Indo-Pacific, a few along the eastern Atlantic and Mediterranean, but none along the coast of the United States.

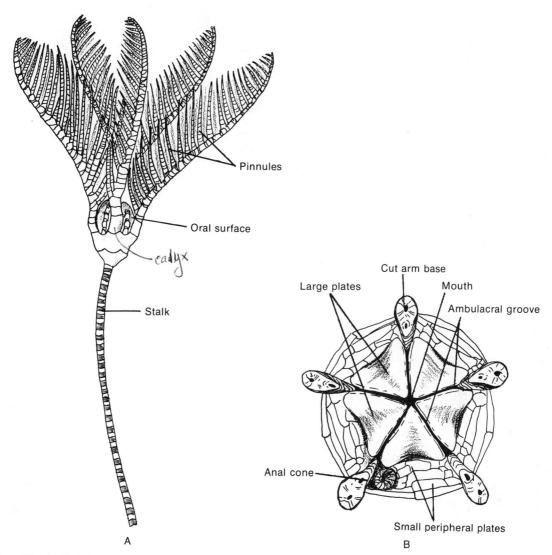

Figure 28.23 *A, Ptilocrinus pinnatus,* a stalked crinoid (or sea lily) with five arms. *B,* Oral surface of a stalked crinoid. (*A* after Clark from Hyman; *B* after Carpenter from Hyman.)

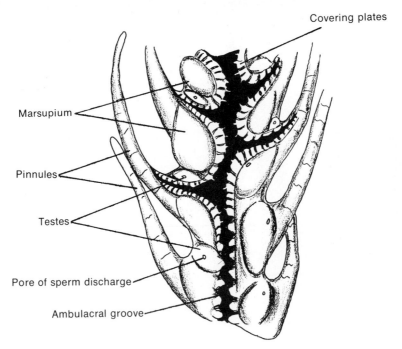

Figure 28.24   Section of arm and pinnule of a crinoid showing ambulacral groove. (After Hyman.)

Crinoids are suspension feeders and this mode of feeding may have been the original function of the water vascular system. On contact, rapid whip-like movements of the podia toss suspended food particles into the ambulacral grooves, where they are entrapped in mucus and conveyed within the ciliated groove down to the mouth. The significance of the branching arms and the pinnules now becomes apparent. These are devices that greatly increase the surface area for collecting food. For example, a 24 cm. Japanese sea lily with 56 arms was estimated to have a total ambulacral groove length of about 80 meters.

The length of the ambulacral groove also reflects the differences in the biological productivity of tropical and cold-water oceans. In general, biological productivity as reflected by plankton density is greater in coastal waters than in the open ocean, and is greater in cold than in warm oceans. These differences are reflected in crinoid structure. In general, cold-water crinoids have fewer and shorter arms than do tropical species; the greater plankton content demands less surface area for suspension feeding.

The anus of crinoids is located on a short elevation to one side of the central oral surface (Fig. 28.23). Feces can thus be more readily swept away without fouling the ambulacral grooves. Podia are the gas exchange organs.

Gametes develop from coelomic epithelium in the arms and pinnules, and spawning takes place by rupture of the body wall. Brooding species retain their eggs on the arms. The larval stage has ciliated bands but no larval arms. The larva eventually settles and attaches, and metamorphosis occurs during attachment, as in asteroids. Sea lilies, of course, remain permanently attached, whereas the crown of feather stars is eventually freed.

28.6   FOSSIL ECHINODERMS

Echinoderms have one of the richest fossil records of any group of animals. The phylum makes its appearance in the Cambrian and is present in every subsequent period. Several extinct classes, as well as crinoids, flourished in great numbers during the Paleozoic. Some of these earlier echinoderms, such as the extinct blastoids and cystoids, had spherical bodies with the ambulacral grooves extending down over the body surface. Significantly, the grooves were bordered with the equivalent of crinoid pinnules (brachioles) (Fig. 28.26).

All of this seems to suggest that these early fossil echinoderms fed in the same manner as do modern crinoids, and that the original function of the water vascular system was feeding. It could scarcely have functioned for locomotion, for the wrong side was directed upward and most of these animals were attached.

From the fossil record and embryonic development we can postulate the following evolutionary history of echinoderms. The ancestors of echinoderms were some group of motile, bilateral coelomates. The ancestral stock took up an attached existence, which resulted in a shift from a bilateral to a more adaptive radial symmetry. Also correlated with an attached mode of existence was the evolution of the calcareous skeleton and of suspension feeding.

We are not certain whether the first echinoderms had arms. They may have, for the oldest known echinoderms, the eocrinoids, possessed arms. We also do not know the evolutionary origin of the water vascular system and podia. Structures seldom arise in evolution *de novo;* they arise as modifications of pre-existing structures. What then could have been the precursor of the water vascular system? Some zoologists have argued that the motile, bilateral ancestors of echinoderms possessed something like a lophophore for feeding and that such a lophophore-like structure became the water vascular system (Fig. 28.27). They suggest that the cavity of each tentacle became a radial water canal. As evidence, they point to the fact that the coelomic cavity of the lophophore is similar to the cavity

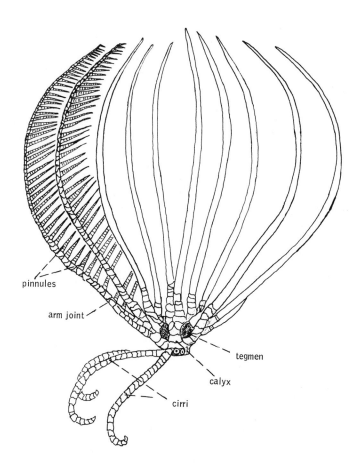

pinnules

arm joint

tegmen

calyx

cirri

Figure 28.25 A Philippine thirty-armed comatulid (or feather star), *Neometra acanthaster.* (After Clark from Hyman.)

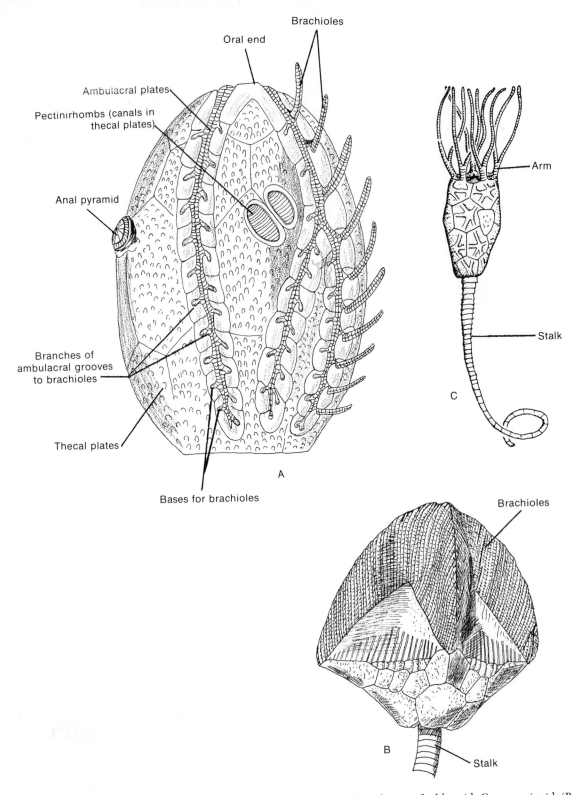

Figure 28.26   Fossil echinoderms. *A*, Lateral view of a cystoid; *B*, lateral view of a blastoid; *C*, an eocrinoid. (*B* after Jackel from Hyman; *C* after Nichols.)

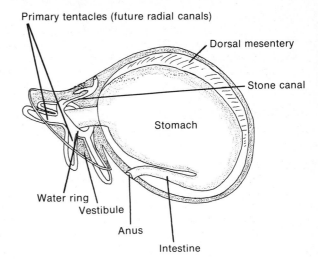

Primary tentacles (future radial canals)

Dorsal mesentery

Stone canal

Stomach

Water ring

Vestibule

Anus

Intestine

Figure 28.27 One of several projected hypothetical ancestors of echinoderms. (After Bury from Hyman.)

of the water vascular system and that the two have similar embryonic origins.

Modern free-moving echinoderms—asteroids, brittle stars, and echinoids—appear late in the fossil record, and we have no fossil forms or any other evidence that can help bridge the great gap between those echinoderms in which the oral surface is directed upward and those in which it is directed downward.

CLASSIFICATION OF THE PHYLUM ECHINODERMATA

Subphylum Echinozoa

Radially symmetrical, commonly globoid echinoderms without arms or arm-like appendages.

CLASS HELICOPLACOIDEA. Extinct Paleozoic echinoderms. Spindle-shaped, with a body wall composed of pleated plates.
CLASS EDRIOASTEROIDEA. Extinct Paleozoic echinoderms. Attached or free; oral surface directed upward.
CLASS OPHIOCISTIOIDEA. Extinct Paleozoic echinoderms. Helmet-shaped test; podia covered with scales.
CLASS HOLOTHUROIDEA. Sea cucumbers. Mouth and anus at opposite ends of cucumber-shaped body. Oral end bearing tentacles derived from podia. Ossicles microscopic.
CLASS ECHINOIDEA. Sea urchins and sand dollars. Body globose or greatly flattened dorsoventrally. Ossicles fused to form a rigid test.

Subphylum Homalozoa

Extinct irregular Paleozoic echinoderms.

CLASS CARPOIDEA. Stalked body bent over so that one side was directed toward substratum.

Subphylum Crinozoa

Radially symmetrical globoid echinoderms with arms or brachioles. Oral surface directed upward.

CLASS CYSTOIDEA. Extinct Paleozoic echinoderms. Body attached directly or by stalk to substratum. Three or five ambulacra extended down over body surface. Ambulacra commonly bordered by brachioles. Body wall perforated by pores, which may have constituted gas exchange system.

CLASS BLASTOIDEA. Extinct Paleozoic echinoderms. Body attached directly or by stalk to substratum. Closely placed brachioles bordered the five ambulacra, which were typically elevated on a ridge, giving the body a pentagonal shape.

CLASS CRINOIDEA. Sea lilies and feather stars. Living and fossil echinoderms dating from the early Paleozoic. Body attached by a stalk (sea lilies) or free (feather stars). Ambulacra located on arms, which are typically branched.

Subphylum Asterozoa

Radially symmetrical, unattached star-shaped echinoderms. Oral surface directed downward.

CLASS STELLEROIDEA

   Subclass Somasteroidea.   Extinct starfish.

   Subclass Asteroidea.[*]   Living starfish. Arms, containing a large coelomic cavity not sharply set off from central disc. Tube feet located within an ambulacral groove.

   Subclass Ophiuroidea.[*]   Brittle stars. Arms, containing a canal-like coelomic cavity, sharply set off from central disc. Ambulacra not constructed as a groove.

ANNOTATED REFERENCES

Detailed accounts of the echinoderms can be found in many of the references listed at the end of Chapter 18. The following works deal with echinoderms alone.

Boolootian, R. A. (Ed.): Physiology of Echinodermata. New York, John Wiley, 1966. An excellent review of echinoderm ecology, behavior, and physiology.

Clark, A. M.: Starfishes and Their Relations. London, British Museum, 1962.

Nichols, D.: Echinoderms. London, Hutchinson University Library, 1966. A good short biology of the phylum.

---

[*] In order to simplify comparisons, starfish and brittle stars were in the earlier discussion assigned the rank of class, an older scheme of classification.

29. Chordates and Hemichordates

The Chordata, the largest of the deuterostome phyla, includes animals with three distinguishing characteristics (see Fig. 29.8): (1) a dorsal hollow **nerve cord,** which arises as an infolding of the surface ectoderm; (2) a dorsal longitudinal skeletal rod, the **notochord,** located beneath the nerve cord; and (3) paired lateral openings, commonly referred to as **gill clefts** or **slits,** through the pharyngeal wall of the gut. Although these three structures are found in the early developmental stages of all chordates, and were probably characteristic of the ancestral chordates, not all persist in the adults. The notochord disappears during development in most vertebrates, and only the hollow nerve cord and the pharyngeal clefts or their derivatives remain in the adult.

Most of the chordates are vertebrates but there are two interesting small subphyla of invertebrate chordates, i.e., chordates lacking a vertebral column or backbone.

29.1 SUBPHYLUM URO-CHORDATA

The members of the subphylum **Urochordata,** called **tunicates** or **ascidians,** are the larger group of invertebrate chordates. Most tunicates are sessile, attached to rocks, shells, pilings, and ship bottoms.

The planktonic larva of urochordates is about 0.7 mm. in length and looks like a tiny fish or tadpole. In fact, it is sometimes called a "tadpole" larva. The finned tail is the locomotor organ, and it contains longitudinal muscle fibers, the nerve cord and the notochord (Fig. 29.1). The notochord prevents a shortening, or telescoping, of the tail when the muscle fibers contract, and converts muscle contraction into lateral undulations of the tail. The anterior mouth, which may not be open during the larval stage, is connected to a large **pharynx** that is perforated by two paired lateral clefts. The clefts lead into another chamber, the **atrium.**

At the end of its planktonic existence, the larva settles to the bottom and becomes attached by anterior adhesive papillae. A radical metamorphosis ensues and the larva develops into the very different adult body form. In the course of the transformation the tail, including the notochord and neural tube, is resorbed and disappears; the pharynx develops into a large food-collecting chamber; and growth on the anterior dorsal sides of the body is so much greater than elsewhere that the mouth ends up almost 180° from its original position (Fig. 29.1).

Adult tunicates have more or less spherical bodies, which range in size from a pinhead to a small potato; in fact, some species actually bear a strong resemblance to potatoes. The body is covered by a thick envelope, or **tunic,** containing **tunicine,** a

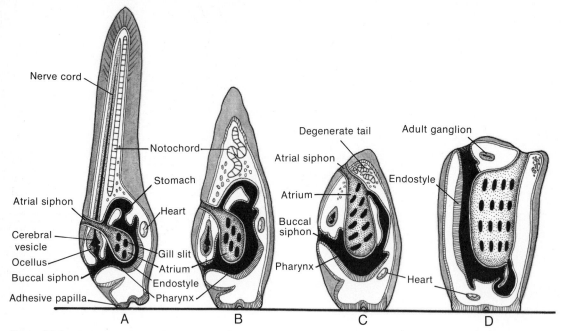

Figure 29.1 *A,* Diagrammatic lateral view of a urochordate tadpole larva, which has just attached to the substratum by the anterior end. *B* and *C,* Metamorphosis. *D,* A young individual just after metamorphosis. (*B* modified after Seeliger from Brien.)

form of cellulose. Cellulose is rarely encountered in animals. The consistency of the tunic may be gelatinous, but more often it is heavy and leathery, even rigid. Many tunicates are drab, but some are bright orange, lavender, green, or other colors.

One end of the urochordate body is attached to the substratum; the opposite end contains two openings, the **buccal** and **atrial siphons** (Fig. 29.2). The buccal siphon opens into a large pharyngeal chamber, which fills the upper half or two-thirds of the body of the animal. The wall of the pharynx is perforated by such a vast number of openings that it resembles a grid. The perforations open into an outer surrounding chamber, the **atrium,** connected to the exterior through the atrial siphon. Through the pharyngeal and atrial chambers flows a current of water, entering through the buccal siphon and leaving through the atrial siphon. The body below the pharynx, called the **abdomen,** contains the **stomach, intestine, heart** and other internal organs (Fig. 29.2).

Nutrition. Tunicates are **filter feeders,** removing plankton from the water stream passing through the pharyngeal chamber.

The current is produced by the beating of cilia which border the margins of the pharyngeal perforations. A deep groove, the **endostyle,** extends down the length of the pharyngeal chamber on the side into which the buccal siphon opens (Fig. 29.2). Mucus secreted by the endostyle is driven by cilia out of the groove and over the inner pharyngeal surface as a film. Plankton is trapped in the film and carried across the pharynx by frontal cilia on the gridlike bars which form its walls. The plankton-laden film eventually collects within a deep gutter on the opposite side of the pharynx. Cilia within the gutter carry the collected material downward to the opening of the esophagus at the bottom of the pharynx (Fig. 29.2).

An enormous amount of water passes through the pharyngeal basket of a tunicate and serves not only for feeding but also for gas exchange. A specimen of *Phalusia* only several centimeters long was estimated to circulate 173 liters of water through the body in 24 hours. The current can be regulated or halted by opening or closing the siphons. Moreover, some small contraction of the body is possible in species which do not have an especially heavy and rigid tunic. Exposed intertidal species may sud-

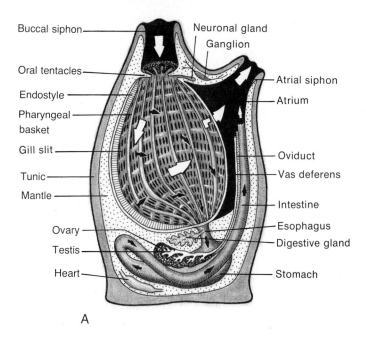

A

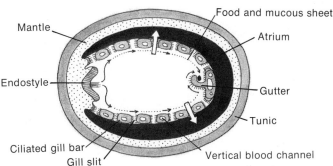

Figure 29.2 Diagrammatic lateral view (A) and cross section (B) of a tunicate showing major internal organs. Large arrows represent the course of the current of water; small arrows that of the food and mucous sheet. The stomach, intestine and other visceral organs are embedded in the mantle.

B

denly eject a spurt of water from the siphons, hence their name, sea squirts.

The gut is U-shaped and a large stomach occupies the bend of the U (Fig. 29.2). An intestine leads back upward and opens into the atrium. Fecal matter is swept away by the water current in the atrial siphon.

Excretion and Internal Transport. There are no special organs for excretion of nitrogenous waste. Most escape by diffusion; some accumulate as inert uric acid crystals in various parts of the body. Tunicates have an open internal transport system, but the blood follows distinct channels, which course through the pharyngeal basket, abdominal organs, and in some species, even the tunic. The heart, located near the stomach, periodically reverses its beat and

the flow of blood, a most unusual phenomenon. Apparently, excitation centers at each end of the heart alternately dominate each other, but the significance of the reversal of flow is not understood.

The blood of some tunicates possesses green cells, which contain large amounts of vanadium. These animals have the ability to remove and concentrate an element which is present in sea water in only trace amounts. The function of the vanadocytes is unknown.

Colonial Tunicates. Most common tunicates are solitary, but a considerable number are colonial. Some colonial tunicates are connected by **stolons** and resemble a vine creeping over the substratum (Fig. 29.3). Others have bodies arranged in clus-

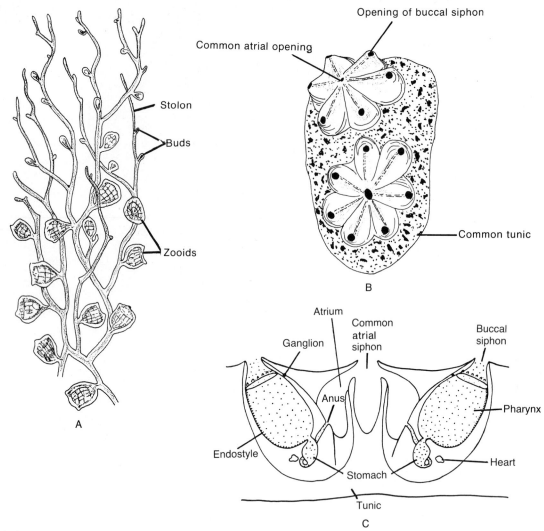

Figure 29.3   Colonial tunicates. *A*, A colony of *Perophora viridis; B*, surface view of *Botryllus schlosseri; C*, vertical section through *Botryllus*. (*A* after Miner; *B* after Milne-Edwards from Yonge; *C* after Delage and Herouard.)

ters; still others are so intimately connected that they are embedded within a common tunic. Such colonial tunicates have the individuals symmetrically arranged and the atrial siphons open into a common cloacal chamber. Some colonial tunicates are unattached and swim using the collective feeding current expelled through a common opening as a water jet for locomotion.

Reproduction. Most tunicates are capable of asexual reproduction by an often extremely complex process of **budding.** All species are **hermaphroditic** with a single abdominal ovary and testis; the oviduct and sperm duct open into the atrium (Fig.

29.2). Fertilization may be external with planktonic development, especially in solitary species. Alternatively, fertilization may take place within the atrium and the eggs brooded there through early embryonic stages. Development eventually leads to the tadpole larva already described.

29.2   CHORDATE METAMERISM

Although members of another subphylum, the **Cephalochordata,** lack a backbone, they do share with vertebrates the condition of **metamerism.** Metamerism appears to have evolved early in chordate history in the line leading to the cephalochordates and vertebrates but after the divergence of the urochordates. As in annelids, segmentation developed as an adaptation for locomotion—but for undulatory swimming rather than burrowing.

The body wall musculature of these animals is arranged laterally to either side of the mid-dorsal notochord. In a theoretical non-metameric chordate, alternate contraction of the muscles on the two sides would cause the body to bow, first one way and then the other, or the tail to move back and forth like a pendulum. These movements would not be especially effective for propulsion in swimming. Breaking up of the body musculature into linearly arranged blocks separated by connective tissue

septa (**myocommata**), makes possible the restriction of muscle contraction to parts of the body. The contracting muscle fibers pull on the septa, which are anchored to the axial skeleton. A wave of contraction affecting the muscle blocks in sequence can pass down the length of the body, alternating with a wave on the opposite side, and throwing the body into undulatory curves (Fig. 29.4). These curves, sweeping from front to back, and especially pronounced in the thinner tail, drive the animal forward. This efficient mode of swimming is still utilized by fish.

As in annelids, it is the musculature that exhibits the primary segmentation. The segmentation of the nervous system and blood vascular systems represents an adjustment of these systems to supply the muscle blocks. Segmentation in chordates does not involve the coelom. The coelom in chordates is not utilized as a localized hydrostatic skeleton; rather it is the notochord against which the muscle blocks are indirectly pulling.

29.3 SUBPHYLUM CEPHALO- CHORDATA

Amphioxus and a related genus of small, superficially fish-shaped chordates constitute the subphylum **Cephalochordata.** Species occur in United States coastal waters south from the Chesapeake (*Amphioxus virginiae*) and Monterey (*A. californiense*) Bays. They usually lie buried in sand with only their anterior end protruding, but they can also swim fairly well.

The body of *Amphioxus* (Figs. 29.5 and 29.6) is elongate, tapering at each end, and compressed from side to side. Small median fins and a pair of lateral finlike **metapleural folds** are present, but apparently they are neither large nor strong enough to keep the animal on an even keel, for *Amphioxus* spirals as it swims. Swimming is accomplished by the contraction of the muscle blocks, or **myomeres**. Shortening of the body is prevented by an unusually long notochord that extends farther anteriorly than in any other chordate, an attribute after which the subphylum is named.

Nutrition. *Amphioxus* is a **filter feeder.**

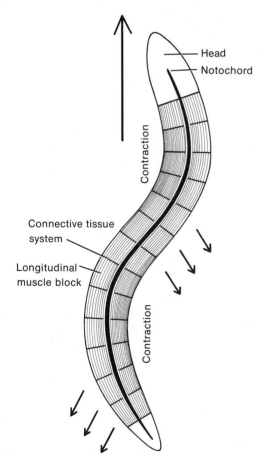

Figure 29.4 Diagrammatic frontal section of a primitive segmented chordate showing pattern of muscle contraction in undulatory swimming. Large arrow indicates direction of movement; small arrows indicate direction of propulsive force generated by wave of muscle contraction sweeping from anterior to posterior.

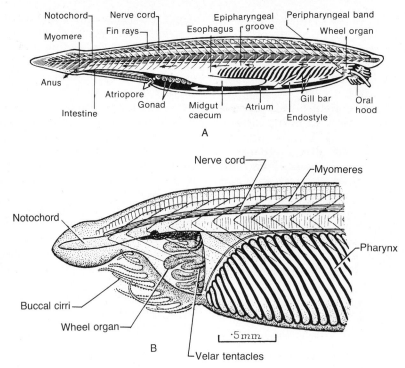

Figure 29.5 *Amphioxus. A,* A diagrammatic lateral view. White arrows represent the course of the current of water; black arrows that of the food. Further details of the course of food particles in the gut are shown in Figure 29.7. *B,* An enlarged view of the anterior end of *Amphioxus,* cleared and stained to show internal structures. (*B* modified after Yonge.)

Water and minute food particles are taken in through the **oral hood,** whose edges bear a series of delicate projections, the **cirri,** that act as a strainer to exclude larger par-

ticles (Fig. 29.5). The inside of the oral hood is lined with bands of cilia, the **wheel organ.** These, together with cilia in the pharynx, produce a current of water that

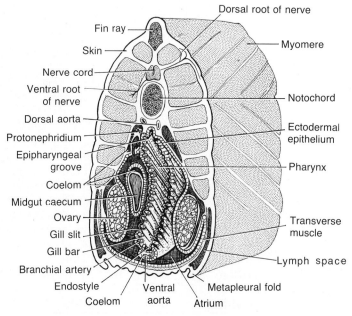

Figure 29.6 A diagrammatic cross section through the posterior part of the pharynx of *Amphioxus.* Branchial arteries extend from the ventral aorta through the gill bars to the dorsal aortas. The portion of the coelom ventral to the endostyle is connected through alternate gill bars with the pair of coelomic canals lying dorsal to the atrium. Other parts of the coelom are associated with the midgut caecum and gonads.

enters the mouth. The mouth proper lies deep within the oral hood surrounded by **velar tentacles.**

Food is entrapped within the pharynx in mucus secreted by an endostyle just as it was in urochordates. Water in the pharynx escapes into an atrium through numerous gill slits. Some gas exchange occurs in the pharynx, but the skin is the major respiratory surface. The pharynx is primarily a food-gathering device. Food particles carried back into the intestinal region are sorted out by complex ciliary currents. Large particles continue posteriorly, but small ones are deflected into the **midgut caecum,** a ventral outgrowth from the floor of the gut, and drift ventrally along the lateral walls of the caecum (Fig. 29.7). Many of these particles are ingested by caecal cells and are digested intracellularly. Undigested material, together with enzymes secreted by certain caecal cells, is carried posteriorly to join the mass of larger food particles. As this mass rotates and is degraded by enzymatic action, small particles are carried forward to the caecum to be engulfed and digested. Material not broken down in the midgut is carried posteriorly. Much of this is undigestible detritus, but some additional digestion and absorption may occur in the posterior part of the intestine. Fecal material is discharged through an **anus,** which, as in vertebrates, lies slightly anterior to the posterior tip of the body.

Internal Transport. Absorbed nutrients are distributed by a circulatory system. A series of veins returns blood from the various parts of the body to a sinus which is located ventral to the posterior part of the pharynx and may be comparable to the posterior part of the vertebrate heart. No muscular heart is present; blood is propelled by the contraction of the arteries. A ventral aorta extends from the sinus forward beneath the pharynx and connects with branchial arteries that extend dorsally through the gill bars to a pair of dorsal aortas (Fig. 29.6). The dorsal aortas, in turn, carry the blood posteriorly to spaces within the tissues. True capillaries are absent, but the general direction of blood flow, i.e., anteriorly in the ventral part of the body and posteriorly in the dorsal part, is similar to that of a vertebrate and different from that of other animals.

The excretory organs are segmentally arranged, ciliated **protonephridia** that lie dorsal to certain gill bars and open into the atrium (Fig. 29.6).

Nervous System. The nervous system of *Amphioxus* consists of a tubular nerve cord located dorsal to the notochord (Figs. 29.5 and 29.6). Its anterior end is differentiated slightly but is not expanded to form a brain. Paired, segmental nerves, consisting of dorsal and ventral roots as in vertebrates, extend into the tissues. The ventral roots go directly into the myomeres, and the dorsal roots pass between myomeres to supply the skin, gut wall and ventral parts of the body. *Amphioxus* is sensitive to light and to chemical and tactile stimuli, but no elaborate sense organs are present. The cirri on the oral hood and a flagellated pit in the skin near the front of the nerve cord appear to be chemoreceptors. Photoreceptive cells, which are partly masked with pigment, lie in the nerve cord. The prominent pigment spot at the anterior end of the cord apparently does not function in light reception.

Reproduction. The sexes are separate in *Amphioxus* and numerous testes or ovaries bulge into the atrial cavity. Actually, they lie within a portion of a highly modified coelom (Fig. 29.6). The gametes are discharged into the atrium upon the rupture of the gonad walls. Fertilization and development are external.

Figure 29.7 Diagram of food particle movements in the intestine and midgut caecum. (From Barrington.)

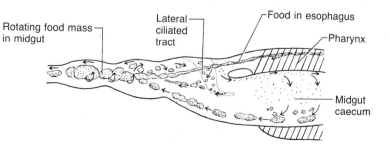

Rotating food mass in midgut

Lateral ciliated tract

Food in esophagus

Pharynx

Midgut caecum

29.4   SUBPHYLUM VERTEBRATA

The Vertebrata is by far the largest and most important of the chordate subphyla, for all but about 2000 of the approximately 41,000 living species of chordates are vertebrates. The eight classes of vertebrates (p. 257) share with the lower chordates the three diagnostic characteristics of the phylum. These are clearly represented at some stage in the life history of the various groups. The dorsal tubular nerve cord has differentiated into a brain and spinal cord, present in the embryos and adults of all species (Fig. 29.8). Embryonic vertebrates have a notochord lying ventral to the nerve cord and extending from the middle of the brain nearly to the posterior end of the body, but a vertebral column replaces the notochord in most adults. All embryonic vertebrates have a series of **pharyngeal pouches** that grow laterally from the walls of the pharynx, but these pouches break through the body surface to form gill slits only in fishes and larval amphibians.

Vertebrates evolved as a group of chordates that became more active and developed more aggressive ways of feeding. Most of the distinctive characteristics of vertebrates are related to these changes in mode of life. Their greater activity is reflected in the replacement of the notochord by a vertebral column, in the continued elaboration of the segmented muscular system seen beginning in *Amphioxus*, in an aggregation of nervous tissue (the brain) and elaborate sense organs at the anterior end of the body, and in the protection of these organs by a brain case or **cranium.** An alternate name for the subphylum, the Craniata, emphasizes this last feature.

Early vertebrates probably continued a filter-feeding mode of life but used muscular movements of their pharynx, and not simply ciliary currents, to draw in water and food. This was certainly more efficient and made possible an increase in size. Larger size, and the eventual evolution of jaws, permitted a yet more active and aggressive mode of life. Respiratory organs (gills or lungs) replaced the general body surface as sites of gas exchange. A muscular **heart** developed which pumps blood effectively through a closed circulatory system. Wastes are excreted by a pair of **kidneys** (Fig. 29.8) composed of numerous tubules quite unlike any invertebrate excretory organ. Vertebrates have become the most successful and dominant group of chordates.

29.5   PHYLUM HEMICHORDATA

We must now examine one last group of deuterostomes, the phylum **Hemichordata,** for the members of this phylum provide some additional evidence linking echinoderms and chordates and must be con-

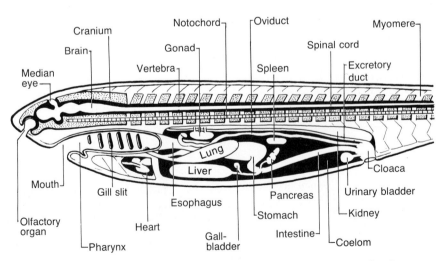

Figure 29.8   A diagrammatic sagittal section through a generalized vertebrate to show the characteristics of vertebrates and the arrangement of the major organs.

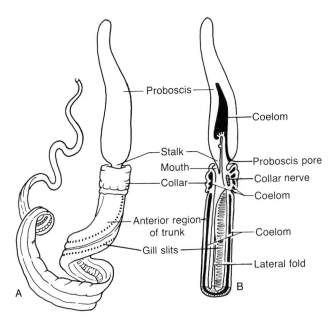

Figure 29.9 Phylum Hemichordata (genus *Saccoglossus*). *A*, External view showing external features (after Bateson). *B*, A diagrammatic section through the anterior part of the body showing some of the internal organs. A lateral fold subdivides the pharynx into a ventral channel along which the sand passes and a dorsal channel containing the gill slits.

sidered in speculating about chordate origins.

Hemichordates are marine worm-like animals, which are found beneath stones and shells, in burrows in sand and mud, and in tubes in a few species. The body is composed of three regions, an anterior **proboscis**, a middle **collar**, and a long posterior **trunk** (Fig. 29.9). The rounded or pointed proboscis is simply the anterior end of the body and in this respect is equivalent to the prostomium of annelids. The shape of the proboscis of some species has resulted in their being called **acorn worms.** At the back of the proboscis on the ventral side is the mouth. It lies just in front of the short band-like collar region. The trunk follows the collar and makes up the greater part of the body.

Hemichordates possess gill clefts. These are present as a line of many perforations on either side of the anterior end of the trunk (Fig. 29.9). A stream of water enters the mouth, passes into the pharynx, and then exits through the gill pores, driven by cilia. Although hemichordates do not contain gills, gas exchange occurs as water passes through the pharyngeal clefts; this appears to be their only function.

Nutrition. Hemichordates feed largely on suspended and deposit material which adheres to mucus on the surface of the proboscis and is driven by cilia into the mouth along with the ventilating current. Within the pharynx the water is separated from the ingested food, which passes into the posterior part of the gut. Some hemichordates (*Balanoglossus*, for example) which live in burrows, ingest large quantities of sand and extract the organic detritus from it. The castings of these animals are often a conspicuous feature of a sand flat exposed at low tide (Fig. 29.10).

Nervous System. Despite the presence of pharyngeal clefts, hemichordates are not chordates, as they were once thought to be, for they have no notochord, nor is there a true dorsal nerve cord. However, within the collar region of some hemichordates, a dense concentration of dorsal epidermal nerve fibers sinks inward and forms a tube-like arrangement (Fig. 29.9); some zoologists have thought this might be the homolog of the chordate nerve cord.

Reproduction. Hemichordates have separate sexes and the numerous paired lateral gonads, each with a separate gonopore, are located within either side of the trunk. The eggs are released in mucous masses and

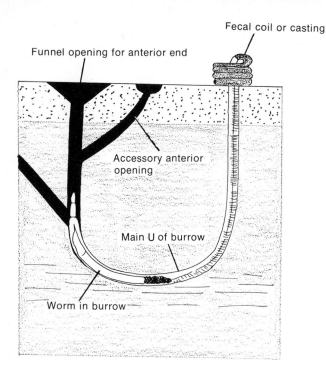

Fecal coil or casting

Funnel opening for anterior end

Accessory anterior opening

Main U of burrow

Worm in burrow

Figure 29.10 Burrow system of the Mediterranean *Balanoglossus clavigerus*. (After Stiasny from Hyman.)

fertilization is external. Development follows the typical deuterostome pattern and leads in many species to a **tornaria larva** (Fig. 29.11), which is strikingly like that of echinoderms. After a planktonic existence the tornaria larva lengthens and settles to the bottom as a young worm.

The preceding discussion has been devoted to the most common hemichordates, members of the class **Enteropneusta.** A small number of species comprises another class, the **Pterobranchia.** These are deep-water hemichordates that live attached within secreted tubes. The collar region bears tentacles and some species lack pharyngeal clefts. Many zoologists consider the pterobranchs to be the most primitive members of the phylum.

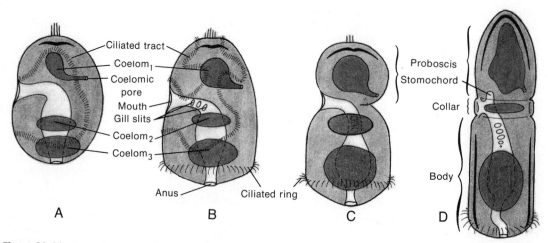

Ciliated tract

Coelom$_1$

Coelomic pore

Mouth

Gill slits

Coelom$_2$

Coelom$_3$

Anus

Ciliated ring

Proboscis

Stomochord

Collar

Body

A   B   C   D

Figure 29.11 Diagrammatic side views of the larval development of a hemichordate. Compare with Figure 28.9. *A*, Early larva; *B*, later larva; *C* and *D*, metamorphosis.

29.6 DEUTEROSTOME RELATIONSHIPS AND CHORDATE ORIGINS

The common pattern in the early embryogeny of echinoderms, hemichordates, and chordates is an important basis for recognizing the deuterostome line of evolution, but there is other evidence of the close evolutionary relationship between these three principal deuterostome phyla. The strikingly similar larval stages clearly indicate a connection between hemichordates and echinoderms. On the other hand, the presence of gill clefts and perhaps the dorsal collar nerve cord relates hemichordates to chordates. Although the precise relationship of the three phyla is unknown, there seems little doubt that they share a common evolutionary history.

The nature of the common ancestral form, as well as the origins of each phylum, is obscure. We speculated about echinoderm origins in the previous chapter. No more is known about chordate origins. An idea elaborated by N. J. Berrill and others, and held by many zoologists, postulates that the ancestral chordates were sessile and that urochordates reflect the primitive condition of the phylum (Fig. 29.12). Pharyngeal clefts were originally an adaptation for filter feeding in the adult, and the notochord and

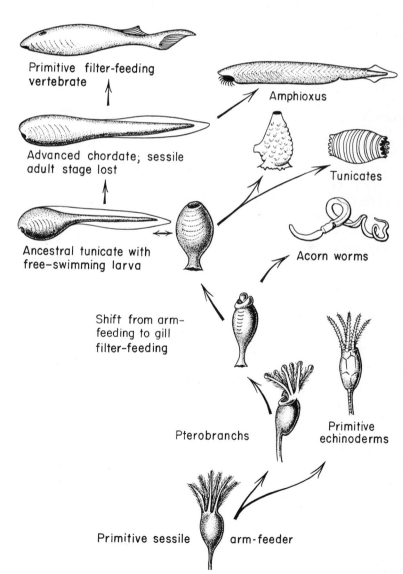

Figure 29.12 A diagram showing Berrill's hypothesis of chordate evolution. (From Romer: Vertebrate Body, 4th ed.)

Primitive filter-feeding vertebrate

Amphioxus

Advanced chordate; sessile adult stage lost

Tunicates

Ancestral tunicate with free-swimming larva

Acorn worms

Shift from arm-feeding to gill filter-feeding

Pterobranchs

Primitive echinoderms

Primitive sessile arm-feeder

finned tail were adaptations for swimming during larval life. According to this theory, the cephalochordates and vertebrates evolved from the urochordate tadpole larva through neoteny and paedogenesis, i.e., prolongation of larval life, precocious sexual development, and suppression of adult features. Metamerism developed as an adaptation for swimming in the neotenous cephalochordate-vertebrate ancestor. The sessile existence of primitive chordates, pterobranch hemichordates, and primitive echinoderms is viewed as a feature resulting from common ancestry.

However, like most theories that attempt to explain evolutionary relationships between classes and phyla, this theory of chordate origins is not perfect. A principal difficulty stems from making the urochordates the ancestral members of the phylum, for they are certainly highly specialized. Indeed, sessility is a specialized condition wherever it occurs in the Animal Kingdom. Larval neoteny is used as a way of getting around this problem and in effect "starting fresh" on the cephalochordate and vertebrate line.

There is also no certainty that the common ancestor of echinoderms and chordates was sessile. Echinoderms appear to have evolved from motile ancestors, and their sessile existence and subsequent radiation may well have been an independent evolutionary event unrelated to the evolution of a sessile existence in urochordates.

Other views of chordate origins assume that the common ancestral deuterostome was a swimming bilateral coelomate. Pharyngeal clefts evolved as a mechanism for gas exchange or filter feeding in the line leading to hemichordates and chordates. The ancestral chordates may have been small swimming bilateral animals having a dorsal nerve tube, a notochord skeleton, and pharyngeal clefts. Whether for gas exchange or filter feeding, pharyngeal clefts would have enabled these ancestral chordates to take up a sessile existence, and some became attached to give rise to the urochordates. In another stock, locomotor adaptations were developed, leading to metamerism and an undulatory mode of swimming. From this group the cephalochordates and vertebrates arose.

ANNOTATED REFERENCES

Accounts of the hemichordates and chordates can be found in many of the references listed at the end of Chapter 18. Those below deal solely with these groups.

Barrington, E.: The Biology of Hemichordates and Protochordates. San Francisco, W. H. Freeman, 1965. A general account emphasizing the behavior, physiology, and reproduction of hemichordates, urochordates, and cephalochordates.

Berrill, N. J.: The Origin of the Vertebrates. London, Oxford University Press, 1955. This work contains Berrill's theory of chordate and vertebrate origins.

Clark, R. B.: Dynamics in Metazoan Evolution. Oxford, Clarendon Press, 1964. A detailed synthesis of the ideas regarding the origins of the coelom and metamerism and their phylogenetic implications.

30. Vertebrates: Fishes

Fishes are aquatic vertebrates well suited for life in the water. There are many groups, each with unique characteristics, but all share many features that adapt them to the aquatic environment. We have considered these when we discussed vertebrate form and function (Chapters 11 to 17), but wish to reemphasize certain points and elaborate upon others at this point.

30.1 AQUATIC ADAPTATIONS

Fishes have a streamlined, or fusiform, body shape that enables them to move through a dense medium, water, with a minimum of effort. The body is supported by a skeleton, but the skeleton is not so strong as in terrestrial vertebrates for the buoyancy of the water also provides considerable support. Trunk muscles are segmented myomeres whose successive activation causes the lateral undulations of the trunk and tail by which the animal swims. As the tail moves from side to side, it pushes diagonally backward against the water, and the water, in turn, has an opposite and equal reaction upon the tail (Fig. 30.1A). This force can be resolved into lateral and forward components. The lateral components cancel each other out, and the forward ones propel the fish. The head has considerable inertia so it does not move from side to side as much as the rest of the body. Primitive fishes have rather heavy bodies and tend to sink, but the forward movement also generates lift forces. Lift is generated at the anterior end of the body by the planing of a head that is somewhat flattened ventrally and by large pectoral fins; it is developed at the posterior end by a **heterocercal tail** (Fig. 30.1B). The vertebral axis turns dorsally into the upper lobe of such a tail and stiffens this part of the tail relative to the lower lobe. As the tail moves from side to side, the lower lobe has a sculling action and generates an upward thrust. The more advanced fishes have evolved a swim bladder, a sac of air dorsal to the body cavity, that gives them a density equal to that of the surrounding water. They do not sink so easily, the pectoral fins are smaller, and the tail is symmetrical. The medial fins, as well as the paired pectoral and pelvic fins, provide stability against rolling, pitching, and yawing (Fig. 30.1C). The paired fins are also used in turning and in changing elevation.

The sense organs of fishes are adapted for receiving stimuli from the water. Movable eyelids and tear glands are absent for the eyes are bathed and cleansed by the surrounding water. Most light refraction is accomplished by a spherical lens; little occurs in the cornea for its refractive index is similar to that of water. An inner ear is present, but fishes lack the external and middle ear found in terrestrial vertebrates. The pressure waves in the water pass easily through the tissues of the fish to the inner ear. Low frequency vibrations and certain water movements are detected by the lateral line system, which has been defined as a

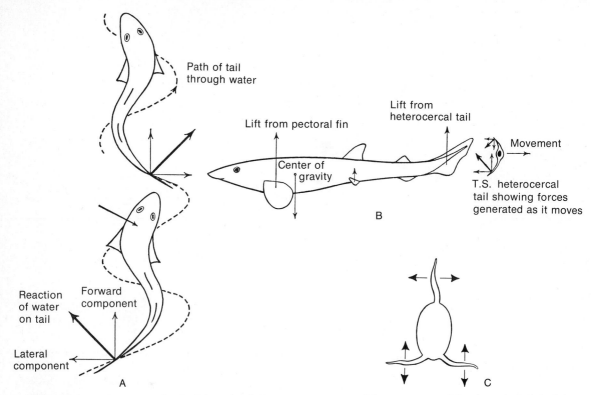

Figure 30.1   Locomotion of a dogfish. *A,* Undulatory movements and forces generated by the tail; *B,* lift of the pectoral fins and heterocercal tail; *C,* cross section of the body showing stabilizing effect of median and paired fins. (Modified from Marshall and Hughes.)

system of distant touch (p. 352). The nose is simply an olfactory organ, and water does not pass through it on the way to the pharynx. A few fishes with lungs do have a connection between nose and mouth cavity.

Gas exchange between the water and blood occurs by diffusion across the gills. Water typically enters the pharynx through the mouth and is discharged through the gill pouches (Fig. 12.10, p. 291). The heart, located just posterior and ventral to the gills, receives blood low in oxygen content, and its contraction drives the blood through the gill capillaries. After leaving the gills, oxygenated blood is distributed to the body under relatively low pressure.

A movable, muscular tongue is absent in fishes. The respiratory current of water carries food posteriorly into the pharynx. Digestion and absorption is similar in all vertebrates, but the ectothermic fishes require less absorptive surface than more active endothermic vertebrates.

Nitrogenous metabolic wastes are eliminated in most fishes by diffusion through

the gills and also by kidneys, which are drained by wolffian ducts. The kidneys also play an important role in water and salt balance. Excess salt swallowed by marine fishes is eliminated by special salt-excreting cells or glands.

Most fishes are oviparous. As the female lays eggs, the male discharges sperm. Fertilization is external, and there is often an aquatic larval stage during development.

30.2   VERTEBRATE BEGINNINGS

Although vertebrate origins are obscure, we have a reasonably complete fossil record of their subsequent evolution. The most primitive fishes are jawless types placed in the class **Agnatha.** This group flourished during the middle Paleozoic era, when it was represented by several orders collectively known as the **ostracoderms** (Fig. 30.2). Most ostracoderms were small, bottom-feeding fishes that were somewhat flattened dorsoventrally (Fig. 30.3). They

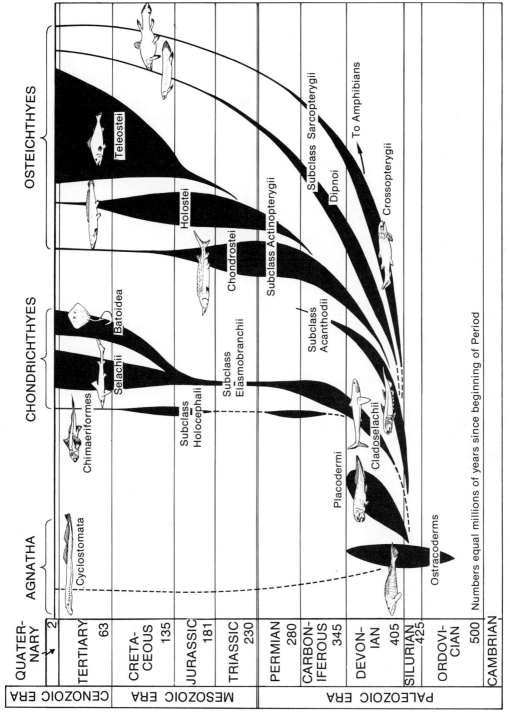

Figure 30.2 An evolutionary tree of fishes. The relationships of the various groups, their relative abundance and their distribution in time are shown.

679

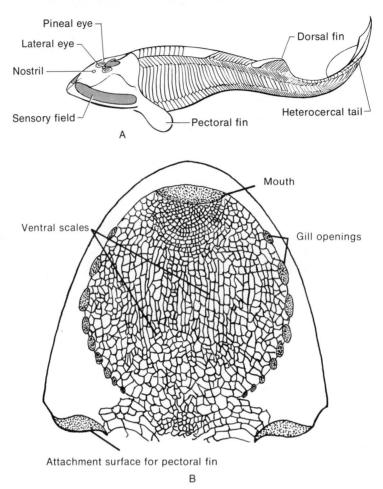

Figure 30.3 *Hemicyclaspis,* a representative ostracoderm of the early Devonian period. This fish was about 20 cm. long. *A,* laterodorsal view; *B,* underside of head. (Modified after Stensiö.)

had an extensive armor of thick bony plates and scales that developed in the dermis of the skin. These were very similar to the **cosmoid scales** of some other primitive fishes (Fig. 30.4), for beneath a thin layer of enamel-like **ganoine** there was a thick layer of dentine-like **cosmine.** The rest of the scale consisted of a layer of **spongy bone** containing many vascular spaces, and a layer of more compact **lamellar bone.** These heavy scales certainly offered some mechanical protection, possibly against aquatic scorpion-like eurypterids of the period; in addition the layers of ganoine and cosmine may have offered some protection against an excessive inflow of water from the freshwater environment in which these fishes probably lived.

Ostracoderms had median fins, and some had paired pectoral fins, but paired fins

were not well developed in most groups. Most had the heterocercal tail characteristic of primitive fishes.

The ostracoderm head was rather unusual. A single **median nostril** was present on the top of the head in the best known species. There was a pair of lateral eyes and a single, median **pineal eye** on the top of the head posterior to the nostril. A median eye is found in many primitive fishes and terrestrial vertebrates, and it is retained in a few living fishes and reptiles. It is a photoreceptor rather than an image-forming eye, and receives stimuli that enable the organism to adjust its physiological activity to the diurnal cycle. Three regions of the head, one dorsomedial and a pair of dorsolateral areas, contained small plates beneath which were enlarged cranial nerves. It is believed that these areas were sensory fields, pos-

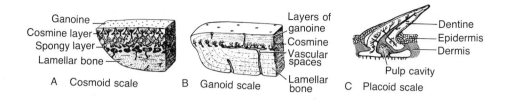

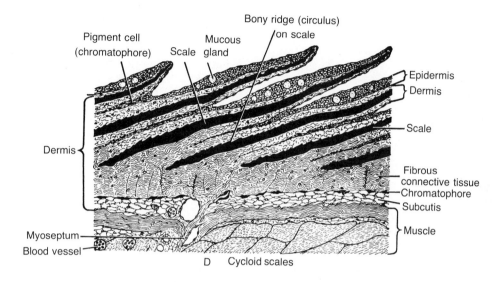

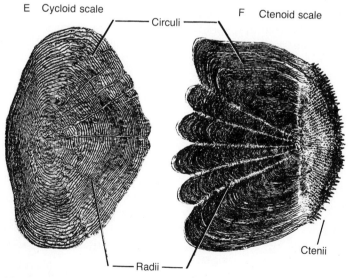

Figure 30.4 Types of fish scales. Vertical sections through (A) a cosmoid scale of the type seen in certain ostracoderms and sarcopterygians, (B) a ganoid scale of the type seen in acanthodians and primitive actinopterygians, (C) a placoid scale of a shark, (D) the skin and cycloid scales of a teleost. Surface view of (E) the cycloid scale of a primitive teleost and (F) the ctenoid scale of an advanced teleost. (A–C from Romer; D, courtesy of General Biological Supply House; E and F from Hubbs and Lagler.)

sibly a part of the lateral line system (p. 352). Much of the ventral surface of the head was covered with small plates forming a flexible floor to the pharynx. Movement of this floor presumably drew water and minute particles of food into the jawless mouth. The water then left the pharynx through as many as 9 pairs of small **gill slits,** but the food particles were somehow trapped in the pharynx. It seems probable that the ancestral vertebrates, like the lower chordates of the present day, were filter-feeders.

30.3 LIVING JAWLESS VERTEBRATES

Ostracoderms became extinct by the end of the Devonian, but the living lampreys and hagfishes of the order **Cyclostomata** are a specialized remnant of the class Agnatha (Figs. 30.5 and 30.6). They are jawless, have more gill slits than other living fishes, lack paired appendages, retain a pineal eye, and have a single median nostril. Besides leading to an olfactory sac, this nostril opens into an **hypophyseal sac** that passes beneath the front of the brain. Much of the pituitary gland of higher vertebrates is derived from an embryonic hypophysis. Cyclostomes differ from ostracoderms in several respects: they have an eel-like shape and a slimy scaleless skin, and they are predators or scavengers. Many lampreys, like the ostra-

coderms, live in fresh water, but some spend their adult life in the ocean and return to fresh water only to reproduce. The hagfishes are exclusively marine.

A familiar example of the group is the sea lamprey, *Petromyzon marinus.* The chief axial support for the body is a **notochord** which persists throughout life and is never replaced by vertebrae. Rudimentary vertebrae are present, however, on each side of the notochord and spinal cord. The brain is encased by a cartilaginous **cranium,** and the gills are supported by a complex, cartilaginous lattice-work known as the **branchial basket,** which may be homologous to the visceral skeleton of other fishes.

The **mouth** lies deep within a **buccal funnel,** a suction-cup mechanism with which the lamprey attaches to other fishes (Fig. 30.5). The mobile **tongue** armed with horny "teeth" rasps away at the prey's flesh, and the lamprey sucks in the blood and bits of tissue. It has special **oral glands** that secrete an anticoagulant which keeps the blood flowing freely. From the mouth cavity, the food enters a specialized **esophagus** that bypasses the pharynx to lead into a straight **intestine.** There is no stomach or spleen. A **liver** is present, but the pancreas is represented only by cells imbedded in the wall of the intestine and in the liver.

The respiratory system consists of seven pairs of gill pouches that connect to a modified pharynx known as the respiratory tube. The respiratory tube is isolated from the

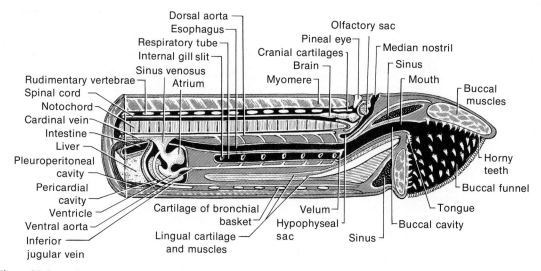

Figure 30.5 A diagrammatic representation of the more important organs found in the anterior part of the lamprey.

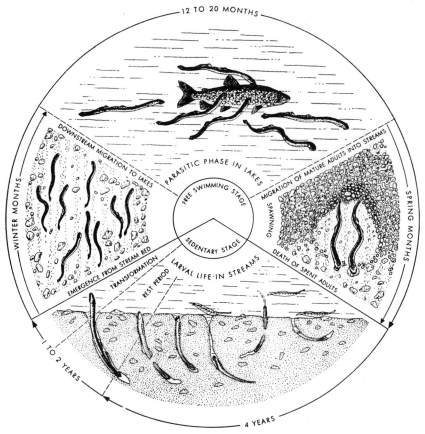

Figure 30.6 The life cycle of the sea lamprey in the Great Lakes. The lamprey spends all but a year or two of its six and one-half to seven and one-half years of life as a larva. (From The Sea Lamprey by Applegate and Moffett. Copyright © by Scientific American, Inc., April, 1955. All rights reserved.)

food passage so the animal can pump water in and out of the gill pouches through the external gill slits while it is feeding. The pumping of the pharyngeal region also changes the pressure in the hypophyseal sac in a way that resembles the compressing and relaxing of the bulb of an eye dropper, and water is thereby circulated across the olfactory sac.

The **kidneys** are drained by **wolffian ducts.** These ducts carry only urine, for sperm or eggs pass from the large **testis** or **ovary** into the coelom. A pair of **genital pores** leads from the coelom into a **uro-genital sinus,** formed by the fused posterior ends of the wolffian ducts, and thence to the cloaca and outside. The absence of genital ducts may be a very primitive feature.

The eggs are laid on the bottom of streams in a shallow nest, which the lampreys make by removing the larger stones (Fig. 30.6).

Fertilization is external and the adults die after spawning.

Developing sea lampreys pass through a larval stage that lasts five to six years. The larva is so different in appearance from adult lampreys that it was originally believed to be a different kind of animal and was named **Ammocoetes.** The ammocoetes larva is eel-shaped but lacks the specialized feeding mechanism of the adult. It lies within burrows in the mud at the bottom of streams and sifts minute food particles from water passing through the pharynx. Like the lower chordates, it has a mucus-producing **endostyle** that aids in trapping food.

Adult lampreys injure and kill many other fishes. The sea lamprey was able to pass the Niagara barrier, presumably through the Welland Canal, and extended its range from Lake Ontario into the other Great Lakes. The lake fishing industry was severely harmed during the late 1940s, and lake trout, which had been an important economic resource, were nearly eliminated. Development of a chemical larvicide that selectively kills larval lampreys gives some hope for a restoration of the fishing industry.

The hagfishes resemble lampreys in major respects, though differing, of course, in certain details. Hags are primarily scavengers feeding upon dead fish along the ocean bottom, but they also attack disabled fish of any sort, including those hooked or netted. They burrow into the fish and eat out the inside, leaving little but a bag of skin and bone. They are a commercial nuisance, but their over-all damage is not great, since they are abundant in only a few localities.

30.4   PRIMITIVE JAWED FISHES

During the Silurian and Devonian periods, other descendants of the ostracoderms became more active and predaceous. A more streamlined body shape, some reduction of the heavy trunk armor, and the presence of paired pelvic as well as pectoral fins suggest greater activity. The presence of jaws enabled these fishes to feed upon a wider variety of food than the jawless ostracoderms.

Jaws evolved as the mouth was displaced posteriorly and became associated with the anterior part of the gill or visceral skeleton (p. 264). One or two visceral arches may have been lost in the process, for jawed fishes have fewer gill slits and arches than jawless ones. However, the most anterior of the remaining visceral arches, known as the **mandibular arch,** became enlarged and, together with dermal bones developed in the skin adjacent to it, formed the jaws. Among the earliest jawed fishes were the "spiny sharks" or **acanthodians** (Fig. 30-2). When bony plates are removed from the cheek and gill region, it is evident that the mandibular arch is a modified gill arch (Fig. 30.7). The second visceral arch, known as the **hyoid arch,** lay close behind the mandibular arch, and in some species appears to have extended as a prop from the brain case to the posterior end of the upper jaw as it does in many contemporary species (Fig. 11.7, p. 264). The remaining visceral arches are **branchial arches** that supported the gills.

The **placoderms** were a second group of early jawed fishes (Figs. 30.2 and 30.8). Many were fresh-water species of modest size but the most spectacular genus was *Dunkleosteus,* a two- to three-meter-long monster of upper Devonian Ohio seas. The head and anterior part of the trunk were encased in heavy bony plates, but the rest of the body was nearly naked. The most conspicuous parts of the jaws were sharp edged plates of dermal bone, but these were attached to a deeper lying mandibular arch.

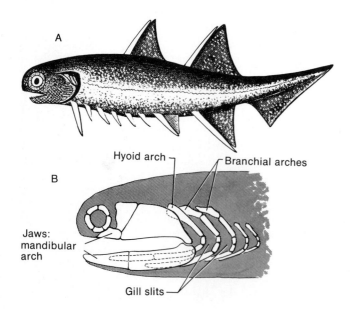

Figure 30.7 *A,* The "spiny shark," *Climatius,* was among the first jawed vertebrates. This fish was about 8 cm. long. After removal of the gill covering and superficial bony scales and plates on the head of a related genus (*Acanthodes, B*), it can be seen that the jaws are modified gill arches. (*A,* after Watson; *B* modified after Watson.)

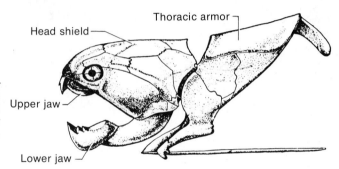

Figure 30.8 A lateral view of the anterior end of the placoderm *Dunkleosteus*. The dermal plates comprising the upper and lower jaws were attached to the mandibular arch. This fish attained a length of 3 meters.

The development of jaws in these early fishes was an important step in the evolution of vertebrates, for jaws enabled them to adapt to many modes of life. The success of jawed fishes probably contributed to the extinction of ostracoderms and to the limitation of cyclostomes to rather specialized ecological niches.

Other jawed fishes can be sorted out into two distinct evolutionary lines (Fig. 30.2). Members of one class, the sharks and skates, have a cartilaginous skeleton, and hence are known as the cartilaginous fishes or **Chondrichthyes.** Species in the other class have at least a partly ossified internal skeleton, hence are known as bony fishes or **Osteichthyes.** These are our most familiar fishes including salmon, minnows, perch, lungfishes, and the like. It is generally agreed that the acanthodians were early members of the Osteichthyes. Although the

placoderms had a partly ossified internal skeleton, other features suggest that they were more closely related to the cartilaginous fishes.

30.5 CHARACTERISTICS OF CARTILAGINOUS FISHES

The early cartilaginous fishes appeared in the Devonian period. Unlike most of their contemporaries, they were a marine group and, with few exceptions, sharks, skates and their allies have remained a marine group. The cartilaginous internal skeleton is sometimes strengthened by the deposition of calcium salts, but there is never any internal ossification. A skeleton composed of cartilage is believed to represent the retention by the adult of the embryonic skeletal ma-

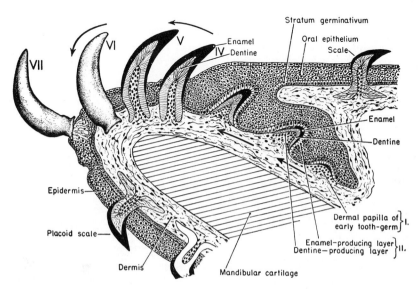

Figure 30.9 A vertical section through the lower jaw of a shark. Teeth, which are structurally very similar to placoid scales, are continually developing, moving to the surface, and sloughing off. (From Rand.)

terial. It is not regarded as the primitive adult condition, for at least parts of the skeleton were ossified in most early fishes. The only bonelike materials present in cartilaginous fishes are teeth and the small, spiny **placoid scales** embedded in the skin (Fig. 30.4), giving it a sandpaper-like quality. Placoid scales are structurally comparable to the enamel-like and dentine-like outer parts of the heavy bony scales of more primitive fishes. The triangular teeth of sharks closely resemble enlarged placoid scales, and have doubtless evolved from them (Fig. 30.9).

Cartilaginous fishes resemble more primitive fishes in being heavy-bodied and denser than water, for they lack air-filled lungs or a swim bladder—features found in bony fishes.

The visceral organs of the dogfish, *Squalus acanthias* (Fig. 30.10), are in many ways more characteristic of primitive fishes than are those of the specialized cyclostomes. The **mouth cavity** is continuous posteriorly with the **pharynx**. A spiracle, containing a vestigial gill, and the gill slits, containing functional gills, open from the pharynx to the body surface. A wide **esophagus** leads from the back of the pharynx to a J-shaped **stomach**. A short, straight **valvular intestine** continues back to the **cloaca.** The valvular intestine receives secretions from the **liver** and **pancreas.** It contains an elaborate spiral fold known as the **spiral valve;** this helical fold serves both to slow the passage of food and to increase the digestive and absorptive surface of the intestine. Excess salt taken in with the food is eliminated by a salt excreting **rectal gland.**

Cartilaginous fishes are unique in their method of reproduction. Part of the male's pelvic fin is modified as a **clasper** with which to transfer sperm to the female reproductive tract (Fig. 30.11). The eggs are fertilized in the upper part of the oviduct, and a horny protective capsule is secreted around them by certain oviducal cells. Skates are **oviparous,** but there is no free larval stage as there is in frogs and many other oviparous animals. The eggs are very heavily laden with yolk, and the embryos develop within the protective capsule. Some sharks are also oviparous, but most have departed from this primitive egg-laying habit. A few are truly **viviparous,** for the fertilized eggs develop in a modified portion of the oviduct known as the **uterus,** an intimate association is established between each embryo's yolk sac and the uterine lining, forming a **yolk sac placenta,** and the embryos have a greater dependence for their nutrient requirements upon the mother than upon food stored in the yolk. However, most sharks, including our common dogfish, are **ovoviviparous;** the eggs also develop within a uterus, but there is a greater dependence upon food stored in the yolk. In some cases, a portion of the nutritional requirements is derived by the absorption of materials secreted by the mother into the uterine fluid, but an intimate placental relationship is not established. It is frequently difficult to make a sharp distinction between viviparous and ovoviviparous reproduction, for the degree of dependence of the embryo for its nutrients upon the mother ranges along a continuum, and in both the young are born at an advanced stage of development as miniature adults.

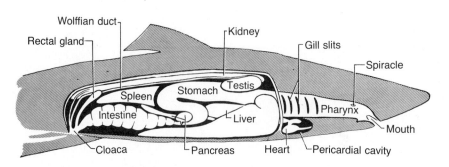

Figure 30.10   The visceral organs of the dogfish.

Figure 30.11   A group of cartilaginous fishes. *A*, A male dogfish, *Squalus acanthias; B*, the sawfish, *Pristis; C*, the stingray, *Dasyatis; D*, the ratfish, *Chimaera*. (*A* modified after Bigelow and Schroeder; *C*, courtesy of Marine Studios; *D* from Romer after Dean.)

30.6 EVOLUTION OF CARTILAGINOUS FISHES

The ancestral Chondrichthyes were essentially sharklike, but in their subsequent evolution the cartilaginous fishes have diverged widely and have become adapted to many modes of life within the aquatic environment. One line of evolution (subclass Holocephali) has led to our present-day, rather rare, deep-water ratfish (*Chimaera*) (Fig. 30.11*D*). In these fishes, the gill slits are covered by an operculum, so there is a common external orifice, and the tail is long and ratlike. The other line of evolution (subclass **Elasmobranchii**) is distinguished by having separate external openings for each gill slit. Elasmobranchs have been far more successful and have diverged into two contemporary orders — **Selachii** (sharks and dogfish) and **Batoidea** (skates and rays).

Sharks. Most selachians are active fishes that feed voraciously with their sharp, triangular-shaped teeth upon other fishes, crustaceans and certain mollusks. The largest sharks, such as the whale shark (*Rhincodon*), which may reach a length of about 15 meters, have minute teeth and feed entirely upon small crustaceans and other organisms that form the drifting plankton. They gulp mouthfuls of water and as the water passes out of the gill slits, the food is kept in their pharynx by a branchial sieve. Whale sharks are the largest living fishes.

Although rare, attacks on man by sharks receive considerable notoriety. In order to obtain reliable data as to their frequency and the circumstances of their occurrence, the Shark Research Panel of the American Institute of Biological Sciences has been maintaining a world-wide "Shark Attack File." There seem to be some 30 unprovoked attacks per year, occurring mostly between latitudes 35°N and 35°S, where the waters are warm and the more dangerous species live. It is well established that sharks are attracted to fresh blood, even in small quantities, coming from injuries to a swimmer or from speared fish a skin diver may be towing. Most sharks circle their intended victim several times before attacking, so if the victim is aware of their presence and he does not attract undue attention by splashing, there is a fair chance of his avoiding or warding off any attack.

Skates and Rays. Most skates and rays are bottom-dwelling fishes that are flattened dorsoventrally and have enormous pectoral fins whose undulations propel the fish along the bottom (Fig. 30.11C). Their mouth is often buried in the sand or mud, and water for respiration enters the pharynx via the pair of enlarged spiracles. A spiracular valve in each one is then closed, and the water is forced out the typical gill slits. Most skates and rays have crushing teeth and feed upon shellfish, but others are adapted for other methods of feeding. The sawfish (*Pristis*) has an elongated, blade-shaped snout armed with toothlike scales. By thrashing about in a shoal of small fishes, it can disable many and eat them at leisure. As in the sharks, the largest members of the group (the devilfish, *Manta*) have reduced teeth and are plankton feeders. Some devilfish have a "wing spread" of 6 meters and can easily upset small boats. Harpooning these is an exciting sport!

30.7 CHARACTERISTICS OF BONY FISHES

While early sharks were becoming dominant in the ocean the bony fishes, class **Osteichthyes,** became dominant in fresh water. They subsequently entered the ocean and became the most successful group there as well.

Bony fishes differ from the cartilaginous fishes in having an ossified internal skeleton and in retaining more of the primitive dermal bony scales and plates. All bony fish also have an **operculum,** a lateral flap of the body wall that extends posteriorly from the head and covers the gill region. The gill pouches lead to an opercular chamber which opens on the body surface at the posterior margin of the operculum (Fig. 12.10, p. 291). An operculum is found only in the holocephalans among the cartilaginous fishes. Male bony fishes never have the pelvic clasper found in all male cartilaginous fishes.

The soft parts of most bony fishes, the perch for example (Fig. 30.12), show a peculiar mixture of primitive and highly specialized characters. Most need not concern us, but one of great interest is the **swim bladder.** In the perch, this is a median membranous sac lying in the dorsal portion of the coelom. The bladder is filled with gases similar to those dissolved in the water (nitrogen, oxygen, carbon dioxide). It functions primarily as a hydrostatic organ, adjusting the density of the body so that the fish can stay at various depths with a minimum of effort. Gases may be secreted into the bladder or absorbed from it, as conditions warrant, through specialized capillary networks in its wall. Under conditions of oxygen deficiency, some fish can utilize the oxygen in the bladder, so the organ may function as a temporary storage site for this gas.

In some bony fishes, the swim bladder is connected to the digestive tract by a pneumatic duct and, in a few, functional **lungs** are present instead. This led many to postulate that the swim bladder was the precursor

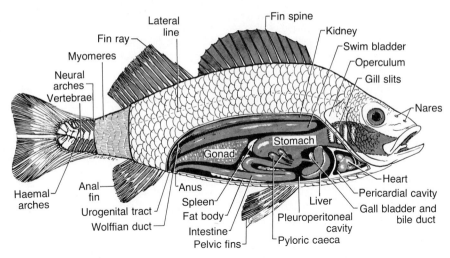

Figure 30.12 The visceral organs of the perch.

of lungs. At present, however, lungs are considered to be the precursor of the swim bladder, for the organ is most lunglike in the most primitive bony fishes.

It is believed that the ancestral bony fishes had lungs similar to those of the living African lungfish (*Protopterus*). In the lungfish a pair of saclike lungs develops as a ventral outgrowth from the posterior part of the pharynx (Fig. 30.13). The lungs enable the fish to survive conditions of stagnant water and drought. The rivers in which the African lungfish live may completely dry up, but the fish can survive curled up within

a mucous cocoon that it secretes around itself in the dried mud. A small opening from the cocoon to the surface of the mud enables the fish to breathe air during this period. The African lungfish has become so dependent upon its lungs that it will die if it cannot occasionally reach the surface to gulp air.

Air breathing probably evolved in fishes as a supplement to gill respiration. Presumably, early bony fishes evolved lungs as an adaptation to the unreliable fresh-water conditions of the Devonian period. Geologic evidence indicates that the Devonian

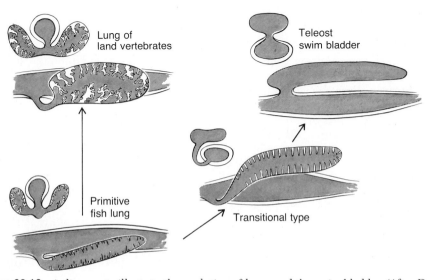

Figure 30.13 A diagram to illustrate the evolution of lungs and the swim bladder. (After Dean.)

was a period of frequent seasonal drought. Bodies of fresh water undoubtedly either became stagnant swamps with a low oxygen content or dried up completely. Only fishes with such an adaptation could survive these conditions. The others became extinct or migrated to the sea, as did many later placoderms and the cartilaginous fishes. Groups of bony fishes that have remained in fresh water throughout their history tended to retain lunglike organs, but those that went to sea no longer needed lungs, for ocean waters are rich in oxygen. Their useless lungs evolved into useful hydrostatic organs. What are presumed to be intermediate stages in this shift can still be seen in certain species. Later, when conditions were more favorable, many salt-water bony fishes re-entered fresh water but retained their swim bladders. The fresh-water perch has such a history. Its ancestors first evolved lungs in a fresh-water environment, then went into the ocean, where the lungs changed into swim bladders; later the fish re-entered fresh water and retained the swim bladders.

30.8   EVOLUTION OF BONY FISHES

Bony fishes can be traced back to the early Devonian and late Silurian (Fig. 30.2). At that time several lines of evolution within the class were already established. Since all share such features as a partly ossified internal skeleton, well-developed bony scales and an operculum, it is assumed that they had a common evolutionary origin, possibly from some ostracoderm group. Yet each evolutionary line is already well established. The acanthodians (p. 684) were characterized by having a prominent supporting spine at the anterior edge of the dorsal, anal and paired fins. Often there were accessory lateral spines between the paired fins. Acanthodians were the earliest of the bony fishes and it is possible that other lines of bony fish evolution were derived from them. They flourished during the Devonian period, but were replaced by more progressive types by the end of the Paleozoic era. Other bony fishes are usually grouped into the subclasses **Actinopterygii** and **Sarcopterygii**.

Actinopterygians. The actinopterygians are the familiar ray-finned fishes such as the perch. Their paired fins are fan-shaped (Fig. 30.14). Skeletal elements enter their base, but most of the fin is supported by numerous dermal rays which evolved from rows of bony scales. Their paired olfactory sacs connect only with the outside and not with the mouth cavity.

Actinopterygian evolution presents a good example of a succession in which early dominant groups became replaced by more successful types. Three infraclasses are recognized and each, in turn, had its day (Fig. 30.2). Currently, the infraclass **Chondrostei** has dwindled to a few species, of which the Nile bichir (*Polypterus*) and the sturgeon (*Scaphirhynchus*) are examples (Fig. 30.14). The infraclass **Holostei** has also dwindled and is represented today by such relict species as the gar (*Lepisosteus*) and bowfin (*Amia*). The infraclass **Teleostei**, in contrast, has been continuously expanding since its origin in the Mesozoic era. It is to this group that the minnows, perch and most familiar fishes belong.

Various evolutionary tendencies can be traced through this succession. The functional lungs of early actinopterygians (still retained in *Polypterus*) became transformed into swim bladders with little respiratory function. Correlated with increased buoyancy and better streamlining, we find that the primitive heterocercal tail of most chondrosteans (*Polypterus* is an exception) became superficially symmetrical in teleosts, but the caudal skeleton still shows indications of the upward tilt of the vertebral column. The caudal fin rays attach to **hypural bones,** which are modifications of the haemal spines that attach to the ventral surface of the tail vertebrae (Fig. 30.15). Such a tail is said to be **homocercal.** Holosteans have an intermediate **abbreviated heterocercal** tail. Early actinopterygians were clothed with thick bony scales of the **ganoid type** (Fig. 30.4B). During subsequent evolution the superficial layers were lost, and the bone was reduced to a thin disc which develops in the dermis of the skin (Fig. 30.4D). Such a scale is termed **cycloid** if its surface is smooth, **ctenoid** if the posterior portion bears minute spiny processes, or ctenii. As the fish grows, increments of bone are added to the scale and these appear as rings, or **circuli.** In some fish, in

Figure 30.14 A group of primitive ray-finned fishes that have survived to the present day. *A*, The Nile bichir, *Polypterus; B,* the shovel-nosed sturgeon, *Scaphirhynchus platorhynchus; C,* the longnose gar, *Lepisosteus oseus; D,* the bowfin, *Amia calva.* (*A* after Dean; *B, C* and *D,* courtesy of the American Museum of Natural History.)

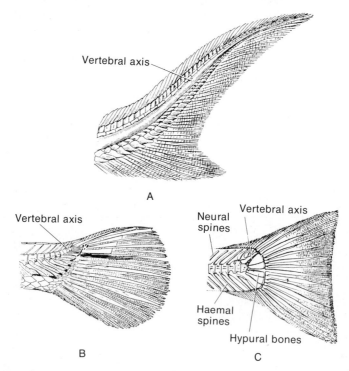

Figure 30.15   Caudal fin types of bony fishes. *A*, The primitive heterocercal tail seen in the sturgeon, *Acipenser;* *B*, the abbreviated heterocercal tail of the garpike, *Lepisosteus; C*, the homocercal tail of a teleost. (*A* and *B* from Jordan; *C* from Romer.)

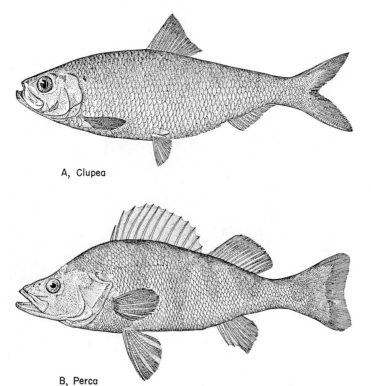

A, Clupea

B, Perca

Figure 30.16   *A*, A primitive teleost, the herring; *B*, an advanced teleost, the perch. Differences in body proportions and fins are apparent. (From Romer: Vertebrate Body, 4th ed.)

which the rate of growth slows down in the winter, the circuli are much closer together and form a check mark. By counting such marks, and not the circuli themselves, estimates of a fish's age can be made.

Teleosts are an exceedingly large and diverse infraclass, for there are 30,000-odd species. The evolutionary relationships between the various groups is far from certain, but certain evolutionary trends are quite apparent. The more primitive teleosts, the tarpons and herrings (Fig. 30.16), are characterized by having elongate, streamlined bodies; a single dorsal fin; pelvic fins lo-

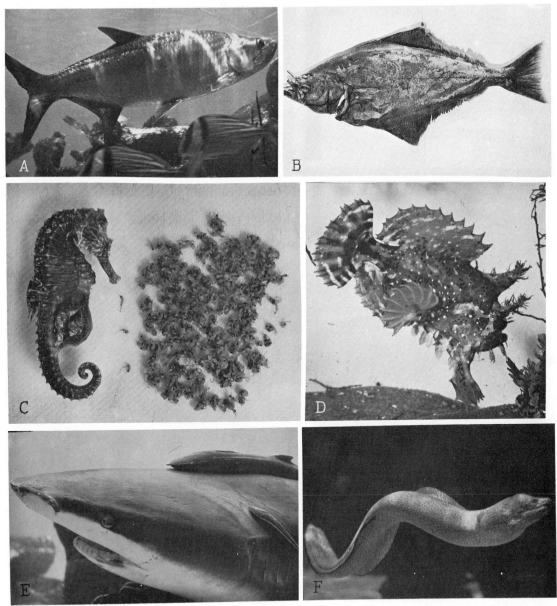

Figure 30.17 Diversity among the teleosts. A, A tarpon, *Tarpon*, one of the more primitive types of teleosts; B, a halibut, *Hippoglossus*, one of the flatfish that feed along the bottom; C, a male sea horse, *Hippocampus*, with the brood pouch in which the female deposits her eggs, young shown at right; D, the sargassum fish, *Histrio*, appears very bizarre out of its natural environment, but is well concealed among seaweed; E, the sharksucker, *Remora*; F, the moray eel, *Gymnothorax*, normally lurks within interstices of coral reefs. (A, C, D, E and F, courtesy of Marine Studios; B, courtesy of the American Museum of Natural History.)

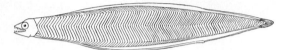

Figure 30.18   The transparent leptocephalus larva of the eel. A similar larval stage occurs during the development of the tarpon. (From Jordan after Eigenmann.)

cated near the posterior part of the trunk; fins supported by flexible and branching bony rays rather than by spines; cycloid scales covering the tail and trunk, but not

extending onto the head; an air, or pneumatic, duct connecting the swim bladder and digestive tract. The most advanced are the spiny-finned teleosts such as the sunfishes and perch. These are often rather short and deep-bodied (from dorsal to ventral) fishes. There is a tendency for the dorsal fin to split into two parts; the anterior being supported by spines, the posterior by flexible bony rays. The pelvic fins have shifted forward to a point beneath the pectoral fins, and spines are present in the an-

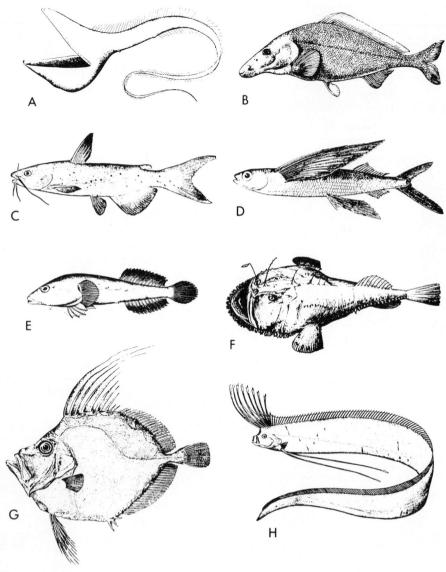

Figure 30.19   Further diversity among the teleosts. *A*, The deep-sea gulper eel, *Saccopharynx*; *B*, *Mormyrus*, one of the primitive fresh-water osteoglossomorphs; *C*, the catfish, *Ictalurus*, one of the ostariophysians; *D*, the flying fish, *Exocoetus*, a specialized top minnow; *E*, the clingfish, *Caularchus*, with pelvic fins modified as a sucking disc; *F*, the angler, *Lophius*, with its tassel-like lure; *G*, the John Dory, *Zeus*; *H*, the oarfish, *Regalecus*. (*C* and *E* from Jordan; others from Romer.)

terior border of these fins. The scales have become ctenoid and have extended onto the head and gill covering. The pneumatic duct is lost.

Teleosts between these two extremes show a mixture of primitive and advanced characteristics and often certain distinctive features of their own. Eels, for example, have long snakelike bodies and have lost their pelvic fins and usually their scales (Fig. 30.17E). A relationship between eels and the more primitive tarpon is suggested, for both have a similar, ribbon-shaped, transparent larval stage known as the **leptocephalus** (Fig. 30.18). Most of our common fresh-water teleosts such as minnows, suckers and catfish (Fig. 30.19C) resemble primitive teleosts in the various features discussed, but are distinctive in having a small set of bones, the **weberian ossicles,** extending from the anterior end of the swim bladder to the inner ear (Fig. 30.20). This mechanism apparently acts as a hydrophone, for the acoustical sensitivity of these fishes is considerably greater than that of any other group. Other groups of teleosts are defined in the synopsis at the end of this chapter.

The more primitive herring and tarpon (Fig. 30.17A), are active, predaceous, streamlined fishes of the open waters. But there have been many interesting departures from these generalized types, for teleosts have spread out into all parts of the aquatic environment and have become adapted to nearly every conceivable ecologic niche. This phenomenon of **adaptive radiation** is seen in all large groups. Apparently, the resources of the environment can be utilized more fully if subgroups become specialized for certain parts of the environment than if all compete with one another in the total environment.

The halibut, soles and flounders have become specialized for a bottom-dwelling life. Like the skates and rays, they are flattened and glide along the bottom with up-and-down undulatory movements. But instead of being flattened dorsoventrally, they are greatly compressed from side to side, and swim turned over on one side (Fig. 30.17B). During larval development, the eye that would be on the "ventral" side migrates to the top surface, but the mouth does not change position. The skates and flounders present a good example of **convergent evolution,** by which animals that are widely separated in the evolutionary scale independently adapt to similar modes of life. They acquire similar adaptive features, in this case a flattened body shape, though in different ways.

Other teleosts have adapted to a life among seaweeds and in coral reefs. The sea horses with their monkey-like, prehensile tails, the sargassum fish with its camouflaging color and weedlike protuberances, and the elongate, snakelike moray eels are examples (Fig. 30.17).

A few teleosts have adapted to life in the ocean depths. Such fish often have light-producing luminescent organs presumably for species recognition, and large mouths and greatly distensible stomachs that enable them to take full advantage of the occasional meal that comes their way (Fig. 30.19A).

Some teleosts live in intimate association with other fishes. The remora has an anterior dorsal fin that is modified as a suction cup and is used to attach to sharks. It feeds upon crumbs of the larger fish's meals, or obtains free rides to favorable feeding grounds. Relationships of this type, in which one organism benefits and the other receives neither benefit nor harm, are known as **commensalism** (Fig. 30.17E).

A few teleosts have become amphibious. The Australian mudskipper frequently hops about on the mud flats of mangrove swamps at low tide in search of food, and may even bask in the sun. It has unusually muscular pectoral fins with which it pulls itself along the land, and it can close its opercular chamber and extract oxygen from the air with its gills.

A number of fishes, both cartilaginous and bony, have evolved (from muscular tissue)

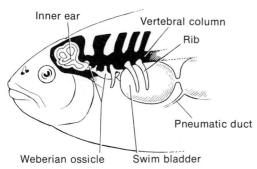

Figure 30.20 A lateral view of the head of a carp to show the connection between inner ear and swim bladder made by the weberian ossicles. (From Portman.)

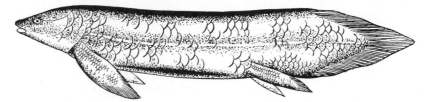

Figure 30.21   The Australian lungfish, *Epiceratodus*. (After Norman.)

electric organs capable of discharging pulses of electrical energy. In such fishes as the electric ray (*Torpedo*) of the North Atlantic and the electric eel (*Electrophorus*) of the Amazon basin the pulses are strong enough to stun prey or predator. *Electrophorus* can put out pulses in excess of 500 volts! In other fishes, the pulses are very weak and their biologic significance was long unknown. Researchers have found that they are part of an electric guidance system. Dr. H. W. Lissmann of the University of Cambridge has found that *Gymnarchus,* a nocturnal fresh-water fish of West Africa, gives off a continuous stream of weak electric pulses that set up an electric field in the surrounding water, resembling a dipole. Objects having an electrical conductivity different from that of the water distort the field, and these changes can be detected by the fish by way of modified portions of its lateral line system. Dr. Lissmann has postulated that electric organs of this type represent an intermediate evolutionary stage between typical muscle with its action potential measured in millivolts and the stunning electric organs. Darwin could not explain how natural selection could favor the evo-

lution of stunning electric organs for, at the lower levels of efficiency, they would be useless. Probably they had a different function when they were first evolving. After a certain threshold of power was reached, the function could change.

Many other fascinating adaptations are found among teleosts, but we must not dwell upon them for these fishes are only a side issue in the total picture of vertebrate evolution. The branch toward the higher vertebrates passed through the less spectacular Sarcopterygians of ancient Devonian swamps.

Sarcopterygians. The sarcopterygians include the lungfishes and crossopterygians (Figs. 30.21 and 30.22). Their paired appendages are typically elongate and lobe-shaped and are supported internally by an axis of flesh and bone. In many species, each of the olfactory sacs connects to the body surface through an **external nostril** and to the front part of the roof of the mouth cavity through an **internal nostril.** Sarcopterygian evolution diverged at an early time into two lines—the lungfishes (order **Dipnoi**) and the crossopterygians (order **Crossopterygii**). The primitive crossopterygians

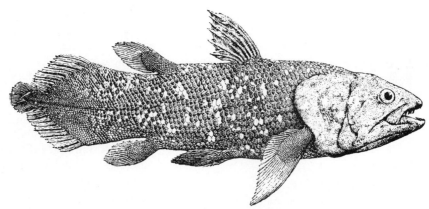

Figure 30.22   *Latimeria*, a living coelacanth found off the coast of the Comoro Islands. (From Millot.)

were the less specialized, having a well-ossified internal skeleton and small conical teeth suited for seizing prey. It is from this group that the amphibians arose. Lungfishes early in their evolution developed specialized crushing tooth plates, which enabled them to feed effectively upon shellfish, and showed tendencies toward reduction of the internal skeleton and paired appendages. In certain other features lungfishes and crossopterygians have paralleled actinopterygian evolution. They evolved symmetrical tails, though of a type that is symmetrical internally as well as externally (**diphycercal**), and their primitive, thick bony scales, which were characterized by having a thick

layer of dentine-like **cosmin** (Fig. 30.4), have tended to thin to the cycloid type.

Both crossopterygians and lungfishes were successful in the fresh waters of the Devonian, but have dwindled to a few relict species today. Lungfishes retained their lungs and have survived in the unstable fresh-water environments of tropical South America, Africa and Australia. It was long believed that all crossopterygians had become extinct, as indeed the primitive fresh-water ones have. However, a few specimens of a somewhat specialized side branch (the coelacanths) have been found in recent years near the Comoro Islands between Africa and Madagascar (Fig. 30.22).

* CLASSIFICATION OF FISHES

CLASS AGNATHA. Jawless fishes. The oldest and most primitive class of vertebrates; all members lack jaws and pelvic fins; many lack pectoral fins.

† *Ostracoderms.* Several orders of small, Paleozoic fishes are collectively called the ostracoderms; most were heavily armored. *Hemicyclaspis.*

Order Cyclostomata. Lampreys, hagfishes. Body eel-shaped; round mouth without jaws; no paired appendages; scaleless slimy skin. The sea lamprey, *Petromyzon.*

† **CLASS PLACODERMI.** Several orders of primitive jawed fish of the Paleozoic; bony head shield movably articulated with trunk shield; probably related to cartilaginous fishes. *Dunkleosteus.*

CLASS CHONDRICHTHYES. Cartilaginous fishes. Skeleton entirely of cartilage; placoid scales usually present; lungs or swim bladder absent; pelvic clasper in male; nearly all are marine.

Subclass Elasmobranchii. Gill slits open independently on the body surface; placoid scales present; lateral line system imbedded in the skin.

† *Order Cladoselachii.* Ancestral sharks with broad-based paired fins. *Cladoselache.*

Order Selachii. Modern sharks with narrow-based paired fins; streamlined bodies; heterocercal tail; most with sharp triangular teeth. The dogfish, *Squalus;* whale shark, *Rhincodon.*

Order Batoidea. Sawfish, skates, rays. Body dorsoventrally flattened; pectoral fins enlarged; trunk and tail reduced; teeth usually in the form of crushing plates. Common skate, *Raja;* devilfish, *Manta;* sawfish, *Pristis.*

Subclass Holocephali. Gill slits covered by an operculum; scales absent; lateral line system in the form of open grooves.

Order Chimaeriformes. The ratfish, *Chimaera.*

CLASS OSTEICHTHYES. Bony fishes. Skeleton includes considerable bone; various types of bony scales, other than placoid, usually present; lungs or swim bladder usually present; abundant in fresh and salt water.

† **Subclass Acanthodii.** Several orders of Paleozoic fishes. Oldest jawed fishes; a prominent spine in front of dorsal, anal and paired fins. The "spiny shark," *Climatius.*

Subclass Actinopterygii. Ray-finned fishes. Paired fins fan-shaped, supported by radiating bony rays or spines; each nasal cavity has an entrance and exit on snout surface.

---

* Based on the classification in A. S. Romer: Vertebrate Paleontology, 3rd ed. Chicago, University of Chicago Press, 1966. The classification is taken down to orders, and in a few cases suborders. All living groups are included, but only the more important extinct groups are mentioned.

† Extinct groups.

INFRACLASS CHONDROSTEI. Two extinct and two living orders of primitive ray-finned fishes. Scales usually ganoid; mouth opening large; tail usually heterocercal.

> **Order Polypteriformes.** The bichir. Body slender; numerous dorsal fins; lungs present; tropical Africa. *Polypterus.*
>
> **Order Acipenseriformes.** Sturgeon and allies. Long sharklike snout; subterminal mouth; several rows of heavy bony scales. Common sturgeon, *Acipenser;* paddlefish, *Polydon.*

INFRACLASS HOLOSTEI. Three extinct and two living orders of intermediate ray-finned fishes. Scales ganoid or cycloid; mouth opening smaller than in chondrosteans; abbreviated heterocercal tail.

> **Order Semionotiformes.** Garpikes. Elongated trunk and snout; scales ganoid. *Lepisosteus.*
>
> **Order Amiiformes.** The bowfin. Snout rounded; scales cycloid; long dorsal fin. *Amia.*

INFRACLASS TELEOSTEI. Advanced ray-finned fishes. Scales cycloid or ctenoid; mouth opening small; tail homocercal; a hydrostatic swim bladder usually present.

> † **Superorder Leptolepimorpha.** One order of primitive herring-like teleosts still retaining some holostean characters, including traces of ganoine on scales.
>
> **Superorder Elpomorpha.** Primitive teleosts with a "leptocephalous" larva.
>
> > **Order Elpoiformes.** Tarpons. Often large, rapidly swimming oceanic fishes; much prized for sport. *Tarpon.*
> >
> > **Order Anguilliformes.** Eels. Body elongate; scales rudimentary or absent; pelvic fins usually absent. American eel, *Anguilla.*
> >
> > **Order Notacanthiformes.** Gulper eels. Deep-sea, somewhat eel-shaped fishes; large mouths. *Saccopharynx.*
>
> **Superorder Clupeomorpha.**
>
> > **Order Clupeiformes.** Herring, sardines, shad and their allies. Primitive teleosts; scales cycloid; head and operculum not scaled; fins without spines; trunk not greatly shortened; single dorsal fin; pelvic fins abdominal; a duct connects swim bladder and digestive tract; most marine. Common herring, *Clupea.*
>
> **Superorder Osteoglossomorpha.**
>
> > **Order Osteoglossiformes.** Primitive fresh-water teleosts of the Old and New World tropics. The elephant-snouted fish, *Mormyrus.*
>
> **Superorder Protacanthopterygii.** Primitive teleosts, some of which are beginning to acquire certain of the advanced features of acanthopterygians.
>
> > **Order Salmoniformes.** Salmon, trout, pike. A fleshy adipose fin is often present on the back. The salmon, *Salmo.*
> >
> > **Order Cetomimiformes.** A few unusual deep-sea fishes.
> >
> > **Order Ctenothrissiformes.** Possibly ancestral spiny-finned teleosts; most are extinct.
> >
> > **Order Gonorhynchiformes.** Milkfishes.
>
> **Superorder Ostariophysi.** A large group of primarily fresh-water teleosts; primitive in most respects, but have webberian ossicles.
>
> > **Order Cypriniformes.** Carp, minnows, suckers and their allies. Primitive ostariophysians; usually retain cycloid scales; most lack spines in fins. Carp, *Cyprinus;* shiner, *Notropis;* sucker, *Catostomus.*
> >
> > **Order Siluriformes.** Catfish. Specialized ostariophysians; scales usually lost; a large spine in dorsal and pectoral fins; long sensory barbels on head. Channel catfish, *Ictalurus.*
>
> **Superorder Atherinomorpha.**
>
> > **Order Atheriniformes.** Top minnows, flying fishes and their allies. Teleosts with a mixture of primitive and advanced characters; scales usually cycloid; scales beginning to spread onto head and operculum; some spines present in fins; pelvic fins abdominal in position. Killifish, *Fundulus;* silverside, *Atherina;* flying fish, *Exocoetus.*
>
> **Superorder Paracanthopterygii.** Several small orders of specialized teleost fishes with advanced characteristics. Scales usually ctenoid; spines usually present in fins; pelvic fins located far forward.

Order Amblyopsiformes. Pirate perch. Anus located far forward under gill region. Pirate perch, *Aphredoderus.*

Order Batrachoidiformes. Toadfishes. Bizarre teleosts with large, somewhat flattened heads and long tapering trunks. *Batrachoides.*

Order Gobiesociformes. Clingfishes. Pelvic fins located far forward and modified as a sucking disc; often found in tidepools. *Caularchus.*

Order Lophiiformes. Anglers. First dorsal spine carries a tassel that dangles as a lure over large mouth; most deep sea. Common angler, *Lophius.*

Order Gadiformes. Cod, haddock. Elongate and subdivided dorsal and anal fins. Cod, *Gadus;* burbot, *Lota.*

Superorder Acanthopterygii. Spiny-finned teleosts. Advanced teleosts; scales ctenoid; head and operculum usually scaled; spines present in fins; trunk often short; often two dorsal fins; pelvic fin thoracic in position; swim bladder without a duct.

Order Beryciformes. Squirrelfish. Moderately deep-bodied fishes; retain 18 or 19 rays in the tail fin in contrast to 17 in more advanced acanthopterygians.

Order Zeiformes. John Dory, *Zeus.*

Order Lampidiformes. Ribbonfish, oarfish. Elongate oceanic fish; some species 6 meters long. The oarfish, *Regalecus.*

Order Gasterosteiformes. Sticklebacks, pipefish, sea horses. Trunk encased in bony armor; small mouth at end of tubular snout. The sea horse, *Hippocampus.*

Order Channiformes. Snakeheads. Small Asiatic coastal fish with lung-like extensions from gill chambers; can live out of water. *Channa.*

Order Synbranchiformes. Eel-like, tropical coastal fish. Swamp eel.

Order Scorpaeniformes. Scorpion fish, sculpins. Enlarged heads and pectoral fins; projecting spines from gill covering. The sculpin, *Acanthocottus.*

Order Dactylopteriformes. Enlarged pectoral fins; glide in a manner similar to the more familiar flying fishes.

Order Pegasiformes. Dragonfishes. Small marine fishes of tropical seas; trunk and tail encased in a bony box or rings. *Pegasus.*

Order Perciformes. The largest order of spiny-finned teleosts; includes perch, bass, mackerel, gobies, barracuda and their allies. A relatively unspecialized group of spiny-finned teleosts. The perch, *Perca.*

Order Pleuronectiformes. Flatfishes. Body highly compressed; fish lie on ocean bottom on one side of body; both eyes on upper side of body. Halibut, *Hippoglossus;* flounder, *Paralichthys.*

Order Tetraodontiformes. Triggerfish, trunkfish, puffers. Strong jaws with a sharp beak; scales often spiny; some inflate by swallowing water. Porcupine fish, *Diodon.*

Subclass Sarcopterygii. Fleshy-finned fishes. Paired fins lobe-shaped with a central axis of flesh and bone; internal nostrils usually present; primitive members had cosmoid scales.

Order Crossopterygii. Crossopterygians. Primitive fleshy-finned fishes, conical teeth present; median eye usually present.

†SUBORDER RHIPIDISTIA. Primitive fresh-water crossopterygians; includes the ancestor of amphibians. *Osteolepis.*

SUBORDER COELACANTHINI. More specialized fresh-water and marine crossopterygians; one living genus. *Latimeria.*

Order Dipnoi. Lungfishes. Specialized fleshy-finned fishes; teeth forming crushing tooth plates; median eye usually absent, three living genera confined to Australia (*Neoceratodus*), Africa (*Protopterus*) and South America (*Lepidosiren*).

ANNOTATED REFERENCES

Vertebrates

The following references contain useful information on many aspects of the biology of fishes and other groups of vertebrates.

Blair, W. F., A. P. Blair, P. Brodkorb, F. R. Cagle, and G. A. Moore: Vertebrates of the United States. 2nd ed. New York, McGraw-Hill Book Company, 1968. A very useful taxonomic treatment of the vertebrates of the United States, apart from Alaska and Hawaii. Keys for all groups are included, and each group and species is discussed briefly.

Colbert, E. H.: Evolution of the Vertebrates. 2nd ed. New York, John Wiley & Sons, Inc. 1969. An excellent and thorough account of vertebrate evolution; all groups, both living and extinct, are considered briefly.

Gregory, W. K.: Evolution Emerging, A Survey of Changing Patterns from Primeval Life to Man. New York, Macmillan, 1951. A valuable and superbly illustrated source book on vertebrate evolution; includes much original material.

Olson, E. E.: Vertebrate Paleozoology. New York, Wiley-Interscience, 1971. A useful reference on vertebrate paleontology; discusses the adaptive nature of many evolutionary changes.

Orr, R. T.: Vertebrate Biology. 3rd ed. Philadelphia, W. B. Saunders Co., 1971. A valuable text and reference on many aspects of vertebrates; a chapter is devoted to each major group of vertebrates and to such general topics as territory, dormancy and population dynamics.

Romer, A. S.: The Vertebrate Story. 4th ed. Chicago, University of Chicago Press, 1959. A very well written and nontechnical account of the evolution of vertebrates.

Romer, A. S.: Notes and Comments on Vertebrate Paleontology. Chicago, University of Chicago Press, 1968. An informal discussion of interesting points and speculations deemed inappropriate for the more formal book on vertebrate paleontology.

Romer, A. S.: Vertebrate Paleontology. 3rd ed. Chicago, University of Chicago Press, 1966. The most recent edition of a standard text and source book.

Young, J. Z.: The Life of Vertebrates. 2nd ed. Oxford, Clarendon Press, 1963. A fascinating account of the evolution and adaptations of vertebrates. One or more chapters are devoted to each class; morphological and physiological perspectives are skillfully interwoven.

Fishes

Breder, C. M., Jr.: Field Book of Marine Fishes of the Atlantic Coast from Labrador to Texas. 2nd ed. New York, S. P. Putnam's Sons, Inc., 1948. A useful guide for the identification of the more common fishes of our Atlantic Coast.

Herald, E. S.: Living Fishes of the World. Garden City, N.Y., Doubleday & Co., 1961. A very well written, nontechnical account of the biology of the various groups of fish. As with other books in this series, the photographs are superb; many are in color.

Hoar, W. S., and D. J. Randall (Eds.): Fish Physiology. New York, Academic Press, 1969–1970. A valuable source book; chapters on various aspects of fish physiology have been written by leading investigators.

Hubbs, C. L., and K. F. Lagler: Fishes of the Great Lakes Region. 2nd ed. Bloomfield Hills, Mich., Cranbrook Institute of Science, 1958. The definitive guide to fishes of the Great Lakes Basin, this book would also be useful, though not infallible, for adjacent parts of the United States.

Lagler, K. F., J. E. Bardach, and R. R. Miller: Ichthyology. New York, John Wiley & Sons, Inc., 1962. A very useful text covering the basic anatomy, physiology, systematics and ecology of fishes.

Marshall, N. B.: The Life of Fishes. Cleveland, World Publishing Co., 1966. A very readable account of the biology of fishes written by a senior investigator at the British Museum of Natural History.

Moy-Thomas, J. A.: Paleozoic Fishes. 2nd ed. by R. S. Miles. Philadelphia, W. B. Saunders Company, 1971. An extensive revision of the classic book on primitive fishes.

Norman, J. R.: A History of Fishes. 2nd ed. Revised by P. H. Greenwood. New York, Hill and Wang, Inc., 1963. A recent edition of a classic on the biology of fishes.

Tee-van, J., H. B. Bigelow, and G. W. Mead (Eds.): Fishes of the Western North Atlantic. New Haven, Sears Foundation for Marine Research, Yale University Press, 1948, 1953, 1963, 1964, 1966. Five parts of this definitive treatise on the taxonomy and biology of Atlantic fishes have been published.

31. Vertebrates: Amphibians and Reptiles

31.1 THE TRANSITION FROM WATER TO LAND

The transition from fresh water to land was a momentous step in vertebrate evolution that opened up vast new areas for exploitation. It was an extremely difficult step because the physical conditions on land are so very different from those in water. Air neither affords so much support nor offers so much resistance as water. The terrestrial environment provides little of the essential body water and salts. Oxygen is more abundant in the air than in water, but it must be extracted from a different medium. The ambient temperature fluctuates much more on the land than in the water. Air and water have different refractive indices.

Successful adaptation to the terrestrial environment necessitated changes throughout the body. Stronger skeletal support and different methods of locomotion evolved. Changes occurred in the equipment for sensory perception and changes in the nervous system were a natural corollary of the more complex muscular system and altered sense organs. An efficient method of obtaining oxygen from the air evolved, as did adaptations to prevent desiccation. The delicate, free-swimming, aquatic larval stage was suppressed, and reproduction upon land became possible. Finally, the ability to maintain a fairly constant and high body temperature was achieved, and terrestrial vertebrates could then be active under a wide range of external temperatures.

In view of the magnitude of these changes, it is not surprising that the transition from water to land was not abrupt but took millions of years and involved the participation of many groups. Indeed, a main theme in the evolution of the terrestrial vertebrates, or **tetrapods,** has been a continual improvement in their adjustment to terrestrial conditions.

The crossopterygians unwittingly made the first steps in this transition. Their lungs, as we have seen, were probably an adaptation to survive conditions of stagnant water or temporary drought. Their relatively strong, lobate paired fins may have enabled them to squirm from one drying and overcrowded swamp to another more favorable one. Crossopterygians were not trying to get onto the land but, in adapting to their own environment, they evolved features that made them viable in a new and different environment. That is, they became **preadapted** to certain terrestrial conditions. Given this preadaptation, an abundance of food (various invertebrates, stranded fishes, plants) upon the land or the shores of swamps, little competition upon the land, and a constriction of favorable fresh water habitats, it is not hard to imagine some of the crossopterygians making the adaptive shift from water to land and becoming the amphibians (Fig. 31.1). No one knows how

Figure 31.1 A restoration of life in a Carboniferous swamp 345 million years ago. The labyrinthodont amphibians were the first terrestrial vertebrates. (Courtesy of the American Museum of Natural History.)

long the transition from crossopterygians to amphibians took, but the first amphibian fossils are found in strata that were formed nearly 50 million years later than those containing the first crossopterygians.

Amphibians, in turn, acquired additional terrestrial features, and reptiles still more. But the pinnacle of terrestrial adaptation is achieved only by the reptiles' descendants — the birds and mammals.

31.2   CHARACTERISTICS OF AMPHIBIANS

During the evolution of the amphibians, many fishlike characteristics of their ancestors were lost, and many features evolved that enabled them to spend considerable time on land. The vertebral column is stronger with more thoroughly ossified vertebrae, and the vertebrae interlock securely by means of overlapping **articular processes** on the vertebral arches (Fig. 11.8, p. 265). These features help provide the necessary support on land. The paired fins of fishes are transformed into **legs,** but the limbs are in a somewhat sprawled position, and the humerus and femur move back and forth in the horizontal plane. Fishlike undulations

of the trunk and tail help to advance the limbs to a position favorable for placing them on the ground, and may help in developing the thrust of the limb on the ground (Fig. 31.2). The footfall pattern seen in amphibians, and even in slower moving higher vertebrates that do not undulate their trunk, follows naturally from the positions assumed by the fins of a fish as a result of the undulations of the trunk. First the left front foot is advanced, and this is followed in turn by the right rear foot, right front foot and left rear foot. During most stages of a stride, just one foot is advanced, leaving a tripod of three feet to support the body.

Important changes have occurred in the respiratory system. Larval amphibians retain gills, but these are lost in adults; **lungs,** supplemented by the skin and buccopharyngeal membrane, become the sites of gas exchange. The loss of bony scales, of course, is a prerequisite for the skin to serve as a respiratory membrane. As explained earlier (p. 292), air is pumped in and out of the lungs by movements of the floor of the

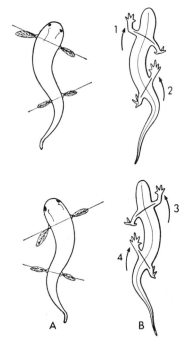

Figure 31.2 A comparison of locomotion in (A) a fish and (B) a salamander. Fish-like undulations of trunk and tail continue in the salamander and help to determine the placement of the limbs upon the ground. Numbers indicate the sequence of limb movements. (A after Marshall and Hughes; B after Romer.)

mouth and pharynx. These are very similar to the movements which in a fish circulate water through the pharynx and across the gills. In the evolution of the respiratory system, we see a good example of the continuity of structure and function which must exist in any evolutionary transition. Lung ventilation by rib and diaphragm movements evolved much later.

The substitution of lungs for gills as the major site of gas exchange has important consequences in the circulatory system. Two types of blood, depleted blood from the body and oxygen-rich blood from the lungs, return to the heart. These two blood streams are kept separate to some extent as they flow through the heart (see p. 311). Depleted blood for the most part is sent to the lungs and skin, and oxygen-rich blood is distributed to the body. Since blood leaving the heart on its way to the body does not first go through the gills, as it would in a fish, there is little loss of blood pressure. Blood pressure in the dorsal aorta of a frog is nearly double that in the dorsal aorta of a dogfish. In this respect, terrestrial vertebrates have a more efficient circulatory system than do fishes.

Catching and swallowing food on land is somewhat more complex than in the water, and amphibians have evolved a muscular **tongue** and **oral glands** that aid in these processes.

Many changes occur in the sensory apparatus, for problems associated with perception on land are quite different from those in the water. The lateral line sensory system is retained by larval amphibians but is lost in the adult. Also, at the time of metamorphosis, **eyelids** and **tear glands** develop, which protect and cleanse the eye in the aerial environment, and an **ear** sensitive to air- or ground-borne vibrations also appears. A number of changes occur in the nervous system correlated with changed patterns of locomotion and changes in the sensory apparatus.

Although well adapted to the terrestrial environment in many ways, amphibians have retained several features from their piscine ancestors that restrict most of them to a rather moist habitat and prevent them from fully exploiting the terrestrial environment. First, most amphibians are unable to prevent a large loss of body water when on land. Since their skin is a thin, moist, vascular membrane, considerable evaporation takes place through it. There has been some shift in the nitrogen metabolism of amphibians so that less of the waste products are removed as highly toxic ammonia, and more are removed as **urea** (Table 31.1). Amphibians do not need as rapid a turnover of water as the fresh water fishes do, yet a fairly large volume is still needed to flush these products from the tissues and body. The structure of their kidney tubules is not yet modified for the reabsorption of a large volume of water from the urine during the process of urine formation.

A second restricting feature is that all amphibians are cold-blooded, or **ectothermic;** their body temperature is close to that of the environment and fluctuates with it. They cannot maintain a constant and rather high body temperature. Since the rate of metabolic processes fluctuates with temperature changes, they cannot be active at low temperatures. This is not too serious for a fish for, in most bodies of water, temperatures, even beneath the ice, remain above freezing. But terrestrial ectotherms living in temperate regions must move during the winter to areas that do not freeze and enter a dormant state known as **hibernation.** Amphibians bury themselves in the mud at the bottom of ponds or burrow into soft ground below the frost line. During hibernation metabolic activities are at a minimum. The only food utilized is that stored within the body; respiration and circulation are very slow. Some tropical amphibians during the hottest and driest parts of the year go into a comparable dormant state known as **aestivation.**

Finally, amphibians are unable to repro-

TABLE 31.1 TYPES OF NITROGEN EXCRETION IN REPRESENTATIVE VERTEBRATES °

Animal	Ammonia	Urea	Uric Acid	Other
Fresh-water minnow, *Cyprinus*	60.0	6.2	0.2	33.6
Bullfrog, *Rana*	3.2	84.0	0.4	12.4
European tortoise, *Testudo*	4.1	22.0	51.9	22.0
Hen	3.0	10.0	87.0	0.0
Adult man	3.5	85.0	2.2	9.3

° Data from Prosser and Brown. Figures in per cent.

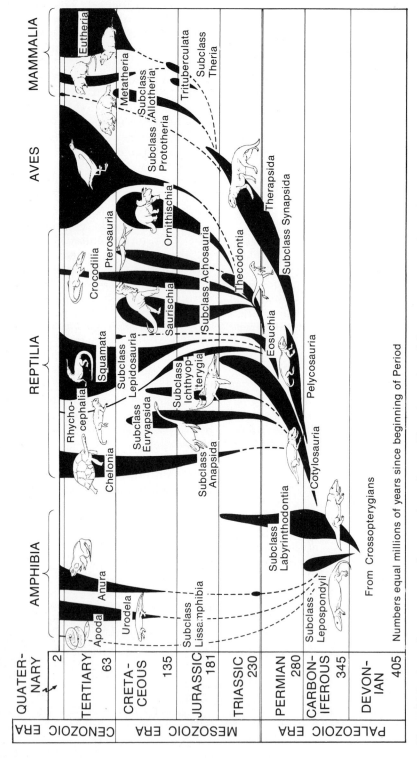

Figure 31.3 An evolutionary tree of terrestrial vertebrates. The relationship of the various groups, their relative abundance and their distribution in time are shown.

duce under truly terrestrial conditions. Most species must return to the water to lay their eggs. Even the terrestrial toad returns to this medium, for it has no means of internal fertilization and sperm cannot be sprayed over eggs upon the land. Neither has it suppressed the free larval stage in development, and these larvae cannot withstand the rigors of the terrestrial environment.

31.3 EVOLUTION OF AMPHIBIANS

The best known of the early amphibians are the **labyrinthodonts,** a group that finally diverged from the crossopterygians during the late Devonian period (Fig. 31.3). An interesting detail they shared with the crossopterygians was a peculiar labyrinthine infolding of the enamel in their teeth. The name labyrinthodont is derived from this feature. All were fairly clumsy, salamander-shaped creatures with rudimentary necks and heavy, muscular tails inherited from their piscine ancestors. Although they had legs and other terrestrial adaptations, many labyrinthodonts must have spent a great deal of time in the water. One early genus had fish-like rays supporting a tail fin. All became extinct during the Triassic. However, the group is important, for it included not only the ancestors of other amphibians but also those of reptiles and, hence, of all higher tetrapods.

Another group of extinct amphibians, the **lepospondyls,** although probably derived from the labyrinthodonts, differed from them in having solid, spool-shaped vertebral centra rather than centra composed of several distinct chunks of bone. The vertebral structure of living amphibians resembles that of this group and it is generally held that the lepospondyls are the ancestors of the present-day amphibians.

Living amphibians are grouped into three orders. Salamanders (order **Urodela**) retain the primitive body form with short legs, long trunk and well developed tail. Caecilians (order **Apoda**) are legless, burrowing, wormlike forms confined to the tropics (Fig. 31.4). Frogs and toads (order **Anura**), with their short trunk and powerful, enlarged hind legs are specialized for a hopping and jumping mode of life (Fig. 31.5). The fossil history of all these groups is fragmentary, but we do have one curious frog fossil from the early Triassic (Fig. 31.6) that is intermediate between specialized contemporary frogs and more primitive amphibians. Its skull shape, shortened trunk, elongated ilia in the pelvic girdle, and slightly enlarged hind legs are suggestive of a frog. But the trunk is not as short as in modern frogs, there is a bit of a tail, and there has been no fusion of the bones of the lower arm and lower leg.

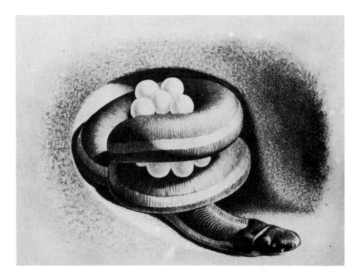

Figure 31.4 A burrowing caecilian, *Ichthyophis,* with her egg mass. (From Romer after the Sarasins.)

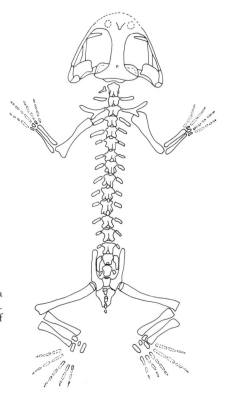

Figure 31.5 Three stages in the leap of a bullfrog. Observe that the pelvic girdle is flexed when the frog is at rest and this causes a characteristic bump in the back (*A*); the hind legs and pelvic girdle extend at take-off (*B*); and the frog lands on its front legs (*C*). (Courtesy of Mr. Earle R. Edmiston.)

Figure 31.6 Restoration of the skeleton of *Triadobatrachus*, a Triassic frog-like amphibian possibly ancestral to modern frogs. (After Estes in Vial (Ed.): Evol. Biol. Anurans, University of Missouri Press, 1973.)

31.4   AMPHIBIAN ADAPTATIONS

Salamanders.   Salamanders are not such familiar amphibians as frogs and toads, for most have secretive habits. They may be found beneath stones and logs in damp woods or beneath stones along the side of streams, and some are entirely aquatic. A rather generalized type is Jefferson's salamander, *Ambystoma jeffersonianum* (Fig. 31.7), of the eastern United States. This species is terrestrial as an adult, but returns to the water in early spring to reproduce. The males deposit sperm in clumps called **spermatophores** on sticks and leaves in the water. Later the females pick these up with their cloacal lips. Fertilization occurs inside the cloaca, and the fertilized eggs are deposited in masses attached to sticks submerged beneath the surface of the water. This habit is adaptive, for these salamanders reproduce very early in the spring, at a time when the surface of the water may freeze over. The larvae of salamanders differ from those of frogs and toads in retaining external gills throughout their larval life and in having true rather than horny teeth.

There have been many special adaptations among salamanders. The most abundant of our American species are woodland types like the red-backed salamander (*Plethodon cinereus*), of the family Plethodontidae. A particularly interesting feature of plethodonts is their complete loss of lungs; gas exchange occurs entirely across the moist membranes lining the mouth and pharynx and the skin. The skin is a more effective respiratory organ than in frogs because the epidermis is very thin and capillaries come close to the surface. Loss of lungs may seem to be a curious adaptation for a terrestrial vertebrate, but it has been postulated that early in their evolution plethodonts became adapted for life in rapid mountain streams. The oxygen content of cold water is rather high, and the low temperature reduces the general rate of metabolism of the animals. Air in the lungs would be disadvantageous under these conditions, for the animals would float and be washed away. Lungs may have been lost in adapting to this habitat. Subsequently many plethodonts entered different environments but never regained the lost lungs.

Several groups of salamanders, including the mudpuppy (*Necturus maculosus*, Fig. 31.8A), have become entirely aquatic. The development of the reproductive organs has

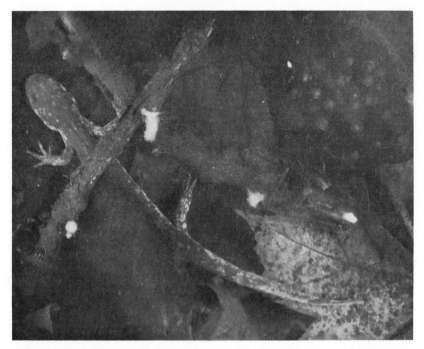

Figure 31.7   Jefferson's salamander, *Ambystoma jeffersonianum*, reproduces in the water. The four white structures attached to sticks are spermatophores. A clump of eggs can be seen in the upper right hand corner.

Figure 31.8 Neotenic salamanders. *A*, The mudpuppy, *Necturus maculosus*, is a permanent larva. *B*, The tiger salamander, *Ambystoma tigrinum*, metamorphoses in most environments, but fails to do so in certain mountain lakes. *C*, The axolotl, or neotenic form of *Ambystoma tigrinum*. (*A* courtesy of Shedd Aquarium, Chicago; *B* and *C* courtesy of the Philadelphia Zoological Society.)

been speeded up in relation to development of other parts of the body. Sexual maturity is achieved in the larval stage and metamorphosis is never completed. This phenomenon is known as **neoteny**. The hormone of the thyroid gland, thyroxine, is necessary for metamorphosis; however, failure of *Necturus* to metamorphose appears to result from the inability of the tissues to respond to thyroxine rather than from an absence of this hormone. Thyroxine is produced, for the thyroid of *Necturus* hastens metamorphosis when transplanted to frog tadpoles. In some other neotenic salamanders, the failure to metamorphose may result from an inhibition of the mechanism that releases thyroxine. The tiger salamander, *Ambystoma tigrinum* (Fig. 31.8*B*), metamorphoses under most conditions, but those living at high altitudes in the Rocky

Mountains fail to do so and remain permanent larvae known as **axolotls.** Apparently, cold inhibits the release of thyroxine, for when axolotls are fed thyroxine or when they are brought to warmer climates, metamorphosis occurs normally.

Frogs and Toads.   Most anurans are amphibious as adults, living near water to which they frequently go to feed or escape danger, but some are more terrestrial in habits, and others have become adapted to an arboreal life. The terms frog, toad, and tree frog or tree toad ordinarily imply amphibious, terrestrial and arboreal modes of life, not natural evolutionary groups. Members of several distinct families of anurans, for example, have become adapted independently to an arboreal life.

Toads have adjusted to a terrestrial life by evolving structures and patterns of behavior that reduce water loss. The epidermis of their skin is more horny and less pervious to water than that of frogs. A thick dry skin reduces cutaneous respiration, but this is compensated for by an increase in the respiratory surface of the lungs. The lining of toad lungs is more complexly folded than that of frogs. Much of the water lost through the kidneys is reabsorbed in the urinary bladder. Toads are crepuscular in habits; they burrow or take shelter by day and come out in the moist evening to feed upon insects.

The chief adaptation to arboreal life has been the evolution of **digital pads** upon the tips of the toes (Fig. 31.9). The surface epithelium of the pads is rough and grips the substratum by friction.

A particularly fascinating aspect of anuran biology is the evolution of methods by which development can proceed elsewhere than in the open water. This has occurred primarily among tropical frogs and probably as a protection against seasonal drought and numerous aquatic enemies such as predaceous insect larvae. A Brazilian tree frog (*Hyla faber*) protects its young by laying its eggs in mud craters which the male builds by circling in shallow water and pushing up mud (Fig. 31.10A). A more

striking means of protection is seen in a small Chilean frog, *Rhinoderma darwinii.* The male of this species stuffs the fertilized eggs into his vocal sacs where they remain until metamorphosis is complete. In the completely aquatic Surinam toad, *Pipa pipa,* of tropical South America, the male pushes the fertilized eggs into the puffy skin on the back of the female. The eggs gradually sink deeper into the skin, and the outer membrane of each one forms a protective cap over its temporary skin pocket. The young emerge as little froglets. The eggs and larvae of all these frogs are equipped with a large supply of yolk, but in other respects the larvae are fairly typical and simply develop in a sheltered environment.

In certain species the vulnerable larval stage is omitted, and the embryo develops directly into a miniature adult. Anurans with **direct development** include the marsupial frog, *Gastrotheca,* and *Eleutherodactylus*—both of the New World tropics. The former carries her eggs in a dorsal brood pouch (Fig. 31.10C); the latter lays eggs in protected damp places such as beneath stones or in the axil of leaves. The jelly layers about the egg of *Eleutherodactylus* help to prevent desiccation; sufficient yolk is stored within the egg for the nutritive

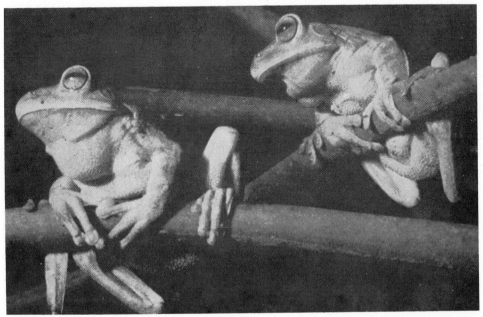

Figure 31.9 The tree frog *Hyla faber* clings to trees by means of its expanded digital pads. (Courtesy of G. A. Lutz, Natural History, Vol. LXX, No. 1.)

A

B

C

Figure 31.10 Adaptations of frogs that protect the larvae from aquatic predators. A, The circular mud crater nests of the Brazilian tree frog, *Hyla faber; B, Pipa pipa*, the female's back is covered with eggs which by the third day have begun the process of sinking into the skin; C, the marsupial frog, *Gastrotheca*, carries eggs in a brood pouch on her back. (A from G. A. Lutz, Natural History, Vol. LXX, No. 1; B from G. Rabb, Natural History, Vol LXX, No. 5; C from E. S. Ross in Orr: Vertebrate Biology, 3rd ed.)

requirements of the embryo; such larval features as horny teeth, gills and opercular fold are vestigial or absent; the fins of the larval tail are expanded, become highly vascular and form an organ for gas exchange; and the period of development is shortened.

Something similar to what has taken place among these frogs today may have occurred among the amphibians which were ancestral to reptiles. If the aquatic larvae of the ancestral amphibians were subjected to a very high predation, any variation that tended toward the suppression of the defenseless larval stage and toward the direct development of their embryos in a less vulnerable environment would have a selective advantage. It is possible that the labyrinthodont amphibians that gave rise to the reptiles developed means of terrestrial reproduction before the adults completely left the water to live on the land.

31.5 CHARACTERISTICS OF REPTILES

Turtles, lizards, snakes, and other reptiles are capable of exploiting the terrestrial environment more completely than amphibians. They have improved upon the means of locomotion, gas exchange and other terrestrial attributes of their amphibian ancestors. In addition, reptiles have made sig-nificant advances over amphibians in being able to conserve body water, reproduce upon the land, and control their body temperature to a limited extent.

A lizard, such as the chuckwalla (*Sauromalus obesus*, Fig. 31.11) of the southwestern United States, is a representative reptile. Its body shape and manner of locomotion are better adapted to land life than to the amphibian condition. The neck is longer and the first two cervical vertebrae are specialized to permit the head to move independently of the rest of the body as the animal feeds. The tail is more slender than in the labyrinthodonts and salamanders. This reflects the decreasing importance of fishlike lateral undulations of the trunk and tail in locomotion, and the increasing importance of the limbs. Well-formed **claws** are borne upon the toes and enable the feet to grip the substratum securely. The more powerful hind legs require a pelvic girdle that is attached more firmly onto the vertebral column. Reptiles typically have two sacral vertebrae, whereas amphibians have only one.

Improved locomotion and increased agility also involve a more elaborate muscular system, nervous system and sense organs. The delicate tympanic membrane is protected by its position deep within a canal, the **external auditory meatus,** and the eye is further protected through the evolu-

Figure 31.11 The chuckwalla (*Sauromalus obesus*) lives in the deserts of the southwestern United States. When unable to shelter, it prevents overheating by panting and placing its body parallel to the sun's rays. (From How Reptiles Regulate Their Body Temperature by C. M. Bogert. Copyright © by Scientific American, Inc. All rights reserved.)

tion of a third, transparent eyelid known as the **nictitating membrane.**

The surface of the skin is covered with dry **horny scales** that minimize water loss by this route. These scales develop through the deposition of **keratin** (a very insoluble and hence waterproofing protein) in the superficial layers of the epidermis. Some lizards and crocodilians also have small plates of dermal bone imbedded in the dermis of the skin beneath many of the horny scales.

The kidney tubules of reptiles are modified in such a way that less water is initially removed from the blood than in amphibians, and much of the water that is removed is later reabsorbed by other parts of the kidney tubule and by the urinary bladder. In reptiles, a large proportion of the nitrogenous waste products is excreted as **uric acid** (Table 31.1). Uric acid is much less toxic and water soluble than ammonia or urea, hence it does not have to be flushed out of the tissues so rapidly nor with such a large volume of water. The urine of animals excreting uric acid typically has a pastelike consistency. The reptilian kidney also differs from that of lower vertebrates in being drained by a duct called the **ureter** instead of by the wolffian duct. The latter becomes a genital duct in males and is lost in females.

The dry horny skin of reptiles reduces cutaneous respiration to a negligible amount, but an increase in the respiratory surface of the lungs not only compensates for this, but also provides for the increased volume of gas exchange necessitated by a general increase in activity. Mechanisms for moving air into and out of the lungs are also more efficient. Instead of pumping air into the lungs by froglike throat movements, reptiles decrease the pressure within their body cavity, and atmospheric pressure drives in air. A subatmospheric pressure is created around the lungs during inspiration by the forward movement of the ribs and the concomitant increase in size of the body cavity. The contraction of abdominal muscles and the elastic recoil of the lungs force out air. Circulatory changes, previously discussed, further separate the oxygenated and unoxygenated blood leaving the heart and make the oxygen supply to the tissues more effective.

Although reptiles are still ectotherms, many can control their body temperature within narrow limits during the daytime by changes in behavior. Studies made on spiny lizards of the southwestern deserts of the United States show that they can maintain their body temperature at about 34° C (93° F) during much of the day. If body temperatures fall below the threshold for normal activity, the lizards lie at right angles to the sun's rays, thereby exposing a maximum body surface to the sun. If body temperatures rise too far, they seek shelter or lie parallel to the sun's rays. The chuckwalla (*Sauromalus*) can also eliminate body heat by panting (Fig. 31.11). Color changes are an additional means of temperature regulation in some lizards. In cold weather, the skin is dark and absorbs a maximum amount of heat; as temperatures rise, pigment migrates to the center of the chromatophores and the skin becomes lighter in color. All these methods are only useful during the daytime. At night, body temperature falls to ambient levels, and in prolonged cold weather reptiles go into a state of dormancy.

Major changes have come about in the method of reproduction. Male reptiles have evolved **copulatory organs** which introduce the sperm directly into the female reproductive tract. Fertilization is internal, and the delicate sperm are not exposed to the external environment. A large quantity of nutritive **yolk** is stored within the egg while it is still in the ovary. As the eggs pass down the oviduct after ovulation, they are fertilized, and additional substances and a **shell** are secreted around each one by certain oviducal cells. **Albumin** and similar materials around the egg provide additional food, ions and water. The leathery or calcareous shell serves for protection against mechanical injury and desiccation, yet it is porous enough to permit gas exchange. Such an egg, which contains, or has the means of providing, all substances necessary for the complete development of the embryo to a miniature adult, is called a **cleidoic egg.** Reptiles lay fewer eggs than lower vertebrates, but the eggs are larger, better equipped, and laid in sheltered situations, so the mortality is low. A collared lizard lays only four to 24 eggs in contrast to the 2000 or 3000 of the leopard frog.

As the embryo develops, it separates from the yolk, which becomes suspended in a **yolk sac** (Fig. 31.12). Protective layers of tissue fold over the embryo. The outermost

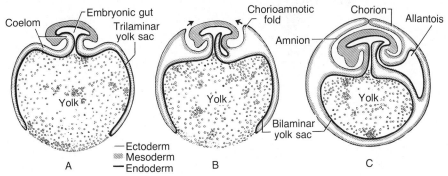

Figure 31.12 Sections of vertebrate embryos to show the extraembryonic membranes. *A*, The trilaminar yolk sac of a large yolk fish embryo consists of all three germ layers. *B*, The chorioamniotic folds of an early embryo of a reptile appear to have evolved from the ectoderm and part of the mesoderm of a trilaminar yolk sac. *C*, A later reptile embryo in which the extraembryonic membranes are complete. Notice that the yolk sac is bilaminar. The albumin and shell, which surround the reptile embryo and extraembryonic membranes, have not been shown.

of these is the **chorion.** An **amnion** lies beneath it and forms around the embryo a fluid-filled chamber, which serves as a protective water cushion and provides an aquatic environment in which the embryo develops. These two membranes appear to be derived phylogenetically from something similar to the superficial layers covering the yolk sac of certain large yolked fish embryos. Another membrane, the **allantois,** is a saclike outgrowth from the embryo's hindgut. It is homologous to the urinary bladder but extends beyond the body wall, passing between the amnion and the chorion. Its highly vascular wall unites with the chorion, and gas exchange with the external environment occurs there. Nitrogenous excretory products, largely in the form of crystals of uric acid, accumulate in the cavity of the allantois. It has been postulated that the evolution of the cleidoic egg was correlated with a shift in nitrogen metabolism. Little water is available to carry off ammonia and urea in the terrestrial environment in which the eggs are laid, and nitrogen cannot be eliminated in gaseous form; consequently much of it must be converted to inert uric acid which can be stored.

Yolk sac, chorion, amnion and allantois are collectively called the **extraembryonic membranes.** These adaptations for terrestrial reproduction are found in the embryos of all reptiles, birds and mammals. These groups of vertebrates are often called **amniotes.** In contrast, the various fish groups and amphibians lacking an amnion are called **anamniotes.**

31.6 EVOLUTION AND ADAPTATIONS OF REPTILES

Stem Reptiles. Having "solved" the essential problems of terrestrial life at a time when there were few competitors upon the land, the reptiles multiplied rapidly, spread into all of the ecologic niches available to them, and became specialized accordingly. Much of their divergence has involved adaptation to different methods of locomotion and feeding. Different feeding patterns have entailed, among other things, modification of the jaw muscles, and this in turn has affected the structure of the temporal region of the skull. Skull morphology, therefore, provides a convenient way to sort out the various lines of reptile evolution. Only a small part of the skull of a primitive tetrapod actually surrounds the brain (Fig. 31.13A). A large space lies lateral to the brain case and posterior to the eyes, and it is largely filled with jaw muscles which extend down to the lower jaw. In ancestral reptiles belonging to the order **Cotylosauria,** as in earlier vertebrates, a solid roof of dermal bone covered these muscles dorsally and laterally (Fig. 31.13A and B). This is referred to as the **anapsid** condition. As a consequence of some brain enlargement, and an enlargement of jaw muscles, various types of openings have evolved in the temporal roof of most later reptile groups. Jaw muscles arise from the brain case and from the periphery of the temporal fenestrae, and they can bulge through the opening when they contract.

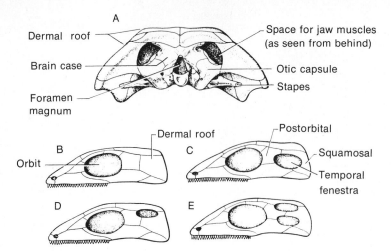

Figure 31.13 Diagrams to show the types of temporal roofs found in reptile skulls. *A*, Posterior view of the skull of a cotylosaur to show the relationship between temporal roof and brain case; *B*, the anapsid condition; *C*, the synapsid condition; *D*, the euryapsid condition; *E*, the diapsid condition. (*A* from Romer after Price; *B–E* from Romer.)

Turtles.   Turtles (order **Chelonia**) are believed to be direct descendants of cotylosaurs (Fig. 31.3), for most have retained the anapsid skull, but they are specialized by being encased in a protective shell composed of bony plates overlaid by horny scales. The bony plates have ossified in the dermis of the skin, but they have also fused with the ribs and some other deeper parts of the skeleton.

The portion of the shell covering the back is known as the **carapace;** the ventral portion, the **plastron.**

Ancestral turtles were stiff-necked creatures, unable to retract their heads, but modern species can withdraw theirs into the shell. This is accomplished by bending the neck in an S-shaped loop in either the vertical plane (North American species such as the red-eared turtle, *Pseudemys scripta elegans*) or in the horizontal plane (Australian side-necked turtle, *Chelodina longicollis*, Fig. 31.14). Sea turtles belong to the former group. They have also adapted to an aquatic mode of life, swimming about by means of oarlike flippers. They come ashore only to lay their eggs in holes which they dig on the beaches.

Marine Blind Alleys.   Sea turtles are not the only reptiles that have returned to the ocean. In the Mesozoic, two lines of reptilian evolution adapted to marine conditions. Plesiosaurs (order **Sauropterygia,** Fig. 31.15) were superficially turtle-shaped (though they lacked the shell), with squat, heavy bodies and long necks. Some species reached a length of 12 meters. They propelled themselves by means of large paddle-shaped appendages. Members of the other line, the ichthyosaurs (order **Ichthyosauria,** Fig. 31.15), were porpoise-like in size and shape, and probably in habits. They moved with fishlike undulations of the trunk. Both groups are characterized by having a temporal opening high up on the side of the skull (Fig. 31.13*D*).

Plesiosaurs could probably get onto the beaches to lay their eggs, but the extreme aquatic adaptation of the ichthyosaurs would preclude their doing so. How then did they reproduce, for cleidoic eggs cannot develop submerged in water? In an unusual fossil, several small ichthyosaurs are lodged in the posterior part of the mother's abdominal cavity, and one individual is part way out the cloaca. These must have been offspring about to be born, for the skeletons of individuals that had been eaten would not remain intact during a passage through the digestive tract. Apparently, these reptiles, like some modern lizards and snakes, were viviparous or ovoviviparous, the eggs being retained in the oviduct until embryonic development was complete.

These marine reptiles flourished during the Mesozoic, competing with the more primitive kinds of fishes. Just why they became extinct near the close of this era is uncertain, but their extinction coincides with the evolution and increase of the teleosts. Possibly they could not compete successfully with these fishes.

Lizard-like Reptiles.   The most abundant of our present-day reptiles are the lizard-

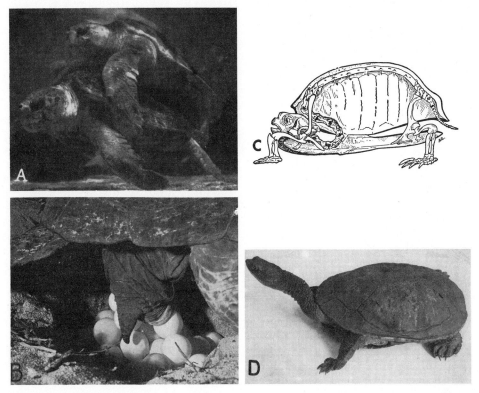

Figure 31.14 *A*, Copulating loggerhead sea turtles, *Caretta caretta; B*, sea turtle laying eggs; *C*, diagram showing retraction of head into the shell by bending the neck in the vertical plane; *D*, the Australian side-necked turtle, *Chelodina longicollis.* (*A*, photograph by Frank Essapian, courtesy of Marine Studios; *B*, LIFE photo by Fritz Goro. © Time, Inc.; *C* from Gregory, William King: Evolution Emerging, Volume II, The Macmillan Co., 1951; *D* courtesy of the New York Zoological Society.)

Figure 31.15 Aquatic reptiles of the Mesozoic era. Plesiosaurs on the left; ichthyosaurs on the right. Plesiosaurs reached a length of 12 meters; ichthyosaurs, a length of about 3 meters. (Courtesy of the Chicago Museum of Natural History.)

like ones of which lizards and snakes are the most familiar examples. The most primitive living member of this group is the tuatara (*Sphenodon*, Fig. 31.16)—the only surviving representative of the order **Rhynchocephalia**. Rhynchocephalians are lizard-like in general appearance, and are characterized by having a skull with two temporal openings (**diapsid** condition); one above and one below the squamosal-postorbital bar (Fig. 31.13*E*). At one time the group was very widespread, but now it is limited to a few small islands off the coast of New Zealand. *Sphenodon* is a surviving "fossil," for it has not changed greatly from species that were living 150 million years ago.

Lizards and snakes, though superficially different from each other, are similar enough in basic structure to be placed in the single order **Squamata**. Lizards (suborder **Lacertilia**) are the older and more primitive. They doubtless evolved from some rhynchocephalian-like ancestor early in the Mesozoic era, but have lost more of the temporal roof including the bar of bone just beneath the lower temporal opening. For the most part lizards are diurnal, terrestrial quadrupeds but, like other successful groups, they have undergone an extensive adaptive radiation (Fig. 31.17).

Several groups have become arboreal and evolved interesting adaptations for climbing. The true chameleon of Africa (not to be confused with the circus chameleon of our Southeast) has a prehensile tail and an odd foot structure in which the toes of each foot are fused together into two groups that oppose each other like the jaws of a pair of pliers. Geckos, in contrast, cling to trees by means of expanded digital pads. Numerous fine ridges on the under surface of the pads increase the friction.

Many lizards, including the horned toads of our Southwest (*Phrynosoma*), burrow to some extent for protection, and some have taken to a burrowing mode of life. Appendages are lost in many burrowing lizards, though vestiges of girdles are present. The eyes may be reduced, and the body form becomes wormlike. The glass snake (*Ophisaurus*), although it burrows only part of the time, is a lizard of this type. The glass snake derives its name from its ability to break off its tail when seized. The tail, which constitutes about two-thirds of the animal's length, fragments into many pieces that writhe about, attracting attention while the lizard moves quietly away. Other lizards also have this ability, though developed to a less spectacular degree. Lost tails are regenerated, but the new tails are supported by a cartilaginous rod rather than by vertebrae.

The only poisonous lizards are the beaded lizards, such as the Gila monster (*Helo-*

Figure 31.16   The tuatara, *Sphenodon*, is one of the most primitive of living reptiles. (Courtesy of the New York Zoological Society.)

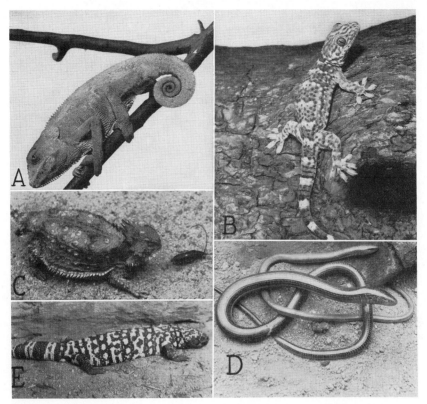

Figure 31.17 Adaptive radiation among lizards. *A*, The Old World chameleon has grasping feet and a prehensile tail with which to climb about the trees. *B*, The gecko climbs by means of digital pads. *C*, The horned-toad, *Phrynosoma*, is a ground-dwelling species that often burrows. *D*, The glass snake, *Ophisaurus*, also burrows. *E*, The Gila monster, *Heloderma*, and a related Mexican species are the only poisonous lizards in the world. (Courtesy of the New York Zoological Society.)

derma) of the Southwestern United States. Modified glands in the floor of the mouth discharge a neurotoxic poison, which is injected into the victim by means of grooved teeth. This is a relatively inefficient method, so the bite is not so dangerous as the bite of most poisonous snakes. Charles Bogert of the American Museum of Natural History reports that eight of 34 bites that have come to his attention were fatal, and he believes that the majority of minor bites are never reported. It is probable that the poison is used for defense rather than for killing prey, for the Gila monster crushes its food with its powerful jaws.

Snakes (suborder **Ophidia**) differ from lizards most notably in being able to swallow animals several times their own diameter (Fig. 31.18). This is made possible by an unusually flexible jaw mechanism, which results in part from the further "erosion" of the temporal roof. Snakes have also lost the squamosal-postorbital bar of bone, and this permits the quadrate bone (the skull bone which in all reptiles and lower vertebrates articulates with the lower jaw) to participate in jaw movements. In a sense, snakes have five jaw joints (Fig. 31.19): (1) the usual one between quadrate and lower jaw, (2) one between quadrate and squamosal, (3) one between squamosal and brain case, (4) one about half-way along the lower jaw, and (5) one at the chin, for the two lower jaws are not united at this point. Other features which characterize snakes are the absence of movable eyelids, of a tympanic membrane and middle ear cavity, and of legs and girdles. The absence of the pectoral girdle is a necessary correlate of swallowing animals larger than the diameter of the body. There are exceptions to these generalizations, for geckos do not have movable eyelids, glass "snakes" lack legs, and some of the more primitive snakes, such as the python, have vestigial hind legs.

Snakes doubtless evolved from some

Figure 31.18   An Indian python (*Python molurus*) swallowing a rabbit. (From Bellairs: The Life of Reptiles, Weidenfeld and Nicolson.)

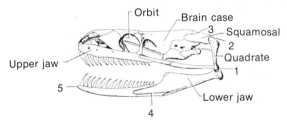

Figure 31.19 Lateral view of the skull of a python to show the five points at which motion can occur when the jaws are opened. (From Romer after M. Smith.)

primitive lizard group, and very probably from burrowing members of that group. The most primitive living snakes are burrowing species, and many details of ophidean anatomy suggest a fossorial ancestry. The absence of a muscular ciliary body and other features of their eyes, for example, indicates that the eyes redeveloped from eyes that had undergone marked retrogressive changes. Their forked tongue, which is often seen darting from the mouth (Fig. 31.20A), is an organ concerned with touch and smelling. Odorous particles adhere to it, the tongue is withdrawn into the mouth, and the tip is projected into a specialized part of the nasal cavity (Jacobson's organ). The great elaboration of such a device would seem to be an adaptation to a

burrowing mode of life in which other senses would be less useful.

In their subsequent evolution some snakes gave up the burrowing habit, adjusted to epigean life, and underwent an extensive adaptive radiation. A pattern of locomotion evolved that is dependent primarily upon undulatory movements of the long trunk. When a snake moves through grass, for example, loops of the trunk form behind the head and move posteriorly. When these loops meet protuberances from the ground, they push upon them and the resultants of these forces move the snake forward.

Of particular interest has been the evolution in several distinct groups of snakes of methods of prey immobilization. If a snake is to swallow an animal larger in diameter than itself, there is a distinct advantage in having the animal quiet, particularly if it is an animal such as a rodent capable of inflicting a severe bite! Boa constrictors, pythons, and many harmless rodent-eating snakes, such as king snakes, entwine their prey in loops of the trunk and gradually suffocate the prey by stopping their respiratory movements. A number of other groups have evolved a poison mechanism consisting of specialized oral glands associated with grooved or hollow hypodermic-like teeth—the **fangs.** Old World vipers and

Figure 31.20 *A,* A coachwhip snake protruding its tongue; *B,* "milking" a rattlesnake to get poison for the production of antivenom. The tongue is a tactile and olfactory organ that is perfectly harmless; it should not be confused with fangs, which are specialized teeth. (*A* courtesy of the New York Zoological Society; *B* courtesy of Ross Allen's Reptile Institute.)

New World pit vipers (rattlesnakes, copperheads, cottonmouths, water moccasins) have a pair of large hollow fangs at the front of the mouth that are articulated to bones of the upper jaw and palate in such a way that they are folded against the roof of the mouth when the mouth is closed and automatically brought forward when the mouth is opened (Fig. 31.20B). The poison of many of the snakes is hemolytic and causes a breakdown of the red blood cells in the animal bitten. Coral snakes belong to a group related to the Old World cobras. Their poison is neurotoxic, and their fangs are a pair of hollow, short, stationary teeth attached to the front of the upper jaw.

Pit vipers differ from other poisonous snakes in having a prominent **sensory pit** on each side of their heads between the nostril and eye. Experiments in which other sense organs have been rendered functionless show that the pit enables these snakes to seek out and strike accurately at objects warmer than their surroundings. The pits probably help the snakes to feed upon small nocturnal mammals.

Dinosaurs and Their Allies. Lizards and snakes are the successful reptiles today, but during the Mesozoic era the land was dominated by another offshoot of primitive rhynchocephalian-like reptiles. These "ruling reptiles" were the **archosaurs**—an assemblage of several orders that shared many features, including a diapsid skull and a tendency to evolve a two-legged gait. Reduced pectoral appendages, enlarged pelvic appendages, and a heavy tail that could act as a counterbalance for the trunk were correlated with this method of locomotion.

Saurischian dinosaurs (order **Saurischia**) evolved from ancestors that were approximately 1 meter long, but later saurischians became giants of the land and swamps. *Tyrannosaurus* (Fig. 31.21A) was the largest terrestrial carnivore that the world has ever seen. It stood about 6 meters high and had large jaws armed with dagger-like teeth 15 cm. long—a truly formidable creature! Other saurischian dinosaurs were herbivores that reverted to a quadruped gait, but their bipedal ancestry is reflected in their retention of a heavy tail and in having hind legs larger than the front ones. *Brontosaurus* (Fig. 31.21B and C) was an enormous herbivore that attained a length of 25 meters and an estimated weight of 45 metric tons.

Its huge size led to the hypothesis that it lived in swamps where it would be supported partly by the buoyancy of the water. A more recent view is that it fed in giraffe-like fashion upon high vegetation. Its limb structure resembled that of elephants, and its tail was not flattened as is usually the case in aquatic vertebrates.

Dinosaurs belonging to the order **Ornithischia** were all herbivores. Some lived in swamps, others in the uplands. Many reverted to a quadruped gait and increased in size, though none was so large as the saurischians. These animals undoubtedly formed much of the diet of carnivores such as *Tyrannosaurus*, and many evolved protective devices such as spiked tails, bony plates on the body and horned skulls. *Stegosaurus* and *Triceratops* (Fig. 31.21C and D) are examples of this group.

The reasons for the evolution of large size are not entirely clear. Within limits, large size has a protective value, but it may also have been a way of achieving a more nearly constant body temperature. Reptiles, being ectothermic, derive a great deal of their body heat during warm weather from the external environment. As mass increases, the relative amount of body surface available for the absorption of heat decreases. An adaptation of this type may have been particularly important for animals that lived in a warm climate and were too big to shelter by burrowing or hiding beneath debris, for it would help to prevent body temperature from reaching a lethal point. As explained earlier, prolonged high tempeature destroys most enzyme systems.

A bipedal gait naturally freed the front legs from use in terrestrial locomotion. The front legs became reduced in many dinosaurs, but in one group of archosaurs they evolved into wings. The wings of the flying reptiles (order **Pterosauria**) consisted of a membrane of skin supported by a greatly elongated fourth finger (Fig. 31.22). The fifth finger was lost, and the others probably were used for clinging to cliffs. The hind legs were very feeble, and the animal must have been nearly helpless on the ground. Certain pterosaurs became very large, one having a wing spread of over 6 meters.

Most of the archosaurs became extinct toward the end of the Mesozoic, but the reason for this is not entirely clear. Perhaps the pterosaurs succumbed in competition

A

B

C

Figure 31.21 Representatives of the main groups of dinosaurs that flourished during the late Mesozoic era. *A, Tyrannosaurus,* a carnivorous saurischian; *B* and *C, Brontosaurus* reconstructed as a swamp dweller or browser of high vegetation; *D, Stegosaurus,* an ornithischian; *E, Triceratops,* another ornithischian. (*C* courtesy of Fauna; others, courtesy of the Chicago Museum of Natural History.)

D

E

Figure 31.22 *Pteranodon*, the largest of the flying reptiles, or pterosaurs, that lived during the late Mesozoic era. (Courtesy of the American Museum of Natural History.)

with birds, which also evolved from primitive bipedal archosaurs. The extinction of the dinosaurs may have resulted from climatic changes. An inability of the specialized herbivores to adapt to changes in vegetation would have led to their death. Their disappearance, in turn, would deprive the huge carnivores of most of their food supply. The factors mentioned may not have wiped out a population but simply have reduced it to a size at which it became very difficult for the species to maintain itself.

Only one group of archosaurs survived this wholesale extinction—the alligators and crocodiles (order **Crocodilia**). Crocodiles have reverted to a quadruped gait (though their hind legs are much longer than the front) and an amphibious mode of life. Only two species occur in the United States—the American alligator, which can be distinguished by its rounded snout, and the American crocodile, which has a much more pointed snout (Fig. 31.23).

Mammal-like Reptiles.   Another line of evolution, which was destined to lead to mammals, diverged from the cotylosaurs millions of years before the advent of archosaurs. Early mammal-like reptiles (order **Pelycosauria**) were very similar to cotylosaurs, differing primarily in having a skull with a temporal opening ventral to the squamosal and postorbital bones (**synapsid** condition, Fig. 31.13C). They were medium-sized, somewhat clumsy, terrestrial quadrupeds with limbs sprawled out at right angles to the body. Some pelycosaurs (*Dimetrodon*, Fig. 31.24A) had bizarre sails upon their backs supported by elongated neural spines. There has been much speculation as to the sail's function, but it did help to maintain a constant ratio of body surface to mass as members of this evolutionary line became larger. It has been suggested that the sail was a primitive thermoregulator, acting as a heat receptor or radiator.

Later mammal-like reptiles (*Lycaenops*, order **Therapsida**, Fig. 31.24B) came to resemble mammals more closely. Their limbs were beneath the body where they could provide better support and move more rapidly back and forth. Their teeth were specialized, like those of mammals, into ones suited for cropping, stabbing, cutting and grinding. The major osteologic character that separated them from mammals was the nature of the jaw joint and the sound-transmitting apparatus. In all nonmammalian vertebrates, the jaw joint lies between the posterior ends of the mandibular arch, which are usually ossified as a **quadrate bone** in the upper jaw and an **articular bone** in the lower jaw. In most terrestrial vertebrates, a **stapes** (a derivative of the fish hyoid arch) transmits air-borne vibrations from a tympanic membrane to the inner ear located within the skull (cf. frog, Fig. 15.10, p. 353). This may have been the case in mammal-like reptiles, although we cannot be sure, for a soft part such as a tympanic membrane is not preserved in fossils. In mammals, the jaw joint is between two dermal bones (the **dentary** of the lower jaw and **squamosal** of the upper jaw)

Figure 31.23 The broad-nosed caiman (*Caiman latirostris*) from South America. The nose, eye and ear (arrow) of a crocodilian can remain above the surface when the rest of the body is submerged. (From Bellairs: The Life of Reptiles, Weidenfeld and Nicolson.)

Figure 31.24 Two mammal-like reptiles. *A, Dimetrodon*, an early member of the group; *B, Lycaenops*, a later mammal-like reptile similar to those that gave rise to mammals. (Courtesy of the American Museum of Natural History.)

that lie just anterior to the quadrate and articular. The mammalian homologues of the quadrate and articular (the **incus** and **malleus,** respectively) are covered by a tympanic membrane and form, with the stapes, a chain of three delicate auditory ossicles that transmit air-borne vibrations from the tympanic membrane to the inner ear (Fig. 15.11). This character had not been achieved by the late therapsids, but adaptations to changes in feeding mechanisms had led to an enlargement of the dentary, which grew posteriorly close to the squamosal, and the quadrate and articular became very small. In a few very late therapsids a dentary-squamosal joint was present concurrently with the quadrate-articular joint. The final change to the mammalian condition was made by the middle of the Mesozoic; shortly afterward the mammal-like reptiles became extinct. Their mammalian descendants remained a rather inconspicuous part of the fauna until the disappearance of the dinosaurs.

CLASSIFICATION OF AMPHIBIANS AND REPTILES

CLASS AMPHIBIA. The amphibians. Ectothermic vertebrates usually having an aquatic larval stage; adults typically terrestrial; skin moist and slimy; scales absent in most groups; microscopic vestiges of bony scales in one living group.

    † **Subclass Labyrinthodontia.** Several orders of extinct and primitive amphibians collectively called labyrinthodonts; vertebral centra of the "arch" type, each usually consisting of two or three arches of bone encasing the notochord.

    † **Subclass Lepospondyli.** Several orders of extinct amphibia with vertebral centra of the "spool" type; each centrum a single structure, spool-shaped, pierced by a longitudinal canal for persistent notochord.

    Subclass Lissamphibia. Modern amphibians, probably evolved from lepospondyls. Skull usually flat and broad; pineal opening lost; brain case poorly ossified; pubis never ossified; carpals and tarsals largely cartilaginous.

        Superorder Salientia. Frogs and their allies. Trunk short; tail reduced or absent; ilia elongated; hind legs elongated.

            † *Order Proanura.* Ancestral frogs of the Triassic. Tail still present; limbs not highly specialized. *Triadobatrachus.*

            Order Anura. Modern frogs and toads. Caudal vertebrae form a urostyle; legs specialized for jumping. The leopard frog, *Rana;* tree frog, *Hyla;* American toad, *Bufo.*

        Superorder Caudata. Tailed amphibians. Trunk long; tail present; legs, if present, not elongated.

            Order Urodela. Salamanders. Tail long; legs usually present. The spotted salamander, *Ambystoma;* redbacked salamander, *Plethodon;* mudpuppy, *Necturus.*

            Order Apoda. Caecilians. Wormlike trunk; limbs absent; tail very short; vestiges of dermal scales in the skin. Confined to the tropics.

CLASS REPTILIA. The reptiles. Ectothermic tetrapods that reproduce upon the land; body covered with horny scales or plates.

    Subclass Anapsida. Primitive reptiles characterized by a solid roof in the temporal region of the skull.

        † *Order Cotylosauria.* The cotylosaurs. Ancestral reptiles retaining a large number of primitive features.

        Order Chelonia. The turtles. Anapsid reptiles with a short trunk encased in a bony shell which in turn is covered by horny plates; teeth absent. Red-eared turtle, *Pseudemys;* green sea turtle, *Chelonia;* side-necked turtle, *Chelodina.*

    † **Subclass Euryapsida.** Ancient marine reptiles that propelled themselves with long paddle-shaped limbs; single temporal opening high on the skull.

        † *Order Sauropterygia.* The plesiosaurs.

    † **Subclass Ichthyopterygia.** Ancient, marine, fishlike reptiles; single temporal opening high on the skull.

        † *Order Ichthyosauria.* The ichthyosaurs.

---

† Extinct.

Subclass Lepidosauria. Primitive reptiles with two temporal openings in the skull (diapsid); pineal foramen usually retained; teeth on palate as well as jaw margins, usually not set in sockets; usually quadrupeds.

> † *Order Eosuchia.* Ancestral lepidosaurians.
>
> *Order Rhynchocephalia.* Primitive lizard-like reptiles. Temporal openings completely bounded by arches of bone; distinctive overhanging beak on upper jaw; now confined to certain islands off New Zealand. The tuatara, *Sphenodon.*
>
> *Order Squamata.* Lizards and snakes. Advanced lepidosaurians in which there has been a loss of one or two arches of bone bordering the temporal openings.
>
>> SUBORDER LACERTILIA. Lizards. Only the lower arch of bone has been lost in the temporal region; most groups retain legs, eyelids and an auditory opening. The chuckwalla, *Sauromalus;* horned toad, *Phrynosoma;* Gila monster, *Heloderma.*
>>
>> SUBORDER OPHIDIA. Snakes. Both arches of bone bounding temporal openings have been lost; jaw mechanism extremely flexible so that food larger than head can be swallowed; pineal foramen, legs and auditory opening are lost; eyes covered by a transparent scale, the spectacle. The garter snake, *Thamnophis;* water snake, *Natrix;* rattlesnake, *Crotalus.*

Subclass Archosauria. Advanced diapsid reptiles. Pineal foramen usually absent; teeth usually restricted to jaw margin, set in sockets; an extra fenestra usually present in front of the eye and/or on the lower jaw; bipedal reptiles, or showing signs of bipedal ancestry.

> † *Order Thecodontia.* Ancestral archosaurs. Show the beginning of archosaur tendencies but not the specializations of later archosaurs.
>
> † *Order Saurischia.* Saurischian dinosaurs. Pelvis three-pronged with pubis located entirely anterior to ischium. *Tyrannosaurus, Brontosaurus.*
>
> † *Order Ornithischia.* Ornithischian dinosaurs. Pelvis birdlike with part of pubis growing posteriorly parallel to ischium; all herbivorous. *Stegasaurus, Triceratops.*
>
> † *Order Pterosauria.* The flying reptiles. Wing a skin membrane supported by an elongated fourth finger. *Pteranodon.*
>
> *Order Crocodilia.* Alligators and crocodiles. Quadrupeds but with hind legs noticeably larger than forelegs; adapted for an aquatic mode of life. The American alligator, *Alligator;* the American crocodile, *Crocodilus.*

Subclass Synapsida. Mammal-like reptiles characterized by a single temporal opening on the lateral surface of the skull.

> *Order Pelycosauria.* Early mammal-like reptiles retaining many primitive features; limbs in primitive tetrapod position with humerus and femur moving in the horizontal plane. *Dimetrodon.*
>
> *Order Therapsida.* Advanced mammal-like reptiles beginning to show many mammalian characteristics; limbs rotated beneath the body with humerus and femur moving in the vertical plane. *Lycaenops.*

ANNOTATED REFERENCES

Many of the general references on vertebrates cited at the end of Chapter 30 contain considerable information on the biology of amphibians and reptiles.

Bellairs, A.: The Life of Reptiles. London, Weidenfeld and Nicolson, 1969. Deals with the anatomy, physiology, development, ecology and evolution of reptiles. The modern successor to Gadow's famous book (see below).

Bishop, S. C.: Handbook of Salamanders of the United States, of Canada and of Lower California. Ithaca, N.Y., Comstock Publishing Co., 1943. A standard reference work on the taxonomy and natural history of salamanders.

Bogert, C. M.: How reptiles regulate their body temperature. Scientific American

---

† Extinct.

200:105 (April) 1959. Many interesting experiments on temperature regulation are discussed.

Carr, A.: Handbook of Turtles of the United States, Canada and Baja California. Ithaca, N.Y., Comstock Publishing Co., 1952. A standard reference work on the taxonomy and natural history of turtles.

Cochran, D. M.: Living Amphibians of the World. New York, Doubleday & Co., 1961. A fascinating and superbly illustrated account of the groups of living amphibians.

Conant, R.: A Field Guide to Reptiles and Amphibians of Eastern North America. Boston, Houghton Mifflin Co., 1958. A very useful guide for the field identification of amphibians and reptiles; similar to the Peterson bird guides.

Gadow, H.: Amphibia and Reptilia. Weinheim, Germany, Engelmann, 1958. A recent reprint of a book originally published as a part of the "Cambridge Natural History" in 1901; it is still a valuable account of the anatomy and evolution of amphibians and reptiles.

Gans, C., A. d'A. Bellairs, and T. S. Parsons (Eds.): Biology of the Reptilia. New York, Academic Press, 1969–1970. The skeleton, sense organs, blood and circulation, and endocrine glands are considered in the first three volumes. Other volumes are in preparation.

Moore, J. (Ed.): Physiology of the Amphibia. New York, Academic Press, 1964. Most aspects of amphibian physiology are covered in this very valuable source book.

Noble, G. K.: The Biology of the Amphibia. New York, Dover Publications, Inc., 1956. This is a recent reprint of an older but still very valuable reference work on the classification, anatomy, physiology and ecology of amphibians.

Porter, K. R.: Herpetology. Philadelphia, W. B. Saunders Co., 1972. An excellent account of most aspects of the biology of amphibians and reptiles.

Schmidt, K. P., and R. F. Inger: Living Reptiles of the World. New York, Doubleday & Co., 1957. A beautifully illustrated account of the natural history of the families of living reptiles.

Smith, H. M.: Handbook of Lizards of the United States and Canada. Ithaca, N.Y., Comstock Publishing Co., 1946. A standard reference work on the taxonomy and natural history of lizards.

Wright, A. H., and A. A. Wright: Handbook of Frogs of the United States and Canada. 3rd ed. Ithaca, N.Y., Comstock Publishing Co., 1949. A standard reference work on the taxonomy and natural history of frogs.

Wright, A. H., and A. A. Wright: Handbook of Snakes of the United States and Canada. Ithaca, N.Y., Comstock Publishing Co., 1957. A standard reference work on the taxonomy and natural history of snakes.

32. Vertebrates: Birds

Reptiles are able to occupy the terrestrial environment more successfully than amphibians largely because they evolved ways of conserving body water, reproducing upon the land and regulating to some degree their body temperature by behavioral patterns. Birds (class **Aves**) and mammals (class **Mammalia**) are descendants of early reptiles, and both groups further adapted to the terrestrial environment. Of particular significance has been the evolution of physiological mechanisms for the maintenance of rather high and constant body temperatures. Birds and mammals are **endothermic,** or warm-blooded, for body heat is generated internally, and body temperatures are regulated by a controlled production and loss of heat (see Section 33.1). Their metabolic processes can proceed at an optimal rate despite the wide range in external temperatures common in the terrestrial environment, and they are typically very active creatures.

Higher metabolic rates require higher rates of exchange of materials with the environment and rapid distribution of these materials within the body. Birds and mammals have met these requirements in somewhat similar ways; their adaptations for increased activity provide interesting examples of convergent evolution, although in other respects they are quite different. Birds evolved from early bipedal archosaurs (Fig. 31.3) and have undergone specializations for flight; mammals evolved from a stock of mammal-like reptiles and have become specialized for terrestrial life.

32.1 PRINCIPLES OF FLIGHT

A group of extinct reptiles, the pterosaurs, and a group of mammals, the bats, have evolved true flight, but neither group has been as successful fliers as have birds. Bird wings are modified pectoral appendages, and the flying surfaces are composed of feathers. A wing is shaped like an airfoil; thick in front and thin and tapering behind, and sometimes cambered so that it is slightly concave on the under surface and convex on the upper surface. As the air stream flows across the wing, the stream moves faster along the upper surface than the lower surface. In accordance with Bernoulli's law in physics, which states that in a fluid stream the pressure is least where the velocity is greatest, this differential in air speed decreases the pressure above the wing relative to the underside. This produces a force acting perpendicular to the wing surface which can be resolved into a lift component perpendicular to the air stream and a drag component parallel to the air stream (Fig. 32.1A). For the bird to fly, the lift force must equal the force of gravity on the bird, and there must be a propulsive force (discussed below) that can overcome the drag force. The somewhat teardrop shape of the wing allows a smooth flow of air across the surface and keeps to a minimum lift-reducing eddies and drag.

Various factors can increase lift. Lift increases in direct proportion to the surface area of the wing. Wing areas differ among different species of bird, and can be varied

727

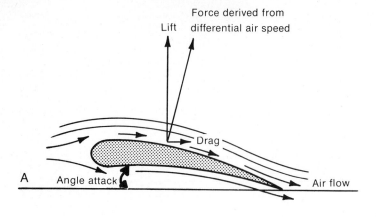

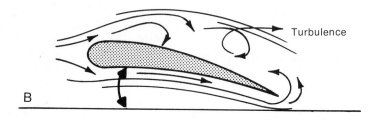

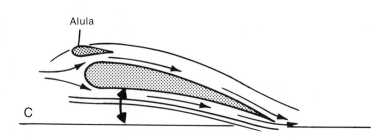

Figure 32.1 The effect of a wing on the airstream. *A*, Air moving more rapidly along the longer upper surface of the wing reduces the pressure on the upper surface and generates a force which can be resolved into lift and drag forces; *B*, if the angle of attack of the wing becomes too great, air swirls into the low pressure area causing turbulence and a reduction in lift; *C*, turbulence can be reduced by raising the alula and forming a slot through which the air moves rapidly and smoothly.

in individual birds by the degree to which the wing is stretched out or unfolded. Lift increases greatly as the speed of airflow across the wing increases, for lift is proportional to the square of the speed. Fast flying birds, such as a swift, can and do have relatively smaller wings than slower flying species. When a bird is flying at low speeds, or during taking off and landing, other mechanisms are used to increase lift. The wing can be tilted so that its anterior edge is considerably higher, a procedure known as increasing the angle of attack (Fig. 32.1*B*). This increases lift, but also tends to

create lift-reducing turbulence above the wing. Separating certain feathers, which produces slots through which the air moves very rapidly, can reduce this turbulence and make it possible for the wing to generate a very high lift. A small group of feathers supported by the first digit and known as the **alula** can produce a slot at the front of the wing (Figs. 32.1*C* and 32.2). Additional slots are often formed along the trailing margin of the wing and at the wing tip. The latter slots reduce the turbulence known as tip vortex. Some birds obtain additional lift on landing by fanning out the tail feathers and

Figure 32.2 Photograph of a red-shouldered hawk landing. Air speed is low and lift is increased by several slots: *1*, alulae; *2*, separation of primary wing feathers; *3*, separation of feathers on leading edge of wing; *4*, auxiliary feathers on upper surface of wing. (From Hertel.)

bending them down. The tail, then, acts both as a brake and as a high-lift, low-speed airfoil.

There are many kinds of flight. The simplest is **gliding,** in which the wings provide some lift and the forward motion comes from falling through the air. Altitude is lost in a glide of this type, but altitude can be maintained or even increased if the bird also soars. Land birds, such as the turkey vulture or the osprey (Fig. 32.3*A*), circle within a rising current of warm air, or above a bluff where air is deflected upward. Birds that engage in **static soaring** of this type have relatively short, broad wings that enable them to maneuver easily in the capricious air currents, yet have enough surface area to provide lift. Flight is slow, and additional lift comes from considerable slotting of the wing, particularly near the tip. Oceanic birds engage in **dynamic soaring,** which makes use of the increase in air speed with increasing elevation above the ocean surface. Friction with the ocean causes air speed to be slowest at the ocean surface (Fig. 32.3*C*). Starting at a high elevation, these birds glide rapidly downward

with the wind. Just above the ocean they wheel into the wind and use the momentum gained in the glide to start to gain altitude. As they gain altitude, they encounter increasingly fast air speeds which in turn generate additional lift. In this way the birds regain their original altitude. Dynamic soarers have long narrow wings.

In the familiar **flapping flight** used by such birds as the pigeon (Fig. 32.3*B*), the wings not only provide the lift, but they also serve as propellers. The up and down movement of the wings relative to the body of the bird is responsible for the forward movement, but the wings do not simply push back against the air as a swimmer would push back against the water. On the downstroke, they move down and forward; on the upstroke, up and back. As a wing moves down, the air pushes up against it and the more flexible posterior margin of the distal part of the wing is twisted up. The distal portion of the wing twists the opposite way on the upstroke. The twisting of the distal portion of the wing gives it a pitch comparable to that of a propeller and this, together with the movements of this part of

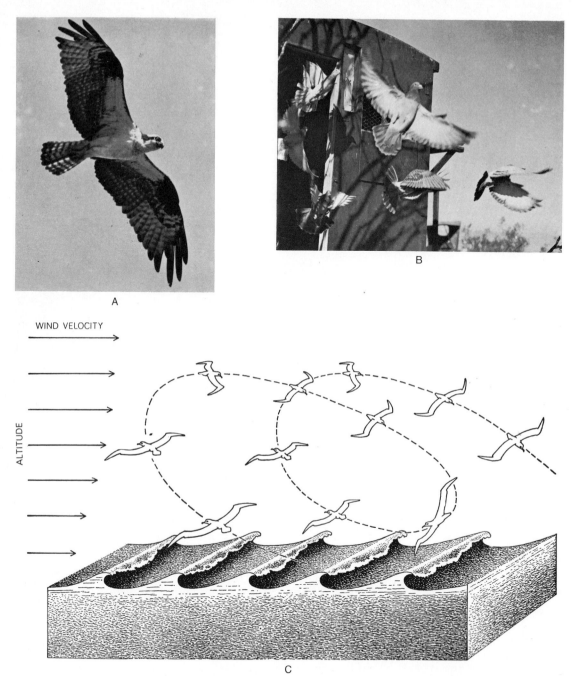

Figure 32.3 Types of flight in birds. *A,* Static soaring of an osprey; *B,* flapping flight of pigeons; *C,* dynamic soaring of oceanic birds. (*A,* photograph by Allan D. Cruickshank, National Audubon Society; *B,* U.S. Army photograph; *C,* from The Soaring Flight of Birds, by C. D. Cone, Jr. Copyright © by Scientific American, Inc. All rights reserved.)

the wing, is responsible for the forward motion. In all types of flight, the tail helps to support and balance the body and is used as a rudder.

Most of the features of bird wings also apply to the wings of airplanes. But a bird's wings and tail have one great advantage over those of an airplane in that they can be varied considerably to adjust to different speeds and types of flight, for different angles of attack, and for many other variables. Even though recently designed planes have wings that can be varied to some extent, any bird is more versatile than any single type of airplane.

32.2 THE ADAPTIVE FEATURES OF BIRDS

There are few features of the anatomy of birds that are not directly or indirectly related to endothermy and flight. They are adapted structurally and functionally to provide a high energy output in a body of low weight.

Scales and Feathers. Birds have retained the horny scales of reptiles on parts of their legs, on their feet and, in modified form, as a covering for their beaks, but the scales that cover the rest of the reptilian body have been transformed into **feathers.** Feathers, like horny scales, are epidermal outgrowths whose cells have accumulated large amounts of keratin and are no longer living. Pigment deposited in these cells during the development of the feather, together with surface modifications that reflect certain light rays, is responsible for the brilliant colors of birds. Although feathers cover a bird, they fan out in most species from localized **feather tracts** rather than growing out uniformly from all of the body surface. Feathers, more than any other single feature, characterize birds, for they are found only in members of this class. They overlap, entrap air and form an insulating layer that reduces loss of body heat and helps to make a high body temperature possible. Those on the tail and wings form the flying surfaces.

The **contour feathers** that cover the body and provide the flying surface consist of a

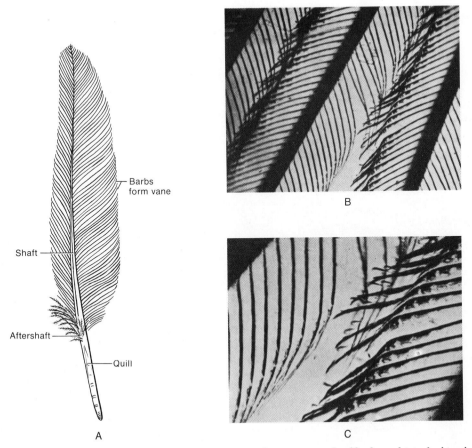

Figure 32.4   Contour feather of a bird. *A,* Entire feather; *B,* photomicrograph of barbs and interlocking barbules; *C,* further enlargement to show hooks and spines on barbules. (*A* modified after Young; *B* and *C* from Welty.)

stiff, central **shaft** bearing numerous parallel side branches, the **barbs**, which collectively form the **vane** (Fig. 32.4). Each barb bears minute hooked branches, **barbules**, along its side, which interlock with the barbules of adjacent barbs to hold the barbs together. If the barbs separate, the bird can preen the feather with its bill until they hook together again; thus, the vane is a strong, light, and easily repaired surface ideal for flight. In birds that have lost the power of flight, such as the ostrich, hooklets are not present upon the barbules, and the feather is very fluffy. The proximal end of the shaft, known as the **quill** is lodged within an epidermal follicle in the skin. Blood vessels enter it during the development of the feather. A small **aftershaft,** bearing a few barbs, may arise from the distal end of the quill.

Other types of feathers include the hairlike **filoplumes**, sometimes visible on plucked fowl, and **down feathers.** Down covers young birds and is found under the contour feathers in the adults of certain species, particularly aquatic ones. It is unusually good insulation, for it has a reduced shaft and long fluffy barbs arising directly from the distal end of the quill. As they preen, many birds spread an oily secretion produced by the **uropygial gland** over their feathers. The gland is located on the back near the tail (Fig. 32.9). It is particularly well developed in water fowl and its secretions waterproof the feathers.

Most birds **molt** once a year, usually after the breeding season. Feathers are lost and replaced in a characteristic sequence for each species. The process is gradual in most cases, and the birds can move about normally during molting, but certain male ducks shed the large flight feathers on their wings so rapidly that they are unable to fly for a while.

Skeleton. Many adaptations for flight are apparent in the skeleton of birds (Fig. 32.5). The bones are very light in weight, for they are hollow and remarkably thin. Extensions

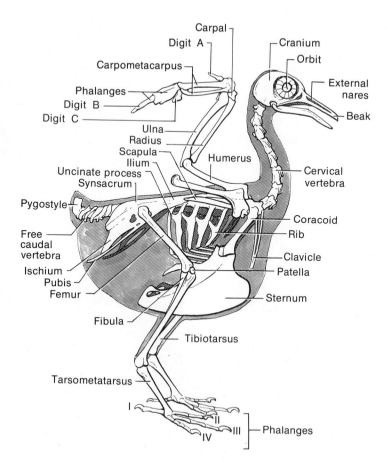

Figure 32.5 Skeleton of a pigeon. The distal part of the right wing has been omitted. (Modified after Heilmann.)

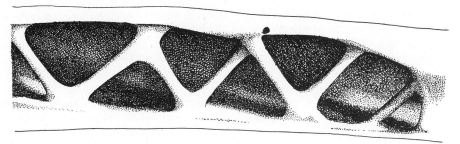

Figure 32.6 Longitudinal section of the metacarpal bone from a vulture's wing. Notice the internal trussing similar to that in an airplane's wing. (From D'Arcy Thompson.)

from the lungs enter the limb bones in many species. Robert Cushman Murphy of the American Museum of Natural History has reported that the skeleton of a frigate bird having a wingspread of over 2 meters weighed only 115 gm., which was less than the weight of its feathers! This is an extreme example, but the skeletons of all birds weigh less relative to their body weight than the skeletons of mammals. The bones are very strong because most of the bone substance is located at the periphery of the bone, where it gives better structural support. A bird bone may be compared to a metal tube, which is more resistant to certain types of stress than a metal rod of equal weight. The rod would be much narrower and could be bent more easily than the tube. Many bird bones are further strengthened by internal struts of bone arranged in a manner similar to the trusses inside the wing of an airplane (Fig. 32.6).

The **skull** is notable for the large size of the cranial region, the large orbits and the toothless beak. The neck region is very long, and the **cervical vertebrae** are articulated in such a way that the head and neck are very mobile. Since the bird's bill is used for feeding, preening, nest building, defense and the like, freedom of movement of the head is very important. The trunk region, in contrast, is shortened, and the **trunk vertebrae** are firmly united to form a strong fulcrum for the action of the wings and a strong point of attachment for the pelvic girdle and hind legs. The hind legs bear the entire weight of the body when the bird is on the ground. In the pigeon, 13 of the more posterior vertebrae (some of the trunk, all of the sacral and some of the caudal vertebrae) are fused together to form a **synsacrum** with which the pelvic girdle is fused. Several

free **caudal vertebrae,** which permit movement of the tail, follow the synsacrum. The terminal caudal vertebrae are fused together as a **pygostyle** and support the large tail feathers.

The last two cervical vertebrae of the pigeon and the thoracic vertebrae bear distinct **ribs.** The thoracic basket is firm yet flexible. Extra firmness is provided by the ossification of the ventral portions of the thoracic ribs (the parts articulating with the sternum and comparable to our costal cartilages) and by posteriorly projecting **uncinate processes** on the dorsal portions of the ribs, which overlap the next posterior ribs. Flexibility, needed in respiratory movements, is made possible by the joints between the dorsal and ventral portions of the ribs. The **sternum,** or breastbone, is greatly expanded and, in all but the flightless birds, has a large midventral **keel** which increases the area available for the attachment of the flight muscles.

The bones of the wing are homologous to those of the pectoral appendage of other tetrapods. The **humerus, radius** and **ulna** can be recognized easily, but the bones of the hand have been greatly modified. Two free **carpals** are present and a **carpometacarpus** (a safety-pin–shaped complex of bone representing the fused carpals and metacarpals of three fingers) lies distal to them. The end of the most anterior finger is represented by a spur-shaped **phalanx** articulated to the proximal end of the carpometacarpus. This supports the alula. The main axis of the hand passes through the next finger, and it has two distinct phalanges articulated to the distal end of the carpometacarpus. Another small, spur-shaped phalanx at the distal end of the carpometacarpus represents the end of the last finger. There is some doubt whether the fingers are homologous to the first three or to the second, third and fourth fingers. The pectoral girdle, which supports the wing, consists of a narrow dorsal **scapula**, a stout **coracoid** extending as a prop from the shoulder joint to the sternum, and a delicate **clavicle,** which unites distally with its mate of the opposite side to form the wishbone.

The legs of birds resemble the hind legs of bipedal archosaurs. The **femur** articulates distally with a reduced **fibula** and a large **tibiotarsus** (fusion of the tibia with certain tarsals). The remaining tarsals and the elongated metatarsals have fused to form a **tarsometatarsus.** The fifth toe has been lost in all birds and the fourth in some species. The first toe is turned posteriorly in the pigeon and many other birds. It serves as a prop and increases the grasping action of the foot when the bird perches. The efficiency of the leg in running on the ground and jumping at take-off is increased by the elongation of the metatarsals, and by the elevation of the heel off the ground. The various fusions of the limb bones reduce the chance of dislocation and injury, for birds' legs must act as shock absorbers when they land. The

pelvic girdle is equally sturdy; the **ilium, ischium** and **pubis** of each side are firmly united with each other and with the vertebral column. The pubes and ischia of the two sides do not unite to form a midventral pelvic symphysis as they do in other tetrapods. This permits a more posterior displacement of the viscera, which, together with the shortened trunk, shifts the center of gravity of the body nearer to the hind legs. The absence of a symphysis also makes possible the laying of large eggs with calcareous shells.

The feet of birds have undergone a variety of modifications as birds have adapted to particular modes of life (Fig. 32.7). The foot and toes become particularly sturdy in ground-dwelling species, and the power of grasping is especially well developed in such perching specialists as our songbirds. In perching birds, the tendons of the foot are so arranged that the weight of the body automatically causes the toes to flex and grasp the perch when the bird alights upon a branch. The woodpeckers have sharp claws, and the fourth toe is turned backward with the first to form a foot ideally suited for clinging onto the sides of trees. Swimming birds have a web stretching be-

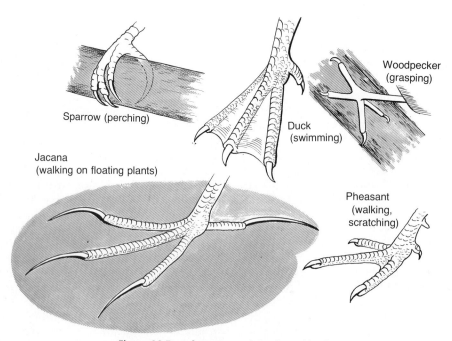

Figure 32.7   Adaptations of the feet of birds.

tween certain of their toes—the three anterior toes in loons, albatrosses, ducks, gulls and many others; all four toes in pelicans, cormorants and their relatives. The marsh-dwelling jacana of the tropics has a foot with exceedingly long toes and claws that enable it to scamper across lily pads and other floating vegetation. Swifts and hummingbirds have very small feet barely strong enough to grasp a perch. These birds spend most of their time on the wing and almost never alight on the ground.

Muscles. The intricate movements of the neck and the support of the body by a single pair of legs entail numerous modifications of the muscular system, but the muscles concerned with flight are of particular interest. A large **pectoralis,** which originates on the sternum and inserts on the ventral surface of the humerus, is responsible for the powerful downstroke of the wings. One might expect that dorsally placed muscles would be responsible for the recovery stroke but, instead, another ventral muscle, the **supracoracoideus** (designated pectoralis minor by some authors), is responsible for the upstroke by virtue of a peculiar pulley-like arrangement of its tendon of insertion (Fig. 32.8). The origin of the supracoracoideus is on the sternum dorsal to the pectoralis. Its tendon passes through a canal in the pectoral girdle near the shoulder joint and inserts on the dorsal surface of the humerus. Muscles within the wing are responsible for its folding and unfolding and the regulation of its shape and angles during flight. Other muscles attach to the follicles of the large flight feathers of the wings and tail and control their positions.

Major Features of the Visceral Organs. Less obvious but no less important adaptations for increased activity and flight are present in many of the internal organs (Fig. 32.9). Increased activity and a high metabolic rate necessitate a large intake of food. The digestive system is compact but it is so effective that, in some of the smaller birds, an amount of food equivalent to 30 per cent of the body weight can be processed each day! Moreover, most of the food that is selected has a high caloric value. Birds eat a variety of insects and other animals and such plant food as fruit and seeds. They do not attempt to eat such bulky, low caloric food as leaves and grass. The bills of birds are modified according to the nature of the food they eat (Fig. 36.1, p. 821). Finches have short heavy bills well suited for picking up and breaking open seeds. The hooked beak of hawks is ideal for tearing apart small animals that they have seized with their powerful talons. Herons use their long sharp bills for spearing fish and frogs, which they deftly flip into their mouths. The length and shape of hummingbirds' bills are correlated with the structure of the flowers from which they extract nectar. Whippoorwills fly about in the evening catching insects with their gaping mouths. Bristle-like feathers at the base of the bill help them to catch their prey. When feeding, the skimmer flies just above the ocean with its elongated lower jaw skimming the surface. Any fish or other organ-

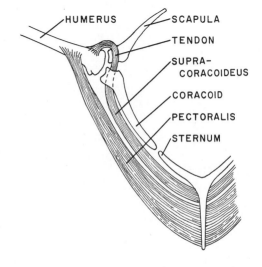

Figure 32.8 A diagrammatic cross section through the shoulder region and sternum showing the arrangement of major flight muscles. (From Welty after Storer.)

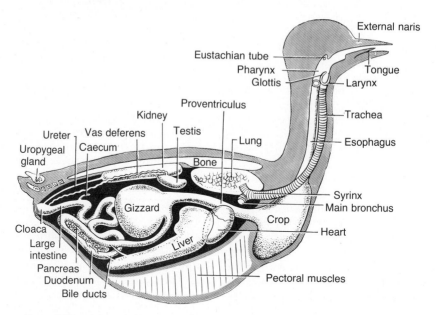

Figure 32.9   A lateral dissection of a pigeon to show the major visceral organs.

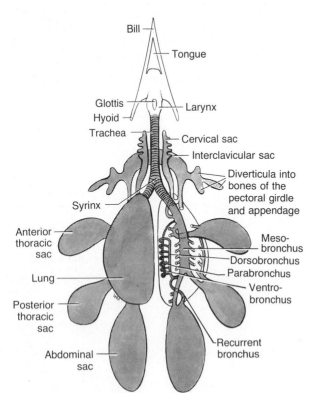

Figure 32.10   The respiratory organs of a bird as seen in a dorsal view. Only one ventrobronchus and one dorsobronchus and their interconnecting parabronchi are shown. (Partly after Welty.)

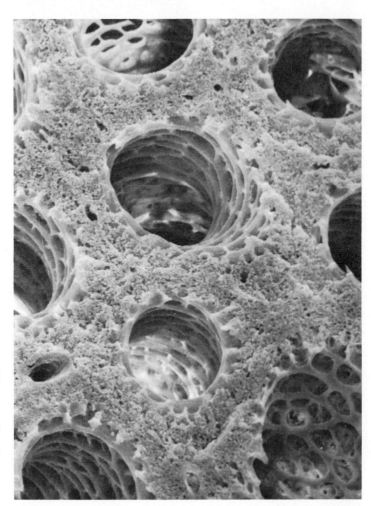

Figure 32.11 Scanning electron micrograph of a section through the lung of a domestic fowl showing parabronchi in cross section and associated air capillaries. (Photograph by H. R. Duncker from How Birds Breathe, by K. Schmidt-Nielsen. Copyright © by Scientific American, Inc. All rights reserved.)

isms that are hit are flicked into its open mouth. The woodcock's long and sensitive bill is adapted for probing for worms in the soft ground. The woodcock can open the tip of its bill slightly to grasp a worm without opening the rest of the mouth!

Food taken into the **mouth** is mixed with a lubricating saliva and passes through the **pharynx** and down the **esophagus** without further treatment, for birds have no teeth. In grain-eating species, such as the pigeon, the lower end of the esophagus is modified to form a **crop** in which the seeds are temporarily stored and softened by the uptake of water. Food is mixed with peptic enzymes in the **proventriculus,** or first part of the stomach, and then passes into the **gizzard,** the highly modified posterior part of the stomach characterized by thick muscular walls and modified glands that

secrete a horny lining. Small stones that have been swallowed are usually found in the gizzard and aid in grinding the food to a pulp and mixing it with the gastric juices. The **intestinal region** is relatively short compared to the intestine of mammals. As in mammals, it is lined with microscopic, finger-like projections, the **villi,** that greatly increase the surface area.

The **lungs** of birds are unique. They are relatively small and compact organs, but lead to an extensive system of **air sacs** that extend into many parts of the body, some even entering the bones (Fig. 32.10). The air sacs connect with a large **mesobronchus** that extends through each lung, and also with **recurrent bronchi** that lead back to the lungs. The larger bronchi connect with secondary bronchi (**dorsobronchi** and **ventrobronchi**) which, in turn, are interconnected by smaller **parabronchi.** Minute branching and anastomosing **air capillaries** radiate from the parabronchi, and it is through their vascular walls that gas exchange with the blood occurs (Fig. 32.11). There are no blind sacs, such as the alveoli of mammal lungs, in bird lungs.

By tracing the flow of certain gases through the lungs and air sacs, and detecting the direction of flow by small probes, Dr. Knut Schmidt-Nielsen and his students have determined the course of air through this complex system. During inspiration, the sternum is lowered, the air sacs expand, and the lungs, surprisingly, are compressed slightly by the contraction of muscles that pull a membranous "diaphragm" against their ventral surface. Air high in oxygen content moves through the trachea and mesobronchi into the posterior air sacs, and stale air, already in the lungs, moves from the lungs into the anterior air sacs. During expiration, the sternum is raised, the air sacs compressed, and the lungs expand. Air high in oxygen content now moves from the posterior sacs into the lungs; stale air in the anterior sacs moves into the mesobronchi, the trachea, then out of the body.

This unusual mechanism of ventilation provides a one-way flow of air through the parabronchi and air capillaries, but it takes two cycles of inspiration and expiration for a given unit of air to move through the system: inspiration 1, fresh air mixed with some stale air remaining in the trachea enters the posterior air sacs; expiration 1, this air enters the lungs where it gives up oxygen and takes on more carbon dioxide; inspiration 2, stale air moves from the lungs to the anterior air sacs; expiration 2, the stale air leaves the anterior sacs and most of it is expelled from the body. The residual carbon dioxide remaining in the trachea mixes with the incoming air, and it prevents too much carbon dioxide being removed from the blood. The amount of carbon dioxide in the blood must be kept above a certain threshold value to maintain the activity of the respiratory center (p. 297), and the proper acid-base balance of the body.

Birds and mammals have evolved quite different ways of providing sufficient gas exchange to sustain a high level of metabolism. Bird lungs are the more efficient for air moves through the lung tissue in only one direction, and the direction is counter to that of the blood flow. The system is analogous to the countercurrent exchange mechanism found in the gills of many fishes (p. 292). The great efficiency of bird lungs permits some species to fly freely at altitudes in excess of 4000 meters.

A mechanism for the production of sounds is associated with the air passages. Membranes are set vibrating by the movement of air; however, the vibratory membranes are not in the larynx at the anterior end of the trachea but in a **syrinx** at its posterior end (Fig. 32.10). Muscles associated with the syrinx vary the pitch of the notes.

Birds resemble mammals in having evolved a double circulation through the heart. The complete separation of venous and arterial blood within the heart, the rapid heart beat (400 to 500 times per minute in a small bird such as a sparrow when it is at rest), and an increase in blood pressure make for a very rapid and efficient circulation. This is of the utmost importance in an endotherm for the tissues need a large supply of food and oxygen, and waste products of metabolism must be removed quickly.

Nitrogenous wastes are removed from the blood by a pair of **kidneys,** which are basically similar to those of reptiles (Fig. 32.9). The high rate of metabolism of birds, however, requires a much greater number of kidney tubules. Indeed, most birds have relatively more tubules than do mammals; a cubic millimeter of tissue from the cortex of a bird's kidney contains 100 to 500 renal corpuscles in contrast to 15 or less in a comparable amount of mammal kidney. Body water is conserved, as in reptiles, by tubular reabsorption and the elimination of most of the nitrogenous waste as uric acid. Birds have lost the urinary bladder, present in ground-dwelling vertebrates, possibly as one means of reducing body weight.

Birds lay cleidoic eggs, as do reptiles; hence their reproductive system is very similar to that of reptiles. Female birds have lost the right ovary and oviduct, probably as another means of reducing weight.

Body salts are generally conserved by terrestrial vertebrates, although any excess can be eliminated by the kidneys. Sea birds; however, have a large salt intake with their food and water and must eliminate more salts than can be disposed of by the kid-

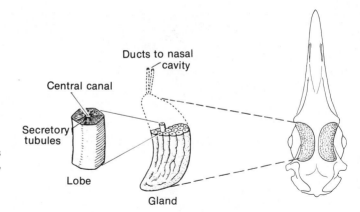

Figure 32.12 The salt-excreting glands of the herring gull. (Modified after Fänge, Schmidt-Nielsen and Osaki.)

neys. This is done by special **salt-excreting glands,** which in the herring gull, for example, are located above the eyes and discharge a concentrated salt solution into the nasal cavities. This solution leaves these cavities through the external nares and drips from the end of the beak (Fig. 32.12). Each gland consists of many lobes, each of which is composed of many vascularized secretory tubules radiating from a central canal.

The sense of smell is less important in vertebrates that spend a considerable part of their life off the ground than it is in terrestrial species, so it is not surprising to find that the olfactory organ and olfactory portions of the brain are reduced in birds. Sight, on the other hand, is very important, and the **eyes** and optic regions of the brain are unusually well developed. The eyes of birds occupy a large portion of the head, and both eyes together are often heavier than the brain. The visual acuity of birds, that is, their ability to distinguish objects as they become smaller and closer together, is several times as great as that of man. The ability to accommodate rapidly is also well developed in birds' eyes, for birds must change quickly from distant to near vision as they maneuver among the branches of a tree or swoop down to the ground from a considerable height. Muscular coordination is also very important in the bird way of life, and the **cerebellum** is correspondingly well developed. The **cerebral hemispheres** are large; however, their size is not a result of the expansion of the surface cortex, as it is in mammals, but rather the enlargement of a more deeply situated mass of gray matter, the **corpus striatum.**

32.3 THE ORIGIN AND EVOLUTION OF BIRDS

One might infer simply from the structure of modern birds that they have evolved from archosaurian reptiles, but we need not stretch our inferences, for three specimens of a fossil bird are known that are clearly intermediate between archosaurs and modern birds. The fossils are preserved with remarkable detail in a fine-grained, lithographic limestone from Jurassic deposits in Bavaria.

Archaeopteryx lithographica (Fig. 32.13A) was about the size of a crow. Its skeleton is reptilian in having toothed jaws, no fusion of trunk or sacral vertebrae, a long tail, and a poorly developed sternum. Birdlike tendencies are evident in the enlarged orbits, some expansion of the brain case, and in the winglike structure of the hand. As in modern birds, the "hand" is elongated and only three "fingers" are present; however, there is little fusion of bones and each finger bears a claw. If the skeleton alone were known, the creature would probably have been regarded as a peculiar archosaur, but it is evident that this was a primitive bird and not a reptile, for there are clear impressions of feathers (Fig. 10.1, p. 236). The feathers would suggest that *Archaeopteryx* was active and warm-blooded. The ratio of its wing surface to its body size, together with the poorly developed sternum, indicates that it was at best a weak flier. These most primitive birds are placed in the subclass **Archaeornithes.**

We, of course, do not know just how flight evolved in the ancestors of *Archaeopteryx,*

Figure 32.13   Extinct birds. *A*, A restoration of *Archaeopteryx*, the earliest known bird; *B*, a restoration of *Hesperornis*, a large diving bird of the Cretaceous. (*A* from Heilmann; *B* courtesy of the American Museum of Natural History.)

but the early stages in the evolution of flight must have been adaptive in some way, even though the "proavis" would not at first have been able to fly. It is possible that the ancestors of birds were becoming more active, and possibly warm-blooded, and feathers may have first been of value in helping to conserve body heat. Given feathers, their enlargement along the posterior margin of the forelimb and tail may have conferred some stability in running rapidly along the ground, or even in rudimentary gliding from low branches. Thus structures originally of value in one way may have attained a certain threshold of size which enabled them to perform a different function. Complex organs such as feathers and wings could have evolved in

Figure 32.14   Representative paleognathous birds. *A*, Ostrich; *B*, a kiwi with its relatively huge egg. (*A* from Grzimek, B., in Natural History, Vol. LXX, No. 1; *B* courtesy of the American Museum of Natural History.)

Figure 32.15   A group of neognathous birds. *A*, Penguins use their modified wings as flippers; *B*, courtship of albatrosses; *C*, the young cormorant has to reach into the throat of its parent to get its food; *D*, an American egret, or heron, a wading bird; *E*, barn owl strikes; *F*, noddy terns on nest. (*A* courtesy of Smithsonian Institute; *B* courtesy of Lt. Col. N. Rankin; *C*, photo by L. W. Walker from National Audubon Society; *D* and *F* courtesy of the American Museum of Natural History; *E* from Payne, R. S., and Drury, W. H., Jr., in Natural History, Vol. LXVII, No. 6.)

this manner and could have been adaptive at all stages of their evolution.

The next group of fossil birds have been found in Cretaceous deposits. These birds had lost the long reptilian tail, had evolved a well-developed sternum and were modern in many other ways. A true pygostyle had not yet evolved, and teeth were present in at least certain species. There are clear indications of teeth in fossils of *Hesperornis* (Fig. 32.13*B*), a large diving species with powerful hind legs and vestigial wings. The nature of the jaws of *Ichthyornis*, which was a tern-sized flying species, is uncertain. Although they are placed in the subclass **Neornithes** along with modern birds, the more primitive nature of these Cretaceous species is recognized by placing them in a distinct superorder — the **Odontognathae.**

All later birds have lost the reptilian teeth, but a few (superorder **Palaeognathae**) retain a somewhat reptilian palate, whereas others (superorder **Neognathae**) have a more specialized palatal structure. Living paleognathous birds are for the most part ground-dwelling, flightless species, such as the ostriches of Africa, the rheas of South America, the cassowaries of Australia and the peculiar kiwi of New Zealand (Fig. 32.14). The legs are well developed and powerful, the wings vestigial, and the feathers are fluffy. Presumably, these birds evolved from flying ancestors but readapted to a terrestrial mode of life in areas where there were an abundant food supply upon the ground and few competitors or enemies. The ancestry of certain of them can be traced back to the early Cenozoic era. A number of large ground-dwelling neognathous birds also lived then, which suggests that there might have been a competition at this time between birds and early mammals for the conquest of the land surface, which had recently, geologically speaking, been vacated by the large reptiles. Mammals won, and only a few ground-dwelling birds survived.

All other birds, including the vast majority of living species, are neognathous types (Fig. 32.15). Twenty-three orders of these birds are recognized (cf. Classification at the end of this chapter). Most can be distinguished by specializations of their bills, wings, tails and feet which reflect their divergence with respect to feeding and method of locomotion. Some, including the loons, ducks and gulls are aquatic as well as good fliers. Albatrosses and petrels are oceanic birds that spend much of their life at sea. It is probable that the penguins evolved from certain diving members of this group. Penguins have lost their ability to fly, and their wings are modified as paddles for swimming under water. The herons, cranes and coots have become specialized for a wading, marsh-dwelling mode of life. Hawks, eagles and owls are birds of prey. The grouse, pheasants and fowl are predominantly terrestrial forms, though they can fly short distances, and the perching and songbirds are well adapted for life in the trees. The songbirds are members of the order **Passeriformes.**

32.4 MIGRATION AND NAVIGATION

An aspect of bird biology of particular interest is the seasonal migration of many species, and their uncanny ability to navigate. Many vertebrates migrate and find their way home if displaced, but birds' power of flight has endowed them with a more spectacular range of movement than in other vertebrates.

The extensive movement of populations back and forth between different areas has certain advantages. The migration of birds from winter quarters in temperate or tropical regions to breeding areas in the north permits them to spread into an area where the days are long in the summer and a large food supply develops for a few months. Birds can establish territories with a minimum of effort, and they have long hours of daylight to obtain food at a time when their population is increasing greatly. As conditions become inclement, the birds return to winter quarters. Migration also prevents predator populations from increasing greatly for the predators of a particular region do not have a sustained food supply if the birds move out at intervals. In addition, migration helps birds avoid harsh climatic extremes and is, in a sense, a homeostatic mechanism.

Although there are advantages to migration, there are also hazards to a long trip.

Many migrants are caught in storms and perish. It is not surprising, then, that not all species of bird migrate. Even within a single species there may be populations that do and others that do not. Barn owls (*Tyto alba*) from the northern part of the United States, for example, tend to migrate, whereas more southern populations are sedentary. Presumably each group of birds has adapted to the particular environmental stresses it encounters. If the advantages of movement outweigh the hazards, the population migrates; if not, it remains more sedentary. Those that migrate are simply using their great capacity of movement to spread out in space biologic functions that in another species occur in different parts of a more restricted range. It is common for many animals to feed and breed in somewhat different areas, even though the areas may not be very far apart.

The stimulus for migration has been a topic of intensive research. It is generally agreed that neuroendocrine changes predispose a bird to migrate. In the spring of the year in the northern hemisphere, the hypothalamus stimulates the pituitary gland to secrete more gonadotropic hormones and the gonads increase in size. There is also a rapid increase in fat deposits, and an increased activity, or restlessness, of the birds. In the fall, the gonad decreases in size, but the birds again become restless and accumulate food reserves for the trip. The interrelation of these factors to each other and to changes in day length are not entirely clear. Once birds are in a migratory condition, favorable weather and other external factors probably trigger the onset of migration.

Most passerine migrants travel at night, stopping to feed and rest during the day. Some may fly several hundred miles during a single night, but then may rest for several days. Many larger birds, including hawks and herons, migrate by day, and ducks and geese may migrate either at night or in the day. The northward advance of birds in the spring averages about 20 to 25 miles per day. Many species tend to follow the advance of certain temperature lines, or isotherms (Fig. 32.16). The length of migration and the route taken are very consistent for each type of bird but vary with the species. The Canada goose winters in the United States from the Great Lakes south, breeds in Canada as far north as the arctic coast, and migrates along a broad front between the two areas. The scarlet tanager winters in parts of South America and breeds in the area from Nova Scotia, southern Quebec and southern Manitoba south to

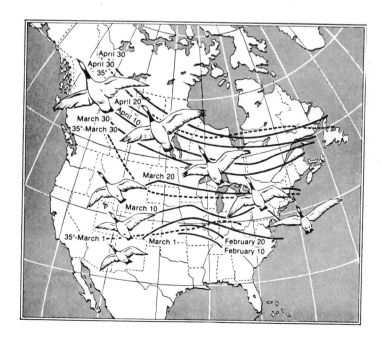

Figure 32.16 The northward migration of the Canada goose keeps pace with spring, following the isotherm of 35° F. (Modified after Lincoln.)

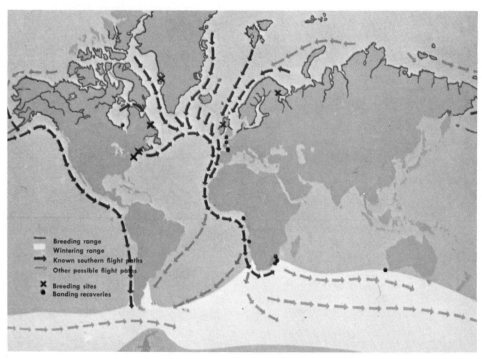

Figure 32.17 Migration routes of different populations of the arctic tern. (From The Natural Science Encyclopedia, Vol. 2. Published by Golden Press. Copyright © 1962 by Western Publishing Company, Inc.)

Sun   Cloud

Figure 32.18 Vanishing point diagrams of the compass course taken by Lesser Black-backed Gulls on sunny and cloudy days. North is toward the top of the diagram and the shortest line represents one bird. The direction of home is shown by the arrow. (From Matthews.)

South Carolina, northern Georgia, northern Alabama and Kansas. In contrast to the Canada goose, it has a narrow migration route, which extends through southern Central America and then across the center of the Gulf of Mexico, passing between Yucatan and Cuba. The longest migration is that of the arctic tern; some of these birds travel 25,000 miles in a year. This species breeds in the Arctic, then migrates to its winter quarters in the South Atlantic (Fig. 32.17).

The season, speed and routes of migration have been carefully described for most migratory species, but how birds navigate and find their way remains one of the intriguing, incompletely solved problems of animal behavior. Obviously, the birds must know where they are going; there must be some feature of the environment that is related to the goal of the bird; and the bird must have some way of perceiving this feature. Theories of navigation based on magnetic fields of the earth, visual landmarks, celestial points of reference and other aspects of the environment have been proposed, but no single one explains all of the facts. The magnetic field theory is weakened by our inability to demonstrate that birds are sensitive to magnetic fields. Many migration routes parallel mountain ranges, coast lines and other major physiographic features, which suggests that visual landmarks are being used. Dr. Griffin, in a study of the related problem of homing, released sea birds (gannets) in unknown territory 100 miles or more inland from their nests and followed their return from an airplane. The birds did not head straight for home, but flew in widening circles over large areas, apparently in an exploratory fashion, until they came into familiar territory, and then they headed directly home. Dr. Matthews of Cambridge University believes that visual landmarks play a minor role in long distance migration and has suggested that birds use the position of the sun, a sense of time, and a knowledge of the position of the sun at different times at their destination to determine their position and find their way. This would be analogous to a mariner using a sextant, a chronometer and a knowledge of latitude and longitude. In one set of experiments Matthews released Lesser Black-backed Gulls south of home on sunny and cloudy days and observed the directions that they took until they vanished from view. The results can be shown in "vanishing point" diagrams (Fig. 32.18) in which the number of birds that took a particular compass course at the vanishing point is shown by the length and direction of lines. Most birds showed an orientation toward home on sunny days, but a disoriented scatter under heavy cloud, which is consistent with the hypothesis of a sun compass.

Studies by Sauer suggest that night migrants may use the star pattern to navigate. Migrating birds caught in the spring and caged so they can see the sky become restless at night and tend to face in a northerly direction. In a planetarium, they orient themselves to the north of the artificial sky regardless of the direction of the true sky. We have learned a great deal about migration, but more observations and experiments will be necessary before the full story is known.

CLASSIFICATION OF BIRDS

CLASS AVES. The birds. Endothermic, typically flying vertebrates covered with feathers.

† **Subclass Archaeornithes.** Ancestral birds retaining many reptilian features including jaws with teeth, long tail, and three unfused fingers, each bearing a claw. *Archaeopteryx.*

Subclass Neornithes. Birds with a reduced number of caudal vertebrae; wing composed of three highly modified fingers partly fused together.

† **Superorder Odontognathae.** Cretaceous birds; some, at least, retaining teeth. *Hesperornis, Ichthyornis.*

---

† Extinct.

Superorder Palaeognathae.   Modern toothless birds with a primitive palate. Most are flightless.

Order Tinamiformes.   Tinamous. Largely a ground-dwelling group, but with weak powers of flight; sternum retains a keel. Mexico to South America, *Tinamus*.

Order Struthioniformes.   Ostriches. Huge flightless birds with small wings; unkeeled sternum; head and neck largely devoid of feathers; large powerful legs with only two toes. Africa and western Asia, *Struthio*.

Order Rheiformes.   Rheas. Large flightless birds with unkeeled sternum; head and neck feathered; heavy legs with three toes. South America, *Rhea*.

Order Casuariformes.   Cassowaries and emus. Large flightless birds with small wings and unkeeled sternum; long hairlike feathers with long aftershaft; heavy legs with three toes. New Guinea and Australia, *Casuarius*.

† **Order Aepyornithiformes.**   Elephant birds. Large flightless birds of Africa and Madagascar; became extinct in historic times; laid largest eggs known, 33 × 23 cm. *Aepyornis*.

† **Order Dinornithiformes.**   Moas. Largest of the flightless birds, attained a height of 3 meters. Became extinct about 700 years ago. New Zealand, *Dinornis*.

Order Apterygiformes.   Kiwis. Modest-sized flightless birds with unkeeled sternum and vestigial wings; four toes on feet; long bill with nostrils near the tip, used in probing soft ground for food. New Zealand, *Apteryx*.

Superorder Neognathae.   Modern birds with a less reptilian palate.

Order Gaviformes.   Loons. Legs located far back on the body; webbed feet; reduced tail; long, compressed and sharply pointed bill; very good divers. The common loon, *Gavia*.

Order Podicipediformes.   Grebes. Legs located far back on the body; lobate toes; reduced tail; very good divers. Eared grebe, *Colymbus*.

Order Procellariiformes.   Albatrosses, shearwaters, fulmars, petrels, tropic birds. Webbed feet; fourth toe vestigial; long narrow wings; tubular nostrils. The petrel, *Oceanodroma*.

Order Sphenisciformes.   Penguins. Flightless oceanic birds with four anteriorly directed toes with a web between three of them; wings modified as paddles; excellent divers. Confined to the southern hemisphere, chiefly Antarctica. The emperor penguin, *Aptenodytes*.

Order Pelecaniformes.   Pelicans, gannets, cormorants, water-turkey, man-o-war bird. Totipalmate swimmers with four toes included in the webbed foot; tendency for the development of a gular sac. The pelican, *Pelecanus*.

Order Ciconiiformes.   Herons, bitterns, storks, ibises, flamingos. Long-legged and long-necked wading birds; feet broad, but usually not webbed; the portion of the head between the eye and nostril (the lores) usually devoid of feathers. Great blue heron, *Ardea*.

Order Anseriformes.   Ducks, geese, swans. Short-legged, web-footed swimming and diving birds; bill usually broad and flat with transverse horny ridges adapted for filtering mud. The mallard, *Anas;* white-fronted goose, *Anser*.

Order Falconiformes.   Vultures, kites, hawks, falcons, eagles. Diurnal birds of prey with strong, hooked bill; sharp, curved talons. Cooper's hawk, *Accipiter;* duck hawk, *Falco*.

Order Galliformes.   Grouse, quails, partridges, pheasants, turkeys, chickens. Seed- and plant-eating, largely ground-dwelling birds; short stout bill; heavy feet with short strong claws, adapted for running and scratching in the ground; wings relatively short. The chicken, *Gallus*.

Order Ralliformes.   Cranes, rails, gallinules, coots. Marsh birds; feet not webbed, but toes sometimes lobed; legs elongate in some groups; lores feathered. Virginia rail, *Rallus*.

† **Order Diatrymiformes.**   Large flightless birds of the early Cenozoic. *Diatryma*.

Order Charadriiformes.   Plovers, woodcock, snipe, sandpipers, stilts, phalaropes, gulls, terns, skimmers, auks, puffins. A diverse group of shore birds. The killdeer, *Charadrius*.

Order Columbiformes. Pigeons, doves. Short slender bill with a fleshy cere at its base overhanging the slitlike nostrils; short legs. The domestic pigeon, *Columba.*

Order Psittaciformes. Parrots. Feet adapted for grasping, with fourth toe capable of being turned back beside the first toe; bill heavy and hooked; often brilliantly colored plumage. Carolina parakeet, *Conurus.*

Order Cuculiformes. Cuckoos, road-runners. Foot zygodactylous with fourth toe permanently reversed beside the first; tail long. The cuckoo, *Cuculus.*

Order Strigiformes. Owls. Nocturnal birds of prey with strong hooked bills; sharp, curved talons; feathers arranged as a facial disc around the large, forwardly turned eyes. The barred owls, *Strix.*

Order Caprimulgiformes. Nighthawks, whippoorwills. Twilight flying birds with small bills, but large mouths surrounded by bristle-like feathers that help net insects; legs and feet small. The whippoorwill, *Caprimulgus.*

Order Apodiformes. Swifts and hummingbirds. Fast flying birds with long narrow wings; legs and feet very small. The chimney swift, *Chaetura.*

Order Coliiformes. The colies of Africa. Small birds with long tails; first and fourth toes can be turned posteriorly.

Order Trogoniformes. Trogons. Short stout bill; small feet; often green, iridescent plumage. The coppery-tailed trogon, *Trogon.*

Order Coraciiformes. Kingfishers. Strong sharp bill; foot syndactylous with third and fourth toes fused at their bases; feathers often forming a crest on the head. The belted kingfisher, *Megaceryle.*

Order Piciformes. Woodpeckers and toucans. Bill chisel-like (woodpeckers) or very large (toucans); zygodactylous foot with fourth toe permanently turned posteriorly. The flicker, *Colaptes.*

Order Passeriformes. The perching birds and songbirds. The largest order of birds, it includes the flycatchers, larks, swallows, crows, jays, chickadees, nuthatches, creepers, wrens, dippers, thrashers, thrushes, robins, bluebirds, kinglets, pipets, waxwings, shrikes, starlings, vireos, wood warblers, weaver finches, blackbirds, orioles, tanagers, finches, sparrows, etc. Foot adapted for perching; three toes in front opposed by one well-developed toe behind. The English sparrow, *Passer.*

ANNOTATED REFERENCES

Many of the general references cited at the end of Chapter 30 contain considerable information on the biology of mammals.

Bent, A. C.: Life Histories of North American Birds. New York, Dover Publications, 1961–1968. A reprinting of Bent's famous multivolume study of the natural history of birds. Originally published between 1919 and 1958 as Bulletins of the U.S. National Museum.

Gilliard, E. T.: Living Birds of the World. New York, Doubleday & Co., 1958. The major groups of birds are summarized and superbly illustrated.

Hertel, H.: Structure, Form, Movement. New York, Reinhold Publishing Corp., 1966. One section of this important book on animal locomotion is devoted to various aspects of bird flight.

Hinde, R. A. (Ed.): Bird Vocalizations. Cambridge, Cambridge University Press, 1969. Many authors discuss the nature, production and biological uses of songs.

Howard, E.: Territory in Bird Life. New York, Atheneum, 1964. A reprint of a classic book on bird behavior.

Marshall, A. J. (Ed.): Biology and Comparative Physiology of Birds. New York, Academic Press, 1960–1961. An important two volume source book on many aspects of the anatomy, physiology, reproduction, migration and behavior of birds.

Matthews, G. V.: Bird Navigation. 2nd ed. Cambridge, Cambridge University Press, 1968. A thorough account is given of the different theories of navigation and homing.

Peterson, R. T.: A Field Guide to the Birds. 2nd ed. Boston, Houghton Mifflin Co., 1947. The standard and widely used guide for the field identification of birds from the Great Plains to the East Coast.

Peterson, R. T.: A Field Guide to Western Birds. Revised ed. Boston, Houghton Mifflin Co., 1961. A companion to the preceding volume, covers the birds from the Pacific Coast to the western parts of the Great Plains.

Pettingill, O. S., Jr.: A Laboratory and Field Manual of Ornithology. 3rd ed. Minneapolis, Burgess Publishing Co., 1956. A manual on the structure, habits and ecology of birds for the serious student of ornithology.

Schmidt-Nielsen, K.: How birds breathe. Scientific American 225:73 (Dec.) 1971. A summary is given of important new findings on the unique lung ventilating mechanism of birds.

Sturkie, P. D.: Avian Physiology. 2nd ed. Ithaca, N.Y., Comstock Publishing Co., 1965. A very important source book; covers most aspects of avian physiology.

Van Tyne, J., and A. J. Berger: Fundamentals of Ornithology. New York, John Wiley & Sons, Inc., 1959. An important text for the serious student; emphasizes the groups of birds.

Welty, J. C.: The Life of Birds. Philadelphia, W. B. Saunders Co., 1962. A comprehensive one-volume work on all aspects of the biology of birds.

33. Vertebrates: Mammals

Mammals (**class Mammalia**) are the familiar hair-covered creatures including cats, mice and men. Most of their adaptive features are related to the evolution of increased activity and greater care of the young. In this they resemble birds, but mammals evolved from reptiles independently of birds and achieved endothermy and greater care of the young in somewhat different ways. **Hair** rather than feathers helps to conserve body heat, and female mammals suckle their young with the secretions of **mammary glands.** Mammals have also evolved a powerful jaw mechanism capable of grinding or cutting up food. A shift in jaw joint is correlated with the changes in jaw mechanics, and mammals differ from all other vertebrates in having a **dentary-squamosal joint** rather than one between the articular and quadrate bones.

33.1 CHARACTERISTICS OF MAMMALS

Temperature Regulation. Their relatively high and constant body temperature permits mammals to be very active animals. Heat loss, or gain, between the organism and its environment is reduced by the hairs that entrap an insulating layer of still air next to the skin. Heat is produced internally by a high level of oxidative metabolism. Body temperature is maintained by physio-

logical controls of the mechanisms for heat loss and production. Important aspects of the control mechanisms can be seen in Figure 33.1, which shows the amount of metabolic work, measured by oxygen consumption, needed to maintain body temperature over a wide range of environmental temperatures. Mammals have a minimum and constant oxygen consumption over a certain range of ambient temperatures that extends downwards from their normal body temperature (37° C., or 98.6° F., in man). This range is called the **thermal neutral zone,** and is bounded by **upper** and **lower critical temperatures.** Little extra metabolic work is required to maintain body temperature within the thermal neutral zone. As ambient temperatures fall slightly below body temperature, heat conservation mechanisms that do not require much energy output are utilized. The hairs are elevated by the contraction of the arrector pili muscles (p. 262) thus increasing the thickness of the insulating layer; blood flow through the skin is reduced by the constriction of cutaneous capillaries; and heat loss by water evaporation is reduced by a decreased activity of the sweat glands. Opposite changes occur so that less heat is conserved as the ambient temperature rises toward body temperature.

Beyond the critical temperatures, however, body temperature can be maintained only by the expenditure of extra metabolic work. The rise in oxygen consumption below the lower critical temperature reflects

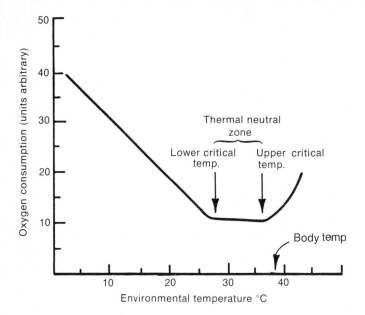

Figure 33.1 A graph to show the relationship between oxygen consumption and environmental temperature in a typical mammal with a body temperature of 37° C. (Modified after Gordon: Animal Function: Principles and Adaptations, Macmillan.)

the extra heat production now needed to maintain body temperature. Shivering, which is an involuntary activity of superficial muscles, is one mechanism of increased heat production, but there is also a general increase in metabolic activity in many parts of the body. The rise in oxygen consumption above the upper critical temperature reflects the added metabolic work needed to dissipate heat when ambient temperature exceeds body temperature. There is now an increase in rate of heart beat and rate of circulation through the skin. Profuse sweating occurs in some mammals and panting in others. Panting, which involves the evaporation of water from the respiratory passages, is an important cooling mechanism in heavily furred mammals that have few sweat glands, and in birds that have none.

Neural mechanisms by which body temperature is regulated are very complex. There are thermal receptors in the skin whereby mammals become aware of environmental temperature changes, and they may modify behavior accordingly. The major thermal mechanism, however, is in the hypothalamus. It responds to slight changes in blood temperature and initiates the changes needed to adjust heat loss and production to the environmental context.

Mammals that live in areas where the environment is rigorous have evolved adaptations that supplement thermal regulation. Arctic mammals, for example, have features that minimize the metabolic cost of living in a cold climate. Typically a thick fur lowers both the lower critical temperature and the slope of the environmental temperature-oxygen consumption curve. The temperature of the appendages, which cannot be insulated as well as the rest of the body, is permitted to fall below that of the body core. Arteries carrying blood to the limbs are sometimes closely intermeshed with the veins returning blood so that a countercurrent exchange mechanism is set up whereby much body heat moves from the arteries to the veins and is not lost. Enough heat must be permitted to enter the appendages, however, to keep them from freezing.

Some arctic and temperate mammals, notably many insectivores, bats and rodents, adjust to winter weather by going into a period of dormancy known as **hibernation.** During this period, they lose considerable control over the regulatory mechanism of body temperature, and their body temperature approaches the ambient temperature. Metabolism is very low during hibernation, yet it is sufficient to sustain life and to keep the body from freezing. There are certain advantages to hibernation for a small endotherm. Small mammals have a relatively higher rate of metabolism than large ones because they have more surface area in proportion to their mass. They lose a great deal of heat through their surface areas and must consume much food just to maintain body temperature. In many regions insects and certain types of plant food are not available in quantity during the winter. If an animal can permit its body temperature to drop, it can get by on less food, or even on the food reserves within its body. Hibernation in mammals should not be confused with hibernation in certain ectotherms (p. 703). Body temperature always fluctuates with ambient temperatures in ectotherms, and when it is low enough the animals naturally are inactive. In a warm-blooded animal, a normally high and constant body temperature is permitted to drop; in a sense, the "thermostat" in the hypothalamus is turned down. Physiologic adjustments are more complex in this situation.

Although mammals in general are endothermic, there are differences among them in their capacity to regulate body temperature. As might be expected, this capacity is less developed among some of the more primitive groups such as the egg-laying monotremes and the pouched marsupials than among higher mammals. The Central American opossum (a marsupial), for example, can nearly stabilize its body temperature over an external temperature range of 10° to 30°C., but loses control of body temperature, and in effect becomes an ectotherm, when external temperatures exceed 30°C. (Fig. 33.3). It can also better maintain a constant body temperature during its period of nocturnal activity than when it is sleeping during the day. Mammals of this type are sometimes described as **heterothermic.**

Metabolic Systems. In order to sustain their high level of metabolism, mammals must be very efficient in obtaining large supplies of food and oxygen, eliminating waste products, and transporting materials throughout the body. The dentition of mammals enables them to obtain and handle a wide variety of foods. Their teeth are not all the same shape, as they are in most reptiles, but are differentiated into various types (Fig. 33.2). Chisel-shaped **incisors** are present at the front of each jaw and are used for nipping and cropping. Next is a single **canine** tooth, which is primitively a long, sharp tooth, useful in attacking and stabbing prey, or in defense. A series of **premolars** and **molars** follow the canine. These teeth tear, crush and grind up the food. In primitive mammals, the premolars are sharper than the molars and have more of a tearing function. A primitive placental mammal, such as an insectivore (Fig. 33.2), has three incisors, one canine, four premolars and three molars in each side of the upper and lower jaw. This can be expressed as a dental formula: $I\frac{3}{3}$, $C\frac{1}{1}$, $Pm\frac{4}{4}$, $M\frac{3}{3}$. No placental mammal has more teeth than this, but the number of teeth is reduced in many groups. Man, for example, has the dental formula of $I\frac{2}{2}$, $C\frac{1}{1}$, $Pm\frac{2}{2}$, $M\frac{3}{3}$. Considerable variation also

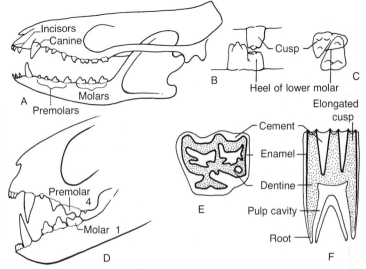

Figure 33.2 Teeth of mammals. *A,* The relatively unspecialized teeth of a primitive insectivore; *B* and *C,* lateral and crown views of the left upper and lower molars of an insectivore to show their occlusion; *D,* the stabbing and cutting teeth of a cat; *E* and *F,* a crown view and a vertical section through the left upper molar of a horse to show its adaptation for crushing and grinding.

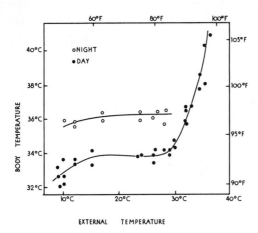

Figure 33.3 A graph showing relationship of body temperature to external temperature in a nocturnally active Central American opossum, *Metachirus.* (From Morrison.)

occurs in the structure of particular teeth in the different groups, as we shall see.

Most mammals do not swallow their food whole but break it up mechanically with their teeth and mix it with saliva which, in addition to lubricating the food, usually contains an enzyme that begins the digestion of carbohydrates. As we have seen (Sections 12.3 and 12.4) digestion is completed in the stomach and intestinal region. Numerous microscopic villi line the small intestine, as they do in birds, and increase the surface area available for absorption.

A greater exchange of oxygen and carbon dioxide is made possible by a many-fold increase in the respiratory surface of the lungs and by improved methods of ventilation. Features that make this possible have been considered earlier (Section 12.11). Of particular interest from an evolutionary viewpoint has been the development of a **secondary palate,** a horizontal partition of bone and flesh in the roof of the mouth that separates the air and food passages in this region (Fig. 33.4). In lower tetrapods the nasal cavities lead directly into the front of the mouth, but in mammals they open more posteriorly into the pharynx. The secondary palate permits nearly continuous breathing, which is a necessity for organisms with a high rate of metabolism. Mammals can manipulate food in their mouths, and breathing need be interrupted only momentarily when the food is swallowed, and in some species not even then (young of opossum,

p. 756). The presence of a secondary palate in late mammal-like reptiles suggests that their rate of metabolism was increasing.

Mammals, like birds, have evolved an efficient system of internal transport of materials between sites of intake, utilization and excretion. Their heart is completely divided internally so there is no mixing of venous and arterial blood. Increased blood pressure also contributes to a more rapid and efficient circulation.

Although mammals eliminate most of their nitrogenous wastes as **urea,** which is more soluble and requires more water for its removal than does the uric acid excreted by some reptiles and birds, approximately 99 per cent of the water that starts down the kidney tubules is later reabsorbed, so that the net loss of water in these animals is minimal. The generally high metabolic rate of mammals results in the formation of a large amount of wastes to be eliminated. An increase in blood pressure, and hence in blood flow through the kidney, and an increase in the number of kidney tubules have enabled mammals to increase the rate of excretion.

Locomotion and Coordination. Mammals also move about with greater agility than lower tetrapods. Their appendages extend directly down to the ground in the vertical plane, instead of out from the body in the horizontal plane as the proximal segment of the limb does in amphibians and most reptiles. This improves the effectiveness of the limbs in support and permits the development of a longer stride and more rapid locomotion. A firmer support is also provided for the pelvic girdle and hind limbs, because most mammals have three sacral vertebrae in contrast to the two of most reptiles. Arboreal species use the tail for balancing, and it plays a major role in the propulsion of aquatic mammals such as the whales, but in most mammals it has lost its primitive role in locomotion and is frequently reduced in size.

The increased speed of locomotion also entailed changes in the neuromuscular apparatus. Shifts in many of the muscles concerned with support and locomotion are correlated with the new limb posture. Moreover, the muscular system of mammals is considerably more elaborate than that of reptiles. This, together with a more highly developed nervous system, permits more varied responses and adjustments to environmental conditions.

Care of the Young. The evolution by reptiles of the cleidoic egg was a successful adjustment to terrestrial reproduction so long as vertebrates were ectothermic. However, embryos that are to become endothermic adults must develop under warm and constant temperatures; birds and mammals cannot lay eggs and then ignore them. Birds incubate their eggs by sitting on them, and one group of primitive mammals, which includes the duckbilled platypus of Australia, does the same. All other mammals are viviparous. The eggs are retained within a specialized region of the female reproductive tract, the **uterus,** and the young are born as miniature adults.

All of the extraembryonic membranes characteristic of reptiles are present in viviparous mammals, but albuminous materials are not ordinarily secreted about the egg. The allantois, or in a few species the yolk sac, unites with the chorion, thereby carrying the fetal blood vessels over to this outermost membrane. The vascularized chorion unites in varying degrees with the uterine lining to form a **placenta,** in which fetal and maternal blood streams come close together, though they remain separated by some layers of tissue (Fig. 14.14). The embryo derives its food and oxygen and eliminates its carbon dioxide and nitrogenous wastes across these membranes.

Different species of mammals are born at different stages of maturity. Certain mice, for example, are extremely **altricial,** being born naked and with closed eyes and plugged ears. Newborn deer and other large herbivores are quite **precocial** and can run about and largely care for themselves. But regardless of maturity at birth, all newborn mammals feed upon milk secreted by specialized **mammary glands** of the female (Fig. 33.5). In such primitive mammals as the platypus (Fig. 33.6A), the milk is discharged onto the hairs and the young lap it up, but in other mammals, nipples or teats are associated with the glands and the young are suckled. When the young finally leave their mother, they are at a relatively advanced stage of development and are equipped to care for themselves.

33.2 PRIMITIVE MAMMALS

Monotremes. The most primitive contemporary mammals are the platypus (*Ornithorhynchus*) and its close relative, the spiny anteater (*Tachyglossus*) (Fig. 33.6A and B). In addition to the egg-laying habit, these mammals retain many other reptilian characteristics, including a cloaca. The ordinal name for the group, **Monotremata,** refers to the presence of a single opening for the discharge of feces, excretory and genital products. In other mammals the cloaca has become divided, and the opening of the intestine, the **anus,** is separate from that of the urogenital ducts.

Monotremes are curious animals that have survived to the present only because they have been isolated from serious com-

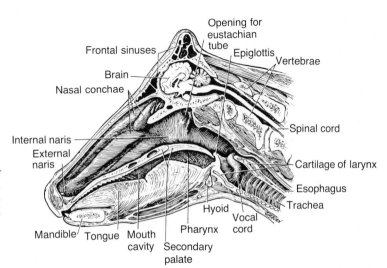

Figure 33.4 A sagittal section of the head of a cow showing the relationship between the digestive and respiratory systems. (Modified after Sisson and Grossman.)

Frontal sinuses
Opening for eustachian tube
Epiglottis
Vertebrae
Brain
Nasal conchae
Spinal cord
Internal naris
External naris
Cartilage of larynx
Esophagus
Trachea
Hyoid
Vocal cord
Mandible
Tongue
Mouth cavity
Pharynx
Secondary palate

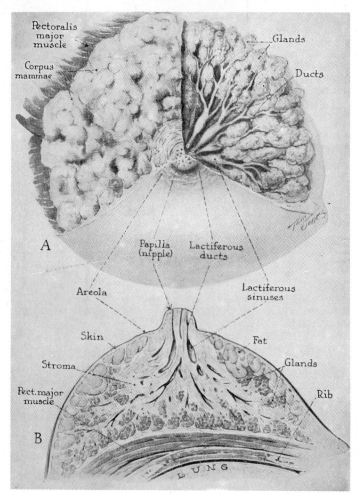

Figure 33.5 *A*, Successive stages of the dissection of the mammary gland of a lactating human being are shown clockwise. *B*, Vertical section through the nipple of a mammary gland. (From King and Showers: Human Anatomy and Physiology, 5th ed.; courtesy of S. H. Camp Co.)

petition in the Australian region. The platypus is a semiaquatic species with webbed feet, short hairs and a bill like a duck's used in grubbing in the mud for food. Spiny anteaters have large claws and a long beak adapted for feeding upon ants and termites. The animal can burrow very effectively with its claws, completely burying itself even in fairly hard ground in a few minutes. Many of its hairs are modified as quills.

When the first skins of the platypus were shipped to Europe in the late eighteenth century, many zoologists viewed them as skillful fakes such as the then current Chinese mermaids (the forepart of a monkey sewn onto the tail of a fish). After the authenticity of the platypus was established, a long controversy ensued as to whether to consider it a reptile or a mammal. Monotremes were finally regarded as mammals, but as such primitive and unusual ones that they are placed in a separate subclass—the **Prototheria.** Some investigators believe that monotremes evolved from mammal-like reptiles earlier than, and independently of, the other mammals. If this is true, mammals have had a polyphyletic rather than a common evolutionary origin (Fig. 31.3). A corollary of such a view is that hair and mammary glands either evolved independently in monotremes and other mammals or were attributes of the mammal-like reptiles.

Marsupials. Other mammals living today are believed to have had a common origin and are placed in the subclass **Theria.** Therian mammals were present in the last half of the Mesozoic era, but they did not become abundant until the extinction of the ruling reptiles. During the Cenozoic, they increased rapidly, radiated widely, and became the dominant terrestrial vertebrates.

Contemporary therians fall into two infraclasses: (1) the **Metatheria,** which includes the opossum, kangaroo and other pouched mammals of the order **Marsupialia** (Fig. 32.6), and (2) the **Eutheria,** or true placental mammals. Both groups are viviparous, though the placental arrangements of marsupials are less effective than those of eutherians. In most marsupials the extra-embryonic membranes, and chiefly the yolk sac, simply absorb a "uterine milk" secreted by the mother. There is no intimate union between the extraembryonic membranes and the uterine lining as there is in most eutherians.

Marsupials are born in what we would regard as a very premature stage. Their front legs, however, are well developed at

Figure 33.6 Monotremes and marsupials. *A,* The duckbilled platypus; *B,* the spiny anteater; *C,* opossum and young; *D,* koala bear; *E,* kangaroo. The platypus and anteater are monotremes; the others are marsupials. (*A* and *B* courtesy of the New York Zoological Society; *C* and *D* courtesy of American Museum of Natural History; *E* from Australian News and Information Bureau.)

birth, and the young pull themselves into a **marsupium,** or pouch, on the belly of the mother, attach to a nipple, and there complete their development. An opossum, for example, is born after only 13 days gestation, but it continues its development in the pouch until it is about 70 days old. A forward extension of the tubular epiglottis of the infant opossum dorsal to its secondary palate completely separates the digestive and respiratory tracts, and breathing and feeding can take place concurrently.

Marsupials were world-wide during the early Cenozoic, but as eutherians began to spread out, marsupials became restricted. They have been most successful in those parts of the world where they have been isolated from competition with eutherians. They are the dominant type of mammal in Australia, have undergone an adaptive radiation and have become specialized for many modes of life. There are carnivorous marsupials such as the Tasmanian wolf, ant-eating types, molelike types, semiarboreal phalangers and koala bears (the original "Teddy-bear"), plains-dwelling kangaroos and rabbit-like bandicoots. In contrast, the only marsupial present in North America is the semiarboreal opossum.

33.3 ADAPTIVE RADIATION OF EUTHERIANS

The eutherians, or placental mammals as they are frequently called, are the most successful mammals in all the parts of the world that they have reached. They have radiated widely and adapted to nearly every conceivable ecologic niche upon the land. Others have readapted successfully to an aquatic mode of life, and some have evolved true flight.

Insectivores. The most primitive eutherians, that is, the stem group from which the other lines of descent evolved, were rather generalized, insect-eating types of the order **Insectivora.** Among modern types are the shrews, moles and the European hedgehog (Fig. 33.13). All are small mammals with an unspecialized limb structure; five clawed toes are retained and the entire foot is placed flat upon the ground, a posture termed **plantigrade** (Fig. 33.7). They have a primitive dentition in which the molar teeth bear sharp cusps well adapted for feeding upon insects and other small invertebrates. The group includes the smallest of all known mammals—one species of shrew which as an adult weighs no more than 2 grams!

Flying Mammals. Bats, order **Chiroptera,** are closely related to this stem group and are sometimes characterized as flying insectivores. As in other flying vertebrates, the pectoral appendages have been transformed into wings (Fig. 33.8). Bat wings are structurally closer to those of pterosaurs than to birds' wings, for the flying surface is a leathery membrane, but the wing of a bat is supported by four elongated fingers (the second to fifth) rather than by a single

Figure 33.7 Insectivores. A, A shrew eats more than its own weight every day; B, a mole in its burrow. (A from Conoway, C. H., in Natural History, Vol. LXVIII, No. 10; B, courtesy of the American Museum of Natural History.)

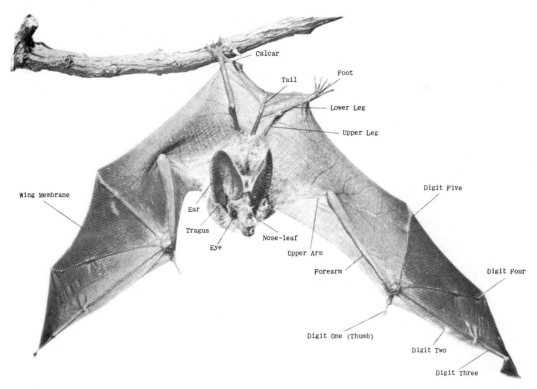

Figure 33.8 The Mexican big-eared bat (*Macrotus mexicanus*), about to take off. (From Walker, E. P.: Mammals of the World. Baltimore, The Johns Hopkins Press, 1964. Photograph by E. P. Walker.)

one as in the pterosaur. The wing membrane attaches onto the hind leg and, in some bats, the tail is included in the membrane. The first finger is free of the wing, bears a small claw and is used for grasping and clinging. The hind legs are small and are of little use upon the ground, but they, too, are effective grasping organs and are used for clinging to a perch from which the bats hang upside down when at rest.

Our familiar bats are insect eaters that fly about at dusk in search of their prey, but bats have diverged considerably in their feeding habits. Among the food that various groups are specialized to take are fruit, pollen and nectar, blood (vampire bats), small mammals and birds, and fish. Fish-eating bats catch their prey near the surface of the water by means of hooked claws on their rather powerful feet.

Bats have evolved a system of echolocation that enables them to avoid obstacles in the dark and also enables many species to find their prey. As early as 1794, Spallanzani observed that a blinded bat could find its way about, but that one in which the ears had been plugged was helpless. However, it was not until the availability of sophisticated electronic apparatus about the time of World War II that Dr. Donald Griffin and others were able to show that bats emitted ultrasonic sounds that bounced off an object and returned as an echo. By analyzing the echoes, many bats can determine the distance, direction, size, and possibly the texture of the object.

There are many differences in method of echolocation among the various groups of bats, which suggests that there has been some independent evolution of mechanisms. As the familiar bats of the family Vespertilionidae fly about at dusk searching for insects, they emit ultrasonic pulses that range from 25,000 to over 100,000 cycles per second, well above our threshold of hearing. Sounds at these frequencies have short wave lengths, hence can produce sharp echoes from small objects. The pulses are frequency modulated and drop about an octave during their 1 to 4 millisecond duration. While the bat is searching, pulses are emitted at the rate of about 10 per second, but the frequency of emission increases considerably, and the duration of the pulses shortens, when a bat detects an insect and homes in on it. The emitted sounds are at very high energy levels, 60 dynes per square cm., which is over twice that in a boiler factory. Tiny muscles in the middle ear contract during sound emission, thereby damping the movement of the auditory ossicles and protecting the inner ear. These muscles relax as the echo returns. Distance appears to be perceived by the time interval between the emitted pulse and the echo. The potential problem of an overlap between the pulse and echo from a close object is avoided in some bats by decreasing the duration of the pulses as they get nearer to the object. In some other cases there is an overlap, and the resolution of the problem is not clear. Directionality appears to be determined by a comparison of the differences in intensity of the echo between the two ears. Bats can perceive meaningful signals in the presence of considerable extraneous noise, but the mechanisms for this are not entirely clear.

Bats are the only mammals to have evolved true flight, but some other mammals can stretch a loose skin fold between their front and hind legs and glide from tree to tree. The colugo (order **Dermoptera**) of the East Indies and Philippines has been observed to glide 136 meters while losing no more than 12 meters in elevation (Fig. 33.9*A*). The flying squirrel is another example of a gliding mammal (Fig. 33.16*A*).

Toothless Mammals. Since primitive mammals were insectivorous, it is not surprising that certain ones became specialized to feed upon ants and termites, which are very abundant in certain regions. The South American anteater, order **Edentata** (Fig. 33.9*B*), is representative of this mode of life. Its large claws enable it to open ant hills, and then it laps up the insects with its long, sticky tongue. In contrast to a primitive insectivore, which crushes its insect food with its teeth, an anteater swallows whole the insects that it eats. Its teeth were not needed for survival and have been lost. The tree sloth and armadillo belong to this same order, though they retain vestiges of teeth.

The pangolins of Africa and Asia (order **Pholidota**) and the aardvark of South Africa (order **Tubulidentata**) are superficially similar to edentates, but this is a result of an independent adaptation to a similar mode of life (Fig. 33.10).

Primates. Members of the order **Primates,** the group to which monkeys and man belong, are also closely related to the primitive insectivorous stock. Indeed, one member of the order, the Oriental tree shrew (*Tupaia*, Fig. 33.11*A*), has at times been considered to be an insectivore. Primates evolved from primitive, semiarboreal insectivores and underwent further specializations for life in the trees. Even those that have secondarily reverted to a terrestrial life bear the stamp of this prior arboreal adaptation. Our flexible limbs and grasping hands are fundamentally adaptations for life in the trees. Claws were transformed into finger- and toenails when grasping hands and feet evolved. The reduction of the olfactory organ and olfactory portion of the brain and the development of stereoscopic, or binocular, vision represent other adaptations of our ancestors to arboreal life. Keen vision and the ability to appreciate depth are very important for animals moving through trees, whereas smell is less important for organisms living some distance from the ground than it is for terrestrial species. Muscular coordination is also very important, and the cerebellum of primates is unusually well developed. The evolution of stereoscopic vision, increased agility, and particularly the influx of a new sort of sensory information gained by the handling of objects with a grasping hand, was accompanied by an extraordinary development of the cerebral hemispheres. The cerebrum is the chief center for the integration of sensory information and the initiation of appropriate motor responses in all mammals, but it is particularly prominent in primates. It is believed that higher mental functions such as conceptual thought could only have evolved in organisms with a grasping hand. In a very real sense, we are a product of the trees.

Primates are often divided into four suborders. The first, suborder **Lemuroidea,** includes the tree shrew, lemurs, lorises, galagos and the peculiar aye-aye. Although fossils of lemurs are found in North America and Eurasia, lemur-like primates are now

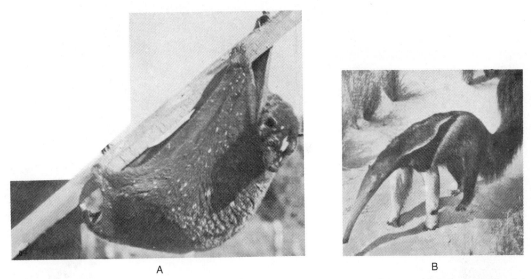

Figure 33.9   *A*, A calugo; *B*, a giant anteater. (*A* from Walker, E. P.: Mammals of the World. Baltimore, The Johns Hopkins Press, 1964. Photograph by J. N. Hamlet. *B*, courtesy of the American Museum of Natural History.)

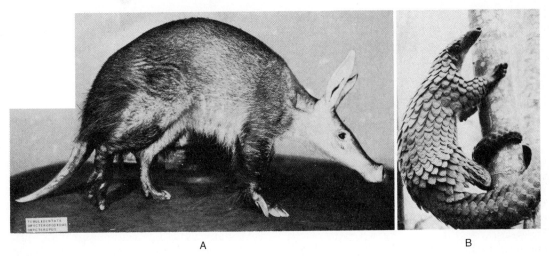

Figure 33.10   (*A*) The African aardvark and (*B*) the African tree pangolin resemble the South American anteater in their adaptations for insect eating, but these common features have evolved in these mammals independently. (*B* from Walker, E. P.: Mammals of the World. Baltimore, The Johns Hopkins Press, 1964. Photograph by Jean-Luc Perret.)

Figure 33.11 A group of primates. *A*, Tree shrew; *B*, lemur; *C*, tarsier; *D*, woolly monkey from South America. (*A* and *C*, courtesy of the American Museum of Natural History; *B*, courtesy of the San Diego Zoo; *D* from Walker, E. P.: Mammals of the World. Baltimore, The Johns Hopkins Press, 1964.)

confined to the Old World tropics; Madagascar has a particularly rich fauna of lemurs. All are rather primitive creatures in which such primate specializations as grasping feet and toenails have begun to appear. However, most lemurs retain a rather long snout, for the nasal region has not been greatly reduced. The suborder **Tarsioidea** includes a single living genus, *Tarsius*, of the East Indies and Philippines. *Tarsius* is a rat-sized animal with large eyes suited for nocturnal vision, and elongated tarsals and digital pads that aid in hopping through the tree tops. It, and the known fossil tarsioids, are too specialized to be the ancestors of other primates, but its flattened face

and forward turned eyes are the sort of advances over lemurs that we would expect to find in the ancestors of the higher primates. Lemuroids and tarsioids are sometimes referred to as the **prosimians.**

Monkeys, the great apes and man, all of whom evolved from certain prosimians, are often collectively called the **anthropoids.** All have a relatively flat face, stereoscopic vision, the capacity to sit on their haunches and examine objects with their hands, and an unusually large brain. However, Old World monkeys, apes and man differ from the New World monkeys in many particulars. New World monkeys retain three premolar teeth, and have evolved a prehensile tail that can be used as a fifth grasping limb in the trees (Fig. 33.11*D*). The two groups of anthropoids have probably had a long independent evolutionary history, which is reflected in the placing of the New World monkeys in the suborder **Platyrrhini** and the Old World species in the suborder **Catarrhini.** Catarrhines are considered more fully in connection with the evolution of man (Chapter 34).

Carnivores. As mammals increased in number and diversity, the opportunity arose for them to feed upon one another. Certain ones became specialized for a flesh-eating mode of life, and these constitute the order **Carnivora.** Familiar living carnivores are the weasels, dogs, raccoons, bears and cats (Fig. 33.12). The shift from an insectivorous to a flesh-eating diet was not difficult. An improvement in the stabbing and shearing action of the teeth and the evolution of a foot structure that enabled them to run fast enough to catch their prey were about all that was necessary. Speed has been increased in most by the development of a longer foot and by standing upon their toes (though not their toe tips) with the rest of the foot raised off the ground in the manner of a sprinter. This **digitigrade** foot posture (Fig. 33.13) gives a longer stride than the primitive **plantigrade** posture, in which the entire foot is placed squarely upon the ground or tree branch.

Most carnivores are semiarboreal or terrestrial, but one branch of the order, which includes the seals, sea lions and walruses, early specialized for exploiting the resources of the sea. In addition to their adaptations as carnivores, which include the large canine tusks of the walrus used in gathering shellfish, these species evolved flippers and other aquatic modifications. When they swim, the large pelvic flippers are turned posteriorly and are moved from side to side like the tail of a fish.

Ungulates. Horses, cows and similar mammals have become highly specialized for a plant diet. This has entailed a considerable change in their dentition, for plant food must be thoroughly ground by the teeth before it can be acted upon by the digestive enzymes. The molars of plant-eating mammals (and those of omnivorous species such as man) have become square, as seen in a surface view. Those of the upper and lower jaws no longer slide vertically across each other to give some cutting action, as do the triangular molars of more primitive mammals, but meet and crush the food between them (Fig. 33.2). A simple squaring of the molars, and to some extent of the premolars, is sufficient for herbivorous mammals that browse upon soft vegetation. But those that feed upon grass and other hard and gritty fare, as do the grazing species, are confronted with the additional problem of the wearing away of the teeth. Two adaptations have occurred: the height of the cusps of the teeth has increased, and cement (a hard material previously found only on the roots of the teeth) has grown up over the surface of the tooth and into the "valleys" between the elongated cusps. More tooth is provided to wear away, and the tooth is more resistant to wear. Teeth of this type are referred to as high-crowned in contrast to the more primitive low-crowned type.

Herbivores constitute the primary food supply of carnivores and protect themselves primarily by the simple expedient of running away. Adaptations for speed have entailed a lengthening of the legs, especially their distal portions, and a relative shortening of the proximal parts of the limbs. The feet are very long, and the animals walk upon their toe tips, a gait termed **unguligrade** (Fig. 33.12). Those toes that no longer reached the ground became vestigial, or disappeared, and the primitive claws on the remaining ones were transformed into hoofs —a characteristic that gives the name ungulate to these mammals. The relatively shorter proximal segment of a limb places the retractor muscles closer to the fulcrum (Fig. 11.17, p. 274); therefore a slight contraction of the muscle can induce a rather extensive movement of the distal end of a limb.

The numerous and varied contemporary ungulates (Fig. 33.14) are grouped into two orders that can be separated on the basis of the type of toe reduction. In the order **Perissodactyla,** the axis of the foot passes through the third toe, and this is always the largest. Ancestral perissodactyls, including the primitive forest-dwelling horses of the early Tertiary, had three well-developed toes (the second, third and fourth) and sometimes a trace of a fourth toe (the fifth). The tapir and rhinoceros, which still

Figure 33.12 Representative carnivores and cetaceans. *A*, Raccoon; *B*, walrus; *C*, the birth of a porpoise; *D*, the whalebone plates of a toothless whale hang down from the roof of the mouth; *E*, weasels in summer pelage. The porpoise and whale are cetaceans; the others are carnivores. (*A, B, D* and *E*, courtesy of the American Museum of Natural History; *C*, courtesy of Marine Studios.)

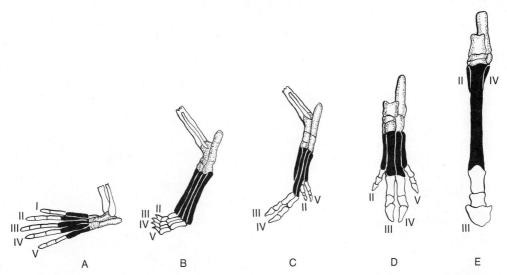

Figure 33.13 Lateral and anterior views of the skeleton of the left hind foot of representative mammals. *A*, The primitive plantigrade foot of a lemur; *B*, the digitigrade foot of a cat; *C* and *D*, the unguligrade foot of a pig, an even-toed ungulate; *E*, the unguligrade foot of a horse, an odd-toed ungulate. The digits are indicated by Roman numerals, the metatarsals are black and the tarsals are stippled.

walk upon soft ground, retain the middle three toes as functional toes, but only the third is left in modern, plains-dwelling horses. Perissodactyls are characterized by having an odd number of toes.

In the order **Artiodactyla,** the axis of the foot passes between the third and fourth toes, which are equal in size and importance. Ancestral artiodactyls had four toes (the second, third, fourth and fifth). Pigs and their allies, which live in a soft ground habitat, retain these four toes, though the second and fifth are reduced in size. Vestiges of the second and fifth toes, the dew claws, are present in some deer, but camels, giraffes, antelope, sheep and cattle retain only the third and fourth toes. Artiodactyls, then, are even-toed ungulates. It is probable that these two orders have had a separate evolutionary origin and owe their points of similarity to parallel evolution.

Subungulates. Subungulates are a group of plant eaters that have certain incipient ungulate tendencies. Elephants (order **Proboscidea,** Fig. 33.15*A*), for example, have five toes, each ending in a hoof-like nail. They also walk to some extent upon their toe tips, but a pad of elastic tissue posterior to the digits supports most of the body weight. Elephants are noted for their enormous size, which must approach the maxi-

mum for a completely terrestrial animal. Though large mammals have a relatively lower metabolic rate than small mammals, the huge mass of elephants necessitates their obtaining large quantities of food. The trunk, which represents the drawn out upper lip and nose, is an effective food-gathering organ. Elephants have a unique dentition in which all of the front teeth are lost except for one pair of incisors, which are modified as tusks. Their premolars, which have come to resemble molars, and their molars are very effective organs for grinding up large quantities of rather coarse plant food. They are high-crowned and so large that there is room for only one in each side of the upper and lower jaws at a time. When it is worn down, a new one replaces it. Unlike reptiles and other lower vertebrates in which there is a continuous replacement of worn-out teeth, mammals have a limited replacement of teeth. Deciduous incisors, canines and premolars are present in young individuals and these are replaced later in life by permanent ones. The molars, which do not develop until after infancy, are not replaced. Elephants, by using up their premolars and molars one at a time, have evolved an interesting way of prolonging total tooth life.

Living elephants are restricted to Africa

Figure 33.14 *A,* Tapir in Rangoon Zoo, Burma; *B,* expedition camel, Kalgan, China; *C,* cattle egret warns the weak-sighted rhinoceros of approaching danger; *D,* hippopotamus. (*A, B* and *D,* courtesy of the American Museum of Natural History; *C* from Natural History, Vol. LXVIII, No. 10, 1959.)

Figure 33.15 Subungulates. The elephant (A) and the manatee (B) are believed to have had a common ancestry. (Courtesy of the American Museum of Natural History.)

and tropical Asia and are only a small remnant of a once worldwide and varied proboscidean population. During the Pleistocene Epoch, mastodons, mammoths and other proboscideans were abundant in North America.

The conies of the Middle East (order **Hyracoidea**), though superficially rabbit-like animals, show an affinity to the elephants in their foot structure and in certain features of their dentition.

A final group of contemporary subungulates are the sea cows or manatees (order **Sirenia**). These animals live in warm coastal waters and feed upon seaweed, grinding it up with molars that are replaced from behind in elephant-like fashion. Sea cows have a powerful, horizontally flattened tail and well-developed pectoral flippers. These features, together with a very mobile and expressive snout and a single pair of pectoral mammary glands, led mariners of long ago to regard them as mermaids.

Rodents and Lagomorphs. Other herbivorous mammals gnaw and, in addition to high-crowned, grinding molars, have an upper and lower pair of enlarged, chisel-like incisor teeth that grow out from the base as fast as they wear away at the tip. Gnawing has been a very successful mode of life; in fact, there are more species, and possibly more individuals, of gnawing mammals, or rodents (order **Rodentia**), than of all other mammals combined. Rodents have undergone their own adaptive radiation and

have evolved specializations for a variety of ecologic niches. Rats, mice and chipmunks live on the ground, gophers and woodchucks burrow, squirrels and porcupines are adept at climbing trees, and muskrats and beavers are semiaquatic (Fig. 33.16).

Rabbits and the related pika of our Western mountains are superficially similar to rodents and were at one time placed in this order. True rodents, however, have only one pair of incisors in each jaw, whereas rabbits have a reduced second pair hidden behind the large pair of upper incisors. It is now believed that rabbits and the pika belong to a separate order, the **Lagomorpha,** and that their resemblance to rodents is a result of parallel evolution.

Whales. Whales, dolphins and porpoises, of the order **Cetacea,** are highly specialized marine mammals that may have evolved from primitive terrestrial carnivores. They have a fish-shaped body, pectoral flippers for steering and balancing, no pelvic flippers, and horizontal flukes on a powerful tail that is moved up and down to propel the animal through the water. Some species have even re-evolved a dorsal fin. Despite these fishlike attributes, cetaceans are air-breathing, viviparous and suckle their young (Fig. 33.12C). Some hair is present in the fetus, but it is vestigial or lost in the adult stage in which its insulating function is performed by a thick layer of blubber.

Figure 33.16 Rodents and lagomorphs. *A*, A flying squirrel; *B*, the pika; *C*, a chipmunk shelling a nut; *D*, a group of beavers. (Courtesy of the American Museum of Natural History.)

Certain species have evolved sonar-like systems that help them to avoid obstacles and find their prey even in muddy river waters that a few enter.

Most cetaceans have a good complement of conical teeth well-suited for feeding upon fish, but the largest whales have lost their teeth and feed upon plankton. With fringed, horny plates (the whalebone) that hang down from the palate (Fig. 33.12*D*), a toothless whale strains these small organisms from water passing through its mouth. The richness of the plankton together with the buoyancy of the water has enabled these whales to attain enormous size. The blue whale, which reaches a length of 30 meters and a weight of 135 metric tons, is the largest animal that has ever existed.

CLASSIFICATION OF MAMMALS

CLASS MAMMALIA. The mammals. Endothermic tetrapods, generally covered with hair; jaw joint between dentary and squamosal bones; three auditory ossicles.

† **Subclass Eotheria.** Two orders of very primitive Triassic and Jurassic mammals, close to the reptile-mammal boundary.

Subclass Prototheria. One order of primitive mammals retaining many reptilian features including the egg-laying habit and cloaca; now confined to the Australian and New Zealand regions.

Order Monotremata. The monotremes. The platypus, *Ornithorhynchus;* spiny anteater, *Tachyglossus.*

† **Subclass Allotheria.** One order of gnawing mammals from the Mesozoic and early Tertiary. The multituberculates.

---

† Extinct.

Subclass Theria. Typical mammals. All living ones are viviparous.

† **INFRACLASS TRITUBERCULATA.** Two orders of Mesozoic mammals with a cusp pattern on their molar teeth which suggests that they were ancestral to higher mammals.

INFRACLASS METATHERIA. Pouched mammals. Young are born at an early stage of development, and complete their development attached to teats which are located in a skin pouch; usually three premolar teeth and four molars in each jaw.

Order Marsupialia. Marsupials. The opossum, *Didelphis*.

INFRACLASS EUTHERIA. Placental mammals. Young develop to a relatively mature stage in the uterus; dental formula never exceeds $I\frac{3}{3}$, $C\frac{1}{1}$, $Pm\frac{4}{4}$, $M\frac{3}{3}$.

Order Insectivora. Insectivores including shrews, moles and hedgehog. Small mammals, usually with long pointed snouts; sharp cusps on molar teeth adapted for insect eating; feet retain five toes and claws. The common shrew, *Sorex*.

Order Dermoptera. The colugo, or flying lemur, of the East Indies and the Philippines. A gliding animal with a lateral fold of skin.

Order Chiroptera. The bats. Pectoral appendages modified as wings; hind legs small and included in wing membranes. The little brown bat, *Myotis*.

Order Primates. The primates. Rather generalized mammals retaining five digits on hands and feet; first digit usually opposable; claws usually replaced by finger- and toenails; eyes typically large and turned forward; often considerable reduction in length of snout.

SUBORDER LEMUROIDEA. The lemurs, lorises, galagos. Primitive primates; snout longer than in other groups; often one or two digits retain claws. The lemur, *Lemur*.

SUBORDER TARSIOIDEA. Short-faced primates; certain tarsal bones elongated. The tarsier, *Tarsius*.

SUBORDER PLATYRRHINI. New World monkeys. Nostrils far apart; three premolars retained; tail often prehensile. The capuchin, *Cebus*.

SUBORDER CATARRHINI. Old World monkeys, apes, man. Nostrils close together; two premolars; tail, if present, never prehensile. Macaque, *Macaca*; chimpanzee, *Pan*; man, *Homo*.

Order Carnivora. The carnivores. Flesh-eating mammals; large canine; certain premolars and molars modified as shearing teeth; claws well developed.

† SUBORDER CREODONTIA. Ancestral carnivores.

SUBORDER FISSIPEDIA. Modern terrestrial carnivores including the dogs, wolves, foxes, raccoons, pandas, bears, weasels, martens, wolverines, badgers, skunks, minks, otters, cats, lions, tigers, mongooses, hyenas. The domestic cat, *Felis*.

SUBORDER PINNIPEDIA. Marine carnivores including the seals, sea lions and walruses. Members of this group have many aquatic specializations such as paddle-like limbs and reduced tail. The harbor seal, *Phoca*.

† **Order Condylarthra.** Ancestral ungulates. Five toes were retained, but each bore a small hoof; except for loss of clavicle, limb skeleton little modified; dentition complete; molars slightly modified for plant eating.

Order Proboscidea. Elephants and related extinct mammoths and mastodons. Massive ungulates retaining five toes, each with a small hoof; two upper incisors elongated as tusks; nose and upper lip modified as a proboscis. African elephant, *Loxodonta*; Indian elephant, *Elephas*.

Order Sirenia. Sea cows. Marine herbivores; pectoral limbs paddle-like; pelvic limbs lost; large horizontally flattened tail used in propulsion. Florida manatee, *Trichechus*.

Order Hydracoidea. Coneys. Small, guinea pig-like herbivores of the Middle East; four toes on front foot, three on hind foot, each with a hoof. *Procavia*.

---

† Extinct.

Order Perissodactyla. Odd-toed ungulates. Axis of support passes through third digit; lateral digits reduced or lost.

SUBORDER HIPPOMORPHA. Horses and their allies. *Equus.*

SUBORDER CERATOMORPHA. Tapirs and rhinoceroses.

Order Artiodactyla. Even-toed ungulates. Axis of support passes between third and fourth toes; first toe lost; second and fifth toes reduced or lost.

SUBORDER SUINA. Pigs, peccaries, hippopotamuses. The pig, *Sus.*

SUBORDER RUMINANTIA. The cud-chewing artiodactyls including camels, llamas, chevrotains, deer, giraffes, pronghorns, antelopes, cattle, sheep and goats. American buffalo, *Bison.*

Order Edentata. New World edentates including sloths, anteaters and armadillos. Teeth reduced or lost; large claws on toes. The armadillo, *Dasypus.*

Order Pholidota. The pangolin, *Manis,* of Africa and southeastern Asia. Teeth lost; long tongue used to feed on insects; body covered with overlapping horny plates.

Order Tubulidentata. The aardvark, *Orycteropus,* of South Africa. Teeth reduced; long tongue used to feed on insects.

Order Cetacea. The whales and their allies. Large marine mammals; pectoral limbs reduced to flippers; pelvic limbs lost; large tail bears horizontal flukes which are used in propulsion.

SUBORDER ODONTOCETI. Toothed whales. The bottlenosed dolphin. *Tursiops.*

SUBORDER MYSTICETI. Whalebone whales. Teeth lost; strain small animals from water with horny whalebone plates that hang down from upper jaws. The blue whale, *Balaenoptera.*

Order Rodentia. The rodents. Gnawing mammals with two pairs of chisel-like incisor teeth. The largest order of mammals, it includes the squirrels, chipmunks, marmots, gophers, beavers, rats, mice, muskrats, lemmings, voles, porcupines, guinea pigs, capybaras and chinchillas. The woodchuck, *Marmota.*

Order Lagomorpha. Hares, rabbits, pikas. Gnawing mammals with two pairs of chisel-like incisors and an extra pair of small upper incisors that lie behind the enlarged first pair. The rabbit, *Lepus.*

ANNOTATED REFERENCES

Many of the general references on vertebrates cited at the end of Chapter 30 contain considerable information on the biology of mammals.

Andersen, H. T. (Ed.): The Biology of Marine Mammals. New York, Academic Press, 1969. Chapters deal with the swimming, diving, echolocation and other aspects of the biology of cetaceans and other marine mammals.

Bourlière, F.: The Natural History of Mammals. 2nd ed. New York, Alfred A. Knopf, Inc., 1956. A fascinating account of the natural history of mammals; originally published in French as Vie et Moeurs des Mammifères.

Burt, W. H., and R. P. Grossenheider: A Field Guide to the Mammals. Boston, Houghton Mifflin Co., 1952. A useful guide, in the style of the Peterson bird guides, for the field identification of mammals.

Hall, E. R., and K. R. Kelson: The Mammals of North America. New York, The Ronald Press Co., 1959. A two-volume monograph on mammals for the serious student.

Sanderson, I. T.: Living Mammals of the World. Garden City, N.Y., Doubleday & Co., 1955. A superbly illustrated account of the major groups of mammals.

Schmidt-Nielsen, K.: Desert Animals. Oxford, Clarendon Press, 1964. The adaptations of camels, the kangaroo rat, man and other animals to desert life are thoroughly analyzed.

Scott, W. B.: A History of Land Mammals in the Western Hemisphere. 2nd ed. New York, Macmillan, 1937. An old but still very valuable source book on the evolution of horses, camels, mastodons and other groups that roamed the New World in ages past.

Slijper, E. J.: Whales. London, Hutchinson & Co., 1962. A valuable source book on the natural history and the specialized anatomy and physiology of cetaceans.

Vaughan, T. A.: Mammalogy. Philadelphia, W. B. Saunders Co., 1972. An excellent textbook with chapters on the origins of mammals, the various groups, ecology, zoogeography, behavior, and various aspects of mammalian physiology.

Walker, E. P., et al.: Mammals of the World. Baltimore, The Johns Hopkins Press, 1964. Each known genus of mammals is discussed and illustrated in the first two volumes of this treatise. A third volume is devoted to a classified bibliography of the literature regarding mammal groups and their anatomy, physiology, ecology, etc.

Wimsatt, W. A. (Ed.): Biology of Bats. New York, Academic Press, 1970. Includes chapters on the evolution, anatomy and physiology of the organ systems, thermoregulation and hibernation, development, echolocation, and ecology of bats.

Young, J. Z.: The Life of Mammals. 2nd ed. Oxford, Clarendon Press, 1957. A very valuable source book emphasizing the anatomy and physiology of mammals.

34. Vertebrates: Catarrhines and Man

During the course of evolution, primates diverged at an early time from primitive insectivores and, as a group, became well adapted for life in the trees. Our grasping hand, reduced sense of smell, short face, stereoscopic vision, and enlarged cerebellum and cerebrum all began as arboreal adaptations. Lemurs, tarsioids, and most monkeys remain quadrupeds in the trees, but some of the higher Old World primates, including the great apes and man, became too large to scamper through the trees on all fours, and evolved different patterns of locomotion.

34.1 OLD WORLD MONKEYS

The great apes, man and the Old World monkeys, all belonging to the suborder **Catarrhini,** probably evolved from tarsioid-like ancestors in early Tertiary times, independently of the New World monkey group (suborder Platyrrhini). Catarrhines are characterized by having their nostrils close together and directed downward (*kata* = down, + *rhino* = nose), only two premolar teeth in contrast to the three of other primates, a partially naked face, and a large brain. Their tail may be long, short, or absent, but it is never prehensile as it is in most New World monkeys.

Contemporary Old World monkeys (fam-

ily **Cercopithecidae**) constitute a large and diverse group including the forest dwelling langurs and green monkeys, the macaques (who are only partially arboreal), and the terrestrial baboons and mandrills. All are quadrupeds, but they tend to sit upright upon **ischial callosities** — hardened and often brilliantly colored skin pads upon their buttocks (Fig. 34.1). Their molar teeth often have two transverse crests that help in grinding the plant food on which this group largely feeds. Many have a primitive social structure; baboons travel in troops and cooperate in obtaining food and protecting the females and young. Their large canines are used in defense.

34.2 LIVING GREAT APES

Contemporary great apes (family **Pongidae**) are the gibbon (*Hylobates*) of Malaysia, the orangutan (*Pongo*) of Borneo and Sumatra, and the chimpanzee (*Pan*) and gorilla (*Gorilla*) of tropical Africa (Fig. 34.1). All are considerably larger than monkeys, lack a tail and have broad chests as in man. They swing gracefully from branch to branch using the arms alternately in a pattern of locomotion known as **brachiating.** Their arms are much longer than their legs. Apes retain the grasping foot of earlier primates, but the hand, with its elongated palm

Figure 34.1   A group of catarrhines. *A*, An Old World monkey (langur) showing ischial callosities; *B*, a baboon; *C*, gorilla walking on all fours; *D*, gibbon brachiating. (*A*, courtesy of Chicago Museum of Natural History; *B*, *C*, and *D* from Campbell: Human Evolution, Aldine Publishing Co.)

and shortened thumb, is used as a hook as they swing through the trees (Fig. 34.2).

The **gibbon** is the smallest and probably the most primitive of all apes. It is the best brachiator and can clear three meters or more with each swing. Apes are erect as they brachiate, and the gibbon also walks erect upon the ground, using its long arms as balancers. The chimpanzee and gorilla are less arboreal than the others, and assume a semi-erect posture on the ground, supporting the front of the body some of the time with their long arms. They walk somewhat awkwardly on the outer edges of their feet and the knuckles of their hands (Fig. 34.1). Gorillas are the largest of the living apes, attaining a height of 1.8 m. and a weight of 230 kg.

Apes are herbivores, feeding upon fruit, young leaves and other plant material. The teeth and jaws are powerful. Prominent bony brow ridges above the orbits help resist the stresses set up in the skull by the powerful jaw mechanisms (Fig. 34.3). The gorilla's skull also has large sagittal and nuchal crests that increase the area available for the attachment of jaw muscles. The molar teeth lack the transverse crests seen in monkeys. The tooth row has a somewhat squarish, or U-shaped appearance, for the molars are in parallel rows and the canine, which is used in defense, is large (Fig. 34.4).

Apes are very intelligent creatures. They live in groups, have a social structure, communicate with each other by primitive sounds and facial expressions, and they can use sticks and other objects as tools to reach food.

34.3   ANCESTRAL APES

Early Oligocene fossil beds of Egypt have yielded fragmentary fossils of primitive catarrhines. *Parapithecus,* known from a fossil mandible, may have been ancestral to all later species; it had two premolars and lacked the dental specialization characteristic of later monkeys or apes. However, the different lines of catarrhine evolution diverged early (Fig. 34.5), for the same beds yield fossils that could be ancestral to present-day Old World monkeys and the great apes.

Propliopithecus, once considered to be an ancestral gibbon, was sufficiently generalized structurally to be ancestral to all later apes. The gibbon, however, probably diverged early from the others, for *Pliopithecus* of later Miocene and Pliocene deposits resembles the gibbon closely in skull features. Other apes may have evolved from *Aegyptopithecus.*

An assemblage of ape remains has been found in Miocene and Pliocene deposits of Europe, South Asia and Africa. Originally assigned to different genera, most are now considered to represent different species of *Dryopithecus.* It is probably from this group that the higher apes and man diverged. Remains of the limb skeleton indicate that the dryopithecines, while clearly apes, were not so specialized for brachiating as modern apes, and the foot structure of certain ones suggests some adaptation toward bipedalism.

East Africa at this time was savanna coun-

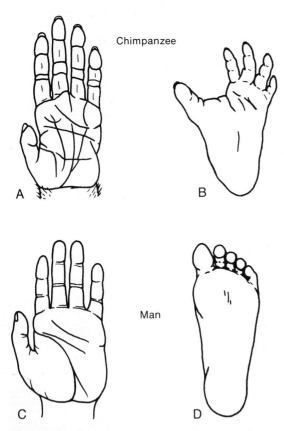

Figure 34.2   The hand of a chimpanzee (A) is adapted for brachiating and its foot (B) for grasping; the hand of man (C) is adapted for grasping and the foot (D) for bipedal locomotion. (A and C modified from Biegert; B and D from Morton.)

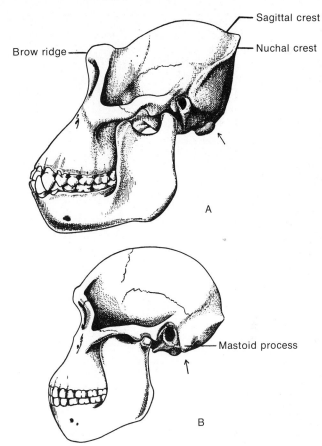

Sagittal crest

Nuchal crest

Brow ridge

A

Mastoid process

B

Figure 34.3 Lateral view of the skull and lower jaw of A, a gorilla; B, *Australopithecus* The arrow indicates the inclination of the foramen magnum. (From LeGros Clark: The Antecedents of Man, Quadrangle Books, Inc.)

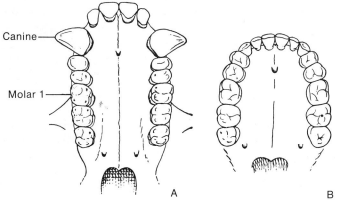

Canine

Molar 1

A

B

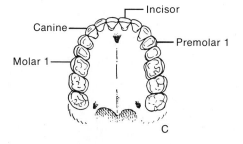

Incisor

Canine

Premolar 1

Molar 1

C

Figure 34.4 Palate and upper teeth of A, gorilla; B, *Australopithecus*; C, modern man. (From LeGros Clark: The Antecedents of Man, Quadrangle Books, Inc.)

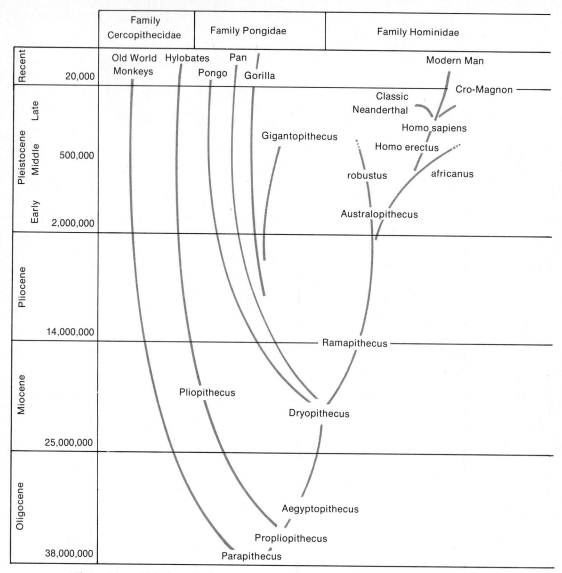

Figure 34.5 A phylogeny of catarrhine primates. The families refer to recent species.

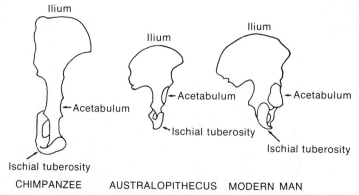

Figure 34.6 Lateral views of the pelvis of a chimpanzee, *Australopithecus*, and modern man. (From C. Coon: The Origin of Races. Copyright © 1962 by Carleton S. Coon. Reprinted by permission of Alfred A. Knopf, Inc.)

try with patches of forest separated by grassland. Many apes remained in the forest country, but some must have lived on the fringes of the forest and entered the grasslands. *Gigantopithecus* may have been such a type. It remained a herbivore but, judging from its huge teeth, specialized on a graminivorous diet of seeds, stems and other coarse material. *Gigantopithecus* was first known from a single molar discovered in 1935 by von Koenigswald as he searched the collections of dragon bones sold by Chinese apothecaries as cure-alls. Many teeth and several jaws are now known from Middle Pleistocene deposits of China and India. If other parts of the body were proportional in size, *Gigantopithecus* must have weighed 300 kg. and stood 2.5 m. tall. Man presumably evolved from other apes that entered the savannas and adapted to an omnivorous, bipedal mode of life.

34.4 CHARACTERISTICS OF MAN

Man is more nearly similar to the chimpanzee and gorilla than to other living apes, but differs enough to be placed in a separate family, the **Hominidae**. Man differs from contemporary apes in being well adapted to a bipedal gait, using the hands not for locomotion but for tool manipulation and manufacture, and having an omnivorous diet that includes both plant and animal material. Many features are correlated with our upright stance. We have a lumbar curve in our back which places our center of gravity over the pelvis and hind legs. Our ilium is broad and flaring, providing a large surface for the attachment of gluteal and other muscles that hold us erect (Fig. 34.6). Our legs are longer and stronger than our arms. The distal ends of our femora are brought close to the midline giving us a knock-kneed appearance, but placing the foot under the projection of the body's center of gravity, thereby enabling us to balance easily on one foot when the other is off the ground (Fig. 34.7). Our foot has lost its primitive grasping ability, for the toes are short and parallel each other (Fig. 34.2). The heel bone is large, the carpals and metacarpals form strong supporting arches, and the great toe, which is closest to the line of support, is enlarged. Our head is balanced upon the top of the vertebral column. The foramen magnum is far under the skull; the nuchal area on the back of the skull, for the attachment of neck muscles, is much reduced; and the mastoid process is enlarged (Figs. 34.3 and 34.10). It has been suggested by some investigators that the loss of body hair was an adaptation for heat loss in primitive man who probably ran down game in savanna country.

Since the hands were no longer needed in locomotion, they could be used in other ways. Man retained and improved upon the grasping ability of the primitive primate hand as he first began to use large bones and stones as clubs and then to make stone tools. Our thumb is longer and the metacarpal portion of our hand shorter than in apes (Fig. 34.2).

A more varied diet is available in savannas than in forests, and early man probably ate insects, lizards, rodents and other small animals as well as plant material. Later he used fire and softened his food by cooking.

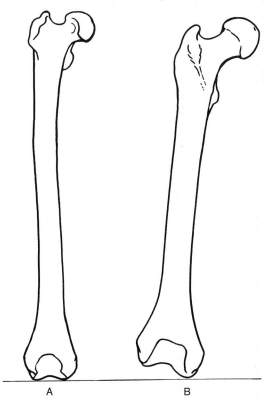

Figure 34.7 Femora of A, an extinct ape (*Dryopithecus*); B, a modern man. When the distal articular surface is in the horizontal plane, the distal end of the human femur inclines medially. (Modified from Campbell: Human Evolution, Aldine Publishing Co.)

The teeth and jaws of man are less massive than in apes, and his face does not protrude as much. Our tooth row is more rounded and the canine small (Fig. 34.4). Man defends himself with his tools and weapons rather than with his teeth.

The brain of early man was not much larger than an ape's, but that of modern man is considerably larger. A large brain evolved quickly, probably during the period when man began to hunt antelope and other larger animals which required more sophisticated weapons, a cooperative social structure, and effective means of communicating and sharing ideas.

34.5   THE APE MEN

Ramapithecus, known from jaw fragments from the Miocene-Pliocene boundary of southern Asia and Africa, is the oldest primate to show man-like characteristics (Fig. 34.5). These fragments indicate a creature having a somewhat rounded dental arch, relatively small canines, and an arched palate. Clearly a change in diet was occurring 14 million years ago, which implies the creature was living some of the time in more open country. A change in locomotion would have had a selective advantage in such a habitat, but we have no direct evidence as to whether the creature was becoming erect.

Many important steps in human evolution must have occurred during the Pliocene, but unfortunately we have no fossil ape-men from this epoch. Raymond Dart, Professor of Anatomy at the University of Witwatersrand, discovered the next stage of human evolution in 1924 when he found a cranial cast and part of the skull of a child in Pleistocene cave deposits of South Africa. He named it *Australopithecus* (Latin *australis*, south, + *pithecus*, ape) *africanus*. Subsequently, a number of australopithecine skulls, pelves and limb bones have been discovered in South and East African deposits ranging from Early to Middle Pleistocene (about 2 million to 500,000 years ago). Particularly fruitful has been Olduvai Gorge, a miniature grand canyon of East Africa explored by Louis and Mary Leakey, for here there is a near continuum of deposits from most of the first half of the Pleistocene.

As new fossils were discovered each was assigned to a different genus, but there is now general agreement that only two ape-men roamed the African savannas during the period—*Australopithecus africanus* and *Australopithecus (Paranthropus) robustus*. They were clearly terrestrial bipeds, although they probably walked with more of a swinging motion than we do. The foramen magnum is far under the skull, the mastoid process is well developed (Fig. 34.3) and the ilium is somewhat wide and flaring (Fig. 34.6). Even parts of a characteristic human foot have been found. *Australopithecus africanus* weighed about 35 kg.; *robustus* was somewhat larger, weighing perhaps 60 kg. The faces of both protruded slightly, and the creatures had no chins, but the brow ridges were not massive and all features of the dentition were very human (Figs. 34.3 and 34.4). Surprisingly, the cranial capacity, which ranged from 450 to 700 ml., was about the size of that of the larger apes (range 340 to 752 ml. in a gorilla), but far below that of modern man (range: 1200 to 2000 ml.).

It seems clear that human limb, dental, and many facial features evolved before there was a great increase in brain size. Darwin and other late 19th century scholars made the opposite assumption, which for a time appeared to be confirmed by the discovery of *Eoanthropus* in an Early Pleistocene gravel pit in Piltdown, England. The fragments suggested a man with a modern cranial capacity and the face and jaws of an ape. As the australopithecines were discovered, *Eoanthropus* appeared more and more as an anachronism. Reinvestigation of the fragments with modern methods of dating revealed a hoax; someone had "salted" the gravel pit with cranial fragments of a Late Pleistocene man and part of the mandible of a contemporary chimpanzee stained to make it look old!

Australopithecus africanus and *robustus* differed somewhat in their dentition and mode of life. The teeth of *robustus* were larger and the jaws more massive. A sagittal crest upon the skull indicates a very powerful chewing mechanism. *Robustus* was without doubt a herbivore, feeding upon tough roots and young shoots along with berries and fruit; *africanus* was a more agile omnivore eating insects and small game along with softer plant material (Fig. 34.8).

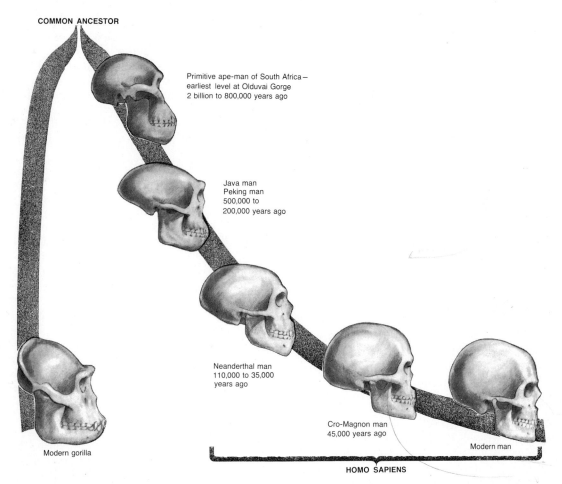

COMMON ANCESTOR

Primitive ape-man of South Africa—
earliest level at Olduvai Gorge
2 billion to 800,000 years ago

Java man
Peking man
500,000 to
200,000 years ago

Neanderthal man
110,000 to 35,000
years ago

Cro-Magnon man
45,000 years ago

Modern man

Modern gorilla

HOMO SAPIENS

Figure 34.8   The evolution of the human skull. Note gradual increase in cranial capacity and reduction of jaw, brow-ridges and other ape-like features. (After Young, L. B. (Ed.): Evolution of Man. Oxford University Press, 1970.)

There is no evidence that *robustus* was a toolmaker, but primitive stone tools of the Oldowan culture are found in the camp-sites of *africanus*. These choppers were little more than rounded stones sharpened by chipping one end (Fig. 34.9), but they, and the stone flakes broken from them, could have been used for killing and skinning small game, for sharpening branches as weapons, and for cracking open bones for the marrow. Many cracked bones and bashed-in skulls have been found at the campsites.

When the australopithecines were first discovered, it was uncertain whether to interpret them as apes or men. The question is largely semantic. Although *robustus* appears to have been a side line that died out by the Middle Pleistocene, we now believe that the line of evolution to more advanced man passed through *africanus*. Hunting and toolmaking probably set up selective forces that favored an increase in brain size in *africanus*. Middle Pleistocene specimens (originally described as *Homo habilis*) are those with the largest cranial capacity.

34.6 HOMO ERECTUS

Long before the australopithecines were discovered in Africa, Haeckel in Germany had postulated the existence of a missing link between apes and modern man. He was presumptuous enough to give a name to his hypothetical creature—*Pithecanthropus alalus*—which literally means ape man without speech. This stirred the imagination of the Dutchman Eugene Dubois, who searched diligently for the missing link in Java, finally discovering in 1894 fossils in a river bank of a primitive man he called *Pithecanthropus erectus*. Shortly afterwards, Davidson Black discovered the remains of a similar man in caves near Peking, which he named *Sinanthropus pekingensis*. Subsequently, other fossils of similar men have been found in the East Indies, China, Africa and Europe. A host of names are associated with them (Algerian man, Heidelberg man), but physical anthropologists regard all of them as representing a single, widespread species now called *Homo erectus*. He is considered to be sufficiently close to modern man (*Homo sapiens*) to be included in the same genus. Differences among the fossils suggest that *Homo erectus* was beginning to show some geographic variation.

The femur discovered by Dubois indicates that *Homo erectus* stood upright and was probably about 1.5 m. tall. Brain size ranged from 750 to 1300 ml., which approaches that of the australopithecines on the one extreme and overlaps that of modern man on the other. Frontal areas of the brain, however, were poorly developed, for *Homo erectus* had a low sloping forehead (Fig. 34.10A). His face was somewhat brutish, protruding slightly, with rather heavy brow ridges and no chin. Teeth, though large, were essentially modern in their configuration.

The oldest remains of *Homo erectus* go back 500,000 years to the Middle Pleistocene. He was still using Oldowan choppers at this time, but by the time he disappeared (about 300,000 years ago) he had acquired a much more sophisticated Acheulian culture. He was chipping flint and other fine-grained stones on all surfaces to fashion hand axes (Fig. 34.9B), and he was producing large stone flakes for a variety of cutting tools. Charred bones indicated the use of fire. Fire, and perhaps crude hide clothing, would have been essential for him to penetrate central Europe and Asia at a time when continental glaciers were advancing. It is clear that he had become a hunter of large game, for campsites contain the bones of bears, horses, and even elephants. Such animals must have been stampeded over cliffs or into bogs to have been killed by primitive man. From a study of early hunting sites in Spain, F. Clark Howell of the University of Chicago has postulated that fire was sometimes used in driving game. All this implies a rather intelligent man, living in groups, with an ability to communicate and to teach the young to make tools and to hunt, and with a knowledge of the seasons and habits of game (Fig. 34.8).

34.7 HOMO SAPIENS

The oldest traces of our own species, *Homo sapiens*, are from deposits 200,000 to

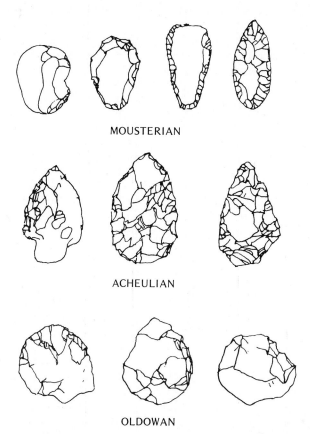

MOUSTERIAN

ACHEULIAN

OLDOWAN

Figure 34.9 Paleolithic tool kits. From bottom to top, these three sets of tools illustrate major stages in the early refinement of stoneworking. (From Keesing and Keesing: New Perspectives in Cultural Anthropology. Copyright © 1971 by Holt, Rinehart, and Winston, Inc. Reprinted by permission of Holt, Rinehart and Winston, Inc.)

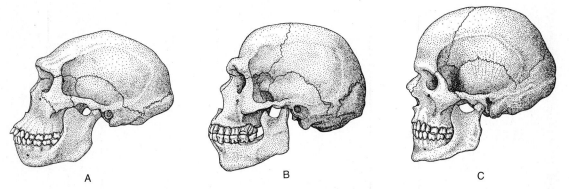

A   B   C

Figure 34.10 Lateral views of the skull and lower jaw of A, *Homo erectus* (Java); B, *Homo sapiens* (Neanderthal); C, *Homo sapiens* (Cro-Magnon). (From T. H. Eaton, Jr.: Evolution. Copyright © 1970 by W. W. Norton & Co., Inc.)

300,000 years old in Swanscombe, England, and Steinheim, Germany. A nearly complete skull can be pieced together from the two remains. The bones are thick, brow ridges are present, and the forehead is somewhat sloping, but all features are within the range of variation of modern man. Acheulian tools are associated with Swanscombe man.

During most of the last glacial stage, western Europe was occupied by a stocky, brutish-looking fellow, called Neanderthal man because his first remains were discovered in the Neander valley (German, *thal* = valley). Originally considered to be a distinct species, Neanderthals are now regarded as an extinct race of *Homo sapiens.* These "classic" Neanderthals had large brains, and they were powerful men with strong jaws, heavy brow ridges, receding chins and short, bandy legs (Fig. 34.10*B*). They often built their fires in hearths in cave floors, and their Mousterian culture included a large tool kit of stone axes, scrapers, borers, knives, spear points, and saw-edged and notched tools probably used in making spear handles and other simple wooden implements. The stones show considerable secondary chipping to refine the shapes and sharpen the edges (Fig. 34.9*C*). The Neanderthals probably had developed a belief in the supernatural and in an afterlife, for they buried their dead with food and tools.

Some physical anthropologists limit the term Neanderthal to these classic fellows of western Europe, but others use the term in a far broader sense for all men living in Europe, Africa, and Asia during the last part of the Pleistocene. All used variants of the Mousterian culture, and some, including Solo man of Java and Rhodesian man of southern Africa, were nearly as brutish in appearance as Neanderthal man. However, many men during this period had more delicate features; they had good chins, high foreheads, and no or weak brow ridges.

It is doubtless from such as these that Cro-Magnon man evolved (Fig. 34.10*C*). Cro-Magnon man abruptly replaced the classic Neanderthal man in western Europe about 35,000 years ago. Physically, he is indistinguishable from many present-day Europeans. He developed a very sophisticated Aurignacian culture, using delicate stone, bone and wooden tools, and he has left fine examples of his painting in caves in France and Spain.

Mankind today can be divided into dozens of local races, that is, populations found in particular parts of the world and sharing certain genes and traits. Examples of local races are the Ainu of northern Japan, the Nordics of northern Europe, the Eskimos of arctic America, and the American Indians. Races differ from one another not in single features but in having different frequencies of many genes and characteristics affecting body proportions, skull shape, degree of skin pigmentation, texture of head hair, abundance of body hair, form of eyelids, thickness of lips, frequency of various blood groups, ability to taste phenylthiocarbamide, and many other anatomical and physiological traits. While certain of these differences, such as skin pigmentation, probably are adaptive, the significance of many is unknown.

Anthropologists believe that the numerous races represent localized differences in larger geographic groups. On the basis of skeletal remains, and the present-day differences in the characteristics and distribution of races, Carleton Coon recognizes five major racial groups: (1) the **Caucasoids,** which include the Nordic, Alpine and Mediterranean races of Europe; the Armenoids and Dinarics of Eastern Europe, the Near East and North Africa; and the Hindus of India; (2) the **Mongoloids,** which include the Chinese, Japanese, Ainu, Eskimos and American Indians; (3) the **Congoids,** or Negroes and Pygmies; (4) the **Capoids,** or Bushmen and Hottentots of Africa; and (5) the **Australoids,** which include the Australian aborigines, Negritos, Tasmanians, and Papuo-melanesians.

Fossil evidence indicates that the early evolution of man occurred in Africa, but by the mid-Pleistocene, man had spread widely through the Old World. It is uncertain at which time the racial groups began to differentiate. Many anthropologists believe that it was in the latter part of the Pleistocene, after the *Homo sapiens* level had been attained, but Professor Coon and others believe that they can find skeletal evidence for the beginning of racial differentiation as far back as *Homo erectus.* If this was the case, then many populations of man in dif-

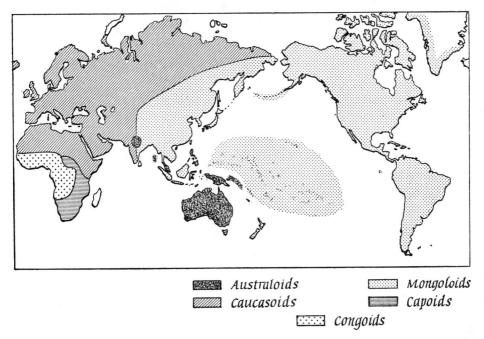

Australoids   Mongoloids
Caucasoids   Capoids
Congoids

Figure 34.11   Probable distribution of the five major racial groups of modern man shortly after the ending of the Pleistocene. (From C. Coon: The Origin of Races. Copyright © 1962 by Carleton S. Coon. Reprinted by permission of Alfred A. Knopf, Inc.)

ferent parts of the world must have crossed the *sapiens* line independently. *Homo erectus* and *sapiens* would then be viewed as a large, interbreeding complex changing in space and time.

Exactly when Mongoloids first crossed the Bering Strait from Asia to the New World is uncertain. Presumably it was during one of the glacial periods of the Pleistocene for at that time much water would have been utilized in forming continental glaciers, the sea level would have been lower, and Siberia and Alaska would have been connected by at least a series of close islands. Geologic evidence indicates that parts of Siberia and Alaska were unglaciated during the ice ages. It is most probable that the crossing was made during the last ice age. The oldest sign of man in the New World is a group of primitive tools found in Venezuela and associated with the remains of extinct mammals such as mastodons and glyptodons (giant anteaters). These have been dated at 16,000 B.C. The earliest Indian sites in North America are in Colorado and Arizona, and are dated at 8800 and 9300 B.C. respectively. The distribution of racial groups shortly after the ending of the Pleistocene is shown in Figure 34.11. Within historic times there has been an increasing movement of peoples and a resultant increase in racial mixture.

ANNOTATED REFERENCES

Binford, S. R., and L. R. Binford: Stone tools and human behavior. Scientific American 220:4 (April) 1969. An Analysis of the social structure of Neanderthal man from a study of the groups of Mousterian tools and their use.

Campbell, B. C.: Human Evolution. Chicago, Aldine Publishing Co., 1966. An excellent account of human evolution with an emphasis on man's unique anatomical adaptations.

Clark, W. E. LeGros: The Antecedents of Man. Edinburgh, Edinburgh University Press, 1959. Also published as a paperback by Harper, 1963. An excellent presentation of primates and their evolution.

Coon, C. S.: The Origin of Races. New York, Alfred A. Knopf, 1962. Describes contemporary human races and traces their origin back to the Middle Pleistocene; all human fossils known to date of publication are described.

Day, M.: Guide to Fossil Man. Cleveland, World Publishing Co., 1965. A tabulation and description of all known human fossils; an excellent guide to the primary literature on human evolution.

Editors of Time-Life Books: The Emergence of Man. New York, Time-Life Books, 1972–1973. A series of authoritative and superbly illustrated books on human evolution written by the editors of Time-Life Books in consultation with leading anthropologists. Volumes on Life Before Man, The Missing Link (Australopithecines) and The First Men (Homo erectus) have been published. Others are in preparation.

Howell, F. C., and the editors of *Life:* Early Man. New York, Time Inc., 1965. A Life Nature Library book; a superbly illustrated account of man's evolution prepared by a leading authority on the subject.

Napir, J.: The antiquity of human walking. Scientific American *216:*4 (April) 1967. An analysis of man's striding gait, its anatomical prerequisites and its probable origin.

Simons, E. L. The earliest apes. Scientific American *217:*6 (Dec.) 1967. A discussion of fossil apes and their interrelationships.

Simons, E. L., and P. C. Ettel: Gigantopithecus. Scientific American *222:*1 (Jan.) 1970. A study of this largest of all apes and its interrelationships and probable mode of life.

Tobias, P. V.: The Brain in Hominid Evolution. New York, Columbia University Press, 1971. A thorough review of brain evolution and its relation to cultural development.

PART FOUR

ANIMALS
AND
THEIR
ENVIRONMENT

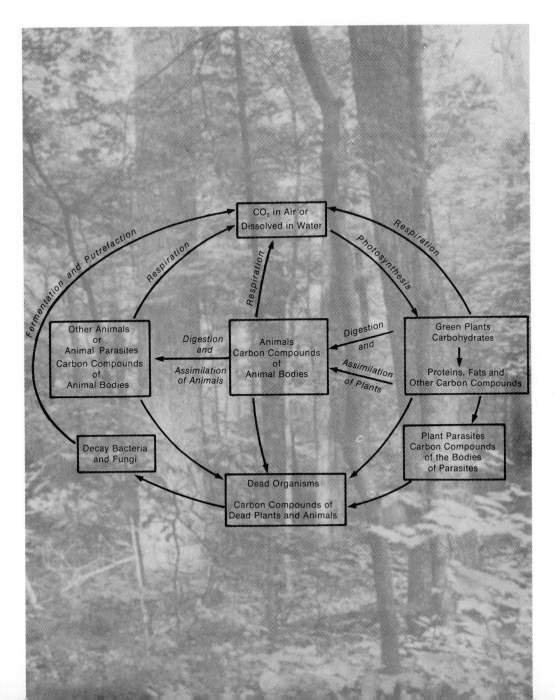

CO₂ in Air or Dissolved in Water

Fermentation and Putrefaction

Respiration

Respiration

Photosynthesis

Respiration

Other Animals
or
Animal Parasites
Carbon Compounds
of
Animal Bodies

Digestion
and
Assimilation
of Animals

Animals
Carbon Compounds
of
Animal Bodies

Digestion
and
Assimilation
of Plants

Green Plants
Carbohydrates

Proteins, Fats and
Other Carbon Compounds

Decay Bacteria
and Fungi

Plant Parasites
Carbon Compounds
of the Bodies
of Parasites

Dead Organisms
Carbon Compounds of
Dead Plants and Animals

35. Behavior

Much of what an animal does can be analyzed in terms of specific behavior patterns—specific coordinated sequences of neuromuscular activity. These typically occur in response to some change in the environment and may result in the movement of the animal's entire body or some part of it. A dog may wag its tail, a bird may sing, and a butterfly may release a volatile sex attractant. Movements of the entire body are involved in more complex behavior patterns such as the stalking movements of a cougar, the courtship movements of a stickleback fish, or the crawling of a snail. Some behavior patterns result in an animal not moving: a pointer freezes when it has detected a pheasant, a possum plays dead in the presence of an enemy, and a young bird remains motionless on the ground in the presence of a predator. All animals exhibit behavior patterns, although in some, such as sponges, clams and bryozoa, the detection of behavior patterns may require extreme patience on the part of the observer. Behavior patterns commonly involve muscles, cilia, flagella, or pseudopodia as effectors, but they may involve responses such as the emission of light, the release of the secretion of an exocrine or endocrine gland, or the generation of electricity in an electric organ.

35.1 BEHAVIORAL ADAPTATIONS

Certain general types of behavior can be detected in many different groups of animals, and the behavior of diverse groups may be influenced by certain common factors. Behavior is one means by which an animal may adapt to changes in the environment. The physiologic basis of behavior resides in the manifold activities of the animal's nervous system. All movements may ultimately be analyzed in terms of the patterns of nervous activity and the patterns of neuromuscular connections that are involved in their production.

A behavior pattern is typically triggered in response to some variation in the environment—some change in the light, temperature, humidity, oxygen, carbon dioxide, pH or texture of its surroundings. Animals may make behavioral responses to changes in their biotic environment; specific behavior patterns may be triggered by the presence of certain other organisms. Predators seek prey; the prey attempts to evade the predator. Changes in an animal's internal environment may also initiate specific behavior patterns. When a cat has not eaten for some time, the concentration of glucose in its blood decreases and the stomach increases its motility. In response to these and other stimuli, the animal becomes restless, moves about and looks for food. When it has eaten, its internal conditions are changed, the restlessness ceases and the animal may either groom itself or sleep.

Behavioral responses are adaptive either for the survival of the individual or the survival of the species. Certain behavioral responses may lead to the death of the individual, but increase the likelihood of survival of the species through the survival of the offspring. Each animal's behavior patterns must enable it to live long enough to reproduce. It must avoid deleterious environments, predators, parasites and competition from members of its own species. It must obtain an adequate supply of raw materials for its biosynthetic processes and an adequate supply of energy for its metabolic machinery. At the appropriate time it may need to locate and mate with another of its kind, and perhaps subsequently guard and educate the young. No two species accomplish all of these ends by the same patterns of behavior. Behavior, indeed, is just as diverse as biological structure and is just as characteristic of a given species as its size, form, color or odor.

Behavior patterns are determined by the capabilities of the organisms's **receptor, effector** and **nervous systems,** which have in turn been determined by evolution. What happens in response to a change in the environmental variable, the stimulus, depends upon how the receptor, effector and nervous systems are interconnected and integrated. Their organization may be relatively direct so that the stimulus triggers a simple set of muscle actions that proceed automatically once started. As muscular action proceeds, it may be constantly monitored by other stimuli that it has generated in the body, which therefore steer and guide it. Alternatively, the stimulus may trigger a complex response that is completely programmed genetically in the central nervous system. The response mechanism in other behav-

ioral systems may involve both learned and genetically programmed components.

35.2   REFLEXES

In the simplest sort of response the stimulus triggers a muscle or a set of muscles which respond in a predictable, unvarying fashion. This behavior is determined only by the presence of the appropriate sense organs connected with a particular set of muscles. In the **knee-jerk** or **stretch reflex** a tap on the tendon below the knee cap stretches the attached muscle. The stretching stimulates the spindle organ (Fig. 35.1). Impulses from this receptor neuron are conducted along its axon to the spinal cord where they pass by a synapse to a motor neuron that stimulates the muscle to contract, producing the knee jerk. The knee-jerk reflex is an example of behavior dependent simply upon the presence of a stretch receptor connected with a specific muscle by a **monosynaptic reflex arc.**

The nature of the reflex response may be altered by the characteristics of the synapses connecting the incoming and outgoing segments of the circuit. There is delay at the synapse, and the stronger the stimulus, the shorter is the delay. This relates the onset of the behavior to the strength of the stimulus. In addition, the phenomenon of **afterdischarge** may appear: the synapse may continue to be active after the stimulus has ceased. A cockroach stimulated to run in response to a touch on its anal cercus may continue to run for some seconds after the stim-

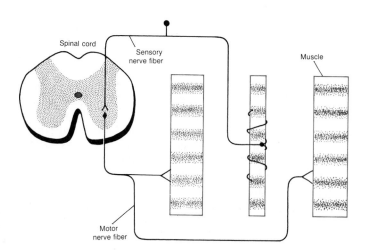

Figure 35.1 A diagram of the anatomy of the stretch reflex, a monosynaptic reflex arc. (After Van der Kloot, W. G.: Behavior. New York, Holt, Rinehart and Winston, Inc., 1968.)

ulus has ceased. The continued running is due in part to after-discharge at a synapse in the ventral nerve cord.

Synapses also exhibit the phenomenon of **temporal summation.** If an investigator applies to a fly's mouth parts a sugar solution that is concentrated enough to stimulate taste hairs but too dilute to evoke a behavioral response, the proboscis is not extended. However, if this stimulation is quickly followed by another weak stimulation (which would not itself evoke a response), proboscis extension does occur.

When two or more neurons synapse with the same postsynaptic neuron, there may be **spatial summation.** Two incoming impulses, one in each presynaptic neuron, each too weak to trigger the discharge of the motor neuron, may add together and be effective.

The reflex arc underlying the flexion reflex involves two synapses—one between the sensory neuron and interneuron and one between interneuron and motor neuron (Fig. 35.2). The interneuron is inhibitory, and when stimulated by sensory input from one muscle, it inhibits activity in the antagonistic muscle. Considerable complexity of behavior can be developed when the number of synapses in the reflex arc is increased. Increasing the strength of the stimulus elicits different responses in the flexion reflex. In experiments with dogs in which the connection between the spinal cord and brain has been cut, a weak noxious stimulus applied to the foot pad results in withdrawal of the foot. An increase in stimulus strength causes a strong flexion of the lower leg. A stronger stimulus elicits flexion of the upper leg as well. This spreading of response with increase in stimulus strength is termed **irradiation.**

35.3 SPONTANEOUS ACTIVITY

Patterns of behavior may occur spontaneously, completely independently of stimuli from the external environment, or the external stimulus may simply serve to release a complex, centrally patterned neural program. The marine lugworm, *Arenicola*, lives in a U-shaped burrow in the mud and keeps the burrow open by rhythmic muscular movements. These arise not in

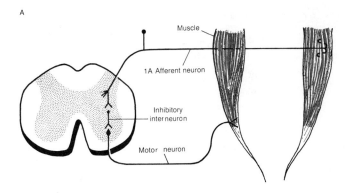

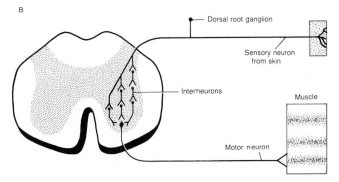

Figure 35.2 *A,* The most direct nerve pathway between a stretch receptor and a motor neuron innervating the antagonist muscle. *B,* A polysynaptic reflex arc between sensory neurons in the skin and motor neurons. (After Van der Kloot, W. G.: Behavior. New York, Holt, Rinehart and Winston, Inc., 1968.)

response to environmental stimuli, but are initiated by a series of spontaneously active clocks in the various ganglia of the worm. The brain is not necessary for these activities to occur, and isolated fragments of the worm show the same kind of movements.

Other behavior programs are built into the central nervous systems of invertebrates and need only be triggered or released by specific stimuli. Wind blowing on special receptors on the head of the locust initiates impulses from the brain to the thoracic ganglion; these in turn send impulses to the motor nerves innervating the wing muscles and result in flight. The pattern of impulses from the thoracic ganglion determines the sequence of the flight muscles. Even when all the muscles have been removed so that there is no feedback from proprioceptors in the wings, normal neural patterns are delivered by the thoracic ganglion.

The singing of crickets requires environmental stimuli only to initiate it. The complex series of muscle contractions needed to produce the various kinds of songs are largely determined in the brain. This inference was drawn from experiments in which electrical stimulation of one area of the cricket's brain triggered normal song rhythms, whereas stimulation of another area produced atypical songs (Fig. 35.3).

Many birds that nest on the ground, such as the graylag goose, retrieve any egg that happens to roll out of the nest by placing their bill on the far side of the egg and rolling the egg back into the nest with side-to-side movements of the bill that prevent the egg from slipping away (Fig. 35.4). When the egg is replaced with a cylinder that does not wobble, the side-to-side compensatory movements cease. Alternatively, if the egg or the cylinder is removed once the retrieving movement has started, the bill is still drawn back to the breast. This is a fixed action pattern that, once elicited by the stimulus (the sight of an egg outside the nest), continues independently. The side-to-side movements are steered by feedback

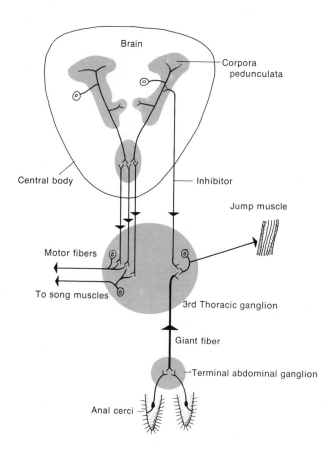

Figure 35.3 The main interactions between ganglia that control singing outbursts in grasshoppers and crickets. Electrical stimulation of a nucleus called the corpora pedunculata in the brain causes a burst of normal song and simultaneously inhibits the motor neurons involved in jumping. The central body in the brain organizes the patterns of the song, and stimulation here produces abnormal noises. Stimulation in the 3rd thoracic ganglion excites motor neurons in a meaningless pattern. The alerting mechanism of the anal cerci, which are sensitive to sounds and air movements, is inhibited during song, thus preventing the animal's alarming itself by its own noise. (After Huber, F.: Invertebrate Nervous Systems, Wiersma, C. A. G. (Ed.). Chicago, University of Chicago Press, 1967.)

Figure 35.4 *A*, Greylag goose retrieving an egg which is outside the nest.
This movement is very stereotyped in
form and used by many ground-nesting
birds *B*, The goose attempts to retrieve
a giant egg in precisely the same fashion.
(From Lorenz, K. Z., and Tinbergen,
N.: Z. Tierpsychol. 2, 1, 1938.)

stimuli coming from the wobbling egg. They simply modify the orientation of the basic pattern.

35.4 STIMULI AND SIGNS

In the terms of electronics, receptors are "narrow-pass filters," transmitting a narrow range of environmental energies. The receptors of different species may pass different portions of the range. The honeybee sees a far different world than we do, for its eye is sensitive to ultraviolet, but not to red light. The flicker-fusion frequency of the human eye is about 50 per second, but that of a bee is as much as 250 per second. Thus, an incandescent lamp that looks like a steady light to us, but is actually flickering 50 or 60 cycles per second, will appear clearly as a flickering light to an insect.

Man and other primates, whose ancestors became adapted during evolution to an arboreal life, came to rely largely on the sense of sight for sensory input. Most other mammals probably obtain more useful information from their sense of smell. Most mammals hunt using the sense of smell, and their prey largely depend on their own sense of smell to detect the predators. Mammals and certain other animals use the sense of smell in defining their territory, in distinguishing friends from enemies, male from female, and young from adult. The importance of pheromones in providing a means of communication from animal to animal utilizing the sense of smell was discussed earlier (p. 402).

Sounds are other major determinants of behavior among the vertebrates. Certain species of frogs have distinctive mating calls; the female of each species will respond only to those mating calls from the males of its species and will ignore those of others. Similarly, the songs of birds provide determinants for a wide range of behavior, mating calls and alarm calls.

There are marked differences in the sensory capacities of animals throughout the animal kingdom. The bat emits sounds with frequencies far beyond the range of the human ear. These sounds bounce off any insect flying in the vicinity, and by **echolocation** the bat is able to home in on the insect and catch it. Certain moths can hear these ultra high-frequency cries of the bat and take evasive action. Porpoises utilize sonar for underwater navigation by echolocation, and spiders can interpret the vibrations produced on their webs by ensnared insects. A rattlesnake can sense heat well enough to strike in the dark at the heat-producing object.

The central nervous systems of many organisms are capable of filtering out environmental stimuli that may be of little value. If an animal were to process all the sensory information it could receive, its central nervous system would be swamped. Certain segments of the retina of the frog's eye are specialized to react to small, dark, convex objects. The principal item in the diet of frogs is flies, and frogs respond more readily to small, dark, circular moving objects near them than to larger, more diffuse or distant objects. Such **sign stimuli** often serve as triggers for fixed action patterns of behavior. When quick action is essential, as in escaping from a predator, a danger sign is more useful than a detailed description of the danger. Alarm signs, whether they are sights or sounds, are usually simple and contrast sharply to the environment. Small birds typically show an immediate flight reaction to animals with large eyes (understandably, considering that their major predators, owls and hawks, have large eyes). Small birds flee any staring eye, and many moths, the prey of birds, have evolved eye-shaped spots on their hind wings. These "eyes" are concealed by the forewings when the moth is at rest, but spring into view (Fig. 35.5) when the moth is disturbed and extends its forewings.

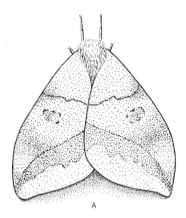

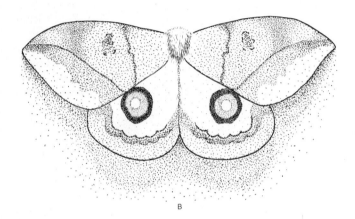

Figure 35.5   The moth *Automeris coresus* (A) at rest, and (B) displaying the vivid eyespots on its hind wings in response to a light touch. (Modified from Blest: Behaviour, 11:257, 1957.)

By varying systematically the characters of a stimulus, the behavioral scientist can discover the specific "sign." In the spring male **sticklebacks** establish territories from which they drive other males. By presenting males with various models, it was found that the sign of the male is a red belly (Fig. 35.6); other aspects of morphology are unimportant.

35.5   METHODS OF STUDYING BEHAVIOR

The neurophysiologic approach to behavior involves analyses of neural function; the neurophysiologist removes neurons, stimulates neurons, and records the activity of neurons. By placing an electrode on the nerve leading from some sense organ and making a recording, the experimenter can determine what kinds of sensory input are available to the animal. By monitoring

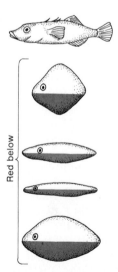

Figure 35.6   A series of models used in tests of aggression with male sticklebacks. The four crude imitations colored red below released more attacks than the accurate model which lacked red. (After Tinbergen, N.: The Study of Instinct. London, Oxford University Press, 1951.)

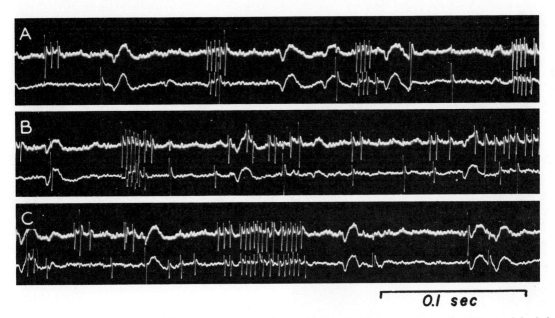

Figure 35.7 Nervous activity of the tympanic nerve of a noctuid moth. The upper trace of activity is of the left "ear" of the moth, the lower trace the right. The response seen is due to the approach of an echolocating (moth-eating) bat. In records (A) and (B) the difference in left and right tympanic activity indicates that the bat is approaching from the left. A burst of activity (C) is the response to a bat "buzz" rather close to the moth. (Courtesy of K. D. Roeder.)

neuronal activity he can determine what stimuli will actually generate action potentials in a given sense organ (Fig. 35.7). In this way, he can distinguish between an animal that is incapable of detecting a particular stimulus and another animal that can detect the stimulus, but simply is not responding under the given test conditions.

Complex patterns of behavior can be produced by electrical stimulation of certain parts of the central nervous system. Although the enormous complexity of the central nervous system makes it difficult to interpret the results of such experiments, the location of the neural pathways involved in the control of certain movements can be established. When an electrode is implanted in the brain of an anesthetized animal and the animal is allowed to wake and move around freely, it can be stimulated successively with impulses of different strengths. Observation of the relation between the area stimulated and the behavior or mood observed in the animal provides some insight into the operation of the brain (Fig. 35.8). Stimulation of a certain portion of the midbrain in a number of mammalian species can be used as a reward

for behavior. Human subjects report that this part of the brain is a pleasure center; that electrical stimulation of the area does not produce sensations of sight, sound, light or odor, but simply of pleasure. Stimulation of this center has been used to a limited extent to relieve suicidal tendencies in extremely manic-depressive individuals. However, a rat which has an electrode implanted in its midbrain that it can activate by pressing a bar at one end of the cage slowly dies of starvation and exhaustion because it neither sleeps nor eats, but simply pleasures itself to death by repeatedly pressing the bar.

In a second method, the investigator makes systematic changes in the sensory input of an animal and observes the changes in output in some segment of its behavioral repertoire. He then makes a diagrammatic representation of the relationship between input and output, usually relying upon feedback loops and other cybernetic phenomena.

In addition to these two methods of studying behavior, which might be described as neurophysiological because they actually concentrate on the neuronal circuitry under-

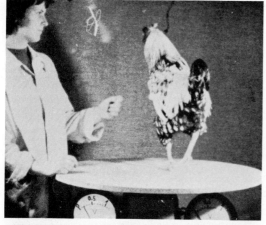

Figure 35.8   Electrical stimulation of attack behavior. The rooster is stimulated by electrodes implanted in a particular part of its brain. It starts to move toward the keeper and, as stimulation is continued, it attacks. (Courtesy of E. von Holst.)

lying behavior, there are two widely used methods of studying behavior itself. The psychological school of behaviorists, led by B. F. Skinner at Harvard, has attempted to establish general principles of behavior and especially of learning. Skinner's studies of training rats and pigeons to do certain tasks, using immediate reward or reinforcement when the task has been accomplished, has been the basis of programmed learning. It is the goal of these studies, carried out in the laboratory, to control all the experimental variables so that the effect of one can be measured.

One of the central concepts that emerged from studies by the comparative psychologists was the stimulus-response theory. This attempted to explain all behavior as a series of reactions to stimuli. Although the importance of stimuli and responses in determining behavior patterns is clearly apparent, animals are more than a collection of sets of stimulus-response reactions. The experimental psychologist confines his subject to carefully defined experimental conditions such as a **maze** or a **shuttle box,** in which the subject learns to move from one side to another to avoid an electric shock, or a **Skinner box,** in which it learns to press a lever or peck at a key to receive a reward of food or drink. The aim of the experimental psychologist is to establish laws of behavior that will describe how an animal's behavior changes after specific kinds of practice, reward or punishment are experienced. In this way the efficacy of one or another experimental situation in producing learning can be measured.

Another system of behavioral studies, founded in the late 1930's by Konrad Lorenz and Niko Tinbergen, is termed **ethology.** The ethologists insisted that animal behavior be studied under conditions that are as nearly natural as possible. Ethologists have explored the rich variety of adaptive behavior patterns shown by birds, fish and insects. They have been especially concerned with analyzing the factors controlling social behavior, including aggressive, reproductive and parental behavior.

Members of the ethologic school insist that the natural behavior of a species be carefully observed before one begins an experimental analysis of the factors involved. The ethologist is greatly concerned with the biological significance of behavior patterns and how these patterns have evolved. Many of the biologically important behavior patterns studied by ethologists appeared to be primarily the product of genetic expression and less the result of learning; initially perhaps the ethologists overemphasized the innate qualities of behavior. Both ethologists and psychologists recognize that the behavior an animal shows at any moment is the result of past selective forces. These determine a particular morphologic, biochemical and neuronal structure in an animal. The functional characteristics of this structure are continually altered by experience, by hormones and by other physiologic changes. The sum of these factors results in the maintenance of the animal in its proper condition, and stimuli from the biotic and the abiotic environment can then elicit particular behavior patterns.

35.6 DETERMINANTS OF BEHAVIOR: MOTIVATION

What an animal does at any given time is determined by the stimuli impinging upon it and by its motivation. The nature of the stimuli to which the animal will respond and the latency of its response depends upon its motivation. Obviously a hungry dog will respond differently from one that has just been fed when both are presented with a dish of food and with some alternative. Physiologic research has revealed further types of interaction between external stimuli and motivation. Each stimulus may evoke a response within the brain by its specific sensory pathways. In addition, it may elicit many less specific responses. In the vertebrate, incoming sensory pathways have sidebranches, or collaterals, extending to a diffuse series of fiber tracts, the **reticular**

formation, which is connected to the brain's higher centers. Impulses from the reticular formation arouse the brain's higher centers into action. Thus, any stimulus may not only evoke a specific response pertaining to itself but may change the animal's state of arousal and responsiveness to other stimuli, both those that are related to the first and those that are unrelated to it.

A given stimulus applied to the same animal at different times does not always evoke the same response. The relationship between stimulus and response may be altered by fatigue, maturation, learning, motivation and even by the season of the year. Some examples of motivated behavior are feeding, drinking, courting, nest-building, territorial defense and care of the young. Perhaps the most striking feature of motivational behavior is that it is highly specific to a particular state and to particular stimuli. In general, a hungry animal will not court and an animal in a highly receptive sexual state will not feed.

Each type of specific motivation causes the animal to orient its behavior towards some specific goal. The pattern begins with a searching phase, usually called **appetitive behavior.** This does not imply that the animal is consciously "searching" or that it necessarily "knows" what the goal is. It simply means that there is a phase of undirected behavior which, in the normal course of events, will lead to a particular source of further stimulation. Only rarely can a hungry fox simply get up and eat; ordinarily it must hunt, and the behavior patterns used in hunting may be many and diverse. It is frequently impossible to identify the nature of appetitive behavior until the actual goal can be observed. An animal searching for food, water or a mate may behave quite similarly, but with each the stimuli required to bring the search to an end are specific. When a fox hunting for food encounters prey, it then switches to an oriented type of behavior; when the food is seized, eating ensues. At this time the various searching patterns give way to a series of responses directed at the goal. These responses, called **consummatory acts,** may be stereotyped, fixed action patterns. Eating is the consummatory act of feeding behavior, drinking relates to thirst, copulation to sexual behavior, and so on. Consummatory acts are normally followed by a period of quiescence during which the animal is no longer responsive to stimuli from the goal; thus, a well fed animal shows no further appetitive behavior even though the appropriate stimuli may be present. Quiescence relates only to one type of behavior; the animal may be actively pursuing some other sort of goal. Frequently responsiveness gradually builds up again so that there is a fairly direct relationship between the threshold level of responsiveness for a given set of goal stimuli and the time that has elapsed since the previous performance of the consummatory act.

The **hypothalamus** exerts a great influence on feeding and drinking behavior, sexual behavior, sleep, maternal behavior and behavioral responses to temperature change. It is generally considered to be the control center for much of what we call motivated behavior. Feeding behavior, for example, is controlled by feeding and satiety centers. Stimulating the **satiety center** in the ventral medial region of the hypothalamus decreases the amount of food eaten. Stimulating the **feeding centers** in the lateral hypothalamus causes the animal to eat (even if it has just had a full meal).

Experiments with goats have demonstrated the role of the hypothalamus in controlling motivated drinking behavior. When a goat (or any other mammal) needs water, the deficit of water in the blood stimulates special cells in the lateral area of the hypothalamus. Impulses from these cells stimulate the pituitary to release antidiuretic hormone. This hormone results in the resorption of more water by the kidneys, and also stimulates the animal to look for water and drink it. When a very small amount of hypertonic salt solution is injected with a fine needle directly into this area of the hypothalamus, the goat drinks enormous quantities of water, even though it had previously satisfied its thirst and needed no water. It will drink water even if it is adulterated with high concentrations of salt or quinine.

Motivation can be measured only indirectly. For example, to determine the motivation towards feeding behavior one can measure the amount of food eaten, how bitter the food can be made before it will be refused by the hungry animal, how fast the animal will run towards food, how strongly it will pull towards food when fitted with a harness, the level of electric shock that will be tolerated by the animal to get to the food, or the rate at which the animal will press a bar to obtain a food reward. Motivational behavior is a response to a need to restore a condition of homeostasis. When homeostatic balance has been restored, there is generally some sort of sensory feedback which signals that the need has been satisfied.

Hormones may alter behavior by stimulating the development of organs employed in behavior, by affecting the early development of the nervous system, by producing changes in peripheral organs concerned with the generation of sensory input to the central nervous system, by influencing special centers in the central nervous system or by exerting general, nonspecific effects on the animal as a whole.

Thyroidectomy, gonadectomy, adrenalectomy and hypophysectomy each enhances the nest-building behavior of rats. Each of the operations leads to a lowered body temperature, which enhances nest-building. One well studied example of hormonally induced changes in peripheral organs affecting behavior is the effect of **prolactin** on parental feeding by doves. "Pigeon milk," on which doves feed their young, is actually sloughed epithelium of the crop sac. Prolactin causes the crop to enlarge just before the young hatch. Sensory stimulation from the crop at this time is an important factor in initiating parental feeding, for the application of a local anesthetic to the crop reduces parental feeding.

The role of hormones in the development and expression of behavior has been demonstrated in many vertebrates by castrating young animals and observing the absence of certain adult or sexual behavior later on. Injecting sex hormones into sexually immature animals will induce sexual behavior prematurely. Sexual behavior in spayed animals is restored by injecting or implanting the appropriate sex hormones.

Sexually mature female cats usually have three estrous cycles a year. When in estrus, a female cat in the presence of a male will elevate the rump, deflect the tail to one side and make treading movements with the hind legs. A cat in *anestrus* will fight if a male approaches too closely.

Figure 35.9 Cryptic resemblance of several different types of insects to leaves. (Courtesy of H. B. Cott.)

35.7 BEHAVIOR PATTERNS

Feeding. All animals show characteristic behavior patterns in finding and consuming food. Even the oyster and other sessile filter feeders which do not hunt their food utilize sense organs, filtering mechanisms and coordinated muscle contractions to obtain particles of food of the correct size. Most animals search actively for specific kinds of food; when the animal's sense organs detect a certain constellation of physical characteristics, the object is treated as food and is eaten. Some insects feed on particular plants only when specific receptors, located on hairs on their mouth parts, are stimulated by a particular chemical compound secreted by the plant. Frogs are visually stimulated by small, dark, moving forms such as flies. The identification of such forms by the frog's nervous system appears to result from the properties of special interneurons in the visual pathway; these fire only when specific types of movement occur across the frog's retina. Young birds initially peck at any small round object which contrasts with the background. By trial and error, their pecking movements become more accurate and discriminating, and they learn not to peck at objects, such as pebbles, which have no nutritive value.

Predatory Behavior. The predator-prey relationship, in which one animal kills another for food, is of special interest because of the interactions between the attack patterns of the predator and the defense patterns of the prey. Lions and leopards show a complex set of movements as they move along downwind of a herd of antelope, creep toward one animal and run it down with a short burst of speed. If the antelope is not caught during this short burst, it usually escapes.

Many birds are attracted to the visual patterns presented by insects, their normal prey. In an evolutionary response to this, a number of larval and adult insects have "eyespot" patterns which are suddenly exposed when the insect is disturbed, creating a "non-insect" pattern (Fig. 35.5). Other insects resemble their environment closely and become immobile when disturbed (Fig. 35.9). Predators are usually attracted by movement, and simply being still is excellent defense for the prey. In a number of arthropods (whip scorpions, many insects,

most millipedes) and some vertebrates (toads, skunks), behavior patterns involving chemical warning and defense devices have evolved as a means of protection against predators (Fig. 35.10).

A common type of predator-prey interaction, termed **protean behavior,** occurs when the prey appears to go through a sequence of possible escape patterns at random, and the predator cannot predict the exact path of the prey's movements. A stickleback fish being chased by a duck follows an erratic path that is very difficult to follow and apparently impossible to predict.

Orientation. The behavioral activities of animals are in large measure directed by sensory information concerning the environment. If a sequence of muscle contractions is to bring about a biologically effective movement, the animal must have information about where it is and where its limbs are. Input from proprioceptors indicates the position of the limbs relative to the body, and input from other sense organs indicates the body's position in space.

The octopus can be trained to attack only when stimulated by certain visual patterns. However, when its gravitational sensory input is cut off by removal of the statocysts, the octopus has great difficulty responding to the visual patterns in the correct way. The central nervous system must integrate its visual and its gravitational information before the proper responses can occur.

Most animals orient themselves in a particular way in the area in which they move about. Amphipods living on the beach use visual stimuli (the sun during the day, the moon at night) as indicators of compass direction to determine their way back to the ocean. This was demonstrated by experiments in which screens and mirrors were used to change the apparent position of the sun or moon. Certain wolf spiders live along river banks and escape from enemies by running across the water. These spiders also use the position of the sun to determine the direction of escape. It appears that the correct escape direction is learned, for when young spiders are moved to a new location, their responses are altered so that they move in the correct direction relative to the water in the new locality.

Mammals frequently mark the limits of their **home range** (the area of their daily movements) or **territory** (the area an animal

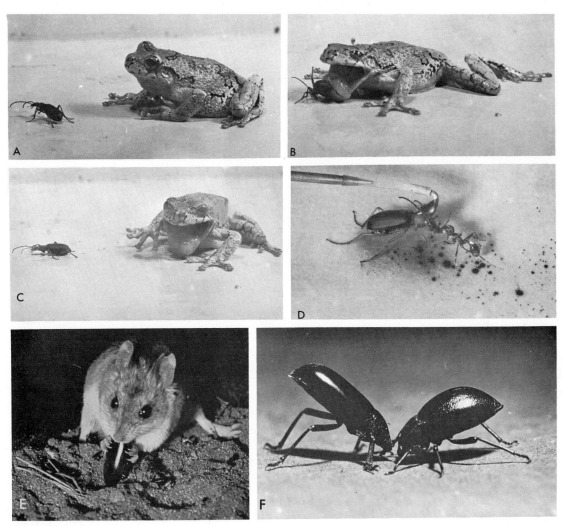

Figure 35.10 Chemical defense systems in arthropods The bombardier beetle produces a noxious chemical spray when attacked by ants (*D*) or a toad (*A*, *B*, and *C*). When the toad tastes the beetle's spray, it rejects the insect (*C*). The beetle *Eleodes* does a headstand as it ejects its defensive spray (animal to the left in *F*); a mimic of *Eleodes*, *Megasica* also does the headstand, although it has no defensive chemical spray (animal to the right in *F*). The grasshopper mouse avoids the spray of *Eleodes* by sticking the beetle's abdomen toward the ground (*E*). (Courtesy of T. Eisner.)

defends) by chemical marks. Some regularly use the secretions of special scent glands to mark the limits of their territory. The familiar sight of a dog at a fire hydrant or shrub is not so much the result of pressure on the dog's bladder as it is an attempt to provide a chemical marker for later orientation.

In addition to those movements involved in daily life, many vertebrates and some arthropods have special behavior patterns of migration, during which they cover relatively great distances, usually moving to a different type of environment. Certain environmental changes bring about specific physiologic changes in the animals; these are followed by the extended periods of locomotor activity that constitute the migration. In locusts, the increased tactile contacts due to crowding bring about physiologic changes which precede the formation of migratory swarms. Spiny lobsters may use tactile stimuli to become entrained in long migrating columns.

Figure 35.11 Defensive display of the ghost crab, *Ocypode*. The bright white mani of the chelipeds are ordinarily held under the body, out of sight. (Courtesy of K. Daumer.)

The migrations of birds have fascinated many investigators. Increasing day length in the spring stimulates the hypothalamus to secrete gonadotropin-releasing hormones (p. 397); these stimulate the pituitary to secrete gonadotropins, which cause the gonads to enlarge and secrete sex hormones, initiating the northward migrations. The length of the migrations and the route taken vary from one species to another. How birds navigate is not completely known. Experiments suggest that some species use the position of the sun or the pattern of stars to navigate; others may use visual landmarks, and still others may be able to detect the magnetic fields of the earth and "fly by compass."

Aggression. Competition for food, mating partners and nesting space may result in aggressive behavior patterns, primarily between members of the same species. When an animal is approached too closely by another, its first response is often a defensive posture (Fig. 35.11). Most arthropods and vertebrates have an **individual distance,** the minimum distance required to separate two animals without a fight ensuing. It can be pictured as a space around an animal into which only sexual partners may enter without eliciting aggressive behavior.

This space is distinct from an animal's **territory,** the total environmental area it will defend. In its territory, an animal obtains the food, nest materials and space needed for the rearing of its young. Since only members of the same species are likely to utilize the same portions of the environment, territorial defense is usually shown only toward members of the same species, and aggressive interactions are especially frequent at territorial boundaries. Such **"boundary encounters"** are usually longer and more complex than fights occurring well within an individual's territory. One animal will quickly retreat from a second when the meeting occurs within the second's territory. At the boundary between territories, neither animal retreats and complex patterns of behavior may be exchanged. The animals may show ambivalent behavior patterns as they are simultaneously motivated to attack and retreat. Under such confusing conditions, the animals often do neither, but instead do something totally unrelated to the situation. Herring gulls, after a series of aggressive postures at a territorial boundary, may either pull grass from the ground, an activity usually associated with nest building, or go through motions similar to grass pulling. Such **displacement activities** may be quite ineffective under the circumstances. Grass pulling has secondarily become part of the gull's aggressive repertoire.

An important characteristic of intraspecific fighting is that it rarely results in any physical damage to the combatants. Through evolutionary selection the acts have become **ritualized** — the movements have a social, communicatory function, and the interactions are mainly exchanges of signals or **displays.** The wrestling of rattlesnakes (Fig. 35.12) produces even less damage to the interacting reptiles than the shows put on

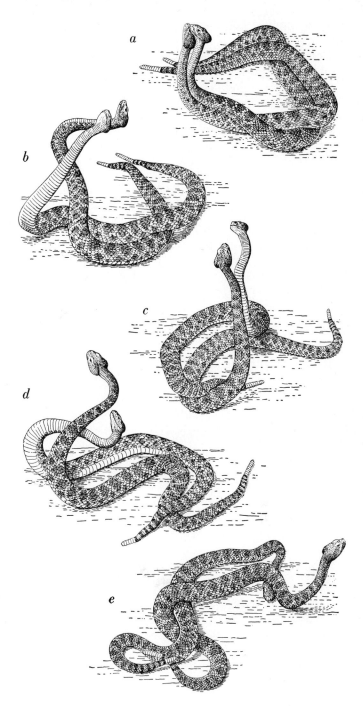

Figure 35.12 Fighting among rattlesnakes. The snakes do not bite one another but wrestle, pushing against each other with their heads (*a* and *b*) or ventral parts of their bodies (*c* and *d*) until one is pinned (*e*). (From The Fighting Behavior of Animals by I. Eibl-Eibesfeldt, Copyright © 1961 by Scientific American, Inc. All rights reserved.)

by "wrestlers" on television. When one individual wins an encounter, the loser frequently assumes a submissive posture which inhibits further attack by the winner. In some species of hermit crabs, exchanges of aggressive movements stop when one or both crabs suddenly squat down on the sand. Fighting may, in some species, continue until damage is done (cat and cock fights are justly famous); equally paired mice will sometimes fight to the point of bloodshed; large male seals regularly lose blood in fights for the possession of a harem of females. The great majority of intraspecific fights, however, do not result in physical damage.

Dominance Hierarchies. In many vertebrates and a few arthropods, the social interactions of the members of a local group are well organized and help to decrease aggressive activity among the members of the group. In the simplest form of such a social hierarchy (the "**peck order**"), one individual, α, is dominant over all others, a second, β, is dominant over all except α, and so on. Dominance is expressed by winning in aggressive interaction, by having first chance at the available food and mating partners and by having first choice of nesting locations. The other animals are subordinate and either assume a submissive posture or retreat from the more dominant individual.

An animal's position in a hierarchy may be a function not only of its size, but also of its sex, individual temperament, immediate physiologic condition and behavioral interactions outside of its social group. A chicken can maintain distinct positions in a number of groups at the same time, but its positions in all groups can be altered by interactions with chickens outside these groups. Thus, when a chicken loses an encounter with another very aggressive chicken outside its usual social group, it may lose subsequent encounters within its social group.

Hierarchies have been observed in a number of invertebrate groups, but the organization within small groups of crayfish appears to be one of the clearest examples of individual recognition-dependent structure. Straight-line hierarchies develop after a small number of fights, and subsequent interactions are limited to exchanges of displays of mild threats.

Communication. Animals often communicate by behavior patterns which mutually affect one another's behavior; typically, the behavior concerns either mating or aggression. One of the classic examples of nonaggressive, nonsexual communication is the means by which honeybees direct other members of the colony to a source of food. When a worker bee has found a source of food, it collects a sample and flies back to the hive. It transmits information to other members of the colony by the kind of "dance" it does on a vertical surface of the hive. If the food is near the hive, the honeybee circles first in one direction and then in the other in a "round dance" (Fig. 35.13). The other worker bees then fly out and search in all directions near the hive. If the food is located at a greater distance from the hive, the bee goes through a half

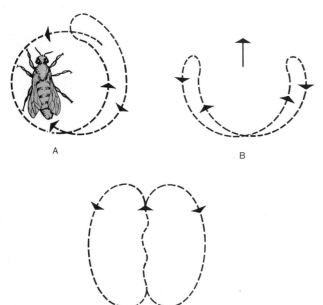

A

B

C

Figure 35.13 Three kinds of communication dances performed by honey bees. *A*, The round dance; *B*, the sickle dance; *C*, waggle dance. (From von Frisch, K., and Lindauer, M.: In McGill, T. (Ed.): Readings on Animal Behavior. New York, Holt, Rinehart and Winston, Inc., 1965.)

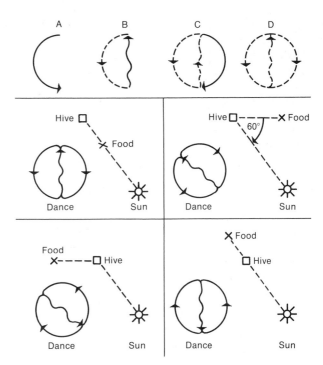

Figure 35.14 Indication of direction by the waggle dance.

circle, then moves in a straight line, waggling its abdomen from side to side, and finally moves in a half circle in the other direction. During the straight portion of the dance, the bee produces a series of sounds. The angle that the straight, waggling portion of the dance makes with the vertical is the same as the angle of flight to the food relative to the sun (Fig. 35.14). Information on the distance to be traveled in that direction is also transmitted by the pattern of the "waggle dance." There is a correlation between the number of waggle dances executed in a given time and the distance of the food from the hive. An even closer correlation, however, exists between distance and the sound pulses the dancing bee produces. Bees can sense the vibrations of the substratum, and sound would seem to be a good way to communicate in the dim interior of the hive.

The distance to be traveled is "calculated" by the bee on the basis of the flight *to* the food source, not the return trip. Moreover, the instructions apparently include corrections for wind and large obstacles. The dancing bee also indicates the richness of the food source by how long and vigorously it repeats the dance. Other bees gain further information about the food source

by the smell of the flowers the bee has recently visited. The more primitive social bees, the stingless bees, lay down an odor trail, a series of chemical marks, to the food source; in addition, the collector bee guides other individuals along the trail.

Fire ants communicate information about food sources by a pheromone (p. 402) produced in a special gland, **Dufor's gland,** located in the abdomen of the worker ant. When one ant has located a source of food, it gathers what it can and starts back toward the nest. As it moves along the ground, it periodically extrudes its stinger and places a small amount of pheromone on the substratum. When another fire ant comes across this trail of chemical dashes it becomes very active, follows the trail and is led to the source of the food.

Other examples of communication between members of a group can be seen in the schooling of fish and the group flight patterns of birds. The importance of visual cues has been established for schooling in some fish, but what controls the sudden group movements of a flock of birds is still a mystery. In both fishes and birds it appears that grouping is a defense against predators. Predators are often able to catch isolated individuals, but are confused by the many movements of the individual animals making up a group.

Courtship and Mating. All bisexual animals possess mechanisms for getting egg and sperm together to continue the existence of the species. For some sessile marine animals, such as clams, urchins and tunicates, courtship and mating are simply the releasing of sperm or eggs into the surrounding waters (where external fertilization takes place) when the animals are chemically stimulated by the presence of a few gametes in the water.

Changes in the physical environment induce (via hormones) physiologic changes in many species which prepare the animals so that subsequent specific stimuli will elicit the appropriate mating patterns. In other species the presence and movements of individuals of the opposite sex initiate the physiologic changes. Mutual exchanges of early courtship movements initiate and synchronize physiologic changes so that the sexes are ready to mate at the same time.

Further precopulatory behavior patterns (or repetition of the ones effective in physiologic priming) serve two important functions: (1) they decrease aggressive tendencies and (2) they establish species and sexual identification. The first function is needed if two animals are to avoid fighting long enough to mate. Where this condition has not been completely fulfilled, special adaptations are necessary. For example, the male praying mantis is smaller than the female and subject to predatory attack by her. However, mating can continue even after the male's head has been bitten off by the female, since the nervous activities controlling copulatory movements are centered in an abdominal ganglion. In fact, removing the head enhances copulatory behavior by eliminating inhibitory impulses originating in the brain.

In some cockroaches and other primitive insects the female feeds on a special secretion on a part of the male's back, which diverts her attention during copulation. The males of certain predatory flies present the females with little packages of food wrapped in silken threads. The males of other species divert the female's attention by presenting her with an empty balloon of silk. The balloon both decreases the female's aggressive tendencies and identifies the sex and species of the male presenting it. A web of silk is used by some pseudoscorpions to form a "hallway" to guide the female. In these species the inter-sex aggressiveness is very great and these offerings by the male play an important role in neutralizing it.

The complex pattern of reproductive behavior of the **stickleback** has been studied intensively as a model (Fig. 35.15). After the male has migrated and staked out his territory, he builds a nest and courts any egg-laden female who enters his territory. In courting he performs a special zigzag dance. The female follows the male and he leads her to the nest, at which point he directs his head toward the entrance. The pointing stimulates the female to enter the nest and she, in turn, stimulates the male to tremble. His trembling stimulates the female to spawn and the male fertilizes the eggs. The presence of the eggs stimulates his sexual behavior. The tendency towards sex and aggression can be measured by presenting to a male on his home territory either another male or a female fish confined in a glass tube. If the fish presented to him is a receptive female with a swollen belly, the male performs the zigzag dance. A count of the number of bites or zigzags per unit time is a measure of the respective strengths of the male's sexual and aggressive tendencies.

The characteristic pattern of singing of a territorial male bird advertises his presence. The song patterns of each species are distinct and presumably function both to keep other males of that species away from a male's territory and to attract a female of the species to the singing male. Certain fishes, frogs and insects also have mating calls which provide effective cues for the discrimination of species. This ethologic isolating mechanism prevents the wasting of genetic material.

The bowerbirds of tropical jungles have a very complex courtship behavior (Fig. 35.16) involving a fixed sequence of sign stimuli flashed from partner to partner. In an arena, a special territory where mating takes place, the male sings and displays his bright colors in special short flight movements. In other species, the male bowerbird's plumage is less brilliantly marked, but he builds and decorates an elaborate display area, thus attracting a female not by his own color, but by the

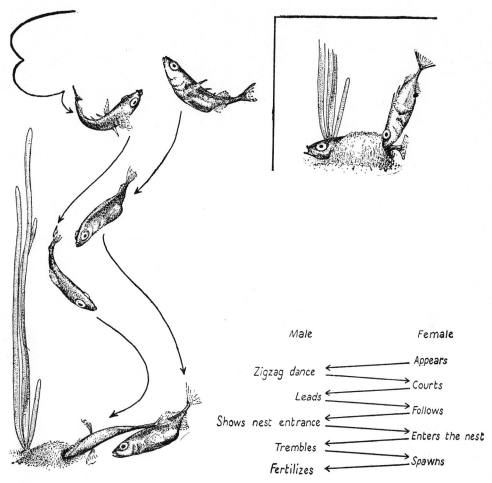

Figure 35.15 Courtship in the three-spined stickleback. The appearance of the female in breeding condition elicits a zigzag dance by a territory holding male. The female follows the male, who leads her to the entrance of the nest he has built. When the female is in the nest, the male's tactile prodding elicits egg laying (inset). (After Tinbergen, N. The Study of Instinct. London, Oxford University Press, 1951.)

form and color of pebbles and berries added to the display area. After mating, the female bowerbird builds the nest and cares for the brood. Most species of birds nest and rear their young in the territory; the male typically shares the tasks of incubating the eggs and feeding the young birds.

Parental Care. Most animals give little or no care to their offspring. This lack of parental behavior is usually correlated with the production of great numbers of eggs and sperm. Some reptiles and insects place their eggs in special burrows or nests where development is completed, but the parents do not aid the young after they hatch. Female crayfish and other crustaceans attach their eggs to their abdominal pleopods and aerate them until the eggs hatch. The young of some scorpions remain with the mother, clinging to her back until after their first molt. Care-giving behavior between one generation and the next is well developed in the social insects and in birds and mammals. Many young birds respond to tactile, acoustic and visual stimuli given by the adult by gaping—stretching upward toward the parent with open mouth. The parent responds to this by feeding the young (some baby birds have brightly colored oral cavities that serve as sign stimuli to the parents).

The extensive exchanges between mother and young in primates often have long-term effects on the behavioral development of

Figure 35.16   Courtship in one species of bower bird. The male builds a bower of sticks (A) and decorates it with pebbles and berries (B). The female bower bird enters the construction (C) and, after the male displays, she sits (D) and the birds mate (E). The female leaves, builds the nest (F) and rears the young birds by herself. (Courtesy of E. T. Gilliard.)

the young. Rhesus monkeys deprived of their normal relationship with mothers or siblings never develop normal social and sexual relationships. Monkeys reared by artificial mothers do not interact normally with other members of the species, nor are they able to mate when mature.

The elements of the maternal-infant relationship were studied by Dr. Harry Harlow using surrogate mothers of cloth and wire (Fig. 35.17). He found that both bodily contact and visual stimuli are important to this relationship. Young monkeys formed strong emotional attachments to cloth mothers, ran to them when frightened by a mechanical teddy bear beating a drum, and would explore strange objects in the room only as long as the "mother" was present as a psychologic refuge. Monkeys without a real or appropriate surrogate mother fled to a corner of the room, clasped their heads and bodies and rocked convulsively back and forth. Their actions resembled the behavior of autistic children.

The young of some primates receive attention from several of the adult females of the troop, which is the social unit consisting of several adult males, three to four adult females and their offspring. Initially the mother will not allow others to hold the infant, but soon it is held and carried by other females, each of which nudges, touches, smells and licks it. If the infant squeals or begins to struggle, another female (usually the mother) comes over and takes it. The touch and color of the fur and the movements of the mother elicit a clinging response in the infant; the infant's color pattern elicits the interest and movements of the female, who then facilitates clinging. These mutual responses of mother and newborn are very important in the development of the social behavior and organization of primate groups.

Figure 35.17 Young rhesus monkey, reared without a real mother, clinging to the surrogate used in raising it. (Photography by Gordon Coster. *In* Love in infant monkeys by H. F. Harlow. Copyright © 1959 by Scientific American, Inc. All rights reserved.)

35.8 CYCLIC BEHAVIOR AND BIOLOGIC CLOCKS

Many animals exhibit behavior patterns that recur with a regular periodicity—each year, each lunar month, each day, or with the tides. Some behavior patterns are regulated by a complex interaction of several types of periods. In recent years there has been much interest in these cycles and in their regulation by some sort of intrinsic "biologic clocks." The organisms in the temperate zones of the world typically show marked seasonal cycles of activity. Most fishes, birds and mammals have a single breeding season in the spring; deer and sheep breed in the fall. Certain animals migrate at specific times to higher latitudes or higher altitudes. Some animals show a pronounced decrease in activity and metabolism during the winter, a phenomenon termed **hibernation.** In the tropics, although there is no sharp change in temperature from one month to the next, a period of heavy rainfall usually alternates with a period of little or no rain. The cycles of plants and animals in such regions are geared to these changes in the environment.

Other cycles of animal and perhaps plant activity reflect the phases of the moon. The most striking ones are those in marine organisms that are tuned to the changes in the tides due to the phases of the moon. The swarming of the Palolo worm at a particular time of the year is governed by a combination of annual, lunar and tidal rhythms (p. 559). The Atlantic fireworm swarms in the surface waters around Bermuda for 55 minutes after sunset on days of the full moon during the three summer months. The grunion, a small pelagic fish of the Pacific coast of the United States, swarms from

April through June on those three or four nights when the spring tide occurs. At precisely the high point of the tide the fish squirm onto the beach, deposit eggs and sperm in the sand and return to the sea in the next wave. By the time the next tide reaches that portion of the beach 15 days later, the young fish have hatched and are ready to enter the sea. The menstrual cycles of man and apes also follow lunar periods.

Organisms (even the various parts of an organism) usually do not function at a constant rate over the entire 24 hours of a day. Repeated sequences of events occur at about 24 hour intervals in the lives of many organisms. These have been termed "circadian rhythms" (*circa* = about; *dies* = day). It is important to note that when the organism is placed under constant conditions, these rhythms have a period of *about* 24 hours, but almost never *exactly* 24 hours. In nature the activity cycle is lined up with the cycle of physical events which determine the length of the day. The daily changes in light are the signals used by most organisms as a "zeitgeber" (*zeit* = time; *geben* = to give).

Some animals are **diurnal**, showing their greatest activity during the day, whereas others are **nocturnal** and are most active during the hours of darkness. Still others are **crepuscular**, having their greatest activity during the twilight hours. Certain insects exhibit diurnal variations in pigmentation and continue to show these when placed in continuous darkness. Some spiders show a diurnal rhythm of the secretion of neurohormones; preparations of nervous tissues isolated from the animal and maintained *in vitro* continue to show this cyclic production of neurohormones. A diurnal cycle of the deposition of glycogen in the liver of the rabbit and mouse has been detected. Many of man's physiologic processes exhibit circadian rhythms. Our body temperature is highest at 4:00 or 5:00 P.M. and lowest (as much as 2° F. lower) at 4:00 or 5:00 A.M. The secretions of a number of hormones, the heart rate and blood pressure, and the rate of excretion of sodium and potassium all show circadian peaks and valleys.

The diurnal cycles of many organisms are firmly entrenched and are not easily changed even by prolonged exposure to constant light, to constant darkness or to abnormal cycles of light and darkness (a cycle having 8 or 10 hours instead of 24, for example).

Many different kinds of animals have evolved biologic clocks by means of which their activities are adapted to the regularly recurring changes in the external physical environment, and perhaps to changes in their internal milieu as well. These clocks, together with other signals from the external environment, indicate to the organism that time of day most appropriate for some activity or physiologic process. Birds and bees have highly developed timing devices and can use them in navigation. Bees are not only able to find their way from hive to feeding ground using the direction of the sun as a reference, but they can make suitable corrections for the sun's position as the day advances; that is, they can correct their celestial navigation for the time of day just as man does.

At one time it was believed that the timing mechanisms of the biologic clocks operated on some sort of "hourglass" principle, that some regularly recurring stimulus such as sunrise triggered a physiologic process requiring a certain length of time for its completion, at which time the signal was sent to the appropriate organ. More recent evidence favors the concept that the timing mechanisms involve some sort of cyclic oscillation. Biologic clocks do require a small but definite amount of biologically useful energy for their operation; they cannot be primarily dependent on enzyme-catalyzed reactions because they show little or no change with temperature. The biologic clock could involve some sort of biophysical mechanism, such as an alteration in the physical state of a complex molecule like a protein, or an alternation of the tension and relaxation of some physical structure.

Snails and other marine organisms whose activities vary with the tide continue to show the cyclic variations in activity when removed to an aquarium and protected from changes in light, temperature and other factors. This persistence of rhythmic changes in activity, coordinated with the cyclic changes in the environment from which the animal was removed, is strong evidence for the existence of biologic clocks.

The keying of these endogenous rhythms to changes in the local environment is obviously of supreme adaptive importance. They enable certain adjustments in physiologic and biochemical processes to occur before the time of the change in the environment. Thus, the organism's activities are synchronized with the various periodicities in the environment. Although the cycles are set and persist under constant conditions, they can be reset if necessary. If animals that breed in the spring are transported from the northern to the southern hemisphere, their cycle eventually shifts over to coincide with the occurrence of spring in their new home. Members of the jet set, flying east or west to widely different time zones, eventually change their pattern of sleeping and wakefulness and their many other circadian rhythms. In short, all cycles are set by local time.

35.9 SOCIAL BEHAVIOR

The behavior of animals must be adjusted not only to factors in the physical environment but also to other individuals, both of the same species and of other species. In the broadest sense, any interaction between two or more individuals constitutes **social behavior.** Usually, however, social relationships imply interactions among members of the same species.

The mere presence of more than one individual does not mean that the behavior is social. Many factors of the physical environment bring animals together in **aggregations,** but interactions, if they occur here, are apt to be circumstantial. A light at night is a stimulus that causes large numbers of moths and other insects to collect around it. The high humidity under a log is a common stimulus causing aggregations of wood lice. A water hole in the dry savanna is a common focus for large numbers of birds and mammals. None of these groups is, strictly speaking, social.

Much of what we recognize as social behavior is based, directly or indirectly, on mating behavior. Except in parthenogenetic species, male and female must come together in a cooperative fashion. Although this might appear to be a simple matter to arrange, there are in fact some major obstacles that must be circumvented. Elaborate patterns of behavior have evolved to overcome these.

First of all, eggs and sperm must be produced at the same time. This means that male and female must find each other even if fertilization in that species is external, as in fish. It also means that both individuals must be in a reproductive state. The synchrony of time and place is accomplished by common stimuli from the external environment and by the stimulatory effect of one sex upon the other. Many fish, birds and mammals are brought into synchronous reproductive readiness by increasing day length, which acts on the pituitary to release gonadotropic hormones, which, in turn, affect behavior. The increasing water temperature in spring induces the hormonal changes in the stickleback which predispose the fish to migrate into shallow water. The visual stimulation provided by the vegetation there releases reproductive behavior.

In many species of animals, family life—a long-term relationship between male, female and young—is a natural consequence of sexual relations. The basis of family relationships is the provision of shelter, food and defense for the young. Parents and young must react to one another in a variety of ways. Many birds will not feed the young unless they gape or beg. Young are stimulated by the sight of the parent's head over the nest, vibration of the nest by the alighting adult or, in the case of the sea gull, by a "mew-call." The gull chick is further stimulated by the parent's bill, a long, thin structure with a red spot at the tip.

Larger groups may consist of several to many families or of individuals no longer tied to a family, as in flocks of birds, schools of fish and herds of animals. There are several adaptive advantages of grouping: A group is more alert to danger than a single individual. What one member may miss another sees, and the whole group takes flight. Often several unrelated species group together. In mixed groups of baboons and impala, the keen-sighted baboons spot visible danger and the keen-scented impala detect hidden danger. A group offers protection against attack. Male musk oxen make a ring around the young and female musk oxen when the group is threatened. A group is capable of communal attack as well as communal defense. The most frequently observed form of communal attack is the mobbing of crows, owls or hawks by groups of smaller birds. Communal hunting has been developed to a fine art by wolves and lions. The origin of a group and its continued cohesiveness depend upon a constant flow of information within the group.

Reproductive fighting and **dominance fighting** are aspects of group behavior that may also have adaptive value.

Reproductive fighting, directed toward reproductive rivals, serves to space out the breeding pairs. The end result is a more even distribution of food, more nesting and breeding space, and the prevention of multiple copulations and their attendant waste. This kind of spacing provides each pair with its own **territory** from which it drives trespassers. The size of a territory may vary from the very small domain of a nesting gull to the extensive hunting territory of a large carnivore. The size of the territory of gulls and many other colonially nesting birds is the diameter of a circle from within which one bird can reach out to peck another without having to stir from its nest. This type of territory is nest-centered.

The territory of most other species is more than the breeding area proper; it is the area which supplies food, and its size is in part a function of the nature of the food. Territorial size may also be determined by topography (i.e., the actual amount of space available) but does not appear to change with variations in the abundance of individuals.

Animals with highly organized societies may have a **colony range** rather than a pair territory. Troops of baboons may claim a range of three to six square miles, but this may overlap the home range of an adjacent troop. The location and extent of ranges is set by group tradition. Territorial boundaries are learned by each generation. As long as food and water are available, troops may never meet. Arboreal species, prevented by the foliage from seeing their neighbors, are spaced out by vocal signals. Howler monkeys, for example, begin each day by howling for about 30 minutes to advertise their position to other troops. When there are shortages, as when only one water hole is available for baboons, the troops may come together, but they do not mix.

Other social animals, exemplified by the prairie dogs, inhabit territories that are subdivided into fixed and stable family units. The resemblance to human social life in this respect is striking. Whereas troops of baboons resemble human nomadic tribal life, prairie dog colonies resemble urban apartment living.

Another way of sharing the available environment may be called **temporal spacing.** It is achieved by the establishment of a dominance hierarchy in a group that occupies common territory. No individual calls an area his own or occupies it for more than a brief period; however, not all spots are equally available to all individuals. The best of everything comes first to dominant members of a group, while members lower on the social ladder must wait their turn. Social order is generally the rule in animals that live in groups. Among vertebrates it attains its highest development in troops of primates. A male achieves a dominant position in the group by his physical condition, age, aggressiveness and fighting ability, and by the rank of his mother. He then commands the immediate surrounding space. Once achieved, his position is held by symbolic threats rather than by actual fighting. These threats are ritualized behavior patterns developed from preparatory movements for fighting, as when a man shakes a clenched fist at his enemy.

The hierarchy is further complicated in baboons by the fact that a number of males may join together in a governing clique. There may be an outside male who can outfight any individual member, but the group combines to stand against the intruder.

Each animal comes to know his rank in the social order, and if challenged by his superior, makes ritualized submission by raising his rump, and the dominant asserts his dominance by mounting. Grooming, which is basically a hygienic measure, has acquired social significance in the more aggressive species of primates. The dominant member is more often groomed than grooming.

The cohesiveness of a primate group and the intensity of the dominance hierarchy are directly related to the level of potential danger to which the species is subjected. Gorillas, for example, do not face the same dangers as baboons, are much less aggressive animals (despite their ferocious appearance), and lead a considerably more relaxed life. There are dominant males who lead, but there is little violence. Baboons and macaques, on the other hand, are subjected to many dangers in their environment; they are, accordingly, very aggressive.

35.10 SOCIAL INSECTS

Among the oldest and most highly developed societies are those of insects. Three hundred million years ago these societies were already in existence. The establishment of social life in these usually short-lived animals was made possible by the prolongation of parental life. First the progeny depended on the parents, then the parents on the progeny.

Various gradations between solitary and social life are evident in existing species. Some solitary insects, such as mosquitoes and dragonflies, drop their eggs anywhere and go about their business. Butterflies and flesh-flies lay their eggs on food suitable for the young. The solitary digger wasps dig a hole, provision it with food such as paralyzed spiders and caterpillars, lay an egg, seal the entrance and depart. The subsocial insects represent a further step toward social habit; they provide a mass of food for each egg, but remain to guard the nest or young. In some species of dung beetles the female collects and rolls a ball of dung, excavates a burrow, drops the ball in, lays eggs and departs. In other species the male assists by guarding the dung balls while the female excavates. In still another species both sexes dig chambers, stock them with dung on which the female lays eggs, then guard them until the young hatch. At this time all disperse.

True social insects forage for food for the colony continuously and the young cooperate in caring for the next generation. The most highly developed and complex of these are the social termites, ants and bees. All of these are matriarchies; all members of the colony are the offspring of a single female and hence all have very similar genotypes.

The greatest diversity is found among ants. The army ants have no permanent nest, but travel in long marauding columns and rest in temporary bivouacs. There are ants that kill off other species, carry off their young and are thenceforth fully dependent upon the "slaves" that develop from these larvae. Other ants collect nectar and feed it to special members of the colony that become living honey casks hanging from the roofs of underground chambers. There are ants that feed on the honeydew produced by aphids, carry these "ant cows" into overwintering quarters and protect them from predators. There are leaf-cutter, or fungus-growing, ants that cut leaves and carry them underground to serve as a substrate for growing pure strains of fungus which is eaten by the ants.

The cohesiveness and organization of insect societies often prompts comparison with human societies, and indeed some notable similarities are readily discernible: (1) Insect societies arise from discrete families (i.e., one female and her offspring) and by swarming frequently send off new colonies. Human societies originally began as discrete families, but this is not repeated phylogenetically. (2) Man has tradition and social heredity; insects, in this case ants, bequeath fungi and real estate (hunting grounds). (3) Man uses tools; insects do not, with the rare exception of the ant that uses its larvae as spinning shuttles to fasten leaves together with larval silk. (4) Man controls the environment of his dwelling; bees and termites also control the temperature of their nests. (5) Man domesticates animals; ants "domesticate" aphids. (6) Man has a language; bees and ants communicate by dances and pheromones.

A basic difference between human and insect societies is in the manner in which the two adapt to different requirements. Man has evolved learning and abstract intelligence whereby he meets different situations. The adaptation provides him with a flexible society. Insects evolved different genotypes and different castes (soldiers, workers and so on), each of which meets a specific need. Their society is, therefore, a rigid one.

35.11 LEARNING

Learning is a relatively long-lasting adaptive change in behavior resulting from experience. It is usually defined by exclusion; that is, it is a modification of behavior that cannot be accounted for by sensory adaptation, central excitatory states, endogenous rhythms, motivational states or maturation. Learning exhibits many forms and is not a unitary phenomenon. Laboratory experiments have shown that the members of almost every animal phylum can undergo learning phenomena, and field observations have shown that learning is important in a variety of natural situations with a wide variety of animals. Learning in different animal species has different characteristics and may involve different mechanisms. There are several categories of learning: habituation, classic conditioning, operant conditioning, trial and error, insight learning and imprinting.

Habituation is perhaps the simplest form of learning, for it does not involve the acquisition of new responses but rather the loss of old ones. An animal gradually stops responding to stimuli that have no relevance to its life, ones that are neither rewarding nor punishing. Birds soon ignore the scarecrow which put them to flight when it was first placed in a field. The taming of animals represents a common form of habituation. The nature of the processes underlying habituation is obscure, but it must be a property of the central nervous system and not of the sense organ.

All other types of learning consist of strengthening responses that are significant to the animal. The simplest of these is **classic reflex conditioning,** or Pavlovian conditioning. Pavlov's classic experiments with conditioning in dogs frequently dealt with the salivation reflex. If food is placed in a dog's mouth, the dog salivates. By placing a tube in the salivary duct Pavlov could collect and measure all the saliva produced in response to a given stimulus. He then gave the dog standard stimuli by puffing known amounts of meat powder into its mouth through a tube. A standard amount of meat powder caused the secretion of a certain amount of saliva. Pavlov then associated the unconditioned stimulus, the meat powder, with another stimulus, the ringing of a bell or the ticking of a metronome. Initially the second stimulus caused no response, but after a number of pairings of the bell and meat powder stimuli, the saliva began to drip from the tube when the bell was rung and before the meat powder was administered. Pavlov called the second stimulus the conditioned stimulus and the response to it, salivation in response to a bell or metronome, the **conditioned reflex.** Pavlov's experiments showed that almost any stimulus could act as a conditioned stimulus provided it did not produce too marked a response itself.

Conditioned reflexes of this type have been observed in many kinds of animals from earthworms to chimpanzees. Birds become conditioned to avoid the black and orange caterpillars of the cinnibar moth, which have a very bad taste. The birds associate the bad taste with the orange and black color pattern and avoid not only the caterpillars but certain wasps and other black and orange colored insects.

In **operant conditioning** a particular act is rewarded (or punished) when it occurs, thus increasing (or decreasing) the probability that the act will be repeated. A central feature of operant conditioning is the concept that the animal must play a role in bringing about the response which is rewarded or punished. In the simplest type, a rat presses a bar in his cage and is rewarded with food. Initially the pressing of the bar occurs by chance, but on successive trials the rat learns that pressing the bar provides him with food. The rat might at first press the bar with its nose or foot, but the experimenter can determine which of these will be rewarded. He can arrange matters so that only when the rat presses the bar with his left front foot will he receive the reward of food. The experimenter can indeed set up his experimental cage in such a way that the rat is punished, perhaps with an electric shock, if he presses the bar with anything other than his left front foot. Thus operant training may be either reward training (food) or escape or avoidance training (avoiding the electric shock). The **trial and error learning** that occurs in nature is probably more like operant conditioning than classic conditioning. In both types of training only the *probability* of the occurrence of a certain act under a given set of conditions is changed.

Some generalizations can be made about associative learning, as exemplified by these conditioning experiments. First, the time relations are critical. If the unconditioned stimulus precedes the conditioned stimulus, there is little or no conditioning. If the conditioned stimulus precedes the unconditioned stimulus by more than a second, little or no conditioning results. Thus, **contiguity** of conditioned and unconditioned stimuli is required for associative learning. Another relevant feature is **repetition.** The more often the conditioned stimulus and the unconditioned stimulus are paired, the stronger is the acquired conditioned response. The amount of saliva produced by Pavlov's dogs in response to the bell or metronome increased with each trial. A rat learning to run a maze makes fewer and fewer mistakes with each repetition. The rate at which an animal learns is described by a **learning curve,** generated by plotting the time taken to complete the task at each trial or the number of errors committed at each trial, against the number of trials. Repetition finally produces a degree of learning beyond which there is no improvement; however, training beyond this point does make the response more resistant to extinction. Extinction is the decay of learning in the absence of reinforcement. Without reinforcement the conditioned response may disappear completely.

Two other concepts associated with conditioning are generalization and discrimination. **Generalization** refers to the common observation that an animal conditioned to one stimulus will also be conditioned to closely related stimuli. The closer the two resemble each other, the better the response will be to the new stimulus. If a dog is conditioned to respond in some way to a 1000 cps tone, it will respond somewhat to a 500 cps tone or to a 1500 cps tone. Its response to a 100 cps or a 2000 cps tone would be poorer. The opposite process is termed **discrimination.**

A type of learning in which an animal learns without reinforcement and later puts the information to good use has been termed **latent learning.** An example of this is exploration. Most animals spend a lot of time exploring new environments, familiarizing themselves with their surroundings. Ants, bees and wasps make **orientation flights** around a nest they have built to learn its position. Some wasps have been shown to be able to learn the essential landmarks around their nest in an orientation flight lasting only nine seconds.

Insight learning, or insight reasoning, is considered to be the highest form of learning. It is defined as the ability to combine two or more isolated experiences to form a new experience tailored to a desired goal. If an animal is blocked from obtaining food he can see and selects an appropriate detour to it on his first try, he has probably used insight to solve the problem. Only monkeys and chimpanzees seem to be able to succeed on their first attempt in situations such as this; even dogs and raccoons usually fail on their first attempt. One of the classic examples of insight learning is that of the chimpanzee who piled up boxes or fitted two bamboo poles together to get some

Figure 35.18 A rhesus monkey performing an oddity-principle test. (From Stone, K. P. (Ed.): Comparative Psychology. Englewood Cliffs, N.J., Prentice-Hall, Inc., 1951.)

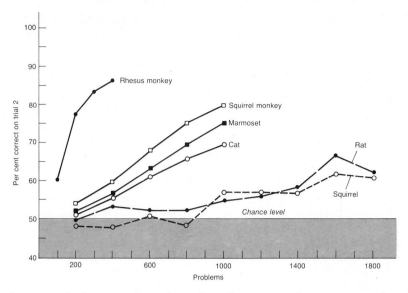

Figure 35.19   The rate at which various mammals can form discrimination learning sets. With each new problem the animal's choice on the first trial has to be random, but if it has learned the principle behind the problems, trial 2 should be correct. Note how long it is before the scores of rats or squirrels on trial 2 become better than chance or 50 per cent. Many monkeys reach almost 100 per cent within 400 problems. (From Warren: The Behavior of Nonhuman Primates. Vol. 1. New York, Academic Press, 1965.)

bananas hung out of reach. The chimpanzee figured out the solution without being taught.

There are several other types of problem-solving tests used in measuring an animal's ability to learn. These are called **learning set tests.** In one, **oddity principle learning,** the animal must pick out the odd member of a set of three no matter what the objects are (Fig. 35.18). In another, **the delayed reaction set,** the animal is allowed to see which of two cups has food hidden under it. Later, after a delay, it has to lift the correct cup. In a third type of **learning set test,** an animal is presented with two dissimilar objects, one of which always has food under it. Repeated trials are necessary before the animal immediately goes to the object concealing the reward. This discrimination problem is repeated with many different pairs of objects. Although each problem is just as difficult as the first, the animal usually performs with increasing accuracy on repeated trials. Finally, when presented with two objects the animal lifts one. If the food is there, it chooses that object on each subsequent presentation. If the food is not there, it lifts the other object and then always chooses that one first on subsequent presen-

tations. The animal finally gets the idea, or forms a learning set, that the food is always under a particular object. If it picks the object correctly the first time, there is no need to look further. The rates at which various mammals form discrimination learning sets vary tremendously (Fig. 35.19).

Imprinting is a form of learning first described in birds but now known to occur in sheep, goats, deer, buffalo and other animals whose young are able to walk around at birth. It is a phenomenon whereby a young animal becomes "attached" to the first moving object it sees (or hears, or smells) and reacts to it as it would toward its mother. Konrad Lorenz, who investigated this phenomenon extensively in the 1930's, described imprinting as a unique learning process whereby the young of precocial birds (those able to walk just after hatching) form an attachment with a mother figure. Normally this is their actual mother, but they will become attached to almost any moving object they see during a brief critical period shortly after hatching (Fig. 35.20). This following response is of considerable adaptive value in keeping the young bird close behind its parent and well within the protective range of the parent.

Figure 35.20 A remote-controlled decoy duck was used by Hess and his colleagues in imprinting experiments. Here both decoy and duckling move about freely rather than on a runway. (From "Imprinting" in Animals by E. H. Hess. Copyright © 1958 by Scientific American, Inc. All rights reserved.)

Lorenz found that the young bird's choice of a mother figure during this imprinting period also affected its choice of a sexual partner when it matured. Young birds reared by a foster mother of another species cease to follow her when they develop and become independent. However, when they became sexually mature, they court and attempt to mate with birds of the foster mother's species.

Imprinting occurs in mammals as well as in birds. Orphan lambs reared by humans follow them about and show little interest in other sheep. Harlow's studies with rhesus monkeys show very clearly that the development of normal sexual and social responses depends on their being reared with their mother or with siblings during infancy. There is little doubt that something similar to imprinting occurs in humans; human infants are extremely sensitive to maternal deprivation from about 18 months to three years of age.

The Physical Basis of Memory and Learning. Learning in some way involves the storage of information in the nervous system and its retrieval on demand. Many of the recent studies of learning have dealt with attempts to discover its physical basis. Somewhere in the nervous system there must be stored a more or less permanent record of what has been learned that can be recalled on future occasions. This record is termed the memory trace or **engram.**

Electrical stimulation of the cerebral cortex of a patient undergoing brain surgery can cause vivid recollection of long-forgotten events. From this it was inferred that the items of memory, the engrams, were filed away somehow in specific places in the brain. Some years ago Karl Lashley investigated the retention of maze learning in rats by removing portions of the cortex after the rats had learned to solve various problems. The essence of Lashley's results was that the extent of the memory removed by the operation was a function of how much of the cortex was removed and not of what specific part of the cortex was removed. Lashley concluded that the cortex was equipotential and that engrams were not sorted out at specific cortical sites but were in some way present throughout its substance. He speculated that memory might be some system of impulses in reverberating circuits.

Other investigations of the role of the cerebral hemispheres in memory have used the "split-brain" technique of R. W. Sperry. The cerebral hemispheres are separated by a cut down the midline which severs the **corpus callosum,** a large band of transverse fibers that passes ventrally and links the two hemispheres (Fig. 35.21). If a subject learns something using one eye (the other eye is

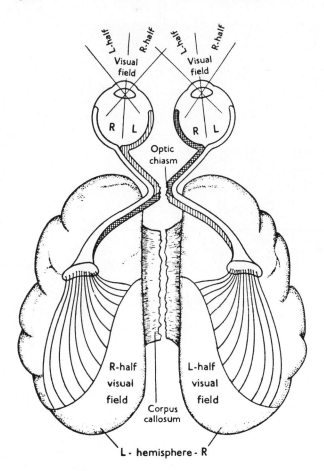

Figure 35.21 Split brain technique. The right and left halves of the brain have been completely separated by cutting the corpus callosum and the optic chiasm. (Redrawn from Sperry, R. W. Science, *133*:1749–1757, 1961.)

covered), there is no problem when the eye coverings are reversed—the subject can make discriminations with its "untrained" eye. Fibers from the left and right halves of the retina cross over in the optic chiasma, and the visual cortex in each cerebral cortex receives information from both eyes. If the optic chiasma is cut, something learned by the right eye is still recognized by the left eye as long as the corpus callosum is intact. However, if the corpus callosum is cut also, half of the brain no longer knows anything learned visually by the other half of the brain following the operation. These split-brain animals are of great interest to psychologists for they are in effect animals with two separate brains. It is possible to train the two halves of the brain to make diametrically opposite discriminations. The left half may learn that the red circle conceals the food, whereas the right half learns that the green circle conceals the food.

These experiments show that one function of the corpus callosum is to transfer engrams (or information used in forming engrams) from one hemisphere to the other and that storage is eventually bilateral. The two cerebral cortices may be linked by fiber tracts lower in the brain, but they apparently are not effective in transferring engrams.

Investigations of the memory system of mammals have shown that memories of recent events, **"short-term" memory,** have properties that differ somewhat from memories of events further in the past, **"long-term" memory.** People suffering from concussion are often unable to recall the events that immediately preceded their accident but have normal recall of events in the past. When their memory returns, events are remembered roughly in the order of their occurrence. This phenomenon has been termed **retrograde amnesia.** It appears that each learning trial sets up a process in the brain which relies on the continuous activity of certain neurons. This **consolidation process** results in the formation of an engram and its storage somewhere in the cortex. Concussion, electroshock and similar treatments can interfere with the consolidation process that forms the engrams, but they have no effect on fully formed engrams. The **hippocampus,** on the lower inner margin of each cerebral hemisphere, is connected to the hypothalamus and to the cortex via the reticular formation of the forebrain. If both hippocampi are damaged, human beings have severe defects in their short-term memory system.

Investigations of the nature of the engram have been pursued actively for many years, and although they have provided a fair amount of evidence about what the engram is *not,* they have given relatively little positive information. The areas of brains known to be the sites of engram storage – the mammalian cerebral cortex and the optic lobes of the octopus – typically have many self-exciting or reverberating circuits. In these, neurons are arranged in a loop so that impulses conducted along the axon eventually reach the dendrite or cell body. One earlier theory suggested that engrams were represented by continuous activity within such reverberating circuits. These loops may well be involved in the consolidation process, but it is unlikely that they could be the basis of long-term storage. Nerve cells are efficient transmitters over brief periods, but the accurate repetition of a very specific pattern of impulses could hardly continue over many minutes, much less days or years. Clear evidence to the contrary was provided by experiments in which rats were cooled down to 0° for periods of one hour. At this temperature all electrical activity in the brain ceased, yet when the animals were rewarmed they retained their memory of events prior to the cooling as well as normal, noncooled rats.

Another theory suggests that memory is not the continuous activity of the special circuit, but that the electrical activity associated with a learning trial "wears a path" which will facilitate transmission along that specific path the next time the events occur.

Several features of the engram suggest that it is some sort of permanent growth process, which in turn suggests that protein synthesis may be involved in some way in the conversion of short-term memory into the long-term engram.

Experiments with both goldfish and mice show that **puromycin,** which inhibits the synthesis of proteins (p. 194) interferes with the consolidation of memory but not with short-term learning. Established long-term memories can be normal in puromycin-treated animals; hence, it is unlikely that the drug affects the mechanism by which memories are recalled. These experiments suggest that short-term memory, memory immediately following training, is stored by neural activity, but this is shortly followed by a storage phase involving protein synthesis. This storage phase, which may last for several days, converts the short-term memory into the stable, long-term memory. This conclusion can be only tentative, for puromycin is not specific and may affect processes that are necessary for the formation of memories but only indirectly connected with it.

The deciphering of the genetic code led some investigators to the theory that memory might have a basis in the synthesis of RNA – that engrams might consist of specific sequences of nucleotides in RNA molecules. There have been reports of the transfer of specific conditioned reflexes in planaria by the administration of RNA extracts. Studies by Hyden and others suggest that there are changes in the structure of RNA in specific neurons that are involved in specific kinds of learning – for example, in the neurons of the vestibular nuclei of rats that have undergone training in a task involving balance.

The homing in of salmon on the specific tributary in which they were born depends primarily on olfactory cues; the salmon smells some distinctive aroma of his home river, remembers it, and subsequently homes in on it. Salmon have been prepared in the laboratory with electrodes in their olfactory lobes to record neuronal activity, and with a catheter in the nasal orifice (Fig. 35.22). When water from the salmon's home river is perfused through the fish's nose, a clear burst of nerve impulses is detected in the olfactory lobe. When water from other rivers is perfused through the nose, little or no electrical activity is de-

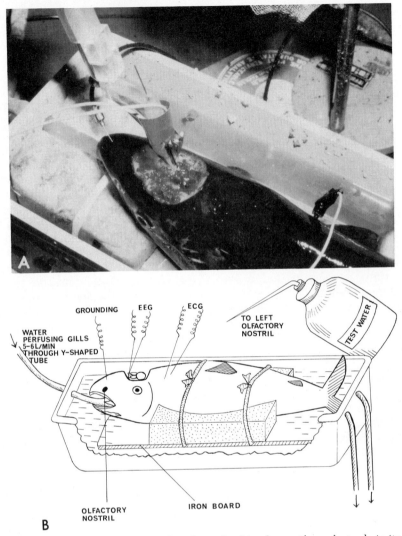

Figure 35.22 Photograph (A) and diagram (B) of a salmon fixed in place with an electrode in its olfactory lobe so that a recording of its action potentials can be made as water from different rivers is placed in its nostril. (Photograph and diagram courtesy of Dr. Kiyoshi Oshima, Kyoto University.)

tected in the electrodes buried in the olfactory lobe. The experimenters prepared an RNA extract of the olfactory lobe of a salmon that homed in on river A and injected that into the olfactory lobe of a salmon that homed in on river B. Within 48 hours the olfactory lobe of this second salmon showed clear bursts of electrical activity, recorded by the indwelling electrode, when water from river A was perfused into the salmon's nostril! This suggests that some sort of memory was transferred from one salmon to another via the RNA extract. There have been similar reports of the transfer of training from one animal to another by the injection of RNA extracts, but negative reports in other instances. There still is no conceptual framework as to how memory might be coded in RNA, in protein, or in any other kind of molecule, but the subject is being attacked experimentally with considerable vigor.

ANNOTATED REFERENCES

The literature relating to the field of animal behavior is exceptionally rich and the titles selected here are only a few of the many excellent sources available.

Bastock, M.: Courtship: A Zoological Study. London, Heinemann Educational, 1967. An interesting book describing courtship and reproductive behavior in a variety of animal types.

Beach, Frank: Hormones and Behavior. New York, Paul B. Hoeber, 1947. A classic study of reproductive behavior.

Carthy, J. D.: The Behavior of Arthropods. New York, Academic Press, 1968. A fascinating summary of studies of mechanisms controlling insect behavior.

Dethier, V. G., and E. Stellar: Animal Behavior. 3rd ed. Englewood Cliffs, N.J., Prentice-Hall, Inc., 1970. A concise survey of the subject.

Deutsch, J. A.: The Structural Basis of Behavior. Cambridge, Cambridge University Press, 1960. An excellent text of neurobiology.

Etkin, William: Social Behavior and Organization Among Vertebrates. Chicago, University of Chicago Press, 1964. A description of social behavior from the ecological and evolutionary points of view.

Gaito, J.: Macromolecules and Behavior. Amsterdam, North-Holland Publishing Co., 1966. Interesting reading about the chemical basis of behavioral phenomena.

Hinde, R. A.: Animal Behavior. 2nd ed. New York, McGraw-Hill Book Co., 1970. A good general text combining ethologic and psychologic viewpoints.

Hirsch, J. (Ed.): Behavior-Genetic Analysis. New York, McGraw-Hill Book Co., 1967. A multi-author account of the inheritance of behavior patterns.

Lorenz, K.: On Aggression. New York, Harcourt, Brace & World, Inc., 1966. A classic volume discussing aggressive behavior.

Manning, A.: An Introduction to Animal Behavior. Reading, Mass., Addison-Wesley, 1967. A shorter introduction to the subject.

Marler, P., and W. J. Hamilton: Mechanisms of Animal Behavior. New York, John Wiley & Sons, Inc., 1966. An excellent general text with detailed accounts of animal behavior.

McGaugh, J. L., N. M. Weinberger, and R. E. Whalen: Psychobiology. San Francisco, W. H. Freeman & Co., 1967. A text by many authors that discusses a wide variety of investigations in animal behavior.

McGill, T. E.: Readings in Animal Behavior. New York, Holt, Rinehart and Winston, 1965. A multi-author text that provides samples of current thought and research on animal behavior.

Morgan, C. T.: Physiological Psychology. 3rd ed. New York, McGraw-Hill Book Co., 1960. An excellent text presenting the physiologic approach to an analysis of behavior.

Schrier, A. M., H. F. Harlow, and F. Stollnitz (Eds.): The Behavior of Nonhuman Primates. New York, Academic Press, 1965. A well-written account of the experimental analysis of primate behavior.

Slucken, W.: Imprinting and Early Learning. London, Methuen & Co., 1964. An account of experiments and observations on early learning.

Thorpe, W. H.: Learning and Instinct in Animals. Harvard University Press, Cambridge, 1956. A classic exposition of theories of learning.

Tinbergen, N.: Social Behavior in Animals. New York, John Wiley & Sons, 1953. A classic book on ethology.

Wells, M. J.: Brain and Behavior in Cephalopods. London, Heinemann Educational, 1962. An interesting description of the experimental analysis of behavior in the octopus.

36. Ecology

As more is learned about any kind of plant or animal, it becomes increasingly clear that each species has undergone adaptations to survive under a particular set of environmental circumstances. Each may show adaptations to wind, sun, moisture, temperature, salinity and other aspects of the physical environment as well as adaptations to specific plants and animals that live in the same region. Studies of the interrelationships of organisms with their physical and biotic environments are termed "ecology." This word is very much in the public consciousness today as we become aware of some of our past and current ecologic malpractices. It is important for everyone to know and appreciate the principles of this aspect of biology so that he can form an intelligent opinion regarding topics such as insecticides, detergents, mercury pollution, sewage disposal, power dams, and their effects on mankind, on human civilization, and on the world we live in.

The Greek *oikos* means "house" or "place to live," and ecology (*oikos logos*) is literally the study of organisms "at home," in their native environment. The term was proposed by the German biologist Ernst Haeckel in 1869, but many of the concepts of ecology antedated the term by a century or more. Ecology is concerned with the biology of groups of organisms and their relations to the environment. The term **autecology** refers to studies of individual organisms, or of populations of single species, and their relations to their environment. The contrasting term, **synecology**, refers to studies of groups of organisms associated to form a functional unit of the environment. Groups of organisms may be associated in three different levels of organization—populations, communities and ecosystems. In ecologic usage, a **population** is a group of individuals of any one kind of organism, a group of individuals of a single species. A community in the ecologic sense, a **biotic community**, includes all of the populations occupying a given defined physical area. The community together with the physical, nonliving environment comprise an **ecosystem**. Thus synecology is concerned with the many relationships within communities and ecosystems. The ecologist deals with such questions as who lives in the shade of whom, who eats whom, who plays a role in the propagation and dispersion of whom, and how energy flows from one individual to the next in a food chain. The ecologist attempts to define and analyze those characteristics of populations which are distinct from the characteristics of individuals and the factors responsible for the aggregation of populations into communities.

36.1 THE CONCEPTS OF RANGES AND LIMITS

Probably no species of plant or animal is found everywhere in the world; some parts of the earth are too hot, too cold, too wet, too dry or too something else for the organism to survive there. The environment may not kill the adult directly, but effectively keeps the species from becoming established if it prevents its reproducing or kills off the egg, embryo or some other stage in the life cycle.

Most species of organisms are not even found in all the regions of the world where they could survive. The existence of barriers prevents their further dispersal and enables us to distinguish the major biogeographic realms (p. 251) characterized by certain assemblages of plants and animals.

Biologists early in the nineteenth century were aware that each species requires certain materials for growth and reproduction, and can be restricted if the environment does not provide a certain minimal amount of each one of these materials. Justus Liebig stated in 1840 what is now known as his **"Law of the Minimum,"** that the rate of growth of each organism is limited by whatever essential nutrient is present in a minimal amount. Liebig, who studied the factors affecting the growth of plants, found that the yield of crops was often limited not by a nutrient required in large amounts, such as water or carbon dioxide, but by something needed only in trace amounts, such as boron or manganese. Liebig's law is strictly applicable only under steady-state conditions, when the inflow of energy and materials equals the outflow. In addition, there may be interactions between factors such that a very high concentration of one nutrient may alter the rate of utilization of another (the rate-limiting one) and hence alter the effective minimal amount required. Certain plants, for example, require less zinc when growing in the shade than when growing in the sunlight.

V. E. Shelford pointed out in 1913 that *too much* of a certain factor would act as a limiting factor just as well as too little of it, and that the distribution of each species is determined by its **range of tolerance** to variations in each environmental factor. Much work has been done to define the limits of tolerance, the limits within which species can exist, and this concept, sometimes called Shelford's **"Law of Tolerance,"** has been helpful in understanding the distribution of organisms.

It has usually been found that certain stages in the life cycle are critical in limiting organisms—seedlings and larvae are usually more sensitive than adult plants and animals. Adult blue crabs, for example, can survive in water with a low salt content and can migrate for some distance up river from the sea, but their larvae cannot, and the species cannot become permanently established there.

Some organisms have very narrow ranges of tolerance to environmental factors; others can survive within much broader limits. Any given organism may have narrow limits of tolerance for one factor and wide limits for another. Ecologists use the prefixes **steno-** and **eury-** to refer to organisms with narrow and wide, respectively, ranges of tolerance to a given factor. A stenothermic organism is one which will tolerate only narrow variations in temperature. The housefly is a eurythermic organism, for it can tolerate temperatures ranging from 5 to 45° C. The adaptation to cold of the antarctic fish *Trematomus bernacchi* is remarkable. It is extremely stenothermic and will tolerate temperatures only between −2° C. and +2° C. At 1.9° C. this fish is immobile from heat prostration!

Temperature is an important limiting factor, as the relative sparseness of life in the desert and arctic demonstrates. Most of the animals that do live in the desert have adapted to the rigors of the environment by living in burrows during the day and coming out to forage only at night. Many animals escape the bitter cold of the northern winter not by migrating south but by burrowing beneath the snow. Measurements made in Alaska show that when the surface temperature is −55° C., the temperature 60 cm. under the snow, at the surface of the soil, is a "warm" −7° C.

Although the ringnecked pheasant has been introduced into the southern states a number of times and the adults survive well, the developing eggs are apparently killed by the high daily temperatures and are unable to complete development.

Light. The amount of light is an important factor in determining the distribution and behavior of both plants and animals. Light is, of course, the ultimate source of energy for life on this planet, yet prolonged direct exposure of cells to light of high intensity may be fatal. Both plants and animals have evolved mechanisms and responses to protect them against too much (or too little) light.

The amount of daylight per day, known as

the **photoperiod,** has a marked influence on the time of flowering of plants, the time of migration of birds, the time of spawning of fish, and the seasonal change of color of certain birds and mammals. The effects of the photoperiod on vertebrates appear to occur via some neurohormonal mechanism involving the hypothalamus, the pituitary and pineal glands. Knowledge of photoperiod phenomena has proved to be of considerable economic importance. Chicken farmers have found that artificial illumination in the hen house, by extending the photoperiod, stimulates the hens to lay more eggs.

Water. Water is a physiologic necessity for all living things but is a limiting factor primarily for land organisms. The total amount of rainfall, its seasonal distribution, the humidity, and the ground supply of water are some of the factors limiting distribution of animals and plants. Some lakes and streams, especially in the western and southwestern United States, periodically become dry or almost dry, and the fish and other aquatic animals are killed. During periods of low water, the water temperature may rise sufficiently to kill off the aquatic forms. Many of the protozoa survive the drying of the puddles in which they normally live by forming thick-walled cysts. Some animals have adapted to desert conditions by digging and living in burrows where the temperature is lower and the humidity is higher than at the surface. Measurements have shown that the temperature in the burrow of a kangaroo rat 60 cm. underground may be only 16° C. when the surface temperature is over 38° C.

An excess of water is fatal to certain animals; earthworms, for example, may be driven from their burrows by heavy rainfall because oxygen is only sparingly soluble in water and they are unable to get enough oxygen when immersed. Knowledge of the limits of water tolerance is helpful in attacking insect and other pests. Wire worms have rather narrow limits of tolerance to water and are most sensitive as larvae and pupae. They can be killed by flooding the infested fields or by planting alfalfa or wheat to dry out the soil below the limit of tolerance of the wire worm larvae.

Other Factors. The supply of oxygen and carbon dioxide is usually not limiting for land organisms except for animals living deep in the soil, on the tops of mountains, or within the bodies of other animals. Animals living in aquatic environments may be limited by the amount of dissolved oxygen present; the oxygen tension in stagnant ponds or in streams fouled by industrial wastes may become so low as to be incompatible with many forms of life. Some parasites have adapted to the low oxygen tension within the host's body by evolving special metabolic pathways by which energy can be released from foodstuffs without the utilization of free oxygen.

The trace elements necessary for plant and animal life are limiting factors in certain parts of the world. The soil in certain parts of Australia, for example, is deficient in copper and cobalt and is unsuitable for raising cattle or sheep. Other trace elements which may be a limiting factor are manganese, zinc, iron, sulfur, selenium and boron.

The amount of carbon dioxide in the air is remarkably constant, but the amount dissolved in water varies widely. An excess of carbon dioxide may be a limiting factor for fish and insect larvae. The hydrogen ion concentration, pH, of water is related physicochemically to the carbon dioxide concentration, and it, too, may be an important limiting factor in aquatic environments.

Water currents are limiting for a number of kinds of animals and plants; the fauna and flora of a still pond and of a rapidly flowing stream are quite different. Winds may have a comparable limiting effect upon land organisms.

The type of soil, the amount of topsoil, its pH, porosity, slope, water-retaining properties, and so on, are limiting factors for a variety of plants, and hence indirectly for animals. The ability of many animals to survive in a given region depends upon the presence of certain plants to provide shelter and cover, as well as food. Grasses, shrubs and trees on land each provide shelter for certain kinds of animals, and seaweeds and fresh-water aquatic plants have a similar role for aquatic animals. Some animals require special shelter for breeding places and the care of the young. In many different kinds of birds, mammals, crustaceans and other animals, each animal or pair establishes a **territory,** a region which supplies food and shelter for it and its offspring, and which it defends vigorously against invasion by other members of the same species (p. 798).

In summary, whether an animal can become established in a given region is the result of a complex interplay of such physical factors as temperature, light, water, winds and salts, and biotic factors such as the plants and other animals in that region which serve as food, compete for food or space, or act as predators or disease organisms.

36.2 STRUCTURAL ADAPTATIONS

In the course of evolution organisms have undergone successive structural adaptations and readaptations as the environment changed or as they migrated to a new environment. As a result, many organisms today have structures or physiologic mechanisms that are useless or even deleterious, but which were advantageous at an earlier time when the organism was adapted to a different environment.

The adaptations of the mouthparts of certain animals to the kind of food eaten are among the most striking that can be cited. The mouthparts of certain insects are adapted for sucking nectar from certain species of plants; other are specialized for sucking blood, for biting or for chewing vegetation. The bills of various kinds of birds (Fig. 36.1) and the teeth of various kinds of mammals may be highly adapted for particular kinds of food.

In many animals, the specialized adaptation to a certain way of life is simply the latest stage in a series of adaptations. For example, both man and the baboon, whose immediate ancestors were tree-dwellers, have returned to the ground and have become readapted to walking rather than to climbing trees.

Readaptation may be a very complicated process. The present-day Australian tree-climbing kangaroos are the descendants of an original ground-dwelling marsupial. From the ground-dwellers evolved forms which, in adaptive radiation, took to the trees and developed limbs adapted to tree climbing (or perhaps the sequence of events was the reverse—first the evolution of specialized limbs, then the adoption of a tree-dwelling mode of life).

Some of these tree-dwellers eventually left the trees and became readapted to ground life, and the hind legs became lengthened, strengthened and adapted for leaping. But finally some of these kangaroos went back to the trees, although now their legs were so highly specialized for leaping that they could not be used for grasping a tree trunk.

Consequently the present-day tree kangaroos must climb like bears, by bracing their feet against the tree trunk. A comparison of the feet of existing Australian marsupials reveals all the stages in this complicated, shifting process of adaptation.

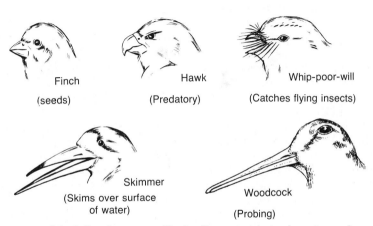

Finch
(seeds)

Hawk
(Predatory)

Whip-poor-will
(Catches flying insects)

Skimmer
(Skims over surface of water)

Woodcock
(Probing)

Figure 36.1 Diagrams of the bills of a variety of birds, illustrating their adaptation to the type of food eaten.

36.3 PHYSIOLOGIC ADAPTATIONS

One of the major struggles among organisms stems from the competition for food, hence a mutation enabling an animal to use a new type of food is extremely advantageous. This may be accomplished in a number of ways—by the evolution of a new digestive or energy-liberating enzyme system, for example. A mutation resulting in a new energy-liberating enzyme enabled the sulfur bacteria to obtain energy from hydrogen sulfide, a substance which is poisonous to almost all other organisms. The evolution of a special enzyme for reducing disulfide bridges (—S—S—) gave the clothes moth its unique ability to digest wool. The protein molecules of wool are cross-linked with many disulfide bridges.

Another type of favorable mutation is one which decreases the growing season of a plant or the total length of time required for an insect to develop. Such mutations enable the organism to survive farther from the equator and open up new areas of living space and new sources of food.

Any mutation that increases a species' limits of temperature tolerance—i.e., makes it more eurythermic—may enable it to live in a new region, at a higher latitude or a higher altitude. Other organisms have solved the problem of living in arctic regions by becoming dormant during the cold season or by developing behavior patterns of migration. Many birds, but only a few mammals, migrate south to avoid the cold northern winter.

A number of kinds of mammals—monotremes, shrews, rodents and bats—hibernate over the winter. It is doubtful whether carnivores such as bears and skunks actually hibernate; they simply sleep for long periods at a time.

In true **hibernation** the body temperature falls to just a degree or two higher than the surrounding air temperature, metabolism decreases greatly, and the heart rate and rate of respiration become very slow. No food is eaten and the animal uses up his stores of body fat, awakening in the spring in an emaciated condition (and probably "as hungry as a bear"). What induces hibernation and what wakes the animal in the spring are not completely clear; changes in the environmental temperature are, of course, important, but changes within the animal body are also involved.

Birds and mammals are unique in possessing mechanisms for controlling body temperature, keeping it constant despite wide fluctuations in the environmental temperature. These thermostated animals are said to be **homoiothermic** ("warm-blooded" is not quite correct; they are really "constant-temperature-blooded"). Fish, amphibians, reptiles and all the invertebrates are **poikilothermic**—their body temperature fluctuates with that of the environment (again, "cold-blooded" is not exactly descriptive; their body temperature is simply determined by the temperature of the environment).

Marine fish are usually adapted to survive within a certain range of pressures and hence at a particular depth. Surface animals are crushed by the terrific pressures of the deep and deep-sea animals usually burst when brought to the surface. The whale, however, is able to withstand great changes in pressure, and can dive to depths of 800 meters without injury. Presumably its lung alveoli collapse when the pressure on the body reaches a certain point and gases are no longer absorbed into the blood.

A man can survive pressures as high as six atmospheres if the pressure is increased and then decreased slowly. The increase in pressure increases the amount of gases dissolved in the blood and body fluids. If the pressure is decreased suddenly, these gases come out of solution and form bubbles throughout the body. Those in the blood impede circulation and bring about the symptoms of diver's disease, or "the bends." The pilot of a jet plane may climb so rapidly that the atmospheric pressure is reduced fast enough to bring bubbles of gas out of solution in his blood and cause a type of the bends.

36.4 COLOR ADAPTATIONS

Adaptations for survival are evident in the color and pattern of plants and animals as well as in their structure and physiologic processes. We can distinguish three types of color adaptation: concealing or **protective coloration**, which enables the organism to blend with its background and be less visible to predators; **warning coloration,**

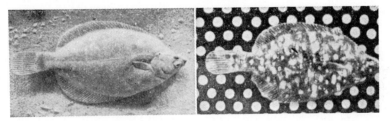

Figure 36.2 An experiment to show the remarkable ability of the flounder to change its color and pattern to conform with its background. Left, a flounder on a uniform, light background; right, the same fish after being placed on a spotted, darker background. (Villee; Biology, 6th ed.)

which consists of bright, conspicuous colors and is assumed by poisonous or unpalatable animals to warn potential predators not to eat them; and **mimicry,** in which the organism resembles some other living or nonliving object—a twig, leaf, stone, or perhaps some other animal which, being poisonous, has warning coloration.

Concealing coloration may serve to hide an animal that wants to escape the notice of potential predators or it may be assumed by a predator in order to be unnoticed by his potential prey. Examples of concealing coloration are legion—the white coats of arctic mammals, and the stripes and spots of tigers, leopards, zebras and giraffes which, though conspicuous in a zoo, blend imperceptibly with the moving pattern of light and dark typical of their native savanna. Some animals—frogs, flounders, chameleons, crabs and others—have a remarkable ability to change their color and pattern as they move from a dark to a light background or from one that is uniform to one that is mottled (Fig. 36.2).

To demonstrate experimentally that concealing coloration does have survival value, F. B. Isely fastened grasshoppers with different body colors to plots of different colored soils—light, dark, grassy, sandy and so on. After these plots had been exposed to the predatory activities of wild birds or chickens for a given length of time, the survivors were tabulated. It was found that there was a significantly higher percentage of survivors among those grasshoppers which matched their background.

H. B. D. Kettlewell of Oxford University made careful field observations of the frequency of capture of light-colored and dark-colored mutants of the moth *Biston betularia* by birds. In the woods of Dorsetshire, where there is little pollution and the tree trunks are light-colored, the light-colored moths

were significantly less likely to be caught and eaten by birds. In woods in the industrial Midlands, where leaves and tree trunks are blackened with soot, the dark-colored moths were significantly less likely to be caught and eaten.

When an animal is equipped with poison fangs, a stinging mechanism or some chemical which gives it a noxious taste, it is probably to its advantage to have the fact widely advertised, and, in fact, many animals so equipped do have warning colors.

A species of European toad with a bright scarlet belly has certain chemicals secreted by its skin glands which make it extremely unpalatable. Whenever a potential predator, such as a stork, swoops over a congregation of toads, they flop on their backs, exposing their scarlet bellies as a warning. The storks and other birds apparently become conditioned by the association of the red color and the bad taste, and avoid the toads assiduously.

Other animals survive by mimicking one of these protectively colored animals; for instance, some harmless, defenseless and palatable animals are identical in shape and color with a poisonous or noxious animal of quite a different family or order, and, being mistaken for it by predators, are left alone. Many tropical insects have evolved this type of protection termed **Batesian mimicry.** If the animal is noxious enough, there will be significant protection afforded to its mimics even if the mimics outnumber their noxious models. The evolution of markedly similar shapes and color patterns by two species, both of which are noxious, is termed **Müllerian mimicry.** It is believed that the similar appearance of the two species increases the probability that predators will learn to avoid that particular shape and color pattern.

The reality of the selective advantage of

color adaptations has been debated. It has been argued that animal vision may be so different from human vision (certain animals may be colorblind, or may be able to see ultraviolet or infrared light) that an animal that appears to be protectively colored to a man may be quite evident to its natural predators. However, many experimental studies, such as the experiments with grasshoppers and moths, have shown that protective coloration does have survival value.

Color and patterns may serve to attract other organisms when that is necessary for survival. The red and blue ischial callosities of monkeys and the gay, extravagant plumage of various birds apparently have an attraction for members of the opposite sex that plays an important role as a prelude to mating. And vividly colored flowers seem to attract the birds and insects whose activities are needed to insure the pollination of the plant or the dispersal of its seed.

36.5 ADAPTATIONS OF SPECIES TO SPECIES

The evolution and adaptation of each species have not occurred in a biologic vacuum, independent of other forms. On the contrary, the adaptation of each species has been influenced markedly by the concurrent adaptations of other species. As a result of this, many types of cross-dependency between species have arisen. Some of the clearest and best understood of these involve insects.

Insects are necessary for the pollination of a great many plants; some plants are so dependent on certain insects that they are unable to survive in a given region unless those particular insects are present. For example, the Smyrna fig could not be grown in California, even though all climatic conditions were favorable, until the fig insect which pollinates the plant was introduced.

Birds, bats, and even snails serve as pollen transporters for some plants, but insects are the prime pollinators. Flowering plants have evolved bright colors and fragrances, presumably to attract insects and birds and ensure pollination. There had been some doubt as to whether insects can detect different odors and colors. The experiments of Karl von Frisch show that

honeybees, at least, are able to differentiate colors and scents and that they are guided in their visits to flowers by these stimuli.

Some of the species-to-species adaptations are so exact that neither form can exist in a region without the other. The yucca plant and the yucca moth, like the fig and fig insect, have evolved to a point of complete interdependence (Fig. 36.3A). The yucca moth, by a series of unlearned acts, goes to a yucca flower, collects some pollen, and takes it to a second flower. There it pushes its ovipositor (egg-laying organ) through the wall of the ovary of the flower and lays an egg. It then carefully places some pollen on the stigma. The yucca plant in this way is sure to be fertilized and produce seeds; the larva of the yucca moth feeds on these yucca seeds. The yucca produces a large number of seeds and is not injured by the loss of the few eaten by the moth larva.

36.6 HABITAT AND ECOLOGIC NICHE

Two basic concepts useful in describing the ecologic relations of organisms are the habitat and the ecologic niche. The **habitat** of an organism is the place where it lives, a physical area, some specific part of the earth's surface, air, soil or water. It may be as large as the ocean or a prairie or as small as the underside of a rotten log or the intestine of a termite, but it is always a tangible, physically demarcated region. More than one animal or plant may live in a particular habitat.

The **ecologic niche** is a more inclusive term that includes not only the physical space occupied by an organism, but its functional role as a member of the community— that is, its trophic position and its position in the gradients of temperature, moisture, pH and other conditions of the environment. The ecologic niche of an organism depends not only on where it lives, but also on what it does—that is, how it transforms energy, how it behaves in response to and modifies its physical and biotic environment, and how it is acted upon by other species. A common analogy is that the habitat is the organism's "address" and the ecologic niche is the organism's "profession," biologically speaking. To define completely

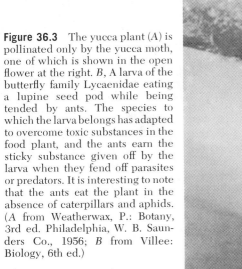

Figure 36.3 The yucca plant (*A*) is pollinated only by the yucca moth, one of which is shown in the open flower at the right. *B*, A larva of the butterfly family Lycaenidae eating a lupine seed pod while being tended by ants. The species to which the larva belongs has adapted to overcome toxic substances in the food plant, and the ants earn the sticky substance given off by the larva when they fend off parasites or predators. It is interesting to note that the ants eat the plant in the absence of caterpillars and aphids. (*A* from Weatherwax, P.: Botany, 3rd ed. Philadelphia, W. B. Saunders Co., 1956; *B* from Villee: Biology, 6th ed.)

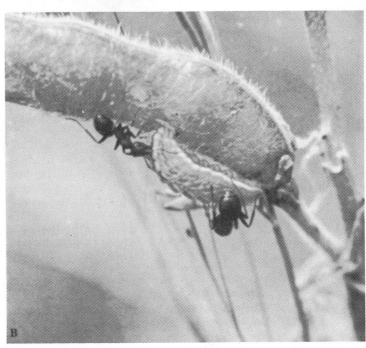

the ecologic niche of any species would require detailed knowledge of a large number of biological characteristics and physical properties of the organism and its environment. Since this is very difficult to obtain, the concept of ecologic niche is used most often to describe differences between species with regard to one or a few of their major features.

Charles Elton, in 1927, was one of the first to use the term in the sense of the functional status of an organism in its community. The term "ecologic niche" has somewhat different meanings to different ecologists. To some it is the ultimate distributional unit within which each species is held by its structural and behavioral limitations. To others it is the functional status of an organism in its community, and these ecologists emphasize the energy relations of the spe-

Figure 36.4 *Notonecta*, the "back-swimmer" (*left*), and *Corixa*, the "water boatman" (*right*), are two aquatic bugs occupying the same habitat—the shallow, vegetation-choked edges of ponds and lakes—but having different ecologic niches.

cies. To others, such as G. E. Hutchinson, the niche is a multidimensional space within which the environment permits an individual or species to survive indefinitely. The ecologic niche is an abstraction that includes all of the physical, chemical, physiologic and biotic factors that an organism needs in order to survive. To describe an animal's ecologic niche we must know what it eats and what eats it, what its activities and range of movements are, and what effects it has on other organisms and on the nonliving parts of the surroundings.

In the shallow waters at the edge of a lake you might find many different kinds of water bugs. They all live in the same place and hence have the same habitat. Some of these water bugs, such as the "back-swimmer" *Notonecta*, are predators, catching and eating other animals of about their size (Fig. 36.4). Other water bugs, such as *Corixa*, feed on dead and decaying organisms. Each has quite a different role in the biologic economy of the lake and each occupies an entirely different ecologic niche.

A single species may occupy somewhat different niches in different regions, depending on such things as the available food supply and the number and kinds of competitors. Animals with distinctly different stages in their life history may occupy different niches in succession. The frog tadpole is a primary consumer, feeding on plants, but an adult frog is a secondary consumer, feeding on insects and other animals. In contrast, young river turtles are secondary consumers, eating snails, worms and insects, whereas the adult turtles are primary consumers and eat green plants such as tape grass.

Two species of organisms that occupy the same or similar ecologic niches in different geographical locations are termed **ecological equivalents.** The array of species present in a given type of community in different biogeographic regions may differ widely. However, similar ecosystems tend to develop wherever there are similar physical habitats; the equivalent functional niches are occupied by whatever biological groups hap-

TABLE 36.1 ECOLOGICAL EQUIVALENTS: SPECIES OCCUPYING COMPARABLE ECOLOGIC NICHES ON NORTH AMERICAN COASTS *

	Grazers on Intertidal Rocks	Fish Feeding on Plankton	Bottom-Dwelling Carnivores
Northeast coast	*Littorina littorea* (periwinkle)	Alewife, Atlantic herring	*Homarus* (lobster)
Gulf coast	*Littorina irrorata*	Menhaden, threadfin	*Menippe* (Stone crab)
Northwest coast	*Littorina danaxis, Littorina scutelata*	Sardine, Pacific herring	*Paralithodes* (King crab)
Tropical coast	*Littorina ziczac*	Anchovy	*Panulirus* (spiny lobster)

* Adapted from Odum, E. P.: Fundamentals of Ecology, 3rd ed. Philadelphia, W. B. Saunders Co., 1971.

pen to be present in the region. Thus, a savanna biome tends to develop wherever the climate permits the development of extensive grasslands, but the species of grass and the species of animals eating the grass may be quite different in different parts of the world. On each of the four continents there are grasslands with large grazing herbivores present. These herbivores are all ecological equivalents. However, in North America the grazing herbivores were bison and prong-horn antelope; in Eurasia, the saga antelope and wild horses; in Africa, other species of antelope and zebra; and in Australia, the large kangaroos. In all four regions these native herbivores have been replaced to a greater or lesser extent by man's domesticated sheep and cattle. As examples of ecological equivalents, the species occupying three marine ecologic niches in four different regions of the coast are listed in Table 36.1. The same kinds of ecologic niches are usually present in similar habitats in different parts of the world. Comparisons of such habitats and analyses of the similarities and differences in the species that are ecological equivalents in these different habitats have been helpful in clarifying the interrelations of these different ecologic niches in any given habitat.

36.7 THE CYCLIC USE OF MATTER

The total mass of all the organisms that have lived on the earth in the past 1.5 billion years is much greater than the mass of carbon and nitrogen atoms present. According to the Law of Conservation of Matter, matter is neither created nor destroyed; obviously the carbon and nitrogen atoms must have been used over and over again in the course of time. The earth neither receives any great amount of matter from other parts of the universe nor does it lose significant amounts of matter to outer space. The atoms of each element, carbon, hydrogen, oxygen, nitrogen and the rest, are taken from the environment, made a part of some cellular component and finally, perhaps by a quite circuitous route involving several other organisms, are returned to the environment to be used over again. An appreciation of the roles of green plants, animals and bacteria in this cyclic use of the elements can be gained from considering the details of the more important cycles.

The Carbon Cycle. There are about six tons of carbon (as carbon dioxide) in the atmosphere over each acre of the earth's surface. Yet each year an acre of plants, such as sugar cane, will extract as much as twenty tons of carbon from the atmosphere and incorporate it into the plant bodies. If there were no way to renew the supply, the green plants would eventually, perhaps in a few centuries, use up the entire atmospheric supply of carbon. Carbon dioxide fixation by bacteria and animals is another, but quantitatively minor, drain on the supply of carbon dioxide. Carbon dioxide is returned to the air by the decarboxylations that occur in cellular respiration. Plants carry on respiration continuously. Plant tissues are eaten by animals which, by respiration, return more carbon dioxide to the air. But respiration alone would be unable to return enough carbon dioxide to the air to balance that withdrawn by photosynthesis. Vast amounts of carbon would accumulate in the compounds making up the dead bodies of plants and animals. The carbon cycle is balanced by the decay bacteria and fungi which cleave the carbon compounds of dead plants and animals and convert the carbon to carbon dioxide (Fig. 36.5).

Investigators at the University of Notre Dame and elsewhere have been able to raise some bacteria-free animals in special incubators. This has been done to study problems such as the synthesis of vitamins by the bacteria normally present in the intestines. In time, these animals become old and die, but their bodies do not decompose. Their constituent atoms have been withdrawn (temporarily) from the carbon cycle.

When the bodies of plants are compressed under water for long periods of time they are not decayed by bacteria but undergo a series of chemical changes to form **peat**, later brown coal or **lignite**, and finally **coal.** The bodies of certain marine plants and animals may undergo somewhat similar changes to form **petroleum.** These processes remove some carbon from the cycle but eventually geologic changes or man's mining and drilling bring the coal and oil to the surface to be burned to carbon dioxide and restored to the cycle.

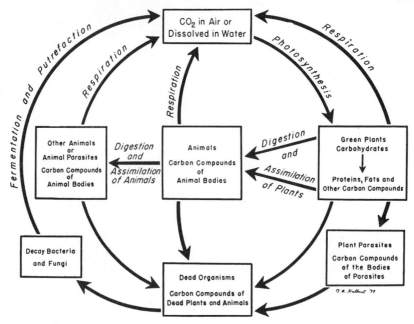

Figure 36.5   The carbon cycle in nature. See text for discussion.

Most of the earth's carbon atoms are present in limestone and marble as carbonates. The rocks are very gradually worn down and the carbonates in time are added to the carbon cycle. But other rocks are forming at the bottom of the sea from the sediments of dead animals and plants, so that the amount of carbon in the carbon cycle remains nearly constant.

The Nitrogen Cycle.   The nitrogen for the synthesis of amino acids and proteins is taken up from the soil and water by plants as nitrates (Fig. 36.6). The nitrates are converted to amino groups and used by the plant cells in the synthesis of amino acids and proteins. Animals may eat the plants and utilize the amino acids from the plant proteins in the synthesis of their own proteins and other cellular constituents. When animals and plants die, the decay bacteria convert the nitrogen of their proteins and other compounds into ammonia. Animals excrete several kinds of nitrogen-containing wastes—urea, uric acid, creatinine and ammonia—and the decay bacteria convert these wastes to ammonia. Most of the ammonia is converted by nitrite bacteria to nitrites and this in turn is converted by nitrate bacteria to nitrates, thus completing the cycle. Denitrifying bacteria convert some of the ammonia to atmospheric nitro-

gen. Atmospheric nitrogen can be "fixed," converted to organic nitrogen compounds such as amino acids, by some blue-green algae (*Nostoc* and *Anabaena*) and by the soil bacteria *Azotobacter* and *Clostridium*.

Other bacteria of the genus *Rhizobium*, although unable to fix atmospheric nitrogen themselves, can do this when in combination with cells from the roots of legumes such as peas and beans. The bacteria invade the roots and stimulate the formation of root nodules, a sort of harmless tumor. The combination of legume cell and bacteria is able to fix atmospheric nitrogen (something neither one can do alone) and for this reason legumes are often planted to restore soil fertility by increasing the content of fixed nitrogen. Nodule bacteria may fix as much as 50 to 100 kilograms of nitrogen per acre per year, and free soil bacteria as much as 12 kilograms per acre per year.

Atmospheric nitrogen can also be fixed by electrical energy, supplied either by lightning or by the electric company. Although 80 per cent of the atmosphere is nitrogen, no animals and only these few plants can utilize it in this form. When the bodies of the nitrogen-fixing bacteria decay, the amino acids are metabolized to ammonia and this in turn is coverted by the nitrite and nitrate bacteria to nitrates to complete the cycle.

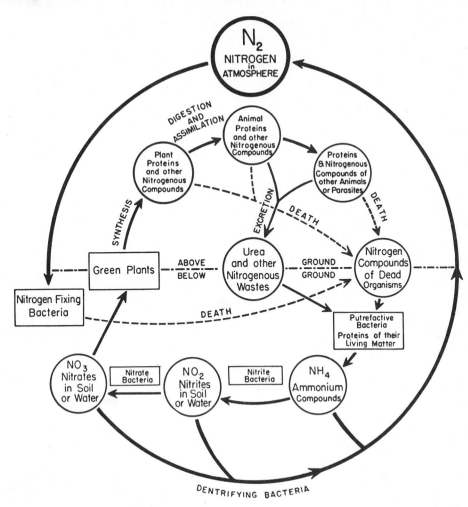

Figure 36.6 The nitrogen cycle in nature. See text for discussion.

The Water Cycle. The great reservoir of water is the ocean. The sun's heat vaporizes water and forms clouds. These, moved by winds, may pass over land, where they are cooled enough to precipitate the water as rain or snow. Some of the precipitated water soaks into the ground, some runs off the surface into streams and goes directly back to the sea. The ground water is returned to the surface by springs, by pumps and by transpiration—the movement of water in plants from roots to leaves. Water inevitably ends up back in the sea, but it may become incorporated into the bodies of several different organisms, one after another, en route. The energy to run the cycle—the heat needed to evaporate water—comes from sunlight.

The Phosphorus Cycle. As water runs over rocks it gradually wears away the surface and carries off a variety of minerals, some in solution and some in suspension. Some of these minerals, such as phosphates, sulfates, calcium, magnesium and others, are necessary for the growth of plants and animals. Phosphorus, an extremely important constituent of all cells, is taken in by plants as inorganic phosphate and converted to a variety of organic phosphates (which are intermediates in the metabolism of carbohydrates, nucleic acids and fats). Animals get their phosphorus as inorganic phosphate in the water they drink or as inorganic plus organic phosphates in the food they eat.

The phosphorus cycle is not completely balanced, for phosphates are being carried into the sediments at the bottom of the sea faster than they are being returned by the actions of marine birds and fish. Sea birds play an important role in returning phosphorus to the cycle by depositing phosphate-rich **guano** on land. Man and other animals, by catching and eating fish, also recover some phosphorus from the sea. In time, geologic upheavals bring some of the sea bottom back to the surface as new mountains are raised, and in this way minerals are recovered from the sea bottom and made available for use once more.

The Energy Cycle.   The cycles of all these types of matter are closed: the atoms are used over and over again. To keep the cycles going does not require new matter but it does require energy, for *the energy cycle is not a closed one.* The Law of the Conservation of Energy, or the First Law of Thermodynamics, states that energy is neither created nor destroyed but only transformed from one kind to another (p. 60). However, the Second Law of Thermodynamics states that whenever energy is transformed from one kind to another, there is an increase in entropy and a decrease in the amount of useful energy. Some energy is degraded into heat and dissipated.

Only a small fraction of the light energy reaching the earth is trapped; considerable areas of the earth have no plants, and plants can utilize in photosynthesis only about 3 per cent of the incident energy. This radiant energy is converted into the potential energy of the chemical bonds of the organic substances made by the plant. When an animal eats a plant (or when bacteria decompose it) and these organic substances are oxidized, the energy liberated is just equal to the amount of energy used in synthesizing the substances (First Law of Thermodynamics), but some of the energy is heat and not useful energy (Second Law of Thermodynamics). If this animal in turn is eaten by another one, a further decrease in useful energy occurs as the second animal oxidizes the organic substances of the first to liberate energy to synthesize its own cellular constituents.

In the successive steps of such a **food chain** (p. 834)—photosynthetic autotroph, herbivorous heterotroph, carnivorous heterotroph, decay bacteria—the number and mass of the organisms in each step is limited by the amount of energy available. Since some energy is lost as heat in each transformation the steps become progressively smaller near the top. This relationship is sometimes called a "food pyramid" or pyramid of numbers, to emphasize that in each successively higher section of the food chain the number (or more precisely the total mass) of the predators decreases.

Eventually, all the energy originally trapped by plants in photosynthesis is converted to heat and dissipated to outer space, and all the carbon of the organic compounds ends up as carbon dioxide. The only important source of energy on earth is sunlight—energy derived from nuclear reactions, largely the conversion of hydrogen to helium, occurring at extremely high temperatures (about $10,000,000°$ C.) in the interior of the sun. When this energy is exhausted and the radiant energy of the sun can no longer support photosynthesis, the carbon cycle will stop, all plants and animals will die and organic carbon will be converted to carbon dioxide. There is no immediate cause for alarm, however; the sun will continue to shine for several billions of years!

36.8   SOLAR RADIATION

Perhaps the most outstanding feature of the earth is the nonuniformity of its physical conditions, which range from Arctic tundra to tropical rain forests. Even the oceans are very patchy, nonuniform places. The earth derives nearly all its energy from the sun, but even the sun's energy is not uniformly distributed over the face of the globe. The solar radiation reaching the surface of the earth varies with the length of the path that the sun's rays take through the atmosphere (whether it is vertical or at an angle), the area of horizontal surface over which is spread a "bundle" of the sun's rays of a given cross-sectional area, the distance of the earth from the sun (which changes seasonally because of the elliptical orbit of the earth around the sun), the amount of water vapor, dust and pollutants in the atmosphere, and the total length of the day (the photoperiod). At higher latitudes the angle of incidence of the sun rays is less than at middle latitudes, and the energy is spread more thinly. The rays must pass through a thicker layer of atmos-

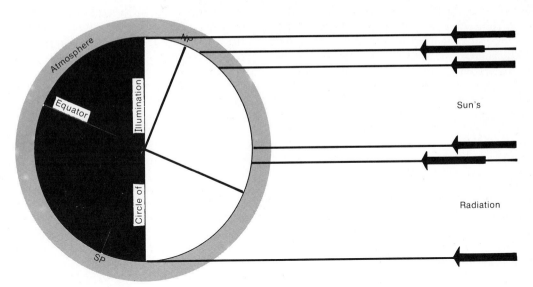

Figure 36.7 Circle of illumination, areas of daylight and darkness, angles of sun's rays at different latitudes and differences in areas affected and thickness of atmosphere penetrated at time of summer solstice. (After Ward, H. B., and Powers, W. E.: Introduction to Weather and Climate. Evanston, Illinois, Northwestern University Press, 1942.)

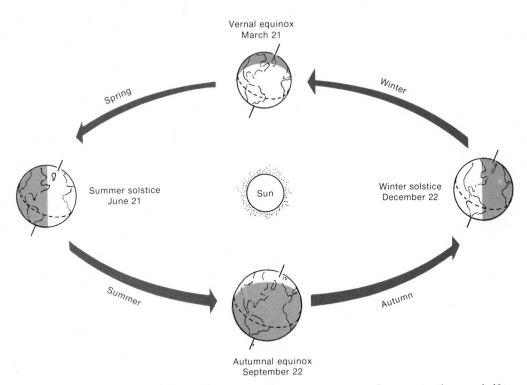

Figure 36.8 The sunlit portions of the Northern Hemisphere are seen to vary from greater than one half in summer to less than one half in winter. The proportion of any latitude that is sunlit is also the proportion of the 24-hour day between sunrise and sunset. (From MacArthur, R. H., and Connell, J. H.: The Biology of Populations. New York, John Wiley & Sons, Inc., 1966.)

phere (Fig. 36.7); consequently, the polar regions receive less radiant energy in the course of a year than do equatorial regions.

The major variations in the amount of incoming solar energy are related to the movements the earth makes with respect to the sun (Fig. 36.8). One complete annual orbit of the earth around the sun requires 365¼ days. The axis of the earth is tilted 23½ degrees relative to its plane of orbit, and therefore the distribution of energy varies throughout the year. The northern hemisphere receives more radiant energy during the period between March 21 and September 21 than in the other half of the year, not only because there are more hours of daylight, but because the angle of incidence of sunlight is more nearly vertical during that period.

The rotation of the earth on its own axis every 24 hours produces day and night and the energy changes associated with these periods. The changes in temperature lag behind the changes in the amount of light energy received. The maximum temperature during the day is usually in mid-afternoon, and the lowest temperature is just before sunrise. The temperature of the soil tends to lag even more than this. The atmosphere changes the distribution of the energy of different wave lengths so that the nature of sunlight actually reaching the earth is different from the sunlight some 100 miles above the atmosphere. Radiant energy of short wave lengths is not absorbed by water or water vapor; it thus passes through the atmosphere with little diminution. Some of the energy of sunlight is absorbed by the earth; some is reflected back as longer wave lengths or heat. Snow, water and light-colored soils reflect heat, whereas bare ground with dark soils absorbs it. This situation has been altered further by man's activities in paving large areas of the earth and in building, plowing, removing trees and causing air pollution. It is estimated that as much as 40 per cent of the heat of the atmosphere is derived from the condensation of water vapor derived from the evaporation of water from the surface of the ocean. The moisture-laden air rises, moves to higher latitudes where it is cooled, and gives up its moisture as clouds or rain. The heat is then absorbed by the wet atmosphere. The atmosphere, heated from below and radiating heat back to the surface of the earth, serves as a heat trap as does the roof of a greenhouse whose glass substitutes for clouds and water vapor.

36.9 ENERGY FLOW

Most of the sun's energy that reaches the earth is eventually lost as heat. A small proportion of the energy of sunlight is absorbed by plants, and a small portion of this is transformed into the potential energy of stored food products. The rest of the energy leaves the plant and becomes part of the earth's general heat loss. All living things except green plants obtain their energy by taking in the products of photosynthesis, carried out by green plants, or the products of chemosynthesis, carried out by microorganisms. Each organism is in a dynamic state and its constituents are constantly being degraded and rebuilt. Thus each organism can be considered as a sort of durable framework through which energy and matter flow in this dynamic state. When carbon or nitrogen atoms enter an organism, they are synthesized into the compounds characteristic for that organism, then are subsequently returned to the environment. The entire earth has a finite amount of carbon, nitrogen and other atoms, and these must constantly be recycled. The degradation of organic molecules by decomposer organisms is of great importance in preventing the catastrophe that would occur if all of the carbon or nitrogen atoms became bound in some life form and could no longer be used.

The various kinds of organisms in nature are balanced with their environment, but for many organisms this balance is a precarious one. Members of the human race have upset the initial balance of nature in an alarming number of instances, and many of these, once undone, are very difficult, if not impossible, to reestablish. The reasons for this will become more apparent when we discuss community successions in the next chapter. In tropical Africa and South America the natives clear portions of the rain forest for fields to grow crops. Without fertilization the cleared area may produce a crop for only a few years. When it becomes unproductive the area is abandoned and more forest is felled to provide new crop land. The abandoned area can probably never again be covered with a mature rain forest. The thin tropical soils have very meager supplies of mineral nutrients; these are leached out of the soil by the heavy rains. The mature rain forest is in precarious balance with the soil and can maintain itself only as long as the balance is not disturbed. Once the balance has been upset, the forest is irretrievably lost. Civilized man in the more technologically advanced countries has done similar things, much more efficiently and with equally devastating results. Many tragic examples can be given of man's blunders in upsetting critical ecological balances. The organisms living in nature are parts of complex interacting communities of many species and are not single, isolated species. The evolution of organisms is determined not simply by the limitations and peculiarities of the temperature, soil, pH, salinity and other factors of the abiotic environment, but also by their relationships with other organisms living in that region. It is true that organisms make the community, but it is equally true that the community makes the organisms.

The carbon, nitrogen and various other cycles in nature operate to conserve the limited amount of usable matter on the earth. In contrast, the amount of energy available is very great and is constantly being renewed in sunlight. The flow of energy is not cyclic but is one-way. By measuring the amount of energy taken up and given off by each kind of organism, the ecologist can determine the functional structure of the organisms living together in a community. From this, he can calculate how much life can be supported in a given area and how many individuals of each species the area can support. As the potential energy of sunlight is transferred from plants and other primary producers through herbivores and their carnivorous predators and parasites, and ultimately, following their deaths, through the decomposer microorganisms, a large portion of the energy is lost at each step as heat. Because of this progressive loss of energy as heat, the total energy flow at each succeeding level is less and less. When an animal eats food, less than 20 per cent of the foodstuff is eventually converted into the flesh of the animal that is doing the eating. The domestic pig is one of the more efficient converters: under the best feeding practices, a pig will convert about 20 per cent of the mass of the food that it eats into pork chops and bacon.

The transfer of energy through a biological community begins when the energy of sunlight is fixed in a green plant by photosynthesis. It is estimated that only 8 per cent of the energy of the sun reaching the planet strikes green plants and that only 3 per cent of this is utilized in photosynthesis. Part of this energy is used by the plant itself to drive the many processes required for maintenance. The amount left over that is stored and expressed as growth represents the **net primary production.**

The net primary production of a field of sugar cane in Hawaii was 190 Kcal. per m.^{2} per day. The average insolation was about 4000 Kcal. per m.^{2} per day. From this we can calculate that the net efficiency of the sugar cane is about 4.8 per cent. Such values can be achieved only by crops under intensive cultivation during a favorable growing season. On an over-all annual basis, sugar cane fields have an efficiency of about 1.9 per cent and tropical forests have an efficiency of about 2 per cent. The stored energy accumulates as living material or biomass. Part is recycled each season by death and decomposition of the organisms; the part that remains alive is called the **standing crop biomass.** This, of course, can vary greatly with the season. In grasslands there is an annual turnover of the biomass, but in forests much of the energy is tied up in wood. The most productive ecosystems on an energy basis are coral reefs and the estuaries of rivers (Fig. 36.9). The least productive are deserts and the open ocean. By and large the production of plant material in each area of the earth has reached an optimum level that is limited only by soil and by the climate. The various kinds of consumers depend upon the production of green plants. Of the net production that is available to herbivores, not all is assimilated. For example, a grasshopper assimilates only about 30 per cent of its food, but some mice assimilate nearly 90 per cent. Most of this goes into the maintenance of the organisms and is eventually lost as heat in the process of respiration. A small residue is stored in the form of new tissue and new individuals; it is this stored energy in the herbivore that is available to the next

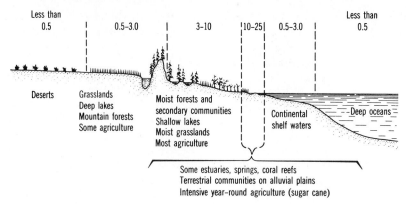

Figure 36.9   The world distribution of primary production, in grams of dry matter per square meter day, as indicated by average daily rates of gross production in major ecosystems. (From Odum, E. P.: Fundamentals of Ecology, 3rd ed. Philadelphia, W. B. Saunders Co., 1971.)

trophic layer, the carnivores. This enormous decrease in biomass at each stage is the basis for the concept of food chains and the pyramidal nature of the successive levels in the food chain.

36.10　FOOD CHAINS AND PYRAMIDS

The number of organisms of each species —or more precisely their total mass—is determined by the rate of flow of energy through the biological part of the ecosystem which includes them.

The transfer of energy from its ultimate source in plants, through a series of organisms each of which eats the preceding and is eaten by the following, is known as a **food chain.** The number of steps in the series is limited to perhaps four or five because of the great decrease in available energy at each step. The percentage of food energy consumed that is converted to new cellular material, and thus is available as food energy for the next animal in the food chain, is known as the percentage efficiency of energy transfer.

The flow of energy in ecosystems, from sunlight through photosynthesis in autotrophic producers, through the tissues of herbivorous primary consumers and the tissues of carnivorous secondary consumers, determines the number and total weight (**biomass**) of organisms at each level in the ecosystem. The flow of energy is greatly reduced at each successive level of nutri-

tion because of the heat losses at each transformation of energy. This decreases the biomass in each level.

Some animals eat but one kind of food and, therefore, are members of a single food chain. Other animals eat many different kinds of food and not only are members of different food chains but may occupy different positions in different food chains. An animal may be a primary consumer in one chain, eating green plants, but a secondary or tertiary consumer in other chains, eating herbivorous animals or other carnivores.

Man is the end of a number of food chains; for example, man eats a fish such as a black bass, which ate little fish which ate small invertebrates which ate algae. The ultimate size of the human population (or the population of any animal) is limited by the length of the food chain, the per cent efficiency of energy transfer at each step in the chain and the amount of light energy falling on the earth.

Since man can do nothing about increasing the amount of incident light energy, and very little about the per cent efficiency of energy transfer, he can increase his supply of food energy only by shortening his food chain, i.e., by eating the primary producers, plants, rather than animals. In overcrowded countries such as India and China men are largely vegetarians because this food chain is shortest and a given area of land can in this way support the greatest number of people. Steak is a luxury in both ecologic and economic terms, but hamburger is just as much an ecologic luxury as steak is.

In addition to predator food chains, such as the man–black bass–minnow–crustacean one, there are parasite food chains. For example, mammals and birds are parasitized by fleas; in the fleas live protozoa which are, in turn, the hosts of bacteria. Since the bacteria might be parasitized by viruses, there could be a five-step parasite food chain.

A third type of food chain is one in which plant material is converted into dead organic matter, **detritus,** before being eaten by animals such as millipedes and earthworms on land, by marine worms and mollusks, or by bacteria and fungi. In a community of organisms in the shallow sea, about 30 per cent of the total energy flows via detritus chains, but in a forest community, with a large biomass of plants and a relatively small biomass of animals, as much as 90 per cent of energy flow may be via detritus pathways. In an intertidal salt marsh, where most of the animals—shellfish, snails and crabs—are detritus eaters, 90 per cent or more of the energy flow is via detritus chains.

Since, in any food chain, there is a loss of energy at each step, it follows that there is a smaller biomass in each successive step. H. T. Odum has calculated that 8100 kilograms of alfalfa plants are required to provide the food for 1000 kilograms of calves, which provide enough food to keep one 12-year-old, 48-kilogram boy alive for one year. Although boys eat many things other than veal, and calves other things besides alfalfa, these numbers illustrate the principle of a food chain. A food chain may be visualized as a **pyramid;** each step in the pyramid is much smaller than the one on which it feeds. Since the predators are usually larger than the ones on which they prey, the pyramid of numbers of individuals in each step of the chain is even more striking than the pyramid of the mass of individuals in successive steps: one boy requires 4.5 calves, which require 20,000,000 alfalfa plants.

36.11 INTERACTIONS BETWEEN SPECIES

Two species may compete for the same space, food or light, or in avoiding predators or disease. They are, in a sense, competing for the same ecologic niche. Competition may result in one species dying off or being forced to change its ecologic niche—to move away or to utilize a different source of food. Careful ecologic studies usually confirm Gause's Rule: that there is only one species in an ecologic niche. The cormorant and the shag are two fish-eating, cliff-nesting sea birds that seemed at first glance to have survived despite the fact that they occupy the same ecologic niche. However, the cormorant feeds on bottom-dwelling fish and shrimp, whereas the shag hunts fish and eels in the upper levels of the sea. Further study showed that these birds typically choose slightly different nesting sites on the cliffs.

Commensalism, the relationship in which two species habitually live together, one of which (the commensal) derives benefit from the association and the other is unharmed, is especially common in the ocean. Practically every worm burrow and shellfish contains some uninvited guests that take advantage of the shelter and possibly of the abundant food provided by the host organism but do it neither good nor harm. Some flatworms live attached to the gills of the horseshoe crab and get their food from the scraps of the crab's meals. They receive shelter and transportation from the host but apparently do not injure it. One of the more startling examples of commensalism is that of a small fish that lives in the posterior end of the digestive tract of the sea cucumber (an echinoderm), entering and leaving it at will. These fish are quickly eaten by other fish if removed from their sheltering host.

If both species gain from an association, but are able to survive without it, the association is termed **protocooperation.** Several kinds of crabs put coelenterates of one sort or another on top of their shells, presumably as camouflage. The coelenterates benefit from the association by obtaining particles of food when the crab captures and eats an animal. Neither crab nor coelenterate is absolutely dependent on the other.

When both species gain from an association and are unable to survive separately, the association is called **mutualism.** It seems probable that interspecific associations begin as commensalism and evolve throughout a stage of protocooperation to one of mutualism. A striking example of a mutualistic association is that of termites and their intestinal flagellates. Termites are famous for their ability to eat wood, yet they have no enzymes to digest it. In their intestines, however, live certain flagellate protozoa that do have the enzymes to digest the cellulose of wood to sugars. Although the flagellates use some of this sugar for their own metabolism, there is enough left over for the termite. Termites cannot survive without their intestinal inhabitants; newly hatched termites instinctively lick the anus of another termite to obtain a supply of flagellates. Since a termite loses all its flagellates along with most of its gut lining at each molt, termites must live in colonies so that a newly molted individual will be able to get a new supply of flagellates from its neighbor. The flagellates also benefit by this arrangement: they are supplied with plenty of food and the proper environment; in fact, they can survive only in the intestines of termites.

In certain types of interspecific associations one of the species is harmed by the other. If one is harmed but the second is unaffected, the relationship is termed **amensalism.** Organisms that produce antibiotics and the species inhibited by the antibiotics are examples of amensalism. The mold *Penicillium* produces **penicillin,** a substance that will inhibit the growth of a variety of bacteria. The mold presumably benefits by having a greater food supply when the competing bacteria have been removed. Man, of course, takes good advantage of this and cultures *Penicillium* and other antibiotic-producing molds in huge quantities to obtain bacteria-inhibiting substances to combat bacterial infections. The use of these bacteria-inhibiting agents has had the unexpected effect of increasing the incidence of fungus-induced diseases in man. These are normally kept in check by the presence of bacteria. When the bacteria are killed off by antibiotics, pathogenic fungi have a golden opportunity to multiply in the host.

We would be quite wrong if we assumed that the host-parasite or predator-prey relationship was invariably harmful to the host or prey *as a species.* This is usually true when such relationships are first set up, but in time, the forces of natural selection tend to decrease the detrimental effects. If the detrimental effects continued, the parasite would eventually kill off all the hosts and, unless it found a new species to parasitize, it would die itself.

Studies of hundreds of different examples of parasite-host and predator-prey interrelations show that in general, where the associations are of long standing, the long-term effect on the host or prey is not very detrimental and may even be beneficial. Conversely, newly acquired predators or parasites are usually quite damaging. The plant parasites and insect pests that are most troublesome to man and his crops are usually those which have recently been carried into some new area and thus have a new group of organisms to attack.

A striking example of the result of upsetting a long-standing predator-prey relationship occurred on the Kaibab plateau, on the north side of the Grand Canyon of the Colorado River. In this area in 1907 there were some 4000 deer and a considerable population of their predators, mountain lions and wolves. When a concerted effort was made to "protect" the deer by killing off the predators, the deer population increased tremendously. By 1925 there were some 100,000 deer on the plateau, far too many for the supply of vegetation. The deer ate everything in reach, grass, tree seedlings and shrubs, and there was marked damage to the vegetation. There was no longer enough vegetation to support the deer population over the winter, and in the next two winters vast numbers of deer starved to death. Finally the deer population fell to about 10,000. The original predator-prey interaction had been maintaining a fairly stable equilibrium, with the number of deer being kept at a level within the available food supply.

The size of the predator population in the wild varies with the size of the population preyed upon. The swings in the size of the predator population lag a bit behind those of the prey.

36.12 INTRASPECIFIC RELATIONS

In addition to the associations between the members of two different species just described, aggregations of animals or plants of a *single* species frequently occur. Some of these aggregations are temporary, for breeding; others are more permanent. Despite the fact that the crowding which accompanies dense aggregations of animals is ecologically undesirable and deleterious, both laboratory experiments and field observations show that such aggregations of individuals may be able to survive when a single individual of the same species placed in the same environment would die. A herd of deer, with many noses and pairs of eyes, is less likely to be surprised by a predator than is a single deer. Wolves hunting in a pack are more likely to make a kill than is a lone wolf. The survival value of intraspecific aggregations is less obvious, but nonetheless real, in some of the lower animals. A group of insects is less likely to dry up and die in a dry environment than is a single insect, and a group of planaria is less likely to be killed by a given dose of ultraviolet light than is a single flatworm. When a dozen goldfish are placed in one bowl and a single one in a second bowl, and the same amount of a toxic agent such as colloidal silver is added to each bowl, the single fish will die, but the group will survive. The explanation for this has proved to be that the slime secreted by the group of fish is enough to precipitate much of the colloidal silver and render it nontoxic, whereas the amount secreted by a single fish is not.

When genes governing a tendency toward aggregation arise in a species and prove to have survival value, natural selection will tend to preserve this inherited behavior pattern. The occurrence of many fish in schools, of birds in flocks and so on are examples of this "unconscious cooperation," which occurs very widely in the animal kingdom.

From such simple animal aggregations there may evolve complex animal societies, composed of specialized types of individuals, such as the colonies of bees, ants and termites. Man is another example of a social animal.

ANNOTATED REFERENCES

Brady, N. C., and H. O. Buckman: The Nature and Properties of Soil. 6th ed. New York, The Macmillan Co., 1960. A classic and complete text of soil science.

Brock, T. D.: The Principles of Microbial Ecology. Englewood Cliffs, N.J., Prentice-Hall, Inc., 1966. Emphasizes the roles of microorganisms in ecological relationships, a subject frequently passed over lightly in general ecologic texts.

Browning, T. O.: Animal Populations. New York, Harper & Row, Publishers, 1963. A discussion of factors regulating the size of animal populations.

Cott, H. B.: Adaptive Coloration in Animals. Oxford, Oxford University Press, 1940. A wonderfully illustrated account of animal camouflage.

Gates, D. M.: Energy Exchange in the Biosphere. New York, Harper & Row, Publishers, 1962. A good summary of the principles regulating the energy environment in which we live.

Kendeigh, S. C.: Animal Ecology. Englewood Cliffs, N.J., Prentice-Hall, Inc., 1961. A concise treatment of the principles of animal ecology.

Lack, D.: The Natural Regulation of Animal Numbers. New York, Oxford University Press, 1954. A concise presentation of the principles of ecologic regulation.

Leopold, Aldo: A Sand County Almanac. New York, Oxford University Press, 1949. An eloquent essay on the relevance of ecology to human affairs.

Odum, E. P.: Fundamentals of Ecology. 3rd ed. Philadelphia, W. B. Saunders Co., 1971. A broad, up-to-date coverage of the field that emphasizes especially the energy relationships in ecology.

Slobodkin, L. B.: Growth and Regulation of Animal Populations. New York, Henry Holt & Co., 1961. An excellent summary of population ecology.

Smith, Robert L.: Ecology and Field Biology. New York, Harper & Row, Publishers, 1966. An excellent general text emphasizing the natural history aspects of the science.

37. Synecology: Communities, Biomes and Life Zones

Each region of the earth—sea, lake, forest, prairie, tundra, desert—is inhabited by a characteristic assemblage of animals and plants which are interrelated in many and diverse ways as competitors, commensals, predators, and so on. The members of each assemblage are not determined by chance but by the total effect of the many interacting physical and biotic factors of the environment. The ecologist refers to the organisms living in a given area as a **biotic community**; this is composed of smaller groups, or **populations**, groups of individuals of any one kind of organism.

The intermeshings of the food chains in any biotic community are very complex and are sometimes called a food web, or "web of life." Some of the interrelated food chains of a deciduous forest in eastern America are indicated in Figure 37.1. The basic principles of the ecologic relations of biotic communities have been elucidated by the study of somewhat simpler communities such as the arctic tundra or desert. The producer organisms of the tundra are lichens, mosses and grasses. Reindeer and caribou feed on the lichens and are preyed upon by wolves and man. Grasses are eaten by the arctic hare and the lemming, which are eaten by the snowy owl and the arctic fox, which is preyed upon by man for its fur.

During the brief arctic summer the food web is enlarged by many insects and by migratory birds which feed upon them.

37.1 POPULATIONS AND THEIR CHARACTERISTICS

A **population** may be defined as a group of organisms of the same species which occupy a given area. It has characteristics which are a function of the whole group and not of the individual members; these are **population density, birth rate, death rate, age distribution, biotic potential, rate of dispersion,** and **growth form.** Although individuals are born and die, individuals do not have birth rates or death rates; these are characteristics of the population as a whole. Modern ecology deals especially with the community and population aspects of the science, and the study of community organization is a particularly active field at present. Population and community relationships are often more important in determining the occurrence and survival of organisms in nature than are the direct effects of physical factors in the environment.

One important attribute of a population is its **density**—the number of individuals per unit area or volume, e.g., human inhabitants

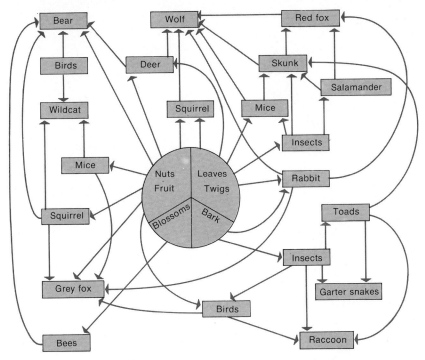

Figure 37.1 Diagram of the interrelationships in a food chain in a deciduous forest in Illinois. (After Shelford, V. E.: Ecological Monographs 21:183–214, 1951.)

per square mile, trees per acre in a forest, millions of diatoms per cubic meter of sea water. This is a measure of the population's success in a given region. Frequently in ecologic studies it is important to know not only the population density, but whether it is changing and, if so, what the rate of change is.

Population density is often difficult to measure in terms of individuals, but measures such as the number of insects caught per hour in a standard trap, the number of sea urchins caught in a standard "sea mop," or the number of birds seen or heard per hour, are usable substitutes. A method that will give good results when used with the proper precautions is that of capturing, let us say, 100 animals, tagging them in some way, and then releasing them. On some subsequent day, another 100 animals are trapped and the proportion of tagged animals is determined. This assumes that animals caught once are neither more likely nor less likely to be caught again, and that both sets of trapped animals are random samples of the population. If the 100 animals caught on the second day include 20 tagged ones, the total population of tagged

and untagged animals in the area of the traps is 500; x/100 = 100/20, hence x = 500.

For many types of ecologic investigations, an estimate of the number of individuals per total area or volume, known as the "crude density," is not sufficiently precise. Only a fraction of that total area may be a suitable habitat for the population, and the size of the individual members of a population may vary tremendously. Ecologists, therefore, calculate an **ecologic density,** defined as the number, or more exactly as the mass, of individuals per area or volume of habitable space. Trapping and tagging experiments might give an estimate of 500 rabbits per square mile, but if only half of that square mile actually consists of areas suitable for rabbits to live in, then the ecologic density will be 1000 rabbits per square mile of rabbit habitat. With species whose members vary greatly in size, such as fish, live weight or some other estimate of the total mass of living fish is a much more satisfactory estimate of density than simply the number of individuals present.

A graph in which the number of organisms or the logarithm of that number is plotted against time is a **population growth**

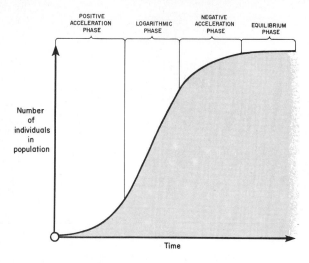

POSITIVE ACCELERATION PHASE **LOGARITHMIC PHASE** **NEGATIVE ACCELERATION PHASE** **EQUILIBRIUM PHASE**

Number of individuals in population

Time

Figure 37.2 A typical sigmoid (S-shaped) growth curve of a population, one in which the total number of individuals is plotted against the time. The absolute units of time and the total number in the population would vary from one species to another, but the shape of the growth curve would be similar for all populations.

curve (Fig. 37.2). Such curves are characteristic of populations, rather than of a single species, and are amazingly similar for populations of almost all organisms from bacteria to man.

Population growth curves have a characteristic shape. When a few individuals enter a previously unoccupied area, growth is slow at first (the positive acceleration phase), then becomes rapid and increases exponentially (the logarithmic phase). The growth rate eventually slows down as environmental resistance gradually increases (the negative acceleration phase) and finally reaches an equilibrium or saturation level. The upper asymptote of the sigmoid curve is termed the "carrying capacity" of the environment.

The birth rate, or natality, of a population is simply the number of new individuals produced per unit time. The **maximum birth rate** is the largest number of organisms that could be produced per unit time under ideal conditions, when there are no limiting factors. This is a constant for a species, determined by physiologic factors such as the number of eggs produced per female per unit time, the proportion of females in the species and so on. The actual birth rate is usually considerably less than this, for not all the eggs laid are able to hatch, not all the larvae or young survive and so on. The size and composition of the population and a variety of environmental conditions affect the actual birth rate. It is difficult to determine the maximum natality, for it is difficult to be sure that all limiting factors have been

removed. However, under experimental conditions, one can get an estimate of this value which is useful in predicting the rate of increase of the population and in providing a yardstick for comparison with the actual birth rate.

The mortality rate of a population refers to the number of individuals dying per unit time. There is a theoretical **minimum mortality,** somewhat analogous to the maximum birth rate, which is the number of deaths which would occur under ideal conditions, deaths due simply to the physiologic changes of old age. This minimum mortality rate is also a constant for a given population. The actual mortality rate will vary depending upon physical factors and on the size, composition and density of the population.

By plotting the number of survivors in a population against time, one gets a **survival curve** (Fig. 37.3). If the units of the time axis are expressed as the percentage of total life span, the survival curves for organisms with very different total life spans can be compared. Civilized man has improved his average life expectancy greatly by modern medical and public health practices, and the curve for human survival approaches the curve for minimal mortality. From such curves one can determine at what stage in the life cycle a particular species is most vulnerable. Reducing or increasing mortality in this vulnerable period will have the greatest effect on the future size of the population. Since the death rate is more variable and

Figure 37.3 Survival curves of four different animals, plotted as number of survivors left at each fraction of the total life span of the species. The total life span for man is about 100 years; the solid curve indicates that about 10 per cent of the babies born die during the first few years of life. Only a small fraction of the human population dies between ages 5 and 45 but after 45 the number of survivors decreases rapidly. Starved fruit flies live only about five days, but almost the entire population lives the same length of time and dies at once. The vast majority of oyster larvae die but the few that become attached to the proper sort of rock or to an old oyster shell survive. The survival curve of hydras is one typical of most animals and plants, in which a relatively constant fraction of the population dies off in each successive time period.

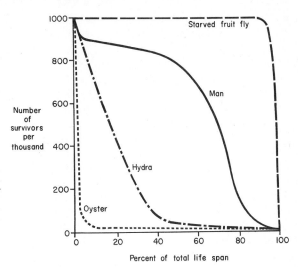

affected to a greater extent by environmental factors than the birth rate, it has a primary role in controlling the size of a population.

It is obvious that populations that differ in the relative numbers of young and old will have different characteristics, different birth and death rates, and different prospects. Death rates typically vary with age, and birth rates are usually proportional to the number of individuals able to reproduce. Three ages can be distinguished in a population in this respect: pre-reproductive, reproductive and post-reproductive. A. J. Lotka has shown from theoretical considerations that a population will tend to become stable and have a constant proportion of individuals of these three ages.

Censuses of the ages of plant or animal populations are of value in predicting population trends. Rapidly growing populations have a high proportion of young forms. The age of fishes can be determined from the growth rings on their scales, and studies of the age ratios of commercial fish catches are of great use in predicting future catches and in preventing overfishing of a region.

The term **biotic potential**, or reproductive potential, refers to the inherent power of a population to increase in numbers when the age ratio is stable and all environmental conditions are optimal. The biotic potential is defined mathematically as the slope of the population growth curve during the logarithmic phase of growth. When environmental conditions are less than optimal, the

rate of population growth is less. The difference between the potential ability of a population to increase and the actual change in the size of the population is a measure of environmental resistance.

Even when a population is growing rapidly in number, each *individual* organism of the reproductive age carries on reproduction at the same rate as at any other time; the increase in numbers is due to increased survival. At a conservative estimate, one man and one woman, with the cooperation of their children and grandchildren, could give rise to 200,000 progeny within a century, and a pair of fruit flies could multiply to give 3368×10^{52} offspring in a year. Since optimal conditions are not maintained, such biologic catastrophes do not occur, but the situations in India, China and elsewhere indicate the tragedy implicit in the tendency toward human overpopulation.

The sum of the physical and biologic factors which prevent a species from reproducing at its maximum rate is termed the **environmental resistance**. Environmental resistance is often low when a species is first introduced into a new territory so that the species increases in number at a fantastic rate, as when the rabbit was introduced into Australia and the English sparrow and Japanese beetle were brought into the United States. But as a species increases in number the environmental resistance to it also increases, in the form of organisms that prey upon it or parasitize it, and the competition between the members of the species for food and living space.

In an essay in 1798 the Englishman Robert Malthus pointed out this tendency for populations to increase in size until checked by the environment. He realized that these same principles control human populations and suggested that wars, famines and pestilences are inevitable and necessary as brakes on population growth. Since Malthus' time man's productive capacity has increased tremendously as has the total human population. But Malthus' basic principle, that there are physical limits to the amount of food that can be produced for any species, remains true. The earth has a finite carrying capacity for human beings just as it does for any other animal. As environmental resistance increases, the rate of increase of the human population will eventually have to decrease. An equilibrium will be reached either by decreasing the birth rate or by increasing the mortality rate.

37.2 POPULATION CYCLES

Once a population becomes established in a certain region and has reached the equilibrium level, the numbers will vary up and down from year to year, depending on variations in environmental resistance or on factors intrinsic to the population. Some of these population variations are completely irregular, but others are regular and cyclic.

One of the best known of these is the regular nine to ten year cycle of abundance and scarcity of the snowshoe hare and the lynx in Canada which can be traced from the records of the number of pelts received by the Hudson's Bay Company. The peak of the hare population comes about a year before the peak of the lynx population (Fig. 37.4). Since the lynx feeds on the hare, it is obvious that the lynx cycle is related to the hare cycle.

Lemmings and voles are small mouselike animals living in the northern tundra region. Every three or four years there is a great increase in the number of lemmings; they eat all the available food in the tundra and then migrate in vast numbers looking for food. They may invade villages in hordes, and finally many reach the sea and drown. The numbers of arctic foxes and snowy owls, which feed on lemmings, increase similarly. When the lemming population decreases, the foxes starve and the owls migrate south—thus there is an invasion of snowy owls in the United States every three or four years.

Although some cycles recur with great regularity, others do not. For example, in the carefully managed forests of Germany the numbers of four species of moths whose caterpillars feed on pine needles were estimated from censuses made each year from 1880 to 1940. The numbers varied from less than one to more than 10,000 per thousand square meters. The cycles of maxima and minima of the four species were quite independent and were irregular

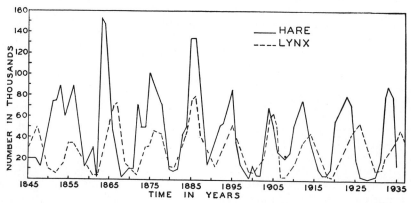

Figure 37.4 Changes in the abundance of the lynx and the snowshoe hare, as indicated by the number of pelts received by the Hudson's Bay Company. This is a classic case of cyclic oscillation in population density. (Redrawn from MacLulich, D. A.: Fluctuations in the numbers of the varying hare (*Lepus americanus*). Univ. Toronto Studies, Biol. series, no. 43, 1937.)

in their frequency and duration. Ecologically speaking, each species was marching to its own tune.

Attempts to explain these vast oscillations in numbers on the basis of climatic changes have been unsuccessful. At one time it was believed that these were caused by sunspots, and the sunspot and lynx cycles do appear to correspond during the early part of the nineteenth century. However, the cycles are of slightly different lengths and by 1920 were completely out of phase, sunspot maxima corresponding to lynx minima. Attempts to correlate these cycles with other periodic weather changes and with cycles of disease organisms have been unsuccessful.

The snowshoe hares die off cyclically even in the absence of predators and in the absence of known disease organisms. The animals apparently die of "shock," characterized by low blood sugar, exhaustion, convulsions and death, symptoms which resemble the "alarm response" induced in laboratory animals subjected to physiologic stress.

This similarity led J. J. Christian (1950) to propose that their death, like the alarm response, is the result of some upset in the adrenal-pituitary system. As population density increases, there is increasing physiologic stress on individual hares, owing to crowding and competition for food. Some individuals are forced into poorer habitats where food is less abundant and predators more abundant. The physiologic stresses stimulate the adrenal medulla to secrete epinephrine, which stimulates the pituitary via the hypothalamus to secrete more ACTH. This, in turn, stimulates the adrenal cortex to produce corticosteroids, an excess or imbalance of which produces the alarm response or physiologic shock.

In the latter part of the winter of a year of peak abundance, with the stress of cold weather, lack of food and the onset of the new reproductive season putting additional demands on the pituitary to secrete gonadotropins, the adrenal-pituitary system fails, becomes unable to maintain its normal control of carbohydrate metabolism, and low blood sugar, convulsions and death ensue. This is a reasonable hypothesis, but the appropriate experiments and observations in the wild needed to test it have not yet been made.

37.3 POPULATION DISPERSION AND TERRITORIALITY

Populations have a tendency to disperse, or spread out in all directions, until some barrier is reached. Within the area, the members of the population may occur at random (this is rarely found), they may be distributed more or less uniformly throughout the area (this occurs when there is competition or antagonism to keep them apart), or, most commonly, they may occur in small groups or clumps.

Aggregation in clumps may increase the competition between the members of the group for food or space, but this is more than counterbalanced by the greater survival power of the group during unfavorable periods. A group of animals has much greater resistance than a single individual to adverse conditions such as desiccation, heat, cold or poisons. The combined effect of the protective mechanisms of the group is effective in countering the adverse environment, whereas that of a single individual is not.

Aggregation may be caused by local habitat differences, weather changes, reproductive urges or social attractions. Such aggregations of individuals may have a definite organization involving social hierarchies of dominant and subordinate individuals arranged in a "peck-order" (p. 800).

Other species of animals are regularly found spaced apart; each member tends to occupy a certain area or **territory** which he defends against intrusion by other members of the same species and sex. Usually a male establishes a territory of his own (perhaps by fighting with other males) and then, by making himself conspicuous, tries to entice a female to share the territory with him.

It has been suggested that territoriality may have survival value for a species in ensuring an adequate amount of food, nesting materials and cover for the young, in protecting the female and young against other males, and in limiting the population to a density that can be supported by the environment. Many species of birds, some mammals, fish, crabs and insects establish such territories, either as regions for gathering food or as nesting areas.

37.4   BIOTIC COMMUNITIES

A biotic community is an assemblage of populations living in a defined area or habitat; it can be either large or small. The interactions of the various kinds of organisms maintain the structure and function of the community and provide the basis for the ecological regulation of community succession. The concept that animals and plants live together in an orderly manner, not strewn haphazardly over the surface of the earth, is an important principle of ecology.

Sometimes adjacent communities are sharply defined and separated from each other; more frequently they blend imperceptibly together. Why certain plants and animals comprise a given community, how they affect each other, and how man can control them to his advantage are some of the major problems of ecologic research.

In trying to control some particular species, it has frequently been found more effective to modify the community rather than to attempt direct control of the species itself. For example, the most effective way to increase the quail population is not to raise and release birds, nor even to kill off predators, but to maintain the particular biotic community in which quail are most successful.

Although each community may contain hundreds or thousands of species of plants and animals, most of these are relatively unimportant, and only a few exert a major control of the community owing to their size, numbers or activities. In land communities these major species are usually plants, for they both produce food and provide shelter for many other species. Many land communities are named for their dominant plants—sagebrush, oak-hickory, pine and so on. Aquatic communities, containing no conspicuous large plants, are usually named for some physical characteristic—stream rapids community, mud flat community and sandy beach community.

In ecologic investigations it is unnecessary (and indeed usually impossible) to consider all of the species present in a community. Usually a study of the major plants which control the community, the larger populations of animals and the fundamental energy relations (the food chains) of the system will define the ecologic relations within the community. For example, in studying a lake one would first investigate the kinds, distribution and abundance of the important producer plants and the physical and chemical factors of the environment which might be limiting. Then the reproductive rates, mortality rates, age distributions and other population characteristics of the important game fish would be determined. A study of the kinds, distribution and abundance of the primary and secondary consumers of the lake, which constitute the food of the game fish, and the nature of other organisms which compete with these fish for food would elucidate the basic food chains in the lake. Quantitative studies of these would reveal the basic energy relationships of the system and show how efficiently the incident light energy is being converted into the desired end product, the flesh of game fish. On the basis of this knowledge the lake could intelligently be managed to increase the production of game fish.

Detailed studies of simpler biotic communities such as those of the arctic or desert, where there are fewer organisms and their interrelations are more evident, have provided a basis for studying and understanding the much more varied and complex forest communities.

A thorough ecologic investigation of a particular region requires that the region be studied at regular intervals throughout the year for a period of several years. The physical, chemical, climatic and other factors of the region are carefully evaluated, and an intensive study is made of a number of carefully delimited areas which are large enough to be representative of the region but small enough to be studied quantitatively. The number and kinds of plants and animals in these study areas are estimated by suitable sampling techniques. Estimates are made periodically throughout the year to determine not only the components of the community at any one time but also their seasonal and annual variations. The biologic and physical data are correlated, the major and minor communities of the region are identified, and the food chains and other important ecologic relationships of the members of the community are analyzed. The particular adaptations of the animals and plants for their respective roles in the community can then be appreciated.

37.5 COMMUNITY SUCCESSION

Any given area tends to have an orderly sequence of communities which change together with the physical conditions and lead finally to a stable mature community or **climax community.** The entire sequence of communities characteristic of a given region is termed a **sere,** and the individual transitional communities are called **seral stages** or seral communities. In successive stages there is not only a change in the species of organisms present but an increase in the number of species and in the total biomass.

These series are so regular in many parts of the world that an ecologist, recognizing the particular seral community present in a given area, can predict the sequence of future changes. The ultimate causes of these successions are not clear. Climate and other physical factors play some role, but the succession is directed in part by the nature of the community itself, for the action of each seral community is to make the area less favorable for itself and more favorable for other species until the stable climax community is reached. Physical factors such as the nature of the soil, the topography and the amount of water may cause the succession of communities to stop short of the expected climax community in what is called an **edaphic climax.**

Occasionally the organisms that man wants to encourage for his own ends — timber, game birds, fresh water game fish — are members of a seral stage in community succession rather than of the climax community. Then the ecologist has the difficult problem of trying to manipulate the community to halt the succession and maintain the desired seral community.

One of the classic studies of ecologic succession was made on the shores of Lake Michigan (Fig. 37.5). As the lake has become smaller it has left successively younger sand dunes, and one can study the stages in ecologic succession as one goes away from the lake. The youngest dunes, nearest the lake, have only grasses and insects; the next older ones have shrubs such as cottonwoods, then evergreens and finally there is a beech-maple climax community, with deep rich soil full of earthworms and snails.

As the lake retreated it also left a series of ponds. The youngest of these contain little rooted vegetation and lots of bass and bluegills. Later the ponds become choked with vegetation and smaller in size as the basins fill. Finally the ponds become marshes and then dry ground, invaded by shrubs and ending in the beech-maple climax forest. Man-made ponds, such as those behind dams, similarly tend to become filled up.

Another dramatic example of community succession began August 7, 1883, when a volcanic explosion occurred on the Indonesian island Krakatoa, causing part of the island to disappear. The remainder was covered with hot volcanic debris to a depth of 60 meters, and all life was obliterated. A year later some grass and a single spider were found. By 1908, 202 species of animals had taken up residence on the island. This increased to 621 species by 1919, and to 880 species by 1934, when there was a young forest on one part of the island.

Ecologic succession can be demonstrated in the laboratory. If a few pieces of dry grass are placed in a beaker of pond water, a population of bacteria will appear in a few days. Next, flagellates appear and eat the bacteria, then ciliated protozoa such as paramecia appear and eat the flagellates. Finally predator protozoa such as *Didinium* will appear and eat the paramecia. The protozoa, present as spores or cysts attached to the grass, emerge in a definite succession of protozoan communities.

Biotic communities show marked **vertical stratification,** determined in large part by vertical differences in physical factors such as temperature, light and oxygen. The operation of such physical factors in determining vertical stratification in lakes and in the ocean is quite evident. In a forest there is a vertical stratification of plant life, from mosses and herbs on the ground, to shrubs, low trees and tall trees. Each of these strata has a distinctive animal population. Even such highly motile animals as birds are restricted, more or less, to certain layers. Some species of birds are found only in shrubs, others only in the tops of tall trees. There are daily and seasonal changes in the populations found in each stratum, and some animals are found first in one, then in another layer as they pass through their life histories. These strata are interrelated in many diverse ways, and most ecologists consider them to be subdivisions of one

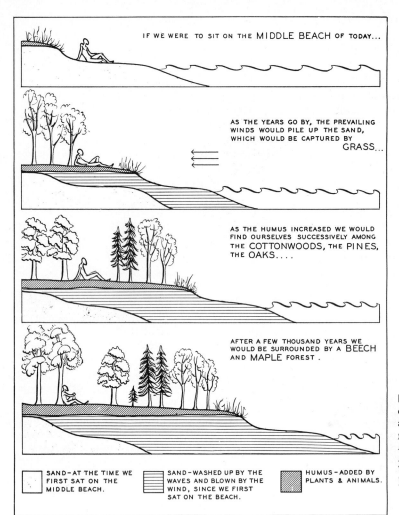

IF WE WERE TO SIT ON THE MIDDLE BEACH OF TODAY...

AS THE YEARS GO BY, THE PREVAILING WINDS WOULD PILE UP THE SAND, WHICH WOULD BE CAPTURED BY GRASS...

AS THE HUMUS INCREASED WE WOULD FIND OURSELVES SUCCESSIVELY AMONG THE COTTONWOODS, THE PINES, THE OAKS....

AFTER A FEW THOUSAND YEARS WE WOULD BE SURROUNDED BY A BEECH AND MAPLE FOREST .

SAND—AT THE TIME WE FIRST SAT ON THE MIDDLE BEACH.

SAND—WASHED UP BY THE WAVES AND BLOWN BY THE WIND, SINCE WE FIRST SAT ON THE BEACH.

HUMUS—ADDED BY PLANTS & ANIMALS.

Figure 37.5 Diagram of the succession of communities with time along the shores of Lake Michigan in northern Indiana. (From Allee, W. C., et al.: Principles of Animal Ecology, Philadelphia, W. B. Saunders Co., 1949. After Buchsbaum.)

large community rather than separate communities. Vertical stratification, by increasing the number of ecologic niches in a given surface area, reduces competition between species and enables more species to coexist in a given area.

37.6 THE CONCEPT OF THE ECOSYSTEM

All the living organisms that inhabit a certain area comprise the biotic community. A larger unit, termed the **ecosystem,** includes the organisms in a given area and the encompassing physical environment. In the ecosystem a flow of energy, derived from organism-environment interactions, leads to a clearly defined trophic structure with biotic diversity and to the cyclic exchange of materials between the living and nonliving parts of the system. From the trophic (nourishment) standpoint, an ecosystem has two components: an **autotrophic** part, in which light energy is captured or "fixed" and used to synthesize complex organic compounds from simple inorganic ones, and a **heterotrophic** part, in which the complex molecules undergo rearrangement, utilization and decomposition. In describing an ecosystem it is convenient to recognize and tabulate the following components: (1) the inorganic substances such as carbon dioxide, water, nitrogen and phosphate that are involved in material cycles; (2) the organic compounds such as proteins, carbohydrates and lipids that are synthesized in the biotic phase; (3) the climate, tempera-

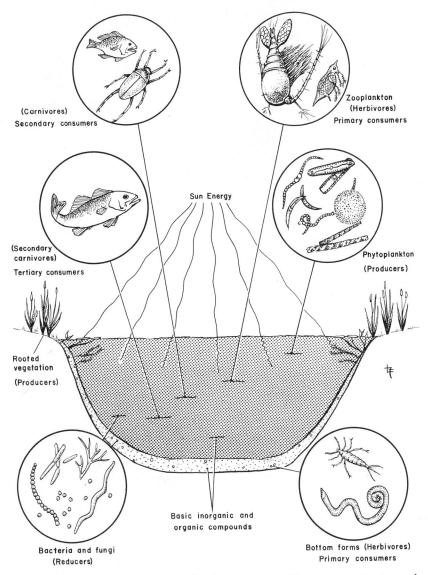

Figure 37.6   A small fresh water pond as an example of an ecosystem. The component parts, producer, consumer and decomposer or reducer organisms, plus the nonliving parts are indicated.

ture and other physical factors; (4) the producers, autotrophic organisms (mostly green plants) which can manufacture complex organic materials from simple inorganic substances; (5) the macroconsumers, or **phagotrophs,** heterotrophic organisms (mostly animals) that ingest other organisms or chunks of organic matter; and (6) the microconsumers, or **saprotrophs,** heterotrophic organisms (mostly fungi and bacteria) which break down the complex compounds of dead organisms, absorb some of the decomposition products and release inorganic nu-

trients that are made available to the producers to complete the various cycles of elements.

The producers, phagotrophs and saprotrophs make up the biomass of the ecosystem —the living weight. In analyzing an ecosystem, the investigator studies the energy circuits present, the food chains, the patterns of biologic diversity in time and space, the nutrient cycles, the development and evolution of the ecosystem and the factors that control the composition of the ecosystem. The ecosystem is the basic functional unit in ecology and includes both the biotic communities and the abiotic environment in a given region, each of which influences the properties of the other and both of which are needed to maintain life on the earth.

A classic example of an ecosystem compact enough to be investigated in quantitative detail is a small lake or pond (Fig. 37.6). The nonliving parts of the lake include the water, dissolved oxygen, carbon dioxide, inorganic salts such as phosphates, nitrates and chlorides of sodium, potassium and calcium, and a multitude of organic compounds. The living part of the lake can be subdivided according to the functions of the organisms, i.e., what they contribute toward keeping the ecosystem operating as a stable, interacting whole. In a lake there are two types of producers, the larger plants growing along the shore or floating in shallow water, and the microscopic floating plants, most of which are algae, that are distributed throughout the water as deep as light will penetrate. These tiny plants, collectively known as **phytoplankton,** are usually not visible unless they are present in great abundance and give the water a greenish tinge. They are usually much more important as food producers for the lake than are the more readily visible plants.

The macroconsumers or phagotrophs include insects and insect larvae, crustacea, fish, and perhaps some fresh water clams. Primary consumers are the plant eaters and secondary consumers are the carnivores that eat the primary consumers. There might be some tertiary consumers that eat the carnivorous secondary consumers.

The ecosystem is completed by saprotrophs or decomposer organisms, bacteria and fungi, which break down the organic compounds of cells from dead producer and consumer organisms either into small organic molecules, which they utilize themselves, or into inorganic substances that can be used as raw materials by green plants.

The term "ecosystem" was proposed by the British ecologist A. G. Tansley in 1935, but the concept of the unity of organisms and environment can be traced back to very early biological literature. The concept of the ecosystem is a very broad one and emphasizes the obligatory relationships, the causal relationships and the interdepend-

ence of biotic and abiotic components. The ecosystem is the level of biological organization particularly suitable for the application of the techniques of systems analysis. An entity may be considered an ecosystem if its major components are present and operate together to achieve some sort of functional stability, even though this may persist for only a short time. Even a temporary pond is a definite ecosystem. It contains characteristic organisms and undergoes characteristic processes even though its existence may be limited to a relatively brief period of time.

The interaction of autotrophic and heterotrophic components is a universal feature of all ecosystems whether they are located on land, in fresh water or in the ocean. Frequently the autotrophic and heterotrophic components are partially separated spatially. The greatest amount of autotrophic metabolism occurs in a "green belt" stratum in which light energy is available. Below this lies a "brown belt" in which the most intense heterotrophic metabolism takes place. In the brown belt organic matter tends to accumulate both in soils and in sediments. The two functions may also be partially separated in time, for there may be a considerable delay in the heterotrophic utilization of the products of autotrophic organisms. In a forest, for example, the products of photosynthesis tend to accumulate in the form of leaves, wood and the food stored in seeds and roots. A relatively long time may elapse before these materials become litter and soil and available to the heterotrophic system.

The two major energy circuits in any ecosystem are the **grazing circuit,** in which animals eat living plants or parts of plants, and the contrasting **organic detritus circuit,** in which dead materials accumulate and are decomposed by bacteria and fungi. From an operational standpoint the living and nonliving parts of ecosystems are tightly interwoven and difficult to separate. Both inorganic compounds and organic compounds not only are found within and without living organisms, but are in a constant state of flux between living and nonliving conditions. A few substances such as ATP are found uniquely inside living cells. In contrast, humic substances, the resistant end products of decomposition, are never found inside living cells, yet are a major and characteristic component of all ecosystems. DNA and chlorophyll may occur both inside and outside organisms, but are nonfunctional when outside the cell. Ecologists can measure the amount of ATP, humus and chlorophyll in a given area or volume to provide an index of the biomass, the decomposition and the production, respectively, in that ecosystem.

The three living components of an ecosystem, the producers, phagotrophs and saprotrophs, are roughly equivalent to plants, animals and bacteria plus fungi, respectively. However, it should be realized that these are functional classifications and that some species of organisms occupy intermediate positions, while others shift their mode of nutrition according to the circumstances of the environment. The heterotrophs can be separated into large and small consumers in an arbitrary fashion.

An excellent way to study an ecosystem is to investigate a small pond, a meadow or an old field. Any area exposed to light—a lawn, a flower-box in the window, or a laboratory aquarium—can be used as long as the physical dimensions and biotic diversity of the area are not so great that observation of the whole is difficult.

37.7   THE HABITAT APPROACH

The subject of ecology can be approached through discussions of the principles and concepts of the science as they apply to different levels of organization, the individual, population, community and ecosystem. Another general approach, the **habitat approach,** describes the distinctive features of the major habitats and their subdivisions, how they are organized, the organisms present in each and the ecologic role of these organisms in that region (i.e., the identity of the major producers, consumers and decomposers).

Four major habitats can be distinguished: **marine, estuarine, fresh water** and **terrestrial.** No plant or animal is found in all four major habitats and, indeed, no animal or plant is found everywhere within any one of these. Every species of animal and plant tends to produce more offspring than can survive within the normal range of the organism. There is strong **population pressure** tending to force the individuals of each

species to spread out and become established in new territories. Competing species, predators, lack of food, adverse climate and the unsuitability of the adjacent regions, perhaps owing to the lack of some requisite physical or chemical factor, all act to counterbalance the population pressure and to prevent the spread of the species. Since all of these factors are subject to change, the range of a species tends to be dynamic rather than static and may change quite suddenly. The spread of a species is prevented by geographic **barriers,** such as oceans, mountains, deserts and large rivers, and is facilitated by **highways,** such as land connections between continents. The present distribution of plants and animals is determined by the barriers and highways that exist now and those that have existed in the geologic past.

The biogeographic realms, discussed on page 251, are regions made up of whole continents, or of large parts of a continent, separated by major geographic barriers and characterized by the presence of certain unique animals and plants. Within these biogeographic realms and established by a complex interaction of climate, other physical factors and biotic factors, are large, distinct, easily differentiated community units called **biomes.** A biome is a large community unit characterized by the kinds of plants and animals present. This may be contrasted to the ecosystem, which is a natural unit of living and nonliving components that interact to form a stable system in which the exchange of materials follows a circular path. Thus an ecosystem might be a small pond or a large area coextensive with a biome, but it includes the *physical* environment as well as the populations of animals, plants and microorganisms.

In each biome the *kind* of climax vegetation is uniform—grasses, conifers, deciduous trees—but the particular *species* of plant may vary in different parts of the biome. The kind of climax vegetation depends upon the physical environment and the two together determine the kind of animals present. The definition of biome includes not only the actual climax community of a region, but also the several intermediate communities that precede the climax community.

There is usually no sharp line of demarcation between adjacent biomes; instead each

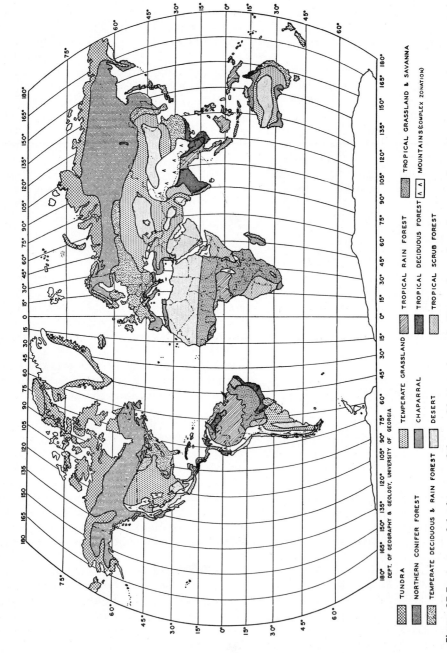

Figure 37.7 A map of the biomes of the world. Note that only the tundra and northern coniferous forest are more or less continuous bands around the world. Other biomes are generally isolated in different biogeographic realms and may be expected to have ecologically equivalent but taxonomically unrelated species. (From Odum, E. P.: Fundamentals of Ecology, 3rd ed. Philadelphia, W. B. Saunders Co., 1971.)

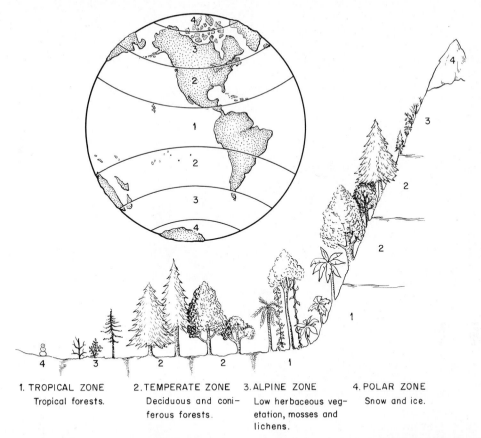

1. TROPICAL ZONE
   Tropical forests.

2. TEMPERATE ZONE
   Deciduous and coni-
   ferous forests.

3. ALPINE ZONE
   Low herbaceous veg-
   etation, mosses and
   lichens.

4. POLAR ZONE
   Snow and ice.

Figure 37.8 Diagram showing correspondence of life zones at successively higher altitudes at the same latitude (1 to 4, right), and at successively higher latitudes at the same altitude (1 to 4, left, and inset).

blends with the next through a fairly broad transition region termed an **ecotone.** There is, for example, an extensive region in northern Canada where the tundra and coniferous forests blend in the tundra-coniferous forest ecotone. The ecotonal community typically consists of some organisms from each of the adjacent biomes plus some that are characteristic of, and perhaps restricted to, the ecotone. There is a tendency (called the **edge effect**) for the ecotone to contain both a greater number of species and a higher population density than either adjacent biome.

Some of the biomes recognized by ecologists are **tundra, coniferous forest, deciduous forest, broad-leaved evergreen subtropical forest, grassland, desert, chaparral** and **tropical rain forest.** These biomes are distributed, though somewhat irregularly, as belts around the world (Fig. 37.7), and as one travels from the equator to the pole he

may traverse tropical rain forests, grassland, desert, deciduous forest, coniferous forest and finally reach the tundra in northern Canada, Alaska or Siberia.

Since climatic conditions at higher altitudes are in many ways similar to those at higher latitudes, there is a similar succession of biomes on the slopes of high mountains (Fig. 37.8). For example, as one goes from the San Joaquin Valley of California into the Sierras, one passes from desert through grassland and chaparral to deciduous forest and coniferous forest, then, above timberline, to a region resembling the tundra of the Arctic.

37.8   THE TUNDRA BIOME

Between the Arctic Ocean and polar ice-caps and the forests to the south lies a band of treeless, wet, arctic grassland called the **tundra** (Fig. 37.9).

Figure 37.9 The tundra biome. *Above:* View of the low tundra near Churchill, Manitoba, in July. Note the numerous ponds. *Below:* View of tundra vegetation showing "lumpy" nature of low tundra and a characteristic tundra bird, the willow ptarmigan. (Lower photo by C. Lynn Haywood.)

Some five million acres of tundra stretch across northern North America, northern Europe and Siberia. The primary characteristics of this region are the low temperatures and a short growing season. The amount of precipitation is rather small but water is usually not a limiting factor because the rate of evaporation is also very low.

The ground usually remains frozen except for the uppermost 10 or 20 cm., which thaw during the brief summer season. The permanently frozen deeper soil layer is called **permafrost.** The rather thin carpet of vegetation includes lichens, mosses, grasses, sedges and a few low shrubs. The animals that have adapted to survive in the tundra are caribou or reindeer, the arctic hare, arctic fox, polar bear, wolves, lemmings, snowy owls, ptarmigans and, during the summer, swarms of flies, mosquitoes and a host of migratory birds.

The caribou and reindeer are highly migratory because there is not enough vegetation produced in any one local area to support them. Although casual inspection might suggest that tundras are rather barren areas, a surprisingly large number of organisms have become adapted to survive the cold. During the long daylight hours of the brief summer, the rate of primary production is quite high. The production from the vegetation on the land, from the plants in the many shallow ponds that dot the landscape and from the phytoplankton in the adjacent Arctic Ocean provides enough food to support a variety of resident mammals and many kinds of migratory birds and insects.

37.9 THE FOREST BIOMES

Several different types of forest biomes can be distinguished. These are generally arranged on a gradient from north to south or from high altitude to lower altitude. Adjacent to the tundra region either at high latitude or high altitude is the **northern coniferous forest** (Fig. 37.10), which stretches across both North America and Eurasia just south of the tundra. This is characterized by spruce, fir and pine trees and by animals such as the snowshoe hare, the lynx and the wolf.

The evergreen conifers provide dense shade throughout the year; this tends to inhibit development of shrubs and a herbaceous undergrowth. The continuous presence of green leaves permits photosynthesis to occur throughout the year despite the low temperature during the winter and results in a fairly high annual rate of primary production.

These coniferous forests are the major source of commercial lumber around the world. After they have fallen, the needles decay very slowly and the soil develops a characteristic condition with relatively little humus. The northern coniferous forest, like the tundra, shows a marked seasonal periodicity, and the populations of animals undergo striking peaks and depressions in numbers.

Along the west coast of North America from Alaska south to central California is a region termed the **moist coniferous forest biome,** characterized by a much greater humidity, somewhat higher temperatures, and smaller seasonal ranges than the classic coniferous forest farther north. There is high rainfall, from 75 to 375 cm. per year, and, in addition, a great deal of moisture is contributed by the frequent fogs. There are forests of Sitka spruce in the northern section, western hemlock, arbor vitae, and Douglas fir in the Puget Sound area, and the coastal redwood, *Sequoia sempervirens*, in California. These forests typically have a luxuriant ground cover of ferns and other herbaceous plants. The potential production of this region is very great, and with careful foresting and replanting, the annual crop of lumber is very high.

Figure 37.10 The coniferous forest biome covers parts of Canada, northern Europe, and Siberia, and extends southward at higher altitudes on the larger mountain ranges. (From Orr, R. T.: Vertebrate Biology, 3rd ed. Philadelphia, W. B. Saunders Co., 1971.)

Figure 37.11 The piñon-juniper biome in Arizona. The small piñon pines and cedars each grow some distance from the neighboring trees, giving an open, parklike appearance to the woodland. (U.S. Forest Service Photo.)

A distinctive subdivision of the northern coniferous forest biome, perhaps distinctive enough to be considered a separate biome, is the pigmy conifer or **piñon-juniper biome** found in west central California and in the Great Basin and Colorado River regions of Nevada, Utah, Colorado, New Mexico and Arizona. This occupies a belt between the desert or grasslands at lower altitudes and the true northern coniferous forest found at higher altitudes where there is more rainfall. In this region the annual rainfall of 25

Figure 37.12 An example of a temperate deciduous forest, Noble County, Ohio. The dominant trees are white and red oaks with an understory of hickory. (Photograph by U.S. Forest Service.)

to 50 cm. is irregularly distributed throughout the year. The small piñon pines and cedars tend to be widely spaced, and the biome has an open, parklike appearance (Fig. 37.11). The pine nuts and cedar berries are eaten by the piñon jay, the gray titmouse and the bush tit.

The **temperate deciduous forest biome** (Fig. 37.12) is found in areas with abundant evenly distributed rainfall (75 to 150 cm. annually) and moderate temperatures with distinct summers and winters. Temperate deciduous forest biomes originally covered eastern North America, all of Europe, parts of Japan and Australia, and the southern portion of South America.

The trees present—beech, maple, oak, hickory and chestnut—lose their leaves during half the year; thus, the contrast between winter and summer is very marked. The undergrowth of shrubs and herbs is generally well developed. The animals originally present in the forest were deer, bears, squirrels, gray foxes, bobcats, wild turkeys and woodpeckers. Much of this forest region has now been replaced by cultivated fields and cities.

In regions of fairly high rainfall but where temperature differences between winter and summer are less marked, as in Florida, the **broad-leaved evergreen subtropical forest biome** is found. The vegetation includes live oaks, magnolias, tamarinds and palm trees, with many vines and epiphytes such as orchids and Spanish moss.

The variety of life reaches its maximum in the **tropical rain forests** (Fig. 37.13), which occupy low lying areas near the equator with annual rainfalls of 200 cm. or more. The thick rain forests, with a tremendous variety of plants and animals, are found in the valleys of the Amazon, Orinoco, Congo and Zambesi rivers and in parts of Central America, Malaya, Borneo and New Guinea.

The extremely dense vegetation makes it difficult to study or even photograph the rain forest biome. The vegetation is vertically stratified with tall trees often covered with vines, creepers, lianas and epiphytes. Under the tall trees is a continuous evergreen carpet, the canopy layer, some 25 to 35 meters tall. The lowest layer is an understory that becomes dense where there is a break in the canopy.

No single species of animal or plant is present in large enough numbers to be dominant. The diversity of species is remarkable; there may be more species of plants and insects in a few acres of tropical rain forest than in all of Europe. The trees of the tropical rain forest are usually evergreen and rather tall. Their roots are often shallow and have swollen bases or flying buttresses.

The tropical rain forest is the ultimate of

Figure 37.13 The rain forest biome: border of a clearing in the Ituri Forest of Nala, The Congo. (Photograph by Herbert Lang. Courtesy of the American Museum of Natural History, New York.)

jungles although the low light intensity at the ground level may result in sparse herbaceous vegetation and actual bare spots in certain areas. Many of the animals live in the upper layers of the vegetation. Among the characteristic animals are monkeys, sloths, termites, ants, anteaters, many reptiles and many brilliantly colored birds — parakeets, toucans and birds of paradise.

37.10   THE GRASSLAND BIOME

The **grassland biome** (Fig. 37.14) is found where rainfall is about 25 to 75 cm. per year, not enough to support a forest, yet more than that of a true desert. Grasslands typically occur in the interiors of continents — the prairies of the western United States, and those of Argentina, Australia, southern Russia and Siberia. Grasslands provide natural pasture for grazing animals, and our principal agricultural food plants have been developed by artificial selection from the grasses.

The mammals of the grassland biome are either grazing or burrowing forms — bison, antelope, zebras, wild horses and asses, rabbits, ground squirrels, prairie dogs and gophers. These characteristically aggregate into herds or colonies; this aggregation probably provides some protection against predators. The birds characteristic of grasslands are prairie chickens, meadow larks and rodent hawks.

The species of grasses present in any given grassland may range from tall species, 150 to 250 cm. in height, to short species of grass that do not exceed 15 cm. in height. Some species of grass grow in clumps or bunches and others spread out and form sods with underground rhizomes. The roots of the several species of grass found in grasslands all penetrate deeply into the soil and the weight of the roots of a healthy plant will be several times the weight of the shoot.

Trees and shrubs may occur in grasslands either as scattered individuals or in belts along the streams and rivers. The soil of grasslands is very rich in humus because of the rapid growth and decay of the individual plants. The grassland soils are well suited for growing cultivated food plants such as corn and wheat which are species of cultivated grasses. The grasslands are also well adapted to serve as natural pastures for cattle, sheep and goats. However, when grasslands are subjected to consistent overgrazing and overplowing, they can be turned into man-made deserts.

There is a broad belt of tropical grassland or **savanna** in Africa lying between the Sahara desert and the tropical rain forest of the Congo basin. Other savannas are found in South America and Australia. Although the annual rainfall is high, as much as 125 cm., a distinct, prolonged dry season prevents the development of a forest. During the dry season there may be extensive fires,

Figure 37.14 A region of short-grass grassland with a herd of bison, originally one of the major grazing animals in the grassland biome of western United States and Canada. The bison in the center is wallowing. (From Odum, E. P.: Fundamentals of Ecology, 3rd ed. Philadelphia, W. B. Saunders Co., 1971.)

Figure 37.15   The savanna biome; characteristic animals of the African grasslands, zebra and wildebeest, Kruger National Park, Transvaal, Republic of South Africa. (From Odum, E. P.: Fundamentals of Ecology, 2nd ed. Philadelphia, W. B. Saunders Co., 1959. Photograph by Herbert Lang.)

which play an important role in the ecology of the region. In this region (Fig. 37.15) are great numbers and great varieties of grazing animals and predators such as lions.

How to make best use of these African grasslands is a problem now facing the new nations of Africa as they work to raise the level of nutrition in their human populations. Many ecologists are of the opinion that it would be better to harvest the native herbivores — antelope, hippopotamuses and wildebeests — on a sustained yield basis rather than try to exterminate them completely and substitute cattle. The diversity of the natural population would mean broader use of all the resources of primary production, and the native species are immune to the many tropical parasites and diseases which plague the cattle that have been introduced.

37.11   THE CHAPARRAL BIOME

In mild, temperate regions of the world with relatively abundant rain in the winter but with very dry summers the climax community includes trees and shrubs with hard, thick evergreen leaves (Fig. 37.16). This type of vegetation is called "chaparral" in California and Mexico, "macchie" around the Mediterranean and "mellee scrub" on Australia's south coast.

The trees and shrubs common in California's chaparral are chamiso and manzanita. Eucalyptus trees introduced from Australia's

Figure 37.16   An example of the chaparral biome in California. The shrubs in the picture are *Eriodictyon tomentosum*. (Photograph by U.S. Forest Service.)

south coast into California's chaparral region have prospered mightily and have replaced to a considerable extent the native woody vegetation in areas near cities.

Mule deer and many kinds of birds live in the chaparral during the rainy season but move north or to higher altitudes to escape the hot, dry summer. Brush rabbits, wood rats, chipmunks, lizards, wren-tits and brown towhees are characteristic animals of the chaparral biome. During the hot, dry season, there is an ever present danger of fire which may sweep rapidly over the chaparral slopes. Following a fire, the shrubs sprout vigorously after the first rains and may reach maximum size within twenty years.

37.12   THE DESERT BIOME

In regions with less than 25 cm. of rain per year, or in certain hot regions where there may be more rainfall but with an uneven distribution in the annual cycle, vegetation is sparse and consists of greasewood, sagebrush or cactus. The individual plants in the desert are typically widely spaced with large bare areas separating them. In the brief rainy season, the California desert becomes carpeted with an amazing variety of wild flowers and grasses, most of which complete their life cycle from seed to seed in a few weeks. The animals present in the desert are reptiles, insects and burrowing rodents such as the kangaroo rat and pocket mouse, both of which are able to live without drinking water by extracting the moisture from the seeds and succulent cactus they eat.

The small amount of rainfall may be due to continued high barometric pressure, as in the Sahara and Australian deserts; a geographical position in the rain shadow of a mountain, as in the western North American deserts; or to high altitude, as in the deserts in Tibet and Bolivia. The only absolute deserts, where little or no rain ever falls, are those of northern Chile and of the central Sahara.

Careful measurements of the amount of dry matter produced for a given area in the course of a year show a clear linear relationship with the amount of rainfall, at least up to 60 cm. per year. This illustrates the primary role of moisture as a limiting factor in the productivity of the desert. Where the soil is favorable, an irrigated desert can be extremely productive because of the large amount of sunlight.

Two types of deserts can be distinguished on the basis of their average temperatures: "hot" deserts, such as that found in Arizona,

Figure 37.17   Two types of desert in western North America, a "cool" desert in Idaho dominated by sagebrush (*above*) and (*below*) a rather luxuriant "hot" desert in Arizona, with giant cactus (Saguaro) and palo verde trees, in addition to creosote bushes and other desert shrubs. In extensive areas of desert country the desert shrubs alone dot the landscape. (Upper photograph by U.S. Forest Service, lower photograph by U.S. Soil Conservation Service.)

characterized by the giant saguaro cactus, palo verde trees and the creosote bush; and "cool" deserts, such as that present in Idaho, dominated by sagebrush (Fig. 37.17).

Certain reptiles and insects are well adapted for survival in deserts because of their thick, impervious integuments and the fact that they excrete dry waste matter. A few species of mammals have become secondarily adapted to the desert by excreting very concentrated urine. They avoid the sun by remaining in their burrows during the day. The camel and the desert birds must have an occasional drink of water but can go for long periods of time using the water stored in the body.

When deserts are irrigated, the large volume of water passing through the irrigation system may lead to the accumulation of salts in the soil as some of the water is evaporated, and this will eventually limit the area's productivity. The water supply itself can fail if the watershed from which it is obtained is not cared for appropriately. The ruins of old irrigation systems and of the civilizations they supported in the deserts of North Africa and the Near East remind us that the irrigated desert will retain its productivity only when the entire system is kept in appropriate balance.

37.13 THE EDGE OF THE SEA: MARSHES AND ESTUARIES

Where the sea meets the land there may be one of several kinds of ecosystems with distinctive characteristics: a rocky shore, a sandy beach, an intertidal mud flat or a tidal estuary containing salt marshes. The word estuary refers to the mouth of a river or a coastal bay where the salinity is less than in the open ocean, intermediate between the sea and fresh water. Most estuaries, particularly those in temperate and arctic regions, undergo marked variations in temperature, salinity and other physical properties in the course of a year. To survive there, estuarine organisms must have a wide range of tolerance to these changes (they must be euryhaline, eurythermic, and so on).

The waters of estuaries are among the most naturally fertile in the world, frequently having a much greater productivity than the adjacent sea or the fresh water up the river. This high productivity is brought about by the action of the tides, which promote a rapid circulation of nutrients and aid in the removal of waste products, and by the presence of many kinds of plants, which provide for an extensive photosynthetic carpet. These include the phytoplankton, the algae living in and on the mud, sand, rocks or other hard surfaces, and the large attached plants, "seaweeds," eel grasses and marsh grasses.

Some of the marsh grass is eaten by insects and other terrestrial herbivores but most of it is converted to detritus and is consumed by the clams, crabs and other marine detritus eaters. Estuaries may have a high productivity of fish, oysters, shrimp and other seafood, which can be tapped by "mariculture," such as the oyster farms of Japan where oysters are grown suspended on rafts hanging from floats. This is an excellent way of obtaining protein foods as a harvest of the natural productivity of the estuaries. The farms must be spaced well apart and protected from pollution.

Estuaries and marshes are high on the list of ecologic regions of the world that are seriously threatened by man's activities. They were long considered to be worthless regions in which waste materials could be dumped. Many have been irretrievably lost by being drained, filled and converted to housing developments or industrial sites. Man is just beginning to appreciate that the best interests of all are served by maintaining estuaries in their natural state and protecting them from waste material and thermal and oil pollution.

37.14 MARINE LIFE ZONES

There has recently been a great upsurge of interest in oceanography in general and in marine ecology in particular as men have appreciated that we have much to learn about the mysterious sea. The oceans, which cover 70 per cent of the earth's surface, constitute one of the great reservoirs of living things and of the essential nutrients needed by both land and marine organisms. It is clear that the total weight of living things (the "biomass") in the ocean far exceeds that of all living things on land and in fresh water.

The seas are continuous one with another and marine organisms are restrained from spreading to all parts of the ocean only by factors such as temperature, salinity and depth. The salinity of the open ocean is about 35 parts per thousand. In the western Baltic it is 12 parts per thousand and 0.6 parts per thousand in the Gulf of Finland owing to the inflow of fresh water. The Red Sea, with no source of fresh water and a high rate of evaporation, reaches 46.5 parts per thousand. The temperature of the oceans ranges from about −2° C. in the polar seas to 32° C. or more in the tropics, but the annual range of variation in any given region is usually no more than 6° C.

The waters of the seas are continually moving in vast currents, such as the Gulf Stream, the North Pacific Current and the Humboldt Current, which circle in a clockwise fashion in the northern hemisphere and counterclockwise in the southern hemisphere. These currents not only influence the distribution of marine forms but also have marked effects on the climates of the adjacent land masses. In addition, there are very slow currents of cold, dense water flowing at great depths from the polar regions toward the equator.

Where the wind consistently moves surface water away from steep coastal slopes, water from the deep is brought to the surface by a process termed **upwelling**. This water is cold and rich in nutrients which have accumulated in the depths. Regions of upwelling typically occur on the western coasts of continents, as in California, Peru and Portugal, and are the most productive of all marine areas. The upwelling produced by the Peru Current has created one of the richest fisheries in the world and supports large populations of seabirds that deposit nitrate- and phosphate-rich **guano** on the headlands and adjacent coastal islands. These upwellings are very important in returning elements such as phosphorus to the surface for recycling.

Although the saltiness of the open ocean is relatively uniform, the concentrations of phosphates, nitrates and other nutrients vary widely in different parts of the sea and at different times of the year, and are usually the major factors limiting the biologic productivity of the seas in a given region.

Like the land, the ocean consists of regions characterized by different physical conditions, and consequently inhabited by specific kinds of plants and animals. All the phyla except Onychophora, and all the classes except amphibians, centipedes, millipedes and insects are well represented

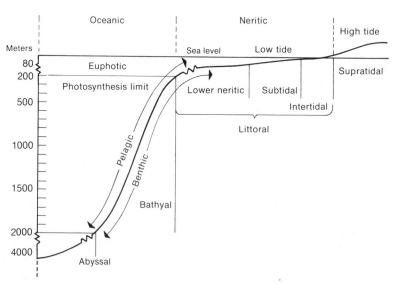

Figure 37.18 Zonation in the sea. (Redrawn from Hedgpeth, J. W.: The Classification of Estuarine and Brackish Waters and the Hydrographic Climate. Rpt. 11. National Research Council Committee on a Treatise on Marine Ecology and Paleoecology, 1951.)

in the oceans. Ctenophores, brachiopods, echinoderms, chaetognaths and a few lesser phyla are found only in the oceans.

A gently sloping continental shelf usually extends some distance offshore; beyond this the ocean floor (the **continental slope**) drops steeply to the abyssal region. The region of shallow water over the continental shelf is called the **neritic zone;** it can be subdivided into **supratidal** (above the high tide mark), **intertidal** (between the high and low tide lines, a region also known as the "littoral") and **subtidal** regions (Fig. 37.18).

The open sea beyond the edge of the continental shelf is the **oceanic zone.** The upper part of the ocean, into which enough light can penetrate to be effective in photosynthesis, is known as the **euphotic zone.** The average lower limit of this is at about 100 meters, but in a few regions of clear tropical water this may extend to twice that depth. The regions of the ocean beneath the euphotic zone are called the **bathyal zone,** over the continental slope, to a depth of perhaps 2000 meters; the depths of the ocean beyond that comprise the **abyssal zone.**

The floor of the ocean is not flat, but is thrown into gigantic ridges and valleys. Some of the ridges rise nearly to the surface (or above it where there are oceanic islands) and some of the valleys lie 10,000 meters below the surface of the sea. Huge underwater avalanches occur from time to time as parts of the ridges tumble into the valleys.

Some organisms are bottom dwellers, called **benthos,** and creep or crawl over the bottom or are **sessile** (attached to it). Others are **pelagic,** living in the open water, and are either active swimmers, **nekton,** or organisms that float with the current, **plankton.** The plankton includes algae, protozoa, small larval forms of a variety of animals, and a few worms. The nekton includes jellyfish, squid, fishes, turtles, seals and whales. Some of the benthic animals —crabs, snails, starfish, certain worms— crawl over the substrate. Clams and worms burrow into the sand, mud or rock of the sea bottom. Sponges, sea anemones, corals, bryozoa, crinoids, oysters, barnacles and tunicates are attached to the substratum.

The intertidal zone is one of the most favorable of all the habitats in the world and many biologists believe that life may have originated here. The abundance of water, light, oxygen, carbon dioxide and minerals makes it extremely salutary for plants. The dense growth of plants, providing food and shelter, makes it an excellent habitat for animals. The plants of the region are primarily a wide variety of algae plus a few grasses. Many of the animals are sessile and are more or less permanently fixed to the sea bottom, though they may be pelagic at some stage of their life cycle. These sessile animals are usually restricted to certain depths of the intertidal zone (Fig. 37.19). There is keen competition among the plants for space, and among the animals for space and food, so the forms living here have had to evolve special adaptations to survive.

The gravitational pulls of sun and moon each cause two bulges of the water on opposite sides of the earth, the high tides. The tides advance westward since the earth is rotating eastward on its axis. Since the earth rotates once a day on its axis, there are two high tides and two low tides per day. The periodicity of the tides is about 12.5 hours; hence the tides are about 50 minutes later on each succeeding day. Twice a month, at full and new moon, the earth, sun and moon are in line; the pulls of the sun and moon on the waters of the earth are additive; and the difference between high and low tide is greater than normal (the spring tides). At the quarter moons, when sun and moon are pulling at right angles, the difference between high and low tide is less than usual (neap tides). The difference in tidal range between high and low tides ranges from about 30 cm. in the open ocean to 15 m. or more in the Bay of Fundy and the English Channel.

Since the intertidal zone is exposed to air twice daily, its inhabitants have had to develop some sort of protection against desiccation. Some animals avoid this by burrowing into the damp sand or rocks until the tide returns; others have developed shells which can be closed, retaining a supply of water inside. Many plants contain jelly-like substances such as agar, which absorb large quantities of water and retain it while the tide is out.

One of the outstanding characteristics of this region is the ever present action of the waves, and the organisms living on a sandy or rocky beach have had to evolve ways of resisting wave action. The many seaweeds

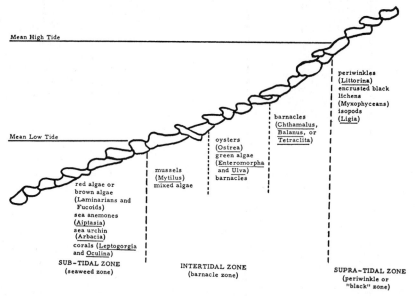

Mean High Tide

periwinkles
(Littorina)
encrusted black
lichens
(Myxophyceans)
isopods
(Ligia)

barnacles
(Chthamalus,
Balanus, or
Tetraclita)

Mean Low Tide

oysters
(Ostrea)
green algae
(Enteromorpha
and Ulva)
barnacles

mussels
(Mytilus)
mixed algae

red algae or
brown algae
(Laminarians and
Fucoids)
sea anemones
(Aiptasia)
sea urchin
(Arbacia)
corals (Leptogorgia
and Oculina)

SUB-TIDAL ZONE
(seaweed zone)

INTERTIDAL ZONE
(barnacle zone)

SUPRA-TIDAL ZONE
(periwinkle or
"black" zone)

Figure 37.19　Distribution of plants and animals on a rocky shore at Beaufort, North Carolina. (From Odum, E. P.: Fundamentals of Ecology, 3rd ed. Philadelphia, W. B. Saunders Co., 1971. Data from Stephenson and Stephenson, 1952.)

have tough pliable bodies, able to bend with the waves without breaking, while the animals either are encased in hard calcareous shells, such as those of mollusks, bryozoa, starfish, barnacles and crabs, or are covered by a strong leathery skin that can bend without breaking, like that of the sea anemone and octopus.

The successive zones of the intertidal region can be seen clearly on most rocky shores (Fig. 37.20). At the uppermost end is a zone of bare rock marking the transition between land and sea. Next is a spray zone with dark patches of algae on which the periwinkles (*Littorina*) graze. Below this is the zone regularly covered by the high tide; rocks in this zone are encrusted with barnacles, limpets and mussels. On the layer of rocks below this there is a cover of rock kelp and Irish moss containing small crabs and snails. Rocky coasts typically have tide pools which contain characteristic assemblages of animals and plants.

The sandy shore may be even more harsh than the rocky shore. It is subject to all the extremes of the latter plus the inconvenience of a constantly shifting substratum. The last makes life on the surface almost impossible; life has retreated below the surface. Zonation on a sandy beach is illustrated in Figure 37.21. It does not, however, conform to a universal

pattern as does that of rocky shores. In the example depicted, the supralittoral zone is inhabited by ghost crabs and beach hoppers. These animals spend most of the daytime hidden in damp burrows. They forage at night. Ghost crabs nightly go to the water to dampen their gill chambers. The intertidal zone is not as rich here as on the rocky shore, but it is the home of ghost shrimps, clams and bristle worms. Lower down the beach are lugworms, trumpet worms and other species of clams. Two interesting inhabitants of this zone are the mole crab and the coquina clam. As waves roll up the beach these two small creatures emerge from the sand, ride the waves up the beach and, as the velocity of the water decreases, burrow quickly into the sand as the waves retreat. Once settled, the crab extends its antennae and the clam its siphon to extract particulate food from the receding waves.

The subtidal zone is also thickly populated, for it has plenty of light and the nutrients required by plants. The absence of the periodic exposure to air and the diminished wave action permit many plants and animals to live here that could not survive in the tidal zone. Here live many species of fish and many single-celled algae; the larger seaweeds, which require a substratum for attachment, are found only in the shallower parts of the region.

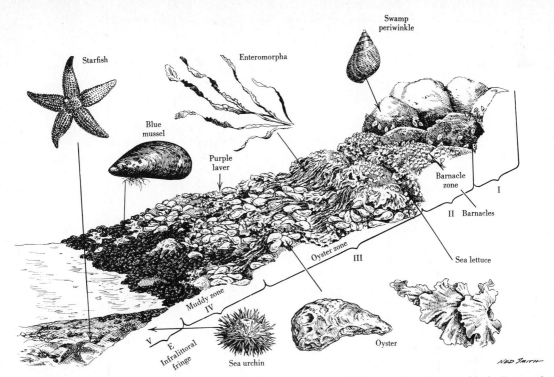

Figure 37.20 Zonation along a rocky shore—mid-Atlantic coast line. (I) Bare rock with some black algae and swamp periwinkle; (II) barnacle zone; (III) oyster zone: oysters, Enteromorpha, sea lettuce, and purple laver; (IV) muddy zone: mussel beds; (V) infralittoral fringe: starfish and so on. Note absence of kelps. (Zonation drawing based on Stephenson, 1952; sketches done from life or specimens From Smith, R. L.: Ecology and Field Biology. New York, Harper & Row, Publishers, 1966.)

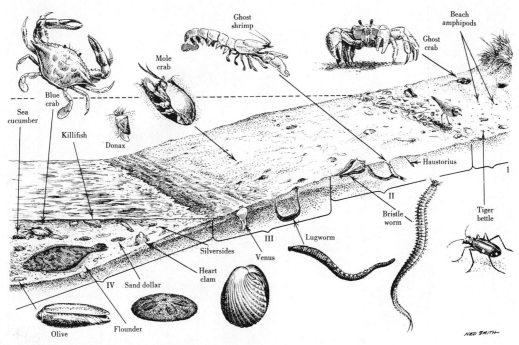

Figure 37.21 Life on a sandy ocean beach along the Atlantic coast. Although strong zonation is absent, organisms still change on a gradient from land to sea (I) Supratidal zone: ghost crabs and sand fleas; (II) flat beach zone: ghost shrimp, bristle worms, clams; (III) intratidal zone: clams, lugworms, mole crabs; (IV) subtidal zone: the dotted line indicates high tide. (From Smith, R. L.: Ecology and Field Biology. New York, Harper & Row, Publishers, 1966.)

The subtidal zone is the home of cockles, razor clams, moon shells and other mollusks. Still farther out are starfish, sand dollars, sea cucumbers, killifish, silversides and flounders.

In this area of the marine environment, as in all others, the **primary producer organisms,** equivalent to the flowering plants on land, are the **phytoplankton** (Fig. 37.22), consisting principally of diatoms and dinoflagellates. It is difficult to appreciate their importance because they are so small. An absolutely minimal estimate would place their density at 12,500,000 individuals per cubic foot. In temperate regions the phytoplankton undergoes two seasonal population explosions or "blooms," one in the spring, the other in late summer or fall. The mechanism is similar to that responsible for "blooms" in lakes. In the wintertime low temperatures and reduced light restrict photosynthesis to a low level; however, these factors do not prevent bacteria and other microorganisms from generating high concentrations of nitrogen, phosphorus and other nutrient elements. When spring brings higher temperatures and more light, photosynthesis accelerates. The nutrient supply is ample because the winter mixing of surface and deep water brings up nutrients that have fallen to and accumulated at the bottom. Within a fortnight, the diatoms multiply 10,000-fold. This prodigious growth accounts for the spring bloom. Soon, however, the nutrients are exhausted. Replacement from lower layers no longer occurs because warming of the surface water keeps it on top and prevents mixing. Nutrients are now locked in the bodies of animals that have eaten the phytoplankton or are slowly falling to the bottom in dead bodies. Whereas temperature and light were the limiting factors during the winter, nutrient level is the limiting factor during the summer, especially since existing phytoplankton is now being consumed by animals.

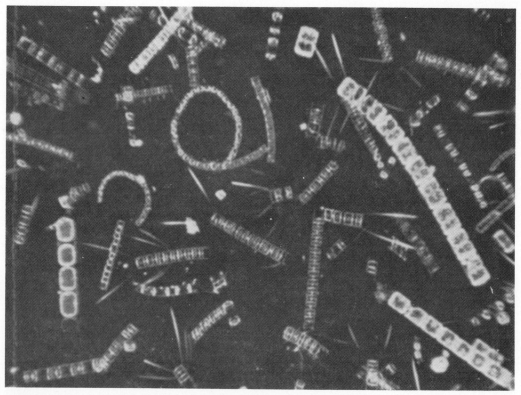

Figure 37.22   Living plants of the plankton (phytoplankton). ×110. Chains of cells of several species of *Chaetoceros* (those with spines), a chain of *Thalassiosira condensata* (at and pointing to bottom right corner), and a chain of *Lauderia borealis* (above the last named). By electronic flash. (From Hardy, A.: The Open Sea. Vol. 1. London, William Collins Sons & Co., Ltd., 1966.)

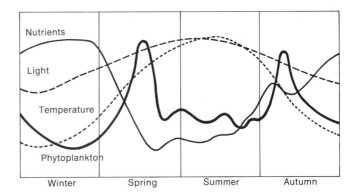

Figure 37.23 The probable mechanisms for phytoplankton "blooms." See text for explanation. (From Odum, E. P.: Fundamentals of Ecology, 3rd ed. Philadelphia, W. B. Saunders Company, 1971.)

Now nutrients begin to accumulate again in lower layers. As fall approaches the upper layers of water begin to cool again. The accompanying density change, together with the autumn equinoctial gales, begins mixing the water again. Water rich in phos-phates and nitrates is brought up from below. Other forms of phytoplankton, especially nitrogen-fixing blue-green algae, now bloom until reduced nutrients or temperature again intercedes (Fig. 37.23).

The other important population in the

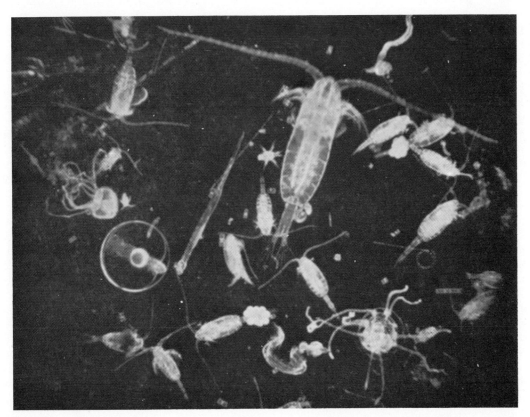

Figure 37.24 Living animals of the plankton (zooplankton). ×16. The copepods *Calanus finmarchicus* (the largest animal) and *Pseudocalanus elongatus* (similar in shape, but much smaller than *Calanus*, and the one with the cluster of eggs); two small anthomedusae with long tentacles; a fish egg (the circular object); a young arrow-worm *Sagitta* (to the right of the fish egg); small nauplius (larval stage) of copepod (close to left side of *Calanus*) and the planktonic tunicate *Oikopleura* (curly objects top right and middle bottom). By electronic flash, but partially narcotized. (From Hardy, A.: The Open Sea. Vol. 1. London, William Collins Sons & Co., Ltd., 1966.)

ocean is the **zooplankton** (Fig. 37.24). The zooplankton comprises all the animals carried passively by moving water. Every major phylum of the animal kingdom is represented, if not as adults, at least as eggs or larval stages. All the organisms are small, but not so small as not to be visible through a 6× hand lens. The beauty and almost limitless variety of forms of these animals beggar description. A sample haul near the Isle of Man gave an average of 4500 animals per cubic meter. The same hauls gave about 727,000 planktonic plants per cubic meter.

The active surface dwellers of the **neritic** zone are many bony fish, large crustacea, turtles, seals, whales and a host of sea birds. All these **consumers** are restricted in their distribution by temperature, salinity and nutrients. There is horizontal zonation and vertical stratification. The greatest numbers of fish (individuals–not species) are found in northern waters and where there are cold upwellings. Only a few species make up the bulk of commercial fisheries. Three fourths of the world's catch consists of herring, cod, haddock, pollock, salmon, flounder, sole, plaice, halibut, mackerel, tuna and bonito groups. The cod and flounder groups are bottom fish. Upon these and other marine organisms feed a great number of birds. Sandpipers, plovers, herons, curlews and others search the supra- and intertidal zones; cormorants, sea ducks, and pelicans, the subtidal zones; petrels and shearwaters, the lower neritic zone farther out to sea.

The **oceanic region** is less rich in species and numbers than the coastal areas, but it has its characteristic species. Many of these are transparent or bluish and since the sediment-free water of the open sea is marvelously transparent, these animals are nearly invisible. Animals that are too thick to be transparent frequently have smooth shiny and silvery bodies which make them invisible by mirroring the water in which they swim. Among the most characteristic animals of the open ocean are the baleen and toothed whales, the only mammals that are truly and absolutely marine. The baleen or whalebone whales live on phytoplankton which they strain from the water. The toothed whales live on nekton. Giant squids comprise part of their diet. The distribution of these whales is correlated directly or indirectly with the distribution of phytoplank-

ton. The flora and fauna of arctic seas differ from those of tropical seas. The boundaries between the two are not sharp and shift with the seasons. The major tropical oceans, the Atlantic and the Indo-Pacific, are separated from each other by regions of cold water. Each contains a large number of unique species of animals and plants.

Above the surface of the open sea are the oceanic birds—the petrels, albatrosses and frigate birds. They are not truly marine because they must come to land to breed. Apart from that they may spend all their lives at sea out of sight of land. Like the whales their distribution may be worldwide but not uniform. They are restricted by the distribution of fish which, in turn, are dependent on the phytoplankton.

Eighty-eight per cent of the ocean is more than 1 mile deep. It is continuous throughout the world except for the deep water of the Arctic Ocean, which is cut off from the rest by a narrow submerged mountain range connecting Greenland, Iceland and Europe. This is the area of great pressure and of perpetual night. Since no photosynthesis is possible, the only source of energy is the constant rain of organic debris, the bodies and waste products of organisms in the surface layers, that falls toward the bottom. The other prerequisite for life, oxygen, gets to the bottom by means of the oceanic circulation discussed earlier.

Life in the deep is poorly known because of the enormous difficulty of observation. The **pelagic** animals are strong swimmers, not easily caught in nets. The scant knowledge available has been gleaned from studies of net hauls and by observation from special undersea craft or via underwater television. In general, the animals of this world are either filter-feeders which sieve out particles before they reach the bottom, grubbers which ingest sediment or predators on the filter-feeders and grubbers.

Most of the fish of the abyssal region are rather small and peculiarly shaped; many are equipped with luminescent organs, which may serve as lures for the forms preyed upon. The majority of the deep-sea creatures are related to shallow-sea forms, and they must have migrated to their present habitat recently (geologically speaking), for none is older than the Mesozoic.

Since the number of members of any one species in these vast depths is small, re-

Figure 37.25 Sexual parasitism in the deep-sea angler fish, *Photocorynus spiniceps*. The small male is permanently attached to the female; he has no independent existence but is nourished by the blood of the female. (From Allee, W. C., et al.: Principles of Animal Ecology, Philadelphia, W. B. Saunders Co., 1949. After Norman.)

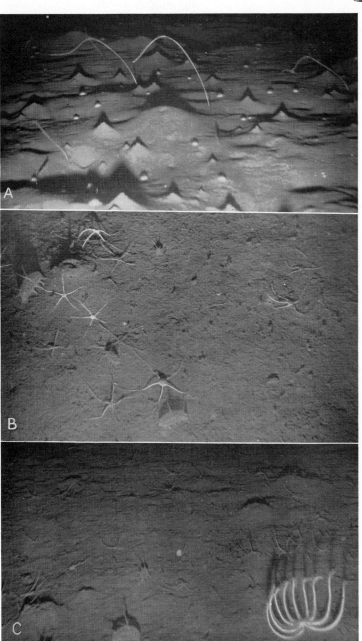

Figure 37.26 Photographs (by means of a special underwater camera called the benthograph) of the ocean bottom at three different depths off southern California (San Diego Trough). *A*, At 95 meters. Note abundant sea urchins (probably *Lytechinus*) appearing as globular, light-colored bodies, and the long curved sea whips (probably *Acanthoptilum*). Burrowing worms have built the conical piles of sediment at the mouth of their burrows. (Emery, 1952.) *B*, At 1200 meters. Vertical photograph of about 4 square meters of bottom composed of green silty mud having a high organic content. Note the numerous brittle stars (Ophiuroidea) and several large sea cucumbers (Holothuroidea). The latter have not been identified as to species as they have never been dredged from the sea and have only been seen in bottom photographs! (Official Navy photo, courtesy U.S. Navy Electronics Laboratory, San Diego.) *C*, At 1400 meters. Note in the right foreground the ten arms of what is probably a comatulid crinoid (a relative of the starfish which is attached to the bottom by a stalklike part). Small worm tubes and brittle stars litter the surface, and two sea cucumbers may be seen in the left foreground. Continual activity of burrowing animals keeps the sea bottom "bumpy." The bottom edge of the picture represents a distance of about 2 meters. (From Emery, K. O.: Submarine photography with the benthograph. Scient. Monthly, 75:3–11, 1952.)

production is more of a problem than in any other region, and some fish have evolved a curious adaptation to ensure that reproduction will occur. At an early age the male becomes attached to and fuses with the head of the female, where he continues to live as a small (2.5 cm.) parasite (Fig. 37.25). In due course he becomes mature and when the female lays her eggs, he releases his sperm into the water to fertilize them.

The bottom of the sea is a soft ooze, made of the organic remains and shells of foraminifera, radiolaria, and other animals and plants. Many invertebrates live on the ocean floor even at great depths (Fig. 37.26). These are usually characterized by thin, almost transparent shells, whereas the related shallow-water forms, exposed to wave action, have thick, hard shells. Apparently even the greatest "deeps" are inhabitated, for tube-dwelling worms have been dredged from depths of 8000 meters, and sea urchins, starfish, bryozoa and brachiopods have been found at depths of 6000 meters. Animals living on these bottom oozes typically have long thin appendages and possess spines or stalks.

37.15   FRESH-WATER LIFE ZONES

Fresh-water habitats can be divided into **standing water**—lakes, ponds and swamps—and **running water**—rivers, creeks and springs—each of which can be further subdivided. The biologic communities of fresh water habitats are in general more familiar than the salt water ones, and many of the animals used as specimens in zoology classes are from fresh water—amebas, hydras, planarias, crayfish and frogs.

Standing water, such as a lake, can be divided (much as the zones of the ocean were distinguished) into the shallow water near the shore (the **littoral zone**), the surface waters away from the shore (the **limnetic zone**), and the deep waters under the limnetic zone (the **profundal zone**).

Aquatic life is probably most prolific in the littoral zone. Within this zone the plant communities form concentric rings around the pond or lake as the depth increases (Fig. 37.27). At the shore proper are the cattails, bulrushes, arrowheads and pickerelweeds—the emergent, firmly rooted vegetation linking water and land environments. Out slightly deeper are the rooted plants with floating leaves such as the water lilies. Still deeper are the fragile thin-stemmed water weeds, rooted but totally submerged. Here also are found diatoms, blue-green algae and green algae. Common green pond scum is one of the latter.

The littoral zone is also the scene of the greatest concentration of animals (Fig. 37.28), distributed in recognizable communities. In or on the bottom are various dragonfly nymphs, crayfish, isopods, worms, snails and clams. Other animals live in or on plants and other objects projecting up from the bottom. These include the climbing dragonfly and damsel fly nymphs, rotifers, flatworms, bryozoa, hydra, snails and others. The zooplankton consists of water fleas such as *Daphnia*, rotifers and ostracods. The larger freely swimming fauna (**nekton**) includes diving beetles and bugs, dipterous larvae (e.g., mosquitoes) and large numbers of many other insects. Among the vertebrates are frogs, salamanders, snakes and turtles. Floating members of the community (**neuston**) include whirligig beetles, water striders and numerous protozoa. Many pond fish (sunfish, top minnows, bass, pike and gar) spend much of their time in the littoral zone.

The limnetic or open-water zone is occupied by many microscopic plants (dinoflagellates, *Euglena*, *Volvox*), many small crustaceans (copepods, cladocera and so on) and many fish.

Deep (profundal) life consists of bacteria, fungi, clams, blood worms (larvae of midges), annelids and other small animals capable of surviving in a region of little light and low oxygen.

As compared to ponds where the littoral zone is large, the water usually shallow and temperature stratification usually absent, lakes have large limnetic and profundal zones, a marked **thermal stratification** and a seasonal cycle of heat and oxygen distribution. In the summertime, the surface water (**epilimnion**) of lakes becomes heated while that below (**hypolimnion**) remains cold. There is no circulatory exchange between

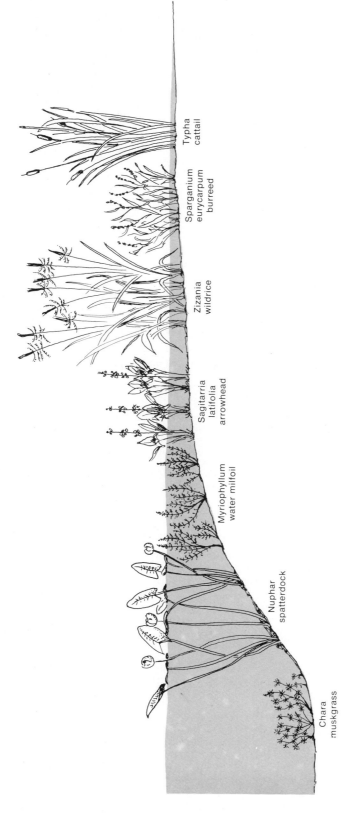

Figure 37.27 Zonation of vegetation about ponds and along river banks. Note the changes in vegetation with water depth. (After Dansereau, P., Biogeography: An Ecological Perspective. New York, The Ronald Press Company, 1959; from Smith, R. L.: Ecology and Field Biology. New York, Harper & Row, Publishers, 1966.)

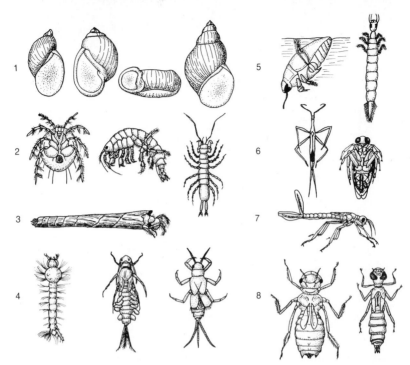

Figure 37.28   Some representative animals of the littoral zone of ponds and lakes. Series 1 to 4 are primarily herbivorous forms (primary consumers); series 5 to 8 are predators (secondary consumers). *1,* Pond snails (*left to right*): *Lymnaea* (pseudosuccinea) *columella; Physa gyrina; Helisoma trivolvis; Campeloma decisum. 2,* Small arthropods living on or near the bottom or associated with plants or detritus (*left to right*): a water mite, or Hydracarina (*Mideopsis*); an amphipod (*Gammarus*); an isopod (*Asellus*). *3,* A pond caddis fly larva (*Triaenodes*), with its thin, light portable case. *4, (Left to right)* A mosquito larva (*Culex pipiens*); a clinging or periphytic mayfly nymph (*Cloeon*); a benthic mayfly nymph (*Caenis*)—note gill covers which protect gills from silt. *5,* A predatory diving beetle, *Dytiscus,* adult and (*right*) larva. *6,* Two predaceous Hemipterans, a water scorpion, *Ranatra* (Nepidae), and (*right*) a back-swimmer, *Notonecta. 7,* A damsel fly nymph, *Lestes* (Odonata-Zygoptera); note three caudal gills. *8,* Two dragonfly nymphs (Odonata-Anisoptera), *Helocordulia,* a long-legged sprawling type (benthos), and (*right*) *Aeschina,* a slender climbing type (periphyton). (After Pennak, Robert W.: Fresh-water Invertebrates of the United States. New York, The Ronald Press Company, 1953.)

upper and lower layers, with the result that the lower layers frequently become deprived of oxygen. Between the two is a region of steep temperature decline (**thermocline**). As the cooler weather of fall approaches, the surface water cools, the temperature is equal at all levels, the water of the whole lake begins to circulate and the deep is again oxygenated. This is the "**fall overturn.**" In winter, as the surface temperature drops to 4° C., the water at the surface becomes less dense, remains at the surface and impedes circulation. The bottom is now warmer than the top. Because bacterial decomposition and respiration are less at low temperatures and cold water holds more oxygen, there is usually no great winter stagnation. The formation of ice may, however, cause oxygen depletion and result in a heavy winterkill of fish. The "**spring overturn**" occurs when the ice melts and the heavier surface water sinks to the bottom (Fig. 37.29).

Moving waters differ in three major aspects from lakes and ponds: current is a controlling and limiting factor; land-water interchange is great because of the small size and depth of moving water systems as compared with lakes; oxygen is almost always in abundant supply except when there is pollution. Temperature extremes tend to be greater than in standing water. Plants and animals living in streams are usually attached to surfaces or, in the case of animals, are exceptionally strong swimmers. Characteristic stream organisms are: caddis fly larvae, blackfly larvae, attached green algae, encrusting diatoms and aquatic mosses.

Fresh water habitats change much more

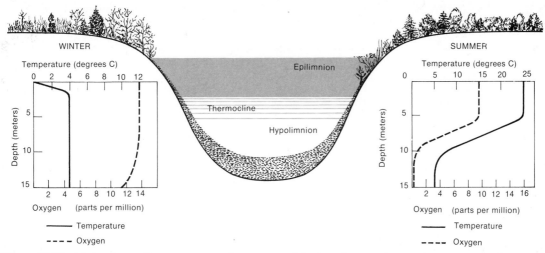

Figure 37.29   Thermal stratification in a north temperate lake (Linsley Pond, Conn.). Summer conditions are shown on the right, winter conditions on the left. Note that in summer the oxygen-rich circulating layer of water, the epilimnion, is separated from the cold oxygen-poor hypolimnion waters by a broad zone, called the thermocline, which is characterized by a rapid change in temperature and oxygen with increasing depth. (From Life in the Depths of a Pond by E. S. Deevey, Jr. Copyright © by Scientific American, Inc. All rights reserved.)

rapidly than other life zones; ponds become swamps, swamps become filled in and converted to dry land, and streams erode their banks and change their course. The kinds of plants and animals present may change markedly and show ecologic successions similar to those on land. The large lakes, such as the Great Lakes, are relatively stable habitats and have more stable populations of plants and animals. Lake Baikal in the Soviet Union is the oldest and deepest lake in the world, formed during the Mesozoic Era and containing many species of fish and other animals found nowhere else.

37.16 THE DYNAMIC BALANCE OF NATURE

The concept of the dynamic state of the cellular constituents was discussed in Chapter 4 and we learned that the protein, fat, carbohydrate and other constituents of the cells are constantly being broken down and resynthesized. A biotic community undergoes an analogous constant reshuffling of its constituent parts; the concept of the dynamic state of biotic communities is an important ecologic principle. Plant and animal populations are constantly subject to changes in their physical and biotic environment and must adapt or die. In addition, communities undergo a number of rhythmic changes—daily, tidal, lunar, seasonal and annual—in the activities or movements of their constituent organisms. These result in periodic changes in the composition of the community as a whole. A population may vary in size but if it outruns its food supply, like the Kaibab deer or the lemmings, equilibrium is quickly restored. Communities of organisms are comparable in many ways to a many-celled organism, and exhibit growth, specialization and interdependence of parts, characteristic form, and even development from immaturity to maturity, old age and death.

ANNOTATED REFERENCES

Aubert de la Rue, E., F. Boulière, and J. P. Harroy: The Tropics. New York, Alfred A. Knopf, 1957. A well illustrated account of the ecologic relations of animals and plants in the tropic rain forest.

Bates, M.: The Forest and the Sea. New York, Random House, 1960. Presents a more detailed discussion of these specific biological communities.

Braun, E. L.: Deciduous Forests of Eastern North America. Philadelphia, The Blakiston Company, 1950. A fine summary of the vegetation present in our deciduous forests.

Carson, R. L.: The Sea Around Us. New York, Oxford University Press, 1951. A beautifully written, nontechnical presentation of certain aspects of marine life.

Carson, R. L.: The Edge of the Sea. Boston, Houghton Mifflin Co., 1955. A companion book describing animal and plant interrelations at the seashore.

Coker, R. E.: Lakes, Streams and Ponds. Chapel Hill, N.C., University of North Carolina Press, 1954. A classic presentation of limnology.

Farb, P.: The Forest. New York, Time-Life Nature Library, 1961. Beautifully illustrated, nontechnical accounts of life in these biomes.

Hardy, A. C.: The Open Sea: The World of Plankton. New York, Houghton Mifflin Co., 1957. An interesting presentation of some of the microscopic forms living in the sea.

Hardy, A. C.: The Open Sea: Fish and Fisheries. New York, Houghton Mifflin Co., 1969. A fascinating discussion of the importance of fish and the seas to world ecology.

Jaeger, E. C.: The North American Deserts. Palo Alto, Stanford University Press, 1957. An interesting presentation of our deserts and the animals that live there.

Leopold, A. C.: The Desert. New York, New York, Time-Life Nature Library, 1961.

Ley, W.: The Poles. New York, Time-Life Nature Library, 1962.

MacArthur, R. H., and J. H. Connell: The Biology of Populations. New York, John Wiley & Sons, 1966. A text that emphasizes the population aspects of ecology.

Malin, J.: The Grasslands of North America. Lawrence, Kan., James Malin, 1956. Presents an interesting discussion of the ecology of the grassland biome that covers much of the midwest.

Odum, E. P.: Fundamentals of Ecology. 3rd ed. Philadelphia, W. B. Saunders Co., 1971. This popular text has an excellent account of each of the major biomes and life zones.

Reid, G.: Ecology of Inland Waters and Estuaries. New York, Reinhold Publishing Corp., 1961. Describes the life zones in rivers and estuaries.

Richard, P. W.: The Tropical Rain Forest. New York, Cambridge University Press, 1952. A classic account of life in the jungle.

Van Dyne, G. (Ed.): The Ecosystem Concept in Natural Resource Management. New York, Academic Press, 1969. An excellent source of material relating to a variety of ecosystems.

Wilson, D. P.: Life of the Shore and Shallow Sea. London, Nicholson and Watson, 1951. A beautifully illustrated description of animals and plants that live in the sea and on the shore.

38. Human Ecology

The dramatic upsurge of popular interest in ecology in recent years has brought to everyone's attention the cycles of water and nutrients, the factors controlling biological productivity, the nature of food chains and the problems of global pollution. The explosive growth of the human population has forced man to consider the ultimate limitations of the earth, rather than simply local limitations. In his famous *Essay on Population* (1798), Thomas Malthus predicted that human populations would increase faster than their food supply. Man is beginning to realize that the earth is simply a big spaceship with finite limits to its productivity and its ability to cope with pollutants. A key question in ecology is whether man is in greater danger of running out of resources or of being overcome by pollution.

The technology to solve many aspects of population and pollution problems is at hand; what is urgently needed is more general appreciation of the nature of the problems by informed laymen and a willingness on everyone's part to apply the principles of ecology for the ultimate benefit of mankind. In the immediate future, this probably depends not so much on advances in the environmental sciences, but rather on the application of economics, law, politics, planning and other areas of humanistic science to ecological problems.

There are many ways in which a knowledge of the principles of ecology can be used to further human society, one of the most important of which is the rational conservation of our natural resources.

Conservation does not mean simply hoarding—not using the resources at all—nor does it imply a simple rationing of our supplies so that some will be left for the future. True conservation implies taking full advantage of our knowledge of ecology and managing our ecosystems so as to establish a balance of harvest and renewal, thus insuring a continuous yield of useful plants, animals and other material. Further, this should at the same time guarantee the preservation of an environment with high quality that offers aesthetic and recreational uses as well as products.

The record of man's squandering of natural resources is indeed a dark one. The slaughter of the bison that once roamed the western plains, the decimation of the whales, the depletion of our supplies of many kinds of fresh water and marine fishes, the extinction of the passenger pigeon and other birds, the razing of thousands of square miles of forest and the burning of even more by careless use of fire, the pollution of streams with sewage and industrial wastes, the careless cultivation of land that has resulted in the complete ruin of many square miles of land and the silting of streams—these are some of the more flagrant examples of natural resources wasted beyond hope of rapid reclamation. Although countermeasures have been taken by state and federal departments of conservation, the chief task at present is to make the population at large realize the urgency and magnitude of the job to be done and to obtain general support for the measures that must be taken. For many aspects of the conservation problem, additional basic ecologic research is needed to determine the possible effects of a particular conservation measure on the ecology of the entire region.

38.1   AGRICULTURE

Wind and water have caused soil erosion through all geologic ages, but unwise farming and forestry practices in the last few centuries greatly increased the rate of erosion in certain parts of the globe. After decades of destructive exploitation of farmlands by planting one crop such as corn or cotton year after year, the soil conservation program sponsored jointly by federal and local agencies and based on sound ecologic principles is beginning to be effective in countering erosion. The rotation of crops, contour farming, the establishment of windbreaks to prevent soil erosion by winds, and the use of proper fertilizers to renew the soil are all measures which have proved effective in maintaining a balanced ecosystem. Successful farming must follow the principles of good land use. It is not conservation to reclaim marginal land for agricultural purposes or to build expensive dams and canals to irrigate land unless the land can produce crops that make the irrigation worthwhile.

If the grasslands of regions with slight rainfall are plowed and planted with wheat, a "dust bowl" will inevitably develop. If the land is kept as grassland and grazed in moderation, the soil will be kept in place, no dust bowl will develop and the land can be used economically year after year. By destroying the grass covering the soil, overgrazing can lead to destructive erosion just as improper plowing does. Overgrazing also leads to the invasion of the grassland by undesirable weeds and desert shrubs that are subsequently difficult to eradicate so that grass may grow again. Poor land use affects not only the unwise farmer but the whole population, which is eventually taxed to pay for rehabilitation of the land.

The ecologists specializing in the management of land have classified land into eight categories on the basis of its slope, kind of soil and natural biotic communities. These categories range from type I, which is excellent for farming and can be cultivated continuously, through three classes that can be used for farming only with special care, and another three classes that are suitable only for permanent pasture or forest, to type VIII, suitable only to be left as it is for game, for recreational and scenic purposes or for watershed protection (Fig. 38.1).

The control of insect pests by chemicals such as DDT must be carried out cautiously, with possible ecologic upsets in mind. Spraying orchards, forests and marshes may destroy not only the pests but also useful insects such as honeybees which

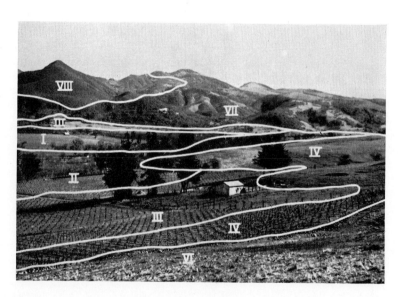

Figure 38.1 Classification of land according to its usefulness. Types I and II may be cultivated continuously; types III and IV are subject to erosion and must be cultivated with great care; types V, VI and VII are suitable for pasture or forests but not for cultivation; type VIII is productive only as a habitat for game. (U.S. Soil Conservation Service photograph.)

pollinate many kinds of fruit trees and crops, and useful insect parasites.

DDT and related chemicals kill other animals in addition to insects; amphibians and reptiles are the most vulnerable vertebrates. The vertebrates are less sensitive than insects, and DDT applied at a level of about one pound per acre is effective in insect control without endangering the vertebrates. However, when applied at a level of 5 to 10 pounds per acre, some of the useful larger animals are killed along with the insects. Some of the newer, stronger insecticides such as parathion have been used without adequate testing of their effects on other animals. Insect pests may actually increase after the use of DDT; the chemical may kill greater numbers of insect enemies of the pest than of the pests themselves. A number of strains of insects resistant to DDT have developed. However, the World Health Organization warns that banning DDT before cheap, safe and effective substitutes are found would be a disaster to world health. The agricultural production of the United States might decrease about 30 per cent if pesticides were banned, and the poor nations of the world would suffer even more from decreased food supplies and increased incidences of insect-borne diseases.

The great increase in agricultural productivity, the "green revolution," has been accomplished with the use of fertilizers, irrigation, pest control, and genetic selection of specific strains of crop plants. This industrialization of agriculture has required the input of considerable amounts of fuel energy. Doubling the crop yield, for example, is achieved only by a ten-fold increase in use of fertilizers, pesticides and fuel energy. As a result, industrialized agriculture is a major cause of both air and water pollution.

38.2 FORESTRY

The management of our forests is an important aspect of applied ecology. Until very recently most of the timber cut came from the accumulated growth of trees over the centuries. The lumber industry must adjust itself to bring the amount harvested into balance with the annual growth of the forests. Careful forest management has been carried on in Europe for many decades but is only beginning in this country. Proper timber management in our national and state forests has been important in demonstrating to the owners of private forests the results that can be obtained in this way. Since, in some regions, the desirable timber trees are members of the climax community, the ecologic problem is simply to find the best way to speed the return of the climax community after the trees have been cut. In other regions, the desirable trees are earlier seral stages of the ecologic succession, and forest management involves establishing means of preventing the succession to the climax community. This is also true of many kinds of animals; most game birds and many of the most valuable game fish are members of, and thrive best in, an early seral stage of their community.

38.3 WILDLIFE MANAGEMENT

The management of our fish and wildlife resources is a field of applied ecology supported by wide public interest, especially by sportsmen's clubs and associations. "Wildlife" used in this connection usually means game and fur-bearing animals. Since the various types of wildlife are adapted to different stages of ecologic succession, their management requires a knowledge of and the proper use of these stages. As the Middle West became more and more intensively farmed, and the original forests and prairies were reduced to small patches, the prairie chickens and ruffed grouse which were adapted to these habitats were greatly decreased in numbers. However, this region has been partially restocked with game birds by introducing ring-necked pheasants and Hungarian partridges, which had become adapted to living in the intensively farmed regions of Europe.

Of the three general methods used to increase the population of game animals—laws restricting the number killed, artificial stocking and the improvement of the habitat—the latter is the most effective. If the game habitats are destroyed or drastically altered, protective laws and artificial stocking of the region are useless. Protective laws must operate to prevent a population from getting too large as well as too small. Deer populations, in the absence of natural predators but subject to a constant, moderate amount of hunting, may increase to a point where they actually ruin the vegetation of the forest. Hunting should be restricted, of course, when populations are small and increased when they are larger. This requires accurate annual estimates of the population density of the game species.

Stocking a region artificially with game animals is effective only if they are being introduced into a new region or into one from which they have been killed off. Beavers, for example, were trapped to extermination in Pennsylvania, but restocking with Canadian beavers has been very successful, and it is estimated that there are some 15 to 20,000 beavers busy building dams in Pennsylvania. These are now an important factor in flood control in that region. The principles of population growth make it clear that if game animals of a certain species are already present, artificially stocking that region with additional members of the species will be futile. Stocking a region with a completely new species must be done cautiously, or the species may succeed so well as to become a pest and upset the biotic community, as has happened with rabbits in Australia and the English sparrow in the United States.

The management of the fish in a pond may be directed toward providing sport for hook and line fishermen or toward raising a crop of food fish and draining the pond at regular intervals to harvest the crop. To provide the best sport fishing a lake or pond should be stocked with a combination of the sport fish and its natural prey; stocking a pond with large-mouth bass plus bluegills gives seven to ten times more bass in three years than does stocking with bass alone. Stocking with fish must be done with care, for if a lake that already has about as many fish occupying a certain ecologic niche as possible is stocked with more of the same kind, there will be a decrease in the rate of growth and the average size of the fish. It has been found that sport fishing with hook and line is not likely to overfish a lake; the lake is more likely to be underfished, and the resulting crowding leads to a decrease in the average size of the fish population.

38.4 AQUACULTURE

The building of dams raises intricate ecologic problems, for dams may be intended for power, for flood control, for the prevention of soil erosion, for irrigation or for the creation of recreational areas. Since no one dam can satisfactorily accomplish all of these objectives, the primary objective must be clearly delineated and the secondary results must be understood. A contrast of two proposals for dealing with the same water-

TABLE 38.1 A COMPARISON OF A SINGLE MAIN RIVER RESERVOIR PLAN WITH A PLAN FOR MULTIPLE SMALLER HEADWATERS RESERVOIRS *

	Main Stream Reservoir	Multiple Headwaters Reservoirs
Number of reservoirs	1	34
Drainage area, square miles	195	190
Flood storage, acre feet	52,000	59,100
Surface water area for recreation, acres	1,950	2,100
Flood pool, acres	3,650	5,100
Bottom farmland inundated, acres	1,850	1,600
Bottom farmland protected, acres	3,371	8,080
Total cost	$6,000,000	$1,983,000

* From Odum, E. P.: Fundamentals of Ecology, 3rd ed. Philadelphia, W. B. Saunders Co., 1971. Data from Peterson, 1952; the costs in 1973 dollars would be greatly inflated!

shed (Table 38.1) shows that the multiple dam plan costs less, destroys a smaller area of productive farmland, impounds more water and is more effective in controlling floods and soil erosion. Since the productivity of a lake is inversely proportional to its depth, the series of small reservoirs would have greater productivity of fish.

To be most effective, the building of a dam must be accompanied by measures to decrease soil erosion upstream or the storage reservoir will fill with silt in a few years.

The management of the fish population in the lakes created by large dams is more difficult than the management of a pond. Sport fishing is usually very good when a dam has first been built, but gradually the silting up of the reservoir and the decrease in productivity change the nature of the fish community from game fish to less desirable catfish and shiners.

The term "ecologic backlash" has come into use to describe unforeseen, detrimental consequences of a project which either cancel out the anticipated gains of a particular modification of the environment or create more problems than were solved. A dam built primarily for hydroelectric power on the Zambesi River in Africa has had many undesirable side effects. The large lake shore created by the dam has greatly increased the habitat for the tsetse fly and has caused a severe outbreak of disease in cattle. Similar dams in other parts of Africa have increased the incidence of sleeping sickness among humans. Because the temperature of the water in the lake is high, the lake does not undergo spring and fall overturns, which mix the waters of lakes in the temperate zone. This greatly decreases the lake's productivity, and the fish catch has not compensated for the lost productivity of the grazing and agricultural land covered by the waters of the lake. The displacement of people from the rich river bank lands covered by the dammed waters to less suitable land has led to increased soil erosion. Some of the displaced people have moved to cities and caused further social upheaval. The regulated flow of water downstream of the dam has proved to be more damaging than the natural flooding which previously inundated the bottom lands annually. Maintenance of the fertility of these lands

will require the input of expensive chemical fertilizers. There is evidence that, similarly, the Aswan dam on the Nile is not an unmixed blessing.

Lake Baikal in Siberia is a very deep and ancient lake that was formed by earth movement during the Mesozoic Era, some 150,000,000 years ago. It contains some 20 to 25 per cent of the world's fresh water supply. It is often called the "Australia" of fresh water lakes because of its long isolation and its unique collection of plants and animals. Three hundred and seventy-six of its 384 species of arthropods and 29 of its 36 species of fish are found nowhere else in the world. The ecology of lake Baikal is now threatened by Soviet industry. Its water is polluted by wastes discharged from an enormous paper factory and the trees of its watershed are being felled for conversion to paper. The soil in the watershed is very thin and soil erosion occurs rapidly after the trees are felled. The soil is washed into the lake, making its originally crystal clear waters turbid and further altering the ecosystem. The region south and east of the lake is desert and with this change in the watershed the desert may spread to the shores of the lake.

The three chief sources of stream pollution are industrial materials, which either are directly toxic themselves or which reduce the oxygen supply in the water; sewage and other materials which decrease the oxygen content of the water and introduce bacteria and other septic organisms (Fig. 38.2); and turbidity caused by soil erosion in the watershed. As the silt settles out downstream it may cover up the spawning grounds of fish and have other direct deleterious effects. Erosion can be prevented by proper soil management, industrial wastes can be prevented by suitable design of the manufacturing process, and properly treated sewage can be emptied into a stream without deranging its ecologic relations. The proper design and operation of "waste management parks" can utilize heated water from power plants (thermal pollution) and certain types of treated and diluted domestic and industrial wastes as energy subsidies for adapted species of fish and other animals. These can ultimately be harvested to provide food and other useful products.

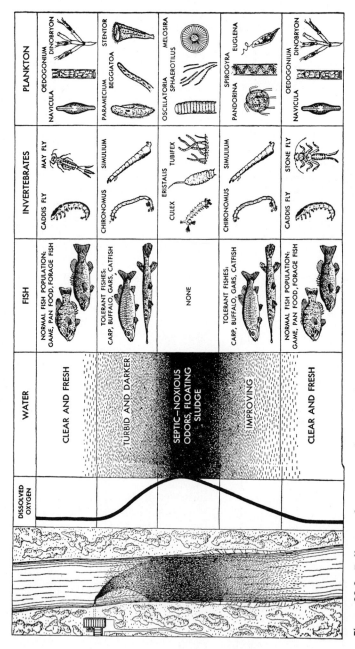

Figure 38.2 Pollution of a stream with untreated sewage and its subsequent recovery. As the amount of oxygen dissolved in the water decreases (*left*), fishes disappear and only organisms able to obtain oxygen from the surface, or those which tolerate low oxygen tensions, are able to survive. When the sewage has been reduced by bacteria, the population of animals and plants in the stream returns to normal. (From Odum, E. P.: Fundamentals of Ecology, 3rd ed. Philadelphia, W. B. Saunders Co., 1971. After Eliassen.)

38.5 MARINE FISHERIES AND ESTUARINE MARICULTURE

The primary productivity of the sea, as measured by the pounds of organic carbon produced per year per acre of surface, is very high. The productivity of the western Atlantic off the coast of North America is 2.5 to 3.5 tons of organic carbon per acre and that of Long Island Sound is 2.5 to 4.5 tons per acre. The productivity of the average forest is about 1 ton per acre; most cultivated land fixes only about ¾ ton of organic carbon per acre; and only the rich, intensively cultivated cornfields of Ohio produce as much as 4 tons per acre. Despite this high productivity, man's actual harvest from the ocean, in terms of pounds of fish caught per acre of surface, is very low. Only the rich fishing grounds of the North Sea produce as much as 15 pounds of fish per acre. The ecologic reasons for this are clear: the fish are secondary or tertiary consumers and are on top of a vast pyramid of producers. There are many organisms competing for the food energy fixed by the algae in addition to the edible fish and crustacea harvested from the sea.

The annual harvest of marine fish over the entire globe was about 18 million tons in 1938 and 55 million tons in 1967. Almost 80 per cent of the total comes from the North Atlantic, the northern and western Pacific and just off the west coasts of Peru and Ecuador in South America. About half of the total harvest is used as human food; the remainder is used as food for pets, poultry and livestock.

Man could undoubtedly recover for his use much more of the biologic productivity of the sea. Although he might be reluctant to eat marine algae himself, they might be filtered from sea water and processed so as to be suitable as feed for cattle or some other gastronomically acceptable animal.

Careful studies by the U.S. Fish and Wildlife Service of the fish population of George's Bank and other commercially fished areas have led to recommendations about the rate of fishing and the size of nets used which ensure that the fish are harvested at an optimal size for greatest yield at present and in the future. These areas, which had been fished so extensively that some of the most desirable species were reduced greatly in numbers, are now beginning to revive under careful management.

The shellfish—oysters, clams, shrimps and lobsters—present somewhat different and more difficult problems, for their habitat is more limited than that of commercial fish and they are more affected by adverse environmental changes. Oysters, whose food consists of algae or detritus of a certain size filtered from the sea water by their gills, are unable to use algae of a different size. Oysters were unable to survive in certain bays of Long Island Sound when ducks were raised in large numbers commercially on the adjacent shore. The wastes from the duck farms were washed into the bays, and the addition of this organic matter changed the community ecology in such a way that the normal food of the oyster, diatoms, was replaced by other algae which could not be used by the oysters. Once an oyster bed has been seriously depleted it may fail to recover even if seeded with oyster larvae, because the larvae require a favorable surface for attachment and the most favorable is the shell of an old oyster. In commercial oyster farms the larvae are provided with artificial sites for attachment. Once they have become attached they may be moved to other waters, even from one ocean to another, to complete their growth in waters that are favorable for feeding although not favorable for the reproduction of the species. The intensive cultivation (mariculture) of oysters, clams and other shellfish in estuaries and other marine environments has provided an important source of protein-rich food in Japan, Indonesia and Australia. It does require a considerable input of energy, at present largely in the form of human hand labor. By using the raft method of oyster culture, rather than simply harvesting the wild population, the production of oyster meat in a given area can be increased five- to ten-fold.

38.6 MINERAL RESOURCES

Because it was assumed that the supply of mineral resources would last for centuries to come and that there was no way of conserving them, little attention has been paid to this aspect of conservation. However, both of these assumptions are quite wrong. The relationship between the resources available and their utilization is expressed by the demographic quotient, Q:

$$Q = \frac{\text{total resources available}}{\text{population density} \times \text{per capita consumption}}$$

The quality of modern life goes down as the quotient goes down, and it is decreasing rapidly at present. Even if the total resources available could be kept constant by recycling, the value of Q decreases as the population density increases and as the per capita consumption increases. The utilization of natural resources in the United States is increasing at a rate of about 10 per cent per year. The demands for iron, copper, lead, tin and even the relatively abundant metal aluminum are such that the country is no longer self-sufficient either in mineral resources or in the fossil fuels needed to extract the new mineral resources. At the moment biotic and mineral resources are more critical and will be exhausted sooner than energy sources. Natural gas may soon be depleted and oil will follow a few decades later; but coal will last for a considerably longer time, hopefully until completely safe methods for producing large amounts of electricity in nuclear reactors have been devised. The limiting factor in the utilization of industrial energy will be pollution rather than supply. The report of the Committee on Pollution of the National Academy of Sciences, entitled "Resources and Man (1969)," made a total of 26 recommendations. These can be summarized in the statement that control of the size of human populations and improved management and recycling of our resources are urgently needed, beginning immediately!

38.7 PUBLIC HEALTH

Many aspects of the field of public health require the application of ecologic principles; the prevention of the spread of diseases carried by animals is an ecologic as well as a medical problem. The most effective way of eliminating malaria, for example, is to eliminate the particular species of mosquito which is the vector of the malaria parasite, yet this must be done without destroying the useful insects of the region. The mosquitoes which transmit malaria in different parts of the world have quite different ecologic niches and, therefore, measures that may be effective in mosquito control in one region may be quite ineffective in another. The malaria of the southeastern United States is transmitted by mosquitoes living in marshes; Italian malarial mosquitoes live in cool running water in the uplands; and Puerto Rican malarial mosquitoes live in brackish (slightly salty) water. Careful ecologic surveys of each region are necessary to formulate the proper measures to control the insects.

The size of the populations of rats, mice and many insect pests increases with the size of cities and the correlated tendency toward the development of slums in the older parts of the town. A survey in England in 1953 revealed that only 0.1 per cent of the houses in towns with less than 25,000 houses were infested with bedbugs, but over 1.0 per cent were infested in towns with more than 100,000 houses! Careful ecologic studies in Baltimore showed that although professional crews of rat trappers might catch as much as half of the rat population, it quickly returned to its former level. Cats proved to be much overrated as rat predators and were not effective in controlling the rat population. However, by changing the essential elements of the rats' habitat, by improving sanitation, thus decreasing the garbage on which the rats fed and the wastes in which they hid, the rat population was reduced to about 10 per cent of its former size. It remained at this lower level because that was the total number of rats which could survive in the altered environment.

38.8 POLLUTION

The ultimate cause of pollution is people, and as the number of people increases there is a corresponding increase in the amount of pollution. As the number of people has increased and the amount of energy used by each person has also increased, the total energy demand has increased at a very high rate. This is reflected in the air pollution caused by the exhaust stacks of industrial plants and electric generating plants burning fossil fuels, and by the exhaust of automobiles transporting people to and from their work and play.

The National Academy of Sciences, in the report by the Committee on Pollution, defined pollution as follows:

Pollution is an undesirable change in the physical, chemical or biological characteristics of our air, land and water that may or will harmfully affect human life or that of desirable species; industrial processes, living conditions and cultural assets; or that may or will waste or deteriorate our raw material resources. Pollutants are residues of the things we make, use and throw away. Pollution increases not only because as people multiply the space available to each person becomes smaller, but also because the demands per person are continually increasing so that each throws away more year by year. As the earth becomes more crowded there is no longer an "away." One person's trash basket is another's living space.*

Pollution is undoubtedly the most important limiting factor for man. It is difficult to put a price tag on the cost of pollution, but it is clear that pollution places a burden on human society (1) from the loss of resources owing to unnecessarily wasteful exploitation, (2) from the cost of abating and controlling pollution and (3) from the cost in human health.

From an ecological viewpoint two types of pollution can be recognized—pollution involving biodegradable pollutants and pollution involving nondegradable pollutants. Biodegradable pollutants such as domestic sewage can be decomposed rapidly by natural processes or by carefully engineered systems, such as a community's sewage treatment plant. Problems arise when the input of degradable pollutants into the environment exceeds the environment's capacity to decompose or disperse them. The problems with disposable sewage result in general from the fact that urban populations have grown much faster than their sewage treatment facilities. Degradable pollutants can be dealt with by a combination of mechanical and biological treatments; but there are limits to the total amount of organic matter that can be decomposed in a given area, and there is an overall limit to the amount of carbon dioxide (one of the decomposition products) that can be released into the air.

The nondegradable pollutants include

metals such as mercury, trace metals, steel and aluminum cans and organic chemicals such as DDT that are degraded only very slowly. Some of these nondegradable pollutants accumulate as they proceed along the food chains. Dealing with these pollutants is a much more difficult and expensive problem. Some of these pollutants, such as aluminum cans, are actually a resource if they are recycled instead of thrown away. Others, such as DDT, can be replaced by more degradable substitutes when they are found.

The design and operation of an efficient waste treatment plant is a very complex subject, beyond the scope of this book, but it is obvious that such operations must be carried out with the best possible application of ecological principles. It is perfectly feasible to produce drinkable water from sewage waste, but this is a very expensive procedure. (However, it is less expensive than producing drinking water from sea water.)

The pollution of the air in industrialized countries such as the United States is a problem of first magnitude. Although the pollution of the air over the entire country is serious enough, the most striking aspect of air pollution is the local concentration of pollutants that may occur over cities such as Los Angeles and New York during a temperature inversion. This occurs when air is trapped under a warm upper layer that prevents a vertical rise of the pollutants. Certain combinations of pollutants may react in the environment to produce additional pollution. Certain components in automobile exhaust can combine in the presence of sunlight to produce even more toxic substances called **photochemical smog.** One of the components of photochemical smog blocks a key reaction in photosynthesis and will kill plants by inhibiting their production of food.

The subject of pesticides and herbicides has been an extremely controversial one in the past decade. Early in this century, when farms were small and diversified and there was plenty of farm labor, the increase in the numbers of insect pests was blocked by simple inorganic salts. The industrialization of agriculture that has occurred within the last few decades required, and was made possible by, broad spectrum poisons such as the organochlorides and organophosphates. Unfortunately these poisons were used much

---

* From Resources and Man (1969), the National Research Council, National Academy of Sciences, Washington, D.C.

too enthusiastically, and since many are persistent and are degraded very slowly, these broad spectrum poisons have accumulated to such an extent in many parts of the environment that they can no longer be used.

Two highly undesirable effects of the saturation of the environment with DDT and other organochlorides are the development of strains of insects that are resistant to the DDT and the accumulation of these toxic materials in the food chain. Most of the problems can be traced to the fact that new insecticides are tested, usually superficially, at the level of the organism against which they are directed and then are used at the ecosystem level without controlled tests of their effects on the entire ecosystem.

It is clear that if man is to raise enough food crops to supply his burgeoning population he must have some way of dealing with the insect pests that are competing for this same supply of food. Chemical insecticides must be used only in appropriate places and amounts and with adequate controls. Certain kinds of insects can be removed by introducing appropriate predators or parasites, by supplying hormonal factors such as juvenile hormone that will prevent the insect from completing its life cycle, by using sex lures and other pheromones that will confuse the reproductive efforts of the insects, or by sterilizing large numbers of male insect pests by chemicals or radiation and then releasing the sterile males into the natural environment.

If all chemical pesticides were banned, the agricultural production of the United States would decrease by 30 per cent or more, according to the United States Department of Agriculture. Some partisans have claimed that 50 million Americans would be in danger of starving to death if chemical pesticides were withheld from agriculture. The poor and undeveloped nations of the world would probably suffer even more from disease and hunger if DDT were banned. In pest control programs on many kinds of crops the organochlorine pesticides are being replaced by organophosphorus and carbamate compounds, which are less resistant and more easily biodegradable. However, larger amounts of these pesticides are usually required to control pests in a given area and they may be more toxic to man than DDT. Maintaining agri-

cultural production with the aid of chemical pesticides and herbicides without damaging the ecosystem irretrievably will be a difficult task.

Recently much interest has centered on the discovery of abnormally large quantities of mercury in both fresh water and marine fish. The ocean contains a great many metals, including gold, silver and platinum as well as mercury. These metals have been washed into the ocean over the course of billions of years and are not primarily the result of industrial pollution. In contrast, the mercury in certain fresh-water ecosystems such as the Great Lakes is primarily the result of industrial pollution, and measures can and should be taken to eliminate further input. Mercury from sea water undoubtedly accumulates in the flesh of all kinds of fish. It is also present in flesh of lobsters, shrimp, clams, oysters and other shellfish. It should be noted that one commonly prescribed medicine, calomel, contains mercury.

Other pollutants in the environment are the various radioactive materials such as strontium-90. Strontium-90 is a beta emitter, having a half-life of 28 years. Strontium is very similar to calcium in its chemical and physical properties and moves with it in its natural cycle. Calcium and strontium are washed out of rocks and move down rivers to the ocean. In the far north radioactive fall-out of strontium-90 has been absorbed by lichens, and the caribou and reindeer that eat the lichens have concentrated the strontium in their flesh. Men who eat the flesh of the animals or drink milk from them may then accumulate strontium in their bones. In other parts of the world strontium-90 has accumulated in the bones of both adults and children who obtain it in cow's milk; the cows obtained it by eating vegetation polluted with strontium-90.

Little strontium-90 has been added to the environment since 1963. If the ban on nuclear testing can be continued, strontium-90, and the related cesium-137, will gradually disappear from the environment. Both are byproducts of fission reactions and enter the environment through fall-out or through faulty waste disposal; they are especially hazardous because they accumulate in successive living organisms along the food chain. Studies in Great Britain showed that strontium-90 was 21 times more concentrated in grass than in soil and some 700 times more concentrated in sheep than in the grass.

Most human beings now live in urban areas, and urban areas not only pollute the air but modify the climate. Because of the absorption of solar radiation by vertical surfaces and the production of heat by the machines in a city, the air temperature in a city will be 1 to 3° F. higher and the humidity will be about 6 per cent lower than in the surrounding countryside. Because of particulate air pollution, cloudiness will be 10 per cent greater and there will be 30 to 100 per cent more fog, the higher value occurring in the winter. Precipitation will be 10 per cent greater, there will be 15 per cent less sunshine and 10 to 30 per cent less ultraviolet radiation in the city than in the countryside.

In 1970, the average population density in the United States was one person to every ten acres of ice-free land area. The average population of the world was at the same level. Even if the anticipated reduction in birth rate occurs, the population of the United States will double in the next 30 years. By the year 2000 there will be only five acres of land for every man, woman and child. Although as little as ⅓ acre can produce enough calories to sustain a person, the kind of optimal quality diet that includes meats, fruits, and vegetables requires at least 1½ acres per person. Another acre per person is needed to produce the paper, wood and cotton needed by each person, and an additional ½ acre is needed for the roads, airports, buildings and other areas not involved in food production. Only about 2 acres per person remain for all the many other uses that man has for land. An affluent, developed nation actually requires more space and resources per person than an undeveloped nation. From this it follows that the key in determining the optimum human population density should be the adequate and usable *pollution-free living space* and not simply the food needed. E. P. Odum, in his book *Fundamentals of Ecology* (1971), has said, in essence, that this earth can feed more warm bodies sustained as so many domestic animals in a polluted feed lot than it can support quality human beings who have a right to a pollution-free environment, a reasonable chance for personal liberty, and a variety of options

for the pursuit of happiness. To obtain this latter goal, at least one-third of all land should remain protected for use as national, state or municipal parks, greenbelts, refuges, wilderness areas and so on. If the land is under private ownership, it should be protected by zoning or other legal means.

38.9 HUMAN ECOLOGY

No great amount of thought is required to realize that the ecologic principles apply to human populations as well as to populations of animals and plants. Human ecology deals not only with the dynamics of human populations but also with the relationship of man to the many physical and biotic factors which impinge upon him. By appreciating that human populations are a part of larger units — of biotic communities and ecosystems — man can deal with his own special problems more intelligently. Man has wittingly and unwittingly exercised a great deal of control over his environment and has modified the communities and ecosystems of which he is a part. However, this control is far from complete, and man must, like other animals and plants, adapt to those situations which he cannot change. By understanding and cooperating with the various cycles of nature, man has a better chance of surviving in the future than if he blindly attempts to change and control them.

The human population is clearly in danger of multiplying beyond the ability of the earth to support it. In the past several centuries, the population of the world has increased tremendously as new territories have been opened for exploitation and as methods of producing food have become more efficient. Most biologists and social scientists believe that the danger of human overpopulation is both great and imminent. It has been amply shown that the Malthusian Principle (that populations have an inherent ability to grow exponentially) is true for organisms generally, and the growth of the human population in the past 300 years follows an exponential curve. The productivity and carrying capacity of the earth for human beings can be maintained and perhaps increased somewhat, but eventually the human biomass must be brought into equilibrium with the space and food available. The human population is probably in greater danger of running out of drinkable water and breathable air than it is of exhausting its food supply! Some limitation of human reproduction is clearly inevitable. It remains to be seen whether man will do this voluntarily or involuntarily.

ANNOTATED REFERENCES

Adams, A.: Eleventh Hour: A Hard Look at Conservation. New York, G. P. Putnam's Sons, 1970. An overview of the conservation movement.

Benarde, M. A.: Our Precarious Habitat, An Integrated Approach to Understanding Man's Effect on His Environment. New York, W. W. Norton & Co., 1970. This presents another viewpoint of the problems arising from man's misuse of the environment.

Carson, R. L.: Silent Spring. Boston, Houghton Mifflin Co., 1962. The beautifully written book that did much to touch off popular concern about preserving our environment.

Cox, G. W. (Ed.): Readings in Conservation Ecology. New York, Appleton-Century-Crofts, 1969. An anthology; an excellent source book for topics related to conservation of natural resources.

Doutt, R. L., and W. W. Kilgore: Pest Control. New York, Academic Press, 1967.

Draybill, H. F. (Ed.): Biological Effects of Pesticides in Mammalian Systems. New York, N.Y. Acad. Sci., 1969. Two excellent books on pests and their control by pesticides.

Ehrlich, P., and A. Ehrlich: Population Resources and Environment: Issues in Human Ecology. San Francisco, W. H. Freeman & Co., 1970. A thorough ecological analysis of man's impact on the productivity of the biosphere, presented by two of the outstanding figures in the field.

Graham, F., Jr.: Since Silent Spring. Boston, Houghton Mifflin Co., 1970. A follow-up, eight years later, of what had happened in popular ecology since the publication of Rachel Carson's book.

Hardin, G. (Ed.): Population, Evolution and Birth Control. San Francisco, W. H. Freeman & Co., 1969. A thoughtful analysis of the present human predicament.

Marine, G.: America The Raped. New York, Simon and Schuster, 1969. Some examples of the outrageous devastation resulting from poor ecologic planning.

McHarg, I.: Design With Nature. Garden City, N.Y., Natural History Press, 1969. An outline of the author's proposals for improving the design of human communities.

Miller, M. W., and G. G. Berg (Eds.): Chemical Fallout. Springfield, Ill., Charles C Thomas, 1969. A collection of two dozen papers discussing pesticides—how they work and their effects on animal and human populations.

Roueche, B.: What's Left. Boston, Little, Brown & Co., 1968. A particularly well written account of the damage to our environment by lack of consideration of ecologic issues.

Index